Matrix Mathematics

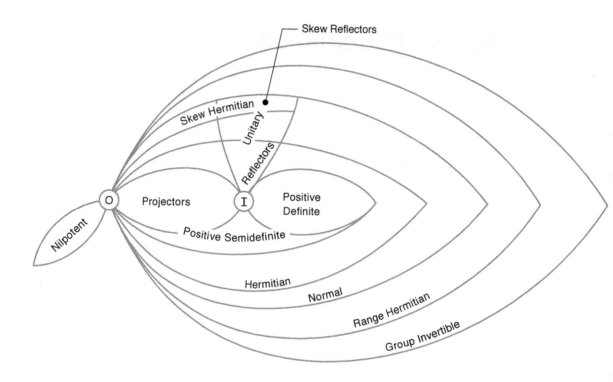

Matrix Mathematics

Theory, Facts, and Formulas

Dennis S. Bernstein

PRINCETON UNIVERSITY PRESS

PRINCETON AND OXFORD

Copyright ©2009 by Princeton University Press

Published by Princeton University Press,
41 William Street, Princeton, New Jersey 08540

In the United Kingdom: Princeton University Press,
6 Oxford Street, Woodstock, Oxfordshire, 0X20 1TW

All Rights Reserved

Library of Congress Cataloging-in-Publication Data

Bernstein, Dennis S., 1954–
        Matrix mathematics: theory, facts, and formulas / Dennis S. Bernstein. – 2nd ed.
            p. cm.
        Includes bibliographical references and index.
        ISBN 978-0-691-13287-7 (hardcover : alk. paper)
        ISBN 978-0-691-14039-1 (pbk. : alk. paper)
        1. Matrices. 2. Linear systems. I. Title.
QA188.B475 2008
512.9'434—dc22

                                                                                2008036257

British Library Cataloging-in-Publication Data is available

This book has been composed in Computer Modern and Helvetica.

The publisher would like to acknowledge the author of this volume for providing the
camera-ready copy from which this book was printed.

Printed on acid-free paper. ∞

press.princeton.edu

Printed in the United States of America

10   9   8   7   6   5   4   3   2   1

To the memory of my parents,

Irma Shorrie (Hirshon) Bernstein and Milton Bernstein,

whose love and guidance are everlasting

To the memory of my parents,

Irma Shorrie (Hirshon) Bernstein and Milton Bernstein,

whose love and guidance are everlasting

... vessels, unable to contain the great light flowing into them, shatter and break. ... the remains of the broken vessels fall ... into the lowest world, where they remain scattered and hidden

— D. W. Menzi and Z. Padeh,
*The Tree of Life, Chayyim Vital's
Introduction to the Kabbalah of
Isaac Luria*, Jason Aaronson,
Northvale, 1999

Thor ... placed the horn to his lips ... He drank with all his might and kept drinking as long as ever he was able; when he paused to look, he could see that the level had sunk a little, ... for the other end lay out in the ocean itself.

— P. A. Munch, *Norse Mythology*,
AMS Press, New York, 1970

Contents

Preface to the Second Edition

This second edition of *Matrix Mathematics* represents a major expansion of the original work. While the total number of pages is increased 57% from 752 to 1181, the increase is actually greater since this edition is typeset in a smaller font to facilitate a manageable physical size.

The second edition expands on the first edition in several ways. For example, the new version includes material on graphs (developed within the framework of relations and partially ordered sets), as well as alternative partial orderings of matrices, such as rank subtractivity, star, and generalized Löwner. This edition also includes additional material on the Kronecker canonical form and matrix pencils; matrix representations of finite groups; zeros of multi-input, multi-output transfer functions; equalities and inequalities for real and complex numbers; bounds on the roots of polynomials; convex functions; and vector and matrix norms.

The additional material as well as works published subsequent to the first edition increased the number of cited works from 820 to 1540, an increase of 87%. To increase the utility of the bibliography, this edition uses the "back reference" feature of LATEX, which indicates where each reference is cited in the text. As in the first edition, the second edition includes an author index. The expansion of the first edition resulted in an increase in the size of the index from 108 pages to 161 pages.

The first edition included 57 problems, while the current edition has 74. These problems represent extensions or generalizations of known results, sometimes motivated by gaps in the literature.

In this edition, I have attempted to correct all errors that appeared in the first edition. As with the first edition, readers are encouraged to contact me about errors or omissions in the current edition, which I will periodically update on my home page.

Acknowledgments

I am grateful to many individuals who kindly provided advice and material for this edition. Some readers alerted me to errors, while others suggested additional material. In other cases I sought out researchers to help me understand the precise nature of interesting results. At the risk of omitting those who were helpful, I am pleased to acknowledge the following: Mark Balas, Jason Bernstein, Sanjay Bhat, Gerald Bourgeois, Adam Brzezinski, Francesco Bullo, Vijay

Chellaboina, Naveena Crasta, Anthony D'Amato, Sever Dragomir, Bojana Drincic, Harry Dym, Matthew Fledderjohn, Haoyun Fu, Masatoshi Fujii, Takayumi Furuta, Steven Gillijns, Rishi Graham, Wassim Haddad, Nicholas Higham, Diederich Hinrichsen, Matthew Holzel, Qing Hui, Masatoshi Ito, Iman Izadi, Pierre Kabamba, Marthe Kassouf, Christopher King, Siddharth Kirtikar, Michael Margliot, Roy Mathias, Peter Mercer, Alex Olshevsky, Paul Otanez, Bela Palancz, Harish Palanthandalam-Madapusi, Fotios Paliogiannis, Isaiah Pantelis, Wei Ren, Ricardo Sanfelice, Mario Santillo, Amit Sanyal, Christoph Schmoeger, Demetrios Serakos, Wasin So, Robert Sullivan, Dogan Sumer, Yongge Tian, Götz Trenkler, Panagiotis Tsiotras, Takeaki Yamazaki, Jin Yan, Masahiro Yanagida, Vera Zeidan, Chenwei Zhang, Fuzhen Zhang, and Qing-Chang Zhong.

As with the first edition, I am especially indebted to my family, who endured four more years of my consistent absence to make this revision a reality. It is clear that any attempt to fully embrace the enormous body of mathematics known as matrix theory is a neverending task. After devoting more than two decades to this project of reassembling the scattered shards, I remain, like Thor, barely able to perceive a dent in the vast knowledge that resides in the hundreds of thousands of pages devoted to this fascinating and incredibly useful subject. Yet, it is my hope that this book will prove to be valuable to everyone who uses matrices, and will inspire interest in a mathematical construction whose secrets and mysteries have no bounds.

Dennis S. Bernstein
Ann Arbor, Michigan
dsbaero@umich.edu
March 2009

Preface to the First Edition

The idea for this book began with the realization that at the heart of the solution to many problems in science, mathematics, and engineering often lies a "matrix fact," that is, an identity, inequality, or property of matrices that is crucial to the solution of the problem. Although there are numerous excellent books on linear algebra and matrix theory, no one book contains all or even most of the vast number of matrix facts that appear throughout the scientific, mathematical, and engineering literature. This book is an attempt to organize many of these facts into a reference source for users of matrix theory in diverse applications areas.

Viewed as an extension of scalar mathematics, matrix mathematics provides the means to manipulate and analyze multidimensional quantities. Matrix mathematics thus provides powerful tools for a broad range of problems in science and engineering. For example, the matrix-based analysis of systems of ordinary differential equations accounts for interaction among all of the state variables. The discretization of partial differential equations by means of finite differences and finite elements yields linear algebraic or differential equations whose matrix structure reflects the nature of physical solutions [1269]. Multivariate probability theory and statistical analysis use matrix methods to represent probability distributions, to compute moments, and to perform linear regression for data analysis [517, 621, 671, 720, 972, 1212]. The study of linear differential equations [709, 710, 746] depends heavily on matrix analysis, while linear systems and control theory are matrix-intensive areas of engineering [3, 68, 146, 150, 319, 321, 356, 379, 381, 456, 515, 631, 764, 877, 890, 960, 1121, 1174, 1182, 1228, 1232, 1243, 1368, 1402, 1490, 1535]. In addition, matrices are widely used in rigid body dynamics [28, 745, 753, 811, 829, 874, 995, 1053, 1095, 1096, 1216, 1231, 1253, 1384], structural mechanics [888, 1015, 1127], computational fluid dynamics [313, 492, 1460], circuit theory [32], queuing and stochastic systems [659, 944, 1061], econometrics [413, 973, 1146], geodesy [1272], game theory [229, 924, 1264], computer graphics [65, 511], computer vision [966], optimization [259, 382, 978], signal processing [720, 1193, 1395], classical and quantum information theory [361, 720, 1069, 1113], communications systems [800, 801], statistics [594, 671, 973, 1146, 1208], statistical mechanics [18, 163, 164, 1406], demography [305, 828], combinatorics, networks, and graph theory [132, 169, 183, 227, 239, 270, 272, 275, 310, 311, 343, 282, 371, 415, 438, 494, 514, 571, 616, 654, 720, 868, 945, 956, 1172, 1421], optics [563, 677, 820], dimensional analysis [658, 1283], and number theory [865].

In all applications involving matrices, computational techniques are essential for obtaining numerical solutions. The development of efficient and reliable algorithms for matrix computations is therefore an important area of research that has been extensively developed [98, 312, 404, 583, 699, 701, 740, 774, 1255, 1256, 1258, 1260, 1347, 1403, 1461, 1465, 1467, 1513]. To facilitate the solution of matrix problems, entire computer packages have been developed using the language of matrices. However, this book is concerned with the analytical properties of matrices rather than their computational aspects.

This book encompasses a broad range of fundamental questions in matrix theory, which, in many cases can be viewed as extensions of related questions in scalar mathematics. A few such questions follow.

What are the basic properties of matrices? How can matrices be characterized, classified, and quantified?

How can a matrix be decomposed into simpler matrices? A matrix decomposition may involve addition, multiplication, and partition. Decomposing a matrix into its fundamental components provides insight into its algebraic and geometric properties. For example, the polar decomposition states that every square matrix can be written as the product of a rotation and a dilation analogous to the polar representation of a complex number.

Given a pair of matrices having certain properties, what can be inferred about the sum, product, and concatenation of these matrices? In particular, if a matrix has a given property, to what extent does that property change or remain unchanged if the matrix is perturbed by another matrix of a certain type by means of addition, multiplication, or concatenation? For example, if a matrix is nonsingular, how large can an additive perturbation to that matrix be without the sum becoming singular?

How can properties of a matrix be determined by means of simple operations? For example, how can the location of the eigenvalues of a matrix be estimated directly in terms of the entries of the matrix?

To what extent do matrices satisfy the formal properties of the real numbers? For example, while $0 \leq a \leq b$ implies that $a^r \leq b^r$ for real numbers a, b and a positive integer r, when does $0 \leq A \leq B$ imply $A^r \leq B^r$ for positive-semidefinite matrices A and B and with the positive-semidefinite ordering?

Questions of these types have occupied matrix theorists for at least a century, with motivation from diverse applications. The existing scope and depth of knowledge are enormous. Taken together, this body of knowledge provides a powerful framework for developing and analyzing models for scientific and engineering applications.

This book is intended to be useful to at least four groups of readers. Since linear algebra is a standard course in the mathematical sciences and engineering, graduate students in these fields can use this book to expand the scope of their linear algebra text. For instructors, many of the facts can be used as exercises to augment standard material in matrix courses. For researchers in the mathematical sciences, including statistics, physics, and engineering, this book can be used as a general reference on matrix theory. Finally, for users of matrices in the applied sciences, this book will provide access to a large body of results in matrix theory. By collecting these results in a single source, it is my hope that this book will prove to be convenient and useful for a broad range of applications. The material in this book is thus intended to complement the large number of classical and modern texts and reference works on linear algebra and matrix theory [11, 384, 516, 554, 555, 572, 600, 719, 812, 897, 964, 981, 988, 1033, 1072, 1078, 1125, 1172, 1225, 1269].

After a review of mathematical preliminaries in Chapter 1, fundamental properties of matrices are described in Chapter 2. Chapter 3 summarizes the major classes of matrices and various matrix transformations. In Chapter 4 we turn to polynomial and rational matrices whose basic properties are essential for understanding the structure of constant matrices. Chapter 5 is concerned with various decompositions of matrices including the Jordan, Schur, and singular value decompositions. Chapter 6 provides a brief treatment of generalized inverses, while Chapter 7 describes the Kronecker and Schur product operations. Chapter 8 is concerned with the properties of positive-semidefinite matrices. A detailed treatment of vector and matrix norms is given in Chapter 9, while formulas for matrix derivatives are given in Chapter 10. Next, Chapter 11 focuses on the matrix exponential and stability theory, which are central to the study of linear differential equations. In Chapter 12 we apply matrix theory to the analysis of linear systems, their state space realizations, and their transfer function representation. This chapter also includes a discussion of the matrix Riccati equation of control theory.

Each chapter provides a core of results with, in many cases, complete proofs. Sections at the end of each chapter provide a collection of Facts organized to correspond to the order of topics in the chapter. These Facts include corollaries and special cases of results presented in the chapter, as well as related results that go beyond the results of the chapter. In some cases the Facts include open problems, illuminating remarks, and hints regarding proofs. The Facts are intended to provide the reader with a useful reference collection of matrix results as well as a gateway to the matrix theory literature.

Acknowledgments

The writing of this book spanned more than a decade and a half, during which time numerous individuals contributed both directly and indirectly. I am grateful for the helpful comments of many people who contributed technical material and insightful suggestions, all of which greatly improved the presentation and content of the book. In addition, numerous individuals generously agreed to read sections or chapters of the book for clarity and accuracy. I wish to thank Jasim Ahmed, Suhail Akhtar, David Bayard, Sanjay Bhat, Tony Bloch, Peter Bullen, Steve Campbell, Agostino Capponi, Ramu Chandra, Jaganath Chandrasekhar, Nalin Chaturvedi, Vijay Chellaboina, Jie Chen, David Clements, Dan Davison,

Dimitris Dimogianopoulos, Jiu Ding, D. Z. Djokovic, R. Scott Erwin, R. W. Fare-brother, Danny Georgiev, Joseph Grcar, Wassim Haddad, Yoram Halevi, Jesse Hoagg, Roger Horn, David Hyland, Iman Izadi, Pierre Kabamba, Vikram Kapila, Fuad Kittaneh, Seth Lacy, Thomas Laffey, Cedric Langbort, Alan Laub, Alexander Leonessa, Kai-Yew Lum, Pertti Makila, Roy Mathias, N. Harris McClamroch, Boris Mordukhovich, Sergei Nersesov, JinHyoung Oh, Concetta Pilotto, Harish Palanthandalum-Madapusi, Michael Piovoso, Leiba Rodman, Phil Roe, Carsten Scherer, Wasin So, Andy Sparks, Edward Tate, Yongge Tian, Panagiotis Tsiotras, Feng Tyan, Ravi Venugopal, Jan Willems, Hong Wong, Vera Zeidan, Xingzhi Zhan, and Fuzhen Zhang for their assistance. Nevertheless, I take full responsibility for any remaining errors, and I encourage readers to alert me to any mistakes, corrections of which will be posted on the web. Solutions to the open problems are also welcome.

Portions of the manuscript were typed by Jill Straehla and Linda Smith at Harris Corporation, and by Debbie Laird, Kathy Stolaruk, and Suzanne Smith at the University of Michigan. John Rogosich of Techsetters, Inc., provided invaluable assistance with LATEX issues, and Jennifer Slater carefully copyedited the entire manuscript. I also thank JinHyoung Oh and Joshua Kang for writing C code to refine the index.

I especially thank Vickie Kearn of Princeton University Press for her wise guidance and constant encouragement. Vickie managed to address all of my concerns and anxieties, and helped me improve the manuscript in many ways.

Finally, I extend my greatest appreciation for the (uncountably) infinite patience of my family, who endured the days, weeks, months, and years that this project consumed. The writing of this book began with toddlers and ended with a teenager and a twenty-year old. We can all be thankful it is finally finished.

Dennis S. Bernstein
Ann Arbor, Michigan
dsbaero@umich.edu
January 2005

Special Symbols

General Notation

π	$3.14159\ldots$
e	$2.71828\ldots$
$\triangleq$	equals by definition
$\lim_{\varepsilon\downarrow 0}$	limit from the right
$\binom{\alpha}{m}$	$\frac{\alpha(\alpha-1)\cdots(\alpha-m+1)}{m!}$
$\binom{n}{m}$	$\frac{n!}{m!(n-m)!}$
$\lfloor a \rfloor$	largest integer less than or equal to a
δ_{ij}	1 if $i = j$, 0 if $i \neq j$ (Kronecker delta)
$\log$	logarithm with base e
$\operatorname{sign}\alpha$	1 if $\alpha > 0$, -1 if $\alpha < 0$, 0 if $\alpha = 0$

Chapter 1

$\{\ \}$	set (p. 2)
$\in$	is an element of (p. 2)
$\notin$	is not an element of (p. 2)
$\varnothing$	empty set (p. 2)
$\{\ \}_{\text{ms}}$	multiset (p. 2)
card	cardinality (p. 2)
$\cap$	intersection (p. 2)
$\cup$	union (p. 2)
$\mathcal{Y}\backslash\mathcal{X}$	complement of $\mathcal{X}$ relative to $\mathcal{Y}$ (p. 2)
$\mathcal{X}^{\sim}$	complement of $\mathcal{X}$ (p. 3)

$\subseteq$	is a subset of (p. 3)
$\subset$	is a proper subset of (p. 3)
$(x_1, \ldots, x_n)$	tuple or n-tuple (p. 3)
$\mathbb{Z}$	integers (p. 3)
$\mathbb{N}$	nonnegative integers (p. 3)
$\mathbb{P}$	positive integers (p. 3)
$\mathbb{R}$	real numbers (p. 3)
$\mathbb{C}$	complex numbers (p. 3)
$\jmath$	$\sqrt{-1}$ (p. 3)
$\overline{z}$	complex conjugate of $z \in \mathbb{C}$ (p. 4)
$\mathrm{Re}\, z$	real part of $z \in \mathbb{C}$ (p. 4)
$\mathrm{Im}\, z$	imaginary part of $z \in \mathbb{C}$ (p. 4)
$|z|$	absolute value of $z \in \mathbb{C}$ (p. 4)
OLHP	open left half plane in $\mathbb{C}$ (p. 4)
CLHP	closed left half plane in $\mathbb{C}$ (p. 4)
ORHP	open right half plane in $\mathbb{C}$ (p. 4)
CRHP	closed right half plane in $\mathbb{C}$ (p. 4)
$\jmath\mathbb{R}$	imaginary numbers (p. 4)
OUD	open unit disk in $\mathbb{C}$ (p. 4)
CUD	closed unit disk in $\mathbb{C}$ (p. 4)
CPP	closed punctured plane in $\mathbb{C}$ (p. 4)
OPP	open punctured plane in $\mathbb{C}$ (p. 4)
$\mathbb{F}$	$\mathbb{R}$ or $\mathbb{C}$ (p. 4)
$f\colon \mathcal{X} \mapsto \mathcal{Y}$	f is a function with domain $\mathcal{X}$ and codomain $\mathcal{Y}$ (p. 4)
$\mathrm{Graph}(f)$	$\{(x, f(x))\colon\ x \in \mathcal{X}\}$ (p. 4)
$f \bullet g$	composition of functions f and g (p. 4)
$f^{-1}(\mathcal{S})$	inverse image of $\mathcal{S}$ (p. 5)
$\mathrm{rev}(\mathcal{R})$	reversal of the relation $\mathcal{R}$ (p. 7)
$\mathcal{R}^{\sim}$	complement of the relation $\mathcal{R}$ (p. 7)
$\mathrm{ref}(\mathcal{R})$	reflexive hull of the relation $\mathcal{R}$ (p. 7)
$\mathrm{sym}(\mathcal{R})$	symmetric hull of the relation $\mathcal{R}$ (p. 7)
$\mathrm{trans}(\mathcal{R})$	transitive hull of the relation $\mathcal{R}$ (p. 7)

equiv($\mathcal{R}$)	equivalence hull of the relation $\mathcal{R}$ (p. 7)
$x \overset{\mathcal{R}}{=} y$	(x, y) is an element of the equivalence relation $\mathcal{R}$ (p. 7)
glb($\mathcal{S}$)	greatest lower bound of $\mathcal{S}$ (p. 8, Definition 1.5.9)
lub($\mathcal{S}$)	least upper bound of $\mathcal{S}$ (p. 8, Definition 1.5.9)
inf($\mathcal{S}$)	infimum of $\mathcal{S}$ (p. 9, Definition 1.5.9)
sup($\mathcal{S}$)	supremum of $\mathcal{S}$ (p. 9, Definition 1.5.9)
rev($\mathcal{G}$)	reversal of the graph $\mathcal{G}$ (p. 9)
$\mathcal{G}^\sim$	complement of the graph $\mathcal{G}$ (p. 9)
ref($\mathcal{G}$)	reflexive hull of the graph $\mathcal{G}$ (p. 9)
sym($\mathcal{G}$)	symmetric hull of the graph $\mathcal{G}$ (p. 9)
trans($\mathcal{G}$)	transitive hull of the graph $\mathcal{G}$ (p. 9)
equiv($\mathcal{G}$)	equivalence hull of the graph $\mathcal{G}$ (p. 9)
indeg(x)	indegree of the node x (p. 10)
outdeg(x)	outdegree of the node x (p. 10)
deg(x)	degree of the node x (p. 10)

Chapter 2

$\mathbb{R}^n$	$\mathbb{R}^{n \times 1}$ (real column vectors) (p. 85)
$\mathbb{C}^n$	$\mathbb{C}^{n \times 1}$ (complex column vectors) (p. 85)
$\mathbb{F}^n$	$\mathbb{R}^n$ or $\mathbb{C}^n$ (p. 85)
$x_{(i)}$	ith component of $x \in \mathbb{F}^n$ (p. 85)
$x \geq\geq y$	$x_{(i)} \geq y_{(i)}$ for all i ($x - y$ is nonnegative) (p. 86)
$x >> y$	$x_{(i)} > y_{(i)}$ for all i ($x - y$ is positive) (p. 86)
$\mathbb{R}^{n \times m}$	$n \times m$ real matrices (p. 86)
$\mathbb{C}^{n \times m}$	$n \times m$ complex matrices (p. 86)
$\mathbb{F}^{n \times m}$	$\mathbb{R}^{n \times m}$ or $\mathbb{C}^{n \times m}$ (p. 86)
row$_i(A)$	ith row of A (p. 87)
col$_i(A)$	ith column of A (p. 87)
$A_{(i,j)}$	(i, j) entry of A (p. 87)

$A \overset{i}{\leftarrow} b$

matrix obtained from $A \in \mathbb{F}^{n \times m}$ by replacing $\mathrm{col}_i(A)$ with $b \in \mathbb{F}^n$ or $\mathrm{row}_i(A)$ with $b \in \mathbb{F}^{1 \times m}$ (p. 87)

$\mathrm{d}_{\max}(A) \triangleq \mathrm{d}_1(A)$

largest diagonal entry of $A \in \mathbb{F}^{n \times n}$ having real diagonal entries (p. 87)

$\mathrm{d}_i(A)$

ith largest diagonal entry of $A \in \mathbb{F}^{n \times n}$ having real diagonal entries (p. 87)

$\mathrm{d}_{\min}(A) \triangleq \mathrm{d}_n(A)$

smallest diagonal entry of $A \in \mathbb{F}^{n \times n}$ having real diagonal entries (p. 87)

$A_{(\mathcal{S}_1, \mathcal{S}_2)}$

submatrix of A formed by retaining the rows of A listed in $\mathcal{S}_1$ and the columns of A listed in $\mathcal{S}_2$ (p. 88)

$A_{(\mathcal{S})}$

$A_{(\mathcal{S}, \mathcal{S})}$ (p. 88)

$A \geq\geq B$

$A_{(i,j)} \geq B_{(i,j)}$ for all i, j ($A - B$ is nonnegative) (p. 88)

$A >> B$

$A_{(i,j)} > B_{(i,j)}$ for all i, j ($A - B$ is positive) (p. 88)

$[A, B]$

commutator $AB - BA$ (p. 89)

$\mathrm{ad}_A(X)$

adjoint operator $[A, X]$ (p. 89)

$x \times y$

cross product of vectors $x, y \in \mathbb{R}^3$ (p. 89)

$K(x)$

cross-product matrix for $x \in \mathbb{R}^3$ (p. 90)

$0_{n \times m}, 0$

$n \times m$ zero matrix (p. 90)

I_n, I

$n \times n$ identity matrix (p. 91)

$\hat{I}_n, \hat{I}$

$n \times n$ reverse permutation matrix

$$\begin{bmatrix} & & 1 \\ & \cdot^{\cdot^{\cdot}} & \\ 1 & & 0 \end{bmatrix} \text{(p. 91)}$$

P_n

$n \times n$ cyclic permutation matrix (p. 91)

N_n, N

$n \times n$ standard nilpotent matrix (p. 92)

$e_{i,n}, e_i$

$\mathrm{col}_i(I_n)$ (p. 92)

$E_{i,j,n \times m}, E_{i,j}$

$e_{i,n} e_{j,m}^{\mathrm{T}}$ (p. 92)

$1_{n \times m}$

$n \times m$ ones matrix (p. 92)

A^{T}

transpose of A (p. 94)

$\mathrm{tr}\, A$

trace of A (p. 94)

$\overline{C}$

complex conjugate of $C \in \mathbb{C}^{n \times m}$ (p. 95)

A^*

$\overline{A}^{\mathrm{T}}$ conjugate transpose of A (p. 95)

$\mathrm{Re}\, A$

real part of $A \in \mathbb{F}^{n \times m}$ (p. 95)

$\operatorname{Im} A$	imaginary part of $A \in \mathbb{F}^{n \times m}$ (p. 95)
$\overline{\mathcal{S}}$	$\{\overline{Z}: \ Z \in \mathcal{S}\}$ or $\{\overline{Z}: \ Z \in \mathcal{S}\}_{\mathrm{ms}}$ (p. 95)
$A^{\hat{\mathrm{T}}}$	$\hat{I} A^{\mathrm{T}} \hat{I}$ reverse transpose of A (p. 96)
$A^{\hat{*}}$	$\hat{I} A^{*} \hat{I}$ reverse complex conjugate transpose of A (p. 96)
$\lvert x \rvert$	absolute value of $x \in \mathbb{F}^n$ (p. 96)
$\lvert A \rvert$	absolute value of $A \in \mathbb{F}^{n \times n}$ (p. 96)
$\operatorname{sign} x$	sign of $x \in \mathbb{R}^n$ (p. 97)
$\operatorname{sign} A$	sign of $A \in \mathbb{R}^{n \times n}$ (p. 97)
$\operatorname{co} \mathcal{S}$	convex hull of $\mathcal{S}$ (p. 98)
$\operatorname{cone} \mathcal{S}$	conical hull of $\mathcal{S}$ (p. 98)
$\operatorname{coco} \mathcal{S}$	convex conical hull of $\mathcal{S}$ (p. 98)
$\operatorname{span} \mathcal{S}$	span of $\mathcal{S}$ (p. 98)
$\operatorname{aff} \mathcal{S}$	affine hull of $\mathcal{S}$ (p. 98)
$\operatorname{dim} \mathcal{S}$	dimension of $\mathcal{S}$ (p. 98)
$\mathcal{S}^{\perp}$	orthogonal complement of $\mathcal{S}$ (p. 99)
$\operatorname{polar} \mathcal{S}$	polar of $\mathcal{S}$ (p. 99)
$\operatorname{dcone} \mathcal{S}$	dual cone of $\mathcal{S}$ (p. 99)
$\mathcal{R}(A)$	range of A (p. 101)
$\mathcal{N}(A)$	null space of A (p. 102)
$\operatorname{rank} A$	rank of A (p. 104)
$\operatorname{def} A$	defect of A (p. 104)
A^{L}	left inverse of A (p. 106)
A^{R}	right inverse of A (p. 106)
A^{-1}	inverse of A (p. 110)
$A^{-\mathrm{T}}$	$\left(A^{\mathrm{T}}\right)^{-1}$ (p. 111)
A^{-*}	$\left(A^{*}\right)^{-1}$ (p. 111)
$\det A$	determinant of A (p. 112)
$A_{[i;j]}$	submatrix $A_{(\{i\}^{\sim},\{j\}^{\sim})}$ of A obtained by deleting $\operatorname{row}_i(A)$ and $\operatorname{col}_j(A)$ (p. 114)
A^{A}	adjugate of A (p. 114)
$A \overset{\mathrm{rs}}{\leq} B$	rank subtractivity partial ordering (p. 129, Fact 2.10.32)

$A \overset{*}{\leq} B$ star partial ordering (p. 130, Fact 2.10.35)

Chapter 3

$\mathrm{diag}(a_1, \ldots, a_n)$ $\begin{bmatrix} a_1 & & 0 \\ & \ddots & \\ 0 & & a_n \end{bmatrix}$ (p. 181)

$\mathrm{revdiag}(a_1, \ldots, a_n)$ $\begin{bmatrix} 0 & & a_1 \\ & \iddots & \\ a_n & & 0 \end{bmatrix}$ (p. 181)

$\mathrm{diag}(A_1, \ldots, A_k)$ block-diagonal matrix $\begin{bmatrix} A_1 & & 0 \\ & \ddots & \\ 0 & & A_k \end{bmatrix}$, where

$A_i \in \mathbb{F}^{n_i \times m_i}$ (p. 181)

J_{2n}, J $\begin{bmatrix} 0 & I_n \\ -I_n & 0 \end{bmatrix}$ (p. 183)

$\mathrm{gl}_{\mathbb{F}}(n)$, $\mathrm{pl}_{\mathbb{C}}(n)$, $\mathrm{sl}_{\mathbb{F}}(n)$,   Lie algebras (p. 185)
$\mathrm{u}(n)$, $\mathrm{su}(n)$, $\mathrm{so}(n)$,
$\mathrm{symp}_{\mathbb{F}}(2n)$, $\mathrm{osymp}_{\mathbb{F}}(2n)$,
$\mathrm{aff}_{\mathbb{F}}(n)$, $\mathrm{se}_{\mathbb{F}}(n)$, $\mathrm{trans}_{\mathbb{F}}(n)$

$\mathcal{S}_1 \approx \mathcal{S}_2$ the groups $\mathcal{S}_1$ and $\mathcal{S}_2$ are isomorphic (p. 186)

$\mathrm{GL}_{\mathbb{F}}(n)$, $\mathrm{PL}_{\mathbb{F}}(n)$, $\mathrm{SL}_{\mathbb{F}}(n)$,   groups (p. 187)
$\mathrm{U}(n)$, $\mathrm{O}(n)$, $\mathrm{U}(n,m)$,
$\mathrm{O}(n,m)$, $\mathrm{SU}(n)$, $\mathrm{SO}(n)$,
$\mathrm{P}(n)$, $\mathrm{A}(n)$, $\mathrm{D}(n)$, $\mathrm{C}(n)$,
$\mathrm{Symp}_{\mathbb{F}}(2n)$, $\mathrm{OSymp}_{\mathbb{F}}(2n)$,
$\mathrm{Aff}_{\mathbb{F}}(n)$, $\mathrm{SE}_{\mathbb{F}}(n)$, $\mathrm{Trans}_{\mathbb{F}}(n)$

$A_{\perp}$ complementary idempotent matrix or
                  projector $I - A$ corresponding to the
                  idempotent matrix or projector A (p. 190)

$\mathrm{ind}\, A$ index of A (p. 190)

$\mathbb{H}$ quaternions (p. 247, Fact 3.24.1)

$\mathrm{Sp}(n)$ symplectic group in $\mathbb{H}$ (p. 249, Fact 3.24.4)

Chapter 4

$\mathbb{F}[s]$ polynomials with coefficients in $\mathbb{F}$ (p. 253)

$\deg p$ degree of $p \in \mathbb{F}[s]$ (p. 253)

$\mathrm{mroots}(p)$ multiset of roots of $p \in \mathbb{F}[s]$ (p. 254)

roots(p)	set of roots of $p \in \mathbb{F}[s]$ (p. 254)
mult$_p(\lambda)$	multiplicity of λ as a root of $p \in \mathbb{F}[s]$ (p. 254)
$\mathbb{F}^{n \times m}[s]$	$n \times m$ matrices with entries in $\mathbb{F}[s]$ ($n \times m$ polynomial matrices with coefficients in $\mathbb{F}$) (p. 256)
rank P	rank of $P \in \mathbb{F}^{n \times m}[s]$ (p. 257)
Szeros(P)	set of Smith zeros of $P \in \mathbb{F}^{n \times m}[s]$ (p. 259)
mSzeros(P)	multiset of Smith zeros of $P \in \mathbb{F}^{n \times m}[s]$ (p. 259)
χ_A	characteristic polynomial of A (p. 262)
$\lambda_{\max}(A) \triangleq \lambda_1(A)$	largest eigenvalue of $A \in \mathbb{F}^{n \times n}$ having real eigenvalues (p. 262)
$\lambda_i(A)$	ith largest eigenvalue of $A \in \mathbb{F}^{n \times n}$ having real eigenvalues (p. 262)
$\lambda_{\min}(A) \triangleq \lambda_n(A)$	smallest eigenvalue of $A \in \mathbb{F}^{n \times n}$ having real eigenvalues (p. 262)
amult$_A(\lambda)$	algebraic multiplicity of $\lambda \in \text{spec}(A)$ (p. 262)
spec(A)	spectrum of A (p. 262)
mspec(A)	multispectrum of A (p. 262)
gmult$_A(\lambda)$	geometric multiplicity of $\lambda \in \text{spec}(A)$ (p. 267)
spabs(A)	spectral abscissa of A (p. 267)
sprad(A)	spectral radius of A (p. 267)
$\nu_-(A), \nu_0(A), \nu_+(A)$	number of eigenvalues of A counting algebraic multiplicity having negative, zero, and positive real part, respectively (p. 267)
In A	inertia of A, that is, $[\nu_-(A) \; \nu_0(A) \; \nu_+(A)]^{\mathrm{T}}$ (p. 267)
sig A	signature of A, that is, $\nu_+(A) - \nu_-(A)$ (p. 267)
μ_A	minimal polynomial of A (p. 269)
$\mathbb{F}(s)$	rational functions with coefficients in $\mathbb{F}$ (SISO rational transfer functions) (p. 271)
$\mathbb{F}_{\text{prop}}(s)$	proper rational functions with coefficients in $\mathbb{F}$ (SISO proper rational transfer functions) (p. 271)
reldeg g	relative degree of $g \in \mathbb{F}_{\text{prop}}(s)$ (p. 271)
$\mathbb{F}^{n \times m}(s)$	$n \times m$ matrices with entries in $\mathbb{F}(s)$ (MIMO rational transfer functions) (p. 271)

Chapter 5

$\mathrm{circ}(a_0, \ldots, a_{n-1})$ circulant matrix of $a_0, \ldots, a_{n-1} \in \mathbb{F}$ (p. 388, Fact 5.16.7)

Chapter 6

A^+ (Moore-Penrose) generalized inverse of A (p. 397)

$D|\mathcal{A}$ Schur complement of D with respect to $\mathcal{A}$ (p. 401)

A^{D} Drazin generalized inverse of A (p. 401)

$A^{\#}$ group generalized inverse of A (p. 403)

Chapter 7

$\mathrm{vec}\,A$ vector formed by stacking columns of A (p. 439)

$\otimes$ Kronecker product (p. 440)

$P_{n,m}$ Kronecker permutation matrix (p. 442)

$\oplus$ Kronecker sum (p. 443)

$A \circ B$ Schur product of A and B (p. 444)

$A^{\circ \alpha}$ Schur power of A, $(A^{\circ \alpha})_{(i,j)} = (A_{(i,j)})^{\alpha}$ (p. 444)

Chapter 8

$\mathbf{H}^n$ $n \times n$ Hermitian matrices (p. 459)

$\mathbf{N}^n$ $n \times n$ positive-semidefinite matrices (p. 459)

$\mathbf{P}^n$ $n \times n$ positive-definite matrices (p. 459)

$A \geq B$ $A - B \in \mathbf{N}^n$ (p. 459)

$A > B$ $A - B \in \mathbf{P}^n$ (p. 459)

$\langle A \rangle$ $(A^*A)^{1/2}$ (p. 474)

$A \# B$ geometric mean of A and B (p. 508, Fact 8.10.43)

$A \#_\alpha B$ generalized geometric mean of A and B (p. 510, Fact 8.10.45)

$A : B$ parallel sum of A and B (p. 581, Fact 8.21.18)

$\mathrm{sh}(A, B)$ shorted operator (p. 582, Fact 8.21.19)

Chapter 9

Chapter 10

$f'(x)$	derivative of f at x (p. 686)
$\dfrac{\mathrm{d}f(x_0)}{\mathrm{d}x_{(i)}}$	$f'(x_0)$ (p. 686)
$f^{(k)}(x)$	kth derivative of f at x (p. 688)
$\dfrac{\mathrm{d}^+ f(x_0)}{\mathrm{d}x_{(i)}}$	right one-sided derivative (p. 688)
$\dfrac{\mathrm{d}^- f(x_0)}{\mathrm{d}x_{(i)}}$	left one-sided derivative (p. 688)
$\mathrm{Sign}(A)$	matrix sign of $A \in \mathbb{C}^{n \times n}$ (p. 690)

Chapter 11

e^A or $\exp(A)$	matrix exponential (p. 707)
$\mathcal{L}$	Laplace transform (p. 710)
$\mathcal{S}_\mathrm{s}(A)$	asymptotically stable subspace of A (p. 729)
$\mathcal{S}_\mathrm{u}(A)$	unstable subspace of A (p. 729)

Chapter 12

$\mathcal{U}(A,C)$	unobservable subspace of (A,C) (p. 800)
$\mathcal{O}(A,C)$	$\begin{bmatrix} C \\ CA \\ CA^2 \\ \vdots \\ CA^{n-1} \end{bmatrix}$ (p. 801)
$\mathcal{C}(A,B)$	controllable subspace of (A,B) (p. 809)
$\mathcal{K}(A,B)$	$\begin{bmatrix} B & AB & A^2B & \cdots & A^{n-1}B \end{bmatrix}$ (p. 809)
$G \sim \left[\begin{array}{c|c} A & B \\ \hline C & D \end{array}\right]$	state space realization of $G \in \mathbb{F}^{l \times m}_{\mathrm{prop}}[s]$ (p. 822)
$\mathcal{H}_{i,j,k}(G)$	Markov block-Hankel matrix $\mathcal{O}_i(A,C)\mathcal{K}_j(A,B)$ (p. 826)
$\mathcal{H}(G)$	Markov block-Hankel matrix $\mathcal{O}(A,C)\mathcal{K}(A,B)$ (p. 827)
$G \overset{\min}{\sim} \left[\begin{array}{c|c} A & B \\ \hline C & D \end{array}\right]$	state space realization of $G \in \mathbb{F}^{l \times m}_{\mathrm{prop}}[s]$ (p. 828)
$\mathcal{H}$	Hamiltonian $\begin{bmatrix} A & \Sigma \\ R_1 & -A^\mathrm{T} \end{bmatrix}$ (p. 853)

Conventions, Notation, and Terminology

The reader is encouraged to review this section in order to ensure correct interpretation of the statements in this book.

When a word is defined, it is italicized.

The definition of a word, phrase, or symbol should always be understood as an "if and only if" statement, although for brevity "only if" is omitted. The symbol $\triangleq$ means equal by definition, where $A \triangleq B$ means that the left-hand expression A is defined to be the right-hand expression B.

A mathematical object defined by a constructive procedure is *well defined* if the constructive procedure produces a uniquely defined object.

Analogous statements are written in parallel using the following style: If n is (even, odd), then $n + 1$ is (odd, even).

The variables i, j, k, l, m, n always denote integers. Hence, $k \geq 0$ denotes a nonnegative integer, $k \geq 1$ denotes a positive integer, and the limit $\lim_{k \to \infty} A^k$ is taken over positive integers.

The imaginary unit $\sqrt{-1}$ is always denoted by dotless $\jmath$.

The letter s always represents a complex scalar. The letter z may or may not represent a complex scalar.

The inequalities $c \leq a \leq d$ and $c \leq b \leq d$ are written simultaneously as

$$c \leq \left\{ \begin{array}{c} a \\ b \end{array} \right\} \leq d.$$

The prefix "non" means "not" in the words nonconstant, nonempty, nonintegral, nonnegative, nonreal, nonsingular, nonsquare, nonunique, and nonzero. In some traditional usage, "non" may mean "not necessarily."

"Unique" means "exactly one."

"Increasing" and "decreasing" indicate strict change for a change in the argument. The word "strict" is superfluous, and thus is omitted. Nonincreasing means nowhere increasing, while nondecreasing means nowhere decreasing.

A set can have a finite or infinite number of elements. A finite set has a finite number of elements.

Multisets can have repeated elements. Hence, $\{x\}_{\text{ms}}$ and $\{x, x\}_{\text{ms}}$ are different. The listed elements α, β, γ of the conventional set $\{\alpha, \beta, \gamma\}$ need not be distinct. For example, $\{\alpha, \beta, \alpha\} = \{\alpha, \beta\}$.

In statements of the form "Let $\text{spec}(A) = \{\lambda_1, \ldots, \lambda_r\}$," the listed elements $\lambda_1, \ldots, \lambda_r$ are assumed to be distinct.

Square brackets are used alternately with parentheses. For example, $f[g(x)]$ denotes $f(g(x))$.

The order in which the elements of the set $\{x_1, \ldots, x_n\}$ and the elements of the multiset $\{x_1, \ldots, x_n\}_{\text{ms}}$ are listed has no significance. The components of the n-tuple $(x_1, \ldots, x_n)$ are ordered.

The notation $(x_i)_{i=1}^{\infty}$ denotes the sequence $(x_1, x_2, \ldots)$. A sequence can be viewed as a tuple with a countably infinite number of components, where the order of the components is relevant and the components need not be distinct.

The composition of functions f and g is denoted by $f \bullet g$. The traditional notation $f \circ g$ is reserved for the Schur product.

$\mathcal{S}_1 \subset \mathcal{S}_2$ means that $\mathcal{S}_1$ is a proper subset of $\mathcal{S}_2$, whereas $\mathcal{S}_1 \subseteq \mathcal{S}_2$ means that $\mathcal{S}_1$ is either a proper subset of $\mathcal{S}_2$ or is equal to $\mathcal{S}_2$. Hence, $\mathcal{S}_1 \subset \mathcal{S}_2$ is equivalent to $\mathcal{S}_1 \subseteq \mathcal{S}_2$ and $\mathcal{S}_1 \neq \mathcal{S}_2$, while $\mathcal{S}_1 \subseteq \mathcal{S}_2$ is equivalent to either $\mathcal{S}_1 \subset \mathcal{S}_2$ or $\mathcal{S}_1 = \mathcal{S}_2$.

The terminology "graph" corresponds to what is commonly called a "simple directed graph," while the terminology "symmetric graph" corresponds to a "simple undirected graph."

The range of $\cos^{-1}$ is $[0, \pi]$, the range of $\sin^{-1}$ is $[-\pi/2, \pi/2]$, the range of $\tan^{-1}$ is $(-\pi/2, \pi/2)$, and the range of $\cot^{-1}$ is $(0, \pi)$.

The *angle between two vectors* is an element of $[0, \pi]$. Therefore, by using $\cos^{-1}$, the inner product of two vectors can be used to compute the angle between two vectors.

$0! \triangleq 1$, $0/0 = (\sin 0)/0 = (1 - \cos 0)/0 = (\sinh 0)/0 \triangleq 1$, and $1/\infty \triangleq 0$.

For all $\alpha \in \mathbb{C}$, $\binom{\alpha}{0} \triangleq 1$. For all $k \in \mathbb{N}$, $\binom{0}{k} \triangleq 1$.

For all square matrices A, $A^0 \triangleq I$. In particular, $0_{n\times n}^0 \triangleq I_n$. With this convention, it is possible to write

$$\sum_{i=0}^{\infty} \alpha^i = \frac{1}{1-\alpha}$$

for all $-1 < \alpha < 1$. Of course, $\lim_{x\downarrow 0} 0^x = 0$, $\lim_{x\downarrow 0} x^0 = 1$, and $\lim_{x\downarrow 0} x^x = 1$.

Neither ∞ nor $-\infty$ is a real number. However, some operations are defined for these objects as extended real numbers, such as $\infty + \infty = \infty$, $\infty\infty = \infty$, and, for all nonzero real numbers α, $\alpha\infty = \text{sign}(\alpha)\infty$. 0∞ and $\infty - \infty$ are not defined. See [71, pp. 14, 15].

Let a and b be real numbers such that $a < b$. A *finite interval* is of the form (a, b), $[a, b)$, $(a, b]$, or $[a, b]$, whereas an *infinite interval* is of the form $(-\infty, a)$, $(-\infty, a]$, (a, ∞), $[a, \infty)$, or $(-\infty, \infty)$. An *interval* is either a finite interval or an infinite interval. An *extended infinite interval* includes either ∞ or $-\infty$. For example, $[-\infty, a)$ and $[-\infty, a]$ include $-\infty$, $(a, \infty]$ and $[a, \infty]$ include ∞, and $[-\infty, \infty]$ includes $-\infty$ and ∞.

The symbol $\mathbb{F}$ denotes either $\mathbb{R}$ or $\mathbb{C}$ consistently in each result. For example, in Theorem 5.6.3, the three appearances of "$\mathbb{F}$" can be read as either all "$\mathbb{C}$" or all "$\mathbb{R}$."

The imaginary numbers are denoted by $_J\mathbb{R}$. Hence, 0 is both a real number and an imaginary number.

The notation $\text{Re}\,A$ and $\text{Im}\,A$ represents the real and imaginary parts of A, respectively. Some books use $\text{Re}\,A$ and $\text{Im}\,A$ to denote the Hermitian and skew-Hermitian matrices $\frac{1}{2}(A + A^*)$ and $\frac{1}{2}(A - A^*)$.

For the scalar ordering "$\leq$," if $x \leq y$, then $x < y$ if and only if $x \neq y$. For the entrywise vector and matrix orderings, $x \leq y$ and $x \neq y$ do not imply that $x < y$.

Operations denoted by superscripts are applied before operations represented by preceding operators. For example, $\text{tr}\,(A+B)^2$ means $\text{tr}\left[(A + B)^2\right]$ and $\text{cl}\,\mathcal{S}^\sim$ means $\text{cl}(\mathcal{S}^\sim)$. This convention simplifies many formulas.

A vector in $\mathbb{F}^n$ is a column vector, which is also a matrix with one column. In mathematics, "vector" generally refers to an abstract vector not resolved in coordinates.

Sets have elements, vectors and sequences have components, and matrices have entries. This terminology has no mathematical consequence.

The notation $x_{(i)}$ represents the ith component of the vector x.

The notation $A_{(i,j)}$ represents the scalar (i,j) entry of A. $A_{i,j}$ or A_{ij} denotes a block or submatrix of A.

All matrices have nonnegative integral dimensions. If a matrix has either zero rows or zero columns, then the matrix is empty.

The entries of a submatrix $\hat{A}$ of a matrix A are the entries of A located in specified rows and columns. $\hat{A}$ is a block of A if $\hat{A}$ is a submatrix of A whose entries are entries of adjacent rows and columns of A. Every matrix is both a submatrix and block of itself.

The determinant of a submatrix is a subdeterminant. Some books use "minor." The determinant of a matrix is also a subdeterminant of the matrix.

The dimension of the null space of a matrix is its defect. Some books use "nullity."

A block of a square matrix is diagonally located if the block is square and the diagonal entries of the block are also diagonal entries of the matrix; otherwise, the block is off-diagonally located. This terminology avoids confusion with a "diagonal block," which is a block that is also a square, diagonal submatrix.

For the partitioned matrix $\left[\begin{smallmatrix} A & B \\ C & D \end{smallmatrix}\right] \in \mathbb{F}^{(n+m)\times(k+l)}$, it can be inferred that $A \in \mathbb{F}^{n\times k}$ and similarly for B, C, and D.

The Schur product of matrices A and B is denoted by $A \circ B$. Matrix multiplication is given priority over Schur multiplication, that is, $A \circ BC$ means $A \circ (BC)$.

The adjugate of $A \in \mathbb{F}^{n\times n}$ is denoted by A^{A}. The traditional notation is $\mathrm{adj}\, A$, while the notation A^{A} is used in [1259]. If $A \in \mathbb{F}$ is a scalar then $A^{\mathrm{A}} = 1$. In particular, $0_{1\times1}^{\mathrm{A}} = 1$. However, for all $n \geq 2$, $0_{n\times n}^{\mathrm{A}} = 0_{n\times n}$.

If $\mathbb{F} = \mathbb{R}$, then $\overline{A}$ becomes A, A^* becomes A^{T}, "Hermitian" becomes "symmetric," "unitary" becomes "orthogonal," "unitarily" becomes "orthogonally," and "congruence" becomes "T-congruence." A square complex matrix A is symmetric if $A^{\mathrm{T}} = A$ and orthogonal if $A^{\mathrm{T}}A = I$.

The diagonal entries of a matrix $A \in \mathbb{F}^{n\times n}$ all of whose diagonal entries are real are ordered as $\mathrm{d}_{\max}(A) = \mathrm{d}_1(A) \geq \mathrm{d}_2(A) \geq \cdots \geq \mathrm{d}_n(A) = \mathrm{d}_{\min}(A)$.

Every $n \times n$ matrix has n eigenvalues. Hence, eigenvalues are counted in accordance with their algebraic multiplicity. The phrase "distinct eigenvalues" ignores algebraic multiplicity.

The eigenvalues of a matrix $A \in \mathbb{F}^{n\times n}$ all of whose eigenvalues are real are ordered as $\lambda_{\max}(A) = \lambda_1(A) \geq \lambda_2(A) \geq \cdots \geq \lambda_n(A) = \lambda_{\min}(A)$.

The inertia of a matrix is written as

$$\operatorname{In} A \triangleq \begin{bmatrix} \nu_-(A) \\ \nu_0(A) \\ \nu_+(A) \end{bmatrix}.$$

Some books use the notation $(\nu(A), \delta(A), \pi(A))$.

For $A \in \mathbb{F}^{n \times n}$, $\operatorname{amult}_A(\lambda)$ is the number of copies of λ in the multispectrum of A, $\operatorname{gmult}_A(\lambda)$ is the number of Jordan blocks of A associated with λ, and $\operatorname{ind}_A(\lambda)$ is the order of the largest Jordan block of A associated with λ. The index of A, denoted by $\operatorname{ind} A = \operatorname{ind}_A(0)$, is the order of the largest Jordan block of A associated with the eigenvalue 0.

The matrix $A \in \mathbb{F}^{n \times n}$ is semisimple if the order of every Jordan block of A is 1, and cyclic if A has exactly one Jordan block associated with each of its eigenvalues. Defective means not semisimple, while derogatory means not cyclic.

An $n \times m$ matrix has exactly $\min\{n, m\}$ singular values, exactly $\operatorname{rank} A$ of which are positive.

The $\min\{n, m\}$ singular values of a matrix $A \in \mathbb{F}^{n \times m}$ are ordered as $\sigma_{\max}(A) \triangleq \sigma_1(A) \geq \sigma_2(A) \geq \cdots \geq \sigma_{\min\{n,m\}}(A)$. If $n = m$, then $\sigma_{\min}(A) \triangleq \sigma_n(A)$. The notation $\sigma_{\min}(A)$ is defined only for square matrices.

Positive-semidefinite and positive-definite matrices are Hermitian.

A square matrix with entries in $\mathbb{F}$ is diagonalizable over $\mathbb{F}$ if and only if it can be transformed into a diagonal matrix whose entries are in $\mathbb{F}$ by means of a similarity transformation whose entries are in $\mathbb{F}$. Therefore, a complex matrix is diagonalizable over $\mathbb{C}$ if and only if all of its eigenvalues are semisimple, whereas a real matrix is diagonalizable over $\mathbb{R}$ if and only if all of its eigenvalues are semisimple and real. The real matrix $\begin{bmatrix} 0 & 1 \\ -1 & 0 \end{bmatrix}$ is diagonalizable over $\mathbb{C}$, although it is not diagonalizable over $\mathbb{R}$. The Hermitian matrix $\begin{bmatrix} 1 & \jmath \\ -\jmath & 2 \end{bmatrix}$ is diagonalizable over $\mathbb{C}$, and also has real eigenvalues.

An idempotent matrix $A \in \mathbb{F}^{n \times n}$ satisfies $A^2 = A$, while a projector is a Hermitian, idempotent matrix. Some books use "projector" for idempotent and "orthogonal projector" for projector. A reflector is a Hermitian, involutory matrix. A projector is a normal matrix each of whose eigenvalues is 1 or 0, while a reflector is a normal matrix each of whose eigenvalues is 1 or -1.

An elementary matrix is a nonsingular matrix formed by adding an outer-product matrix to the identity matrix. An elementary reflector is a reflector exactly one of whose eigenvalues is -1. An elementary projector is a projector exactly one of whose eigenvalues is 0. Elementary reflectors are elementary matrices. However, elementary projectors are not elementary matrices since elementary projectors are singular.

A range-Hermitian matrix is a square matrix whose range is equal to the range of its complex conjugate transpose. These matrices are also called "EP" matrices.

The polynomials 1 and $s^3 + 5s^2 - 4$ are monic. The zero polynomial is not monic.

The rank of a polynomial matrix P is the maximum rank of $P(s)$ over $\mathbb{C}$. This quantity is also called the normal rank. We denote this quantity by $\operatorname{rank} P$ as distinct from $\operatorname{rank} P(s)$, which denotes the rank of the matrix $P(s)$.

The rank of a rational transfer function G is the maximum rank of $G(s)$ over $\mathbb{C}$ excluding poles of the entries of G. This quantity is also called the normal rank. We denote this quantity by $\operatorname{rank} G$ as distinct from $\operatorname{rank} G(s)$, which denotes the rank of the matrix $G(s)$.

The symbol $\oplus$ denotes the Kronecker sum. Some books use $\oplus$ to denote the direct sum of matrices or subspaces.

The notation $|A|$ represents the matrix obtained by replacing every entry of A by its absolute value.

The notation $\langle A \rangle$ represents the matrix $(A^*A)^{1/2}$. Some books use $|A|$ to denote this matrix.

The Hölder norms for vectors and matrices are denoted by $\| \cdot \|_p$. The matrix norm induced by $\| \cdot \|_q$ on the domain and $\| \cdot \|_p$ on the codomain is denoted by $\| \cdot \|_{p,q}$.

The Schatten norms for matrices are denoted by $\| \cdot \|_{\sigma p}$, and the Frobenius norm is denoted by $\| \cdot \|_F$. Hence, $\| \cdot \|_{\sigma \infty} = \| \cdot \|_{2,2} = \sigma_{\max}(\cdot)$, $\| \cdot \|_{\sigma 2} = \| \cdot \|_F$, and $\| \cdot \|_{\sigma 1} = \operatorname{tr} \langle \cdot \rangle$.

Terminology Relating to Inequalities

Let "$\leq$" be a partial ordering, let X be a set, and consider the inequality

$$f(x) \leq g(x) \text{ for all } x \in X. \tag{1}$$

Inequality (1) is *sharp* if there exists $x_0 \in X$ such that $f(x_0) = g(x_0)$.

The inequality

$$f(x) \leq f(y) \text{ for all } x \leq y \tag{2}$$

is a monotonicity result.

The inequality

$$f(x) \leq p(x) \leq g(x) \text{ for all } x \in X, \tag{3}$$

where p is not identically equal to either f or g on X, is an *interpolation* or *refinement* of (1). The inequality

$$g(x) \leq \alpha f(x) \text{ for all } x \in X, \tag{4}$$

where $\alpha > 1$, is a *reversal* of (1).

Defining $h(x) \triangleq g(x) - f(x)$, it follows that (1) is equivalent to

$$h(x) \geq 0 \text{ for all } x \in X. \tag{5}$$

Now, suppose that h has a global minimizer $x_0 \in X$. Then, (5) implies that

$$0 \leq h(x_0) = \min_{x \in X} h(x) \leq h(y) \text{ for all } y \in X. \tag{6}$$

Consequently, inequalities are often expressed equivalently in terms of optimization problems, and vice versa.

Many inequalities are based on a single function that is either monotonic or convex.

Matrix Mathematics

Chapter One

Preliminaries

In this chapter we review some basic terminology and results concerning logic, sets, functions, and related concepts. This material is used throughout the book.

1.1 Logic

Every *statement* is either true or false, but not both. Let A and B be statements. The *negation* of A is the statement (not A), the *both* of A and B is the statement (A and B), and the *either* of A and B is the statement (A or B). The statement (A or B) does not contradict (A and B), that is, the word "or" is inclusive. Exclusive "or" is indicated by the phrase "but not both."

The statements "A and B or C" and "A or B and C" are ambiguous. We therefore write "A and either B or C" and "either A or both B and C."

Let A and B be statements. The *implication* statement "if A is satisfied, then B is satisfied" or, equivalently, "A implies B" is written as $A \implies B$, while $A \iff B$ is equivalent to $[(A \implies B) \text{ and } (A \impliedby B)]$. Of course, $A \impliedby B$ means $B \implies A$. A *tautology* is a statement that is true regardless of whether the component statements are true or false. For example, the statement "(A and B) implies A" is a tautology. A *contradiction* is a statement that is false regardless of whether the component statements are true or false. For example, the statement "A implies (not)A" is a contradiction.

Suppose that $A \iff B$. Then, A is satisfied *if and only if* B is satisfied. The implication $A \implies B$ (the "only if" part) is *necessity*, while $B \implies A$ (the "if" part) is *sufficiency*. The *converse* statement of $A \implies B$ is $B \implies A$. The statement $A \implies B$ is equivalent to its *contrapositive* statement (not B) $\implies$ (not A).

A *theorem* is a significant statement, while a *proposition* is a theorem of less significance. The primary role of a *lemma* is to support the proof of a theorem or proposition. Furthermore, a *corollary* is a consequence of a theorem or proposition. Finally, a *fact* is either a theorem, proposition, lemma, or corollary. Theorems, propositions, lemmas, corollaries, and facts are provably true statements.

Suppose that $A' \implies A \implies B \implies B'$. Then, $A' \implies B'$ is a corollary of $A \implies B$.

Let A, B, and C be statements, and assume that $A \implies B$. Then, $A \implies B$ is a *strengthening* of the statement $(A$ and $C) \implies B$. If, in addition, $A \implies C$, then the statement $(A$ and $C) \implies B$ has a *redundant assumption*.

1.2 Sets

A *set* $\{x, y, \ldots\}$ is a collection of elements. A set may have a finite or infinite number of elements. A *finite set* has a finite number of elements.

Let $\mathfrak{X}$ be a set. Then,

$$x \in \mathfrak{X} \tag{1.2.1}$$

means that x is an *element* of $\mathfrak{X}$. If w is not an element of $\mathfrak{X}$, then we write

$$w \notin \mathfrak{X}. \tag{1.2.2}$$

The statement "$x \in \mathfrak{X}$" is either true or false, but not both. The statement "$\mathfrak{X} \notin \mathfrak{X}$" is true by convention, and thus no set can be an element of itself. Therefore, there does not exist a set that contains every set. The set with no elements, denoted by $\varnothing$, is the *empty set*. If $\mathfrak{X} \neq \varnothing$, then $\mathfrak{X}$ is *nonempty*.

A set cannot have repeated elements. For example, $\{x, x\} = \{x\}$. However, a *multiset* is a collection of elements that allows for repetition. The multiset consisting of two copies of x is written as $\{x, x\}_{\text{ms}}$. However, we do not assume that the listed elements x, y of the conventional set $\{x, y\}$ are distinct. The number of distinct elements of the set $\mathcal{S}$ or not-necessarily-distinct elements of the multiset $\mathcal{S}$ is the *cardinality* of $\mathcal{S}$, which is denoted by $\text{card}(\mathcal{S})$.

There are two basic types of mathematical statements for quantifiers. An *existential statement* is of the form

$$\text{there exists } x \in \mathfrak{X} \text{ such that statement } Z \text{ is satisfied,} \tag{1.2.3}$$

while a *universal statement* has the structure

$$\text{for all } x \in \mathfrak{X}, \text{ it follows that statement } Z \text{ is satisfied,} \tag{1.2.4}$$

or, equivalently,

$$\text{statement } Z \text{ is satisfied for all } x \in \mathfrak{X}. \tag{1.2.5}$$

Let $\mathfrak{X}$ and $\mathcal{Y}$ be sets. The *intersection* of $\mathfrak{X}$ and $\mathcal{Y}$ is the set of common elements of $\mathfrak{X}$ and $\mathcal{Y}$ given by

$$\mathfrak{X} \cap \mathcal{Y} \triangleq \{x\colon\ x \in \mathfrak{X} \text{ and } x \in \mathcal{Y}\} = \{x \in \mathfrak{X}\colon\ x \in \mathcal{Y}\} \tag{1.2.6}$$

$$= \{x \in \mathcal{Y}\colon\ x \in \mathfrak{X}\} = \mathcal{Y} \cap \mathfrak{X}, \tag{1.2.7}$$

while the set of elements in either $\mathfrak{X}$ or $\mathcal{Y}$ (the *union* of $\mathfrak{X}$ and $\mathcal{Y}$) is

$$\mathfrak{X} \cup \mathcal{Y} \triangleq \{x\colon\ x \in \mathfrak{X} \text{ or } x \in \mathcal{Y}\} = \mathcal{Y} \cup \mathfrak{X}. \tag{1.2.8}$$

The *complement* of $\mathfrak{X}$ *relative* to $\mathcal{Y}$ is

$$\mathcal{Y} \backslash \mathfrak{X} \triangleq \{x \in \mathcal{Y}\colon\ x \notin \mathfrak{X}\}. \tag{1.2.9}$$

If $\mathcal{Y}$ is specified, then the *complement* of $\mathcal{X}$ is

$$\mathcal{X}^{\sim} \triangleq \mathcal{Y}\backslash\mathcal{X}. \tag{1.2.10}$$

If $x \in \mathcal{X}$ implies that $x \in \mathcal{Y}$, then $\mathcal{X}$ is *contained* in $\mathcal{Y}$ ($\mathcal{X}$ is a *subset* of $\mathcal{Y}$), which is written as

$$\mathcal{X} \subseteq \mathcal{Y}. \tag{1.2.11}$$

The statement $\mathcal{X} = \mathcal{Y}$ is equivalent to the validity of both $\mathcal{X} \subseteq \mathcal{Y}$ and $\mathcal{Y} \subseteq \mathcal{X}$. If $\mathcal{X} \subseteq \mathcal{Y}$ and $\mathcal{X} \neq \mathcal{Y}$, then $\mathcal{X}$ is a *proper subset* of $\mathcal{Y}$ and we write $\mathcal{X} \subset \mathcal{Y}$. The sets $\mathcal{X}$ and $\mathcal{Y}$ are *disjoint* if $\mathcal{X} \cap \mathcal{Y} = \varnothing$. A *partition* of $\mathcal{X}$ is a set of pairwise-disjoint and nonempty subsets of $\mathcal{X}$ whose union is equal to $\mathcal{X}$.

The operations "$\cap$," "$\cup$," and "$\backslash$" and the relations "$\subset$" and "$\subseteq$" extend directly to multisets. For example,

$$\{x, x\}_{\mathrm{ms}} \cup \{x\}_{\mathrm{ms}} = \{x, x, x\}_{\mathrm{ms}}. \tag{1.2.12}$$

By ignoring repetitions, a multiset can be converted to a set, while a set can be viewed as a multiset with distinct elements.

The *Cartesian product* $\mathcal{X}_1 \times \cdots \times \mathcal{X}_n$ of sets $\mathcal{X}_1, \ldots, \mathcal{X}_n$ is the set consisting of *tuples* of the form $(x_1, \ldots, x_n)$, where $x_i \in \mathcal{X}_i$ for all $i \in \{1, \ldots, n\}$. A tuple with n components is an *n-tuple*. Note that the components of an n-tuple are ordered but need not be distinct.

By replacing the logical operations "$\Longrightarrow$," "and," "or," and "not" by "$\subseteq$," "$\cup$," "$\cap$," and "$\sim$," respectively, statements about statements A and B can be transformed into statements about sets $\mathcal{A}$ and $\mathcal{B}$, and vice versa. For example, the tautology

$$A \text{ and } (B \text{ or } C) \iff (A \text{ and } B) \text{ or } (A \text{ and } C)$$

is equivalent to

$$\mathcal{A} \cap (\mathcal{B} \cup \mathcal{C}) = (\mathcal{A} \cap \mathcal{B}) \cup (\mathcal{A} \cap \mathcal{C}).$$

1.3 Integers, Real Numbers, and Complex Numbers

The symbols $\mathbb{Z}$, $\mathbb{N}$, and $\mathbb{P}$ denote the sets of integers, nonnegative integers, and positive integers, respectively. The symbols $\mathbb{R}$ and $\mathbb{C}$ denote the real and complex number fields, respectively, whose elements are *scalars*. Define

$$\jmath \triangleq \sqrt{-1}.$$

Let $x \in \mathbb{C}$. Then, $x = y + \jmath z$, where $y, z \in \mathbb{R}$. Define the *complex conjugate* $\overline{x}$ of x by

$$\overline{x} \triangleq y - \jmath z \tag{1.3.1}$$

and the real part $\mathrm{Re}\, x$ of x and the imaginary part $\mathrm{Im}\, x$ of x by

$$\mathrm{Re}\, x \triangleq \tfrac{1}{2}(x + \overline{x}) = y \tag{1.3.2}$$

and

$$\mathrm{Im}\, x \triangleq \tfrac{1}{\jmath 2}(x - \overline{x}) = z. \tag{1.3.3}$$

Furthermore, the *absolute value* $|x|$ of x is defined by

$$|x| \triangleq \sqrt{y^2 + z^2}. \tag{1.3.4}$$

The *closed left half plane* (CLHP), *open left half plane* (OLHP), *closed right half plane* (CRHP), and *open right half plane* (ORHP) are the subsets of $\mathbb{C}$ defined by

$$\text{OLHP} \triangleq \{x \in \mathbb{C}: \ \operatorname{Re} x < 0\}, \tag{1.3.5}$$

$$\text{CLHP} \triangleq \{x \in \mathbb{C}: \ \operatorname{Re} x \leq 0\}, \tag{1.3.6}$$

$$\text{ORHP} \triangleq \{x \in \mathbb{C}: \ \operatorname{Re} x > 0\}, \tag{1.3.7}$$

$$\text{CRHP} \triangleq \{x \in \mathbb{C}: \ \operatorname{Re} x \geq 0\}. \tag{1.3.8}$$

The imaginary numbers are represented by $j\mathbb{R}$. Note that 0 is both a real number and an imaginary number.

Next, we define the *open unit disk* (OUD) and the *closed unit disk* (CUD) by

$$\text{OUD} \triangleq \{x \in \mathbb{C}: \ |x| < 1\} \tag{1.3.9}$$

and

$$\text{CUD} \triangleq \{x \in \mathbb{C}: \ |x| \leq 1\}. \tag{1.3.10}$$

The complements of the open unit disk and the closed unit disk are given, respectively, by the *closed punctured plane* (CPP) and the *open punctured plane*, which are defined by

$$\text{CPP} \triangleq \{x \in \mathbb{C}: \ |x| \geq 1\} \tag{1.3.11}$$

and

$$\text{OPP} \triangleq \{x \in \mathbb{C}: \ |x| > 1\}. \tag{1.3.12}$$

Since $\mathbb{R}$ is a proper subset of $\mathbb{C}$, we state many results for $\mathbb{C}$. In other cases, we treat $\mathbb{R}$ and $\mathbb{C}$ separately. To do this efficiently, we use the symbol $\mathbb{F}$ to consistently denote either $\mathbb{R}$ or $\mathbb{C}$.

1.4 Functions

Let $\mathcal{X}$ and $\mathcal{Y}$ be sets. Then, a *function* f that maps $\mathcal{X}$ into $\mathcal{Y}$ is a rule $f\colon \mathcal{X} \mapsto \mathcal{Y}$ that assigns a unique element $f(x)$ (the *image* of x) of $\mathcal{Y}$ to each element x of $\mathcal{X}$. Equivalently, a function $f\colon \mathcal{X} \mapsto \mathcal{Y}$ can be viewed as a subset $\mathcal{F}$ of $\mathcal{X} \times \mathcal{Y}$ such that, for all $x \in \mathcal{X}$, it follows that there exists $y \in \mathcal{Y}$ such that $(x, y) \in \mathcal{F}$ and such that, if $(x, y_1), (x, y_2) \in \mathcal{F}$, then $y_1 = y_2$. In this case, $\mathcal{F} = \text{Graph}(f) \triangleq \{(x, f(x))\colon \ x \in \mathcal{X}\}$. The set $\mathcal{X}$ is the *domain* of f, while the set $\mathcal{Y}$ is the *codomain* of f. If $f\colon \mathcal{X} \mapsto \mathcal{X}$, then f is a function on $\mathcal{X}$. For $\mathcal{X}_1 \subseteq \mathcal{X}$, it is convenient to define $f(\mathcal{X}_1) \triangleq \{f(x)\colon \ x \in \mathcal{X}_1\}$. The set $f(\mathcal{X})$, which is denoted by $\mathcal{R}(f)$, is the *range* of f. If, in addition, $\mathcal{Z}$ is a set and $g\colon f(\mathcal{X}) \mapsto \mathcal{Z}$, then $g \bullet f\colon \mathcal{X} \mapsto \mathcal{Z}$ (the *composition* of g and f) is the function $(g \bullet f)(x) \triangleq g[f(x)]$. If $x_1, x_2 \in \mathcal{X}$ and $f(x_1) = f(x_2)$ implies that $x_1 = x_2$, then f

is *one-to-one*; if $\mathcal{R}(f) = \mathcal{Y}$, then f is *onto*. The function $I_{\mathcal{X}}\colon \mathcal{X} \mapsto \mathcal{X}$ defined by $I_{\mathcal{X}}(x) \triangleq x$ for all $x \in \mathcal{X}$ is the *identity* on $\mathcal{X}$. Finally, $x \in \mathcal{X}$ is a *fixed point* of the function $f\colon \mathcal{X} \mapsto \mathcal{X}$ if $f(x) = x$.

The following result shows that function composition is associative.

Proposition 1.4.1. Let $\mathcal{X}$, $\mathcal{Y}$, $\mathcal{Z}$, and $\mathcal{W}$ be sets, and let $f\colon \mathcal{X} \mapsto \mathcal{Y}$, $g\colon \mathcal{Y} \mapsto \mathcal{Z}$, $h\colon \mathcal{Z} \mapsto \mathcal{W}$. Then,

$$h \bullet (g \bullet f) = (h \bullet g) \bullet f. \tag{1.4.1}$$

Hence, we write $h \bullet g \bullet f$ for $h \bullet (g \bullet f)$ and $(h \bullet g) \bullet f$.

Let $\mathcal{X}$ be a set, and let $\hat{\mathcal{X}}$ be a partition of $\mathcal{X}$. Furthermore, let $f\colon \hat{\mathcal{X}} \mapsto \mathcal{X}$, where, for all $\mathcal{S} \in \hat{\mathcal{X}}$, it follows that $f(\mathcal{S}) \in \mathcal{S}$. Then, f is a *canonical mapping*, and $f(\mathcal{S})$ is a *canonical form*. That is, for all components $\mathcal{S}$ of the partition $\hat{\mathcal{X}}$ of $\mathcal{X}$, it follows that the function f assigns an element of $\mathcal{S}$ to the set $\mathcal{S}$.

Let $f\colon \mathcal{X} \mapsto \mathcal{Y}$. Then, f is *left invertible* if there exists a function $g\colon \mathcal{Y} \mapsto \mathcal{X}$ (a *left inverse* of f) such that $g \bullet f = I_{\mathcal{X}}$, whereas f is *right invertible* if there exists a function $h\colon \mathcal{Y} \mapsto \mathcal{X}$ (a *right inverse* of f) such that $f \bullet h = I_{\mathcal{Y}}$. In addition, the function $f\colon \mathcal{X} \mapsto \mathcal{Y}$ is *invertible* if there exists a function $f^{-1}\colon \mathcal{Y} \mapsto \mathcal{X}$ (the *inverse* of f) such that $f^{-1} \bullet f = I_{\mathcal{X}}$ and $f \bullet f^{-1} = I_{\mathcal{Y}}$. The *inverse image* $f^{-1}(\mathcal{S})$ of $\mathcal{S} \subseteq \mathcal{Y}$ is defined by

$$f^{-1}(\mathcal{S}) \triangleq \{x \in \mathcal{X}\colon\ f(x) \in \mathcal{S}\}. \tag{1.4.2}$$

Note that the set $f^{-1}(\mathcal{S})$ can be defined whether or not f is invertible. In fact, $f^{-1}[f(\mathcal{X})] = \mathcal{X}$.

Theorem 1.4.2. Let $\mathcal{X}$ and $\mathcal{Y}$ be sets, and let $f\colon \mathcal{X} \mapsto \mathcal{Y}$. Then, the following statements hold:

i) f is left invertible if and only if f is one-to-one.

ii) f is right invertible if and only if f is onto.

Furthermore, the following statements are equivalent:

iii) f is invertible.

iv) f has a unique inverse.

v) f is one-to-one and onto.

vi) f is left invertible and right invertible.

vii) f has a unique left inverse.

viii) f has a unique right inverse.

Proof. To prove *i)*, suppose that f is left invertible with left inverse $g\colon \mathcal{Y} \mapsto \mathcal{X}$. Furthermore, suppose that $x_1, x_2 \in \mathcal{X}$ satisfy $f(x_1) = f(x_2)$. Then, $x_1 = g[f(x_1)] = g[f(x_2)] = x_2$, which shows that f is one-to-one. Conversely, suppose that f is one-to-one so that, for all $y \in \mathcal{R}(f)$, there exists a unique $x \in \mathcal{X}$ such that $f(x) = y$.

Hence, define the function $g\colon \mathcal{Y} \mapsto \mathcal{X}$ by $g(y) \triangleq x$ for all $y = f(x) \in \mathcal{R}(f)$ and by $g(y)$ arbitrary for all $y \in \mathcal{Y}\backslash\mathcal{R}(f)$. Consequently, $g[f(x)] = x$ for all $x \in \mathcal{X}$, which shows that g is a left inverse of f.

To prove ii), suppose that f is right invertible with right inverse $g\colon \mathcal{Y} \mapsto \mathcal{X}$. Then, for all $y \in \mathcal{Y}$, it follows that $f[g(y)] = y$, which shows that f is onto. Conversely, suppose that f is onto so that, for all $y \in \mathcal{Y}$, there exists at least one $x \in \mathcal{X}$ such that $f(x) = y$. Selecting one such x arbitrarily, define $g\colon \mathcal{Y} \mapsto \mathcal{X}$ by $g(y) \triangleq x$. Consequently, $f[g(y)] = y$ for all $y \in \mathcal{Y}$, which shows that g is a right inverse of f. $\qquad\square$

Definition 1.4.3. Let $\mathcal{I} \subset \mathbb{R}$ be a finite or infinite interval, and let $f\colon \mathcal{I} \mapsto \mathbb{R}$. Then, f is *convex* if, for all $\alpha \in [0, 1]$ and for all $x, y \in \mathcal{I}$, it follows that

$$f[\alpha x + (1 - \alpha)y] \le \alpha f(x) + (1 - \alpha)f(y). \qquad (1.4.3)$$

Furthermore, f is *strictly convex* if, for all $\alpha \in (0, 1)$ and for all distinct $x, y \in \mathcal{I}$, it follows that

$$f[\alpha x + (1 - \alpha)y] < \alpha f(x) + (1 - \alpha)f(y).$$

A more general definition of convexity is given by Definition 8.6.14.

1.5 Relations

Let $\mathcal{X}$, $\mathcal{X}_1$, and $\mathcal{X}_2$ be sets. A *relation* $\mathcal{R}$ on $\mathcal{X}_1 \times \mathcal{X}_2$ is a subset of $\mathcal{X}_1 \times \mathcal{X}_2$. A *relation* $\mathcal{R}$ on $\mathcal{X}$ is a relation on $\mathcal{X} \times \mathcal{X}$. Likewise, a *multirelation* $\mathcal{R}$ on $\mathcal{X}_1 \times \mathcal{X}_2$ is a multisubset of $\mathcal{X}_1 \times \mathcal{X}_2$, while a *multirelation* $\mathcal{R}$ on $\mathcal{X}$ is a multirelation on $\mathcal{X} \times \mathcal{X}$.

Let $\mathcal{X}$ be a set, and let $\mathcal{R}_1$ and $\mathcal{R}_2$ be relations on $\mathcal{X}$. Then, $\mathcal{R}_1 \cap \mathcal{R}_2$, $\mathcal{R}_1\backslash\mathcal{R}_2$, and $\mathcal{R}_1 \cup \mathcal{R}_2$ are relations on $\mathcal{X}$. Furthermore, if $\mathcal{R}$ is a relation on $\mathcal{X}$ and $\mathcal{X}_0 \subseteq \mathcal{X}$, then we define $\mathcal{R}|_{\mathcal{X}_0} \triangleq \mathcal{R} \cap (\mathcal{X}_0 \times \mathcal{X}_0)$, which is a relation on $\mathcal{X}_0$.

The following result shows that relations can be viewed as generalizations of functions.

Proposition 1.5.1. Let $\mathcal{X}_1$ and $\mathcal{X}_2$ be sets, and let $\mathcal{R}$ be a relation on $\mathcal{X}_1 \times \mathcal{X}_2$. Then, there exists a function $f\colon \mathcal{X}_1 \mapsto \mathcal{X}_2$ such that $\mathcal{R} = \text{Graph}(f)$ if and only if, for all $x \in \mathcal{X}_1$, there exists a unique $y \in \mathcal{X}_2$ such that $(x, y) \in \mathcal{R}$. In this case, $f(x) = y$.

Definition 1.5.2. Let $\mathcal{R}$ be a relation on the set $\mathcal{X}$. Then, the following terminology is defined:

 i) $\mathcal{R}$ is *reflexive* if, for all $x \in \mathcal{X}$, it follows that $(x, x) \in \mathcal{R}$.

 ii) $\mathcal{R}$ is *symmetric* if, for all $(x_1, x_2) \in \mathcal{R}$, it follows that $(x_2, x_1) \in \mathcal{R}$.

 iii) $\mathcal{R}$ is *transitive* if, for all $(x_1, x_2) \in \mathcal{R}$ and $(x_2, x_3) \in \mathcal{R}$, it follows that $(x_1, x_3) \in \mathcal{R}$.

 iv) $\mathcal{R}$ is an *equivalence relation* if $\mathcal{R}$ is reflexive, symmetric, and transitive.

Proposition 1.5.3. Let $\mathcal{R}_1$ and $\mathcal{R}_2$ be relations on the set $\mathcal{X}$. If $\mathcal{R}_1$ and $\mathcal{R}_2$ are (reflexive, symmetric) relations, then so are $\mathcal{R}_1 \cap \mathcal{R}_2$ and $\mathcal{R}_1 \cup \mathcal{R}_2$. If $\mathcal{R}_1$ and $\mathcal{R}_2$ are (transitive, equivalence) relations, then so is $\mathcal{R}_1 \cap \mathcal{R}_2$.

Definition 1.5.4. Let $\mathcal{R}$ be a relation on the set $\mathcal{X}$. Then, the following terminology is defined:

i) The *complement* $\mathcal{R}^\sim$ of $\mathcal{R}$ is the relation $\mathcal{R}^\sim \triangleq (\mathcal{X} \times \mathcal{X}) \backslash \mathcal{R}$.

ii) The *support* $\mathrm{supp}(\mathcal{R})$ of $\mathcal{R}$ is the smallest subset $\mathcal{X}_0$ of $\mathcal{X}$ such that $\mathcal{R}$ is a relation on $\mathcal{X}_0$.

iii) The *reversal* $\mathrm{rev}(\mathcal{R})$ of $\mathcal{R}$ is the relation $\mathrm{rev}(\mathcal{R}) \triangleq \{(y, x) \colon (x, y) \in \mathcal{R}\}$.

iv) The *shortcut* $\mathrm{shortcut}(\mathcal{R})$ of $\mathcal{R}$ is the relation $\mathrm{shortcut}(\mathcal{R}) \triangleq \{(x, y) \in \mathcal{X} \times \mathcal{X} \colon x$ and y are distinct and there exist $k \geq 1$ and $x_1, \ldots, x_k \in \mathcal{X}$ such that $(x, x_1), (x_1, x_2), \ldots, (x_k, y) \in \mathcal{R}\}$.

v) The *reflexive hull* $\mathrm{ref}(\mathcal{R})$ of $\mathcal{R}$ is the smallest reflexive relation on $\mathcal{X}$ that contains $\mathcal{R}$.

vi) The *symmetric hull* $\mathrm{sym}(\mathcal{R})$ of $\mathcal{R}$ is the smallest symmetric relation on $\mathcal{X}$ that contains $\mathcal{R}$.

vii) The *transitive hull* $\mathrm{trans}(\mathcal{R})$ of $\mathcal{R}$ is the smallest transitive relation on $\mathcal{X}$ that contains $\mathcal{R}$.

viii) The *equivalence hull* $\mathrm{equiv}(\mathcal{R})$ of $\mathcal{R}$ is the smallest equivalence relation on $\mathcal{X}$ that contains $\mathcal{R}$.

Proposition 1.5.5. Let $\mathcal{R}$ be a relation on the set $\mathcal{X}$. Then, the following statements hold:

i) $\mathrm{ref}(\mathcal{R}) = \mathcal{R} \cup \{(x, x) \colon x \in \mathcal{X}\}$.

ii) $\mathrm{sym}(\mathcal{R}) = \mathcal{R} \cup \mathrm{rev}(\mathcal{R})$.

iii) $\mathrm{trans}(\mathcal{R}) = \mathcal{R} \cup \mathrm{shortcut}(\mathcal{R})$.

iv) $\mathrm{equiv}(\mathcal{R}) = \mathcal{R} \cup \mathrm{ref}(\mathcal{R}) \cup \mathrm{sym}(\mathcal{R}) \cup \mathrm{trans}(\mathcal{R})$.

v) $\mathrm{equiv}(\mathcal{R}) = \mathcal{R} \cup \mathrm{ref}(\mathcal{R}) \cup \mathrm{rev}(\mathcal{R}) \cup \mathrm{shortcut}(\mathcal{R})$.

Furthermore, the following statements hold:

vi) $\mathcal{R}$ is reflexive if and only if $\mathcal{R} = \mathrm{ref}(\mathcal{R})$.

vii) $\mathcal{R}$ is symmetric if and only if $\mathcal{R} = \mathrm{rev}(\mathcal{R})$.

viii) $\mathcal{R}$ is transitive if and only if $\mathcal{R} = \mathrm{trans}(\mathcal{R})$.

ix) $\mathcal{R}$ is an equivalence relation if and only if $\mathcal{R} = \mathrm{equiv}(\mathcal{R})$.

For an equivalence relation $\mathcal{R}$ on the set $\mathcal{X}$, $(x_1, x_2) \in \mathcal{R}$ is denoted by $x_1 \overset{\mathcal{R}}{=} x_2$. If $\mathcal{R}$ is an equivalence relation and $x \in \mathcal{X}$, then the subset $\mathcal{E}_x \triangleq \{y \in \mathcal{X} \colon y \overset{\mathcal{R}}{=} x\}$ of $\mathcal{X}$ is the *equivalence class of x induced by $\mathcal{R}$*.

Theorem 1.5.6. Let $\mathcal{R}$ be an equivalence relation on a set $\mathcal{X}$. Then, the set $\{\mathcal{E}_x \colon x \in \mathcal{X}\}$ of equivalence classes induced by $\mathcal{R}$ is a partition of $\mathcal{X}$.

Proof. Since $\mathcal{X} = \bigcup_{x \in \mathcal{X}} \mathcal{E}_x$, it suffices to show that if $x, y \in \mathcal{X}$, then either $\mathcal{E}_x = \mathcal{E}_y$ or $\mathcal{E}_x \cap \mathcal{E}_y = \varnothing$. Hence, let $x, y \in \mathcal{X}$, and suppose that $\mathcal{E}_x$ and $\mathcal{E}_y$ are not disjoint so that there exists $z \in \mathcal{E}_x \cap \mathcal{E}_y$. Thus, $(x, z) \in \mathcal{R}$ and $(z, y) \in \mathcal{R}$. Now, let $w \in \mathcal{E}_x$. Then, $(w, x) \in \mathcal{R}$, $(x, z) \in \mathcal{R}$, and $(z, y) \in \mathcal{R}$ imply that $(w, y) \in \mathcal{R}$. Hence, $w \in \mathcal{E}_y$, which implies that $\mathcal{E}_x \subseteq \mathcal{E}_y$. By a similar argument, $\mathcal{E}_y \subseteq \mathcal{E}_x$. Consequently, $\mathcal{E}_x = \mathcal{E}_y$. $\qquad\square$

The following result, which is the converse of Theorem 1.5.6, shows that a partition of a set $\mathcal{X}$ defines an equivalence relation on $\mathcal{X}$.

Theorem 1.5.7. Let $\mathcal{X}$ be a set, consider a partition of $\mathcal{X}$, and define the relation $\mathcal{R}$ on $\mathcal{X}$ by $(x, y) \in \mathcal{R}$ if and only if x and y belong to the same partition subset of $\mathcal{X}$. Then, $\mathcal{R}$ is an equivalence relation on $\mathcal{X}$.

Definition 1.5.8. Let $\mathcal{R}$ be a relation on the set $\mathcal{X}$. Then, the following terminology is defined:

  i) $\mathcal{R}$ is *antisymmetric* if $(x_1, x_2) \in \mathcal{R}$ and $(x_2, x_1) \in \mathcal{R}$ imply that $x_1 = x_2$.

  ii) $\mathcal{R}$ is a *partial ordering* on $\mathcal{X}$ if $\mathcal{R}$ is reflexive, antisymmetric, and transitive.

Let $\mathcal{R}$ be a partial ordering on $\mathcal{X}$. Then, $(x_1, x_2) \in \mathcal{R}$ is denoted by $x_1 \overset{\mathcal{R}}{\leq} x_2$. If $x_1 \overset{\mathcal{R}}{\leq} x_2$ and $x_2 \overset{\mathcal{R}}{\leq} x_1$, then, since $\mathcal{R}$ is antisymmetric, it follows that $x_1 = x_2$. Furthermore, if $x_1 \overset{\mathcal{R}}{\leq} x_2$ and $x_2 \overset{\mathcal{R}}{\leq} x_3$, then, since $\mathcal{R}$ is transitive, it follows that $x_1 \overset{\mathcal{R}}{\leq} x_3$.

Definition 1.5.9. Let "$\overset{\mathcal{R}}{\leq}$" be a partial ordering on $\mathcal{X}$. Then, the following terminology is defined:

  i) Let $\mathcal{S} \subseteq \mathcal{X}$. Then, $y \in \mathcal{X}$ is a *lower bound* for $\mathcal{S}$ if, for all $x \in \mathcal{S}$, it follows that $y \overset{\mathcal{R}}{\leq} x$.

  ii) Let $\mathcal{S} \subseteq \mathcal{X}$. Then, $y \in \mathcal{X}$ is an *upper bound* for $\mathcal{S}$ if, for all $x \in \mathcal{S}$, it follows that $x \overset{\mathcal{R}}{\leq} y$.

  iii) Let $\mathcal{S} \subseteq \mathcal{X}$. Then, $y \in \mathcal{X}$ is the *least upper bound* lub($\mathcal{S}$) for $\mathcal{S}$ if y is an upper bound for $\mathcal{S}$ and, for all upper bounds $x \in \mathcal{X}$ for $\mathcal{S}$, it follows that $y \overset{\mathcal{R}}{\leq} x$. In this case, we write $y = \text{lub}(\mathcal{S})$.

  iv) Let $\mathcal{S} \subseteq \mathcal{X}$. Then, $y \in \mathcal{X}$ is the *greatest lower bound* for $\mathcal{S}$ if y is a lower bound for $\mathcal{S}$ and, for all lower bounds $x \in \mathcal{X}$ for $\mathcal{S}$, it follows that $x \overset{\mathcal{R}}{\leq} y$. In this case, we write $y = \text{glb}(\mathcal{S})$.

  v) $\overset{\mathcal{R}}{\leq}$ is a *lattice* on $\mathcal{X}$ if, for all distinct $x, y \in \mathcal{X}$, the set $\{x, y\}$ has a least upper bound and a greatest lower bound.

  vi) $\mathcal{R}$ is a *total ordering* on $\mathcal{X}$ if, for all $x, y \in \mathcal{X}$, it follows that either $(x, y) \in \mathcal{R}$ or $(y, x) \in \mathcal{R}$.

For a subset $\mathcal{S}$ of the real numbers, it is traditional to write $\inf \mathcal{S}$ and $\sup \mathcal{S}$ for $\mathrm{glb}(\mathcal{S})$ and $\mathrm{lub}(\mathcal{S})$, respectively, where "inf" and "sup" denote infimum and supremum, respectively.

1.6 Graphs

Let $\mathcal{X}$ be a finite, nonempty set, and let $\mathcal{R}$ be a relation on $\mathcal{X}$. Then, the pair $\mathcal{G} = (\mathcal{X}, \mathcal{R})$ is a *graph*. The elements of $\mathcal{X}$ are the *nodes* of $\mathcal{G}$, while the elements of $\mathcal{R}$ are the *arcs* of $\mathcal{G}$. If $\mathcal{R}$ is a multirelation on $\mathcal{X}$, then $\mathcal{G} = (\mathcal{X}, \mathcal{R})$ is a *multigraph*.

The graph $\mathcal{G} = (\mathcal{X}, \mathcal{R})$ can be visualized as a set of points in the plane representing the nodes in $\mathcal{X}$ connected by the arcs in $\mathcal{R}$. Specifically, the arc $(x, y) \in \mathcal{R}$ from x to y can be visualized as a directed line segment or curve connecting node x to node y. The direction of an arc can be denoted by an arrow head. For example, consider a graph that represents a city with streets (arcs) connecting houses (nodes). Then, a symmetric relation is a street plan with no one-way streets, whereas an antisymmetric relation is a street plan with no two-way streets.

Definition 1.6.1. Let $\mathcal{G} = (\mathcal{X}, \mathcal{R})$ be a graph. Then, the following terminology is defined:

i) The arc $(x, x) \in \mathcal{R}$ is a *self-loop*.

ii) The *reversal* of $(x, y) \in \mathcal{R}$ is (y, x).

iii) If $x, y \in \mathcal{X}$ and $(x, y) \in \mathcal{R}$, then y is the *head* of (x, y) and x is the *tail* of (x, y).

iv) If $x, y \in \mathcal{X}$ and $(x, y) \in \mathcal{R}$, then x is a *parent* of y, and y is a *child* of x.

v) If $x, y \in \mathcal{X}$ and either $(x, y) \in \mathcal{R}$ or $(y, x) \in \mathcal{R}$, then x and y are *adjacent*.

vi) If $x \in \mathcal{X}$ has no parent, then x is a *root*.

vii) If $x \in \mathcal{X}$ has no child, then x is a *leaf*.

Suppose that $(x, x) \in \mathcal{R}$. Then, x is both the head and the tail of (x, x), and thus x is a parent and child of itself. Consequently, x is neither a root nor a leaf. Furthermore, x is adjacent to itself.

Definition 1.6.2. Let $\mathcal{G} = (\mathcal{X}, \mathcal{R})$ be a graph. Then, the following terminology is defined:

i) The *reversal* of $\mathcal{G}$ is the graph $\mathrm{rev}(\mathcal{G}) \triangleq (\mathcal{X}, \mathrm{rev}(\mathcal{R}))$.

ii) The *complement* of $\mathcal{G}$ is the graph $\mathcal{G}^{\sim} \triangleq (\mathcal{X}, \mathcal{R}^{\sim})$.

iii) The *reflexive hull* of $\mathcal{G}$ is the graph $\mathrm{ref}(\mathcal{G}) \triangleq (\mathcal{X}, \mathrm{ref}(\mathcal{R}))$.

iv) The *symmetric hull* of $\mathcal{G}$ is the graph $\mathrm{sym}(\mathcal{G}) \triangleq (\mathcal{X}, \mathrm{sym}(\mathcal{R}))$.

v) The *transitive hull* of $\mathcal{G}$ is the graph $\mathrm{trans}(\mathcal{G}) \triangleq (\mathcal{X}, \mathrm{trans}(\mathcal{R}))$.

vi) The *equivalence hull* of $\mathcal{G}$ is the graph $\mathrm{equiv}(\mathcal{G}) \triangleq (\mathcal{X}, \mathrm{equiv}(\mathcal{R}))$.

vii) $\mathcal{G}$ is *reflexive* if $\mathcal{R}$ is reflexive.

viii) $\mathcal{G}$ is *symmetric* if $\mathcal{R}$ is symmetric. In this case, the arcs (x, y) and (y, x) in $\mathcal{R}$ are denoted by the subset $\{x, y\}$ of $\mathcal{X}$, called an *edge*. For the self-loop (x, x), the corresponding edge is the subset $\{x\}$.

ix) $\mathcal{G}$ is *transitive* if $\mathcal{R}$ is transitive.

x) $\mathcal{G}$ is an *equivalence graph* if $\mathcal{R}$ is an equivalence relation.

xi) $\mathcal{G}$ is *antisymmetric* if $\mathcal{R}$ is antisymmetric.

xii) $\mathcal{G}$ is *partially ordered* if $\mathcal{R}$ is a partial ordering on $\mathcal{X}$.

xiii) $\mathcal{G}$ is *totally ordered* if $\mathcal{R}$ is a total ordering on $\mathcal{X}$.

xiv) $\mathcal{G}$ is a *tournament* if $\mathcal{G}$ has no self-loops, is antisymmetric, and $\operatorname{sym}(\mathcal{R}) = \mathcal{X} \times \mathcal{X}$.

Note that a symmetric graph can include self-loops, whereas a reflexive graph has a self-loop at every node.

Definition 1.6.3. Let $\mathcal{G} = (\mathcal{X}, \mathcal{R})$ be a graph. Then, the following terminology is defined:

i) The graph $\mathcal{G}' = (\mathcal{X}', \mathcal{R}')$ is a *subgraph* of $\mathcal{G}$ if $\mathcal{X}' \subseteq \mathcal{X}$ and $\mathcal{R}' \subseteq \mathcal{R}$.

ii) The subgraph $\mathcal{G}' = (\mathcal{X}', \mathcal{R}')$ of $\mathcal{G}$ is a *spanning subgraph* of $\mathcal{G}$ if $\operatorname{supp}(\mathcal{R}) = \operatorname{supp}(\mathcal{R}')$.

iii) For $x, y \in \mathcal{X}$, a *walk* in $\mathcal{G}$ from x to y is an n-tuple of arcs of the form $(x, y) \in \mathcal{R}$ for $n = 1$ and $\left((x, x_1), (x_1, x_2), \ldots, (x_{n-1}, y) \right) \in \mathcal{R}^n$ for $n \geq 2$. The *length* of the walk is n. The nodes $x, x_1, \ldots, x_{n-1}, y$ are the *nodes* of the walk. Furthermore, if $n \geq 2$, then the nodes $x_1, \ldots, x_{n-1}$ are the *intermediate nodes* of the walk.

iv) $\mathcal{G}$ is *connected* if, for all distinct $x, y \in \mathcal{X}$, there exists a walk in $\mathcal{G}$ from x to y.

v) For $x, y \in \mathcal{X}$, a *trail* in $\mathcal{G}$ from x to y is a walk in $\mathcal{G}$ from x to y whose arcs are distinct and such that, if (w, z) is an arc of the walk, then (z, w) is not an arc of the walk.

vi) For $x, y \in \mathcal{X}$, a *path* in $\mathcal{G}$ from x to y is a trail in $\mathcal{G}$ from x to y whose intermediate nodes (if any) are distinct.

vii) $\mathcal{G}$ is *traceable* if $\mathcal{G}$ has a path such that every node in $\mathcal{X}$ is a node of the path. Such a path is called a *Hamiltonian path*.

viii) For $x \in \mathcal{X}$, a *cycle* in $\mathcal{G}$ at x is a path in $\mathcal{G}$ from x to x whose length is greater than 1.

ix) $\mathcal{G}$ is *acyclic* if $\mathcal{G}$ has no cycles.

x) The *period* of $\mathcal{G}$ is the greatest common divisor of the lengths of the cycles in $\mathcal{G}$. Furthermore, $\mathcal{G}$ is *aperiodic* if the period of $\mathcal{G}$ is 1.

xi) $\mathcal{G}$ is *Hamiltonian* if $\mathcal{G}$ has a cycle such that every node in $\mathcal{X}$ is a node of the cycle. Such a cycle is called a *Hamiltonian cycle*.

xii) $\mathcal{G}$ is a *tree* if $\mathcal{G}$ has exactly one root x and, for all $y \in \mathcal{X}$ such that $y \neq x$, there exists a unique path from x to y.

xiii) $\mathcal{G}$ is a *forest* if $\mathcal{G}$ is a union of trees.

xiv) $\mathcal{G}$ is a *chain* if $\mathcal{G}$ is a tree and has exactly one leaf.

xv) The *indegree* of $x \in \mathcal{X}$ is $\operatorname{indeg}(x) \triangleq \operatorname{card}\{y \in \mathcal{X}: y \text{ is a parent of } x\}$.

xvi) The *outdegree* of $x \in \mathcal{X}$ is $\operatorname{outdeg}(x) \triangleq \operatorname{card}\{y \in \mathcal{X}: y \text{ is a child of } x\}$.

xvii) If $\mathcal{G}$ is symmetric, then the *degree* of $x \in \mathcal{X}$ is $\deg(x) \triangleq \operatorname{indeg}(x) = \operatorname{outdeg}(x)$.

xviii) If $\mathcal{X}_0 \subseteq \mathcal{X}$, then,
$$\mathcal{G}|_{\mathcal{X}_0} \triangleq (\mathcal{X}_0, \mathcal{R}|_{\mathcal{X}_0}).$$

xix) If $\mathcal{G}' = (\mathcal{X}', \mathcal{R}')$ is a graph, then $\mathcal{G} \cup \mathcal{G}' \triangleq (\mathcal{X} \cup \mathcal{X}', \mathcal{R} \cup \mathcal{R}')$ and $\mathcal{G} \cap \mathcal{G}' \triangleq (\mathcal{X} \cap \mathcal{X}', \mathcal{R} \cap \mathcal{R}')$.

xx) Let $\mathcal{X} = \mathcal{X}_1 \cup \mathcal{X}_2$, where $\mathcal{X}_1$ and $\mathcal{X}_2$ are nonempty and disjoint, and assume that $\mathcal{X} = \operatorname{supp}(\mathcal{G})$. Then, $(\mathcal{X}_1, \mathcal{X}_2)$ is a *directed cut* of $\mathcal{G}$ if, for all $x_1 \in \mathcal{X}_1$ and $x_2 \in \mathcal{X}_2$, there does not exist a walk from x_1 to x_2.

Note that a graph that is acyclic cannot be aperiodic.

Let $\mathcal{G} = (\mathcal{X}, \mathcal{R})$ be a graph, and let $w: \mathcal{X} \times \mathcal{X} \mapsto [0, \infty)$, where $w(x, y) > 0$ if $(x, y) \in \mathcal{R}$ and $w(x, y) = 0$ if $(x, y) \notin \mathcal{R}$. For each arc $(x, y) \in \mathcal{R}$, $w(x, y)$ is the *weight* associated with the arc (x, y), and the triple $\mathcal{G} = (\mathcal{X}, \mathcal{R}, w)$ is a *weighted graph*. Every graph can be viewed as a weighted graph by defining $w[(x, y)] \triangleq 1$ for all $(x, y) \in \mathcal{R}$ and $w[(x, y)] \triangleq 0$ for all $(x, y) \notin \mathcal{R}$. The graph $\mathcal{G}' = (\mathcal{X}', \mathcal{R}', w')$ is a *weighted subgraph* of $\mathcal{G}$ if $\mathcal{X} \subseteq \mathcal{X}'$, $\mathcal{R}'$ is a relation on $\mathcal{X}'$, $\mathcal{R}' \subseteq \mathcal{R}$, and w' is the restriction of w to $\mathcal{R}'$. Finally, if $\mathcal{G}$ is symmetric, then w is defined on edges $\{x, y\}$ of $\mathcal{G}$.

1.7 Facts on Logic, Sets, Functions, and Relations

Fact 1.7.1. Let A and B be statements. Then, the following statements hold:

i) $\operatorname{not}(A \text{ or } B) \Longleftrightarrow [(\operatorname{not} A) \text{ and } (\operatorname{not} B)]$.

ii) $\operatorname{not}(A \text{ and } B) \Longleftrightarrow (\operatorname{not} A) \text{ or } (\operatorname{not} B)$.

iii) $(A \text{ or } B) \Longleftrightarrow [(\operatorname{not} A) \Longrightarrow B]$.

iv) $(A \Longrightarrow B) \Longleftrightarrow [(\operatorname{not} A) \text{ or } B]$.

v) $[A \text{ and } (\operatorname{not} B)] \Longleftrightarrow [\operatorname{not}(A \Longrightarrow B)]$.

Remark: Each statement is a tautology.

Remark: Statements *i*) and *ii*) are *De Morgan's laws*. See [233, p. 24].

Fact 1.7.2. The following statements are equivalent:

i) $A \Longrightarrow (B \text{ or } C)$.

ii) $[A \text{ and } (\text{not } B)] \Longrightarrow C$.

Remark: The statement that *i*) and *ii*) are equivalent is a tautology.

Fact 1.7.3. The following statements are equivalent:

i) $A \Longleftrightarrow B$.

ii) $[A \text{ or } (\text{not } B)] \text{ and } (\text{not } [A \text{ and } (\text{not } B)])$.

Remark: The statement that *i*) and *ii*) are equivalent is a tautology.

Fact 1.7.4. The following statements are equivalent:

i) Not [for all x, there exists y such that statement Z is satisfied].

ii) There exists x such that, for all y, statement Z is not satisfied.

Fact 1.7.5. Let $\mathcal{A}$, $\mathcal{B}$, and $\mathcal{C}$ be sets, and assume that each of these sets has a finite number of elements. Then,

$$\text{card}(\mathcal{A} \cup \mathcal{B}) = \text{card}(\mathcal{A}) + \text{card}(\mathcal{B}) - \text{card}(\mathcal{A} \cap \mathcal{B})$$

and

$$\begin{aligned}\text{card}(\mathcal{A} \cup \mathcal{B} \cup \mathcal{C}) = {}&\text{card}(\mathcal{A}) + \text{card}(\mathcal{B}) + \text{card}(\mathcal{C}) \\ &- \text{card}(\mathcal{A} \cap \mathcal{B}) - \text{card}(\mathcal{A} \cap \mathcal{C}) - \text{card}(\mathcal{B} \cap \mathcal{C}) \\ &+ \text{card}(\mathcal{A} \cap \mathcal{B} \cap \mathcal{C}).\end{aligned}$$

Remark: This result is the *inclusion-exclusion principle*. See [181, p. 82] or [1249, pp. 64–67].

Fact 1.7.6. Let $\mathcal{A}, \mathcal{B}, \mathcal{C}$ be subsets of a set $\mathfrak{X}$. Then, the following statements hold:

i) $\mathcal{A} \cap \mathcal{A} = \mathcal{A} \cup \mathcal{A} = \mathcal{A}$.

ii) $(\mathcal{A} \cup \mathcal{B})^{\sim} = \mathcal{A}^{\sim} \cap \mathcal{B}^{\sim}$.

iii) $(\mathcal{A} \cap \mathcal{B})^{\sim} = \mathcal{A}^{\sim} \cup \mathcal{B}^{\sim}$.

iv) $\mathcal{A} = (\mathcal{A} \backslash \mathcal{B}) \cup (\mathcal{A} \cap \mathcal{B})$.

v) $[\mathcal{A} \backslash (\mathcal{A} \cap \mathcal{B})] \cup \mathcal{B} = \mathcal{A} \cup \mathcal{B}$.

vi) $(\mathcal{A} \cup \mathcal{B}) \backslash (\mathcal{A} \cap \mathcal{B}) = (\mathcal{A} \cap \mathcal{B}^{\sim}) \cup (\mathcal{A}^{\sim} \cap \mathcal{B})$.

vii) $\mathcal{A} \cap (\mathcal{B} \cup \mathcal{C}) = (\mathcal{A} \cap \mathcal{B}) \cup (\mathcal{A} \cap \mathcal{C})$.

viii) $\mathcal{A} \cup (\mathcal{B} \cap \mathcal{C}) = (\mathcal{A} \cup \mathcal{B}) \cap (\mathcal{A} \cup \mathcal{C})$.

ix) $(\mathcal{A} \backslash \mathcal{B}) \backslash \mathcal{C} = \mathcal{A} \backslash (\mathcal{B} \cup \mathcal{C})$.

x) $(\mathcal{A} \cap \mathcal{B}) \backslash \mathcal{C} = (\mathcal{A} \backslash \mathcal{C}) \cap (\mathcal{B} \backslash \mathcal{C})$.

xi) $(\mathcal{A} \cap \mathcal{B}) \backslash (\mathcal{C} \cap \mathcal{B}) = (\mathcal{A} \backslash \mathcal{C}) \cap \mathcal{B}$.

xii) $(A \cup B) \backslash C = (A \backslash C) \cup (B \backslash C) = [A \backslash (B \cup C)] \cup (B \backslash C).$

xiii) $(A \cup B) \backslash (C \cap B) = (A \backslash B) \cup (B \backslash C).$

xiv) $(A \cup B) \cap (A \cup B^{\sim}) = A.$

xv) $(A \cup B) \cap (A^{\sim} \cup B) \cap (A \cup B^{\sim}) = A \cap B.$

Fact 1.7.7. Define the relation $\mathcal{R}$ on $\mathbb{R} \times \mathbb{R}$ by

$$\mathcal{R} \triangleq \{((x_1, y_1), (x_2, y_2)) \in (\mathbb{R} \times \mathbb{R}) \times (\mathbb{R} \times \mathbb{R}) : x_1 \le x_2 \text{ and } y_1 \le y_2\}.$$

Then, $\mathcal{R}$ is a partial ordering.

Fact 1.7.8. Define the relation $\mathcal{L}$ on $\mathbb{R} \times \mathbb{R}$ by

$$\mathcal{L} \triangleq \{((x_1, y_1), (x_2, y_2)) \in (\mathbb{R} \times \mathbb{R}) \times (\mathbb{R} \times \mathbb{R}) :$$
$$x_1 \le x_2 \text{ and, if } x_1 = x_2, \text{ then } y_1 \le y_2\}.$$

Then, $\mathcal{L}$ is a total ordering on $\mathbb{R} \times \mathbb{R}$.

Remark: Denoting this total ordering by "$\overset{d}{\le}$," note that $(1, 4) \overset{d}{\le} (2, 3)$ and $(1, 4) \overset{d}{\le} (1, 5)$.

Remark: This ordering is the *lexicographic ordering* or *dictionary ordering*, where 'book' $\overset{d}{\le}$ 'box'. Note that the ordering of words in a dictionary is reflexive, antisymmetric, and transitive, and that every pair of words can be ordered.

Remark: See Fact 2.9.31.

Fact 1.7.9. Let $f \colon \mathcal{X} \mapsto \mathcal{Y}$, and assume that f is invertible. Then,

$$(f^{-1})^{-1} = f.$$

Fact 1.7.10. Let $f \colon \mathcal{X} \mapsto \mathcal{Y}$ and $g \colon \mathcal{Y} \mapsto \mathcal{Z}$, and assume that f and g are invertible. Then, $g \bullet f$ is invertible and

$$(g \bullet f)^{-1} = f^{-1} \bullet g^{-1}.$$

Fact 1.7.11. Let $f \colon \mathcal{X} \mapsto \mathcal{Y}$, and let $A, B \subseteq \mathcal{X}$. Then, the following statements hold:

i) If $A \subseteq B$, then $f(A) \subseteq f(B)$.

ii) $f(A \cup B) = f(A) \cup f(B)$.

iii) $f(A \cap B) \subseteq f(A) \cap f(B)$.

Fact 1.7.12. Let $f \colon \mathcal{X} \mapsto \mathcal{Y}$, and let $A, B \subseteq \mathcal{Y}$. Then, the following statements hold:

i) $f[f^{-1}(A)] \subseteq A \subseteq f^{-1}[f(A)]$.

ii) $f^{-1}(A \cup B) = f^{-1}(A) \cup f^{-1}(B)$.

iii) $f^{-1}(A \cap B) = f^{-1}(A) \cap f^{-1}(B)$.

iv) $f^{-1}(A \backslash B) = f^{-1}(A) \backslash f^{-1}(B)$.

Fact 1.7.13. Let $\mathcal{X}$ and $\mathcal{Y}$ be finite sets, and let $f\colon \mathcal{X} \mapsto \mathcal{Y}$. Then, the following statements hold:

i) If $\text{card}(\mathcal{X}) < \text{card}(\mathcal{Y})$, then f is not onto.

ii) If $\text{card}(\mathcal{Y}) < \text{card}(\mathcal{X})$, then f is not one-to-one.

iii) If f is one-to-one and onto, then $\text{card}(\mathcal{X}) = \text{card}(\mathcal{Y})$.

Now, assume in addition that $\text{card}(\mathcal{X}) = \text{card}(\mathcal{Y})$. Then, the followings statements are equivalent:

iv) f is one-to-one.

v) f is onto.

vi) $\text{card}[f(\mathcal{X})] = \text{card}(\mathcal{X})$.

Remark: See Fact 1.8.1.

Fact 1.7.14. Let $S_1, \ldots, S_m$ be finite sets, and let

$$n \triangleq \sum_{i=1}^{m} \text{card}(S_i).$$

Then,

$$\left\lceil \frac{n}{m} \right\rceil \leq \max_{i=1,\ldots,m} \text{card}(S_i).$$

In addition, if $m < n$, then there exists $i \in \{1, \ldots, m\}$ such that $\text{card}(S_i) \geq 2$.

Remark: $\lceil x \rceil$ is the smallest integer greater than or equal to x.

Remark: This result is the *pigeonhole principle*.

Fact 1.7.15. Let $f\colon \mathcal{X} \mapsto \mathcal{Y}$. Then, the following statements are equivalent:

i) f is one-to-one.

ii) For all $A \subseteq \mathcal{X}$ and $B \subseteq \mathcal{X}$, it follows that $f(A \cap B) = f(A) \cap f(B)$.

iii) For all $A \subseteq \mathcal{X}$, it follows that $f^{-1}[f(A)] = A$.

iv) For all disjoint $A \subseteq \mathcal{X}$ and $B \subseteq \mathcal{X}$, it follows that $f(A)$ and $f(B)$ are disjoint.

v) For all $A \subseteq \mathcal{X}$ and $B \subseteq \mathcal{X}$ such that $A \subseteq B$, it follows that $f(A \backslash B) = f(A) \backslash f(B)$.

Proof: See [71, pp. 44, 45].

Fact 1.7.16. Let $f\colon \mathcal{X} \mapsto \mathcal{Y}$. Then, the following statements are equivalent:

i) f is onto.

ii) For all $A \subseteq \mathcal{X}$, it follows that $f[f^{-1}(A)] = A$.

Fact 1.7.17. Let $f\colon X \mapsto Y$, and let $g\colon Y \mapsto Z$. Then, the following statements hold:

i) If f and g are one-to-one, then $f \bullet g$ is one-to-one.

ii) If f and g are onto, then $f \bullet g$ is onto.

Remark: A matrix version of this result is given by Fact 2.10.3.

Fact 1.7.18. Let $f\colon X \mapsto Y$, let $g\colon Y \mapsto X$, and assume that f and g are one-to-one. Then, there exists $h\colon X \mapsto Y$ such that h is one-to-one and onto.

Proof: See [1049, pp. 16, 17].

Remark: This result is the *Schroeder-Bernstein theorem.*

Fact 1.7.19. Let X be a set, and let $\mathfrak{X}$ denote the class of subsets of X. Then, "$\subset$" and "$\subseteq$" are transitive relations on $\mathfrak{X}$, and "$\subseteq$" is a partial ordering on $\mathfrak{X}$.

1.8 Facts on Graphs

Fact 1.8.1. Let $\mathcal{G} = (X, \mathcal{R})$ be a graph. Then, the following statements hold:

i) $\mathcal{R}$ is the graph of a function on X if and only if every node in X has exactly one child.

Furthermore, the following statements are equivalent:

ii) $\mathcal{R}$ is the graph of a one-to-one function on X.

iii) $\mathcal{R}$ is the graph of an onto function on X.

iv) $\mathcal{R}$ is the graph of a one-to-one and onto function on X.

v) Every node in X has exactly one child and not more than one parent.

vi) Every node in X has exactly one child and at least one parent.

vii) Every node in X has exactly one child and exactly one parent.

Remark: See Fact 1.7.13.

Fact 1.8.2. Let $\mathcal{G} = (X, \mathcal{R})$ be a graph, and assume that $\mathcal{R}$ is the graph of a function $f\colon X \mapsto X$. Then, either f is the identity function or $\mathcal{G}$ has a cycle.

Fact 1.8.3. Let $\mathcal{G} = (X, \mathcal{R})$ be a graph, and assume that $\mathcal{G}$ has a Hamiltonian cycle. Then, $\mathcal{G}$ has no roots and no leaves.

Fact 1.8.4. Let $\mathcal{G} = (X, \mathcal{R})$ be a graph. Then, $\mathcal{G}$ has either a root or a cycle.

Fact 1.8.5. Let $\mathcal{G} = (X, \mathcal{R})$ be a symmetric graph. Then, the following statements are equivalent:

i) $\mathcal{G}$ is a forest.

ii) $\mathcal{G}$ has no cycles.

iii) No pair of nodes in $\mathfrak{X}$ is connected by more than one path.

Furthermore, the following statements are equivalent:

iv) $\mathcal{G}$ is a tree.

v) $\mathcal{G}$ is a connected forest.

vi) $\mathcal{G}$ is connected and has no cycles.

vii) $\mathcal{G}$ is connected and has $\mathrm{card}(\mathfrak{X}) - 1$ edges.

viii) $\mathcal{G}$ has no cycles and has $\mathrm{card}(\mathfrak{X}) - 1$ edges.

ix) Every pair of nodes in $\mathfrak{X}$ is connected by exactly one path.

Fact 1.8.6. Let $\mathcal{G} = (\mathfrak{X}, \mathcal{R})$ be a tournament. Then, $\mathcal{G}$ has a Hamiltonian path. Furthermore, every Hamiltonian path is a Hamiltonian cycle if and only if $\mathcal{G}$ is connected.

Fact 1.8.7. Let $\mathcal{G} = (\mathfrak{X}, \mathcal{R})$ be a symmetric graph, where $\mathfrak{X} \subset \mathbb{R}^2$, assume that $n \triangleq \mathrm{card}(\mathfrak{X}) \geq 3$, assume that $\mathcal{G}$ is connected, and assume that the edges in $\mathcal{R}$ can be represented by line segments that lie in the same plane and that are either disjoint or intersect at a node. Furthermore, let m denote the number of edges of $\mathcal{G}$, and let f denote the number of disjoint regions in $\mathbb{R}^2$ whose boundaries are the edges of $\mathcal{G}$. Then,

$$n - m + f = 2.$$

Furthermore,

$$f \leq 2(n - 2),$$

and thus

$$m \leq 3(n - 2).$$

Remark: The equality gives the *Euler characteristic* for a planar graph. A similar result holds for the surfaces of convex polyhedra in three dimensions, such as the tetrahedron, cube (hexahedron), octahedron, dodecahedron, and icosahedron. See [1152].

1.9 Facts on Binomial Identities and Sums

Fact 1.9.1. The following statements hold:

i) Let $0 \leq k \leq n$. Then,

$$\binom{n}{k} = \binom{n}{n-k}.$$

ii) Let $1 \leq k \leq n$. Then,

$$k\binom{n}{k} = n\binom{n-1}{k-1}.$$

iii) Let $2 \leq k \leq n$. Then,

$$k(k-1)\binom{n}{k} = n(n-1)\binom{n-2}{k-2}.$$

iv) Let $0 \leq k < n$. Then,

$$(n-k)\binom{n}{k} = n\binom{n-1}{k}.$$

v) Let $1 \leq k \leq n$. Then,

$$\binom{n}{k} = \binom{n-1}{k} + \binom{n-1}{k-1}.$$

vi) Let $0 \leq m \leq k \leq n$. Then,

$$\binom{n}{k}\binom{k}{m} = \binom{n}{m}\binom{n-m}{k-m}.$$

vii) Let $m, n \geq 0$. Then,

$$\sum_{i=0}^{m}\binom{n+i}{n} = \binom{n+m+1}{m}.$$

viii) Let $k \geq 0$ and $n \geq 1$. Then,

$$\sum_{i=0}^{n-1}\frac{(k+i)!}{i!} = k!\binom{k+n}{k+1}.$$

ix) Let $0 \leq k \leq n$. Then,

$$\sum_{i=k}^{n}\binom{i}{k} = \binom{n+1}{k+1}.$$

x) Let $n, m \geq 0$, and let $0 \leq k \leq \min\{n, m\}$. Then,

$$\sum_{i=0}^{k}\binom{n}{i}\binom{m}{k-i} = \binom{n+m}{k}.$$

xi) Let $n \geq 0$. Then,

$$\sum_{i=1}^{n}\binom{n}{i}\binom{n}{i-1} = \binom{2n}{n+1}.$$

xii) Let $0 \leq k \leq n$. Then,

$$\sum_{i=0}^{n-k}\binom{n}{i}\binom{n}{k+i} = \frac{(2n)!}{(n-k)!(n+k)!}.$$

xiii) Let $0 \leq k \leq n/2$. Then,

$$\sum_{i=k}^{n-k}\binom{i}{k}\binom{n-i}{k} = \binom{n+1}{2k+1}.$$

xiv) Let $1 \leq k \leq n/2$. Then,

$$\sum_{i=0}^{k}2^{2i}\binom{n}{k-i}\binom{n-k+i}{2i} = \binom{2n}{2k}.$$

xv) Let $1 \le k \le (n-1)/2$. Then,

$$\sum_{i=0}^{k} 2^{2i+1} \binom{n}{k-i} \binom{n-k+i}{2i+1} = \binom{2n}{2k+1}.$$

xvi) Let $n \ge 0$. Then,

$$\sum_{i=0}^{n} \binom{n}{i}^2 = \binom{2n}{n}.$$

xvii) Let $n \ge 1$. Then,

$$\sum_{i=0}^{n} i\binom{n}{i}^2 = n\binom{2n-1}{n-1}.$$

xviii) For all $x, y \in \mathbb{C}$ and $n \ge 0$,

$$(x+y)^n = \sum_{i=0}^{n} \binom{n}{i} x^{n-i} y^i.$$

xix) Let $n \ge 0$. Then,

$$\sum_{i=0}^{n} \binom{n}{i} = 2^n.$$

xx) Let $n \ge 0$. Then,

$$\sum_{i=0}^{n} \frac{1}{i+1} \binom{n}{i} = \frac{2^{n+1}-1}{n+1}.$$

xxi) Let $n \ge 0$. Then,

$$\sum_{i=0}^{n} \binom{2n+1}{i} = \sum_{i=0}^{2n} \binom{2n}{i} = 4^n.$$

xxii) Let $n > 1$. Then,

$$\sum_{i=0}^{n-1} (n-i)^2 \binom{2n}{i} = 4^{n-1} n.$$

xxiii) Let $n \ge 0$. Then,

$$\sum_{i=0}^{\lfloor n/2 \rfloor} \binom{n}{2i} = 2^{n-1}.$$

xxiv) Let $n \ge 0$. Then,

$$\sum_{i=0}^{\lfloor (n-1)/2 \rfloor} \binom{n}{2i+1} = 2^{n-1}.$$

xxv) Let $n \ge 0$. Then,

$$\sum_{i=0}^{\lfloor n/2 \rfloor} (-1)^i \binom{n}{2i} = 2^{n/2} \cos \frac{n\pi}{4}.$$

xxvi) Let $n \ge 0$. Then,

$$\sum_{i=0}^{\lfloor (n-1)/2 \rfloor} (-1)^i \binom{n}{2i+1} = 2^{n/2} \sin \frac{n\pi}{4}.$$

xxvii) Let $n \geq 1$. Then,

$$\sum_{i=1}^{n} i \binom{n}{i} = n 2^{n-1}.$$

xxviii) Let $n \geq 1$. Then,

$$\sum_{i=0}^{n} \binom{n}{2i} = 2^{n-1}.$$

xxix) Let $0 \leq k < n$. Then,

$$\sum_{i=0}^{k} (-1)^i \binom{n}{i} = (-1)^k \binom{n-1}{k}.$$

xxx) Let $n \geq 1$. Then,

$$\sum_{i=0}^{n} (-1)^i \binom{n}{i} = 0.$$

xxxi) Let $n \geq 1$. Then,

$$\sum_{i=0}^{n} \frac{2^i}{i+1} = \frac{2^n}{n+1} \sum_{i=0}^{n} \frac{1}{\binom{n}{i}}.$$

Proof: See [181, pp. 64–68, 78], [340], [598, pp. 1, 2], [686, pp. 2–10, 74], and [1207]. Statement *xxxi)* is given in [242, p. 55].

Remark: Statement *x)* is *Vandermonde's identity.*

Remark: For all $\alpha \in \mathbb{C}$, $\binom{\alpha}{0} \triangleq 1$. For all $k \in \mathbb{N}$, $\binom{0}{k} \triangleq 1$.

Fact 1.9.2. The following inequalities hold:

i) Let $n \geq 2$. Then,

$$\frac{4^n}{n+1} < \binom{2n}{n} < 4^n.$$

ii) Let $n \geq 7$. Then,

$$\left(\frac{n}{3}\right)^n < n! < \left(\frac{n}{2}\right)^n.$$

iii) Let $1 \leq k \leq n$. Then,

$$\left(\frac{n}{k}\right)^k \leq \binom{n}{k} \leq \min\left\{\frac{n^k}{k!}, \left(\frac{ne}{k}\right)^k\right\}.$$

iv) Let $0 \leq k \leq n$. Then,

$$(n+1)^k \binom{n}{k} \leq n^k \binom{n+1}{k}.$$

v) Let $1 \leq k \leq n-1$. Then,

$$\sum_{i=1}^{k} i(i+1) \binom{2n}{k-i} < \frac{2^{2n-2} k(k+1)}{n}.$$

vi) Let $1 \le k \le n$. Then,

$$n^k \le k^{k/2}(k+1)^{(k-1)/2}\binom{n}{k}.$$

Proof: Statements *i*) and *ii*) are given in [242, p. 210]. Statement *iv*) is given in [686, p. 111]. Statement *vi*) is given in [464].

Fact 1.9.3. Let n be a positive integer. Then,

$$\sum_{i=1}^{n} i = \tfrac{1}{2}n(n+1),$$

$$\sum_{i=1}^{n} (2i-1) = n^2,$$

$$\sum_{i=1}^{n} i^2 = \tfrac{1}{6}n(n+1)(2n+1),$$

$$\sum_{i=1}^{n} i^3 = \tfrac{1}{4}n^2(n+1)^2 = \left(\sum_{i=1}^{n} i\right)^2,$$

$$\sum_{i=1}^{n} i^4 = \tfrac{1}{30}n(n+1)(2n+1)(3n^2+3n-1),$$

$$\sum_{i=1}^{n} i^5 = \tfrac{1}{12}n^2(n+1)^2(2n^2+2n-1).$$

Remark: See Fact 1.17.9 and [686, p. 11].

Remark: A matrix approach to sums of powers of integers is given in [453].

Fact 1.9.4. Let $n \ge 2$. Then,

$$n(\sqrt[n]{n+1}-1) < \sum_{i=1}^{n} \frac{1}{i} < 1 + n\left(1 - \frac{1}{\sqrt[n]{n}}\right).$$

Proof: See [686, pp. 158, 161].

Fact 1.9.5. Let n be a positive integer. Then,

$$\log(n+1) < \sum_{i=1}^{n} \frac{1}{i}.$$

Furthermore,

$$\gamma \triangleq \lim_{n\to\infty}\left[\left(\sum_{i=1}^{n}\frac{1}{i}\right) - \log(n+1)\right] = \lim_{n\to\infty}\left[\left(\sum_{i=1}^{n}\frac{1}{i}\right) - \log n\right],$$

where $\gamma \approx 0.57721$. In addition,

$$\left(\sum_{i=1}^{n}\frac{1}{i}\right) - \log(n+1) < \gamma < \left(\sum_{i=1}^{n}\frac{1}{i}\right) - \log n,$$

where the (leftmost, rightmost) term is (an increasing, a decreasing) function of n. Finally,

$$\lim_{n \to \infty} \frac{\sum_{i=1}^{n} \frac{1}{i}}{\log n} = 1.$$

Remark: γ is the *Euler constant*. See [409, 674].

Fact 1.9.6. The following statements hold:

$$\sum_{i=0}^{\infty} \frac{1}{i!} = e,$$

$$\sum_{i=1}^{\infty} \frac{1}{i^2} = \frac{\pi^2}{6},$$

$$\sum_{i=1}^{\infty} \frac{1}{i^4} = \frac{\pi^4}{90},$$

$$\sum_{i=1}^{\infty} \frac{1}{i^6} = \frac{\pi^6}{945},$$

$$\sum_{i=1}^{\infty} \frac{1}{(2i-1)^2} = \frac{\pi^2}{8},$$

$$\sum_{i=1}^{\infty} \frac{1}{(2i-1)^4} = \frac{\pi^4}{96},$$

$$\sum_{i=1}^{\infty} \frac{1}{(2i-1)^6} = \frac{\pi^6}{960},$$

$$\sum_{i=1}^{\infty} (-1)^{i+1} \frac{1}{i^2} = \frac{\pi^2}{12},$$

$$\sum_{i=1}^{\infty} (-1)^{i+1} \frac{1}{i^4} = \frac{7\pi^4}{720},$$

$$\sum_{i=1}^{\infty} (-1)^{i+1} \frac{1}{i^6} = \frac{31\pi^6}{30240},$$

$$\sum_{i=1}^{\infty} (-1)^{i+1} \frac{1}{2i-1} = \frac{\pi}{4},$$

$$\sum_{i=1}^{\infty} (-1)^{i+1} \frac{1}{(2i-1)^3} = \frac{5\pi^5}{1536},$$

$$\sum_{i=1}^{\infty} (-1)^{i+1} \frac{1}{(2i-1)^5} = \frac{61\pi^7}{184320}.$$

Fact 1.9.7. The following statements hold:

$$\sum_{i=1}^{\infty} \frac{1}{i^i} = \int_0^1 \frac{1}{x^x}\, dx \approx 1.29128$$

and

$$\sum_{i=1}^{\infty} (-1)^{i+1} \frac{1}{i^i} = \int_0^1 x^x \, \mathrm{d}x \approx 0.78343.$$

Proof: See [242, pp. 4, 44].

Fact 1.9.8. The following statements hold:

$$\sum_{i=1}^{\infty} \frac{1}{\binom{2i}{i}} = \frac{1}{3} + \frac{2\pi}{9\sqrt{3}},$$

$$\sum_{i=1}^{\infty} \frac{i}{\binom{2i}{i}} = \frac{2}{3} + \frac{2\pi}{9\sqrt{3}},$$

$$\sum_{i=1}^{\infty} \frac{i^2}{\binom{2i}{i}} = \frac{4}{3} + \frac{10\pi}{27\sqrt{3}},$$

$$\sum_{i=1}^{\infty} \frac{1}{i\binom{2i}{i}} = \frac{\pi}{3\sqrt{3}},$$

$$\sum_{i=1}^{\infty} \frac{1}{i^2\binom{2i}{i}} = \frac{\pi^2}{18},$$

$$\sum_{i=1}^{\infty} \frac{2-i}{\binom{2i}{i}} = \frac{2\pi}{9\sqrt{3}},$$

$$\sum_{i=0}^{\infty} \frac{25i-3}{2^{i-1}\binom{3i}{i}} = \pi.$$

Proof: See [242, pp. 20, 25, 26].

Fact 1.9.9. The following statements hold:

$$\prod_{i=2}^{\infty} \frac{i^2-1}{i^2+1} = \frac{1}{2}\prod_{i=2}^{\infty} \frac{i^2}{i^2+1} = \frac{\pi}{\sinh\pi} \approx 0.2720,$$

$$\prod_{i=2}^{\infty} \frac{i^2-1}{i^2} = \frac{1}{2},$$

$$\prod_{i=2}^{\infty} \frac{i^3-1}{i^3+1} = \frac{2}{3},$$

$$\prod_{i=2}^{\infty} \frac{i^4-1}{i^4+1} = \frac{\pi\sinh\pi}{\cosh(\sqrt{2}\pi) - \cos(\sqrt{2}\pi)} \approx 0.8480.$$

Proof: See [242, pp. 4, 5].

Fact 1.9.10. The following statements hold for all real numbers x:

$$\sin x = x \prod_{i=1}^{\infty} \left(1 - \frac{x^2}{i^2 \pi^2}\right),$$

$$\cos x = \prod_{i=1}^{\infty} \left(1 - \frac{4x^2}{(2i-1)^2 \pi^2}\right),$$

$$\sinh x = x \prod_{i=1}^{\infty} \left(1 + \frac{x^2}{i^2 \pi^2}\right),$$

$$\cosh x = \prod_{i=1}^{\infty} \left(1 + \frac{4x^2}{(2i-1)^2 \pi^2}\right),$$

$$\sin x = x \prod_{i=1}^{\infty} \cos \frac{x}{2^i}.$$

Fact 1.9.11. For $i \in \{1, 2, \ldots\}$, let p_i denote the ith prime number, where $p_1 = 2$. Then,

$$\frac{\pi^2}{6} = \prod_{i=1}^{\infty} \frac{1}{1 - p_i^{-2}} \approx 1.6449.$$

Remark: This equality is the *Euler product formula* for $\zeta(2)$, where ζ is the zeta function.

1.10 Facts on Convex Functions

Fact 1.10.1. Let $\mathfrak{I}$ be a finite or infinite interval, and let $f: \mathfrak{I} \mapsto \mathbb{R}$. Then, in each case below, f is convex:

i) $\mathfrak{I} = (0, \infty)$, $f(x) = -\log x$.

ii) $\mathfrak{I} = (0, \infty)$, $f(x) = x \log x$.

iii) $\mathfrak{I} = (0, \infty)$, $f(x) = x^p$, where $p < 0$.

iv) $\mathfrak{I} = [0, \infty)$, $f(x) = -x^p$, where $p \in (0, 1)$.

v) $\mathfrak{I} = [0, \infty)$, $f(x) = x^p$, where $p \in (1, \infty)$.

vi) $\mathfrak{I} = [0, \infty)$, $f(x) = (1 + x^p)^{1/p}$, where $p \in (1, \infty)$.

vii) $\mathfrak{I} = \mathbb{R}$, $f(x) = \frac{a^x - b^x}{c^x - d^x}$, where $0 < d < c < b < a$.

viii) $\mathfrak{I} = \mathbb{R}$, $f(x) = \log \frac{a^x - b^x}{c^x - d^x}$, where $0 < d < c < b < a$ and $ad \geq bc$.

Proof: Statements *vii)* and *viii)* are given in [242, p. 39].

Fact 1.10.2. Let $\mathfrak{I} \subseteq (0, \infty)$ be a finite or infinite interval, let $f: \mathfrak{I} \mapsto \mathbb{R}$, and define $g: \mathfrak{I} \mapsto \mathbb{R}$ by $g(x) = xf(1/x)$. Then, f is (convex, strictly convex) if and only if g is (convex, strictly convex).

Proof: See [1066, p. 13].

Fact 1.10.3. Let $f: \mathbb{R} \mapsto \mathbb{R}$, assume that f is convex, and assume that there exists $\alpha \in \mathbb{R}$ such that, for all $x \in \mathbb{R}$, $f(x) \leq \alpha$. Then, f is constant.

Proof: See [1066, p. 35].

Fact 1.10.4. Let $\mathfrak{I} \subseteq \mathbb{R}$ be a finite or infinite interval, let $f: \mathfrak{I} \mapsto \mathbb{R}$, and assume that f is continuous. Then, the following statements are equivalent:

i) f is convex.

ii) For all $n \in \mathbb{P}$, $x_1, \ldots, x_n \in \mathfrak{I}$, and $\alpha_1, \ldots, \alpha_n \in [0, 1]$ such that $\sum_{i=1}^{n} \alpha_i = 1$, it follows that

$$f\left(\sum_{i=1}^{n} \alpha_i x_i\right) \leq \sum_{i=1}^{n} \alpha_i f(x_i).$$

Remark: This result is *Jensen's inequality*.

Remark: Setting $f(x) = x^p$ yields Fact 1.17.36, whereas setting $f(x) = \log x$ for $x \in (0, \infty)$ yields the arithmetic-mean–geometric-mean inequality given by Fact 1.17.14.

Remark: See Fact 10.11.7.

Fact 1.10.5. Let $[a, b] \subset \mathbb{R}$, let $f: [a, b] \mapsto \mathbb{R}$ be convex, and let $x, y \in [a, b]$. Then,

$$\tfrac{1}{2}[f(x) + f(y)] - f[\tfrac{1}{2}(x + y)] \leq \tfrac{1}{2}[f(a) + f(b)] - f[\tfrac{1}{2}(a + b)].$$

Remark: This result is *Niculescu's inequality*. See [102, p. 13].

Fact 1.10.6. Let $\mathfrak{I} \subseteq \mathbb{R}$ be a finite or infinite interval, let $f: \mathfrak{I} \mapsto \mathbb{R}$. Then, the following statements are equivalent:

i) f is convex.

ii) f is continuous, and, for all $x, y \in \mathfrak{I}$,

$$\tfrac{2}{3}(f[\tfrac{1}{2}(x+y)]+f[\tfrac{1}{2}(y+z)]+f[\tfrac{1}{2}(x+z)]) \leq \tfrac{1}{3}[f(x)+f(y)+f(z)]+f[\tfrac{1}{3}(x+y+z).$$

Remark: This result is *Popoviciu's inequality*. See [1066, p. 12].

Remark: For the case of a scalar argument and $f(x) = |x|$, this result implies Hlawka's inequality given by Fact 9.7.4. See Fact 1.20.2 and [1068].

Problem: Extend this result so that it yields Hlawka's inequality for vector arguments.

Fact 1.10.7. Let $[a, b] \subset \mathbb{R}$, let $f: [a, b] \mapsto \mathbb{R}$, and assume that f is convex. Then,

$$f[\tfrac{1}{2}(a + b)] \leq \tfrac{1}{b-a} \int_a^b f(x)\, \mathrm{d}x \leq \tfrac{1}{2}[f(a) + f(b)].$$

Proof: See [1066, pp. 50–53] and [1186, 1188].

Remark: This result is the *Hermite-Hadamard inequality*.

1.11 Facts on Scalar Identities and Inequalities in One Variable

Fact 1.11.1. Let x and α be real numbers, and assume that $x \geq -1$. Then, the following statements hold:

i) If $\alpha \leq 0$, then
$$1 + \alpha x \leq (1 + x)^\alpha.$$

Furthermore, equality holds if and only if either $x = 0$ or $\alpha = 0$.

ii) If $\alpha \in [0, 1]$, then
$$(1 + x)^\alpha \leq 1 + \alpha x.$$

Furthermore, equality holds if and only if either $x = 0$, $\alpha = 0$, or $\alpha = 1$.

iii) If $\alpha \geq 1$, then
$$1 + \alpha x \leq (1 + x)^\alpha.$$

Furthermore, equality holds if and only if either $x = 0$ or $\alpha = 1$.

Proof: See [36], [280, p. 4], and [1035, p. 65]. Alternatively, the result follows from Fact 1.11.26. See [1481].

Remark: These results are *Bernoulli's inequality*. An equivalent version is given by Fact 1.11.2.

Remark: The proof of *i)* and *iii)* in [36] is based on the fact that, for $x \geq -1$, the function $f(x) \triangleq \frac{(1+x)^\alpha - 1}{x}$ for $x \neq 0$ and $f(0) \triangleq \alpha$, is increasing.

Fact 1.11.2. Let x be a nonnegative number, and let α be a real number. If $\alpha \in [0, 1]$, then
$$\alpha + x^\alpha \leq 1 + \alpha x,$$

whereas, if either $\alpha \leq 0$ or $\alpha \geq 1$, then
$$1 + \alpha x \leq \alpha + x^\alpha.$$

Proof: Set $y = x + 1$ in Fact 1.11.1. Alternatively, for the case $\alpha \in [0, 1]$, set $y = 1$ in the right-hand inequality in Fact 1.12.21. For the case $\alpha \geq 1$, note that $f(x) \triangleq \alpha + x^\alpha - 1 - \alpha x$ satisfies $f(1) = 0$, $f'(1) = 0$, and, for all $x \geq 0$, $f''(x) = \alpha(\alpha - 1)x^{\alpha-2} > 0$.

Remark: This result is equivalent to Bernoulli's inequality. See Fact 1.11.1.

Remark: For $\alpha \in [0, 1]$ a matrix version is given by Fact 8.9.43.

Problem: Compare the second inequality to Fact 1.12.22 with $y = 1$.

Fact 1.11.3. Let x and α be real numbers, assume that either $\alpha \leq 0$ or $\alpha \geq 1$, and assume that $x \in [0, 1]$. Then,
$$(1 + x)^\alpha \leq 1 + (2^\alpha - 1)x.$$

Furthermore, equality holds if and only if either $\alpha = 0$, $\alpha = 1$, $x = 0$, or $x = 1$.

Proof: See [36].

Fact 1.11.4. Let $x \in (0, 1)$, and let k be a positive integer. Then,

$$(1 - x)^k < \frac{1}{1 + kx}.$$

Proof: See [686, p. 137].

Fact 1.11.5. Let x be a nonnegative number. Then,

$$8x < x^4 + 9,$$
$$3x^2 \leq x^3 + 4,$$
$$4x^2 < x^4 + x^3 + x + 1,$$
$$8x^2 < x^4 + x^3 + 4x + 4,$$
$$3x^5 < x^{11} + x^4 + 1.$$

Now, let n be a positive integer. Then,

$$(2n + 1)x^n \leq \sum_{i=1}^{2n} x^i.$$

Proof: See [686, pp. 117, 123, 152, 153, 155].

Fact 1.11.6. Let x be a positive number. Then,

$$1 + \tfrac{1}{2}x - \tfrac{1}{8}x^2 < \sqrt{1 + x} < 1 + \tfrac{1}{2}x - \tfrac{1}{8}x^2 + \tfrac{1}{16}x^3.$$

Proof: See [805, p. 55].

Fact 1.11.7. Let $x \in (0, 1)$. Then,

$$\frac{1}{2 - x} < x^x < x^2 - x + 1.$$

Proof: See [686, p. 164].

Fact 1.11.8. Let $x, p \in [1, \infty)$. Then,

$$x^{1/p}(x - 1) < px(x^{1/p} - 1).$$

Furthermore, equality holds if and only if either $p = 1$ or $x = 1$.

Proof: See [544, p. 194].

Fact 1.11.9. If $p \in [\sqrt{2}, 2)$, then, for all $x \in (0, 1)$, it follows that

$$\left[\frac{1 - x^p}{p(1 - x)} \right]^2 \leq \tfrac{1}{2}(1 + x^{p-1}).$$

Furthermore, if $p \in (1, \sqrt{2})$, then there exists $x \in (0, 1)$, such that

$$\tfrac{1}{2}(1 + x^{p-1}) < \left[\frac{1 - x^p}{p(1 - x)} \right]^2.$$

Proof: See [210].

Fact 1.11.10. Let $x, p \in [1, \infty)$. Then,

$$(p-1)^{p-1}(x^p-1)^p \leq p^p(x-1)(x^p-x)^{p-1}x^{p-1}.$$

Furthermore, equality holds if and only if either $p = 1$ or $x = 1$.

Proof: See [544, p. 194].

Fact 1.11.11. Let $x \in [1, \infty)$, and let $p, q \in (1, \infty)$ satisfy $1/p + 1/q = 1$. Then,

$$px^{1/q} \leq 1 + (p-1)x.$$

Furthermore, equality holds if and only if $x = 1$.

Proof: See [544, p. 194].

Fact 1.11.12. Let $x \in [1, \infty)$, and let $p, q \in (1, \infty)$ satisfy $1/p + 1/q = 1$. Then,

$$x - 1 \leq p^{1/p}q^{1/q}(x^{1/p}-1)^{1/p}(x^{1/q}-1)^{1/q}x^{2/(pq)}.$$

Furthermore, equality holds if and only if $x = 1$.

Proof: See [544, p. 195].

Fact 1.11.13. Let x be a real number, and let $p, q \in (1, \infty)$ satisfy $1/p + 1/q = 1$. Then,

$$\tfrac{1}{p}e^{px} + \tfrac{1}{q}e^{-qx} \leq e^{p^2q^2x^2/8}.$$

Proof: See [893, p. 260].

Fact 1.11.14. Let x and y be positive numbers. If $x \in (0, 1]$ and $y \in [0, x]$, then

$$\left(1 + \frac{1}{x}\right)^y \leq 1 + \frac{y}{x}.$$

Equality holds if and only if either $y = 0$ or $x = y = 1$. If $x \in (0, 1)$, then

$$\left(1 + \frac{1}{x}\right)^x < 2.$$

If $x > 1$ and $y \in [1, x]$, then

$$1 + \frac{y}{x} \leq \left(1 + \frac{1}{x}\right)^y < 1 + \frac{y}{x} + \frac{y^2}{x^2}.$$

The left-hand inequality is an equality if and only if $y = 1$. Finally, if $x > 1$, then

$$2 < \left(1 + \frac{1}{x}\right)^x < 3.$$

Proof: See [686, p. 137].

Fact 1.11.15. Let x be a nonnegative number, and let p and q be real numbers such that $0 < p \leq q$. Then,

$$e^x\left(1 + \frac{1}{p}\right)^{-x} \leq \left(1 + \frac{x}{p}\right)^p \leq \left(1 + \frac{x}{q}\right)^q \leq e^x.$$

Furthermore, if $p < q$, then equality holds if and only if $x = 0$. Finally,

$$\lim_{q \to \infty} \left(1 + \frac{x}{q}\right)^q = e^x.$$

Proof: See [280, pp. 7, 8].

Remark: For $q \to \infty$, $(1 + 1/q)^q = e + O(1/q)$, whereas $(1 + 1/q)^q[1 + 1/(2q)] = e + O(1/q^2)$. See [853].

Fact 1.11.16. Let x be a positive number. Then,

$$\sqrt{\frac{x}{x+1}}\, e < \left(1 + \frac{1}{x}\right)^x < \frac{2x+1}{2x+2}\, e$$

and

$$\sqrt{1 + \frac{1}{x}}\, e^{-1/[12x(x+1)]} < \frac{2x+2}{2x+1}\, e^{1/[6(2x+1)^2]}$$

$$< \frac{e}{\left(1 + \frac{1}{x}\right)^x}$$

$$< \sqrt{1 + \frac{1}{x}}\, e^{-1/[3(2x+1)^2]}.$$

Proof: See [1190].

Fact 1.11.17. e is given by

$$\lim_{q \to \infty} \left(\frac{q+1}{q-1}\right)^{q/2} = e$$

and

$$\lim_{q \to \infty} \left[\frac{q^q}{(q-1)^{q-1}} - \frac{(q-1)^{q-1}}{(q-2)^{q-2}}\right] = e.$$

Proof: These expressions are given in [1187] and [853], respectively.

Fact 1.11.18. Let x be a positive number. Then,

$$\left(1 + \frac{1}{x + \frac{1}{5}}\right)^{1/2} < \left(1 + \frac{2}{3x+1}\right)^{3/4}$$

$$< \left(1 + \frac{1}{\frac{5}{4}x + \frac{1}{3}}\right)^{5/8}$$

$$< \frac{e}{\left(1 + \frac{1}{x}\right)^x}$$

$$< \left(1 + \frac{1}{x + \frac{1}{6}}\right)^{1/2}.$$

Proof: See [946].

Fact 1.11.19. Let n be a positive integer. If $n \geq 3$, then

$$n! < 2^{n(n-1)/2}.$$

If $n \geq 6$, then

$$\left(\frac{n}{3}\right)^n < n! < \left(\frac{n}{2}\right)^n.$$

Proof: See [686, p. 137].

Fact 1.11.20. Let $n \geq 3$. Then,

$$e\left(\frac{n}{e}\right)^n < \sqrt{2\pi n}\left(\frac{n}{e}\right)^n < n! < \sqrt{\frac{n}{n-1}}\sqrt{2\pi n}\left(\frac{n}{e}\right)^n < \left(\frac{n+1}{2}\right)^n < e\left(\frac{n}{2}\right)^n.$$

Proof: See [1190].

Remark: The second expression is *Stirling's formula* for $n!$.

Remark: The fourth inequality requires $n > 2$.

Remark: $\sqrt{2\pi} < e < 2\sqrt{\pi}$.

Fact 1.11.21. Let x and a be positive numbers. Then,

$$\log x \leq ax - \log a - 1.$$

In particular,

$$\log x \leq \frac{x}{e}.$$

Fact 1.11.22. Let x be a positive number. Then,

$$\frac{x-1}{x} \leq \log x \leq x - 1.$$

Furthermore, equality holds if and only if $x = 1$.

Fact 1.11.23. Let x be a positive number such that $x > 1$. Then,

$$\frac{1}{x^2+1} \leq \frac{\log x}{x^2-1} \leq \frac{1}{2x}.$$

Furthermore,

$$\lim_{x\downarrow 1}\frac{\log x}{x^2-1} = \frac{1}{2}.$$

Remark: In this case, the indeterminate $0/0$ is $1/2$ rather than the convention 1.

Fact 1.11.24. Let x be a positive number. Then,

$$\frac{2|x-1|}{x+1} \leq |\log x| \leq \frac{|x-1|(1+x^{1/3})}{x+x^{1/3}} \leq \frac{|x-1|}{\sqrt{x}}.$$

Furthermore, equality holds if and only if $x = 1$.

Proof: See [280, p. 8].

Fact 1.11.25. If $x \in (0,1]$, then

$$\frac{x-1}{x} \le \frac{x^2-1}{2x} \le \frac{x-1}{\sqrt{x}} \le \frac{(x-1)(1+x^{1/3})}{x+x^{1/3}} \le \log x \le \frac{2(x-1)}{x+1} \le \frac{x^2-1}{x^2+1} \le x-1.$$

If $x \ge 1$, then

$$\frac{x-1}{x} \le \frac{x^2-1}{x^2+1} \le \frac{2(x-1)}{x+1} \le \log x \le \frac{(x-1)(1+x^{1/3})}{x+x^{1/3}} \le \frac{x-1}{\sqrt{x}} \le \frac{x^2-1}{2x} \le x-1.$$

Furthermore, equality holds in all cases if and only if $x = 1$.

Proof: See [280, p. 8] and [640].

Fact 1.11.26. Let x be a positive number, and let p and q be real numbers such that $0 < p \le q$. Then,

$$\log x \le \frac{x^p-1}{p} \le \frac{x^q-1}{q} \le x^q \log x.$$

In particular,

$$\log x \le 2(\sqrt{x}-1) \le x-1.$$

Furthermore, equality holds in the second inequality if and only if either $p = q$ or $x = 1$. Finally,

$$\lim_{p\downarrow 0} \frac{x^p-1}{p} = \log x.$$

Proof: See [36, 1481] and [280, p. 8].

Remark: See Proposition 8.6.4 and Fact 8.13.1.

Fact 1.11.27. Let $x > 0$. Then,

$$x - \tfrac{1}{2}x^2 + \tfrac{1}{3}x^3 - \tfrac{1}{4}x^4 < \log(1+x) < x - \tfrac{1}{2}x^2 + \tfrac{1}{3}x^3.$$

Proof: See [805, p. 55].

Fact 1.11.28. Let $x > 1$. Then,

$$\frac{x-1}{\log x} < \left(\frac{x^{1/2}+x^{1/4}+1}{3}\right)^2 < \left(\frac{x^{1/3}+1}{2}\right)^3.$$

Proof: See [778].

Fact 1.11.29. Let x be a real number. Then, the following statements hold:

i) If $x \in [0, \pi/2]$, then

$$\left.\begin{array}{c} x\cos x \\[2pt] \frac{2}{\pi}x \le \frac{2}{\pi}x + \frac{1}{\pi^3}x(\pi^2-4x^2) \\[2pt] \dfrac{x}{\sqrt{(1-4/\pi^2)x^2+1}} \end{array}\right\} \le \sin x \le \left\{\begin{array}{c} \frac{2}{\pi}x + \frac{\pi-2}{\pi^3}x(\pi^2-4x^2) \\[2pt] x \le \tan x \\[2pt] 1. \end{array}\right.$$

ii) If $x \in (0, \pi/2]$, then

$$\cot^2 x < \frac{1}{x^2} < 1 + \cot^2 x.$$

iii) If $x \in (0, \pi)$, then

$$\tfrac{1}{\pi}x(\pi - x) \le \sin x \le \tfrac{4}{\pi^2}x(\pi - x).$$

iv) If $x \in [-4, 4]$, then

$$\cos x \le \frac{\sin x}{x} \le 1.$$

v) If $x \in [-\pi/2, \pi/2]$ and $p \in [0, 3]$, then

$$\cos x \le \left(\frac{\sin x}{x}\right)^p \le 1.$$

vi) If $x \ne 0$, then

$$x - \tfrac{1}{6}x^3 < \sin x < x - \tfrac{1}{6}x^3 + \tfrac{1}{120}x^5.$$

vii) If $x \ne 0$, then

$$1 - \tfrac{1}{2}x^2 < \cos x < 1 - \tfrac{1}{2}x^2 + \tfrac{1}{24}x^4.$$

viii) If $x \ge \sqrt{3}$, then

$$1 + x\cos\frac{\pi}{x} < (x + 1)\cos\frac{\pi}{x + 1}.$$

ix) If $x \in [0, \pi/2)$,

$$\frac{4x}{\pi - 2x} \le \pi\tan x.$$

x) If $x \in [0, \pi/2)$, then

$$2 \le \tfrac{16}{\pi^4}x^3\tan x + 2 \le \left(\frac{\sin x}{x}\right)^2 + \frac{\tan x}{x} \le \tfrac{8}{45}x^3\tan x + 2.$$

xi) If $x \in (0, \pi/2)$, then

$$3x < 2\sin x + \tan x.$$

xii) For all $x > 0$,

$$3\sin x < (2 + \cos x)x.$$

xiii) If $x \in [0, \pi/2]$,

$$2\log\sec x \le (\sin x)\tan x.$$

xiv) If $x \in (0, 1)$, then

$$\sin^{-1} x < \frac{x}{1 - x^2}.$$

xv) If $x > 0$, then

$$\left.\begin{array}{c} \dfrac{x}{x^2 + 1} \\[2mm] \dfrac{3x}{1 + 2\sqrt{x^2 + 1}} \end{array}\right\} < \tan^{-1} x.$$

xvi) If $x \in (0, \pi/2)$, then

$$\sinh x < 2\tan x.$$

xvii) If $x \in \mathbb{R}$, then

$$1 \le \frac{\sinh x}{x} \le \cosh x \le \left(\frac{\sinh x}{x}\right)^3.$$

xviii) If $x > 0$ and $p \ge 3$, then

$$\cosh x < \left(\frac{\sinh x}{x}\right)^p.$$

xix) If $x > 0$, then

$$2 \leq \tfrac{8}{45}x^3 \tan x + 2 \leq \left(\frac{\sinh x}{x}\right)^2 + \frac{\tanh x}{x}.$$

xx) If $x > 0$, then

$$\frac{\sinh x}{\sqrt{\sinh^2 x + \cosh^2 x}} < \tanh x < x < \sinh x < \tfrac{1}{2}\sinh 2x.$$

Proof: Statements *i*), *iv*), *viii*), *ix*), and *xiii*) are given in [279, pp. 250, 251]. For *i*), see also [805, p. 75] and [928]. Statement *ii*) follows from $\sin x < x < \tan x$ in statement *i*). Statement *iii*) is given in [805, p. 72]. Statement *v*) is given in [1537]. Statements *vi*) and *vii*) are given in [805, p. 68]. Statement *x*) is given in [36, 1466]. See also [280, p. 9], [893, pp. 270–271], and [1536, 1537]. Statement *xi*) is *Huygens's inequality*. See [805, p. 71] and [893, p. 266]. Statement *xii*) is given in [805, p. 71] and [893, p. 266]. Statement *xiv*) is given in [893, p. 271]. Statements *xv*) and *xvi*) are given in [805, pp. 70, 75]. Statement *xvii*) is given in [279, pp. 131] and [691, p. 71]. Statements *xviii*) and *xix*) are given in [1537]. Statement *xx*) is given in [805, p. 74].

Remark: The inequality $2/\pi \leq (\sin x)/x$ is *Jordan's inequality*. See [928].

Fact 1.11.30. The following statements hold:

i) If $x \in \mathbb{R}$, then
$$\frac{1 - x^2}{1 + x^2} \leq \frac{\sin \pi x}{\pi x}.$$

ii) If $|x| \geq 1$, then
$$\frac{1 - x^2}{1 + x^2} + \frac{(1 - x)^2}{x(1 + x^2)} \leq \frac{\sin \pi x}{\pi x}.$$

iii) If $x \in (0, 1)$, then
$$\frac{(1 - x^2)(4 - x^2)(9 - x^2)}{x^6 - 2x^4 + 13x^2 + 36} \leq \frac{\sin \pi x}{\pi x} \leq \frac{1 - x^2}{\sqrt{1 + 3x^4}}.$$

Proof: See [928].

Fact 1.11.31. Let n be a positive integer, and let r be a positive number. Then,

$$\frac{n}{n + 1} \leq \left[\frac{(n + 1)\sum_{i=1}^{n} i^r}{n \sum_{i=1}^{n+1} i^r}\right]^{1/r} \leq \frac{\sqrt[n]{n!}}{\sqrt[n+1]{(n + 1)!}}.$$

Proof: See [4].

Remark: The left-hand inequality is *Alzer's inequality*, while the right-hand inequality is *Martins's inequality*.

1.12 Facts on Scalar Identities and Inequalities in Two Variables

Fact 1.12.1. Let m and n be positive integers. Then,
$$(m^2 - n^2)^2 + (2mn)^2 = (m^2 + n^2)^2.$$
In particular, if $m = 2$ and $n = 1$, then
$$3^2 + 4^2 = 5^2,$$
whereas, if $m = 3$ and $n = 2$, then
$$5^2 + 12^2 = 13^2.$$
Furthermore, if $m = 4$ and $n = 1$, then
$$8^2 + 15^2 = 17^2,$$
whereas, if $m = 4$ and $n = 3$, then
$$7^2 + 24^2 = 25^2.$$

Remark: This result characterizes all *Pythagorean triples* within an integer multiple.

Fact 1.12.2. The following integer statements hold:

i) $3^3 + 4^3 + 5^3 = 6^3$.

ii) $1^3 + 12^3 = 9^3 + 10^3$.

iii) $10^2 + 11^2 + 12^2 = 13^2 + 14^2$.

iv) $21^2 + 22^2 + 23^2 + 24^2 = 25^2 + 26^2 + 27^2$.

Remark: The cube of a positive integer cannot be the sum of the cubes of two positive integers. See Fact 1.13.18.

Fact 1.12.3. Let $x, y \in \mathbb{R}$. Then,
$$x^2 - y^2 = (x - y)(x + y),$$
$$x^3 - y^3 = (x - y)(x^2 + xy + y^2),$$
$$x^3 + y^3 = (x + y)(x^2 - xy + y^2),$$
$$x^4 - y^4 = (x - y)(x + y)(x^2 + y^2),$$
$$x^4 + x^2 y^2 + y^4 = (x^2 + xy + y^2)(x^2 - xy + y^2),$$
$$x^4 + (x + y)^4 + y^4 = 2(x^2 + xy + y^2)^2,$$
$$x^5 - y^5 = (x - y)(x^4 + x^3 y + x^2 y^2 + xy^3 + y^4),$$
$$x^5 + y^5 = (x + y)(x^4 - x^3 y + x^2 y^2 - xy^3 + y^4),$$
$$x^6 - y^6 = (x - y)(x + y)(x^2 + xy + y^2)(x^2 - xy + y^2).$$

Fact 1.12.4. Let x and y be real numbers. Then,

$$xy \le \tfrac{1}{4}(x+y)^2 \le \tfrac{1}{2}(x^2+y^2).$$

If, in addition, x and y are positive, then

$$2 \le \frac{x}{y} + \frac{y}{x}$$

and

$$\frac{2}{\frac{1}{x} + \frac{1}{y}} \le \sqrt{xy} \le \tfrac{1}{2}(x+y).$$

Remark: See Fact 8.10.7.

Fact 1.12.5. Let x and y be positive numbers, and assume that $0 < x < y$. Then,

$$\frac{(x-y)^2}{8y} < \frac{(x-y)^2}{4(x+y)} < \tfrac{1}{2}(x+y) - \sqrt{xy} < \frac{(x-y)^2}{8x}.$$

Proof: See [140, p. 231] and [470, p. 183].

Fact 1.12.6. Let x and y be real numbers, and let $\alpha \in [0,1]$. Then,

$$\sqrt{\alpha}x + \sqrt{1-\alpha}y \le (x^2+y^2)^{1/2}.$$

Furthermore, equality holds if and only if one of the following conditions holds:

i) $x = y = 0$.

ii) $x = 0$, $y > 0$, and $\alpha = 0$.

iii) $x > 0$, $y = 0$, and $\alpha = 1$.

iv) $x > 0$, $y > 0$, and $\alpha = \frac{x^2}{x^2+y^2}$.

Fact 1.12.7. Let α be a real number. Then,

$$0 \le x^2 + \alpha xy + y^2$$

for all real numbers x, y if and only if $\alpha \in [-2, 2]$.

Fact 1.12.8. Let x and y be nonnegative numbers. Then,

$$9xy^2 \le 3x^3 + 7y^3,$$
$$27x^2y \le 4(x+y)^3,$$
$$6xy^2 \le x^3 + y^6 + 8,$$
$$x^2y + y^2x \le x^3 + y^3,$$
$$x^3y + y^3x \le x^4 + y^4,$$
$$x^4y + y^4x \le x^5 + y^5,$$
$$5x^6y^6 \le 2x^{15} + 3y^{10},$$
$$8(x^3y + y^3x) \le (x+y)^4,$$
$$4x^2y \le x^4 + x^3y + y^2 + xy,$$
$$4x^2y \le x^4 + x^3y^2 + y^2 + x,$$

$$12xy \leq 4x^2y + 4y^2x + 4x + y,$$
$$6x^2y^2 \leq x^4 + 2x^3y + 2y^3x + y^4,$$
$$9xy \leq (x^2 + x + 1)(y^2 + y + 1),$$
$$4(x^2y + y^2x) \leq 2(x^2 + y^2)^2 + x^2 + y^2,$$
$$2(x^2y + y^2x + x^2y^2) \leq 2(x^4 + y^4) + x^2 + y^2.$$

Proof: See Fact 1.17.8, [470, p. 183], [686, pp. 117, 120, 123, 124, 150, 153, 155].

Fact 1.12.9. Let x and y be real numbers. Then,

$$x^3y + y^3x \leq x^4 + y^4,$$
$$4xy(x - y)^2 \leq (x^2 - y^2)^2,$$
$$2x + 2xy \leq x^2y^2 + x^2 + 2,$$
$$3(x + y + xy) \leq (x + y + 1)^2.$$

Proof: See [686, p. 117].

Fact 1.12.10. Let x and y be real numbers. Then,

$$2|(x + y)(1 - xy)| \leq (1 + x^2)(1 + y^2).$$

Proof: See [470, p. 185].

Fact 1.12.11. Let x and y be real numbers, and assume that $xy(x + y) \geq 0$. Then,

$$(x^2 + y^2)(x^3 + y^3) \leq (x + y)(x^4 + y^4).$$

Proof: See [470, p. 183].

Fact 1.12.12. Let x and y be real numbers. Then,

$$[x^2 + y^2 + (x + y)^2]^2 = 2[x^4 + y^4 + (x + y)^4].$$

Therefore,

$$\tfrac{1}{2}(x^2 + y^2)^2 \leq x^4 + y^4 + (x + y)^4$$

and

$$x^4 + y^4 \leq \tfrac{1}{2}[x^2 + y^2 + (x + y)^2]^2.$$

Remark: This result is *Candido's identity*. See [27].

Fact 1.12.13. Let x and y be real numbers. Then,

$$54x^2y^2(x + y)^2 \leq [x^2 + y^2 + (x + y)^2]^3.$$

Equivalently,

$$[x^2y^2(x + y)^2]^{1/3} \leq \tfrac{1}{\sqrt[3]{2}}\tfrac{1}{3}[x^2 + y^2 + (x + y)^2]^3.$$

Remark: This result interpolates the arithmetic-mean–geometric-mean inequality due to the factor $1/\sqrt[3]{2}$.

Remark: This inequality is used in Fact 4.10.3.

Fact 1.12.14. Let x and y be real numbers, and let $p \in [1, \infty)$. Then,

$$(p-1)(x-y)^2 + [\tfrac{1}{2}(x+y)]^2 \le [\tfrac{1}{2}(|x|^p + |y|^p)]^{2/p}.$$

Proof: See [556, p. 148].

Fact 1.12.15. Let x and y be complex numbers. If $p \in [1, 2]$, then

$$[|x|^2 + (p-1)|y|^2]^{1/2} \le [\tfrac{1}{2}(|x+y|^p + |x-y|^p)]^{1/p}.$$

If $p \in [2, \infty]$, then

$$[\tfrac{1}{2}(|x+y|^p + |x-y|^p)]^{1/p} \le [|x|^2 + (p-1)|y|^2]^{1/2}.$$

Proof: See Fact 9.9.35.

Fact 1.12.16. Let x and y be real numbers, let p and q be real numbers, and assume that $1 \le p \le q$. Then,

$$[\tfrac{1}{2}(|x + \tfrac{y}{\sqrt{q-1}}|^q + |x - \tfrac{y}{\sqrt{q-1}}|^q)]^{1/q} \le [\tfrac{1}{2}(|x + \tfrac{y}{\sqrt{p-1}}|^p + |x - \tfrac{y}{\sqrt{p-1}}|^p)]^{1/p}.$$

Proof: See [556, p. 206].

Remark: This result is the scalar version of Bonami's inequality. See Fact 9.7.20.

Fact 1.12.17. Let x and y be positive numbers, and let n be a positive integer. Then,

$$(n+1)(xy^n)^{1/(n+1)} < x + ny.$$

Proof: See [893, p. 252].

Fact 1.12.18. Let x and y be positive numbers such that $x < y$, and let n be a positive integer. Then,

$$(n+1)(y-x)x^n < y^{n+1} - x^{n+1} < (n+1)(y-x)y^n.$$

Proof: See [893, p. 248].

Fact 1.12.19. Let $[a, b] \subset \mathbb{R}$, and let $x, y \in [a, b]$. Then,

$$|x| + |y| - |x+y| \le |a| + |b| - |a+b|.$$

Proof: Use Fact 1.10.5.

Fact 1.12.20. Let $[a, b] \subset (0, \infty)$, and let $x, y \in [a, b]$. Then,

$$\sqrt{\frac{x}{y}} + \sqrt{\frac{y}{x}} \le \sqrt{\frac{a}{b}} + \sqrt{\frac{b}{a}}.$$

Proof: Use Fact 1.10.5.

Fact 1.12.21. Let x and y be nonnegative numbers, and let $\alpha \in [0, 1]$. Then,

$$[\alpha x^{-1} + (1-\alpha)y^{-1}]^{-1} \le x^\alpha y^{1-\alpha} \le \alpha x + (1-\alpha)y.$$

Remark: The right-hand inequality follows from the concavity of the logarithm function.

Remark: The left-hand inequality is the scalar *Young inequality*. See Fact 8.10.46, Fact 8.12.27, and Fact 8.12.28.

Fact 1.12.22. Let x and y be distinct positive numbers, and let $\alpha \in [0,1]$. Then,

$$\alpha x + (1-\alpha)y \le \gamma x^\alpha y^{1-\alpha},$$

where $\gamma > 0$ is defined by

$$\gamma \triangleq \frac{(h-1)h^{1/(h-1)}}{e \log h}$$

and $h \triangleq \max\{y/x, x/y\}$. In particular,

$$\sqrt{xy} \le \tfrac{1}{2}(x+y) \le \gamma \sqrt{xy}.$$

Remark: This result is the *reverse Young inequality*. See Fact 1.12.21. The case $\alpha = 1/2$ is the *reverse arithmetic-mean–geometric mean inequality*. See Fact 1.17.19.

Remark: $\gamma = S(1,h)$ is *Specht's ratio*. See Fact 1.17.19 and Fact 11.14.22.

Remark: This result is due to Tominaga. See [528].

Fact 1.12.23. Let x and y be positive numbers. Then,

$$1 < x^y + y^x.$$

Proof: See [470, p. 184] or [805, p. 75].

Fact 1.12.24. Let x and y be positive numbers. Then,

$$(x+y)\log[\tfrac{1}{2}(x+y)] \le x\log x + y\log y.$$

Proof: This result follows from the fact that $f(x) = x\log x$ is convex on $(0,\infty)$. See [805, p. 62].

Fact 1.12.25. Let x be a positive number, and let y be a real number. Then,

$$y - \frac{e^{y-1}}{x} \le \log x.$$

Furthermore, equality holds if $x = y = 1$.

Fact 1.12.26. Let x and y be real numbers, and let $\alpha \in [0,1]$. Then,

$$[\alpha e^{-x} + (1-\alpha)e^{-y}]^{-1} \le e^{\alpha x + (1-\alpha)y} \le \alpha e^x + (1-\alpha)e^y.$$

Proof: Replace x and y by e^x and e^y, respectively, in Fact 1.12.21.

Remark: The right-hand inequality follows from the convexity of the exponential function.

Fact 1.12.27. Let x and y be real numbers, and assume that $x \ne y$. Then,

$$e^{(x+y)/2} \le \frac{e^x - e^y}{x-y} \le \tfrac{1}{2}(e^x + e^y).$$

Proof: See [26].

Remark: See Fact 1.12.37.

Fact 1.12.28. Let x and y be real numbers. Then,

$$2 - y - e^{-x-y} \leq 1 + x \leq y + e^{x-y}.$$

Furthermore, equality holds on the left if and only if $x = -y$, and on the right if and only if $x = y$. In particular,

$$2 - e^{-x} \leq 1 + x \leq e^x.$$

Fact 1.12.29. Let x and y be real numbers. Then, the following statements hold:

i) If $0 \leq x \leq y \leq \pi/2$, then

$$\frac{x}{y} \leq \frac{\sin x}{\sin y} \leq \frac{\pi}{2}\left(\frac{x}{y}\right).$$

ii) If either $x, y \in [0, 1]$ or $x, y \in [1, \pi/2]$, then

$$(\tan x) \tan y \leq (\tan 1) \tan xy.$$

iii) If $x, y \in [0, 1]$, then

$$(\sin^{-1} x) \sin^{-1} y \leq \tfrac{1}{2} \sin^{-1} xy.$$

iv) If $y \in (0, \pi/2]$ and $x \in [0, y]$, then

$$\left(\frac{\sin y}{y}\right) x \leq \sin x \leq \sin\left[y\left(\frac{x}{y}\right)^{y \cot y}\right].$$

v) If $x, y \in [0, \pi]$ are distinct, then

$$\tfrac{1}{2}(\sin x + \sin y) < \frac{\cos x - \cos y}{y - x} < \sin[\tfrac{1}{2}(x + y)].$$

vi) If $0 \leq x < y < \pi/2$, then

$$\frac{1}{\cos^2 x} < \frac{\tan x - \tan y}{x - y} < \frac{1}{\cos^2 y}.$$

vii) If x and y are positive numbers, then

$$(\sinh x) \sinh xy \leq xy \sinh(x + xy).$$

viii) If $0 < y < x < \pi/2$, then

$$\frac{\sin x}{\sin y} < \frac{x}{y} < \frac{\tan x}{\tan y}.$$

Proof: Statements *i)–iii)* are given in [279, pp. 250, 251]. Statement *iv)* is given in [1066, p. 26]. Statement *v)* is a consequence of the Hermite-Hadamard inequality given by Fact 1.10.6. See [1066, p. 51]. Statement *vi)* follows from the mean value theorem and monotonicity of the cosine function. See [893, p. 264]. Statement *vii)* is given in [691, p. 71]. Statement *viii)* is given in [893, p. 267].

Remark: $(\sin 0)/0 = (\sinh 0)/0 = 1$.

Fact 1.12.30. Let x and y be positive numbers. If $p \in [1, \infty)$, then

$$x^p + y^p \le (x + y)^p.$$

Furthermore, if $p \in [0, 1)$, then

$$(x + y)^p \le x^p + y^p.$$

Proof: For the first statement, set $p = 1$ in Fact 1.17.35. For the second statement, set $q = 1$ in Fact 1.17.35.

Fact 1.12.31. Let x, y, p, q be nonnegative numbers. Then,

$$x^p y^q + x^q y^p \le x^{p+q} + y^{p+q}.$$

Furthermore, equality holds if and only if either $pq = 0$ or $x = y$.

Proof: See [686, p. 96].

Fact 1.12.32. Let x and y be nonnegative numbers, and let $p, q \in (1, \infty)$ satisfy $1/p + 1/q = 1$. Then,

$$xy \le \frac{x^p}{p} + \frac{y^q}{q}.$$

Furthermore, equality holds if and only if $x^p = y^q$.

Proof: See [440, p. 12] or [441, p. 10].

Remark: This result is *Young's inequality.* An extension is given by Fact 1.17.31. Matrix versions are given by Fact 8.12.12 and Fact 9.14.22.

Remark: $1/p + 1/q = 1$ is equivalent to $(p - 1)(q - 1) = 1$.

Fact 1.12.33. Let x and y be positive numbers, and let p and q be real numbers such that $0 \le p \le q$. Then,

$$\frac{x^p + y^p}{(xy)^{p/2}} \le \frac{x^q + y^q}{(xy)^{q/2}}.$$

Remark: See Fact 8.8.9.

Fact 1.12.34. Let x and y be positive numbers, and let p and q be nonzero real numbers such that $p \le q$. Then,

$$\left(\frac{x^p + y^p}{2} \right)^{1/p} \le \left(\frac{x^q + y^q}{2} \right)^{1/q}.$$

Furthermore, equality holds if and only if either $p = q$ or $x = y$. Finally,

$$\sqrt{xy} = \lim_{p \to 0} \left(\frac{x^p + y^p}{2} \right)^{1/p}.$$

Hence, if $p < 0 < q$, then

$$\left(\frac{x^p + y^p}{2} \right)^{1/p} \le \sqrt{xy} \le \left(\frac{x^q + y^q}{2} \right)^{1/q}$$

where equality holds if and only if $x = y$.

Proof: See [823, pp. 63–65] and [941].

Remark: This result is a *power mean inequality*. Letting $q = 1$ yields the arithmetic-mean–geometric-mean inequality $\sqrt{xy} \leq \frac{1}{2}(x+y)$.

Fact 1.12.35. Let x and y be positive numbers, assume that $x \leq y$, let p and q be nonzero real numbers such that $0 < p \leq q \leq 1$, and define $f \colon [0, \infty) \mapsto \mathbb{R}$ by

$$f(t) \triangleq \begin{cases} \frac{p(t^q - 1)}{q(t^p - 1)}, & t \neq 1, \\ 1, & t = 1. \end{cases}$$

Then, $f(x) \leq f(y)$.

Proof: See [651].

Fact 1.12.36. Let x and y be positive numbers, and let p and q be nonzero real numbers such that $p \leq q$. Then,

$$\frac{x^p + y^p}{x^{p-1} + y^{p-1}} \leq \frac{x^q + y^q}{x^{q-1} + y^{q-1}}.$$

Furthermore, equality holds if and only if either $x = y$ or $p = q$.

Proof: See [102, p. 23].

Remark: The quantity $\frac{x^p + y^p}{x^{p-1} + y^{p-1}}$ is the *Lehmer mean*.

Fact 1.12.37. Let x and y be positive numbers such that $x < y$, and define

$$G \triangleq \sqrt{xy}, \quad L \triangleq \frac{y - x}{\log y - \log x}, \quad I \triangleq \frac{1}{e} \left(\frac{x^x}{y^y} \right)^{1/(y-x)}, \quad A \triangleq \frac{1}{2}(x + y).$$

Then,

$$x < G < L < I < A < y,$$

$$G < \sqrt{GA} < \sqrt[3]{G^2 A} < \sqrt[3]{\frac{1}{4}(G+A)^2 G} < L < \left. \begin{cases} \frac{1}{3}(2G + A) < \frac{1}{3}(G + 2A) \\ \sqrt{LA} < \frac{1}{2}(L + A) \end{cases} \right\} < I < A,$$

and

$$G + \frac{(x-y)^2 (x+3y)(y+3x)}{8(x+y)(x^2 + 6xy + y^2)} \leq A.$$

Now, let p and q be real numbers such that $1/3 \leq p < 1 < q$. Then,

$$L < \left(\frac{x^p + y^p}{2} \right)^{1/p} < A < \left(\frac{x^q + y^q}{2} \right)^{1/q}.$$

Proof: See [941, 1185, 1267] and [686, p. 106]. The inequality $L < \frac{1}{3}(2G + A)$ is *Polya's inequality*. See [1066, p. 53]. The inequality $\frac{1}{3}(G + 2A) < I$ is due to Sandor. See [102, p. 24].

Remark: These inequalities refine the arithmetic-mean–geometric-mean inequality Fact 1.17.14.

Remark: L is the *logarithmic mean*. Note that $L = \int_0^1 x^t y^{1-t} \, dt$.

Remark: I is the *identric mean*. See [1267].

Remark: See Fact 1.12.26 and Fact 1.17.26.

Fact 1.12.38. Let x and y be positive numbers, and define

$$L \triangleq \frac{y-x}{\log y - \log x}, \quad H_p \triangleq \left(\frac{x^p + (xy)^{p/2} + y^p}{3} \right)^{1/p}, \quad M_p \triangleq \left(\frac{x^p + y^p}{2} \right)^{1/p}.$$

If p, q are positive numbers such that $p < q$, then

$$M_p < M_q$$

and

$$H_p < H_q.$$

Now, let p, q, r be positive numbers such that $0.5283 \approx (\log 3)/(3 \log 2) \leq p \leq 3q/2$ and $1/3 < r < [(\log 2)/\log 3]p \approx 0.6309p$. Then,

$$L < H_{1/2} < M_{1/3} < M_r < H_p < M_q.$$

In particular, if $r \leq (\log 2)/\log 3 \approx 0.6309$ and $q \geq 2/3 \approx 0.6667$, then

$$\left(\frac{x^r + y^r}{2} \right)^{1/r} < \frac{x + \sqrt{xy} + y}{3} < \left(\frac{x^q + y^q}{2} \right)^{1/q}.$$

Finally, if $1/2 \leq p \leq 3q/2$, then

$$\frac{y-x}{\log y - \log x} < \left(\frac{x^p + (xy)^{p/2} + y^p}{3} \right)^{1/p} < \left(\frac{x^q + y^q}{2} \right)^{1/q}.$$

Proof: See [281, p. 350] and [619, 778].

Remark: The center term is the *Heron mean*.

Fact 1.12.39. Let x and y be distinct positive numbers, and let $\alpha \in [0, 1]$. Then,

$$\sqrt{xy} \leq \tfrac{1}{2}(x^{1-\alpha}y^\alpha + x^\alpha y^{1-\alpha}) \leq \tfrac{1}{2}(x + y).$$

Furthermore,

$$\tfrac{1}{2}(x^{1-\alpha}y^\alpha + x^\alpha y^{1-\alpha}) \leq \frac{y-x}{\log y - \log x}$$

if and only if $\alpha \in [\tfrac{1}{2}(1 - 1/\sqrt{3}), \tfrac{1}{2}(1 + 1/\sqrt{3})]$, whereas

$$\frac{y-x}{\log y - \log x} \leq \tfrac{1}{2}(x^{1-\alpha}y^\alpha + x^\alpha y^{1-\alpha})$$

if and only if $\alpha \in [0, \tfrac{1}{2}(1 - 1/\sqrt{3})] \cup [\tfrac{1}{2}(1 + 1/\sqrt{3})]$.

Proof: See [447].

Remark: The first string of inequalities refines the arithmetic-mean–geometric-mean inequality Fact 1.17.14. The center term is the *Heinz mean*. Monotonicity is considered in Fact 1.18.1, while matrix extensions are given by Fact 9.9.49.

Fact 1.12.40. Let x and y be positive numbers. Then,

$$\left(\frac{x}{y}\right)^y \le \left(\frac{x+1}{y+1}\right)^{y+1}.$$

Furthermore, equality holds if and only if $x = y$.

Proof: See [893, p. 267].

Fact 1.12.41. Let x and y be real numbers. If either $0 < x < y < 1$ or $1 < x < y$, then

$$\frac{y^x}{x^y} < \frac{y}{x}$$

and

$$\frac{y^y}{x^x} < \left(\frac{y}{x}\right)^{xy}.$$

If $0 < x < 1 < y$, then both inequalities are reversed. If either $0 < x < 1 < y$ or $0 < x < y < e$, then

$$1 < \left(\frac{y \log x}{x \log y}\right)\left(\frac{y^x - 1}{x^y - 1}\right) < \frac{y^x}{x^y}.$$

If $e < x < y$, then both inequalities are reversed.

Proof: See [1132].

Fact 1.12.42. Let x and y be real numbers, let p and q be odd positive integers, assume that $p \le q$, and define $\alpha \triangleq p/q$. Then,

$$|x^\alpha - y^\alpha| \le 2^{1-\alpha}|x - y|^\alpha.$$

In particular, if $k \ge 1$, then

$$|x - y|^{2k+1} \le 2^{2k}|x^{2k+1} - y^{2k+1}|.$$

Now, assume in addition that x and y are nonnegative. If $r \ge 1$, then

$$|x - y|^r \le |x^r - y^r|.$$

Proof: See [713, 752].

Remark: Matrix versions of these results are given in [713]. Applications to nonlinear control appear in [752, 1133].

Problem: Unify the first and third inequalities.

1.13 Facts on Scalar Identities and Inequalities in Three Variables

Fact 1.13.1. Let x, y, z be real numbers. Then,

$$|x| + |y| + |z| \le |x + y - z| + |y + z - x| + |z + x - y|$$

and

$$\frac{|x+y|}{1 + |x+y|} \le \frac{|x|}{1 + |x|} + \frac{|y|}{1 + |y|}.$$

Proof: See [470, pp. 181, 183].

Problem: Extend these results to $\mathbb{C}$ and vector arguments.

Remark: Equality holds in the first result if x, y, z represent the lengths of the sides of a triangle. See Fact 1.13.17.

Fact 1.13.2. Let x, y, z be real numbers. Then,

$$2[(x-y)(x-z) + (y-z)(y-x) + (z-x)(z-y)] = (x-y)^2 + (y-z)^2 + (z-x)^2.$$

Proof: See [140, pp. 242, 402].

Fact 1.13.3. Let x, y, z be real numbers. Then,

$$(x+y)z \leq \tfrac{1}{2}(x^2 + y^2) + z^2.$$

Proof: See [140, p. 230].

Fact 1.13.4. Let x, y, z be real numbers. Then,

$$(\tfrac{1}{2}x + \tfrac{1}{3}y + \tfrac{1}{6}z)^2 \leq \tfrac{1}{2}x^2 + \tfrac{1}{3}y^2 + \tfrac{1}{6}z^2.$$

Proof: See [686, p. 129].

Fact 1.13.5. Let x, y be nonnegative numbers, and let z be a positive number. Then,

$$x + y \leq z^y x + z^{-x} y.$$

Proof: See [686, p. 163].

Fact 1.13.6. Let x, y, z be nonnegative numbers. Then,

$$\sqrt[3]{xyz} \leq \tfrac{1}{3}(\sqrt{xy} + \sqrt{yz} + \sqrt{zx}) \leq \tfrac{1}{6}(x+y+z) + \tfrac{1}{2}\sqrt[3]{xyz} \leq \tfrac{1}{3}(x+y+z).$$

Proof: The first inequality is given by Fact 1.17.21, while the second inequality is given in [1067].

Fact 1.13.7. Let x, y, z be nonnegative numbers. Then,

$$\begin{aligned}
xy + yz + zx &\leq (\sqrt{xy} + \sqrt{yz} + \sqrt{zx})^2 \\
&\leq 3(xy + yz + zx) \\
&\leq (x+y+z)^2 \\
&\leq 3(x^2 + y^2 + z^2),
\end{aligned}$$

$$4(xy + yz) \leq (x+y+z)^2,$$

$$2(x+y+z) \leq x^2 + y^2 + z^2 + 3,$$

$$2(xy + yz - zx) \leq x^2 + y^2 + z^2,$$

$$5xy + 3yz + 7zx \leq 6x^2 + 4y^2 + 5z^2.$$

Proof: See Fact 1.17.7 and [686, pp. 117, 126].

Fact 1.13.8. Let x, y, z be nonnegative numbers. Then,

$$12xy + 6xyz \le 6x^2 + y^2(z+2)(2z+3),$$

$$(x + y - z)(y + z - x)(z + x - y) \le xyz,$$

$$8xyz \le (x+y)(y+z)(z+x),$$

$$6xyz \le x^2y^2 + y^2z^2 + z^2x^2 + x^2 + y^2 + z^2,$$

$$15xyz \le x^3 + y^3 + z^3 + 2(x^2y + y^2z + z^2x + xy^2 + yz^2 + zx^2),$$

$$15xyz + x^3 + y^3 + z^3 \le 2(x + y + z)(x^2 + y^2 + z^2),$$

$$16xyz \le (x+1)(y+1)(x+z)(y+z),$$

$$27xyz \le (x^2 + x + 1)(y^2 + y + 1)(z^2 + z + 1),$$

$$4xyz \le x^2y^2z^2 + xy + yz + zx,$$

$$x^2y + y^2z + z^2x \le x^3 + y^3 + z^3,$$

$$x^2(z + y - x) + y^2(z + x - y) + z^2(x + y - z)$$
$$\le 3xyz$$
$$\le xy^2 + yz^2 + zx^2$$
$$\le x^3 + y^3 + z^3,$$

$$27xyz \le 3(x + y + z)(xy + yz + zx)$$
$$\le (x + y + z)^3$$
$$\le 3(x + y + z)(x^2 + y^2 + z^2)$$
$$\le 9(x^3 + y^3 + z^3).$$

Proof: See Fact 1.13.11, [470, pp. 166, 169, 179, 182], [686, pp. 117, 120, 152], and [893, pp. 247, 257].

Remark: Note the factorization

$$x^3 + y^3 + z^3 - 3xyz = (x + y + z)(x^2 + y^2 + z^2 - xy - yz - zx),$$

where both sides are nonnegative due to the arithmetic-mean–geometric-mean inequality.

Remark: For positive x, y, z, the inequality $9xyz \le (x + y + z)(xy + yz + zx)$ is given by Fact 1.17.16.

Remark: For positive x, y, z, the inequality $3xyz \le xy^2 + yz^2 + zx^2$ is given by Fact 1.17.17.

Fact 1.13.9. Let x, y, z be nonnegative numbers. Then,

$$\left.\begin{array}{r} xyz(x+y+z) \\ 2xyz|x+y-z| \\ 2xyz|x-y+z| \\ 2xyz|-x+y+z| \end{array}\right\} \le \left\{\begin{array}{c} x^2y^2 + y^2z^2 + z^2x^2 \\ 3xyz(x+y+z) \end{array}\right\}$$

$$\le (xy + yz + zx)^2$$

$$\le 3(x^2y^2 + y^2z^2 + z^2x^2)$$

$$\le (x^2 + y^2 + z^2)^2$$

$$\le (x + y + z)(x^3 + y^3 + z^3)$$

$$\le \left\{\begin{array}{c} 3(x^4 + y^4 + z^4) \\ (x + y + z)^4 \end{array}\right\}$$

$$\le 27(x^4 + y^4 + z^4),$$

$$x^2y^2 + y^2z^2 + z^2x^2 \le \tfrac{1}{2}[x^4 + y^4 + z^4 + xyz(x+y+z)]$$

$$\le x^4 + y^4 + z^4$$

$$\le (x^2 + y^2 + z^2)^2,$$

$$xyz(x+y+z) \le x^3y + y^3z + z^3x \le x^4 + y^4 + z^4,$$

$$\left.\begin{array}{r} 2xyz|x+y-z| \\ 2xyz|x-y+z| \\ 2xyz|-x+y+z| \end{array}\right\} \le 3(x^3y + y^3z + z^3x) \le (x^2 + y^2 + z^2)^2,$$

$$(x^2 + y^2 + z^2)(x^3 + y^3 + z^3) \le 3(x^5 + y^5 + z^5).$$

Furthermore,

$$\frac{1}{3}(x + y + z) \le \frac{x^3}{x^2 + xy + y^2} + \frac{y^3}{y^2 + yz + z^2} + \frac{z^3}{z^2 + zx + x^2}.$$

Proof: See [470, pp. 170, 180], [686, pp. 106, 108, 149], [893, pp. 247, 257], Fact 1.17.2, Fact 1.17.4, and Fact 1.17.22.

Remark: The inequality $2xyz(x + y - z) \le x^2y^2 + y^2z^2 + z^2x^2$ follows from $(xy - yz - zx)^2$, and thus is valid for all real x, y, z. See [470, p. 194].

Remark: The inequality $3xyz(x + y + z) \le (xy + yz + zx)^2$ follows from Newton's inequality. See Fact 1.17.11.

Fact 1.13.10. Let x, y, z be nonnegative numbers. Then,

$$9x^2y^2z^2 \le (x^2y + y^2z + z^2x)(xy^2 + yz^2 + zx^2),$$

$$27x^2y^2z^2 \le 3xyz(x + y + z)(xy + yz + zx)$$

$$\le \begin{Bmatrix} xyz(x + y + z)^3 \\ (xy + yz + zx)^3 \end{Bmatrix}$$

$$\le \tfrac{27}{64}(x + y)^2(y + z)^2(z + x)^2$$

$$\le \tfrac{9}{64}[(x + y)^6 + (y + z)^6 + (z + x)^6]$$

$$\le \tfrac{1}{27}(x + y + z)^6$$

$$\le 9(x^6 + y^6 + z^6),$$

$$432xy^2z^3 \le (x + y + z)^6,$$

$$3x^2y^2z^2 \le \begin{Bmatrix} x^3yz^2 + x^2y^3z + xy^2z^3 \\ xy^3z^2 + x^2yz^3 + x^3y^2z \end{Bmatrix} \le x^2y^4 + y^2z^4 + z^2x^4,$$

$$9(x^2 + yz)(y^2 + zx)(z^2 + xy) \le 8(x^3 + y^3 + z^3)^2,$$

$$3xyz(x^3 + y^3 + z^3) \le (xy + yz + zx)(x^4 + y^4 + z^4),$$

$$2(x^3y^3 + y^3z^3 + z^3x^3) \le x^6 + y^6 + z^6 + 3x^2y^2z^2,$$

$$xyz(x + y + z)(x^3 + y^3 + z^3) \le (xy + yz + zx)(x^5 + y^5 + z^5),$$

$$(xy + yz + zx)x^2y^2z^2 \le x^8 + y^8 + z^8,$$

$$(xy + yz + zx)^2(xyz^2 + x^2yz + xy^2z) \le 3(y^2z^2 + z^2x^2 + x^2y^2)^2,$$

$$(xyz + 1)^3 \le (x^3 + 1)(y^3 + 1)(z^3 + 1).$$

Finally, if $\alpha \in [3/7, 7/3]$, then

$$(\alpha + 1)^6(xy + yz + zx)^3 \le 27(\alpha x + y)^2(\alpha y + z)^2(\alpha z + x)^2.$$

In particular,

$$64(xy + yz + zx)^3 \le (x + y)^2(y + z)^2(z + x)^2$$

and

$$27(xy + yz + zx)^3 \le (2x + y)^2(2y + z)^2(2z + x)^2.$$

Proof: See [140, p. 229], [279, p. 244], [334, p. 114], [470, pp. 179, 182], [686, pp. 105, 134, 150, 155, 169], [893, pp. 247, 252, 257], [1066, p. 14], [1408], Fact 1.13.11, Fact 1.13.21, Fact 1.17.2, Fact 1.17.4, and Fact 1.17.8. For the last inequality, see [66].

Remark: The inequality $(xy + yz + zx)^2(xyz^2 + x^2yz + xy^2z) \le 3(y^2z^2 + z^2x^2 + x^2y^2)^2$ is due to Klamkin. See Fact 2.20.11 and [1408].

Remark: Extensions of several of these inequalities are given in [768].

Fact 1.13.11. Let x, y, z be positive numbers. Then,

$$6 \le \frac{9}{2} + \frac{x}{y+z} + \frac{y}{z+x} + \frac{z}{x+y} \le \frac{x+y}{z} + \frac{y+z}{x} + \frac{z+x}{y}.$$

Proof: See [102, pp. 33, 34].

Fact 1.13.12. Let x, y, z be real numbers. Then,

$$2xyz \le x^2 + y^2 z^2$$

and

$$3x^2 y^2 z^2 \le x^4 y^2 + x^2 y^4 + z^6.$$

Proof: See [686, p. 117] and [157, p. 78].

Fact 1.13.13. Let x, y, z be positive numbers, and assume that $x < y + z$. Then,

$$\frac{x}{1+x} < \frac{y}{1+y} + \frac{z}{1+z}.$$

Proof: See [893, p. 44].

Fact 1.13.14. Let x, y, z be nonnegative numbers. Then,

$$xy(x+y) + yz(y+z) + zx(z+x) \le x^3 + y^3 + z^3 + 3xyz.$$

Proof: See [686, p. 98].

Fact 1.13.15. Let x, y, z be nonnegative numbers, and assume that $x + y < z$. Then,

$$2(x+y)^2 z \le x^3 + y^3 + z^3 + 3xyz.$$

Proof: See [686, p. 98].

Fact 1.13.16. Let x, y, z be nonnegative numbers, and assume that $z < x + y$. Then,

$$2(x+y)z^2 \le x^3 + y^3 + z^3 + 3xyz.$$

Proof: See [686, p. 100].

Fact 1.13.17. Let x, y, z be positive numbers. Then, the following statements are equivalent:

i) x, y, z represent the lengths of the sides of a triangle.

ii) $z < x + y$, $x < y + z$, and $y < z + x$.

iii) $(x + y - z)(y + z - x)(z + x - y) > 0$.

iv) $x > |y - z|$, $y > |z - x|$, and $z > |x - y|$.

v) $|y - z| < x < y + z$.

vi) There exist positive numbers a, b, c such that $x = a + b$, $y = b + c$, and $z = c + a$.

vii) $2(x^4 + y^4 + z^4) < (x^2 + y^2 + z^2)^2$.

In this case, a, b, c in v) are given by

$$a = \tfrac{1}{2}(z + x - y), \quad b = \tfrac{1}{2}(x + y - z), \quad c = \tfrac{1}{2}(y + z - x).$$

Proof: See [470, p. 164]. Statements v) and vii) are given in [686, p. 125].

Remark: See Fact 8.9.5.

Fact 1.13.18. Let $n \geq 2$ be an integer, let x, y, z be positive numbers, and assume that $x^n + y^n = z^n$. Then, x, y, z represent the lengths of the sides of a triangle.

Proof: See [686, p. 112].

Remark: A lengthy proof shows that, for all $n \geq 3$, the equation $x^n + y^n = z^n$ has no solution in positive integers. This result was stated without proof by Fermat.

Fact 1.13.19. Let x, y, z be positive numbers that represent the lengths of the sides of a triangle. Then, $1/(x + y)$, $1/(y + z)$, and $1/(z + x)$ represent the lengths of the sides of a triangle.

Proof: See [893, p. 44].

Remark: See Fact 1.13.17 and Fact 1.13.20.

Fact 1.13.20. Let x, y, z be positive numbers that represent the lengths of the sides of a triangle. Then, $\sqrt{x}$, $\sqrt{y}$, and $\sqrt{z}$, represent the lengths of the sides of a triangle.

Proof: See [686, p. 99].

Remark: See Fact 1.13.17 and Fact 1.13.19.

Fact 1.13.21. Let x, y, z be positive numbers that represent the lengths of the sides of a triangle. Then,

$$3(xy + yz + zx) < (x + y + z)^2 < 4(xy + yz + zx),$$

$$2(x^2 + y^2 + z^2) < (x + y + z)^2 < 3(x^2 + y^2 + z^2),$$

$$\tfrac{1}{4}(x + y + z)^2 \leq \left\{ \begin{array}{c} xy + yz + zx \\ \tfrac{1}{3}(x + y + z)^2 \end{array} \right\} \leq x^2 + y^2 + z^2 \leq 2(xy + yz + zx),$$

$$3 < \frac{2x}{y + z} + \frac{2y}{z + x} + \frac{2z}{x + y} < 4,$$

$$x(y^2 + z^2) + y(z^2 + x^2) + z(x^2 + y^2) \leq 3xyz + x^3 + y^3 + z^3,$$

$$\tfrac{1}{4}(x + y + z)^3 \leq (x + y)(y + z)(z + x) \leq \tfrac{8}{27}(x + y + z)^3,$$

$$\tfrac{13}{27}(x + y + z)^3 \leq (x^2 + y^2 + z^2)(x + y + z) + 4xyz \leq \tfrac{1}{2}(x + y + z)^3,$$

$$xyz(x + y + z) \leq x^2 y^2 + y^2 z^2 + z^2 x^2 \leq x^3 y + y^3 z + z^3 x,$$

$$xyz \leq \tfrac{1}{8}(x + y)(y + z)(z + x).$$

If, in addition, the triangle is isosceles, then

$$3(xy + yz + zx) < (x + y + z)^2 < \tfrac{16}{5}(xy + yz + zx),$$

$$\tfrac{8}{3}(x^2 + y^2 + z^2) < (x + y + z)^2 < 3(x^2 + y^2 + z^2),$$

$$\tfrac{9}{32}(x + y + z)^3 \le (x + y)(y + z)(z + x) \le \tfrac{8}{27}(x + y + z)^3.$$

Proof: The first string is given in [893, p. 42]. In the second string, the lower bound is given in [470, p. 179], while the upper bound, which holds for all positive x, y, z, is given in Fact 1.13.8. Both the first and second strings are given in [996, p. 199]. In the third string, the upper leftmost inequality follows from Fact 1.13.21; the upper inequality second from the left follows from Fact 1.13.7 whether or not x, y, z represent the lengths of the sides of a triangle; the rightmost inequality is given in [470, p. 179]; the lower leftmost inequality is immediate; and the lower inequality second from the left follows from Fact 1.17.2. The fourth string is given in [893, pp. 267]. The fifth string is given in [470, p. 183]. This result can be written as [470, p. 186]

$$3 \le \frac{x}{y + z - x} + \frac{y}{z + x - y} + \frac{z}{x + y - z}.$$

The sixth string is given in [996, p. 199]. The seventh string is given in [1445]. In the eighth string, the left-hand inequality holds for all positive x, y, z. See Fact 1.13.9. The right-hand inequality, which is given in [470, p. 183], orders and interpolates two upper bounds for $xyz(x + y + z)$ given in Fact 1.13.9. The ninth string is given in [996, p. 201]. The inequalities for the case of an obtuse triangle are given in [240] and [996, p. 199].

Remark: In the fourth string, the lower left inequality is *Nesbitt's inequality*. See [470, p. 163].

Remark: See Fact 1.13.17 and Fact 2.20.11.

Fact 1.13.22. Let x, y, z represent the lengths of the sides of a triangle. Then,

$$\frac{9}{x + y + z} \le \frac{1}{x} + \frac{1}{y} + \frac{1}{z} \le \frac{1}{x + y - z} + \frac{1}{x + z - y} + \frac{1}{y + z - x}.$$

Proof: The lower bound, which holds for all x, y, z, follows from Fact 1.13.21. The upper bound is given in [996, p. 72].

Remark: The upper bound is *Walker's inequality*.

Fact 1.13.23. Let x, y, z be positive numbers such that $x + y + z = 1$. Then,

$$\frac{25}{1 + 48xyz} \le \frac{1}{x} + \frac{1}{y} + \frac{1}{z}.$$

Proof: See [1504].

Fact 1.13.24. Let x, y, z be positive numbers that represent the lengths of the sides of a triangle. Then,

$$\left| \frac{x}{y} + \frac{y}{z} + \frac{z}{x} - \left(\frac{y}{x} + \frac{z}{y} + \frac{x}{z} \right) \right| < 1.$$

Proof: See [470, p. 181].

Fact 1.13.25. Let x, y, z be positive numbers that represent the lengths of the sides of a triangle. Then,

$$\left| \frac{x-y}{x+y} + \frac{y-z}{y+z} + \frac{z-x}{z+x} \right| < \frac{1}{8}.$$

Proof: See [470, p. 183].

Fact 1.13.26. Let x, y, z be real numbers. Then,

$$\frac{|x-z|}{\sqrt{1+x^2}\sqrt{1+z^2}} \le \frac{|x-y|}{\sqrt{1+x^2}\sqrt{1+y^2}} + \frac{|y-z|}{\sqrt{1+y^2}\sqrt{1+z^2}}.$$

Proof: See [470, p. 184].

1.14 Facts on Scalar Identities and Inequalities in Four Variables

Fact 1.14.1. Let w, x, y, z be nonnegative numbers. Then,

$$\sqrt{wx} + \sqrt{yz} \le \sqrt{(w+y)(x+z)}$$

and

$$6\sqrt[4]{wxyz} \le \sqrt{(w+x)(y+z)} + \sqrt{(w+y)(x+z)} + \sqrt{(w+z)(x+y)}.$$

Proof: Use Fact 1.12.4, and see [686, p. 120].

Fact 1.14.2. Let w, x, y, z be nonnegative numbers. Then,

$$4(wx + xy + yz + zw) \le (w+x+y+z)^2,$$

$$8(wx + xy + yz + zw + wy + xz) \le 3(w+x+y+z)^2,$$

$$16(wxy + xyz + yzw + zwx) \le (w+x+y+z)^3,$$

$$256wxyz \le 16(w+x+y+z)(wxy + xyz + yzw + zwx)$$
$$\le (w+x+y+z)^4$$
$$\le 16(w+x+y+z)(w^3 + x^3 + y^3 + z^3),$$

$$4wxyz \le w^2xy + xyz^2 + y^2zw + zwx^2 = (wx + yz)(wy + xz),$$

$$4wxyz \le wx^2z + xy^2w + yz^2x + zw^2y,$$

$$8wxyz \le (wx + yz)(w+x)(y+z),$$

$$(wx + wy + wz + xy + xz + yz)^2 \le 6(w^2x^2 + w^2y^2 + w^2z^2 + x^2y^2 + x^2z^2 + y^2z^2),$$

$$4(wxy + xyz + yzw + zwx)^2 \le (w^2 + x^2 + y^2 + z^2)^3,$$

$$81wxyz \le (w^2 + w + 1)(x^2 + x + 1)(y^2 + y + 1)(z^2 + z + 1),$$

$$w^3 x^3 y^3 + x^3 y^3 z^3 + y^3 z^3 w^3 + z^3 w^3 x^3 \le (wxy + xyz + yzw + zwx)^3$$
$$\le 16(w^3 x^3 y^3 + x^3 y^3 z^3 + y^3 z^3 w^3 + z^3 w^3 x^3),$$

$$\frac{16}{3(w + x + y + z)} \le \frac{1}{w + x + y} + \frac{1}{x + y + z} + \frac{1}{y + z + w} + \frac{1}{z + w + x}.$$

Proof: See [470, p. 179], [686, pp. 120, 123, 124, 134, 144, 161], [819], Fact 1.17.22, and Fact 1.17.20.

Remark: The inequality $(w + x + y + z)^3 \le 16(w^3 + x^3 + y^3 + z^3)$ is given by Fact 1.17.2.

Remark: The inequality $16wxyz \le (w + x + y + z)(wxy + xyz + yzw + zwx)$ is given by Fact 1.17.16.

Remark: The inequality $4wxyz \le w^2 xy + xyz^2 + y^2 zw + zwx^2$ follows from Fact 1.17.17 with $n = 2$.

Remark: The inequality $4wxyz \le wx^2 z + xy^2 w + yz^2 x + zw^2 y$ is given by Fact 1.17.17.

Fact 1.14.3. Let w, x, y, z be real numbers. Then,
$$4wxyz \le w^2 x^2 + x^2 y^2 + y^2 w^2 + z^4$$
and
$$(wxyz + 1)^3 \le (w^3 + 1)(x^3 + 1)(y^3 + 1)(z^3 + 1).$$

Proof: See [157, p. 78] and [686, p. 134].

Fact 1.14.4. Let w, x, y, z be real numbers. Then,
$$(w^2 + x^2)(y^2 + z^2) = (wz + xy)^2 + (wy - xz)^2$$
$$= (wz - xy)^2 + (wy + xz)^2.$$

Hence,
$$\left.\begin{array}{l} (wz + xy)^2 \\ (wy - xz)^2 \\ (wz - xy)^2 \\ (wy + xz)^2 \end{array}\right\} \le (w^2 + x^2)(y^2 + z^2).$$

Remark: The equality is a statement of the fact that, for complex numbers z_1, z_2, $|z_1|^2 |z_2|^2 = |z_1 z_2|^2 = |\operatorname{Re}(z_1 z_2)|^2 + |\operatorname{Im}(z_1 z_2)|^2$. See [354, p. 77].

Fact 1.14.5. Let w, x, y, z be real numbers. Then,
$$w^4 + x^4 + y^4 + z^4 - 4wxyz = (w^2 - x^2)^2 + (y^2 - z^2)^2 + 2(wx - yz)^2.$$

Remark: This result yields the arithmetic-mean–geometric-mean inequality for four variables. See [140, pp. 226, 367].

1.15 Facts on Scalar Identities and Inequalities in Six Variables

Fact 1.15.1. Let x, y, z, u, v, w be real numbers. Then,

$$x^6 + y^6 + z^6 + u^6 + v^6 + w^6 - 6xyzuvw$$

$$= \tfrac{1}{2}(x^2 + y^2 + z^2)^2[(x^2 - y^2)^2 + (y^2 - z^2)^2 + (z^2 - x^2)^2]$$

$$+ \tfrac{1}{2}(u^2 + v^2 + w^2)^2[(u^2 - v^2)^2 + (v^2 - w^2)^2 + (w^2 - u^2)^2]$$

$$+ 3(xyz - uvw)^2.$$

Remark: This result yields the arithmetic-mean–geometric-mean inequality for six variables. See [140, p. 226].

1.16 Facts on Scalar Identities and Inequalities in Eight Variables

Fact 1.16.1. Let $x_1, x_2, x_3, x_4, y_1, y_2, y_3, y_4$ be real numbers. Then,

$$(x_1^2 + x_2^2 + x_3^2 + x_4^2)(y_1^2 + y_2^2 + y_3^2 + y_4^2)$$

$$= (x_1 y_1 - x_2 y_2 - x_3 y_3 - x_4 y_4)^2 + (x_1 y_2 + x_2 y_1 + x_3 y_4 - x_4 y_3)^2$$

$$+ (x_1 y_3 - x_2 y_4 + x_3 y_1 + x_4 y_2)^2 + (x_1 y_4 + x_2 y_3 - x_3 y_2 + x_4 y_1)^2.$$

Hence,

$$\left.\begin{array}{c} (x_1 y_1 - x_2 y_2 - x_3 y_3 - x_4 y_4)^2 + (x_1 y_2 + x_2 y_1 + x_3 y_4 - x_4 y_3)^2 \\ + (x_1 y_3 - x_2 y_4 + x_3 y_1 + x_4 y_2)^2 \\ (x_1 y_1 - x_2 y_2 - x_3 y_3 - x_4 y_4)^2 + (x_1 y_2 + x_2 y_1 + x_3 y_4 - x_4 y_3)^2 \\ + (x_1 y_4 + x_2 y_3 - x_3 y_2 + x_4 y_1)^2 \\ (x_1 y_1 - x_2 y_2 - x_3 y_3 - x_4 y_4)^2 + (x_1 y_3 - x_2 y_4 + x_3 y_1 + x_4 y_2)^2 \\ + (x_1 y_4 + x_2 y_3 - x_3 y_2 + x_4 y_1)^2 \\ (x_1 y_2 + x_2 y_1 + x_3 y_4 - x_4 y_3)^2 + (x_1 y_3 - x_2 y_4 + x_3 y_1 + x_4 y_2)^2 \\ + (x_1 y_4 + x_2 y_3 - x_3 y_2 + x_4 y_1)^2 \end{array}\right\}$$

$$\leq (x_1^2 + x_2^2 + x_3^2 + x_4^2)(y_1^2 + y_2^2 + y_3^2 + y_4^2).$$

Remark: The equality represents a relationship between a pair of quaternions. An analogous equality holds for two sets of eight variables representing a pair of octonions. See [354, p. 77].

1.17 Facts on Scalar Identities and Inequalities in n Variables

Fact 1.17.1. Let $x_1, \ldots, x_n$ be real numbers, and let k be a positive integer. Then,

$$\left(\sum_{i=1}^{n} x_i\right)^k = \sum_{i_1 + \cdots + i_n = k} \frac{k!}{i_1! \cdots i_n!} x_1^{i_1} \cdots x_n^{i_n}.$$

Remark: This result is the *multinomial theorem*.

Fact 1.17.2. Let $x_1, \ldots, x_n$ be nonnegative numbers, and let k be a positive integer. Then,

$$\sum_{i=1}^{n} x_i^k \leq \left(\sum_{i=1}^{n} x_i\right)^k \leq n^{k-1} \sum_{i=1}^{n} x_i^k.$$

Furthermore, equality holds in the second inequality if and only if $x_1 = \cdots = x_n$.

Remark: The case $n = 4$, $k = 3$ is given by the inequality $(w + x + y + z)^3 \leq 16(w^3 + x^3 + y^3 + z^3)$ of Fact 1.14.2.

Fact 1.17.3. Let $x_1, \ldots, x_n$ be nonnegative numbers. Then,

$$\left(\sum_{i=1}^{n} x_i\right)^2 \leq n \sum_{i=1}^{n} x_i^2.$$

Furthermore, equality holds if and only if $x_1 = \cdots = x_n$.

Remark: This result is equivalent to *i*) of Fact 9.8.12 with $m = 1$.

Fact 1.17.4. Let $x_1, \ldots, x_n$ be nonnegative numbers, and let k be a positive integer. Then,

$$\sum_{i=1}^{n} x_i^k \leq \left(\sum_{i=1}^{n} x_i\right)\left(\sum_{i=1}^{n} x_i^{k-1}\right) \leq n \sum_{i=1}^{n} x_i^k.$$

Proof: See [893, pp. 257, 258].

Fact 1.17.5. Let $x_1, \ldots, x_n$ be nonnegative numbers, and let $p, q \in [1, \infty)$, where $p \leq q$. Then,

$$\left(\sum_{i=1}^{n} x_i^q\right)^{1/q} \leq \left(\sum_{i=1}^{n} x_i^p\right)^{1/p} \leq n^{1/p - 1/q}\left(\sum_{i=1}^{n} x_i^q\right)^{1/q}.$$

Equivalently,

$$\sum_{i=1}^{n} x_i^q \leq \left(\sum_{i=1}^{n} x_i^p\right)^{q/p} \leq n^{q/p - 1} \sum_{i=1}^{n} x_i^q.$$

Proof: See Fact 9.7.29.

Remark: Setting $p = 1$ and $q = k$ yields Fact 1.17.2.

Fact 1.17.6. Let $x_1, \ldots, x_n$ be nonnegative numbers. Then,

$$\left(\sum_{i=1}^{n} x_i^3\right)^2 \leq \left(\sum_{i=1}^{n} x_i^2\right)^3 \leq n\left(\sum_{i=1}^{n} x_i^3\right)^2.$$

Proof: Set $p = 2$ and $q = 3$ in Fact 1.17.5 and square all terms.

Fact 1.17.7. Let $x_1, \ldots, x_n$ be nonnegative numbers. For $n = 2$,

$$2(x_1 x_2 + x_2 x_1) \leq (x_1 + x_2)^2.$$

For $n = 3$,

$$3(x_1x_2 + x_2x_3 + x_3x_1) \leq (x_1 + x_2 + x_3)^2.$$

If $n \geq 4$, then

$$4(x_1x_2 + x_2x_3 + \cdots + x_nx_1) \leq \left(\sum_{i=1}^{n} x_i\right)^2.$$

Proof: See [686, p. 144]. The cases $n = 2, 3, 4$ are given by Fact 1.12.4, Fact 1.13.7, and Fact 1.14.2.

Problem: Is 4 the best constant for $n \geq 5$?

Fact 1.17.8. Let $x_1, \ldots, x_n$ be nonnegative numbers. Then,

$$\left(\sum_{i=1}^{n} x_i\right)\left(\sum_{i=1}^{n} x_i^3\right) \leq \left(\sum_{i=1}^{n} x_i^5\right)\left(\sum_{i=1}^{n} \frac{1}{x_i}\right).$$

Proof: See [686, p. 150].

Fact 1.17.9. Let $x_1, \ldots, x_n$ be positive numbers, and assume that, for all $i \in \{1, \ldots, n-1\}$, $x_i < x_{i+1} \leq x_i + 1$. Then,

$$\sum_{i=1}^{n} x_i^3 \leq \left(\sum_{i=1}^{n} x_i\right)^2.$$

Proof: See [470, p. 183].

Remark: Equality holds in Fact 1.9.3.

Fact 1.17.10. Let $x_1, \ldots, x_n$ be complex numbers, define $E_0 \triangleq 1$, and, for $1 \leq k \leq n$, define

$$E_k \triangleq \sum_{i_1 < \cdots < i_k} \prod_{j=1}^{k} x_{i_j}.$$

Furthermore, for each positive integer k define

$$\mu_k \triangleq \sum_{i=1}^{n} x_i^k.$$

Then, for all $k \in \{1, \ldots, n\}$,

$$kE_k = \sum_{i=1}^{k} (-1)^{i-1} E_{k-i} \mu_i.$$

In particular,

$$E_1 = \mu_1,$$
$$2E_2 = E_1\mu_1 - \mu_2,$$
$$3E_3 = E_2\mu_2 - E_1\mu_2 + \mu_3.$$

Furthermore,

$$E_1 = \mu_1,$$
$$E_2 = \tfrac{1}{2}(\mu_1^2 - \mu_2),$$
$$E_3 = \tfrac{1}{6}(\mu_1^3 - 3\mu_1\mu_2 + 2\mu_3)$$

and

$$\mu_1 = E_1,$$
$$\mu_2 = E_1^2 - 2E_2,$$
$$\mu_3 = E_1^3 - 3E_1E_2 + 3E_3.$$

Remark: This result is *Newton's identity.* An application to roots of polynomials is given by Fact 4.8.2.

Remark: E_k is the kth *elementary symmetric polynomial.*

Remark: See Fact 1.17.11.

Fact 1.17.11. Let $x_1, \ldots, x_n$ be complex numbers, let k be a positive integer such that $1 < k < n$, and define

$$S_k \triangleq \binom{n}{k}^{-1} \sum_{i_1 < \cdots < i_k} \prod_{j=1}^{k} x_{i_j}.$$

Then,

$$S_{k-1}S_{k+1} \leq S_k^2.$$

Remark: This result is *Newton's inequality.* The case $n = 3$, $k = 2$ is given by Fact 1.13.9.

Remark: S_k is the kth *elementary symmetric mean.*

Remark: See Fact 1.17.10.

Fact 1.17.12. Let $x_1, \ldots, x_n$ be real numbers, and define

$$\overline{x} \triangleq \frac{1}{n} \sum_{j=1}^{n} x_j$$

and

$$\sigma \triangleq \sqrt{\frac{1}{n} \sum_{j=1}^{n} (x_j - \overline{x})^2} = \sqrt{\left(\frac{1}{n} \sum_{j=1}^{n} x_j^2\right) - \overline{x}^2}.$$

Then, for all $i \in \{1, \ldots, n\}$,

$$|x_i - \overline{x}| \leq \sqrt{n-1}\,\sigma.$$

Equality holds if and only if all of the elements of $\{x_1, \ldots, x_n\}_{\mathrm{ms}} \backslash \{x_i\}$ are equal. In addition,

$$\frac{\sigma}{\sqrt{n-1}} \leq \max\{x_1, \ldots, x_n\} - \overline{x} \leq \sqrt{n-1}\,\sigma.$$

Equality holds in either the left-hand inequality or the right-hand inequality if and only if all of the elements of $\{x_1,\ldots,x_n\}_{\mathrm{ms}}\backslash\{\max\{x_1,\ldots,x_n\}\}$ are equal. Finally,

$$\frac{\sigma}{\sqrt{n-1}} \le \bar{x} - \min\{x_1,\ldots,x_n\} \le \sqrt{n-1}\sigma.$$

Equality holds in either the left-hand inequality or the right-hand inequality if and only if all of the elements of $\{x_1,\ldots,x_n\}_{\mathrm{ms}}\backslash\{\min\{x_1,\ldots,x_n\}\}$ are equal.

Proof: The first result is the *Laguerre-Samuelson inequality*. See [588, 751, 776, 1070, 1168, 1364]. The lower bounds in the second and third strings are given in [1483]. See also [1168].

Remark: A vector extension of the Laguerre-Samuelson inequality is given by Fact 8.9.36. An application to eigenvalue bounds is given by Fact 5.11.45.

Fact 1.17.13. Let $x_1,\ldots,x_n$ be real numbers, and let α, δ, and p be positive numbers. If $p \ge 1$, then

$$\left(\frac{\alpha}{\alpha+n}\right)^{p-1}\delta^p \le \left|\delta - \sum_{i=1}^n x_i\right|^p + \alpha^{p-1}\sum_{i=1}^n |x_i|^p.$$

In particular,

$$\frac{\alpha\delta^2}{\alpha+n} \le \left(\delta - \sum_{i=1}^n x_i\right)^2 + \alpha\sum_{i=1}^n x_i^2.$$

Furthermore, if $p \le 1$, $x_1,\ldots,x_n$ are nonnegative, and $\sum_{i=1}^n x_i \le \delta$, then

$$\left|\delta - \sum_{i=1}^n x_i\right|^p + \alpha^{p-1}\sum_{i=1}^n |x_i|^p \le \left(\frac{\alpha}{\alpha+n}\right)^{p-1}\delta^p.$$

Finally, equality holds in all cases if and only if $x_1 = \cdots = x_n = \delta/(\alpha+n)$.

Proof: See [1284].

Remark: This result is *Wang's inequality*. The special case $p = 2$ is *Hua's inequality*. Generalizations are given by Fact 9.7.8 and Fact 9.7.9.

Fact 1.17.14. Let $x_1,\ldots,x_n$ be nonnegative numbers. Then,

$$\left(\prod_{i=1}^n x_i\right)^{1/n} \le \frac{1}{n}\sum_{i=1}^n x_i.$$

Furthermore, equality holds if and only if $x_1 = x_2 = \cdots = x_n$.

Remark: This result is the *arithmetic-mean–geometric-mean inequality*. Several proofs are given in [281]. See also [322]. Bounds for the difference between these quantities are given in [30, 303, 1376].

Fact 1.17.15. Let $x_1,\ldots,x_n$ be positive numbers. Then,

$$\frac{n}{\frac{1}{x_1}+\cdots+\frac{1}{x_n}} \le \sqrt[n]{x_1\cdots x_n} \le \frac{1}{n}(x_1+\cdots+x_n) \le \sqrt{\frac{1}{n}\sum_{i=1}^n x_i^2}.$$

Furthermore, equality holds in each inequality if and only if $x_1 = x_2 = \cdots = x_n$.

Remark: The lower bound for the geometric mean is the *harmonic mean*, while the first and third terms constitute the *arithmetic-mean–harmonic-mean inequality*. See Fact 1.17.38.

Remark: The upper bound for the arithmetic mean is the *quadratic mean*. See [627] and Fact 1.17.32.

Fact 1.17.16. Let $x_1, \ldots, x_n$ be positive numbers. Then,

$$\frac{n^2}{x_1 + \cdots + x_n} \le \frac{1}{x_1} + \cdots + \frac{1}{x_n}.$$

Proof: Use Fact 1.17.15. See also [686, p. 130].

Remark: The case $n = 3$ yields the inequality $9xyz \le (x + y + z)(xy + yz + zx)$ of Fact 1.13.8.

Remark: The case $n = 4$ yields the inequality $16wxyz \le (w + x + y + z)(wxy + xyz + yzw + zwx)$ of Fact 1.14.2.

Fact 1.17.17. Let $x_1, \ldots, x_n$ be positive numbers. Then,

$$n \le \frac{x_1}{x_2} + \frac{x_2}{x_3} + \cdots + \frac{x_{n-1}}{x_n} + \frac{x_n}{x_1}.$$

Remark: The case $n = 3$ yields the inequality $3xyz \le xy^2 + yz^2 + zx^2$ of Fact 1.13.8.

Remark: The case $n = 4$ yields the inequality $4wxyz \le wx^2z + xy^2w + yz^2x + zw^2y$ of Fact 1.14.2.

Fact 1.17.18. Let $x_1, \ldots, x_n$ be nonnegative numbers. Then,

$$\left(\prod_{i=1}^{n} x_i\right)^{1/n} \le \frac{1}{n}\sum_{i=1}^{n} x_i \le \left(\prod_{i=1}^{n} x_i\right)^{1/n} + \frac{1}{n}\sum_{i<j} |x_i - x_j|.$$

Proof: See [470, p. 186].

Fact 1.17.19. Let $x_1, \ldots, x_n$ be positive numbers contained in $[a, b]$, where $a > 0$. Then,

$$\left(\prod_{i=1}^{n} x_i\right)^{1/n} \le \frac{1}{n}\sum_{i=1}^{n} x_i \le \gamma \left(\prod_{i=1}^{n} x_i\right)^{1/n},$$

where γ is defined by

$$\gamma \triangleq \frac{(h-1)h^{1/(h-1)}}{e \log h}$$

and $h \triangleq b/a$.

Remark: The right-hand inequality is a *reverse arithmetic-mean–geometric mean inequality*; see [524, 529, 1505]. This result is due to Specht. For the case $n = 2$, see Fact 1.12.22.

Remark: $\gamma = S(1, h)$ is Specht's ratio. See Fact 1.12.22 and Fact 11.14.22.

Remark: Matrix extensions are considered in [21, 832].

Fact 1.17.20. Let $x_1, \ldots, x_n$ be positive numbers, and let k satisfy $1 \leq k \leq n$. Then,

$$\left(\binom{n}{k}^{-1} \sum_{i_1 < \cdots < i_k} \prod_{j=1}^{k} x_{i_j} \right)^{1/k} \leq \tfrac{1}{n} \sum_{i=1}^{n} x_i.$$

Equivalently,

$$\sum_{i_1 < \cdots < i_k} \prod_{j=1}^{k} x_{i_j} \leq \binom{n}{k} \left(\tfrac{1}{n} \sum_{i=1}^{n} x_i \right)^{k}.$$

Proof: This result follows from the fact that the kth elementary symmetric function is Schur concave. See [556, p. 102, Exercise 7.11].

Remark: Equality holds if $k = 1$. The case $n = k$ is the arithmetic-mean–geometric-mean inequality. The case $n = 3$, $k = 2$ yields the third inequality in Fact 1.13.7. The cases $n = 4$, $k = 3$ and $n = 4$, $k = 2$ are given in Fact 1.14.2.

Fact 1.17.21. Let $x_1, \ldots, x_n$ be positive numbers, and let k and k' satisfy $1 \leq k \leq k' \leq n$. Then,

$$\left(\prod_{i=1}^{n} x_i \right)^{1/n} \leq \binom{n}{k'}^{-1} \sum_{i_1 < \cdots < i'_k} \prod_{j=1}^{k'} x_{i_j}^{1/k'} \leq \binom{n}{k}^{-1} \sum_{i_1 < \cdots < i_k} \prod_{j=1}^{k} x_{i_j}^{1/k} \leq \tfrac{1}{n} \sum_{i=1}^{n} x_i.$$

Proof: See [556, p. 23] and [819].

Remark: This result is an interpolation of the arithmetic-mean–geometric-mean inequality. An alternative interpolation is given by Fact 1.17.25.

Remark: If $k = 1$, then the right-hand inequality is an equality. If $k = n$, then the left-hand inequality is an equality. The case $n = 3$ and $k = 2$ is given by Fact 1.13.6.

Fact 1.17.22. Let $x_1, \ldots, x_n$ be nonnegative numbers, and let k be a positive integer such that $1 \leq k \leq n$. Then,

$$\left(\sum_{i_1 < \cdots < i_k} \prod_{j=1}^{k} x_{i_j} \right)^{k} \leq \binom{n}{k}^{k-1} \sum_{i_1 < \cdots < i_k} \prod_{j=1}^{k} x_{i_j}^{k}.$$

Remark: Equality holds if $k = 1$ or $k = n$. The case $n = 3$, $k = 2$ is given by Fact 1.13.9. The cases $n = 4$, $k = 3$ and $n = 4$, $k = 2$ are given by Fact 1.14.2.

Fact 1.17.23. Let $x_1, \ldots, x_n$ be positive numbers, and let k satisfy $1 \leq k \leq n$. Then,

$$\left(\prod_{i=1}^{n} x_i \right)^{1/n} \leq \binom{n}{k}^{-1} \sum_{i_1 < \cdots < i_k} \prod_{j=1}^{k} x_{i_j}^{1/k} \leq \left(\binom{n}{k}^{-1} \sum_{i_1 < \cdots < i_k} \prod_{j=1}^{k} x_{i_j} \right)^{1/k} \leq \tfrac{1}{n} \sum_{i=1}^{n} x_i.$$

Proof: Use Fact 1.17.22 to merge Fact 1.17.20 and Fact 1.17.21.

Fact 1.17.24. Let $x_1, \ldots, x_n$ be positive numbers, and let k and k' satisfy $1 \le k \le k' \le n$. Then,

$$\left(\prod_{i=1}^{n} x_i \right)^{1/n} \le \left(\binom{n}{k'}^{-1} \sum_{i_1 < \cdots < i'_k} \prod_{j=1}^{k'} x_{i_j} \right)^{1/k'} \le \left(\binom{n}{k}^{-1} \sum_{i_1 < \cdots < i_k} \prod_{j=1}^{k} x_{i_j} \right)^{1/k} \le \tfrac{1}{n} \sum_{i=1}^{n} x_i.$$

Proof: See [819].

Fact 1.17.25. Let $x_1, \ldots, x_n$ be positive numbers, let $\alpha_1, \ldots, \alpha_n$ be nonnegative numbers, and assume that $\sum_{i=1}^{n} \alpha_i = 1$. Then,

$$\left(\prod_{i=1}^{n} x_i \right)^{1/n} \le \tfrac{1}{n!} \sum \prod_{j=1}^{n} x_{i_j}^{\alpha_j} \le \tfrac{1}{n} \sum_{i=1}^{n} x_i,$$

where the summation is taken over all $n!$ permutations $\{i_1, \ldots, i_n\}$ of $\{1, \ldots, n\}$.

Proof: See [556, p. 100].

Remark: This result is a consequence of *Muirhead's theorem*, which states that the middle expression is a Schur convex function of the exponents. See Fact 2.21.5.

Fact 1.17.26. Let $x_1, \ldots, x_n$ be positive numbers. Then,

$$\left(\prod_{i=1}^{n} x_i \right)^{1/n} < \tfrac{1}{n} \left(\frac{x_2 - x_1}{\log x_2 - \log x_1} + \frac{x_3 - x_2}{\log x_3 - \log x_2} + \cdots + \frac{x_1 - x_n}{\log x_1 - \log x_n} \right) < \tfrac{1}{n} \sum_{i=1}^{n} x_i.$$

Proof: See [102, p. 44].

Remark: This result is due to Bencze.

Remark: This result extends Fact 1.12.37 to n variables. See also [1500].

Fact 1.17.27. Let $x_1, \ldots, x_n$ be positive numbers contained in $[a, b]$, where $a > 0$. Then,

$$\frac{a}{2n^2} \sum_{i<j} (\log x_i - \log x_j)^2 \le \tfrac{1}{n} \sum_{i=1}^{n} x_i - \left(\prod_{i=1}^{n} x_i \right)^{1/n} \le \frac{b}{2n^2} \sum_{i<j} (\log x_i - \log x_j)^2.$$

Proof: See [1066, p. 86] or [1067].

Fact 1.17.28. Let $x_1, \ldots, x_n$ be nonnegative numbers contained in $(0, 1/2]$. Furthermore, define

$$A \triangleq \tfrac{1}{n} \sum_{i=1}^{n} x_i, \qquad G \triangleq \prod_{i=1}^{n} x_i^{1/n}, \qquad H \triangleq \frac{n}{\displaystyle\sum_{i=1}^{n} \frac{1}{x_i}}$$

and

$$A' \triangleq \tfrac{1}{n} \sum_{i=1}^{n} (1 - x_i), \qquad G' \triangleq \prod_{i=1}^{n} (1 - x_i)^{1/n}, \qquad H' \triangleq \frac{n}{\displaystyle\sum_{i=1}^{n} \frac{1}{1 - x_i}}.$$

Then, the following statements hold:

    i) $A'/G' \leq A/G$. Furthermore, equality holds if and only if $x_1 = \cdots = x_n$.

    ii) $A' - G' \leq A - G$. Furthermore, equality holds if and only if $x_1 = \cdots = x_n$.

    iii) $A^n - G^n \leq A'^n - G'^n$. Furthermore, equality holds for $n = 1$ and $n = 2$, and, for $n \geq 3$, if and only if $x_1 = \cdots = x_n$.

    iv) $G'/H' \leq G/H$.

Proof: See [1169]. For a proof of *iv)*, see [1189].

Remark: Result *i)* is due to Fan. See [1189].

    Fact 1.17.29. Let $x_1, \ldots, x_n$ be positive numbers, and, for all $k \in \{1, \ldots, n\}$, define

$$A_k \triangleq \frac{1}{k} \sum_{i=1}^{k} x_i, \qquad G_k \triangleq \prod_{i=1}^{k} x_i^{1/k}.$$

Then,

$$1 = \left(\frac{A_1}{G_1}\right)^1 \leq \left(\frac{A_2}{G_2}\right)^2 \leq \cdots \leq \left(\frac{A_n}{G_n}\right)^n$$

and

$$0 = 1(A_1 - G_1) \leq 2(A_2 - G_2) \leq \cdots \leq n(A_n - G_n).$$

Proof: See [1066, p. 13].

Remark: The first result is due to Popoviciu, while the second result is due to Rado.

    Fact 1.17.30. Let $x_1, \ldots, x_n$ be positive numbers, let p be a real number, and define

$$M_p \triangleq \begin{cases} \left(\prod_{i=1}^{n} x_i\right)^{1/n}, & p = 0, \\ \left(\frac{1}{n} \sum_{i=1}^{n} x_i^p\right)^{1/p}, & p \neq 0. \end{cases}$$

Now, let p and q be real numbers such that $p \leq q$. Then,

$$M_p \leq M_q$$

and

$$\lim_{r \to -\infty} M_r = \min\{x_1, \ldots, x_n\} \leq \lim_{r \to 0} M_r = M_0 \leq \lim_{r \to \infty} M_r = \max\{x_1, \ldots, x_n\}.$$

Finally, $p < q$ and at least two of the numbers $x_1, \ldots, x_n$ are distinct if and only if

$$M_p < M_q.$$

Proof: See [279, p. 210] and [988, p. 105]. If p and q are nonzero and $p \leq q$, then

$$\left(\sum_{i=1}^{n} x_i^p\right)^{1/p} \leq \left(\frac{1}{n}\right)^{1/q - 1/p} \left(\sum_{i=1}^{n} x_i^q\right)^{1/q},$$

which is a reverse form of Fact 1.17.35.

Proof: To verify the limit, take the log of both sides and use l'Hôpital's rule.

Remark: This result is a *power mean inequality*. $M_0 \leq M_1$ is the arithmetic-mean–geometric-mean inequality given by Fact 1.17.14.

Remark: A matrix application of this result is given by Fact 8.12.1.

Fact 1.17.31. Let $x_1, \ldots, x_n$ be nonnegative numbers, let $\alpha_1, \ldots, \alpha_n$ be nonnegative numbers, and assume that $\sum_{i=1}^{n} \alpha_i = 1$. Then,

$$\prod_{i=1}^{n} x_i \leq \sum_{i=1}^{n} \alpha_i x_i^{1/\alpha_i}.$$

Furthermore, equality holds if and only if $x_1 = x_2 = \cdots = x_n$.

Proof: See [459].

Remark: This result is a generalization of Young's inequality. See Fact 1.12.32. Matrix versions are given by Fact 8.12.12 and Fact 9.14.22.

Remark: This result is equivalent to Fact 1.17.32.

Fact 1.17.32. Let $x_1, \ldots, x_n$ be positive numbers, let $\alpha_1, \ldots, \alpha_n$ be nonnegative numbers, and assume that $\sum_{i=1}^{n} \alpha_i = 1$. Then,

$$\frac{1}{\sum_{i=1}^{n} \frac{\alpha_i}{x_i}} \leq \prod_{i=1}^{n} x_i^{\alpha_i} \leq \sum_{i=1}^{n} \alpha_i x_i.$$

Now, let r be a real number, define

$$M_r \triangleq \left(\sum_{i=1}^{n} \alpha_i x_i^r \right)^{1/r}.$$

and let p and q be real numbers such that $p \leq q$. Then,

$$M_p \leq M_q$$

and

$$\lim_{r \to -\infty} M_r = \min\{x_1, \ldots, x_n\} \leq \lim_{r \to 0} M_r = M_0 \leq \lim_{r \to \infty} M_r = \max\{x_1, \ldots, x_n\}.$$

Furthermore, equality holds if and only if $x_1 = x_2 = \cdots = x_n$.

Proof: Since $f(x) = -\log x$ is convex, it follows that

$$\log \prod_{i=1}^{n} x_i^{\alpha_i} = \sum_{i=1}^{n} \alpha_i \log x_i \leq \log \sum_{i=1}^{n} \alpha_i x_i.$$

To prove the second statement, define $f \colon [0, \infty)^n \mapsto [0, \infty)$ by $f(\mu_1, \ldots, \mu_n) \triangleq \sum_{i=1}^{n} \alpha_i \mu_i - \prod_{i=1}^{n} \mu_i^{\alpha_i}$. Note that $f(\mu, \ldots, \mu) = 0$ for all $\mu \geq 0$. If $x_1, \ldots, x_n$ minimizes f, then $\partial f / \partial \mu_i (x_1, \ldots, x_n) = 0$ for all $i \in \{1, \ldots, n\}$, which implies that $x_1 = x_2 = \cdots = x_n$.

Remark: This result is the *weighted arithmetic-mean–geometric-mean* inequality. Setting $\alpha_1 = \cdots = \alpha_n = 1/n$ yields Fact 1.17.14.

Remark: This result is equivalent to Fact 1.17.31.

Remark: See [1066, p. 11].

Fact 1.17.33. Let $x_1, \ldots, x_n$ be positive numbers, let $\alpha_1, \ldots, \alpha_n$ be nonnegative numbers, and assume that $\sum_{i=1}^n \alpha_i = 1$. Then,

$$e^{2\beta} \prod_{i=1}^n x_i^{\alpha_i} \le \sum_{i=1}^n \alpha_i x_i.$$

where

$$\beta \triangleq 1 - \frac{\sum_{i=1}^n \alpha_i \sqrt{x_i}}{\sqrt{\sum_{i=1}^n \alpha_i x_i}}.$$

Proof: See [17].

Remark: This result is a refinement of the weighted arithmetic-mean–geometric-mean inequality. See Fact 1.17.32.

Remark: The convexity of $f(x) = x^2$ implies that $\beta \ge 0$.

Fact 1.17.34. Let $x_1, \ldots, x_n$ be nonnegative numbers. Then,

$$1 + \left(\prod_{i=1}^n x_i \right)^{1/n} \le \left[\prod_{i=1}^n (1 + x_i) \right]^{1/n}.$$

Furthermore, equality holds if and only if $x_1 = x_2 = \cdots = x_n$.

Proof: Use Fact 1.17.14. See [242, p. 210].

Remark: This inequality is used to prove Corollary 8.4.15.

Fact 1.17.35. Let $x_1, \ldots, x_n$ be nonnegative numbers, and let p, q be positive numbers such that $p \le q$. Then,

$$\left(\sum_{i=1}^n x_i^q \right)^{1/q} \le \left(\sum_{i=1}^n x_i^p \right)^{1/p}.$$

Furthermore, the inequality is strict if and only if $p < q$ and at least two of the numbers $x_1, \ldots, x_n$ are nonzero.

Proof: See Proposition 9.1.5.

Remark: This result is the *power-sum inequality*. See [279, p. 213]. This result implies that the Hölder norm is a monotonic function of the exponent.

Fact 1.17.36. Let $x_1, \ldots, x_n$ be positive numbers, and let $\alpha_1, \ldots, \alpha_n \in [0, 1]$ be such that $\sum_{i=1}^n \alpha_i = 1$. If $p \le 0$ or $p \ge 1$, then

$$\left(\sum_{i=1}^n \alpha_i x_i \right)^p \le \sum_{i=1}^n \alpha_i x_i^p.$$

Alternatively, if $p \in [0, 1]$, then

$$\sum_{i=1}^n \alpha_i x_i^p \le \left(\sum_{i=1}^n \alpha_i x_i \right)^p.$$

Finally, equality in both cases holds if and only if either $p = 0$ or $p = 1$ or $x_1 = \cdots = x_n$.

Remark: This result is a consequence of Jensen's inequality given by Fact 1.10.4.

Fact 1.17.37. Let $0 < x_1 < \cdots < x_n$, and let $\alpha_1, \ldots, \alpha_n \geq 0$ satisfy $\sum_{i=1}^{n} \alpha_i = 1$. Then,

$$1 \leq \left(\sum_{i=1}^{n} \alpha_i x_i \right) \left(\sum_{i=1}^{n} \frac{\alpha_i}{x_i} \right) \leq \frac{(x_1 + x_n)^2}{4 x_1 x_n}.$$

Remark: This result is the *Kantorovich inequality*. See Fact 8.15.10 and [952].

Remark: See Fact 1.17.38.

Fact 1.17.38. Let $x_1, \ldots, x_n$ be positive numbers, and define $\alpha \triangleq \min_{i=1,\ldots,n} x_i$ and $\beta \triangleq \max_{i=1,\ldots,n} x_i$. Then,

$$1 \leq \left(\frac{1}{n} \sum_{i=1}^{n} x_i \right) \left(\frac{1}{n} \sum_{i=1}^{n} \frac{1}{x_i} \right) \leq \frac{(\alpha + \beta)^2}{4\alpha\beta}.$$

Proof: Use Fact 1.17.37 or Fact 1.18.21. See [440, p. 94] or [441, p. 119].

Remark: The left-hand inequality is the arithmetic-mean–harmonic-mean inequality. See Fact 1.17.15. The right-hand inequality is *Schweitzer's inequality*. See [1428, 1443] for historical details.

Remark: A matrix extension is given by Fact 8.10.32.

Fact 1.17.39. Let $x_1, \ldots, x_n$ be positive numbers, and let p and q be positive numbers. Then,

$$\left(\frac{1}{n} \sum_{i=1}^{n} x_i^p \right) \left(\frac{1}{n} \sum_{i=1}^{n} x_i^q \right) \leq \frac{1}{n} \sum_{i=1}^{n} x_i^{p+q}.$$

In particular, if $p \in [0, 1]$, Then,

$$\left(\frac{1}{n} \sum_{i=1}^{n} x_i^p \right) \left(\frac{1}{n} \sum_{i=1}^{n} x_i^{1-p} \right) \leq \frac{1}{n} \sum_{i=1}^{n} x_i^p.$$

Proof: See [1432].

Remark: These inequalities are interpolated in [1432].

Fact 1.17.40. Let $x_1, \ldots, x_n$ be positive numbers. Then,

$$\frac{1}{n} \sum_{k=1}^{n} \left(\prod_{i=1}^{k} x_i \right)^{1/k} \leq \left[\prod_{k=1}^{n} \left(\frac{1}{k} \sum_{i=1}^{k} x_i \right) \right]^{1/k}.$$

Furthermore, equality holds if and only if $x_1 = \cdots = x_n$.

Remark: This result can be expressed as $\frac{1}{n}(z_1 + \cdots + z_n) \leq \sqrt[n]{y_1 \cdots y_n}$, where $z_k \triangleq \sqrt[k]{x_1 \cdots x_k} \leq y_k \triangleq \frac{1}{k}(x_1 + \cdots + x_k)$.

Remark: This result is the *mixed arithmetic-geometric mean inequality*. This result is due to Nanjundiah. See [344, 1008].

Fact 1.17.41. Let $x_1, \ldots, x_n$ be positive numbers, where $n \geq 2$. Then,

$$\sum_{k=1}^{n} \left(\prod_{i=1}^{k} x_i \right)^{1/k} \leq \frac{n}{\sqrt[n]{n!}} \sum_{k=1}^{n} x_k \leq e^{(n-1)/n} \sum_{k=1}^{n} x_k \leq e \sum_{k=1}^{n} x_k.$$

Furthermore, equality holds in all of these inequalities if and only if $x_1 = \cdots = x_n = 0$.

Remark: The inequality $\frac{n}{\sqrt[n]{n!}} < e^{(n-1)/n}$, which is equivalent to $e(n/e)^n < n!$, follows from Fact 1.11.20.

Remark: This result is a finite version of *Carleman's inequality*. See [344] and [556, p. 22].

Fact 1.17.42. Let $x_1, \ldots, x_n$ be positive numbers, not all of which are zero. Then,

$$\left(\sum_{i=1}^{n} x_i \right)^4 < (2\tan^{-1} n)^2 \left(\sum_{i=1}^{n} x_i^2 \right) \sum_{i=1}^{n} i^2 x_i^2 < \pi^2 \left(\sum_{i=1}^{n} x_i^2 \right) \sum_{i=1}^{n} i^2 x_i^2.$$

Furthermore,

$$\left(\sum_{i=1}^{n} x_i \right)^2 < \frac{\pi^2}{6} \sum_{i=1}^{n} i^2 x_i^2.$$

Proof: See [158] or [894, p. 18].

Remark: The first and third terms in the first inequality constitute a finite version of the *Carlson inequality*. The last inequality follows from the Cauchy-Schwarz inequality. See [470, p. 175].

Fact 1.17.43. Let $x_1, \ldots, x_n$ be nonnegative numbers, and let $p > 1$. Then,

$$\sum_{k=1}^{n} \left(\frac{1}{k} \sum_{i=1}^{k} x_i \right)^p \leq \left(\frac{p}{p-1} \right)^p \sum_{k=1}^{n} x_k^p.$$

Proof: See [873].

Remark: This result is the *Hardy inequality*. See [344, 873].

Fact 1.17.44. Let $x_1, \ldots, x_n$ be nonnegative numbers, and let $p > 1$. Then,

$$\sum_{k=1}^{n} \left(\sum_{i=k}^{n} \frac{x_i}{i} \right)^p \leq p^p \sum_{k=1}^{n} x_k^p.$$

Proof: See [873].

Remark: This result is the *Copson inequality*.

Fact 1.17.45. Let $x_1, \ldots, x_n$, α, and β be positive numbers, let p and q be real numbers, and assume that one of the following conditions is satisfied:

  i) $p \in (-\infty, 1] \backslash \{0\}$ and $(n-1)\alpha \leq \beta$.

  ii) $p \geq 1$ and $(n^p - 1)\alpha \leq \beta$.

Then,

$$\frac{n}{(\alpha + \beta)^{1/p}} \le \sum_{i=1}^{n} \left(\frac{x_i^q}{\alpha x_i^q + \beta \prod_{k=1}^{n} x_k^{q/n}} \right)^{1/p}.$$

Proof: See [1498].

Fact 1.17.46. Let $x_1, \ldots, x_n$ be nonnegative numbers, and assume that $\sum_{i=1}^{n} x_i = 1$. Then,

$$0 \le \log n - \sum_{i=1}^{n} x_i \log \frac{1}{x_i} \le \frac{1}{2}(n^2 - n) \max_{i,j=1,\ldots,n} |x_i - x_j|^2.$$

Furthermore, $\sum_{i=1}^{n} x_i \log \frac{1}{x_i} = 0$ if and only if $x_i = 1$ for some i, while $\sum_{i=1}^{n} x_i \log \frac{1}{x_i} = \log n$ if and only if $x_1 = \cdots = x_n = 1/n$.

Proof: See [443].

Remark: Define $0 \log \frac{1}{0} \triangleq 0$.

Remark: Alternative entropy bounds involving $\max_{i,j=1,\ldots,n} x_i/x_j$ are given in [444].

Fact 1.17.47. Let $x_1, \ldots, x_n$ be positive numbers, and assume that $\sum_{i=1}^{n} x_i = 1$. Then,

$$0 \le \log n - \sum_{i=1}^{n} x_i \log \frac{1}{x_i} \le \left(n \sum_{i=1}^{n} x_i^2 \right) - 1 \le \left(\sum_{i=1}^{n} x_i^3 \right)^{1/2} \left[\left(\sum_{i=1}^{n} \frac{1}{x_i} \right) - n^2 \right]^{1/2}.$$

Consequently,

$$\log n + 1 - n \sum_{i=1}^{n} x_i^2 \le \sum_{i=1}^{n} x_i \log \frac{1}{x_i} \le \log n.$$

Proof: See [443, 1007].

Remark: It follows from Fact 1.17.38 that $n^2 \le \sum_{i=1}^{n} \frac{1}{x_i}$.

Fact 1.17.48. Let $x_1, \ldots, x_n$ be positive numbers, assume that $\sum_{i=1}^{n} x_i = 1$, and define $a \triangleq \min_{i=1,\ldots,n} x_i$ and $b \triangleq \max_{i=1,\ldots,n} x_i$. Then,

$$0 \le \log n - \sum_{i=1}^{n} x_i \log \frac{1}{x_i} \le \frac{1}{n} \lfloor \frac{n^2}{4} \rfloor (b-a) \log \frac{b}{a} \le \frac{1}{n} \lfloor \frac{n^2}{4} \rfloor \frac{(b-a)^2}{\sqrt{ab}}.$$

Proof: See [445].

Remark: This result is based on Fact 1.18.18.

Remark: See Fact 2.21.6.

Fact 1.17.49. Let $x_1, \ldots, x_n$ be nonnegative numbers. Then,

$$\frac{e^2}{4} \sum_{i=1}^{n} x_i^2 \le \prod_{i=1}^{n} e^{x_i}.$$

Furthermore, equality holds for $n = 1$ and $x_1 = 2$.

Proof: See [1131].

1.18 Facts on Scalar Identities and Inequalities in $2n$ Variables

Fact 1.18.1. Let $x_1, \ldots, x_n$ and $y_1, \ldots, y_n$ be nonnegative numbers, let $\alpha, \beta \in \mathbb{R}$, and assume that either $0 \le \beta \le \alpha \le \frac{1}{2}$ or $\frac{1}{2} \le \alpha \le \beta \le 1$. Then,

$$\sum_{i=1}^n x_i^{1-\alpha} y_i^{\alpha} \sum_{i=1}^n x_i^{\alpha} y_i^{1-\alpha} \le \sum_{i=1}^n x_i^{1-\beta} y_i^{\beta} \sum_{i=1}^n x_i^{\beta} y_i^{1-\beta}.$$

Furthermore, if x and y are nonnegative numbers, then

$$x^{1-\alpha} y^{\alpha} + x^{\alpha} y^{1-\alpha} \le x^{1-\beta} y^{\beta} + x^{\beta} y^{1-\beta}.$$

Remark: This monotonicity inequality is due to Callebaut. See [1420].

Fact 1.18.2. Let $x_1, \ldots, x_n$ and $y_1, \ldots, y_n$ be real numbers. Furthermore, let $x_{[1]}, \ldots, x_{[n]}$ denote a rearrangement of $x_1, \ldots, x_n$ such that $x_{[1]} \ge \cdots \ge x_{[n]}$. Then,

$$\sum_{i=1}^n (x_{[i]} - y_{[i]})^2 \le \sum_{i=1}^n (x_{[i]} - y_i)^2.$$

Proof: See [470, p. 180].

Fact 1.18.3. Let $x_1, \ldots, x_n$ and $y_1, \ldots, y_n$ be real numbers, and assume that $x_1 \le \cdots \le x_n$ and $y_1 \le \cdots \le y_n$. Furthermore, let $x_{[1]}, \ldots, x_{[n]}$ denote a rearrangement of $x_1, \ldots, x_n$ such that $x_{[1]} \ge \cdots \ge x_{[n]}$. Then,

$$n \sum_{i=1}^n x_{[i]} y_{[n-i+1]} \le \left(\sum_{i=1}^n x_i \right) \left(\sum_{i=1}^n y_i \right) \le n \sum_{i=1}^n x_{[i]} y_{[i]}.$$

Furthermore, each inequality is an equality if and only if either $x_1 = \cdots = x_n$ or $y_1 = \cdots = y_n$.

Proof: See [686, pp. 148, 149].

Remark: This result is *Chebyshev's inequality.*

Fact 1.18.4. Let $x_1, \ldots, x_n$ and $y_1, \ldots, y_n$ be real numbers. Furthermore, let $x_{[1]}, \ldots, x_{[n]}$ denote a rearrangement of $x_1, \ldots, x_n$ such that $x_{[1]} \ge \cdots \ge x_{[n]}$. Then,

$$\sum_{i=1}^n x_{[i]} y_{[n-i+1]} \le \sum_{i=1}^n x_i y_i \le \sum_{i=1}^n x_{[i]} y_{[i]}.$$

Proof: See [240, p. 127] and [996, p. 141].

Remark: This result is the *Hardy-Littlewood rearrangement inequality.*

Remark: See Fact 8.19.19.

Fact 1.18.5. Let $x_1, \ldots, x_n$ be nonnegative numbers, and let $y_1, \ldots, y_n$ be real numbers. Furthermore, let $y_{[1]}, \ldots, y_{[n]}$ denote a rearrangement of $y_1, \ldots, y_n$ such that $y_{[1]} \geq \cdots \geq y_{[n]}$. Then, for all $k \in \{1, \ldots, n\}$, it follows that

$$\sum_{i=1}^{k} x_{[i]} y_i \leq \sum_{i=1}^{k} x_{[i]} y_{[i]}$$

and

$$\sum_{i=1}^{k} x_{[i]} y_{[n-i+1]} \leq \sum_{i=1}^{k} x_i y_i.$$

Now, assume in addition that $y_1, \ldots, y_n$ are nonnegative numbers. Then, for all $k \in \{1, \ldots, n\}$, it follows that

$$\sum_{i=1}^{k} x_{[i]} y_{[n-i+1]} \leq \sum_{i=1}^{k} x_i y_i \leq \sum_{i=1}^{k} x_{[i]} y_i \leq \sum_{i=1}^{k} x_{[i]} y_{[i]}.$$

Proof: See [389, 862] and [996, p. 141].

Remark: This result is an extension of the *Hardy-Littlewood rearrangement inequality.*

Fact 1.18.6. Let $x_1, \ldots, x_n$ and $y_1, \ldots, y_n$ be positive numbers, and let p, q be positive numbers such that, for all $i \in \{1, \ldots, n\}$,

$$q \leq \frac{x_i}{y_i} \leq p.$$

Furthermore, let $x_{[1]}, \ldots, x_{[n]}$ denote a rearrangement of $x_1, \ldots, x_n$ such that $x_{[1]} \geq \cdots \geq x_{[n]}$. Then,

$$\sum_{i=1}^{n} x_{[i]} y_{[i]} \leq \frac{p+q}{2\sqrt{pq}} \sum_{i=1}^{n} x_i y_i.$$

Proof: See [255].

Remark: This result is a reverse rearrangement inequality.

Remark: Equality holds for $x_1 = 2$, $x_2 = 1$, $y_1 = 1/2$, $y_2 = 2$, $q = 1$, and $p = 4$. Consequently, if $q = \min_{i=1,\ldots,n} x_i/y_i$ and $p = \max_{i=1,\ldots,n} x_i/y_i$, then the coefficient $\frac{p+q}{2\sqrt{pq}}$ is the best possible.

Fact 1.18.7. Let $x_1, \ldots, x_n$ and $y_1, \ldots, y_n$ be nonnegative numbers, and assume that $x_1 \geq \cdots \geq x_n$ and $y_1 \geq \cdots \geq y_n$. Then,

$$\prod_{i=1}^{n} (x_i^2 + y_i^2) \leq \prod_{i=1}^{n} (x_i^2 + y_{n-i+1}^2).$$

Remark: See Fact 8.13.11.

Fact 1.18.8. Let $x_1, \ldots, x_n$ and $y_1, \ldots, y_n$ be complex numbers. Then,

$$\left| \sum_{i=1}^{n} x_i y_i \right|^2 = \sum_{i=1}^{n} |x_i|^2 \sum_{i=1}^{n} |y_i|^2 - \sum_{i<j} |\overline{x}_i y_j - \overline{x}_j y_i|^2.$$

Remark: This result is the *Lagrange identity*. For the complex case, see [440, p. 6] or [441, p. 3]. For the real case, see [1354, 322].

Fact 1.18.9. Let $x_1, \ldots, x_n$ and $y_1, \ldots, y_n$ be real numbers. Then,

$$\sum_{i=1}^{n} x_i y_i \leq \left(\sum_{i=1}^{n} x_i^2 \right)^{1/2} \left(\sum_{i=1}^{n} y_i^2 \right)^{1/2}.$$

Furthermore, equality holds if and only if $\begin{bmatrix} x_1 & \cdots & x_n \end{bmatrix}^{\mathrm{T}}$ and $\begin{bmatrix} y_1 & \cdots & y_n \end{bmatrix}^{\mathrm{T}}$ are linearly dependent.

Remark: This result is the *Cauchy-Schwarz inequality*.

Fact 1.18.10. Let $x_1, \ldots, x_n$ and $y_1, \ldots, y_n$ be real numbers, and assume that $x_1 \leq \cdots \leq x_n$ and $y_1 \leq \cdots \leq y_n$. Then,

$$\left(\sum_{i=1}^{n} x_i \right) \left(\sum_{i=1}^{n} y_i \right) \leq n \sum_{i=1}^{n} x_i y_i.$$

Proof: See [71, p. 27].

Fact 1.18.11. Let $x_1, \ldots, x_n$ and $y_1, \ldots, y_n$ be nonnegative numbers, and let $\alpha \in [0, 1]$. Then,

$$\sum_{i=1}^{n} x_i^{\alpha} y_i^{1-\alpha} \leq \left(\sum_{i=1}^{n} x_i \right)^{\alpha} \left(\sum_{i=1}^{n} y_i \right)^{1-\alpha}.$$

Now, let $p, q \in [1, \infty]$ satisfy $1/p + 1/q = 1$. Then, equivalently,

$$\sum_{i=1}^{n} x_i y_i \leq \left(\sum_{i=1}^{n} x_i^p \right)^{1/p} \left(\sum_{i=1}^{n} y_i^q \right)^{1/q}.$$

Furthermore, equality holds if and only if $\begin{bmatrix} x_1^p & \cdots & x_n^p \end{bmatrix}^{\mathrm{T}}$ and $\begin{bmatrix} y_1^q & \cdots & y_n^q \end{bmatrix}^{\mathrm{T}}$ are linearly dependent.

Remark: This result is *Hölder's inequality*.

Remark: Note the relationship between the *conjugate parameters* p, q and the *barycentric coordinates* $\alpha, 1 - \alpha$. See Fact 8.22.52.

Remark: See Fact 9.7.34.

Fact 1.18.12. Let $x_1, \ldots, x_n$ and $y_1, \ldots, y_n$ be complex numbers, let p, q, r be positive numbers, and assume that $1/p + 1/q = 1/r$. If $p \in (0, 1)$, $q < 0$, and $r = 1$, then

$$\left(\sum_{i=1}^{n} |x_i|^p \right)^{1/p} \left(\sum_{i=1}^{n} |y_i|^q \right)^{1/q} \leq \sum_{i=1}^{n} |x_i y_i|.$$

Furthermore, if $p, q, r > 0$, then

$$\left(\sum_{i=1}^{n} |x_i y_i|^r \right)^{1/r} \leq \left(\sum_{i=1}^{n} |x_i|^p \right)^{1/p} \left(\sum_{i=1}^{n} |y_i|^q \right)^{1/q}.$$

Proof: See [1066, p. 19].

Remark: This result is the *Rogers-Hölder inequality*.

Remark: Extensions of this result for negative values of p, q, and r are considered in [1066, p. 19].

Remark: See Proposition 9.1.6.

Fact 1.18.13. Let $x_1, \ldots, x_n$ and $y_1, \ldots, y_n$ be nonnegative numbers, and let $p, q \in [1, \infty]$ satisfy $1/p + 1/q = 1$. Then,

$$\sum_{i=1}^{n} \sum_{j=1}^{n} \frac{x_i y_j}{i+j-1} \le \frac{\pi}{\sin(\pi/p)} \left(\sum_{i=1}^{n} x_i^p \right)^{1/p} \left(\sum_{i=1}^{n} y_i^q \right)^{1/q}.$$

In particular,

$$\sum_{i=1}^{n} \sum_{j=1}^{n} \frac{x_i y_j}{i+j-1} \le \pi \left(\sum_{i=1}^{n} x_i^2 \right)^{1/2} \left(\sum_{i=1}^{n} y_i^2 \right)^{1/2}.$$

Proof: See [556, p. 66] or [873].

Remark: This result is the *Hardy-Hilbert inequality*.

Remark: It follows from Fact 1.18.11 that

$$\sum_{i=1}^{n} \sum_{j=1}^{n} x_i y_j \le n \left(\sum_{i=1}^{n} x_i^p \right)^{1/p} \left(\sum_{i=1}^{n} y_i^q \right)^{1/q}.$$

Fact 1.18.14. Let $x_1, \ldots, x_n$ and $y_1, \ldots, y_n$ be nonnegative numbers, and let $p, q \in [1, \infty]$ satisfy $1/p + 1/q = 1$. Then,

$$\sum_{i=1}^{n} \sum_{j=1}^{n} \frac{x_i y_j}{\max\{i, j\}} \le pq \left(\sum_{i=1}^{n} x_i^p \right)^{1/p} \left(\sum_{i=1}^{n} y_i^q \right)^{1/q}.$$

Furthermore,

$$\sum_{i=2}^{n} \sum_{j=2}^{n} \frac{x_i y_j}{\log ij} \le \frac{\pi}{\sin(\pi/p)} \left(\sum_{i=2}^{n} i^{p-1} x_i^p \right)^{1/p} \left(\sum_{i=2}^{n} i^{q-1} y_i^q \right)^{1/q}.$$

In particular,

$$\sum_{i=2}^{n} \sum_{j=2}^{n} \frac{x_i y_j}{\log ij} \le \pi \left(\sum_{i=2}^{n} i x_i^2 \right)^{1/2} \left(\sum_{i=2}^{n} i y_i^2 \right)^{1/2}.$$

Proof: For the first result, see [99]. For the second result see [1507].

Remark: Related inequalities are given in [1508].

Fact 1.18.15. Let $x_1, \ldots, x_n$ and $y_1, \ldots, y_n$ be nonnegative numbers, and assume that, for all $i \in \{1, \ldots, n\}$, $x_i + y_i > 0$. Then,

$$\left(\sum_{i=1}^{n} x_i y_i \right)^2 \le \sum_{i=1}^{n} (x_i^2 + y_i^2) \sum_{i=1}^{n} \frac{x_i^2 y_i^2}{x_i^2 + y_i^2} \le \sum_{i=1}^{n} x_i^2 \sum_{i=1}^{n} y_i^2.$$

Proof: See [440, p. 37], [441, p. 51], or [1420].

Remark: This interpolation of the Cauchy-Schwarz inequality is *Milne's inequality*.

Fact 1.18.16. Let $x_1, \ldots, x_n$ and $y_1, \ldots, y_n$ be nonnegative numbers, and let $\alpha \in [0, 1]$. Then,

$$\left(\sum_{i=1}^n x_i y_i\right)^2 \le \sum_{i=1}^n x_i^{1+\alpha} y_i^{1-\alpha} \sum_{i=1}^n x_i^{1-\alpha} y_i^{1+\alpha} \le \sum_{i=1}^n x_i^2 \sum_{i=1}^n y_i^2.$$

Proof: See [440, p. 43], [441, p. 51], or [1420].

Remark: This interpolation of the Cauchy-Schwarz inequality is *Callebaut's inequality*.

Fact 1.18.17. Let $x_1, \ldots, x_{2n}$ and $y_1, \ldots, y_{2n}$ be real numbers. Then,

$$\left(\sum_{i=1}^{2n} x_i y_i\right)^2 \le \left(\sum_{i=1}^{2n} x_i y_i\right)^2 + \left[\sum_{i=1}^n (x_i y_{n+i} - x_{n+i} y_i)\right]^2 \le \sum_{i=1}^{2n} x_i^2 \sum_{i=1}^{2n} y_i^2.$$

Proof: See [440, p. 41] or [441, p. 49].

Remark: This interpolation of the Cauchy-Schwarz inequality is *McLaughlin's inequality*.

Fact 1.18.18. Let $x_1, \ldots, x_n$ and $y_1, \ldots, y_n$ be nonnegative numbers, and define $a \triangleq \min_{i=1,\ldots,n} x_i$, and $b \triangleq \max_{i=1,\ldots,n} x_i$, $c \triangleq \min_{i=1,\ldots,n} y_i$, and $d \triangleq \max_{i=1,\ldots,n} y_i$. Then,

$$\left|\sum_{i=1}^n x_i y_i - \frac{1}{n}\sum_{i=1}^n x_i \sum_{i=1}^n y_i\right| \le \lfloor\tfrac{n}{2}\rfloor(1 - \tfrac{1}{n}\lfloor\tfrac{n}{2}\rfloor)(b-a)(d-c).$$

Proof: See [445].

Remark: This result is used in Fact 1.17.46.

Fact 1.18.19. Let $x_1, \ldots, x_n$ and $y_1, \ldots, y_n$ be positive numbers, and assume that $\sum_{i=2}^n x_i^2 < x_1^2$. Then,

$$\left(x_1^2 - \sum_{i=2}^n x_i^2\right)\left(y_1^2 - \sum_{i=2}^n y_i^2\right) \le \left(x_1 y_1 - \sum_{i=2}^n x_i y_i\right)^2.$$

Remark: This result is *Aczels's inequality*. See [279, p. 16]. Extensions are given in [1496] and Fact 9.7.4.

Fact 1.18.20. Let $x_1, \ldots, x_n$ be real numbers, and let $z_1, \ldots, z_n$ be complex numbers. Then,

$$\left|\sum_{i=1}^n x_i z_i\right|^2 \le \tfrac{1}{2}\sum_{i=1}^n x_i^2\left(\sum_{i=1}^n |z_i|^2 + \left|\sum_{i=1}^n z_i^2\right|\right) \le \sum_{i=1}^n x_i^2 \sum_{i=1}^n |z_i|^2.$$

Proof: See [440, p. 40] or [441, p. 48].

Remark: Conditions for equality in the left-hand inequality are given in [440, p. 40] or [441, p. 48].

Remark: This interpolation of the Cauchy-Schwarz inequality is *De Bruijn's inequality.*

Fact 1.18.21. Let $x_1, \ldots, x_n$ and $y_1, \ldots, y_n$ be positive numbers, and define $\alpha \triangleq \min_{i=1,\ldots,n} x_i/y_i$ and $\beta \triangleq \max_{i=1,\ldots,n} x_i/y_i$. Then,

$$\left(\sum_{i=1}^n x_i y_i\right)^2 \leq \sum_{i=1}^n x_i^2 \sum_{i=1}^n y_i^2 \leq \tfrac{(\alpha+\beta)^2}{4\alpha\beta} \left(\sum_{i=1}^n x_i y_i\right)^2.$$

Equivalently, let $a \triangleq \min_{i=1,\ldots,n} x_i$, $A \triangleq \max_{i=1,\ldots,n} x_i$, $b \triangleq \min_{i=1,\ldots,n} y_i$, and $B \triangleq \max_{i=1,\ldots,n} y_i$. Then,

$$\left(\sum_{i=1}^n x_i y_i\right)^2 \leq \sum_{i=1}^n x_i^2 \sum_{i=1}^n y_i^2 \leq \frac{(ab+AB)^2}{4abAB} \left(\sum_{i=1}^n x_i y_i\right)^2.$$

Proof: See [440, p. 73] or [441, p. 92].

Remark: This reversal of the Cauchy-Schwarz inequality is the *Polya-Szego inequality.*

Fact 1.18.22. Let $x_1, \ldots, x_n$ and $y_1, \ldots, y_n$ be positive numbers, let $a \triangleq \min_{i=1,\ldots,n} x_i$, $A \triangleq \max_{i=1,\ldots,n} x_i$, $b \triangleq \min_{i=1,\ldots,n} y_i$, and $B \triangleq \max_{i=1,\ldots,n} y_i$, let p, q be positive numbers, and assume that $1/p + 1/q = 1$. Then,

$$\sum_{i=1}^n x_i y_i \leq \left(\sum_{i=1}^n x_i^p\right)^{1/p} \left(\sum_{i=1}^n y_i^q\right)^{1/q} \leq \gamma \sum_{i=1}^n x_i y_i,$$

where
$$\gamma \triangleq \frac{A^p B^q - a^p b^q}{[p(AbB^q - aBb^q)]^{1/p}[q(aBA^p - Aba^p)]^{1/q}}.$$

Proof: See [1428].

Remark: The right-hand inequality, which is a reversal of Hölder's inequality, is the *Diaz-Goldman-Metcalf inequality.*

Remark: Setting $p = q = 1/2$ yields Fact 1.18.21.

Remark: The case in which $1/p + 1/q = 1/r$ is discussed in [1428].

Fact 1.18.23. Let $x_1, \ldots, x_n$ and $y_1, \ldots, y_n$ be nonnegative numbers, and define $m_x \triangleq \min_{i=1,\ldots,n} x_i$, $m_y \triangleq \min_{i=1,\ldots,n} y_i$, $M_x \triangleq \max_{i=1,\ldots,n} x_i$, and $M_y \triangleq \max_{i=1,\ldots,n} y_i$. Then,

$$\left(\sum_{i=1}^n x_i y_i\right)^2 \leq \sum_{i=1}^n x_i^2 \sum_{i=1}^n y_i^2 \leq \left(\sum_{i=1}^n x_i y_i\right)^2 + \tfrac{n^2}{3}(M_x M_y - m_x m_y)^2.$$

Proof: See [770].

Remark: The right-hand inequality, which is a reversal of the Cauchy-Schwarz inequality, is *Ozeki's inequality.*

Fact 1.18.24. Let $x_1, \ldots, x_n$ and $y_1, \ldots, y_n$ be nonnegative numbers, and assume that, for all $i \in \{1, \ldots, n\}$, $x_i + y_i > 0$. Then,

$$\sum_{i=1}^n \frac{x_i y_i}{x_i + y_i} \sum_{i=1}^n (x_i + y_i) \le \sum_{i=1}^n x_i \sum_{i=1}^n y_i.$$

Proof: See [440, p. 36] or [441, p. 42].

Remark: For positive numbers x and y, define the *harmonic mean* $H(x, y)$ of x and y by

$$H(x, y) \triangleq \frac{2}{\frac{1}{x} + \frac{1}{y}}.$$

Then, this result is equivalent to

$$\sum_{i=1}^n H(x_i, y_i) \le H\left(\sum_{i=1}^n x_i, \sum_{i=1}^n y_i\right).$$

See [440, p. 37] or [441, p. 43]. The factor of 2 appearing on the right-hand side in [440, 441] is not needed.

Remark: This result is *Dragomir's inequality*.

Remark: Letting α, β be positive numbers and defining the arithmetic mean $A(\alpha, \beta) \triangleq \frac{1}{2}(\alpha + \beta)$, it follows that

$$\frac{(\alpha + \beta)^2}{4\alpha\beta} = \frac{A(\alpha, \beta)}{H(\alpha, \beta)}.$$

For details, see [1443].

Fact 1.18.25. Let $x_1, \ldots, x_n$ and $y_1, \ldots, y_n$ be nonnegative numbers. If $p \in (0, 1]$, then

$$\left[\sum_{i=1}^n (x_i + y_i)^p\right]^{1/p} \ge \left(\sum_{i=1}^n x_i^p\right)^{1/p} + \left(\sum_{i=1}^n y_i^p\right)^{1/p}.$$

If $p \ge 1$, then

$$\left[\sum_{i=1}^n (x_i + y_i)^p\right]^{1/p} \le \left(\sum_{i=1}^n x_i^p\right)^{1/p} + \left(\sum_{i=1}^n y_i^p\right)^{1/p}.$$

Furthermore, equality holds if and only if either $p = 1$ or $\begin{bmatrix} x_1 & \cdots & x_n \end{bmatrix}^{\mathrm{T}}$ and $\begin{bmatrix} y_1 & \cdots & y_n \end{bmatrix}^{\mathrm{T}}$ are linearly dependent.

Proof: See [267].

Remark: This result is *Minkowski's inequality*.

Fact 1.18.26. Let $x_1, \ldots, x_n$ and $y_1, \ldots, y_n$ be nonnegative numbers, let $\alpha_1, \ldots, \alpha_n$ be nonnegative numbers, and assume that $\sum_{i=1}^n \alpha_i = 1$. Then,

$$x_1^{\alpha_1} \cdots x_n^{\alpha_n} + y_1^{\alpha_1} \cdots y_n^{\alpha_n} \le (x_1 + y_1)^{\alpha_1} \cdots (x_n + y_n)^{\alpha_n}.$$

Proof: See [805, p. 64].

Fact 1.18.27. Let $x_1, \ldots, x_n, y_1, \ldots, y_n \in (-1, 1)$, and let m be a positive integer. Then,

$$\left[\sum_{i=1}^{n} \frac{1}{(1 - x_i y_i)^m} \right]^2 \leq \left[\sum_{i=1}^{n} \frac{1}{(1 - x_i^2)^m} \right] \left[\sum_{i=1}^{n} \frac{1}{(1 - y_i^2)^m} \right].$$

Proof: See [440, p. 19] or [441, p. 19].

Fact 1.18.28. Let $x_1, \ldots, x_n$ and $y_1, \ldots, y_n$ be nonnegative numbers, and assume that $\sum_{i=1}^{n} x_i$ and $\sum_{i=1}^{n} y_i$ are nonzero. Then,

$$\left(\frac{\sum_{i=1}^{n} x_i}{\sum_{i=1}^{n} y_i} \right)^{\sum_{i=1}^{n} x_i} \prod_{i=1}^{n} y_i^{x_i} \leq \prod_{i=1}^{n} x_i^{x_i}.$$

Furthermore, equality holds if and only if there exists $\alpha > 0$ such that, for all $i \in \{1, \ldots, n\}$, $x_i = \alpha y_i$.

Proof: See [134].

Fact 1.18.29. Let $x_1, \ldots, x_n$ and $y_1, \ldots, y_n$ be nonnegative numbers, and assume that $\sum_{i=1}^{n} x_i = \sum_{i=1}^{n} y_i$. Then,

$$\prod_{i=1}^{n} y_i^{x_i} \leq \prod_{i=1}^{n} x_i^{x_i}.$$

In particular,

$$\left(\frac{1}{n} \sum_{i=1}^{n} x_i \right)^{\sum_{i=1}^{n} x_i} \leq \prod_{i=1}^{n} x_i^{x_i}.$$

Proof: See Fact 1.18.28 and [1190].

Fact 1.18.30. Let $x_1, \ldots, x_n$ and $y_1, \ldots, y_n$ be positive numbers. Then,

$$\sum_{i=1}^{n} x_i \log \frac{\sum_{j=1}^{n} x_j}{\sum_{j=1}^{n} y_j} \leq \sum_{i=1}^{n} x_i \log \frac{x_i}{y_i}.$$

If $\sum_{i=1}^{n} x_i = 1$, then

$$\sum_{i=1}^{n} x_i \log \frac{1}{x_i} \leq \sum_{i=1}^{n} x_i \log \frac{1}{y_i} + \log \sum_{i=1}^{n} y_i.$$

On the other hand, if $\sum_{i=1}^{n} x_i = \sum_{i=1}^{n} y_i$, then

$$0 \leq \sum_{i=1}^{n} x_i \log \frac{1}{y_i} + \log \sum_{i=1}^{n} y_i.$$

Finally, if $\sum_{i=1}^{n} x_i = \sum_{i=1}^{n} y_i = 1$, then

$$\sum_{i=1}^{n} x_i \log \frac{1}{x_i} \leq \sum_{i=1}^{n} x_i \log \frac{1}{y_i},$$

or, equivalently,

$$0 \le \sum_{i=1}^{n} x_i \log \frac{x_i}{y_i}.$$

Proof: See [1007].

Remark: $\sum_{i=1}^{n} x_i \log \frac{1}{x_i}$ is the *entropy*.

Remark: A refined upper bound and positive lower bound for $\sum_{i=1}^{n} x_i \log \frac{x_i}{y_i}$ are given in [640].

Remark: See Fact 2.21.6.

Remark: Related results are given in [1215, p. 276].

1.19 Facts on Scalar Identities and Inequalities in $3n$ Variables

Fact 1.19.1. Let $x_1, \ldots, x_n, y_1, \ldots, y_n, z_1, \ldots, z_n$ be real numbers. Then,

$$\left(\sum_{i=1}^{n} x_i y_i z_i \right)^4 \le \left(\sum_{i=1}^{n} x_i^4 \right) \left(\sum_{i=1}^{n} y_i^2 \right)^2 \left(\sum_{i=1}^{n} z_i^4 \right).$$

Proof: See [71, p. 27].

Fact 1.19.2. Let $x_1, \ldots, x_n, y_1, \ldots, y_n, z_1, \ldots, z_n$ be complex numbers. Then,

$$\left| \sum_{i=1}^{n} x_i \overline{z_i} \sum_{i=1}^{n} z_i \overline{y_i} \right| \le \frac{1}{2} \left(\sqrt{ \sum_{i=1}^{n} |x_i|^2 \sum_{i=1}^{n} |y_i|^2 } + \left| \sum_{i=1}^{n} x_i \overline{y_i} \right| \right) \sum_{i=1}^{n} |z_i|^2.$$

Proof: See [527].

Remark: This extension of the Cauchy-Schwarz inequality is *Buzano's inequality*.

Remark: See *xv*) of Fact 9.7.4.

1.20 Facts on Scalar Identities and Inequalities in Complex Variables

Fact 1.20.1. Let z be a complex number with complex conjugate $\overline{z}$, real part $\operatorname{Re} z$, and imaginary part $\operatorname{Im} z$. Then, the following statements hold:

i) $-|z| \le \operatorname{Re} z \le |\operatorname{Re} z| \le |z|$.

ii) $-|z| \le \operatorname{Im} z \le |\operatorname{Im} z| \le |z|$.

iii) $0 \le |z| = |-z| = |\overline{z}|$.

iv) $\operatorname{Re} z = |\operatorname{Re} z| = |z|$ if and only if $\operatorname{Re} z \ge 0$ and $\operatorname{Im} z = 0$.

v) $\operatorname{Im} z = |\operatorname{Im} z| = |z|$ if and only if $\operatorname{Im} z \ge 0$ and $\operatorname{Re} z = 0$.

vi) If $z \ne 0$, then $\overline{z^{-1}} = \overline{z}^{-1}$.

vii) If $z \neq 0$, then $z^{-1} = \overline{z}/|z|^2$.

viii) If $z \neq 0$, then $|z^{-1}| = 1/|z|$.

ix) If $|z| = 1$, then $z^{-1} = \overline{z}$.

x) If $z \neq 0$, then $\operatorname{Re} z^{-1} = (\operatorname{Re} z)/|z|^2$.

xi) $\operatorname{Re} z \neq 0$ if and only if $\operatorname{Re} z^{-1} \neq 0$.

xii) If $\operatorname{Re} z \neq 0$, then $|z| = \sqrt{(\operatorname{Re} z)/(\operatorname{Re} z^{-1})}$.

xiii) $|z^2| = |z|^2 = z\overline{z}$.

xiv) $z^2 \geq 0$ if and only if $\operatorname{Im} z = 0$.

xv) $z^2 \leq 0$ if and only if $\operatorname{Re} z = 0$.

xvi) $z^2 + \overline{z}^2 + 4(\operatorname{Im} z)^2 = 2|z|^2$.

xvii) $z^2 + \overline{z}^2 + 2|z|^2 = 4(\operatorname{Re} z)^2$.

xviii) $z^2 + \overline{z}^2 + 2(\operatorname{Im} z)^2 = 2(\operatorname{Re} z)^2$.

xix) $z^2 + \overline{z}^2 \leq \left\{ \begin{matrix} |z^2 + \overline{z}^2| \\ (\operatorname{Re} z)^2 \end{matrix} \right\} \leq 2|z|^2$.

xx) $z^2 + \overline{z}^2 = |z^2 + \overline{z}^2| = (\operatorname{Re} z)^2 = 2|z|^2$ if and only if $\operatorname{Im} z = 0$.

xxi) Let n be a positive integer. If $z \neq 1$, then

$$\frac{1 - z^n}{1 - z} = \sum_{i=0}^{n-1} z^i = 1 + z + \cdots + z^{n-1}.$$

Furthermore,

$$\lim_{z \to 1} \frac{1 - z^n}{1 - z} = n.$$

Remark: A matrix version of *i*) is given in [1302].

Fact 1.20.2. Let z_1 and z_2 be complex numbers. Then, the following statements hold:

i) $|z_1 z_2| = |z_1| |z_2|$.

ii) If $z_2 \neq 0$, then $|z_1/z_2| = |z_1|/|z_2|$.

iii) $\big||z_1| - |z_2|\big| \leq |z_1 + z_2| \leq |z_1| + |z_2|$.

iv) $|z_1 + z_2| = |z_1| + |z_2|$ if and only if $\operatorname{Re}(z_1 \overline{z_2}) = |z_1||z_2|$.

v) $|z_1 + z_2| = |z_1| + |z_2|$ if and only if there exists $\alpha \geq 0$ such that either $z_1 = \alpha z_2$ or $z_2 = \alpha z_1$, that is, if and only if z_1 and z_2 have the same phase angle.

vi) $\big||z_1| - |z_2|\big| \leq |z_1 - z_2|$.

vii) $\big||z_1| - |z_2|\big| = |z_1 - z_2|$ if and only if there exists $\alpha \geq 0$ such that either $z_1 = \alpha z_2$ or $z_2 = \alpha z_1$, that is, if and only if z_1 and z_2 have the same phase angle.

viii) $|1+\overline{z_1}z_2|^2 = (1-|z_1|)^2(1-|z_2|)^2+|z_1+z_2|^2 = (1+|z_1|^2)(1+|z_2|^2)-|z_1-z_2|^2.$

ix) $|z_1 - z_2|^2 \leq (1 + |z_1|^2)(1 + |z_2|^2).$

x) $\frac{1}{2}|z_1 - z_2 + |\frac{z_2}{z_1}|z_1 - |\frac{z_1}{z_2}|z_2| = \frac{1}{2}(|z_1| + |z_2|)\left|\frac{z_1}{|z_1|} - \frac{z_2}{|z_2|}\right| \leq |z_1 - z_2|.$

xi) $2\operatorname{Re}(z_1 z_2) \leq |z_1|^2 + |z_2|^2.$

xii) $2\operatorname{Re}(z_1 z_2) = |z_1|^2 + |z_2|^2$ if and only if $z_1 = \overline{z_2}.$

xiii) $\frac{1}{2}(|z_1 + z_2|^2 + |z_1 - z_2|^2) = |z_1|^2 + |z_2|^2.$

xiv) $z_1\overline{z_2} = \frac{1}{4}(|z_1 + z_2|^2 - |z_1 - z_2|^2 + \jmath|z_1 + \jmath z_2|^2 - \jmath|z_1 - \jmath z_2|^2).$

xv) If $a, b \in \mathbb{C}$, $|a| \neq |b|$, and $z_2 = az_1 + b\overline{z_1}$, then
$$z_1 = \frac{\overline{a}z_2 - b\overline{z_2}}{|a|^2 - |b|^2}.$$

xvi) If $p \geq 1$, then
$$|z_1 + z_2|^p \leq 2^{p-1}(|z_1|^p + |z_2|^p).$$

xvii) If $p \geq 2$, then
$$2(|z_1|^p + |z_2|^p) \leq |z_1 + z_2|^p + |z_1 - z_2|^p \leq 2^{p-1}(|z_1|^p + |z_2|^p).$$

xviii) If $p \geq 2$, $q > 0$, and $1/p + 1/q = 1$, then
$$2(|z_1|^p + |z_2|^p)^{q-1} \leq |z_1 + z_2|^q + |z_1 - z_2|^q.$$

xix) If $p \in (1, 2]$, $q > 0$, and $1/p + 1/q = 1$, then
$$|z_1 + z_2|^q + |z_1 - z_2|^q \leq 2(|z_1|^p + |z_2|^p)^{q-1}.$$

xx) Let n be a positive integer. If $z_1 \neq z_2$, then
$$\frac{z_1^n - z_2^n}{z_1 - z_2} = z_1^{n-1} + z_2 z_1^{n-2} + \cdots + z_2^{n-1}.$$

Furthermore,
$$\lim_{z_2 \to z_1} \frac{z_1^n - z_2^n}{z_1 - z_2} = nz_1^{n-1}.$$

Now, let z_1, z_2, and z_3 be complex numbers. Then, the following statements hold:

xxi) $|z_1 + z_2|^2 + |z_2 + z_3|^2 + |z_3 + z_1|^2 = |z_1|^2 + |z_2|^2 + |z_3|^2 + |z_1 + z_2 + z_2|^2.$

xxii) $|z_1 + z_2| + |z_2 + z_3| + |z_3 + z_1| \leq |z_1| + |z_2| + |z_3| + |z_1 + z_2 + z_2|.$

xxiii) $4(|z_1|^2 + |z_2|^2 + |z_3|^2) \leq |z_1 + z_2 + z_3|^2 + |z_1 + z_2 - z_3|^2 + |z_1 - z_2 + z_3|^2 + |z_1 - z_2 - z_3|^2.$

xxiv) If z_1, z_2, z_3 are nonzero and $z_1^7 + z_2^7 + z_3^7 = 0$, then $|z_1| = |z_2| = |z_3|.$

Finally, for $i \in \{1, \ldots, n\}$, let $z_i = r_i e^{\jmath\phi_i}$ be complex numbers, where $r_i \geq 0$ and $\phi_i \in \mathbb{R}$, and assume that there exist $\theta_1, \theta_2 \in \mathbb{R}$ such that $0 < \theta_2 - \theta_1 < \pi$ and such that, for all $i \in \{1, \ldots, n\}$, $\theta_1 \leq \phi_i \leq \theta_2$. Then, the following inequality holds:

xxv) $\cos[\frac{1}{2}(\theta_2 - \theta_1)]\sum_{i=1}^n |z_i| \leq |\sum_{i=1}^n z_i|.$

Remark: Matrix versions of *i)*, *iii)*, *v)*–*vii)* are given in [1302]. Result *viii)* is given in [61, p. 19] and [1502]. Result *x)* is the *Dunkl-Williams inequality*. See [440, p.

43] or [441, p. 52] and *ii*) of Fact 9.7.4. Result *xiii*) is the parallelogram law; see [462] and Fact 9.7.4. Result *xiv*) is the *polarization identity*; see [376, p. 54], [1057, p. 276], and Fact 9.7.4. Result *xv*) is given in [754]. Result *xvi*) is given in [713]. Results *xvii*)–*xix*) are due to Clarkson; see [713], [1035, p. 536], and Fact 9.9.34. Result *xxi*) is given in [61, p. 19]. Result *xxii*) is *Hlawka's inequality*. See Fact 1.10.6 and Fact 9.7.4. Result *xxiii*) is given in [462]. Result *xxiv*) is given in [61, pp. 186, 187]. Result *xxv*) is due to Petrovich; see [442].

Remark: The absolute value $|z| = |x + \jmath y|$, where x and y are real, is identical to the Euclidean norm $\left\| \begin{bmatrix} x \\ y \end{bmatrix} \right\|_2$. Therefore, each result in Section 9.7 for the Euclidean norm on $\mathbb{R}^2$ can be recast in terms of complex numbers.

Problem: Compare the lower bounds for $|z_1 - z_2|$ given by *iv*) and *vii*).

Fact 1.20.3. Let a, b, c be complex numbers, and assume that $a \neq 0$. Then, $z \in \mathbb{C}$ satisfies

$$az^2 + bz + c = 0$$

if and only if

$$z = \frac{1}{2a}(y - b),$$

where

$$y = \pm \tfrac{1}{\sqrt{2}} \left(\sqrt{|\Delta| + \operatorname{Re} \Delta} + \jmath \operatorname{sign}(\operatorname{Im} \Delta) \sqrt{|\Delta| + \operatorname{Re} \Delta} \right)$$

and

$$\Delta \triangleq b^2 - 4ac.$$

If, in addition, a, b, c are real, then $z \in \mathbb{C}$ satisfies

$$az^2 + bz + c = 0$$

if and only if

$$z = \frac{1}{2a}(-b \pm \sqrt{b^2 - 4ac}).$$

Proof: See [61, pp. 15, 16].

Fact 1.20.4. Let $z, z_1, \ldots, z_n$ be complex numbers. Then,

$$\frac{1}{n} \sum_{i=1}^{n} |z - z_i|^2 = \left| z - \frac{1}{n} \sum_{i=1}^{n} z_i \right|^2 + \frac{1}{n} \sum_{1 \le i < j \le n} |z_i - z_j|^2.$$

Proof: See [61, pp. 146].

Fact 1.20.5. let z_1 and z_2 be complex numbers. Then,

$$\frac{|z_1 - z_2| - \big||z_1| - |z_2|\big|}{\min\{|z_1|, |z_2|\}} \leq \left|\frac{z_1}{|z_1|} - \frac{z_2}{|z_2|}\right|$$

$$\leq \left\{ \begin{array}{c} \dfrac{|z_1 - z_2| + \big||z_1| - |z_2|\big|}{\max\{|z_1|, |z_2|\}} \\[3mm] \dfrac{2|z_1 - z_2|}{|z_1| + |z_2|} \end{array} \right\}$$

$$\leq \left\{ \begin{array}{c} \dfrac{2|z_1 - z_2|}{\max\{|z_1|, |z_2|\}} \\[3mm] \dfrac{2(|z_1 - z_2| + \big||z_1| - |z_2|\big|)}{|z_1| + |z_2|} \end{array} \right\}$$

$$\leq \frac{4|z_1 - z_2|}{|z_1| + |z_2|}.$$

Proof: See Fact 9.7.10.

Remark: The second and lower third terms constitute the Dunkl-Williams inequality given by Fact 1.20.2.

Fact 1.20.6. Let z be a complex number. Then, the following statements hold:

i) $0 < |e^z| \leq e^{|z|}$.

ii) $|e^z| = e^{|z|}$ if and only if $\operatorname{Im} z = 0$ and $\operatorname{Re} z \geq 0$.

iii) $|e^z| = 1$ if and only if $\operatorname{Re} z = 0$.

iv) $\big||e^z| - 1\big| \leq |e^z - 1| \leq e^{|z|} - 1$.

v) If $|z| < \log 2$, then $|e^z - 1| \leq e^{|z|} - 1 < 1$.

vi) $e^z = e^{\operatorname{Re} z}[\cos(\operatorname{Im} z) + \jmath \sin(\operatorname{Im} z)]$.

vii) $\operatorname{Re} e^z = 0$ if and only if $\operatorname{Im} z$ is an odd integer multiple of $\pm \pi/2$.

viii) $\operatorname{Im} e^z = 0$ if and only if $\operatorname{Im} z$ is an integer multiple of $\pm \pi$.

ix) If z is nonzero, then $|z^{\jmath}| < e^{\pi}$.

Furthermore, let θ_1 and θ_2 be real numbers. Then, the following statements hold:

x) $|e^{\jmath\theta_1} - e^{\jmath\theta_2}| \leq |\theta_1 - \theta_2|$.

xi) $|e^{\jmath\theta_1} - e^{\jmath\theta_2}| = |\theta_1 - \theta_2|$ if and only if $\theta_1 = \theta_2$.

Finally, let r_1 and r_2 be nonnegative numbers, at least one of which is positive. Then, the following statement holds:

xii) $|e^{\jmath\theta_1} - e^{\jmath\theta_2}| \leq \frac{2|r_1 e^{\jmath\theta_1} - r_2 e^{\jmath\theta_2}|}{r_1 + r_2}$.

Proof: Statement xii) is given in [701, p. 218].

Remark: A matrix version of x) is given by Fact 11.16.13.

Fact 1.20.7. Let z be a complex number. Then, for all nonzero $z \in \mathbb{C}$, there exist infinitely many $s \in \mathbb{C}$ such that $e^s = z$. Specifically, let $z = re^{\jmath\phi}$, where $r > 0$ and $\phi \in \mathbb{R}$. Then, for all $k \in \mathbb{Z}$, $s = \log r + \jmath(\phi + 2\pi k)$ satisfies $e^s = z$, where $\log r$ is the positive logarithm of r. In particular, for all odd integers k, $e^{\pm\jmath\pi k} = -1$, while, for all even integers k, $e^{\pm\jmath\pi k} = 1$. To obtain a single-valued definition of log, let $z \in \mathbb{C}$ be nonzero, and write z uniquely as $z = re^{\jmath\phi}$, where $r > 0$ and $\phi \in (-\pi, \pi]$. Then, the *principal branch* of the log function $\log z \in \mathbb{C}$ is defined as

$$\log z \triangleq \log r + \jmath\phi.$$

The principal branch of the log function

$$\log: \mathbb{C}\backslash\{0\} \mapsto \{z: \operatorname{Re} z \neq 0 \text{ and } -\pi < \operatorname{Im} z \leq \pi\}$$

has the following properties:

i) If $z \in \mathbb{C}$ is nonzero, then
$$e^{\log z} = z.$$

ii) Let $z = re^{\jmath\phi} \in \mathbb{C}$, where $r \geq 0$ and $\phi \in (-\pi, \pi]$, and assume that $r\sin\phi \in (-\pi, \pi]$. Then,
$$\log e^z = z.$$

iii) Let $z_1 = r_1 e^{\jmath\phi_1}$ and $z_2 = r_2 e^{\jmath\phi_2}$, where $r_1, r_2 > 0$ and $\phi_1, \phi_2 \in (-\pi, \pi]$, and assume that $\phi_1 + \phi_2 \in (-\pi, \pi]$. Then,
$$\log z_1 z_2 = \log z_1 + \log z_2.$$

Now, define $\mathcal{D} \triangleq \{z \in \mathbb{C}: |z - 1| < 1\}$. Then, the following statements hold:

iv) For all $z \in \mathcal{D}$, $\log z$ is given by the convergent series
$$\log z = \sum_{i=1}^{\infty} \frac{(-1)^{i+1}}{i}(z-1)^i.$$

v) If $z \in \mathcal{D}$, then
$$\log e^z = z.$$

vi) If $z_1, z_2 \in \mathcal{D}$, then
$$\log z_1 z_2 = \log z_1 + \log z_2.$$

vi) If $|z| < 1$, then
$$|\log(1 + z)| \leq -\log(1 - |z|)$$

and
$$\frac{|z|}{1 + |z|} \leq |\log(1 + z)| \leq \frac{|z|(1 + |z|)}{|1 + z|}.$$

Remark: Let $z = re^{\jmath\theta} \in \mathbb{C}$ satisfy $|z - 1| < 1$. Then, $-\pi/2 < \theta < \pi/2$. Furthermore, $\log z = (\log r) + \jmath\theta$, and thus $-\pi/2 < \operatorname{Im}\log z < \pi/2$. Consequently, the infinite series in iv) gives the principal log of z.

Fact 1.20.8. The following infinite series converge for the given values of the complex argument z:

i) For all $z \in \mathbb{C}$, $\quad \sin z = z - \frac{1}{3!}z^3 + \frac{1}{5!}z^5 - \frac{1}{7!}z^7 + \cdots$.

ii) For all $z \in \mathbb{C}$,
$$\cos z = 1 - \tfrac{1}{2!}z^2 + \tfrac{1}{4!}z^4 - \tfrac{1}{6!}z^6 + \cdots .$$

iii) For all $|z| < \pi/2$,
$$\tan z = z + \tfrac{1}{3}z^3 + \tfrac{2}{15}z^5 + \tfrac{17}{315}z^7 + \tfrac{62}{2835}z^9 + \cdots .$$

iv) For all $z \in \mathbb{C}$,
$$e^z = 1 + z + \tfrac{1}{2!}z^2 + \tfrac{1}{3!}z^3 + \tfrac{1}{4!}z^4 + \cdots .$$

v) For all nonzero $z \in \mathbb{C}$ such that $|z - 1| \leq 1$,
$$\log z = -\left[1 - z + \tfrac{1}{2}(1 - z)^2 + \tfrac{1}{3}(1 - z)^3 + \tfrac{1}{4}(1 - z)^4 + \cdots \right].$$

vi) For all $z \in \text{CUD}\backslash\{1\}$,
$$\log(1 - z) = -\left(z + \tfrac{1}{2}z^2 + \tfrac{1}{3}z^3 + \tfrac{1}{4}z^4 + \cdots \right).$$

vii) For all $z \in \text{CUD}\backslash\{-1\}$,
$$\log(1 + z) = z - \tfrac{1}{2}z^2 + \tfrac{1}{3}z^3 - \tfrac{1}{4}z^4 + \cdots .$$

viii) For all $z \in \text{CUD}\backslash\{-1, 1\}$,
$$\log \frac{1 + z}{1 - z} = 2(z + \tfrac{1}{3}z^3 + \tfrac{1}{5}z^5 + \cdots).$$

ix) For all $z \in \mathbb{C}$ such that $\text{Re}\, z > 0$,
$$\log z = \sum_{i=0}^{\infty} \frac{2}{2i + 1} \left(\frac{z - 1}{z + 1} \right)^{2i+1} .$$

x) For all $z \in \mathbb{C}$,
$$\sinh z = \sin \jmath z = z + \tfrac{1}{3!}z^3 + \tfrac{1}{5!}z^5 + \tfrac{1}{7!}z^7 + \cdots .$$

xi) For all $z \in \mathbb{C}$,
$$\cosh z = \cos \jmath z = 1 + \tfrac{1}{2!}z^2 + \tfrac{1}{4!}z^4 + \tfrac{1}{6!}z^6 + \cdots .$$

xii) For all $|z| < \pi/2$,
$$\tanh z = \tan \jmath z = z - \tfrac{1}{3}z^3 + \tfrac{2}{15}z^5 - \tfrac{17}{315}z^7 + \tfrac{62}{2835}z^9 - \cdots .$$

xiii) For all $\alpha \in \mathbb{C}$ and $|z| \leq 1$ such that either $|z| < 1$ or both $\text{Re}\, \alpha > -1$ and $|z| \neq -1$,
$$(1 + z)^\alpha = 1 + \alpha z + \frac{\alpha(\alpha - 1)}{2!}z^2 + \frac{\alpha(\alpha - 1)(\alpha - 2)}{3!}z^3 + \frac{\alpha(\alpha - 1)(\alpha - 2)(\alpha - 3)}{4!}z^4 + \cdots$$
$$= \binom{\alpha}{0} + \binom{\alpha}{1}z + \binom{\alpha}{2}z^2 + \binom{\alpha}{3}z^3 + \binom{\alpha}{4}z^4 + \cdots .$$

xiv) For all $\alpha \in \mathbb{C}$ and $|z| < 1$,
$$\frac{1}{(1 - z)^{\alpha+1}} = \binom{\alpha}{0} + \binom{1+\alpha}{1}z + \binom{2+\alpha}{2}z^2 + \binom{3+\alpha}{3}z^3 + \binom{4+\alpha}{4}z^4 + \cdots .$$

xv) For all $|z| < 1$,
$$(1 - z)^{-1} = 1 + z + z^2 + z^3 + z^4 + \cdots .$$

Proof: See [772, pp. 11, 12]. For $x \in \mathbb{R}$ such that $|x| < 1$, it follows that

$$\frac{d}{dx} \log(1 - x) = \frac{-1}{1 - x} = -(1 + x + x^2 + \cdots).$$

Integrating yields

$$\log(1 - x) = -(x + \tfrac{1}{2}x^2 + \tfrac{1}{3}x^3 + \cdots).$$

Using analytic continuation to replace $x \in \mathbb{R}$ satisfying $|x| < 1$ with $z \in \mathbb{C}$ satisfying $|z| < 1$ yields *vi*).

Remark: *vii*) is *Mercator's series*, while *viii*) and *ix*) are equivalent forms of *Gregory's series*. See [701, p. 273].

Remark: The coefficients in *iii*) can be expressed in terms of Bernoulli numbers. See [772, p. 129].

Remark: *xiii*) is the *binomial series*.

1.21 Facts on Trigonometric and Hyperbolic Identities

Fact 1.21.1. Let x be a real number such that the expressions below are defined. Then, the following statements hold:

i) $\sin x = \frac{1}{j2}(e^{jx} - e^{-jx})$.

ii) $\cos x = \frac{1}{2}(e^{jx} + e^{-jx})$.

iii) $\sin(x + y) = (\sin x)(\cos y) + (\cos x)\sin y$.

iv) $\sin(x - y) = (\sin x)(\cos y) - (\cos x)\sin y$.

v) $\cos(x + y) = (\cos x)(\cos y) - (\sin x)\sin y$.

vi) $\cos(x - y) = (\cos x)(\cos y) + (\sin x)\sin y$.

vii) $(\sin x)\sin y = \frac{1}{2}[\cos(x - y) - \cos(x + y)]$.

viii) $(\sin x)\cos y = \frac{1}{2}[\sin(x + y) + \sin(x - y)]$.

ix) $(\cos x)\cos y = \frac{1}{2}[\cos(x + y) + \cos(x - y)]$.

x) $\sin^2 x - \sin^2 y = [\sin(x + y)]\sin(x - y)$.

xi) $\cos^2 x - \sin^2 y = [\cos(x + y)]\cos(x - y)$.

xii) $\cos^2 x - \cos^2 y = [\sin(x + y)]\sin(y - x)$.

xiii) $\sin x + \sin y = 2[\sin \frac{1}{2}(x + y)]\cos \frac{1}{2}(x - y)$.

xiv) $\sin x - \sin y = 2[\sin \frac{1}{2}(x - y)]\cos \frac{1}{2}(x + y)$.

xv) $\cos x + \cos y = 2[\cos \frac{1}{2}(x + y)]\cos \frac{1}{2}(x - y)$.

xvi) $\cos x - \cos y = 2[\sin \frac{1}{2}(x + y)]\sin \frac{1}{2}(y - x)$.

xvii) $\tan(x + y) = \frac{(\tan x) + \tan y}{1 - (\tan x)\tan y}$.

xviii) $\tan(x - y) = \frac{(\tan x) - \tan y}{1 + (\tan x)\tan y}$.

xix) $\tan x + \tan y = \frac{\sin(x+y)}{(\cos x)\cos y}$.

xx) $\tan x - \tan y = \frac{\sin(x-y)}{(\cos x)\cos y}$.

xxi) $\sin x = 2(\sin \frac{x}{2})\cos \frac{x}{2}$.

xxii) $\cos x = 2(\cos^2 \frac{x}{2}) - 1$.

xxiii) $\sin 2x = 2(\sin x)\cos x$.

xxiv) $\cos 2x = 2(\cos^2 x) - 1$.

xxv) $\tan 2x = \frac{2\tan x}{1-\tan^2 x}$.

xxvi) $\sin 3x = 3(\sin x) - 4\sin^3 x$.

xxvii) $\cos 3x = 4(\cos^3 x) - 3\cos x$.

xxviii) $\tan 3x = \frac{3(\tan x)-\tan^3 x}{1-3\tan^2 x}$.

xxix) $\sin^2 x = \frac{1}{2}(1 - \cos 2x)$.

xxx) $\cos^2 x = \frac{1}{2}(1 + \cos 2x)$.

xxxi) $\tan^2 x = \frac{1-\cos 2x}{1+\cos 2x}$.

xxxii) $\sin x = \frac{2\tan \frac{x}{2}}{1+\tan^2 \frac{x}{2}}$.

xxxiii) $\cos x = \frac{1-\tan^2 \frac{x}{2}}{1+\tan^2 \frac{x}{2}}$.

xxxiv) $\tan x = \frac{\sin 2x}{1+\cos 2x} = \frac{1-\cos 2x}{\sin 2x} = \frac{2\tan \frac{x}{2}}{1-\tan^2 \frac{x}{2}}$.

xxxv) $\tan \frac{1}{2}x = \frac{\sin x}{1+\cos x} = \frac{1-\cos x}{\sin x}$.

xxxvi) $\sin^2 \frac{x}{2} = \frac{1}{2}(1 - \cos x)$.

xxxvii) $\cos^2 \frac{x}{2} = \frac{1}{2}(1 + \cos x)$.

xxxviii) For all $t \geq 0$ and $\alpha \in (0, 1)$,

$$\int_0^\infty \frac{tx^{\alpha-1}}{t+x}\, \mathrm{d}x = \frac{t^\alpha \pi}{\sin \alpha\pi}.$$

Remark: See [772, pp. 114–116]. The last result is given in [1540, p. 448, formula 589]. See also [556, p. 69].

Fact 1.21.2. Let x be a real number. Then, the following statements hold:

i) If $x \in [-1, 1]$, then $\sin^{-1} x + \sin^{-1}(-x) = 0$.

ii) If $x \in [-1, 1]$, then $\cos^{-1} x + \cos^{-1}(-x) = \pi$.

iii) $\tan^{-1} x + \tan^{-1}(-x) = 0$.

iv) $\cot^{-1} x + \cot^{-1}(-x) = \pi$.

v) If $x \in [-1, 1]$, then $\sin^{-1} x + \cos^{-1} x = \frac{\pi}{2}$.

vi) $\tan^{-1} x + \cot^{-1} x = \frac{\pi}{2}$.

vii) If $x \in [0, 1]$, then $\sin^{-1} x = \cos^{-1} \sqrt{1 - x^2}$.

viii) If $x^2 < 1$, then $\sin^{-1} x = \tan^{-1} \frac{x}{\sqrt{1-x^2}}$.

ix) If $x \in (0, 1]$, then $\sin^{-1} x = \cot^{-1} \frac{\sqrt{1-x^2}}{x}$.

x) If $x \in [-1, 0)$, then $\pi + \sin^{-1} x = \cot^{-1} \frac{\sqrt{1-x^2}}{x}$.

xi) If $x \in [0, 1]$, then $\cos^{-1} x = \sin^{-1} \sqrt{1 - x^2}$.

xii) If $x \in [-1, 0]$, then $\cos^{-1} x + \sin^{-1} \sqrt{1 - x^2} = \pi$.

xiii) If $x \in (0, 1]$, then $\cos^{-1} x = \tan^{-1} \frac{\sqrt{1-x^2}}{x}$.

xiv) If $x \in [-1, 0)$, then $\cos^{-1} x = \pi + \cot^{-1} \frac{\sqrt{1-x^2}}{x}$.

xv) If $x \in [-1, 1)$, then $\cos^{-1} x = \cot^{-1} \frac{x}{\sqrt{1-x^2}}$.

xvi) $\tan^{-1} x = \sin^{-1} \frac{x}{\sqrt{1+x^2}}$.

xvii) If $x \geq 0$, then $\tan^{-1} x = \cos^{-1} \frac{1}{\sqrt{1+x^2}}$.

xviii) If $x > 0$, then $\tan^{-1} x = \cot^{-1} \frac{1}{x}$.

xix) If $x < 0$, then $\pi + \tan^{-1} x + \cot^{-1} \frac{1}{x} = 0$.

xx) If $x > 0$, then $\tan^{-1} x = \cos^{-1} \frac{1}{\sqrt{1+x^2}}$.

xxi) If $x < 0$, then $\tan^{-1} x + \cos^{-1} \frac{1}{\sqrt{1+x^2}} = \pi$.

xxii) $\cot^{-1} x = \cos^{-1} \frac{x}{\sqrt{1+x^2}}$.

xxiii) If $x \in [-1, 1]$, then $\sin^{-1} x = 2 \tan^{-1} \frac{x}{1+\sqrt{1-x^2}}$.

xxiv) If $x \in (-1, 1]$, then $\cos^{-1} x = 2 \tan^{-1} \frac{\sqrt{1-x^2}}{1+x}$.

xxv) $\tan^{-1} 1 + \tan^{-1} 2 + \tan^{-1} 3 = \pi$.

xxvi) $4 \tan^{-1} \frac{1}{5} = \frac{\pi}{4} + \tan^{-1} \frac{1}{239}$.

xxvii) $\tan^{-1} x = 2 \tan^{-1} \frac{x}{1+\sqrt{1+x^2}}$.

xxviii) If $y \in \mathbb{R}$ and $xy < 1$, then $\tan^{-1} x + \tan^{-1} y = \tan^{-1} \frac{xy}{1-xy}$.

Remark: The range of $\cos^{-1}$ is $[0, \pi]$, the range of $\sin^{-1}$ is $[-\pi/2, \pi/2]$, the range of $\tan^{-1}$ is $(-\pi/2, \pi/2)$, and the range of $\cot^{-1}$ is $(0, \pi)$.

Remark: A geometric proof of *xxv)* is given in [835].

Fact 1.21.3. Let x be a real number such that the expressions below are defined. Then, the following statements hold:

i) $\sinh x = \frac{1}{2}(e^x - e^{-x})$.

ii) $\cosh x = \frac{1}{2}(e^x + e^{-x})$.

iii) $\tanh x = \frac{\sinh x}{\cosh x}$.

iv) $\sin jx = j\sinh x$.

v) $\cos jx = j\cosh x$.

vi) $\tan jx = j\tanh x$.

vii) $\sinh jx = j\sin x$.

viii) $\cosh jx = j\cos x$.

ix) $\tanh jx = j\tan x$.

x) $\sinh(x+y) = (\sinh x)(\cosh y) + (\cosh x)\sinh y$.

xi) $\cosh(x+y) = (\cosh x)(\cosh y) + (\sinh x)\sinh y$.

xii) $\tanh(x+y) = \frac{(\tanh x)+\tanh y}{1+(\tanh x)\tanh y}$.

Remark: See [772, pp. 117–119].

Fact 1.21.4. Let $z = x + jy$, where z is a complex number and x and y are real numbers. Then, the following statements hold:

i) $\sin z = (\sin x)(\cosh y) + j(\cos x)\sinh y$.

ii) $\cos z = (\cos x)(\cosh y) - j(\sin x)\sinh y$.

iii) $\tan z = \frac{(\sin 2x)+j\sinh 2y}{(\cos 2x)+\cosh 2y}$.

1.22 Notes

Much of the preliminary material in this chapter can be found in [1057]. A related treatment of mathematical preliminaries is given in [1157]. An extensive introduction to logic and mathematical fundamentals is given in [233]. In [233], the notation "$A \to B$" is used to denote an implication, which is called a *disjunction*, while "$A \implies B$" indicates a tautology.

An extensive treatment of partially ordered sets is given in [1210]. Lattices are discussed in [233].

A graph that is not necessarily symmetric is traditionally called a *directed graph* or *digraph*. In this context, a connected graph is traditionally called *strongly connected*.

Alternative terminology for "one-to-one" and "onto" is *injective* and *surjective*, respectively, while a function that is injective and surjective is *bijective*.

Reference works on inequalities include [166, 279, 280, 281, 348, 653, 988, 996, 1035, 1252]. Recommended texts on complex variables include [744, 1058, 1093].

Chapter Two

Basic Matrix Properties

In this chapter we provide a detailed treatment of the basic properties of matrices such as range, null space, rank, and invertibility. We also consider properties of convex sets, cones, and subspaces.

2.1 Matrix Algebra

The set $\mathbb{F}^n$ consists of *vectors* x of the form

$$x = \begin{bmatrix} x_{(1)} \\ \vdots \\ x_{(n)} \end{bmatrix}, \tag{2.1.1}$$

where $x_{(1)}, \ldots, x_{(n)} \in \mathbb{F}$ are the *components* of x, and $\mathbb{F}$ represents either $\mathbb{R}$ or $\mathbb{C}$. Hence, the elements of $\mathbb{F}^n$ are *column vectors*. Since $\mathbb{F}^1 = \mathbb{F}$, it follows that every scalar is also a vector. If $x \in \mathbb{R}^n$ and every component of x is nonnegative, then x is *nonnegative*, while, if every component of x is positive, then x is *positive*.

Definition 2.1.1. Let $x, y \in \mathbb{R}^n$, and assume that $x_{(1)} \geq \cdots \geq x_{(n)}$ and $y_{(1)} \geq \cdots \geq y_{(n)}$. Then, the following terminology is defined:

  i) y *weakly majorizes* x if, for all $k \in \{1, \ldots, n\}$, it follows that

$$\sum_{i=1}^{k} x_{(i)} \leq \sum_{i=1}^{k} y_{(i)}. \tag{2.1.2}$$

  ii) y *strongly majorizes* x if y weakly majorizes x and

$$\sum_{i=1}^{n} x_{(i)} = \sum_{i=1}^{n} y_{(i)}. \tag{2.1.3}$$

Now, assume in addition that x and y are nonnegative. Then, the following terminology is defined:

  iii) y *weakly log majorizes* x if, for all $k \in \{1, \ldots, n\}$, it follows that

$$\prod_{i=1}^{k} x_{(i)} \leq \prod_{i=1}^{k} y_{(i)}. \tag{2.1.4}$$

iv) y strongly log majorizes x if *y* weakly log majorizes *x* and

$$\prod_{i=1}^{n} x_{(i)} = \prod_{i=1}^{n} y_{(i)}.\tag{2.1.5}$$

Clearly, if *y* strongly majorizes *x*, then *y* weakly majorizes *x*, and, if *y* strongly log majorizes *x*, then *y* weakly log majorizes *x*. Fact 2.21.12 states that, if *y* weakly log majorizes *x*, then *y* weakly majorizes *x*. Finally, in the notation of Definition 2.1.1, if *y* majorizes *x*, then $x_{(1)} \le y_{(1)}$, while, if *y* strongly majorizes *x*, then $y_{(n)} \le x_{(n)}$.

Definition 2.1.2. Let $S \subseteq \mathbb{R}^n$, and let $f: S \mapsto \mathbb{R}$. Then, *f* is *Schur convex* if, for all $x, y \in S$ such that *y* strongly majorizes *x*, it follows that $f(x) \le f(y)$. Furthermore, *f* is *Schur concave* if $-f$ is Schur convex.

If $\alpha \in \mathbb{F}$ and $x \in \mathbb{F}^n$, then $\alpha x \in \mathbb{F}^n$ is given by

$$\alpha x = \begin{bmatrix} \alpha x_{(1)} \\ \vdots \\ \alpha x_{(n)} \end{bmatrix}.\tag{2.1.6}$$

If $x, y \in \mathbb{F}^n$, then *x* and *y* are *linearly dependent* if there exists $\alpha \in \mathbb{F}$ such that either $x = \alpha y$ or $y = \alpha x$. Linear dependence for a set of two or more vectors is defined in Section 2.3. Furthermore, vectors add component by component, that is, if $x, y \in \mathbb{F}^n$, then

$$x + y = \begin{bmatrix} x_{(1)} + y_{(1)} \\ \vdots \\ x_{(n)} + y_{(n)} \end{bmatrix}.\tag{2.1.7}$$

Thus, if $\alpha, \beta \in \mathbb{F}$, then the *linear combination* $\alpha x + \beta y$ is given by

$$\alpha x + \beta y = \begin{bmatrix} \alpha x_{(1)} + \beta y_{(1)} \\ \vdots \\ \alpha x_{(n)} + \beta y_{(n)} \end{bmatrix}.\tag{2.1.8}$$

If $x \in \mathbb{R}^n$ and *x* is nonnegative, then we write $x \geq\geq 0$, while, if *x* is positive, then we write $x >> 0$. If $x, y \in \mathbb{R}^n$, then $x \geq\geq y$ means that $x - y \geq\geq 0$, while $x >> y$ means that $x - y >> 0$.

The vectors $x_1, \ldots, x_m \in \mathbb{F}^n$ placed side by side form the *matrix*

$$A \triangleq \begin{bmatrix} x_1 & \cdots & x_m \end{bmatrix},\tag{2.1.9}$$

which has *n rows* and *m columns*. The components of the vectors $x_1, \ldots, x_m$ are the *entries* of *A*. We write $A \in \mathbb{F}^{n \times m}$ and say that *A* has *size* $n \times m$. Since $\mathbb{F}^n = \mathbb{F}^{n \times 1}$, it follows that every vector is also a matrix. Note that $\mathbb{F}^{1 \times 1} = \mathbb{F}^1 = \mathbb{F}$. If $n = m$, then *n* is the *order* of *A*, and *A* is *square*. The *i*th row of *A* and the *j*th column of

A are denoted by $\text{row}_i(A)$ and $\text{col}_j(A)$, respectively. Hence,

$$A = \begin{bmatrix} \text{row}_1(A) \\ \vdots \\ \text{row}_n(A) \end{bmatrix} = \begin{bmatrix} \text{col}_1(A) & \cdots & \text{col}_m(A) \end{bmatrix}. \tag{2.1.10}$$

The entry $x_{j(i)}$ of A in both the ith row of A and the jth column of A is denoted by $A_{(i,j)}$. Therefore, $x \in \mathbb{F}^n$ can be written as

$$x = \begin{bmatrix} x_{(1)} \\ \vdots \\ x_{(n)} \end{bmatrix} = \begin{bmatrix} x_{(1,1)} \\ \vdots \\ x_{(n,1)} \end{bmatrix}. \tag{2.1.11}$$

Let $A \in \mathbb{F}^{n \times m}$. For $b \in \mathbb{F}^n$, the matrix obtained from A by replacing $\text{col}_i(A)$ with b is denoted by

$$A \overset{i}{\leftarrow} b. \tag{2.1.12}$$

Likewise, for $b \in \mathbb{F}^{1 \times m}$, the matrix obtained from A by replacing $\text{row}_i(A)$ with b is denoted by (2.1.12).

Let $A \in \mathbb{F}^{n \times m}$, and let $l \triangleq \min\{n, m\}$. Then, the entries $A_{(i,i)}$ for all $i \in \{1, \ldots, l\}$ and $A_{(i,j)}$ for all $i \neq j$ are the *diagonal entries* and *off-diagonal entries* of A, respectively. Moreover, for all $i \in \{1, \ldots, l-1\}$, the entries $A_{(i,i+1)}$ and $A_{(i+1,i)}$ are the *superdiagonal entries* and *subdiagonal entries* of A, respectively. In addition, the entries $A_{(i,l+1-i)}$ for all $i \in \{1, \ldots, l\}$ are the *reverse-diagonal entries* of A. If the diagonal entries $A_{(1,1)}, \ldots, A_{(l,l)}$ of A are real, then the diagonal entries of A are labeled from largest to smallest as

$$\text{d}_1(A) \geq \cdots \geq \text{d}_l(A), \tag{2.1.13}$$

and we define

$$\text{d}_{\max}(A) \triangleq \text{d}_1(A), \quad \text{d}_{\min}(A) \triangleq \text{d}_l(A). \tag{2.1.14}$$

Partitioned matrices are of the form

$$\begin{bmatrix} A_{11} & \cdots & A_{1l} \\ \vdots & \ddots & \vdots \\ A_{k1} & \cdots & A_{kl} \end{bmatrix}, \tag{2.1.15}$$

where, for all $i \in \{1, \ldots, k\}$ and $j \in \{1, \ldots, l\}$, the *block* A_{ij} of A is a matrix of size $n_i \times m_j$. If $n_i = m_j$ and the diagonal entries of A_{ij} lie on the diagonal of A, then the square matrix A_{ij} is a *diagonally located block*; otherwise, A_{ij} is an *off-diagonally located block*.

Let $A \in \mathbb{F}^{n \times m}$. Then, a *submatrix* of A is formed by deleting rows and columns of A. By convention, A is a submatrix of A. If A is a partitioned matrix, then every block of A is a submatrix of A. A block is thus a submatrix whose entries are entries of adjacent rows and adjacent columns. A submatrix can be specified in terms of the rows and columns that are retained. If like-numbered rows and columns of A are retained, then the resulting square submatrix of A is a *principal submatrix* of A. Every diagonally located block is a principal submatrix. Finally, if rows and columns $1, \ldots, j$ of A are retained, then the resulting $j \times j$ submatrix of A is a *leading principal submatrix* of A.

Let $A \in \mathbb{F}^{n \times m}$, and let $\mathcal{S}_1$ and $\mathcal{S}_2$ be subsets of $\{1, \ldots, n\}$ and $\{1, \ldots, m\}$, respectively. Then, $A_{(\mathcal{S}_1, \mathcal{S}_2)}$ is the $\mathrm{card}(\mathcal{S}_1) \times \mathrm{card}(\mathcal{S}_2)$ submatrix of A formed by retaining the rows of A listed in $\mathcal{S}_1$ and the columns of A listed in $\mathcal{S}_2$. Therefore, $A_{(\mathcal{S}_1^\sim, \mathcal{S}_2^\sim)}$ is the $[n - \mathrm{card}(\mathcal{S}_1)] \times [m - \mathrm{card}(\mathcal{S}_2)]$ submatrix of A formed by deleting the rows of A listed in $\mathcal{S}_1$ and the columns of A listed in $\mathcal{S}_2$. If $\mathcal{S} \subseteq \{1, \ldots, \min\{n, m\}\}$, then we define $A_{(\mathcal{S})} \triangleq A_{(\mathcal{S}, \mathcal{S})}$, which is a principal submatrix of A.

Let $A, B \in \mathbb{F}^{n \times m}$. Then, A and B add entry by entry, that is, for all $i \in \{1, \ldots, n\}$ and $j \in \{1, \ldots, m\}$, $(A + B)_{(i,j)} = A_{(i,j)} + B_{(i,j)}$. Furthermore, for all $i \in \{1, \ldots, n\}$ and $j \in \{1, \ldots, m\}$, it follows that, for all $\alpha \in \mathbb{F}$, $(\alpha A)_{(i,j)} = \alpha A_{(i,j)}$. Hence, for all $\alpha, \beta \in \mathbb{F}$, $(\alpha A + \beta B)_{(i,j)} = \alpha A_{(i,j)} + \beta B_{(i,j)}$. If $A, B \in \mathbb{F}^{n \times m}$, then A and B are *linearly dependent* if there exists $\alpha \in \mathbb{F}$ such that either $A = \alpha B$ or $B = \alpha A$.

Let $A \in \mathbb{R}^{n \times m}$. If every entry of A is nonnegative, then A is *nonnegative*, which is written as $A \geq\geq 0$. If every entry of A is positive, then A is *positive*, which is written as $A >> 0$. If $A, B \in \mathbb{R}^{n \times m}$, then $A \geq\geq B$ means that $A - B \geq\geq 0$, while $A >> B$ means that $A - B >> 0$.

Let $z \in \mathbb{F}^{1 \times n}$ and $y \in \mathbb{F}^n = \mathbb{F}^{n \times 1}$. Then, the scalar $zy \in \mathbb{F}$ is defined by

$$zy \triangleq \sum_{i=1}^n z_{(1,i)} y_{(i)}. \tag{2.1.16}$$

Now, let $A \in \mathbb{F}^{n \times m}$ and $x \in \mathbb{F}^m$. Then, the matrix-vector product Ax is defined by

$$Ax \triangleq \begin{bmatrix} \mathrm{row}_1(A)x \\ \vdots \\ \mathrm{row}_n(A)x \end{bmatrix}. \tag{2.1.17}$$

It can be seen that Ax is a linear combination of the columns of A, that is,

$$Ax = \sum_{i=1}^m x_{(i)} \mathrm{col}_i(A). \tag{2.1.18}$$

The matrix A can be associated with the function $f: \mathbb{F}^m \mapsto \mathbb{F}^n$ defined by $f(x) \triangleq Ax$ for all $x \in \mathbb{F}^m$. The function $f: \mathbb{F}^m \mapsto \mathbb{F}^n$ is *linear* since, for all $\alpha, \beta \in \mathbb{F}$ and $x, y \in \mathbb{F}^m$, it follows that

$$f(\alpha x + \beta y) = \alpha Ax + \beta Ay. \tag{2.1.19}$$

The function $f: \mathbb{F}^m \mapsto \mathbb{F}^n$ defined by

$$f(x) \triangleq Ax + z, \tag{2.1.20}$$

where $z \in \mathbb{F}^n$, is *affine*.

Theorem 2.1.3. Let $A \in \mathbb{F}^{n \times m}$ and $B \in \mathbb{F}^{m \times l}$, and define $f: \mathbb{F}^m \mapsto \mathbb{F}^n$ and $g: \mathbb{F}^l \mapsto \mathbb{F}^m$ by $f(x) \triangleq Ax$ and $g(y) \triangleq By$. Furthermore, define the composition

$h \triangleq f \bullet g \colon \mathbb{F}^l \mapsto \mathbb{F}^n$. Then, for all $y \in \mathbb{R}^l$,

$$h(y) = f[g(y)] = A(By) = (AB)y, \tag{2.1.21}$$

where, for all $i \in \{1, \ldots, n\}$ and $j \in \{1, \ldots, l\}$, $AB \in \mathbb{F}^{n \times l}$ is defined by

$$(AB)_{(i,j)} \triangleq \sum_{k=1}^{m} A_{(i,k)} B_{(k,j)}. \tag{2.1.22}$$

Hence, we write ABy for $(AB)y$ and $A(By)$.

Let $A \in \mathbb{F}^{n \times m}$ and $B \in \mathbb{F}^{m \times l}$. Then, $AB \in \mathbb{F}^{n \times l}$ is the *product* of A and B. The matrices A and B are *conformable*, and the product (2.1.22) defines *matrix multiplication*.

Let $A \in \mathbb{F}^{n \times m}$ and $B \in \mathbb{F}^{m \times l}$. Then, AB can be written as

$$AB = \begin{bmatrix} A\mathrm{col}_1(B) & \cdots & A\mathrm{col}_l(B) \end{bmatrix} = \begin{bmatrix} \mathrm{row}_1(A)B \\ \vdots \\ \mathrm{row}_n(A)B \end{bmatrix}. \tag{2.1.23}$$

Thus, for all $i \in \{1, \ldots, n\}$ and $j \in \{1, \ldots, l\}$,

$$(AB)_{(i,j)} = \mathrm{row}_i(A)\mathrm{col}_j(B), \tag{2.1.24}$$

$$\mathrm{col}_j(AB) = A\mathrm{col}_j(B), \tag{2.1.25}$$

$$\mathrm{row}_i(AB) = \mathrm{row}_i(A)B. \tag{2.1.26}$$

For conformable matrices A, B, C, the associative and distributive equalities

$$(AB)C = A(BC), \tag{2.1.27}$$

$$A(B + C) = AB + AC, \tag{2.1.28}$$

$$(A + B)C = AC + BC \tag{2.1.29}$$

are valid. Hence, we write ABC for $(AB)C$ and $A(BC)$. Note that (2.1.27) is a special case of (1.4.1).

Let $A, B \in \mathbb{F}^{n \times n}$. Then, the *commutator* $[A, B] \in \mathbb{F}^{n \times n}$ of A and B is the matrix

$$[A, B] \triangleq AB - BA. \tag{2.1.30}$$

The *adjoint operator* $\mathrm{ad}_A \colon \mathbb{F}^{n \times n} \mapsto \mathbb{F}^{n \times n}$ is defined by

$$\mathrm{ad}_A(X) \triangleq [A, X]. \tag{2.1.31}$$

Let $x, y \in \mathbb{R}^3$. Then, the *cross product* $x \times y \in \mathbb{R}^3$ of x and y is defined by

$$x \times y \triangleq \begin{bmatrix} x_{(2)}y_{(3)} - x_{(3)}y_{(2)} \\ x_{(3)}y_{(1)} - x_{(1)}y_{(3)} \\ x_{(1)}y_{(2)} - x_{(2)}y_{(1)} \end{bmatrix}. \tag{2.1.32}$$

Furthermore, the 3×3 *cross-product matrix* is defined by

$$K(x) \triangleq \begin{bmatrix} 0 & -x_{(3)} & x_{(2)} \\ x_{(3)} & 0 & -x_{(1)} \\ -x_{(2)} & x_{(1)} & 0 \end{bmatrix}. \tag{2.1.33}$$

Note that

$$x \times y = K(x)y. \tag{2.1.34}$$

Multiplication of partitioned matrices is analogous to matrix multiplication with scalar entries. For example, for matrices with conformable blocks,

$$\begin{bmatrix} A & B \end{bmatrix} \begin{bmatrix} C \\ D \end{bmatrix} = AC + BD, \tag{2.1.35}$$

$$\begin{bmatrix} A \\ B \end{bmatrix} C = \begin{bmatrix} AC \\ BC \end{bmatrix}, \tag{2.1.36}$$

$$\begin{bmatrix} A \\ B \end{bmatrix} \begin{bmatrix} C & D \end{bmatrix} = \begin{bmatrix} AC & AD \\ BC & BD \end{bmatrix}, \tag{2.1.37}$$

$$\begin{bmatrix} A & B \\ C & D \end{bmatrix} \begin{bmatrix} E & F \\ G & H \end{bmatrix} = \begin{bmatrix} AE + BG & AF + BH \\ CE + DG & CF + DH \end{bmatrix}. \tag{2.1.38}$$

The $n \times m$ *zero matrix*, all of whose entries are zero, is written as $0_{n \times m}$. If the dimensions are unambiguous, then we write just 0. Let $x \in \mathbb{F}^m$ and $A \in \mathbb{F}^{n \times m}$. Then, the zero matrix satisfies

$$0_{k \times m} x = 0_k, \tag{2.1.39}$$

$$A 0_{m \times l} = 0_{n \times l}, \tag{2.1.40}$$

$$0_{k \times n} A = 0_{k \times m}. \tag{2.1.41}$$

Another special matrix is the *empty matrix*. For $n \in \mathbb{N}$, the $0 \times n$ empty matrix, which is written as $0_{0 \times n}$, has zero rows and n columns, while the $n \times 0$ empty matrix, which is written as $0_{n \times 0}$, has n rows and zero columns. For $A \in \mathbb{F}^{n \times m}$, where $n, m \in \mathbb{N}$, the empty matrix satisfies the multiplication rules

$$0_{0 \times n} A = 0_{0 \times m} \tag{2.1.42}$$

and

$$A 0_{m \times 0} = 0_{n \times 0}. \tag{2.1.43}$$

Although empty matrices have no entries, it is useful to define the product

$$0_{n \times 0} 0_{0 \times m} \triangleq 0_{n \times m}. \tag{2.1.44}$$

Also, we define

$$I_0 \triangleq \hat{I}_0 \triangleq 0_{0 \times 0}. \tag{2.1.45}$$

For $n, m \in \mathbb{N}$, we define $\mathbb{F}^{0 \times m} \triangleq \{0_{0 \times m}\}$, $\mathbb{F}^{n \times 0} \triangleq \{0_{n \times 0}\}$, and $\mathbb{F}^0 \triangleq \mathbb{F}^{0 \times 1}$. Note that

$$\begin{bmatrix} 0_{n \times 0} & 0_{n \times m} \\ 0_{0 \times 0} & 0_{0 \times m} \end{bmatrix} = 0_{n \times m}. \tag{2.1.46}$$

The empty matrix can be viewed as a useful device for matrices just as 0 is for real numbers and $\varnothing$ is for sets.

The $n \times n$ *identity matrix*, which has 1's on the diagonal and 0's elsewhere, is denoted by I_n or just I. Let $x \in \mathbb{F}^n$ and $A \in \mathbb{F}^{n \times m}$. Then, the identity matrix satisfies

$$I_n x = x \qquad (2.1.47)$$

and

$$A I_m = I_n A = A. \qquad (2.1.48)$$

Let $A \in \mathbb{F}^{n \times n}$. Then, $A^2 \triangleq AA$ and, for all $k \geq 1$, $A^k \triangleq AA^{k-1}$. We use the convention $A^0 \triangleq I$ even if A is the zero matrix.

The $n \times n$ *reverse permutation matrix*, which has 1's on the reverse diagonal and 0's elsewhere, is denoted by $\hat{I}_n$ or just $\hat{I}$. In particular, $\hat{I}_1 \triangleq 1$. Multiplication of $x \in \mathbb{F}^n$ by $\hat{I}_n$ reverses the components of x. Likewise, left multiplication of $A \in \mathbb{F}^{n \times m}$ by $\hat{I}_n$ reverses the rows of A, while right multiplication of A by $\hat{I}_m$ reverses the columns of A. Note that

$$\hat{I}_n^2 = I_n. \qquad (2.1.49)$$

The $n \times n$ *cyclic permutation matrix* P_n is defined by $P_1 \triangleq 1$ and, for $n > 1$,

$$P_n \triangleq \begin{bmatrix} 0 & 1 & 0 & \cdots & 0 & 0 \\ 0 & 0 & 1 & \ddots & 0 & 0 \\ 0 & 0 & 0 & \ddots & 0 & 0 \\ \vdots & \ddots & \ddots & \ddots & \ddots & \vdots \\ 0 & 0 & 0 & \ddots & 0 & 1 \\ 1 & 0 & 0 & \cdots & 0 & 0 \end{bmatrix}.$$

Note that $P_1 = \hat{I}_1$, $P_2 = \hat{I}_2$, and

$$P_n^n = I_n. \qquad (2.1.50)$$

The $n \times n$ *standard nilpotent matrix* N_n, or just N, is defined by $N_0 \triangleq 0_{0 \times 0}$, $N_1 \triangleq 0$, and, for $n > 1$,

$$N_n \triangleq \begin{bmatrix} 0 & 1 & 0 & \cdots & 0 & 0 \\ 0 & 0 & 1 & \ddots & 0 & 0 \\ 0 & 0 & 0 & \ddots & 0 & 0 \\ \vdots & \ddots & \ddots & \ddots & \ddots & \vdots \\ 0 & 0 & 0 & \ddots & 0 & 1 \\ 0 & 0 & 0 & \cdots & 0 & 0 \end{bmatrix}.$$

Note that

$$N_n^n = 0. \tag{2.1.51}$$

2.2 Transpose and Inner Product

A fundamental vector and matrix operation is the transpose. If $x \in \mathbb{F}^n$, then the *transpose* x^{T} of x is defined to be the row vector

$$x^{\mathrm{T}} \triangleq \left[\begin{array}{ccc} x_{(1)} & \cdots & x_{(n)} \end{array}\right] \in \mathbb{F}^{1 \times n}. \tag{2.2.1}$$

Similarly, if $x = \left[\begin{array}{ccc} x_{(1,1)} & \cdots & x_{(1,n)} \end{array}\right] \in \mathbb{F}^{1 \times n}$, then

$$x^{\mathrm{T}} = \left[\begin{array}{c} x_{(1,1)} \\ \vdots \\ x_{(1,n)} \end{array}\right] \in \mathbb{F}^{n \times 1}. \tag{2.2.2}$$

Let $x, y \in \mathbb{F}^n$. Then, $x^{\mathrm{T}}y \in \mathbb{F}$ is a scalar, and

$$x^{\mathrm{T}}y = y^{\mathrm{T}}x = \sum_{i=1}^{n} x_{(i)}y_{(i)}. \tag{2.2.3}$$

Note that

$$x^{\mathrm{T}}x = \sum_{i=1}^{n} x_{(i)}^2. \tag{2.2.4}$$

The vector $e_{i,n} \in \mathbb{R}^n$, or just e_i, has 1 as its ith component and 0's elsewhere. Thus,

$$e_{i,n} = \mathrm{col}_i(I_n). \tag{2.2.5}$$

Let $A \in \mathbb{F}^{n \times m}$. Then, $e_i^{\mathrm{T}}A = \mathrm{row}_i(A)$ and $Ae_i = \mathrm{col}_i(A)$. Furthermore, the (i, j) entry of A can be written as

$$A_{(i,j)} = e_i^{\mathrm{T}}Ae_j. \tag{2.2.6}$$

The $n \times m$ matrix $E_{i,j,n \times m} \in \mathbb{R}^{n \times m}$, or just $E_{i,j}$, has 1 as its (i, j) entry and 0's elsewhere. Thus,

$$E_{i,j,n \times m} = e_{i,n}e_{j,m}^{\mathrm{T}}. \tag{2.2.7}$$

Note that $E_{i,1,n \times 1} = e_{i,n}$ and

$$I_n = E_{1,1} + \cdots + E_{n,n} = \sum_{i=1}^{n} e_i e_i^{\mathrm{T}}. \tag{2.2.8}$$

Finally, the $n \times m$ *ones matrix*, all of whose entries are 1, is written as $1_{n \times m}$ or just 1. Thus,

$$1_{n \times m} = \sum_{i,j=1}^{n,m} E_{i,j,n \times m}. \tag{2.2.9}$$

Note that

$$1_{n\times 1} = \sum_{i=1}^{n} e_{i,n} = \begin{bmatrix} 1 \\ \vdots \\ 1 \end{bmatrix} \tag{2.2.10}$$

and

$$1_{n\times m} = 1_{n\times 1} 1_{1\times m}. \tag{2.2.11}$$

Lemma 2.2.1. Let $x \in \mathbb{R}$. Then, $x^{\mathrm{T}}x = 0$ if and only if $x = 0$.

Let $x, y \in \mathbb{R}^n$. Then, $x^{\mathrm{T}}y \in \mathbb{R}$ is the *inner product* of x and y. Furthermore, x and y are *orthogonal* if $x^{\mathrm{T}}y = 0$. If x and y are nonzero, then the *angle* $\theta \in [0, \pi]$ between x and y is defined by

$$\theta \triangleq \cos^{-1} \frac{x^{\mathrm{T}}y}{\sqrt{x^{\mathrm{T}}xy^{\mathrm{T}}y}}. \tag{2.2.12}$$

Note that x and y are orthogonal if and only if $\theta = \pi/2$.

Let $x \in \mathbb{C}^n$. Then, $x = y + \jmath z$, where $y, z \in \mathbb{R}^n$. Therefore, the transpose x^{T} of x is given by

$$x^{\mathrm{T}} = y^{\mathrm{T}} + \jmath z^{\mathrm{T}}. \tag{2.2.13}$$

The *complex conjugate* $\bar{x}$ of x is defined by

$$\bar{x} \triangleq y - \jmath z, \tag{2.2.14}$$

while the *complex conjugate transpose* x^* of x is defined by

$$x^* \triangleq \bar{x}^{\mathrm{T}} = y^{\mathrm{T}} - \jmath z^{\mathrm{T}}. \tag{2.2.15}$$

The vectors y and z are the *real* and *imaginary* parts $\mathrm{Re}\, x$ and $\mathrm{Im}\, x$ of x, respectively, which are defined by

$$\mathrm{Re}\, x \triangleq \tfrac{1}{2}(x + \bar{x}) = y \tag{2.2.16}$$

and

$$\mathrm{Im}\, x \triangleq \tfrac{1}{\jmath 2}(x - \bar{x}) = z. \tag{2.2.17}$$

Note that

$$x^*x = \sum_{i=1}^{n} \bar{x}_{(i)} x_{(i)} = \sum_{i=1}^{n} |x_{(i)}|^2 = \sum_{i=1}^{n} \left[y_{(i)}^2 + z_{(i)}^2 \right]. \tag{2.2.18}$$

If $w, x \in \mathbb{C}^n$, then $w^{\mathrm{T}}x = x^{\mathrm{T}}w$.

Lemma 2.2.2. Let $x \in \mathbb{C}^n$. Then, $x^*x = 0$ if and only if $x = 0$.

Let $x, y \in \mathbb{C}^n$. Then, $x^*y \in \mathbb{C}$ is the *inner product* of x and y, which is given by

$$x^*y = \sum_{i=1}^{n} \bar{x}_{(i)} y_{(i)}. \tag{2.2.19}$$

Furthermore, x and y are *orthogonal* if $x^*y = 0$. If x and y are nonzero, then the *angle* $\theta \in [0, \pi]$ between x and y is defined by

$$\theta \triangleq \cos^{-1} \frac{x^*y}{\sqrt{x^*xy^*y}}. \tag{2.2.20}$$

Note that x and y are orthogonal if and only if $\theta = \pi/2$. It follows from the Cauchy-Schwarz inequality given by Corollary 9.1.7 that the arguments of $\cos^{-1}$ in (2.2.12) and (2.2.20) are elements of the interval $[-1, 1]$.

Let $A \in \mathbb{F}^{n \times m}$. Then, the *transpose* $A^{\mathrm{T}} \in \mathbb{F}^{m \times n}$ of A is defined by

$$A^{\mathrm{T}} \triangleq \begin{bmatrix} [\mathrm{row}_1(A)]^{\mathrm{T}} & \cdots & [\mathrm{row}_n(A)]^{\mathrm{T}} \end{bmatrix} = \begin{bmatrix} [\mathrm{col}_1(A)]^{\mathrm{T}} \\ \vdots \\ [\mathrm{col}_m(A)]^{\mathrm{T}} \end{bmatrix}, \tag{2.2.21}$$

that is, $\mathrm{col}_i(A^{\mathrm{T}}) = [\mathrm{row}_i(A)]^{\mathrm{T}}$ for all $i \in \{1, \ldots, n\}$ and $\mathrm{row}_i(A^{\mathrm{T}}) = [\mathrm{col}_i(A)]^{\mathrm{T}}$ for all $i \in \{1, \ldots, m\}$. Hence, $(A^{\mathrm{T}})_{(i,j)} = A_{(j,i)}$ and $(A^{\mathrm{T}})^{\mathrm{T}} = A$. If $B \in \mathbb{F}^{m \times l}$, then

$$(AB)^{\mathrm{T}} = B^{\mathrm{T}}A^{\mathrm{T}}. \tag{2.2.22}$$

In particular, if $x \in \mathbb{F}^m$, then

$$(Ax)^{\mathrm{T}} = x^{\mathrm{T}}A^{\mathrm{T}}, \tag{2.2.23}$$

while, if, in addition, $y \in \mathbb{F}^n$, then $y^{\mathrm{T}}Ax$ is a scalar and

$$y^{\mathrm{T}}Ax = (y^{\mathrm{T}}Ax)^{\mathrm{T}} = x^{\mathrm{T}}A^{\mathrm{T}}y. \tag{2.2.24}$$

If $B \in \mathbb{F}^{n \times m}$, then, for all $\alpha, \beta \in \mathbb{F}$,

$$(\alpha A + \beta B)^{\mathrm{T}} = \alpha A^{\mathrm{T}} + \beta B^{\mathrm{T}}. \tag{2.2.25}$$

Let $x \in \mathbb{F}^n$ and $y \in \mathbb{F}^m$. Then, the matrix $xy^{\mathrm{T}} \in \mathbb{F}^{n \times m}$ is the *outer product* of x and y. The outer product xy^{T} is nonzero if and only if both x and y are nonzero.

The *trace* of a square matrix $A \in \mathbb{F}^{n \times n}$, denoted by $\mathrm{tr}\, A$, is defined to be the sum of its diagonal entries, that is,

$$\mathrm{tr}\, A \triangleq \sum_{i=1}^n A_{(i,i)}. \tag{2.2.26}$$

Note that

$$\mathrm{tr}\, A = \mathrm{tr}\, A^{\mathrm{T}}. \tag{2.2.27}$$

Let $A \in \mathbb{F}^{n \times m}$ and $B \in \mathbb{F}^{m \times n}$. Then, AB and BA are square,

$$\mathrm{tr}\, AB = \mathrm{tr}\, BA = \mathrm{tr}\, A^{\mathrm{T}}B^{\mathrm{T}} = \mathrm{tr}\, B^{\mathrm{T}}A^{\mathrm{T}} = \sum_{i,j=1}^{n,m} A_{(i,j)}B_{(j,i)}, \tag{2.2.28}$$

and

$$\mathrm{tr}\, AA^{\mathrm{T}} = \mathrm{tr}\, A^{\mathrm{T}}A = \sum_{i,j=1}^{n,m} A_{(i,j)}^2. \tag{2.2.29}$$

Furthermore, if $n = m$, then, for all $\alpha, \beta \in \mathbb{F}$,

$$\operatorname{tr}(\alpha A + \beta B) = \alpha \operatorname{tr} A + \beta \operatorname{tr} B. \tag{2.2.30}$$

Lemma 2.2.3. Let $A \in \mathbb{R}^{n \times m}$. Then, $\operatorname{tr} A^{\mathrm{T}} A = 0$ if and only if $A = 0$.

Let $A, B \in \mathbb{R}^{n \times m}$. Then, the *inner product* of A and B is $\operatorname{tr} A^{\mathrm{T}} B$. Furthermore, A is *orthogonal* to B if $\operatorname{tr} A^{\mathrm{T}} B = 0$.

Let $C \in \mathbb{C}^{n \times m}$. Then, $C = A + \jmath B$, where $A, B \in \mathbb{R}^{n \times m}$. Therefore, the transpose C^{T} of C is given by

$$C^{\mathrm{T}} = A^{\mathrm{T}} + \jmath B^{\mathrm{T}}. \tag{2.2.31}$$

The *complex conjugate* $\overline{C}$ of C is

$$\overline{C} \triangleq A - \jmath B, \tag{2.2.32}$$

while the *complex conjugate transpose* C^* of C is

$$C^* \triangleq \overline{C}^{\mathrm{T}} = A^{\mathrm{T}} - \jmath B^{\mathrm{T}}. \tag{2.2.33}$$

Note that $\overline{C} = C$ if and only if $B = 0$, and that

$$\left(C^{\mathrm{T}}\right)^{\mathrm{T}} = \overline{\overline{C}} = (C^*)^* = C. \tag{2.2.34}$$

The matrices A and B are the real and imaginary parts $\operatorname{Re} C$ and $\operatorname{Im} C$ of C, respectively, which are denoted by

$$\operatorname{Re} C \triangleq \tfrac{1}{2}(C + \overline{C}) = A \tag{2.2.35}$$

and

$$\operatorname{Im} C \triangleq \tfrac{1}{\jmath 2}(C - \overline{C}) = B. \tag{2.2.36}$$

If C is square, then

$$\operatorname{tr} C = \operatorname{tr} A + \jmath \operatorname{tr} B \tag{2.2.37}$$

and

$$\operatorname{tr} C = \operatorname{tr} C^{\mathrm{T}} = \overline{\operatorname{tr} \overline{C}} = \overline{\operatorname{tr} C^*}. \tag{2.2.38}$$

If $\mathcal{S} \subseteq \mathbb{C}^{n \times m}$, then

$$\overline{\mathcal{S}} \triangleq \{\overline{A} : \ A \in \mathcal{S}\}. \tag{2.2.39}$$

If $\mathcal{S}$ is a multiset with elements in $\mathbb{C}^{n \times m}$, then

$$\overline{\mathcal{S}} = \{\overline{A} : \ A \in \mathcal{S}\}_{\mathrm{ms}}. \tag{2.2.40}$$

Let $A \in \mathbb{F}^{n \times n}$. Then, for all $k \in \mathbb{N}$,

$$A^{k\mathrm{T}} \triangleq (A^k)^{\mathrm{T}} = \left(A^{\mathrm{T}}\right)^k, \tag{2.2.41}$$

$$\overline{A^k} = \overline{A}^k, \tag{2.2.42}$$

and

$$A^{k*} \triangleq (A^k)^* = (A^*)^k. \tag{2.2.43}$$

Lemma 2.2.4. Let $A \in \mathbb{C}^{n \times m}$. Then, $\operatorname{tr} A^* A = 0$ if and only if $A = 0$.

Let $A, B \in \mathbb{C}^{n \times m}$. Then, the *inner product* of A and B is tr A^*B. Furthermore, A is *orthogonal* to B if tr $A^*B = 0$.

If $A, B \in \mathbb{C}^{n \times m}$, then, for all $\alpha, \beta \in \mathbb{C}$,

$$(\alpha A + \beta B)^* = \overline{\alpha} A^* + \overline{\beta} B^*, \tag{2.2.44}$$

while, if $A \in \mathbb{C}^{n \times m}$ and $B \in \mathbb{C}^{m \times l}$, then

$$\overline{AB} = \overline{A}\,\overline{B} \tag{2.2.45}$$

and

$$(AB)^* = B^*A^*. \tag{2.2.46}$$

In particular, if $A \in \mathbb{C}^{n \times m}$ and $x \in \mathbb{C}^m$, then

$$(Ax)^* = x^*A^*, \tag{2.2.47}$$

while, if, in addition, $y \in \mathbb{C}^n$, then

$$y^*Ax = (y^*Ax)^{\mathrm{T}} = x^{\mathrm{T}}A^{\mathrm{T}}\overline{y} \tag{2.2.48}$$

and

$$(y^*Ax)^* = \left(\overline{y^*Ax}\right)^{\mathrm{T}} = \left(y^{\mathrm{T}}\overline{A}\overline{x}\right)^{\mathrm{T}} = x^*A^*y. \tag{2.2.49}$$

For $A \in \mathbb{F}^{n \times m}$, define the *reverse transpose* of A by

$$A^{\hat{\mathrm{T}}} \triangleq \hat{I}_m A^{\mathrm{T}} \hat{I}_n \tag{2.2.50}$$

and the *reverse complex conjugate transpose* of A by

$$A^{\hat{*}} \triangleq \hat{I}_m A^* \hat{I}_n. \tag{2.2.51}$$

For example,

$$\begin{bmatrix} 1 & 2 & 3 \\ 4 & 5 & 6 \end{bmatrix}^{\hat{\mathrm{T}}} = \begin{bmatrix} 6 & 3 \\ 5 & 2 \\ 4 & 1 \end{bmatrix}. \tag{2.2.52}$$

In general,

$$(A^*)^{\hat{*}} = (A^{\hat{*}})^* = (A^{\mathrm{T}})^{\hat{\mathrm{T}}} = \left(A^{\hat{\mathrm{T}}}\right)^{\mathrm{T}} = \hat{I}_n A \hat{I}_m \tag{2.2.53}$$

and

$$(A^{\hat{*}})^{\hat{*}} = \left(A^{\hat{\mathrm{T}}}\right)^{\hat{\mathrm{T}}} = A. \tag{2.2.54}$$

Note that, if $B \in \mathbb{F}^{m \times l}$, then

$$(AB)^{\hat{*}} = B^{\hat{*}} A^{\hat{*}} \tag{2.2.55}$$

and

$$(AB)^{\hat{\mathrm{T}}} = B^{\hat{\mathrm{T}}} A^{\hat{\mathrm{T}}}. \tag{2.2.56}$$

Let $x \in \mathbb{F}^m$ and $A \in \mathbb{F}^{n \times m}$. Then, every component of x and every entry of A can be replaced by its absolute value to obtain $|x| \in \mathbb{R}^m$ and $|A| \in \mathbb{R}^{n \times m}$, where, for all $i \in \{1, \ldots, n\}$,

$$|x|_{(i)} \triangleq |x_{(i)}|, \tag{2.2.57}$$

and, for all $i \in \{1, \ldots, n\}$ and $j \in \{1, \ldots, m\}$,

$$|A|_{(i,j)} \triangleq |A_{(i,j)}|. \qquad (2.2.58)$$

Note that

$$|Ax| \leq\leq |A||x|. \qquad (2.2.59)$$

Furthermore, if $B \in \mathbb{F}^{m \times l}$, then

$$|AB| \leq\leq |A||B|. \qquad (2.2.60)$$

For $x \in \mathbb{R}^n$ and $A \in \mathbb{R}^{n \times m}$, every component of x and every entry of A can be replaced by its sign to obtain $\text{sign}\, x \in \mathbb{R}^n$ and $\text{sign}\, A \in \mathbb{R}^{n \times m}$ defined by

$$(\text{sign}\, x)_{(i)} \triangleq \text{sign}\, x_{(i)} \qquad (2.2.61)$$

for all $i \in \{1, \ldots, n\}$, and

$$(\text{sign}\, A)_{(i,j)} \triangleq \text{sign}\, A_{(i,j)} \qquad (2.2.62)$$

for all $i \in \{1, \ldots, n\}$ and $j \in \{1, \ldots, m\}$.

2.3 Convex Sets, Cones, and Subspaces

The definitions in this section are stated for subsets of $\mathbb{F}^n$. All of these definitions apply to subsets of $\mathbb{F}^{n \times m}$.

Let $\mathcal{S} \subseteq \mathbb{F}^n$. If $\alpha \in \mathbb{F}$, then $\alpha \mathcal{S} \triangleq \{\alpha x : x \in \mathcal{S}\}$. We write $-\mathcal{S}$ for $(-1)\mathcal{S}$. The set $\mathcal{S}$ is *symmetric* if $\mathcal{S} = -\mathcal{S}$, that is, $x \in \mathcal{S}$ if and only if $-x \in \mathcal{S}$.

For $\mathcal{S}_1, \mathcal{S}_2 \subseteq \mathbb{F}^n$, define $\mathcal{S}_1 + \mathcal{S}_2 \triangleq \{x + y : x \in \mathcal{S}_1 \text{ and } y \in \mathcal{S}_2\}$. In particular, for $y \in \mathbb{F}^n$ and $\mathcal{S} \subseteq \mathbb{F}^n$, define $y + \mathcal{S} \triangleq \{y\} + \mathcal{S} = \{y + x : x \in \mathcal{S}\}$ and $\mathcal{S} + y \triangleq \mathcal{S} + \{y\} = \{y\} + \mathcal{S}$. Note that, for all $\alpha_1, \alpha_2 \in \mathbb{F}$, $(\alpha + \beta)\mathcal{S} \subseteq \alpha\mathcal{S} + \beta\mathcal{S}$. Trivially, $\mathcal{S} + \varnothing = \varnothing$.

If $x, y \in \mathbb{F}^n$ and $\alpha \in [0, 1]$, then $\alpha x + (1 - \alpha)y$ is a *convex combination* of x and y with *barycentric coordinates* α and $1 - \alpha$. The set $\mathcal{S} \subseteq \mathbb{F}^n$ is *convex* if, for all $x, y \in \mathcal{S}$, every convex combination of x and y is an element of $\mathcal{S}$. Trivially, the empty set is convex.

Let $\mathcal{S} \subseteq \mathbb{F}^n$. Then, $\mathcal{S}$ is a *cone* if, for all $x \in \mathcal{S}$ and all $\alpha > 0$, the vector αx is an element of $\mathcal{S}$. Now, assume in addition that $\mathcal{S}$ is a cone. Then, $\mathcal{S}$ is *pointed* if $0 \in \mathcal{S}$, while $\mathcal{S}$ is *blunt* if $0 \notin \mathcal{S}$. Furthermore, $\mathcal{S}$ is *one-sided* if $x, -x \in \mathcal{S}$ implies that $x = 0$. Hence, $\mathcal{S}$ is one-sided if and only if $\mathcal{S} \cap -\mathcal{S} \subseteq \{0\}$. Furthermore, $\mathcal{S}$ is a *convex cone* if it is convex. Trivially, the empty set is a convex cone.

Let $\mathcal{S} \subseteq \mathbb{F}^n$. Then, $\mathcal{S}$ is a *subspace* if, for all $x, y \in \mathcal{S}$ and $\alpha, \beta \in \mathbb{F}$, the vector $\alpha x + \beta y$ is an element of $\mathcal{S}$. Note that, if $\{x_1, \ldots, x_r\} \subset \mathbb{F}^n$, then the set $\{\sum_{i=1}^r \alpha_i x_i : \alpha_1, \ldots, \alpha_r \in \mathbb{F}\}$ is a subspace. In addition, $\mathcal{S}$ is an *affine subspace* if there exists a vector $z \in \mathbb{F}^n$ such that $\mathcal{S} + z$ is a subspace. Affine subspaces

$\mathcal{S}_1, \mathcal{S}_2 \subseteq \mathbb{F}^n$ are *parallel* if there exists a vector $z \in \mathbb{F}^n$ such that $\mathcal{S}_1 + z = \mathcal{S}_2$. If $\mathcal{S}$ is an affine subspace, then there exists a unique subspace parallel to $\mathcal{S}$. Trivially, the empty set is a subspace and an affine subspace.

Let $\mathcal{S} \subseteq \mathbb{F}^n$. The *convex hull* of $\mathcal{S}$, denoted by $\mathrm{co}\, \mathcal{S}$, is the smallest convex set containing $\mathcal{S}$. Hence, $\mathrm{co}\, \mathcal{S}$ is the intersection of all convex subsets of $\mathbb{F}^n$ that contain $\mathcal{S}$. The *conical hull* of $\mathcal{S}$, denoted by $\mathrm{cone}\, \mathcal{S}$, is the smallest cone in $\mathbb{F}^n$ containing $\mathcal{S}$, while the *convex conical hull* of $\mathcal{S}$, denoted by $\mathrm{coco}\, \mathcal{S}$, is the smallest convex cone in $\mathbb{F}^n$ containing $\mathcal{S}$. If $\mathcal{S}$ has a finite number of elements, then $\mathrm{co}\, \mathcal{S}$ is a *polytope* and $\mathrm{coco}\, \mathcal{S}$ is a *polyhedral convex cone*. The *span* of $\mathcal{S}$, denoted by $\mathrm{span}\, \mathcal{S}$, is the smallest subspace in $\mathbb{F}^n$ containing $\mathcal{S}$, while, if $\mathcal{S}$ is nonempty, then the *affine hull* of $\mathcal{S}$, denoted by $\mathrm{aff}\, \mathcal{S}$, is the smallest affine subspace in $\mathbb{F}^n$ containing $\mathcal{S}$. Note that $\mathcal{S}$ is convex if and only if $\mathcal{S} = \mathrm{co}\, \mathcal{S}$, while similar statements hold for $\mathrm{cone}\, \mathcal{S}$, $\mathrm{coco}\, \mathcal{S}$, $\mathrm{span}\, \mathcal{S}$, and $\mathrm{aff}\, \mathcal{S}$. Trivially, $\mathrm{co}\, \varnothing = \mathrm{cone}\, \varnothing = \mathrm{coco}\, \varnothing = \mathrm{span}\, \varnothing = \mathrm{aff}\, \varnothing = \varnothing$.

Let $x_1, \ldots, x_r \in \mathbb{F}^n$. Then, $x_1, \ldots, x_r$ are *linearly independent* if $\alpha_1, \ldots, \alpha_r \in \mathbb{F}$ and

$$\sum_{i=1}^{r} \alpha_i x_i = 0 \tag{2.3.1}$$

imply that $\alpha_1 = \alpha_2 = \cdots = \alpha_r = 0$. Clearly, $x_1, \ldots, x_r$ is linearly independent if and only if $\overline{x_1}, \ldots, \overline{x_r}$ are linearly independent. If $x_1, \ldots, x_r$ are not linearly independent, then $x_1, \ldots, x_r$ are *linearly dependent*. Note that $0_{n \times 1}$ is linearly dependent.

Let $\mathcal{S} \subseteq \mathbb{F}^n$, and assume that $\mathcal{S}$ is not empty. If $\mathcal{S}$ is not equal to $\{0_{n \times 1}\}$, then there exist $r \geq 1$ vectors $x_1, \ldots, x_r \in \mathbb{F}^n$ such that $x_1, \ldots, x_r$ are linearly independent over $\mathbb{F}$ and such that $\mathrm{span}\{x_1, \ldots, x_r\} = \mathcal{S}$. The set of vectors $\{x_1, \ldots, x_r\}$ is a *basis* for $\mathcal{S}$. The positive integer r, which is the *dimension* $\dim \mathcal{S}$ of $\mathcal{S}$, is uniquely defined. For nonnegative n, we define $\dim\{0_{n \times 1}\} = 0$. If $\mathcal{S}$ is an affine subspace, then the *dimension* $\dim \mathcal{S}$ of $\mathcal{S}$ is the dimension of the subspace parallel to $\mathcal{S}$. If $\mathcal{S}$ is not an affine subspace, then the *dimension* $\dim \mathcal{S}$ of $\mathcal{S}$ is the dimension of $\mathrm{aff}\, \mathcal{S}$. We define $\dim \varnothing \triangleq -\infty$.

Let $x_1, \ldots, x_{n+1} \in \mathbb{R}^n$, and define $\mathcal{S} \triangleq \mathrm{co}\{x_1, \ldots, x_{n+1}\}$. The set $\mathcal{S}$ is a *simplex* if $\dim \mathcal{S} = n$.

The following result is the *subspace dimension theorem*.

Theorem 2.3.1. Let $\mathcal{S}_1, \mathcal{S}_2 \subseteq \mathbb{F}^n$ be subspaces. Then,

$$\dim(\mathcal{S}_1 + \mathcal{S}_2) + \dim(\mathcal{S}_1 \cap \mathcal{S}_2) = \dim \mathcal{S}_1 + \dim \mathcal{S}_2. \tag{2.3.2}$$

Proof. See [645, p. 227]. $\qquad\qquad\qquad\qquad\qquad\qquad\qquad\qquad\qquad\qquad\qquad\qquad\square$

Let $\mathcal{S}_1, \mathcal{S}_2 \subseteq \mathbb{F}^n$ be subspaces. Then, $\mathcal{S}_1$ and $\mathcal{S}_2$ are *complementary* if $\mathcal{S}_1 + \mathcal{S}_2 = \mathbb{F}^n$ and $\mathcal{S}_1 \cap \mathcal{S}_2 = \{0\}$. In this case, we say that $\mathcal{S}_1$ is complementary to $\mathcal{S}_2$, or vice versa.

Corollary 2.3.2. Let $S_1, S_2 \subseteq \mathbb{F}^n$ be subspaces, and consider the following conditions:

i) $\dim(S_1 + S_2) = n$.

ii) $S_1 \cap S_2 = \{0\}$.

iii) $\dim S_1 + \dim S_2 = n$.

iv) S_1 and S_2 are complementary subspaces.

Then,

$$[i), ii)] \iff [i), iii)] \iff [ii), iii)] \iff [i), ii), iii)] \iff [iv)].$$

Let $S \subseteq \mathbb{F}^n$ be nonempty. Then, the *orthogonal complement* $S^\perp$ of S is defined by

$$S^\perp \triangleq \{x \in \mathbb{F}^n: \ x^*y = 0 \text{ for all } y \in S\}. \tag{2.3.3}$$

The orthogonal complement $S^\perp$ of S is a subspace even if S is not.

Let $y \in \mathbb{F}^n$ be nonzero. Then, the subspace $\{y\}^\perp$, whose dimension is $n-1$, is a *hyperplane*. Furthermore, S is an *affine hyperplane* if there exists a vector $z \in \mathbb{F}^n$ such that $S + z$ is a hyperplane. The set $\{x \in \mathbb{F}^n: \ \operatorname{Re} x^*y \le 0\}$ is a *closed half space*, while the set $\{x \in \mathbb{F}^n: \ \operatorname{Re} x^*y < 0\}$ is an *open half space*. Finally, S is an *affine (closed, open) half space* if there exists a vector $z \in \mathbb{F}^n$ such that $S + z$ is a (closed, open) half space.

Let $S \subseteq \mathbb{F}^n$. Then,

$$\operatorname{polar} S \triangleq \{x \in \mathbb{F}^n: \ \operatorname{Re} x^*y \le 1 \text{ for all } y \in S\} \tag{2.3.4}$$

is the *polar* of S. Note that $\operatorname{polar} S$ is a convex set. Furthermore,

$$\operatorname{polar} S = \operatorname{polar} \operatorname{co} S. \tag{2.3.5}$$

Let $S \subseteq \mathbb{F}^n$. Then,

$$\operatorname{dcone} S \triangleq \{x \in \mathbb{F}^n: \ \operatorname{Re} x^*y \le 0 \text{ for all } y \in S\} \tag{2.3.6}$$

is the *dual cone* of S. Note that $\operatorname{dcone} S$ is a pointed convex cone. Furthermore,

$$\operatorname{dcone} S = \operatorname{dcone} \operatorname{cone} S = \operatorname{dcone} \operatorname{coco} S. \tag{2.3.7}$$

Let $S_1, S_2 \subseteq \mathbb{F}^n$ be subspaces. Then, S_1 and S_2 are *orthogonally complementary* if S_1 and S_2 are complementary and $x^*y = 0$ for all $x \in S_1$ and $y \in S_2$.

Proposition 2.3.3. Let $S_1, S_2 \subseteq \mathbb{F}^n$ be subspaces. Then, S_1 and S_2 are orthogonally complementary if and only if $S_1 = S_2^\perp$.

For the next result, note that "$\subset$" indicates proper inclusion.

Lemma 2.3.4. Let $S_1, S_2 \subseteq \mathbb{F}^n$ be subspaces such that $S_1 \subseteq S_2$. Then, $S_1 \subset S_2$ if and only if $\dim S_1 < \dim S_2$. Equivalently, $S_1 = S_2$ if and only if $\dim S_1 = \dim S_2$.

The following result provides constructive characterizations of $\operatorname{co} \mathcal{S}$, $\operatorname{cone} \mathcal{S}$, $\operatorname{coco} \mathcal{S}$, $\operatorname{span} \mathcal{S}$, and $\operatorname{aff} \mathcal{S}$.

Theorem 2.3.5. Let $\mathcal{S} \subseteq \mathbb{R}^n$ be nonempty. Then,

$$\operatorname{co} \mathcal{S} = \bigcup_{k \in \mathbb{P}} \left\{ \sum_{i=1}^{k} \alpha_i x_i \colon \ \alpha_i \geq 0, \ x_i \in \mathcal{S}, \text{ and } \sum_{i=1}^{k} \alpha_i = 1 \right\} \tag{2.3.8}$$

$$= \left\{ \sum_{i=1}^{n+1} \alpha_i x_i \colon \ \alpha_i \geq 0, \ x_i \in \mathcal{S}, \text{ and } \sum_{i=1}^{n+1} \alpha_i = 1 \right\}, \tag{2.3.9}$$

$$\operatorname{cone} \mathcal{S} = \{ \alpha x \colon \ x \in \mathcal{S} \text{ and } \alpha > 0 \}, \tag{2.3.10}$$

$$\operatorname{coco} \mathcal{S} = \bigcup_{k \in \mathbb{P}} \left\{ \sum_{i=1}^{k} \alpha_i x_i \colon \ \alpha_i \geq 0, \ x_i \in \mathcal{S}, \text{ and } \sum_{i=1}^{k} \alpha_i > 0 \right\} \tag{2.3.11}$$

$$= \left\{ \sum_{i=1}^{n+1} \alpha_i x_i \colon \ \alpha_i \geq 0, \ x_i \in \mathcal{S}, \text{ and } \sum_{i=1}^{n} \alpha_i > 0 \right\}, \tag{2.3.12}$$

$$\operatorname{span} \mathcal{S} = \bigcup_{k \in \mathbb{P}} \left\{ \sum_{i=1}^{k} \alpha_i x_i \colon \ \alpha_i \in \mathbb{R} \text{ and } x_i \in \mathcal{S} \right\} \tag{2.3.13}$$

$$= \left\{ \sum_{i=1}^{n} \alpha_i x_i \colon \ \alpha_i \in \mathbb{R} \text{ and } x_i \in \mathcal{S} \right\}, \tag{2.3.14}$$

$$\operatorname{aff} \mathcal{S} = \bigcup_{k \in \mathbb{P}} \left\{ \sum_{i=1}^{k} \alpha_i x_i \colon \ \alpha_i \in \mathbb{R}, \ x_i \in \mathcal{S}, \text{ and } \sum_{i=1}^{k} \alpha_i = 1 \right\} \tag{2.3.15}$$

$$= \left\{ \sum_{i=1}^{n+1} \alpha_i x_i \colon \ \alpha_i \in \mathbb{R}, \ x_i \in \mathcal{S}, \text{ and } \sum_{i=1}^{n+1} \alpha_i = 1 \right\}. \tag{2.3.16}$$

Now, let $\mathcal{S} \subseteq \mathbb{C}^n$. Then,

$$\operatorname{co} \mathcal{S} = \bigcup_{k \in \mathbb{P}} \left\{ \sum_{i=1}^{k} \alpha_i x_i \colon \ \alpha_i \geq 0, \ x_i \in \mathcal{S}, \text{ and } \sum_{i=1}^{k} \alpha_i = 1 \right\} \tag{2.3.17}$$

$$= \left\{ \sum_{i=1}^{2n+1} \alpha_i x_i \colon \ \alpha_i \geq 0, \ x_i \in \mathcal{S}, \text{ and } \sum_{i=1}^{2n+1} \alpha_i = 1 \right\}, \tag{2.3.18}$$

$$\operatorname{cone} \mathcal{S} = \{ \alpha x \colon \ x \in \mathcal{S} \text{ and } \alpha > 0 \}, \tag{2.3.19}$$

$$\operatorname{coco} \mathcal{S} = \bigcup_{k \in \mathbb{P}} \left\{ \sum_{i=1}^{k} \alpha_i x_i \colon \ \alpha_i \geq 0, \ x_i \in \mathcal{S}, \text{ and } \sum_{i=1}^{k} \alpha_i > 0 \right\} \tag{2.3.20}$$

$$= \left\{ \sum_{i=1}^{2n+1} \alpha_i x_i \colon \ \alpha_i \geq 0, \ x_i \in \mathcal{S}, \text{ and } \sum_{i=1}^{2n} \alpha_i > 0 \right\}, \tag{2.3.21}$$

$$\operatorname{span} \mathcal{S} = \bigcup_{k \in \mathbb{P}} \left\{ \sum_{i=1}^{k} \alpha_i x_i \colon \ \alpha_i \in \mathbb{C} \text{ and } x_i \in \mathcal{S} \right\} \qquad (2.3.22)$$

$$= \left\{ \sum_{i=1}^{n} \alpha_i x_i \colon \ \alpha_i \in \mathbb{C} \text{ and } x_i \in \mathcal{S} \right\}, \qquad (2.3.23)$$

$$\operatorname{aff} \mathcal{S} = \bigcup_{k \in \mathbb{P}} \left\{ \sum_{i=1}^{k} \alpha_i x_i \colon \ \alpha_i \in \mathbb{C}, \ x_i \in \mathcal{S}, \text{ and } \sum_{i=1}^{k} \alpha_i = 1 \right\} \qquad (2.3.24)$$

$$= \left\{ \sum_{i=1}^{n+1} \alpha_i x_i \colon \ \alpha_i \in \mathbb{C}, \ x_i \in \mathcal{S}, \text{ and } \sum_{i=1}^{n+1} \alpha_i = 1 \right\}. \qquad (2.3.25)$$

Proof. Result (2.3.8) is immediate, while (2.3.9) is proved in [904, p. 17]. Furthermore, (2.3.10) is immediate. Next, note that, since $\operatorname{coco} \mathcal{S} = \operatorname{co} \operatorname{cone} \mathcal{S}$, it follows that (2.3.8) and (2.3.10) imply (2.3.12) with n replaced by $n+1$. However, every element of $\operatorname{coco} \mathcal{S}$ lies in the convex hull of $n+1$ points one of which is the origin. It thus follows that we can set $x_{n+1} = 0$, which yields (2.3.12). Similar arguments yield (2.3.14). Finally, note that all vectors of the form $x_1 + \beta(x_2 - x_1)$, where $x_1, x_2 \in \mathcal{S}$ and $\beta \in \mathbb{R}$, are elements of $\operatorname{aff} \mathcal{S}$. Forming the convex hull of these vectors yields (2.3.16). $\qquad \square$

The following result shows that cones can be used to induce relations on $\mathbb{F}^n$.

Proposition 2.3.6. Let $\mathcal{S} \subseteq \mathbb{F}^n$ be a cone and, for $x, y \in \mathbb{F}^n$, let $x \leq y$ denote the relation $y - x \in \mathcal{S}$. Then, the following statements hold:

i) "$\leq$" is reflexive if and only if $\mathcal{S}$ is a pointed cone.

ii) "$\leq$" is antisymmetric if and only if $\mathcal{S}$ is a one-sided cone.

iii) "$\leq$" is symmetric if and only if $\mathcal{S}$ is a symmetric cone.

iv) "$\leq$" is transitive if and only if $\mathcal{S}$ is a convex cone.

Proof. The proofs of *i)*, *ii)*, and *iii)* are immediate. To prove *iv)*, suppose that "$\leq$" is transitive, and let $x, y \in \mathcal{S}$ so that $0 \leq \alpha x \leq \alpha x + (1 - \alpha)y$ for all $\alpha \in (0, 1]$. Hence, $\alpha x + (1 - \alpha)y \in \mathcal{S}$ for all $\alpha \in (0, 1]$, and thus $\mathcal{S}$ is convex. Conversely, suppose that $\mathcal{S}$ is a convex cone, and assume that $x \leq y$ and $y \leq z$. Then, $y - x \in \mathcal{S}$ and $z - y \in \mathcal{S}$ imply that $z - x = 2\left[\frac{1}{2}(y - x) + \frac{1}{2}(z - y) \right] \in \mathcal{S}$. Hence, $x \leq z$, and thus "$\leq$" is transitive. $\qquad \square$

2.4 Range and Null Space

Two key features of a matrix $A \in \mathbb{F}^{n \times m}$ are its range and null space, denoted by $\mathcal{R}(A)$ and $\mathcal{N}(A)$, respectively. The *range* of A is defined by

$$\mathcal{R}(A) \triangleq \{ Ax \colon x \in \mathbb{F}^m \}. \qquad (2.4.1)$$

Note that, for all nonnegative n and m, it follows that $\mathcal{R}(0_{n \times 0}) = \{0_{n \times 1}\}$ and $\mathcal{R}(0_{0 \times m}) = \{0_{0 \times 1}\}$. Letting α_i denote $x_{(i)}$, it can be seen that

$$\mathcal{R}(A) = \left\{ \sum_{i=1}^{m} \alpha_i \mathrm{col}_i(A): \; \alpha_1, \dots, \alpha_m \in \mathbb{F} \right\}, \tag{2.4.2}$$

which shows that $\mathcal{R}(A)$ is a subspace of $\mathbb{F}^n$. It thus follows from Theorem 2.3.5 that

$$\mathcal{R}(A) = \mathrm{span}\,\{\mathrm{col}_1(A), \dots, \mathrm{col}_m(A)\}. \tag{2.4.3}$$

By viewing A as a function from $\mathbb{F}^m$ into $\mathbb{F}^n$, we can write $\mathcal{R}(A) = A\mathbb{F}^m$.

The *null space* of $A \in \mathbb{F}^{n \times m}$ is defined by

$$\mathcal{N}(A) \triangleq \{x \in \mathbb{F}^m: \; Ax = 0\}. \tag{2.4.4}$$

Note that $\mathcal{N}(0_{n \times 0}) = \mathbb{F}^0 = \{0_{0 \times 1}\}$ and $\mathcal{N}(0_{0 \times m}) = \mathbb{F}^m$. Equivalently,

$$\mathcal{N}(A) = \left\{ x \in \mathbb{F}^m: \; x^{\mathrm{T}}[\mathrm{row}_i(A)]^{\mathrm{T}} = 0 \text{ for all } i \in \{1, \dots, n\} \right\} \tag{2.4.5}$$

$$= \left\{ [\mathrm{row}_1(A)]^{\mathrm{T}}, \dots, [\mathrm{row}_n(A)]^{\mathrm{T}} \right\}^{\perp}, \tag{2.4.6}$$

which shows that $\mathcal{N}(A)$ is a subspace of $\mathbb{F}^m$. Note that, if $\alpha \in \mathbb{F}$ is nonzero, then $\mathcal{R}(\alpha A) = \mathcal{R}(A)$ and $\mathcal{N}(\alpha A) = \mathcal{N}(A)$. Finally, if $\mathbb{F} = \mathbb{C}$, then $\mathcal{R}(A)$ and $\mathcal{R}(\overline{A})$ are not necessarily identical. For example, let $A \triangleq \begin{bmatrix} \jmath \\ 1 \end{bmatrix}$.

Let $A \in \mathbb{F}^{n \times n}$, and let $\mathcal{S} \subseteq \mathbb{F}^n$ be a subspace. Then, $\mathcal{S}$ is an *invariant subspace* of A if $A\mathcal{S} \subseteq \mathcal{S}$. Note that $A\mathcal{R}(A) \subseteq A\mathbb{F}^n = \mathcal{R}(A)$ and $A\mathcal{N}(A) = \{0_n\} \subseteq \mathcal{N}(A)$. Hence, $\mathcal{R}(A)$ and $\mathcal{N}(A)$ are invariant subspaces of A.

If $A \in \mathbb{F}^{n \times m}$ and $B \in \mathbb{F}^{m \times l}$, then it is easy to see that

$$\mathcal{R}(AB) = A\mathcal{R}(B). \tag{2.4.7}$$

Hence, the following result is not surprising.

Lemma 2.4.1. Let $A \in \mathbb{F}^{n \times m}$, $B \in \mathbb{F}^{m \times l}$, and $C \in \mathbb{F}^{k \times n}$. Then,

$$\mathcal{R}(AB) \subseteq \mathcal{R}(A) \tag{2.4.8}$$

and

$$\mathcal{N}(A) \subseteq \mathcal{N}(CA). \tag{2.4.9}$$

Proof. Since $\mathcal{R}(B) \subseteq \mathbb{F}^m$, it follows that $\mathcal{R}(AB) = A\mathcal{R}(B) \subseteq A\mathbb{F}^m = \mathcal{R}(A)$. Furthermore, $y \in \mathcal{N}(A)$ implies that $Ay = 0$, and thus $CAy = 0$. $\square$

Corollary 2.4.2. Let $A \in \mathbb{F}^{n \times n}$, and let $k \geq 1$. Then,

$$\mathcal{R}(A^k) \subseteq \mathcal{R}(A) \tag{2.4.10}$$

and

$$\mathcal{N}(A) \subseteq \mathcal{N}(A^k). \tag{2.4.11}$$

Although $\mathcal{R}(AB) \subseteq \mathcal{R}(A)$ for arbitrary conformable matrices A, B, we now show that equality holds in the special case $B = A^*$. This result, along with others, is the subject of the following basic theorem.

Theorem 2.4.3. Let $A \in \mathbb{F}^{n \times m}$. Then, the following statements hold:

i) $\mathcal{R}(A)^\perp = \mathcal{N}(A^*)$.

ii) $\mathcal{R}(A) = \mathcal{R}(AA^*)$.

iii) $\mathcal{N}(A) = \mathcal{N}(A^*A)$.

Proof. To prove *i)*, we first show that $\mathcal{R}(A)^\perp \subseteq \mathcal{N}(A^*)$. Let $x \in \mathcal{R}(A)^\perp$. Then, $x^*z = 0$ for all $z \in \mathcal{R}(A)$. Hence, $x^*Ay = 0$ for all $y \in \mathbb{R}^m$. Equivalently, $y^*A^*x = 0$ for all $y \in \mathbb{R}^m$. Letting $y = A^*x$, it follows that $x^*AA^*x = 0$. Now, Lemma 2.2.2 implies that $A^*x = 0$. Thus, $x \in \mathcal{N}(A^*)$. Conversely, let us show that $\mathcal{N}(A^*) \subseteq \mathcal{R}(A)^\perp$. Letting $x \in \mathcal{N}(A^*)$, it follows that $A^*x = 0$, and, hence, $y^*A^*x = 0$ for all $y \in \mathbb{R}^m$. Equivalently, $x^*Ay = 0$ for all $y \in \mathbb{R}^m$. Hence, $x^*z = 0$ for all $z \in \mathcal{R}(A)$. Thus, $x \in \mathcal{R}(A)^\perp$, which proves *i)*.

To prove *ii)*, note that Lemma 2.4.1 with $B = A^*$ implies that $\mathcal{R}(AA^*) \subseteq \mathcal{R}(A)$. To show that $\mathcal{R}(A) \subseteq \mathcal{R}(AA^*)$, let $x \in \mathcal{R}(A)$, and suppose that $x \notin \mathcal{R}(AA^*)$. Then, it follows from Proposition 2.3.3 that $x = x_1 + x_2$, where $x_1 \in \mathcal{R}(AA^*)$ and $x_2 \in \mathcal{R}(AA^*)^\perp$ with $x_2 \neq 0$. Thus, $x_2^*AA^*y = 0$ for all $y \in \mathbb{R}^n$, and setting $y = x_2$ yields $x_2^*AA^*x_2 = 0$. Hence, Lemma 2.2.2 implies that $A^*x_2 = 0$, so that, by *i)*, $x_2 \in \mathcal{N}(A^*) = \mathcal{R}(A)^\perp$. Since $x \in \mathcal{R}(A)$, it follows that $0 = x_2^*x = x_2^*x_1 + x_2^*x_2$. However, $x_2^*x_1 = 0$ so that $x_2^*x_2 = 0$ and $x_2 = 0$, which is a contradiction. This proves *ii)*.

To prove *iii)*, note that *ii)* with A replaced by A^* implies that $\mathcal{R}(A^*A)^\perp = \mathcal{R}(A^*)^\perp$. Furthermore, replacing A by A^* in *i)* yields $\mathcal{R}(A^*)^\perp = \mathcal{N}(A)$. Hence, $\mathcal{N}(A) = \mathcal{R}(A^*A)^\perp$. Now, *i)* with A replaced by A^*A implies that $\mathcal{R}(A^*A)^\perp = \mathcal{N}(A^*A)$. Hence, $\mathcal{N}(A) = \mathcal{N}(A^*A)$, which proves *iii)*. $\qquad\square$

Result *i)* of Theorem 2.4.3 can be written equivalently as

$$\mathcal{N}(A)^\perp = \mathcal{R}(A^*), \tag{2.4.12}$$

$$\mathcal{N}(A) = \mathcal{R}(A^*)^\perp, \tag{2.4.13}$$

$$\mathcal{N}(A^*)^\perp = \mathcal{R}(A), \tag{2.4.14}$$

while replacing A by A^* in *ii)* and *iii)* of Theorem 2.4.3 yields

$$\mathcal{R}(A^*) = \mathcal{R}(A^*A), \tag{2.4.15}$$

$$\mathcal{N}(A^*) = \mathcal{N}(AA^*). \tag{2.4.16}$$

Using *ii)* of Theorem 2.4.3 and (2.4.15), it follows that

$$\mathcal{R}(AA^*A) = A\mathcal{R}(A^*A) = A\mathcal{R}(A^*) = \mathcal{R}(AA^*) = \mathcal{R}(A). \tag{2.4.17}$$

Letting $A \triangleq \begin{bmatrix} 1 & \jmath \end{bmatrix}$ shows that $\mathcal{R}(A)$ and $\mathcal{R}(AA^{\mathrm{T}})$ may be different.

2.5 Rank and Defect

The *rank* of $A \in \mathbb{F}^{n \times m}$ is defined by

$$\operatorname{rank} A \triangleq \dim \mathcal{R}(A). \tag{2.5.1}$$

It can be seen that the rank of A is equal to the number of linearly independent columns of A over $\mathbb{F}$. For example, if $\mathbb{F} = \mathbb{C}$, then $\operatorname{rank} \begin{bmatrix} 1 & \jmath \end{bmatrix} = 1$, while, if either $\mathbb{F} = \mathbb{R}$ or $\mathbb{F} = \mathbb{C}$, then $\operatorname{rank} \begin{bmatrix} 1 & 1 \end{bmatrix} = 1$. Furthermore, $\operatorname{rank} A = \operatorname{rank} \overline{A}$, $\operatorname{rank} A^{\mathrm{T}} = \operatorname{rank} A^*$, $\operatorname{rank} A \leq m$, and $\operatorname{rank} A^{\mathrm{T}} \leq n$. If $\operatorname{rank} A = m$, then A has *full column rank*, while, if $\operatorname{rank} A^{\mathrm{T}} = n$, then A has *full row rank*. If A has either full column rank or full row rank, then A has *full rank*. Note that, for all nonnegative n and m, it follows that $\operatorname{rank}(0_{n \times m}) = \operatorname{rank}(0_{n \times 0}) = \operatorname{rank}(0_{0 \times m}) = 0$. Finally, the *defect* of A is

$$\operatorname{def} A \triangleq \dim \mathcal{N}(A). \tag{2.5.2}$$

The following result follows from Theorem 2.4.3.

Corollary 2.5.1. Let $A \in \mathbb{F}^{n \times m}$. Then, the following statements hold:

i) $\operatorname{rank} A^* + \operatorname{def} A = m$.

ii) $\operatorname{rank} A = \operatorname{rank} AA^*$.

iii) $\operatorname{def} A = \operatorname{def} A^*A$.

Proof. It follows from (2.4.12) and Proposition 2.3.2 that $\operatorname{rank} A^* = \dim \mathcal{R}(A^*) = \dim \mathcal{N}(A)^{\perp} = m - \dim \mathcal{N}(A) = m - \operatorname{def} A$, which proves *i)*. Results *ii)* and *iii)* follow from *ii)* and *iii)* of Theorem 2.4.3. $\qquad \square$

Replacing A by A^* in Corollary 2.5.1 yields

$$\operatorname{rank} A + \operatorname{def} A^* = n, \tag{2.5.3}$$

$$\operatorname{rank} A^* = \operatorname{rank} A^*A, \tag{2.5.4}$$

$$\operatorname{def} A^* = \operatorname{def} AA^*. \tag{2.5.5}$$

Furthermore, note that

$$\operatorname{def} A = \operatorname{def} \overline{A} \tag{2.5.6}$$

and

$$\operatorname{def} A^{\mathrm{T}} = \operatorname{def} A^*. \tag{2.5.7}$$

Lemma 2.5.2. Let $A \in \mathbb{F}^{n \times m}$ and $B \in \mathbb{F}^{m \times l}$. Then,

$$\operatorname{rank} AB \leq \min\{\operatorname{rank} A, \operatorname{rank} B\}. \tag{2.5.8}$$

Proof. Since, by Lemma 2.4.1, $\mathcal{R}(AB) \subseteq \mathcal{R}(A)$, it follows that $\operatorname{rank} AB \leq \operatorname{rank} A$. Next, suppose that $\operatorname{rank} B < \operatorname{rank} AB$. Let $\{y_1, \ldots, y_r\} \subset \mathbb{F}^n$ be a basis for $\mathcal{R}(AB)$, where $r \triangleq \operatorname{rank} AB$, and, since $y_i \in A\mathcal{R}(B)$ for all $i \in \{1, \ldots, r\}$, let $x_i \in \mathcal{R}(B)$ be such that $y_i = Ax_i$ for all $i \in \{1, \ldots, r\}$. Since $\operatorname{rank} B < r$, it follows that $x_1, \ldots, x_r$ are linearly dependent. Hence, there exist $\alpha_1, \ldots, \alpha_r \in \mathbb{F}$, not all zero, such that $\sum_{i=1}^{r} \alpha_i x_i = 0$, which implies that $\sum_{i=1}^{r} \alpha_i A x_i = \sum_{i=1}^{r} \alpha_i y_i = 0$. Thus, $y_1, \ldots, y_r$ are linearly dependent, which is a contradiction. $\qquad \square$

Corollary 2.5.3. Let $A \in \mathbb{F}^{n \times m}$. Then,

$$\operatorname{rank} A = \operatorname{rank} A^* \tag{2.5.9}$$

and

$$\operatorname{def} A = \operatorname{def} A^* + m - n. \tag{2.5.10}$$

Therefore,

$$\operatorname{rank} A = \operatorname{rank} A^*A.$$

If, in addition, $n = m$, then

$$\operatorname{def} A = \operatorname{def} A^*. \tag{2.5.11}$$

Proof. It follows from (2.5.8) with $B = A^*$ that $\operatorname{rank} AA^* \leq \operatorname{rank} A^*$. Furthermore, $ii)$ of Corollary 2.5.1 implies that $\operatorname{rank} A = \operatorname{rank} AA^*$. Hence, $\operatorname{rank} A \leq \operatorname{rank} A^*$. Interchanging A and A^* and repeating this argument yields $\operatorname{rank} A^* \leq \operatorname{rank} A$. Hence, $\operatorname{rank} A = \operatorname{rank} A^*$. Next, using $i)$ of Corollary 2.5.1, (2.5.9), and (2.5.3) it follows that $\operatorname{def} A = m - \operatorname{rank} A^* = m - \operatorname{rank} A = m - (n - \operatorname{def} A^*)$, which proves (2.5.10). $\qquad \square$

Corollary 2.5.4. Let $A \in \mathbb{F}^{n \times m}$. Then,

$$\operatorname{rank} A \leq \min\{m, n\}. \tag{2.5.12}$$

Proof. By definition, $\operatorname{rank} A \leq m$, while it follows from (2.5.9) that $\operatorname{rank} A = \operatorname{rank} A^* \leq n$. $\qquad \square$

The *dimension theorem* is given by (2.5.13) in the following result.

Corollary 2.5.5. Let $A \in \mathbb{F}^{n \times m}$. Then,

$$\operatorname{rank} A + \operatorname{def} A = m \tag{2.5.13}$$

and

$$\operatorname{rank} A = \operatorname{rank} A^*A. \tag{2.5.14}$$

Proof. The result (2.5.13) follows from $i)$ of Corollary 2.5.1 and (2.5.9), while (2.5.14) follows from (2.5.4) and (2.5.9). $\qquad \square$

The following result follows from the subspace dimension theorem and the dimension theorem.

Corollary 2.5.6. Let $A \in \mathbb{F}^{n \times m}$. Then,

$$\dim[\mathcal{R}(A) + \mathcal{N}(A)] + \dim[\mathcal{R}(A) \cap \mathcal{N}(A)] = m. \tag{2.5.15}$$

Corollary 2.5.7. Let $A \in \mathbb{F}^{n \times n}$ and $k \geq 1$. Then,

$$\operatorname{rank} A^k \leq \operatorname{rank} A \tag{2.5.16}$$

and

$$\operatorname{def} A \leq \operatorname{def} A^k. \tag{2.5.17}$$

Proposition 2.5.8. Let $A \in \mathbb{F}^{n \times n}$. If rank $A^2 = $ rank A, then, for all $k \geq 1$, rank $A^k = $ rank A. Equivalently, if def $A^2 = $ def A, then, for all $k \geq 1$, def $A^k = $ def A.

Proof. Since rank $A^2 = $ rank A and $\mathcal{R}(A^2) \subseteq \mathcal{R}(A)$, it follows from Lemma 2.3.4 that $\mathcal{R}(A^2) = \mathcal{R}(A)$. Hence, $\mathcal{R}(A^3) = A\mathcal{R}(A^2) = A\mathcal{R}(A) = \mathcal{R}(A^2)$. Thus, rank $A^3 = $ rank A. Similar arguments yield rank $A^k = $ rank A for all $k \geq 1$. $\square$

We now prove *Sylvester's inequality*, which provides a lower bound for the rank of the product of two matrices.

Proposition 2.5.9. Let $A \in \mathbb{F}^{n \times m}$ and $B \in \mathbb{F}^{m \times l}$. Then,

$$\text{rank } A + \text{rank } B \leq m + \text{rank } AB. \qquad (2.5.18)$$

Proof. Using (2.5.8) to obtain the second inequality below, it follows that

$$
\begin{aligned}
\text{rank } A + \text{rank } B &= \text{rank} \begin{bmatrix} 0 & A \\ B & 0 \end{bmatrix} \\
&\leq \text{rank} \begin{bmatrix} 0 & A \\ B & I \end{bmatrix} \\
&= \text{rank} \begin{bmatrix} I & A \\ 0 & I \end{bmatrix} \begin{bmatrix} -AB & 0 \\ B & I \end{bmatrix} \\
&\leq \text{rank} \begin{bmatrix} -AB & 0 \\ B & I \end{bmatrix} \\
&\leq \text{rank} \begin{bmatrix} -AB & 0 \end{bmatrix} + \text{rank} \begin{bmatrix} B & I \end{bmatrix} \\
&= \text{rank } AB + m. \qquad\qquad \square
\end{aligned}
$$

Combining (2.5.8) with (2.5.18) yields the following result.

Corollary 2.5.10. Let $A \in \mathbb{F}^{n \times m}$ and $B \in \mathbb{F}^{m \times l}$. Then,

$$\text{rank } A + \text{rank } B - m \leq \text{rank } AB \leq \min\{\text{rank } A, \text{rank } B\}. \qquad (2.5.19)$$

2.6 Invertibility

Let $A \in \mathbb{F}^{n \times m}$. Then, A is *left invertible* if there exists a matrix $A^{\mathrm{L}} \in \mathbb{F}^{m \times n}$ such that $A^{\mathrm{L}} A = I_m$, while A is *right invertible* if there exists a matrix $A^{\mathrm{R}} \in \mathbb{F}^{m \times n}$ such that $A A^{\mathrm{R}} = I_n$. These definitions are consistent with the definitions of left and right invertibility given in Chapter 1 applied to the function $f \colon \mathbb{F}^m \mapsto \mathbb{F}^n$ given by $f(x) = Ax$. Note that A^{L} (when it exists) and A^* are the same size, and likewise for A^{R}.

Theorem 2.6.1. Let $A \in \mathbb{F}^{n \times m}$. Then, the following statements are equivalent:

i) A is left invertible.

ii) A is one-to-one.

iii) $\operatorname{def} A = 0$.

iv) $\operatorname{rank} A = m$.

v) A has full column rank.

The following statements are also equivalent:

vi) A is right invertible.

vii) A is onto.

viii) $\operatorname{def} A = m - n$.

ix) $\operatorname{rank} A = n$.

x) A has full row rank.

Proposition 2.6.2. Let $A \in \mathbb{F}^{n \times m}$. Then, the following statements are equivalent:

i) A has a unique left inverse.

ii) A has a unique right inverse.

iii) $\operatorname{rank} A = n = m$.

Proof. To prove that *i)* implies *iii)*, suppose that $\operatorname{rank} A = m < n$ so that A is left invertible but nonsquare. Then, it follows from the dimension theorem Corollary 2.5.5 that $\operatorname{def} A^{\mathrm{T}} = n - m > 0$. Hence, there exist infinitely many matrices $A^{\mathrm{L}} \in \mathbb{F}^{m \times n}$ such that $A^{\mathrm{L}} A = I_m$. Conversely, suppose that $B \in \mathbb{F}^{n \times n}$ and $C \in \mathbb{F}^{n \times n}$ are left inverses of A. Then, $(B - C)A = 0$, and it follows from Sylvester's inequality Proposition 2.5.9 that $B = C$. $\qquad\square$

The following result shows that the rank and defect of a matrix are not affected by either left multiplication by a left invertible matrix or right multiplication by a right invertible matrix.

Proposition 2.6.3. Let $A \in \mathbb{F}^{n \times m}$, and let $C \in \mathbb{F}^{k \times n}$ be left invertible and $B \in \mathbb{F}^{m \times l}$ be right invertible. Then,

$$\mathcal{R}(A) = \mathcal{R}(AB) \tag{2.6.1}$$

and

$$\mathcal{N}(A) = \mathcal{N}(CA). \tag{2.6.2}$$

Furthermore,

$$\operatorname{rank} A = \operatorname{rank} CA = \operatorname{rank} AB \tag{2.6.3}$$

and

$$\text{def } A = \text{def } CA = \text{def } AB + m - l. \tag{2.6.4}$$

Proof. Let C^L be a left inverse of C. Using both inequalities in (2.5.19) and the fact that $\text{rank } A \leq n$, it follows that

$$\text{rank } A = \text{rank } A + \text{rank } C^L C - n \leq \text{rank } C^L CA \leq \text{rank } CA \leq \text{rank } A,$$

which implies that $\text{rank } A = \text{rank } CA$. Next, (2.5.13) and (2.6.3) imply that $m - \text{def } A = m - \text{def } CA = l - \text{def } AB$, which yields (2.6.4). $\qquad\square$

As shown in Proposition 2.6.2, left and right inverses of nonsquare matrices are not unique. For example, the matrix $A = \begin{bmatrix} 0 \\ 1 \end{bmatrix}$ is left invertible and has left inverses $\begin{bmatrix} 0 & 1 \end{bmatrix}$ and $\begin{bmatrix} 1 & 1 \end{bmatrix}$. In spite of this nonuniqueness, however, left inverses are useful for solving equations of the form $Ax = b$, where $A \in \mathbb{F}^{n \times m}$, $x \in \mathbb{F}^m$, and $b \in \mathbb{F}^n$. If A is left invertible, then one can formally (although not rigorously) solve $Ax = b$ by noting that $x = A^L Ax = A^L b$, where $A^L \in \mathbb{R}^{m \times n}$ is a left inverse of A. However, it is necessary to determine beforehand whether or not there actually exists a vector x satisfying $Ax = b$. For example, if $A = \begin{bmatrix} 0 \\ 1 \end{bmatrix}$ and $b = \begin{bmatrix} 1 \\ 0 \end{bmatrix}$, then A is left invertible although there does not exist a vector x satisfying $Ax = b$. The following result addresses the various possibilities that can arise. One interesting feature of this result is that, if there exists a solution of $Ax = b$ and A is left invertible, then the solution is unique even if A does not have a unique left inverse, which is the case if and only if $m < n$. For this result, $\begin{bmatrix} A & b \end{bmatrix}$ denotes the $n \times (m+1)$ partitioned matrix formed from A and b. Note that $\text{rank } A \leq \text{rank } \begin{bmatrix} A & b \end{bmatrix} \leq m + 1$, while $\text{rank } A = \text{rank } \begin{bmatrix} A & b \end{bmatrix}$ is equivalent to $b \in \mathcal{R}(A)$.

Theorem 2.6.4. Let $A \in \mathbb{F}^{n \times m}$ and $b \in \mathbb{F}^n$. Then, the following statements hold:

i) $Ax = b$ has no solution if and only if $\text{rank } A < \text{rank } \begin{bmatrix} A & b \end{bmatrix}$.

ii) $Ax = b$ has at least one solution if and only if $\text{rank } A = \text{rank } \begin{bmatrix} A & b \end{bmatrix}$.

iii) The following statements are equivalent:

    a) $Ax = b$ has a unique solution.

    b) $\text{rank } A = \text{rank } \begin{bmatrix} A & b \end{bmatrix} = m$.

    c) $Ax = b$ has at least one solution, and A is left invertible.

    In this case, let $A^L \in \mathbb{F}^{m \times n}$ be a left inverse of A. Then, the unique solution of $Ax = b$ is given by $x = A^L b$.

iv) Assume that $Ax = b$ has at least one solution. Then, $Ax = b$ has a unique solution if and only if A is left invertible. In this case, let $A^L \in \mathbb{F}^{m \times n}$ be a left inverse of A. Then, the unique solution of $Ax = b$ is given by $x = A^L b$.

v) The following statements are equivalent:

    a) $Ax = b$ has infinitely many solutions.

    b) $\text{rank } A = \text{rank } \begin{bmatrix} A & b \end{bmatrix} < m$.

    c) $Ax = b$ has at least one solution, and A is not left invertible.

    In this case, let $\hat{x} \in \mathbb{F}^m$ satisfy $A\hat{x} = b$. Then, the set of solutions of $Ax = b$

is given by $\hat{x} + \mathcal{N}(A)$.

vi) Assume that rank $A = n$. Then, $Ax = b$ has at least one solution. Now, let $A^{\mathrm{R}} \in \mathbb{F}^{m \times n}$ be a right inverse of A. Then, $x = A^{\mathrm{R}}b$ is a solution of $Ax = b$. Furthermore, if $\hat{x} \in \mathbb{F}^n$ satisfies $A\hat{x} = b$, then the set of solutions of $Ax = b$ is given by $\hat{x} + \mathcal{N}(A)$. Finally, $n = m$ if and only if $x = A^{\mathrm{R}}b$ is the unique solution of $Ax = b$. In this case, $A^{\mathrm{R}} = A^{-1}$.

Proof. To prove *i)*, note that rank $A <$ rank $\begin{bmatrix} A & b \end{bmatrix}$ is equivalent to the statement that b is not a linear combination of columns of A, that is, $Ax = b$ does not have a solution $x \in \mathbb{F}^m$. Statement *ii)* is the contrapositive of *i)*.

To prove $a) \implies c)$ of *iii)*, let $\hat{x}$ satisfy $A\hat{x} = b$, and suppose that A is not left invertible. Then, it follows from Theorem 2.6.1 that $\mathcal{N}(A)$ is a nonzero subspace. Since every element of the affine subspace $\hat{x} + \mathcal{N}(A)$ satisfies $Ax = b$, it follows that $Ax = b$ does not have a unique solution. Conversely, let A^{L} be a left inverse of A, and let $x, \hat{x} \in \mathbb{F}^m$ be solutions of $Ax = b$. Then, $A(x - \hat{x}) = 0$, which implies that $x - \hat{x} = A^{\mathrm{L}}A(x - \hat{x}) = 0$. Consequently, $Ax = b$ has a unique solution. The equivalence of *b)* and *c)* follows from *i)* and Theorem 2.6.1.

Statement *iv)* is a consequence of *iii)*.

To prove the last statement of *v)*, first let $x = \hat{x} + x_0$, where $x_0 \in \mathcal{N}(A)$. Then, $Ax = A\hat{x} + Ax_0 = b$. Therefore, every element of $\hat{x} + \mathcal{N}(A)$ satisfies $Ax = b$. To prove the reverse inclusion, let x satisfy $Ax = b$. Then, $A(x - \hat{x}) = 0$, which implies that $x - \hat{x} \in \mathcal{N}(A)$, that is, $x \in \hat{x} + \mathcal{N}(A)$. Hence, $\hat{x} + \mathcal{N}(A)$ is the set of solutions of $Ax = b$.

To prove *vi)*, note that, since rank $A = n$, it follows that rank $A =$ rank $\begin{bmatrix} A & b \end{bmatrix}$, and thus it follows from *ii)* that $Ax = b$ has at least one solution. In fact, $AA^{\mathrm{R}}b = b$. Now, let $\hat{x}$ be a solution of $Ax = b$. Then, if $x \in \mathbb{F}^m$ satisfies $Ax = b$, then $A(x - \hat{x}) = 0$, and thus $x \in \hat{x} + \mathcal{N}(A)$. To prove the reverse inclusion, let $x \in \hat{x} + \mathcal{N}(A)$. Then, $A(x - \hat{x}) = 0$, and thus $Ax = A\hat{x} = b$. Hence, $\hat{x} + \mathcal{N}(A)$ is the set of solutions of $Ax = b$. To prove the last statement, note that, if $n = m$, then A has a unique right inverse given by A^{-1}, and thus $x = A^{-1}b$ is the unique solution of $Ax = b$. Conversely, suppose that $Ax = b$ has a unique solution $\hat{x}$. Then, since $\hat{x} + \mathcal{N}(A)$ is the set of solutions of $Ax = b$, it follows that $\mathcal{N}(A) = \{0\}$, which implies that def $A = 0$. Hence, rank $A = m$, and thus $n = m$. $\qquad\square$

The set of solutions of $Ax = b$ is further characterized by Proposition 6.1.7, while connections to least squares solutions are discussed in Fact 9.15.4.

Let $A \in \mathbb{F}^{n \times m}$. Proposition 2.6.2 considers the uniqueness of left and right inverses of A, but does not consider the case in which a matrix is both a left inverse and a right inverse of A. Consequently, we say that A is *nonsingular* if there exists a matrix $B \in \mathbb{F}^{m \times n}$, the *inverse* of A, such that $BA = I_m$ and $AB = I_n$, that is, B is both a left and right inverse of A.

Proposition 2.6.5. Let $A \in \mathbb{F}^{n \times m}$. Then, the following statements are equivalent:

  i) A is nonsingular

  ii) rank $A = n = m$.

In this case, A has a unique inverse.

Proof. If A is nonsingular, then, since B is both left and right invertible, it follows from Theorem 2.6.1 that rank $A = m$ and rank $A = n$. Hence, *ii)* holds. Conversely, it follows from Theorem 2.6.1 that A has both a left inverse B and a right inverse C. Then, $B = BI_n = BAC = I_nC = C$. Hence, B is also a right inverse of A. Thus, A is nonsingular. In fact, the same argument shows that A has a unique inverse. $\qquad\square$

The following result can be viewed as a specialization of Theorem 1.4.2 to the function $f \colon \mathbb{F}^n \mapsto \mathbb{F}^n$, where $f(x) = Ax$.

Corollary 2.6.6. Let $A \in \mathbb{F}^{n \times n}$. Then, the following statements are equivalent:

  i) A is nonsingular.

  ii) A has a unique inverse.

  iii) A is one-to-one.

  iv) A is onto.

  v) A is left invertible.

  vi) A is right invertible.

  vii) A has a unique left inverse.

 viii) A has a unique right inverse.

  ix) rank $A = n$.

  x) def $A = 0$.

Let $A \in \mathbb{F}^{n \times n}$ be nonsingular. Then, the inverse of A, denoted by A^{-1}, is a unique $n \times n$ matrix with entries in $\mathbb{F}$. If A is not nonsingular, then A is *singular*.

The following result is a specialization of Theorem 2.6.4 to the case $n = m$.

Corollary 2.6.7. Let $A \in \mathbb{F}^{n \times n}$ and $b \in \mathbb{F}^n$. Then, the following statements hold:

  i) A is nonsingular if and only if there exists a unique vector $x \in \mathbb{F}^n$ satisfying $Ax = b$. In this case, $x = A^{-1}b$.

  ii) A is singular and rank $A = \text{rank} \begin{bmatrix} A & b \end{bmatrix}$ if and only if there exist infinitely many $x \in \mathbb{R}^n$ satisfying $Ax = b$. In this case, let $\hat{x} \in \mathbb{F}^m$ satisfy $A\hat{x} = b$. Then, the set of solutions of $Ax = b$ is given by $\hat{x} + \mathcal{N}(A)$.

Proposition 2.6.8. Let $A \in \mathbb{F}^{n \times n}$. Then, the following statements are equivalent:

   i) A is nonsingular.

  ii) $\overline{A}$ is nonsingular.

 iii) A^{T} is nonsingular.

 iv) A^* is nonsingular.

In this case,

$$(\overline{A})^{-1} = \overline{A^{-1}}, \tag{2.6.5}$$

$$\left(A^{\mathrm{T}}\right)^{-1} = \left(A^{-1}\right)^{\mathrm{T}}, \tag{2.6.6}$$

$$\left(A^*\right)^{-1} = \left(A^{-1}\right)^*. \tag{2.6.7}$$

Proof. Since $AA^{-1} = I$, it follows that $\left(A^{-1}\right)^* A^* = I$. Hence, $\left(A^{-1}\right)^* = (A^*)^{-1}$. $\square$

We thus use $A^{-\mathrm{T}}$ to denote $\left(A^{\mathrm{T}}\right)^{-1}$ or $\left(A^{-1}\right)^{\mathrm{T}}$ and A^{-*} to denote $(A^*)^{-1}$ or $\left(A^{-1}\right)^*$.

Proposition 2.6.9. Let $A, B \in \mathbb{F}^{n \times n}$ be nonsingular. Then,

$$(AB)^{-1} = B^{-1}A^{-1}, \tag{2.6.8}$$

$$(AB)^{-\mathrm{T}} = A^{-\mathrm{T}}B^{-\mathrm{T}}, \tag{2.6.9}$$

$$(AB)^{-*} = A^{-*}B^{-*}. \tag{2.6.10}$$

Proof. Note that $ABB^{-1}A^{-1} = AIA^{-1} = I$, which shows that $B^{-1}A^{-1}$ is the inverse of AB. Similarly, $(AB)^* A^{-*}B^{-*} = B^*A^*A^{-*}B^{-*} = B^*IB^{-*} = I$, which shows that $A^{-*}B^{-*}$ is the inverse of $(AB)^*$. $\square$

For a nonsingular matrix $A \in \mathbb{F}^{n \times n}$ and $r \in \mathbb{Z}$ we write

$$A^{-r} \triangleq (A^r)^{-1} = \left(A^{-1}\right)^r, \tag{2.6.11}$$

$$A^{-r\mathrm{T}} \triangleq (A^r)^{-\mathrm{T}} = \left(A^{-\mathrm{T}}\right)^r = (A^{-r})^{\mathrm{T}} = \left(A^{\mathrm{T}}\right)^{-r}, \tag{2.6.12}$$

$$A^{-r*} \triangleq (A^r)^{-*} = (A^{-*})^r = (A^{-r})^* = (A^*)^{-r}. \tag{2.6.13}$$

For example, $A^{-2*} = (A^{-*})^2$.

2.7 The Determinant

One of the most useful quantities associated with a square matrix is its determinant. In this section we develop some basic results pertaining to the determinant of a matrix.

The *determinant* of $A \in \mathbb{F}^{n \times n}$ is defined by

$$\det A \triangleq \sum_{\sigma} (-1)^{N_\sigma} \prod_{i=1}^{n} A_{(i,\sigma(i))}, \qquad (2.7.1)$$

where the sum is taken over all $n!$ permutations $\sigma = (\sigma(1), \dots, \sigma(n))$ of the column indices $1, \dots, n$, and where N_σ is the minimal number of pairwise transpositions needed to transform $\sigma(1), \dots, \sigma(n)$ to $1, \dots, n$. The following result is an immediate consequence of this definition.

Proposition 2.7.1. Let $A \in \mathbb{F}^{n \times n}$. Then,

$$\det A^{\mathrm{T}} = \det A, \qquad (2.7.2)$$

$$\det \overline{A} = \overline{\det A}, \qquad (2.7.3)$$

$$\det A^* = \overline{\det A}, \qquad (2.7.4)$$

and, for all $\alpha \in \mathbb{F}$,

$$\det \alpha A = \alpha^n \det A. \qquad (2.7.5)$$

If, in addition, $B \in \mathbb{F}^{m \times n}$ and $C \in \mathbb{F}^{m \times m}$, then

$$\det \begin{bmatrix} A & 0 \\ B & C \end{bmatrix} = (\det A)(\det C). \qquad (2.7.6)$$

The following observations are immediate consequences of the definition of the determinant.

Proposition 2.7.2. Let $A, B \in \mathbb{F}^{n \times n}$. Then, the following statements hold:

i) If every off-diagonal entry of A is zero, then

$$\det A = \prod_{i=1}^{n} A_{(i,i)}. \qquad (2.7.7)$$

In particular, $\det I_n = 1$.

ii) If A has a row or column consisting entirely of 0's, then $\det A = 0$.

iii) If A has two identical rows or two identical columns, then $\det A = 0$.

iv) If $x \in \mathbb{F}^n$ and $i \in \{1, \dots, n\}$, then

$$\det\left(A + x e_i^{\mathrm{T}}\right) = \det A + \det\left(A \overset{i}{\leftarrow} x\right). \qquad (2.7.8)$$

v) If $x \in \mathbb{F}^{1 \times n}$ and $i \in \{1, \dots, n\}$, then

$$\det(A + e_i x) = \det A + \det\left(A \overset{i}{\leftarrow} x\right). \qquad (2.7.9)$$

vi) If B is equal to A except that, for some $i \in \{1, \dots, n\}$ and $\alpha \in \mathbb{F}$, either $\mathrm{col}_i(B) = \alpha \mathrm{col}_i(A)$ or $\mathrm{row}_i(B) = \alpha \mathrm{row}_i(A)$, then $\det B = \alpha \det A$.

vii) If B is formed from A by interchanging two rows or two columns of A, then $\det B = -\det A$.

viii) If B is formed from A by adding a multiple of a (row, column) of A to another (row, column) of A, then $\det B = \det A$.

Statements *vi*)–*viii*) correspond, respectively, to multiplying the matrix A on the left or right by matrices of the form

$$I_n + (\alpha - 1)E_{i,i} = \begin{bmatrix} I_{i-1} & 0 & 0 \\ 0 & \alpha & 0 \\ 0 & 0 & I_{n-i} \end{bmatrix}, \tag{2.7.10}$$

$$I_n + E_{i,j} + E_{j,i} - E_{i,i} - E_{j,j} = \begin{bmatrix} I_{i-1} & 0 & 0 & 0 & 0 \\ 0 & 0 & 0 & 1 & 0 \\ 0 & 0 & I_{j-i-1} & 0 & 0 \\ 0 & 1 & 0 & 0 & 0 \\ 0 & 0 & 0 & 0 & I_{n-j} \end{bmatrix}, \tag{2.7.11}$$

where $i \neq j$, and

$$I_n + \beta E_{i,j} = \begin{bmatrix} I_{i-1} & 0 & 0 & 0 & 0 \\ 0 & 1 & 0 & \beta & 0 \\ 0 & 0 & I_{j-i-1} & 0 & 0 \\ 0 & 0 & 0 & 1 & 0 \\ 0 & 0 & 0 & 0 & I_{n-j} \end{bmatrix}, \tag{2.7.12}$$

where $\beta \in \mathbb{F}$ and $i \neq j$. The matrices in (2.7.11) and (2.7.12) illustrate the case $i < j$. Since $I + (\alpha-1)E_{i,i} = I + (\alpha-1)e_i e_i^{\mathrm{T}}$, $I + E_{i,j} + E_{j,i} - E_{i,i} - E_{j,j} = I - (e_i - e_j)(e_i - e_j)^{\mathrm{T}}$, and $I + \beta E_{i,j} = I + \beta e_i e_j^{\mathrm{T}}$, it follows that all of these matrices are of the form $I - xy^{\mathrm{T}}$. In terms of Definition 3.1.1, (2.7.10) is an elementary matrix if and only if $\alpha \neq 0$, (2.7.11) is an elementary matrix, and (2.7.12) is an elementary matrix if and only if either $i \neq j$ or $\beta \neq -1$.

Proposition 2.7.3. Let $A, B \in \mathbb{F}^{n \times n}$. Then,

$$\det AB = \det BA = (\det A)(\det B). \tag{2.7.13}$$

Proof. First note the equality

$$\begin{bmatrix} A & 0 \\ I & B \end{bmatrix} = \begin{bmatrix} I & A \\ 0 & I \end{bmatrix}\begin{bmatrix} -AB & 0 \\ 0 & I \end{bmatrix}\begin{bmatrix} I & 0 \\ B & I \end{bmatrix}\begin{bmatrix} 0 & I \\ I & 0 \end{bmatrix}.$$

The first and third matrices on the right-hand side of this equality add multiples of rows and columns of $\begin{bmatrix} -AB & 0 \\ 0 & I \end{bmatrix}$ to other rows and columns of $\begin{bmatrix} -AB & 0 \\ 0 & I \end{bmatrix}$. As already noted, these operations do not affect the determinant of $\begin{bmatrix} -AB & 0 \\ 0 & I \end{bmatrix}$. In addition, the fourth matrix on the right-hand side of this equality interchanges n pairs of columns of $\begin{bmatrix} 0 & A \\ B & I \end{bmatrix}$. Using (2.7.5), (2.7.6), and the fact that every interchange of a pair of columns of $\begin{bmatrix} 0 & A \\ B & I \end{bmatrix}$ entails a factor of -1, it thus follows that $(\det A)(\det B) = \det \begin{bmatrix} A & 0 \\ I & B \end{bmatrix} = (-1)^n \det \begin{bmatrix} -AB & 0 \\ 0 & I \end{bmatrix} = (-1)^n \det(-AB) = \det AB$. $\square$

Corollary 2.7.4. Let $A \in \mathbb{F}^{n \times n}$ be nonsingular. Then, $\det A \neq 0$ and

$$\det A^{-1} = (\det A)^{-1}. \tag{2.7.14}$$

Proof. Since $AA^{-1} = I_n$, it follows that $\det AA^{-1} = (\det A)(\det A^{-1}) = 1$. Hence, $\det A \neq 0$. In addition, $\det A^{-1} = 1/\det A$. $\square$

Let $A \in \mathbb{F}^{n \times m}$. The determinant of a square submatrix of A is a *subdeterminant* of A. By convention, the determinant of A is a subdeterminant of A. The determinant of a $j \times j$ (principal, leading principal) submatrix of A is a $j \times j$ (*principal, leading principal*) *subdeterminant* of A.

Let $A \in \mathbb{F}^{n \times n}$. Then, the *cofactor* of $A_{(i,j)}$, denoted by $A_{[i;j]}$, is the $(n-1) \times (n-1)$ submatrix of A obtained by deleting the ith row and jth column of A. In other words,

$$A_{[i;j]} \triangleq A_{(\{i\}^\sim, \{j\}^\sim)}. \tag{2.7.15}$$

The following result provides a cofactor expansion of $\det A$.

Proposition 2.7.5. Let $A \in \mathbb{F}^{n \times n}$. Then, for all $i \in \{1, \ldots, n\}$,

$$\sum_{k=1}^{n} (-1)^{i+k} A_{(i,k)} \det A_{[i;k]} = \det A. \tag{2.7.16}$$

Furthermore, for all $i, j \in \{1, \ldots, n\}$ such that $j \neq i$,

$$\sum_{k=1}^{n} (-1)^{i+k} A_{(j,k)} \det A_{[i;k]} = 0. \tag{2.7.17}$$

Proof. Identity (2.7.16) is an equivalent recursive form of the definition $\det A$, while the right-hand side of (2.7.17) is equal to $\det B$, where B is obtained from A by replacing $\text{row}_i(A)$ by $\text{row}_j(A)$. As already noted, $\det B = 0$. $\square$

Let $A \in \mathbb{F}^{n \times n}$, where $n \geq 2$. To simplify (2.7.16) and (2.7.17) it is useful to define the *adjugate* of A, denoted by $A^{\mathrm{A}} \in \mathbb{F}^{n \times n}$, where, for all $i, j \in \{1, \ldots, n\}$,

$$\left(A^{\mathrm{A}} \right)_{(i,j)} \triangleq (-1)^{i+j} \det A_{[j;i]} = \det(A \overset{i}{\leftarrow} e_j). \tag{2.7.18}$$

Then, (2.7.16) implies that, for all $i \in \{1, \ldots, n\}$,

$$\sum_{k=1}^{n} A_{(i,k)} (A^{\mathrm{A}})_{(k,i)} = \left(AA^{\mathrm{A}} \right)_{(i,i)} = \left(A^{\mathrm{A}}A \right)_{(i,i)} = \det A, \tag{2.7.19}$$

while (2.7.17) implies that, for all $i, j \in \{1, \ldots, n\}$ such that $j \neq i$,

$$\sum_{k=1}^{n} A_{(i,k)} (A^{\mathrm{A}})_{(k,j)} = \left(AA^{\mathrm{A}} \right)_{(i,j)} = \left(A^{\mathrm{A}}A \right)_{(i,j)} = 0. \tag{2.7.20}$$

Thus,

$$AA^{\mathrm{A}} = A^{\mathrm{A}}A = (\det A)I. \tag{2.7.21}$$

Consequently, if $\det A \neq 0$, then

$$A^{-1} = \frac{1}{\det A} A^{\mathrm{A}}, \tag{2.7.22}$$

whereas, if $\det A = 0$, then

$$AA^{\mathrm{A}} = A^{\mathrm{A}}A = 0. \tag{2.7.23}$$

For a scalar $A \in \mathbb{F}$, we define $A^{\mathrm{A}} \triangleq 1$.

The following result provides the converse of Corollary 2.7.4 by using (2.7.22) to construct A^{-1} in terms of $(n-1) \times (n-1)$ subdeterminants of A.

Corollary 2.7.6. Let $A \in \mathbb{F}^{n \times n}$. Then, A is nonsingular if and only if $\det A \neq 0$. In this case, for all $i, j \in \{1, \ldots, n\}$, the (i,j) entry of A^{-1} is given by

$$\left(A^{-1}\right)_{(i,j)} = (-1)^{i+j} \frac{\det A_{[j;i]}}{\det A}. \tag{2.7.24}$$

Finally, the following result uses the nonsingularity of submatrices to characterize the rank of a matrix.

Proposition 2.7.7. Let $A \in \mathbb{F}^{n \times m}$. Then, rank A is the largest order of all nonsingular submatrices of A.

2.8 Partitioned Matrices

Partitioned matrices were used to state or prove several results in this chapter including Proposition 2.5.9, Theorem 2.6.4, Proposition 2.7.1, and Proposition 2.7.3. In this section we give several useful equalities for partitioned matrices.

Proposition 2.8.1. Let $A_{ij} \in \mathbb{F}^{n_i \times m_j}$ for all $i \in \{1, \ldots, k\}$ and $j \in \{1, \ldots, l\}$. Then,

$$\begin{bmatrix} A_{11} & \cdots & A_{1l} \\ \vdots & \ddots & \vdots \\ A_{k1} & \cdots & A_{kl} \end{bmatrix}^{\mathrm{T}} = \begin{bmatrix} A_{11}^{\mathrm{T}} & \cdots & A_{k1}^{\mathrm{T}} \\ \vdots & \ddots & \vdots \\ A_{1l}^{\mathrm{T}} & \cdots & A_{kl}^{\mathrm{T}} \end{bmatrix} \tag{2.8.1}$$

and

$$\begin{bmatrix} A_{11} & \cdots & A_{1l} \\ \vdots & \ddots & \vdots \\ A_{k1} & \cdots & A_{kl} \end{bmatrix}^{*} = \begin{bmatrix} A_{11}^{*} & \cdots & A_{k1}^{*} \\ \vdots & \ddots & \vdots \\ A_{1l}^{*} & \cdots & A_{kl}^{*} \end{bmatrix}. \tag{2.8.2}$$

If, in addition, $k = l$ and $n_i = m_i$ for all $i \in \{1, \ldots, m\}$, then

$$\mathrm{tr} \begin{bmatrix} A_{11} & \cdots & A_{1k} \\ \vdots & \ddots & \vdots \\ A_{k1} & \cdots & A_{kk} \end{bmatrix} = \sum_{i=1}^{k} \mathrm{tr}\, A_{ii} \tag{2.8.3}$$

and

$$\det \begin{bmatrix} A_{11} & A_{12} & \cdots & A_{1k} \\ 0 & A_{22} & \cdots & A_{2k} \\ \vdots & \ddots & \ddots & \vdots \\ 0 & 0 & \cdots & A_{kk} \end{bmatrix} = \prod_{i=1}^{k} \det A_{ii}. \tag{2.8.4}$$

Lemma 2.8.2. Let $B \in \mathbb{F}^{n \times m}$ and $C \in \mathbb{F}^{m \times n}$. Then,

$$\begin{bmatrix} I & B \\ 0 & I \end{bmatrix}^{-1} = \begin{bmatrix} I & -B \\ 0 & I \end{bmatrix} \tag{2.8.5}$$

and

$$\begin{bmatrix} I & 0 \\ C & I \end{bmatrix}^{-1} = \begin{bmatrix} I & 0 \\ -C & I \end{bmatrix}. \tag{2.8.6}$$

Let $A \in \mathbb{F}^{n \times n}$ and $D \in \mathbb{F}^{m \times m}$ be nonsingular. Then,

$$\begin{bmatrix} A & 0 \\ 0 & D \end{bmatrix}^{-1} = \begin{bmatrix} A^{-1} & 0 \\ 0 & D^{-1} \end{bmatrix}. \tag{2.8.7}$$

Proposition 2.8.3. Let $A \in \mathbb{F}^{n \times n}$, $B \in \mathbb{F}^{n \times m}$, $C \in \mathbb{F}^{l \times n}$, and $D \in \mathbb{F}^{l \times m}$, and assume that A is nonsingular. Then,

$$\begin{bmatrix} A & B \\ C & D \end{bmatrix} = \begin{bmatrix} I & 0 \\ CA^{-1} & I \end{bmatrix} \begin{bmatrix} A & 0 \\ 0 & D - CA^{-1}B \end{bmatrix} \begin{bmatrix} I & A^{-1}B \\ 0 & I \end{bmatrix} \tag{2.8.8}$$

and

$$\operatorname{rank} \begin{bmatrix} A & B \\ C & D \end{bmatrix} = n + \operatorname{rank}(D - CA^{-1}B). \tag{2.8.9}$$

If, furthermore, $l = m$, then

$$\det \begin{bmatrix} A & B \\ C & D \end{bmatrix} = (\det A) \det(D - CA^{-1}B). \tag{2.8.10}$$

Proposition 2.8.4. Let $A \in \mathbb{F}^{n \times m}$, $B \in \mathbb{F}^{n \times l}$, $C \in \mathbb{F}^{l \times m}$, and $D \in \mathbb{F}^{l \times l}$, and assume that D is nonsingular. Then,

$$\begin{bmatrix} A & B \\ C & D \end{bmatrix} = \begin{bmatrix} I & BD^{-1} \\ 0 & I \end{bmatrix} \begin{bmatrix} A - BD^{-1}C & 0 \\ 0 & D \end{bmatrix} \begin{bmatrix} I & 0 \\ D^{-1}C & I \end{bmatrix} \tag{2.8.11}$$

and

$$\operatorname{rank} \begin{bmatrix} A & B \\ C & D \end{bmatrix} = l + \operatorname{rank}(A - BD^{-1}C). \tag{2.8.12}$$

If, furthermore, $n = m$, then

$$\det \begin{bmatrix} A & B \\ C & D \end{bmatrix} = (\det D) \det(A - BD^{-1}C). \tag{2.8.13}$$

Corollary 2.8.5. Let $A \in \mathbb{F}^{n \times m}$ and $B \in \mathbb{F}^{m \times n}$. Then,

$$\begin{bmatrix} I_n & A \\ B & I_m \end{bmatrix} = \begin{bmatrix} I_n & 0 \\ B & I_m \end{bmatrix} \begin{bmatrix} I_n & 0 \\ 0 & I_m - BA \end{bmatrix} \begin{bmatrix} I_n & A \\ 0 & I_m \end{bmatrix}$$

$$= \begin{bmatrix} I_n & A \\ 0 & I_m \end{bmatrix} \begin{bmatrix} I_n - AB & 0 \\ 0 & I_m \end{bmatrix} \begin{bmatrix} I_n & 0 \\ B & I_m \end{bmatrix}.$$

Hence,

$$\operatorname{rank} \begin{bmatrix} I_n & A \\ B & I_m \end{bmatrix} = n + \operatorname{rank}(I_m - BA) = m + \operatorname{rank}(I_n - AB)$$

and

$$\det \begin{bmatrix} I_n & A \\ B & I_m \end{bmatrix} = \det(I_m - BA) = \det(I_n - AB). \qquad (2.8.14)$$

Hence, $I_n + AB$ is nonsingular if and only if $I_m + BA$ is nonsingular.

Lemma 2.8.6. Let $A \in \mathbb{F}^{n \times n}$, $B \in \mathbb{F}^{n \times m}$, $C \in \mathbb{F}^{m \times n}$, and $D \in \mathbb{F}^{m \times m}$. If A and D are nonsingular, then

$$(\det A) \det(D - CA^{-1}B) = (\det D) \det(A - BD^{-1}C), \qquad (2.8.15)$$

and thus $D - CA^{-1}B$ is nonsingular if and only if $A - BD^{-1}C$ is nonsingular.

Proposition 2.8.7. Let $A \in \mathbb{F}^{n \times n}$, $B \in \mathbb{F}^{n \times m}$, $C \in \mathbb{F}^{m \times n}$, and $D \in \mathbb{F}^{m \times m}$. If A and $D - CA^{-1}B$ are nonsingular, then

$$\begin{bmatrix} A & B \\ C & D \end{bmatrix}^{-1}$$
$$= \begin{bmatrix} A^{-1} + A^{-1}B(D - CA^{-1}B)^{-1}CA^{-1} & -A^{-1}B(D - CA^{-1}B)^{-1} \\ -(D - CA^{-1}B)^{-1}CA^{-1} & (D - CA^{-1}B)^{-1} \end{bmatrix}. \qquad (2.8.16)$$

If D and $A - BD^{-1}C$ are nonsingular, then

$$\begin{bmatrix} A & B \\ C & D \end{bmatrix}^{-1}$$
$$= \begin{bmatrix} (A - BD^{-1}C)^{-1} & -(A - BD^{-1}C)^{-1}BD^{-1} \\ -D^{-1}C(A - BD^{-1}C)^{-1} & D^{-1} + D^{-1}C(A - BD^{-1}C)^{-1}BD^{-1} \end{bmatrix}. \qquad (2.8.17)$$

If A, D, and $D - CA^{-1}B$ are nonsingular, then $A - BD^{-1}C$ is nonsingular, and

$$\begin{bmatrix} A & B \\ C & D \end{bmatrix}^{-1}$$
$$= \begin{bmatrix} (A - BD^{-1}C)^{-1} & -(A - BD^{-1}C)^{-1}BD^{-1} \\ -(D - CA^{-1}B)^{-1}CA^{-1} & (D - CA^{-1}B)^{-1} \end{bmatrix}. \qquad (2.8.18)$$

The following result is the *matrix inversion lemma.*

Corollary 2.8.8. Let $A \in \mathbb{F}^{n \times n}$, $B \in \mathbb{F}^{n \times m}$, $C \in \mathbb{F}^{m \times n}$, and $D \in \mathbb{F}^{m \times m}$. If A, $D - CA^{-1}B$, and D are nonsingular, then $A - BD^{-1}C$ is nonsingular,

$$(A - BD^{-1}C)^{-1} = A^{-1} + A^{-1}B(D - CA^{-1}B)^{-1}CA^{-1}, \qquad (2.8.19)$$

and

$$C(A - BD^{-1}C)^{-1}A = D(D - CA^{-1}B)^{-1}C. \qquad (2.8.20)$$

If A and $I - CA^{-1}B$ are nonsingular, then $A - BC$ is nonsingular, and

$$(A - BC)^{-1} = A^{-1} + A^{-1}B(I - CA^{-1}B)^{-1}CA^{-1}. \qquad (2.8.21)$$

If $D - CB$, and D are nonsingular, then $I - BD^{-1}C$ is nonsingular, and

$$\left(I - BD^{-1}C\right)^{-1} = I + B(D - CB)^{-1}C. \tag{2.8.22}$$

If $I - CB$ is nonsingular, then $I - BC$ is nonsingular, and

$$(I - BC)^{-1} = I + B(I - CB)^{-1}C. \tag{2.8.23}$$

Corollary 2.8.9. Let $A, B, C, D \in \mathbb{F}^{n \times n}$. If A, B, $C - DB^{-1}A$, and $D - CA^{-1}B$ are nonsingular, then

$$\begin{bmatrix} A & B \\ C & D \end{bmatrix}^{-1} = \begin{bmatrix} A^{-1} - \left(C - DB^{-1}A\right)^{-1}CA^{-1} & \left(C - DB^{-1}A\right)^{-1} \\ -\left(D - CA^{-1}B\right)^{-1}CA^{-1} & \left(D - CA^{-1}B\right)^{-1} \end{bmatrix}. \tag{2.8.24}$$

If A, C, $B - AC^{-1}D$, and $D - CA^{-1}B$ are nonsingular, then

$$\begin{bmatrix} A & B \\ C & D \end{bmatrix}^{-1} = \begin{bmatrix} A^{-1} - A^{-1}B\left(B - AC^{-1}D\right)^{-1} & -A^{-1}B\left(D - CA^{-1}B\right)^{-1} \\ \left(B - AC^{-1}D\right)^{-1} & \left(D - CA^{-1}B\right)^{-1} \end{bmatrix}. \tag{2.8.25}$$

If A, B, C, $B - AC^{-1}D$, and $D - CA^{-1}B$ are nonsingular, then $C - DB^{-1}A$ is nonsingular, and

$$\begin{bmatrix} A & B \\ C & D \end{bmatrix}^{-1} = \begin{bmatrix} A^{-1} - A^{-1}B\left(B - AC^{-1}D\right)^{-1} & \left(C - DB^{-1}A\right)^{-1} \\ \left(B - AC^{-1}D\right)^{-1} & \left(D - CA^{-1}B\right)^{-1} \end{bmatrix}. \tag{2.8.26}$$

If B, D, $A - BD^{-1}C$, and $C - DB^{-1}A$ are nonsingular, then

$$\begin{bmatrix} A & B \\ C & D \end{bmatrix}^{-1} = \begin{bmatrix} \left(A - BD^{-1}C\right)^{-1} & \left(C - DB^{-1}A\right)^{-1} \\ -D^{-1}C\left(A - BD^{-1}C\right)^{-1} & D^{-1} - D^{-1}C\left(C - DB^{-1}A\right)^{-1} \end{bmatrix}. \tag{2.8.27}$$

If C, D, $A - BD^{-1}C$, and $B - AC^{-1}D$ are nonsingular, then

$$\begin{bmatrix} A & B \\ C & D \end{bmatrix}^{-1} = \begin{bmatrix} \left(A - BD^{-1}C\right)^{-1} & -\left(A - BD^{-1}C\right)^{-1}BD^{-1} \\ \left(B - AC^{-1}D\right)^{-1} & D^{-1} - \left(B - AC^{-1}D\right)^{-1}BD^{-1} \end{bmatrix}. \tag{2.8.28}$$

If B, C, D, $A - BD^{-1}C$, and $C - DB^{-1}A$ are nonsingular, then $B - AC^{-1}D$ is nonsingular, and

$$\begin{bmatrix} A & B \\ C & D \end{bmatrix}^{-1} = \begin{bmatrix} \left(A - BD^{-1}C\right)^{-1} & \left(C - DB^{-1}A\right) \\ \left(B - AC^{-1}D\right)^{-1} & D^{-1} - D^{-1}C\left(C - DB^{-1}A\right)^{-1} \end{bmatrix}. \tag{2.8.29}$$

Finally, if A, B, C, D, $A - BD^{-1}C$, and $B - AC^{-1}D$, are nonsingular, then $C - DB^{-1}A$ and $D - CA^{-1}B$ are nonsingular, and

$$\begin{bmatrix} A & B \\ C & D \end{bmatrix}^{-1} = \begin{bmatrix} \left(A - BD^{-1}C\right)^{-1} & \left(C - DB^{-1}A\right)^{-1} \\ \left(B - AC^{-1}D\right)^{-1} & \left(D - CA^{-1}B\right)^{-1} \end{bmatrix}. \tag{2.8.30}$$

Corollary 2.8.10. Let $A, B \in \mathbb{F}^{n \times n}$, and assume that A and $I - A^{-1}B$ are nonsingular. Then, $A - B$ is nonsingular, and

$$(A - B)^{-1} = A^{-1} + A^{-1}B\left(I - A^{-1}B\right)^{-1}A^{-1}. \tag{2.8.31}$$

If, in addition, B is nonsingular, then

$$(A - B)^{-1} = A^{-1} + A^{-1}\left(B^{-1} - A^{-1}\right)^{-1}A^{-1}. \tag{2.8.32}$$

2.9 Facts on Polars, Cones, Dual Cones, Convex Hulls, and Subspaces

Fact 2.9.1. Let $\mathcal{S} \subseteq \mathbb{F}^n$, let $\alpha, \beta \in \mathbb{F}$, and assume that at least one of the following statements holds:

i) $\operatorname{card}(\mathcal{S}) \leq 1$.

ii) $\mathcal{S}$ is a cone, and α and β are nonnegative numbers.

iii) $\mathcal{S}$ is convex, and α and β are nonnegative numbers such that $\alpha + \beta = 1$.

iv) $\mathcal{S}$ is an affine subspace, and $\alpha + \beta \neq 0$.

Then,
$$\alpha\mathcal{S} + \beta\mathcal{S} = (\alpha + \beta)\mathcal{S}.$$

Remark: The inclusion "$\supseteq$" holds for all $\mathcal{S} \subseteq \mathbb{F}^n$ and $\alpha, \beta \in \mathbb{F}$.

Fact 2.9.2. Let $\mathcal{S}_1, \mathcal{S}_2 \subseteq \mathbb{F}^n$, and assume that $\mathcal{S}_1$ and $\mathcal{S}_2$ are convex. Then, $\mathcal{S}_1 + \mathcal{S}_2$ is convex.

Fact 2.9.3. Let $\mathcal{S} \subseteq \mathbb{F}^n$. Then, the following statements hold:

i) $\operatorname{coco}\mathcal{S} = \operatorname{co}\operatorname{cone}\mathcal{S} = \operatorname{cone}\operatorname{co}\mathcal{S}$.

ii) $\mathcal{S}^{\perp\perp} = \operatorname{span}\mathcal{S} = \operatorname{coco}(\mathcal{S} \cup -\mathcal{S})$.

iii) $\mathcal{S} \subseteq \operatorname{co}\mathcal{S} \subseteq (\operatorname{aff}\mathcal{S} \cap \operatorname{coco}\mathcal{S}) \subseteq \left\{ \begin{matrix} \operatorname{aff}\mathcal{S} \\ \operatorname{coco}\mathcal{S} \end{matrix} \right\} \subseteq \operatorname{span}\mathcal{S}$.

iv) $\mathcal{S} \subseteq (\operatorname{co}\mathcal{S} \cap \operatorname{cone}\mathcal{S}) \subseteq \left\{ \begin{matrix} \operatorname{co}\mathcal{S} \\ \operatorname{cone}\mathcal{S} \end{matrix} \right\} \subseteq \operatorname{coco}\mathcal{S} \subseteq \operatorname{span}\mathcal{S}$.

v) $\operatorname{dcone}\operatorname{dcone}\mathcal{S} = \operatorname{cl}\operatorname{coco}\mathcal{S}$.

Proof: For *v)*, see [243, p. 54].

Remark: See [180, p. 52]. Note that "pointed" in [180] means one-sided.

Fact 2.9.4. Let $\mathcal{S}, \mathcal{S}_1, \mathcal{S}_2 \subseteq \mathbb{F}^n$. Then, the following statements hold:

i) $\operatorname{polar}\mathcal{S}$ is a closed, convex set containing the origin.

ii) $\operatorname{polar}\mathbb{F}^n = \{0\}$, and $\operatorname{polar}\{0\} = \mathbb{F}^n$.

iii) If $\alpha > 0$, then $\operatorname{polar}\alpha\mathcal{S} = \frac{1}{\alpha}\operatorname{polar}\mathcal{S}$.

iv) $\mathcal{S} \subseteq \operatorname{polar}\operatorname{polar}\mathcal{S}$.

v) If $\mathcal{S}$ is nonempty, then polar polar polar $\mathcal{S}$ = polar $\mathcal{S}$.

vi) If $\mathcal{S}$ is nonempty, then polar polar $\mathcal{S}$ = cl co($\mathcal{S} \cup \{0\}$).

vii) If $0 \in \mathcal{S}$ and $\mathcal{S}$ is closed and convex, then polar polar $\mathcal{S}$ = $\mathcal{S}$.

viii) If $\mathcal{S}_1 \subseteq \mathcal{S}_2$, then polar $\mathcal{S}_2 \subseteq$ polar $\mathcal{S}_1$.

ix) polar($\mathcal{S}_1 \cup \mathcal{S}_2$) = (polar $\mathcal{S}_1$) $\cap$ (polar $\mathcal{S}_2$).

x) If $\mathcal{S}$ is a convex cone, then polar $\mathcal{S}$ = dcone $\mathcal{S}$.

Proof: See [157, pp. 143–147].

Fact 2.9.5. Let $\mathcal{S}_1, \mathcal{S}_2 \subseteq \mathbb{F}^n$, and assume that $\mathcal{S}_1$ and $\mathcal{S}_2$ are cones. Then,
$$\text{dcone}(\mathcal{S}_1 + \mathcal{S}_2) = (\text{dcone } \mathcal{S}_1) \cap (\text{dcone } \mathcal{S}_1).$$
If, in addition, $\mathcal{S}_1$ and $\mathcal{S}_2$ are closed and convex, then
$$\text{dcone}(\mathcal{S}_1 \cap \mathcal{S}_2) = \text{cl}[(\text{dcone } \mathcal{S}_1) + (\text{dcone } \mathcal{S}_2)].$$
Proof: See [157, p. 147] or [243, pp. 58, 59].

Fact 2.9.6. Let $\mathcal{S} \subset \mathbb{F}^n$. Then, the following statements hold:

i) $\mathcal{S}$ is an affine hyperplane if and only if there exist a nonzero vector $y \in \mathbb{F}^n$ and $\alpha \in \mathbb{R}$ such that $\mathcal{S} = \{x: \ \text{Re}\, x^*y = \alpha\}$.

ii) $\mathcal{S}$ is an affine closed half space if and only if there exist a nonzero vector $y \in \mathbb{F}^n$ and $\alpha \in \mathbb{R}$ such that $\mathcal{S} = \{x \in \mathbb{F}^n: \ \text{Re}\, x^*y \le \alpha\}$.

iii) $\mathcal{S}$ is an affine open half space if and only if there exist a nonzero vector $y \in \mathbb{F}^n$ and $\alpha \in \mathbb{R}$ such that $\mathcal{S} = \{x \in \mathbb{F}^n: \ \text{Re}\, x^*y \le \alpha\}$.

Proof: Let $z \in \mathbb{F}^n$ satisfy $z^*y = \alpha$. Then, $\{x: \ x^*y = \alpha\} = \{y\}^\perp + z$.

Fact 2.9.7. Let $x_1, \ldots, x_k \in \mathbb{F}^n$. Then,
$$\text{aff}\, \{x_1, \ldots, x_k\} = x_1 + \text{span}\, \{x_2 - x_1, \ldots, x_k - x_1\}.$$
Remark: See Fact 10.8.12.

Fact 2.9.8. Let $\mathcal{S} \subseteq \mathbb{F}^n$, and assume that $\mathcal{S}$ is an affine subspace. Then, $\mathcal{S}$ is a subspace if and only if $0 \in \mathcal{S}$.

Fact 2.9.9. Let $\mathcal{S}_1, \mathcal{S}_2 \subseteq \mathbb{F}^n$ be (cones, convex sets, convex cones, subspaces). Then, so are $\mathcal{S}_1 \cap \mathcal{S}_2$ and $\mathcal{S}_1 + \mathcal{S}_2$.

Fact 2.9.10. Let $\mathcal{S}_1, \mathcal{S}_2 \subseteq \mathbb{F}^n$ be pointed convex cones. Then,
$$\text{co}(\mathcal{S}_1 \cup \mathcal{S}_2) = \mathcal{S}_1 + \mathcal{S}_2.$$

Fact 2.9.11. Let $\mathcal{S}_1, \mathcal{S}_2 \subseteq \mathbb{F}^n$ be subspaces. Then,
$$\mathcal{S}_1 \cup \mathcal{S}_2 \subseteq \text{co}(\mathcal{S}_1 \cup \mathcal{S}_2) = \mathcal{S}_1 + \mathcal{S}_2.$$
Furthermore, if $\mathcal{S}_3 \subseteq \mathbb{F}^n$ is a subspace such that $\mathcal{S}_1 \cup \mathcal{S}_2 \subseteq \mathcal{S}_3$, then $\mathcal{S}_1 + \mathcal{S}_2 \subseteq \mathcal{S}_3$.

Finally, $S_1 \cup S_2$ is a subspace if and only if either $S_1 \subseteq S_2$ or $S_2 \subseteq S_1$. In this case,

$$S_1 \cup S_2 = S_1 + S_2.$$

Remark: $S_1 + S_2$ is the smallest subspace containing $S_1 \cup S_2$. See [1208, p. 81].

Fact 2.9.12. Let $S_1, S_2 \subseteq \mathbb{F}^n$. Then,

$$(\operatorname{span} S_1) \cup (\operatorname{span} S_2) \subseteq \operatorname{span}(S_1 \cup S_2)$$

and

$$\operatorname{span}(S_1 \cap S_2) \subseteq (\operatorname{span} S_1) \cap (\operatorname{span} S_2).$$

Proof: See [1215, p. 11].

Fact 2.9.13. Let $S_1, S_2 \subseteq \mathbb{F}^n$ be subspaces. Then,

$$\operatorname{span}(S_1 \cup S_2) = S_1 + S_2.$$

Therefore, $S_1 + S_2$ is the smallest subspace that contains $S_1 \cup S_2$.

Fact 2.9.14. Let $S_1, S_2 \subseteq \mathbb{F}^n$ be subspaces. Then, the following statements are equivalent:

i) $S_1 \subseteq S_2$

ii) $S_2^\perp \subseteq S_1^\perp$.

iii) For all $x \in S_1$ and $y \in S_2^\perp$, $x^* y = 0$.

Furthermore, $S_1 \subset S_2$ if and only if $S_2^\perp \subset S_1^\perp$.

Fact 2.9.15. Let $S_1, S_2 \subseteq \mathbb{F}^n$. Then,

$$S_1^\perp \cap S_2^\perp \subseteq (S_1 + S_2)^\perp.$$

Problem: Determine necessary and sufficient conditions under which equality holds.

Fact 2.9.16. Let $S_1, S_2 \subseteq \mathbb{F}^n$ be subspaces. Then,

$$(S_1 \cap S_2)^\perp = S_1^\perp + S_2^\perp$$

and

$$(S_1 + S_2)^\perp = S_1^\perp \cap S_2^\perp.$$

Fact 2.9.17. Let $S_1, S_2, S_3 \subseteq \mathbb{F}^n$ be subspaces. Then,

$$S_1 + (S_2 \cap S_3) \subseteq (S_1 + S_2) \cap (S_1 + S_3)$$

and

$$S_1 \cap (S_2 + S_3) \supseteq (S_1 \cap S_2) + (S_1 \cap S_3).$$

Fact 2.9.18. Let $S_1, S_2 \subseteq \mathbb{F}^n$ be subspaces. Then, S_1, S_2 are complementary subspaces if and only if $S_1^\perp, S_2^\perp$ are complementary subspaces.

Remark: See Fact 3.12.1.

Fact 2.9.19. Let $\mathcal{S}_1, \mathcal{S}_2 \subseteq \mathbb{F}^n$ be nonzero subspaces, and define $\theta \in [0, \pi/2]$ by

$$\cos\theta = \max\{|x^*y| : (x,y) \in \mathcal{S}_1 \times \mathcal{S}_2 \text{ and } x^*x = y^*y = 1\}.$$

Then,

$$\cos\theta = \max\{|x^*y| : (x,y) \in \mathcal{S}_1^\perp \times \mathcal{S}_2^\perp \text{ and } x^*x = y^*y = 1\}.$$

Furthermore, $\theta = 0$ if and only if $\mathcal{S}_1 \cap \mathcal{S}_2 = \{0\}$, and $\theta = \pi/2$ if and only if $\mathcal{S}_1 = \mathcal{S}_2^\perp$.

Remark: See [551, 765].

Remark: θ is a *principal angle*. See Fact 5.9.31, Fact 5.11.39, and Fact 5.12.17.

Fact 2.9.20. Let $\mathcal{S}_1, \mathcal{S}_2 \subseteq \mathbb{F}^n$ be subspaces, and assume that $\mathcal{S}_1 \cap \mathcal{S}_2 = \{0\}$. Then,

$$\dim \mathcal{S}_1 + \dim \mathcal{S}_2 \leq n.$$

Fact 2.9.21. Let $\mathcal{S}_1, \mathcal{S}_2 \subseteq \mathbb{F}^n$ be subspaces. Then,

$$\dim(\mathcal{S}_1 \cap \mathcal{S}_2) \leq \min\{\dim \mathcal{S}_1, \dim \mathcal{S}_2\}$$
$$\leq \left\{ \begin{matrix} \dim \mathcal{S}_1 \\ \dim \mathcal{S}_2 \end{matrix} \right\}$$
$$\leq \max\{\dim \mathcal{S}_1, \dim \mathcal{S}_2\}$$
$$\leq \dim(\mathcal{S}_1 + \mathcal{S}_2)$$
$$\leq \min\{\dim \mathcal{S}_1 + \dim \mathcal{S}_2, n\}.$$

Fact 2.9.22. Let $\mathcal{S}_1, \mathcal{S}_2, \mathcal{S}_3 \subseteq \mathbb{F}^n$ be subspaces. Then,

$$\dim(\mathcal{S}_1 + \mathcal{S}_2 + \mathcal{S}_3) + \max\{\dim(\mathcal{S}_1 \cap \mathcal{S}_2), \dim(\mathcal{S}_1 \cap \mathcal{S}_3), \dim(\mathcal{S}_2 \cap \mathcal{S}_3)\}$$
$$\leq \dim \mathcal{S}_1 + \dim \mathcal{S}_2 + \dim \mathcal{S}_3.$$

Proof: See [400, p. 124].

Remark: Setting $\mathcal{S}_3 = \{0\}$ yields a weaker version of Theorem 2.3.1.

Fact 2.9.23. Let $\mathcal{S}_1, \ldots, \mathcal{S}_k \subseteq \mathbb{F}^n$ be subspaces having the same dimension. Then, there exists a subspace $\hat{\mathcal{S}} \subseteq \mathbb{F}^n$ such that, for all $i \in \{1, \ldots, k\}$, $\hat{\mathcal{S}}$ and $\mathcal{S}_i$ are complementary.

Proof: See [644, pp. 78, 79, 259, 260].

Fact 2.9.24. Let $\mathcal{S} \subseteq \mathbb{F}^n$ be a subspace. Then, for all $m \geq \dim \mathcal{S}$, there exists a matrix $A \in \mathbb{F}^{n \times m}$ such that $\mathcal{S} = \mathcal{R}(A)$.

Fact 2.9.25. Let $A \in \mathbb{F}^{n \times n}$, let $\mathcal{S} \subseteq \mathbb{F}^n$, assume that $\mathcal{S}$ is a subspace, let $k \triangleq \dim \mathcal{S}$, let $S \in \mathbb{F}^{n \times k}$, and assume that $\mathcal{R}(S) = \mathcal{S}$. Then, $\mathcal{S}$ is an invariant subspace of A if and only if there exists a matrix $M \in \mathbb{F}^{k \times k}$ such that $AS = SM$.

Proof: Set $B = I$ in Fact 5.13.1. See [897, p. 99].

Fact 2.9.26. Let $\mathcal{S} \subseteq \mathbb{F}^m$, and let $A \in \mathbb{F}^{n \times m}$. Then,

$$\text{cone}\, A\mathcal{S} = A\,\text{cone}\, \mathcal{S},$$
$$\text{co}\, A\mathcal{S} = A\,\text{co}\, \mathcal{S},$$
$$\text{span}\, A\mathcal{S} = A\,\text{span}\, \mathcal{S},$$
$$\text{aff}\, A\mathcal{S} = A\,\text{aff}\, \mathcal{S}.$$

Hence, if $\mathcal{S}$ is a (cone, convex set, subspace, affine subspace), then so is $A\mathcal{S}$. Now, assume in addition that A is left invertible, and let $A^{\mathrm{L}} \in \mathbb{F}^{m \times n}$ be a left inverse of A. Then,

$$\text{cone}\, \mathcal{S} = A^{\mathrm{L}}\,\text{cone}\, A\mathcal{S},$$
$$\text{co}\, \mathcal{S} = A^{\mathrm{L}}\,\text{co}\, A\mathcal{S},$$
$$\text{span}\, \mathcal{S} = A^{\mathrm{L}}\,\text{span}\, A\mathcal{S},$$
$$\text{aff}\, \mathcal{S} = A^{\mathrm{L}}\,\text{aff}\, A\mathcal{S}.$$

Hence, if $A\mathcal{S}$ is a (cone, convex set, subspace, affine subspace), then so is $\mathcal{S}$.

Fact 2.9.27. Let $\mathcal{S} \subseteq \mathbb{F}^n$, and let $A \in \mathbb{F}^{n \times m}$. Then, the following statements hold:

i) If A is right invertible and A^{R} is a right inverse of A, then

$$(A\mathcal{S})^{\perp} \subseteq A^{\mathrm{R}*}\mathcal{S}^{\perp}.$$

ii) If A is left invertible and A^{L} is a left inverse of A, then

$$A\mathcal{S}^{\perp} \subseteq (A^{\mathrm{L}*}\mathcal{S})^{\perp}.$$

iii) If $n = m$ and A is nonsingular, then

$$(A\mathcal{S})^{\perp} = A^{-*}\mathcal{S}^{\perp}.$$

Proof: The third statement is an immediate consequence of the first two statements.

Fact 2.9.28. Let $A \in \mathbb{F}^{n \times m}$, and let $\mathcal{S}_1 \subseteq \mathbb{R}^m$ and $\mathcal{S}_2 \subseteq \mathbb{F}^n$ be subspaces. Then, the following statements are equivalent:

i) $A\mathcal{S}_1 \subseteq \mathcal{S}_2$.

ii) $A^*\mathcal{S}_2^{\perp} \subseteq \mathcal{S}_1^{\perp}$.

Proof: See [319, p. 12].

Fact 2.9.29. Let $\mathcal{S}_1, \mathcal{S}_2 \subseteq \mathbb{F}^m$ be subspaces, and let $A \in \mathbb{F}^{n \times m}$. Then, the following statements hold:

i) $A(\mathcal{S}_1 \cup \mathcal{S}_2) = A\mathcal{S}_1 \cup A\mathcal{S}_2$.

ii) $A(\mathcal{S}_1 \cap \mathcal{S}_2) \subseteq A\mathcal{S}_1 \cap A\mathcal{S}_2$.

iii) $A(\mathcal{S}_1 + \mathcal{S}_2) = A\mathcal{S}_1 + A\mathcal{S}_2$.

If, in addition, A is left invertible, then the following statement holds:

iv) $A(\mathcal{S}_1 \cap \mathcal{S}_2) = A\mathcal{S}_1 \cap A\mathcal{S}_2$.

Proof: See Fact 1.7.11, Fact 1.7.15, and [319, p. 12].

Fact 2.9.30. Let $\mathcal{S}, \mathcal{S}_1, \mathcal{S}_2 \subseteq \mathbb{F}^n$ be subspaces, let $A \in \mathbb{F}^{n \times m}$, and define $f: \mathbb{F}^m \mapsto \mathbb{F}^n$ by $f(x) \triangleq Ax$. Then, the following statements hold:

i) $f[f^{-1}(\mathcal{S})] \subseteq \mathcal{S} \subseteq f^{-1}[f(\mathcal{S})]$.

ii) $[f^{-1}(\mathcal{S})]^\perp = A^* \mathcal{S}^\perp$.

iii) $f^{-1}(\mathcal{S}_1 \cup \mathcal{S}_2) = f^{-1}(\mathcal{S}_1) \cup f^{-1}(\mathcal{S}_2)$.

iv) $f^{-1}(\mathcal{S}_1 \cap \mathcal{S}_2) = f^{-1}(\mathcal{S}_1) \cap f^{-1}(\mathcal{S}_2)$.

v) $f^{-1}(\mathcal{S}_1 + \mathcal{S}_2) \supseteq f^{-1}(\mathcal{S}_1) + f^{-1}(\mathcal{S}_2)$.

Proof: See Fact 1.7.12 and [319, p. 12].

Problem: For a subspace $\mathcal{S} \subseteq \mathbb{F}^n$, $A \in \mathbb{F}^{n \times m}$, and $f(x) \triangleq Ax$, determine $B \in \mathbb{F}^{m \times n}$ such that $f^{-1}(\mathcal{S}) = B\mathcal{S}$, that is, $AB\mathcal{S} \subseteq \mathcal{S}$ and $B\mathcal{S}$ is maximal.

Fact 2.9.31. Define the convex pointed cone $\mathcal{S} \subset \mathbb{R}^2$ by

$$\mathcal{S} \triangleq \{(x_1, x_2) \in [0, \infty) \times \mathbb{R}: \text{ if } x_1 = 0, \text{ then } x_2 \geq 0\},$$

that is,

$$\mathcal{S} = ([0, \infty) \times \mathbb{R}) \backslash [\{0\} \times (-\infty, 0)].$$

Furthermore, for $x, y \in \mathbb{R}^2$, define $x \overset{\mathrm{d}}{\leq} y$ if and only if $y - x \in \mathcal{S}$. Then, "$\overset{\mathrm{d}}{\leq}$" is a total ordering on $\mathbb{R}^2$.

Remark: "$\overset{\mathrm{d}}{\leq}$" is the lexicographic or dictionary ordering. See Fact 1.7.8.

Remark: See [157, p. 161].

2.10 Facts on Range, Null Space, Rank, and Defect

Fact 2.10.1. Let $A \in \mathbb{F}^{n \times n}$. Then,

$$\mathcal{N}(A) \subseteq \mathcal{R}(I - A)$$

and

$$\mathcal{N}(I - A) \subseteq \mathcal{R}(A).$$

Remark: See Fact 3.12.3.

Fact 2.10.2. Let $A \in \mathbb{F}^{n \times m}$. Then, the following statements hold:

i) If $B \in \mathbb{F}^{m \times l}$ and $\operatorname{rank} B = m$, then $\mathcal{R}(A) = \mathcal{R}(AB)$.

ii) If $C \in \mathbb{F}^{k \times n}$ and $\operatorname{rank} C = n$, then $\mathcal{N}(A) = \mathcal{N}(CA)$.

iii) If $S \in \mathbb{F}^{m \times m}$ and S is nonsingular, then $\mathcal{N}(A) = S\mathcal{N}(AS)$.

Remark: See Lemma 2.4.1.

Fact 2.10.3. Let $A \in \mathbb{F}^{n \times m}$ and $B \in \mathbb{F}^{m \times l}$. Then, the following statements hold:

   $i)$ If A and B are right invertible, then so is AB.

   $ii)$ If A and B are left invertible, then so is AB.

   $iii)$ If $n = m = l$ and A and B are nonsingular, then so is AB.

Proof: This result follows from either Corollary 2.5.10 or Proposition 2.6.3.

Remark: See Fact 1.7.17.

Fact 2.10.4. Let $\mathcal{S} \subseteq \mathbb{F}^m$, assume that $\mathcal{S}$ is an affine subspace, and let $A \in \mathbb{F}^{n \times m}$. Then, the following statements hold:

   $i)$ $\operatorname{rank} A + \dim \mathcal{S} - m \leq \dim A\mathcal{S} \leq \min\{\operatorname{rank} A, \dim \mathcal{S}\}$.

   $ii)$ $\dim(A\mathcal{S}) + \dim[\mathcal{N}(A) \cap \mathcal{S}] = \dim \mathcal{S}$.

   $iii)$ $\dim A\mathcal{S} \leq \dim \mathcal{S}$.

   $iv)$ If A is left invertible, then $\dim A\mathcal{S} = \dim \mathcal{S}$.

Proof: For $ii)$, see [1157, p. 413]. For $iii)$, note that $\dim A\mathcal{S} \leq \dim \mathcal{S} = \dim A^L A\mathcal{S} \leq \dim A\mathcal{S}$.

Remark: See Fact 2.9.26 and Fact 10.8.17.

Fact 2.10.5. Let $A \in \mathbb{F}^{n \times m}$ and $B \in \mathbb{F}^{1 \times m}$. Then, $\mathcal{N}(A) \subseteq \mathcal{N}(B)$ if and only if there exists a vector $\lambda \in \mathbb{F}^n$ such that $B = \lambda^* A$.

Fact 2.10.6. Let $A \in \mathbb{F}^{n \times m}$ and $b \in \mathbb{F}^n$. Then, there exists a vector $x \in \mathbb{F}^n$ satisfying $Ax = b$ if and only if $b^* \lambda = 0$ for all $\lambda \in \mathcal{N}(A^*)$.

Proof: Assume that $A^* \lambda = 0$ implies that $b^* \lambda = 0$. Then, $\mathcal{N}(A^*) \subseteq \mathcal{N}(b^*)$. Hence, $b \in \mathcal{R}(b) \subseteq \mathcal{R}(A)$.

Fact 2.10.7. Let $A \in \mathbb{F}^{n \times m}$ and $B \in \mathbb{F}^{l \times m}$. Then, $\mathcal{N}(B) \subseteq \mathcal{N}(A)$ if and only if there exists a matrix $C \in \mathbb{F}^{n \times l}$ such that $A = CB$. Now, let $A \in \mathbb{F}^{n \times m}$ and $B \in \mathbb{F}^{n \times l}$. Then, $\mathcal{R}(A) \subseteq \mathcal{R}(B)$ if and only if there exists a matrix $C \in \mathbb{F}^{l \times m}$ such that $A = BC$.

Fact 2.10.8. Let $A, B \in \mathbb{F}^{n \times m}$, and let $C \in \mathbb{F}^{m \times l}$ be right invertible. Then, $\mathcal{R}(A) \subseteq \mathcal{R}(B)$ if and only if $\mathcal{R}(AC) \subseteq \mathcal{R}(BC)$. Furthermore, $\mathcal{R}(A) = \mathcal{R}(B)$ if and only if $\mathcal{R}(AC) = \mathcal{R}(BC)$.

Proof: Since C is right invertible, it follows that $\mathcal{R}(A) = \mathcal{R}(AC)$.

Fact 2.10.9. Let $A, B \in \mathbb{F}^{n \times n}$, and assume there exists $\alpha \in \mathbb{F}$ such that $\alpha A + B$ is nonsingular. Then, $\mathcal{N}(A) \cap \mathcal{N}(B) = \{0\}$.

Remark: The converse is not true. Let $A \triangleq \left[\begin{smallmatrix} 1 & 0 \\ 2 & 0 \end{smallmatrix}\right]$ and $B \triangleq \left[\begin{smallmatrix} 0 & 1 \\ 0 & 2 \end{smallmatrix}\right]$.

Fact 2.10.10. Let $A, B \in \mathbb{F}^{n \times m}$, and let $\alpha \in \mathbb{F}$ be nonzero. Then,

$$\mathcal{N}(A) \cap \mathcal{N}(B) = \mathcal{N}(A) \cap \mathcal{N}(A + \alpha B) = \mathcal{N}(\alpha A + B) \cap \mathcal{N}(B).$$

Remark: See Fact 2.11.3.

Fact 2.10.11. Let $x \in \mathbb{F}^n$ and $y \in \mathbb{F}^m$. If either $x = 0$ or $y \neq 0$, then
$$\mathcal{R}(xy^{\mathrm{T}}) = \mathcal{R}(x) = \mathrm{span}\,\{x\}.$$
Furthermore, if either $x \neq 0$ or $y = 0$, then
$$\mathcal{N}(xy^{\mathrm{T}}) = \mathcal{N}(y^{\mathrm{T}}) = \{\overline{y}\}^{\perp}.$$

Fact 2.10.12. Let $A \in \mathbb{F}^{n \times m}$ and $B \in \mathbb{F}^{m \times l}$. Then, $\mathrm{rank}\,AB = \mathrm{rank}\,A$ if and only if $\mathcal{R}(AB) = \mathcal{R}(A)$.

Proof: If $\mathcal{R}(AB) \subset \mathcal{R}(A)$ (note proper inclusion), then Lemma 2.3.4 implies that $\mathrm{rank}\,AB < \mathrm{rank}\,A$.

Fact 2.10.13. Let $A \in \mathbb{F}^{n \times m}$, $B \in \mathbb{F}^{m \times l}$, and $C \in \mathbb{F}^{l \times k}$. If $\mathrm{rank}\,AB = \mathrm{rank}\,B$, then $\mathrm{rank}\,ABC = \mathrm{rank}\,BC$.

Proof: $\mathrm{rank}\,B^{\mathrm{T}}A^{\mathrm{T}} = \mathrm{rank}\,B^{\mathrm{T}}$ implies that $\mathcal{R}(C^{\mathrm{T}}B^{\mathrm{T}}A^{\mathrm{T}}) = \mathcal{R}(C^{\mathrm{T}}B^{\mathrm{T}})$.

Fact 2.10.14. Let $A \in \mathbb{F}^{n \times m}$ and $B \in \mathbb{F}^{m \times l}$. Then, the following statements hold:

i) $\mathrm{rank}\,AB + \mathrm{def}\,A = \dim[\mathcal{N}(A) + \mathcal{R}(B)]$.

ii) $\mathrm{rank}\,AB + \dim[\mathcal{N}(A) \cap \mathcal{R}(B)] = \mathrm{rank}\,B$.

iii) $\mathrm{rank}\,AB + \dim[\mathcal{N}(A^*) \cap \mathcal{R}(B^*)] = \mathrm{rank}\,A$.

iv) $\mathrm{def}\,AB + \mathrm{rank}\,A + \dim[\mathcal{N}(A) + \mathcal{R}(B)] = l + m$.

v) $\mathrm{def}\,AB = \mathrm{def}\,B + \dim[\mathcal{N}(A) \cap \mathcal{R}(B)]$.

vi) $\mathrm{def}\,AB + m = \mathrm{def}\,A + \dim[\mathcal{N}(A^*) \cap \mathcal{R}(B^*)] + l$.

Remark: $\mathrm{rank}\,B - \mathrm{rank}\,AB = \dim[\mathcal{N}(A) \cap \mathcal{R}(B)] \leq \dim \mathcal{N}(A) = m - \mathrm{rank}\,A$ yields (2.5.18).

Fact 2.10.15. Let $A \in \mathbb{F}^{n \times m}$ and $B \in \mathbb{F}^{m \times l}$. Then,
$$\max\{\mathrm{def}\,A + l - m, \mathrm{def}\,B\} \leq \mathrm{def}\,AB \leq \mathrm{def}\,A + \mathrm{def}\,B.$$
If, in addition, $m = l$, then
$$\max\{\mathrm{def}\,A, \mathrm{def}\,B\} \leq \mathrm{def}\,AB.$$

Remark: The first inequality is *Sylvester's law of nullity*.

Fact 2.10.16. Let $A \in \mathbb{F}^{n \times m}$ and $B \in \mathbb{F}^{n \times p}$. Then, there exists a matrix $X \in \mathbb{F}^{m \times p}$ satisfying $AX = B$ and $\mathrm{rank}\,X = q$ if and only if
$$\mathrm{rank}\,B \leq q \leq \min\{m + \mathrm{rank}\,B - \mathrm{rank}\,A, p\}.$$

Proof: See [1386].

Fact 2.10.17. The following statements hold:

i) $\operatorname{rank} A \geq 0$ for all $A \in \mathbb{F}^{n \times m}$.

ii) $\operatorname{rank} A = 0$ if and only if $A = 0$.

iii) $\operatorname{rank} \alpha A = (\operatorname{sign} |\alpha|) \operatorname{rank} A$ for all $\alpha \in \mathbb{F}$ and $A \in \mathbb{F}^{n \times m}$.

iv) $\operatorname{rank}(A + B) \leq \operatorname{rank} A + \operatorname{rank} B$ for all $A, B \in \mathbb{F}^{n \times m}$.

Remark: Compare these conditions to the properties of a matrix norm given by Definition 9.2.1.

Fact 2.10.18. Let $n, m, k \in \mathbb{P}$. Then, $\operatorname{rank} 1_{n \times m} = 1$ and $1_{n \times n}^k = n^{k-1} 1_{n \times n}$.

Fact 2.10.19. Let $A \in \mathbb{F}^{n \times m}$. Then, $\operatorname{rank} A = 1$ if and only if there exist vectors $x \in \mathbb{F}^n$ and $y \in \mathbb{F}^m$ such that $x \neq 0$, $y \neq 0$, and $A = xy^{\mathrm{T}}$. In this case, $\operatorname{tr} A = y^{\mathrm{T}} x$.

Remark: See Fact 4.10.1.

Fact 2.10.20. Let $A \in \mathbb{F}^{n \times n}$, $k \geq 1$, and $l \in \mathbb{N}$. Then, the following statements hold:

i) $\mathcal{R}\left[(AA^*)^k\right] = \mathcal{R}\left[(AA^*)^l A\right]$.

ii) $\mathcal{N}\left[(A^*A)^k\right] = \mathcal{N}\left[A(A^*A)^l\right]$.

iii) $\operatorname{rank}(AA^*)^k = \operatorname{rank}(AA^*)^l A$.

iv) $\operatorname{def}(A^*A)^k = \operatorname{def} A(A^*A)^l$.

Fact 2.10.21. Let $A \in \mathbb{F}^{n \times m}$, and let $B \in \mathbb{F}^{m \times p}$. Then,

$$\operatorname{rank} AB = \operatorname{rank} A^*AB = \operatorname{rank} ABB^*.$$

Proof: See [1215, p. 37].

Fact 2.10.22. Let $A \in \mathbb{F}^{n \times n}$. Then,

$$2\operatorname{rank} A^2 \leq \operatorname{rank} A + \operatorname{rank} A^3.$$

Proof: See [400, p. 126] and consider a Jordan block of A.

Fact 2.10.23. Let $A \in \mathbb{F}^{n \times n}$. Then,

$$\operatorname{rank} A + \operatorname{rank}(A - A^3) = \operatorname{rank}(A + A^2) + \operatorname{rank}(A - A^2).$$

Consequently,

$$\operatorname{rank} A \leq \operatorname{rank}(A + A^2) + \operatorname{rank}(A - A^2),$$

and A is tripotent if and only if

$$\operatorname{rank} A = \operatorname{rank}(A + A^2) + \operatorname{rank}(A - A^2).$$

Proof: See [1340].

Remark: This result is due to Anderson and Styan.

Fact 2.10.24. Let $x, y \in \mathbb{F}^n$. Then,

$$\mathcal{R}(xy^{\mathrm{T}} + yx^{\mathrm{T}}) = \mathcal{R}([\begin{array}{cc} x & y \end{array}]),$$
$$\mathcal{N}(xy^{\mathrm{T}} + yx^{\mathrm{T}}) = \{x\}^{\perp} \cap \{y\}^{\perp},$$
$$\operatorname{rank}(xy^{\mathrm{T}} + yx^{\mathrm{T}}) \leq 2.$$

Furthermore, $\operatorname{rank}(xy^{\mathrm{T}} + yx^{\mathrm{T}}) = 1$ if and only if there exists $\alpha \in \mathbb{F}$ such that $x = \alpha y \neq 0$.

Remark: $xy^{\mathrm{T}} + yx^{\mathrm{T}}$ is a *doublet*. See [382, pp. 539, 540].

Fact 2.10.25. Let $A \in \mathbb{F}^{n \times m}$, $x \in \mathbb{F}^n$, and $y \in \mathbb{F}^m$. Then,

$$(\operatorname{rank} A) - 1 \leq \operatorname{rank}(A + xy^*) \leq (\operatorname{rank} A) + 1.$$

Remark: See Fact 6.4.2.

Fact 2.10.26. Let $A \triangleq \left[\begin{smallmatrix} 1 & 0 \\ 0 & 0 \end{smallmatrix}\right]$ and $B \triangleq \left[\begin{smallmatrix} 0 & 1 \\ 0 & 0 \end{smallmatrix}\right]$. Then, $\operatorname{rank} AB = 1$ and $\operatorname{rank} BA = 0$.

Remark: See Fact 3.7.30.

Fact 2.10.27. Let $A, B \in \mathbb{F}^{n \times m}$. Then,

$$|\operatorname{rank} A - \operatorname{rank} B| \leq \left\{ \begin{array}{c} \operatorname{rank}(A + B) \\ \operatorname{rank}(A - B) \end{array} \right\} \leq \operatorname{rank} A + \operatorname{rank} B.$$

If, in addition, $\operatorname{rank} B \leq k$, then

$$(\operatorname{rank} A) - k \leq \left\{ \begin{array}{c} \operatorname{rank}(A + B) \\ \operatorname{rank}(A - B) \end{array} \right\} \leq (\operatorname{rank} A) + k.$$

Fact 2.10.28. Let $A, B \in \mathbb{F}^{n \times m}$. Then, the following statements are equivalent:

i) $\operatorname{rank}(A + B) = \operatorname{rank} A + \operatorname{rank} B$.

ii) $\mathcal{R}(A) \cap \mathcal{R}(B) = \{0\}$ and $\mathcal{R}(A^{\mathrm{T}}) \cap \mathcal{R}(B^{\mathrm{T}}) = \{0\}$.

Proof: See [288].

Remark: See Fact 2.10.29.

Fact 2.10.29. Let $A, B \in \mathbb{F}^{n \times m}$, and assume that $A^*B = 0$ and $BA^* = 0$. Then,

$$\operatorname{rank}(A + B) = \operatorname{rank} A + \operatorname{rank} B.$$

Proof: Use Fact 2.11.4 and Proposition 6.1.6. See [347].

Remark: See Fact 2.10.28.

Fact 2.10.30. Let $A, B \in \mathbb{F}^{n \times m}$. Then, the following statements are equivalent:

i) $\operatorname{rank}(B - A) = \operatorname{rank} B - \operatorname{rank} A$.

ii) There exists $M \in \mathbb{F}^{m \times n}$ such that $A = BMB$ and $M = MBM$.

iii) There exists $M \in \mathbb{F}^{m \times n}$ such that $B = BMB$, $MA = 0$, and $AM = 0$.

iv) There exists $M \in \mathbb{F}^{m \times n}$ such that $A = AMA$, $MB = 0$, and $BM = 0$.

Proof: See [347].

Fact 2.10.31. Let $A, B, C \in \mathbb{F}^{n \times m}$, and assume that

$$\operatorname{rank}(B - A) = \operatorname{rank} B - \operatorname{rank} A$$

and

$$\operatorname{rank}(C - B) = \operatorname{rank} C - \operatorname{rank} B.$$

Then,

$$\operatorname{rank}(C - A) = \operatorname{rank} C - \operatorname{rank} A.$$

Proof: $\operatorname{rank}(C - A) \leq \operatorname{rank}(C - B) + \operatorname{rank}(B - A) = \operatorname{rank} C - \operatorname{rank} A$. Furthermore, $\operatorname{rank} C \leq \operatorname{rank}(C - A) + \operatorname{rank} A$, and thus $\operatorname{rank}(C - A) \geq \operatorname{rank} C - \operatorname{rank} A$. Alternatively, use Fact 2.10.30.

Remark: This result is due to [664].

Fact 2.10.32. Let $A, B \in \mathbb{F}^{n \times m}$, and define

$$A \overset{\mathrm{rs}}{\leq} B$$

if and only if

$$\operatorname{rank}(B - A) = \operatorname{rank} B - \operatorname{rank} A.$$

Then, "$\overset{\mathrm{rs}}{\leq}$" is a partial ordering on $\mathbb{F}^{n \times m}$.

Proof: Use Fact 2.10.31.

Remark: The relation "$\overset{\mathrm{rs}}{\leq}$" is the *rank subtractivity partial ordering.*

Remark: See Fact 8.20.4.

Fact 2.10.33. Let $A, B \in \mathbb{F}^{n \times m}$, and assume that the following conditions hold:

i) $A^*A = A^*B$.

ii) $AA^* = BA^*$.

iii) $B^*B = B^*A$.

iv) $BB^* = AB^*$.

Then, $A = B$.

Proof: See [669].

Fact 2.10.34. Let $A, B, C \in \mathbb{F}^{n \times m}$, and assume that the following conditions hold:

i) $A^*A = A^*B$.

ii) $AA^* = BA^*$.

iii) $B^*B = B^*C$.

iv) $BB^* = CB^*$.

Then, the following conditions hold:

v) $A^*A = A^*C$.

vi) $AA^* = CA^*$.

Proof: See [669].

Fact 2.10.35. Let $A, B \in \mathbb{F}^{n \times m}$. Then,

$$A \overset{*}{\leq} B$$

if and only if

$$A^*A = A^*B$$

and

$$AA^* = BA^*.$$

Then, "$\overset{*}{\leq}$" is a partial ordering on $\mathbb{F}^{n \times m}$.

Proof: Use Fact 2.10.33 and Fact 2.10.34.

Remark: The relation "$\overset{*}{\leq}$" is the *star partial ordering*. See [114, 669].

Remark: See Fact 8.20.6.

Fact 2.10.36. Let $A, B \in \mathbb{F}^{n \times n}$, and assume that $A \overset{*}{\leq} B$ and $AB = BA$. Then, $A^2 \overset{*}{\leq} B^2$.

Proof: See [109].

Remark: See Fact 8.20.4.

2.11 Facts on the Range, Rank, Null Space, and Defect of Partitioned Matrices

Fact 2.11.1. Let $A \in \mathbb{F}^{n \times m}$ and $B \in \mathbb{F}^{n \times l}$. Then,

$$\mathcal{R}(\begin{bmatrix} A & B \end{bmatrix}) = \mathcal{R}(A) + \mathcal{R}(B).$$

Consequently,

$$\operatorname{rank} \begin{bmatrix} A & B \end{bmatrix} = \dim[\mathcal{R}(A) + \mathcal{R}(B)].$$

Furthermore, the followings statements are equivalent:

i) $\operatorname{rank} \begin{bmatrix} A & B \end{bmatrix} = n$.

ii) $\operatorname{def} \begin{bmatrix} A^* \\ B^* \end{bmatrix} = 0$.

iii) $\mathcal{N}(A^*) \cap \mathcal{N}(B^*) = \{0\}$.

Fact 2.11.2. Let $A \in \mathbb{F}^{n \times m}$ and $B \in \mathbb{F}^{l \times m}$. Then,

$$\text{rank} \begin{bmatrix} A \\ B \end{bmatrix} = \dim [\mathcal{R}(A^*) + \mathcal{R}(B^*)].$$

Proof: Use Fact 2.11.1.

Fact 2.11.3. Let $A \in \mathbb{F}^{n \times m}$ and $B \in \mathbb{F}^{l \times m}$. Then,

$$\mathcal{N}\left(\begin{bmatrix} A \\ B \end{bmatrix} \right) = \mathcal{N}(A) \cap \mathcal{N}(B).$$

Consequently,

$$\text{def} \begin{bmatrix} A \\ B \end{bmatrix} = \dim[\mathcal{N}(A) \cap \mathcal{N}(B)].$$

Furthermore, the followings statements are equivalent:

i) $\text{rank} \begin{bmatrix} A \\ B \end{bmatrix} = m.$

ii) $\text{def} \begin{bmatrix} A \\ B \end{bmatrix} = 0.$

iii) $\mathcal{N}(A) \cap \mathcal{N}(B) = \{0\}.$

Remark: See Fact 2.10.10.

Fact 2.11.4. Let $A, B \in \mathbb{F}^{n \times m}$. Then, the following statements are equivalent:

i) $\text{rank}(A + B) = \text{rank}\, A + \text{rank}\, B.$

ii) $\text{rank} \begin{bmatrix} A & B \end{bmatrix} = \text{rank} \begin{bmatrix} A \\ B \end{bmatrix} = \text{rank}\, A + \text{rank}\, B.$

iii) $\dim[\mathcal{R}(A) \cap \mathcal{R}(B)] = \dim[\mathcal{R}(A^*) \cap \mathcal{R}(B^*)] = 0.$

iv) $\mathcal{R}(A) \cap \mathcal{R}(B) = \mathcal{R}(A^*) \cap \mathcal{R}(B^*) = \{0\}.$

v) There exists a matrix $C \in \mathbb{F}^{m \times n}$ such that $ACA = A$, $CB = 0$, and $BC = 0$.

Proof: See [347, 993].

Remark: Additional conditions are given by Fact 6.4.35 under the assumption that $A + B$ is nonsingular.

Fact 2.11.5. Let $A \in \mathbb{F}^{n \times m}$ and $B \in \mathbb{F}^{n \times l}$. Then,

$$\mathcal{R}(A) = \mathcal{R}(B)$$

if and only if

$$\text{rank}\, A = \text{rank}\, B = \text{rank} \begin{bmatrix} A & B \end{bmatrix}.$$

Fact 2.11.6. Let $A \in \mathbb{F}^{n \times m}$, and let $A_0 \in \mathbb{F}^{k \times l}$ be a submatrix of A. Then,

$$\text{rank}\, A_0 \leq \text{rank}\, A.$$

Fact 2.11.7. Let $A \in \mathbb{F}^{n \times m}$, $B \in \mathbb{F}^{k \times m}$, $C \in \mathbb{F}^{m \times l}$, and $D \in \mathbb{F}^{m \times p}$, and assume that

$$\mathrm{rank} \begin{bmatrix} A \\ B \end{bmatrix} = \mathrm{rank}\, A$$

and

$$\mathrm{rank} \begin{bmatrix} C & D \end{bmatrix} = \mathrm{rank}\, C.$$

Then,

$$\mathrm{rank} \begin{bmatrix} A \\ B \end{bmatrix} \begin{bmatrix} C & D \end{bmatrix} = \mathrm{rank}\, AC.$$

Proof: Use i) of Fact 2.10.14.

Fact 2.11.8. Let $A \in \mathbb{F}^{n \times m}$ and $B \in \mathbb{F}^{n \times l}$. Then,

$$\max\{\mathrm{rank}\, A, \mathrm{rank}\, B\} \le \mathrm{rank} \begin{bmatrix} A & B \end{bmatrix}$$
$$= \mathrm{rank}\, A + \mathrm{rank}\, B - \dim[\mathcal{R}(A) \cap \mathcal{R}(B)]$$
$$\le \mathrm{rank}\, A + \mathrm{rank}\, B$$

and

$$\mathrm{def}\, A + \mathrm{def}\, B \le \mathrm{def} \begin{bmatrix} A & B \end{bmatrix}$$
$$= \mathrm{def}\, A + \mathrm{def}\, B + \dim[\mathcal{R}(A) \cap \mathcal{R}(B)]$$
$$\le \min\{l + \mathrm{def}\, A, m + \mathrm{def}\, B\}.$$

If, in addition, $A^*B = 0$, then

$$\mathrm{rank} \begin{bmatrix} A & B \end{bmatrix} = \mathrm{rank}\, A + \mathrm{rank}\, B$$

and

$$\mathrm{def} \begin{bmatrix} A & B \end{bmatrix} = \mathrm{def}\, A + \mathrm{def}\, B.$$

Proof: To prove the first equality, use Theorem 2.3.1 and Fact 2.11.1. For the case $A^*B = 0$, note that

$$\mathrm{rank} \begin{bmatrix} A & B \end{bmatrix} = \mathrm{rank} \begin{bmatrix} A^* \\ B^* \end{bmatrix} \begin{bmatrix} A & B \end{bmatrix} = \begin{bmatrix} A^*A & 0 \\ 0 & B^*B \end{bmatrix}$$
$$= \mathrm{rank}\, A^*A + \mathrm{rank}\, B^*B = \mathrm{rank}\, A + \mathrm{rank}\, B.$$

Remark: See Fact 6.5.6 and Fact 6.4.49.

Fact 2.11.9. Let $A \in \mathbb{F}^{n \times m}$ and $B \in \mathbb{F}^{n \times l}$. Then,

$$\mathrm{rank} \begin{bmatrix} A & B \end{bmatrix} + \dim[\mathcal{R}(A) \cap \mathcal{R}(B)] = \mathrm{rank}\, A + \mathrm{rank}\, B.$$

Proof: Use Theorem 2.3.1 and Fact 2.11.1.

Fact 2.11.10. Let $A \in \mathbb{F}^{n \times m}$ and $B \in \mathbb{F}^{l \times m}$. Then,

$$\mathrm{rank} \begin{bmatrix} A \\ B \end{bmatrix} + \dim[\mathcal{R}(A^*) \cap \mathcal{R}(B^*)] = \mathrm{rank}\, A + \mathrm{rank}\, B.$$

Proof: Use Fact 2.11.9.

Fact 2.11.11. Let $A \in \mathbb{F}^{n \times m}$ and $B \in \mathbb{F}^{l \times m}$. Then,

$$\max\{\operatorname{rank} A, \operatorname{rank} B\} \leq \operatorname{rank} \left[\begin{array}{c} A \\ B \end{array} \right]$$
$$= \operatorname{rank} A + \operatorname{rank} B - \dim[\mathcal{R}(A^*) \cap \mathcal{R}(B^*)]$$
$$\leq \operatorname{rank} A + \operatorname{rank} B$$

and

$$\operatorname{def} A - \operatorname{rank} B \leq \operatorname{def} A - \operatorname{rank} B + \dim[\mathcal{R}(A^*) \cap \mathcal{R}(B^*)]$$
$$= \operatorname{def} \left[\begin{array}{c} A \\ B \end{array} \right]$$
$$\leq \min\{\operatorname{def} A, \operatorname{def} B\}.$$

If, in addition, $AB^* = 0$, then

$$\operatorname{rank} \left[\begin{array}{c} A \\ B \end{array} \right] = \operatorname{rank} A + \operatorname{rank} B$$

and

$$\operatorname{def} \left[\begin{array}{c} A \\ B \end{array} \right] = \operatorname{def} A - \operatorname{rank} B.$$

Proof: Use Fact 2.11.8 and Fact 2.9.21.

Remark: See Fact 6.5.6.

Fact 2.11.12. Let $A, B \in \mathbb{F}^{n \times m}$. Then,

$$\left. \begin{array}{c} \max\{\operatorname{rank} A, \operatorname{rank} B\} \\ \\ \operatorname{rank}(A + B) \end{array} \right\} \leq \left\{ \begin{array}{c} \operatorname{rank} \left[\begin{array}{cc} A & B \end{array} \right] \\ \\ \operatorname{rank} \left[\begin{array}{c} A \\ B \end{array} \right] \end{array} \right\} \leq \operatorname{rank} A + \operatorname{rank} B$$

and

$$\operatorname{def} A - \operatorname{rank} B \leq \left\{ \begin{array}{c} \operatorname{def} \left[\begin{array}{cc} A & B \end{array} \right] - m \\ \\ \operatorname{def} \left[\begin{array}{c} A \\ B \end{array} \right] \end{array} \right\} \leq \left\{ \begin{array}{c} \min\{\operatorname{def} A, \operatorname{def} B\} \\ \\ \operatorname{def}(A + B). \end{array} \right.$$

Proof: $\operatorname{rank}(A + B) = \operatorname{rank} \left[\begin{array}{cc} A & B \end{array} \right] \left[\begin{array}{c} I \\ I \end{array} \right] \leq \operatorname{rank} \left[\begin{array}{cc} A & B \end{array} \right]$, and $\operatorname{rank}(A + B) = \operatorname{rank} \left[\begin{array}{cc} I & I \end{array} \right] \left[\begin{array}{c} A \\ B \end{array} \right] \leq \operatorname{rank} \left[\begin{array}{c} A \\ B \end{array} \right]$.

Fact 2.11.13. Let $A \in \mathbb{F}^{n \times m}$, $B \in \mathbb{F}^{l \times k}$, and $C \in \mathbb{F}^{l \times m}$. Then,

$$\operatorname{rank} A + \operatorname{rank} B = \operatorname{rank} \left[\begin{array}{cc} A & 0 \\ 0 & B \end{array} \right] \leq \operatorname{rank} \left[\begin{array}{cc} A & 0 \\ C & B \end{array} \right]$$

and

$$\operatorname{rank} A + \operatorname{rank} B = \operatorname{rank} \left[\begin{array}{cc} 0 & A \\ B & 0 \end{array} \right] \leq \operatorname{rank} \left[\begin{array}{cc} 0 & A \\ B & C \end{array} \right].$$

Finally, let $D \in \mathbb{F}^{k \times m}$ and $E \in \mathbb{F}^{l \times n}$. Then,

$$\operatorname{rank} A + \operatorname{rank} B = \operatorname{rank} \begin{bmatrix} A & 0 \\ BD + EA & B \end{bmatrix} = \operatorname{rank} \begin{bmatrix} 0 & A \\ B & BD + EA \end{bmatrix}.$$

Fact 2.11.14. Let $A \in \mathbb{F}^{n \times m}$, $B \in \mathbb{F}^{m \times l}$, and $C \in \mathbb{F}^{l \times k}$. Then,

$$\operatorname{rank} AB + \operatorname{rank} BC \leq \operatorname{rank} \begin{bmatrix} 0 & AB \\ BC & B \end{bmatrix} = \operatorname{rank} B + \operatorname{rank} ABC.$$

Consequently,

$$\operatorname{rank} AB + \operatorname{rank} BC - \operatorname{rank} B \leq \operatorname{rank} ABC.$$

Furthermore, the following statements are equivalent:

i) $\operatorname{rank} \begin{bmatrix} 0 & AB \\ BC & B \end{bmatrix} = \operatorname{rank} AB + \operatorname{rank} BC.$

ii) $\operatorname{rank} AB + \operatorname{rank} BC - \operatorname{rank} B = \operatorname{rank} ABC.$

iii) There exist $X \in \mathbb{F}^{k \times l}$ and $Y \in \mathbb{F}^{m \times n}$ such that

$$BCX + YAB = B.$$

Remark: This result is the *Frobenius inequality*.

Proof: Use Fact 2.11.13 and $\begin{bmatrix} 0 & AB \\ BC & B \end{bmatrix} = \begin{bmatrix} I & A \\ 0 & I \end{bmatrix} \begin{bmatrix} -ABC & 0 \\ 0 & B \end{bmatrix} \begin{bmatrix} I & 0 \\ C & I \end{bmatrix}$. The last statement follows from Fact 5.10.21. See [1339, 1340].

Remark: See Fact 6.5.15 for the case of equality.

Fact 2.11.15. Let $A, B \in \mathbb{F}^{n \times m}$. Then,

$$\operatorname{rank} \begin{bmatrix} A & B \end{bmatrix} + \operatorname{rank} \begin{bmatrix} A \\ B \end{bmatrix} \leq \operatorname{rank} \begin{bmatrix} 0 & A & B \\ A & A & 0 \\ B & 0 & B \end{bmatrix}$$

$$= \operatorname{rank} A + \operatorname{rank} B + \operatorname{rank}(A + B).$$

Proof: Use the Frobenius inequality with $A \triangleq C^{\mathrm{T}} \triangleq \begin{bmatrix} I & I \end{bmatrix}$ and with B replaced by $\begin{bmatrix} A & 0 \\ 0 & B \end{bmatrix}$.

Fact 2.11.16. Let $A \in \mathbb{F}^{n \times m}$, $B \in \mathbb{F}^{n \times l}$, and $C \in \mathbb{F}^{n \times k}$. Then,

$$\operatorname{rank} \begin{bmatrix} A & B & C \end{bmatrix} \leq \operatorname{rank} \begin{bmatrix} A & B \end{bmatrix} + \operatorname{rank} \begin{bmatrix} B & C \end{bmatrix} - \operatorname{rank} B$$

$$\leq \operatorname{rank} \begin{bmatrix} A & B \end{bmatrix} + \operatorname{rank} C$$

$$\leq \operatorname{rank} A + \operatorname{rank} B + \operatorname{rank} C.$$

Proof: See [962].

Fact 2.11.17. Let $A \in \mathbb{F}^{n \times m}$ and $B \in \mathbb{F}^{k \times l}$, and assume that B is a submatrix of A. Then,

$$k + l - \operatorname{rank} B \leq n + m - \operatorname{rank} A.$$

Proof: See [138].

Fact 2.11.18. Let $A \in \mathbb{F}^{n \times m}$ and $B \in \mathbb{F}^{m \times n}$. Then,

$$
\begin{bmatrix} I_n & I_n - AB \\ B & 0 \end{bmatrix} = \begin{bmatrix} I_n & A \\ 0 & I_m \end{bmatrix} \begin{bmatrix} 0 & I_n - AB \\ B & 0 \end{bmatrix} \begin{bmatrix} I_n & 0 \\ I_n & I_n \end{bmatrix}
$$

$$
= \begin{bmatrix} I_n & 0 \\ B & I_m \end{bmatrix} \begin{bmatrix} I_n & 0 \\ 0 & BAB - B \end{bmatrix} \begin{bmatrix} I_n & I_n - AB \\ 0 & I_m \end{bmatrix}.
$$

Hence,

$$
\text{rank} \begin{bmatrix} I_n & I_n - AB \\ B & 0 \end{bmatrix} = \text{rank } B + \text{rank}(I_n - AB) = n + \text{rank}(BAB - B).
$$

Remark: See Fact 2.14.7.

Fact 2.11.19. Let $A \in \mathbb{F}^{n \times m}$ and $B \in \mathbb{F}^{m \times n}$. Then,

$$
\begin{bmatrix} A & AB \\ BA & B \end{bmatrix} = \begin{bmatrix} I_n & 0 \\ B & I_m \end{bmatrix} \begin{bmatrix} A & 0 \\ 0 & B - BAB \end{bmatrix} \begin{bmatrix} I_m & B \\ 0 & I_n \end{bmatrix}
$$

$$
= \begin{bmatrix} I_n & A \\ 0 & I_m \end{bmatrix} \begin{bmatrix} A - ABA & 0 \\ 0 & B \end{bmatrix} \begin{bmatrix} I_m & 0 \\ A & I_n \end{bmatrix}.
$$

Hence,

$$
\text{rank} \begin{bmatrix} A & AB \\ BA & B \end{bmatrix} = \text{rank } A + \text{rank}(B - BAB) = \text{rank } B + \text{rank}(A - ABA).
$$

Remark: See Fact 2.14.10.

Fact 2.11.20. Let $\left[\begin{smallmatrix} A & B \\ C & D \end{smallmatrix}\right] \in \mathbb{F}^{(n_1+n_2) \times (m_1+m_2)}$, assume that $\left[\begin{smallmatrix} A & B \\ C & D \end{smallmatrix}\right]$ is nonsingular, and define $\left[\begin{smallmatrix} E & F \\ G & H \end{smallmatrix}\right] \in \mathbb{F}^{(m_1+m_2) \times (n_1+n_2)}$ by

$$
\begin{bmatrix} E & F \\ G & H \end{bmatrix} \triangleq \begin{bmatrix} A & B \\ C & D \end{bmatrix}^{-1}.
$$

Then,

$$
\text{def } A = \text{def } H,
$$
$$
\text{def } B = \text{def } F,
$$
$$
\text{def } C = \text{def } G,
$$
$$
\text{def } D = \text{def } E.
$$

More generally, if U and V are complementary submatrices of a matrix and its inverse, then $\text{def } U = \text{def } V$.

Proof: See [1273, 1398] and [1399, p. 38].

Remark: U and V are *complementary submatrices* if the row numbers not used to create U are the column numbers used to create V, and the column numbers not used to create U are the row numbers used to create V.

Remark: Note the sizes of the matrix blocks, which differ from the sizes in Fact 2.14.28.

Remark: This result is the *nullity theorem*. A history of this result is given in [1273]. See Fact 3.19.1.

Fact 2.11.21. Let $A \in \mathbb{F}^{n \times n}$, assume that A is nonsingular, and let $\mathcal{S}_1, \mathcal{S}_2 \subseteq \{1, \ldots, n\}$. Then,

$$\operatorname{rank}(A^{-1})_{(\mathcal{S}_1, \mathcal{S}_2)} = \operatorname{rank} A_{(\mathcal{S}_2^\sim, \mathcal{S}_1^\sim)} + \operatorname{card}(\mathcal{S}_1) + \operatorname{card}(\mathcal{S}_2) - n.$$

Proof: See [1399, p. 40].

Remark: See Fact 2.11.22 and Fact 2.13.6.

Fact 2.11.22. Let $A \in \mathbb{F}^{n \times n}$, assume that A is nonsingular, and let $\mathcal{S} \subseteq \{1, \ldots, n\}$. Then,
$$\operatorname{rank}(A^{-1})_{(\mathcal{S}, \mathcal{S}\sim)} = \operatorname{rank} A_{(\mathcal{S}, \mathcal{S}\sim)}.$$

Proof: Apply Fact 2.11.21 with $\mathcal{S}_2 = \mathcal{S}_1^\sim$.

2.12 Facts on the Inner Product, Outer Product, Trace, and Matrix Powers

Fact 2.12.1. Let $x, y, z \in \mathbb{F}^n$, and assume that $x^*x = y^*y = z^*z = 1$. Then,
$$\sqrt{1 - |x^*y|^2} \leq \sqrt{1 - |x^*z|^2} + \sqrt{1 - |z^*y|^2}.$$

Equality holds if and only if there exists $\alpha \in \mathbb{F}$ such that either $z = \alpha x$ or $z = \alpha y$.

Proof: See [1526, p. 155].

Remark: See Fact 3.11.24.

Fact 2.12.2. Let $x, y \in \mathbb{F}^n$. Then, $x^*x = y^*y$ and $\operatorname{Im} x^*y = 0$ if and only if $x - y$ is orthogonal to $x + y$.

Fact 2.12.3. Let $x, y \in \mathbb{R}^n$. Then, $xx^{\mathrm{T}} = yy^{\mathrm{T}}$ if and only if either $x = y$ or $x = -y$.

Fact 2.12.4. Let $x, y \in \mathbb{R}^n$. Then, $xy^{\mathrm{T}} = yx^{\mathrm{T}}$ if and only if x and y are linearly dependent.

Fact 2.12.5. Let $x, y \in \mathbb{R}^n$. Then, $xy^{\mathrm{T}} = -yx^{\mathrm{T}}$ if and only if either $x = 0$ or $y = 0$.

Proof: If $x_{(i)} \neq 0$ and $y_{(j)} \neq 0$, then $x_{(j)} = y_{(i)} = 0$ and $0 \neq x_{(i)}y_{(j)} \neq x_{(j)}y_{(i)} = 0$.

Fact 2.12.6. Let $x, y \in \mathbb{R}^n$. Then, $yx^{\mathrm{T}} + xy^{\mathrm{T}} = y^{\mathrm{T}}yxx^{\mathrm{T}}$ if and only if either $x = 0$ or $y = \frac{1}{2}y^{\mathrm{T}}yx$.

Fact 2.12.7. Let $x, y \in \mathbb{F}^n$. Then,

$$(xy^*)^r = (y^*x)^{r-1}xy^*.$$

Fact 2.12.8. Let $x_1, \ldots, x_k \in \mathbb{F}^n$, and let $y_1, \ldots, y_k \in \mathbb{F}^m$. Then, the following statements are equivalent:

i) $x_1, \ldots, x_k$ are linearly independent, and $y_1, \ldots, y_k$ are linearly independent.

ii) $\mathcal{R}\left(\sum_{i=1}^{k} x_i y_i^{\mathrm{T}}\right) = k$.

Proof: See [382, p. 537].

Fact 2.12.9. Let $A, B, C \in \mathbb{R}^{2 \times 2}$. Then,

$$\operatorname{tr}(ABC + ACB) + (\operatorname{tr} A)(\operatorname{tr} B)\operatorname{tr} C$$
$$= (\operatorname{tr} A)\operatorname{tr} BC + (\operatorname{tr} B)\operatorname{tr} AC + (\operatorname{tr} C)\operatorname{tr} AB.$$

Proof: See [275, p. 330].

Remark: See Fact 4.9.3.

Fact 2.12.10. Let $A \in \mathbb{F}^{n \times m}$ and $B \in \mathbb{F}^{l \times k}$. Then,

$$A E_{i,j,m \times l} B = \operatorname{col}_i(A)\operatorname{row}_j(B).$$

Fact 2.12.11. Let $A \in \mathbb{F}^{n \times m}$, $B \in \mathbb{F}^{m \times l}$, and $C \in \mathbb{F}^{l \times n}$. Then,

$$\operatorname{tr} ABC = \sum_{i=1}^{n} \operatorname{row}_i(A) B \operatorname{col}_i(C).$$

Fact 2.12.12. Let $A \in \mathbb{F}^{n \times m}$. Then, the following statements are equivalent:

i) $A = 0$.

ii) $Ax = 0$ for all $x \in \mathbb{F}^m$.

iii) $\operatorname{tr} AA^* = 0$.

Fact 2.12.13. Let $A \in \mathbb{F}^{n \times n}$ and $k \geq 1$. Then,

$$\operatorname{Re} \operatorname{tr} A^{2k} \leq \operatorname{tr} A^k A^{k*} \leq \operatorname{tr}(AA^*)^k.$$

Remark: To prove the left-hand inequality, consider $\operatorname{tr}(A^k - A^{k*})(A^{k*} - A^k)$. For the right-hand inequality when $k = 2$, consider $\operatorname{tr}(AA^* - A^*A)^2$.

Fact 2.12.14. Let $A \in \mathbb{F}^{n \times n}$. Then, $\operatorname{tr} A^k = 0$ for all $k \in \{1, \ldots, n\}$ if and only if $A^n = 0$.

Proof: For sufficiency, Fact 4.10.8 implies that $\operatorname{spec}(A) = \{0\}$, and thus the Jordan form of A is a block-diagonal matrix each of whose diagonally located blocks is a standard nilpotent matrix. For necessity, see [1526, p. 112].

Fact 2.12.15. Let $A \in \mathbb{F}^{n \times n}$, and assume that $\operatorname{tr} A = 0$. If $A^2 = A$, then $A = 0$. If $A^k = A$, where $k \geq 4$ and $2 \leq n < p$, where p is the smallest prime divisor of $k - 1$, then $A = 0$.

Proof: See [352].

Fact 2.12.16. Let $A, B \in \mathbb{F}^{n \times n}$. Then,

$$\operatorname{Re}\operatorname{tr} AB \leq \tfrac{1}{2}\operatorname{tr}(AA^* + BB^*).$$

Proof: See [748].

Remark: See Fact 8.12.20.

Fact 2.12.17. Let $A, B \in \mathbb{F}^{n \times n}$, and assume that $AB = 0$. Then, for all $k \geq 1$,

$$\operatorname{tr}(A + B)^k = \operatorname{tr} A^k + \operatorname{tr} B^k.$$

Fact 2.12.18. Let $A \in \mathbb{R}^{n \times n}$, let $x, y \in \mathbb{R}^n$, and let $k \geq 1$. Then,

$$\left(A + xy^{\mathrm{T}}\right)^k = A^k + B\hat{I}_k C^{\mathrm{T}},$$

where

$$B \triangleq \begin{bmatrix} x & Ax & \cdots & A^{k-1}x \end{bmatrix}$$

and

$$C \triangleq \begin{bmatrix} y & (A^{\mathrm{T}} + yx^{\mathrm{T}})y & \cdots & (A^{\mathrm{T}} + yx^{\mathrm{T}})^k y \end{bmatrix}.$$

Proof: See [196].

Fact 2.12.19. Let $A, B \in \mathbb{F}^{n \times n}$. Then, the following statements hold:

i) $AB + BA = \tfrac{1}{2}[(A + B)^2 - (A - B)^2].$

ii) $(A + B)(A - B) = A^2 - B^2 - [A, B].$

iii) $(A - B)(A + B) = A^2 - B^2 + [A, B].$

iv) $A^2 - B^2 = \tfrac{1}{2}[(A + B)(A - B) + (A - B)(A + B)].$

Fact 2.12.20. Let $A, B \in \mathbb{F}^{n \times n}$, and let k be a positive integer. Then,

$$A^k - B^k = \sum_{i=0}^{k-1} A^i(A - B)B^{k-1-i} = \sum_{i=1}^{k} A^{k-i}(A - B)B^{i-1}.$$

Fact 2.12.21. Let $A \in \mathbb{F}^{n \times n}$, $B \in \mathbb{F}^{n \times m}$, and $C \in \mathbb{F}^{m \times m}$, and let $k \geq 1$. Then,

$$\begin{bmatrix} A & B \\ 0 & C \end{bmatrix}^k = \begin{bmatrix} A^k & \sum_{i=1}^{k} A^{k-i}BC^{i-1} \\ 0 & C^k \end{bmatrix}.$$

Fact 2.12.22. Let $A, B \in \mathbb{F}^{n \times n}$, and define $\mathcal{A} \triangleq \begin{bmatrix} A & A \\ A & A \end{bmatrix}$ and $\mathcal{B} \triangleq \begin{bmatrix} B & -B \\ -B & B \end{bmatrix}$. Then,

$$\mathcal{A}\mathcal{B} = \mathcal{B}\mathcal{A} = 0.$$

Fact 2.12.23. A cube root of I_2 is given by

$$\begin{bmatrix} -\tfrac{1}{2} & \tfrac{\sqrt{3}}{2} \\ \tfrac{-\sqrt{3}}{2} & -\tfrac{1}{2} \end{bmatrix}^3 = \begin{bmatrix} -1 & -1 \\ 1 & 0 \end{bmatrix}^3 = I_2.$$

Fact 2.12.24. Let n be an integer, and define

$$\begin{bmatrix} a_n \\ b_n \\ c_n \end{bmatrix} \triangleq \begin{bmatrix} 63 & 104 & -68 \\ 64 & 104 & -67 \\ 80 & 131 & -85 \end{bmatrix}^n \begin{bmatrix} 1 \\ 2 \\ 2 \end{bmatrix}.$$

Then,

$$\sum_{n=0}^{\infty} a_n x^n = \frac{1 + 53x + 9x^2}{1 - 82x - 82x^2 + x^3},$$

$$\sum_{n=0}^{\infty} b_n x^n = \frac{2 - 26x - 12x^2}{1 - 82x - 82x^2 + x^3},$$

$$\sum_{n=0}^{\infty} c_n x^n = \frac{2 + 8x - 10x^2}{1 - 82x - 82x^2 + x^3},$$

and

$$a_n^3 + b_n^3 = c_n^3 + (-1)^n.$$

Remark: This result is due to Ramanujan. See [647].

Remark: The last equality holds for all integers, not necessarily positive.

2.13 Facts on the Determinant

Fact 2.13.1. Let n be a positive integer. Then,

$$\det \hat{I}_n = (-1)^{\lfloor n/2 \rfloor} = (-1)^{n(n-1)/2}.$$

Consequently, $\hat{I}_n$ is an (even, odd) permutation matrix if and only if $\frac{1}{2}n(n-1)$ is (even, odd).

Proof: Since $\hat{I}_n$ is a permutation matrix, its determinant reflects whether it permutes the components of a vector in an odd or even manner, which reflects the parity of pairwise component swaps that it performs on a vector. The total number of swaps performed by a permutation matrix is given by the sum over all rows of the number of 1's in successive rows that are to the left of the 1 in each row. For $\hat{I}_n$, this number is given by $(n-1) + (n-2) + \cdots + 2 + 1 = n(n-1)/2$. See [803, pp. 29–32].

Remark: See Fact 3.11.16 and Fact 3.23.2.

Fact 2.13.2. Let n be a positive integer. Then,

$$\det P_n = (-1)^{n-1}.$$

Consequently, P_n is an (even, odd) permutation matrix if and only if n is (odd, even).

Remark: See Fact 3.11.16 and Fact 5.16.7.

Fact 2.13.3. $\det(I_n + \alpha 1_{n \times n}) = 1 + \alpha n$.

Fact 2.13.4. Let $A \in \mathbb{F}^{n \times m}$, let $B \in \mathbb{F}^{m \times n}$, and assume that $m < n$. Then, $\det AB = 0$.

Fact 2.13.5. Let $A \in \mathbb{F}^{n \times m}$, let $B \in \mathbb{F}^{m \times n}$, and assume that $n \leq m$. Then,

$$\det AB = \sum_{1 \leq i_1 < \cdots < i_n \leq m} \det A_{(\{1,\ldots,n\},\{i_1,\ldots,i_n\})} \det B_{(\{i_1,\ldots,i_n\},\{1,\ldots,n\})}.$$

Proof: See [459, p. 102].

Remark: $\det AB$ is equal to the sum of all $\binom{m}{n}$ products of pairs of subdeterminants of A and B formed by choosing n columns of A and the corresponding n rows of B.

Remark: This equality is a special case of the Binet-Cauchy formula given by Fact 7.5.17. The special case $n = m$ is given by Proposition 2.7.1.

Remark: Determinantal and minor equalities are given in [276, 905].

Remark: See Fact 2.14.8.

Fact 2.13.6. Let $A \in \mathbb{F}^{n \times n}$, assume that A is nonsingular, let $\mathcal{S}_1, \mathcal{S}_2 \subseteq \{1,\ldots,n\}$, and assume that $\text{card}(\mathcal{S}_1) = \text{card}(\mathcal{S}_2)$. Then,

$$\left| \det (A^{-1})_{(\mathcal{S}_1,\mathcal{S}_2)} \right| = \frac{\left| \det A_{(\mathcal{S}_{\tilde{1}},\mathcal{S}_{\tilde{2}})} \right|}{|\det A|}.$$

Proof: See [1399, p. 38].

Remark: In the case $\text{card}(\mathcal{S}_1) = \text{card}(\mathcal{S}_2) = 1$, this result yields the absolute value of (2.7.24).

Remark: See Fact 2.11.21.

Fact 2.13.7. Let $A \in \mathbb{F}^{n \times n}$, assume that A is nonsingular, and let $b \in \mathbb{F}^n$. Then, the solution $x \in \mathbb{F}^n$ of $Ax = b$ is given by

$$x = \begin{bmatrix} \dfrac{\det \left(A \overset{1}{\leftarrow} b \right)}{\det A} \\ \vdots \\ \dfrac{\det \left(A \overset{n}{\leftarrow} b \right)}{\det A} \end{bmatrix}.$$

Proof: Note that $A\left(I \overset{i}{\leftarrow} x \right) = A \overset{i}{\leftarrow} b$. Since $\det \left(I \overset{i}{\leftarrow} x \right) = x_{(i)}$, it follows that $(\det A)x_{(i)} = \det \left(A \overset{i}{\leftarrow} b \right)$.

Remark: This equality is *Cramer's rule*. See Fact 2.13.8 for extensions to non-square A.

Fact 2.13.8. Let $A \in \mathbb{F}^{n \times m}$ be right invertible, and let $b \in \mathbb{F}^n$. Then, a solution $x \in \mathbb{F}^m$ of $Ax = b$ is given by

$$x_{(i)} = \frac{\det\left[\left(A \overset{i}{\leftarrow} b\right)A^*\right] - \det\left[\left(A \overset{i}{\leftarrow} 0\right)A^*\right]}{\det(AA^*)},$$

for all $i \in \{1, \ldots, m\}$.

Proof: See [887].

Remark: This result is a generalization of Cramer's rule. See Fact 2.13.7. Extensions to generalized inverses are given in [182, 777, 880] and [1430, Chapter 3].

Fact 2.13.9. Let $A \in \mathbb{F}^{n \times n}$, and assume that either $A_{(i,j)} = 0$ for all i, j such that $i + j < n + 1$ or $A_{(i,j)} = 0$ for all i, j such that $i + j > n + 1$. Then,

$$\det A = (-1)^{\lfloor n/2 \rfloor} \prod_{i=1}^{n} A_{(i,n+1-i)}.$$

Remark: A is *lower reverse triangular*.

Fact 2.13.10. Let $a_1, \ldots, a_n \in \mathbb{F}$. Then,

$$\det \begin{bmatrix} 1 + a_1 & a_2 & \cdots & a_n \\ a_1 & 1 + a_2 & \cdots & a_n \\ \vdots & \vdots & \ddots & \vdots \\ a_1 & a_2 & \cdots & 1 + a_n \end{bmatrix} = 1 + \sum_{i=1}^{n} a_i.$$

Fact 2.13.11. Let $a_1, \ldots, a_n \in \mathbb{F}$ be nonzero. Then,

$$\det \begin{bmatrix} \frac{1+a_1}{a_1} & 1 & \cdots & 1 \\ 1 & \frac{1+a_2}{a_2} & \cdots & 1 \\ \vdots & \vdots & \ddots & \vdots \\ 1 & 1 & \cdots & \frac{1+a_n}{a_n} \end{bmatrix} = \frac{1 + \sum_{i=1}^{n} a_i}{\prod_{i=1}^{n} a_i}.$$

Fact 2.13.12. Let $a, b, c_1, \ldots, c_n \in \mathbb{F}$, define $A \in \mathbb{F}^{n \times n}$ by

$$A \triangleq \begin{bmatrix} c_1 & a & a & \cdots & a \\ b & c_2 & a & \cdots & a \\ b & b & c_3 & \ddots & a \\ \vdots & \vdots & \ddots & \ddots & \vdots \\ b & b & b & \cdots & c_n \end{bmatrix},$$

and let $p(x) = (c_1 - x)(c_2 - x) \cdots (c_n - x)$ and $p_i(x) = p(x)/(c_i - x)$ for all $i \in$

$\{1, \ldots, n\}$. Then,

$$\det A = \begin{cases} \dfrac{bp(a) - ap(b)}{b - a}, & b \neq a, \\ a\displaystyle\sum_{i=1}^{n-1} p_i(a) + c_n p_n(a), & b = a. \end{cases}$$

Proof: See [1523, p. 10].

Fact 2.13.13. Let $a, b \in \mathbb{F}$, and define $A, B \in \mathbb{F}^{n \times n}$ by

$$A \triangleq (a - b)I_n + b1_{n \times n} = \begin{bmatrix} a & b & b & \cdots & b \\ b & a & b & \cdots & b \\ b & b & a & \ddots & b \\ \vdots & \vdots & \ddots & \ddots & \vdots \\ b & b & b & \cdots & a \end{bmatrix}$$

and

$$B \triangleq aI_n + b1_{n \times n} = \begin{bmatrix} a+b & b & b & \cdots & b \\ b & a+b & b & \cdots & b \\ b & b & a+b & \ddots & b \\ \vdots & \vdots & \ddots & \ddots & \vdots \\ b & b & b & \cdots & a+b \end{bmatrix}.$$

Then,

$$\det A = (a - b)^{n-1}[a + b(n - 1)]$$

and, if $\det A \neq 0$, then

$$A^{-1} = \frac{1}{a - b}I_n + \frac{b}{(b - a)[a + b(n - 1)]}1_{n \times n}.$$

Furthermore,

$$\det B = a^{n-1}(a + nb)$$

and, if $\det B \neq 0$, then

$$B^{-1} = \frac{1}{a}\left(I_n - \frac{b}{a + nb}1_{n \times n}\right).$$

Remark: See Fact 2.14.26, Fact 4.10.16, and Fact 8.9.35.

Remark: The matrix $aI_n + b1_{n \times n}$ arises in combinatorics. See [273, 275].

Fact 2.13.14. Let $A \in \mathbb{F}^{n \times n}$, and define $\gamma \triangleq \max_{i,j=1,\ldots,n} |A_{(i,j)}|$. Then,

$$|\det A| \leq \gamma^n n^{n/2}.$$

Proof: This result is a consequence of the arithmetic-mean–geometric-mean inequality Fact 1.17.14 and Schur's inequality Fact 8.18.5. See [459, p. 200].

Remark: See Fact 8.13.35.

Fact 2.13.15. Let $A \in \mathbb{R}^{n \times n}$, and, for $i = 1, \ldots, n$, let α_i denote the sum of the positive components in $\text{row}_i(A)$ and let β_i denote the sum of the positive components in $\text{row}_i(-A)$. Then,

$$|\det A| \le \prod_{i=1}^{n} \max\{\alpha_i, \beta_i\} - \prod_{i=1}^{n} \min\{\alpha_i, \beta_i\}.$$

Proof: See [789].

Remark: This result is an extension of a result due to Schinzel.

Fact 2.13.16. For $i = 1, \ldots, 4$, let $A_i, B_i \in \mathbb{F}^{2 \times 2}$, where $\det A_i = \det B_i = 1$. Furthermore, define $\mathcal{A}, \mathcal{B}, \mathcal{C}, \mathcal{D} \in \mathbb{F}^{4 \times 4}$, where, for $i, j = 1, \ldots, 4$,

$$\mathcal{A}_{(i,j)} = \text{tr } A_i A_j,$$
$$\mathcal{B}_{(i,j)} = \text{tr } B_i B_j,$$
$$\mathcal{C}_{(i,j)} = \text{tr } A_i B_j,$$
$$\mathcal{D}_{(i,j)} = \text{tr } A_i B_j^{-1}.$$

Then,

$$\det \mathcal{C} + \det \mathcal{D} = 0$$

and

$$(\det \mathcal{A})(\det \mathcal{B}) = (\det \mathcal{C})^2.$$

Remark: These equalities are due to Magnus. See [756].

Fact 2.13.17. Let $\mathcal{I} \subseteq \mathbb{R}$ be a finite or infinite interval, and let $f: \mathcal{I} \mapsto \mathbb{R}$. Then, the following statements are equivalent:

i) f is convex.

ii) For all distinct $x, y, z \in \mathcal{I}$,

$$\frac{\det \begin{bmatrix} 1 & x & f(x) \\ 1 & y & f(y) \\ 1 & z & f(z) \end{bmatrix}}{\det \begin{bmatrix} 1 & x & x^2 \\ 1 & y & y^2 \\ 1 & z & z^2 \end{bmatrix}} \ge 0.$$

iii) For all $x, y, z \in \mathcal{I}$ such that $x < y < z$,

$$\det \begin{bmatrix} 1 & x & f(x) \\ 1 & y & f(y) \\ 1 & z & f(z) \end{bmatrix} \ge 0.$$

Proof: See [1066, p. 21].

2.14 Facts on the Determinant of Partitioned Matrices

Fact 2.14.1. Let $A \in \mathbb{F}^{n \times n}$, let A_0 be the $k \times k$ leading principal submatrix of A, and let $B \in \mathbb{F}^{(n-k) \times (n-k)}$, where, for all $i, j \in \{1, \ldots, n-k\}$, $B_{(i,j)}$ is the determinant of the submatrix of A comprised of rows $1, \ldots, k$ and $k+i$ and columns $1, \ldots, k$ and $k+j$. Then,

$$\det B = (\det A_0)^{n-k-1} \det A.$$

If, in addition, A_0 is nonsingular, then

$$\det A = \frac{\det B}{(\det A_0)^{n-k-1}}.$$

Remark: If $k = n-1$, then $B = \det A$.

Remark: This result is *Sylvester's identity.*

Fact 2.14.2. Let $A \in \mathbb{F}^{n \times n}$, $x, y \in \mathbb{F}^n$, and $a \in \mathbb{F}$. Then,

$$\det \begin{bmatrix} A & x \\ y^{\mathrm{T}} & a \end{bmatrix} = a(\det A) - y^{\mathrm{T}} A^{\mathrm{A}} x.$$

Hence,

$$\det \begin{bmatrix} A & x \\ y^{\mathrm{T}} & a \end{bmatrix} = \begin{cases} (\det A)(a - y^{\mathrm{T}} A^{-1} x), & \det A \neq 0, \\ a \det(A - a^{-1} x y^{\mathrm{T}}), & a \neq 0, \\ -y^{\mathrm{T}} A^{\mathrm{A}} x, & a = 0 \text{ or } \det A = 0. \end{cases}$$

In particular,

$$\det \begin{bmatrix} A & Ax \\ y^{\mathrm{T}} A & y^{\mathrm{T}} Ax \end{bmatrix} = 0.$$

Finally,

$$\det(A + xy^{\mathrm{T}}) = (\det A) + y^{\mathrm{T}} A^{\mathrm{A}} x = -\det \begin{bmatrix} A & x \\ y^{\mathrm{T}} & -1 \end{bmatrix}.$$

If, in addition, A is nonsingular, then

$$\det(A + xy^{\mathrm{T}}) = (\det A)(1 + y^{\mathrm{T}} A^{-1} x).$$

Remark: See Fact 2.14.3, Fact 2.16.2, and Fact 2.16.4.

Fact 2.14.3. Let $A \in \mathbb{F}^{n \times n}$, $b \in \mathbb{F}^n$, and $a \in \mathbb{F}$. Then,

$$\det \begin{bmatrix} A & b \\ b^* & a \end{bmatrix} = a(\det A) - b^* A^{\mathrm{A}} b.$$

In particular,

$$\det \begin{bmatrix} A & b \\ b^* & a \end{bmatrix} = \begin{cases} (\det A)(a - b^* A^{-1} b), & \det A \neq 0, \\ a \det(A - a^{-1} b b^*), & a \neq 0, \\ -b^* A^{\mathrm{A}} b, & a = 0. \end{cases}$$

Remark: This equality is a specialization of Fact 2.14.2 with $x = b$ and $y = \bar{b}$.

Remark: See Fact 8.15.5.

Fact 2.14.4. Let $A \in \mathbb{F}^{n \times n}$. Then,

$$\mathrm{rank} \begin{bmatrix} A & A \\ A & A \end{bmatrix} = \mathrm{rank} \begin{bmatrix} A & -A \\ -A & A \end{bmatrix} = \mathrm{rank}\, A,$$

$$\mathrm{rank} \begin{bmatrix} A & A \\ -A & A \end{bmatrix} = 2\mathrm{rank}\, A,$$

$$\det \begin{bmatrix} A & A \\ A & A \end{bmatrix} = \det \begin{bmatrix} A & -A \\ -A & A \end{bmatrix} = 0,$$

$$\det \begin{bmatrix} A & A \\ -A & A \end{bmatrix} = 2^n (\det A)^2.$$

Remark: See Fact 2.14.5.

Fact 2.14.5. Let $a, b, c, d \in \mathbb{F}$, let $A \in \mathbb{F}^{n \times n}$, and define $\mathcal{A} \triangleq \begin{bmatrix} aA & bA \\ cA & dA \end{bmatrix}$. Then,

$$\mathrm{rank}\, \mathcal{A} = \left(\mathrm{rank} \begin{bmatrix} a & b \\ c & d \end{bmatrix} \right) \mathrm{rank}\, A$$

and

$$\det \mathcal{A} = (ad - bc)^n (\det A)^2.$$

Remark: See Fact 2.14.4.

Proof: See Proposition 7.1.11 and Fact 7.4.24.

Fact 2.14.6. $\det \begin{bmatrix} 0 & I_n \\ I_m & 0 \end{bmatrix} = (-1)^{nm}$.

Fact 2.14.7. Let $A, B \in \mathbb{F}^{n \times n}$. Then,

$$\det \begin{bmatrix} I_n & I_n - AB \\ B & 0 \end{bmatrix} = \det \begin{bmatrix} 0 & I_n - AB \\ B & 0 \end{bmatrix} = \det(BAB - B).$$

Remark: See Fact 2.11.18 and Fact 2.14.6.

Fact 2.14.8. Let $A \in \mathbb{F}^{n \times m}$, let $B \in \mathbb{F}^{m \times n}$, and assume that $n \leq m$. Then,

$$\det AB = (-1)^{(n+1)m} \det \begin{bmatrix} A & 0_{n \times n} \\ -I_m & B \end{bmatrix}.$$

Proof: See [459].

Remark: See Fact 2.13.5.

Fact 2.14.9. Let A, B, C, D be conformable matrices with entries in $\mathbb{F}$. Then,

$$\begin{bmatrix} A & AB \\ C & D \end{bmatrix} = \begin{bmatrix} I & 0 \\ C & I \end{bmatrix} \begin{bmatrix} A & 0 \\ C - CA & D - CB \end{bmatrix} \begin{bmatrix} I & B \\ 0 & I \end{bmatrix},$$

$$\det \begin{bmatrix} A & AB \\ C & D \end{bmatrix} = (\det A) \det(D - CB),$$

$$\begin{bmatrix} A & B \\ CA & D \end{bmatrix} = \begin{bmatrix} I & 0 \\ C & I \end{bmatrix} \begin{bmatrix} A & B - AB \\ 0 & D - CB \end{bmatrix} \begin{bmatrix} I & B \\ 0 & I \end{bmatrix},$$

$$\det \begin{bmatrix} A & B \\ CA & D \end{bmatrix} = (\det A) \det(D - CB),$$

$$\begin{bmatrix} A & BD \\ C & D \end{bmatrix} = \begin{bmatrix} I & B \\ 0 & I \end{bmatrix} \begin{bmatrix} A - BC & 0 \\ C - DC & D \end{bmatrix} \begin{bmatrix} I & 0 \\ C & I \end{bmatrix},$$

$$\det \begin{bmatrix} A & BD \\ C & D \end{bmatrix} = \det(A - BC) \det D,$$

$$\begin{bmatrix} A & B \\ DC & D \end{bmatrix} = \begin{bmatrix} I & B \\ 0 & I \end{bmatrix} \begin{bmatrix} A - BC & B - BD \\ 0 & D \end{bmatrix} \begin{bmatrix} I & 0 \\ C & I \end{bmatrix},$$

$$\det \begin{bmatrix} A & B \\ DC & D \end{bmatrix} = \det(A - BC) \det D.$$

Remark: See Fact 6.5.25.

Fact 2.14.10. Let $A, B \in \mathbb{F}^{n \times n}$. Then,

$$\det \begin{bmatrix} A & AB \\ BA & B \end{bmatrix} = (\det A) \det(B - BAB) = (\det B) \det(A - ABA).$$

Proof: See Fact 2.11.19 and Fact 2.14.7.

Fact 2.14.11. Let $A_1, A_2, B_1, B_2 \in \mathbb{F}^{n \times m}$, and define $\mathcal{A} \triangleq \begin{bmatrix} A_1 & A_2 \\ A_2 & A_1 \end{bmatrix}$ and $\mathcal{B} \triangleq \begin{bmatrix} B_1 & B_2 \\ B_2 & B_1 \end{bmatrix}$. Then,

$$\operatorname{rank} \begin{bmatrix} \mathcal{A} & \mathcal{B} \\ \mathcal{B} & \mathcal{A} \end{bmatrix} = \sum_{i=1}^{4} \operatorname{rank} C_i,$$

where $C_1 \triangleq A_1 + A_2 + B_1 + B_2$, $C_2 \triangleq A_1 + A_2 - B_1 - B_2$, $C_3 \triangleq A_1 - A_2 + B_1 - B_2$, and $C_4 \triangleq A_1 - A_2 - B_1 + B_2$. If, in addition, $n = m$, then

$$\det \begin{bmatrix} \mathcal{A} & \mathcal{B} \\ \mathcal{B} & \mathcal{A} \end{bmatrix} = \prod_{i=1}^{4} \det C_i.$$

Proof: See [1337].

Remark: See Fact 3.24.8.

Fact 2.14.12. Let $A, B, C, D \in \mathbb{F}^{n \times n}$, and assume that $\operatorname{rank} \begin{bmatrix} A & B \\ C & D \end{bmatrix} = n$. Then,

$$\det \begin{bmatrix} \det A & \det B \\ \det C & \det D \end{bmatrix} = 0.$$

Fact 2.14.13. Let $A, B, C, D \in \mathbb{F}^{n \times n}$. Then,

$$\det \begin{bmatrix} A & B \\ C & D \end{bmatrix} = \begin{cases} \det(DA - CB), & AB = BA, \\ \det(AD - CB), & AC = CA, \\ \det(AD - BC), & DC = CD, \\ \det(DA - BC), & DB = BD. \end{cases}$$

Remark: These equalities are *Schur's formulas*. See [150, p. 11].

Proof: If A is nonsingular, then

$$\det \begin{bmatrix} A & B \\ C & D \end{bmatrix} = (\det A) \det(D - CA^{-1}B) = \det(DA - CA^{-1}BA)$$

$$= \det(DA - CB).$$

Alternatively, note the equality

$$\begin{bmatrix} A & B \\ C & D \end{bmatrix} = \begin{bmatrix} A & 0 \\ C & DA - CB \end{bmatrix} \begin{bmatrix} I & BA^{-1} \\ 0 & A^{-1} \end{bmatrix}.$$

If A is singular, then replace A by $A + \varepsilon I$ and use continuity.

Problem: Find a direct proof for the case in which A is singular.

Fact 2.14.14. Let $A, B, C, D \in \mathbb{F}^{n \times n}$. Then,

$$\det \begin{bmatrix} A & B \\ C & D \end{bmatrix} = \begin{cases} \det(AD^{\mathrm{T}} - B^{\mathrm{T}}C^{\mathrm{T}}), & AB = BA^{\mathrm{T}}, \\ \det(AD^{\mathrm{T}} - BC), & DC = CD^{\mathrm{T}}, \\ \det(A^{\mathrm{T}}D - CB), & A^{\mathrm{T}}C = CA, \\ \det(A^{\mathrm{T}}D - C^{\mathrm{T}}B^{\mathrm{T}}), & D^{\mathrm{T}}B = BD. \end{cases}$$

Proof: Define the nonsingular matrix $A_\varepsilon \triangleq A + \varepsilon I$, which satisfies $A_\varepsilon B = BA_\varepsilon^{\mathrm{T}}$. Then,

$$\det \begin{bmatrix} A_\varepsilon & B \\ C & D \end{bmatrix} = (\det A_\varepsilon) \det(D - CA_\varepsilon^{-1}B)$$

$$= \det(DA_\varepsilon^{\mathrm{T}} - CA_\varepsilon^{-1}BA_\varepsilon^{\mathrm{T}}) = \det(DA_\varepsilon^{\mathrm{T}} - CB).$$

Fact 2.14.15. Let $A, B, C, D \in \mathbb{F}^{n \times n}$. Then,

$$\det\left[\begin{array}{cc} A & B \\ C & D \end{array}\right] = \begin{cases} (-1)^{\operatorname{rank} C}\det\left(A^{\mathrm{T}}D + C^{\mathrm{T}}B\right), & A^{\mathrm{T}}C = -C^{\mathrm{T}}A, \\ (-1)^{n+\operatorname{rank} A}\det\left(A^{\mathrm{T}}D + C^{\mathrm{T}}B\right), & A^{\mathrm{T}}C = -C^{\mathrm{T}}A, \\ (-1)^{\operatorname{rank} B}\det\left(A^{\mathrm{T}}D + C^{\mathrm{T}}B\right), & B^{\mathrm{T}}D = -D^{\mathrm{T}}B, \\ (-1)^{n+\operatorname{rank} D}\det\left(A^{\mathrm{T}}D + C^{\mathrm{T}}B\right), & B^{\mathrm{T}}D = -D^{\mathrm{T}}B, \\ (-1)^{\operatorname{rank} B}\det\left(AD^{\mathrm{T}} + BC^{\mathrm{T}}\right), & AB^{\mathrm{T}} = -BA^{\mathrm{T}}, \\ (-1)^{n+\operatorname{rank} A}\det\left(AD^{\mathrm{T}} + BC^{\mathrm{T}}\right), & AB^{\mathrm{T}} = -BA^{\mathrm{T}}, \\ (-1)^{\operatorname{rank} C}\det\left(AD^{\mathrm{T}} + BC^{\mathrm{T}}\right), & CD^{\mathrm{T}} = -DC^{\mathrm{T}}, \\ (-1)^{n+\operatorname{rank} D}\det\left(AD^{\mathrm{T}} + BC^{\mathrm{T}}\right), & CD^{\mathrm{T}} = -DC^{\mathrm{T}}. \end{cases}$$

Proof: See [985, 1439].

Remark: This result is due to Callan. See [1439].

Remark: If $A^{\mathrm{T}}C = -C^{\mathrm{T}}A$ and $\operatorname{rank} A + \operatorname{rank} C + n$ is odd, then $\left[\begin{smallmatrix} A & B \\ C & D \end{smallmatrix}\right]$ is singular.

Fact 2.14.16. Let $A, B, C, D \in \mathbb{F}^{n \times n}$. Then,

$$\det\left[\begin{array}{cc} A & B \\ C & D \end{array}\right] = \begin{cases} \det\left(AD^{\mathrm{T}} - BC^{\mathrm{T}}\right), & AB^{\mathrm{T}} = BA^{\mathrm{T}}, \\ \det\left(AD^{\mathrm{T}} - BC^{\mathrm{T}}\right), & DC^{\mathrm{T}} = CD^{\mathrm{T}}, \\ \det\left(A^{\mathrm{T}}D - C^{\mathrm{T}}B\right), & A^{\mathrm{T}}C = C^{\mathrm{T}}A, \\ \det\left(A^{\mathrm{T}}D - C^{\mathrm{T}}B\right), & D^{\mathrm{T}}B = B^{\mathrm{T}}D. \end{cases}$$

Proof: See [985].

Fact 2.14.17. Let $A \in \mathbb{F}^{n \times m}$, $B \in \mathbb{F}^{n \times l}$, $C \in \mathbb{F}^{k \times m}$, and $D \in \mathbb{F}^{k \times l}$, and assume that $n + k = m + l$. If $AC^{\mathrm{T}} + BD^{\mathrm{T}} = 0$, then

$$\det\left[\begin{array}{cc} A & B \\ C & D \end{array}\right]^2 = \det\left(AA^{\mathrm{T}} + BB^{\mathrm{T}}\right)\det\left(CC^{\mathrm{T}} + DD^{\mathrm{T}}\right).$$

Alternatively, if $A^{\mathrm{T}}B + C^{\mathrm{T}}D = 0$, then

$$\det\left[\begin{array}{cc} A & B \\ C & D \end{array}\right]^2 = \det\left(A^{\mathrm{T}}A + C^{\mathrm{T}}C\right)\det\left(B^{\mathrm{T}}B + D^{\mathrm{T}}D\right).$$

Proof: Form $\left[\begin{smallmatrix} A & B \\ C & D \end{smallmatrix}\right]\left[\begin{smallmatrix} A & B \\ C & D \end{smallmatrix}\right]^{\mathrm{T}}$ and $\left[\begin{smallmatrix} A & B \\ C & D \end{smallmatrix}\right]^{\mathrm{T}}\left[\begin{smallmatrix} A & B \\ C & D \end{smallmatrix}\right]$.

Fact 2.14.18. Let $A \in \mathbb{F}^{n \times m}$, $B \in \mathbb{F}^{n \times m}$, $C \in \mathbb{F}^{k \times m}$, and $D \in \mathbb{F}^{k \times m}$, and assume that $n + k = 2m$. If $AD^{\mathrm{T}} + BC^{\mathrm{T}} = 0$, then

$$\det\left[\begin{array}{cc} A & B \\ C & D \end{array}\right]^2 = (-1)^m\det\left(AB^{\mathrm{T}} + BA^{\mathrm{T}}\right)\det\left(CD^{\mathrm{T}} + DC^{\mathrm{T}}\right).$$

Alternatively, if $AB^{\mathrm{T}} + BA^{\mathrm{T}} = 0$ or $CD^{\mathrm{T}} + DC^{\mathrm{T}} = 0$, then

$$\det \begin{bmatrix} A & B \\ C & D \end{bmatrix}^2 = (-1)^{m^2 + nk} \det \left(AD^{\mathrm{T}} + BC^{\mathrm{T}} \right)^2.$$

Proof: Form $\begin{bmatrix} A & B \\ C & D \end{bmatrix} \begin{bmatrix} B^{\mathrm{T}} & D^{\mathrm{T}} \\ A^{\mathrm{T}} & C^{\mathrm{T}} \end{bmatrix}$ and $\begin{bmatrix} A & B \\ C & D \end{bmatrix} \begin{bmatrix} D^{\mathrm{T}} & B^{\mathrm{T}} \\ C^{\mathrm{T}} & A^{\mathrm{T}} \end{bmatrix}$. See [1439].

Fact 2.14.19. Let $A \in \mathbb{F}^{n \times m}$, $B \in \mathbb{F}^{n \times l}$, $C \in \mathbb{F}^{n \times m}$, and $D \in \mathbb{F}^{n \times l}$, and assume that $m + l = 2n$. If $A^{\mathrm{T}}D + C^{\mathrm{T}}B = 0$, then

$$\det \begin{bmatrix} A & B \\ C & D \end{bmatrix}^2 = (-1)^n \det \left(C^{\mathrm{T}}A + A^{\mathrm{T}}C \right) \det \left(D^{\mathrm{T}}B + B^{\mathrm{T}}D \right).$$

Alternatively, if $B^{\mathrm{T}}D + D^{\mathrm{T}}B = 0$ or $A^{\mathrm{T}}C + C^{\mathrm{T}}A = 0$, then

$$\det \begin{bmatrix} A & B \\ C & D \end{bmatrix}^2 = (-1)^{n^2 + ml} \det \left(A^{\mathrm{T}}D + C^{\mathrm{T}}B \right)^2.$$

Proof: Form $\begin{bmatrix} C^{\mathrm{T}} & A^{\mathrm{T}} \\ D^{\mathrm{T}} & B^{\mathrm{T}} \end{bmatrix} \begin{bmatrix} A & B \\ C & D \end{bmatrix}$ and $\begin{bmatrix} D^{\mathrm{T}} & B^{\mathrm{T}} \\ C^{\mathrm{T}} & A^{\mathrm{T}} \end{bmatrix} \begin{bmatrix} A & B \\ C & D \end{bmatrix}$.

Fact 2.14.20. Let $A \in \mathbb{F}^{n \times n}$, $B \in \mathbb{F}^{n \times k}$, $C \in \mathbb{F}^{k \times n}$, and $D \in \mathbb{F}^{k \times k}$. If $AB + BD = 0$ or $CA + DC = 0$, then

$$\det \begin{bmatrix} A & B \\ C & D \end{bmatrix}^2 = \det \left(A^2 + BC \right) \det \left(CB + D^2 \right).$$

Alternatively, if $A^2 + BC = 0$ or $CB + D^2 = 0$, then

$$\det \begin{bmatrix} A & B \\ C & D \end{bmatrix}^2 = (-1)^{nk} \det (AB + BD) \det (CA + DC).$$

Proof: Form $\begin{bmatrix} A & B \\ C & D \end{bmatrix}^2$ and $\begin{bmatrix} A & B \\ C & D \end{bmatrix} \begin{bmatrix} B & A \\ D & C \end{bmatrix}$.

Fact 2.14.21. Let $A \in \mathbb{F}^{n \times m}$, $B \in \mathbb{F}^{n \times n}$, $C \in \mathbb{F}^{m \times m}$, and $D \in \mathbb{F}^{m \times n}$. If $AD + B^2 = 0$ or $C^2 + DA = 0$, then

$$\det \begin{bmatrix} A & B \\ C & D \end{bmatrix}^2 = (-1)^{nm} \det (AC + BA) \det (CD + DB).$$

Alternatively, if $AC + BA = 0$ or $CD + DB = 0$, then

$$\det \begin{bmatrix} A & B \\ C & D \end{bmatrix}^2 = \det \left(AD + B^2 \right) \det \left(C^2 + DA \right).$$

Proof: Form $\begin{bmatrix} A & B \\ C & D \end{bmatrix} \begin{bmatrix} C & D \\ A & B \end{bmatrix}$ and $\begin{bmatrix} A & B \\ C & D \end{bmatrix} \begin{bmatrix} D & C \\ B & A \end{bmatrix}$.

Fact 2.14.22. Let $A \in \mathbb{F}^{n \times m}$, $B \in \mathbb{F}^{n \times l}$, $C \in \mathbb{F}^{k \times m}$, and $D \in \mathbb{F}^{k \times l}$, and assume that $n + k = m + l$. If $AC^* + BD^* = 0$, then

$$\left| \det \begin{bmatrix} A & B \\ C & D \end{bmatrix} \right|^2 = \det(AA^* + BB^*) \det(CC^* + DD^*).$$

Alternatively, if $A^*B + C^*D = 0$, then

$$\left| \det \begin{bmatrix} A & B \\ C & D \end{bmatrix} \right|^2 = \det(A^*A + C^*C)\det(B^*B + D^*D).$$

Proof: Form $\begin{bmatrix} A & B \\ C & D \end{bmatrix}\begin{bmatrix} A & B \\ C & D \end{bmatrix}^*$ and $\begin{bmatrix} A & B \\ C & D \end{bmatrix}^*\begin{bmatrix} A & B \\ C & D \end{bmatrix}$.

Remark: See Fact 8.13.28.

Fact 2.14.23. Let $A \in \mathbb{F}^{n \times m}$, $B \in \mathbb{F}^{n \times m}$, $C \in \mathbb{F}^{k \times m}$, and $D \in \mathbb{F}^{k \times m}$, and assume that $n + k = 2m$. If $AD^* + BC^* = 0$, then

$$\left| \det \begin{bmatrix} A & B \\ C & D \end{bmatrix} \right|^2 = (-1)^m \det(AB^* + BA^*)\det(CD^* + DC^*).$$

Alternatively, if $AB^* + BA^* = 0$ or $CD^* + DC^* = 0$, then

$$\left| \det \begin{bmatrix} A & B \\ C & D \end{bmatrix} \right|^2 = (-1)^{m^2+nk} |\det(AD^* + BC^*)|^2.$$

Proof: Form $\begin{bmatrix} A & B \\ C & D \end{bmatrix}\begin{bmatrix} B^* & D^* \\ A^* & C^* \end{bmatrix}$ and $\begin{bmatrix} A & B \\ C & D \end{bmatrix}\begin{bmatrix} D^* & B^* \\ C^* & A^* \end{bmatrix}$.

Remark: If $m^2 + nk$ is odd, then $\begin{bmatrix} A & B \\ C & D \end{bmatrix}$ is singular.

Fact 2.14.24. Let $A \in \mathbb{F}^{n \times m}$, $B \in \mathbb{F}^{n \times l}$, $C \in \mathbb{F}^{n \times m}$, and $D \in \mathbb{F}^{n \times l}$, and assume that $m + l = 2n$. If $A^*D + C^*B = 0$, then

$$\left| \det \begin{bmatrix} A & B \\ C & D \end{bmatrix} \right|^2 = (-1)^m \det(C^*A + A^*C)\det(D^*B + B^*D).$$

Alternatively, if $D^*B + B^*D = 0$ or $C^*A + A^*C = 0$, then

$$\left| \det \begin{bmatrix} A & B \\ C & D \end{bmatrix} \right|^2 = (-1)^{n^2+ml} |\det(A^*D + C^*B)|^2.$$

Proof: Form $\begin{bmatrix} C^* & A^* \\ D^* & B^* \end{bmatrix}\begin{bmatrix} A & B \\ C & D \end{bmatrix}$ and $\begin{bmatrix} D^* & B^* \\ C^* & A^* \end{bmatrix}\begin{bmatrix} A & B \\ C & D \end{bmatrix}$.

Remark: If $n^2 + ml$ is odd, then $\begin{bmatrix} A & B \\ C & D \end{bmatrix}$ is singular.

Fact 2.14.25. Let $A \in \mathbb{F}^{n \times m}$ and $B \in \mathbb{F}^{n \times l}$. Then,

$$\det \begin{bmatrix} A^*A & A^*B \\ B^*A & B^*B \end{bmatrix} = \begin{cases} \det(A^*A)\det[B^*B - B^*A(A^*A)^{-1}A^*B], & \operatorname{rank} A = m, \\ \det(B^*B)\det[A^*A - A^*B(B^*B)^{-1}B^*A], & \operatorname{rank} B = l, \\ 0, & n < m + l. \end{cases}$$

If, in addition, $m + l = n$, then

$$\det \begin{bmatrix} A^*A & A^*B \\ B^*A & B^*B \end{bmatrix} = \det(AA^* + BB^*).$$

Remark: See Fact 6.5.27.

Fact 2.14.26. Let $A, B \in \mathbb{F}^{n \times n}$, and define $\mathcal{A} \in \mathbb{F}^{kn \times kn}$ by

$$\mathcal{A} \triangleq \begin{bmatrix} A & B & B & \cdots & B \\ B & A & B & \cdots & B \\ B & B & A & \ddots & B \\ \vdots & \vdots & \ddots & \ddots & \vdots \\ B & B & B & \cdots & A \end{bmatrix}.$$

Then,

$$\det \mathcal{A} = [\det(A - B)]^{k-1} \det[A + (k-1)B].$$

If $k = 2$, then

$$\det \begin{bmatrix} A & B \\ B & A \end{bmatrix} = \det[(A + B)(A - B)] = \det(A^2 - B^2 - [A, B]).$$

Proof: See [587].

Remark: For $k = 2$, the result follows from Fact 4.10.26.

Remark: See Fact 2.13.13.

Fact 2.14.27. Let $A \in \mathbb{F}^{n \times n}$, $B \in \mathbb{F}^{n \times m}$, $C \in \mathbb{F}^{m \times n}$, and $D \in \mathbb{F}^{m \times m}$, and define $M \triangleq \left[\begin{smallmatrix} A & B \\ C & D \end{smallmatrix}\right] \in \mathbb{F}^{(n+m) \times (n+m)}$. Furthermore, let $\left[\begin{smallmatrix} A' & B' \\ C' & D' \end{smallmatrix}\right] \triangleq M^{\mathrm{A}}$, where $A' \in \mathbb{F}^{n \times n}$ and $D' \in \mathbb{F}^{m \times m}$. Then,

$$\det D' = (\det M)^{m-1} \det A$$

and

$$\det A' = (\det M)^{n-1} \det D.$$

Proof: See [1215, p. 297].

Remark: See Fact 2.14.28.

Fact 2.14.28. Let $A \in \mathbb{F}^{n \times n}$, $B \in \mathbb{F}^{n \times m}$, $C \in \mathbb{F}^{m \times n}$, and $D \in \mathbb{F}^{m \times m}$, define $M \triangleq \left[\begin{smallmatrix} A & B \\ C & D \end{smallmatrix}\right] \in \mathbb{F}^{(n+m) \times (n+m)}$, and assume that M is nonsingular. Furthermore, let $\left[\begin{smallmatrix} A' & B' \\ C' & D' \end{smallmatrix}\right] \triangleq M^{-1}$, where $A' \in \mathbb{F}^{n \times n}$ and $D' \in \mathbb{F}^{m \times m}$. Then,

$$\det D' = \frac{\det A}{\det M}$$

and

$$\det A' = \frac{\det D}{\det M}.$$

Consequently, A is nonsingular if and only if D' is nonsingular, and D is nonsingular if and only if A' is nonsingular.

Proof: Use $M\left[\begin{smallmatrix} I & B' \\ 0 & D' \end{smallmatrix}\right] = \left[\begin{smallmatrix} A & 0 \\ C & I \end{smallmatrix}\right]$. See [1219].

Remark: This equality is a special case of *Jacobi's identity*. See [728, p. 21].

Remark: See Fact 2.14.27 and Fact 3.11.19.

2.15 Facts on Left and Right Inverses

Fact 2.15.1. Let $A \in \mathbb{F}^{n \times m}$. Then, the following statements hold:

i) If $A^{\mathrm{L}} \in \mathbb{F}^{m \times n}$ is a left inverse of A, then $\overline{A^{\mathrm{L}}} \in \mathbb{F}^{m \times n}$ is a left inverse of $\overline{A}$.

ii) If $A^{\mathrm{L}} \in \mathbb{F}^{m \times n}$ is a left inverse of A, then $A^{\mathrm{LT}} \in \mathbb{F}^{n \times m}$ is a right inverse of A^{T}.

iii) If $A^{\mathrm{L}} \in \mathbb{F}^{m \times n}$ is a left inverse of A, then $A^{\mathrm{L}*} \in \mathbb{F}^{n \times m}$ is a right inverse of A^{*}.

iv) If $A^{\mathrm{R}} \in \mathbb{F}^{m \times n}$ is a right inverse of A, then $\overline{A^{\mathrm{R}}} \in \mathbb{F}^{m \times n}$ is a right inverse of $\overline{A}$.

v) If $A^{\mathrm{R}} \in \mathbb{F}^{m \times n}$ is a right inverse of A, then $A^{\mathrm{RT}} \in \mathbb{F}^{n \times m}$ is a left inverse of A^{T}.

vi) If $A^{\mathrm{R}} \in \mathbb{F}^{m \times n}$ is a right inverse of A, then $A^{\mathrm{R}*} \in \mathbb{F}^{n \times m}$ is a left inverse of A^{*}.

Furthermore, the following statements are equivalent:

vii) A is left invertible.

viii) $\overline{A}$ is left invertible.

ix) A^{T} is right invertible.

x) A^{*} is right invertible.

Finally, the following statements are equivalent:

xi) A is right invertible.

xii) $\overline{A}$ is right invertible.

xiii) A^{T} is left invertible.

xiv) A^{*} is left invertible.

Fact 2.15.2. Let $A \in \mathbb{F}^{n \times m}$. If $\operatorname{rank} A = m$, then $(A^{*}A)^{-1}A^{*}$ is a left inverse of A. If $\operatorname{rank} A = n$, then $A^{*}(AA^{*})^{-1}$ is a right inverse of A.

Remark: See Fact 3.7.25, Fact 3.7.26, and Fact 3.13.6.

Fact 2.15.3. Let $A \in \mathbb{F}^{n \times m}$, and assume that $\operatorname{rank} A = m$. Then, $A^{\mathrm{L}} \in \mathbb{F}^{m \times n}$ is a left inverse of A if and only if there exists a matrix $B \in \mathbb{F}^{m \times n}$ such that BA is nonsingular and
$$A^{\mathrm{L}} = (BA)^{-1}B.$$
Proof: For necessity, let $B = A^{\mathrm{L}}$.

Fact 2.15.4. Let $A \in \mathbb{F}^{n \times m}$, and assume that $\operatorname{rank} A = n$. Then, $A^{\mathrm{R}} \in \mathbb{F}^{m \times n}$ is a right inverse of A if and only if there exists a matrix $B \in \mathbb{F}^{m \times n}$ such that AB is nonsingular and
$$A^{\mathrm{R}} = B(AB)^{-1}.$$
Proof: For necessity, let $B = A^{\mathrm{R}}$.

Fact 2.15.5. Let $A \in \mathbb{F}^{n \times m}$ and $B \in \mathbb{F}^{m \times l}$, and assume that A and B are left invertible. Then, AB is left invertible. If, in addition, A^L is a left inverse of A and B^L is a left inverse of B, then $B^L A^L$ is a left inverse of AB.

Fact 2.15.6. Let $A \in \mathbb{F}^{n \times m}$ and $B \in \mathbb{F}^{m \times l}$, and assume that A and B are right invertible. Then, AB is right invertible. If, in addition, A^R is a right inverse of A and B^R is a right inverse of B, then $B^R A^R$ is a right inverse of AB.

2.16 Facts on the Adjugate and Inverses

Fact 2.16.1. Let $x, y \in \mathbb{F}^n$. Then,

$$\left(I + xy^T\right)^A = \left(1 + y^T x\right)I - xy^T$$

and

$$\det\left(I + xy^T\right) = \det\left(I + yx^T\right) = 1 + x^T y = 1 + y^T x.$$

If, in addition, $x^T y \neq -1$, then

$$\left(I + xy^T\right)^{-1} = I - \left(1 + x^T y\right)^{-1} xy^T.$$

Fact 2.16.2. Let $A \in \mathbb{F}^{n \times n}$, $x, y \in \mathbb{F}^n$, and $a \in \mathbb{F}$. Then,

$$\begin{bmatrix} A & x \\ y^T & a \end{bmatrix} = \begin{cases} \begin{bmatrix} I & 0 \\ y^T A^{-1} & 1 \end{bmatrix} \begin{bmatrix} A & 0 \\ 0 & a - y^T A^{-1} x \end{bmatrix} \begin{bmatrix} I & A^{-1} x \\ 0 & 1 \end{bmatrix}, & \det A \neq 0, \\[3ex] \begin{bmatrix} I & 0 \\ y^T & 1 \end{bmatrix} \begin{bmatrix} A & 0 \\ y^T - y^T A & a - y^T A^{-1} x \end{bmatrix} \begin{bmatrix} I & A^{-1} x \\ 0 & 1 \end{bmatrix}, & \det A \neq 0, \\[3ex] \begin{bmatrix} I & a^{-1} x \\ 0 & 1 \end{bmatrix} \begin{bmatrix} A - a^{-1} xy^T & 0 \\ 0 & a \end{bmatrix} \begin{bmatrix} I & 0 \\ a^{-1} y^T & 1 \end{bmatrix}, & a \neq 0. \end{cases}$$

Remark: The second factorization follows from Fact 6.5.25 in the case in which A is nonsingular. Fact 6.5.25 provides a factorization for the case in which A is singular and $a = 0$, but with the additional assumption that $x \in \mathcal{R}(A)$.

Fact 2.16.3. Let $A \in \mathbb{F}^{n \times n}$, assume that A is nonsingular, and let $x, y \in \mathbb{F}^n$. Then,
$$\det\left(A + xy^T\right) = \left(1 + y^T A^{-1} x\right)\det A$$

and

$$\left(A + xy^T\right)^A = \left(1 + y^T A^{-1} x\right)(\det A)I - A^A xy^T.$$

Furthermore, the following statements are equivalent:

 i) $\det\left(A + xy^T\right) \neq 0$

 ii) $y^T A^{-1} x \neq -1$.

 iii) $\begin{bmatrix} A & x \\ y^T & -1 \end{bmatrix}$ is nonsingular.

In this case,

$$\left(A + xy^{\mathrm{T}}\right)^{-1} = A^{-1} - \left(1 + y^{\mathrm{T}}A^{-1}x\right)^{-1}A^{-1}xy^{\mathrm{T}}A^{-1}.$$

Remark: See Fact 2.16.2 and Fact 2.14.2.

Remark: The last equality, which is a special case of the matrix inversion lemma given by Corollary 2.8.8, is the *Sherman-Morrison-Woodbury formula.*

Fact 2.16.4. Let $A \in \mathbb{F}^{n \times n}$, let $x, y \in \mathbb{F}^n$, and let $a \in \mathbb{F}$. Then,

$$\left[\begin{array}{cc} A & x \\ y^{\mathrm{T}} & a \end{array}\right]^{\mathrm{A}} = \left[\begin{array}{cc} (a+1)A^{\mathrm{A}} - \left(A + xy^{\mathrm{T}}\right)^{\mathrm{A}} & -A^{\mathrm{A}}x \\ -y^{\mathrm{T}}A^{\mathrm{A}} & \det A \end{array}\right].$$

Now, assume in addition that $\left[\begin{smallmatrix} A & x \\ y^{\mathrm{T}} & a \end{smallmatrix}\right]$ is nonsingular. Then,

$$\left[\begin{array}{cc} A & x \\ y^{\mathrm{T}} & a \end{array}\right]^{-1}$$

$$= \begin{cases} \dfrac{1}{(\det A)(a - y^{\mathrm{T}}A^{-1}x)} \left[\begin{array}{cc} (a - y^{\mathrm{T}}A^{-1}x)A^{-1} + A^{-1}xy^{\mathrm{T}}A^{-1} & -A^{-1}x \\ -y^{\mathrm{T}}A^{-1} & 1 \end{array}\right], & \det A \neq 0, \\[20pt] \dfrac{1}{a\det(A - a^{-1}xy^{\mathrm{T}})} \left[\begin{array}{cc} (a+1)A^{\mathrm{A}} - \left(A + xy^{\mathrm{T}}\right)^{\mathrm{A}} & -A^{\mathrm{A}}x \\ -y^{\mathrm{T}}A^{\mathrm{A}} & \det A \end{array}\right], & a \neq 0, \\[20pt] \dfrac{1}{-y^{\mathrm{T}}A^{\mathrm{A}}x} \left[\begin{array}{cc} (a+1)A^{\mathrm{A}} - \left(A + xy^{\mathrm{T}}\right)^{\mathrm{A}} & -A^{\mathrm{A}}x \\ -y^{\mathrm{T}}A^{\mathrm{A}} & \det A \end{array}\right], & a = 0. \end{cases}$$

Proof: Use Fact 2.14.2, and see [468, 704].

Fact 2.16.5. Let $A \in \mathbb{F}^{n \times n}$. Then, the following statements hold:

i) $\left(\overline{A}\right)^{\mathrm{A}} = \overline{A^{\mathrm{A}}}$.

ii) $\left(A^{\mathrm{T}}\right)^{\mathrm{A}} = \left(A^{\mathrm{A}}\right)^{\mathrm{T}}$.

iii) $\left(A^*\right)^{\mathrm{A}} = \left(A^{\mathrm{A}}\right)^*$.

iv) If $\alpha \in \mathbb{F}$, then $(\alpha A)^{\mathrm{A}} = \alpha^{n-1}A^{\mathrm{A}}$.

v) $\det A^{\mathrm{A}} = (\det A)^{n-1}$.

vi) $\left(A^{\mathrm{A}}\right)^{\mathrm{A}} = (\det A)^{n-2}A$.

vii) $\det\left(A^{\mathrm{A}}\right)^{\mathrm{A}} = (\det A)^{(n-1)^2}$.

viii) If A is nonsingular, then $(A^{-1})^{\mathrm{A}} = (A^{\mathrm{A}})^{-1}$.

Proof: See [704].

Remark: With $0/0 \triangleq 1$ in vi), all of these results hold in the degenerate case $n = 1$.

Fact 2.16.6. Let $A \in \mathbb{F}^{n \times n}$. Then,

$$\det(A + 1_{n \times n}) - \det A = 1_{1 \times n}^{\mathrm{T}} A^{\mathrm{A}} 1 = \sum_{i=1}^{n} \det\left(A \overset{i}{\leftarrow} 1_{n \times 1}\right).$$

Proof: See [226].

Remark: See Fact 2.14.2, Fact 2.16.9, and Fact 10.12.10.

Fact 2.16.7. Let $A \in \mathbb{F}^{n \times n}$, and assume that A is singular. Then,

$$\mathcal{R}(A) \subseteq \mathcal{N}(A^{\mathrm{A}}).$$

Hence,

$$\operatorname{rank} A \leq \operatorname{def} A^{\mathrm{A}}$$

and

$$\operatorname{rank} A + \operatorname{rank} A^{\mathrm{A}} \leq n.$$

Furthermore, $\mathcal{R}(A) = \mathcal{N}(A^{\mathrm{A}})$ if and only if $\operatorname{rank} A = n - 1$.

Fact 2.16.8. Let $A \in \mathbb{F}^{n \times n}$. Then, the following statements hold:

i) $\operatorname{rank} A^{\mathrm{A}} = n$ if and only if $\operatorname{rank} A = n$.

ii) $\operatorname{rank} A^{\mathrm{A}} = 1$ if and only if $\operatorname{rank} A = n - 1$.

iii) $A^{\mathrm{A}} = 0$ if and only if $\operatorname{rank} A \leq n - 2$.

Proof: See [1125, p. 12].

Remark: See Fact 4.10.9.

Remark: Fact 6.3.6 provides an expression for A^{A} in the case $\operatorname{rank} A^{\mathrm{A}} = 1$.

Fact 2.16.9. Let $A, B \in \mathbb{F}^{n \times n}$. Then,

$$\left(A^{\mathrm{A}} B\right)_{(i,j)} = \det\left[A \overset{i}{\leftarrow} \operatorname{col}_j(B)\right].$$

Remark: See Fact 10.12.10.

Fact 2.16.10. Let $A, B \in \mathbb{F}^{n \times n}$. Then, the following statements hold:

i) $(AB)^{\mathrm{A}} = B^{\mathrm{A}} A^{\mathrm{A}}$.

ii) If B is nonsingular, then $\left(BAB^{-1}\right)^{\mathrm{A}} = BA^{\mathrm{A}} B^{-1}$.

iii) If $AB = BA$, then $A^{\mathrm{A}} B = BA^{\mathrm{A}}$, $AB^{\mathrm{A}} = B^{\mathrm{A}} A$, and $A^{\mathrm{A}} B^{\mathrm{A}} = B^{\mathrm{A}} A^{\mathrm{A}}$.

Fact 2.16.11. Let $A, B, C, D \in \mathbb{F}^{n \times n}$ and $ABCD = I$. Then, $ABCD = DABC = CDAB = BCDA$.

Fact 2.16.12. Let $A = \left[\begin{smallmatrix} a & b \\ c & d \end{smallmatrix}\right] \in \mathbb{F}^{2 \times 2}$, where $ad - bc \neq 0$. Then,

$$A^{-1} = (ad - bc)^{-1} \begin{bmatrix} d & -b \\ -c & a \end{bmatrix}.$$

Furthermore, if $A = \begin{bmatrix} a & b & c \\ d & e & f \\ g & h & i \end{bmatrix} \in \mathbb{F}^{3 \times 3}$ and $\beta = a(ei - fh) - b(di - fg) + c(dh - eg) \neq 0$, then

$$A^{-1} = \beta^{-1} \begin{bmatrix} ei - fh & -(bi - ch) & bf - ce \\ -(di - fg) & ai - cg & -(af - cd) \\ dh - eg & -(ah - bg) & ae - bd \end{bmatrix}.$$

Fact 2.16.13. Let $A, B \in \mathbb{F}^{n \times n}$, and assume that $A + B$ is nonsingular. Then,

$$A(A + B)^{-1}B = B(A + B)^{-1}A = A - A(A + B)^{-1}A = B - B(A + B)^{-1}B.$$

Fact 2.16.14. Let $A, B \in \mathbb{F}^{n \times n}$, and assume that A and B are nonsingular. Then,

$$A^{-1} + B^{-1} = A^{-1}(A + B)B^{-1}.$$

Furthermore, $A^{-1} + B^{-1}$ is nonsingular if and only if $A + B$ is nonsingular. In this case,

$$\begin{aligned}
\left(A^{-1} + B^{-1}\right)^{-1} &= A(A + B)^{-1}B \\
&= B(A + B)^{-1}A \\
&= A - A(A + B)^{-1}A \\
&= B - B(A + B)^{-1}B.
\end{aligned}$$

Fact 2.16.15. Let $A, B \in \mathbb{F}^{n \times n}$, and assume that A and B are nonsingular. Then,

$$A - B = A(B^{-1} - A^{-1})B.$$

Therefore,

$$\operatorname{rank}(A - B) = \operatorname{rank}(A^{-1} - B^{-1}).$$

In particular, $A - B$ is nonsingular if and only if $A^{-1} - B^{-1}$ is nonsingular. In this case,

$$\left(A^{-1} - B^{-1}\right)^{-1} = A - A(A - B)^{-1}A.$$

Fact 2.16.16. Let $A \in \mathbb{F}^{n \times m}$ and $B \in \mathbb{F}^{m \times n}$, and assume that $I + AB$ is nonsingular. Then, $I + BA$ is nonsingular and

$$(I_n + AB)^{-1}A = A(I_m + BA)^{-1}.$$

Remark: This result is the *push-through identity*. Furthermore,

$$(I + AB)^{-1} = I - (I + AB)^{-1}AB.$$

Fact 2.16.17. Let $A, B \in \mathbb{F}^{n \times n}$, and assume that $I + BA$ is nonsingular. Then,

$$(I + AB)^{-1} = I - A(I + BA)^{-1}B.$$

Fact 2.16.18. Let $A \in \mathbb{F}^{n \times n}$, and assume that A and $A + I$ are nonsingular. Then,

$$(A + I)^{-1} + \left(A^{-1} + I\right)^{-1} = (A + I)^{-1} + (A + I)^{-1}A = I.$$

Fact 2.16.19. Let $A \in \mathbb{F}^{n \times m}$. Then,

$$(I + AA^*)^{-1} = I - A(I + A^*A)^{-1}A^*.$$

Fact 2.16.20. Let $A \in \mathbb{F}^{n \times n}$, assume that A is nonsingular, let $B \in \mathbb{F}^{n \times m}$, let $C \in \mathbb{F}^{m \times n}$, and assume that $A + BC$ and $I + CA^{-1}B$ are nonsingular. Then,

$$(A + BC)^{-1}B = A^{-1}B(I + CA^{-1}B)^{-1}.$$

In particular, if $A + BB^*$ and $I + B^*A^{-1}B$ are nonsingular, then

$$(A + BB^*)^{-1}B = A^{-1}B(I + B^*A^{-1}B)^{-1}.$$

Fact 2.16.21. Let $A \in \mathbb{F}^{n \times n}$, $B \in \mathbb{F}^{n \times m}$, $C \in \mathbb{F}^{l \times n}$, and $D \in \mathbb{F}^{m \times l}$, and assume that A and $A + BDC$ are nonsingular. Then,

$$
\begin{aligned}
(A + BDC)^{-1} &= A^{-1} - (I + A^{-1}BDC)^{-1}A^{-1}BDCA^{-1} \\
&= A^{-1} - A^{-1}(I + BDCA^{-1})^{-1}BDCA^{-1} \\
&= A^{-1} - A^{-1}B(I + DCA^{-1}B)^{-1}DCA^{-1} \\
&= A^{-1} - A^{-1}BD(I + CA^{-1}BD)^{-1}CA^{-1} \\
&= A^{-1} - A^{-1}BDC(I + A^{-1}BDC)^{-1}A^{-1} \\
&= A^{-1} - A^{-1}BDCA^{-1}(I + BDCA^{-1})^{-1}.
\end{aligned}
$$

Proof: See [684].

Remark: The third equality generalizes the matrix inversion lemma given by Corollary 2.8.8 in the form

$$(A + BDC)^{-1} = A^{-1} - A^{-1}B(D^{-1} + CA^{-1}B)^{-1}CA^{-1}$$

since D need not be square or invertible.

Fact 2.16.22. Let $A \in \mathbb{F}^{n \times m}$, let $C, D \in \mathbb{F}^{n \times m}$, and assume that $I + DB$ is nonsingular. Then,

$$I + AC - (A + B)(I + DB)^{-1}(D + C) = (I - AD)(I + BD)^{-1}(I - BC).$$

Proof: See [1502].

Remark: See Fact 2.16.23 and Fact 8.11.21.

Fact 2.16.23. Let $A, B, C \in \mathbb{F}^{n \times m}$. Then,

$$I + AC^* - (A + B)(I + B^*B)^{-1}(B + C)^* = (I - AB^*)(I + BB^*)^{-1}(I - BC^*).$$

Proof: Set $D = B^*$ and replace C by C^* in Fact 2.16.22.

Fact 2.16.24. Let $A, B \in \mathbb{F}^{n \times n}$, and assume that B is nonsingular. Then,

$$A = B[I + B^{-1}(A - B)].$$

Fact 2.16.25. Let $A \in \mathbb{F}^{n \times n}$, assume that A is either upper triangular or lower triangular, let D denote the diagonal part of A, and assume that D is nonsingular. Then,

$$A^{-1} = \sum_{i=0}^{n}(I - D^{-1}A)^i D^{-1}.$$

Remark: Using the Schur product notation, $D = I \circ A$.

Fact 2.16.26. Let $A, B \in \mathbb{F}^{n \times n}$, and assume that A and $A+B$ are nonsingular. Then, for all $k \in \mathbb{N}$,

$$(A + B)^{-1} = \sum_{i=0}^{k} A^{-1}(-BA^{-1})^i + (-A^{-1}B)^{k+1}(A + B)^{-1}$$

$$= \sum_{i=0}^{k} A^{-1}(-BA^{-1})^i + A^{-1}(-BA^{-1})^{k+1}(I + BA^{-1})^{-1}.$$

Fact 2.16.27. Let $A, B \in \mathbb{F}^{n \times n}$ and $\alpha \in \mathbb{F}$, and assume that A, B, $\alpha A^{-1} + (1 - \alpha)B^{-1}$, and $\alpha B + (1 - \alpha)A$ are nonsingular. Then,

$$\alpha A + (1 - \alpha)B - \left[\alpha A^{-1} + (1 - \alpha)B^{-1}\right]^{-1}$$
$$= \alpha(1 - \alpha)(A - B)[\alpha B + (1 - \alpha)A]^{-1}(A - B).$$

Remark: This equality is relevant to *iv*) of Proposition 8.6.17.

Fact 2.16.28. Let $A \in \mathbb{F}^{n \times n}$, assume that A is nonsingular, and define $A_0 \triangleq I_n$. Furthermore, for all $k \in \{1, \ldots, n\}$, let

$$\alpha_k = \tfrac{1}{k} \operatorname{tr} AA_{k-1},$$

and, for all $k \in \{1, \ldots, n-1\}$, let

$$A_k = AA_{k-1} - \alpha_k I.$$

Then,

$$A^{-1} = \tfrac{1}{\alpha_n} A_{n-1}.$$

Remark: This result is due to Frame. See [174, p. 99].

Fact 2.16.29. Let $A \in \mathbb{F}^{n \times n}$, assume that A is nonsingular, and define $(B_i)_{i=1}^{\infty}$ by

$$B_{i+1} \triangleq 2B_i - B_i AB_i,$$

where $B_0 \in \mathbb{F}^{n \times n}$ satisfies $\operatorname{sprad}(I - B_0 A) < 1$. Then,

$$B_i \to A^{-1}$$

as $i \to \infty$.

Proof: See [148, p. 167].

Remark: This sequence is given by a Newton-Raphson algorithm.

Remark: See Fact 6.3.34 for the case in which A is singular or nonsquare.

Fact 2.16.30. Let $A \in \mathbb{F}^{n \times n}$, and assume that A is nonsingular. Then, $A + A^{-*}$ is nonsingular.

Proof: Note that $AA^* + I$ is positive definite.

2.17 Facts on the Inverse of Partitioned Matrices

Fact 2.17.1. Let $A \in \mathbb{F}^{n \times n}$, $B \in \mathbb{F}^{n \times m}$, $C \in \mathbb{F}^{m \times n}$, and $D \in \mathbb{F}^{m \times m}$, and assume that A and D are nonsingular. Then,

$$\begin{bmatrix} A & B \\ 0 & D \end{bmatrix}^{-1} = \begin{bmatrix} A^{-1} & -A^{-1}BD^{-1} \\ 0 & D^{-1} \end{bmatrix}$$

and

$$\begin{bmatrix} A & 0 \\ C & D \end{bmatrix}^{-1} = \begin{bmatrix} A^{-1} & 0 \\ -D^{-1}CA^{-1} & D^{-1} \end{bmatrix}.$$

Fact 2.17.2. Let $A \in \mathbb{F}^{n \times n}$, $B \in \mathbb{F}^{m \times m}$, and $C \in \mathbb{F}^{m \times n}$. Then,

$$\det \begin{bmatrix} 0 & A \\ B & C \end{bmatrix} = \det \begin{bmatrix} C & B \\ A & 0 \end{bmatrix} = (-1)^{nm}(\det A)(\det B).$$

If, in addition, A and B are nonsingular, then

$$\begin{bmatrix} 0 & A \\ B & C \end{bmatrix}^{-1} = \begin{bmatrix} -B^{-1}CA^{-1} & B^{-1} \\ A^{-1} & 0 \end{bmatrix}$$

and

$$\begin{bmatrix} C & B \\ A & 0 \end{bmatrix}^{-1} = \begin{bmatrix} 0 & A^{-1} \\ B^{-1} & -B^{-1}CA^{-1} \end{bmatrix}.$$

Fact 2.17.3. Let $A \in \mathbb{F}^{n \times n}$, $B \in \mathbb{F}^{n \times m}$, and $C \in \mathbb{F}^{m \times m}$, and assume that C is nonsingular. Then,

$$\begin{bmatrix} A & B \\ B^{\mathrm{T}} & C \end{bmatrix} = \begin{bmatrix} A - BC^{-1}B^{\mathrm{T}} & B \\ 0 & C \end{bmatrix} \begin{bmatrix} I & 0 \\ C^{-1}B^{\mathrm{T}} & I \end{bmatrix}.$$

If, in addition, $A - BC^{-1}B^{\mathrm{T}}$ is nonsingular, then $\begin{bmatrix} A & B \\ B^{\mathrm{T}} & C \end{bmatrix}$ is nonsingular and

$$\begin{bmatrix} A & B \\ B^{\mathrm{T}} & C \end{bmatrix}^{-1}$$
$$= \begin{bmatrix} (A - BC^{-1}B^{\mathrm{T}})^{-1} & -(A - BC^{-1}B^{\mathrm{T}})^{-1}BC^{-1} \\ -C^{-1}B^{\mathrm{T}}(A - BC^{-1}B^{\mathrm{T}})^{-1} & C^{-1}B^{\mathrm{T}}(A - BC^{-1}B^{\mathrm{T}})^{-1}BC^{-1} + C^{-1} \end{bmatrix}.$$

Fact 2.17.4. Let $A, B \in \mathbb{F}^{n \times n}$. Then,

$$\det \begin{bmatrix} I & A \\ B & I \end{bmatrix} = \det(I - AB) = \det(I - BA).$$

If $\det(I - BA) \neq 0$, then

$$\begin{bmatrix} I & A \\ B & I \end{bmatrix}^{-1} = \begin{bmatrix} I + A(I - BA)^{-1}B & -A(I - BA)^{-1} \\ -(I - BA)^{-1}B & (I - BA)^{-1} \end{bmatrix}$$

$$= \begin{bmatrix} (I - AB)^{-1} & -(I - AB)^{-1}A \\ -B(I - AB)^{-1} & I + B(I - AB)^{-1}A \end{bmatrix}.$$

Fact 2.17.5. Let $A, B \in \mathbb{F}^{n \times m}$. Then,

$$\begin{bmatrix} A & B \\ B & A \end{bmatrix} = \frac{1}{2} \begin{bmatrix} I & I \\ I & -I \end{bmatrix} \begin{bmatrix} A + B & 0 \\ 0 & A - B \end{bmatrix} \begin{bmatrix} I & I \\ I & -I \end{bmatrix}.$$

Therefore,

$$\operatorname{rank} \begin{bmatrix} A & B \\ B & A \end{bmatrix} = \operatorname{rank}(A + B) + \operatorname{rank}(A - B).$$

Now, assume in addition that $n = m$. Then,

$$\det \begin{bmatrix} A & B \\ B & A \end{bmatrix} = \det[(A + B)(A - B)] = \det(A^2 - B^2 - [A, B]).$$

Hence, $\begin{bmatrix} A & B \\ B & A \end{bmatrix}$ is nonsingular if and only if $A + B$ and $A - B$ are nonsingular. In this case,

$$\begin{bmatrix} A & B \\ B & A \end{bmatrix}^{-1} = \frac{1}{2} \begin{bmatrix} (A + B)^{-1} + (A - B)^{-1} & (A + B)^{-1} - (A - B)^{-1} \\ (A + B)^{-1} - (A - B)^{-1} & (A + B)^{-1} + (A - B)^{-1} \end{bmatrix},$$

$$(A + B)^{-1} = \frac{1}{2} \begin{bmatrix} I & I \end{bmatrix} \begin{bmatrix} A & B \\ B & A \end{bmatrix}^{-1} \begin{bmatrix} I \\ I \end{bmatrix},$$

and

$$(A - B)^{-1} = \frac{1}{2} \begin{bmatrix} I & -I \end{bmatrix} \begin{bmatrix} A & B \\ B & A \end{bmatrix}^{-1} \begin{bmatrix} I \\ -I \end{bmatrix}.$$

Remark: See Fact 6.5.1.

Fact 2.17.6. Let $A_1, \ldots, A_k \in \mathbb{F}^{n \times n}$, and assume that the $kn \times kn$ partitioned matrix below is nonsingular. Then, $A_1 + \cdots + A_k$ is nonsingular, and

$$(A_1 + \cdots + A_k)^{-1} = \frac{1}{k} \begin{bmatrix} I_n & \cdots & I_n \end{bmatrix} \begin{bmatrix} A_1 & A_2 & \cdots & A_k \\ A_k & A_1 & \cdots & A_{k-1} \\ \vdots & \vdots & \ddots & \vdots \\ A_2 & A_3 & \cdots & A_1 \end{bmatrix}^{-1} \begin{bmatrix} I_m \\ \vdots \\ I_m \end{bmatrix}.$$

Proof: See [1314].

Remark: The partitioned matrix is *block circulant*. See Fact 6.5.2 and Fact 6.6.1.

Fact 2.17.7. Let $\mathcal{A} \triangleq \left[\begin{smallmatrix} A & B \\ 0_{m\times m} & C \end{smallmatrix}\right]$, where $A \in \mathbb{F}^{n\times m}$, $B \in \mathbb{F}^{n\times n}$, and $C \in \mathbb{F}^{m\times n}$, and assume that CA is nonsingular. Furthermore, define $P \triangleq A(CA)^{-1}C$ and $P_\perp \triangleq I - P$. Then, $\mathcal{A}$ is nonsingular if and only if $P + P_\perp B P_\perp$ is nonsingular. In this case,

$$\mathcal{A}^{-1} = \left[\begin{array}{cc} (CA)^{-1}(C - CBD) & -(CA)^{-1}CB(A - DBA)(CA)^{-1} \\ D & (A - DBA)(CA)^{-1} \end{array} \right],$$

where $D \triangleq (P + P_\perp B P_\perp)^{-1}P_\perp$.

Proof: See [656].

Fact 2.17.8. Let $A \in \mathbb{F}^{n\times m}$ and $B \in \mathbb{F}^{n\times(n-m)}$, and assume that $\left[\begin{array}{cc} A & B \end{array} \right]$ is nonsingular and $A^*B = 0$. Then,

$$\left[\begin{array}{cc} A & B \end{array} \right]^{-1} = \left[\begin{array}{c} (A^*A)^{-1}A^* \\ (B^*B)^{-1}B^* \end{array} \right].$$

Remark: See Fact 6.5.18.

Problem: Find an expression for $\left[\begin{array}{cc} A & B \end{array} \right]^{-1}$ without assuming $A^*B = 0$.

Fact 2.17.9. Let $A \in \mathbb{F}^{n\times m}$, $B \in \mathbb{F}^{n\times l}$, and $C \in \mathbb{F}^{m\times l}$. Then,

$$\left[\begin{array}{ccc} I_n & A & B \\ 0 & I_m & C \\ 0 & 0 & I_l \end{array} \right]^{-1} = \left[\begin{array}{ccc} I_n & -A & AC - B \\ 0 & I_m & -C \\ 0 & 0 & I_l \end{array} \right].$$

Fact 2.17.10. Let $A \in \mathbb{F}^{n\times n}$, and assume that A is nonsingular. Then, $X = A^{-1}$ is the unique matrix satisfying

$$\mathrm{rank} \left[\begin{array}{cc} A & I \\ I & X \end{array} \right] = \mathrm{rank}\, A.$$

Remark: See Fact 6.3.29 and Fact 6.6.2.

Proof: See [496].

2.18 Facts on Commutators

Fact 2.18.1. Let $A, B \in \mathbb{F}^{2\times 2}$. Then,

$$[A, B]^2 = \tfrac{1}{2}(\mathrm{tr}\,[A, B]^2)I_2.$$

Remark: See [512, 513].

Fact 2.18.2. Let $A, B \in \mathbb{F}^{n\times n}$. Then,

$$\mathrm{tr}\,[A, B]^3 = 3\,\mathrm{tr}(A^2B^2AB - B^2A^2BA) = -3\,\mathrm{tr}(AB^2A[A, B]).$$

Fact 2.18.3. Let $A, B \in \mathbb{F}^{n\times n}$, assume that $[A, B] = 0$, and let $k, l \in \mathbb{N}$. Then, $[A^k, B^l] = 0$.

Fact 2.18.4. Let $A, B, C \in \mathbb{F}^{n \times n}$. Then, the following statements hold:

i) $[A, A] = 0$.

ii) $[A, B] = [-A, -B] = -[B, A]$.

iii) $[A, B + C] = [A, B] + [A, C]$.

iv) $[\alpha A, B] = [A, \alpha B] = \alpha[A, B]$ for all $\alpha \in \mathbb{F}$.

v) $[A, [B, C]] + [B, [C, A]] + [C, [A, B]] = 0$.

vi) $[A, B]^{\mathrm{T}} = [B^{\mathrm{T}}, A^{\mathrm{T}}] = -[A^{\mathrm{T}}, B^{\mathrm{T}}]$.

vii) $\mathrm{tr}\,[A, B] = 0$.

viii) $\mathrm{tr}\,A^k[A, B] = \mathrm{tr}\,B^k[A, B] = 0$ for all $k \geq 1$.

ix) $[[A, B], B - A] = [[B, A], A - B]$.

x) $[A, [A, B]] = -[A, [B, A]]$.

Remark: *v)* is the *Jacobi identity*.

Fact 2.18.5. Let $A, B \in \mathbb{F}^{n \times n}$. Then, for all $X \in \mathbb{F}^{n \times n}$,

$$\mathrm{ad}_{[A,B]} = [\mathrm{ad}_A, \mathrm{ad}_B],$$

that is,

$$\mathrm{ad}_{[A,B]}(X) = \mathrm{ad}_A[\mathrm{ad}_B(X)] - \mathrm{ad}_B[\mathrm{ad}_A(X)]$$

or, equivalently,

$$[[A, B], X] = [A, [B, X]] - [B, [A, X]].$$

Fact 2.18.6. Let $A \in \mathbb{F}^{n \times n}$ and, for all $X \in \mathbb{F}^{n \times n}$, define

$$\mathrm{ad}_A^k(X) \triangleq \begin{cases} \mathrm{ad}_A(X), & k = 1, \\ \mathrm{ad}_A^{k-1}[\mathrm{ad}_A(X)], & k \geq 2. \end{cases}$$

Then, for all $X \in \mathbb{F}^{n \times n}$ and $k \geq 1$,

$$\mathrm{ad}_A^2(X) = [A, [A, X]] - [[A, X], A]$$

and

$$\mathrm{ad}_A^k(X) = \sum_{i=0}^{k} (-1)^{k-i} \binom{k}{i} A^i X A^{k-i}.$$

Proof: For the last equality, see [1125, pp. 176, 207].

Remark: The proof of Proposition 11.4.7 is based on $g\left(e^{t\,\mathrm{ad}_A} e^{t\,\mathrm{ad}_B}\right)$, where $g(z) \triangleq (\log z)/(z - 1)$. See [1192, p. 35].

Remark: See Fact 11.14.4.

Fact 2.18.7. Let $A, B \in \mathbb{F}^{n \times n}$, and assume that $[A, B] = A$. Then, A is singular.

Proof: If A is nonsingular, then $\mathrm{tr}\,B = \mathrm{tr}\,ABA^{-1} = \mathrm{tr}\,B + n$.

Fact 2.18.8. Let $A, B \in \mathbb{R}^{n \times n}$ be such that $AB = BA$. Then, there exists a matrix $C \in \mathbb{R}^{n \times n}$ such that $A^2 + B^2 = C^2$.

Proof: See [425].

Remark: This result applies to real matrices only.

Fact 2.18.9. Let $A \in \mathbb{F}^{n \times n}$. Then,

$$n \leq \dim \{X \in \mathbb{F}^{n \times n} \colon AX = XA\}$$

and

$$\dim \{[A, X] \colon X \in \mathbb{F}^{n \times n}\} \leq n^2 - n.$$

Proof: See [400, pp. 125, 142, 493, 537].

Remark: The first set is the *centralizer* or *commutant* of A. See Fact 7.5.2.

Remark: These quantities are the defect and rank, respectively, of the operator $f \colon \mathbb{F}^{n \times n} \mapsto \mathbb{F}^{n \times n}$ defined by $f(X) \triangleq AX - XA$. See Fact 7.5.2.

Remark: See Fact 5.14.21 and Fact 5.14.23.

Fact 2.18.10. Let $A \in \mathbb{F}^{n \times n}$. Then, there exists $\alpha \in \mathbb{F}$ such that $A = \alpha I$ if and only if, for all $X \in \mathbb{F}^{n \times n}$, $AX = XA$.

Proof: To prove sufficiency, note that $A^{\mathrm{T}} \oplus -A = 0$. Hence, $\{0\} = \mathrm{spec}(A^{\mathrm{T}} \oplus -A) = \{\lambda - \mu \colon \lambda, \mu \in \mathrm{spec}(A)\}$. Therefore, $\mathrm{spec}(A) = \{\alpha\}$, and thus $A = \alpha I + N$, where N is nilpotent. Consequently, for all $X \in \mathbb{F}^{n \times n}$, $NX = XN$. Setting $X = N^*$, it follows that N is normal. Hence, $N = 0$.

Remark: This result determines the center subgroup of $\mathrm{GL}(n)$.

Fact 2.18.11. Define $\mathcal{S} \subseteq \mathbb{F}^{n \times n}$ by

$$\mathcal{S} \triangleq \{[X, Y] \colon X, Y \in \mathbb{F}^{n \times n}\}.$$

Then, $\mathcal{S}$ is a subspace. Furthermore,

$$\mathcal{S} = \{Z \in \mathbb{F}^{n \times n} \colon \mathrm{tr}\, Z = 0\}$$

and

$$\dim \mathcal{S} = n^2 - 1.$$

Proof: See [400, pp. 125, 493]. Alternatively, note that $\mathrm{tr} \colon \mathbb{F}^{n^2} \mapsto \mathbb{F}$ is onto, and use Corollary 2.5.5.

Fact 2.18.12. Let $A, B, C, D \in \mathbb{F}^{n \times n}$. Then, there exist $E, F \in \mathbb{F}^{n \times n}$ such that

$$[E, F] = [A, B] + [C, D].$$

Proof: This result follows from Fact 2.18.11.

Problem: Construct E and F.

2.19 Facts on Complex Matrices

Fact 2.19.1. Let $a, b \in \mathbb{R}$. Then, $\begin{bmatrix} a & b \\ -b & a \end{bmatrix}$ is a representation of the complex number $a + \jmath b$ that preserves addition, multiplication, and inversion of complex numbers. In particular, if $a^2 + b^2 \neq 0$, then

$$\begin{bmatrix} a & b \\ -b & a \end{bmatrix}^{-1} = \begin{bmatrix} \frac{a}{a^2+b^2} & \frac{-b}{a^2+b^2} \\ \frac{b}{a^2+b^2} & \frac{a}{a^2+b^2} \end{bmatrix}$$

and

$$(a + \jmath b)^{-1} = \frac{a}{a^2 + b^2} - \jmath \frac{b}{a^2 + b^2}.$$

Remark: $\begin{bmatrix} a & b \\ -b & a \end{bmatrix}$ is a *rotation-dilation*. See Fact 3.24.6.

Fact 2.19.2. Let $\nu, \omega \in \mathbb{R}$. Then,

$$\begin{bmatrix} \nu & \omega \\ -\omega & \nu \end{bmatrix} = \frac{1}{\sqrt{2}} \begin{bmatrix} 1 & 1 \\ \jmath & -\jmath \end{bmatrix} \begin{bmatrix} \nu + \jmath\omega & 0 \\ 0 & \nu - \jmath\omega \end{bmatrix} \frac{1}{\sqrt{2}} \begin{bmatrix} 1 & 1 \\ \jmath & -\jmath \end{bmatrix}^*$$

$$= \frac{1}{\sqrt{2}} \begin{bmatrix} 1 & \jmath \\ \jmath & 1 \end{bmatrix} \begin{bmatrix} \nu + \jmath\omega & 0 \\ 0 & \nu - \jmath\omega \end{bmatrix} \frac{1}{\sqrt{2}} \begin{bmatrix} 1 & \jmath \\ \jmath & 1 \end{bmatrix}^*$$

$$= \frac{1}{\sqrt{2}} \begin{bmatrix} 1 & -\jmath \\ \jmath & -1 \end{bmatrix} \begin{bmatrix} \nu + \jmath\omega & 0 \\ 0 & \nu - \jmath\omega \end{bmatrix} \frac{1}{\sqrt{2}} \begin{bmatrix} 1 & -\jmath \\ \jmath & -1 \end{bmatrix}$$

and

$$\begin{bmatrix} \nu & \omega \\ -\omega & \nu \end{bmatrix}^{-1} = \frac{1}{\nu^2 + \omega^2} \begin{bmatrix} \nu & -\omega \\ \omega & \nu \end{bmatrix}.$$

Remark: See Fact 2.19.1.

Remark: All three transformations are unitary. The third transformation is also Hermitian.

Fact 2.19.3. Let $A, B \in \mathbb{R}^{n \times m}$. Then,

$$\begin{bmatrix} A & B \\ -B & A \end{bmatrix} = \frac{1}{2} \begin{bmatrix} I & I \\ \jmath I & -\jmath I \end{bmatrix} \begin{bmatrix} A + \jmath B & 0 \\ 0 & A - \jmath B \end{bmatrix} \begin{bmatrix} I & -\jmath I \\ I & \jmath I \end{bmatrix}$$

$$= \frac{1}{2} \begin{bmatrix} I & \jmath I \\ -\jmath I & -I \end{bmatrix} \begin{bmatrix} A - \jmath B & 0 \\ 0 & A + \jmath B \end{bmatrix} \begin{bmatrix} I & \jmath I \\ -\jmath I & -I \end{bmatrix}$$

$$= \begin{bmatrix} I & 0 \\ \jmath I & I \end{bmatrix} \begin{bmatrix} A + \jmath B & B \\ 0 & A - \jmath B \end{bmatrix} \begin{bmatrix} I & 0 \\ -\jmath I & I \end{bmatrix}.$$

Consequently,

$$\begin{bmatrix} A + \jmath B & 0 \\ 0 & A - \jmath B \end{bmatrix} = \frac{1}{2} \begin{bmatrix} I & -\jmath I \\ I & \jmath I \end{bmatrix} \begin{bmatrix} A & B \\ -B & A \end{bmatrix} \begin{bmatrix} I & I \\ \jmath I & -\jmath I \end{bmatrix},$$

and thus

$$A + \jmath B = \frac{1}{2} \begin{bmatrix} I & -\jmath I \end{bmatrix} \begin{bmatrix} A & B \\ -B & A \end{bmatrix} \begin{bmatrix} I \\ \jmath I \end{bmatrix}.$$

Furthermore,

$$\text{rank}(A + \jmath B) = \text{rank}(A - \jmath B) = \tfrac{1}{2}\text{rank}\begin{bmatrix} A & B \\ -B & A \end{bmatrix}.$$

Now, assume in addition that $n = m$. Then,

$$\det\begin{bmatrix} A & B \\ -B & A \end{bmatrix} = \det(A + \jmath B)\det(A - \jmath B)$$

$$= |\det(A + \jmath B)|^2$$

$$= \det\left[A^2 + B^2 + \jmath(AB - BA)\right]$$

$$\geq 0.$$

Hence, $\begin{bmatrix} A & B \\ -B & A \end{bmatrix}$ is nonsingular if and only if $A + \jmath B$ is nonsingular. If A is nonsingular, then

$$\det\begin{bmatrix} A & B \\ -B & A \end{bmatrix} = \det\left(A^2 + ABA^{-1}B\right).$$

If $AB = BA$, then

$$\det\begin{bmatrix} A & B \\ -B & A \end{bmatrix} = \det\left(A^2 + B^2\right).$$

Proof: If A is nonsingular, then use

$$\begin{bmatrix} A & B \\ -B & A \end{bmatrix} = \begin{bmatrix} A & 0 \\ 0 & A \end{bmatrix}\begin{bmatrix} I & A^{-1}B \\ -A^{-1}B & I \end{bmatrix}$$

and

$$\det\begin{bmatrix} I & A^{-1}B \\ -A^{-1}B & I \end{bmatrix} = \det\left[I + \left(A^{-1}B\right)^2\right].$$

Remark: See Fact 4.10.27 and [82, 1313].

Fact 2.19.4. Let $A, B \in \mathbb{R}^{n \times m}$. Then, $\begin{bmatrix} A & B \\ -B & A \end{bmatrix}$ and $\begin{bmatrix} A & -B \\ B & A \end{bmatrix}$ are representations of the complex matrices $A + \jmath B$ and $\overline{A + \jmath B}$, respectively. Furthermore, $\begin{bmatrix} A^\mathrm{T} & B^\mathrm{T} \\ -B^\mathrm{T} & A^\mathrm{T} \end{bmatrix}$ and $\begin{bmatrix} A^\mathrm{T} & -B^\mathrm{T} \\ B^\mathrm{T} & A^\mathrm{T} \end{bmatrix}$ are representations of the complex matrices $(A + \jmath B)^\mathrm{T}$ and $(A + \jmath B)^*$, respectively.

Fact 2.19.5. Let $A, B \in \mathbb{R}^{n \times m}$ and $C, D \in \mathbb{R}^{m \times l}$. Then, for all $\alpha, \beta \in \mathbb{R}$, $\begin{bmatrix} A & B \\ -B & A \end{bmatrix}$, $\begin{bmatrix} C & D \\ -D & C \end{bmatrix}$, and $\begin{bmatrix} \alpha A + \beta C & \alpha B + \beta D \\ -(\alpha B + \beta D) & \alpha A + \beta C \end{bmatrix} = \alpha\begin{bmatrix} A & B \\ -B & A \end{bmatrix} + \beta\begin{bmatrix} C & D \\ -D & C \end{bmatrix}$ are representations of the complex matrices $A + \jmath B$, $C + \jmath D$, and $\alpha(A + \jmath B) + \beta(C + \jmath D)$, respectively.

Fact 2.19.6. Let $A, B \in \mathbb{R}^{n \times m}$ and $C, D \in \mathbb{R}^{m \times l}$. Then, $\begin{bmatrix} A & B \\ -B & A \end{bmatrix}$, $\begin{bmatrix} C & D \\ -D & C \end{bmatrix}$, and $\begin{bmatrix} AC - BD & AD + BC \\ -(AD + BC) & AC - BD \end{bmatrix} = \begin{bmatrix} A & B \\ -B & A \end{bmatrix}\begin{bmatrix} C & D \\ -D & C \end{bmatrix}$ are representations of the complex matrices $A + \jmath B$, $C + \jmath D$, and $(A + \jmath B)(C + \jmath D)$, respectively.

Fact 2.19.7. Let $A, B \in \mathbb{R}^{n \times n}$. Then, $A + \jmath B$ is nonsingular if and only if $\begin{bmatrix} A & B \\ -B & A \end{bmatrix}$ is nonsingular. In this case,

$$(A + \jmath B)^{-1} = \tfrac{1}{2}\begin{bmatrix} I & -\jmath I \end{bmatrix}\begin{bmatrix} A & B \\ -B & A \end{bmatrix}^{-1}\begin{bmatrix} I \\ \jmath I \end{bmatrix}.$$

If A is nonsingular, then $A + \jmath B$ is nonsingular if and only if $A + BA^{-1}B$ is nonsingular. In this case,

$$\begin{bmatrix} A & B \\ -B & A \end{bmatrix}^{-1} = \begin{bmatrix} (A + BA^{-1}B)^{-1} & -A^{-1}B(A + BA^{-1}B)^{-1} \\ A^{-1}B(A + BA^{-1}B)^{-1} & (A + BA^{-1}B)^{-1} \end{bmatrix}$$

and

$$(A + \jmath B)^{-1} = (A + BA^{-1}B)^{-1} - \jmath A^{-1}B(A + BA^{-1}B)^{-1}.$$

Alternatively, if B is nonsingular. Then, $A + \jmath B$ is nonsingular if and only if $B + AB^{-1}A$ is nonsingular. In this case,

$$\begin{bmatrix} A & B \\ -B & A \end{bmatrix}^{-1} = \begin{bmatrix} B^{-1}A(B + AB^{-1}A)^{-1} & -(B + AB^{-1}A)^{-1} \\ (B + AB^{-1}A)^{-1} & B^{-1}A(B + AB^{-1}A)^{-1} \end{bmatrix}$$

and

$$(A + \jmath B)^{-1} = B^{-1}A(B + AB^{-1}A)^{-1} - \jmath(B + AB^{-1}A)^{-1}.$$

Remark: See Fact 3.11.10, Fact 6.5.1, and [1314].

Fact 2.19.8. Let $A \in \mathbb{F}^{n \times n}$. Then,

$$\det(I + A\overline{A}) \geq 0.$$

Proof: See [426].

Fact 2.19.9. Let $A, B \in \mathbb{F}^{n \times n}$. Then,

$$\det \begin{bmatrix} A & B \\ -\overline{B} & \overline{A} \end{bmatrix} \geq 0.$$

If, in addition, A is nonsingular, then

$$\det \begin{bmatrix} A & B \\ -\overline{B} & \overline{A} \end{bmatrix} = |\det A|^2 \det\left(I + \overline{A^{-1}B}A^{-1}B\right).$$

Proof: See [1525].

Remark: Fact 2.19.8 implies that $\det\left(I + \overline{A^{-1}B}A^{-1}B\right) \geq 0$.

Fact 2.19.10. Let $A, B \in \mathbb{R}^{n \times n}$, and define $C \in \mathbb{R}^{2n \times 2n}$ by $C \triangleq \begin{bmatrix} C_{11} & C_{12} & \cdots \\ C_{21} & \cdots & \\ \vdots & & \end{bmatrix}$, where $C_{ij} \triangleq \begin{bmatrix} A_{(i,j)} & B_{(i,j)} \\ -B_{(i,j)} & A_{(i,j)} \end{bmatrix} \in \mathbb{R}^{2 \times 2}$ for all $i, j \in \{1, \ldots, n\}$. Then,

$$\det C = |\det(A + \jmath B)|^2.$$

Proof: Note that

$$C = A \otimes I_2 + B \otimes J_2 = P_{2,n}(I_2 \otimes A + J_2 \otimes B)P_{2,n} = P_{2,n} \begin{bmatrix} A & B \\ -B & A \end{bmatrix} P_{2,n}.$$

See [261].

2.20 Facts on Geometry

Fact 2.20.1. The points $x, y, z \in \mathbb{R}^2$ lie on one line if and only if

$$\det \begin{bmatrix} x & y & z \\ 1 & 1 & 1 \end{bmatrix} = 0.$$

Fact 2.20.2. The points $w, x, y, z \in \mathbb{R}^3$ lie in one plane if and only if

$$\det \begin{bmatrix} w & x & y & z \\ 1 & 1 & 1 & 1 \end{bmatrix} = 0.$$

Fact 2.20.3. Let $x_1, \ldots, x_n \in \mathbb{R}^n$. Then,

$$\text{rank} \begin{bmatrix} 1 & \cdots & 1 \\ x_1 & \cdots & x_n \end{bmatrix} = \text{rank} \begin{bmatrix} 1 & 0 & \cdots & 0 \\ x_1 & x_2 - x_1 & \cdots & x_n - x_1 \end{bmatrix}.$$

Hence,

$$\text{rank} \begin{bmatrix} 1 & \cdots & 1 \\ x_1 & \cdots & x_n \end{bmatrix} = n$$

if and only if

$$\text{rank} \begin{bmatrix} x_2 - x_1 & \cdots & x_n - x_1 \end{bmatrix} = n - 1.$$

In this case,

$$\text{aff} \{x_1, \ldots, x_n\} = x_1 + \text{span} \{x_2 - x_1, \ldots, x_n - x_1\},$$

and thus aff $\{x_1, \ldots, x_n\}$ is an affine hyperplane. Finally,

$$\text{aff} \{x_1, \ldots, x_n\} = \{x \in \mathbb{R}^n \colon \det \begin{bmatrix} 1 & 1 & \cdots & 1 \\ x & x_1 & \cdots & x_n \end{bmatrix} = 0\}.$$

Proof: See [1215, p. 31].

Remark: See Fact 2.20.4.

Fact 2.20.4. Let $x_1, \ldots, x_{n+1} \in \mathbb{R}^n$. Then, the following statements are equivalent:

i) co $\{x_1, \ldots, x_{n+1}\}$ is a simplex.

ii) co $\{x_1, \ldots, x_{n+1}\}$ has nonempty interior.

iii) aff $\{x_1, \ldots, x_{n+1}\} = \mathbb{R}^n$.

iv) span $\{x_2 - x_1, \ldots, x_{n+1} - x_1\} = \mathbb{R}^n$.

v) $\begin{bmatrix} 1 & \cdots & 1 \\ x_1 & \cdots & x_{n+1} \end{bmatrix}$ is nonsingular.

Proof: The equivalence of *i)* and *ii)* follows from Fact 10.8.9. The equivalence of *i)* and *iv)* follows from Fact 2.9.7. Finally, the equivalence of *iv)* and *v)* follows from

$$\begin{bmatrix} 1 & \cdots & 1 \\ x_1 & \cdots & x_{n+1} \end{bmatrix} = \begin{bmatrix} 1 & 0 & \cdots & 0 \\ x_1 & x_2 - x_1 & \cdots & x_{n+1} - x_1 \end{bmatrix} \begin{bmatrix} 1 & 1 & 1 & \cdots & 1 \\ 0 & 1 & 0 & \cdots & 0 \\ 0 & 0 & 1 & \cdots & 0 \\ \vdots & \vdots & \ddots & \ddots & \vdots \\ 0 & 0 & \cdots & \cdots & 1 \end{bmatrix}.$$

Remark: See Fact 2.20.3 and Fact 10.8.12.

Fact 2.20.5. Let z_1, z_2, z be complex numbers, and assume that $z_1 \neq z_2$. Then, the following statements are equivalent:

i) z lies on the line passing through z_1 and z_2.

ii) $\frac{z-z_1}{z_2-z_1}$ is real.

iii) $\det \begin{bmatrix} z - z_1 & \overline{z} - \overline{z_1} \\ z_2 - z_1 & \overline{z_2} - \overline{z_1} \end{bmatrix} = 0.$

iv) $\det \begin{bmatrix} z & \overline{z} & 1 \\ z_1 & \overline{z_1} & 1 \\ z_2 & \overline{z_2} & 1 \end{bmatrix} = 0.$

Furthermore, the following statements are equivalent:

v) z lies on the line segment connecting z_1 and z_2.

vi) $\frac{z-z_1}{z_2-z_1}$ is a positive number.

vii) There exists $\phi \in (-\pi, \pi]$ such that $|z - z_1|e^{J\phi} = |z_2 - z_1|e^{J\phi}$.

Proof: See [61, pp. 54–56].

Fact 2.20.6. Let z_1, z_2, z_3 be distinct complex numbers. Then, the following statements are equivalent:

i) z_1, z_2, z_3 are the vertices of an equilateral triangle.

ii) $|z_1 - z_2| = |z_2 - z_3| = |z_3 - z_1|$.

iii) $z_1^2 + z_2^2 + z_3^2 = z_1 z_2 + z_2 z_3 + z_3 z_1$.

iv) $\frac{z_2 - z_1}{z_3 - z_2} = \frac{z_3 - z_2}{z_1 - z_2}$.

Proof: See [61, pp. 70, 71] and [893, p. 316].

Fact 2.20.7. Let $S \subset \mathbb{R}^2$ denote the triangle with vertices $\begin{bmatrix} 0 \\ 0 \end{bmatrix}, \begin{bmatrix} x_1 \\ y_1 \end{bmatrix}, \begin{bmatrix} x_2 \\ y_2 \end{bmatrix} \in \mathbb{R}^2$. Then,

$$\text{area}(S) = \tfrac{1}{2} \left| \det \begin{bmatrix} x_1 & x_2 \\ y_1 & y_2 \end{bmatrix} \right|.$$

Fact 2.20.8. Let $S \subset \mathbb{R}^2$ denote the triangle with vertices $\begin{bmatrix} x_1 \\ y_1 \end{bmatrix}, \begin{bmatrix} x_2 \\ y_2 \end{bmatrix}, \begin{bmatrix} x_3 \\ y_3 \end{bmatrix} \in \mathbb{R}^2$. Then,

$$\text{area}(S) = \tfrac{1}{2} \left| \det \begin{bmatrix} 1 & 1 & 1 \\ x_1 & x_2 & x_3 \\ y_1 & y_2 & y_3 \end{bmatrix} \right|.$$

Proof: See [1215, p. 32].

Fact 2.20.9. Let z_1, z_2, z_3 be complex numbers. Then, the area of the triangle $\mathcal{S}$ formed by z_1, z_2, z_3 is given by

$$\text{area}(\mathcal{S}) = \tfrac{1}{4}\left|\det\begin{bmatrix} z_1 & \overline{z_1} & 1 \\ z_2 & \overline{z_2} & 1 \\ z_3 & \overline{z_3} & 1 \end{bmatrix}\right|.$$

Proof: See [61, p. 79].

Fact 2.20.10. Let $\mathcal{S} \subset \mathbb{R}^3$ denote the triangle with vertices $x, y, z \in \mathbb{R}^3$. Then,

$$\text{area}(\mathcal{S}) = \tfrac{1}{2}\sqrt{[(y-x)\times(z-x)]^{\mathrm{T}}[(y-x)\times(z-x)]} = \tfrac{1}{2}\|(y-x)\times(z-x)\|_2.$$

Furthermore,

$$\text{area}(\mathcal{S}) = \tfrac{1}{2}\|y-x\|_2\|z-x\|_2\sin\theta,$$

where $\theta \in (0, \pi)$ is the angle between $y - x$ and $z - x$. Now, assume in addition that $x = 0$. Then,

$$\text{area}(\mathcal{S}) = \tfrac{1}{2}\sqrt{[y\times z]^{\mathrm{T}}[y\times z]} = \tfrac{1}{2}\|y\times z\|_2.$$

Furthermore,

$$\text{area}(\mathcal{S}) = \tfrac{1}{2}\|y\|_2\|z\|_2\sin\theta,$$

where $\theta \in (0, \pi)$ is the angle between y and z.

Remark: The connection between the norm of the cross product of two vectors and the angle between the vectors is given by *xxviii)* of Fact 3.10.1.

Fact 2.20.11. Let $\mathcal{S} \subset \mathbb{R}^2$ denote a triangle whose sides have lengths a, b, and c, let A, B, C denote the angles of the triangle opposite the sides having lengths a, b, and c, respectively, define the semiperimeter $s \triangleq \tfrac{1}{2}(a+b+c)$, let r denote the radius of the largest inscribed circle, and let R denote the radius of the smallest circumscribed circle. Then, the following statements hold:

i) $A + B + C = \pi$.

ii) $a^2 + b^2 = c^2 + 2ab\cos C$.

iii) $\frac{\sin A}{a} = \frac{\sin B}{b} = \frac{\sin C}{c}$.

iv) $\text{area}(\mathcal{S}) = \tfrac{1}{2}ab\sin C = \frac{c^2}{2}\frac{(\sin A)\sin B}{\sin C}$.

v) $\text{area}(\mathcal{S}) = \sqrt{s(s-a)(s-b)(s-c)} = rs = \frac{abc}{4R}$.

vi) $\text{area}(\mathcal{S}) \le \frac{\sqrt{3}}{12}(a^2 + b^2 + c^2)$.

vii) If $\mathcal{S}$ is equilateral, then $\text{area}(\mathcal{S}) = \frac{\sqrt{3}}{4}a^2$ and $R = 2r = \frac{\sqrt{3}}{3}a$.

viii) a, b, c are the roots of the cubic equation

$$x^3 - 2sx^2 + (s^2 + r^2 + 4rR)x - 4srR = 0.$$

That is,

$$a + b + c = 2s, \quad ab + bc + ca = s^2 + r^2 + 4rR, \quad abc = 4rRs.$$

ix) a, b, c satisfy

$$a^2 + b^2 + c^2 = 2(s^2 - r^2 - 4rR)$$

and

$$a^3 + b^3 + c^3 = 2s(s^2 - 3r^2 - 6rR).$$

x) If r_1, r_2, r_3 denote the altitudes of the triangle, then

$$\frac{1}{r} = \frac{1}{r_1} + \frac{1}{r_2} + \frac{1}{r_3}.$$

xi) $r \le \frac{1}{2}\left(\frac{2}{1+\sqrt{5}}\right)^{5/2}(a+b) \approx 0.15(a+b)$. If, in addition, $\mathcal{S}$ is equilateral, then
$r = \frac{\sqrt{3}}{12}(a+b) \approx 0.14(a+b)$.

Furthermore, the following statements hold:

xii) $2 \le \frac{a}{b} + \frac{b}{a} \le \frac{R}{r}$.

xiii) $2 \le \frac{2}{3}\left(\frac{a}{b} + \frac{b}{c} + \frac{c}{a}\right) \le \frac{a}{b} + \frac{b}{c} + \frac{c}{a} - 1 \le \frac{1}{2}\left(1 + \frac{a^2}{bc} + \frac{b^2}{ca} + \frac{c^2}{ab}\right) \le \frac{R}{r}$.

xiv) $1 \le \frac{2a^2}{2a^2-(b-c)^2}\frac{2b^2}{2b^2-(c-a)^2}\frac{2c^2}{2c^2-(a-b)^2} \le \frac{R}{2r}$.

xv) $\frac{a}{2}\frac{4r-R}{R} \le \sqrt{(s-b)(s-c)} \le \frac{a}{2}$.

xvi) A triangle $\mathcal{S}$ with values area($\mathcal{S}$), r, and R exists if and only if

$$r\sqrt{2R^2 + 10rR - r^2 - 2(R-2r)\sqrt{R(R-2r)}}$$
$$\le \text{area}(\mathcal{S}) \le r\sqrt{2R^2 + 10rR - r^2 + 2(R-2r)\sqrt{R(R-2r)}}.$$

xvii) Let $\theta \triangleq \min\{|A - B|, |A - C|, |B - C|\}_{\text{ms}}$. Then,

$$r\sqrt{2R^2 + 10rR - r^2 - 2(R-2r)\sqrt{R(R-2r)}\cos\theta}$$
$$\le \text{area}(\mathcal{S}) \le r\sqrt{2R^2 + 10rR - r^2 + 2(R-2r)\sqrt{R(R-2r)}\cos\theta}.$$

xviii) area($\mathcal{S}$) $\le (R + \frac{1}{2}r)^2$.

xix) area($\mathcal{S}$) $\le \frac{1}{\sqrt{3}}(R+r)^2$.

xx) area($\mathcal{S}$) $\le \frac{3\sqrt{3}}{25}(R+3r)^2$.

xxi) $3\sqrt{3}r^2 \le \text{area}(\mathcal{S}) \le 2rR + (3\sqrt{3} - 4)r^2$.

xxii) $r\sqrt{16rR - 5r^2} \le \text{area}(\mathcal{S}) \le r\sqrt{4R^2 + 4rR + 3r^2}$.

xxiii) For all $n \ge 0$, $a^n + b^n + c^n \le 2^{n+1}R^n + 2^n(3^{1+n/2} - 2^{1+n})r^n$.

xxiv) A triangle $\mathcal{S}$ with values $u = \cos A$, $v = \cos B$, and $v = \cos C$ exists if and only if $u + v + w \ge 1$, $uvw \ge -1$, and $u^2 + v^2 + w^2 + 2uvw = 1$.

xxv) If P is a point inside $\mathcal{S}$ and d_1, d_2, d_3 are the distances from P to each of the sides, then

$$\sqrt{d_1} + \sqrt{d_2} + \sqrt{d_3} \le \sqrt{\frac{a^2+b^2+c^2}{2R}}.$$

In particular,

$$18R^2 \le a^2 + b^2 + c^2.$$

xxvi) $4r^2[8R^2 - (a^2 + b^2 + c^2)] \le R^2(R^2 - 4r^2)$.

xxvii) $abc \le 3\sqrt{3}R^3$.

xxviii) The triangle $\mathcal{S}$ is similar to the triangle $\mathcal{S}'$ with sides of length a', b', c' if and only if

$$\sqrt{aa'} + \sqrt{bb'} + \sqrt{cc'} = \sqrt{(a+b+c)(a'+b'+c')}.$$

xxix) $(\sin \frac{1}{2}A)(\sin \frac{1}{2}B)(\sin \frac{1}{2}C) < (\sin \frac{1}{2}\sqrt[3]{ABC})^3 < \frac{1}{8}$.

xxx) $(\cos \frac{1}{2}A)(\cos \frac{1}{2}B)(\cos \frac{1}{2}C) < [\sin \frac{1}{2}\sqrt[3]{(\pi - A)(\pi - B)(\pi - C)}]^3$.

xxxi) $(\tan \frac{1}{2}\sqrt[3]{ABC})^3 < (\tan \frac{1}{2}A)(\tan \frac{1}{2}B)(\tan \frac{1}{2}C)$.

xxxii) $1 \le \tan^2(\frac{1}{2}A) + \tan^2(\frac{1}{2}B) + \tan^2(\frac{1}{2}C)$.

xxxiii) $\frac{\pi}{3}(a+b+c) \le Aa + Bb + Cc \le \frac{\pi - \min\{A,B,C\}}{2}(a+b+c)$.

xxxiv) If x, y, z are positive numbers, then

$$x \sin A + y \sin B + z \sin C \le \tfrac{1}{2}(xy + yz + zx)\sqrt{\frac{1}{xy} + \frac{1}{yz} + \frac{1}{zx}}$$

$$\le \frac{\sqrt{3}}{2}\left(\frac{yz}{x} + \frac{zx}{y} + \frac{xy}{z}\right).$$

xxxv) $\sin A + \sin B + \sin C \le \frac{3\sqrt{3}}{2}$.

Proof: Results *i*)–*v*) are classical. The first expression for area($\mathcal{S}$) in *v*) is *Heron's formula*. Statements *ii*) and *iii*) are the *cosine rule* and *sine rule*, respectively. See [1540, p. 319]. Statement *vi*) is due to Weitzenbock. See [61, p. 145] and [470, p. 170]. The expression for area($\mathcal{S}$) in *vii*) follows from *v*) and provides the case of equality in *vi*). Statements *viii*) and *ix*) are given in [61, pp. 110, 111]. Statement *xi*) is given in [105]. Statements *xii*) and *xiii*) are given in [1408]. Statement *xiv*) is due to [1124]. See [470, p. 174]. Statement *xv*) is given in [1175]. Statement *xvi*), which is due to Ramus, is the *fundamental triangle inequality*. See [1036]. The interpolation of *xvi*) given by *xvii*) is given in [1497]. The bounds *xviii*)–*xx*) are given in [1499]. The bounds *xxi*) and *xxii*) are due to Blundon. See [1191]. Statement *xxiii*) is given in [1191]. Statement *xxiv*) is given in [637]. Statement *xxv*) is given in [893, pp. 255, 256]. Statement *xxvi*) follows from [61, p. 189]. Statement *xxvii*) follows from [61, p. 144]. Statement *xxviii*) is given in [470, p. 183]. Necessity is immediate. Statements *xxix*)–*xxxi*) are given in [1067]. Statement *xxxii*) is given in [140, p. 231]. Statement *xxxiii*) is given in [996, p. 203]. The first inequality in statement *xxxiv*) is *Klamkin's inequality*. The first and third terms constitute *Vasic's inequality*. See [1408]. Statement *xxxiv*) follows from statement *xxxii*) with $x = y = z = 1$.

Remark: $2r \le R$ in *xii*) is *Euler's inequality*. The interpolation is *Bandila's inequality*. The inequality consisting of the second and fifth terms in *xiii*) is due to Zhang and Song. See [1408].

Remark: The bound *xxi*) is *Mircea's inequality*, while *xxii*) is due to Carliz and Leuenberger. See [1499].

Remark: Additional inequalities for the sides and angles of a triangle are given in Fact 1.13.21, [248], and [996, pp. 192–203].

Remark: The second inequality in *xxxiv*) is given in Fact 1.13.10.

Fact 2.20.12. Let a be a complex number, let $b \in (0, |a|^2)$, and define

$$\mathcal{S} \triangleq \{z \in \mathbb{C} \colon |z|^2 - \bar{a}z - a\bar{z} + b = 0\}.$$

Then, $\mathcal{S}$ is the circle with center at a and radius $\sqrt{|a|^2 - b}$. That is,

$$\mathcal{S} = \{z \in \mathbb{C} \colon |z - a| = \sqrt{|a|^2 - b}\}.$$

Proof: See [61, p. 84, 85].

Fact 2.20.13. Let $\mathcal{S} \subset \mathbb{R}^2$ be a convex quadrilateral whose sides have lengths a, b, c, d, define the semiperimeter $s \triangleq \frac{1}{2}(a + b + c + d)$, let A, B, C, D denote the angles of $\mathcal{S}$ labeled consecutively, and define $\theta \triangleq \frac{1}{2}(A + C) = \pi - \frac{1}{2}(B + D)$. Then,

$$\text{area}(\mathcal{S}) = \sqrt{(s - a)(s - b)(s - c)(s - d) - abcd \cos^2 \theta}.$$

Now, let p, q be the lengths of the diagonals of $\mathcal{S}$. Then,

$$pq \leq ac + bc$$

and

$$\text{area}(\mathcal{S}) = \sqrt{(s - a)(s - b)(s - c)(s - d) - \tfrac{1}{4}(ac + bd + pq)(ac + bd - pq)}.$$

If the quadrilateral has an inscribed circle that contacts all four sides of the quadrilateral, then

$$\text{area}(\mathcal{S}) = \sqrt{abcd} = \sqrt{p^2 q^2 - (ac - bd)^2}.$$

Finally, all of the vertices of $\mathcal{S}$ lie on a circle if and only if

$$pq = ac + bc.$$

In this case,

$$\text{area}(\mathcal{S}) = \sqrt{(s - a)(s - b)(s - c)(s - d)}$$

and

$$\text{area}(\mathcal{S}) = \frac{1}{4R} \sqrt{(ad + bc)(ac + bd)(ab + cd)},$$

where R is the radius of the circumscribed circle.

Proof: See [62, pp. 37, 38], Wikipedia, PlanetMath, and MathWorld.

Remark: $pq \leq ac + bc$ is *Ptolemy's inequality*, which holds for nonconvex quadrilaterals. The equality case is *Ptolemy's theorem*. See [61, p. 130].

Remark: The fourth expression for area($\mathcal{S}$) is *Brahmagupta's formula*. The limiting case $d = 0$ yields Heron's formula. See Fact 2.20.11.

Remark: For each quadrilateral, there exists a quadrilateral with the same side lengths and whose vertices lie on a circle. The area of the latter quadrilateral is maximum over all quadrilaterals with the same side lengths. See [1109].

Remark: See Fact 9.7.5.

Problem: For which quadrilaterals does there exist a quadrilateral with the same side lengths and whose sides are tangent to an inscribed circle?

Fact 2.20.14. Let $\mathcal{S} \subset \mathbb{R}^2$ denote the polygon with vertices $\begin{bmatrix} x_1 \\ y_1 \end{bmatrix}, \ldots, \begin{bmatrix} x_n \\ y_n \end{bmatrix} \in \mathbb{R}^2$ arranged in counterclockwise order, and assume that the interior of the polygon is either empty or simply connected. Then,

$$\text{area}(\mathcal{S}) = \tfrac{1}{2}\det \begin{bmatrix} x_1 & x_2 \\ y_1 & y_2 \end{bmatrix} + \tfrac{1}{2}\det \begin{bmatrix} x_2 & x_3 \\ y_2 & y_3 \end{bmatrix} + \cdots$$

$$+ \tfrac{1}{2}\det \begin{bmatrix} x_{n-1} & x_n \\ y_{n-1} & y_n \end{bmatrix} + \tfrac{1}{2}\det \begin{bmatrix} x_n & x_1 \\ y_n & y_1 \end{bmatrix}.$$

Remark: The polygon need not be convex, while "counterclockwise" is determined with respect to a point in the interior of the polygon. *Simply connected* means that the polygon has no holes. See [1268].

Remark: See [61, p. 100].

Remark: See Fact 9.7.5.

Fact 2.20.15. Let $\mathcal{S} \subset \mathbb{R}^3$ denote the tetrahedron with vertices $x, y, z, w \in \mathbb{R}^3$. Then,
$$\text{volume}(\mathcal{S}) = \tfrac{1}{6}\left|(x - w)^{\mathrm{T}}[(y - w) \times (z - w)]\right|.$$

Proof: The volume of the unit simplex $\mathcal{S} \subset \mathbb{R}^3$ with vertices $(0,0,0), (1,0,0)$, $(0,1,0), (0,0,1)$ is $1/6$. Now, Fact 2.20.18 implies that the volume of $A\mathcal{S}$ is $(1/6)|\det A|$.

Remark: The connection between the *signed volume* of a simplex and the determinant is discussed in [903, pp. 32, 33].

Fact 2.20.16. Let $\mathcal{S} \subset \mathbb{R}^3$ denote the parallelepiped with vertices $x, y, z, x + y, x + z, y + z, x + y + z \in \mathbb{R}^3$. Then,

$$\text{volume}(\mathcal{S}) = |\det \begin{bmatrix} x & y & z \end{bmatrix}|.$$

Fact 2.20.17. Let $A \in \mathbb{R}^{n \times m}$, assume that $\text{rank}\, A = m$, and let $\mathcal{S} \subset \mathbb{R}^n$ denote the parallelepiped in $\mathbb{R}^n$ with a vertex at 0 and generated by the m columns of A, that is,

$$\mathcal{S} = \left\{ \sum_{i=1}^m \alpha_i \text{col}_i(A) \colon 0 \le \alpha_i \le 1 \text{ for all } i \in \{1, \ldots, m\} \right\}.$$

Then,
$$\text{volume}(\mathcal{S}) = \left[\det\left(A^{\mathrm{T}}A\right)\right]^{1/2}.$$

If, in addition, $m = n$, then
$$\text{volume}(\mathcal{S}) = |\det A|.$$

Remark: volume$(\mathcal{S})$ denotes the m-dimensional volume of $\mathcal{S}$. If $m = 2$, then volume$(\mathcal{S})$ is the area of a parallelogram. See [459, p. 202].

Fact 2.20.18. Let $S \subset \mathbb{R}^n$ and $A \in \mathbb{R}^{n \times n}$. Then,

$$\text{volume}(AS) = |\det A| \text{volume}(S).$$

Remark: See [1023, p. 468].

Fact 2.20.19. Let $S \subset \mathbb{R}^n$ be a simplex, and assume that S is inscribed in a sphere of radius R. Then,

$$\text{volume}(S) \le \sqrt{\frac{(n+1)^{n+1}}{n^n}} \frac{R^n}{n!}.$$

Furthermore, equality holds if and only if S is a regular polytope.

Proof: See [1407].

Remark: See [495, p. 66-13].

Fact 2.20.20. Let $x_1, \ldots, x_{n+1} \in \mathbb{R}^n$, define

$$S \triangleq \text{co}\{x_1, \ldots, x_{n+1}\},$$

and define $A \in \mathbb{R}^{(n+2) \times (n+2)}$ by

$$A \triangleq \begin{bmatrix} 0 & 1 & 1 & \cdots & 1 \\ 1 & 0 & \|x_1 - x_2\|_2^2 & \cdots & \|x_1 - x_{n+1}\|_2^2 \\ 1 & \|x_2 - x_1\|_2^2 & 0 & \cdots & \|x_2 - x_{n+1}\|_2^2 \\ \vdots & \vdots & \vdots & \ddots & \vdots \\ 1 & \|x_{n+1} - x_1\|_2^2 & \|x_{n+1} - x_2\|_2^2 & \cdots & 0 \end{bmatrix}.$$

Then, the n-dimensional volume of S is given by

$$\text{vol}(S) = \frac{\sqrt{|\det A|}}{2^{n-1} n!}.$$

Proof: See [236, pp. 97–99] and [242, pp. 234, 235].

Remark: $\det A$ is the *Cayley-Menger determinant*.

Remark: In the case $n = 2$, this result yields Heron's formula for the area of a triangle. See Fact 2.20.11.

Fact 2.20.21. Let S denote the spherical triangle on the surface of the unit sphere whose vertices are $x, y, z \in \mathbb{R}^3$, and let A, B, C denote the angles of S located at the points x, y, z, respectively. Furthermore, let a, b, c denote the planar angles subtended by the pairs (y, z), (x, z), (x, y), respectively, or, equivalently, a, b, c denote the sides of the spherical triangle opposite A, B, C, respectively. Finally, define the solid angle Ω to be the area of S. Then,

$$\Omega = A + B + C - \pi.$$

Furthermore,

$$\tan\frac{\Omega}{2} = \frac{|[\begin{array}{ccc} x & y & z \end{array}]|}{1 + x^\mathrm{T}y + x^\mathrm{T}z + y^\mathrm{T}z}.$$

Equivalently,

$$\tan\frac{\Omega}{2} = \frac{\sqrt{1 - \cos^2 a - \cos^2 b - \cos^2 c + 2(\cos a)(\cos b)\cos c}}{1 + \cos a + \cos b + \cos c}.$$

Finally,

$$\tan\frac{\Omega}{4} = \sqrt{(\tan\tfrac{s}{2})(\tan\tfrac{s-a}{2})(\tan\tfrac{s-b}{2})\tan\tfrac{s-c}{2}}.$$

Proof: See [474] and [1540, pp. 368–371].

Remark: Spherical triangles are discussed in [490, pp. 253–260], [775, Chapter 2], [1459, pp. 904–907], and [1470, pp. 26–29]. A linear algebraic approach is given in [131].

Fact 2.20.22. Let $\mathcal{S}$ denote a circular cap on the surface of the unit sphere, where the angle subtended by cross sections of the cone with apex at the center of the sphere is 2θ. Furthermore, define the solid angle Ω to be the area of $\mathcal{S}$. Then,

$$\Omega = 2\pi(1 - \cos\theta).$$

Fact 2.20.23. Let $\mathcal{S}$ denote a region on the surface of the unit sphere subtended by the sides of a right rectangular pyramid with apex at the center of the sphere, where the subtended planar angles of the edges of the pyramid are θ and ϕ. Furthermore, define the solid angle Ω to be the area of $\mathcal{S}$. Then,

$$\Omega = 4\sin^{-1}\left[(\sin\tfrac{\theta}{2})\sin\tfrac{\phi}{2}\right].$$

2.21 Facts on Majorization

Fact 2.21.1. Let $x \in \mathbb{R}^n$, where $x_{(1)} \geq \cdots \geq x_{(n)} \geq 0$, and assume that $\sum_{i=1}^{n} x_{(i)} = 1$. Then, $e_{1,n}$ strongly majorizes x, and x strongly majorizes $\frac{1}{n}1_{n\times 1}$.

Proof: See [996, p. 95].

Remark: See Fact 2.21.2.

Fact 2.21.2. Let $x, y, z \in \mathbb{R}^n$, assume that $x_{(1)} \geq \cdots \geq x_{(n)}$, $y_{(1)} \geq \cdots \geq y_{(n)}$, and $z_{(1)} \geq \cdots \geq z_{(n)} \geq 0$, and assume that y weakly majorizes x. Then,

$$x^\mathrm{T}z \leq y^\mathrm{T}z.$$

Proof: See [996, p. 95].

Remark: See Fact 2.21.3.

Fact 2.21.3. Let $x, y, z \in \mathbb{R}^n$, assume that $x_{(1)} \geq \cdots \geq x_{(n)}$, $y_{(1)} \geq \cdots \geq y_{(n)}$, and $z_{(1)} \geq \cdots \geq z_{(n)}$, and assume that y strongly majorizes x. Then,

$$x^\mathrm{T}z \leq y^\mathrm{T}z.$$

Proof: See [996, p. 92].

Fact 2.21.4. Let $a < b$, let $f \colon (a,b)^n \mapsto \mathbb{R}$, and assume that f is C^1. Then, f is Schur convex if and only if f is symmetric and, for all $x \in (a,b)^n$,

$$(x_{(1)} - x_{(2)}) \left(\frac{\partial f(x)}{\partial x_{(1)}} - \frac{\partial f(x)}{\partial x_{(2)}} \right) \geq 0.$$

Proof: See [996, p. 57].

Remark: f is symmetric means that $f(Ax) = f(x)$ for all $x \in (a,b)^n$ and for every permutation matrix $A \in \mathbb{R}^{n \times n}$.

Remark: See [801].

Fact 2.21.5. Let $x, y \in \mathbb{R}^n$, assume that $x_{(1)} \geq \cdots \geq x_{(n)} \geq 0$ and $y_{(1)} \geq \cdots \geq y_{(n)} \geq 0$, assume that y strongly majorizes x, and let $p_1, \ldots, p_n$ be nonnegative numbers. Then,

$$\sum \prod_{j=1}^{n} p_{i_j}^{x_{(j)}} \leq \tfrac{1}{n!} \sum \prod_{j=1}^{n} p_{i_j}^{y_{(j)}}$$

where the summation is taken over all $n!$ permutations $\{i_1, \ldots, i_n\}$ of $\{1, \ldots, n\}$.

Proof: See [556, p. 99] and [996, p. 88].

Remark: This result is *Muirhead's theorem*, which is based on a function that is *Schur convex*. An immediate consequence is an interpolated version of the arithmetic-mean–geometric-mean inequality. See Fact 1.17.25.

Fact 2.21.6. Let $x, y \in \mathbb{R}^n$, where $x_{(1)} \geq \cdots \geq x_{(n)} \geq 0$ and $y_{(1)} \geq \cdots \geq y_{(n)} \geq 0$, assume that y strongly majorizes x, and assume that $\sum_{i=1}^{n} x_{(i)} = 1$. Then,

$$\sum_{i=1}^{n} y_i \log \tfrac{1}{y_{(i)}} \leq \sum_{i=1}^{n} x_i \log \tfrac{1}{x_{(i)}} \leq \log n.$$

Proof: See [556, p. 102] and [996, pp. 71, 405].

Remark: For $x_{(1)}, x_{(2)} > 0$, note that $(x_{(1)} - x_{(2)}) \log(x_{(1)}/x_{(2)}) \geq 0$. Hence, it follows from Fact 2.21.4 that the entropy function is *Schur concave*.

Remark: Entropy bounds are given in Fact 1.17.46, Fact 1.17.47, and Fact 1.17.48.

Fact 2.21.7. Let $x, y \in \mathbb{R}^n$, where $x_{(1)} \geq \cdots \geq x_{(n)}$ and $y_{(1)} \geq \cdots \geq y_{(n)}$, assume that y strongly majorizes x, let $f \colon [\min\{x_{(n)}, y_{(n)}\}, y_{(1)}] \mapsto \mathbb{R}$, assume that f is convex, and let $\{i_1, \ldots, i_n\} = \{j_1, \ldots, j_n\} = \{1, \ldots, n\}$ be such that $f(x_{(i_1)}) \geq \cdots \geq f(x_{(i_n)})$ and $f(y_{(i_1)}) \geq \cdots \geq f(y_{(i_n)})$. Then, $\begin{bmatrix} f(y_{(j_1)}) & \cdots & f(y_{(j_n)}) \end{bmatrix}^{\mathrm{T}}$ weakly majorizes $\begin{bmatrix} f(x_{(i_1)}) & \cdots & f(x_{(i_n)}) \end{bmatrix}^{\mathrm{T}}$.

Proof: See [201, p. 42], [730, p. 173], or [996, p. 116].

Fact 2.21.8. Let $x, y \in \mathbb{R}^n$, where $x_{(1)} \geq \cdots \geq x_{(n)} \geq 0$ and $y_{(1)} \geq \cdots \geq y_{(n)} \geq 0$, assume that y strongly log majorizes x, let $f \colon [0, \infty) \mapsto \mathbb{R}$, assume that $g \colon \mathbb{R} \mapsto \mathbb{R}$ defined by $g(z) \triangleq f(e^z)$ is convex, and let $\{i_1, \ldots, i_n\} = \{j_1, \ldots, j_n\} = \{1, \ldots, n\}$ be such that $f(x_{(i_1)}) \geq \cdots \geq f(x_{(i_n)})$ and $f(y_{(j_1)}) \geq \cdots \geq f(y_{(j_n)})$. Then, $\begin{bmatrix} f(y_{(j_1)}) & \cdots & f(y_{(j_n)}) \end{bmatrix}^{\mathrm{T}}$ weakly majorizes $\begin{bmatrix} f(x_{(i_1)}) & \cdots & f(x_{(i_n)}) \end{bmatrix}^{\mathrm{T}}$.

Proof: Apply Fact 2.21.7.

Fact 2.21.9. Let $x, y \in \mathbb{R}^n$, where $x_{(1)} \geq \cdots \geq x_{(n)}$ and $y_{(1)} \geq \cdots \geq y_{(n)}$, assume that y weakly majorizes x, let $f \colon [\min\{x_{(n)}, y_{(n)}\}, y_{(1)}] \mapsto \mathbb{R}$, assume that f is convex and increasing, and let $\{i_1, \ldots, i_n\} = \{j_1, \ldots, j_n\} = \{1, \ldots, n\}$ be such that $f(x_{(i_1)}) \geq \cdots \geq f(x_{(i_n)})$ and $f(y_{(j_1)}) \geq \cdots \geq f(y_{(j_n)})$. Then, $\begin{bmatrix} f(y_{(j_1)}) & \cdots & f(y_{(j_n)}) \end{bmatrix}^{\mathrm{T}}$ weakly majorizes $\begin{bmatrix} f(x_{(i_1)}) & \cdots & f(x_{(i_n)}) \end{bmatrix}^{\mathrm{T}}$.

Proof: See [201, p. 42], [730, p. 173], or [996, p. 116].

Remark: See Fact 2.21.10.

Fact 2.21.10. Let $x, y \in \mathbb{R}^n$, where $x_{(1)} \geq \cdots \geq x_{(n)} \geq 0$ and $y_{(1)} \geq \cdots \geq y_{(n)} \geq 0$, assume that y strongly majorizes x, and let $r \geq 1$. Then, $\begin{bmatrix} y_{(1)}^r & \cdots & y_{(n)}^r \end{bmatrix}^{\mathrm{T}}$ weakly majorizes $\begin{bmatrix} x_{(1)}^r & \cdots & x_{(n)}^r \end{bmatrix}^{\mathrm{T}}$.

Proof: Use Fact 2.21.10.

Remark: Using the Schur power (see Section 7.3), the conclusion can be stated as the fact that $y^{\circ r}$ weakly majorizes $x^{\circ r}$.

Fact 2.21.11. Let $x, y \in \mathbb{R}^n$, where $x_{(1)} \geq \cdots \geq x_{(n)} \geq 0$ and $y_{(1)} \geq \cdots \geq y_{(n)} \geq 0$, assume that y weakly log majorizes x, let $f \colon [0, \infty) \mapsto \mathbb{R}$, assume that $g \colon \mathbb{R} \mapsto \mathbb{R}$ defined by $g(z) \triangleq f(e^z)$ is convex and increasing, and let $\{i_1, \ldots, i_n\} = \{j_1, \ldots, j_n\} = \{1, \ldots, n\}$ be such that $f(x_{(i_1)}) \geq \cdots \geq f(x_{(i_n)})$ and $f(y_{(j_1)}) \geq \cdots \geq f(y_{(j_n)})$. Then, $\begin{bmatrix} f(y_{(j_1)}) & \cdots & f(y_{(j_n)}) \end{bmatrix}^{\mathrm{T}}$ weakly majorizes $\begin{bmatrix} f(x_{(i_1)}) & \cdots & f(x_{(i_n)}) \end{bmatrix}^{\mathrm{T}}$.

Proof: Use Fact 2.21.9.

Fact 2.21.12. Let $x, y \in \mathbb{R}^n$, where $x_{(1)} \geq \cdots \geq x_{(n)} \geq 0$ and $y_{(1)} \geq \cdots \geq y_{(n)} \geq 0$, and assume that y weakly log majorizes x. Then, y weakly majorizes x.

Proof: Use Fact 2.21.11 with $f(t) = t$. See [1521, p. 19].

Fact 2.21.13. Let $x, y \in \mathbb{R}^n$, where $x_{(1)} \geq \cdots \geq x_{(n)} \geq 0$ and $y_{(1)} \geq \cdots \geq y_{(n)} \geq 0$, assume that y weakly majorizes x, and let $p \in [1, \infty)$. Then, for all $k \in \{1, \ldots, n\}$,

$$\left(\sum_{i=1}^{k} x_{(i)}^p \right)^{1/p} \leq \left(\sum_{i=1}^{k} y_{(i)}^p \right)^{1/p}.$$

Proof: Use Fact 2.21.9. See [996, p. 96].

Remark: $\phi(x) \triangleq \left(\sum_{i=1}^{k} x_{(i)}^p \right)^{1/p}$ is a *symmetric gauge function*. See Fact 9.8.42.

2.22 Notes

The theory of determinants is discussed in [1050, 1379]. A graph-theoretic interpretation is given in [272, Chapter 4]. Applications to physics are described in [1405, 1406]. Contributors to the development of this subject are highlighted in [595]. The empty matrix is discussed in [390, 1059], [1157, pp. 462–464], and [1266, p. 3]. Recent versions of Matlab follow the properties of the empty matrix given in this chapter [694, pp. 305, 306]. Convexity is the subject of [184, 243, 259, 463, 904, 1161, 1266, 1388, 1446]. Convex optimization theory is developed in [180, 259]. In [243] the dual cone is called the *polar cone*.

The development of rank properties is based on [993]. Theorem 2.6.4 is based on [1072]. The term "subdeterminant" is used in [1108] and is equivalent to *minor*. The notation A^{A} for adjugate is used in [1259]. Numerous papers on basic topics in matrix theory and linear algebra are collected in [300, 301]. A geometric interpretation of $\mathcal{N}(A)$, $\mathcal{R}(A)$, $\mathcal{N}(A^*)$, and $\mathcal{R}(A^{\mathrm{T}})$ is given in [1270]. Some reflections on matrix theory are given in [1290, 1307]. Applications of the matrix inversion lemma are discussed in [634]. Some historical notes on the determinant and inverse of partitioned matrices as well as the matrix inversion lemma are given in [684].

Results on majorization are extensively developed in [996, 998].

Chapter Three

Matrix Classes and Transformations

This chapter presents definitions of various types of matrices as well as transformations for analyzing matrices.

3.1 Matrix Classes

In this section we categorize various types of matrices based on their algebraic and structural properties.

The following definition introduces various types of square matrices.

Definition 3.1.1. For $A \in \mathbb{F}^{n \times n}$ define the following types of matrices:

i) A is *group invertible* if $\mathcal{R}(A) = \mathcal{R}(A^2)$.

ii) A is *involutory* if $A^2 = I$.

iii) A is *skew involutory* if $A^2 = -I$.

iv) A is *idempotent* if $A^2 = A$.

v) A is *skew idempotent* if $A^2 = -A$.

vi) A is *tripotent* if $A^3 = A$.

vii) A is *nilpotent* if there exists $k \in \mathbb{P}$ such that $A^k = 0$.

viii) A is *unipotent* if $A - I$ is nilpotent.

ix) A is *range Hermitian* if $\mathcal{R}(A) = \mathcal{R}(A^*)$.

x) A is *range symmetric* if $\mathcal{R}(A) = \mathcal{R}(A^{\mathrm{T}})$.

xi) A is *Hermitian* if $A = A^*$.

xii) A is *symmetric* if $A = A^{\mathrm{T}}$.

xiii) A is *skew Hermitian* if $A = -A^*$.

xiv) A is *skew symmetric* if $A = -A^{\mathrm{T}}$.

xv) A is *normal* if $AA^* = A^*A$.

xvi) A is *positive semidefinite* ($A \geq 0$) if A is Hermitian and $x^*Ax \geq 0$ for all

$x \in \mathbb{F}^n$.

xvii) A is *negative semidefinite* $(A \leq 0)$ if $-A$ is positive semidefinite.

xviii) A is *positive definite* $(A > 0)$ if A is Hermitian and $x^*Ax > 0$ for all $x \in \mathbb{F}^n$ such that $x \neq 0$.

xix) A is *negative definite* $(A < 0)$ if $-A$ is positive definite.

xx) A is *semidissipative* if $A + A^*$ is negative semidefinite.

xxi) A is *dissipative* if $A + A^*$ is negative definite.

xxii) A is *unitary* if $A^*A = I$.

xxiii) A is *shifted unitary* if $A + A^* = 2A^*A$.

xxiv) A is *orthogonal* if $A^{\mathrm{T}}A = I$.

xxv) A is *shifted orthogonal* if $A + A^{\mathrm{T}} = 2A^{\mathrm{T}}A$.

xxvi) A is a *projector* if A is Hermitian and idempotent.

xxvii) A is a *reflector* if A is Hermitian and unitary.

xxviii) A is a *skew reflector* if A is skew Hermitian and unitary.

xxix) A is an *elementary projector* if there exists a nonzero vector $x \in \mathbb{F}^n$ such that $A = I - (x^*x)^{-1}xx^*$.

xxx) A is an *elementary reflector* if there exists a nonzero vector $x \in \mathbb{F}^n$ such that $A = I - 2(x^*x)^{-1}xx^*$.

xxxi) A is an *elementary matrix* if there exist vectors $x, y \in \mathbb{F}^n$ such that $A = I - xy^{\mathrm{T}}$ and $x^{\mathrm{T}}y \neq 1$.

xxxii) A is *reverse Hermitian* if $A = A^{\hat{*}}$.

xxxiii) A is *reverse symmetric* if $A = A^{\hat{\mathrm{T}}}$.

xxxiv) A is a *permutation matrix* if each row of A and each column of A possess one 1 and zeros otherwise. A is an *(even, odd) permutation matrix* if A is a permutation matrix and $(\det A = 1, \det A = -1)$. A is a *transposition matrix* if it is a permutation matrix and A has exactly two off-diagonal entries that are nonzero.

xxxv) A is *reducible* if either both $n = 1$ and $A = 0$ or both $n \geq 2$ and there exist $k \geq 1$ and a permutation matrix $S \in \mathbb{R}^{n \times n}$ such that $SAS^{\mathrm{T}} = \begin{bmatrix} B & C \\ 0_{k \times (n-k)} & D \end{bmatrix}$, where $B \in \mathbb{F}^{(n-k) \times (n-k)}$, $C \in \mathbb{F}^{(n-k) \times k}$, and $D \in \mathbb{F}^{k \times k}$.

xxxvi) A is *irreducible* if A is not reducible.

Let $A \in \mathbb{F}^{n \times n}$ be Hermitian. Then, the function $f \colon \mathbb{F}^n \mapsto \mathbb{R}$ defined by

$$f(x) \triangleq x^*Ax \tag{3.1.1}$$

is a *quadratic form*.

The following definition considers matrices that are not necessarily square.

Definition 3.1.2. For $A \in \mathbb{F}^{n \times m}$ define the following types of matrices:

i) A is *semicontractive* if $I_n - AA^*$ is positive semidefinite.

ii) A is *contractive* if $I_n - AA^*$ is positive definite.

iii) A is *left inner* if $A^*A = I_m$.

iv) A is *right inner* if $AA^* = I_n$.

v) A is *centrohermitian* if $A = \hat{I}_n \overline{A} \hat{I}_m$.

vi) A is *centrosymmetric* if $A = \hat{I}_n A \hat{I}_m$.

vii) A is an *outer-product matrix* if there exist $x \in \mathbb{F}^n$ and $y \in \mathbb{F}^m$ such that $A = xy^{\mathrm{T}}$.

The following definition introduces various types of structured and patterned matrices.

Definition 3.1.3. For $A \in \mathbb{F}^{n \times m}$ define the following types of matrices:

i) A is *diagonal* if $A_{(i,j)} = 0$ for all $i \neq j$. If $n = m$, then

$$A = \mathrm{diag}\big(A_{(1,1)}, \ldots, A_{(n,n)}\big).$$

ii) A is *tridiagonal* if $A_{(i,j)} = 0$ for all $|i - j| > 1$.

iii) A is *reverse diagonal* if $A_{(i,j)} = 0$ for all $i + j \neq \min\{n, m\} + 1$. If $n = m$, then

$$A = \mathrm{revdiag}\big(A_{(1,n)}, \ldots, A_{(n,1)}\big).$$

iv) A is (*upper triangular, strictly upper triangular*) if $A_{(i,j)} = 0$ for all ($i > j, i \geq j$).

v) A is (*lower triangular, strictly lower triangular*) if $A_{(i,j)} = 0$ for all ($i < j, i \leq j$).

vi) A is (*upper Hessenberg, lower Hessenberg*) if $A_{(i,j)} = 0$ for all ($i > j+1, i < j+1$).

vii) A is *Toeplitz* if $A_{(i,j)} = A_{(k,l)}$ for all $k - i = l - j$, that is,

$$A = \begin{bmatrix} a & b & c & \cdots \\ d & a & b & \ddots \\ e & d & a & \ddots \\ \vdots & \ddots & \ddots & \ddots \end{bmatrix}.$$

viii) A is *Hankel* if $A_{(i,j)} = A_{(k,l)}$ for all $i + j = k + l$, that is,

$$A = \begin{bmatrix} a & b & c & \cdots \\ b & c & d & \iddots \\ c & d & e & \iddots \\ \vdots & \iddots & \iddots & \iddots \end{bmatrix}.$$

ix) A is *block diagonal* if

$$A = \begin{bmatrix} A_1 & & 0 \\ & \ddots & \\ 0 & & A_k \end{bmatrix} = \mathrm{diag}(A_1, \ldots, A_k),$$

where $A_i \in \mathbb{F}^{n_i \times m_i}$ for all $i \in \{1, \ldots, k\}$.

x) A is *upper block triangular* if

$$A = \begin{bmatrix} A_{11} & A_{12} & \cdots & A_{1k} \\ 0 & A_{22} & \cdots & A_{2k} \\ \vdots & \ddots & \ddots & \vdots \\ 0 & 0 & \cdots & A_{kk} \end{bmatrix},$$

where $A_{ij} \in \mathbb{F}^{n_i \times n_j}$ for all $i, j \in \{1, \ldots, k\}$.

xi) A is *lower block triangular* if

$$A = \begin{bmatrix} A_{11} & 0 & \cdots & 0 \\ A_{21} & A_{22} & \ddots & 0 \\ \vdots & \vdots & \ddots & \vdots \\ A_{k1} & A_{k2} & \cdots & A_{kk} \end{bmatrix},$$

where $A_{ij} \in \mathbb{F}^{n_i \times n_j}$ for all $i, j \in \{1, \ldots, k\}$.

xii) A is *block Toeplitz* if $A_{(i,j)} = A_{(k,l)}$ for all $k - i = l - j$, that is,

$$A = \begin{bmatrix} A_1 & A_2 & A_3 & \cdots \\ A_4 & A_1 & A_2 & \ddots \\ A_5 & A_4 & A_1 & \ddots \\ \vdots & \ddots & \ddots & \ddots \end{bmatrix},$$

where $A_i \in \mathbb{F}^{n_i \times m_i}$.

xiii) A is *block Hankel* if $A_{(i,j)} = A_{(k,l)}$ for all $i + j = k + l$, that is,

$$A = \begin{bmatrix} A_1 & A_2 & A_3 & \cdots \\ A_2 & A_3 & A_4 & \iddots \\ A_3 & A_4 & A_5 & \iddots \\ \vdots & \iddots & \iddots & \iddots \end{bmatrix},$$

where $A_i \in \mathbb{F}^{n_i \times m_i}$.

Definition 3.1.4. For $A \in \mathbb{R}^{n \times m}$ define the following types of matrices:

i) A is *nonnegative* ($A \geq\geq 0$) if $A_{(i,j)} \geq 0$ for all $i \in \{1, \ldots, n\}$ and $j \in \{1, \ldots, m\}$.

ii) A is *row stochastic* if A is nonnegative and $A 1_{m \times 1} = 1_{n \times 1}$.

iii) A is *column stochastic* if A is nonnegative and $1_{1 \times n} A = 1_{1 \times m}$.

iv) A is *doubly stochastic* if A is both row stochastic and column stochastic.

v) A is *positive* $(A \gg 0)$ if $A_{(i,j)} > 0$ for all $i \in \{1, \dots, n\}$ and $j \in \{1, \dots, m\}$.

Now, assume in addition that $n = m$. Then, define the following types of matrices:

vi) A is *almost nonnegative* if $A_{(i,j)} \geq 0$ for all $i, j \in \{1, \dots, n\}$ such that $i \neq j$.

vii) A is a *Z-matrix* if $-A$ is almost nonnegative.

Define the *unit imaginary matrix* $J_{2n} \in \mathbb{R}^{2n \times 2n}$ (or just J) by

$$J_{2n} \triangleq \begin{bmatrix} 0 & I_n \\ -I_n & 0 \end{bmatrix}. \tag{3.1.2}$$

In particular,

$$J_2 = \begin{bmatrix} 0 & 1 \\ -1 & 0 \end{bmatrix}. \tag{3.1.3}$$

Note that J_{2n} is skew symmetric and orthogonal, that is,

$$J_{2n}^{\mathrm{T}} = -J_{2n} = J_{2n}^{-1}. \tag{3.1.4}$$

Hence, J_{2n} is skew involutory and a skew reflector.

The following definition introduces structured matrices of even order. Note that $\mathbb{F}$ can represent either $\mathbb{R}$ or $\mathbb{C}$, although A^{T} does not become A^* in the latter case.

Definition 3.1.5. For $A \in \mathbb{F}^{2n \times 2n}$ define the following types of matrices:

i) A is *Hamiltonian* if $J^{-1}A^{\mathrm{T}}J = -A$.

ii) A is *symplectic* if A is nonsingular and $J^{-1}A^{\mathrm{T}}J = A^{-1}$.

Proposition 3.1.6. Let $A \in \mathbb{F}^{n \times n}$. Then, the following statements hold:

i) If A is Hermitian, skew Hermitian, or unitary, then A is normal.

ii) If A is nonsingular or normal, then A is range Hermitian.

iii) If A is range Hermitian, idempotent, or tripotent, then A is group invertible.

iv) If A is a reflector, then A is tripotent.

v) If A is a permutation matrix, then A is orthogonal.

Proof. *i)* is immediate. To prove *ii)*, note that, if A is nonsingular, then $\mathcal{R}(A) = \mathcal{R}(A^*) = \mathbb{F}^n$, and thus A is range Hermitian. If A is normal, then it follows from Theorem 2.4.3 that $\mathcal{R}(A) = \mathcal{R}(AA^*) = \mathcal{R}(A^*A) = \mathcal{R}(A^*)$, which proves that A is range Hermitian. To prove *iii)*, note that, if A is range Hermitian, then $\mathcal{R}(A) = \mathcal{R}(AA^*) = A\mathcal{R}(A^*) = A\mathcal{R}(A) = \mathcal{R}(A^2)$, while, if A is idempotent, then $\mathcal{R}(A) = \mathcal{R}(A^2)$. If A is tripotent, then $\mathcal{R}(A) = \mathcal{R}(A^3) = A^2\mathcal{R}(A) \subseteq \mathcal{R}(A^2) = A\mathcal{R}(A) \subseteq \mathcal{R}(A)$. Hence, $\mathcal{R}(A) = \mathcal{R}(A^2)$. $\qquad\square$

Proposition 3.1.7. Let $\mathcal{A} \in \mathbb{F}^{2n \times 2n}$. Then, $\mathcal{A}$ is Hamiltonian if and only if there exist matrices $A, B, C \in \mathbb{F}^{n \times n}$ such that B and C are symmetric and

$$\mathcal{A} = \begin{bmatrix} A & B \\ C & -A^{\mathrm{T}} \end{bmatrix}. \tag{3.1.5}$$

3.2 Matrices Related to Graphs

Definition 3.2.1. Let $\mathcal{G} = (\mathcal{X}, \mathcal{R})$ be a graph, where $\mathcal{X} = \{x_1, \ldots, x_n\}$. Then, the following terminology is defined:

i) The *adjacency matrix* $A \in \mathbb{R}^{n \times n}$ of $\mathcal{G}$ is given by $A_{(i,j)} = 1$ if $(x_j, x_i) \in \mathcal{R}$ and $A_{(i,j)} = 0$ if $(x_j, x_i) \notin \mathcal{R}$, for all $i, j \in \{1, \ldots, n\}$.

ii) The *inbound Laplacian matrix* $L_{\mathrm{in}} \in \mathbb{R}^{n \times n}$ of $\mathcal{G}$ is given by $L_{\mathrm{in}(i,i)} = \sum_{j=1, j \neq i}^{n} A_{(i,j)}$, for all $i \in \{1, \ldots, n\}$, and $L_{\mathrm{in}(i,j)} = -A_{(i,j)}$, for all distinct $i, j \in \{1, \ldots, n\}$.

iii) The *outbound Laplacian matrix* $L_{\mathrm{out}} \in \mathbb{R}^{n \times n}$ of $\mathcal{G}$ is given by $L_{\mathrm{out}(i,i)} = \sum_{j=1, j \neq i}^{n} A_{(j,i)}$, for all $i \in \{1, \ldots, n\}$, and $L_{\mathrm{out}(i,j)} = -A_{(i,j)}$, for all distinct $i, j \in \{1, \ldots, n\}$.

iv) The *indegree matrix* $D_{\mathrm{in}} \in \mathbb{R}^{n \times n}$ is the diagonal matrix such that $D_{\mathrm{in}(i,i)} = \mathrm{indeg}(x_i)$, for all $i \in \{1, \ldots, n\}$.

v) The *outdegree matrix* $D_{\mathrm{out}} \in \mathbb{R}^{n \times n}$ is the diagonal matrix such that $D_{\mathrm{out}(i,i)} = \mathrm{outdeg}(x_i)$, for all $i \in \{1, \ldots, n\}$.

vi) Assume that $\mathcal{G}$ has no self-loops, and let $\mathcal{R} = \{a_1, \ldots, a_m\}$. Then, the *incidence matrix* $B \in \mathbb{R}^{n \times m}$ of $\mathcal{G}$ is given by $B_{(i,j)} = 1$ if i is the tail of a_j, $B_{(i,j)} = -1$ if i is the head of a_j, and $B_{(i,j)} = 0$ otherwise, for all $i \in \{1, \ldots, n\}$ and $j \in \{1, \ldots, m\}$.

vii) If $\mathcal{G}$ is symmetric, then the *Laplacian matrix* of $\mathcal{G}$ is given by $L \triangleq L_{\mathrm{in}} = L_{\mathrm{out}}$.

viii) If $\mathcal{G}$ is symmetric, then the *degree matrix* $D \in \mathbb{R}^{n \times n}$ of $\mathcal{G}$ is given by $D \triangleq D_{\mathrm{in}} = D_{\mathrm{out}}$.

ix) If $\mathcal{G} = (\mathcal{X}, \mathcal{R}, w)$ is a weighted graph, then the *adjacency matrix* $A \in \mathbb{R}^{n \times n}$ of $\mathcal{G}$ is given by $A_{(i,j)} = w[(x_j, x_i)]$ if $(x_j, x_i) \in \mathcal{R}$ and $A_{(i,j)} = 0$ if $(x_j, x_i) \notin \mathcal{R}$, for all $i, j \in \{1, \ldots, n\}$.

Note that the adjacency matrix is nonnegative, while the inbound Laplacian, outbound Laplacian, and Laplacian matrices are Z-matrices. Furthermore, note that the inbound Laplacian, outbound Laplacian, and Laplacian matrices are unaffected by the presence of self-loops. However, the indegree and outdegree matrices account for self-loops. It can be seen that, for the arc (x_k, x_l), the ith column of the incidence matrix B is given by $\mathrm{col}_i(B) = e_l - e_k$. Finally, if $\mathcal{G}$ is a symmetric graph, then A and L are symmetric.

Theorem 3.2.2. Let $\mathcal{G} = (\mathcal{X}, \mathcal{R})$ be a graph, where $\mathcal{X} = \{x_1, \ldots, x_n\}$, and let L_{in}, L_{out}, D_{in}, D_{out}, and A denote the inbound Laplacian, outbound Laplacian, indegree, outdegree, and adjacency matrices of $\mathcal{G}$, respectively. Then,

$$L_{\text{in}} = D_{\text{in}} - A \tag{3.2.1}$$

and

$$L_{\text{out}} = D_{\text{out}} - A. \tag{3.2.2}$$

Theorem 3.2.3. Let $\mathcal{G} = (\mathcal{X}, \mathcal{R})$ be a symmetric graph, where $\mathcal{X} = \{x_1, \ldots, x_n\}$, and let A, L, D, and B denote the adjacency, Laplacian, degree, and incidence matrices of $\mathcal{G}$, respectively. Then,

$$L = D - A. \tag{3.2.3}$$

Now, assume in addition that $\mathcal{G}$ has no self-loops. Then,

$$L = \tfrac{1}{2} B B^{\text{T}}. \tag{3.2.4}$$

Definition 3.2.4. Let $M \in \mathbb{F}^{n \times n}$, and let $\mathcal{X} = \{x_1, \ldots, x_n\}$. Then, the *graph of M* is $\mathcal{G}(M) \triangleq (\mathcal{X}, \mathcal{R})$, where, for all $i, j \in \{1, \ldots, n\}$, $(x_j, x_i) \in \mathcal{R}$ if and only if $M_{(i,j)} \neq 0$.

Proposition 3.2.5. Let $M \in \mathbb{F}^{n \times n}$. Then, the adjacency matrix A of $\mathcal{G}(M)$ is given by

$$A = \text{sign}\, |M|. \tag{3.2.5}$$

3.3 Lie Algebras and Groups

In this section we introduce Lie algebras and groups. Lie groups are discussed in Section 11.5. In the following definition, note that the coefficients α and β are required to be real when $\mathbb{F} = \mathbb{C}$.

Definition 3.3.1. Let $\mathcal{S} \subseteq \mathbb{F}^{n \times n}$. Then, $\mathcal{S}$ is a *Lie algebra* if the following conditions are satisfied:

i) If $A, B \in \mathcal{S}$ and $\alpha, \beta \in \mathbb{R}$, then $\alpha A + \beta B \in \mathcal{S}$.

ii) If $A, B \in \mathcal{S}$, then $[A, B] \in \mathcal{S}$.

Note that, if $\mathbb{F} = \mathbb{R}$, then statement *i*) is equivalent to the statement that $\mathcal{S}$ is a subspace. However, if $\mathbb{F} = \mathbb{C}$ and $\mathcal{S}$ contains matrices that are not real, then $\mathcal{S}$ is not a subspace.

Proposition 3.3.2. The following sets are Lie algebras:

i) $\text{gl}_{\mathbb{F}}(n) \triangleq \mathbb{F}^{n \times n}$.

ii) $\text{pl}_{\mathbb{C}}(n) \triangleq \{A \in \mathbb{C}^{n \times n}: \ \text{tr}\, A \in \mathbb{R}\}$.

iii) $\text{sl}_{\mathbb{F}}(n) \triangleq \{A \in \mathbb{F}^{n \times n}: \ \text{tr}\, A = 0\}$.

iv) $\mathrm{u}(n) \triangleq \{A \in \mathbb{C}^{n \times n}\colon A \text{ is skew Hermitian}\}.$

v) $\mathrm{su}(n) \triangleq \{A \in \mathbb{C}^{n \times n}\colon A \text{ is skew Hermitian and tr } A = 0\}.$

vi) $\mathrm{so}(n) \triangleq \{A \in \mathbb{R}^{n \times n}\colon A \text{ is skew symmetric}\}.$

vii) $\mathrm{su}(n,m) \triangleq \{A \in \mathbb{C}^{(n+m) \times (n+m)}\colon \operatorname{diag}(I_n,-I_m)A^*\operatorname{diag}(I_n,-I_m) = -A \text{ and}$
$\text{tr } A = 0\}.$

viii) $\mathrm{so}(n,m) \triangleq \{A \in \mathbb{R}^{(n+m) \times (n+m)}\colon \operatorname{diag}(I_n,-I_m)A^{\mathrm{T}}\operatorname{diag}(I_n,-I_m) = -A\}.$

ix) $\mathrm{symp}_{\mathbb{F}}(2n) \triangleq \{A \in \mathbb{F}^{2n \times 2n}\colon A \text{ is Hamiltonian}\}.$

x) $\mathrm{osymp}_{\mathbb{C}}(2n) \triangleq \mathrm{su}(2n) \cap \mathrm{symp}_{\mathbb{C}}(2n).$

xi) $\mathrm{osymp}_{\mathbb{R}}(2n) \triangleq \mathrm{so}(2n) \cap \mathrm{symp}_{\mathbb{R}}(2n).$

xii) $\mathrm{aff}_{\mathbb{F}}(n) \triangleq \left\{ \begin{bmatrix} A & b \\ 0 & 0 \end{bmatrix}\colon A \in \mathrm{gl}_{\mathbb{F}}(n),\ b \in \mathbb{F}^n \right\}.$

xiii) $\mathrm{se}_{\mathbb{C}}(n) \triangleq \left\{ \begin{bmatrix} A & b \\ 0 & 0 \end{bmatrix}\colon A \in \mathrm{su}(n),\ b \in \mathbb{C}^n \right\}.$

xiv) $\mathrm{se}_{\mathbb{R}}(n) \triangleq \left\{ \begin{bmatrix} A & b \\ 0 & 0 \end{bmatrix}\colon A \in \mathrm{so}(n),\ b \in \mathbb{R}^n \right\}.$

xv) $\mathrm{trans}_{\mathbb{F}}(n) \triangleq \left\{ \begin{bmatrix} 0 & b \\ 0 & 0 \end{bmatrix}\colon b \in \mathbb{F}^n \right\}.$

Definition 3.3.3. Let $\mathcal{S} \subset \mathbb{F}^{n \times n}$. Then, $\mathcal{S}$ is a *group* if the following conditions are satisfied:

i) If $A \in \mathcal{S}$, then A is nonsingular.

ii) If $A \in \mathcal{S}$, then $A^{-1} \in \mathcal{S}$.

iii) If $A, B \in \mathcal{S}$, then $AB \in \mathcal{S}$.

Now, assume in addition that $\mathcal{S}$ is group. Then, $\mathcal{S}$ is an *Abelian group* if the following condition is satisfied:

iv) For all $A, B \in \mathcal{S}$, $[A, B] = 0$.

Furthermore, if $\mathcal{S}_1 \subset \mathcal{S}$, and $\mathcal{S}_1$ is a group, then $\mathcal{S}_1$ is a *subgroup* of $\mathcal{S}$. Finally, $\mathcal{S}$ is a *finite group* if $\mathcal{S}$ has a finite number of elements.

Definition 3.3.4. Let $\mathcal{S}_1 \subset \mathbb{F}^{n_1 \times n_1}$ and $\mathcal{S}_2 \subset \mathbb{F}^{n_1 \times n_1}$ be groups. Then, $\mathcal{S}_1$ and $\mathcal{S}_2$ are *isomorphic* if there exists a one-to-one and onto function $\phi\colon \mathcal{S}_1 \mapsto \mathcal{S}_2$ such that, for all $A, B \in \mathcal{S}_1$, $\phi(AB) = \phi(A)\phi(B)$. In this case, $\mathcal{S}_1 \approx \mathcal{S}_2$, and ϕ is an *isomorphism*.

Proposition 3.3.5. Let $\mathcal{S}_1 \subset \mathbb{F}^{n_1 \times n_1}$ and $\mathcal{S}_2 \subset \mathbb{F}^{n_1 \times n_1}$ be groups, and assume that $\mathcal{S}_1$ and $\mathcal{S}_2$ are isomorphic with isomorphism $\phi\colon \mathcal{S}_1 \mapsto \mathcal{S}_2$. Then, $\phi(I_{n_1}) = I_{n_2}$, and, for all $A \in \mathcal{S}_1$, $\phi(A^{-1}) = [\phi(A)]^{-1}$.

Note that, if $\mathcal{S} \subset \mathbb{F}^{n \times n}$ is a group, then $I_n \in \mathcal{S}$.

The following result lists classical groups that arise in physics and engineering. For example, $O(1,3)$ is the *Lorentz group* [1192, p. 16], [1217, p. 126]. The special orthogonal group $SO(n)$ consists of the orthogonal matrices whose determinant is 1. In particular, each matrix in $SO(2)$ and $SO(3)$ is a *rotation matrix*. Furthermore, $P(n)$, $A(n)$, $D(n)$, and $C(n)$ are the $n \times n$ *permutation group*, *alternating group*, *dihedral group*, and *cyclic group*, respectively.

Proposition 3.3.6. The following sets are groups:

i) $GL_{\mathbb{F}}(n) \triangleq \{A \in \mathbb{F}^{n \times n} : \det A \neq 0\}$.

ii) $PL_{\mathbb{F}}(n) \triangleq \{A \in \mathbb{F}^{n \times n} : \det A > 0\}$.

iii) $SL_{\mathbb{F}}(n) \triangleq \{A \in \mathbb{F}^{n \times n} : \det A = 1\}$.

iv) $U(n) \triangleq \{A \in \mathbb{C}^{n \times n} : A \text{ is unitary}\}$.

v) $O(n) \triangleq \{A \in \mathbb{R}^{n \times n} : A \text{ is orthogonal}\}$.

vi) $SU(n) \triangleq \{A \in U(n) : \det A = 1\}$.

vii) $SO(n) \triangleq \{A \in O(n) : \det A = 1\}$.

viii) $P(n) \triangleq \{A \in \mathbb{R}^{n \times n} : A \text{ is a permutation matrix}\}$.

ix) $A(n) \triangleq \{A \in P(n) : A \text{ is an even permutation matrix}\}$.

x) $D(2) \triangleq \{I_2, -I_2, \hat{I}_2, -\hat{I}_2\}$.

xi) $D(n) \triangleq \{I_n, P_n, P_n^2, \ldots, P_n^{n-1}, \hat{I}_n, \hat{I}_n P_n, \hat{I}_n P_n^2, \ldots, \hat{I}_n P_n^{n-1}\}$, where $n \geq 3$.

xii) $C(n) \triangleq \{I_n, P_n, P_n^2, \ldots, P_n^{n-1}\}$.

xiii) $U(n,m) \triangleq \{A \in \mathbb{C}^{(n+m) \times (n+m)} : A^* \text{diag}(I_n, -I_m)A = \text{diag}(I_n, -I_m)\}$.

xiv) $O(n,m) \triangleq \{A \in \mathbb{R}^{(n+m) \times (n+m)} : A^T \text{diag}(I_n, -I_m)A = \text{diag}(I_n, -I_m)\}$.

xv) $SU(n,m) \triangleq \{A \in U(n,m) : \det A = 1\}$.

xvi) $SO(n,m) \triangleq \{A \in O(n,m) : \det A = 1\}$.

xvii) $\text{Symp}_{\mathbb{F}}(2n) \triangleq \{A \in \mathbb{F}^{2n \times 2n} : A \text{ is symplectic}\}$.

xviii) $\text{OSymp}_{\mathbb{C}}(2n) \triangleq U(2n) \cap \text{Symp}_{\mathbb{C}}(2n)$.

xix) $\text{OSymp}_{\mathbb{R}}(2n) \triangleq O(2n) \cap \text{Symp}_{\mathbb{R}}(2n)$.

xx) $\text{Aff}_{\mathbb{F}}(n) \triangleq \left\{ \begin{bmatrix} A & b \\ 0 & 1 \end{bmatrix} : A \in GL_{\mathbb{F}}(n),\ b \in \mathbb{F}^n \right\}$.

xxi) $\text{SE}_{\mathbb{C}}(n) \triangleq \left\{ \begin{bmatrix} A & b \\ 0 & 1 \end{bmatrix} : A \in SU(n),\ b \in \mathbb{C}^n \right\}$.

xxii) $\text{SE}_{\mathbb{R}}(n) \triangleq \left\{ \begin{bmatrix} A & b \\ 0 & 1 \end{bmatrix} : A \in SO(n),\ b \in \mathbb{R}^n \right\}$.

xxiii) $\text{Trans}_{\mathbb{F}}(n) \triangleq \left\{ \begin{bmatrix} I & b \\ 0 & 1 \end{bmatrix} : b \in \mathbb{F}^n \right\}$.

3.4 Matrix Transformations

The following results use groups to define equivalence relations.

Proposition 3.4.1. Let $\mathcal{S}_1 \subset \mathbb{F}^{n \times n}$ and $\mathcal{S}_2 \subset \mathbb{F}^{m \times m}$ be groups, and let $\mathcal{M} \subseteq \mathbb{F}^{n \times m}$. Then, the subset of $\mathcal{M} \times \mathcal{M}$ defined by

$$\mathcal{R} \triangleq \{(A, B) \in \mathcal{M} \times \mathcal{M}: \text{ there exist } S_1 \in \mathcal{S}_1 \text{ and } S_2 \in \mathcal{S}_2 \text{ such that } A = S_1 B S_2\}$$

is an equivalence relation on $\mathcal{M}$.

Proposition 3.4.2. Let $\mathcal{S} \subset \mathbb{F}^{n \times n}$ be a group, and let $\mathcal{M} \subseteq \mathbb{F}^{n \times n}$. Then, the following subsets of $\mathcal{M} \times \mathcal{M}$ are equivalence relations:

i) $\mathcal{R} \triangleq \{(A, B) \in \mathcal{M} \times \mathcal{M}: \text{ there exists } S \in \mathcal{S} \text{ such that } A = SBS^{-1}\}$.

ii) $\mathcal{R} \triangleq \{(A, B) \in \mathcal{M} \times \mathcal{M}: \text{ there exists } S \in \mathcal{S} \text{ such that } A = SBS^*\}$.

iii) $\mathcal{R} \triangleq \{(A, B) \in \mathcal{M} \times \mathcal{M}: \text{ there exists } S \in \mathcal{S} \text{ such that } A = SBS^{\mathrm{T}}\}$.

If, in addition, $\mathcal{S}$ is an Abelian group, then the following subset $\mathcal{M} \times \mathcal{M}$ is an equivalence relation:

iv) $\mathcal{R} \triangleq \{(A, B) \in \mathcal{M} \times \mathcal{M}: \text{ there exists } S \in \mathcal{S} \text{ such that } A = SBS\}$.

Various transformations can be employed for analyzing matrices. Propositions 3.4.1 and 3.4.2 imply that these transformations define equivalence relations.

Definition 3.4.3. Let $A, B \in \mathbb{F}^{n \times m}$. Then, the following terminology is defined:

i) A and B are *left equivalent* if there exists a nonsingular matrix $S_1 \in \mathbb{F}^{n \times n}$ such that $A = S_1 B$.

ii) A and B are *right equivalent* if there exists a nonsingular matrix $S_2 \in \mathbb{F}^{m \times m}$ such that $A = B S_2$.

iii) A and B are *biequivalent* if there exist nonsingular matrices $S_1 \in \mathbb{F}^{n \times n}$ and $S_2 \in \mathbb{F}^{m \times m}$ such that $A = S_1 B S_2$.

iv) A and B are *unitarily left equivalent* if there exists a unitary matrix $S_1 \in \mathbb{F}^{n \times n}$ such that $A = S_1 B$.

v) A and B are *unitarily right equivalent* if there exists a unitary matrix $S_2 \in \mathbb{F}^{m \times m}$ such that $A = B S_2$.

vi) A and B are *unitarily biequivalent* if there exist unitary matrices $S_1 \in \mathbb{F}^{n \times n}$ and $S_2 \in \mathbb{F}^{m \times m}$ such that $A = S_1 B S_2$.

Definition 3.4.4. Let $A, B \in \mathbb{F}^{n \times n}$. Then, the following terminology is defined:

i) A and B are *similar* if there exists a nonsingular matrix $S \in \mathbb{F}^{n \times n}$ such that $A = SBS^{-1}$.

ii) A and B are *congruent* if there exists a nonsingular matrix $S \in \mathbb{F}^{n \times n}$ such that $A = SBS^*$.

iii) A and B are *T-congruent* if there exists a nonsingular matrix $S \in \mathbb{F}^{n \times n}$ such that $A = SBS^{\mathrm{T}}$.

iv) A and B are *unitarily similar* if there exists a unitary matrix $S \in \mathbb{F}^{n \times n}$ such that $A = SBS^* = SBS^{-1}$.

The transformations that appear in Definition 3.4.3 and Definition 3.4.4 are called *left equivalence, right equivalence, biequivalence, unitary left equivalence, unitary right equivalence, unitary biequivalence, similarity, congruence, T-congruence, and unitary similarity* transformations, respectively. The following results summarize some matrix properties that are preserved under left equivalence, right equivalence, biequivalence, similarity, congruence, and unitary similarity.

Proposition 3.4.5. Let $A, B \in \mathbb{F}^{n \times n}$. If A and B are similar, then the following statements hold:

i) A and B are biequivalent.

ii) $\operatorname{tr} A = \operatorname{tr} B$.

iii) $\det A = \det B$.

iv) A^k and B^k are similar for all $k \geq 1$.

v) A^{k*} and B^{k*} are similar for all $k \geq 1$.

vi) A is nonsingular if and only if B is; in this case, A^{-k} and B^{-k} are similar for all $k \geq 1$.

vii) A is (group invertible, involutory, skew involutory, idempotent, tripotent, nilpotent) if and only if B is.

If A and B are congruent, then the following statements hold:

viii) A and B are biequivalent.

ix) A^* and B^* are congruent.

x) A is nonsingular if and only if B is; in this case, A^{-1} and B^{-1} are congruent.

xi) A is (range Hermitian, Hermitian, skew Hermitian, positive semidefinite, positive definite) if and only if B is.

If A and B are unitarily similar, then the following statements hold:

xii) A and B are similar.

xiii) A and B are congruent.

xiv) A is (range Hermitian, group invertible, normal, Hermitian, skew Hermitian, positive semidefinite, positive definite, unitary, involutory, skew involutory, idempotent, tripotent, nilpotent) if and only if B is.

3.5 Projectors, Idempotent Matrices, and Subspaces

The following result shows that a unique projector can be associated with each subspace.

Proposition 3.5.1. Let $\mathcal{S} \subseteq \mathbb{F}^n$ be a subspace. Then, there exists a unique projector $A \in \mathbb{F}^{n \times n}$ such that $\mathcal{S} = \mathcal{R}(A)$. Furthermore, $x \in \mathcal{S}$ if and only if $x = Ax$.

Proof. See [1023, p. 386] and Fact 3.13.15. $\qquad\square$

For a subspace $\mathcal{S} \subseteq \mathbb{F}^n$, the matrix $A \in \mathbb{F}^{n \times n}$ given by Proposition 3.5.1 is the *projector onto* $\mathcal{S}$. If, in addition, $\mathcal{S}' \subseteq \mathbb{F}^n$, then $A\mathcal{S}'$ is the *projection* of $\mathcal{S}'$ *onto* $\mathcal{S}$.

Let $A \in \mathbb{F}^{n \times n}$ be a projector. Then, the *complementary projector* $A_\perp$ is the projector defined by

$$A_\perp \triangleq I - A. \tag{3.5.1}$$

Proposition 3.5.2. Let $\mathcal{S} \subseteq \mathbb{F}^n$ be a subspace, and let $A \in \mathbb{F}^{n \times n}$ be the projector onto $\mathcal{S}$. Then, $A_\perp$ is the projector onto $\mathcal{S}^\perp$. Furthermore,

$$\mathcal{R}(A)^\perp = \mathcal{N}(A) = \mathcal{R}(A_\perp) = \mathcal{S}^\perp. \tag{3.5.2}$$

The following result shows that a unique idempotent matrix can be associated with each pair of complementary subspaces.

Proposition 3.5.3. Let $\mathcal{S}_1, \mathcal{S}_2 \subseteq \mathbb{F}^n$ be complementary subspaces. Then, there exists a unique idempotent matrix $A \in \mathbb{F}^{n \times n}$ such that $\mathcal{R}(A) = \mathcal{S}_1$ and $\mathcal{N}(A) = \mathcal{S}_2$.

Proof. See [186, p. 118] or [1023, p. 386]. $\qquad\square$

For complementary subspaces $\mathcal{S}_1, \mathcal{S}_2 \subseteq \mathbb{F}^n$, the unique idempotent matrix $A \in \mathbb{F}^{n \times n}$ given by Proposition 3.5.3 is the *idempotent matrix onto* $\mathcal{S}_1 = \mathcal{R}(A)$ *along* $\mathcal{S}_2 = \mathcal{N}(A)$.

For an idempotent matrix $A \in \mathbb{F}^{n \times n}$, the *complementary idempotent matrix* $A_\perp$ defined by (3.5.1) is also idempotent.

Proposition 3.5.4. Let $\mathcal{S}_1, \mathcal{S}_2 \subseteq \mathbb{F}^n$ be complementary subspaces, and let $A \in \mathbb{F}^{n \times n}$ be the idempotent matrix onto $\mathcal{S}_1 = \mathcal{R}(A)$ along $\mathcal{S}_2 = \mathcal{N}(A)$. Then, $\mathcal{R}(A_\perp) = \mathcal{S}_2$ and $\mathcal{N}(A_\perp) = \mathcal{S}_1$, that is, $A_\perp$ is the idempotent matrix onto $\mathcal{S}_2$ along $\mathcal{S}_1$.

Definition 3.5.5. The *index of A*, denoted by $\operatorname{ind} A$, is the smallest nonnegative integer k such that

$$\mathcal{R}(A^k) = \mathcal{R}(A^{k+1}). \tag{3.5.3}$$

Proposition 3.5.6. Let $A \in \mathbb{F}^{n \times n}$. Then, A is nonsingular if and only if $\operatorname{ind} A = 0$. Furthermore, A is group invertible if and only if $\operatorname{ind} A \leq 1$.

Note that $\operatorname{ind} 0_{n \times n} = 1$.

Proposition 3.5.7. Let $A \in \mathbb{F}^{n \times n}$, and let $k \geq 1$. Then, $\operatorname{ind} A \leq k$ if and only if $\mathcal{R}(A^k)$ and $\mathcal{N}(A^k)$ are complementary subspaces.

Fact 3.6.3 states that the null space and range of a range-Hermitian matrix are orthogonally complementary subspaces. Furthermore, Proposition 3.1.6 states that every range-Hermitian matrix is group invertible. Hence, the null space and range of a group-invertible matrix are complementary subspaces. The following corollary of Proposition 3.5.7 shows that the converse is true. Note that every idempotent matrix is group invertible.

Corollary 3.5.8. Let $A \in \mathbb{F}^{n \times n}$. Then, A is group invertible if and only if $\mathcal{R}(A)$ and $\mathcal{N}(A)$ are complementary subspaces.

For a group-invertible matrix $A \in \mathbb{F}^{n \times n}$, the following result shows how to construct the idempotent matrix onto $\mathcal{R}(A)$ along $\mathcal{N}(A)$.

Proposition 3.5.9. Let $A \in \mathbb{F}^{n \times n}$, and let $r \triangleq \operatorname{rank} A$. Then, A is group invertible if and only if there exist matrices $B \in \mathbb{F}^{n \times r}$ and $C \in \mathbb{F}^{r \times n}$ such that $A = BC$ and $\operatorname{rank} B = \operatorname{rank} C = r$. In this case, the idempotent matrix $P \triangleq B(CB)^{-1}C$ is the idempotent matrix onto $\mathcal{R}(A)$ along $\mathcal{N}(A)$.

Proof. See [1023, p. 634].   □

An alternative expression for the idempotent matrix onto $\mathcal{R}(A)$ along $\mathcal{N}(A)$ is given by Proposition 6.2.3.

3.6 Facts on Group-Invertible and Range-Hermitian Matrices

Fact 3.6.1. Let $A \in \mathbb{F}^{n \times n}$. Then, the following statements are equivalent:

i) A is group invertible.

ii) A^* is group invertible.

iii) A^{T} is group invertible.

iv) $\overline{A}$ is group invertible.

v) $\mathcal{R}(A) = \mathcal{R}(A^2)$.

vi) $\mathcal{N}(A) = \mathcal{N}(A^2)$.

vii) $\mathcal{N}(A) \cap \mathcal{R}(A) = \{0\}$.

viii) $\mathcal{N}(A) + \mathcal{R}(A) = \mathbb{F}^n$.

ix) A and A^2 are left equivalent.

x) A and A^2 are right equivalent.

xi) $\operatorname{ind} A \leq 1$.

xii) $\text{rank}\, A = \text{rank}\, A^2$.

xiii) $\text{def}\, A = \text{def}\, A^2$.

xiv) $\text{def}\, A = \text{amult}_A(0)$.

Remark: See Corollary 3.5.8, Proposition 3.5.9, and Corollary 5.5.9.

Fact 3.6.2. Let $A \in \mathbb{F}^{n \times n}$. Then, $\text{ind}\, A \leq k$ if and only if A^k is group invertible.

Fact 3.6.3. Let $A \in \mathbb{F}^{n \times n}$. Then, the following statements are equivalent:

i) A is range Hermitian.

ii) A^* is range Hermitian.

iii) $\mathcal{R}(A) = \mathcal{R}(A^*)$.

iv) $\mathcal{R}(A) \subseteq \mathcal{R}(A^*)$.

v) $\mathcal{R}(A^*) \subseteq \mathcal{R}(A)$.

vi) $\mathcal{N}(A) = \mathcal{N}(A^*)$.

vii) A and A^* are right equivalent.

viii) $\mathcal{R}(A)^\perp = \mathcal{N}(A)$.

ix) $\mathcal{N}(A)^\perp = \mathcal{R}(A)$.

x) $\mathcal{R}(A)$ and $\mathcal{N}(A)$ are orthogonally complementary subspaces.

xi) $\text{rank}\, A = \text{rank} \begin{bmatrix} A & A^* \end{bmatrix}$.

Proof: See [331, 1309].

Remark: Using Fact 3.13.15, Proposition 3.5.2, and Proposition 6.1.6, *vi)* is equivalent to $A^+A = I - (I - A^+A) = AA^+$. See Fact 6.3.9, Fact 6.3.10, and Fact 6.4.13.

Fact 3.6.4. Let $A \in \mathbb{F}^{n \times n}$, and assume that $A^2 = A^*$. Then, A is range Hermitian.

Proof: See [118].

Remark: A is a *generalized projector*.

Fact 3.6.5. Let $A, B \in \mathbb{F}^{n \times n}$, and assume that A and B are range Hermitian. Then,
$$\text{rank}\, AB = \text{rank}\, BA.$$

Proof: See [126].

3.7 Facts on Normal, Hermitian, and Skew-Hermitian Matrices

Fact 3.7.1. Let $A \in \mathbb{F}^{n \times n}$, assume that A is nonsingular, and assume that A is (normal, Hermitian, skew Hermitian, unitary). Then, so is A^{-1}.

Fact 3.7.2. Let $A \in \mathbb{F}^{n \times m}$. Then, $AA^{\mathrm{T}} \in \mathbb{F}^{n \times n}$ and $A^{\mathrm{T}}A \in \mathbb{F}^{m \times m}$ are symmetric.

Fact 3.7.3. Let $\alpha \in \mathbb{R}$ and $A \in \mathbb{R}^{n \times n}$. Then, the matrix equation $\alpha A + A^{\mathrm{T}} = 0$ has a nonzero solution A if and only if $\alpha = 1$ or $\alpha = -1$.

Fact 3.7.4. Let $A \in \mathbb{F}^{n \times n}$, assume that A is Hermitian, and let $k \geq 1$. Then, $\mathcal{R}(A) = \mathcal{R}(A^k)$ and $\mathcal{N}(A) = \mathcal{N}(A^k)$.

Fact 3.7.5. Let $A \in \mathbb{R}^{n \times n}$. Then, the following statements hold:

i) $x^{\mathrm{T}}Ax = 0$ for all $x \in \mathbb{R}^n$ if and only if A is skew symmetric.

ii) A is symmetric and $x^{\mathrm{T}}Ax = 0$ for all $x \in \mathbb{R}^n$ if and only if $A = 0$.

Fact 3.7.6. Let $A \in \mathbb{C}^{n \times n}$. Then, the following statements hold:

i) x^*Ax is real for all $x \in \mathbb{C}^n$ if and only if A is Hermitian.

ii) x^*Ax is imaginary for all $x \in \mathbb{C}^n$ if and only if A is skew Hermitian.

iii) $x^*Ax = 0$ for all $x \in \mathbb{C}^n$ if and only if $A = 0$.

Fact 3.7.7. Let $A \in \mathbb{R}^{n \times n}$. Then, the following statements are equivalent:

i) $x^*Ax > 0$ for all nonzero $x \in \mathbb{C}^n$.

ii) $x^{\mathrm{T}}Ax > 0$ for all nonzero $x \in \mathbb{R}^n$.

Fact 3.7.8. Let $A \in \mathbb{F}^{n \times n}$, and assume that A is block diagonal. Then, A is (normal, Hermitian, skew Hermitian) if and only if every diagonally located block has the same property.

Fact 3.7.9. Let $A \in \mathbb{C}^{n \times n}$. Then, the following statements hold:

i) A is Hermitian if and only if $\jmath A$ is skew Hermitian.

ii) A is skew Hermitian if and only if $\jmath A$ is Hermitian.

iii) A is Hermitian if and only if $\mathrm{Re}\,A$ is symmetric and $\mathrm{Im}\,A$ is skew symmetric.

iv) A is skew Hermitian if and only if $\mathrm{Re}\,A$ is skew symmetric and $\mathrm{Im}\,A$ is symmetric.

v) A is positive semidefinite if and only if $\mathrm{Re}\,A$ is positive semidefinite.

vi) A is positive definite if and only if $\mathrm{Re}\,A$ is positive definite.

vii) A is symmetric if and only if $\left[\begin{smallmatrix} 0 & A \\ A & 0 \end{smallmatrix}\right]$ is symmetric.

viii) A is Hermitian if and only if $\left[\begin{smallmatrix} 0 & A \\ A & 0 \end{smallmatrix}\right]$ is Hermitian.

ix) A is symmetric if and only if $\left[\begin{smallmatrix} 0 & A \\ -A & 0 \end{smallmatrix}\right]$ is skew symmetric.

x) A is Hermitian if and only if $\left[\begin{smallmatrix} 0 & A \\ -A & 0 \end{smallmatrix}\right]$ is skew Hermitian.

Remark: *x)* is a real analogue of *i)* since $\left[\begin{smallmatrix} 0 & A \\ -A & 0 \end{smallmatrix}\right] = I_2 \otimes A$, and I_2 is a real representation of $\jmath$.

Fact 3.7.10. Let $A \in \mathbb{F}^{n \times n}$. Then, the following statements hold:

i) If A is (normal, unitary, Hermitian, positive semidefinite, positive definite), then so is A^{A}.

ii) If A is skew Hermitian and n is odd, then A^{A} is Hermitian.

iii) If A is skew Hermitian and n is even, then A^{A} is skew Hermitian.

iv) If A is diagonal, then so is A^{A}, and, for all $i \in \{1, \ldots, n\}$,

$$\left(A^{\mathrm{A}}\right)_{(i,i)} = \prod_{\substack{j=1 \\ j \neq i}}^{n} A_{(j,j)}.$$

Proof: Use Fact 2.16.10.

Remark: See Fact 5.14.4.

Fact 3.7.11. Let $A \in \mathbb{F}^{n \times n}$, assume that n is even, let $x \in \mathbb{F}^n$, and let $\alpha \in \mathbb{F}$. Then,
$$\det(A + \alpha x x^*) = \det A.$$

Proof: Use Fact 2.16.3 and Fact 3.7.10.

Fact 3.7.12. Let $A \in \mathbb{F}^{n \times n}$. Then, the following statements are equivalent:

i) A is normal.

ii) $A^2 A^* = A A^* A$.

iii) $A A^* A = A^* A^2$.

iv) $\operatorname{tr}(A A^*)^2 = \operatorname{tr} A^2 A^{2*}$.

v) There exists $k \geq 1$ such that
$$\operatorname{tr}(A A^*)^k = \operatorname{tr} A^k A^{k*}.$$

vi) There exist $k, l \in \mathbb{P}$ such that
$$\operatorname{tr}(A A^*)^{kl} = \operatorname{tr}\left(A^k A^{k*}\right)^l.$$

vii) A is range Hermitian, and $A A^* A^2 = A^2 A^* A$.

viii) $A A^* - A^* A$ is positive semidefinite.

ix) $[A, A^* A] = 0$.

x) $[A, [A, A^*]] = 0$.

Proof: See [119, 331, 465, 467, 603, 1239].

Remark: See Fact 3.11.5, Fact 5.14.14, Fact 5.15.4, Fact 6.3.15, Fact 6.6.11, Fact 8.9.28, Fact 8.12.5, Fact 8.18.5, Fact 11.15.4, and Fact 11.16.14.

Fact 3.7.13. Let $A \in \mathbb{F}^{n \times n}$. Then, the following statements are equivalent:

i) A is Hermitian.

ii) $A^2 = A^*A$.

iii) $A^2 = AA^*$.

iv) $A^{*2} = A^*A$.

v) $A^{*2} = AA^*$.

vi) There exists $\alpha \in \mathbb{F}$ such that $A^2 = \alpha A^*A + (1 - \alpha)AA^*$.

vii) There exists $\alpha \in \mathbb{F}$ such that $A^{*2} = \alpha A^*A + (1 - \alpha)AA^*$.

viii) $\operatorname{tr} A^2 = \operatorname{tr} A^*A$.

ix) $\operatorname{tr} A^2 = \operatorname{tr} AA^*$.

x) $\operatorname{tr} A^{*2} = \operatorname{tr} A^*A$.

xi) $\operatorname{tr} A^{*2} = \operatorname{tr} AA^*$.

If, in addition, $\mathbb{F} = \mathbb{R}$, then the following condition is equivalent to *i)–xi)*:

xii) There exist $\alpha, \beta \in \mathbb{R}$ such that
$$\alpha A^2 + (1 - \alpha)A^{\mathrm{T}2} = \beta A^{\mathrm{T}}A + (1 - \beta)AA^{\mathrm{T}}.$$

Proof: To prove that *viii)* implies *i)*, use the Schur decomposition Theorem 5.4.1 to replace A with $D + S$, where D is diagonal and S is strictly upper triangular. Then, $\operatorname{tr} D^*D + \operatorname{tr} S^*S = \operatorname{tr} D^2 \leq \operatorname{tr} D^*D$. Hence, $S = 0$, and thus $\operatorname{tr} D^*D = \operatorname{tr} D^2$, which implies that D is real. See [119, 881].

Remark: See Fact 3.13.1.

Remark: Fact 9.11.3 states that, for all $A \in \mathbb{F}^{n \times n}$, $|\operatorname{tr} A^2| \leq \operatorname{tr} A^*A$.

Fact 3.7.14. Let $A \in \mathbb{F}^{n \times n}$, let $\alpha, \beta \in \mathbb{F}$, and assume that $\alpha \neq 0$. Then, the following statements are equivalent:

i) A is normal.

ii) $\alpha A + \beta I$ is normal.

Now, assume, in addition, that $\alpha, \beta \in \mathbb{R}$. Then, the following statements are equivalent:

iii) A is Hermitian.

iv) $\alpha A + \beta I$ is Hermitian.

Remark: The function $f(A) = \alpha A + \beta I$ is an *affine mapping*.

Fact 3.7.15. Let $A \in \mathbb{R}^{n \times n}$, assume that A is skew symmetric, and let $\alpha > 0$. Then, $-A^2$ is positive semidefinite, $\det A \geq 0$, and $\det(\alpha I + A) > 0$. If, in addition, n is odd, then $\det A = 0$.

Fact 3.7.16. Let $A \in \mathbb{F}^{n \times n}$, and assume that A is skew Hermitian. If n is even, then $\det A \geq 0$. If n is odd, then $\det A$ is imaginary.

Proof: The first statement follows from Proposition 5.5.20.

Fact 3.7.17. Let $x, y \in \mathbb{F}^n$, and define

$$A \triangleq \begin{bmatrix} x & y \end{bmatrix}.$$

Then,

$$xy^* - yx^* = AJ_2A^*.$$

Furthermore, $xy^* - yx^*$ is skew Hermitian and has rank 0 or 2.

Fact 3.7.18. Let $x, y \in \mathbb{F}^n$. Then, the following statements hold:

i) xy^{T} is idempotent if and only if either $xy^{\mathrm{T}} = 0$ or $x^{\mathrm{T}}y = 1$.

ii) xy^{T} is Hermitian if and only if there exists $\alpha \in \mathbb{R}$ such that either $y = \alpha \bar{x}$ or $x = \alpha \bar{y}$.

Fact 3.7.19. Let $x, y \in \mathbb{F}^n$, and define $A \triangleq I - xy^{\mathrm{T}}$. Then, the following statements hold:

i) $\det A = 1 - x^{\mathrm{T}}y$.

ii) A is nonsingular if and only if $x^{\mathrm{T}}y \neq 1$.

iii) A is nonsingular if and only if A is elementary.

iv) $\mathrm{rank}\, A = n - 1$ if and only if $x^{\mathrm{T}}y = 1$.

v) A is Hermitian if and only if there exists $\alpha \in \mathbb{R}$ such that either $y = \alpha \bar{x}$ or $x = \alpha \bar{y}$.

vi) A is positive semidefinite if and only if A is Hermitian and $x^{\mathrm{T}}y \leq 1$.

vii) A is positive definite if and only if A is Hermitian and $x^{\mathrm{T}}y < 1$.

viii) A is idempotent if and only if either $xy^{\mathrm{T}} = 0$ or $x^{\mathrm{T}}y = 1$.

ix) A is orthogonal if and only if either $x = 0$ or $y = \frac{1}{2}y^{\mathrm{T}}yx$.

x) A is involutory if and only if $x^{\mathrm{T}}y = 2$.

xi) A is a projector if and only if either $y = 0$ or $x = x^*xy$.

xii) A is a reflector if and only if either $y = 0$ or $2x = x^*xy$.

xiii) A is an elementary projector if and only if $x \neq 0$ and $y = (x^*x)^{-1}x$.

xiv) A is an elementary reflector if and only if $x \neq 0$ and $y = 2(x^*x)^{-1}x$.

Remark: See Fact 3.13.9.

Fact 3.7.20. Let $x, y \in \mathbb{F}^n$ satisfy $x^{\mathrm{T}}y \neq 1$. Then, $I - xy^{\mathrm{T}}$ is nonsingular and

$$\left(I - xy^{\mathrm{T}}\right)^{-1} = I - \frac{1}{x^{\mathrm{T}}y - 1}xy^{\mathrm{T}}.$$

Remark: The inverse of an elementary matrix is an elementary matrix.

Fact 3.7.21. Let $A \in \mathbb{F}^{n \times n}$, and assume that A is Hermitian. Then, $\det A$ is real.

Fact 3.7.22. Let $A \in \mathbb{F}^{n \times n}$, and assume that A is Hermitian. Then,

$$(\operatorname{tr} A)^2 \leq (\operatorname{rank} A) \operatorname{tr} A^2.$$

Furthermore, equality holds if and only if there exists $\alpha \in \mathbb{R}$ such that $A^2 = \alpha A$.

Remark: See Fact 5.11.10 and Fact 9.13.12.

Fact 3.7.23. Let $A \in \mathbb{R}^{n \times n}$, and assume that A is skew symmetric. Then, $\operatorname{tr} A = 0$. If, in addition, $B \in \mathbb{R}^{n \times n}$ is symmetric, then $\operatorname{tr} AB = 0$.

Fact 3.7.24. Let $A \in \mathbb{F}^{n \times n}$, and assume that A is skew Hermitian. Then, $\operatorname{Re} \operatorname{tr} A = 0$. If, in addition, $B \in \mathbb{F}^{n \times n}$ is Hermitian, then $\operatorname{Re} \operatorname{tr} AB = 0$.

Fact 3.7.25. Let $A \in \mathbb{F}^{n \times m}$. Then, A^*A is positive semidefinite. Furthermore, A^*A is positive definite if and only if A is left invertible. In this case, $A^{\mathrm{L}} \in \mathbb{F}^{m \times n}$ defined by

$$A^{\mathrm{L}} \triangleq (A^*A)^{-1}A^*$$

is a left inverse of A.

Remark: See Fact 2.15.2, Fact 3.7.26, and Fact 3.13.6.

Fact 3.7.26. Let $A \in \mathbb{F}^{n \times m}$. Then, AA^* is positive semidefinite. Furthermore, AA^* is positive definite if and only if A is right invertible. In this case, $A^{\mathrm{R}} \in \mathbb{F}^{m \times n}$ defined by

$$A^{\mathrm{R}} \triangleq A^*(AA^*)^{-1}$$

is a right inverse of A.

Remark: See Fact 2.15.2, Fact 3.13.6, and Fact 3.7.25.

Fact 3.7.27. Let $A \in \mathbb{F}^{n \times m}$. Then, A^*A, AA^*, and $\left[\begin{smallmatrix} 0 & A^* \\ A & 0 \end{smallmatrix} \right]$ are Hermitian, and $\left[\begin{smallmatrix} 0 & A^* \\ -A & 0 \end{smallmatrix} \right]$ is skew Hermitian. Now, assume in addition that $n = m$. Then, $A + A^*$, $\jmath(A - A^*)$, and $\frac{1}{\jmath 2}(A - A^*)$ are Hermitian, while $A - A^*$ is skew Hermitian. Finally,

$$A = \tfrac{1}{2}(A + A^*) + \tfrac{1}{2}(A - A^*)$$

and

$$A = \tfrac{1}{2}(A + A^*) + \jmath[\tfrac{1}{\jmath 2}(A - A^*)].$$

Remark: The last two equalities are Cartesian decompositions.

Fact 3.7.28. Let $A \in \mathbb{F}^{n \times n}$. Then, there exist a unique Hermitian matrix $B \in \mathbb{F}^{n \times n}$ and a unique skew-Hermitian matrix $C \in \mathbb{F}^{n \times n}$ such that $A = B + C$. Specifically, if $A = \hat{B} + \jmath\hat{C}$, where $\hat{B}, \hat{C} \in \mathbb{R}^{n \times n}$, then $\hat{B}$ and $\hat{C}$ are given by

$$B = \tfrac{1}{2}(A + A^*) = \tfrac{1}{2}(\hat{B} + \hat{B}^{\mathrm{T}}) + \jmath\tfrac{1}{2}(\hat{C} - \hat{C}^{\mathrm{T}})$$

and

$$C = \tfrac{1}{2}(A - A^*) = \tfrac{1}{2}(\hat{B} - \hat{B}^{\mathrm{T}}) + \jmath\tfrac{1}{2}(\hat{C} + \hat{C}^{\mathrm{T}}).$$

Furthermore, A is normal if and only if $BC = CB$.

Remark: See Fact 11.13.9.

Fact 3.7.29. Let $A \in \mathbb{F}^{n \times n}$. Then, there exist unique Hermitian matrices $B, C \in \mathbb{C}^{n \times n}$ such that $A = B + \jmath C$. Specifically, if $A = \hat{B} + \jmath \hat{C}$, where $\hat{B}, \hat{C} \in \mathbb{R}^{n \times n}$, then $\hat{B}$ and $\hat{C}$ are given by

$$B = \tfrac{1}{2}(A + A^*) = \tfrac{1}{2}(\hat{B} + \hat{B}^{\mathrm{T}}) + \jmath\tfrac{1}{2}(\hat{C} - \hat{C}^{\mathrm{T}})$$

and

$$C = \tfrac{1}{\jmath 2}(A - A^*) = \tfrac{1}{2}(\hat{C} + \hat{C}^{\mathrm{T}}) - \jmath\tfrac{1}{2}(\hat{B} - \hat{B}^{\mathrm{T}}).$$

Furthermore, A is normal if and only if $BC = CB$.

Remark: This result is the *Cartesian decomposition*.

Fact 3.7.30. Let $A, B \in \mathbb{C}^{n \times n}$, assume that A is either Hermitian or skew Hermitian, and assume that B is either Hermitian or skew Hermitian. Then,

$$\operatorname{rank} AB = \operatorname{rank} BA.$$

Proof: AB and $(AB)^* = BA$ have the same singular values. See Fact 5.11.19.

Remark: See Fact 2.10.26.

Fact 3.7.31. Let $A, B \in \mathbb{R}^{3 \times 3}$, and assume that A and B are skew symmetric. Then,

$$\operatorname{tr} AB^3 = \tfrac{1}{2}(\operatorname{tr} AB)(\operatorname{tr} B^2)$$

and

$$\operatorname{tr} A^3 B^3 = \tfrac{1}{4}(\operatorname{tr} A^2)(\operatorname{tr} AB)(\operatorname{tr} B^2) + \tfrac{1}{3}(\operatorname{tr} A^3)(\operatorname{tr} B^3).$$

Proof: See [82].

Fact 3.7.32. Let $A \in \mathbb{F}^{n \times n}$ and $k \geq 1$. If A is (normal, Hermitian, unitary, involutory, positive semidefinite, positive definite, idempotent, nilpotent), then so is A^k. If A is (skew Hermitian, skew involutory), then so is A^{2k+1}. If A is Hermitian, then A^{2k} is positive semidefinite. If A is tripotent, then so is A^{3k}.

Fact 3.7.33. Let $a, b, c, d, e, f \in \mathbb{R}$, and define the skew-symmetric matrix $A \in \mathbb{R}^{4 \times 4}$ given by

$$A \triangleq \begin{bmatrix} 0 & a & b & c \\ -a & 0 & d & e \\ -b & -d & 0 & f \\ -c & -e & -f & 0 \end{bmatrix}.$$

Then,

$$\det A = (af - be + cd)^2.$$

Proof: See [1215, p. 63].

Remark: See Fact 4.8.14 and Fact 4.10.4.

Fact 3.7.34. Let $A \in \mathbb{R}^{2n \times 2n}$, and assume that A is skew symmetric. Then, there exists a nonsingular matrix $S \in \mathbb{R}^{2n \times 2n}$ such that $S^{\mathrm{T}} A S = J_{2n}$.

Proof: See [106, p. 231].

Fact 3.7.35. Let $A \in \mathbb{R}^{n \times n}$, and assume that A is positive definite. Then,

$$\mathcal{E} \triangleq \{x \in \mathbb{R}^n: \ x^{\mathrm{T}}Ax \le 1\}$$

is a hyperellipsoid. Furthermore, the volume V of $\mathcal{E}$ is given by

$$V = \frac{\alpha(n)}{\sqrt{\det A}},$$

where

$$\alpha(n) \triangleq \begin{cases} \pi^{n/2}/(n/2)!, & n \text{ even}, \\[2mm] 2^n \pi^{(n-1)/2}[(n-1)/2]!/n!, & n \text{ odd}. \end{cases}$$

In particular, the area of the ellipse $\{x \in \mathbb{R}^2: \ x^{\mathrm{T}}Ax \le 1\}$ is $\pi/\det A$.

Remark: $\alpha(n)$ is the volume of the unit n-*dimensional hypersphere*.

Remark: See [824, p. 36].

3.8 Facts on Commutators

Fact 3.8.1. Let $A, B \in \mathbb{F}^{n \times n}$. If either A and B are Hermitian or A and B are skew Hermitian, then $[A, B]$ is skew Hermitian. Furthermore, if A is Hermitian and B is skew Hermitian, or vice versa, then $[A, B]$ is Hermitian.

Fact 3.8.2. Let $A \in \mathbb{F}^{n \times n}$. Then, the following statements are equivalent:

i) $\operatorname{tr} A = 0$.

ii) There exist matrices $B, C \in \mathbb{F}^{n \times n}$ such that B is Hermitian, $\operatorname{tr} C = 0$, and $A = [B, C]$.

iii) There exist matrices $B, C \in \mathbb{F}^{n \times n}$ such that $A = [B, C]$.

Proof: See [549] and Fact 5.9.20. If every diagonal entry of A is zero, then let $B \triangleq \operatorname{diag}(1, \ldots, n)$, $C_{(i,i)} \triangleq 0$, and, for $i \ne j$, $C_{(i,j)} \triangleq A_{(i,j)}/(i - j)$. See [1523, p. 110]. See also [1125, p. 172].

Fact 3.8.3. Let $A \in \mathbb{F}^{n \times n}$. Then, the following statements are equivalent:

i) A is Hermitian, and $\operatorname{tr} A = 0$.

ii) There exists a nonsingular matrix $B \in \mathbb{F}^{n \times n}$ such that $A = [B, B^*]$.

iii) There exist a Hermitian matrix $B \in \mathbb{F}^{n \times n}$ and a skew-Hermitian matrix $C \in \mathbb{F}^{n \times n}$ such that $A = [B, C]$.

iv) There exist a skew-Hermitian matrix $B \in \mathbb{F}^{n \times n}$ and a Hermitian matrix $C \in \mathbb{F}^{n \times n}$ such that $A = [B, C]$.

Proof: See [549] and [1297].

Fact 3.8.4. Let $A \in \mathbb{F}^{n \times n}$. Then, the following statements are equivalent:

i) A is skew Hermitian, and $\operatorname{tr} A = 0$.

ii) There exists a nonsingular matrix $B \in \mathbb{F}^{n \times n}$ such that $A = [\jmath B, B^*]$.

iii) If $A \in \mathbb{C}^{n \times n}$ is skew Hermitian, then there exist Hermitian matrices $B, C \in \mathbb{F}^{n \times n}$ such that $A = [B, C]$.

Proof: See [549] or use Fact 3.8.3.

Fact 3.8.5. Let $A \in \mathbb{F}^{n \times n}$, and assume that A is skew symmetric. Then, there exist symmetric matrices $B, C \in \mathbb{F}^{n \times n}$ such that $A = [B, C]$.

Proof: Use Fact 5.15.24. See [1125, pp. 83, 89].

Remark: "Symmetric" is correct for $\mathbb{F} = \mathbb{C}$.

Fact 3.8.6. Let $A \in \mathbb{F}^{n \times n}$, and assume that $[A, [A, A^*]] = 0$. Then, A is normal.

Remark: See [1523, p. 32].

Fact 3.8.7. Let $A \in \mathbb{F}^{n \times n}$. Then, there exist $B, C \in \mathbb{F}^{n \times n}$ such that B is normal, C is Hermitian, and
$$A = B + [C, B].$$

Remark: See [450].

3.9 Facts on Linear Interpolation

Fact 3.9.1. Let $y \in \mathbb{F}^n$ and $x \in \mathbb{F}^m$. Then, there exists a matrix $A \in \mathbb{F}^{n \times m}$ such that $y = Ax$ if and only if either $y = 0$ or $x \neq 0$. If $y = 0$, then one such matrix is $A = 0$. If $x \neq 0$, then one such matrix is
$$A = (x^*x)^{-1} y x^*.$$

Remark: This is a linear interpolation problem. See [795].

Fact 3.9.2. Let $x, y \in \mathbb{F}^n$, and assume that $x \neq 0$. Then, there exists a Hermitian matrix $A \in \mathbb{F}^{n \times n}$ such that $y = Ax$ if and only if x^*y is real. One such matrix is
$$A = (x^*x)^{-1} [yx^* + xy^* - x^*yI].$$

Now, assume in addition that x and y are real. Then,
$$\sigma_{\max}(A) = \frac{\|x\|_2}{\|y\|_2} = \min\{\sigma_{\max}(B) \colon B \in \mathbb{R}^{n \times n} \text{ is symmetric and } y = Bx\}.$$

Proof: The last statement is given in [1236].

Fact 3.9.3. Let $x, y \in \mathbb{F}^n$, and assume that $x \neq 0$. Then, there exists a positive-definite matrix $A \in \mathbb{F}^{n \times n}$ such that $y = Ax$ if and only if x^*y is real and positive. One such matrix is

$$A = I + (x^*y)^{-1}yy^* - (x^*x)^{-1}xx^*.$$

Proof: To show that A is positive definite, note that the elementary projector $I - (x^*x)^{-1}xx^*$ is positive semidefinite and $\text{rank}[I - (x^*x)^{-1}xx^*] = n - 1$. Since $(x^*y)^{-1}yy^*$ is positive semidefinite, it follows that $\mathcal{N}(A) \subseteq \mathcal{N}[I - (x^*x)^{-1}xx^*]$. Next, since $x^*y > 0$, it follows that $y^*x \neq 0$ and $y \neq 0$, and thus $x \notin \mathcal{N}(A)$. Consequently, $\mathcal{N}(A) \subset \mathcal{N}[I - (x^*x)^{-1}xx^*]$ (note proper inclusion), and thus $\text{def } A < 1$. Hence, A is nonsingular.

Fact 3.9.4. Let $x, y \in \mathbb{F}^n$. Then, there exists a skew-Hermitian matrix $A \in \mathbb{F}^{n \times n}$ such that $y = Ax$ if and only if either $y = 0$ or $x \neq 0$ and $x^*y = 0$. If $x \neq 0$ and $x^*y = 0$, then one such matrix is

$$A = (x^*x)^{-1}(yx^* - xy^*).$$

Proof: See [949].

Fact 3.9.5. Let $x, y \in \mathbb{R}^n$. Then, there exists an orthogonal matrix $A \in \mathbb{R}^{n \times n}$ such that $Ax = y$ if and only if $x^\mathrm{T}x = y^\mathrm{T}y$.

Remark: One such matrix is given by a product of n plane rotations given by Fact 5.15.16. Another matrix is given by the product of elementary reflectors given by Fact 5.15.15. For $n = 3$, one such matrix is given by Fact 3.11.29, while another is given by the exponential of a skew-symmetric matrix given by Fact 11.11.7. See Fact 3.14.4.

Remark: See Fact 9.15.10.

Problem: Extend this result to $\mathbb{C}^n$.

Fact 3.9.6. Let $x, y \in \mathbb{R}^n$, where $x_{(1)} \geq \cdots \geq x_{(n)}$ and $y_{(1)} \geq \cdots \geq y_{(n)}$. Then, the following statements are equivalent:

i) y strongly majorizes x.

ii) $x \in \text{co}\{Ay \colon A \in \text{P}(n)\}$.

iii) There exists a doubly stochastic matrix $A \in \mathbb{R}^{n \times n}$ such that $y = Ax$.

Proof: See [201, p. 33], [728, p. 197], and [996, p. 22].

Remark: The equivalence of *i)* and *ii)* is due to Rado. See [996, p. 113]. The equivalence of *i)* and *iii)* is the *Hardy-Littlewood-Polya theorem*.

Remark: See Fact 3.11.3 and Fact 8.18.8.

3.10 Facts on the Cross Product

Fact 3.10.1. Let $x, y, z, w \in \mathbb{R}^3$, and define the cross-product matrix $K(x) \in \mathbb{R}^{3 \times 3}$ by

$$K(x) \triangleq \begin{bmatrix} 0 & -x_{(3)} & x_{(2)} \\ x_{(3)} & 0 & -x_{(1)} \\ -x_{(2)} & x_{(1)} & 0 \end{bmatrix}.$$

Then, the following statements hold:

i) $x \times x = K(x)x = 0$.

ii) $x^{\mathrm{T}}K(x) = 0$.

iii) $K^{\mathrm{T}}(x) = -K(x)$.

iv) $K^2(x) = xx^{\mathrm{T}} - (x^{\mathrm{T}}x)I$.

v) $\operatorname{tr} K^{\mathrm{T}}(x)K(x) = -\operatorname{tr} K^2(x) = 2x^{\mathrm{T}}x$.

vi) $K^3(x) = -(x^{\mathrm{T}}x)K(x)$.

vii) $[I - K(x)]^{-1} = I + (1 + x^{\mathrm{T}}x)^{-1}[K(x) + K^2(x)]$.

viii) $[I + \frac{1}{2}K(x)][I - \frac{1}{2}K(x)]^{-1} = I + \frac{4}{4 + x^{\mathrm{T}}x}[K(x) + \frac{1}{2}K^2(x)]$.

ix) Define

$$H(x) \triangleq \tfrac{1}{2}[\tfrac{1}{2}(1 - x^{\mathrm{T}}x)I + xx^{\mathrm{T}} + K(x)].$$

Then,
$$H(x)H^{\mathrm{T}}(x) = \tfrac{1}{16}(1 + x^{\mathrm{T}}x)^2 I.$$

x) For all $\alpha, \beta \in \mathbb{R}$, $K(\alpha x + \beta y) = \alpha K(x) + \beta K(y)$.

xi) $x \times y = -(y \times x) = K(x)y = -K(y)x = K^{\mathrm{T}}(y)x$.

xii) If $x \times y \neq 0$, then $\mathcal{N}[(x \times y)^{\mathrm{T}}] = \{x \times y\}^{\perp} = \mathcal{R}([\; x \quad y \;])$.

xiii) $K(x \times y) = K[K(x)y] = [K(x), K(y)]$.

xiv) $K(x \times y) = yx^{\mathrm{T}} - xy^{\mathrm{T}} = [\; x \quad y \;]\begin{bmatrix} -y^{\mathrm{T}} \\ x^{\mathrm{T}} \end{bmatrix} = -[\; x \quad y \;]J_2[\; x \quad y \;]^{\mathrm{T}}$.

xv) $(x \times y) \times x = (x^{\mathrm{T}}xI - xx^{\mathrm{T}})y$.

xvi) $K[(x \times y) \times x] = (x^{\mathrm{T}}x)K(y) - (x^{\mathrm{T}}y)K(x)$.

xvii) $(x \times y)^{\mathrm{T}}(x \times y) = \det[\; x \quad y \quad x \times y \;]$.

xviii) $(x \times y)^{\mathrm{T}}z = x^{\mathrm{T}}(y \times z) = \det[\; x \quad y \quad z \;]$.

xix) $x \times (y \times z) = (x^{\mathrm{T}}z)y - (x^{\mathrm{T}}y)z$.

xx) $(x \times y) \times z = (x^{\mathrm{T}}z)y - (y^{\mathrm{T}}z)x$.

xxi) $K[(x \times y) \times z] = (x^{\mathrm{T}}z)K(y) - (y^{\mathrm{T}}z)K(x)$.

xxii) $K[x \times (y \times z)] = (x^{\mathrm{T}}z)K(y) - (x^{\mathrm{T}}y)K(z)$.

xxiii) $(x \times y)^{\mathrm{T}}(x \times y) = x^{\mathrm{T}}xy^{\mathrm{T}}y - (x^{\mathrm{T}}y)^2$.

xxiv) $K(x)K(y) = yx^{\mathrm{T}} - x^{\mathrm{T}}yI_3$.

xxv) $K(x)K(y)K(x) = -(x^{\mathrm{T}}y)K(x)$.

xxvi) $K^2(x)K(y) + K(y)K^2(x) = -(x^{\mathrm{T}}x)K(y) - (x^{\mathrm{T}}y)K(x)$.

xxvii) $K^2(x)K^2(y) - K^2(y)K^2(x) = -(x^{\mathrm{T}}y)K(x \times y)$.

xxviii) $\sqrt{(x \times y)^{\mathrm{T}}(x \times y)} = \sqrt{x^{\mathrm{T}}xy^{\mathrm{T}}y}\sin\theta$, where θ is the angle between x and y.

xxix) $2xx^{\mathrm{T}}K(y) = (x \times y)x^{\mathrm{T}} + x(x \times y)^{\mathrm{T}} + x^{\mathrm{T}}xK(y) - x^{\mathrm{T}}yK(x)$.

xxx) If $\|x\|_2 = \|y\|_2 = \|z\|_2 = 1$, then
$$1 + 2(x^{\mathrm{T}}y)(y^{\mathrm{T}}z)(z^{\mathrm{T}}x) = [x^{\mathrm{T}}(y \times z)]^2 + (x^{\mathrm{T}}y)^2 + (y^{\mathrm{T}}z)^2 + (z^{\mathrm{T}}x)^2.$$

xxxi) $K(x)K(z)(x^{\mathrm{T}}wy - x^{\mathrm{T}}yw) = K(x)K(w)x^{\mathrm{T}}zy$.

xxxii) $(x \times y)^{\mathrm{T}}(z \times w) = x^{\mathrm{T}}zy^{\mathrm{T}}w - x^{\mathrm{T}}wy^{\mathrm{T}}z = \det\begin{bmatrix} x^{\mathrm{T}}z & x^{\mathrm{T}}w \\ y^{\mathrm{T}}z & y^{\mathrm{T}}w \end{bmatrix}$.

xxxiii) $(x \times y) \times (z \times w) = x^{\mathrm{T}}(y \times w)z - x^{\mathrm{T}}(y \times z)w = x^{\mathrm{T}}(z \times w)y - y^{\mathrm{T}}(z \times w)x$.

xxxiv) $x \times [y \times (z \times w)] = (y^{\mathrm{T}}w)(x \times z) - (y^{\mathrm{T}}z)(x \times w)$.

xxxv) $x \times [y \times (y \times x)] = y \times [x \times (y \times x)] = (y^{\mathrm{T}}x)(x \times y)$.

xxxvi) Let $A \in \mathbb{R}^{3 \times 3}$. Then,
$$A^{\mathrm{T}}K(Ax)A = (\det A)K(x),$$
and thus
$$A^{\mathrm{T}}(Ax \times Ay) = (\det A)(x \times y).$$

xxxvii) Let $A \in \mathbb{R}^{3 \times 3}$, and assume that A is orthogonal. Then,
$$K(Ax)A = (\det A)AK(x),$$
and thus
$$Ax \times Ay = (\det A)A(x \times y).$$

xxxviii) Let $A \in \mathbb{R}^{3 \times 3}$, and assume that A is orthogonal and $\det A = 1$. Then,
$$K(Ax)A = AK(x),$$
and thus
$$Ax \times Ay = A(x \times y).$$

xxxix) $\begin{bmatrix} x & y & z \end{bmatrix}^{\mathrm{A}} = \begin{bmatrix} y \times z & z \times x & x \times y \end{bmatrix}^{\mathrm{T}}$.

xl) $\det\begin{bmatrix} K(x) & y \\ -y^{\mathrm{T}} & 0 \end{bmatrix} = (x^{\mathrm{T}}y)^2$.

xli) $\begin{bmatrix} K(x) & y \\ -y^{\mathrm{T}} & 0 \end{bmatrix}^{\mathrm{A}} = -x^{\mathrm{T}}y\begin{bmatrix} K(y) & x \\ -x^{\mathrm{T}} & 0 \end{bmatrix}$.

xlii) If $x^{\mathrm{T}}y \neq 0$, then
$$\begin{bmatrix} K(x) & y \\ -y^{\mathrm{T}} & 0 \end{bmatrix}^{-1} = \frac{-1}{x^{\mathrm{T}}y}\begin{bmatrix} K(y) & x \\ -x^{\mathrm{T}} & 0 \end{bmatrix}.$$

xliii) If $x \neq 0$, then $K^+(x) = (x^{\mathrm{T}}x)^{-1}K(x)$.

xliv) If $x^\mathrm{T}y = 0$ and $x^\mathrm{T}x + y^\mathrm{T}y \neq 0$, then

$$\begin{bmatrix} K(x) & y \\ -y^\mathrm{T} & 0 \end{bmatrix}^+ = \frac{-1}{x^\mathrm{T}x + y^\mathrm{T}y} \begin{bmatrix} K(x) & y \\ -y^\mathrm{T} & 0 \end{bmatrix}.$$

Proof: Results *vii*), *viii*), and *xxv*)–*xxvii*) are given in [767, p. 363]. Result *ix*) is given in [1374]. Statement *xxiii*) is equivalent to the fact that $\sin^2\theta + \cos^2\theta = 1$. Statement *xxix*) arises in quaternion multiplication. See Fact 3.11.32. Statement *xxx*) is due to Crasta. Statement *xxxi*) is a consequence of a result given in [586, p. 58]. Using *xviii*),

$$e_i^\mathrm{T}A^\mathrm{T}(Ax \times Ay) = \det \begin{bmatrix} Ax & Ay & Ae_i \end{bmatrix} = (\det A)e_i^\mathrm{T}(x \times y)$$

for all $i \in \{1, 2, 3\}$, which proves *xxxvi*). Result *xxxix*) is given in [1351]. Results *xl*)–*xliv*) are proved in [1366]. See [420, 487, 767, 1085, 1223, 1293, 1359].

Remark: Cross products of complex vectors are considered in [613].

Remark: A cross product can be defined on $\mathbb{R}^7$. See [490, pp. 297–299].

Remark: An extension of the cross product to higher dimensions is given by the outer product in Clifford algebras. See Fact 9.7.5 and [357, 435, 569, 620, 689, 690, 895, 959].

Remark: See Fact 11.11.11.

Remark: If $\theta \in (0, \pi)$, then statement *xxviii*) gives twice the area of the triangle with vertices 0, x, and y. See Fact 2.20.10.

Problem: Extend these equalities to complex vectors and matrices.

Fact 3.10.2. Let $A \in \mathbb{R}^{3\times3}$, assume that A is orthogonal, let $B \in \mathbb{C}^{3\times3}$, and assume that B is symmetric. Then,

$$\sum_{i=1}^3 (Ae_i) \times (BAe_i) = 0.$$

Proof: For $i = 1, 2, 3$, multiply by $e_i^\mathrm{T}A^\mathrm{T}$.

Fact 3.10.3. Let α_1, α_2, and α_3 be distinct positive numbers, let $A \in \mathbb{R}^{3\times3}$, assume that A is orthogonal, and assume that

$$\sum_{i=1}^3 \alpha_i e_i \times Ae_i = 0.$$

Then,

$$A \in \{I, \mathrm{diag}(1, -1, -1), \mathrm{diag}(-1, 1, -1), \mathrm{diag}(-1, -1, 1)\}.$$

Remark: This result characterizes equilibria for a dynamical system on SO(3). See [314].

3.11 Facts on Unitary and Shifted-Unitary Matrices

Fact 3.11.1. Let $S_1, S_2 \subseteq \mathbb{F}^n$, assume that S_1 and S_2 are subspaces, and assume that $\dim S_1 \leq \dim S_2$. Then, there exists a unitary matrix $A \in \mathbb{F}^{n \times n}$ such that $AS_1 \subseteq S_2$.

Fact 3.11.2. Let $S_1, S_2 \subseteq \mathbb{F}^n$, assume that S_1 and S_2 are subspaces, and assume that $\dim S_1 + \dim S_2 \leq n$. Then, there exists a unitary matrix $A \in \mathbb{F}^{n \times n}$ such that $AS_1 \subseteq S_2^{\perp}$.

Proof: Use Fact 3.11.1.

Fact 3.11.3. Let $A \in \mathbb{R}^{n \times n}$. Then, the following statements are equivalent:

i) A is a doubly stochastic matrix.

ii) $A \in \operatorname{co} P(n)$.

Proof: See [728, p. 527].

Remark: See Fact 3.9.6 and Fact 4.11.6.

Remark: This result is due to Birkhoff.

Fact 3.11.4. Let $A \in \mathbb{F}^{n \times n}$, and assume that A is unitary. Then, the following statements hold:

i) $A = A^{-*}$.

ii) $A^{\mathrm{T}} = \overline{A}^{-1} = \overline{A}^{*}$.

iii) $\overline{A} = A^{-\mathrm{T}} = \overline{A}^{-*}$.

iv) $A^* = A^{-1}$.

Fact 3.11.5. Let $A \in \mathbb{F}^{n \times n}$, and assume that A is nonsingular. Then, the following statements are equivalent:

i) A is normal.

ii) $A^{-1}A^*$ is unitary.

iii) $[A, A^*] = 0$.

iv) $[A, A^{-*}] = 0$.

v) $[A^{-1}, A^{-*}] = 0$.

Proof: See [603].

Remark: See Fact 3.7.12, Fact 5.15.4, Fact 6.3.15, and Fact 6.6.11.

Fact 3.11.6. Let $A \in \mathbb{F}^{n \times m}$. If A is (left inner, right inner), then A is (left invertible, right invertible) and A^* is a (left inverse, right inverse) of A.

Fact 3.11.7. Let $x, y \in \mathbb{F}^n$, let $A \in \mathbb{F}^{n \times n}$, and assume that A is unitary. Then, $x^*y = 0$ if and only if $(Ax)^*Ay = 0$.

Fact 3.11.8. Let $A \in \mathbb{F}^{n \times n}$, and assume that A is block diagonal. Then, A is (unitary, shifted unitary) if and only if every diagonally located block has the same property.

Fact 3.11.9. Let $A \in \mathbb{F}^{n \times n}$, and assume that A is unitary. Then, $\frac{1}{\sqrt{2}}\begin{bmatrix} A & -A \\ A & A \end{bmatrix}$ is unitary.

Fact 3.11.10. Let $A, B \in \mathbb{R}^{n \times n}$. Then, $A + \jmath B$ is (Hermitian, skew Hermitian, unitary) if and only if $\begin{bmatrix} A & B \\ -B & A \end{bmatrix}$ is (symmetric, skew symmetric, orthogonal).

Remark: See Fact 2.19.7.

Fact 3.11.11. Let $A \in \mathbb{F}^{n \times n}$, and assume that A is unitary. Then,

$$|\operatorname{Re} \operatorname{tr} A| \leq n,$$
$$|\operatorname{Im} \operatorname{tr} A| \leq n,$$

and

$$|\operatorname{tr} A| \leq n.$$

Remark: The third inequality does not follow from the first two inequalities.

Fact 3.11.12. Let $A \in \mathbb{R}^{n \times n}$, and assume that A is orthogonal. Then,

$$-1_{n \times n} \leq\leq A \leq\leq 1_{n \times n},$$

that is, for all $i, j = 1, \ldots, n$, $|A_{(i,j)}| \leq 1$. Hence,

$$|\operatorname{tr} A| \leq n.$$

Furthermore, the following statements are equivalent:

i) $A = I$.

ii) $I \circ A = I$.

iii) $\operatorname{tr} A = n$.

Finally, if n is odd and $\det A = 1$, then

$$2 - n \leq \operatorname{tr} A \leq n.$$

Remark: See Fact 3.11.13.

Remark: $I \circ A$ is the diagonal matrix whose diagonal entries are the same as the diagonal entries of A.

Fact 3.11.13. Let $A \in \mathbb{R}^{n \times n}$, assume that A is orthogonal, let $B \in \mathbb{R}^{n \times n}$, and assume that B is diagonal and positive definite. Then,

$$-B1_{n \times n} \leq\leq BA \leq\leq B1_{n \times n}$$

and

$$-\operatorname{tr} B \leq \operatorname{tr} BA \leq \operatorname{tr} B.$$

Furthermore, the following statements are equivalent:

i) $BA = B$.

$ii)$ $I \circ (BA) = B.$

$iii)$ $\text{tr}\, BA = \text{tr}\, B.$

Remark: See Fact 3.11.12.

Fact 3.11.14. Let $x \in \mathbb{C}^n$, where $n \geq 2$. Then, the following statements are equivalent:

$i)$ There exists a unitary matrix $A \in \mathbb{C}^{n \times n}$ such that

$$x = \begin{bmatrix} A_{(1,1)} \\ \vdots \\ A_{(n,n)} \end{bmatrix}.$$

$ii)$ For all $j \in \{1, \ldots, n\}$, $|x_{(j)}| \leq 1$ and

$$2(1 - |x_{(j)}|) + \sum_{i=1}^{n} |x_{(i)}| \leq n.$$

Proof: See [1370].

Remark: This result is equivalent to the Schur-Horn theorem given by Fact 8.18.10.

Remark: The inequalities in $ii)$ define a polytope.

Fact 3.11.15. Let $A \in \mathbb{C}^{n \times n}$, and assume that A is unitary. Then, $|\det A| = 1$.

Fact 3.11.16. Let $A \in \mathbb{R}^{n \times n}$, and assume that A is orthogonal. Then, either $\det A = 1$ or $\det A = -1$. Consequently, if A is a permutation matrix, then A is either an even permutation matrix or an odd permutation matrix. Finally, if A is a transposition matrix, then A is odd.

Remark: See Fact 2.13.1 and Fact 3.23.2.

Fact 3.11.17. Let $A, B \in \text{SO}(3)$. Then,

$$\det(A + B) \geq 0.$$

Proof: See [1038].

Fact 3.11.18. Let $A \in \mathbb{F}^{n \times n}$, and assume that A is unitary. Then,

$$|\det(I + A)| \leq 2^n.$$

If, in addition, A is real, then

$$0 \leq \det(I + A) \leq 2^n.$$

Fact 3.11.19. Let $M \triangleq \begin{bmatrix} A & B \\ C & D \end{bmatrix} \in \mathbb{F}^{(n+m) \times (n+m)}$, and assume that M is unitary. Then,

$$\det A = (\det M)\overline{\det D}.$$

Proof: Let $\begin{bmatrix} \hat{A} & \hat{B} \\ \hat{C} & \hat{D} \end{bmatrix} \triangleq A^{-1}$, and take the determinant of $A \begin{bmatrix} I & \hat{B} \\ 0 & \hat{D} \end{bmatrix} = \begin{bmatrix} A & 0 \\ C & I \end{bmatrix}$. See [13] or [1219].

Remark: See Fact 2.14.28 and Fact 2.14.7.

Fact 3.11.20. Let $A \in \mathbb{F}^{n \times n}$, assume that A is unitary, and let $x \in \mathbb{F}^n$ be such that $x^*x = 1$ and $Ax = -x$. Then, the following statements hold:

 i) $\det(A + I) = 0$.

 ii) $A + 2xx^*$ is unitary.

 iii) $A = (A + 2xx^*)(I_n - 2xx^*) = (I_n - 2xx^*)(A + 2xx^*)$.

 iv) $\det(A + 2xx^*) = -\det A$.

Fact 3.11.21. The following statements hold:

 i) If $A \in \mathbb{F}^{n \times n}$ is Hermitian, then $A + \jmath I$ is nonsingular, $B \triangleq (\jmath I - A)(\jmath I + A)^{-1}$ is unitary, and $I + B = \jmath 2(\jmath I + A)^{-1}$.

 ii) If $B \in \mathbb{F}^{n \times n}$ is unitary and $\lambda \in \mathbb{C}$ is such that $|\lambda| = 1$ and $I + \lambda B$ is nonsingular, then $A \triangleq \jmath(I - \lambda B)(I + \lambda B)^{-1}$ is Hermitian and $\jmath I + A = \jmath 2(I + \lambda B)^{-1}$.

 iii) If $A \in \mathbb{F}^{n \times n}$ is Hermitian, then there exists a unique unitary matrix $B \in \mathbb{F}^{n \times n}$ such that $I + B$ is nonsingular and $A = \jmath(I - B)(I + B)^{-1}$. In fact, $B = (\jmath I - A)(\jmath I + A)^{-1}$.

 iv) If $B \in \mathbb{F}^{n \times n}$ is unitary and $\lambda \in \mathbb{C}$ is such that $|\lambda| = 1$ and $I + \lambda B$ is nonsingular, then there exists a unique Hermitian matrix $A \in \mathbb{F}^{n \times n}$ such that $\lambda B = (\jmath I - A)(\jmath I + A)^{-1}$. In fact, $A = \jmath(I - \lambda B)(I + \lambda B)^{-1}$.

Proof: See [521, pp. 168, 169].

Remark: The linear fractional transformation $f(s) = (\jmath - s)/(\jmath + s)$ maps the upper half plane of $\mathbb{C}$ onto the unit disk in $\mathbb{C}$, and the real line onto the unit circle in $\mathbb{C}$.

Remark: $\mathcal{C}(A) \triangleq (A - I)(A + I)^{-1} = I - 2(A + I)^{-1}$ is the *Cayley transform* of A. See Fact 3.11.29, Fact 3.11.22, Fact 3.11.23, Fact 3.20.12, Fact 8.9.31, and Fact 11.21.9.

Fact 3.11.22. The following statements hold:

 i) If $A \in \mathbb{F}^{n \times n}$ is skew Hermitian, then $I + A$ is nonsingular, $B \triangleq (I - A)(I + A)^{-1}$ is unitary, and $I + B = 2(I + A)^{-1}$. If, in addition, $\operatorname{mspec}(A) = \overline{\operatorname{mspec}(A)}$, then $\det B = 1$.

 ii) If $B \in \mathbb{F}^{n \times n}$ is unitary and $\lambda \in \mathbb{C}$ is such that $|\lambda| = 1$ and $I + \lambda B$ is nonsingular, then $A \triangleq (I + \lambda B)^{-1}(I - \lambda B)$ is skew Hermitian and $I + A = 2(I + \lambda B)^{-1}$.

 iii) If $A \in \mathbb{F}^{n \times n}$ is skew Hermitian, then there exists a unique unitary matrix $B \in \mathbb{F}^{n \times n}$ such that $I + B$ is nonsingular and $A = (I + B)^{-1}(I - B)$. In fact, $B = (I - A)(I + A)^{-1}$.

 iv) If B is unitary and $\lambda \in \mathbb{C}$ is such that $|\lambda| = 1$ and $I + \lambda B$ is nonsingular, then there exists a unique skew-Hermitian matrix $A \in \mathbb{F}^{n \times n}$ such that $B = \overline{\lambda}(I - A)(I + A)^{-1}$. In fact, $A = (I + \lambda B)^{-1}(I - \lambda B)$.

Proof: See [521, p. 184] and [730, p. 440].

Fact 3.11.23. The following statements hold:

i) If $A \in \mathbb{R}^{n \times n}$ is skew symmetric, then $I + A$ is nonsingular, $B \triangleq (I - A)(I + A)^{-1}$ is orthogonal, $I + B = 2(I + A)^{-1}$, and $\det B = 1$.

ii) If $B \in \mathbb{R}^{n \times n}$ is orthogonal, $C \in \mathbb{R}^{n \times n}$ is diagonal with diagonally located entries ± 1, and $I + CB$ is nonsingular, then $A \triangleq (I + CB)^{-1}(I - CB)$ is skew symmetric, $I + A = 2(I + CB)^{-1}$, and $\det CB = 1$.

iii) If $A \in \mathbb{R}^{n \times n}$ is skew symmetric, then there exists a unique orthogonal matrix $B \in \mathbb{R}^{n \times n}$ such that $I + B$ is nonsingular and $A = (I + B)^{-1}(I - B)$. In fact, $B = (I - A)(I + A)^{-1}$.

iv) If $B \in \mathbb{R}^{n \times n}$ is orthogonal and $C \in \mathbb{R}^{n \times n}$ is diagonal with diagonally located entries ± 1, then there exists a unique skew-symmetric matrix $A \in \mathbb{R}^{n \times n}$ such that $CB = (I - A)(I + A)^{-1}$. In fact, $A = (I + CB)^{-1}(I - CB)$.

Remark: The last statement is due to Hsu. See [1125, p. 101].

Remark: The Cayley transform is a one-to-one and onto map from the set of skew-symmetric matrices to the set of orthogonal matrices whose spectrum does not include -1.

Fact 3.11.24. Furthermore, if $A, B \in \mathbb{F}^{n \times n}$ are unitary, then

$$\sqrt{1 - \left|\tfrac{1}{n} \operatorname{tr} AB\right|^2} \leq \sqrt{1 - \left|\tfrac{1}{n} \operatorname{tr} A\right|^2} + \sqrt{1 - \left|\tfrac{1}{n} \operatorname{tr} B\right|^2}.$$

Proof: See [1425].

Remark: See Fact 2.12.1.

Fact 3.11.25. If $A \in \mathbb{F}^{n \times n}$ is shifted unitary, then $B \triangleq 2A - I$ is unitary. Conversely, If $B \in \mathbb{F}^{n \times n}$ is unitary, then $A \triangleq \tfrac{1}{2}(B + I)$ is shifted unitary.

Remark: The affine mapping $f(A) \triangleq 2A - I$ from the shifted-unitary matrices to the unitary matrices is one-to-one and onto. See Fact 3.14.1 and Fact 3.15.2.

Remark: See Fact 3.7.14 and Fact 3.13.13.

Fact 3.11.26. If $A \in \mathbb{F}^{n \times n}$ is shifted unitary, then A is normal. Hence, the following statements are equivalent:

i) A is shifted unitary.

ii) $A + A^* = 2A^*A$.

iii) $A + A^* = 2AA^*$.

Proof: By Fact 3.11.25 there exists a unitary matrix B such that $A = \tfrac{1}{2}(B + I)$. Since B is normal, it follows from Fact 3.7.14 that A is normal.

Fact 3.11.27. Let $\theta \in \mathbb{R}$, and define the orthogonal matrix

$$A(\theta) \triangleq \begin{bmatrix} \cos\theta & \sin\theta \\ -\sin\theta & \cos\theta \end{bmatrix}.$$

Now, let $\theta_1, \theta_2 \in \mathbb{R}$. Then,

$$A(\theta_1)A(\theta_2) = A(\theta_1 + \theta_2).$$

Consequently,

$$\cos(\theta_1 + \theta_2) = (\cos\theta_1)\cos\theta_2 - (\sin\theta_1)\sin\theta_2,$$
$$\sin(\theta_1 + \theta_2) = (\cos\theta_1)\sin\theta_2 + (\sin\theta_1)\cos\theta_2.$$

Furthermore,

$$\mathrm{SO}(2) = \{A(\theta) \colon \ \theta \in \mathbb{R}\}$$

and

$$\mathrm{O}(2)\backslash\mathrm{SO}(2) = \left\{ \begin{bmatrix} \cos\theta & \sin\theta \\ \sin\theta & -\cos\theta \end{bmatrix} \colon \ \theta \in \mathbb{R} \right\}.$$

Remark: See Proposition 3.3.6 and Fact 11.11.3.

Fact 3.11.28. Let $A \in \mathbb{R}^{3\times3}$. Then, $A \in \mathrm{O}(3)\backslash\mathrm{SO}(3)$ if and only if $-A \in \mathrm{SO}(3)$.

Fact 3.11.29. Let $z \in \mathbb{R}^3$, assume that $\|z\|_2 = 1$, let $\theta \in \mathbb{R}$, and define $A_z(\theta) \in \mathbb{R}^{3\times3}$ by

$$A_z(\theta) \triangleq (\cos\theta)I + (\sin\theta)K(z) + (1 - \cos\theta)zz^{\mathrm{T}}.$$

Then, the following statements hold:

i) $A_z(\theta)$ is given by

$$A_z(\theta) = I + (\sin\theta)K(z) + (1 - \cos\theta)K^2(z).$$

ii) $A_z(\theta)$ is a rotation matrix, that is, $A_z(\theta)$ is orthogonal and $\det A_z(\theta) = 1$.

iii) $A_z(\theta) = I$ if and only if $|\theta/(2\pi)|$ is an even integer.

iv) $A_z(\theta) \neq I$ if and only if $|\theta/(2\pi)|$ is not an even integer.

v) Both $A_z(\theta) \neq I$ and $A_z^2(\theta) = I$ if and only if $|\theta/(2\pi)|$ is an odd integer. In this case, $A_z(\theta) = -I + 2zz^{\mathrm{T}}$.

vi) $A_{-z}(2\pi - \theta) = A_z(\theta)$.

vii) $A_z^{-1}(\theta) = A_z^{\mathrm{T}}(\theta) = A_z(-\theta) = A_{-z}(\theta)$.

viii) Let $x, y \in \mathbb{R}^3$, assume that $\|x\|_2 = \|y\|_2 \neq 0$, and let $\theta \in [0, \pi]$ denote the angle between x and y. Furthermore, if $\theta \in (0, \pi)$, then let $z \in \mathbb{R}^3$ be given by

$$z = \frac{1}{\|x \times y\|_2} x \times y,$$

whereas, if $\theta \in \{0, \pi\}$, then let $z \in \{x\}^\perp$ be such that $\|z\|_2 = 1$. Then,

$$y = A_z(\theta)x.$$

ix) Let $x \in \mathbb{R}^3$. Then, $A_z(\theta)x = x$ if and only if either $\theta = 0$ or $z \times x = 0$.

x) $A_z(\theta)$ satisfies

$$A_z(\theta) = [I - B_z(\theta)][I + B_z(\theta)]^{-1},$$

where

$$B_z(\theta) \triangleq -\tan(\tfrac{1}{2}\theta)K(z).$$

Consequently,

$$[I + A_z(\theta)]^{-1}[I - A_z(\theta)] = -\tan(\tfrac{1}{2}\theta)K(z).$$

Now, let $\phi \in [0, \pi]$, and let $w \in \mathbb{R}^3$, where $\|w\|_2 = 1$. Then, the following statements hold:

xi) The commutator $[A_z(\theta), A_w(\phi)]$ is given by

$$\begin{aligned}[A_z(\theta), A_w(\phi)] = {}& (\sin\theta)(\sin\phi)K(z \times w) \\ &+ (\sin\theta)(1 - \cos\phi)[K(z), ww^{\mathrm{T}}] \\ &+ (\sin\phi)(1 - \cos\theta)[zz^{\mathrm{T}}, K(w)] \\ &+ (1 - \cos\theta)(1 - \cos\phi)w^{\mathrm{T}}z(zw^{\mathrm{T}} - wz^{\mathrm{T}}).\end{aligned}$$

xii) $[A_z(\theta), A_w(\phi)] = 0$ if and only if at least one of the following statements holds:

a) $z \times w = 0$.

b) Either $A_z(\theta) = I$ or $A_w(\phi) = I$.

c) $A_z^2(\theta) = I$, $A_w^2(\phi) = I$, and $w^{\mathrm{T}}z = 0$.

Proof: The expression for $A_z(\theta)$ in terms of $B_z(\theta)$ in x) is derived in [12]. The expression involving $B_z(\theta)$ is derived in [1033, pp. 244, 245]. Result xii) is due to Bhat.

Remark: If x and y in $viii$) are linearly independent, then θ is given by

$$\theta = \cos^{-1}\frac{x^{\mathrm{T}}y}{\|x\|_2\|y\|_2}.$$

Furthermore, it follows from $xxviii$) of Fact 3.10.1 that

$$\sin\theta = \frac{\|x \times y\|_2}{\|x\|_2\|y\|_2}.$$

Remark: $A_z(\theta)$ can be written as

$$\begin{aligned}A_z(\theta) &= (\cos\theta)I + \frac{1}{\|x\|_2^2}(yx^{\mathrm{T}} - xy^{\mathrm{T}}) + \frac{1 - \cos\theta}{\|x \times y\|_2^2}(x \times y)(x \times y)^{\mathrm{T}} \\ &= \frac{x^{\mathrm{T}}y}{x^{\mathrm{T}}x}I + \frac{1}{x^{\mathrm{T}}x}(yx^{\mathrm{T}} - xy^{\mathrm{T}}) + \frac{1 - \cos\theta}{(x^{\mathrm{T}}x\sin\theta)^2}(x \times y)(x \times y)^{\mathrm{T}} \\ &= \frac{x^{\mathrm{T}}y}{x^{\mathrm{T}}x}I + \frac{1}{x^{\mathrm{T}}x}(yx^{\mathrm{T}} - xy^{\mathrm{T}}) + \frac{\tan(\tfrac{1}{2}\theta)}{(x^{\mathrm{T}}x)^2\sin\theta}(x \times y)(x \times y)^{\mathrm{T}} \\ &= \frac{x^{\mathrm{T}}y}{x^{\mathrm{T}}x}I + \frac{1}{x^{\mathrm{T}}x}(yx^{\mathrm{T}} - xy^{\mathrm{T}}) + \frac{1}{(x^{\mathrm{T}}x)^2(1 + \cos\theta)}(x \times y)(x \times y)^{\mathrm{T}} \\ &= \frac{x^{\mathrm{T}}y}{x^{\mathrm{T}}x}I + \frac{1}{x^{\mathrm{T}}x}(yx^{\mathrm{T}} - xy^{\mathrm{T}}) + \frac{1}{x^{\mathrm{T}}x(x^{\mathrm{T}}x + x^{\mathrm{T}}y)}(x \times y)(x \times y)^{\mathrm{T}}.\end{aligned}$$

As a check, note that

$$A_z(\theta)x = (\cos\theta)x + \frac{1}{\|x\|_2^2}(x^\mathrm{T}xy - y^\mathrm{T}xx) + \frac{1-\cos\theta}{\|x \times y\|_2^2}(x \times y)(x \times y)^\mathrm{T}x$$

$$= \frac{x^\mathrm{T}y}{\|x\|_2^2}x + \frac{1}{\|x\|_2^2}(x^\mathrm{T}xy - y^\mathrm{T}xx)$$

$$= y.$$

Furthermore, $B_z(\theta)$ can be written as

$$B_z(\theta) = \frac{1}{x^\mathrm{T}x + x^\mathrm{T}y}(xy^\mathrm{T} - yx^\mathrm{T}).$$

These expressions satisfy

$$A_z(\theta) + B_z(\theta) + A_z(\theta)B_z(\theta) = I.$$

Remark: In *viii*), the matrix $A_z(\theta)$ represents a right-hand rule rotation of the nonzero vector x through the angle θ around z to yield the vector y, which has the same length as x. The cases $x = y$ and $x = -y$ correspond, respectively, to $\theta = 0$ and $\theta = \pi$; in these cases, the vector z representing the axis of rotation is not unique.

Remark: See Fact 11.11.6.

Remark: *viii*) is a linear interpolation result. See Fact 3.9.5, Fact 11.11.7, and [139, 795].

Remark: Extensions of the Cayley transform are discussed in [1375].

Fact 3.11.30. Let $A \in \mathbb{R}^{3\times 3}$, and let $z \triangleq \left[\begin{smallmatrix} b \\ c \\ d \end{smallmatrix}\right]$, where $b^2 + c^2 + d^2 = 1$. Then, $A \in \mathrm{SO}(3)$, and A rotates every vector in $\mathbb{R}^3$ by the angle π about z if and only if

$$A = \begin{bmatrix} 2b^2 - 1 & 2bc & 2bd \\ 2bc & 2c^2 - 1 & 2cd \\ 2bd & 2cd & 2d^2 - 1 \end{bmatrix}.$$

Proof: This formula follows from the last expression for A in Fact 3.11.31 with $\theta = \pi$. See [365, p. 30].

Remark: A is a reflector.

Problem: Solve for b, c, and d in terms of the entries of A.

Fact 3.11.31. Let $A \in \mathbb{R}^{3\times 3}$. Then, $A \in \mathrm{SO}(3)$ if and only if there exist real numbers a, b, c, d such that $a^2 + b^2 + c^2 + d^2 = 1$ and

$$A = \begin{bmatrix} a^2 + b^2 - c^2 - d^2 & 2(bc - ad) & 2(ac + bd) \\ 2(ad + bc) & a^2 - b^2 + c^2 - d^2 & 2(cd - ab) \\ 2(bd - ac) & 2(ab + cd) & a^2 - b^2 - c^2 + d^2 \end{bmatrix}.$$

In this case,

$$a = \pm\tfrac{1}{2}\sqrt{1 + \mathrm{tr}\, A}.$$

If, in addition, $a \neq 0$, then b, c, and d are given by

$$b = \frac{A_{(3,2)} - A_{(2,3)}}{4a}, \quad c = \frac{A_{(1,3)} - A_{(3,1)}}{4a}, \quad d = \frac{A_{(2,1)} - A_{(1,2)}}{4a}.$$

Now, define $v \triangleq \begin{bmatrix} b & c & d \end{bmatrix}^{\mathrm{T}}$. Then, A represents a rotation about the unit-length vector $z \triangleq (\csc \frac{\theta}{2})v$ through the angle $\theta \in [0, 2\pi]$ that satisfies

$$a = \cos \tfrac{\theta}{2},$$

where the direction of rotation is determined by the right-hand rule. Therefore,

$$\theta \triangleq 2\cos^{-1} a.$$

If $a \in [0, 1]$, then

$$\theta = 2\cos^{-1}(\tfrac{1}{2}\sqrt{1 + \operatorname{tr} A}) = \cos^{-1}(\tfrac{1}{2}[(\operatorname{tr} A) - 1]),$$

whereas, if $a \in [-1, 0]$, then

$$\theta = 2\cos^{-1}(-\tfrac{1}{2}\sqrt{1 + \operatorname{tr} A}) = \pi + \cos^{-1}(\tfrac{1}{2}[1 - \operatorname{tr} A]).$$

In particular, $a = 1$ if and only if $\theta = 0$; $a = 0$ if and only if $\theta = \pi$; and $a = -1$ if and only if $\theta = 2\pi$. Furthermore,

$$A = (2a^2 - 1)I_n + 2aK(v) + 2vv^{\mathrm{T}}$$

$$= (\cos\theta)I + (\sin\theta)K(z) + (1 - \cos\theta)zz^{\mathrm{T}}$$

$$= I + (\sin\theta)K(z) + (1 - \cos\theta)K^2(z).$$

Furthermore,

$$A - A^{\mathrm{T}} = 4aK(v) = 2(\sin\theta)K(z),$$

and thus

$$2a\sin\tfrac{\theta}{2} = \sin\theta.$$

If $\theta = 0$ or $\theta = 2\pi$, then $v = z = 0$, whereas, if $\theta = \pi$, then

$$K^2(z) = \tfrac{1}{2}(A - I).$$

Conversely, let $\theta \in \mathbb{R}$, let $z \in \mathbb{R}^3$, assume that $z^{\mathrm{T}}z = 1$, and define

$$\begin{bmatrix} a \\ b \\ c \\ d \end{bmatrix} = \begin{bmatrix} \cos\frac{\theta}{2} \\ (\sin\frac{\theta}{2})z \end{bmatrix}.$$

Then, A represents a rotation about the unit-length vector z through the angle θ, where the direction of rotation is determined by the right-hand rule. In this case, A is given by

$$A = \begin{bmatrix} z_{(1)}^2 + (z_{(2)}^2 + z_{(3)}^2)\cos\theta & z_{(1)}z_{(2)}(1 - \cos\theta) - z_{(3)}\sin\theta & z_{(1)}z_{(3)}(1 - \cos\theta) + z_{(2)}\sin\theta \\ z_{(1)}z_{(2)}(1 - \cos\theta) + z_{(3)}\sin\theta & z_{(2)}^2 + (z_{(1)}^2 + z_{(3)}^2)\cos\theta & z_{(2)}z_{(3)}(1 - \cos\theta) - z_{(1)}\sin\theta \\ z_{(1)}z_{(3)}(1 - \cos\theta) - z_{(2)}\sin\theta & z_{(2)}z_{(3)}(1 - \cos\theta) + z_{(1)}\sin\theta & z_{(3)}^2 + (z_{(1)}^2 + z_{(2)}^2)\cos\theta \end{bmatrix}.$$

Proof: See [490, p. 162], [569, p. 22], [1216, p. 19], and use Fact 3.11.29.

Remark: This result is due to Rodrigues. A history of Rodrigues's contributions is given in [29].

Remark: The numbers a, b, c, d, which are *Euler parameters*, are elements of S^3, which is the sphere in $\mathbb{R}^4$. The elements of S^3 can be viewed as unit quaternions, thus giving S^3 a group structure. See Fact 3.23.9. Conversely, a, b, c, d can be expressed in terms of the entries of a 3×3 orthogonal matrix, which are the *direction cosines*. See [156, pp. 384–387]. See also Fact 3.24.1.

Remark: Replacing a by $-a$ in A but keeping b, c, d unchanged yields the transpose of A.

Remark: Note that A is unchanged when a, b, c, d are replaced by $-a, -b, -c, -d$. Conversely, given the direction cosines of a rotation matrix A, there exist exactly two distinct quadruples (a, b, c, d) of Euler parameters that parameterize A. Therefore, the Euler parameters, which parameterize the unit sphere S^3 in $\mathbb{R}^4$, provide a *double cover* of SO(3). See [994, p. 304] and Fact 3.24.1.

Remark: Sp(1) is a double cover of SO(3), Sp(1) × Sp(1) is a double cover of SO(4), Sp(2) is a double cover of SO(5), and SU(4) is a double cover of SO(3). For each n, SO(n) is double covered by the *spin group* Spin(n). See [370, p. 141], [1287, p. 130], and [1470, pp. 42–47]. Sp(2) is defined in Fact 3.24.4.

Remark: Rotation matrices in $\mathbb{R}^{2 \times 2}$ are discussed in [1227].

Remark: See Fact 8.9.27 and Fact 11.15.10.

Remark: Extensions to $n \times n$ matrices are considered in [552].

Fact 3.11.32. Let $\theta_1, \theta_2 \in \mathbb{R}$, let $z_1, z_2 \in \mathbb{R}^3$, assume that $z_1^T z_1 = z_2^T z_2 = 1$, and, for $i = 1, 2$, let $A_i \in \mathbb{R}^{3 \times 3}$ be the rotation matrix that represents the rotation about the unit-length vector z_i through the angle θ_i, where the direction of rotation is determined by the right-hand rule. Then, $A_3 \triangleq A_2 A_1$ represents the rotation about the unit-length vector z_3 through the angle θ_3, where the direction of rotation is determined by the right-hand rule, and where θ_3 and z_3 are given by

$$\cos \tfrac{\theta_3}{2} = (\cos \tfrac{\theta_2}) \cos \tfrac{\theta_1}{2} - (\sin \tfrac{\theta_2}{2}) \sin \tfrac{\theta_1}{2} z_2^T z_1$$

and

$$z_3 = (\csc \tfrac{\theta_3}{2})[(\sin \tfrac{\theta_2}{2})(\cos \tfrac{\theta_1}{2}) z_2 + (\cos \tfrac{\theta_2}{2})(\sin \tfrac{\theta_1}{2}) z_1 + (\sin \tfrac{\theta_2}{2})(\sin \tfrac{\theta_1}{2})(z_2 \times z_1)]$$

$$= \frac{\cot \tfrac{\theta_3}{2}}{1 - z_2^T z_1 (\tan \tfrac{\theta_2}{2}) \tan \tfrac{\theta_1}{2}}[(\tan \tfrac{\theta_2}{2}) z_2 + (\tan \tfrac{\theta_1}{2}) z_1 + (\tan \tfrac{\theta_2}{2})(\tan \tfrac{\theta_1}{2})(z_2 \times z_1)].$$

Proof: See [569, pp. 22–24].

Remark: These expressions are *Rodrigues's formulas*, which are identical to the quaternion multiplication formula given by

$$\begin{bmatrix} a_3 \\ b_3 \\ c_3 \\ d_3 \end{bmatrix} = \begin{bmatrix} \cos \tfrac{\theta_3}{2} \\ (\sin \tfrac{\theta_3}{2}) z_3 \end{bmatrix} = \begin{bmatrix} a_1 a_2 - z_2^T z_1 \\ a_1 z_2 + a_2 z_1 + z_2 \times z_1 \end{bmatrix}$$

with

$$\begin{bmatrix} a_2 \\ b_2 \\ c_2 \\ d_2 \end{bmatrix} = \begin{bmatrix} \cos\frac{\theta_2}{2} \\ (\sin\frac{\theta_2}{2})z_2 \end{bmatrix}, \qquad \begin{bmatrix} a_1 \\ b_1 \\ c_1 \\ d_1 \end{bmatrix} = \begin{bmatrix} \cos\frac{\theta_1}{2} \\ (\sin\frac{\theta_1}{2})z_1 \end{bmatrix}.$$

in Fact 3.24.1. See [29].

Fact 3.11.33. Let $x, y, z \in \mathbb{R}^2$. If x is rotated according to the right-hand rule through an angle $\theta \in \mathbb{R}$ about y, then the resulting vector $\hat{x} \in \mathbb{R}^2$ is given by

$$\hat{x} = \begin{bmatrix} \cos\theta & -\sin\theta \\ \sin\theta & \cos\theta \end{bmatrix} x + \begin{bmatrix} y_{(1)}(1 - \cos\theta) + y_{(2)}\sin\theta \\ y_{(2)}(1 - \cos\theta) + y_{(1)}\sin\theta \end{bmatrix}.$$

If x is reflected across the line passing through 0 and z and parallel to the line passing through 0 and y, then the resulting vector $\hat{x} \in \mathbb{R}^2$ is given by

$$\hat{x} = \begin{bmatrix} y_{(1)}^2 - y_{(2)}^2 & 2y_{(1)}y_{(2)} \\ 2y_{(1)}y_{(2)} & y_{(2)}^2 - y_{(1)}^2 \end{bmatrix} x + \begin{bmatrix} -z_{(1)}\left(y_{(1)}^2 - y_{(2)}^2 - 1\right) - 2z_{(2)}y_{(1)}y_{(2)} \\ -z_{(2)}\left(y_{(1)}^2 - y_{(2)}^2 - 1\right) - 2z_{(1)}y_{(1)}y_{(2)} \end{bmatrix}.$$

Remark: These *affine planar transformations* are used in computer graphics. See [65, 511, 1122].

Remark: See Fact 3.11.34 and Fact 3.11.29.

Fact 3.11.34. Let $x, y \in \mathbb{R}^3$, and assume that $y^{\mathrm{T}}y = 1$. If x is rotated according to the right-hand rule through an angle $\theta \in \mathbb{R}$ about the line passing through 0 and y, then the resulting vector $\hat{x} \in \mathbb{R}^3$ is given by

$$\hat{x} = x + (\sin\theta)(y \times x) + (1 - \cos\theta)[y \times (y \times x)].$$

Proof: See [25].

Remark: See Fact 3.11.33 and Fact 3.11.29.

3.12 Facts on Idempotent Matrices

Fact 3.12.1. Let $S_1, S_2 \subseteq \mathbb{F}^n$ be complementary subspaces, and let $A \in \mathbb{F}^{n \times n}$ be the idempotent matrix onto S_1 along S_2. Then, A^* is the idempotent matrix onto $S_2^\perp$ along $S_1^\perp$, and $A_\perp^*$ is the idempotent matrix onto $S_1^\perp$ along $S_2^\perp$.

Remark: See Fact 2.9.18.

Fact 3.12.2. Let $A \in \mathbb{F}^{n \times n}$. Then, A is idempotent if and only if there exists a positive integer k such that $A^{k+1} = A^k$.

Fact 3.12.3. Let $A \in \mathbb{F}^{n \times n}$. Then, the following statements are equivalent:

i) A is idempotent.

ii) $\mathcal{N}(A) = \mathcal{R}(A_\perp)$.

iii) $\mathcal{R}(A) = \mathcal{N}(A_\perp)$.

In this case, the following statements hold:

iv) A is the idempotent matrix onto $\mathcal{R}(A)$ along $\mathcal{N}(A)$.

v) $A_\perp$ is the idempotent matrix onto $\mathcal{N}(A)$ along $\mathcal{R}(A)$.

vi) A^* is the idempotent matrix onto $\mathcal{N}(A)^\perp$ along $\mathcal{R}(A)^\perp$.

vii) $A_\perp^*$ is the idempotent matrix onto $\mathcal{R}(A)^\perp$ along $\mathcal{N}(A)^\perp$.

Proof: See [671, p. 146].

Remark: See Fact 2.10.1 and Fact 5.12.18.

Fact 3.12.4. Let $A \in \mathbb{F}^{n \times n}$, and assume that A is idempotent. Then,

$$\mathcal{R}(I - AA^*) = \mathcal{R}(2I - A - A^*).$$

Proof: See [1319].

Fact 3.12.5. Let $A \in \mathbb{F}^{n \times n}$. Then, A is idempotent if and only if $-A$ is skew idempotent.

Fact 3.12.6. Let $A \in \mathbb{F}^{n \times n}$. Then, A is idempotent and rank $A = 1$ if and only if there exist vectors $x, y \in \mathbb{F}^n$ such that $y^\mathrm{T}x = 1$ and $A = xy^\mathrm{T}$.

Fact 3.12.7. Let $A \in \mathbb{F}^{n \times n}$, and assume that A is idempotent. Then, $A^\mathrm{T}, \overline{A}$, and A^* are idempotent.

Fact 3.12.8. Let $A \in \mathbb{F}^{n \times n}$, and assume that A is idempotent and skew Hermitian. Then, $A = 0$.

Fact 3.12.9. Let $A \in \mathbb{F}^{n \times n}$. Then, A is idempotent if and only if rank $A +$ rank$(I - A) = n$.

Fact 3.12.10. Let $A \in \mathbb{F}^{n \times m}$. If $A^\mathrm{L} \in \mathbb{F}^{m \times n}$ is a left inverse of A, then AA^L is idempotent and rank $A^\mathrm{L} = $ rank A. Furthermore, if $A^\mathrm{R} \in \mathbb{F}^{m \times n}$ is a right inverse of A, then $A^\mathrm{R}A$ is idempotent and rank $A^\mathrm{R} = $ rank A.

Fact 3.12.11. Let $A \in \mathbb{F}^{n \times n}$, and assume that A is nonsingular and idempotent. Then, $A = I_n$.

Fact 3.12.12. Let $A \in \mathbb{F}^{n \times n}$, and assume that A is idempotent. Then, so is $A_\perp \triangleq I - A$, and, furthermore, $AA_\perp = A_\perp A = 0$.

Fact 3.12.13. Let $A \in \mathbb{F}^{n \times n}$, and assume that A is idempotent. Then,

$$\det(I + A) = 2^{\mathrm{tr}\, A}$$

and

$$(I + A)^{-1} = I - \tfrac{1}{2}A.$$

Fact 3.12.14. Let $A \in \mathbb{F}^{n \times n}$ and $\alpha \in \mathbb{F}$, where $\alpha \neq 0$. Then, the matrices

$$\begin{bmatrix} A & A^* \\ A^* & A \end{bmatrix}, \quad \begin{bmatrix} A & \alpha^{-1}A \\ \alpha(I - A) & I - A \end{bmatrix}, \quad \begin{bmatrix} A & \alpha^{-1}A \\ -\alpha A & -A \end{bmatrix}$$

are, respectively, normal, idempotent, and nilpotent.

Fact 3.12.15. Let $A, B \in \mathbb{F}^{n \times n}$, and assume that A and B are idempotent. Then,

$$\mathcal{R}([A, B]) = \mathcal{R}(A - B) \cap \mathcal{R}(A_\perp - B)$$

and

$$\mathcal{N}([A, B]) = \mathcal{N}(A - B) \cap \mathcal{N}(A_\perp - B).$$

Proof: See [1458].

Fact 3.12.16. Let $A \in \mathbb{F}^{n \times n}$, and assume that A is nilpotent. Then, there exist idempotent matrices $B, C \in \mathbb{F}^{n \times n}$ such that $A = [B, C]$.

Proof: See [449].

Remark: A necessary and sufficient condition for a matrix to be a commutator of a pair of idempotents is given in [449].

Remark: See Fact 9.9.9 for the case of projectors.

Fact 3.12.17. Let $A, B \in \mathbb{F}^{n \times n}$, assume that A and B are idempotent, and define $A_\perp \triangleq I - A$ and $B_\perp \triangleq I - B$. Then, the following statements hold:

i) $(A - B)^2 + (A_\perp - B)^2 = I$.

ii) $[A, B] = [B, A_\perp] = [B_\perp, A] = [A_\perp, B_\perp]$.

iii) $A - B = AB_\perp - A_\perp B$.

iv) $AB_\perp + BA_\perp = AB_\perp A + A_\perp BA_\perp$.

v) $A[A, B] = [A, B]A_\perp$.

vi) $B[A, B] = [A, B]B_\perp$.

Proof: See [1071].

Fact 3.12.18. Let $A, B \in \mathbb{R}^{n \times n}$. Then, the following statements hold:

i) Assume that $A^3 = -A$ and $B = I + A + A^2$. Then, $B^4 = I$, $B^{-1} = I - A + A^2$, $B^3 - B^2 + B - I = 0$, $A = \frac{1}{2}(B - B^3)$, and $I + A^2$ is idempotent.

ii) Assume that $B^3 - B^2 + B - I = 0$ and $A = \frac{1}{2}(B - B^3)$. Then, $A^3 = -A$ and $B = I + A + A^2$.

iii) Assume that $B^4 = I$ and $A = \frac{1}{2}(B - B^{-1})$. Then, $A^3 = -A$, and $\frac{1}{4}(I + B + B^2 + B^3)$ is idempotent.

Remark: The geometric meaning of these results is discussed in [487, pp. 153, 212–214, 242].

Fact 3.12.19. Let $A \in \mathbb{F}^{n \times n}$, $B \in \mathbb{F}^{n \times m}$, and $C \in \mathbb{F}^{l \times n}$, and assume that A is idempotent, $\mathrm{rank}\begin{bmatrix} C^* & B \end{bmatrix} = n$, and $CB = 0$. Then,

$$\mathrm{rank}\, CAB = \mathrm{rank}\, CA + \mathrm{rank}\, AB - \mathrm{rank}\, A.$$

Proof: See [1339].

Remark: See Fact 3.12.20.

Fact 3.12.20. $A \triangleq \begin{bmatrix} A_{11} & A_{12} \\ A_{21} & A_{22} \end{bmatrix} \in \mathbb{F}^{(n+m) \times (n+m)}$, and assume that A is idempotent. Then,

$$\mathrm{rank}\, A = \mathrm{rank}\begin{bmatrix} A_{12} \\ A_{22} \end{bmatrix} + \mathrm{rank}\begin{bmatrix} A_{11} & A_{12} \end{bmatrix} - \mathrm{rank}\, A_{12}$$

$$= \mathrm{rank}\begin{bmatrix} A_{11} \\ A_{21} \end{bmatrix} + \mathrm{rank}\begin{bmatrix} A_{21} & A_{22} \end{bmatrix} - \mathrm{rank}\, A_{21}.$$

Proof: See [1339] and Fact 3.12.19.

Remark: See Fact 3.13.12 and Fact 6.5.13.

Fact 3.12.21. Let $A \in \mathbb{F}^{n \times m}$ and $B \in \mathbb{F}^{m \times n}$, and assume that AB is nonsingular. Then, $B(AB)^{-1}A$ is idempotent.

Fact 3.12.22. Let $A, B \in \mathbb{F}^{n \times n}$, assume that A and B are idempotent, and let $\alpha, \beta \in \mathbb{F}$ be nonzero and satisfy $\alpha + \beta \neq 0$. Then,

$$\begin{aligned}
\mathrm{rank}(A + B) &= \mathrm{rank}(\alpha A + \beta B) \\
&= \mathrm{rank}\, A + \mathrm{rank}(A_\perp B A_\perp) \\
&= n - \dim[\mathcal{N}(A_\perp B) \cap \mathcal{N}(A)] \\
&= \mathrm{rank}\begin{bmatrix} 0 & A & B \\ A & 0 & 0 \\ B & 0 & 2B \end{bmatrix} - \mathrm{rank}\, A - \mathrm{rank}\, B \\
&= \mathrm{rank}\begin{bmatrix} A & B \\ B & 0 \end{bmatrix} - \mathrm{rank}\, B = \mathrm{rank}\begin{bmatrix} B & A \\ A & 0 \end{bmatrix} - \mathrm{rank}\, A \\
&= \mathrm{rank}(B_\perp A B_\perp) + \mathrm{rank}\, B = \mathrm{rank}(A_\perp B A_\perp) + \mathrm{rank}\, A \\
&= \mathrm{rank}(A + A_\perp B) = \mathrm{rank}(A + B A_\perp) \\
&= \mathrm{rank}(B + B_\perp A) = \mathrm{rank}(B + A B_\perp) \\
&= \mathrm{rank}(I - A_\perp B_\perp) = \mathrm{rank}(I - B_\perp A_\perp) \\
&= \mathrm{rank}\begin{bmatrix} A B_\perp & B \end{bmatrix} = \mathrm{rank}\begin{bmatrix} B A_\perp & A \end{bmatrix} \\
&= \mathrm{rank}\begin{bmatrix} B_\perp A \\ B \end{bmatrix} = \mathrm{rank}\begin{bmatrix} A_\perp B \\ A \end{bmatrix} \\
&= \mathrm{rank}\, A + \mathrm{rank}\, B - n + \mathrm{rank}\begin{bmatrix} A_\perp & A_\perp B_\perp \\ B_\perp A_\perp & B_\perp \end{bmatrix}.
\end{aligned}$$

Furthermore, the following statements hold:

i) If $AB = 0$, then

$$\operatorname{rank}(A + B) = \operatorname{rank}(BA_\perp) + \operatorname{rank} A$$
$$= \operatorname{rank}(B_\perp A) + \operatorname{rank} B.$$

ii) If $BA = 0$, then

$$\operatorname{rank}(A + B) = \operatorname{rank}(AB_\perp) + \operatorname{rank} B$$
$$= \operatorname{rank}(A_\perp B) + \operatorname{rank} A.$$

iii) If $AB = BA$, then

$$\operatorname{rank}(A + B) = \operatorname{rank}(AB_\perp) + \operatorname{rank} B$$
$$= \operatorname{rank}(BA_\perp) + \operatorname{rank} A.$$

iv) $A + B$ is idempotent if and only if $AB = BA = 0$.

v) $A + B = I$ if and only if $AB = BA = 0$ and $\operatorname{rank} [A, B] = \operatorname{rank} A + \operatorname{rank} B = n$.

Remark: See Fact 6.4.36.

Proof: See [611, 859, 860, 1338, 1341]. To prove necessity in *iv*) note that $AB + BA = 0$ implies $AB + ABA = ABA + BA = 0$, which implies that $AB - BA = 0$, and hence $AB = 0$. See [645, p. 250] and [671, p. 435].

Fact 3.12.23. Let $A \in \mathbb{F}^{n \times n}$, let $r \triangleq \operatorname{rank} A$, and let $B \in \mathbb{F}^{n \times r}$ and $C \in \mathbb{F}^{r \times n}$ satisfy $A = BC$. Then, A is idempotent if and only if $CB = I$.

Proof: See [1430, p. 16].

Remark: $A = BC$ is a full-rank factorization.

Fact 3.12.24. Let $A, B \in \mathbb{F}^{n \times n}$, assume that A and B are idempotent, and let $C \in \mathbb{F}^{n \times m}$. Then,

$$\operatorname{rank}(AC - CB) = \operatorname{rank}(AC - ACB) + \operatorname{rank}(ACB - CB)$$

$$= \operatorname{rank} \begin{bmatrix} AC \\ B \end{bmatrix} + \operatorname{rank} \begin{bmatrix} CB & A \end{bmatrix} - \operatorname{rank} A - \operatorname{rank} B.$$

Proof: See [1313].

Fact 3.12.25. Let $A, B \in \mathbb{F}^{n \times n}$, and assume that A and B are idempotent. Then, the following statements are equivalent:

i) $A + B$ is nonsingular.

ii) There exist $\alpha, \beta \in \mathbb{F}$ such that $\alpha + \beta \neq 0$ and $\alpha A + \beta B$ is nonsingular.

iii) For all nonzero $\alpha, \beta \in \mathbb{F}$ such that $\alpha + \beta \neq 0$, $\alpha A + \beta B$ is nonsingular.

Proof: See [107, 857, 1341].

Fact 3.12.26. Let $A, B \in \mathbb{F}^{n \times n}$, and assume that A and B are idempotent. Then,

$$\operatorname{rank}(A - B) = \operatorname{rank} \begin{bmatrix} 0 & A & B \\ A & 0 & 0 \\ B & 0 & 0 \end{bmatrix} - \operatorname{rank} A - \operatorname{rank} B$$

$$= \operatorname{rank} \begin{bmatrix} A \\ B \end{bmatrix} + \operatorname{rank} \begin{bmatrix} A & B \end{bmatrix} - \operatorname{rank} A - \operatorname{rank} B$$

$$= n - \dim[\mathcal{N}(A) \cap \mathcal{N}(B)] - \dim[\mathcal{R}(A) \cap \mathcal{R}(B)]$$

$$= \operatorname{rank}(AB_\perp) + \operatorname{rank}(A_\perp B)$$

$$\leq \operatorname{rank}(A + B)$$

$$\leq \operatorname{rank} A + \operatorname{rank} B.$$

Furthermore, if either $AB = 0$ or $BA = 0$, then

$$\operatorname{rank}(A - B) = \operatorname{rank}(A + B) = \operatorname{rank} A + \operatorname{rank} B.$$

Proof: See [611, 860, 1338, 1341]. The inequality $\operatorname{rank}(A - B) \leq \operatorname{rank}(A + B)$ follows from Fact 2.11.13 and the block 3×3 expressions in this result and in Fact 3.12.22. To prove the last statement in the case that $AB = 0$, first note that $\operatorname{rank} A + \operatorname{rank} B = \operatorname{rank}(A - B)$, which yields $\operatorname{rank}(A - B) \leq \operatorname{rank}(A + B) \leq \operatorname{rank} A + \operatorname{rank} B = \operatorname{rank}(A - B)$.

Remark: See Fact 6.4.36.

Fact 3.12.27. Let $A, B \in \mathbb{F}^{n \times n}$, and assume that A and B are idempotent. Then, the following statements are equivalent:

$i)$ $A - B$ is idempotent.

$ii)$ $\operatorname{rank}(A_\perp + B) + \operatorname{rank}(A - B) = n$.

$iii)$ $ABA = B$.

$iv)$ $\operatorname{rank}(A - B) = \operatorname{rank} A - \operatorname{rank} B$.

$v)$ $\mathcal{R}(B) \subseteq \mathcal{R}(A)$ and $\mathcal{R}(B^*) \subseteq \mathcal{R}(A^*)$.

Proof: See [1340].

Remark: This result is due to Hartwig and Styan.

Fact 3.12.28. Let $A, B \in \mathbb{F}^{n \times n}$, and assume that A and B are idempotent. Then, the following statements are equivalent:

$i)$ $A - B$ is nonsingular.

$ii)$ $I - AB$ is nonsingular, and there exist $\alpha, \beta \in \mathbb{F}$ such that $\alpha + \beta \neq 0$ and $\alpha A + \beta B$ is nonsingular.

$iii)$ $I - AB$ is nonsingular, and $\alpha A + \beta B$ is nonsingular for all $\alpha, \beta \in \mathbb{F}$ such that $\alpha + \beta \neq 0$.

$iv)$ $I - AB$ and $A + A_\perp B$ are nonsingular.

$v)$ $I - AB$ and $A + B$ are nonsingular.

vi) $\mathcal{R}(A) + \mathcal{R}(B) = \mathbb{F}^n$ and $\mathcal{R}(A^*) + \mathcal{R}(B^*) = \mathbb{F}^n$.

vii) $\mathcal{R}(A) + \mathcal{R}(B) = \mathbb{F}^n$ and $\mathcal{N}(A) + \mathcal{N}(B) = \mathbb{F}^n$.

viii) $\mathcal{R}(A) \cap \mathcal{R}(B) = \{0\}$ and $\mathcal{N}(A) \cap \mathcal{N}(B) = \{0\}$.

ix) $\operatorname{rank}\left[\begin{smallmatrix} A \\ B \end{smallmatrix}\right] = \operatorname{rank}\left[\begin{array}{cc} A & B \end{array}\right] = \operatorname{rank} A + \operatorname{rank} B = n$.

Proof: See [107, 611, 858, 860, 1338].

Fact 3.12.29. Let $A, B \in \mathbb{F}^{n \times n}$, assume that A and B are idempotent. Then, the following statements hold:

i) $\mathcal{R}(A) \cap \mathcal{R}(B) \subseteq \mathcal{R}(AB)$.

ii) $\mathcal{N}(B) + [\mathcal{N}(A) \cap \mathcal{R}(B)] \subseteq \mathcal{N}(AB) \subseteq \mathcal{R}(I - AB) \subseteq \mathcal{N}(A) + \mathcal{N}(B)$.

iii) If $AB = BA$, then AB is the idempotent matrix onto $\mathcal{R}(A) \cap \mathcal{R}(B)$ along $\mathcal{N}(A) + \mathcal{N}(B)$.

Furthermore, the following statements are equivalent:

iv) $AB = BA$.

v) $\operatorname{rank} AB = \operatorname{rank} BA$, and AB is the idempotent matrix onto $\mathcal{R}(A) \cap \mathcal{R}(B)$ along $\mathcal{N}(A) + \mathcal{N}(B)$.

vi) $\operatorname{rank} AB = \operatorname{rank} BA$, and $A + B - AB$ is the idempotent matrix onto $\mathcal{R}(A) + \mathcal{R}(B)$ along $\mathcal{N}(A) \cap \mathcal{N}(B)$.

In addition, the following statements are equivalent:

vii) AB is idempotent.

viii) $\mathcal{R}(AB) \subseteq \mathcal{R}(B) + [\mathcal{N}(A) \cap \mathcal{N}(B)]$.

ix) $\mathcal{R}(AB) = \mathcal{R}(A) \cap (\mathcal{R}(B) + [\mathcal{N}(A) \cap \mathcal{N}(B)])$.

x) $\mathcal{N}(B) + [\mathcal{N}(A) \cap \mathcal{R}(B)] = \mathcal{R}(I - AB)$.

Finally, the following statements hold:

xi) $A - B$ is idempotent if and only if B is the idempotent matrix onto $\mathcal{R}(A) \cap \mathcal{R}(B)$ along $\mathcal{N}(A) + \mathcal{N}(B)$.

xii) $A + B$ is idempotent if and only if A is the idempotent matrix onto $\mathcal{R}(A) \cap \mathcal{N}(B)$ along $\mathcal{N}(A) + \mathcal{R}(B)$.

Proof: See [550, p. 53] and [610].

Remark: See Fact 5.12.19.

Fact 3.12.30. Let $A, B \in \mathbb{F}^{n \times n}$, assume that A and B are idempotent, and assume that $AB = BA$. Then, the following statements are equivalent:

i) $A - B$ is nonsingular.

ii) $(A - B)^2 = I$.

iii) $A + B = I$.

Proof: See [611].

Fact 3.12.31. Let $A, B \in \mathbb{F}^{n \times n}$, and assume that A and B are idempotent. Then,

$$\mathrm{rank}\,[A, B] = \mathrm{rank}(A - B) + \mathrm{rank}(A_\perp - B) - n$$

$$= \mathrm{rank}(A - B) + \mathrm{rank}\,AB + \mathrm{rank}\,BA - \mathrm{rank}\,A - \mathrm{rank}\,B.$$

Furthermore, the following statements hold:

i) $AB = BA$ if and only if $\mathcal{R}(AB) = \mathcal{R}(BA)$ and $\mathcal{R}[(AB)^*] = \mathcal{R}[(BA)^*]$.

ii) $AB = BA$ if and only if

$$\mathrm{rank}(A - B) + \mathrm{rank}(A_\perp - B) = n.$$

iii) $[A, B]$ is nonsingular if and only if $A - B$ and $A_\perp - B$ are nonsingular.

iv) $\max\{\mathrm{rank}\,AB, \mathrm{rank}\,BA\} \le \mathrm{rank}(AB + BA)$.

v) $AB + BA = 0$ if and only if $AB = BA = 0$.

vi) $AB + BA$ is nonsingular if and only if $A + B$ and $A_\perp - B$ are nonsingular.

vii) $\mathrm{rank}(AB + BA) = \mathrm{rank}(\alpha AB + \beta BA)$.

viii) $A_\perp - B$ is nonsingular if and only if $\mathrm{rank}\,A = \mathrm{rank}\,B = \mathrm{rank}\,AB = \mathrm{rank}\,BA$. In this case, A and B are similar.

ix) $\mathrm{rank}(A + B) + \mathrm{rank}(AB - BA) = \mathrm{rank}(A - B) + \mathrm{rank}(AB + BA)$.

x) $\mathrm{rank}(AB - BA) \le \mathrm{rank}(AB + BA)$.

Proof: See [860].

Fact 3.12.32. Let $A, B \in \mathbb{F}^{n \times n}$, assume that A and B are idempotent, and assume that $A - B$ is nonsingular. Then, $A + B$ is nonsingular. Now, define $F, G \in \mathbb{F}^{n \times n}$ by

$$F \triangleq A(A - B)^{-1} = (A - B)^{-1}(I - B)$$

and

$$G \triangleq (A - B)^{-1}A = (I - A)(A - B)^{-1}.$$

Then, F and G are idempotent. In particular, F is the idempotent matrix onto $\mathcal{R}(A)$ along $\mathcal{N}(B)$, and G^* is the idempotent matrix onto $\mathcal{R}(A^*)$ along $\mathcal{R}(B^*)$. Furthermore,

$$FB = AG = 0,$$

$$(A - B)^{-1} = F - G_\perp,$$

$$(A - B)^{-1} = (A + B)^{-1}(A - B)(A + B)^{-1},$$

$$(A + B)^{-1} = I - G_\perp F - GF_\perp,$$

$$(A + B)^{-1} = (A - B)^{-1}(A + B)(A - B)^{-1}.$$

Proof: See [860].

Remark: See [860] for an explicit expression for $(A + B)^{-1}$ in the case that $A - B$ is singular.

Remark: See Proposition 3.5.3.

Fact 3.12.33. If $A \in \mathbb{F}^{n \times m}$ and $B \in \mathbb{F}^{n \times (n-m)}$, assume that $[A \ B]$ is nonsingular, and define

$$P \triangleq \begin{bmatrix} A & 0 \end{bmatrix} \begin{bmatrix} A & B \end{bmatrix}^{-1}$$

and

$$Q \triangleq \begin{bmatrix} 0 & B \end{bmatrix} \begin{bmatrix} A & B \end{bmatrix}^{-1}.$$

Then, the following statements hold:

i) P and Q are idempotent.

ii) $P + Q = I_n$.

iii) $PQ = 0$.

iv) $P \begin{bmatrix} A & 0 \end{bmatrix} = \begin{bmatrix} A & 0 \end{bmatrix}$.

v) $Q \begin{bmatrix} 0 & B \end{bmatrix} = \begin{bmatrix} 0 & B \end{bmatrix}$.

vi) $\mathcal{R}(P) = \mathcal{R}(A)$ and $\mathcal{N}(P) = \mathcal{R}(B)$.

vii) $\mathcal{R}(Q) = \mathcal{R}(B)$ and $\mathcal{N}(Q) = \mathcal{R}(A)$.

viii) $A^*B = 0$ if and only if P and Q are projectors. In this case, $P = A(A^*A)^{-1}A^*$ and $Q = B(B^*B)^{-1}B^*$.

ix) $\mathcal{R}(A)$ and $\mathcal{R}(B)$ are complementary subspaces.

x) P is the idempotent matrix onto $\mathcal{R}(A)$ along $\mathcal{R}(B)$.

xi) Q is the idempotent matrix onto $\mathcal{R}(B)$ along $\mathcal{R}(A)$.

Proof: See [1533] or [1534, pp. 74, 75].

Remark: See Fact 3.13.24, Fact 6.4.21, and Fact 6.4.22.

3.13 Facts on Projectors

Fact 3.13.1. Let $A \in \mathbb{F}^{n \times n}$. Then, the following statements are equivalent:

i) A is a projector.

ii) $A = AA^*$.

iii) $A = A^*A$.

iv) A is idempotent and normal.

v) A and A^*A are idempotent.

vi) $AA^*A = A$, and A is idempotent.

vii) A and $\frac{1}{2}(A + A^*)$ are idempotent.

viii) A is idempotent, and $AA^* + A^*A = A + A^*$.

ix) A is tripotent, and $A^2 = A^*$.

x) $AA^* = A^*AA^*$.

xi) A is idempotent, and $\operatorname{rank} A + \operatorname{rank}(I - A^*A) = n$.

xii) A is idempotent, and, for all $x \in \mathbb{F}^n$, $x^*Ax \geq 0$.

Remark: See Fact 3.13.2, Fact 3.13.3, and Fact 6.3.26.

Remark: The matrix $A = \left[\begin{smallmatrix} 1/2 & 1/2 \\ 0 & 0 \end{smallmatrix}\right]$ satisfies $\operatorname{tr} A = \operatorname{tr} A^*A$ but is not a projector. See Fact 3.7.13.

Fact 3.13.2. Let $A \in \mathbb{F}^{n \times n}$, and assume that A is Hermitian. Then, the following statements are equivalent:

i) A is a projector.

ii) $\operatorname{rank} A = \operatorname{tr} A = \operatorname{tr} A^2$.

Proof: See [1215, p. 55].

Remark: See Fact 3.13.1 and Fact 3.13.3.

Fact 3.13.3. Let $A \in \mathbb{F}^{n \times n}$, and assume that A is idempotent. Then, the following statements are equivalent:

i) A is a projector.

ii) $AA^*A = A$.

iii) A is Hermitian.

iv) A is normal.

v) A is range Hermitian.

Proof: See [1367].

Remark: See Fact 3.13.1 and Fact 3.13.2.

Fact 3.13.4. Let $A \in \mathbb{F}^{n \times n}$, and assume that A is a projector. Then, A is positive semidefinite.

Fact 3.13.5. Let $A \in \mathbb{F}^{n \times n}$, assume that A is a projector, and let $x \in \mathbb{F}^n$. Then, $x \in \mathcal{R}(A)$ if and only if $x = Ax$.

Fact 3.13.6. Let $A \in \mathbb{F}^{n \times m}$. If $\operatorname{rank} A = m$, then $B \triangleq A(A^*A)^{-1}A^*$ is a projector and $\operatorname{rank} B = m$. If $\operatorname{rank} A = n$, then $B \triangleq A^*(AA^*)^{-1}A$ is a projector and $\operatorname{rank} B = n$.

Remark: See Fact 2.15.2, Fact 3.7.25, and Fact 3.7.26.

Fact 3.13.7. Let $x \in \mathbb{F}^n$ be nonzero, and define the elementary projector $A \triangleq I - (x^*x)^{-1}xx^*$. Then, the following statements hold:

i) $\operatorname{rank} A = n - 1$.

ii) $\mathcal{N}(A) = \operatorname{span}\{x\}$.

iii) $\mathcal{R}(A) = \{x\}^{\perp}$.

iv) $2A - I$ is the elementary reflector $I - 2(x^*x)^{-1}xx^*$.

Remark: If $y \in \mathbb{F}^n$, then Ay is the *projection* of y on $\{x\}^{\perp}$.

Fact 3.13.8. Let $n > 1$, let $\mathcal{S} \subset \mathbb{F}^n$, and assume that $\mathcal{S}$ is a hyperplane. Then, there exists a unique elementary projector $A \in \mathbb{F}^{n \times n}$ such that $\mathcal{R}(A) = \mathcal{S}$ and $\mathcal{N}(A) = \mathcal{S}^{\perp}$. Furthermore, if $x \in \mathbb{F}^n$ is nonzero and $\mathcal{S} \triangleq \{x\}^{\perp}$, then $A = I - (x^*x)^{-1}xx^*$.

Fact 3.13.9. Let $A \in \mathbb{F}^{n \times n}$. Then, A is a projector and $\operatorname{rank} A = n - 1$ if and only if there exists a nonzero vector $x \in \mathcal{N}(A)$ such that

$$A = I - (x^*x)^{-1}xx^*.$$

In this case, it follows that, for all $y \in \mathbb{F}^n$,

$$y^*y - y^*Ay = \frac{|y^*x|^2}{x^*x}.$$

Furthermore, for $y \in \mathbb{F}^n$, the following statements are equivalent:

i) $y^*Ay = y^*y$.

ii) $y^*x = 0$.

iii) $Ay = y$.

Remark: See Fact 3.7.19.

Fact 3.13.10. Let $A \in \mathbb{F}^{n \times n}$, assume that A is a projector, and let $x \in \mathbb{F}^n$. Then,
$$x^*Ax \le x^*x.$$

Furthermore, the following statements are equivalent:

i) $x^*Ax = x^*x$.

ii) $Ax = x$.

iii) $x \in \mathcal{R}(A)$.

Fact 3.13.11. Let $A \in \mathbb{F}^{n \times n}$, and assume that A is idempotent. Then, A is a projector if and only if, for all $x \in \mathbb{F}^n$, $x^*Ax \le x^*x$.

Proof: See [1125, p. 105].

Fact 3.13.12. $A \triangleq \begin{bmatrix} A_{11} & A_{12} \\ A_{12}^* & A_{22} \end{bmatrix} \in \mathbb{F}^{(n+m) \times (n+m)}$, and assume that A is a projector. Then,
$$\operatorname{rank} A = \operatorname{rank} A_{11} + \operatorname{rank} A_{22} - \operatorname{rank} A_{12}.$$

Proof: See [1340] and Fact 3.12.20.

Remark: See Fact 3.12.20 and Fact 6.5.13.

Fact 3.13.13. Let $A \in \mathbb{F}^{n \times n}$, and assume that A satisfies two out of the three properties (Hermitian, shifted unitary, idempotent). Then, A satisfies the remaining property. Furthermore, these matrices are the projectors.

Proof: If A is idempotent and shifted unitary, then $(2A - I)^{-1} = 2A - I = (2A^* - I)^{-1}$. Hence, A is Hermitian.

Remark: The condition $A + A^* = 2AA^*$ is considered in Fact 3.11.25.

Remark: See Fact 3.14.2 and Fact 3.14.6.

Fact 3.13.14. Let $A \in \mathbb{F}^{n \times n}$, let $B \in \mathbb{F}^{n \times m}$, assume that A is a projector, and assume that $\mathcal{R}(AB) = \mathcal{R}(B)$. Then, $AB = B$.

Proof: $0 = \mathcal{R}(A_\perp AB) = A_\perp \mathcal{R}(AB) = A_\perp \mathcal{R}(B) = \mathcal{R}(A_\perp B)$. Hence, $A_\perp B = 0$. Consequently, $B = (A + A_\perp)B = AB$.

Remark: See Fact 6.4.19.

Fact 3.13.15. Let $A, B \in \mathbb{F}^{n \times n}$, and assume that A and B are projectors. Then, $\mathcal{R}(A) = \mathcal{R}(B)$ if and only if $A = B$.

Remark: See Proposition 3.5.1.

Fact 3.13.16. Let $A, B \in \mathbb{F}^{n \times n}$, assume that A and B are projectors, and assume that $\operatorname{rank} A = \operatorname{rank} B$. Then, there exists a reflector $S \in \mathbb{F}^{n \times n}$ such that $A = SBS$. If, in addition, $A + B - I$ is nonsingular, then one such reflector is given by $S = \langle A + B - I \rangle (A + B - I)^{-1}$.

Proof: See [335].

Remark: The notation $\langle \cdot \rangle$ is defined in Chapter 8.

Fact 3.13.17. Let $A, B \in \mathbb{F}^{n \times n}$, and assume that A and B are projectors. Then, the following statements are equivalent:

 i) $\mathcal{R}(A) \subseteq \mathcal{R}(B)$.

 ii) $A \leq B$.

 iii) $AB = A$.

 iv) $BA = A$.

 v) $B - A$ is a projector

Proof: See [1215, pp. 24, 169].

Remark: See Fact 9.8.3.

Fact 3.13.18. Let $A, B \in \mathbb{F}^{n \times n}$, and assume that A and B are projectors. Then,
$$\mathcal{R}(I - AB) = \mathcal{N}(A) + \mathcal{N}(B)$$
and
$$\mathcal{R}(A + A_\perp B) = \mathcal{R}(A) + \mathcal{R}(B).$$

Proof: See [608, 1360].

Fact 3.13.19. Let $A, B \in \mathbb{F}^{n \times n}$, and assume that A and B are projectors. Then, the following statements are equivalent:

i) $AB = 0$.

ii) $BA = 0$.

iii) $\mathcal{R}(A) = \mathcal{R}(B)^{\perp}$.

iv) $A + B$ is a projector.

In this case, $\mathcal{R}(A + B) = \mathcal{R}(A) + \mathcal{R}(B)$.

Proof: See [544, pp. 42–44].

Remark: See [551].

Fact 3.13.20. Let $A, B \in \mathbb{F}^{n \times n}$, and assume that A and B are projectors. Then, the following statements are equivalent:

i) AB is a projector.

ii) $AB = BA$.

iii) AB is idempotent.

iv) AB is Hermitian.

v) AB is normal.

vi) AB is range Hermitian.

In this case, the following statements hold:

vii) $\mathcal{R}(AB) = \mathcal{R}(A) \cap \mathcal{R}(B)$.

viii) AB is the projector onto $\mathcal{R}(A) \cap \mathcal{R}(B)$.

ix) $A + A_{\perp}B$ is a projector.

x) $A + A_{\perp}B$ is the projector onto $\mathcal{R}(A) + \mathcal{R}(B)$.

Proof: See [544, pp. 42–44] and [1353, 1457].

Remark: See Fact 5.12.16 and Fact 6.4.26.

Problem: If $A + A_{\perp}B$ is a projector, then does it follow that A and B commute?

Fact 3.13.21. Let $A, B \in \mathbb{F}^{n \times n}$, and assume that A and B are projectors. Then, AB is group invertible.

Proof: $\mathcal{N}(BA) \subseteq \mathcal{N}(BABA) \subseteq \mathcal{N}(ABABA) = \mathcal{N}(ABAABA) = \mathcal{N}(ABA) = \mathcal{N}(ABBA) = \mathcal{N}(BA)$.

Remark: See [1457].

Fact 3.13.22. Let $A, B \in \mathbb{F}^{n \times n}$, and assume that A and B are projectors. Then, the rank of the $ln \times ln$ matrix given below is

$$
\mathrm{rank} \begin{bmatrix} A+B & AB & & & \\ AB & A+B & \ddots & & \\ & \ddots & \ddots & \ddots & \\ & & \ddots & A+B & AB \\ & & & AB & A+B \end{bmatrix} = l\,\mathrm{rank}(A+B).
$$

Proof: See [1341].

Fact 3.13.23. Let $A, B \in \mathbb{F}^{n \times n}$, and assume that A and B are projectors. Then,

$$
\mathrm{rank}(A+B) = \mathrm{rank}\,A + \mathrm{rank}\,B - n + \mathrm{rank}(A_\perp + B_\perp),
$$

$$
\mathrm{rank}\begin{bmatrix} A & B \end{bmatrix} = \mathrm{rank}\,A + \mathrm{rank}\,B - n + \mathrm{rank}\begin{bmatrix} A_\perp & B_\perp \end{bmatrix},
$$

$$
\mathrm{rank}\,[A, B] = 2\big(\mathrm{rank}\begin{bmatrix} A & B \end{bmatrix} + \mathrm{rank}\,AB - \mathrm{rank}\,A - \mathrm{rank}\,B\big).
$$

Proof: See [1338, 1341].

Fact 3.13.24. Let $A, B \in \mathbb{F}^{n \times n}$, and assume that A and B are projectors. Then, the following statements are equivalent:

i) $A - B$ is nonsingular.

ii) $\mathrm{rank}\begin{bmatrix} A & B \end{bmatrix} = \mathrm{rank}\,A + \mathrm{rank}\,B = n.$

iii) $\mathcal{R}(A)$ and $\mathcal{R}(B)$ are complementary subspaces.

Now, assume in addition that *i)–iii)* hold. Then, the following statements hold:

iv) $I - BA$ is nonsingular.

v) $A + B - AB$ is nonsingular.

vi) The idempotent matrix $M \in \mathbb{F}^{n \times n}$ onto $\mathcal{R}(B)$ along $\mathcal{R}(A)$ is given by

$$
\begin{aligned}
M &= (I - BA)^{-1}B(I - BA) \\
&= B(I - AB)^{-1}(I - BA) \\
&= (I - AB)^{-1}(I - A) \\
&= A(A + B - AB)^{-1}.
\end{aligned}
$$

vii) M satisfies

$$
M + M^* = (B - A)^{-1} + I,
$$

that is,

$$
(B - A)^{-1} = M + M^* - I = M - M_\perp^*.
$$

Proof: See Fact 5.12.17 and [6, 277, 551, 602, 765, 1142]. The uniqueness of M follows from Proposition 3.5.3, while *vii)* follows from Fact 5.12.18.

Remark: See Fact 3.12.33, Fact 5.12.18, Fact 6.4.21, and Fact 6.4.22.

Fact 3.13.25. Let $A_1, \ldots, A_r \in \mathbb{F}^{n \times n}$, assume that $A_1, \ldots, A_r$ are Hermitian, and consider the following statements:

i) $A_1, \ldots, A_r$ are projectors.

ii) $\sum_{i=1}^r A_i$ is a projector.

iii) For all distinct $i, j \in \{1, \ldots, r\}$, it follows that $A_i A_j = 0$.

iv) $\operatorname{rank} \sum_{i=1}^r A_i = \sum_{i=1}^r \operatorname{rank} A_i$.

Then, if at least one of the pairs of conditions $[i),ii)]$ $[i),iii)]$, $[ii),iii)]$, $[ii),iv)]$ holds, then $i)$–$iv)$ hold. In particular, if $A_1, \ldots, A_r$ are projectors and $\sum_{i=1}^r A_i = I$, then, for all distinct $i, j \in \{1, \ldots, r\}$, it follows that $A_i A_j = 0$.

Proof: See [1208, pp. 400–402].

Remark: The last statement is *Cochran's theorem*.

3.14 Facts on Reflectors

Fact 3.14.1. If $A \in \mathbb{F}^{n \times n}$ is a projector, then $B \triangleq 2A - I$ is a reflector. Conversely, if $B \in \mathbb{F}^{n \times n}$ is a reflector, then $A \triangleq \frac{1}{2}(B + I)$ is a projector.

Remark: See Fact 3.15.2.

Remark: The affine mapping $f(A) \triangleq 2A - I$ from the projectors to the reflectors is one-to-one and onto. See Fact 3.11.25 and Fact 3.15.2.

Fact 3.14.2. Let $A \in \mathbb{F}^{n \times n}$, and assume that A satisfies two out of the three properties (Hermitian, unitary, involutory). Then, A also satisfies the remaining property. Furthermore, these matrices are the reflectors.

Remark: See Fact 3.13.13 and Fact 3.14.6.

Fact 3.14.3. Let $x \in \mathbb{F}^n$ be nonzero, and define the elementary reflector $A \triangleq I - 2(x^*x)^{-1}xx^*$. Then, the following statements hold:

i) $\det A = -1$.

ii) If $y \in \mathbb{F}^n$, then Ay is the reflection of y across $\{x\}^\perp$.

iii) $Ax = -x$.

iv) $\frac{1}{2}(A + I)$ is the elementary projector $I - (x^*x)^{-1}xx^*$.

Fact 3.14.4. Let $x, y \in \mathbb{F}^n$. Then, there exists a unique elementary reflector $A \in \mathbb{F}^{n \times n}$ such that $Ax = y$ if and only if x^*y is real and $x^*x = y^*y$. If, in addition, $x \neq y$, then A is given by

$$A = I - 2[(x - y)^*(x - y)]^{-1}(x - y)(x - y)^*.$$

Remark: This result is the *reflection theorem*. See [572, pp. 16–18] and [1157, p. 357]. See Fact 3.9.5.

Fact 3.14.5. Let $n > 1$, let $\mathcal{S} \subset \mathbb{F}^n$, and assume that $\mathcal{S}$ is a hyperplane. Then, there exists a unique elementary reflector $A \in \mathbb{F}^{n \times n}$ such that, for all $y = y_1 + y_2 \in \mathbb{F}^n$, where $y_1 \in \mathcal{S}$ and $y_2 = \mathcal{S}^{\perp}$, it follows that $Ay = y_1 - y_2$. Furthermore, if $\mathcal{S} = \{x\}^{\perp}$, then $A = I - 2(x^*x)^{-1}xx^*$.

Fact 3.14.6. Let $A \in \mathbb{F}^{n \times n}$, and assume that A satisfies two out of the three properties (skew Hermitian, unitary, skew involutory). Then, A also satisfies the remaining property. Furthermore, these matrices are the skew reflectors.

Remark: See Fact 3.13.13, Fact 3.14.2, and Fact 3.14.7.

Fact 3.14.7. Let $A \in \mathbb{C}^{n \times n}$. Then, A is a reflector if and only if $\jmath A$ is a skew reflector.

Remark: The mapping $f(A) \triangleq \jmath A$ relates Fact 3.14.2 to Fact 3.14.6.

Problem: When A is real and n is even, determine a real transformation between the reflectors and the skew reflectors.

Fact 3.14.8. Let $A \in \mathbb{F}^{n \times n}$. Then, the following statements are equivalent:

i) A is a reflector.

ii) $A = AA^* + A^* - I$.

iii) $A = \frac{1}{2}(A + I)(A^* + I) - I$.

3.15 Facts on Involutory Matrices

Fact 3.15.1. Let $A \in \mathbb{F}^{n \times n}$, and assume that A is involutory. Then, either $\det A = 1$ or $\det A = -1$.

Fact 3.15.2. If $A \in \mathbb{F}^{n \times n}$ is idempotent, then $B \triangleq 2A - I$ is involutory. Conversely, if $B \in \mathbb{F}^{n \times n}$ is involutory, then $A_1 \triangleq \frac{1}{2}(I + B)$ and $A_2 \triangleq \frac{1}{2}(I - B)$ are idempotent.

Remark: See Fact 3.14.1.

Remark: The affine mapping $f(A) \triangleq 2A - I$ from the idempotent matrices to the involutory matrices is one-to-one and onto. See Fact 3.11.25 and Fact 3.14.1.

Fact 3.15.3. Let $A \in \mathbb{F}^{n \times n}$. Then, A is involutory if and only if

$$(A + I)(A - I) = 0.$$

Fact 3.15.4. Let $n \geq 1$. Then, $\hat{I}_n$ is involutory.

Fact 3.15.5. Let $n \geq 1$. Then,

$$P_n \hat{I}_n P_n = \hat{I}_n.$$

Consequently, $P_n \hat{I}_n$ is involutory.

Remark: This identity plays a role in defining the generators for the dihedral group $D(n)$. See [803, p. 169].

Fact 3.15.6. Let $A, B \in \mathbb{F}^{n \times n}$, and assume that A and B are involutory. Then,

$$\mathcal{R}([A, B]) = \mathcal{R}(A - B) \cap \mathcal{R}(A + B)$$

and

$$\mathcal{N}([A, B]) = \mathcal{N}(A - B) \cap \mathcal{N}(A + B).$$

Proof: See [1324].

Fact 3.15.7. Let $A \in \mathbb{F}^{n \times m}$, let $B \in \mathbb{F}^{m \times n}$, and define

$$C \triangleq \begin{bmatrix} I - BA & B \\ 2A - ABA & AB - I \end{bmatrix}.$$

Then, C is involutory.

Proof: See [1023, p. 113].

Fact 3.15.8. Let $A \in \mathbb{R}^{n \times n}$, and assume that A is skew involutory. Then, n is even.

3.16 Facts on Tripotent Matrices

Fact 3.16.1. Let $A \in \mathbb{F}^{n \times n}$, and assume that A is tripotent. Then, A^2 is idempotent.

Remark: The converse is false. A counterexample is $\begin{bmatrix} 0 & 1 \\ 0 & 0 \end{bmatrix}$.

Fact 3.16.2. Let $A \in \mathbb{F}^{n \times n}$. Then, A is nonsingular and tripotent if and only if A is involutory.

Fact 3.16.3. Let $A \in \mathbb{F}^{n \times n}$, and assume that A is Hermitian. Then, A is tripotent if and only if

$$\operatorname{rank} A = \operatorname{rank}(A + A^2) + \operatorname{rank}(A - A^2).$$

Proof: See [1215, p. 176].

Fact 3.16.4. Let $A \in \mathbb{R}^{n \times n}$ be tripotent. Then,

$$\operatorname{rank} A = \operatorname{rank} A^2 = \operatorname{tr} A^2.$$

Fact 3.16.5. If $A, B \in \mathbb{F}^{n \times n}$ are idempotent and $AB = 0$, then $A + BA_\perp$ is idempotent and $C \triangleq A - B$ is tripotent. Conversely, if $C \in \mathbb{F}^{n \times n}$ is tripotent, then $A \triangleq \frac{1}{2}(C^2 + C)$ and $B \triangleq \frac{1}{2}(C^2 - C)$ are idempotent and satisfy $C = A - B$ and $AB = BA = 0$.

Proof: See [1012, p. 114].

3.17 Facts on Nilpotent Matrices

Fact 3.17.1. Let $A \in \mathbb{F}^{n \times n}$. Then, the following statements are equivalent:

i) $\mathcal{R}(A) = \mathcal{N}(A)$.

ii) A is similar to a block-diagonal matrix each of whose diagonal blocks is N_2.

Proof: To prove that *i)* $\implies$ *ii)*, let $S \in \mathbb{F}^{n \times n}$ transform A into its Jordan form. Then, it follows from Fact 2.10.2 that $\mathcal{R}(SAS^{-1}) = S\mathcal{R}(AS^{-1}) = S\mathcal{R}(A) = S\mathcal{N}(A) = S\mathcal{N}(AS^{-1}S) = \mathcal{N}(AS^{-1}) = \mathcal{N}(SAS^{-1})$. The only Jordan block J that satisfies $\mathcal{R}(J) = \mathcal{N}(J)$ is $J = N_2$. Using $\mathcal{R}(N_2) = \mathcal{N}(N_2)$ and reversing these steps yields the converse result.

Remark: The fact that n is even follows from $\operatorname{rank} A + \operatorname{def} A = n$ and $\operatorname{rank} A = \operatorname{def} A$.

Remark: See Fact 3.17.2 and Fact 3.17.3.

Fact 3.17.2. Let $A \in \mathbb{F}^{n \times n}$. Then, the following statements are equivalent:

i) $\mathcal{N}(A) \subseteq \mathcal{R}(A)$.

ii) A is similar to a block-diagonal matrix each of whose diagonal blocks is either nonsingular or N_2.

Remark: See Fact 3.17.1 and Fact 3.17.3.

Fact 3.17.3. Let $A \in \mathbb{F}^{n \times n}$. Then, the following statements are equivalent:

i) $\mathcal{R}(A) \subseteq \mathcal{N}(A)$.

ii) A is similar to a block-diagonal matrix each of whose diagonal blocks is either zero or N_2.

Remark: See Fact 3.17.1 and Fact 3.17.2.

Fact 3.17.4. Let $n \in \mathbb{P}$ and $k \in \{0, \ldots, n\}$. Then, $\operatorname{rank} N_n^k = n - k$.

Fact 3.17.5. Let $A \in \mathbb{R}^{n \times n}$. Then, $\operatorname{rank} A^k$ is a nonincreasing function of $k \geq 1$. Furthermore, if there exists $k \in \{1, \ldots, n\}$ such that $\operatorname{rank} A^{k+1} = \operatorname{rank} A^k$, then $\operatorname{rank} A^l = \operatorname{rank} A^k$ for all $l \geq k$. Finally, if A is nilpotent and $A^l \neq 0$, then $\operatorname{rank} A^{k+1} < \operatorname{rank} A^k$ for all $k \in \{1, \ldots, l\}$.

Fact 3.17.6. Let $A \in \mathbb{F}^{n \times n}$. Then, A is nilpotent if and only if, for all $k \in \{1, \ldots, n\}$, $\operatorname{tr} A^k = 0$.

Proof: See [1125, p. 103] or use Fact 4.8.2 with $p = \chi_A$ and $\mu_1 = \cdots = \mu_n = 0$.

Fact 3.17.7. Let $\lambda \in \mathbb{F}$ and $n, k \in \mathbb{P}$. Then,

$$(\lambda I_n + N_n)^k = \begin{cases} \lambda^k I_n + \binom{k}{1}\lambda^{k-1}N_n + \cdots + \binom{k}{k}N_n^k, & k < n-1, \\ \\ \lambda^k I_n + \binom{k}{1}\lambda^{k-1}N_n + \cdots + \binom{k}{n-1}\lambda^{k-n+1}N_n^{n-1}, & k \geq n-1, \end{cases}$$

that is, for $k \geq n - 1$,

$$\begin{bmatrix} \lambda & 1 & \cdots & 0 & 0 \\ 0 & \lambda & \ddots & 0 & 0 \\ \vdots & \ddots & \ddots & \ddots & \vdots \\ 0 & 0 & \ddots & \lambda & 1 \\ 0 & 0 & \cdots & 0 & \lambda \end{bmatrix}^k = \begin{bmatrix} \lambda^k & \binom{k}{1}\lambda^{k-1} & \cdots & \binom{k}{n-2}\lambda^{k-n+1} & \binom{k}{n-1}\lambda^{k-n+1} \\ 0 & \lambda^k & \ddots & \binom{k}{n-3}\lambda^{k-n+2} & \binom{k}{n-2}\lambda^{k-n+2} \\ \vdots & \ddots & \ddots & \ddots & \vdots \\ 0 & 0 & \ddots & \lambda^k & \binom{k}{1}\lambda^{k-1} \\ 0 & 0 & \cdots & 0 & \lambda^k \end{bmatrix}.$$

Fact 3.17.8. Let $A \in \mathbb{R}^{n \times n}$, assume that A is nilpotent, and let $k \geq 1$ be such that $A^k = 0$. Then,
$$\det(I - A) = 1$$
and
$$(I - A)^{-1} = \sum_{i=0}^{k-1} A^i.$$

Fact 3.17.9. Let $A, B \in \mathbb{F}^{n \times n}$, assume that B is nilpotent, and assume that $AB = BA$. Then, $\det(A + B) = \det A$.

Proof: Use Fact 5.17.4.

Fact 3.17.10. Let $A, B \in \mathbb{R}^{n \times n}$, assume that A and B are nilpotent, and assume that $AB = BA$. Then, $A + B$ is nilpotent.

Proof: If $A^k = B^l = 0$, then $(A + B)^{k+l} = 0$.

Fact 3.17.11. Let $A, B \in \mathbb{F}^{n \times n}$, and assume that A and B are either both upper triangular or both lower triangular. Then,
$$[A, B]^n = 0.$$

Hence, $[A, B]$ is nilpotent.

Remark: See [512, 513].

Remark: See Fact 5.17.6.

Fact 3.17.12. Let $A, B \in \mathbb{F}^{n \times n}$, and assume that $[A, [A, B]] = 0$. Then, $[A, B]$ is nilpotent.

Remark: This result is due to Jacobson. See [505] or [728, p. 98].

Fact 3.17.13. Let $A, B \in \mathbb{F}^{n \times n}$, and assume that there exist $k \in \mathbb{P}$ and nonzero $\alpha \in \mathbb{R}$ such that $[A^k, B] = \alpha A$. Then, A is nilpotent.

Proof: For all $l \in \mathbb{N}$, $A^{k+l}B - A^l B A^k = \alpha A^{l+1}$, and thus $\operatorname{tr} A^{l+1} = 0$. The result now follows from Fact 3.17.6.

Remark: See [1174].

3.18 Facts on Hankel and Toeplitz Matrices

Fact 3.18.1. Let $A \in \mathbb{F}^{n \times m}$. Then, the following statements hold:

i) If A is Toeplitz, then $\hat{I}A$ and $A\hat{I}$ are Hankel.

ii) If A is Hankel, then $\hat{I}A$ and $A\hat{I}$ are Toeplitz.

iii) A is Toeplitz if and only if $\hat{I}A\hat{I}$ is Toeplitz.

iv) A is Hankel if and only if $\hat{I}A\hat{I}$ is Hankel.

Fact 3.18.2. Let $A \in \mathbb{F}^{n \times n}$, assume that A is Hankel, and consider the following conditions:

i) A is Hermitian.

ii) A is real.

iii) A is symmetric.

Then, i) $\Longrightarrow$ ii) $\Longrightarrow$ iii).

Fact 3.18.3. Let $A \in \mathbb{F}^{n \times n}$, and assume that A is a partitioned matrix, each of whose blocks is a $k \times k$ (circulant, Hankel, Toeplitz) matrix. Then, A is similar to a block-(circulant, Hankel, Toeplitz) matrix.

Proof: See [144].

Fact 3.18.4. For all $i, j \in \{1, \ldots, n\}$, define $A \in \mathbb{R}^{n \times n}$ by

$$A_{(i,j)} \triangleq \frac{1}{i + j - 1}.$$

Then, A is Hankel, positive definite, and

$$\det A = \frac{[1!2! \cdots (n-1)!]^4}{1!2! \cdots (2n-1)!}.$$

Furthermore, for all $i, j \in \{1, \ldots, n\}$, A^{-1} has integer entries given by

$$\left(A^{-1}\right)_{(i,j)} = (-1)^{i+j}(i+j-1)\binom{n+i-1}{n-j}\binom{n+j-1}{n-i}\binom{i+j-2}{i-1}^2.$$

Finally, for large n,

$$\det A \approx 2^{-2n^2}.$$

Remark: A is the *Hilbert matrix*, which is a Cauchy matrix. See [699, p. 513], Fact 1.12.37, Fact 3.22.9, Fact 3.22.10, and Fact 12.21.18.

Remark: See [333].

Fact 3.18.5. Let $A \in \mathbb{F}^{n \times n}$, and assume that A is Toeplitz. Then, A is reverse symmetric.

Fact 3.18.6. Let $A \in \mathbb{F}^{n \times n}$. Then, A is Toeplitz if and only if there exist $a_0, \ldots, a_n \in \mathbb{F}$ and $b_1, \ldots, b_n \in \mathbb{F}$ such that

$$A = \sum_{i=1}^{n} b_i N_n^{iT} + \sum_{i=0}^{n} a_i N_n^i.$$

Fact 3.18.7. Let $A \in \mathbb{F}^{n \times n}$, let $k \geq 1$, and assume that A is (lower triangular, strictly lower triangular, upper triangular, strictly upper triangular). Then, so is A^k. If, in addition, A is Toeplitz, then so is A^k.

Remark: If A is Toeplitz, then A^2 is not necessarily Toeplitz.

Remark: See Fact 11.13.1.

Fact 3.18.8. Let $n \geq 2$ and $m \geq 2$, and define $A \in \mathbb{F}^{n \times m}$ by

$$A \triangleq \begin{bmatrix} \sin\theta & \sin 2\theta & \sin 3\theta & \cdots & & \sin m\theta \\ \sin 2\theta & \sin 3\theta & \cdot^{\cdot} & \cdot^{\cdot} & & \cdot^{\cdot} \\ \sin 3\theta & \cdot^{\cdot} & \cdot^{\cdot} & \cdot^{\cdot} & & \cdot^{\cdot} \\ \vdots & \cdot^{\cdot} & \cdot^{\cdot} & \cdot^{\cdot} & & \cdot^{\cdot} \\ \sin n\theta & \cdot^{\cdot} & & \cdot^{\cdot} & \cdot^{\cdot} & \sin(m+n-1)\theta \end{bmatrix}.$$

Then,
$$\operatorname{rank} A = 2.$$

Proof: Use Proposition 12.9.11.

Fact 3.18.9. Let $A \in \mathbb{F}^{n \times n}$ be the tridiagonal, Toeplitz matrix

$$A \triangleq \begin{bmatrix} b & c & 0 & \cdots & 0 & 0 \\ a & b & c & \cdots & 0 & 0 \\ 0 & a & b & \ddots & 0 & 0 \\ \vdots & \vdots & \ddots & \ddots & \ddots & \vdots \\ 0 & 0 & 0 & \ddots & b & c \\ 0 & 0 & 0 & \cdots & a & b \end{bmatrix},$$

and define
$$\alpha \triangleq \tfrac{1}{2}(b + \sqrt{b^2 - 4ac}), \qquad \beta \triangleq \tfrac{1}{2}(b - \sqrt{b^2 - 4ac}).$$

Then,

$$\det A = \begin{cases} b^n, & ac = 0, \\ (n+1)(b/2)^n, & b^2 = 4ac, \\ (\alpha^{n+1} - \beta^{n+1})/(\alpha - \beta), & b^2 \neq 4ac. \end{cases}$$

Proof: See [1526, pp. 101, 102].

Remark: See Fact 3.19.2 and Fact 5.11.43.

Fact 3.18.10. Let $A \in \mathbb{R}^{n \times n}$ be the symmetric, tridiagonal, Toeplitz matrix

$$A \triangleq \begin{bmatrix} 2 & -1 & 0 & \cdots & 0 & 0 \\ -1 & 2 & -1 & \cdots & 0 & 0 \\ 0 & -1 & 2 & \ddots & 0 & 0 \\ \vdots & \vdots & \ddots & \ddots & \ddots & \vdots \\ 0 & 0 & 0 & \ddots & 2 & -1 \\ 0 & 0 & 0 & \cdots & -1 & 1 \end{bmatrix}.$$

Then,

$$A^{-1} = \begin{bmatrix} 1 & 1 & 1 & \cdots & 1 & 1 \\ 1 & 2 & 2 & \cdots & 2 & 2 \\ 1 & 2 & 3 & \cdots & 3 & 3 \\ \vdots & \vdots & \vdots & \ddots & \vdots & \vdots \\ 1 & 2 & 3 & \cdots & n-1 & n-1 \\ 1 & 2 & 3 & \cdots & n-1 & n \end{bmatrix}.$$

Proof: See [1215, p. 182], where the (n, n) entry of A is incorrect.

Remark: See Fact 3.19.4.

Fact 3.18.11. Let $A \in \mathbb{R}^{n \times n}$ be the Toeplitz matrix

$$A \triangleq \begin{bmatrix} 1 & a & a^2 & \cdots & a^{n-2} & a^{n-1} \\ b & 1 & a & \cdots & a^{n-1} & a^{n-2} \\ b^2 & b & 1 & \ddots & a^{n-2} & a^{n-3} \\ \vdots & \vdots & \ddots & \ddots & \ddots & \vdots \\ b^{n-2} & b^{n-3} & b^{n-4} & \ddots & 1 & a \\ b^{n-1} & b^{n-2} & b^{n-3} & \cdots & b & 1 \end{bmatrix}.$$

Then, A is nonsingular if and only if $ab \neq 1$. In this case,

$$A^{-1} = \begin{bmatrix} c & -ac & 0 & \cdots & 0 & 0 \\ -bc & (ab+1)c & -ac & \cdots & 0 & 0 \\ 0 & -bc & (ab+1)c & \ddots & 0 & 0 \\ \vdots & \vdots & & \ddots & \ddots & \vdots \\ 0 & 0 & 0 & \ddots & (ab+1)c & -ac \\ 0 & 0 & 0 & \cdots & -bc & c \end{bmatrix},$$

where $c \triangleq (1 - ab)^{-1}$. Now, assume in addition that $a = b$. Then, A is nonsingular if and only if $|a| < 1$. In this case, A is positive definite.

Remark: See [1208, pp. 348, 349].

3.19 Facts on Tridiagonal Matrices

Fact 3.19.1. Let $A \in \mathbb{F}^{n \times n}$, assume that A is nonsingular, and let $l \in \{0, \ldots, n\}$ and $k \in \{1, \ldots, n\}$. Then, the following statements are equivalent:

i) Every submatrix B of A whose entries are entries of A located above the lth superdiagonal of A satisfies $\operatorname{rank} B \leq k - 1$.

ii) Every submatrix C of A whose entries are entries of A^{-1} located above the lth subdiagonal of A^{-1} satisfies $\operatorname{rank} C \leq l + k - 1$.

Specifically, the following statements hold:

iii) A is lower triangular if and only if A^{-1} is lower triangular.

iv) A is diagonal if and only if A^{-1} is diagonal.

v) A is lower Hessenberg if and only if every submatrix C of A^{-1} whose entries are entries of A^{-1} located on or above the diagonal of A^{-1} satisfies $\operatorname{rank} C \leq 1$.

vi) A is tridiagonal if and only if every submatrix C of A^{-1} whose entries are entries of A^{-1} located on or above the diagonal of A^{-1} satisfies $\operatorname{rank} C \leq 1$ and every submatrix C of A^{-1} whose entries are entries of A^{-1} located on or below the diagonal of A^{-1} satisfies $\operatorname{rank} C \leq 1$.

Proof: See [1273].

Remark: The 0th subdiagonal and the 0th superdiagonal are the diagonal.

Remark: Statement *iii)* corresponds to $l = 0$ and $k = 1$, *iv)* corresponds to $l = 0$ and $k = 1$ applied to A and A^{T}, *v)* corresponds to $l = 1$ and $k = 1$, and *vi)* corresponds to $l = 1$ and $k = 1$ applied to A and A^{T}.

Remark: See Fact 2.11.20.

Remark: Extensions to generalized inverses are considered in [135, 1159].

Fact 3.19.2. Let $A \in \mathbb{F}^{n \times n}$ be the tridiagonal, Toeplitz matrix

$$A \triangleq \begin{bmatrix} a + b & ab & 0 & \cdots & 0 & 0 \\ 1 & a + b & ab & \cdots & 0 & 0 \\ 0 & 1 & a + b & \ddots & 0 & 0 \\ \vdots & \vdots & \ddots & \ddots & \ddots & \vdots \\ 0 & 0 & 0 & \ddots & a + b & ab \\ 0 & 0 & 0 & \cdots & 1 & a + b \end{bmatrix}.$$

Then,

$$\det A = \begin{cases} (n + 1)a^n, & a = b, \\ \dfrac{a^{n+1} - b^{n+1}}{a - b}, & a \neq b. \end{cases}$$

Proof: See [865, pp. 401, 621].

Fact 3.19.3. Let $A \in \mathbb{R}^{n \times n}$, assume that A is tridiagonal with positive diagonal entries, and assume that, for all $i \in \{2, \ldots, n\}$,

$$A_{(i,i-1)} A_{(i-1,i)} < \tfrac{1}{4} \left(\cos \tfrac{\pi}{n+1} \right)^{-2} A_{(i,i)} A_{(i-1,i-1)}.$$

Then, $\det A > 0$. If, in addition, A is symmetric, then A is positive definite.

Proof: See [788].

Remark: Related results are given in [332].

Remark: See Fact 8.8.18.

Fact 3.19.4. Let $A \in \mathbb{R}^{n \times n}$, assume that A is tridiagonal, assume that every entry of the superdiagonal and subdiagonal of A is nonzero, assume that every leading principal subdeterminant of A and every trailing principal subdeterminant of A is nonzero. Then, every entry of A^{-1} is nonzero.

Proof: See [718].

Fact 3.19.5. Let $A \in \mathbb{F}^{n \times n}$, assume that A is nonsingular, and assume that $A_{(2,2)}, \ldots, A_{(n-1,n-1)}$ are nonzero. Then, A^{-1} is tridiagonal if and only if, for all $i, j \in \{1, \ldots, n\}$ such that $|i - j| \geq 2$, and for all k satisfying $\min\{i, j\} < k < \max\{i, j\}$, it follows that

$$A_{(i,j)} = \frac{A_{(i,k)} A_{(k,j)}}{A_{(k,k)}}.$$

Proof: See [151].

3.20 Facts on Hamiltonian and Symplectic Matrices

Fact 3.20.1. Let $A \in \mathbb{F}^{2n \times 2n}$. Then, A is Hamiltonian if and only if $JA = (JA)^{\mathrm{T}}$. Furthermore, A is symplectic if and only if $A^{\mathrm{T}}JA = J$.

Fact 3.20.2. Assume that $n \in \mathbb{P}$ is even, let $A \in \mathbb{F}^{n \times n}$, and assume that A is Hamiltonian and symplectic. Then, A is skew involutory.

Remark: See Fact 3.20.3.

Fact 3.20.3. The following statements hold:

i) I_{2n} is orthogonal, shifted orthogonal, a projector, a reflector, and symplectic.

ii) J_{2n} is skew symmetric, orthogonal, skew involutory, a skew reflector, symplectic, and Hamiltonian.

iii) $\hat{I}_{2n}$ is symmetric, orthogonal, involutory, shifted orthogonal, a projector, a reflector, and Hamiltonian.

Remark: See Fact 3.20.2 and Fact 5.9.27.

Fact 3.20.4. Let $A \in \mathbb{F}^{2n \times 2n}$, assume that A is Hamiltonian, and let $S \in \mathbb{F}^{2n \times 2n}$ be symplectic. Then, SAS^{-1} is Hamiltonian.

Fact 3.20.5. Let $A \in \mathbb{F}^{2n \times 2n}$, and assume that A is Hamiltonian and nonsingular. Then, A^{-1} is Hamiltonian.

Fact 3.20.6. Let $\mathcal{A} \in \mathbb{F}^{2n \times 2n}$. Then, $\mathcal{A}$ is Hamiltonian if and only if there exist $A, B, C, D \in \mathbb{F}^{n \times n}$ such that B and C are symmetric and

$$\mathcal{A} = \left[\begin{array}{cc} A & B \\ C & -A^{\mathrm{T}} \end{array} \right].$$

Remark: See Fact 4.9.24.

Fact 3.20.7. Let $A \in \mathbb{F}^{2n \times 2n}$, and assume that A is Hamiltonian. Then, $\operatorname{tr} A = 0$.

Fact 3.20.8. Let $\mathcal{A} \in \mathbb{F}^{2n \times 2n}$. Then, $\mathcal{A}$ is skew symmetric and Hamiltonian if and only if there exist a skew-symmetric matrix $A \in \mathbb{F}^{n \times n}$ and a symmetric matrix $B \in \mathbb{F}^{n \times n}$ such that

$$\mathcal{A} = \left[\begin{array}{cc} A & B \\ -B & A \end{array} \right].$$

Fact 3.20.9. Let $\mathcal{A} \triangleq \left[\begin{smallmatrix} A & B \\ C & D \end{smallmatrix} \right] \in \mathbb{F}^{2n \times 2n}$, where $A, B, C, D \in \mathbb{F}^{n \times n}$. Then, $\mathcal{A}$ is symplectic if and only if $A^{\mathrm{T}}C$ and $B^{\mathrm{T}}D$ are symmetric and $A^{\mathrm{T}}D - C^{\mathrm{T}}B = I$.

Fact 3.20.10. Let $A \in \mathbb{F}^{2n \times 2n}$, and assume that A is symplectic. Then, $\det A = 1$.

Proof: Using Fact 2.14.16 and Fact 3.20.9 it follows that $\det \mathcal{A} = \det(A^{\mathrm{T}}D - C^{\mathrm{T}}B) = \det I = 1$. See also [106, p. 27], [433], [639, p. 8], or [1217, p. 128].

Fact 3.20.11. Let $A \in \mathbb{F}^{2 \times 2}$. Then, A is symplectic if and only if $\det A = 1$. Hence, $\mathrm{SL}_{\mathbb{F}}(2) = \mathrm{Symp}_{\mathbb{F}}(2)$.

Fact 3.20.12. The following statements hold:

i) If $A \in \mathbb{F}^{2n \times 2n}$ is Hamiltonian and $A + I$ is nonsingular, then $B \triangleq (A - I)(A + I)^{-1}$ is symplectic, $I - B$ is nonsingular, and $(I - B)^{-1} = \frac{1}{2}(A + I)$.

ii) If $B \in \mathbb{F}^{2n \times 2n}$ is symplectic and $I - B$ is nonsingular, then $A = (I + B)(I - B)^{-1}$ is Hamiltonian, $A + I$ is nonsingular, and $(A + I)^{-1} = \frac{1}{2}(I - B)$.

iii) If $A \in \mathbb{F}^{2n \times 2n}$ is Hamiltonian, then there exists a unique symplectic matrix $B \in \mathbb{F}^{2n \times 2n}$ such that $I - B$ is nonsingular and $A = (I + B)(I - B)^{-1}$. In fact, $B = (A - I)(A + I)^{-1}$.

iv) If $B \in \mathbb{F}^{2n \times 2n}$ is symplectic and $I - B$ is nonsingular, then there exists a unique Hamiltonian matrix $A \in \mathbb{F}^{2n \times 2n}$ such that $B = (A - I)(A + I)^{-1}$. In fact, $A = (I + B)(I - B)^{-1}$.

Remark: See Fact 3.11.21, Fact 3.11.22, and Fact 3.11.23.

Fact 3.20.13. Let $\mathcal{A} \in \mathbb{R}^{2n \times 2n}$. Then, $\mathcal{A} \in \mathrm{osymp}_{\mathbb{R}}(2n)$ if and only if there exist $A, B \in \mathbb{R}^{n \times n}$ such that A is skew symmetric, B is symmetric, and $\mathcal{A} = \begin{bmatrix} A & B \\ -B & A \end{bmatrix}$.

Proof: See [403].

Remark: $\mathrm{OSymp}_{\mathbb{R}}(2n)$ is the *orthosymplectic group*.

3.21 Facts on Matrices Related to Graphs

Fact 3.21.1. Let $\mathcal{G} = (\{x_1, \ldots, x_n\}, \mathcal{R})$ be a graph without self-loops, assume that $\mathcal{G}$ is antisymmetric, let $A \in \mathbb{R}^{n \times n}$ denote the adjacency matrix of $\mathcal{G}$, let $L_{\mathrm{in}} \in \mathbb{R}^{n \times n}$ and $L_{\mathrm{out}} \in \mathbb{R}^{n \times n}$ denote the inbound and outbound Laplacians of $\mathcal{G}$, respectively, let D_{in} and D_{out} be the indegree and outdegree matrices of $\mathcal{G}$, respectively, and let A_{sym}, D_{sym}, and L_{sym} denote the adjacency, degree, and Laplacian matrices, respectively, of $\mathrm{sym}(\mathcal{G})$. Then,

$$D_{\mathrm{sym}} = D_{\mathrm{in}} + D_{\mathrm{out}},$$

$$A_{\mathrm{sym}} = A + A^{\mathrm{T}},$$

and

$$L_{\mathrm{sym}} = L_{\mathrm{in}} + L_{\mathrm{out}}^{\mathrm{T}} = L_{\mathrm{in}}^{\mathrm{T}} + L_{\mathrm{out}} = D_{\mathrm{sym}} - A_{\mathrm{sym}}.$$

Fact 3.21.2. Let $\mathcal{G} = (\{x_1, \ldots, x_n\}, \mathcal{R})$ be a graph, let $A \in \mathbb{R}^{n \times n}$ be the adjacency matrix of $\mathcal{G}$, and let D_{out} be the outdegree matrix of $\mathcal{G}$. Then,

$$F \triangleq (I + D_{\mathrm{out}})^{-1}(I + A)$$

is row stochastic. If, in addition, every node has a self-loop, then

$$F_0 \triangleq D_{\mathrm{out}}^{-1} A$$

is row stochastic.

3.22 Facts on Triangular, Irreducible, Cauchy, Dissipative, Contractive, and Centrosymmetric Matrices

Fact 3.22.1. Let $A \in \mathbb{F}^{n \times n}$, and assume that A is either upper triangular or lower triangular. Then,

$$\det A = \prod_{i=1}^{n} A_{(i,i)}.$$

Remark: See Fact 4.10.10.

Fact 3.22.2. Let $A, B \in \mathbb{F}^n$, and assume that A and B are (upper triangular, lower triangular). Then, AB is (upper triangular, lower triangular). If, in addition, either A or B is (strictly upper triangular, strictly lower triangular), then AB is (strictly upper triangular, strictly lower triangular).

Remark: See Fact 3.23.12.

Fact 3.22.3. Let $A \in \mathbb{F}^{n \times n}$, and assume that there exists $i \in \{1, \ldots, n\}$ such that either $\mathrm{row}_i(A) = 0$ or $\mathrm{col}_i(A) = 0$. Then, A is reducible.

Fact 3.22.4. Let $A \in \mathbb{F}^{n \times n}$, and assume that A is reducible. Then, A has at least $n - 1$ entries that are equal to zero.

Fact 3.22.5. Let $A \in \mathbb{R}^{n \times n}$, and assume that A is a permutation matrix. Then, A is irreducible if and only if there exists a permutation matrix $S \in \mathbb{R}^{n \times n}$ such that SAS^{-1} is the $n \times n$ cyclic permutation matrix P_n.

Proof: See [1215, p. 177].

Fact 3.22.6. Let $A \in \mathbb{F}^{n \times n}$. Then, A is reducible if and only if $|A|$ is reducible. Furthermore, A is irreducible if and only if $|A|$ is irreducible.

Fact 3.22.7. Let $A \in \mathbb{F}^{n \times m}$. Then, A is (semicontractive, contractive) if and only if A^* is.

Fact 3.22.8. Let $A \in \mathbb{F}^{n \times n}$, and assume that A is dissipative. Then, A is nonsingular.

Proof: Suppose that A is singular, and let $x \in \mathcal{N}(A)$. Then, $x^*(A + A^*)x = 0$.

Remark: If $A + A^*$ is nonsingular, then A is not necessarily nonsingular. Consider $A = \left[\begin{smallmatrix} 0 & 1 \\ 0 & 0 \end{smallmatrix} \right]$.

Fact 3.22.9. Let $a_1, \ldots, a_n, b_1, \ldots, b_n \in \mathbb{R}$, assume that, for all $i, j \in \{1, \ldots, n\}$, $a_i + b_j \neq 0$, and, for all $i, j \in \{1, \ldots, n\}$, define $A \in \mathbb{R}^{n \times n}$ by

$$A_{(i,j)} \triangleq \frac{1}{a_i + b_j}.$$

Then,

$$\det A = \frac{\prod\limits_{1 \leq i < j \leq n} (a_j - a_i)(b_j - b_i)}{\prod\limits_{1 \leq i,j \leq n} (a_i + b_j)}.$$

Now, assume in addition that $a_1, \ldots, a_n$ are distinct and $b_1, \ldots, b_n$ are distinct. Then, A is nonsingular and

$$\left(A^{-1}\right)_{(i,j)} = \frac{\prod\limits_{1 \leq k \leq n} (a_j + b_k)(a_k + b_i)}{(a_j + b_i) \prod\limits_{\substack{1 \leq k \leq n \\ k \neq j}} (a_j - a_k) \prod\limits_{\substack{1 \leq k \leq n \\ k \neq i}} (b_i - b_k)}.$$

Furthermore,

$$1_{1 \times n} A^{-1} 1_{n \times 1} = \sum_{i=1}^n (a_i + b_i).$$

Remark: A is a *Cauchy matrix*. See [203], [699, p. 515], Fact 3.18.4, Fact 3.22.10, and Fact 12.21.18.

Fact 3.22.10. Let $x_1, \ldots, x_n$ be distinct positive numbers, let $y_1, \ldots, y_n$ be distinct positive numbers, and let $A \in \mathbb{R}^{n \times n}$, where, for all $i, j \in \{1, \ldots, n\}$,

$$A_{(i,j)} \triangleq \frac{1}{x_i + y_j}.$$

Then, A is nonsingular.

Proof: See [879].

Remark: A is a Cauchy matrix. See Fact 3.18.4, Fact 3.22.9, and Fact 12.21.18.

Fact 3.22.11. Let $A \in \mathbb{F}^{n \times m}$. Then, A is centrosymmetric if and only if $A^{\mathrm{T}} = A^{\hat{\mathrm{T}}}$. Furthermore, A is centrohermitian if and only if $A^* = A^{\hat{*}}$.

Fact 3.22.12. Let $A \in \mathbb{F}^{n \times m}$ and $B \in \mathbb{F}^{m \times l}$. If A and B are both (centrohermitian, centrosymmetric), then so is AB.

Proof: See [703].

Remark: Centrosymmetric and centrohermitian matrices are discussed in [908, 1444].

3.23 Facts on Groups

Fact 3.23.1. The following subsets of $\mathbb{R}$ are groups:

i) $\{x \in \mathbb{R} \colon x \neq 0\}$.

ii) $\{x \in \mathbb{R} \colon x > 0\}$.

iii) $\{x \in \mathbb{R} \colon x \neq 0 \text{ and } x \text{ is rational}\}$.

iv) $\{x \in \mathbb{R} \colon x > 0 \text{ and } x \text{ is rational}\}$.

v) $\{-1, 1\} = \mathrm{O}(1)$.

vi) $\{1\} = \mathrm{SO}(1) = \mathrm{SU}(1)$.

Fact 3.23.2. Let $A \in \mathrm{P}(n)$. Then, there exist transposition matrices $T_1, \ldots, T_k \in \mathbb{R}^{n \times n}$ such that

$$A = T_1 \cdots T_k.$$

Furthermore, the following statements hold:

i) $\det A = (-1)^k$.

ii) A is an even permutation matrix if and only if k is even.

iii) A is an odd permutation matrix if and only if k is odd.

Remark: Every permutation of n objects can be realized as a finite sequence of pairwise transpositions. See [457, pp. 106, 107] or [510, p. 82].

Example:

$$P_3 = \begin{bmatrix} 0 & 1 & 0 \\ 0 & 0 & 1 \\ 1 & 0 & 0 \end{bmatrix} = \begin{bmatrix} 0 & 0 & 1 \\ 0 & 1 & 0 \\ 1 & 0 & 0 \end{bmatrix} \begin{bmatrix} 1 & 0 & 0 \\ 0 & 0 & 1 \\ 0 & 1 & 0 \end{bmatrix} = \begin{bmatrix} 1 & 0 & 0 \\ 0 & 0 & 1 \\ 0 & 1 & 0 \end{bmatrix} \begin{bmatrix} 0 & 1 & 0 \\ 1 & 0 & 0 \\ 0 & 0 & 1 \end{bmatrix},$$

which represents a 3-cycle.

Remark: As the above example shows, factorization in terms of transpositions is not unique. However, Fact 5.16.8 shows that every permutation can be written uniquely up to a relabelling as a product of disjoint cycles.

Remark: See Fact 2.13.1 and Fact 3.11.16.

Fact 3.23.3. For all $n \geq 2$, the following statements hold:

i) $\mathrm{card}[\mathrm{P}(n)] = n!$.

ii) $\mathrm{card}[\mathrm{A}(n)] = \frac{1}{2}n!$.

iii) $\mathrm{card}[\mathrm{D}(n)] = 2n$.

iv) $\mathrm{card}[\mathrm{C}(n)] = n$.

In addition, the following statements hold:

v) $\mathrm{A}(2) \subset \mathrm{C}(2) = \mathrm{P}(2) \subset \mathrm{D}(2)$.

vii) $\mathrm{A}(3) = \mathrm{C}(3) \subset \mathrm{P}(3) = \mathrm{D}(3)$.

vi) If $n \geq 4$, then
$$\mathrm{A}(n) \subset \mathrm{SO}(n) \subset \mathrm{O}(n)$$

and
$$\left.\begin{matrix} \mathrm{C}(n) \subset \mathrm{D}(n) \\ \mathrm{A}(n) \end{matrix}\right\} \subset \mathrm{P}(n) \subset \mathrm{O}(n).$$

ix) If $n \geq 5$ and n is odd, then
$$\mathrm{C}(n) \subset \left\{\begin{matrix} \mathrm{D}(n) \\ \mathrm{A}(n) \end{matrix}\right\} \subset \mathrm{P}(n) \subset \mathrm{O}(n).$$

x) If $n \geq 4$ and $n(n-1)/2$ is even, then
$$\mathrm{C}(n) \subset \mathrm{D}(n) \subset \mathrm{A}(n) \subset \mathrm{P}(n) \subset \mathrm{O}(n).$$

Proof: Use Fact 2.13.1 and Fact 2.13.2.

Fact 3.23.4. Let k be a positive integer, and define $R_k \in \mathbb{R}^{2 \times 2}$ by
$$R_k \triangleq \begin{bmatrix} \cos \frac{2\pi}{k} & \sin \frac{2\pi}{k} \\ -\sin \frac{2\pi}{k} & \cos \frac{2\pi}{k} \end{bmatrix},$$

note that $R_1 = I_2$, $R_2 = -I_2$, and $R_k^k = I_2$, and define
$$O_k(2) \triangleq \{I, R_k, \ldots, R_k^{k-1}, \hat{I}_2, \hat{I}_2 R_k, \ldots, \hat{I}_2 R_k^{k-1}\}$$

and
$$\mathrm{SO}_k(2) \triangleq \{I, R_k, \ldots, R_k^{k-1}\}.$$

Furthermore, define
$$\mathrm{SU}_k(1) \triangleq \{1, e^{\jmath 2\pi/k}, e^{\jmath 4\pi/k}, \ldots, e^{\jmath 2(k-1)\pi/k}\}.$$

Then, the following statements hold:

i) $\mathrm{C}(1) = \mathrm{SU}_1(1) = \{1\} \approx \mathrm{SO}_1(2) = \{I_2\}$.

ii) $O_1(2) = P(2) = C(2) = \{I_2, \hat{I}_2\} \approx SO_2(2) = \{I_2, -\hat{I}_2\} \approx \{I_2, -I_2\}$
$\approx SU_2(1) = \{1, -1\}.$

iii) $O_2(2) = D(2) = \{I_2, -I_2, \hat{I}_2, -\hat{I}_2\}.$

iv) $C(3) = \{I_3, P_3, P_3^2\} \approx SO_3(2) = \{I_2, R_3, R_3^2\} \approx SU_3(1)$
$= \{1, -\frac{1}{2} + \frac{\sqrt{3}}{2}\jmath, -\frac{1}{2} - \frac{\sqrt{3}}{2}\jmath\}.$

v) $P(3) = \{I_3, \hat{I}_3, P_3, P_3\hat{I}_3, \hat{I}_3P_3, \hat{I}_3P_3\hat{I}_3\}.$

vi) $P(3) \approx \left\{ \begin{bmatrix} 1 & 0 \\ 0 & 1 \end{bmatrix}, \begin{bmatrix} 0 & 1 \\ -1 & -1 \end{bmatrix}, \begin{bmatrix} -1 & -1 \\ 1 & 0 \end{bmatrix}, \begin{bmatrix} 0 & 1 \\ 1 & 0 \end{bmatrix}, \begin{bmatrix} -1 & -1 \\ 0 & 1 \end{bmatrix}, \begin{bmatrix} 1 & 0 \\ -1 & -1 \end{bmatrix} \right\}.$

vii) $C(4) = \{I_4, P_4, P_4^2, P_4^3\} \approx SO_4(2) = \{I_2, R_4, R_4^2, R_4^3\} \approx SU_4(1)$
$= \{1, -1, \jmath, -\jmath\}.$

viii) $A(4) \approx \{I_3, D_1, D_2, D_3, P_3, D_1P_3, D_2P_3, D_3P_3, P_3^2, D_1P_3^2, D_2P_3^2, D_3P_3^2\},$
where $D_1 \triangleq \mathrm{diag}(1, -1, -1)$, $D_2 \triangleq \mathrm{diag}(-1, 1, -1)\}$, and $D_3 \triangleq$
$\mathrm{diag}(-1, -1, 1).$

ix) $P(4) \approx \{A \in SO(3): |A| \in P(3)\}.$

x) $C(6) \approx \left\{ I_5, \begin{bmatrix} P_2 & 0 \\ 0 & I_3 \end{bmatrix}, \begin{bmatrix} I_2 & 0 \\ 0 & P_3 \end{bmatrix}, \begin{bmatrix} I_2 & 0 \\ 0 & P_3^2 \end{bmatrix}, \begin{bmatrix} P_2 & 0 \\ 0 & P_3 \end{bmatrix}, \begin{bmatrix} P_2 & 0 \\ 0 & P_3^2 \end{bmatrix} \right\}.$

xi) For all $k \geq 2$, $O_k(2) \approx D(k)$.

xii) For all $k \geq 2$, $C(k) \approx SO_k(2) \approx SU_k(1)$.

xiii) $\mathrm{card}[O_k(2)] = 2k$.

xiv) $\mathrm{card}[SO_k(2)] = k$.

Remark: $P(k)$, $A(k)$, $D(k)$, and $C(k)$ are matrix representations of *symmetry groups*, which are groups of transformations that map a set onto itself. Specifically, $P(k)$, $A(k)$, $D(k) \approx O_k(2)$, and $C(k) \approx SO_k(2)$ are matrix representations of the *symmetric group* S_k, the *alternating group* A_k, the *dihedral group* D_k, and the *cyclic group* C_k, respectively, all of which can be viewed as abstract groups having matrix representations. An *abstract group* is a collection of objects (not necessarily matrices) that satisfy the properties of a group as defined by Definition 3.3.3. The elements of S_n permute n-tuples arbitrarily, while the elements of A_n permute n-tuples evenly. Consequently, $A_n \approx A(n)$, where $A(n)$ is the group of $n \times n$ even permutation matrices. See Fact 5.16.8 for the decomposition of a permutation matrix in terms of cyclic permutation matrices. The elements of $SO_k(2)$ perform counterclockwise rotations of planar figures by the angle $2\pi/k$ about a line perpendicular to the plane and passing through 0, while the elements of $O_k(2)$ perform the rotations of $SO_k(2)$ and reflect planar figures across the line $y = x$. See [457, pp. 41, 845]. Matrix representations of groups are discussed in [324, 533, 617, 646, 721, 1218].

Remark: Every finite subgroup of $O(2)$ is a representation of either D_k or C_k for some k. Furthermore, every finite subgroup of $SO(3)$ is a representation of either D_k or C_k for some k or A_4, S_4, or A_5. The symmetry groups A_4, S_4, and A_5 are represented by bijective transformations of regular solids. Specifically, A_4 is represented by the *tetrahedral group*, which consists of $4!/2 = 12$ rotation matrices that map a regular tetrahedron onto itself; S_4 is represented by the *octahedral group*, which consists of $4! = 24$ rotation matrices that map an octahedron or a cube onto itself; and A_5 is represented by the *icosahedral group*, which consists of $5!/2 = 60$ rotation matrices that map a regular icosahedron or a regular dodecahedron onto itself. See [78, p. 184], [354, p. 32], [585, pp. 176–193], [618, pp. 9–23], [1178, p. 69], [1218, pp. 35–43], or [1287, pp. 45–47].

Remark: The dihedral group D_2 is called the *Klein four group*.

Remark: The permutation group S_k is not Abelian for all $k \geq 3$. The alternating group A_3 is Abelian, whereas A_k is not Abelian for all $k \geq 4$. For all $k \geq 5$, the alternating group A_k has no normal subgroups other than the identity and the group itself [803, p. 145]. This result is essential to the classical result of Abel and Galois that there exist polynomials of order 5 and greater whose roots cannot be expressed in terms of radicals involving the coefficients. Two such polynomials are $p(x) = x^5 - x - 1$ and $p(x) = x^5 - 16x + 2$. See [78, p. 574] and [457, pp. 32, 625–639].

Remark: Statement *viii*) is given in [585, p. 180].

Remark: The 24 elements of the octahedral group representing S_4 or $P(4)$ are given in *ix*) by the 3×3 signed permutation matrices with determinant 1, where a *signed permutation matrix* has exactly one nonzero entry, which is either 1 or −1, in each row and column.

Remark: Statement *x*) shows that $C_6 \approx C_2 \times C_3$. See [803, p. 169].

Fact 3.23.5. The following statements hold:

i) There exists exactly one isomorphically distinct group consisting of one element. A representation is $\{I_n\}$.

ii) There exists exactly one isomorphically distinct group consisting of two elements, namely, the cyclic group C_2, which is isomorphic to the permutation group S_2. Representations of C_2 are given by $P(2)$, $C(2)$, $O_1(2)$, $SO_2(2)$, and $SU_2(1) = \{1, -1\}$.

iii) There exists exactly one isomorphically distinct group consisting of three elements, namely, the cyclic group C_3, which is isomorphic to the alternating group A_3. Representations of C_3 are given by $A(3)$, $C(3)$, $SO_3(2)$, and $SU_3(1)$.

iv) There exist exactly two isomorphically distinct groups consisting of four elements, namely, the cyclic group C_4 and the dihedral group D_2. Representations of C_4 are given by $C(4)$, $SO_4(2)$, and $SU_4(1) = \{1, -1, j, -j\}$. A representation of D_2 is given by $O_2(2)$.

v) There exists exactly one isomorphically distinct group consisting of five elements, namely, the cyclic group C_5. Representations of C_5 are given by $C(5)$, $SO_5(2)$, and $SU_5(1)$.

vi) There exist exactly two isomorphically distinct groups consisting of six elements, namely, the cyclic group C_6 and the dihedral group D_3, which is isomorphic to S_3. Representations of C_6 are given by $C(6)$, $SO_6(2)$, and $SU_6(1)$. Representations of D_3 are given by $P(3)$ and $O_3(2)$.

vii) There exists exactly one isomorphically distinct group consisting of seven elements, namely, the cyclic group C_7. Representations of C_7 are given by $C(7)$, $SO_7(2)$, and $SU_7(1)$.

viii) There exist exactly five isomorphically distinct groups consisting of eight elements, namely, C_8, $D_2 \times C_2$, $C_4 \times C_2$, D_4, and the quaternion group $\{\pm 1, \pm \hat{\imath}, \pm \hat{\jmath}, \pm \hat{k}\}$. Representations of C_8 are given by $C(8)$, $SO_8(2)$, and $SU_8(1)$. A representation of D_4 is given by $O_4(2)$. Representations of the quaternion group are given by *ii)* of Fact 3.24.3 and *v)* of Fact 3.24.6.

Proof: See [569, pp. 4–7] and [803, pp. 168–172].

Remark: $SU_k(1)$ is defined in Fact 3.23.4.

Remark: There are more than 250 isomorphically distinct groups consisting of 63 elements. See [803, p. 168].

Remark: The collection of *finite simple groups*, that is, finite groups that contain no normal subgroups other than the identity and the group itself, consists of 18 countably infinite sets and 26 groups, called *sporadic groups*. The largest sporadic group is the *monster group* M, which includes 19 sporadic groups as proper subgroups and satisfies $\mathrm{card}(M) \approx 8 \times 10^{53}$.

Fact 3.23.6. Let $S_1 \subset \mathbb{F}^{n \times n}$ and $S_2 \subset \mathbb{F}^{n \times n}$ be finite groups, and assume that S_1 is a subgroup of S_2. Then, $\mathrm{card}(S_1)$ divides $\mathrm{card}(S_2)$.

Proof: See [803, pp. 77, 95].

Remark: This result is due to Lagrange.

Fact 3.23.7. Let $S \subset \mathbb{F}^{n \times n}$, and assume that S is a group. Then, $\{A^{\mathrm{T}}: A \in S\}$ and $\{\overline{A}: A \in S\}$ are groups.

Fact 3.23.8. Let $P \in \mathbb{F}^{n \times n}$, and define $S \triangleq \{A \in \mathrm{GL}_{\mathbb{F}}(n): A^{\mathrm{T}}PA = P\}$. Then, S is a group. If, in addition, P is nonsingular and skew symmetric, then, for every matrix $P \in S$, it follows that $\det P = 1$.

Proof: See [349].

Remark: If $\mathbb{F} = \mathbb{R}$, n is even, and $P = J_n$, then $S = \mathrm{Symp}_{\mathbb{R}}(n)$.

Remark: Weaker conditions on P such that $\det P = 1$ for all $P \in S$ are given in [349].

Fact 3.23.9. Let n be a nonnegative integer, and define

$$S^n \triangleq \{x \in \mathbb{R}^{n+1}: x^{\mathrm{T}}x = 1\},$$

which is the unit sphere in $\mathbb{R}^{n+1}$. Then, the following statements hold:

i) $S^0 = \{-1, 1\} = O(1)$.

ii) $U(1) = \{e^{\jmath\theta}: \theta \in [0, 2\pi)\} \approx SO(2)$.

iii) $S^1 = \{\begin{bmatrix} \cos\theta & \sin\theta \end{bmatrix}^{\mathrm{T}} \in \mathbb{R}^2: \theta \in [0, 2\pi)\} = \{\begin{bmatrix} \mathrm{Re}\, z & \mathrm{Im}\, z \end{bmatrix}^{\mathrm{T}}: z \in U(1)\}$.

iv) $SU(2) = \{\begin{bmatrix} z & w \\ -\overline{w} & \overline{z} \end{bmatrix} \in \mathbb{C}^{2 \times 2}: z, w \in \mathbb{C}$ and $|z|^2 + |w|^2 = 1\} \approx Sp(1)$.

v) $S^3 = \{ \begin{bmatrix} \mathrm{Re}\,z & \mathrm{Im}\,z & \mathrm{Re}\,w & \mathrm{Im}\,w \end{bmatrix}^{\mathrm{T}} \in \mathbb{R}^4 \colon z, w \in \mathbb{C} \text{ and } |z|^2 + |w|^2 = 1 \}.$

Proof: See [1287, p. 40].

Remark: $\mathrm{Sp}(1) \subset \mathbb{H}^{1 \times 1}$ is the group of unit quaternions. See Fact 3.24.1 and Fact 3.24.4.

Remark: A group operation can be defined on S^n if and only if $n = 0$, 1, or 3. See [1287, p. 40].

Fact 3.23.10. The groups $\mathrm{U}(n)$ and $\mathrm{O}(2n) \cap \mathrm{Symp}_{\mathbb{R}}(2n)$ are isomorphic. In particular, $\mathrm{U}(1)$ and $\mathrm{O}(2) \cap \mathrm{Symp}_{\mathbb{R}}(2) = \mathrm{SO}(2)$ are isomorphic.

Proof: See [100].

Fact 3.23.11. The following subsets of $\mathbb{F}^{n \times n}$ are Lie algebras:

i) $\mathrm{ut}(n) \triangleq \{ A \in \mathrm{gl}_{\mathbb{F}}(n) \colon A \text{ is upper triangular} \}.$

ii) $\mathrm{sut}(n) \triangleq \{ A \in \mathrm{gl}_{\mathbb{F}}(n) \colon A \text{ is strictly upper triangular} \}.$

iii) $\{ 0_{n \times n} \}.$

Fact 3.23.12. The following subsets of $\mathbb{F}^{n \times n}$ are groups:

i) $\mathrm{UT}(n) \triangleq \{ A \in \mathrm{GL}_{\mathbb{F}}(n) \colon A \text{ is upper triangular} \}.$

ii) $\mathrm{UT}_+(n) \triangleq \{ A \in \mathrm{UT}(n) \colon A_{(i,i)} > 0 \text{ for all } i \in \{1, \ldots, n\} \}.$

iii) $\mathrm{UT}_{\pm 1}(n) \triangleq \{ A \in \mathrm{UT}(n) \colon A_{(i,i)} = \pm 1 \text{ for all } i \in \{1, \ldots, n\} \}.$

iv) $\mathrm{SUT}(n) \triangleq \{ A \in \mathrm{UT}(n) \colon A_{(i,i)} = 1 \text{ for all } i \in \{1, \ldots, n\} \}.$

v) $\{ I_n \}.$

Remark: The matrices in $\mathrm{SUT}(n)$ are unipotent. See Fact 5.15.5.

Remark: $\mathrm{SUT}(3)$ for $\mathbb{F} = \mathbb{R}$ is the *Heisenberg group*.

Remark: See Fact 3.22.2.

3.24 Facts on Quaternions

Fact 3.24.1. Let $\hat{\imath}, \hat{\jmath}, \hat{k}$ satisfy

$$\hat{\imath}^2 = \hat{\jmath}^2 = \hat{k}^2 = -1,$$

$$\hat{\imath}\hat{\jmath} = \hat{k} = -\hat{\jmath}\hat{\imath},$$

$$\hat{\jmath}\hat{k} = \hat{\imath} = -\hat{k}\hat{\jmath},$$

$$\hat{k}\hat{\imath} = \hat{\jmath} = -\hat{\imath}\hat{k},$$

and define

$$\mathbb{H} \triangleq \{ a + b\hat{\imath} + c\hat{\jmath} + d\hat{k} \colon a, b, c, d \in \mathbb{R} \}.$$

Furthermore, for $a, b, c, d \in \mathbb{R}$, define $q \triangleq a + b\hat{\imath} + c\hat{\jmath} + d\hat{k}$, $\overline{q} \triangleq a - b\hat{\imath} - c\hat{\jmath} - d\hat{k}$, and $|q| \triangleq \sqrt{q\overline{q}} = \sqrt{a^2 + b^2 + c^2 + d^2} = |\overline{q}|$. Then,

$$qI_4 = U\Omega(q)U,$$

where

$$\Omega(q) \triangleq \begin{bmatrix} a & -b & -c & -d \\ b & a & -d & c \\ c & d & a & -b \\ d & -c & b & a \end{bmatrix}$$

and

$$U \triangleq \frac{1}{2} \begin{bmatrix} 1 & \hat{\imath} & \hat{\jmath} & k \\ -\hat{\imath} & 1 & \hat{k} & -\hat{\jmath} \\ -\hat{\jmath} & -\hat{k} & 1 & \hat{\imath} \\ -\hat{k} & \hat{\jmath} & -\hat{\imath} & 1 \end{bmatrix}$$

satisfies $U^2 = I_4$. In addition,

$$\det \Omega(q) = (a^2 + b^2 + c^2 + c^2)^2.$$

Furthermore, if $|q| = 1$, then $\begin{bmatrix} a & -b & -c & -d \\ b & a & -d & c \\ c & d & a & -b \\ d & -c & b & a \end{bmatrix}$ is orthogonal. Next, for $i = 1, 2$, let $a_i, b_i, c_i, d_i \in \mathbb{R}$, define $q_i \triangleq a_i + b_i\hat{\imath} + c_i\hat{\jmath} + d_i\hat{k}$, and define

$$q_3 \triangleq q_2 q_1 = a_3 + b_3\hat{\imath} + c_3\hat{\jmath} + d_3\hat{k}.$$

Then,

$$\overline{q_3} = \overline{q_2}\,\overline{q_1},$$

$$|q_3| = |q_2 q_1| = |q_1 q_2| = |q_1 \overline{q_2}| = |\overline{q_1} q_2| = |\overline{q_1}\,\overline{q_2}| = |q_1||q_2|,$$

$$\Omega(q_3) = \Omega(q_2)\Omega(q_1),$$

and

$$\begin{bmatrix} a_3 \\ b_3 \\ c_3 \\ d_3 \end{bmatrix} = \Omega(q_2) \begin{bmatrix} a_1 \\ b_1 \\ c_1 \\ d_1 \end{bmatrix}.$$

Next, for $i = 1, 2$, define $v_i \triangleq \begin{bmatrix} b_i & c_i & d_i \end{bmatrix}^{\mathrm{T}}$. Then,

$$\begin{bmatrix} a_3 \\ b_3 \\ c_3 \\ d_3 \end{bmatrix} = \begin{bmatrix} a_2 a_1 - v_2^{\mathrm{T}} v_1 \\ a_1 v_2 + a_2 v_1 + v_2 \times v_1 \end{bmatrix}.$$

Remark: q is a *quaternion*. See [490, pp. 287–294]. Note the analogy between $\hat{\imath}, \hat{\jmath}, \hat{k}$ and the unit vectors in $\mathbb{R}^3$ under cross-product multiplication. See [106, p. 119].

Remark: The group Sp(1) of unit-length quaternions is isomorphic to SU(2). See [370, p. 30], [1287, p. 40], Fact 3.20.11, and Fact 3.24.4.

Remark: The unit-length quaternions, whose coefficients comprise the unit sphere $S^3 \subset \mathbb{R}^4$ and are called *Euler parameters*, provide a double cover of SO(3) as shown by Fact 3.11.31. See [156, p. 380] and [28, 354, 874, 1226].

Remark: An equivalent formulation of quaternion multiplication is given by Rodrigues's formulas. See Fact 3.11.32.

Remark: Determinants of matrices with quaternion entries are discussed in [83] and [1287, p. 31].

Remark: The *Clifford algebras* include the *quaternion algebra* $\mathbb{H}$ and the *octonion algebra* $\mathbb{O}$, which involves the *Cayley numbers*. See [490, pp. 295–300]. These ideas from the basis for *geometric algebra*. See [1248, p. 100] and [101, 354, 357, 372, 421, 435, 436, 490, 620, 622, 652, 688, 689, 690, 702, 855, 895, 959, 1046, 1125, 1216, 1281, 1287, 1311].

Fact 3.24.2. Let $a, b, c, d \in \mathbb{R}$, and let $q \triangleq a + b\hat{\imath} + c\hat{\jmath} + d\hat{k} \in \mathbb{H}$. Then,

$$q = a + b\hat{\imath} + (c + d\hat{\imath})\hat{\jmath}.$$

Remark: For all $q \in \mathbb{H}$, there exist $z, w \in \mathbb{C}$ such that $q = z + w\hat{\jmath}$, where we interpret $\mathbb{C}$ as $\{a + b\hat{\imath} : a, b \in \mathbb{R}\}$. This observation is analogous to the fact that, for all $z \in \mathbb{C}$, there exist $a, b \in \mathbb{R}$ such that $z = a + b\jmath$, where $\jmath \triangleq \sqrt{-1}$. See [1287, p. 10].

Fact 3.24.3. The following sets are groups:

i) $Q \triangleq \{\pm 1, \pm\hat{\imath}, \pm\hat{\jmath}, \pm\hat{k}\}$.

ii) $\mathrm{GL}_\mathbb{H}(1) \triangleq \mathbb{H}\backslash\{0\} = \{a + b\hat{\imath} + c\hat{\jmath} + d\hat{k} : a, b, c, d \in \mathbb{R} \text{ and } a^2 + b^2 + c^2 + d^2 > 0\}$.

iii) $\mathrm{Sp}(1) \triangleq \{a + b\hat{\imath} + c\hat{\jmath} + d\hat{k} : a, b, c, d \in \mathbb{R} \text{ and } a^2 + b^2 + c^2 + d^2 = 1\}$.

iv) $Q_\mathbb{R} \triangleq \left\{ \pm I_4, \pm \begin{bmatrix} 0 & -1 & 0 & 0 \\ 1 & 0 & 0 & 0 \\ 0 & 0 & 0 & -1 \\ 0 & 0 & 1 & 0 \end{bmatrix}, \pm \begin{bmatrix} 0 & 0 & -1 & 0 \\ 0 & 0 & 0 & 1 \\ 1 & 0 & 0 & 0 \\ 0 & -1 & 0 & 0 \end{bmatrix}, \pm \begin{bmatrix} 0 & 0 & 0 & -1 \\ 0 & 0 & -1 & 0 \\ 0 & 1 & 0 & 0 \\ 1 & 0 & 0 & 0 \end{bmatrix} \right\}$.

v) $\mathrm{GL}_{\mathbb{H},\mathbb{R}}(1) \triangleq \left\{ \begin{bmatrix} a & -b & -c & -d \\ b & a & -d & c \\ c & d & a & -b \\ d & -c & b & a \end{bmatrix} : a^2 + b^2 + c^2 + d^2 > 0 \right\}$.

vi) $\mathrm{GL}'_{\mathbb{H},\mathbb{R}}(1) \triangleq \left\{ \begin{bmatrix} a & -b & -c & -d \\ b & a & -d & c \\ c & d & a & -b \\ d & -c & b & a \end{bmatrix} : a^2 + b^2 + c^2 + d^2 = 1 \right\}$.

Furthermore, Q and $Q_\mathbb{R}$ are isomorphic, $\mathrm{GL}_\mathbb{H}(1)$ and $\mathrm{GL}_{\mathbb{H},\mathbb{R}}(1)$ are isomorphic, $\mathrm{Sp}(1)$ and $\mathrm{GL}'_{\mathbb{H},\mathbb{R}}(1)$ are isomorphic, and $\mathrm{GL}'_{\mathbb{H},\mathbb{R}}(1) \subset \mathrm{SO}(4) \cap \mathrm{Symp}_\mathbb{R}(4)$.

Remark: J_4 is an element of $\mathrm{Symp}_\mathbb{R}(4) \cap \mathrm{SO}(4)$ but is not contained in $\mathrm{GL}'_{\mathbb{H},\mathbb{R}}(1)$.

Remark: See Fact 3.24.1.

Fact 3.24.4. Define

$$\mathrm{Sp}(n) \triangleq \{A \in \mathbb{H}^{n \times n} : A^*A = I\},$$

where $\mathbb{H}$ is the quaternion algebra, $A^* \triangleq \overline{A}^\mathrm{T}$, and, for $q = a + b\hat{\imath} + c\hat{\jmath} + d\hat{k} \in \mathbb{H}$, $\bar{q} \triangleq a - b\hat{\imath} - c\hat{\jmath} - d\hat{k}$. Then, the groups $\mathrm{Sp}(n)$ and $\mathrm{U}(2n) \cap \mathrm{Symp}_\mathbb{C}(2n)$ are isomorphic.

In particular, $\mathrm{Sp}(1)$ and $\mathrm{U}(2) \cap \mathrm{Symp}_{\mathbb{C}}(2) = \mathrm{SU}(2)$ are isomorphic.

Proof: See [100].

Remark: $\mathrm{U}(n)$ and $\mathrm{O}(2n) \cap \mathrm{Symp}_{\mathbb{R}}(2n)$ are isomorphic.

Remark: See Fact 3.24.3.

Fact 3.24.5. Let n be a positive integer. Then, $\mathrm{SO}(2n) \cap \mathrm{Symp}_{\mathbb{R}}(2n)$ is a matrix group whose Lie algebra is $\mathrm{so}(2n) \cap \mathrm{symp}_{\mathbb{R}}(2n)$. Furthermore, $A \in \mathrm{SO}(2n) \cap \mathrm{Symp}_{\mathbb{R}}(2n)$ if and only if $A \in \mathrm{Symp}_{\mathbb{R}}(2n)$ and $AJ_{2n} = J_{2n}A$. Finally, $A \in \mathrm{so}(2n) \cap \mathrm{symp}_{\mathbb{R}}(2n)$ if and only if $A \in \mathrm{symp}_{\mathbb{R}}(2n)$ and $AJ_{2n} = J_{2n}A$.

Proof: See [198].

Fact 3.24.6. Define $Q_0, Q_1, Q_2, Q_3 \in \mathbb{C}^{2\times 2}$ by

$$Q_0 \triangleq I_2, \quad Q_1 \triangleq \begin{bmatrix} 0 & -1 \\ 1 & 0 \end{bmatrix}, \quad Q_2 \triangleq \begin{bmatrix} -\jmath & 0 \\ 0 & \jmath \end{bmatrix}, \quad Q_3 \triangleq \begin{bmatrix} 0 & -\jmath \\ -\jmath & 0 \end{bmatrix}.$$

Then, the following statements hold:

i) $Q_0^* = Q_0$ and $Q_i^* = -Q_i$ for all $i \in \{1,2,3\}$.

ii) $Q_0^2 = Q_0$ and $Q_i^2 = -Q_0$ for all $i \in \{1,2,3\}$.

iii) $Q_iQ_j = -Q_jQ_i$ for all $1 \le i < j \le 3$.

iv) $Q_1Q_2 = Q_3$, $Q_2Q_3 = Q_1$, and $Q_3Q_1 = Q_2$.

v) $\{\pm Q_0, \pm Q_1, \pm Q_2, \pm Q_3\}$ is a group.

For $\beta \triangleq \begin{bmatrix} \beta_0 & \beta_1 & \beta_2 & \beta_3 \end{bmatrix}^{\mathrm{T}} \in \mathbb{R}^4$ define

$$Q(\beta) \triangleq \sum_{i=0}^{3} \beta_i Q_i = \begin{bmatrix} \beta_0 + \beta_1\jmath & -(\beta_2 + \beta_3\jmath) \\ \beta_2 - \beta_3\jmath & \beta_0 - \beta_1\jmath \end{bmatrix}.$$

Then,

$$Q(\beta)Q^*(\beta) = \beta^{\mathrm{T}}\beta I_2$$

and

$$\det Q(\beta) = \beta^{\mathrm{T}}\beta.$$

Hence, if $\beta^{\mathrm{T}}\beta = 1$, then $Q(\beta)$ is unitary. Furthermore, the complex matrices Q_0, Q_1, Q_2, Q_3, and $Q(\beta)$ have the real representations

$$\mathcal{Q}_0 = I_4, \qquad \mathcal{Q}_1 = \begin{bmatrix} -J_2 & 0 \\ 0 & -J_2 \end{bmatrix},$$

$$\mathcal{Q}_2 = \begin{bmatrix} 0 & 0 & -1 & 0 \\ 0 & 0 & 0 & 1 \\ 1 & 0 & 0 & 0 \\ 0 & -1 & 0 & 0 \end{bmatrix}, \qquad \mathcal{Q}_3 = \begin{bmatrix} 0 & 0 & 0 & -1 \\ 0 & 0 & -1 & 0 \\ 0 & 1 & 0 & 0 \\ 1 & 0 & 0 & 0 \end{bmatrix},$$

$$\mathcal{Q}(\beta) = \begin{bmatrix} \beta_0 & -\beta_1 & -\beta_2 & -\beta_3 \\ \beta_1 & \beta_0 & -\beta_3 & \beta_2 \\ \beta_2 & \beta_3 & \beta_0 & -\beta_1 \\ \beta_3 & -\beta_2 & \beta_1 & \beta_0 \end{bmatrix}.$$

Hence,

$$\mathcal{Q}(\beta)\mathcal{Q}^{\mathrm{T}}(\beta) = \beta^{\mathrm{T}}\beta I_4$$

and

$$\det \mathcal{Q}(\beta) = \left(\beta^{\mathrm{T}}\beta\right)^2.$$

Remark: Q_0, Q_1, Q_2, Q_3 represent the quaternions $1, \hat{\imath}, \hat{\jmath}, \hat{k}$. See Fact 3.24.1. An alternative representation is given by the *Pauli spin matrices* given by $\sigma_0 = I_2, \sigma_1 = \jmath Q_3, \sigma_2 = \jmath Q_1, \sigma_3 = \jmath Q_2$. See [652, pp. 143–144], [799].

Remark: For applications of quaternions, see [28, 622, 652, 874].

Remark: $\mathcal{Q}(\beta)$ has the form $\left[\begin{smallmatrix} A & B \\ -B & A \end{smallmatrix}\right]$, where A and $\hat{I}B$ are rotation-dilations. See Fact 2.19.1.

Fact 3.24.7. Let $A, B, C, D \in \mathbb{R}^{n \times m}$, define $\hat{\imath}, \hat{\jmath}, \hat{k}$ as in Fact 3.24.1, and let $Q \triangleq A + \hat{\imath}B + \hat{\jmath}C + \hat{k}D$. Then,

$$\mathrm{diag}(Q, Q) = U_n^* \left[\begin{array}{cc} A + \hat{\imath}B & -C - \hat{\imath}D \\ C - \hat{\imath}D & A - \hat{\imath}B \end{array}\right] U_m,$$

where

$$U_n \triangleq \frac{1}{\sqrt{2}} \left[\begin{array}{cc} I_n & -\hat{\imath}I_n \\ -\hat{\jmath}I_n & \hat{k}I_n \end{array}\right].$$

Furthermore, $U_n U_n^* = I_{2n}$.

Proof: See [1336, 1337].

Remark: When $n = m$, this equality uses a similarity transformation to construct a complex representation of quaternions.

Remark: The complex conjugate U_n^* is constructed as in Fact 3.24.7.

Fact 3.24.8. Let $A, B, C, D \in \mathbb{R}^{n \times n}$, define $\hat{\imath}, \hat{\jmath}, \hat{k}$ as in Fact 3.24.1, and let $Q \triangleq A + \hat{\imath}B + \hat{\jmath}C + \hat{k}D$. Then,

$$\mathrm{diag}(Q, Q, Q, Q) = U_n \left[\begin{array}{cccc} A & -B & -C & -D \\ B & A & -D & C \\ C & D & A & -B \\ D & -C & B & A \end{array}\right] U_m,$$

where

$$U_n \triangleq \frac{1}{2} \left[\begin{array}{cccc} I_n & \hat{\imath}I_n & \hat{\jmath}I_n & \hat{k}I_n \\ -\hat{\imath}I_n & I_n & \hat{k}I_n & -\hat{\jmath}I_n \\ -\hat{\jmath}I_n & -\hat{k}I_n & I_n & \hat{\imath}I_n \\ -\hat{k}I_n & \hat{\jmath}I_n & -\hat{\imath}I_n & I_n \end{array}\right].$$

Furthermore, $U_n^* = U_n$ and $U_n^2 = I_{4n}$.

Proof: See [1336, 1337]. See also [83, 261, 483, 614, 1524].

Remark: When $n = m$, this equality uses a similarity transformation to construct a real representation of quaternions. See Fact 2.14.11.

Remark: The complex conjugate U_n^* is constructed by replacing $\hat{\imath}, \hat{\jmath}, \hat{k}$ by $-\hat{\imath}, -\hat{\jmath}, -\hat{k}$, respectively, in U_n^{T}.

Fact 3.24.9. Let $A \in \mathbb{C}^{2 \times 2}$. Then, A is unitary if and only if there exist $\theta \in \mathbb{R}$ and $\beta \in \mathbb{R}^4$ such that $A = e^{\jmath\theta}Q(\beta)$, where $Q(\beta)$ is defined in Fact 3.24.6.

Proof: See [1157, p. 228].

3.25 Notes

In the literature on generalized inverses, range-Hermitian matrices are traditionally called *EP matrices*. Elementary reflectors are traditionally called *Householder matrices* or *Householder reflections*.

An alternative term for irreducible is *indecomposable*, see [988, p. 147].

Left equivalence, right equivalence, and biequivalence are treated in [1157]. Each of the groups defined in Proposition 3.3.6 is a *Lie group*; see Definition 11.6.1. Elementary treatments of Lie algebras and Lie groups are given in [78, 80, 106, 370, 472, 486, 567, 568, 743, 1104, 1176, 1216], while an advanced treatment appears in [1400]. Some additional groups of structured and patterned matrices are given in [969]. Applications of group theory are discussed in [803].

Almost nonnegative matrices are called *ML-matrices* in [1215, p. 208], *essentially nonnegative matrices* in [186, 194, 632], and *Metzler matrices* in [1228, p. 402].

The terminology "idempotent" and "projector" is not standardized in the literature. Some writers use "projector," "oblique projector," or "projection" [550] for idempotent, and "orthogonal projector" or "orthoprojector" for projector.

Matrices with set-valued entries are discussed in [565]. Matrices with entries having physical dimensions are discussed in [658, 1089].

Chapter Four

Polynomial Matrices and Rational Transfer Functions

In this chapter we consider matrices whose entries are polynomials or rational functions. The decomposition of polynomial matrices in terms of the Smith form provides the foundation for developing canonical forms in Chapter 5. In this chapter we also present some basic properties of eigenvalues and eigenvectors as well as the minimal and characteristic polynomials of a square matrix. Finally, we consider the extension of the Smith form to the Smith-McMillan form for rational transfer functions.

4.1 Polynomials

A function $p \colon \mathbb{C} \mapsto \mathbb{C}$ of the form

$$p(s) = \beta_k s^k + \beta_{k-1} s^{k-1} + \cdots + \beta_1 s + \beta_0, \qquad (4.1.1)$$

where $k \in \mathbb{N}$ and $\beta_0, \ldots, \beta_k \in \mathbb{F}$, is a *polynomial*. The set of polynomials is denoted by $\mathbb{F}[s]$. If the coefficient $\beta_k \in \mathbb{F}$ is nonzero, then the *degree* of p, denoted by $\deg p$, is k. If, in addition, $\beta_k = 1$, then p is *monic*. If $k = 0$, then p is *constant*. The degree of a nonzero constant polynomial is zero, while the degree of the zero polynomial is defined to be $-\infty$.

Let p_1 and p_2 be polynomials. Then,

$$\deg p_1 p_2 = \deg p_1 + \deg p_2. \qquad (4.1.2)$$

If $p_1 = 0$ or $p_2 = 0$, then $\deg p_1 p_2 = \deg p_1 + \deg p_2 = -\infty$. If p_2 is a nonzero constant, then $\deg p_2 = 0$, and thus $\deg p_1 p_2 = \deg p_1$. Furthermore,

$$\deg(p_1 + p_2) \le \max\{\deg p_1, \deg p_2\}. \qquad (4.1.3)$$

Therefore, $\deg(p_1 + p_2) = \max\{\deg p_1, \deg p_2\}$ if and only if either *i)* $\deg p_1 \ne \deg p_2$ or *ii)* $p_1 = p_2 = 0$ or *iii)* $r \triangleq \deg p_1 = \deg p_2 \ne -\infty$ and the sum of the coefficients of s^r in p_1 and p_2 is not zero. Equivalently, $\deg(p_1 + p_2) < \max\{\deg p_1, \deg p_2\}$ if and only if $r \triangleq \deg p_1 = \deg p_2 \ne -\infty$ and the sum of the coefficients of s^r in p_1 and p_2 is zero.

Let $p \in \mathbb{F}[s]$ be a polynomial of degree $k \geq 1$. Then, it follows from the *fundamental theorem of algebra* that p has k possibly repeated complex roots $\lambda_1, \ldots, \lambda_k$ and thus can be factored as

$$p(s) = \beta \prod_{i=1}^{k} (s - \lambda_i), \tag{4.1.4}$$

where $\beta \in \mathbb{F}$. The multiplicity of a root $\lambda \in \mathbb{C}$ of p is denoted by $\mathrm{mult}_p(\lambda)$. If λ is not a root of p, then $\mathrm{mult}_p(\lambda) = 0$. The multiset consisting of the roots of p including multiplicity is $\mathrm{mroots}(p) = \{\lambda_1, \ldots, \lambda_k\}_{\mathrm{ms}}$, while the set of roots of p ignoring multiplicity is $\mathrm{roots}(p) = \{\hat{\lambda}_1, \ldots, \hat{\lambda}_l\}$, where $\sum_{i=1}^{l} \mathrm{mult}_p(\hat{\lambda}_i) = k$. If $\mathbb{F} = \mathbb{R}$, then the multiplicity of a root λ_i whose imaginary part is nonzero is equal to the multiplicity of its complex conjugate $\bar{\lambda}_i$. Hence, $\mathrm{mroots}(p)$ is *self-conjugate*, that is, $\mathrm{mroots}(p) = \overline{\mathrm{mroots}(p)}$.

Let $p \in \mathbb{F}[s]$. If $p(-s) = p(s)$ for all $s \in \mathbb{C}$, then p is *even*, while, if $p(-s) = -p(s)$ for all $s \in \mathbb{C}$, then p is *odd*. If p is either odd or even, then $\mathrm{mroots}(p) = -\mathrm{mroots}(p)$. If $p \in \mathbb{R}[s]$ and there exists a polynomial $q \in \mathbb{R}[s]$ such that $p(s) = q(s)q(-s)$ for all $s \in \mathbb{C}$, then p has a *spectral factorization*. If p has a spectral factorization, then p is even and $\deg p$ is an even integer.

Proposition 4.1.1. Let $p \in \mathbb{R}[s]$. Then, the following statements are equivalent:

i) p has a spectral factorization.

ii) p is even, and every imaginary root of p has even multiplicity.

iii) p is even, and $p(\jmath\omega) \geq 0$ for all $\omega \in \mathbb{R}$.

Proof. The equivalence of *i)* and *ii)* is immediate. To prove that *i)* $\implies$ *iii)*, note that, for all $\omega \in \mathbb{R}$,

$$p(\jmath\omega) = q(\jmath\omega)q(-\jmath\omega) = |q(\jmath\omega)|^2 \geq 0.$$

Conversely, to prove *iii)* $\implies$ *i)* write $p = p_1 p_2$, where every root of p_1 is imaginary and none of the roots of p_2 are imaginary. Now, let z be a root of p_2. Then, $-z$, $\bar{z}$, and $-\bar{z}$ are also roots of p_2 with the same multiplicity as z. Hence, there exists a polynomial $p_{20} \in \mathbb{R}[s]$ such that $p_2(s) = p_{20}(s)p_{20}(-s)$ for all $s \in \mathbb{C}$.

Next, assuming that p has at least one imaginary root, write $p_1(s) = \prod_{i=1}^{k} (s^2 + \omega_i^2)^{m_i}$, where $0 \leq \omega_1 < \cdots < \omega_k$ and $m_i \triangleq \mathrm{mult}_p(\jmath\omega_i)$. Let ω_{i_0} denote the smallest element of the set $\{\omega_1, \ldots, \omega_k\}$ such that m_i is odd. Then, it follows that $p_1(\jmath\omega) = \prod_{i=1}^{k} (\omega_i^2 - \omega^2)^{m_i} < 0$ for all $\omega \in (\omega_{i_0}, \omega_{i_0+1})$, where $\omega_{k+1} \triangleq \infty$. However, note that $p_1(\jmath\omega) = p(\jmath\omega)/p_2(\jmath\omega) = p(\jmath\omega)/|p_{20}(\jmath\omega)|^2 \geq 0$ for all $\omega \in \mathbb{R}$, which is a contradiction. Therefore, m_i is even for all $i \in \{1, \ldots, k\}$, and thus $p_1(s) = p_{10}(s)p_{10}(-s)$ for all $s \in \mathbb{C}$, where $p_{10}(s) \triangleq \prod_{i=1}^{k} (s^2 + \omega_i^2)^{m_i/2}$. Consequently, $p(s) = p_{10}(s)p_{20}(s)p_{10}(-s)p_{20}(-s)$ for all $s \in \mathbb{C}$. Finally, if p has no imaginary roots, then $p_1 = 1$, and $p(s) = p_{20}(s)p_{20}(-s)$ for all $s \in \mathbb{C}$. $\square$

The following division algorithm is essential to the study of polynomials.

Lemma 4.1.2. Let $p_1, p_2 \in \mathbb{F}[s]$, and assume that p_2 is not the zero polynomial. Then, there exist unique polynomials $q, r \in \mathbb{F}[s]$ such that $\deg r < \deg p_2$ and

$$p_1 = qp_2 + r. \tag{4.1.5}$$

Proof. Define $n \triangleq \deg p_1$ and $m \triangleq \deg p_2$. If $n < m$, then $q = 0$ and $r = p_1$. Hence, $\deg r = \deg p_1 = n < m = \deg p_2$.

Now, assume that $n \geq m \geq 0$, and write $p_1(s) = \beta_n s^n + \cdots + \beta_0$ and $p_2(s) = \gamma_m s^m + \cdots + \gamma_0$. If $n = 0$, then $m = 0$, $\gamma_0 \neq 0$, $q = \beta_0/\gamma_0$, and $r = 0$. Hence, $-\infty = \deg r < 0 = \deg p_2$.

If $n = 1$, then either $m = 0$ or $m = 1$. If $m = 0$, then $p_2(s) = \gamma_0 \neq 0$, and (4.1.5) is satisfied with $q(s) = p_1(s)/\gamma_0$ and $r = 0$, in which case $-\infty = \deg r < 0 = \deg p_2$. If $m = 1$, then (4.1.5) is satisfied with $q(s) = \beta_1/\gamma_1$ and $r(s) = \beta_0 - \beta_1 \gamma_0/\gamma_1$. Hence, $\deg r \leq 0 < 1 = \deg p_2$.

Now, suppose that $n = 2$. Then, $\hat{p}_1(s) = p_1(s) - (\beta_2/\gamma_m)s^{2-m}p_2(s)$ has degree 1. Applying (4.1.5) with p_1 replaced by $\hat{p}_1$, it follows that there exist polynomials $q_1, r_1 \in \mathbb{F}[s]$ such that $\hat{p}_1 = q_1 p_2 + r_1$ and such that $\deg r_1 < \deg p_2$. It thus follows that $p_1(s) = q_1(s)p_2(s) + r_1(s) + (\beta_2/\gamma_m)s^{2-m}p_2(s) = q(s)p_2(s) + r(s)$, where $q(s) = q_1(s) + (\beta_2/\gamma_m)s^{n-m}$ and $r = r_1$, which verifies (4.1.5). Similar arguments apply to successively larger values of n.

To prove uniqueness, suppose there exist polynomials $\hat{q}$ and $\hat{r}$ such that $\deg \hat{r} < \deg p_2$ and $p_1 = \hat{q}p_2 + \hat{r}$. Then, it follows that $(\hat{q} - q)p_2 = r - \hat{r}$. Next, note that $\deg(r - \hat{r}) < \deg p_2$. If $\hat{q} \neq q$, then $\deg p_2 \leq \deg[(\hat{q} - q)p_2]$ so that $\deg(r - \hat{r}) < \deg[(\hat{q} - q)p_2]$, which is a contradiction. Thus, $\hat{q} = q$, and, hence, $r = \hat{r}$. $\square$

In Lemma 4.1.2, q is the *quotient* of p_1 and p_2, while r is the *remainder*. If $r = 0$, then p_2 *divides* p_1, or, equivalently, p_1 is a *multiple* of p_2. Note that, if $p_2(s) = s - \alpha$, where $\alpha \in \mathbb{F}$, then r is constant and is given by $r(s) = p_1(\alpha)$.

If a polynomial $p_3 \in \mathbb{F}[s]$ divides two polynomials $p_1, p_2 \in \mathbb{F}[s]$, then p_3 is a *common divisor* of p_1 and p_2. Given polynomials $p_1, p_2 \in \mathbb{F}[s]$, there exists a unique monic polynomial $p_3 \in \mathbb{F}[s]$, the *greatest common divisor* of p_1 and p_2, such that p_3 is a common divisor of p_1 and p_2 and such that every common divisor of p_1 and p_2 divides p_3. In addition, there exist polynomials $q_1, q_2 \in \mathbb{F}[s]$ such that the greatest common divisor p_3 of p_1 and p_2 is given by $p_3 = q_1 p_1 + q_2 p_2$. See [1108, p. 113] for proofs of these results. Finally, p_1 and p_2 are *coprime* if their greatest common divisor is $p_3 = 1$, while a polynomial $p \in \mathbb{F}[s]$ is *irreducible* if there do not exist nonconstant polynomials $p_1, p_2 \in \mathbb{F}[s]$ such that $p = p_1 p_2$. For example, if $\mathbb{F} = \mathbb{R}$, then $p(s) = s^2 + s + 1$ is irreducible.

If a polynomial $p_3 \in \mathbb{F}[s]$ is a multiple of two polynomials $p_1, p_2 \in \mathbb{F}[s]$, then p_3 is a *common multiple* of p_1 and p_2. Given nonzero polynomials p_1 and p_2, there exists (see [1108, p. 113]) a unique monic polynomial $p_3 \in \mathbb{F}[s]$ that is a common multiple of p_1 and p_2 and that divides every common multiple of p_1 and p_2. The polynomial p_3 is the *least common multiple* of p_1 and p_2.

The polynomial $p \in \mathbb{F}[s]$ given by (4.1.1) can be evaluated with a square matrix argument $A \in \mathbb{F}^{n \times n}$ by defining

$$p(A) \triangleq \beta_k A^k + \beta_{k-1} A^{k-1} + \cdots + \beta_1 A + \beta_0 I. \tag{4.1.6}$$

4.2 Polynomial Matrices

The set $\mathbb{F}^{n \times m}[s]$ of *polynomial matrices* consists of matrix functions $P \colon \mathbb{C} \mapsto \mathbb{C}^{n \times m}$ whose entries are elements of $\mathbb{F}[s]$. A polynomial matrix $P \in \mathbb{F}^{n \times m}[s]$ can thus be written as

$$P(s) = s^k B_k + s^{k-1} B_{k-1} + \cdots + s B_1 + B_0, \tag{4.2.1}$$

where $B_0, \ldots, B_k \in \mathbb{F}^{n \times m}$. If B_k is nonzero, then the *degree* of P, denoted by $\deg P$, is k, whereas, if $P = 0$, then $\deg P = -\infty$. If $n = m$ and B_k is nonsingular, then P is *regular*, while, if $B_k = I$, then P is *monic*.

The following result, which generalizes Lemma 4.1.2, provides a division algorithm for polynomial matrices.

Lemma 4.2.1. Let $P_1, P_2 \in \mathbb{F}^{n \times n}[s]$, where P_2 is regular. Then, there exist unique polynomial matrices $Q, R, \hat{Q}, \hat{R} \in \mathbb{F}^{n \times n}[s]$ such that $\deg R < \deg P_2$, $\deg \hat{R} < \deg P_2$,

$$P_1 = Q P_2 + R, \tag{4.2.2}$$

and

$$P_1 = P_2 \hat{Q} + \hat{R}. \tag{4.2.3}$$

Proof. See [573, p. 90] or [1108, pp. 134–135]. □

If $R = 0$, then P_2 *right divides* P_1, while, if $\hat{R} = 0$, then P_2 *left divides* P_1.

Let the polynomial matrix $P \in \mathbb{F}^{n \times m}[s]$ be given by (4.2.1). Then, P can be evaluated with a square matrix argument in two different ways, either from the right or from the left. For $A \in \mathbb{C}^{m \times m}$ define

$$P_{\mathrm{R}}(A) \triangleq B_k A^k + B_{k-1} A^{k-1} + \cdots + B_1 A + B_0, \tag{4.2.4}$$

while, for $A \in \mathbb{C}^{n \times n}$, define

$$P_{\mathrm{L}}(A) \triangleq A^k B_k + A^{k-1} B_{k-1} + \cdots + A B_1 + B_0. \tag{4.2.5}$$

$P_{\mathrm{R}}(A)$ and $P_{\mathrm{L}}(A)$ are *matrix polynomials*.

If $n = m$, then $P_{\mathrm{R}}(A)$ and $P_{\mathrm{L}}(A)$ can be evaluated for all $A \in \mathbb{F}^{n \times n}$, although these matrices may be different.

The following result is useful.

Lemma 4.2.2. Let $Q, \hat{Q} \in \mathbb{F}^{n \times n}[s]$ and $A \in \mathbb{F}^{n \times n}$. Furthermore, define $P, \hat{P} \in \mathbb{F}^{n \times n}[s]$ by $P(s) \triangleq Q(s)(sI - A)$ and $\hat{P}(s) \triangleq (sI - A)\hat{Q}(s)$. Then, $P_{\mathrm{R}}(A) = 0$ and $\hat{P}_{\mathrm{L}}(A) = 0$.

Let $p \in \mathbb{F}[s]$ be given by (4.1.1), and define $P(s) \triangleq p(s)I_n = s^k \beta_k I_n + s^{k-1}\beta_{k-1}I_n + \cdots + s\beta_1 I_n + \beta_0 I_n \in \mathbb{F}^{n \times n}[s]$. For $A \in \mathbb{C}^{n \times n}$ it follows that $p(A) = P(A) = P_{\mathrm{R}}(A) = P_{\mathrm{L}}(A)$.

The following result specializes Lemma 4.2.1 to the case of polynomial matrix divisors of degree 1.

Corollary 4.2.3. Let $P \in \mathbb{F}^{n \times n}[s]$ and $A \in \mathbb{F}^{n \times n}$. Then, there exist unique polynomial matrices $Q, \hat{Q} \in \mathbb{F}^{n \times n}[s]$ and unique matrices $R, \hat{R} \in \mathbb{F}^{n \times n}$ such that

$$P(s) = Q(s)(sI - A) + R \tag{4.2.6}$$

and

$$P(s) = (sI - A)\hat{Q}(s) + \hat{R}. \tag{4.2.7}$$

Furthermore, $R = P_{\mathrm{R}}(A)$ and $\hat{R} = P_{\mathrm{L}}(A)$.

Proof. In Lemma 4.2.1 set $P_1 = P$ and $P_2(s) = sI - A$. Since $\deg P_2 = 1$, it follows that $\deg R = \deg \hat{R} = 0$, and thus R and $\hat{R}$ are constant. Finally, the last statement follows from Lemma 4.2.2. $\square$

Definition 4.2.4. Let $P \in \mathbb{F}^{n \times m}[s]$. Then, $\operatorname{rank} P$ is defined by

$$\operatorname{rank} P \triangleq \max_{s \in \mathbb{C}} \operatorname{rank} P(s). \tag{4.2.8}$$

Let $P \in \mathbb{F}^{n \times n}[s]$. Then, $P(s) \in \mathbb{C}^{n \times n}$ for all $s \in \mathbb{C}$. Furthermore, $\det P$ is a polynomial in s, that is, $\det P \in \mathbb{F}[s]$.

Definition 4.2.5. Let $P \in \mathbb{F}^{n \times n}[s]$. Then, P is *nonsingular* if $\det P$ is not the zero polynomial; otherwise, P is *singular*.

Proposition 4.2.6. Let $P \in \mathbb{F}^{n \times n}[s]$, and assume that P is regular. Then, P is nonsingular.

Let $P \in \mathbb{F}^{n \times n}[s]$. If P is nonsingular, then the *inverse* P^{-1} of P can be constructed according to (2.7.22). In general, the entries of P^{-1} are rational functions of s (see Definition 4.7.1). For example, if $P(s) = \begin{bmatrix} s+2 & s+1 \\ s-2 & s-1 \end{bmatrix}$, then $P^{-1}(s) = \frac{1}{2s}\begin{bmatrix} s-1 & -s-1 \\ -s+2 & s+2 \end{bmatrix}$. In certain cases, P^{-1} is also a polynomial matrix. For example, if $P(s) = \begin{bmatrix} s & 1 \\ s^2+s-1 & s+1 \end{bmatrix}$, then $P^{-1}(s) = \begin{bmatrix} s+1 & -1 \\ -s^2-s+1 & s \end{bmatrix}$.

The following result is an extension of Proposition 2.7.7 from constant matrices to polynomial matrices.

Proposition 4.2.7. Let $P \in \mathbb{F}^{n \times m}[s]$. Then, $\operatorname{rank} P$ is the order of the largest nonsingular polynomial matrix that is a submatrix of P.

Proof. For all $s \in \mathbb{C}$ it follows from Proposition 2.7.7 that $\operatorname{rank} P(s)$ is the order of the largest nonsingular submatrix of $P(s)$. Now, let $s_0 \in \mathbb{C}$ be such that $\operatorname{rank} P(s_0) = \operatorname{rank} P$. Then, $P(s_0)$ has a nonsingular submatrix of maximal order $\operatorname{rank} P$. Therefore, P has a nonsingular polynomial submatrix of maximal order $\operatorname{rank} P$. $\qquad\qquad\square$

A polynomial matrix can be transformed by performing elementary row and column operations of the following types:

i) Multiply a row or a column by a nonzero constant.

ii) Interchange two rows or two columns.

iii) Add a polynomial multiple of one (row, column) to another (row, column).

These operations correspond respectively to left multiplication or right multiplication by the elementary matrices

$$I_n + (\alpha - 1)E_{i,i} = \begin{bmatrix} I_{i-1} & 0 & 0 \\ 0 & \alpha & 0 \\ 0 & 0 & I_{n-i} \end{bmatrix}, \tag{4.2.9}$$

where $\alpha \in \mathbb{F}$ is nonzero,

$$I_n + E_{i,j} + E_{j,i} - E_{i,i} - E_{j,j} = \begin{bmatrix} I_{i-1} & 0 & 0 & 0 & 0 \\ 0 & 0 & 0 & 1 & 0 \\ 0 & 0 & I_{j-i-1} & 0 & 0 \\ 0 & 1 & 0 & 0 & 0 \\ 0 & 0 & 0 & 0 & I_{n-j} \end{bmatrix}, \tag{4.2.10}$$

where $i \neq j$, and the *elementary polynomial matrix*

$$I_n + pE_{i,j} = \begin{bmatrix} I_{i-1} & 0 & 0 & 0 & 0 \\ 0 & 1 & 0 & p & 0 \\ 0 & 0 & I_{j-i-1} & 0 & 0 \\ 0 & 0 & 0 & 1 & 0 \\ 0 & 0 & 0 & 0 & I_{n-j} \end{bmatrix}, \tag{4.2.11}$$

where $i \neq j$ and $p \in \mathbb{F}[s]$. The matrices shown in (4.2.10) and (4.2.11) illustrate the case $i < j$. Applying these operations sequentially corresponds to forming products of elementary matrices and elementary polynomial matrices. Note that the elementary polynomial matrix $I + pE_{i,j}$ is nonsingular, and that $(I + pE_{i,j})^{-1} = I - pE_{i,j}$. Therefore, the inverse of an elementary polynomial matrix is an elementary polynomial matrix.

4.3 The Smith Decomposition and Similarity Invariants

Definition 4.3.1. Let $P \in \mathbb{F}^{n \times n}[s]$. Then, P is *unimodular* if P is the product of elementary matrices and elementary polynomial matrices.

The following result provides a canonical form, known as the *Smith form*, for polynomial matrices under unimodular transformation.

Theorem 4.3.2. Let $P \in \mathbb{F}^{n \times m}[s]$, and let $r \triangleq \operatorname{rank} P$. Then, there exist unimodular matrices $S_1 \in \mathbb{F}^{n \times n}[s]$ and $S_2 \in \mathbb{F}^{m \times m}[s]$ and monic polynomials $p_1, \ldots, p_r \in \mathbb{F}[s]$ such that p_i divides p_{i+1} for all $i \in \{1, \ldots, r-1\}$ and such that

$$P = S_1 \begin{bmatrix} p_1 & & & 0 \\ & \ddots & & \\ & & p_r & \\ 0 & & & 0_{(n-r) \times (m-r)} \end{bmatrix} S_2. \qquad (4.3.1)$$

Furthermore, for all $i \in \{1, \ldots, r\}$, let Δ_i denote the monic greatest common divisor of all $i \times i$ subdeterminants of P. Then, p_i is uniquely determined by

$$\Delta_i = p_1 \cdots p_i. \qquad (4.3.2)$$

Proof. This result is obtained by sequentially applying elementary row and column operations to P. For details, see [809, pp. 390–392] or [1108, pp. 125–128]. □

Definition 4.3.3. The monic polynomials $p_1, \ldots, p_r \in \mathbb{F}[s]$ of the Smith form (4.3.1) of $P \in \mathbb{F}^{n \times m}[s]$ are the *Smith polynomials* of P. The *Smith zeros* of P are the roots of $p_1, \ldots, p_r$. Let

$$\operatorname{Szeros}(P) \triangleq \operatorname{roots}(p_r) \qquad (4.3.3)$$

and

$$\operatorname{mSzeros}(P) \triangleq \bigcup_{i=1}^{r} \operatorname{mroots}(p_i). \qquad (4.3.4)$$

Proposition 4.3.4. Let $P \in \mathbb{R}^{n \times m}[s]$, and assume there exist unimodular matrices $S_1 \in \mathbb{F}^{n \times n}[s]$ and $S_2 \in \mathbb{F}^{m \times m}[s]$ and monic polynomials $p_1, \ldots, p_r \in \mathbb{F}[s]$ satisfying (4.3.1). Then, $\operatorname{rank} P = r$.

Proposition 4.3.5. Let $P \in \mathbb{F}^{n \times m}[s]$, and let $r \triangleq \operatorname{rank} P$. Then, r is the largest order of all nonsingular submatrices of P.

Proof. Let r_0 denote the largest order of all nonsingular submatrices of P, and let $P_0 \in \mathbb{F}^{r_0 \times r_0}[s]$ be a nonsingular submatrix of P. First, assume that $r < r_0$. Then, there exists $s_0 \in \mathbb{C}$ such that $\operatorname{rank} P(s_0) = \operatorname{rank} P_0(s_0) = r_0$. Thus, $r = \operatorname{rank} P = \max_{s \in \mathbb{C}} \operatorname{rank} P(s) \geq \operatorname{rank} P(s_0) = r_0$, which is a contradiction. Next, assume that $r > r_0$. Then, it follows from (4.3.1) that there exists $s_0 \in \mathbb{C}$ such that $\operatorname{rank} P(s_0) = r$. Consequently, $P(s_0)$ has a nonsingular $r \times r$ submatrix. Let $\hat{P}_0 \in \mathbb{F}^{r \times r}[s]$ denote the corresponding submatrix of P. Thus, $\hat{P}_0$ is nonsingular, which implies that P has a nonsingular submatrix whose order is greater than r_0, which is a contradiction. Consequently, $r = r_0$. □

Proposition 4.3.6. Let $P \in \mathbb{F}^{n \times m}[s]$, and let $\mathcal{S} \subset \mathbb{C}$ be a finite set. Then,

$$\operatorname{rank} P = \max_{s \in \mathbb{C} \backslash \mathcal{S}} \operatorname{rank} P(s). \tag{4.3.5}$$

Proposition 4.3.7. Let $P \in \mathbb{F}^{n \times n}[s]$. Then, the following statements are equivalent:

i) P is unimodular.

ii) $\det P$ is a nonzero constant.

iii) The Smith form of P is the identity matrix.

iv) P is nonsingular, and P^{-1} is a polynomial matrix.

v) P is nonsingular, and P^{-1} is unimodular.

Proof. To prove that *i)* $\Longrightarrow$ *ii)*, note that every elementary matrix and every elementary polynomial matrix has a constant nonzero determinant. Since P is a product of elementary matrices and elementary polynomial matrices, its determinant is a constant.

To prove that *ii)* $\Longrightarrow$ *iii)*, note that it follows from (4.3.1) that $\operatorname{rank} P = n$ and $\det P = (\det S_1)(\det S_2)p_1 \cdots p_n$, where $S_1, S_2 \in \mathbb{F}^{n \times n}$ are unimodular and $p_1, \ldots, p_n$ are monic polynomials. From the result *i)* $\Longrightarrow$ *ii)*, it follows that $\det S_1$ and $\det S_2$ are nonzero constants. Since $\det P$ is a nonzero constant, it follows that $p_1 \cdots p_n = \det P/[(\det S_1)(\det S_2)]$ is a nonzero constant. Since $p_1, \ldots, p_n$ are monic polynomials, it follows that $p_1 = \cdots = p_n = 1$.

Next, to prove *iii)* $\Longrightarrow$ *iv)*, note that P is unimodular, and thus it follows that $\det P$ is a nonzero constant. Furthermore, since P^{A} is a polynomial matrix, it follows that $P^{-1} = (\det P)^{-1}P^{\mathrm{A}}$ is a polynomial matrix.

To prove that *iv)* $\Longrightarrow$ *v)*, note that $\det P^{-1}$ is a polynomial. Since $\det P$ is a polynomial and $\det P^{-1} = 1/\det P$ it follows that $\det P$ is a nonzero constant. Hence, P is unimodular, and thus $P^{-1} = (\det P)^{-1}P^{\mathrm{A}}$ is unimodular.

Finally, to prove *v)* $\Longrightarrow$ *i)*, note that $\det P^{-1}$ is a nonzero constant, and thus $P = [\det P^{-1}]^{-1}[P^{-1}]^{\mathrm{A}}$ is a polynomial matrix. Furthermore, since $\det P = 1/\det P^{-1}$, it follows that $\det P$ is a nonzero constant. Hence, P is unimodular. $\square$

Proposition 4.3.8. Let $A_1, B_1, A_2, B_2 \in \mathbb{F}^{n \times n}$, where A_2 is nonsingular, and define the polynomial matrices $P_1, P_2 \in \mathbb{F}^{n \times n}[s]$ by $P_1(s) \triangleq sA_1 + B_1$ and $P_2(s) \triangleq sA_2 + B_2$. Then, P_1 and P_2 have the same Smith polynomials if and only if there exist nonsingular matrices $S_1, S_2 \in \mathbb{F}^{n \times n}$ such that $P_2 = S_1 P_1 S_2$.

Proof. The sufficiency result is immediate. To prove necessity, note that it follows from Theorem 4.3.2 that there exist unimodular matrices $T_1, T_2 \in \mathbb{F}^{n \times n}[s]$ such that $P_2 = T_2 P_1 T_1$. Now, since P_2 is regular, it follows from Lemma 4.2.1 that there exist polynomial matrices $Q, \hat{Q} \in \mathbb{F}^{n \times n}[s]$ and constant matrices $R, \hat{R} \in \mathbb{F}^{n \times n}$

such that $T_1 = QP_2 + R$ and $T_2 = P_2\hat{Q} + \hat{R}$. Next, we have

$$
\begin{aligned}
P_2 &= T_2 P_1 T_1 \\
&= (P_2\hat{Q} + \hat{R})P_1 T_1 \\
&= \hat{R}P_1 T_1 + P_2\hat{Q}T_2^{-1}P_2 \\
&= \hat{R}P_1(QP_2 + R) + P_2\hat{Q}T_2^{-1}P_2 \\
&= \hat{R}P_1 R + (T_2 - P_2\hat{Q})P_1 Q P_2 + P_2\hat{Q}T_2^{-1}P_2 \\
&= \hat{R}P_1 R + T_2 P_1 Q P_2 + P_2\left(-\hat{Q}P_1 Q + \hat{Q}T_2^{-1}\right)P_2 \\
&= \hat{R}P_1 R + P_2\left(T_1^{-1}Q - \hat{Q}P_1 Q + \hat{Q}T_2^{-1}\right)P_2.
\end{aligned}
$$

Since P_2 is regular and has degree 1, it follows that, if $T_1^{-1}Q - \hat{Q}P_1 Q + \hat{Q}T_2^{-1}$ is not zero, then $\deg P_2\left(T_1^{-1}Q - \hat{Q}P_1 Q + \hat{Q}T_2^{-1}\right)P_2 \geq 2$. However, since P_2 and $\hat{R}P_1 R$ have degree less than 2, it follows that $T_1^{-1}Q - \hat{Q}P_1 Q + \hat{Q}T_2^{-1} = 0$. Hence, $P_2 = \hat{R}P_1 R$.

Next, to show that $\hat{R}$ and R are nonsingular, note that, for all $s \in \mathbb{C}$,

$$
P_2(s) = \hat{R}P_1(s)R = s\hat{R}A_1 R + \hat{R}B_1 R,
$$

which implies that $A_2 = S_1 A_1 S_2$, where $S_1 = \hat{R}$ and $S_2 = R$. Since A_2 is nonsingular, it follows that S_1 and S_2 are nonsingular. $\qquad\square$

Definition 4.3.9. Let $A \in \mathbb{F}^{n \times n}$. Then, the *similarity invariants* of A are the Smith polynomials of $sI - A$.

The following result provides necessary and sufficient conditions for two matrices to be similar.

Theorem 4.3.10. Let $A, B \in \mathbb{F}^{n \times n}$. Then, A and B are similar if and only if they have the same similarity invariants.

Proof. To prove necessity, assume that A and B are similar. Then, the matrices $sI - A$ and $sI - B$ have the same Smith form and thus the same similarity invariants. To prove sufficiency, it follows from Proposition 4.3.8 that there exist nonsingular matrices $S_1, S_2 \in \mathbb{F}^{n \times n}$ such that $sI - A = S_1(sI - B)S_2$. Thus, $S_1 = S_2^{-1}$, and, hence, $A = S_1 B S_1^{-1}$. $\qquad\square$

Corollary 4.3.11. Let $A \in \mathbb{F}^{n \times n}$. Then, A and A^{T} are similar.

An improved form of Corollary 4.3.11 is given by Corollary 5.3.8.

4.4 Eigenvalues

Let $A \in \mathbb{F}^{n \times n}$. Then, the polynomial matrix $sI - A \in \mathbb{F}^{n \times n}[s]$ is monic and has degree 1.

Definition 4.4.1. Let $A \in \mathbb{F}^{n \times n}$. Then, the *characteristic polynomial* of A is the polynomial $\chi_A \in \mathbb{F}[s]$ given by

$$\chi_A(s) \triangleq \det(sI - A). \tag{4.4.1}$$

Since $sI - A$ is a polynomial matrix, its determinant is the product of its Smith polynomials, that is, the similarity invariants of A.

Proposition 4.4.2. Let $A \in \mathbb{F}^{n \times n}$, and let $p_1, \ldots, p_n \in \mathbb{F}[s]$ denote the similarity invariants of A. Then,

$$\chi_A = \prod_{i=1}^{n} p_i. \tag{4.4.2}$$

Proposition 4.4.3. Let $A \in \mathbb{F}^{n \times n}$. Then, χ_A is monic and $\deg \chi_A = n$.

Let $A \in \mathbb{F}^{n \times n}$, and write the characteristic polynomial of A as

$$\chi_A(s) = s^n + \beta_{n-1} s^{n-1} + \cdots + \beta_1 s + \beta_0, \tag{4.4.3}$$

where $\beta_0, \ldots, \beta_{n-1} \in \mathbb{F}$. The *eigenvalues* of A are the n possibly repeated roots $\lambda_1, \ldots, \lambda_n \in \mathbb{C}$ of χ_A, that is, the solutions of the *characteristic equation*

$$\chi_A(s) = 0. \tag{4.4.4}$$

It is often convenient to denote the eigenvalues of A by $\lambda_1(A), \ldots, \lambda_n(A)$ or just $\lambda_1, \ldots, \lambda_n$. This notation may be ambiguous, however, since it does not uniquely specify which eigenvalue is denoted by λ_i. If, however, every eigenvalue of A is real, then we employ the notational convention

$$\lambda_1 \geq \cdots \geq \lambda_n, \tag{4.4.5}$$

and we define

$$\lambda_{\max}(A) \triangleq \lambda_1, \quad \lambda_{\min}(A) \triangleq \lambda_n. \tag{4.4.6}$$

Definition 4.4.4. Let $A \in \mathbb{F}^{n \times n}$. The *algebraic multiplicity* of an eigenvalue λ of A, denoted by $\mathrm{amult}_A(\lambda)$, is the algebraic multiplicity of λ as a root of χ_A, that is,

$$\mathrm{amult}_A(\lambda) \triangleq \mathrm{mult}_{\chi_A}(\lambda). \tag{4.4.7}$$

The multiset consisting of the eigenvalues of A including their algebraic multiplicity, denoted by $\mathrm{mspec}(A)$, is the *multispectrum* of A, that is,

$$\mathrm{mspec}(A) \triangleq \mathrm{mroots}(\chi_A). \tag{4.4.8}$$

Ignoring algebraic multiplicity, $\mathrm{spec}(A)$ denotes the *spectrum* of A, that is,

$$\mathrm{spec}(A) \triangleq \mathrm{roots}(\chi_A). \tag{4.4.9}$$

Note that

$$\mathrm{Szeros}(sI - A) = \mathrm{spec}(A) \tag{4.4.10}$$

and
$$\mathrm{mSzeros}(sI - A) = \mathrm{mspec}(A). \tag{4.4.11}$$

If $\lambda \notin \mathrm{spec}(A)$, then $\lambda \notin \mathrm{roots}(\chi_A)$, and thus $\mathrm{amult}_A(\lambda) = \mathrm{mult}_{\chi_A}(\lambda) = 0$.

Let $A \in \mathbb{F}^{n \times n}$ and $\mathrm{mroots}(\chi_A) = \{\lambda_1, \dots, \lambda_n\}_{\mathrm{ms}}$. Then,
$$\chi_A(s) = \prod_{i=1}^{n}(s - \lambda_i). \tag{4.4.12}$$

If $\mathbb{F} = \mathbb{R}$, then $\chi_A(s)$ has real coefficients, and thus the eigenvalues of A occur in complex conjugate pairs, that is, $\overline{\mathrm{mroots}(\chi_A)} = \mathrm{mroots}(\chi_A)$. Now, let $\mathrm{spec}(A) = \{\lambda_1, \dots, \lambda_r\}$, and, for all $i \in \{1, \dots, r\}$, let n_i denote the algebraic multiplicity of λ_i. Then,
$$\chi_A(s) = \prod_{i=1}^{r}(s - \lambda_i)^{n_i}. \tag{4.4.13}$$

The following result gives some basic properties of the spectrum of a matrix.

Proposition 4.4.5. Let $A, B \in \mathbb{F}^{n \times n}$. Then, the following statements hold:

i) $\chi_{A^{\mathrm{T}}} = \chi_A$.

ii) For all $s \in \mathbb{C}$, $\chi_{-A}(s) = (-1)^n \chi_A(-s)$.

iii) $\mathrm{mspec}(A^{\mathrm{T}}) = \mathrm{mspec}(A)$.

iv) $\mathrm{mspec}(\overline{A}) = \overline{\mathrm{mspec}(A)}$.

v) $\mathrm{mspec}(A^*) = \overline{\mathrm{mspec}(A)}$.

vi) $0 \in \mathrm{spec}(A)$ if and only if $\det A = 0$.

vii) If $k \in \mathbb{N}$ or if A is nonsingular and $k \in \mathbb{Z}$, then
$$\mathrm{mspec}(A^k) = \{\lambda^k \colon \ \lambda \in \mathrm{mspec}(A)\}_{\mathrm{ms}}. \tag{4.4.14}$$

viii) If $\alpha \in \mathbb{F}$, then $\chi_{\alpha A + I}(s) = \chi_A(s - \alpha)$.

ix) If $\alpha \in \mathbb{F}$, then $\mathrm{mspec}(\alpha I + A) = \alpha + \mathrm{mspec}(A)$.

x) If $\alpha \in \mathbb{F}$, then $\mathrm{mspec}(\alpha A) = \alpha \mathrm{mspec}(A)$.

xi) If A is Hermitian, then $\mathrm{spec}(A) \subset \mathbb{R}$.

xii) If A and B are similar, then $\chi_A = \chi_B$ and $\mathrm{mspec}(A) = \mathrm{mspec}(B)$.

Proof. To prove *i)*, note that
$$\det(sI - A^{\mathrm{T}}) = \det(sI - A)^{\mathrm{T}} = \det(sI - A).$$
To prove *ii)*, note that
$$\chi_{-A}(s) = \det(sI + A) = (-1)^n \det(-sI - A) = (-1)^n \chi_A(-s).$$

Next, *iii*) follows from *i*). Next, *iv*) follows from

$$\det(sI - \overline{A}) = \det(\overline{s}I - A) = \overline{\det(\overline{s}I - A)},$$

while *v*) follows from *iii*) and *iv*).

Next, *vi*) follows from the fact that $\chi_A(0) = (-1)^n \det A$. To prove "$\supseteq$" in *vii*), note that, if $\lambda \in \operatorname{spec}(A)$ and $x \in \mathbb{C}^n$ is an eigenvector of A associated with λ (see Section 4.5), then $A^2x = A(Ax) = A(\lambda x) = \lambda Ax = \lambda^2 x$. Similarly, if A is nonsingular, then $Ax = \lambda x$ implies that $A^{-1}x = \lambda^{-1}x$, and thus $A^{-2}x = \lambda^{-2}x$. Similar arguments apply to arbitrary $k \in \mathbb{Z}$. The reverse inclusion follows from the Jordan decomposition given by Theorem 5.3.3.

To prove *viii*), note that

$$\chi_{\alpha I + A}(s) = \det[sI - (\alpha I + A)] = \det[(s - \alpha)I - A] = \chi_A(s - \alpha).$$

Statement *ix*) follows immediately.

Statement *x*) is true for $\alpha = 0$. For $\alpha \neq 0$, it follows that

$$\chi_{\alpha A}(s) = \det(sI - \alpha A) = \alpha^n \det[(s/\alpha)I - A] = \alpha^n \chi_A(s/\alpha).$$

To prove *xi*), assume that $A = A^*$, let $\lambda \in \operatorname{spec}(A)$, and let $x \in \mathbb{C}^n$ be an eigenvector of A associated with λ. Then, $\lambda = x^*Ax/x^*x$, which is real. Finally, *xii*) is immediate. $\qquad\square$

The following result characterizes the coefficients of χ_A in terms of the eigenvalues of A.

Proposition 4.4.6. Let $A \in \mathbb{F}^{n \times n}$, let $\operatorname{mspec}(A) = \{\lambda_1, \ldots, \lambda_n\}_{\mathrm{ms}}$, and, for all $i \in \{1, \ldots, n\}$, let γ_i denote the sum of all $i \times i$ principal subdeterminants of A. Then, for all $i \in \{1, \ldots, n-1\}$,

$$\gamma_i = \sum_{1 \leq j_1 < \cdots < j_i \leq n} \lambda_{j_1} \cdots \lambda_{j_i}. \tag{4.4.15}$$

Furthermore, for all $i \in \{0, \ldots, n-1\}$, the coefficient β_i of s^i in (4.4.3) is given by

$$\beta_i = (-1)^{n-i} \gamma_{n-i}. \tag{4.4.16}$$

In particular,

$$\beta_{n-1} = -\operatorname{tr} A = -\sum_{i=1}^{n} \lambda_i, \tag{4.4.17}$$

$$\beta_{n-2} = \tfrac{1}{2}\left[(\operatorname{tr} A)^2 - \operatorname{tr} A^2\right] = \sum_{1 \leq j_1 < j_2 \leq n} \lambda_{j_1}\lambda_{j_2}, \tag{4.4.18}$$

$$\beta_1 = (-1)^{n-1}\operatorname{tr} A^{\mathrm{A}} = (-1)^{n-1} \sum_{1 \leq j_1 < \cdots < j_{n-1} \leq n} \lambda_{j_1} \cdots \lambda_{j_{n-1}} = (-1)^{n-1} \sum_{i=1}^{n} \det A_{[i;i]}, \tag{4.4.19}$$

$$\beta_0 = (-1)^n \det A = (-1)^n \prod_{i=1}^{n} \lambda_i. \tag{4.4.20}$$

Proof. The expression for γ_i given by (4.4.15) follows from the factored form of $\chi_A(s)$ given by (4.4.12), while the expression for β_i given by (4.4.16) follows by examining the cofactor expansion (2.7.16) of $\det(sI - A)$. For details, see [1023, p. 495]. Equation (4.4.17) follows from (4.4.16) and the fact that the $(n-1) \times (n-1)$ principal subdeterminants of A are the diagonal entries $A_{(i,i)}$. Using

$$\sum_{i=1}^{n} \lambda_i^2 = \left(\sum_{i=1}^{n} \lambda_i \right)^2 - 2 \sum \lambda_{j_1} \lambda_{j_2},$$

where the third summation is taken over all pairs of elements of $\mathrm{mspec}(A)$, and (4.4.17) yields (4.4.18). Next, if A is nonsingular, then $\chi_{A^{-1}}(s) = (-s)^n (\det A^{-1}) \chi_A(1/s)$. Using (4.4.3) with s replaced by $1/s$ and (4.4.17), it follows that $\operatorname{tr} A^{-1} = (-1)^{n-1} (\det A^{-1}) \beta_1$, and, hence, (4.4.19) is satisfied. Using continuity for the case in which A is singular yields (4.4.19) for arbitrary A. Finally, $\beta_0 = \chi_A(0) = \det(0I - A) = (-1)^n \det A$, which verifies (4.4.20). $\qquad\square$

From the definition of the adjugate of a matrix it follows that $(sI - A)^{\mathrm{A}} \in \mathbb{F}^{n \times n}[s]$ is a monic polynomial matrix of degree $n-1$ of the form

$$(sI - A)^{\mathrm{A}} = s^{n-1}I + s^{n-2}B_{n-2} + \cdots + sB_1 + B_0, \tag{4.4.21}$$

where $B_0, B_1, \ldots, B_{n-2} \in \mathbb{F}^{n \times n}$. Since $(sI - A)^{\mathrm{A}}$ is regular, it follows from Proposition 4.2.6 that $(sI - A)^{\mathrm{A}}$ is a nonsingular polynomial matrix. The matrix $(sI - A)^{-1}$ is the *resolvent* of A, which is given by

$$(sI - A)^{-1} = \frac{1}{\chi_A(s)} (sI - A)^{\mathrm{A}}. \tag{4.4.22}$$

Therefore,

$$(sI - A)^{-1} = \frac{s^{n-1}}{\chi_A(s)} I + \frac{s^{n-2}}{\chi_A(s)} B_{n-2} + \cdots + \frac{s}{\chi_A(s)} B_1 + \frac{1}{\chi_A(s)} B_0. \tag{4.4.23}$$

The next result is the *Cayley-Hamilton theorem*, which shows that every matrix is a "root" of its characteristic polynomial.

Theorem 4.4.7. Let $A \in \mathbb{F}^{n \times n}$. Then,

$$\chi_A(A) = 0. \tag{4.4.24}$$

Proof. Define $P, Q \in \mathbb{F}^{n \times n}[s]$ by $P(s) \triangleq \chi_A(s)I$ and $Q(s) \triangleq (sI - A)^{\mathrm{A}}$. Then, (4.4.22) implies that $P(s) = Q(s)(sI - A)$. It thus follows from Lemma 4.2.2 that $P_{\mathrm{R}}(A) = 0$. Furthermore, $\chi_A(A) = P(A) = P_{\mathrm{R}}(A)$. Hence, $\chi_A(A) = 0$. $\qquad\square$

In the notation of (4.4.13), it follows from Theorem 4.4.7 that

$$\prod_{i=1}^{r} (\lambda_i I - A)^{n_i} = 0. \tag{4.4.25}$$

Lemma 4.4.8. Let $A \in \mathbb{F}^{n \times n}$. Then,

$$\frac{\mathrm{d}}{\mathrm{d}s}\chi_A(s) = \mathrm{tr}\left[(sI - A)^{\mathrm{A}}\right] = \sum_{i=1}^{n} \det\left(sI - A_{[i;i]}\right). \tag{4.4.26}$$

Proof. It follows from (4.4.19) that $\frac{\mathrm{d}}{\mathrm{d}s}\chi_A(s)\big|_{s=0} = \beta_1 = (-1)^{n-1}\mathrm{tr}\,A^{\mathrm{A}}$. Hence,

$$\frac{\mathrm{d}}{\mathrm{d}s}\chi_A(s) = \frac{\mathrm{d}}{\mathrm{d}z}\det[(s+z)I - A]\bigg|_{z=0} = \frac{\mathrm{d}}{\mathrm{d}z}\det[zI - (-sI + A)]\bigg|_{z=0}$$

$$= (-1)^{n-1}\mathrm{tr}\left[(-sI + A)^{\mathrm{A}}\right] = \mathrm{tr}\left[(sI - A)^{\mathrm{A}}\right]. \qquad \square$$

The following result, known as *Leverrier's algorithm*, provides a recursive formula for the coefficients $\beta_0, \ldots, \beta_{n-1}$ of χ_A and $B_0, \ldots, B_{n-2}$ of $(sI - A)^{\mathrm{A}}$.

Proposition 4.4.9. Let $A \in \mathbb{F}^{n \times n}$, let χ_A be given by (4.4.3), and let $(sI - A)^{\mathrm{A}}$ be given by (4.4.21). Then, $\beta_{n-1}, \ldots, \beta_0$ and $B_{n-2}, \ldots, B_0$ are given by

$$\beta_k = \frac{1}{k-n}\mathrm{tr}\,AB_k, \quad k = n-1, \ldots, 0, \tag{4.4.27}$$

$$B_{k-1} = AB_k + \beta_k I, \quad k = n-1, \ldots, 1, \tag{4.4.28}$$

where $B_{n-1} = I$.

Proof. Since $(sI - A)(sI - A)^{\mathrm{A}} = \chi_A(s)I$, it follows that

$$s^n I + s^{n-1}(B_{n-2} - A) + s^{n-2}(B_{n-3} - AB_{n-2}) + \cdots + s(B_0 - AB_1) - AB_0$$
$$= (s^n + \beta_{n-1}s^{n-1} + \cdots + \beta_1 s + \beta_0)I.$$

Equating coefficients of powers of s yields (4.4.28) along with $-AB_0 = \beta_0 I$. Taking the trace of this last equality yields $\beta_0 = -\frac{1}{n}\mathrm{tr}\,AB_0$, which confirms (4.4.27) for $k = 0$. Next, using (4.4.26) and (4.4.21), it follows that

$$\frac{\mathrm{d}}{\mathrm{d}s}\chi_A(s) = \sum_{k=1}^{n} k\beta_k s^{k-1} = \sum_{k=1}^{n} (\mathrm{tr}\,B_{k-1})s^{k-1},$$

where $B_{n-1} \triangleq I_n$ and $\beta_n \triangleq 1$. Equating powers of s, it follows that $k\beta_k = \mathrm{tr}\,B_{k-1}$ for all $k \in \{1, \ldots, n\}$. Now, (4.4.28) implies that $k\beta_k = \mathrm{tr}(AB_k + \beta_k I)$ for all $k \in \{1, \ldots, n-1\}$, which implies (4.4.27). $\qquad \square$

Proposition 4.4.10. Let $A \in \mathbb{F}^{n \times m}$ and $B \in \mathbb{F}^{m \times n}$, and assume that $m \leq n$. Then,

$$\chi_{AB}(s) = s^{n-m}\chi_{BA}(s). \tag{4.4.29}$$

Consequently, $\quad \mathrm{mspec}(AB) = \mathrm{mspec}(BA) \cup \{0, \ldots, 0\}_{\mathrm{ms}}, \tag{4.4.30}$

where the multiset $\{0, \ldots, 0\}_{\mathrm{ms}}$ contains $n - m$ 0's.

Proof. First note that

$$\begin{bmatrix} 0_{m \times m} & 0_{m \times n} \\ A & AB \end{bmatrix} = \begin{bmatrix} I_m & -B \\ 0_{n \times m} & I_n \end{bmatrix} \begin{bmatrix} BA & 0_{m \times n} \\ A & 0_{n \times n} \end{bmatrix} \begin{bmatrix} I_m & B \\ 0_{n \times m} & I_n \end{bmatrix},$$

which shows that $\begin{bmatrix} 0_{m \times m} & 0_{m \times n} \\ A & AB \end{bmatrix}$ and $\begin{bmatrix} BA & 0_{m \times n} \\ A & 0_{n \times n} \end{bmatrix}$ are similar. It thus follows from $xi)$ of Proposition 4.4.5 that $s^m \chi_{AB}(s) = s^n \chi_{BA}(s)$, which implies (4.4.29). Finally, (4.4.30) follows immediately from (4.4.29). $\square$

If $n = m$, then Proposition 4.4.10 specializes to the following result.

Corollary 4.4.11. Let $A, B \in \mathbb{F}^{n \times n}$. Then,

$$\chi_{AB} = \chi_{BA}. \tag{4.4.31}$$

Consequently,

$$\mathrm{mspec}(AB) = \mathrm{mspec}(BA). \tag{4.4.32}$$

We define the *spectral abscissa* of $A \in \mathbb{F}^{n \times n}$ by

$$\mathrm{spabs}(A) \triangleq \max\{\mathrm{Re}\,\lambda: \ \lambda \in \mathrm{spec}(A)\} \tag{4.4.33}$$

and the *spectral radius* of $A \in \mathbb{F}^{n \times n}$ by

$$\mathrm{sprad}(A) \triangleq \max\{|\lambda|: \ \lambda \in \mathrm{spec}(A)\}. \tag{4.4.34}$$

Let $A \in \mathbb{F}^{n \times n}$. Then, $\nu_-(A), \nu_0(A)$, and $\nu_+(A)$ denote the number of eigenvalues of A counting algebraic multiplicity having, respectively, negative, zero, and positive real part. Define the *inertia* of A by

$$\mathrm{In}\, A \triangleq \begin{bmatrix} \nu_-(A) \\ \nu_0(A) \\ \nu_+(A) \end{bmatrix} \tag{4.4.35}$$

and the *signature* of A by

$$\mathrm{sig}\, A \triangleq \nu_+(A) - \nu_-(A). \tag{4.4.36}$$

Note that $\mathrm{spabs}(A) < 0$ if and only if $\nu_-(A) = n$, while $\mathrm{spabs}(A) = 0$ if and only if $\nu_+(A) = 0$.

4.5 Eigenvectors

Let $A \in \mathbb{F}^{n \times n}$, and let $\lambda \in \mathbb{C}$ be an eigenvalue of A. Then, $\chi_A(\lambda) = \det(\lambda I - A) = 0$, and thus $\lambda I - A \in \mathbb{C}^{n \times n}$ is singular. Furthermore, $\mathcal{N}(\lambda I - A)$ is a nontrivial subspace of $\mathbb{C}^n$, that is, $\mathrm{def}(\lambda I - A) > 0$. If $x \in \mathcal{N}(\lambda I - A)$, that is, $Ax = \lambda x$, and $x \neq 0$, then x is an *eigenvector of A associated with* λ. By definition, all eigenvectors are nonzero. Note that, if A and λ are real, then there exists a real eigenvector associated with λ.

Definition 4.5.1. The *geometric multiplicity* of $\lambda \in \mathrm{spec}(A)$, denoted by $\mathrm{gmult}_A(\lambda)$, is the number of linearly independent eigenvectors associated with λ, that is,

$$\mathrm{gmult}_A(\lambda) \triangleq \mathrm{def}(\lambda I - A). \tag{4.5.1}$$

By convention, if $\lambda \notin \mathrm{spec}(A)$, then $\mathrm{gmult}_A(\lambda) \triangleq 0$.

Proposition 4.5.2. Let $A \in \mathbb{F}^{n \times n}$, and let $\lambda \in \operatorname{spec}(A)$. Then, the following statements hold:

i) $\operatorname{rank}(\lambda I - A) + \operatorname{gmult}_A(\lambda) = n$.

ii) $\operatorname{def} A = \operatorname{gmult}_A(0)$.

iii) $\operatorname{rank} A + \operatorname{gmult}_A(0) = n$.

The spectral properties of normal matrices deserve special attention.

Lemma 4.5.3. Let $A \in \mathbb{F}^{n \times n}$ be normal, let $\lambda \in \operatorname{spec}(A)$, and let $x \in \mathbb{C}^n$ be an eigenvector of A associated with λ. Then, x is an eigenvector of A^* associated with $\overline{\lambda} \in \operatorname{spec}(A^*)$.

Proof. Since $\lambda \in \operatorname{spec}(A)$, statement *v)* of Proposition 4.4.5 implies that $\overline{\lambda} \in \operatorname{spec}(A^*)$. Next, since x and λ satisfy $Ax = \lambda x$, $x^* A^* = \overline{\lambda} x^*$, and $AA^* = A^*A$, it follows that

$$(A^* x - \overline{\lambda} x)^*(A^* x - \overline{\lambda} x) = x^* A A^* x - \overline{\lambda} x^* A x - \lambda x^* A^* x + \lambda \overline{\lambda} x^* x$$
$$= x^* A^* A x - \lambda \overline{\lambda} x^* x - \lambda \overline{\lambda} x^* x + \lambda \overline{\lambda} x^* x$$
$$= \lambda \overline{\lambda} x^* x - \lambda \overline{\lambda} x^* x = 0.$$

Hence, $A^* x = \overline{\lambda} x$. $\qquad \square$

Proposition 4.5.4. Let $A \in \mathbb{F}^{n \times n}$. Then, eigenvectors associated with distinct eigenvalues of A are linearly independent. If, in addition, A is normal, then these eigenvectors are mutually orthogonal.

Proof. Let $\lambda_1, \lambda_2 \in \operatorname{spec}(A)$ be distinct with associated eigenvectors $x_1, x_2 \in \mathbb{C}^n$. Suppose that x_1 and x_2 are linearly dependent, that is, $x_1 = \alpha x_2$, where $\alpha \in \mathbb{C}$ and $\alpha \neq 0$. Then, $Ax_1 = \lambda_1 x_1 = \lambda_1 \alpha x_2$, while also $Ax_1 = A\alpha x_2 = \alpha \lambda_2 x_2$. Hence, $\alpha(\lambda_1 - \lambda_2) x_2 = 0$, which contradicts $\alpha \neq 0$. Since pairwise linear independence does not imply the linear independence of larger sets, next, let $\lambda_1, \lambda_2, \lambda_3 \in \operatorname{spec}(A)$ be distinct with associated eigenvectors $x_1, x_2, x_3 \in \mathbb{C}^n$. Suppose that x_1, x_2, x_3 are linearly dependent. In this case, there exist $a_1, a_2, a_3 \in \mathbb{C}$, not all zero, such that $a_1 x_1 + a_2 x_2 + a_3 x_3 = 0$. If $a_1 = 0$, then $a_2 x_2 + a_3 x_3 = 0$. However, $\lambda_2 \neq \lambda_3$ implies that x_2 and x_3 are linearly independent, which in turn implies that $a_2 = 0$ and $a_3 = 0$. Since a_1, a_2, a_3 are not all zero, it follows that $a_1 \neq 0$. Therefore, $x_1 = \alpha x_2 + \beta x_3$, where $\alpha \triangleq -a_2/a_1$ and $\beta \triangleq -a_3/a_1$ are not both zero. Thus, $Ax_1 = A(\alpha x_2 + \beta x_3) = \alpha Ax_2 + \beta Ax_3 = \alpha \lambda_2 x_2 + \beta \lambda_3 x_3$. However, $Ax_1 = \lambda_1 x_1 = \lambda_1(\alpha x_2 + \beta x_3) = \alpha \lambda_1 x_2 + \beta \lambda_1 x_3$. Subtracting these relations yields $0 = \alpha(\lambda_1 - \lambda_2)x_2 + \beta(\lambda_1 - \lambda_3)x_3$. Since x_2 and x_3 are linearly independent, it follows that $\alpha(\lambda_1 - \lambda_2) = 0$ and $\beta(\lambda_1 - \lambda_3) = 0$. Since α and β are not both zero, it follows that $\lambda_1 = \lambda_2$ or $\lambda_1 = \lambda_3$, which contradicts the assumption that $\lambda_1, \lambda_2, \lambda_3$ are distinct. The same arguments apply to sets of four or more eigenvectors.

Now, suppose that A is normal, and let $\lambda_1, \lambda_2 \in \operatorname{spec}(A)$ be distinct eigenvalues with associated eigenvectors $x_1, x_2 \in \mathbb{C}^n$. Then, by Lemma 4.5.3, $Ax_1 = \lambda_1 x_1$ implies that $A^* x_1 = \overline{\lambda}_1 x_1$. Consequently, $x_1^* A = \lambda_1 x_1^*$, which implies that $x_1^* A x_2 = \lambda_1 x_1^* x_2$. Furthermore, $x_1^* A x_2 = \lambda_2 x_1^* x_2$. It thus follows that $0 = (\lambda_1 - \lambda_2) x_1^* x_2$.

Hence, $\lambda_1 \neq \lambda_2$ implies that $x_1^* x_2 = 0$. $\qquad\square$

If $A \in \mathbb{R}^{n \times n}$ is symmetric, then Lemma 4.5.3 is not needed and the proof of Proposition 4.5.4 is simpler. In this case, it follows from x) of Proposition 4.4.5 that $\lambda_1, \lambda_2 \in \operatorname{spec}(A)$ are real, and thus associated eigenvectors $x_1 \in \mathcal{N}(\lambda_1 I - A)$ and $x_2 \in \mathcal{N}(\lambda_2 I - A)$ can be chosen to be real. Hence, $Ax_1 = \lambda_1 x_1$ and $Ax_2 = \lambda_2 x_2$ imply that $x_2^{\mathrm{T}} A x_1 = \lambda_1 x_2^{\mathrm{T}} x_1$ and $x_1^{\mathrm{T}} A x_2 = \lambda_2 x_1^{\mathrm{T}} x_2$. Since $x_1^{\mathrm{T}} A x_2 = x_2^{\mathrm{T}} A^{\mathrm{T}} x_1 = x_2^{\mathrm{T}} A x_1$ and $x_1^{\mathrm{T}} x_2 = x_2^{\mathrm{T}} x_1$, it follows that $(\lambda_1 - \lambda_2) x_1^{\mathrm{T}} x_2 = 0$. Since $\lambda_1 \neq \lambda_2$, it follows that $x_1^{\mathrm{T}} x_2 = 0$.

4.6 The Minimal Polynomial

Theorem 4.4.7 showed that every square matrix $A \in \mathbb{F}^{n \times n}$ is a root of its characteristic polynomial. However, there may be polynomials of degree less than n having A as a root. In fact, the following result shows that there exists a unique monic polynomial that has A as a root and that divides all polynomials that have A as a root.

Theorem 4.6.1. Let $A \in \mathbb{F}^{n \times n}$. Then, there exists a unique monic polynomial $\mu_A \in \mathbb{F}[s]$ of minimal degree such that $\mu_A(A) = 0$. Furthermore, $\deg \mu_A \leq n$, and μ_A divides every polynomial $p \in \mathbb{F}[s]$ satisfying $p(A) = 0$.

Proof. Since $\chi_A(A) = 0$ and $\deg \chi_A = n$, it follows that there exists a minimal positive integer $n_0 \leq n$ such that there exists a monic polynomial $p_0 \in \mathbb{F}[s]$ satisfying $p_0(A) = 0$ and $\deg p_0 = n_0$. Let $p \in \mathbb{F}[s]$ satisfy $p(A) = 0$. Then, by Lemma 4.1.2, there exist polynomials $q, r \in \mathbb{F}[s]$ such that $p = q p_0 + r$ and $\deg r < \deg p_0$. However, $p(A) = p_0(A) = 0$ implies that $r(A) = 0$. If $r \neq 0$, then r can be normalized to obtain a monic polynomial of degree less than n_0, which contradicts the definition n_0. Hence, $r = 0$, which implies that p_0 divides p. This proves existence.

Now, suppose there exist two monic polynomials $p_0, \hat{p}_0 \in \mathbb{F}[s]$ of degree n_0 and such that $p_0(A) = \hat{p}_0(A) = 0$. By the previous argument, p_0 divides $\hat{p}_0$, and vice versa. Therefore, p_0 is a constant multiple of $\hat{p}_0$. Since p_0 and $\hat{p}_0$ are both monic, it follows that $p_0 = \hat{p}_0$. This proves uniqueness. Denote this polynomial by μ_A. $\qquad\square$

The monic polynomial μ_A of smallest degree having A as a root is the *minimal polynomial* of A.

The following result relates the characteristic polynomial and minimal polynomial of $A \in \mathbb{F}^{n \times n}$ to the similarity invariants of A. Note that $\operatorname{rank}(sI - A) = n$, so that A has n similarity invariants $p_1, \ldots, p_n \in \mathbb{F}[s]$. In this case, (4.3.1) becomes

$$sI - A = S_1(s) \begin{bmatrix} p_1(s) & & 0 \\ & \ddots & \\ 0 & & p_n(s) \end{bmatrix} S_2(s), \qquad (4.6.1)$$

where $S_1, S_2 \in \mathbb{F}^{n \times n}[s]$ are unimodular and p_i divides p_{i+1} for all $i \in \{1, \ldots, n-1\}$.

Proposition 4.6.2. Let $A \in \mathbb{F}^{n \times n}$, and let $p_1, \ldots, p_n \in \mathbb{F}[s]$ be the similarity invariants of A, where p_i divides p_{i+1} for all $i \in \{1, \ldots, n-1\}$. Then,

$$\chi_A = \prod_{i=1}^{n} p_i \tag{4.6.2}$$

and

$$\mu_A = p_n. \tag{4.6.3}$$

Proof. Using Theorem 4.3.2 and (4.6.1), it follows that

$$\chi_A(s) = \det(sI - A) = [\det S_1(s)] [\det S_2(s)] \prod_{i=1}^{n} p_i(s).$$

Since S_1 and S_2 are unimodular and χ_A and $p_1, \ldots, p_n$ are monic, it follows that $[\det S_1(s)][\det S_2(s)] = 1$, which proves (4.6.2).

To prove (4.6.3), first note that it follows from Theorem 4.3.2 that $\chi_A = \Delta_{n-1} p_n$, where $\Delta_{n-1} \in \mathbb{F}[s]$ is the greatest common divisor of all $(n-1) \times (n-1)$ subdeterminants of $sI - A$. Since the $(n-1) \times (n-1)$ subdeterminants of $sI - A$ are the entries of $\pm(sI - A)^{\mathrm{A}}$, it follows that Δ_{n-1} divides every entry of $(sI - A)^{\mathrm{A}}$. Hence, there exists a polynomial matrix $P \in \mathbb{F}^{n \times n}[s]$ such that $(sI - A)^{\mathrm{A}} = \Delta_{n-1}(s)P(s)$. Furthermore, since $(sI - A)^{\mathrm{A}}(sI - A) = \chi_A(s)I$, it follows that $\Delta_{n-1}(s)P(s)(sI - A) = \chi_A(s)I = \Delta_{n-1}(s)p_n(s)I$, and thus $P(s)(sI - A) = p_n(s)I$. Lemma 4.2.2 now implies that $p_n(A) = 0$.

Since $p_n(A) = 0$, it follows from Theorem 4.6.1 that μ_A divides p_n. Hence, let $q \in \mathbb{F}[s]$ be the monic polynomial satisfying $p_n = q\mu_A$. Furthermore, since $\mu_A(A) = 0$, it follows from Corollary 4.2.3 that there exists a polynomial matrix $Q \in \mathbb{F}^{n \times n}[s]$ such that $\mu_A(s)I = Q(s)(sI - A)$. Thus, $P(s)(sI - A) = p_n(s)I = q(s)\mu_A(s)I = q(s)Q(s)(sI - A)$, which implies that $P = qQ$. Thus, q divides every entry of P. However, since P is obtained by dividing $(sI - A)^{\mathrm{A}}$ by the greatest common divisor of all of its entries, it follows that the greatest common divisor of the entries of P is 1. Hence, $q = 1$, which implies that $p_n = \mu_A$, which proves (4.6.3). $\square$

Proposition 4.6.2 shows that μ_A divides χ_A, which is also a consequence of Theorem 4.4.7 and Theorem 4.6.1. Proposition 4.6.2 also shows that $\mu_A = \chi_A$ if and only if $p_1 = \cdots = p_{n-1} = 1$, that is, if and only if $p_n = \chi_A$ is the only nonconstant similarity invariant of A. Note that, in general, it follows from (4.6.2) that $\sum_{i=1}^{n} \deg p_i = n$.

Finally, note that the similarity invariants of the $n \times n$ identity matrix I_n are given by $p_i(s) = s - 1$ for all $i \in \{1, \ldots, n\}$. Thus, $\chi_{I_n}(s) = (s-1)^n$ and $\mu_{I_n}(s) = s - 1$.

Proposition 4.6.3. Let $A \in \mathbb{F}^{n \times n}$, and assume that A and B are similar. Then,

$$\mu_A = \mu_B. \tag{4.6.4}$$

4.7 Rational Transfer Functions and the Smith-McMillan Decomposition

We now turn our attention to rational functions.

Definition 4.7.1. The set $\mathbb{F}(s)$ of *rational functions* consists of functions $g\colon \mathbb{C}\backslash\mathcal{S} \mapsto \mathbb{C}$, where $g(s) = p(s)/q(s)$, $p, q \in \mathbb{F}[s]$, $q \neq 0$, and $\mathcal{S} \triangleq \text{roots}(q)$. The rational function g is *strictly proper, proper, exactly proper, improper*, respectively, if $\deg p < \deg q$, $\deg p \leq \deg q$, $\deg p = \deg q$, $\deg p > \deg q$. If p and q are coprime, then the *zeros* of g are the elements of $\text{mroots}(p)$, while the *poles* of g are the elements of $\text{mroots}(q)$. The set of proper rational functions is denoted by $\mathbb{F}_{\text{prop}}(s)$. The *relative degree* of $g \in \mathbb{F}_{\text{prop}}(s)$, denoted by $\text{reldeg}\, g$, is $\deg q - \deg p$.

Definition 4.7.2. The set $\mathbb{F}^{l\times m}(s)$ of *rational transfer functions* consists of matrices whose entries are elements of $\mathbb{F}(s)$. The rational transfer function $G \in \mathbb{F}^{l\times m}(s)$ is *strictly proper* if every entry of G is strictly proper, *proper* if every entry of G is proper, *exactly proper* if every entry of G is proper and at least one entry of G is exactly proper, and *improper* if at least one entry of G is improper. The set of proper rational transfer functions is denoted by $\mathbb{F}^{l\times m}_{\text{prop}}(s)$.

Definition 4.7.3. Let $G \in \mathbb{F}^{l\times m}_{\text{prop}}(s)$. Then, the *relative degree* of G, denoted by $\text{reldeg}\, G$, is defined by

$$\text{reldeg}\, G \triangleq \min_{\substack{i=1,\ldots,l \\ j=1,\ldots,m}} \text{reldeg}\, G_{(i,j)}. \tag{4.7.1}$$

By writing $(sI - A)^{-1}$ as

$$(sI - A)^{-1} = \frac{1}{\chi_A(s)}(sI - A)^{\mathrm{A}}, \tag{4.7.2}$$

it follows from (4.4.21) that $(sI - A)^{-1}$ is a strictly proper rational transfer function. In fact, for all $i \in \{1, \ldots, n\}$,

$$\text{reldeg}\, \left[(sI - A)^{-1}\right]_{(i,i)} = 1, \tag{4.7.3}$$

and thus

$$\text{reldeg}\, (sI - A)^{-1} = 1. \tag{4.7.4}$$

The following definition is an extension of Definition 4.2.4 to rational transfer functions.

Definition 4.7.4. Let $G \in \mathbb{F}^{l\times m}(s)$, and, for all $i \in \{1, \ldots, l\}$ and $j \in \{1, \ldots, m\}$, let $G_{(i,j)} = p_{ij}/q_{ij}$, where $q_{ij} \neq 0$, and $p_{ij}, q_{ij} \in \mathbb{F}[s]$ are coprime. Then, the *poles* of G are the elements of the set

$$\text{poles}(G) \triangleq \bigcup_{i,j=1}^{l,m} \text{roots}(q_{ij}), \tag{4.7.5}$$

and the *blocking zeros* of G are the elements of the set

$$\text{bzeros}(G) \triangleq \bigcap_{i,j=1}^{l,m} \text{roots}(p_{ij}). \tag{4.7.6}$$

Finally, the rank of G is the nonnegative integer

$$\text{rank}\,G \triangleq \max_{s \in \mathbb{C} \backslash \text{poles}(G)} \text{rank}\,G(s). \tag{4.7.7}$$

The following result provides a canonical form, known as the *Smith-McMillan form*, for rational transfer functions under unimodular transformation.

Theorem 4.7.5. Let $G \in \mathbb{F}^{l \times m}(s)$, and let $r \triangleq \text{rank}\,G$. Then, there exist unimodular matrices $S_1 \in \mathbb{F}^{l \times l}[s]$ and $S_2 \in \mathbb{F}^{m \times m}[s]$ and monic polynomials $p_1, \ldots, p_r, q_1, \ldots, q_r \in \mathbb{F}[s]$ such that p_i and q_i are coprime for all $i \in \{1, \ldots, r\}$, p_i divides p_{i+1} for all $i \in \{1, \ldots, r-1\}$, q_{i+1} divides q_i for all $i \in \{1, \ldots, r-1\}$, and

$$G = S_1 \begin{bmatrix} p_1/q_1 & & & \\ & \ddots & & 0_{r \times (m-r)} \\ & & p_r/q_r & \\ & 0_{(l-r) \times r} & & 0_{(l-r) \times (m-r)} \end{bmatrix} S_2. \tag{4.7.8}$$

Proof. Let n_{ij}/d_{ij} denote the (i,j) entry of G, where $n_{ij}, d_{ij} \in \mathbb{F}[s]$ are coprime, and let $d \in \mathbb{F}[s]$ denote the least common multiple of d_{ij} for all $i \in \{1, \ldots, l\}$ and $j \in \{1, \ldots, m\}$. From Theorem 4.3.2 it follows that the polynomial matrix dG has the Smith form $\text{diag}(\hat{p}_1, \ldots, \hat{p}_r, 0, \ldots, 0)$, where $\hat{p}_1, \ldots, \hat{p}_r \in \mathbb{F}[s]$ and $\hat{p}_i$ divides $\hat{p}_{i+1}$ for all $i \in \{1, \ldots, r-1\}$. Now, divide this Smith form by d and express every rational function $\hat{p}_i/d$ in coprime form p_i/q_i so that p_i divides p_{i+1} for all $i \in \{1, \ldots, r-1\}$ and q_{i+1} divides q_i for all $i \in \{1, \ldots, r-1\}$. $\qquad\square$

Proposition 4.7.6. Let $G \in \mathbb{F}^{l \times m}(s)$, and assume that there exist unimodular matrices $S_1 \in \mathbb{F}^{l \times l}[s]$ and $S_2 \in \mathbb{F}^{m \times m}[s]$ and monic polynomials $p_1, \ldots, p_r, q_1, \ldots, q_r \in \mathbb{F}[s]$ such that p_i and q_i are coprime for all $i \in \{1, \ldots, r\}$ and such that (4.7.8) holds. Then, $\text{rank}\,G = r$.

Proposition 4.7.7. Let $G \in \mathbb{F}^{n \times m}[s]$, and let $r \triangleq \text{rank}\,G$. Then, r is the largest order of all nonsingular submatrices of G.

Proposition 4.7.8. Let $G \in \mathbb{F}^{n \times m}(s)$, and let $\mathcal{S} \subset \mathbb{C}$ be a finite set such that $\text{poles}(G) \subseteq \mathcal{S}$. Then,

$$\text{rank}\,G = \max_{s \in \mathbb{C} \backslash \mathcal{S}} \text{rank}\,G(s). \tag{4.7.9}$$

Let $g_1, \ldots, g_r \in \mathbb{F}^n(s)$. Then, $g_1, \ldots, g_r$ are *linearly independent* if $\alpha_1, \ldots, \alpha_r \in \mathbb{F}[s]$ and $\sum_{n=1}^r \alpha_i g_i = 0$ imply that $\alpha_1 = \cdots = \alpha_r = 0$. Equivalently, $g_1, \ldots, g_r$ are *linearly independent* if $\alpha_1, \ldots, \alpha_r \in \mathbb{F}(s)$ and $\sum_{n=1}^r \alpha_i g_i = 0$ imply that $\alpha_1 = \cdots = \alpha_r = 0$. In other words, the coefficients α_i can be either polynomials or rational functions.

Proposition 4.7.9. Let $G \in \mathbb{F}^{l \times m}(s)$. Then, $\operatorname{rank} G$ is equal to the number of linearly independent columns of G.

Since $G \in \mathbb{F}^{l \times m}[s] \subset \mathbb{F}^{l \times m}(s)$, Proposition 4.7.9 applies to polynomial matrices.

Definition 4.7.10. Let $G \in \mathbb{F}^{l \times m}(s)$, assume that $G \neq 0$, let $r \triangleq \operatorname{rank} G$, and let $p_1, \ldots, p_r, q_1, \ldots, q_r \in \mathbb{F}[s]$ be given by Theorem 4.7.5. Then, the *McMillan degree* $\operatorname{Mcdeg} G$ of G is defined by

$$\operatorname{Mcdeg} G \triangleq \sum_{i=1}^{r} \deg q_i. \tag{4.7.10}$$

Furthermore, the *transmission zeros* of G are the elements of the set

$$\operatorname{tzeros}(G) \triangleq \operatorname{roots}(p_r). \tag{4.7.11}$$

Proposition 4.7.11. Let $G \in \mathbb{F}^{l \times m}(s)$, assume that $G \neq 0$, and assume that G has the Smith-McMillan form (4.7.8). Then,

$$\operatorname{poles}(G) = \operatorname{roots}(q_1) \tag{4.7.12}$$

and

$$\operatorname{bzeros}(G) = \operatorname{roots}(p_1). \tag{4.7.13}$$

Note that

$$\operatorname{bzeros}(G) \subseteq \operatorname{tzeros}(G). \tag{4.7.14}$$

Furthermore, we define the multisets

$$\operatorname{mpoles}(G) \triangleq \bigcup_{i=1}^{r} \operatorname{mroots}(q_i), \tag{4.7.15}$$

$$\operatorname{mtzeros}(G) \triangleq \bigcup_{i=1}^{r} \operatorname{mroots}(p_i), \tag{4.7.16}$$

$$\operatorname{mbzeros}(G) \triangleq \operatorname{mroots}(p_1). \tag{4.7.17}$$

Note that

$$\operatorname{mbzeros}(G) \subseteq \operatorname{mtzeros}(G). \tag{4.7.18}$$

If $G = 0$, then these multisets as well as the sets $\operatorname{poles}(G)$, $\operatorname{tzeros}(G)$, and $\operatorname{bzeros}(G)$ are empty.

Proposition 4.7.12. Let $G \in \mathbb{F}_{\operatorname{prop}}^{l \times m}(s)$, assume that $G \neq 0$, let $z \in \mathbb{C}$, and assume that z is not a pole of G. Then, z is a transmission zero of G if and only if $\operatorname{rank} G(z) < \operatorname{rank} G$. Furthermore, z is a blocking zero of G if and only if $G(z) = 0$.

The following example shows that a pole of G can also be a transmission zero of G.

Example 4.7.13. Define $G \in \mathbb{R}^{2 \times 2}_{\text{prop}}(s)$ by

$$G(s) = \left[\begin{array}{cc} \frac{1}{(s+1)^2} & \frac{1}{(s+1)(s+2)} \\ \frac{1}{(s+1)(s+2)} & \frac{s+3}{(s+2)^2} \end{array} \right].$$

Then, rank $G = 2$. Furthermore,

$$G(s) = S_1(s) \left[\begin{array}{cc} \frac{1}{(s+1)^2(s+2)^2} & 0 \\ 0 & s+2 \end{array} \right] S_2(s),$$

where $S_1, S_2 \in \mathbb{R}^{2 \times 2}[s]$ are the unimodular matrices

$$S_1(s) = \left[\begin{array}{cc} (s+2)(s^3+4s^2+5s+1) & 1 \\ (s+1)(s^3+5s^2+8s+3) & 1 \end{array} \right]$$

and

$$S_2(s) = \left[\begin{array}{cc} -(s+2) & (s+1)(s^2+3s+1) \\ 1 & -s(s+2) \end{array} \right].$$

Hence, the McMillan degree of G is 4, the poles of G are -1 and -2, the transmission zero of G is -2, and G has no blocking zeros. Note that -2 is both a pole and a transmission zero of G. Note also that, although G is strictly proper, the Smith-McMillan form of G is improper.

Let $G \in \mathbb{F}^{l \times m}_{\text{prop}}(s)$. A factorization of G of the form

$$G(s) = N(s)D^{-1}(s), \tag{4.7.19}$$

where $N \in \mathbb{F}^{l \times m}[s]$ and $D \in \mathbb{F}^{m \times m}[s]$, is a *right polynomial fraction description* of G. We say that N and D are *right coprime* if every $R \in \mathbb{F}^{m \times m}[s]$ that right divides both N and D is unimodular. In this case, (4.7.19) is a *coprime right polynomial fraction description* of G.

Theorem 4.7.14. Let $N \in \mathbb{F}^{l \times m}[s]$ and $D \in \mathbb{F}^{m \times m}[s]$. Then, the following statements are equivalent:

i) N and D are right coprime.

ii) There exist $X \in \mathbb{F}^{m \times l}[s]$ and $Y \in \mathbb{F}^{m \times m}[s]$ such that

$$XN + YD = I. \tag{4.7.20}$$

iii) For all $s \in \mathbb{C}$,

$$\text{rank} \left[\begin{array}{c} N(s) \\ D(s) \end{array} \right] = m. \tag{4.7.21}$$

Proof. See [1179, p. 297]. □

Equation (4.7.20) is the *Bezout identity*.

The following result shows that all coprime right polynomial fraction descriptions of a proper rational transfer function G are related by a unimodular transformation.

Proposition 4.7.15. Let $G \in \mathbb{F}^{l \times m}_{\text{prop}}(s)$, let $N, \hat{N} \in \mathbb{F}^{l \times m}[s]$, let $D, \hat{D} \in \mathbb{F}^{m \times m}[s]$, and assume that $G = ND^{-1} = \hat{N}\hat{D}^{-1}$. Then, there exists a unimodular matrix $R \in \mathbb{F}^{m \times m}[s]$ such that $N = \hat{N}R$ and $D = \hat{D}R$.

Proof. See [1179, p. 298]. □

The following result uses the Smith-McMillan form to show that every proper rational transfer function has a coprime right polynomial fraction description.

Proposition 4.7.16. Let $G \in \mathbb{F}^{l \times m}_{\text{prop}}(s)$. Then, G has a coprime right polynomial fraction description. If, in addition, $G(s) = N(s)D^{-1}(s)$, where $N \in \mathbb{F}^{l \times m}[s]$ and $D \in \mathbb{F}^{m \times m}[s]$, is a coprime right polynomial fraction description of G, then

$$\text{Szeros}(N) = \text{tzeros}(G) \tag{4.7.22}$$

and

$$\text{Szeros}(D) = \text{poles}(G). \tag{4.7.23}$$

Proof. Note that (4.7.8) can be written as

$$G = S_1 \begin{bmatrix} p_1/q_1 & & & 0 \\ & \ddots & & \\ & & p_r/q_r & \\ 0 & & & 0_{(l-r)\times(m-r)} \end{bmatrix} S_2$$

$$= S_1 \begin{bmatrix} p_1 & & & 0 \\ & \ddots & & \\ & & p_r & \\ 0 & & & 0_{(l-r)\times(m-r)} \end{bmatrix} \begin{bmatrix} q_1 & & & 0 \\ & \ddots & & \\ & & q_r & \\ 0 & & & I_{m-r} \end{bmatrix}^{-1} S_2$$

$$= S_1 \begin{bmatrix} p_1 & & & 0 \\ & \ddots & & \\ & & p_r & \\ 0 & & & 0_{(l-r)\times(m-r)} \end{bmatrix} \left(S_2^{-1} \begin{bmatrix} q_1 & & & 0 \\ & \ddots & & \\ & & q_r & \\ 0 & & & I_{m-r} \end{bmatrix} \right)^{-1},$$

which, by Theorem 4.7.14, is a right coprime polynomial fraction description of G. The last statement follows from Theorem 4.7.5 and Proposition 4.7.15. □

4.8 Facts on Polynomials and Rational Functions

Fact 4.8.1. Let $p \in \mathbb{R}[s]$ be monic, and define $q(s) \triangleq s^n p(1/s)$, where $n \triangleq$ deg p. If $0 \notin \text{roots}(p)$, then $\deg(q) = n$ and

$$\text{mroots}(q) = \{1/\lambda\colon \lambda \in \text{mroots}(p)\}_{\text{ms}}.$$

If $0 \in \text{roots}(p)$ with multiplicity r, then $\deg(q) = n - r$ and

$$\text{mroots}(q) = \{1/\lambda\colon \lambda \neq 0 \text{ and } \lambda \in \text{mroots}(p)\}_{\text{ms}}.$$

Remark: See Fact 11.17.4 and Fact 11.17.5.

Fact 4.8.2. Let $p \in \mathbb{F}^n[s]$ be given by

$$p(s) = s^n + \beta_{n-1}s^{n-1} + \cdots + \beta_1 s + \beta_0,$$

let $\beta_n \triangleq 1$, let $\text{mroots}(p) = \{\lambda_1, \ldots, \lambda_n\}_{\text{ms}}$, and define $\mu_1, \ldots, \mu_n$ by

$$\mu_i \triangleq \lambda_1^i + \cdots + \lambda_n^i.$$

Then, for all $k \in \{1, \ldots, n\}$,

$$k\beta_{n-k} + \mu_1\beta_{n-k+1} + \mu_2\beta_{n-k+2} + \cdots + \mu_k\beta_n = 0.$$

That is,

$$\begin{bmatrix} n & \mu_1 & \mu_2 & \mu_3 & \mu_4 & \cdots & \mu_n \\ 0 & n-1 & \mu_1 & \mu_2 & \mu_3 & \cdots & \mu_{n-1} \\ \vdots & \ddots & \ddots & \ddots & \ddots & \ddots & \vdots \\ \vdots & \ddots & \ddots & \ddots & \ddots & \ddots & \vdots \\ 0 & 0 & \cdots & 0 & 2 & \mu_1 & \mu_2 \\ 0 & 0 & \cdots & 0 & 0 & 1 & \mu_1 \end{bmatrix} \begin{bmatrix} \beta_0 \\ \beta_1 \\ \vdots \\ \beta_{n-1} \\ \beta_n \end{bmatrix} = 0.$$

Consequently, $\beta_1, \ldots, \beta_{n-1}$ are uniquely determined by $\mu_1, \ldots, \mu_n$. In particular,

$$\beta_{n-1} = -\mu_1,$$
$$\beta_{n-2} = \tfrac{1}{2}(\mu_1^2 - \mu_2),$$
$$\beta_3 = \tfrac{1}{6}(-\mu_1^3 + 3\mu_1\mu_2 - 2\mu_3).$$

Proof: See [728, p. 44] and [1027, p. 9].

Remark: These equations are a consequence of Newton's identities given by Fact 1.17.11. Note that, for $i = 0, \ldots, n$, it follows that $\beta_i = (-1)^{n-i}E_{n-i}$, where E_i is the ith elementary symmetric polynomial of the roots of p.

Fact 4.8.3. Let $p, q \in \mathbb{F}[s]$ be monic. Then, p and q are coprime if and only if their least common multiple is pq.

Fact 4.8.4. Let $p, q \in \mathbb{F}[s]$, where $p(s) = a_n s^n + \cdots + a_1 s + a_0$, $q(s) = b_m s^m + \cdots + b_1 s + b_0$, $\deg p = n$, and $\deg q = m$. Furthermore, define the Toeplitz matrices $[p]^{(m)} \in \mathbb{F}^{m \times (n+m)}$ and $[q]^{(n)} \in \mathbb{F}^{n \times (n+m)}$ by

$$[p]^{(m)} \triangleq \begin{bmatrix} a_n & a_{n-1} & \cdots & a_1 & a_0 & 0 & 0 & \cdots & 0 \\ 0 & a_n & a_{n-1} & \cdots & a_1 & a_0 & 0 & \cdots & 0 \\ \vdots & \ddots & \ddots & \ddots & \cdots & \ddots & \ddots & \ddots & \vdots \end{bmatrix}$$

and

$$[q]^{(n)} \triangleq \begin{bmatrix} b_m & b_{m-1} & \cdots & b_1 & b_0 & 0 & 0 & \cdots & 0 \\ 0 & b_m & b_{m-1} & \cdots & b_1 & b_0 & 0 & \cdots & 0 \\ \vdots & \ddots & \ddots & \ddots & \cdots & \ddots & \ddots & \ddots & \vdots \end{bmatrix}.$$

Then, p and q are coprime if and only if

$$\det \begin{bmatrix} [p]^{(m)} \\ [q]^{(n)} \end{bmatrix} \neq 0.$$

Proof: See [494, p. 162] or [1125, pp. 187–191].

Remark: $\begin{bmatrix} A \\ B \end{bmatrix}$ is the *Sylvester matrix*, and $\det \begin{bmatrix} A \\ B \end{bmatrix}$ is the *resultant* of p and q.

Remark: The form $\begin{bmatrix} [p]^{(m)} \\ [q]^{(n)} \end{bmatrix}$ appears in [1125, pp. 187–191]. This result is given in [494, p. 162] in terms of $\begin{bmatrix} \hat{I}[p]^{(m)} \\ \hat{I}[q]^{(n)} \end{bmatrix} \hat{I}$ and in [1540, p. 85] in terms of $\begin{bmatrix} [p]^{(m)} \\ \hat{I}[q]^{(n)} \end{bmatrix}$. Interweaving the rows of $[p]^{(m)}$ and $[q]^{(n)}$ and taking the transpose yields a *step-down matrix* [397].

Fact 4.8.5. Let $p_1, \ldots, p_n \in \mathbb{F}[s]$, and let $d \in \mathbb{F}[s]$ be the greatest common divisor of $p_1, \ldots, p_n$. Then, there exist polynomials $q_1, \ldots, q_n \in \mathbb{F}[s]$ such that

$$d = \sum_{i=1}^{n} q_i p_i.$$

In addition, $p_1, \ldots, p_n$ are coprime if and only if there exist polynomials $q_1, \ldots, q_n \in \mathbb{F}[s]$ such that

$$1 = \sum_{i=1}^{n} q_i p_i.$$

Proof: See [521, p. 16].

Remark: The polynomial d is given by the *Bezout equation*.

Fact 4.8.6. Let $p, q \in \mathbb{F}[s]$, where $p(s) = a_n s^n + \cdots + a_1 s + a_0$ and $q(s) = b_n s^n + \cdots + b_1 s + b_0$, and define $[p]^{(n)}, [q]^{(n)} \in \mathbb{F}^{n \times 2n}$ as in Fact 4.8.4. Furthermore, define

$$R(p, q) \triangleq \begin{bmatrix} [p]^{(n)} \\ [q]^{(n)} \end{bmatrix} = \begin{bmatrix} A_1 & A_2 \\ B_1 & B_2 \end{bmatrix},$$

where $A_1, A_2, B_1, B_2 \in \mathbb{F}^{n \times n}$, and define $\hat{p}(s) \triangleq s^n p(-s)$ and $\hat{q}(s) \triangleq s^n q(-s)$. Then,

$$\begin{bmatrix} A_1 & A_2 \\ B_1 & B_2 \end{bmatrix} = \begin{bmatrix} \hat{p}(N_n^{\mathrm{T}}) & p(N_n) \\ \hat{q}(N_n^{\mathrm{T}}) & q(N_n) \end{bmatrix},$$

$$A_1 B_1 = B_1 A_1,$$

$$A_2 B_2 = B_2 A_2,$$

$$A_1 B_2 + A_2 B_1 = B_1 A_2 + B_2 A_1.$$

Therefore,

$$\begin{bmatrix} I & 0 \\ -B_1 & A_1 \end{bmatrix} \begin{bmatrix} A_1 & A_2 \\ B_1 & B_2 \end{bmatrix} = \begin{bmatrix} A_1 & A_2 \\ 0 & A_1 B_2 - B_1 A_2 \end{bmatrix},$$

$$\begin{bmatrix} -B_2 & A_2 \\ 0 & I \end{bmatrix} \begin{bmatrix} A_1 & A_2 \\ B_1 & B_2 \end{bmatrix} = \begin{bmatrix} A_2 B_1 - B_2 A_1 & 0 \\ B_1 & B_2 \end{bmatrix},$$

and

$$\det R(p,q) = \det(A_1 B_2 - B_1 A_2) = \det(B_2 A_1 - A_2 B_1).$$

Now, define $B(p,q) \in \mathbb{F}^{n \times n}$ by

$$B(p,q) \triangleq (A_1 B_2 - B_1 A_2)\hat{I}.$$

Then, the following statements hold:

i) For all $s, \hat{s} \in \mathbb{C}$,

$$p(s)q(\hat{s}) - q(s)p(\hat{s}) = (s - \hat{s}) \begin{bmatrix} 1 \\ s \\ \vdots \\ s^{n-1} \end{bmatrix}^{\mathrm{T}} B(p,q) \begin{bmatrix} 1 \\ \hat{s} \\ \vdots \\ \hat{s}^{n-1} \end{bmatrix}.$$

ii) $B(p,q) = (B_2 A_1 - A_2 B_1)\hat{I} = \hat{I}(A_1^{\mathrm{T}} B_2^{\mathrm{T}} - B_1^{\mathrm{T}} A_2^{\mathrm{T}}) = \hat{I}(B_1^{\mathrm{T}} A_2^{\mathrm{T}} - A_1^{\mathrm{T}} B_2^{\mathrm{T}}).$

iii) $\begin{bmatrix} 0 & B(p,q) \\ -B(p,q) & 0 \end{bmatrix} = QR^{\mathrm{T}}(p,q)QR(p,q)Q$, where $Q \triangleq \begin{bmatrix} 0 & \hat{I} \\ -\hat{I} & 0 \end{bmatrix}.$

iv) $|\det B(p,q)| = |\det R(p,q)| = |\det q[C(p)]|.$

v) $B(p,q)$ and $\hat{B}(p,q)$ are symmetric.

vi) $B(p,q)$ is a linear function of (p,q).

vii) $B(p,q) = -B(q,p).$

Now, assume in addition that $\deg q \leq \deg p = n$ and p is monic. Then, the following statements hold:

viii) def $B(p,q)$ equals the degree of the greatest common divisor of p and q.

ix) p and q are coprime if and only if $B(p,q)$ is nonsingular.

x) If $B(p,q)$ is nonsingular, then $[B(p,q)]^{-1}$ is Hankel. In fact,

$$[B(p,q)]^{-1} = H_{n,n}(b/p),$$

where $a, b \in \mathbb{F}[s]$ satisfy the Bezout equation $ap + bq = 1$ and $H_{n,n}$ is defined in Fact 4.8.8.

xi) If $q = q_1 q_2$, where $q_1, q_2 \in \mathbb{F}[s]$, then

$$B(p, q) = B(p, q_1) q_2 [C(p)] = q_1 [C^{\mathrm{T}}(p)] B(p, q_2).$$

xii) $B(p, q) = B(p, q) C(p) = C^{\mathrm{T}}(p) B(p, q).$

xiii) $B(p, q) = B(p, 1) q [C(p)] = q [C^{\mathrm{T}}(p)] B(p, 1)$, where $B(p, 1)$ is the Hankel matrix

$$B(p, 1) = \begin{bmatrix} a_1 & a_2 & \cdots & a_{n-1} & 1 \\ a_2 & a_3 & \cdot^{\cdot^{\cdot}} & 1 & 0 \\ \vdots & \cdot^{\cdot^{\cdot}} & \cdot^{\cdot^{\cdot}} & \cdot^{\cdot^{\cdot}} & \vdots \\ a_{n-1} & 1 & \cdot^{\cdot^{\cdot}} & 0 & 0 \\ 1 & 0 & \cdots & 0 & 0 \end{bmatrix}.$$

In particular, for $n = 3$ and $q(s) = s$, it follows that

$$\begin{bmatrix} -a_0 & 0 & 0 \\ 0 & a_2 & 1 \\ 0 & 1 & 0 \end{bmatrix} = \begin{bmatrix} a_1 & a_2 & 1 \\ a_2 & 1 & 0 \\ 1 & 0 & 0 \end{bmatrix} \begin{bmatrix} 0 & 1 & 0 \\ 0 & 0 & 1 \\ -a_0 & -a_1 & -a_2 \end{bmatrix}.$$

xiv) If A_2 is nonsingular, then

$$\begin{bmatrix} A_1 & A_2 \\ B_1 & B_2 \end{bmatrix} = \begin{bmatrix} 0 & I \\ A_2^{-1}\hat{I} & B_2 A_2^{-1} \end{bmatrix} \begin{bmatrix} B(p, q) & 0 \\ 0 & I \end{bmatrix} \begin{bmatrix} I & 0 \\ A_1 & A_2 \end{bmatrix}.$$

xv) If p has distinct roots $\lambda_1, \ldots, \lambda_n$, then

$$V^{\mathrm{T}}(\lambda_1, \ldots, \lambda_n) B(p, q) V(\lambda_1, \ldots, \lambda_n) = \mathrm{diag}[q(\lambda_1) p'(\lambda_1), \ldots, q(\lambda_n) p'(\lambda_n)].$$

Proof: See [494, pp. 164–167], [521, pp. 200–207], and [681]. To prove *ii)*, note that A_1, A_2, B_1, B_2 are square and Toeplitz, and thus reverse symmetric, that is, $A_1 = A_1^{\hat{\mathrm{T}}}$. See Fact 3.18.5.

Remark: $B(p, q)$ is the *Bezout matrix* of p and q. See [149, 680, 741, 1389, 1478], [1125, p. 189], and Fact 5.15.24.

Remark: *xiii)* is the *Barnett factorization*. See [142, 1389]. The definitions of $B(p, q)$ and *ii)* are the *Gohberg-Semencul formulas*. See [521, p. 206].

Remark: It follows from continuity that the expressions for $\det R(p, q)$ are valid whether or not A_1 or B_2 is singular. See Fact 2.14.13.

Remark: The inverse of a Hankel matrix is a Bezout matrix. See [494, p. 174].

Fact 4.8.7. Let $p, q \in \mathbb{F}[s]$, where $p(s) = \alpha_1 s + \alpha_0$ and $q(s) = s^2 + \beta_1 s + \beta_0$. Then, p and q are coprime if and only if $\alpha_0^2 + \alpha_1^2 \beta_0 \neq \alpha_0 \alpha_1 \beta_1$.

Proof: Use Fact 4.8.6.

Fact 4.8.8. Let $p, q \in \mathbb{F}[s]$, assume that q is monic, assume that $\deg p < \deg q = n$, and define $B(p, q)$ as in Fact 4.8.6. Furthermore, define $g \in \mathbb{F}(s)$ by

$$g(s) \triangleq \frac{p(s)}{q(s)} = \sum_{i=1}^{\infty} \frac{h_i}{s^i}.$$

Finally, define the Hankel matrix $H_{i,j}(g) \in \mathbb{R}^{i \times j}$ by

$$H_{i,j}(g) = \begin{bmatrix} h_1 & h_2 & h_{k+3} & \cdots & h_j \\ h_{k+2} & h_{k+3} & \cdots & \cdots & \vdots \\ h_{k+3} & \cdots & \cdots & \cdots & \vdots \\ \vdots & \cdots & \cdots & \cdots & \vdots \\ \vdots & \cdots & \cdots & \cdots & \vdots \\ h_i & \cdots & \cdots & \cdots & h_{j+i-1} \end{bmatrix}.$$

Then, the following statements are equivalent:

  $i)$ p and q are coprime.

  $ii)$ $H_{n,n}(g)$ is nonsingular.

  $iii)$ For all $i,j \geq n$, rank $H_{i,j}(g) = n$.

  $iv)$ There exist $i,j \geq n$ such that rank $H_{i,j}(g) = n$.

Furthermore, the following statements hold:

  $v)$ If p and q are coprime, then $[H_{n,n}(g)]^{-1} = B(q,a)$, where $a, b \in \mathbb{F}[s]$ satisfy the Bezout equation $ap + bq = 1$.

  $vi)$ $B(q,p) = B(q,1)H_{n,n}(g)B(q,1)$.

  $vii)$ $B(q,p)$ and $H_{n,n}(g)$ are congruent.

  $viii)$ In $B(q,p) = $ In $H_{n,n}(g)$.

  $ix)$ det $H_{n,n}(g) = $ det $B(q,p)$.

Proof: See [521, pp. 215–221].

Remark: See Proposition 12.9.11.

Fact 4.8.9. Let $q \in \mathbb{R}[s]$, define $g \in \mathbb{F}(s)$ by $g \triangleq q'/q$, and define $B(q,q')$ as in Fact 4.8.6. Then, the following statements hold:

  $i)$ The number of distinct roots of q is rank $B(q,q')$.

  $ii)$ q has n distinct roots if and only if $B(q,q')$ is nonsingular.

  $iii)$ The number of distinct real roots of q is sig $B(q,q')$.

  $iv)$ q has n distinct, real roots if and only if $B(q,q')$ is positive definite.

  $v)$ The number of distinct complex roots of q is $2\nu_-[B(q,q')]$.

  $vi)$ q has n distinct, complex roots if and only if n is even and $\nu_-[B(q,q')] = n/2$.

  $vii)$ q has n real roots if and only if $B(q,q')$ is positive semidefinite.

Proof: See [521, p. 252].

Remark: $q'(s) \triangleq (d/ds)q(s)$.

Fact 4.8.10. Let $q \in \mathbb{F}[s]$, where $q(s) = \sum_{i=0}^n b_i s^i$, and define

$$\text{coeff}(q) \triangleq \begin{bmatrix} b_n \\ \vdots \\ b_0 \end{bmatrix}.$$

Now, let $p \in \mathbb{F}[s]$, where $p(s) = \sum_{i=0}^n a_i s^i$. Then,

$$\text{coeff}(pq) = A\text{coeff}(q),$$

where $A \in \mathbb{F}^{2n \times (n+1)}$ is the Toeplitz matrix

$$A = \begin{bmatrix} a_n & 0 & 0 & \cdots & 0 \\ a_{n-1} & a_n & 0 & \cdots & 0 \\ \vdots & \ddots & \ddots & \ddots & \vdots \\ a_1 & \ddots & \ddots & \ddots & \vdots \\ a_0 & a_1 & \ddots & \ddots & a_n \\ 0 & a_0 & \ddots & \ddots & a_{n-1} \\ \vdots & \vdots & \ddots & \ddots & \vdots \\ 0 & 0 & \cdots & a_0 & a_1 \end{bmatrix}.$$

In particular, if $n = 3$, then

$$A = \begin{bmatrix} a_2 & 0 & 0 \\ a_1 & a_2 & 0 \\ a_0 & a_1 & a_2 \\ 0 & a_0 & a_1 \end{bmatrix}.$$

Fact 4.8.11. Let $\lambda_1, \ldots, \lambda_n \in \mathbb{C}$ be distinct and, for all $i \in \{1, \ldots, n\}$, define

$$p_i(s) \triangleq \prod_{\substack{j=1 \\ j \neq i}}^n \frac{s - \lambda_i}{\lambda_i - \lambda_j}.$$

Then, for all $i \in \{1, \ldots, n\}$,

$$p_i(\lambda_j) = \begin{cases} 1, & i = j, \\ 0, & i \neq j. \end{cases}$$

Remark: This equality is the *Lagrange interpolation formula.*

Fact 4.8.12. Let $A \in \mathbb{F}^{n \times n}$, and assume that $\det(I + A) \neq 0$. Then, there exists $p \in \mathbb{F}[s]$ such that $\deg p \leq n - 1$ and $(I + A)^{-1} = p(A)$.

Remark: See Fact 4.8.12.

Fact 4.8.13. Let $A \in \mathbb{F}^{n \times n}$, let $q \in \mathbb{F}[s]$, and assume that $q(A)$ is nonsingular. Then, there exists $p \in \mathbb{F}[s]$ such that $\deg p \leq n - 1$ and $[q(A)]^{-1} = p(A)$.

Proof: See Fact 5.14.23.

Fact 4.8.14. Let $A \in \mathbb{R}^{n \times n}$, assume that A is skew symmetric, and let the components of $x_A \in \mathbb{R}^{n(n-1)/2}$ be the entries $A_{(i,j)}$ for all $i > j$. Then, there exists a polynomial function $p \colon \mathbb{R}^{n(n-1)/2} \mapsto \mathbb{R}$ such that, for all $\alpha \in \mathbb{R}$ and $x \in \mathbb{R}^{n(n-1)/2}$,

$$p(\alpha x) = \alpha^{n/2} p(x)$$

and

$$\det A = p^2(x_A).$$

In particular,

$$\det \begin{bmatrix} 0 & a \\ -a & 0 \end{bmatrix} = a^2$$

and

$$\det \begin{bmatrix} 0 & a & b & c \\ -a & 0 & d & e \\ -b & -d & 0 & f \\ -c & -e & -f & 0 \end{bmatrix} = (af - be + cd)^2.$$

Proof: See [903, p. 224] and [1125, pp. 125–127].

Remark: The polynomial p is the *Pfaffian*, and this result is *Pfaff's theorem*.

Remark: An extension to the product of a pair of skew-symmetric matrices is given in [446].

Remark: See Fact 3.7.33.

Fact 4.8.15. Let $G \in \mathbb{F}^{n \times m}(s)$, and let $G_{(i,j)} = n_{ij}/d_{ij}$, where $n_{ij} \in \mathbb{F}[s]$ and $d_{ij} \in \mathbb{F}[s]$ are coprime for all $i \in \{1, \ldots, n\}$ and $j \in \{1, \ldots, m\}$. Then, q_1 given by the Smith-McMillan form is the least common multiple of $d_{11}, d_{12}, \ldots, d_{nm}$.

Fact 4.8.16. Let $G \in \mathbb{F}^{n \times m}(s)$, assume that $\operatorname{rank} G = m$, and let $\lambda \in \mathbb{C}$, where λ is not a pole of G. Then, λ is a transmission zero of G if and only if there exists a vector $u \in \mathbb{C}^m$ such that $G(\lambda)u = 0$. Furthermore, if G is square, then λ is a transmission zero of G if and only if $\det G(\lambda) = 0$.

Fact 4.8.17. Let $G \in \mathbb{F}^{n \times m}(s)$, let $\omega \in \mathbb{R}$, and assume that $\jmath\omega$ is not a pole of G. Then,

$$\operatorname{Im} G(-\jmath\omega) = -\operatorname{Im} G(\jmath\omega).$$

4.9 Facts on the Characteristic and Minimal Polynomials

Fact 4.9.1. Let $A = \begin{bmatrix} a & b \\ c & d \end{bmatrix} \in \mathbb{R}^{2 \times 2}$. Then, the following statements hold:

i) $\operatorname{mspec}(A) = \left\{ \frac{1}{2}\left[a + d \pm \sqrt{(a-d)^2 + 4bc} \right] \right\}_{\mathrm{ms}}$

$$= \left\{ \frac{1}{2}\left[\operatorname{tr} A \pm \sqrt{(\operatorname{tr} A)^2 - 4\det A} \right] \right\}_{\mathrm{ms}}.$$

ii) $\chi_A(s) = s^2 - (\operatorname{tr} A)s + \det A$.

iii) $\det A = \frac{1}{2}\left[(\operatorname{tr} A)^2 - \operatorname{tr} A^2 \right]$.

iv) $(sI - A)^{\mathrm{A}} = sI + A - (\operatorname{tr} A)I$.

v) $A^{-1} = (\det A)^{-1}[(\operatorname{tr} A)I - A]$.

vi) $A^{\mathrm{A}} = (\operatorname{tr} A)I - A$.

vii) $\operatorname{tr} A^{-1} = \operatorname{tr} A/\det A$.

Fact 4.9.2. Let $A \in \mathbb{R}^{3\times3}$. Then, the following statements hold:

i) $\chi_A(s) = s^3 - (\operatorname{tr} A)s^2 + (\operatorname{tr} A^{\mathrm{A}})s - \det A$.

ii) $\operatorname{tr} A^{\mathrm{A}} = \frac{1}{2}[(\operatorname{tr} A)^2 - \operatorname{tr} A^2]$.

iii) $\det A = \frac{1}{3}\operatorname{tr} A^3 - \frac{1}{2}(\operatorname{tr} A)\operatorname{tr} A^2 + \frac{1}{6}(\operatorname{tr} A)^3$.

iv) $(sI - A)^{\mathrm{A}} = s^2 I + s[A - (\operatorname{tr} A)I] + A^2 - (\operatorname{tr} A)A + \frac{1}{2}[(\operatorname{tr} A)^2 - \operatorname{tr} A^2]I$.

Remark: See Fact 7.5.17.

Fact 4.9.3. Let $A, B \in \mathbb{F}^{2\times2}$. Then,

$$AB + BA - (\operatorname{tr} A)B - (\operatorname{tr} B)A + [(\operatorname{tr} A)(\operatorname{tr} B) - \operatorname{tr} AB]I = 0.$$

Furthermore,

$$\det(A + B) - \det A - \det B = (\operatorname{tr} A)(\operatorname{tr} B) - \operatorname{tr} AB.$$

Proof: Apply the Cayley-Hamilton theorem to $A + xB$, differentiate with respect to x, and set $x = 0$. For the second equality, evaluate the Cayley-Hamilton theorem with $A + B$. See [512, 513, 915, 1156] or [1217, p. 37].

Remark: This equality is a *polarized Cayley-Hamilton theorem*. See [81].

Fact 4.9.4. Let $A, B, C \in \mathbb{F}^{2\times2}$. Then,

$$\begin{aligned}
2ABC = \; & (\operatorname{tr} A)BC + (\operatorname{tr} B)AC + (\operatorname{tr} C)AB \\
& - (\operatorname{tr} AC)B + [(\operatorname{tr} AB) - (\operatorname{tr} A)(\operatorname{tr} B)]C \\
& + [(\operatorname{tr} BC) - (\operatorname{tr} B)(\operatorname{tr} C)]A \\
& - [(\operatorname{tr} ACB) - (\operatorname{tr} AC)(\operatorname{tr} B)]I.
\end{aligned}$$

Remark: This equality is a *polarized Cayley-Hamilton theorem*. See [81].

Remark: An analogous formula exists for the product of six 3×3 matrices. See [81].

Fact 4.9.5. Let $A, B, C \in \mathbb{F}^{3\times3}$, and assume that $\operatorname{tr} A = \operatorname{tr} A = \operatorname{tr} C = 0$. Then,

$$4\operatorname{tr}(A^2 B^2) + 2\operatorname{tr}[(AB)^2] = \operatorname{tr}(A^2)\operatorname{tr}(B^2) + 2[\operatorname{tr}(AB)]^2$$

and

$$6\operatorname{tr}(A^2 B^2 AB) + 6\operatorname{tr}(B^2 A^2 BA) + 2\operatorname{tr}(AB)\operatorname{tr}[(AB)^2] + 2\operatorname{tr}(A^3)\operatorname{tr}(B^3)$$

$$= 2\operatorname{tr}(AB)\operatorname{tr}(A^2 B^2) + \operatorname{tr}(A^2)\operatorname{tr}(AB)\operatorname{tr}(B^2) + 2[\operatorname{tr}(AB)]^3 + 6\operatorname{tr}(A^2 B)\operatorname{tr}(AB^2).$$

Proof: See [84].

Fact 4.9.6. Let $A, B, C \in \mathbb{F}^{3\times3}$. Then,

$$\sum [A'B'C' - (\operatorname{tr} A')B'C' + (\operatorname{tr} A')(\operatorname{tr} B')C' - (\operatorname{tr} A'B')C']$$
$$- [(\operatorname{tr} A)(\operatorname{tr} B)\operatorname{tr} C - (\operatorname{tr} A)\operatorname{tr} BC - (\operatorname{tr} B)\operatorname{tr} CA - (\operatorname{tr} C)\operatorname{tr} AB + \operatorname{tr} ABC$$
$$+ \operatorname{tr} CBA]I = 0,$$

where the sum is taken over all six permutations A', B', C' of A, B, C.

Remark: This equality is a *polarized Cayley-Hamilton theorem*. See [82, 915, 1156].

Fact 4.9.7. Let $A, B \in \mathbb{F}^{n\times n}$, assume that A and B commute, and define $f \colon \mathbb{C}^2 \mapsto \mathbb{C}$ by $f(r, s) \triangleq \det(rA - sB)$. Then, $f(B, A) = 0$.

Remark: This result is the *generalized Cayley-Hamilton theorem*. See [364, 700].

Fact 4.9.8. Let $A \in \mathbb{F}^{n\times n}$, let $\chi_A(s) = s^n + \beta_{n-1}s^{n-1} + \cdots + \beta_0$, and let $\operatorname{mspec}(A) = \{\lambda_1, \ldots, \lambda_n\}_{\mathrm{ms}}$. Then,

$$A^{\mathrm{A}} = (-1)^{n-1}\left(A^{n-1} + \beta_{n-1}A^{n-2} + \cdots + \beta_1 I\right).$$

Furthermore,

$$\operatorname{tr} A^{\mathrm{A}} = (-1)^{n-1}\chi'_A(0) = (-1)^{n-1}\beta_1 = \sum_{1\le j_1 < \cdots < j_{n-1}\le n} \lambda_{j_1}\cdots\lambda_{j_{n-1}} = \sum_{i=1}^{n} \det A_{[i;i]}.$$

Proof: Use $A^{-1}\chi_A(A) = 0$. The second equality follows from (4.4.19) or Lemma 4.4.8.

Remark: See Fact 4.10.9.

Fact 4.9.9. Let $A \in \mathbb{F}^{n\times n}$, assume that A is nonsingular, and let $\chi_A(s) = s^n + \beta_{n-1}s^{n-1} + \cdots + \beta_0$. Then,

$$\chi_{A^{-1}}(s) = \frac{1}{\det A}(-s)^n \chi_A(1/s)$$

$$= s^n + (\beta_1/\beta_0)s^{n-1} + \cdots + (\beta_{n-1}/\beta_0)s + 1/\beta_0.$$

Remark: See Fact 5.16.2.

Fact 4.9.10. Let $A \in \mathbb{F}^{n\times n}$, and assume that either A and $-A$ are similar or A^{T} and $-A$ are similar. Then,

$$\chi_A(s) = (-1)^n \chi_A(-s).$$

Furthermore, if n is even, then χ_A is even, whereas, if n is odd, then χ_A is odd.

Remark: A and A^{T} are similar. See Corollary 4.3.11 and Corollary 5.3.8.

Fact 4.9.11. Let $A \in \mathbb{F}^{n\times n}$. Then, for all $s \in \mathbb{C}$,

$$(sI - A)^{\mathrm{A}} = \chi_A(s)(sI - A)^{-1} = \sum_{i=0}^{n-1} \chi_A^{[i]}(s)A^i,$$

where

$$\chi_A(s) = s^n + \beta_{n-1}s^{n-1} + \cdots + \beta_1 s + \beta_0$$

and, for all $i \in \{0, \ldots, n-1\}$, the polynomial $\chi_A^{[i]}$ is defined by

$$\chi_A^{[i]}(s) \triangleq s^{n-i} + \beta_{n-1}s^{n-1-i} + \cdots + \beta_{i+1}.$$

Note that

$$\chi_A^{[n-1]}(s) = s + \beta_{n-1}, \quad \chi_A^{[n]}(s) = 1,$$

and that, for all $i \in \{0, \ldots, n-1\}$ and with $\chi_A^{[0]} \triangleq \chi_A$, the polynomials $\chi_A^{[i]}$ satisfy the recursion

$$s\chi_A^{[i+1]}(s) = \chi_A^{[i]}(s) - \beta_i.$$

Proof: See [1490, p. 31].

Fact 4.9.12. Define $A \in \mathbb{F}^{(n+1)\times(n+1)}[s]$ by

$$A(s) \triangleq \begin{bmatrix} a_n & a_{n-1} & a_{n-2} & \cdots & a_1 & a_0 \\ -1 & s & 0 & \cdots & 0 & 0 \\ 0 & -1 & s & \ddots & 0 & 0 \\ \vdots & \ddots & \ddots & \ddots & \ddots & \vdots \\ 0 & 0 & 0 & \ddots & s & 0 \\ 0 & 0 & 0 & \cdots & -1 & s \end{bmatrix}.$$

Then,

$$\det A(s) = \sum_{i=0}^{n} a_i s^i.$$

Proof: See [272, p. 95].

Fact 4.9.13. Let $A \in \mathbb{R}^{n\times n}$, and assume that A is skew symmetric. If n is even, then χ_A is even, whereas, if n is odd, then χ_A is odd.

Fact 4.9.14. Let $A \in \mathbb{F}^{n\times n}$, and assume that A is skew Hermitian. Then, for all $s \in \mathbb{C}$,

$$\chi_A(-s) = (-1)^n \overline{p(\bar{s})}.$$

Fact 4.9.15. Let $A \in \mathbb{F}^{n\times n}$. Then, $\chi_{\mathcal{A}}$ is even for the matrices $\mathcal{A} \in \mathbb{F}^{2n\times 2n}$ given by $\begin{bmatrix} 0 & A \\ A^* & 0 \end{bmatrix}$, $\begin{bmatrix} A & 0 \\ 0 & -A \end{bmatrix}$, and $\begin{bmatrix} A & 0 \\ 0 & -A^* \end{bmatrix}$.

Fact 4.9.16. Let $A, B \in \mathbb{F}^{n\times n}$, and define $\mathcal{A} \triangleq \begin{bmatrix} 0 & A \\ B & 0 \end{bmatrix}$. Then,

$$\chi_{\mathcal{A}}(s) = \chi_{AB}(s^2) = \chi_{BA}(s^2).$$

Consequently, $\chi_{\mathcal{A}}$ is even.

Proof: Use Fact 2.14.13 and Proposition 4.4.10.

Fact 4.9.17. Let $x, y, z, w \in \mathbb{F}^n$, and define $A \triangleq xy^{\mathrm{T}}$ and $B \triangleq xy^{\mathrm{T}} + zw^{\mathrm{T}}$. Then,

$$\chi_A(s) = s^{n-1}(s - x^{\mathrm{T}}y)$$

and

$$\chi_B(s) = s^{n-2}[s^2 - (x^{\mathrm{T}}y + z^{\mathrm{T}}w)s + x^{\mathrm{T}}yz^{\mathrm{T}}w - y^{\mathrm{T}}zx^{\mathrm{T}}w].$$

Remark: See Fact 5.11.13.

Fact 4.9.18. Let $x, y \in \mathbb{F}^{n-1}$, and define $A \in \mathbb{F}^{n \times n}$ by

$$A \triangleq \begin{bmatrix} 0 & x^{\mathrm{T}} \\ y & 0 \end{bmatrix}.$$

Then,

$$\chi_A(s) = s^{n-1}(s^2 - y^{\mathrm{T}}x).$$

Proof: See [1365].

Fact 4.9.19. Let $x, y, z, w \in \mathbb{F}^{n-1}$, and define $A \in \mathbb{F}^{n \times n}$ by

$$A \triangleq \begin{bmatrix} 1 & x^{\mathrm{T}} \\ y & zw^{\mathrm{T}} \end{bmatrix}.$$

Then,

$$\chi_A(s) = s^{n-3}\left[s^3 - \left(1 + w^{\mathrm{T}}z\right)s^2 + \left(w^{\mathrm{T}}z - x^{\mathrm{T}}y\right)s + w^{\mathrm{T}}zx^{\mathrm{T}}y - x^{\mathrm{T}}zw^{\mathrm{T}}y\right].$$

Proof: See [419].

Remark: Extensions are given in [1365].

Fact 4.9.20. Let $x \in \mathbb{R}^3$, and define $\theta \triangleq \sqrt{x^{\mathrm{T}}x}$. Then,

$$\chi_{K(x)}(s) = s^3 + \theta^2 s.$$

Hence,

$$\mathrm{mspec}[K(x)] = \{0, \jmath\theta, -\jmath\theta\}_{\mathrm{ms}}.$$

Now, assume in addition that $x \neq 0$. Then, x is an eigenvector corresponding to the eigenvalue 0, that is, $K(x)x = 0$. Furthermore, if either $x_{(1)} \neq 0$ or $x_{(2)} \neq 0$, then

$$\begin{bmatrix} x_{(1)}x_{(3)} + \jmath\theta x_{(2)} \\ x_{(2)}x_{(3)} - \jmath\theta x_{(1)} \\ -x_{(1)}^2 - x_{(2)}^2 \end{bmatrix}$$

is an eigenvector corresponding to the eigenvalue $\jmath\theta$. Finally, if $x_{(1)} = x_{(2)} = 0$, then $\begin{bmatrix} \jmath \\ 1 \\ 0 \end{bmatrix}$ is an eigenvector corresponding to the eigenvalue $\jmath\theta$.

Remark: See Fact 11.11.6.

Fact 4.9.21. Let $a, b \in \mathbb{R}^3$, where $a = \begin{bmatrix} a_1 & a_2 & a_3 \end{bmatrix}^{\mathrm{T}}$ and $b = \begin{bmatrix} b_1 & b_2 & b_3 \end{bmatrix}^{\mathrm{T}}$, and define the skew-symmetric matrix $A \in \mathbb{R}^{4 \times 4}$ by

$$A \triangleq \begin{bmatrix} K(a) & b \\ -b^{\mathrm{T}} & 0 \end{bmatrix}.$$

Then, the following statements hold:

i) $\det A = \left(a^{\mathrm{T}}b\right)^2$.

ii) $\chi_A(s) = s^4 + \left(a^{\mathrm{T}}a + b^{\mathrm{T}}b\right)s^2 + \left(a^{\mathrm{T}}b\right)^2$.

iii) $A^{\mathrm{A}} = -a^{\mathrm{T}}b \begin{bmatrix} K(b) & a \\ -a^{\mathrm{T}} & 0 \end{bmatrix}.$

iv) If $\det A \neq 0$, then $A^{-1} = -\left(a^{\mathrm{T}}b\right)^{-1} \begin{bmatrix} K(b) & a \\ -a^{\mathrm{T}} & 0 \end{bmatrix}.$

v) If $\det A = 0$, then

$$A^3 = -\left(a^{\mathrm{T}}a + b^{\mathrm{T}}b\right)^2 A$$

and

$$A^+ = -\left(a^{\mathrm{T}}a + b^{\mathrm{T}}b\right)^{-2} A.$$

Proof: See [1366].

Remark: See Fact 4.10.4 and Fact 11.11.18.

Fact 4.9.22. Let $A \in \mathbb{R}^{2n \times 2n}$, and assume that A is Hamiltonian. Then, χ_A is even, and thus $\mathrm{mspec}(A) = -\mathrm{mspec}(A)$.

Remark: See Fact 5.9.26.

Fact 4.9.23. Let $A, B, C \in \mathbb{R}^{n \times n}$, and define

$$\mathcal{A} \triangleq \begin{bmatrix} A & B \\ C & -A^{\mathrm{T}} \end{bmatrix}.$$

If B and C are symmetric, then $\mathcal{A}$ is Hamiltonian. If B and C are skew symmetric, then $\chi_{\mathcal{A}}$ is even, although $\mathcal{A}$ is not necessarily Hamiltonian.

Proof: For the second result replace J_{2n} by $\begin{bmatrix} 0 & I_n \\ I_n & 0 \end{bmatrix}$.

Fact 4.9.24. Let $A \in \mathbb{R}^{n \times n}$, $R \in \mathbb{R}^{n \times n}$, and $B \in \mathbb{R}^{n \times m}$, and define $\mathcal{A} \in \mathbb{R}^{2n \times 2n}$ by

$$\mathcal{A} \triangleq \begin{bmatrix} A & BB^{\mathrm{T}} \\ R & -A^{\mathrm{T}} \end{bmatrix}.$$

Then, for all $s \notin \mathrm{spec}(A)$,

$$\chi_{\mathcal{A}}(s) = (-1)^n \chi_A(s)\chi_A(-s)\det\left[I + B^{\mathrm{T}}\left(-sI - A^{\mathrm{T}}\right)^{-1}R(sI - A)^{-1}B\right].$$

Now, assume in addition that R is symmetric. Then, $\mathcal{A}$ is Hamiltonian, and $\chi_{\mathcal{A}}$ is even. If, in addition, R is positive semidefinite, then $(-1)^n \chi_{\mathcal{A}}$ has a spectral factorization.

Proof: Using (2.8.10) and (2.8.14), it follows that, for all $\pm s \notin \mathrm{spec}(A)$,

$$\chi_{\mathcal{A}}(s) = \det(sI - A)\det\left[sI + A^{\mathrm{T}} - R(sI - A)^{-1}BB^{\mathrm{T}}\right]$$

$$= (-1)^n \chi_A(s)\chi_A(-s)\det\left[I - B^{\mathrm{T}}\left(sI + A^{\mathrm{T}}\right)^{-1}R(sI - A)^{-1}B\right].$$

To prove the second statement, note that, for all $\omega \in \mathbb{R}$ such that $\jmath\omega \notin \mathrm{spec}(A)$, it follows that

$$\chi_{\mathcal{A}}(\jmath\omega) = (-1)^n \chi_A(\jmath\omega)\overline{\chi_A(\jmath\omega)}\det\left[I + B^{\mathrm{T}}(\jmath\omega I - A)^{-*}R(\jmath\omega I - A)^{-1}B\right].$$

Thus, $(-1)^n \chi_{\mathcal{A}}(\jmath\omega) \geq 0$. By continuity, $(-1)^n \chi_{\mathcal{A}}(\jmath\omega) \geq 0$ for all $\omega \in \mathbb{R}$. Now, Proposition 4.1.1 implies that $(-1)^n \chi_{\mathcal{A}}$ has a spectral factorization.

Remark: Not all Hamiltonian matrices $\mathcal{A} \in \mathbb{R}^{2n \times 2n}$ have the property that $(-1)^n \chi_{\mathcal{A}}$ has a spectral factorization. Consider $\begin{bmatrix} 0 & 0 & 1 & 0 \\ 0 & 0 & 0 & 1 \\ -1 & 0 & 0 & 0 \\ 0 & -3 & 0 & 0 \end{bmatrix}$, whose spectrum is $\{\jmath, -\jmath, \sqrt{3}\jmath, -\sqrt{3}\jmath\}$.

Remark: This result is closely related to Proposition 12.17.8.

Remark: See Fact 3.20.6.

Fact 4.9.25. Let $A \in \mathbb{F}^{n \times n}$. Then, $\mu_A = \chi_A$ if and only if there exists a unique monic polynomial $p \in \mathbb{F}[s]$ of degree n such that $p(A) = 0$.

Proof: To prove necessity, note that if $\hat{p} \neq p$ is monic, of degree n, and satisfies $\hat{p}(A) = 0$, then $p - \hat{p}$ is nonzero, has degree less than n, and satisfies $(p - \hat{p})(A) = 0$. Conversely, if $\mu_A \neq \chi_A$, then $\mu_A + \chi_A$ is monic, has degree n, and satisfies $(\mu_A + \chi_A)(A) = 0$.

Fact 4.9.26. Let $A \in \mathbb{F}^{n \times n}$, and let $p \in \mathbb{F}[s]$. Then, μ_A divides p if and only if $\mathrm{spec}(A) \subseteq \mathrm{roots}(p)$ and, for all $\lambda \in \mathrm{spec}(A)$, $\mathrm{ind}_A(\lambda) \leq \mathrm{mult}_p(\lambda)$.

Fact 4.9.27. Let $A \in \mathbb{F}^{n \times n}$, let $\mathrm{mspec}(A) = \{\lambda_1, \ldots, \lambda_n\}_{\mathrm{ms}}$, and let $p \in \mathbb{F}[s]$. Then, the following statements hold:

i) $\mathrm{mspec}[p(A)] = \{p(\lambda_1), \ldots, p(\lambda_n)\}_{\mathrm{ms}}$.

ii) $\mathrm{roots}(p) \cap \mathrm{spec}(A) = \varnothing$ if and only if $p(A)$ is nonsingular.

iii) μ_A divides p if and only if $p(A) = 0$.

4.10 Facts on the Spectrum

Fact 4.10.1. Let $A \in \mathbb{F}^{n \times n}$. Then, $\mathrm{rank}\, A = 1$ if and only if $\mathrm{gmult}_A(0) = n-1$. In this case, $\mathrm{mspec}(A) = \{\mathrm{tr}\, A, 0, \ldots, 0\}_{\mathrm{ms}}$.

Proof: Use Proposition 4.5.2.

Remark: See Fact 2.10.19.

Fact 4.10.2. Let $A \in \mathbb{R}^{n \times n}$, and assume that A is row stochastic. Then, $1 \in \mathrm{spec}(A)$.

Fact 4.10.3. Let $A \in \mathbb{F}^{3 \times 3}$, assume that A is symmetric, let $\lambda_1, \lambda_2, \lambda_3 \in \mathbb{R}$ denote the eigenvalues of A, where $\lambda_1 \geq \lambda_2 \geq \lambda_3$, and define

$$p = \tfrac{1}{6} \mathrm{tr}\, [A - \tfrac{1}{3}(\mathrm{tr}\, A)I]^2$$

and

$$q = \tfrac{1}{2} \det [A - \tfrac{1}{3}(\mathrm{tr}\, A)I].$$

Then, the following statements hold:

i) $0 \leq |q| \leq p^{3/2}$.

ii) $p = 0$ if and only if $\lambda_1 = \lambda_2 = \lambda_3 = \tfrac{1}{3} \mathrm{tr}\, A$.

iii) $p > 0$ if and only if

$$\lambda_1 = \tfrac{1}{3}\operatorname{tr} A + 2\sqrt{p}\cos\phi,$$
$$\lambda_2 = \tfrac{1}{3}\operatorname{tr} A + \sqrt{3p}\sin\phi - \sqrt{p}\cos\phi,$$
$$\lambda_3 = \tfrac{1}{3}\operatorname{tr} A - \sqrt{3p}\sin\phi - \sqrt{p}\cos\phi,$$

where $\phi \in [0, \pi/3]$ is given by

$$\phi = \tfrac{1}{3}\cos^{-1}\frac{q}{p^{3/2}}.$$

iv) $\phi = 0$ if and only if $q = p^{3/2} > 0$. In this case,

$$\lambda_1 = \tfrac{1}{3}\operatorname{tr} A + 2\sqrt{p},$$
$$\lambda_2 = \lambda_3 = \tfrac{1}{3}\operatorname{tr} A - \sqrt{p}.$$

v) $\phi = \pi/6$ if and only if $p > 0$ and $q = 0$. In this case, $\sin\phi = 1/2$, $\cos\phi = \sqrt{3}/2$, and

$$\lambda_1 = \tfrac{1}{3}\operatorname{tr} A + \sqrt{3p},$$
$$\lambda_2 = \tfrac{1}{3}\operatorname{tr} A,$$
$$\lambda_3 = \tfrac{1}{3}\operatorname{tr} A - \sqrt{3p}.$$

vi) $\phi = \pi/3$ if and only if $q = -p^{3/2} < 0$. In this case, $\sin\phi = \sqrt{3}/2$, $\cos\phi = 1/2$, and

$$\lambda_1 = \lambda_2 = \tfrac{1}{3}\operatorname{tr} A + \sqrt{p},$$
$$\lambda_3 = \tfrac{1}{3}\operatorname{tr} A - 2\sqrt{p}.$$

Proof: See [1234].

Remark: This result is based on *Cardano's trigonometric solution* for the roots of a cubic polynomial. See [238, 1234].

Remark: The inequality $q^2 \le p^3$ follows from Fact 1.12.13.

Fact 4.10.4. Let $a, b, c, d, \omega \in \mathbb{R}$, and define the skew-symmetric matrix $A \in \mathbb{R}^{4\times4}$ given by

$$A \triangleq \begin{bmatrix} 0 & \omega & a & b \\ -\omega & 0 & c & d \\ -a & -c & 0 & \omega \\ -b & -d & -\omega & 0 \end{bmatrix}.$$

Then,

$$\chi_A(s) = s^4 + (2\omega^2 + a^2 + b^2 + c^2 + d^2)s^2 + \left[\omega^2 - (ad - bc)\right]^2$$

and

$$\det A = \left[\omega^2 - (ad - bc)\right]^2.$$

Hence, A is singular if and only if $bc \le ad$ and $\omega = \sqrt{ad - bc}$. Furthermore, A has a repeated eigenvalue if and only if either *i)* A is singular or *ii)* $a = -d$ and $b = c$. In case *i)*, A has the repeated eigenvalue 0, while, in case *ii)*, A has the repeated eigenvalues $\jmath\sqrt{\omega^2 + a^2 + b^2}$ and $-\jmath\sqrt{\omega^2 + a^2 + b^2}$. Finally, cases *i)* and *ii)* cannot occur simultaneously.

Remark: See Fact 3.7.33, Fact 4.9.21, Fact 11.11.16, and Fact 11.11.18.

Fact 4.10.5. Define $A, B \in \mathbb{R}^{n \times n}$ by

$$A \triangleq \begin{bmatrix} 1 & -2 & 0 & \cdots & 0 & 0 \\ 0 & 1 & -2 & \ddots & 0 & 0 \\ 0 & 0 & 1 & \ddots & 0 & 0 \\ \vdots & \vdots & \ddots & \ddots & \ddots & \vdots \\ 0 & 0 & 0 & \ddots & 1 & -2 \\ 0 & 0 & 0 & \cdots & 0 & 1 \end{bmatrix}$$

and

$$B \triangleq \begin{bmatrix} 1 & -2 & 0 & \cdots & 0 & 0 \\ 0 & 1 & -2 & \ddots & 0 & 0 \\ 0 & 0 & 1 & \ddots & 0 & 0 \\ \vdots & \vdots & \ddots & \ddots & \ddots & \vdots \\ 0 & 0 & 0 & \ddots & 1 & -2 \\ \alpha & 0 & 0 & \cdots & 0 & 1 \end{bmatrix},$$

where $\alpha \triangleq -1/2^{n-1}$. Then,

$$\mathrm{spec}(A) = \{1\}$$

and

$$\det B = 0.$$

Fact 4.10.6. Let $A \in \mathbb{F}^{n \times n}$. Then,

$$|\mathrm{spabs}(A)| \leq \mathrm{sprad}(A).$$

Fact 4.10.7. Let $A \in \mathbb{F}^{n \times n}$, assume that A is nonsingular, and assume that $\mathrm{sprad}(I - A) < 1$. Then,

$$A^{-1} = \sum_{k=0}^{\infty} (I - A)^k.$$

Fact 4.10.8. Let $A \in \mathbb{F}^{n \times n}$ and $B \in \mathbb{F}^{m \times m}$. If $\mathrm{tr}\, A^k = \mathrm{tr}\, B^k$ for all $k \in \{1, \ldots, \max\{m, n\}\}$, then A and B have the same nonzero eigenvalues with the same algebraic multiplicity. Now, assume in addition that $n = m$. Then, $\mathrm{tr}\, A^k = \mathrm{tr}\, B^k$ for all $k \in \{1, \ldots, n\}$ if and only if $\mathrm{mspec}(A) = \mathrm{mspec}(B)$.

Proof: Use *Newton's identities*. See Fact 4.8.2.

Remark: This result yields Proposition 4.4.10 since $\mathrm{tr}\,(AB)^k = \mathrm{tr}\,(BA)^k$ for all $k \geq 1$ and for all nonsquare matrices A and B.

Remark: Setting $B = 0_{n \times n}$ yields necessity in Fact 2.12.14.

Fact 4.10.9. Let $A \in \mathbb{F}^{n \times n}$, and let $\operatorname{mspec}(A) = \{\lambda_1, \ldots, \lambda_n\}_{\mathrm{ms}}$. Then,

$$\operatorname{mspec}(A^{\mathrm{A}}) = \begin{cases} \left\{ \dfrac{\det A}{\lambda_1}, \ldots, \dfrac{\det A}{\lambda_n} \right\}_{\mathrm{ms}}, & \operatorname{rank} A = n, \\[2ex] \left\{ \displaystyle\sum_{i=1}^{n} \det A_{[i;i]}, 0, \ldots, 0 \right\}_{\mathrm{ms}}, & \operatorname{rank} A = n - 1, \\[2ex] \{0\}, & \operatorname{rank} A \leq n - 2. \end{cases}$$

Remark: If $\operatorname{rank} A = n - 1$ and $\lambda_n = 0$, then it follows from (4.4.19) that

$$\sum_{i=1}^{n} \det A_{[i;i]} = \lambda_1 \cdots \lambda_{n-1}.$$

Remark: See Fact 2.16.8, Fact 4.9.8, and Fact 5.11.36.

Fact 4.10.10. Let $A \in \mathbb{F}^{n \times n}$, and assume that A is either upper triangular or lower triangular. Then,

$$\chi_A(s) = \prod_{i=1}^{n} (s - A_{(i,i)}).$$

Consequently,

$$\operatorname{mspec}(A) = \{A_{(1,1)}, \ldots, A_{(n,n)}\}_{\mathrm{ms}}.$$

Remark: See Fact 3.22.1.

Fact 4.10.11. Let $A \in \mathbb{F}^{n \times n}$, $B \in \mathbb{F}^{n \times m}$, and $C \in \mathbb{F}^{m \times m}$, and let $p \in \mathbb{F}[s]$. Then,

$$p\left(\begin{bmatrix} A & B \\ 0 & C \end{bmatrix} \right) = \begin{bmatrix} p(A) & \hat{B} \\ 0 & p(C) \end{bmatrix},$$

where $\hat{B} \in \mathbb{F}^{n \times m}$.

Fact 4.10.12. Let $A_1 \in \mathbb{F}^{n \times n}$, $A_{12} \in \mathbb{F}^{n \times m}$, and $A_2 \in \mathbb{F}^{m \times m}$, and define $A \in \mathbb{F}^{(n+m) \times (n+m)}$ by

$$A \triangleq \begin{bmatrix} A_1 & A_{12} \\ 0 & A_2 \end{bmatrix}.$$

Then,

$$\chi_A = \chi_{A_1} \chi_{A_2}.$$

Furthermore,

$$\chi_{A_1}(A) = \begin{bmatrix} 0 & B_1 \\ 0 & \chi_{A_1}(A_2) \end{bmatrix}$$

and

$$\chi_{A_2}(A) = \begin{bmatrix} \chi_{A_2}(A_1) & B_2 \\ 0 & 0 \end{bmatrix},$$

where $B_1, B_2 \in \mathbb{F}^{n \times m}$. Therefore,

$$\mathcal{R}[\chi_{A_2}(A)] \subseteq \mathcal{R}\left(\begin{bmatrix} I_n \\ 0 \end{bmatrix} \right) \subseteq \mathcal{N}[\chi_{A_1}(A)]$$

and

$$\chi_{A_2}(A_1)B_1 + B_2\chi_{A_1}(A_2) = 0.$$

Hence,
$$\chi_A(A) = \chi_{A_1}(A)\chi_{A_2}(A) = \chi_{A_2}(A)\chi_{A_1}(A) = 0.$$

Fact 4.10.13. Let $A_1 \in \mathbb{F}^{n \times n}$, $A_{12} \in \mathbb{F}^{n \times m}$, and $A_2 \in \mathbb{F}^{m \times m}$, assume that $\mathrm{spec}(A_1)$ and $\mathrm{spec}(A_2)$ are disjoint, and define $A \in \mathbb{F}^{(n+m) \times (n+m)}$ by
$$A \triangleq \begin{bmatrix} A_1 & A_{12} \\ 0 & A_2 \end{bmatrix}.$$

Furthermore, let $\mu_1, \mu_2 \in \mathbb{F}[s]$ be such that
$$\mu_A = \mu_1\mu_2,$$
$$\mathrm{roots}(\mu_1) = \mathrm{spec}(A_1),$$
$$\mathrm{roots}(\mu_2) = \mathrm{spec}(A_2).$$

Then,
$$\mu_1(A) = \begin{bmatrix} 0 & B_1 \\ 0 & \mu_1(A_2) \end{bmatrix}$$

and
$$\mu_2(A) = \begin{bmatrix} \mu_2(A_1) & B_2 \\ 0 & 0 \end{bmatrix},$$

where $B_1, B_2 \in \mathbb{F}^{n \times m}$. Therefore,
$$\mathcal{R}[\mu_2(A)] \subseteq \mathcal{R}\left(\begin{bmatrix} I_n \\ 0 \end{bmatrix}\right) \subseteq \mathcal{N}[\mu_1(A)]$$

and
$$\mu_2(A_1)B_1 + B_2\mu_1(A_2) = 0.$$

Hence,
$$\mu_A(A) = \mu_1(A)\mu_2(A) = \mu_2(A)\mu_1(A) = 0.$$

Fact 4.10.14. Let $A_1, A_2, A_3, A_4, B_1, B_2 \in \mathbb{F}^{n \times n}$, and define $A \in \mathbb{F}^{4n \times 4n}$ by
$$A \triangleq \begin{bmatrix} A_1 & B_1 & 0 & 0 \\ 0 & A_2 & 0 & 0 \\ 0 & 0 & A_3 & 0 \\ 0 & 0 & B_2 & A_4 \end{bmatrix}.$$

Then,
$$\mathrm{mspec}(A) = \bigcup_{i=1}^{4} \mathrm{mspec}(A_i).$$

Fact 4.10.15. Let $A \in \mathbb{F}^{n \times m}$ and $B \in \mathbb{F}^{m \times n}$, and assume that $m < n$. Then,
$$\mathrm{mspec}(I_n + AB) = \mathrm{mspec}(I_m + BA) \cup \{1, \ldots, 1\}_{\mathrm{ms}}.$$

Fact 4.10.16. Let $a, b \in \mathbb{F}$, and define the symmetric, Toeplitz matrix $A \in \mathbb{F}^{n \times n}$ by

$$
A \triangleq \begin{bmatrix}
a & b & b & \cdots & b \\
b & a & b & \cdots & b \\
b & b & a & \cdots & b \\
\vdots & \vdots & \vdots & \ddots & \vdots \\
b & b & b & \cdots & a
\end{bmatrix}.
$$

Then,

$$
\mathrm{mspec}(A) = \{a + (n-1)b, a - b, \ldots, a - b\}_{\mathrm{ms}},
$$

$$
A 1_{n \times 1} = [a + (n-1)b] 1_{n \times 1},
$$

and

$$
A^2 + a_1 A + a_0 I = 0,
$$

where $a_1 \triangleq -2a + (2 - n)b$ and $a_0 \triangleq a^2 + (n-2)ab + (1-n)b^2$. Finally,

$$
\mathrm{mspec}(a I_n + b 1_{n \times n}) = \{a + nb, a, \ldots, a\}_{\mathrm{ms}}.
$$

Remark: See Fact 2.13.13 and Fact 8.9.35.

Remark: For the remaining eigenvectors of A, see [1215, pp. 149, 317].

Fact 4.10.17. Let $A \in \mathbb{F}^{n \times n}$. Then,

$$
\mathrm{spec}(A) \subset \bigcup_{i=1}^{n} \left\{ s \in \mathbb{C} \colon |s - A_{(i,i)}| \leq \sum_{\substack{j=1 \\ j \neq i}}^{n} |A_{(i,j)}| \right\}.
$$

Remark: This result is the *Gershgorin circle theorem*. See [274, 1404] for a proof and related results.

Remark: This result yields Corollary 9.4.5 for $\| \cdot \|_{\mathrm{col}}$ and $\| \cdot \|_{\mathrm{row}}$.

Fact 4.10.18. Let $A \in \mathbb{F}^{n \times n}$, and assume that, for all $i \in \{1, \ldots, n\}$,

$$
\sum_{\substack{j=1 \\ j \neq i}}^{n} |A_{(i,j)}| < |A_{(i,i)}|.
$$

Then, A is nonsingular.

Proof: Apply the Gershgorin circle theorem.

Remark: This result is the *diagonal dominance theorem*, and A is *diagonally dominant*. See [1204] for a history of this result.

Remark: For related results, see Fact 4.10.20 and [469, 1045, 1134].

Fact 4.10.19. Let $A \in \mathbb{F}^{n \times n}$, assume that, for all $i \in \{1, \ldots, n\}$, $A_{(i,i)} \neq 0$, and assume that

$$\alpha_i \triangleq \frac{\sum_{j=1, j \neq i}^{n} |A_{(i,j)}|}{|A_{(i,i)}|} < 1.$$

Then,

$$|A_{(1,1)}| \prod_{i=2}^{n} (|A_{(i,i)}| - l_i + L_i) \leq |\det A|,$$

where

$$l_i \triangleq \sum_{j=1}^{i-1} \alpha_j |A_{(i,j)}|, \qquad L_i \triangleq \left| \frac{A_{(i,1)}}{A_{(1,1)}} \right| \sum_{j=i+1}^{n} |A_{(i,j)}|.$$

Proof: See [260].

Remark: Note that, for all $i \in \{1, \ldots, n\}$,

$$l_i = \sum_{j=1}^{i-1} \alpha_j |A_{(i,j)}| \leq \sum_{j=1, j \neq i}^{n} \alpha_j |A_{(i,j)}| \leq \sum_{j=1, j \neq i}^{n} |A_{(i,j)}| = \alpha_i |A_{(i,i)}| < |A_{(i,i)}|.$$

Hence, the lower bound for $|\det A|$ is positive.

Fact 4.10.20. Let $A \in \mathbb{F}^{n \times n}$, and, for all $i \in \{1, \ldots, n\}$, define

$$r_i \triangleq \sum_{\substack{j=1 \\ j \neq i}}^{n} |A_{(i,j)}|, \qquad c_i \triangleq \sum_{\substack{j=1 \\ j \neq i}}^{n} |A_{(j,i)}|.$$

Furthermore, assume that at least one of the following conditions is satisfied:

  i) For all distinct $i, j \in \{1, \ldots, n\}$, $r_i c_j < |A_{(i,i)} A_{(j,j)}|$.

  ii) A is irreducible, for all $i \in \{1, \ldots, n\}$ it follows that $r_i \leq |A_{(i,i)}|$, and there exists $i \in \{1, \ldots, n\}$ such that $r_i < |A_{(i,i)}|$.

  iii) There exist positive integers $k_1, \ldots, k_n$ such that $\sum_{i=1}^{n} (1 + k_i)^{-1} \leq 1$ and such that, for all $i \in \{1, \ldots, n\}$, $k_i \max_{j=1, \ldots, n, j \neq i} |A_{(i,j)}| < |A_{(i,i)}|$.

  iv) There exists $\alpha \in [0, 1]$ such that, for all $i \in \{1, \ldots, n\}$, $r_i^\alpha c_i^{1-\alpha} < |A_{(i,i)}|$.

Then, A is nonsingular.

Proof: See [104].

Remark: All three conditions yield stronger results than Fact 4.10.18.

Fact 4.10.21. Let $A \in \mathbb{R}^{n \times n}$, assume that A is symmetric, and, for all $i \in \{1, \ldots, n\}$, define

$$\alpha_i \triangleq \sum_{\substack{j=1 \\ j \neq i}}^{n} |A_{(i,j)}|.$$

Then,

$$\text{spec}(A) \subset \bigcup_{i=1}^{n} [A_{(i,i)} - \alpha_i, A_{(i,i)} + \alpha_i].$$

Furthermore, for $i = 1, \ldots, n$, define

$$\beta_i \triangleq \max\{0, \max_{\substack{j=1,\ldots,n \\ j \neq i}} A_{(i,j)}\}$$

and

$$\gamma_i \triangleq \min\{0, \min_{\substack{j=1,\ldots,n \\ j \neq i}} A_{(i,j)}\}.$$

Then,

$$\text{spec}(A) \subset \bigcup_{i=1}^n \left[\left(\sum_{j=1}^n A_{(i,j)} \right) - n\beta_i, \left(\sum_{j=1}^n A_{(i,j)} \right) - n\gamma_i \right].$$

Proof: The first statement is the specialization of the Gershgorin circle theorem to real, symmetric matrices. See Fact 4.10.17. The second result is given in [141].

Fact 4.10.22. Let $A \in \mathbb{F}^{n \times n}$. Then,

$$\text{spec}(A) \subset \bigcup_{\substack{i,j=1 \\ i \neq j}}^n \left\{ s \in \mathbb{C} : |s - A_{(i,i)}||s - A_{(j,j)}| \leq \sum_{\substack{k=1 \\ k \neq i}}^n |A_{(i,k)}| \sum_{\substack{k=1 \\ k \neq j}}^n |A_{(j,k)}| \right\}.$$

Remark: The inclusion region is the *ovals of Cassini*. This result is due to Brauer. See [728, p. 380].

Fact 4.10.23. Let $A \in \mathbb{F}^{n \times n}$, and let λ_n denote the eigenvalue of A of smallest absolute value. Then,

$$|\lambda_n| \leq \max_{i=1,\ldots,n} |\text{tr } A^i|^{1/i}.$$

Furthermore,

$$\text{sprad}(A) \leq \max_{i=1,\ldots,2n-1} |\text{tr } A^i|^{1/i}$$

and

$$\text{sprad}(A) \leq \frac{5}{n} \max_{i=1,\ldots,n} |\text{tr } A^i|^{1/i}.$$

Remark: These results are *Turan's inequalities*. See [1035, p. 657].

Fact 4.10.24. Let $A \in \mathbb{F}^{n \times n}$, and, for $j = 1, \ldots, n$, define $b_j \triangleq \sum_{i=1}^n |A_{(i,j)}|$. Then,

$$\sum_{j=1}^n |A_{(j,j)}|/b_j \leq \text{rank } A.$$

Proof: See [1125, p. 67].

Remark: Interpret $0/0$ as 0.

Remark: See Fact 4.10.18.

Fact 4.10.25. Let $A_1, \ldots, A_r \in \mathbb{F}^{n \times n}$, assume that $A_1, \ldots, A_r$ are normal, and let $A \in \text{co}\{A_1, \ldots, A_r\}$. Then,

$$\text{spec}(A) \subseteq \text{co} \bigcup_{i=1,\ldots,r} \text{spec}(A_i).$$

Proof: See [1433].

Remark: See Fact 8.14.7.

Fact 4.10.26. Let $A, B \in \mathbb{R}^{n \times n}$. Then,

$$\mathrm{mspec}\left(\begin{bmatrix} A & B \\ B & A \end{bmatrix}\right) = \mathrm{mspec}(A + B) \cup \mathrm{mspec}(A - B).$$

Proof: See [1215, p. 93].

Remark: See Fact 2.14.26.

Fact 4.10.27. Let $A, B \in \mathbb{R}^{n \times n}$. Then,

$$\mathrm{mspec}\left(\begin{bmatrix} A & B \\ -B & A \end{bmatrix}\right) = \mathrm{mspec}(A + \jmath B) \cup \mathrm{mspec}(A - \jmath B).$$

Now, assume in addition that A is symmetric and B is skew symmetric. Then, $\begin{bmatrix} A & B \\ B^\mathrm{T} & A \end{bmatrix}$ is symmetric, $A + \jmath B$ is Hermitian, and

$$\mathrm{mspec}\left(\begin{bmatrix} A & B \\ B^\mathrm{T} & A \end{bmatrix}\right) = \mathrm{mspec}(A + \jmath B) \cup \mathrm{mspec}(A + \jmath B).$$

Remark: See Fact 2.19.3 and Fact 8.15.7.

Fact 4.10.28. Let $A \in \mathbb{F}^{n \times n}$, $B \in \mathbb{F}^{m \times m}$, and $C \in \mathbb{F}^{n \times m}$, assume that A and B are Hermitian, and define $\mathcal{A}_0 \triangleq \begin{bmatrix} A & 0 \\ 0 & B \end{bmatrix}$ and $\mathcal{A} \triangleq \begin{bmatrix} A & C \\ C^* & B \end{bmatrix}$. Furthermore, define

$$\eta \triangleq \min_{\substack{i=1,\ldots,n \\ j=1,\ldots,m}} |\lambda_i(A) - \lambda_j(B)|.$$

Then, for all $i \in \{1, \ldots, n+m\}$,

$$|\lambda_i(\mathcal{A}) - \lambda_i(\mathcal{A}_0)| \leq \frac{2\sigma_{\max}^2(C)}{\eta + \sqrt{\eta^2 + 4\sigma_{\max}(C)}}.$$

Proof: See [204, pp. 142–146] or [919].

Fact 4.10.29. Let $A \in \mathbb{R}^{n \times n}$, let $b, c \in \mathbb{R}^n$, define $p \in \mathbb{R}[s]$ by $p(s) \triangleq c^\mathrm{T}(sI - A)^\mathrm{A}b$, assume that p and $\det(sI - A)$ are coprime, define $A_\alpha \triangleq A + \alpha bc^\mathrm{T}$ for $\alpha \in [0, \infty)$, and let $\lambda \colon [0, \infty) \to \mathbb{C}$ be a continuous function such that $\lambda(\alpha) \in \mathrm{spec}(A_\alpha)$ for all $\alpha \in [0, \infty)$. Then, either $\lim_{\alpha \to \infty} |\lambda(\alpha)| = \infty$ or $\lim_{\alpha \to \infty} \lambda(\alpha) \in \mathrm{roots}(p)$.

Remark: This result is a consequence of *root locus* analysis from classical control theory, which determines asymptotic pole locations under high-gain feedback.

Fact 4.10.30. Let $A \in \mathbb{F}^{n \times n}$, where $n \geq 2$, and assume that there exist $\alpha \in [0, \infty)$ and $B \in \mathbb{F}^{n \times n}$ such that $A = \alpha I - B$ and $\mathrm{sprad}(B) \leq \alpha$. Then,

$$\mathrm{spec}(A) \subset \{0\} \cup \mathrm{ORHP}.$$

If, in addition, $\mathrm{sprad}(B) < \alpha$, then

$$\mathrm{spec}(A) \subset \mathrm{ORHP},$$

and thus A is nonsingular.

Proof: Let $\lambda \in \mathrm{spec}(A)$. Then, there exists $\mu \in \mathrm{spec}(B)$ such that $\lambda = \alpha - \mu$. Hence, $\mathrm{Re}\,\lambda = \alpha - \mathrm{Re}\,\mu$. Since $\mathrm{Re}\,\mu \leq |\mathrm{Re}\,\mu| \leq |\mu| \leq \mathrm{sprad}(B)$, it follows that

$\operatorname{Re}\lambda \geq \alpha - |\operatorname{Re}\mu| \geq \alpha - |\mu| \geq \alpha - \operatorname{sprad}(B) \geq 0$. Hence, $\operatorname{Re}\lambda \geq 0$. Now, suppose that $\operatorname{Re}\lambda = 0$. Then, since $\alpha - \lambda = \mu \in \operatorname{spec}(B)$, it follows that $\alpha^2 + |\lambda|^2 \leq [\operatorname{sprad}(B)]^2 \leq \alpha^2$. Hence, $\lambda = 0$. By a similar argument, if $\operatorname{sprad}(B) < \alpha$, then $\operatorname{Re}\lambda > 0$.

Remark: Converses of these statements hold when B is nonnegative. See Fact 4.11.8.

4.11 Facts on Graphs and Nonnegative Matrices

Fact 4.11.1. Let $\mathcal{G} = (\mathcal{X}, \mathcal{R})$ be a graph, where $\mathcal{X} = \{x_1, \ldots, x_n\}$, and let A be the adjacency matrix of $\mathcal{G}$. Then, the following statements hold:

i) The number of distinct walks from x_i to x_j of length $k \geq 1$ is $(A^k)_{(j,i)}$.

ii) Let k be an integer such that $1 \leq k \leq n-1$. Then, for distinct $x_i, x_j \in \mathcal{X}$, the number of distinct walks from x_i to x_j whose length is less than or equal to k is $[(I+A)^k]_{(j,i)}$.

Fact 4.11.2. Let $A \in \mathbb{F}^{n \times n}$, and consider $\mathcal{G}(A) = (\mathcal{X}, \mathcal{R})$, where $\mathcal{X} = \{x_1, \ldots, x_n\}$. Then, the following statements are equivalent:

i) $\mathcal{G}(A)$ is connected.

ii) There exists $k \geq 1$ such that $(I + |A|)^{k-1}$ is positive.

iii) $(I + |A|)^{n-1}$ is positive.

Proof: See [728, pp. 358, 359].

Fact 4.11.3. Let $\mathcal{G} = (\{x_1, \ldots, x_n\}, \mathcal{R})$ be a graph, and let $A \in \mathbb{R}^{n \times n}$ be the adjacency matrix of $\mathcal{G}$. Then, the following statements are equivalent:

i) $\mathcal{G}$ is connected.

ii) $\mathcal{G}$ has no directed cuts.

iii) A is irreducible.

iv) $\sum_{i=0}^{n-1} A^i$ is positive.

v) $(I + A)^{n-1}$ is positive.

If, in addition, every node has a self-loop, then the following condition is equivalent to i)–v):

vi) A^{n-1} is positive.

Furthermore, the following statements are equivalent:

vii) $\mathcal{G}$ is not connected.

viii) $\mathcal{G}$ has a directed cut.

ix) A is reducible.

x) $\sum_{i=0}^{n-1} A^i$ has at least one entry that is zero.

xi) $(I + A)^{n-1}$ has at least one entry that is zero.

If, in addition, every node has a self-loop, then the following condition is equivalent to *vii)*–*xi)*:

xii) A^{n-1} has at least one zero entry.

Finally, suppose that A is reducible and there exist $k \geq 1$ and a permutation matrix $S \in \mathbb{R}^{n \times n}$ such that $SAS^{\mathrm{T}} = \begin{bmatrix} B & C \\ 0_{k \times (n-k)} & D \end{bmatrix}$, where $B \in \mathbb{F}^{(n-k) \times (n-k)}$, $C \in \mathbb{F}^{(n-k) \times k}$, and $D \in \mathbb{F}^{k \times k}$. Then, $(\{x_{i_1}, \ldots, x_{i_{n-k}}\}, \{x_{i_{n-k+1}}, \ldots, x_{i_n}\})$ is a directed cut, where $\begin{bmatrix} i_1 & \cdots & i_n \end{bmatrix}^{\mathrm{T}} = S \begin{bmatrix} 1 & \cdots & n \end{bmatrix}^{\mathrm{T}}$.

Proof: See [728, p. 362] and [1177, p. 9-3].

Fact 4.11.4. Let $A \in \mathbb{R}^{n \times n}$, where $n \geq 2$, and assume that A is nonnegative. Then, the following statements hold:

i) $\mathrm{sprad}(A)$ is an eigenvalue of A, and there exists a nonnegative eigenvector $x \in \mathbb{R}^n$ associated with $\mathrm{sprad}(A)$.

ii) If $x \in \mathbb{R}^n$ is a positive eigenvector of A associated with the eigenvalue $\lambda \in \mathbb{C}$, then $\lambda = \mathrm{sprad}(A)$.

Furthermore, the following statements are equivalent:

iii) A is irreducible.

iv) $(I + A)^{n-1}$ is positive.

v) $\mathcal{G}(A)$ is connected.

vi) A has exactly one positive eigenvector whose components sum to 1, and A has no other nonnegative eigenvector whose components sum to 1.

If A is irreducible, then the following statements hold:

vii) $\mathrm{sprad}(A) > 0$.

viii) $\mathrm{sprad}(A)$ is a simple eigenvalue of A.

ix) A has exactly one positive eigenvector $x \in \mathbb{R}^n$ whose components sum to 1. Furthermore, $Ax = \mathrm{sprad}(A)x$.

x) If $x \in \mathbb{R}^n$ is a positive eigenvector of A associated with the eigenvalue $\lambda \in \mathbb{C}$, then $\lambda = \mathrm{sprad}(A)$.

xi) Assume that $\{\lambda_1, \ldots, \lambda_k\}_{\mathrm{ms}} = \{\lambda \in \mathrm{mspec}(A) : |\lambda| = \mathrm{sprad}(A)\}_{\mathrm{ms}}$. Then, $\lambda_1, \ldots, \lambda_k$ are distinct, and
$$\{\lambda_1, \ldots, \lambda_k\} = \{e^{j2\pi i/k} \mathrm{sprad}(A) : \ i = 1, \ldots, k\}.$$
Furthermore,
$$\mathrm{mspec}(A) = e^{j2\pi/k} \mathrm{mspec}(A).$$

xii) If at least one diagonal entry of A is positive, then $\mathrm{sprad}(A)$ is the only eigenvalue of A whose absolute value is $\mathrm{sprad}(A)$.

xiii) If A has at least m positive diagonal entries, then A^{2n-m-1} is positive.

xiv) If $x, y \in \mathbb{R}^n$ are positive and satisfy $Ax = \mathrm{sprad}(A)x$ and $A^{\mathrm{T}}y = \mathrm{sprad}(A)y$, then

$$\lim_{k \to \infty} \frac{1}{k} \sum_{i=1}^{k} ([\mathrm{sprad}(A)]^{-1}A)^i = \frac{1}{x^{\mathrm{T}}y}xy^{\mathrm{T}}.$$

In addition, the following statements are equivalent:

xv) There exists $k \geq 1$ such that A^k is positive.

xvi) A is irreducible and $|\lambda| < \mathrm{sprad}(A)$ for all $\lambda \in \mathrm{spec}(A)\backslash\{\mathrm{sprad}(A)\}$.

xvii) A^{n^2-2n+2} is positive.

xviii) A is irreducible and $\mathcal{G}(A)$ is aperiodic.

A is *primitive* if *xv*)–*xviii*) are satisfied. If A is primitive, then the following statements hold:

xix) For all $k \in \mathbb{P}$, A^k is primitive.

xx) If $k \in \mathbb{P}$ and A^k is positive, then, for all $l \geq k$, A^l is positive.

xxi) There exists a positive integer $k \leq (n-1)n^n$ such that A^k is positive.

xxii) If $x, y \in \mathbb{R}^n$ are positive and satisfy $Ax = \mathrm{sprad}(A)x$ and $A^{\mathrm{T}}y = \mathrm{sprad}(A)y$, then

$$\lim_{k \to \infty} ([\mathrm{sprad}(A)]^{-1}A)^k = \frac{1}{x^{\mathrm{T}}y}xy^{\mathrm{T}}.$$

xxiii) If $x_0 \in \mathbb{R}^n$ is nonzero and nonnegative and $x, y \in \mathbb{R}^n$ are positive and satisfy $Ax = \mathrm{sprad}(A)x$ and $A^{\mathrm{T}}y = \mathrm{sprad}(A)y$, then

$$\lim_{k \to \infty} \frac{A^k x_0 - [\mathrm{sprad}(A)]^k y^{\mathrm{T}} x_0 x}{\|A^k x_0\|_2} = 0.$$

xxiv) $\mathrm{sprad}(A) = \lim_{k \to \infty} (\mathrm{tr}\ A^k)^{1/k}$.

Proof: See [18, pp. 45–49], [137, p. 17], [185, pp. 26–28, 32, 55], [494, Chapter 4], [728, pp. 507–518, 524, 525], and [1177, p. 9-3]. For *xxiii*), see [1224] and [1403, p. 49].

Remark: This result is the *Perron-Frobenius theorem*.

Remark: Statement *xvii*) is due to Wielandt. See [1125, p. 157].

Remark: Statement *xviii*) is given in [1177, p. 9-3].

Remark: See Fact 6.6.21 and Fact 11.18.20.

Example: $\left[\begin{smallmatrix}1&1\\0&0\end{smallmatrix}\right]$ and $\left[\begin{smallmatrix}1&1\\0&1\end{smallmatrix}\right]$ are reducible. $\left[\begin{smallmatrix}0&1\\1&0\end{smallmatrix}\right]$ is irreducible but not primitive. $\left[\begin{smallmatrix}1&1\\1&0\end{smallmatrix}\right]$ is primitive. $\left[\begin{smallmatrix}1&1\\1&1\end{smallmatrix}\right]$ is positive and thus primitive.

Remark: For an arbitrary nonzero and nonnegative initial condition, the state $x_k = A^k x_0$ of the difference equation $x_{k+1} = Ax_k$ approaches a distribution given by the eigenvector associated with the positive eigenvalue of maximum absolute value. In demography, this eigenvector is interpreted as the *stable age distribution*. See [828, pp. 47, 63].

Example: Let x and y be positive numbers such that $x + y < 1$, and define

$$A \triangleq \begin{bmatrix} x & y & 1-x-y \\ 1-x-y & x & y \\ y & 1-x-y & x \end{bmatrix}.$$

Then, $A 1_{3 \times 1} = A^{\mathrm{T}} 1_{3 \times 1} = 1_{3 \times 1}$, and thus $\lim_{k \to \infty} A^k = \frac{1}{3} 1_{3 \times 3}$. See [242, p. 213].

Fact 4.11.5. Let $A \in \mathbb{R}^{n \times n}$, where $n \geq 2$, and assume that A is nonnegative. Then, there exists a perturbation matrix $S \in \mathbb{R}^{n \times n}$ such that SAS^{T} is upper block triangular and such that every diagonally located block is either zero or irreducible.

Proof: If A is either zero or irreducible, then the result holds with $S = I$. If A is either nonzero or reducible, then there exists an $n \times n$ perturbation matrix S such that SAS^{T} is upper block triangular with square, nonnegative, diagonally located blocks B and C. If each matrix B and C is either zero or irreducible, then the result holds. If not, then either B or C can be transformed as needed.

Example: $\begin{bmatrix} 1 & 1 \\ 0 & 0 \end{bmatrix}$ and $\begin{bmatrix} 1 & 1 \\ 0 & 1 \end{bmatrix}$ are reducible, and every diagonally located block is either zero or irreducible.

Remark: This result gives the *Frobenius normal form.* See [271, p. 27-6]. Note that in [271, p. 27-5], all scalar matrices are defined to be irreducible, whereas, in [728, p. 360], the scalar zero matrix is defined to be reducible, which is the convention used here.

Fact 4.11.6. Let $A \in \mathbb{R}^{n \times n}$, where $n \geq 2$, and assume that A is row stochastic. Then, $\mathrm{sprad}(A) = 1$.

Proof: Since $1_{n \times 1}$ is an eigenvector of A associated with the eigenvalue 1, the result follows from *ii)* of Fact 4.11.4. Alternatively, note that $\|A\|_{\infty,\infty} = \|A\|_{\mathrm{row}} = 1$. Since $1 \in \mathrm{spec}(A)$ and $\|\cdot\|_{\infty,\infty}$ is an induced norm, it follows from Corollary 9.4.5 that $1 \leq \mathrm{sprad}(A) \leq \|A\|_{\infty,\infty} = 1$.

Remark: It follows from Fact 3.11.3 that, if $A \in \mathrm{co}\,\mathrm{P}(n)$, then $\mathrm{sprad}(A) = 1$.

Remark: See Fact 11.21.11.

Fact 4.11.7. Let $\mathcal{G} = (\mathcal{X}, \mathcal{R})$ be a graph, where $\mathcal{X} = \{x_1, \ldots, x_n\}$, and let $L_{\mathrm{out}} \in \mathbb{R}^{n \times n}$ denote the outbound Laplacian of $\mathcal{G}$. Then, the following statements hold:

  i) $0 \in \mathrm{spec}(L_{\mathrm{out}}) \subset \{0\} \cup \mathrm{ORHP}$.

  ii) $L_{\mathrm{out}} 1_{n \times 1} = 0$.

  iii) 0 is a semisimple eigenvalue of L_{out}.

  iv) If $\mathcal{G}$ is connected, then 0 is a simple eigenvalue of L_{out}.

  v) The following statements are equivalent:

     a) There exists a node $x \in \mathrm{supp}(\mathcal{G})$ such that, for every node $y \in \mathrm{supp}(\mathcal{G})$, there exists a walk from y to x.

     b) 0 is a simple eigenvalue of L_{out}.

vi) $L_{\text{out}} + L_{\text{out}}^{\text{T}}$ is positive semidefinite.

vii) L_{out} is symmetric if and only if $\mathcal{G}$ is symmetric.

Now, assume in addition that $\mathcal{G}$ is symmetric, and let $L \in \mathbb{R}^{n \times n}$ denote the Laplacian of $\mathcal{G}$. Then, the following statements hold:

viii) L is positive semidefinite.

ix) $0 \in \text{spec}(L) \subset \{0\} \cup [0, \infty)$.

x) $\mathcal{G}$ is connected if and only if $\text{rank}\, L = n - 1$.

xi) If $\mathcal{G}$ is connected, then 0 is a simple eigenvalue of L.

xii) 0 is a simple eigenvalue of L if and only if $\mathcal{G}$ has a spanning subgraph that is a tree.

Proof: See [282, p. 40, 77]. For the last statement, see [1018, p. 147].

Remark: See Fact 11.19.7.

Remark: Statement *v)* means that $\mathcal{G}$ has at least one globally reachable node. See [282, p. 40].

Fact 4.11.8. Let $A \in \mathbb{R}^{n \times n}$, where $n \geq 2$, and assume that A is a Z-matrix. Then, the following statements are equivalent:

i) There exist $\alpha \in (0, \infty)$ and $B \in \mathbb{R}^{n \times n}$ such that $A = \alpha I - B$, B is nonnegative, and $\text{sprad}(B) \leq \alpha$.

ii) $\text{spec}(A) \subset \text{ORHP} \cup \{0\}$.

iii) $\text{spec}(A) \subset \text{CRHP}$.

iv) If $\lambda \in \text{spec}(A)$ is real, then $\lambda \geq 0$.

v) Every principal subdeterminant of A is nonnegative.

vi) For every diagonal, positive-definite matrix $D \in \mathbb{R}^{n \times n}$, it follows that $A + D$ is nonsingular.

A is an *M-matrix* if A is a Z-matrix and *i)–vi)* hold. In addition, the following statements are equivalent:

vii) There exist $\alpha \in (0, \infty)$ and $B \in \mathbb{R}^{n \times n}$ such that $A = \alpha I - B$, B is nonnegative, and $\text{sprad}(B) < \alpha$.

viii) $\text{spec}(A) \subset \text{ORHP}$.

Proof: The result *i)* $\Longrightarrow$ *ii)* follows from Fact 4.10.30, while *ii)* $\Longrightarrow$ *iii)* is immediate. To prove that *iii)* $\Longrightarrow$ *i)*, let $\alpha \in (0, \infty)$ be sufficiently large that $B \triangleq \alpha I - A$ is nonnegative. Hence, for every $\mu \in \text{spec}(B)$, it follows that $\lambda \triangleq \alpha - \mu \in \text{spec}(A)$. Since $\text{Re}\,\lambda \geq 0$, it follows that every $\mu \in \text{spec}(B)$ satisfies $\text{Re}\,\mu \leq \alpha$. Since B is nonnegative, it follows from *i)* of Fact 4.11.4 that $\text{sprad}(B)$ is an eigenvalue of B. Hence, setting $\mu = \text{sprad}(B)$ implies that $\text{sprad}(B) \leq \alpha$. Conditions *iv)* and *v)* are proved in [186, pp. 149, 150]. Finally, the argument used to prove that *i)* $\Longrightarrow$ *ii)* shows in addition that *vii)* $\Longrightarrow$ *viii)*.

Example: $A = \begin{bmatrix} 0 & -1 \\ 0 & 0 \end{bmatrix} = I - \begin{bmatrix} 1 & 1 \\ 0 & 1 \end{bmatrix})$ is an M-matrix.

Remark: A is a *nonsingular M-matrix* if *vii*) and *viii*) hold. See Fact 11.19.5.

Remark: See Fact 11.19.3.

Fact 4.11.9. Let $A \in \mathbb{R}^{n \times n}$, where $n \geq 2$. If A is a Z-matrix, then every principal submatrix of A is also a Z-matrix. Furthermore, if A is an M-matrix, then every principal submatrix of A is also an M-matrix.

Proof: See [730, p. 114].

Fact 4.11.10. Let $A \in \mathbb{R}^{n \times n}$, where $n \geq 2$, and assume that A is a nonsingular M-matrix, B is a Z-matrix, and $A \leq\leq B$. Then, the following statements hold:

  i) $\text{tr}(A^{-1}A^{\text{T}}) \leq n$.

  ii) $\text{tr}(A^{-1}A^{\text{T}}) = n$ if and only if A is symmetric.

  iii) B is a nonsingular M-matrix.

  iv) $0 \leq B^{-1} \leq A^{-1}$.

  v) $0 < \det A \leq \det B$.

Proof: See [730, pp. 117, 370].

Fact 4.11.11. Let $A \in \mathbb{R}^{n \times n}$, where $n \geq 2$, assume that A is a Z-matrix, and define
$$\tau(A) \triangleq \min\{\text{Re}\,\lambda \colon \lambda \in \text{spec}(A)\}.$$
Then, the following statements hold:

  i) $\tau(A) \in \text{spec}(A)$.

  ii) $\min_{i=1,\ldots,n} \sum_{j=1}^{n} A_{(i,j)} \leq \tau(A)$.

Now, assume in addition that A is an M-matrix. Then, the following statements hold:

  iii) If A is nonsingular, then $\tau(A) = 1/\text{sprad}(A^{-1})$.

  iv) $[\tau(A)]^n \leq \det A$.

  v) If $B \in \mathbb{R}^{n \times n}$, B is an M-matrix, and $B \leq\leq A$, then $\tau(B) \leq \tau(A)$.

Proof: See [730, pp. 128–131].

Remark: $\tau(A)$ is the *minimum eigenvalue* of A.

Remark: See Fact 7.6.15.

Fact 4.11.12. Let $A \in \mathbb{R}^{n \times n}$, where $n \geq 2$, and assume that A is an M-matrix. Then, the following statements hold:

  i) There exists a nonzero nonnegative vector $x \in \mathbb{R}^n$ such that Ax is nonnegative.

  ii) If A is irreducible, then there exists a positive vector $x \in \mathbb{R}^n$ such that Ax is nonnegative.

Now, assume in addition that A is singular. Then, the following statements hold:

iii) $\operatorname{rank} A = n - 1$.

iv) There exists a positive vector $x \in \mathbb{R}^n$ such that $Ax = 0$.

v) A is group invertible.

vi) Every principal submatrix of A of order less than n and greater than 1 is a nonsingular M-matrix.

vii) If $x \in \mathbb{R}^n$ and Ax is nonnegative, then $Ax = 0$.

Proof: To prove the first statement, it follows from Fact 4.11.8 that there exist $\alpha \in (0, \infty)$ and $B \in \mathbb{R}^{n \times n}$ such that $A = \alpha I - B$, B is nonnegative, and $\operatorname{sprad}(B) \leq \alpha$. Consequently, it follows from *ii*) of Fact 4.11.4 that there exists a nonzero nonnegative vector $x \in \mathbb{R}^n$ such that $Bx = \operatorname{sprad}(B)x$. Therefore, $Ax = [\alpha - \operatorname{sprad}(B)]x$ is nonnegative. Statements *iii*)–*vii*) are given in [186, p. 156].

Fact 4.11.13. Let $A \triangleq \left[\begin{smallmatrix} 1 & 1 \\ 1 & 0 \end{smallmatrix}\right]$. Then, $\chi_A(s) = s^2 - s - 1$ and $\operatorname{spec}(A) = \{\alpha, \beta\}$, where $\alpha \triangleq \frac{1}{2}(1 + \sqrt{5}) \approx 1.61803$ and $\beta \triangleq \frac{1}{2}(1 - \sqrt{5}) \approx -0.61803$ satisfy

$$\alpha - 1 = 1/\alpha, \qquad \beta - 1 = 1/\beta.$$

Furthermore, $\left[\begin{smallmatrix} \alpha \\ 1 \end{smallmatrix}\right]$ is an eigenvector of A associated with α. Now, for $k \geq 0$, consider the difference equation

$$x_{k+1} = Ax_k.$$

Then, for all $k \geq 0$,

$$x_k = A^k x_0$$

and

$$x_{k+2(1)} = x_{k+1(1)} + x_{k(1)}.$$

Furthermore, if x_0 is positive, then

$$\lim_{k \to \infty} \frac{x_{k(1)}}{x_{k(2)}} = \alpha.$$

In particular, if $x_0 \triangleq \left[\begin{smallmatrix} 1 \\ 1 \end{smallmatrix}\right]$, then, for all $k \geq 0$,

$$x_k = \begin{bmatrix} F_{k+2} \\ F_{k+1} \end{bmatrix},$$

where $F_1 \triangleq F_2 \triangleq 1$ and, for all $k \geq 1$, F_k is given by

$$F_k = \frac{1}{\sqrt{5}}(\alpha^k - \beta^k)$$

and satisfies

$$F_{k+2} = F_{k+1} + F_k.$$

Furthermore,

$$\frac{1}{1 - x - x^2} = F_1 x + F_2 x^2 + \cdots$$

and

$$A^k = \begin{bmatrix} F_{k+1} & F_k \\ F_k & F_{k-1} \end{bmatrix}.$$

On the other hand, if $x_0 \triangleq \left[\begin{smallmatrix} 3 \\ 1 \end{smallmatrix}\right]$, then, for all $k \geq 0$,

$$x_k = \left[\begin{array}{c} L_{k+2} \\ L_{k+1} \end{array} \right],$$

where $L_1 \triangleq 1$, $L_2 \triangleq 3$, and, for all $k \geq 1$, L_k is given by

$$L_k = \alpha^k + \beta^k$$

and satisfies

$$L_{k+2} = L_{k+1} + L_k.$$

Moreover,

$$\lim_{k \to \infty} \frac{F_{k+1}}{F_k} = \frac{L_{k+1}}{L_k} = \alpha.$$

In addition,

$$\alpha = \sqrt{1 + \sqrt{1 + \sqrt{1 + \sqrt{1 + \cdots}}}}.$$

Finally, for all $k \geq 1$,

$$F_{k+1} = \det \left[\begin{array}{ccccccc} 1 & \jmath & 0 & \cdots & 0 & 0 \\ \jmath & 1 & \jmath & \cdots & 0 & 0 \\ 0 & \jmath & 1 & \ddots & 0 & 0 \\ \vdots & \vdots & \ddots & \ddots & \ddots & 0 \\ 0 & 0 & 0 & \ddots & 1 & \jmath \\ 0 & 0 & 0 & \cdots & \jmath & 1 \end{array} \right] = \det \left[\begin{array}{ccccccc} 1 & 1 & 0 & \cdots & 0 & 0 \\ -1 & 1 & 1 & \cdots & 0 & 0 \\ 0 & -1 & 1 & \ddots & 0 & 0 \\ \vdots & \vdots & \ddots & \ddots & \ddots & 0 \\ 0 & 0 & 0 & \ddots & 1 & 1 \\ 0 & 0 & 0 & \cdots & -1 & 1 \end{array} \right],$$

where both matrices are of size $k \times k$.

Proof: Use the last statement of Fact 4.11.4.

Remark: F_k is the kth *Fibonacci number*, L_k is the kth *Lucas number*, and α is the *golden ratio*. See [865, pp. 6–8, 239–241, 362, 363] and Fact 12.23.4. The expressions for F_k and L_k involving powers of α and β are *Binet's formulas*. See [181, p. 125]. The iterated square root equality is given in [490, p. 24]. The determinant equalities are given in [286] and [1146, p. 515].

Remark: $1/(1 - x - x^2)$ is a *generating function* for the Fibonacci numbers. See [1441].

Fact 4.11.14. Consider the nonnegative companion matrix $A \in \mathbb{R}^{n \times n}$ defined by

$$A \triangleq \left[\begin{array}{cccccc} 0 & 1 & 0 & \cdots & 0 & 0 \\ 0 & 0 & 1 & \ddots & 0 & 0 \\ 0 & 0 & 0 & \ddots & 0 & 0 \\ \vdots & \vdots & \vdots & \ddots & \ddots & \vdots \\ 0 & 0 & 0 & \cdots & 0 & 1 \\ 1/n & 1/n & 1/n & \cdots & 1/n & 1/n \end{array} \right].$$

Then, A is irreducible, 1 is a simple eigenvalue of A with associated eigenvector $1_{n\times 1}$, and $|\lambda| < 1$ for all $\lambda \in \mathrm{spec}(A)\backslash\{1\}$. Furthermore, if $x \in \mathbb{R}^n$, then

$$\lim_{k\to\infty} A^k x = \left[\frac{2}{n(n+1)}\sum_{i=1}^{n} ix_{(i-1)}\right]1_{n\times 1}.$$

Proof: See [644, pp. 82, 83, 263–266].

Remark: This result follows from Fact 4.11.4.

Fact 4.11.15. Let $A \in \mathbb{R}^{n\times m}$ and $b \in \mathbb{R}^m$. Then, the following statements are equivalent:

i) If $x \in \mathbb{R}^m$ and $Ax \geq\geq 0$, then $b^{\mathrm{T}}x \geq 0$.

ii) There exists a vector $y \in \mathbb{R}^n$ such that $y \geq\geq 0$ and $A^{\mathrm{T}}y = b$.

Equivalently, exactly one of the following two statements is satisfied:

iii) There exists a vector $x \in \mathbb{R}^m$ such that $Ax \geq\geq 0$ and $b^{\mathrm{T}}x < 0$.

iv) There exists a vector $y \in \mathbb{R}^n$ such that $y \geq\geq 0$ and $A^{\mathrm{T}}y = b$.

Proof: See [161, p. 47] or [243, p. 24].

Remark: This result is the *Farkas theorem*.

Fact 4.11.16. Let $A \in \mathbb{R}^{n\times m}$. Then, the following statements are equivalent:

i) There exists a vector $x \in \mathbb{R}^m$ such that $Ax >> 0$.

ii) If $y \in \mathbb{R}^n$ is nonzero and $y \geq\geq 0$, then $A^{\mathrm{T}}y \neq 0$.

Equivalently, exactly one of the following two statements is satisfied:

iii) There exists a vector $x \in \mathbb{R}^m$ such that $Ax >> 0$.

iv) There exists a nonzero vector $y \in \mathbb{R}^n$ such that $y \geq\geq 0$ and $A^{\mathrm{T}}y = 0$.

Proof: See [161, p. 47] or [243, p. 23].

Remark: This result is *Gordan's theorem*.

Fact 4.11.17. Let $A \in \mathbb{C}^{n\times n}$, and define $|A| \in \mathbb{R}^{n\times n}$ by $|A|_{(i,j)} \triangleq |A_{(i,j)}|$ for all $i,j \in \{1,\ldots,n\}$. Then,

$$\mathrm{sprad}(A) \leq \mathrm{sprad}(|A|).$$

Proof: See [1023, p. 619].

Fact 4.11.18. Let $A \in \mathbb{R}^{n\times n}$, assume that A is nonnegative, and let $\alpha \in [0,1]$. Then,

$$\mathrm{sprad}(A) \leq \mathrm{sprad}\left[\alpha A + (1-\alpha)A^{\mathrm{T}}\right].$$

Proof: See [134].

Fact 4.11.19. Let $A, B \in \mathbb{R}^{n \times n}$, where $0 \leq\leq A \leq\leq B$. Then,

$$\mathrm{sprad}(A) \leq \mathrm{sprad}(B).$$

In particular, $B_0 \in \mathbb{R}^{m \times m}$ is a principal submatrix of B, then

$$\mathrm{sprad}(B_0) \leq \mathrm{sprad}(B).$$

If, in addition, $A \neq B$ and $A + B$ is irreducible, then

$$\mathrm{sprad}(A) < \mathrm{sprad}(B).$$

Hence, if $\mathrm{sprad}(A) = \mathrm{sprad}(B)$ and $A + B$ is irreducible, then $A = B$.

Proof: See [174, p. 27]. See also [459, pp. 500, 501].

Fact 4.11.20. Let $A, B \in \mathbb{R}^{n \times n}$, assume that B is diagonal, assume that A and $A + B$ are nonnegative, and let $\alpha \in [0, 1]$. Then,

$$\mathrm{sprad}[\alpha A + (1 - \alpha)B] \leq \alpha \, \mathrm{sprad}(A) + (1 - \alpha) \, \mathrm{sprad}(A + B).$$

Proof: See [1177, p. 9-5].

Fact 4.11.21. Let $A \in \mathbb{R}^{n \times n}$, assume that $A >> 0$, and let $\lambda \in \mathrm{spec}(A) \backslash \{\mathrm{sprad}(A)\}$. Then,

$$|\lambda| \leq \frac{A_{\max} - A_{\min}}{A_{\max} + A_{\min}} \, \mathrm{sprad}(A),$$

where

$$A_{\max} \triangleq \max \{ A_{(i,j)} \colon \ i, j = 1, \ldots, n \}$$

and

$$A_{\min} \triangleq \min \{ A_{(i,j)} \colon \ i, j = 1, \ldots, n \}.$$

Remark: This result is *Hopf's theorem*.

Remark: The equality case is discussed in [706].

Fact 4.11.22. Let $A \in \mathbb{R}^{n \times n}$, assume that A is nonnegative and irreducible, and let $x, y \in \mathbb{R}^n$, where $x > 0$ and $y > 0$ satisfy $Ax = \mathrm{sprad}(A)x$ and $A^{\mathrm{T}}y = \mathrm{sprad}(A)y$. Then,

$$\lim_{l \to \infty} \frac{1}{l} \sum_{k=1}^{l} \left[\frac{1}{\mathrm{sprad}(A)} A \right]^k = xy^{\mathrm{T}}.$$

If, in addition, A is primitive, then

$$\lim_{k \to \infty} \left[\frac{1}{\mathrm{sprad}(A)} A \right]^k = xy^{\mathrm{T}}.$$

Proof: See [459, p. 503] and [728, p. 516].

Fact 4.11.23. Let $A \in \mathbb{R}^{n \times n}$, assume that A is nonnegative, and let k and m be positive integers. Then,

$$\left[\mathrm{tr} \, A^k \right]^m \leq n^{m-1} \mathrm{tr} \, A^{km}.$$

Proof: See [885].

Remark: This result is the *JLL inequality*.

4.12 Notes

Much of the development in this chapter is based on [1108]. Additional discussions of the Smith and Smith-McMillan forms are given in [809] and [1535]. The proofs of Lemma 4.4.8 and Leverrier's algorithm Proposition 4.4.9 are based on [1157, pp. 432, 433], where it is called the *Souriau-Frame algorithm.* Alternative proofs of Leverrier's algorithm are given in [147, 739]. The proof of Theorem 4.6.1 is based on [728]. Polynomial-based approaches to linear algebra are given in [283, 521], while polynomial matrices and rational transfer functions are studied in [573, 1402].

The term *normal rank* is often used to refer to what we call the rank of a rational transfer function.

Chapter Five

Matrix Decompositions

In this chapter we present several matrix decompositions, namely, the Smith, multicompanion, elementary multicompanion, hypercompanion, Jordan, Schur, and singular value decompositions.

5.1 Smith Form

The first decomposition involves rectangular matrices subject to a biequivalence transformation. This result is the specialization of the Smith decomposition given by Theorem 4.3.2 to constant matrices.

Theorem 5.1.1. Let $A \in \mathbb{F}^{n \times m}$ and $r \triangleq \operatorname{rank} A$. Then, there exist nonsingular matrices $S_1 \in \mathbb{F}^{n \times n}$ and $S_2 \in \mathbb{F}^{m \times m}$ such that

$$A = S_1 \begin{bmatrix} I_r & 0_{r \times (m-r)} \\ 0_{(n-r) \times r} & 0_{(n-r) \times (m-r)} \end{bmatrix} S_2. \tag{5.1.1}$$

Corollary 5.1.2. Let $A, B \in \mathbb{F}^{n \times m}$. Then, A and B are biequivalent if and only if A and B have the same Smith form.

Proposition 5.1.3. Let $A, B \in \mathbb{F}^{n \times m}$. Then, the following statements hold:

  i) A and B are left equivalent if and only if $\mathcal{N}(A) = \mathcal{N}(B)$.

 ii) A and B are right equivalent if and only if $\mathcal{R}(A) = \mathcal{R}(B)$.

iii) A and B are biequivalent if and only if $\operatorname{rank} A = \operatorname{rank} B$.

Proof. The proof of necessity is immediate in i)–iii). Sufficiency in iii) follows from Corollary 5.1.2. For sufficiency in i) and ii), see [1157, pp. 179–181]. $\qquad \square$

5.2 Multicompanion Form

For the monic polynomial $p(s) = s^n + \beta_{n-1} s^{n-1} + \cdots + \beta_1 s + \beta_0 \in \mathbb{F}[s]$ of degree $n \geq 1$, the *companion matrix* $C(p) \in \mathbb{F}^{n \times n}$ associated with p is defined to be

$$C(p) \triangleq \begin{bmatrix} 0 & 1 & 0 & \cdots & 0 & 0 \\ 0 & 0 & 1 & \ddots & 0 & 0 \\ 0 & 0 & 0 & \ddots & 0 & 0 \\ \vdots & \vdots & \vdots & \ddots & \ddots & \vdots \\ 0 & 0 & 0 & \cdots & 0 & 1 \\ -\beta_0 & -\beta_1 & -\beta_2 & \cdots & -\beta_{n-2} & -\beta_{n-1} \end{bmatrix}. \tag{5.2.1}$$

If $n = 1$, then $p(s) = s + \beta_0$ and $C(p) = -\beta_0$. Furthermore, if $n = 0$ and $p = 1$, then we define $C(p) \triangleq 0_{0 \times 0}$. Note that, if $n \geq 1$, then $\text{tr}\, C(p) = -\beta_{n-1}$ and $\det C(p) = (-1)^n \beta_0 = (-1)^n p(0)$.

It is easy to see that the characteristic polynomial of the companion matrix $C(p)$ associated with p is p. For example, let $n = 3$ so that

$$C(p) = \begin{bmatrix} 0 & 1 & 0 \\ 0 & 0 & 1 \\ -\beta_0 & -\beta_1 & -\beta_2 \end{bmatrix}, \tag{5.2.2}$$

and thus

$$sI - C(p) = \begin{bmatrix} s & -1 & 0 \\ 0 & s & -1 \\ \beta_0 & \beta_1 & s + \beta_2 \end{bmatrix}. \tag{5.2.3}$$

Adding s times the second column and s^2 times the third column to the first column leaves the determinant of $sI - C(p)$ unchanged and yields

$$\begin{bmatrix} 0 & -1 & 0 \\ 0 & s & -1 \\ p(s) & \beta_1 & s + \beta_2 \end{bmatrix}. \tag{5.2.4}$$

Hence, $\chi_{C(p)} = p$. If $n = 0$ and $p = 1$, then we define $\chi_{C(p)} \triangleq \chi_{0_{0 \times 0}} = 1$. The following result shows that companion matrices have the same characteristic polynomial and the same minimal polynomial.

Proposition 5.2.1. Let $p \in \mathbb{F}[s]$ be a monic polynomial having degree n. Then, there exist unimodular matrices $S_1, S_2 \in \mathbb{F}^{n \times n}[s]$ such that

$$sI - C(p) = S_1(s) \begin{bmatrix} I_{n-1} & 0_{(n-1) \times 1} \\ 0_{1 \times (n-1)} & p(s) \end{bmatrix} S_2(s). \tag{5.2.5}$$

Furthermore,

$$\chi_{C(p)} = \mu_{C(p)} = p. \tag{5.2.6}$$

Proof. Since $\chi_{C(p)} = p$, it follows that $\text{rank}[sI - C(p)] = n$. Next, since $\det\big([sI - C(p)]_{[n;1]}\big) = (-1)^{n-1}$, it follows that $\Delta_{n-1} = 1$, where Δ_{n-1} is the greatest common divisor (which is monic by definition) of all $(n-1) \times (n-1)$ subdeterminants of $sI - C(p)$. Furthermore, since Δ_{i-1} divides Δ_i for all $i \in \{2, \ldots, n-1\}$, it follows that $\Delta_1 = \cdots = \Delta_{n-2} = 1$. Consequently, $p_1 = \cdots = p_{n-1} = 1$. Since, by

Proposition 4.6.2, $\chi_{C(p)} = \prod_{i=1}^{n} p_i = p_n$ and $\mu_{C(p)} = p_n$, it follows that $\chi_{C(p)} = \mu_{C(p)} = p$. $\qquad \square$

Next, we consider block-diagonal matrices all of whose diagonally located blocks are companion matrices.

Lemma 5.2.2. Let $p_1, \ldots, p_n \in \mathbb{F}[s]$ be monic polynomials such that p_i divides p_{i+1} for all $i \in \{1, \ldots, n-1\}$ and $n = \sum_{i=1}^{n} \deg p_i$. Furthermore, define $C \triangleq \operatorname{diag}[C(p_1), \ldots, C(p_n)] \in \mathbb{F}^{n \times n}$. Then, there exist unimodular matrices $S_1, S_2 \in \mathbb{F}^{n \times n}[s]$ such that

$$sI - C = S_1(s) \begin{bmatrix} p_1(s) & & 0 \\ & \ddots & \\ 0 & & p_n(s) \end{bmatrix} S_2(s). \qquad (5.2.7)$$

Proof. Letting $k_i = \deg p_i$, Proposition 5.2.1 implies that the Smith form of $sI_{k_i} - C(p_i)$ is $0_{0 \times 0}$ if $k_i = 0$ and $\operatorname{diag}(I_{k_i-1}, p_i)$ if $k_i \geq 1$. Note that $p_1 = \cdots = p_{n_0} = 1$, where $n_0 \triangleq \sum_{i=1}^{n} \max\{0, k_i - 1\}$. By combining these Smith forms and rearranging diagonal entries, it follows that there exist unimodular matrices $S_1, S_2 \in \mathbb{F}^{n \times n}[s]$ such that

$$sI - C = \begin{bmatrix} sI_{k_1} - C(p_1) & & \\ & \ddots & \\ & & sI_{k_n} - C(p_n) \end{bmatrix}$$

$$= S_1(s) \begin{bmatrix} p_1(s) & & 0 \\ & \ddots & \\ 0 & & p_n(s) \end{bmatrix} S_2(s).$$

Since p_i divides p_{i+1} for all $i \in \{1, \ldots, n-1\}$, it follows that this diagonal matrix is the Smith form of $sI - C$. $\qquad \square$

The following result uses Lemma 5.2.2 to construct a canonical form, known as the *multicompanion form*, for square matrices under a similarity transformation.

Theorem 5.2.3. Let $A \in \mathbb{F}^{n \times n}$, and let $p_1, \ldots, p_n \in \mathbb{F}[s]$ denote the similarity invariants of A, where p_i divides p_{i+1} for all $i \in \{1, \ldots, n-1\}$. Then, there exists a nonsingular matrix $S \in \mathbb{F}^{n \times n}$ such that

$$A = S \begin{bmatrix} C(p_1) & & 0 \\ & \ddots & \\ 0 & & C(p_n) \end{bmatrix} S^{-1}. \qquad (5.2.8)$$

Proof. Lemma 5.2.2 implies that the $n \times n$ matrix $sI - C$, where $C \triangleq \operatorname{diag}[C(p_1), \ldots, C(p_n)]$, has the Smith form $\operatorname{diag}(p_1, \ldots, p_n)$. Now, since $sI - A$ has the same similarity invariants as C, it follows from Theorem 4.3.10 that A and C are similar. $\qquad \square$

Corollary 5.2.4. Let $A \in \mathbb{F}^{n \times n}$. Then, $\mu_A = \chi_A$ if and only if A is similar to $C(\chi_A)$.

Proof. Suppose that $\mu_A = \chi_A$. Then, it follows from Proposition 4.6.2 that $p_i = 1$ for all $i \in \{1, \ldots, n-1\}$ and $p_n = \chi_A$ is the only nonconstant similarity invariant of A. Thus, $C(p_i) = 0_{0 \times 0}$ for all $i \in \{1, \ldots, n-1\}$, and it follows from Theorem 5.2.3 that A is similar to $C(\chi_A)$. The converse follows from (5.2.6), $xi)$ of Proposition 4.4.5, and Proposition 4.6.3. $\qquad \square$

Corollary 5.2.5. Let $A \in \mathbb{F}^{n \times n}$ be a companion matrix. Then, $A = C(\chi_A)$ and $\mu_A = \chi_A$.

Note that, if $A = I_n$, then the similarity invariants of A are $p_i(s) = s - 1$ for all $i \in \{1, \ldots, n\}$. Thus, $C(p_i) = 1$ for all $i \in \{1, \ldots, n\}$, as expected.

Corollary 5.2.6. Let $A, B \in \mathbb{F}^{n \times n}$. Then, the following statements are equivalent:

i) A and B are similar.

ii) A and B have the same similarity invariants.

iii) A and B have the same multicompanion form.

The multicompanion form given by Theorem 5.2.3 provides a canonical form for A in terms of a block-diagonal matrix of companion matrices. As shown below, however, the multicompanion form is only one such decomposition. The goal of the remainder of this section is to obtain an additional canonical form by applying a similarity transformation to the multicompanion form.

To begin, note that, if A_i is similar to B_i for all $i \in \{1, \ldots, r\}$, then $\mathrm{diag}(A_1, \ldots, A_r)$ is similar to $\mathrm{diag}(B_1, \ldots, B_r)$. Therefore, it follows from Corollary 5.2.6 that, if $sI - A_i$ and $sI - B_i$ have the same Smith form for all $i \in \{1, \ldots, r\}$, then $sI - \mathrm{diag}(A_1, \ldots, A_r)$ and $sI - \mathrm{diag}(B_1, \ldots, B_r)$ have the same Smith form. The following lemma is needed.

Lemma 5.2.7. Let $A = \mathrm{diag}(A_1, A_2)$, where $A_i \in \mathbb{F}^{n_i \times n_i}$ for $i = 1, 2$. Then, μ_A is the least common multiple of μ_{A_1} and μ_{A_2}. In particular, if μ_{A_1} and μ_{A_2} are coprime, then $\mu_A = \mu_{A_1} \mu_{A_2}$.

Proof. Since $0 = \mu_A(A) = \mathrm{diag}[\mu_A(A_1), \mu_A(A_2)]$, it follows that $\mu_A(A_1) = 0$ and $\mu_A(A_2) = 0$. Therefore, Theorem 4.6.1 implies that μ_{A_1} and μ_{A_2} both divide μ_A. Consequently, the least common multiple q of μ_{A_1} and μ_{A_2} also divides μ_A. Since $q(A_1) = 0$ and $q(A_2) = 0$, it follows that $q(A) = 0$. Therefore, μ_A divides q. Hence, $q = \mu_A$. If, in addition, μ_{A_1} and μ_{A_2} are coprime, then $\mu_A = \mu_{A_1} \mu_{A_2}$. $\qquad \square$

Proposition 5.2.8. Let $p \in \mathbb{F}[s]$ be a monic polynomial of positive degree n, and let $p = p_1 \cdots p_r$, where $p_1, \ldots, p_r \in \mathbb{F}[s]$ are monic and pairwise coprime polynomials. Then, the matrices $C(p)$ and $\mathrm{diag}[C(p_1), \ldots, C(p_r)]$ are similar.

Proof. Let $\hat{p}_2 = p_2 \cdots p_r$ and $\hat{C} \triangleq \mathrm{diag}[C(p_1), C(\hat{p}_2)]$. Since p_1 and $\hat{p}_2$ are coprime, it follows from Lemma 5.2.7 that $\mu_{\hat{C}} = \mu_{C(p_1)}\mu_{C(\hat{p}_2)}$. Furthermore, $\chi_{\hat{C}} = \chi_{C(p_1)}\chi_{C(\hat{p}_2)} = \mu_{\hat{C}}$. Hence, Corollary 5.2.4 implies that $\hat{C}$ is similar to $C(\chi_{\hat{C}})$. However, $\chi_{\hat{C}} = p_1 \cdots p_r = p$, so that $\hat{C}$ is similar to $C(p)$. If $r > 2$, then the same argument can be used to decompose $C(\hat{p}_2)$ to show that $C(p)$ is similar to $\mathrm{diag}[C(p_1), \ldots, C(p_r)]$. $\qquad\square$

Proposition 5.2.8 can be used to decompose every companion block of a multicompanion form into smaller companion matrices. This procedure can be carried out for every companion block whose characteristic polynomial has coprime factors. For example, suppose that $A \in \mathbb{R}^{10 \times 10}$ has the similarity invariants $p_i(s) = 1$ for all $i \in \{1, \ldots, 7\}$, $p_8(s) = (s+1)^2$, $p_9(s) = (s+1)^2(s+2)$, and $p_{10}(s) = (s+1)^2(s+2)(s^2+3)$, so that, by Theorem 5.2.3, the multicompanion form of A is $\mathrm{diag}[C(p_8), C(p_9), C(p_{10})]$, where $C(p_8) \in \mathbb{R}^{2 \times 2}$, $C(p_9) \in \mathbb{R}^{3 \times 3}$, and $C(p_{10}) \in \mathbb{R}^{5 \times 5}$. According to Proposition 5.2.8, the companion matrices $C(p_9)$ and $C(p_{10})$ can be further decomposed. For example, $C(p_9)$ is similar to $\mathrm{diag}[C(p_{9,1}), C(p_{9,2})]$, where $p_{9,1}(s) = (s+1)^2$ and $p_{9,2}(s) = s+2$ are coprime. Furthermore, $C(p_{10})$ is similar to four different diagonal matrices, three of which have two companion blocks while the fourth has three companion blocks. Since $p_8(s) = (s+1)^2$ does not have nonconstant coprime factors, however, it follows that the companion matrix $C(p_8)$ cannot be decomposed into smaller companion matrices.

The largest number of companion blocks achievable by similarity transformation is obtained by factoring every similarity invariant into *elementary divisors*, which are powers of irreducible polynomials that are nonconstant, monic, and pairwise coprime. In the above example, this factorization is given by $p_9(s) = p_{9,1}(s)p_{9,2}(s)$, where $p_{9,1}(s) = (s+1)^2$ and $p_{9,2}(s) = s+2$, and by $p_{10} = p_{10,1}p_{10,2}p_{10,3}$, where $p_{10,1}(s) = (s+1)^2$, $p_{10,2}(s) = s+2$, and $p_{10,3}(s) = s^2+3$. The elementary divisors of A are thus $(s+1)^2$, $(s+1)^2$, $s+2$, $(s+1)^2$, $s+2$, and s^2+3, which yields six companion blocks. Viewing $A \in \mathbb{C}^{n \times n}$ we can further factor $p_{10,3}(s) = (s+\jmath\sqrt{3})(s-\jmath\sqrt{3})$, which yields a total of seven companion blocks. From Proposition 5.2.8 and Theorem 5.2.3 we obtain the *elementary multicompanion form*, which provides another canonical form for A.

Theorem 5.2.9. Let $A \in \mathbb{F}^{n \times n}$, and let $q_1^{l_1}, \ldots, q_h^{l_h} \in \mathbb{F}[s]$ be the elementary divisors of A, where $l_1, \ldots, l_h$ are positive integers. Then, there exists a nonsingular matrix $S \in \mathbb{F}^{n \times n}$ such that

$$A = S \begin{bmatrix} C\!\left(q_1^{l_1}\right) & & 0 \\ & \ddots & \\ 0 & & C\!\left(q_h^{l_h}\right) \end{bmatrix} S^{-1}. \qquad (5.2.9)$$

5.3 Hypercompanion Form and Jordan Form

We now present an alternative form of the companion blocks of the elementary multicompanion form (5.2.9). To do this we define the *hypercompanion matrix* $\mathcal{H}_l(q)$ associated with the elementary divisor $q^l \in \mathbb{F}[s]$, where l is a positive integer, as follows. For $q(s) = s - \lambda \in \mathbb{C}[s]$, define the $l \times l$ Toeplitz hypercompanion matrix

$$
\mathcal{H}_l(q) \triangleq \lambda I_l + N_l =
\begin{bmatrix}
\lambda & 1 & 0 & & & & \\
0 & \lambda & 1 & & 0 & & \\
& & \ddots & \ddots & & & \\
& & & \ddots & 1 & 0 \\
& 0 & & & \lambda & 1 \\
& & & & & 0 & \lambda
\end{bmatrix},
\tag{5.3.1}
$$

whereas, for $q(s) = s^2 - \beta_1 s - \beta_0 \in \mathbb{R}[s]$, define the $2l \times 2l$ real, tridiagonal hypercompanion matrix

$$
\mathcal{H}_l(q) \triangleq
\begin{bmatrix}
0 & 1 & & & & & \\
\beta_0 & \beta_1 & 1 & & & 0 & \\
& 0 & 0 & 1 & & & \\
& & \beta_0 & \beta_1 & 1 & & \\
& & & \ddots & \ddots & \ddots & \\
0 & & & & \ddots & 0 & 1 \\
& & & & & \beta_0 & \beta_1
\end{bmatrix}.
\tag{5.3.2}
$$

The following result shows that the hypercompanion matrix $\mathcal{H}_l(q)$ is similar to the companion matrix $C(q^l)$ associated with the elementary divisor q^l of $\mathcal{H}_l(q)$.

Lemma 5.3.1. Let $l \in \mathbb{P}$, and let $q(s) = s - \lambda \in \mathbb{C}[s]$ or $q(s) = s^2 - \beta_1 s - \beta_0 \in \mathbb{R}[s]$. Then, q^l is the only elementary divisor of $\mathcal{H}_l(q)$, and $\mathcal{H}_l(q)$ is similar to $C(q^l)$.

Proof. Let k denote the order of $\mathcal{H}_l(q)$. Then, $\chi_{\mathcal{H}_l(q)} = q^l$ and $\det\left([sI - \mathcal{H}_l(q)]_{[k;1]}\right) = (-1)^{k-1}$. Hence, as in the proof of Proposition 5.2.1, it follows that $\chi_{\mathcal{H}_l(q)} = \mu_{\mathcal{H}_l(q)}$. Corollary 5.2.4 now implies that $\mathcal{H}_l(q)$ is similar to $C(q^l)$. $\qquad \square$

Proposition 5.2.8 and Lemma 5.3.1 yield the following canonical form, which is known as the *hypercompanion form*.

Theorem 5.3.2. Let $A \in \mathbb{F}^{n \times n}$, and let $q_1^{l_1}, \ldots, q_h^{l_h} \in \mathbb{F}[s]$ be the elementary divisors of A, where $l_1, \ldots, l_h$ are positive integers. Then, there exists a nonsingular matrix $S \in \mathbb{F}^{n \times n}$ such that

$$
A = S
\begin{bmatrix}
\mathcal{H}_{l_1}(q_1) & & 0 \\
& \ddots & \\
0 & & \mathcal{H}_{l_h}(q_h)
\end{bmatrix}
S^{-1}.
\tag{5.3.3}
$$

Next, consider Theorem 5.3.2 with $\mathbb{F} = \mathbb{C}$. In this case, every elementary divisor $q_i^{l_i}$ is of the form $(s - \lambda_i)^{l_i}$, where $\lambda_i \in \mathbb{C}$. Furthermore, $S \in \mathbb{C}^{n \times n}$, and the hypercompanion form (5.3.3) is a block-diagonal matrix whose diagonally located blocks are of the form (5.3.1). The hypercompanion form (5.3.3) with every diagonally located block of the form (5.3.1) is the *Jordan form*, as given by the following result.

Theorem 5.3.3. Let $A \in \mathbb{C}^{n \times n}$, and let $q_1^{l_1}, \ldots, q_h^{l_h} \in \mathbb{C}[s]$ be the elementary divisors of A, where $l_1, \ldots, l_h$ are positive integers and each of the polynomials $q_1, \ldots, q_h \in \mathbb{C}[s]$ has degree 1. Then, there exists a nonsingular matrix $S \in \mathbb{C}^{n \times n}$ such that

$$A = S \begin{bmatrix} \mathcal{H}_{l_1}(q_1) & & 0 \\ & \ddots & \\ 0 & & \mathcal{H}_{l_h}(q_h) \end{bmatrix} S^{-1}. \tag{5.3.4}$$

Corollary 5.3.4. Let $p \in \mathbb{F}[s]$, let $\lambda_1, \ldots, \lambda_r$ denote the distinct roots of p, and, for $i = 1, \ldots, r$, let $l_i \triangleq m_p(\lambda_i)$ and $p_i(s) \triangleq s - \lambda_i$. Then, $C(p)$ is similar to $\mathrm{diag}[\mathcal{H}_{l_1}(p_1), \ldots, \mathcal{H}_{l_r}(p_r)]$.

To illustrate the structure of the Jordan form, let $l_i = 3$ and $q_i(s) = s - \lambda_i$, where $\lambda_i \in \mathbb{C}$. Then, $\mathcal{H}_{l_i}(q_i)$ is the 3×3 matrix

$$\mathcal{H}_{l_i}(q_i) = \lambda_i I_3 + N_3 = \begin{bmatrix} \lambda_i & 1 & 0 \\ 0 & \lambda_i & 1 \\ 0 & 0 & \lambda_i \end{bmatrix} \tag{5.3.5}$$

so that $\mathrm{mspec}[\mathcal{H}_{l_i}(q_i)] = \{\lambda_i, \lambda_i, \lambda_i\}_{\mathrm{ms}}$. If $\mathcal{H}_{l_i}(q_i)$ is the only diagonally located block of the Jordan form associated with the eigenvalue λ_i, then the algebraic multiplicity of λ_i is equal to 3, while its geometric multiplicity is equal to 1.

Now, consider Theorem 5.3.2 with $\mathbb{F} = \mathbb{R}$. In this case, every elementary divisor $q_i^{l_i}$ is either of the form $(s - \lambda_i)^{l_i}$ or of the form $(s^2 - \beta_{1i} s - \beta_{0i})^{l_i}$, where $\beta_{0i}, \beta_{1i} \in \mathbb{R}$. Furthermore, $S \in \mathbb{R}^{n \times n}$, and the hypercompanion form (5.3.3) is a block-diagonal matrix whose diagonally located blocks are real matrices of the form (5.3.1) or (5.3.2). In this case, (5.3.3) is the *real hypercompanion form*.

Applying an additional real similarity transformation to each diagonally located block of the real hypercompanion form yields the *real Jordan form*. To do this, define the *real Jordan matrix* $\mathcal{J}_l(q)$ for the positive integer l as follows. For $q(s) = s - \lambda \in \mathbb{F}[s]$ define $\mathcal{J}_l(q) \triangleq \mathcal{H}_l(q)$, whereas, if $q(s) = s^2 - \beta_1 s - \beta_0 \in \mathbb{F}[s]$ is irreducible with a nonreal root $\lambda = \nu + \jmath\omega$, then define the $2l \times 2l$ upper Hessenberg matrix

$$\mathfrak{J}_l(q) \triangleq \begin{bmatrix} \nu & \omega & 1 & 0 & & & & \\ -\omega & \nu & 0 & 1 & \ddots & & 0 & \\ & & \nu & \omega & 1 & \ddots & & \\ & & -\omega & \nu & 0 & \ddots & \ddots & \\ & & & & \ddots & \ddots & 1 & 0 \\ & & & & & \ddots & 0 & 1 \\ & 0 & & & & & \nu & \omega \\ & & & & & & -\omega & \nu \end{bmatrix}. \tag{5.3.6}$$

Theorem 5.3.5. Let $A \in \mathbb{R}^{n \times n}$, and let $q_1^{l_1}, \dots, q_h^{l_h} \in \mathbb{R}[s]$, where $l_1, \dots, l_h$ are positive integers, are the elementary divisors of A. Then, there exists a nonsingular matrix $S \in \mathbb{R}^{n \times n}$ such that

$$A = S \begin{bmatrix} \mathfrak{J}_{l_1}(q_1) & & 0 \\ & \ddots & \\ 0 & & \mathfrak{J}_{l_h}(q_h) \end{bmatrix} S^{-1}. \tag{5.3.7}$$

Proof. For the irreducible quadratic $q(s) = s^2 - \beta_1 s - \beta_0 \in \mathbb{R}[s]$, we show that $\mathfrak{J}_l(q)$ and $\mathcal{H}_l(q)$ are similar. Writing $q(s) = (s - \lambda)(s - \overline{\lambda})$, it follows from Theorem 5.3.3 that $\mathcal{H}_l(q) \in \mathbb{R}^{2l \times 2l}$ is similar to $\text{diag}(\lambda I_l + N_l, \overline{\lambda} I_l + N_l)$. Next, by using a permutation similarity transformation, it follows that $\mathcal{H}_l(q)$ is similar to

$$\begin{bmatrix} \lambda & 0 & 1 & 0 & & & & \\ 0 & \overline{\lambda} & 0 & 1 & 0 & & 0 & \\ & 0 & \lambda & 0 & 1 & 0 & & \\ & & 0 & \overline{\lambda} & 0 & 1 & & \\ & & & & \ddots & \ddots & \ddots & \\ & & & & & \ddots & 1 & 0 \\ & & & & & \ddots & 0 & 1 \\ & 0 & & & & & \lambda & 0 \\ & & & & & & 0 & \overline{\lambda} \end{bmatrix},$$

Finally, applying the similarity transformation $S \triangleq \text{diag}(\hat{S}, \dots, \hat{S})$ to the above matrix, where $\hat{S} \triangleq \begin{bmatrix} -\jmath & -\jmath \\ 1 & -1 \end{bmatrix}$ and $\hat{S}^{-1} = \frac{1}{2} \begin{bmatrix} \jmath & 1 \\ \jmath & -1 \end{bmatrix}$, yields $\mathfrak{J}_l(q)$. $\square$

Example 5.3.6. Let $A, B \in \mathbb{R}^{4 \times 4}$ and $C \in \mathbb{C}^{4 \times 4}$ be given by

$$A = \begin{bmatrix} 0 & 1 & 0 & 0 \\ 0 & 0 & 1 & 0 \\ 0 & 0 & 0 & 1 \\ -16 & 0 & -8 & 0 \end{bmatrix},$$

$$B = \begin{bmatrix} 0 & 1 & 0 & 0 \\ -4 & 0 & 1 & 0 \\ 0 & 0 & 0 & 1 \\ 0 & 0 & -4 & 0 \end{bmatrix},$$

and

$$C = \begin{bmatrix} \jmath 2 & 1 & 0 & 0 \\ 0 & 2\jmath & 0 & 0 \\ 0 & 0 & -\jmath 2 & 1 \\ 0 & 0 & 0 & -\jmath 2 \end{bmatrix}.$$

Then, A is in companion form, B is in real hypercompanion form, and C is in Jordan form. Furthermore, A, B, and C are similar.

Example 5.3.7. Let $A, B \in \mathbb{R}^{6\times 6}$ and $C \in \mathbb{C}^{6\times 6}$ be given by

$$A = \begin{bmatrix} 0 & 1 & 0 & 0 & 0 & 0 \\ 0 & 0 & 1 & 0 & 0 & 0 \\ 0 & 0 & 0 & 1 & 0 & 0 \\ 0 & 0 & 0 & 0 & 1 & 0 \\ 0 & 0 & 0 & 0 & 0 & 1 \\ -27 & 54 & -63 & 44 & -21 & 6 \end{bmatrix},$$

$$B = \begin{bmatrix} 0 & 1 & 0 & 0 & 0 & 0 \\ -3 & 2 & 1 & 0 & 0 & 0 \\ 0 & 0 & 0 & 1 & 0 & 0 \\ 0 & 0 & -3 & 2 & 1 & 0 \\ 0 & 0 & 0 & 0 & 0 & 1 \\ 0 & 0 & 0 & 0 & -3 & 2 \end{bmatrix},$$

and

$$C = \begin{bmatrix} 1+\jmath\sqrt{2} & 1 & 0 & 0 & 0 & 0 \\ 0 & 1+\jmath\sqrt{2} & 1 & 0 & 0 & 0 \\ 0 & 0 & 1+\jmath\sqrt{2} & 0 & 0 & 0 \\ 0 & 0 & 0 & 1-\jmath\sqrt{2} & 1 & 0 \\ 0 & 0 & 0 & 0 & 1-\jmath\sqrt{2} & 1 \\ 0 & 0 & 0 & 0 & 0 & 1-\jmath\sqrt{2} \end{bmatrix}.$$

Then, A is in companion form, B is in real hypercompanion form, and C is in Jordan form. Furthermore, A, B, and C are similar.

The next result shows that every matrix is similar to its transpose by means of a symmetric similarity transformation. This result, which improves Corollary 4.3.11, is due to Frobenius.

Corollary 5.3.8. Let $A \in \mathbb{F}^{n\times n}$. Then, there exists a symmetric, nonsingular matrix $S \in \mathbb{F}^{n\times n}$ such that $A = SA^{\mathrm{T}}S^{-1}$.

Proof. It follows from Theorem 5.3.3 that there exists a nonsingular matrix $\hat{S} \in \mathbb{C}^{n\times n}$ such that $A = \hat{S}B\hat{S}^{-1}$, where $B = \mathrm{diag}(B_1, \ldots, B_r)$ is the Jordan form of A, and $B_i \in \mathbb{C}^{n_i\times n_i}$ for all $i \in \{1, \ldots, r\}$. Now, define the symmetric nonsingular matrix $S \triangleq \hat{S}\tilde{I}\hat{S}^{\mathrm{T}}$, where $\tilde{I} \triangleq \operatorname{diag}\left(\hat{I}_{n_1},\ldots,\hat{I}_{n_r}\right)$ is symmetric and involutory. Furthermore, note that $\hat{I}_{n_i}B_i\hat{I}_{n_i} = B_i^{\mathrm{T}}$ for all $i \in \{1,\ldots,r\}$ so that $\tilde{I}B\tilde{I} = B^{\mathrm{T}}$, and thus $\tilde{I}B^{\mathrm{T}}\tilde{I} = B$. Hence, it follows that

$$SA^{\mathrm{T}}S^{-1} = S\hat{S}^{-\mathrm{T}}B^{\mathrm{T}}\hat{S}^{\mathrm{T}}S^{-1} = \hat{S}\tilde{I}\hat{S}^{\mathrm{T}}\hat{S}^{-\mathrm{T}}B^{\mathrm{T}}\hat{S}^{\mathrm{T}}\hat{S}^{-\mathrm{T}}\tilde{I}\hat{S}^{-1}$$
$$= \hat{S}\tilde{I}B^{\mathrm{T}}\tilde{I}\hat{S}^{-1} = \hat{S}B\hat{S}^{-1} = A.$$

If A is real, then a similar argument based on the real Jordan form shows that S can be chosen to be real. $\qquad\square$

An extension of Corollary 5.3.8 to the case in which A is normal is given by Fact 5.9.11.

Corollary 5.3.9. Let $A \in \mathbb{F}^{n\times n}$. Then, there exist symmetric matrices $S_1, S_2 \in \mathbb{F}^{n\times n}$ such that S_2 is nonsingular and $A = S_1 S_2$.

Proof. From Corollary 5.3.8 it follows that there exists a symmetric, nonsingular matrix $S \in \mathbb{F}^{n\times n}$ such that $A = SA^{\mathrm{T}}S^{-1}$. Now, let $S_1 \triangleq SA^{\mathrm{T}}$ and $S_2 \triangleq S^{-1}$. Note that S_2 is symmetric and nonsingular. Furthermore, $S_1^{\mathrm{T}} = AS = SA^{\mathrm{T}} = S_1$, which shows that S_1 is symmetric. $\qquad\square$

Note that Corollary 5.3.8 follows from Corollary 5.3.9. If $A = S_1 S_2$, where S_1, S_2 are symmetric and S_2 is nonsingular, then $A = S_2^{-1}S_2 S_1 S_2 = S_2^{-1}A^{\mathrm{T}}S_2$.

5.4 Schur Decomposition

The *Schur decomposition* uses a unitary similarity transformation to transform an arbitrary square matrix into an upper triangular matrix.

Theorem 5.4.1. Let $A \in \mathbb{C}^{n\times n}$. Then, there exist a unitary matrix $S \in \mathbb{C}^{n\times n}$ and an upper triangular matrix $B \in \mathbb{C}^{n\times n}$ such that

$$A = SBS^*. \tag{5.4.1}$$

Proof. Let $\lambda_1 \in \mathbb{C}$ be an eigenvalue of A with associated eigenvector $x \in \mathbb{C}^n$ chosen such that $x^*x = 1$. Furthermore, let $S_1 \triangleq \begin{bmatrix} x & \hat{S}_1 \end{bmatrix} \in \mathbb{C}^{n\times n}$ be unitary, where $\hat{S}_1 \in \mathbb{C}^{n\times(n-1)}$ satisfies $\hat{S}_1^*\hat{S}_1 = I_{n-1}$ and $x^*\hat{S}_1 = 0_{1\times(n-1)}$. Then, $S_1 e_1 = x$, and

$$\operatorname{col}_1(S_1^{-1}AS_1) = S_1^{-1}Ax = \lambda_1 S_1^{-1}x = \lambda_1 e_1.$$

Consequently,

$$A = S_1 \begin{bmatrix} \lambda_1 & C_1 \\ 0_{(n-1)\times 1} & A_1 \end{bmatrix} S_1^{-1},$$

where $C_1 \in \mathbb{C}^{1\times(n-1)}$ and $A_1 \in \mathbb{C}^{(n-1)\times(n-1)}$. Next, let $S_{20} \in \mathbb{C}^{(n-1)\times(n-1)}$ be a unitary matrix such that

$$A_1 = S_{20} \begin{bmatrix} \lambda_2 & C_2 \\ 0_{(n-2)\times 1} & A_2 \end{bmatrix} S_{20}^{-1},$$

where $C_2 \in \mathbb{C}^{1 \times (n-2)}$ and $A_2 \in \mathbb{C}^{(n-2) \times (n-2)}$. Hence,

$$A = S_1 S_2 \begin{bmatrix} \lambda_1 & C_{11} & C_{12} \\ 0 & \lambda_2 & C_2 \\ 0 & 0 & A_2 \end{bmatrix} S_2^{-1} S_1,$$

where $C_1 = \begin{bmatrix} C_{11} & C_{12} \end{bmatrix}$, $C_{11} \in \mathbb{C}$, and $S_2 \triangleq \begin{bmatrix} 1 & 0 \\ 0 & S_{20} \end{bmatrix}$ is unitary. Proceeding in a similar manner yields (5.4.1) with $S \triangleq S_1 S_2 \cdots S_{n-1}$, where $S_1, \ldots, S_{n-1} \in \mathbb{C}^{n \times n}$ are unitary. $\square$

Since A and B in (5.4.1) are similar and B is upper triangular, it follows from Fact 4.10.10 that A and B have the same eigenvalues with the same algebraic multiplicities.

The *real Schur decomposition* uses a real orthogonal similarity transformation to transform a real matrix into an upper Hessenberg matrix with real 1×1 and 2×2 diagonally located blocks.

Corollary 5.4.2. Let $A \in \mathbb{R}^{n \times n}$, and let $\mathrm{mspec}(A) = \{\lambda_1, \ldots, \lambda_r\}_{\mathrm{ms}} \cup \{\nu_1 + \jmath\omega_1, \nu_1 - \jmath\omega_1, \ldots, \nu_l + \jmath\omega_l, \nu_l - \jmath\omega_l\}_{\mathrm{ms}}$, where $\lambda_1, \ldots, \lambda_r \in \mathbb{R}$ and, for all $i \in \{1, \ldots, l\}$, $\nu_i, \omega_i \in \mathbb{R}$ and $\omega_i \neq 0$. Then, there exists an orthogonal matrix $S \in \mathbb{R}^{n \times n}$ such that

$$A = SBS^{\mathrm{T}}, \tag{5.4.2}$$

where B is upper block triangular and the diagonally located blocks $B_1, \ldots, B_r \in \mathbb{R}$ and $\hat{B}_1, \ldots, \hat{B}_l \in \mathbb{R}^{2 \times 2}$ of B satisfy $B_i \triangleq [\lambda_i]$ for all $i \in \{1, \ldots, r\}$ and $\mathrm{spec}(\hat{B}_i) = \{\nu_i + \jmath\omega_i, \nu_i - \jmath\omega_i\}$ for all $i \in \{1, \ldots, l\}$.

Proof. The proof is analogous to the proof of Theorem 5.3.5. See also [728, p. 82]. $\square$

Corollary 5.4.3. Let $A \in \mathbb{R}^{n \times n}$, and assume that the spectrum of A is real. Then, there exist an orthogonal matrix $S \in \mathbb{R}^{n \times n}$ and an upper triangular matrix $B \in \mathbb{R}^{n \times n}$ such that

$$A = SBS^{\mathrm{T}}. \tag{5.4.3}$$

The Schur decomposition reveals the structure of range-Hermitian matrices and thus, as a special case, normal matrices.

Corollary 5.4.4. Let $A \in \mathbb{F}^{n \times n}$, and define $r \triangleq \mathrm{rank}\, A$. Then, A is range Hermitian if and only if there exist a unitary matrix $S \in \mathbb{F}^{n \times n}$ and a nonsingular matrix $B \in \mathbb{F}^{r \times r}$ such that

$$A = S \begin{bmatrix} B & 0 \\ 0 & 0 \end{bmatrix} S^*. \tag{5.4.4}$$

In addition, A is normal if and only if there exist a unitary matrix $S \in \mathbb{C}^{n \times n}$ and a diagonal matrix $B \in \mathbb{C}^{r \times r}$ such that (5.4.4) is satisfied.

Proof. Suppose that A is range Hermitian, and let $A = S\hat{B}S^*$, where $\hat{B}$ is upper triangular and $S \in \mathbb{F}^{n \times n}$ is unitary. Assume that A is singular, and choose S such that $\hat{B}_{(j,j)} = \hat{B}_{(j+1,j+1)} = \cdots = \hat{B}_{(n,n)} = 0$ and such that all other diagonal entries of $\hat{B}$ are nonzero. Thus, $\text{row}_n(\hat{B}) = 0$, which implies that $e_n \in \mathcal{R}(\hat{B})^\perp$. Since A is range Hermitian, it follows that $\mathcal{R}(\hat{B}) = \mathcal{R}(\hat{B}^*)$, and thus $e_n \in \mathcal{R}(\hat{B}^*)^\perp$. Therefore, it follows from (2.4.13) that $e_n \in \mathcal{N}(\hat{B})$, which implies that $\text{col}_n(\hat{B}) = 0$. If, in addition, $\hat{B}_{(n-1,n-1)} = 0$, then $\text{col}_{n-1}(\hat{B}) = 0$. Repeating this argument shows that $\hat{B}$ has the form $\begin{bmatrix} B & 0 \\ 0 & 0 \end{bmatrix}$, where $B \in \mathbb{F}^{r \times r}$ is nonsingular.

Now, suppose that A is normal, and let $A = S\hat{B}S^*$, where $\hat{B} \in \mathbb{C}^{n \times n}$ is upper triangular and $S \in \mathbb{C}^{n \times n}$ is unitary. Since A is normal, it follows that $AA^* = A^*A$, which implies that $\hat{B}\hat{B}^* = \hat{B}^*\hat{B}$. Since $\hat{B}$ is upper triangular, it follows that $(\hat{B}^*\hat{B})_{(1,1)} = \hat{B}_{(1,1)}\overline{\hat{B}_{(1,1)}}$, whereas $(\hat{B}\hat{B}^*)_{(1,1)} = \text{row}_1(\hat{B})[\text{row}_1(\hat{B})]^* = \sum_{i=1}^n \hat{B}_{(1,i)}\overline{\hat{B}_{(1,i)}}$. Since $(\hat{B}^*\hat{B})_{(1,1)} = (\hat{B}\hat{B}^*)_{(1,1)}$, it follows that $\hat{B}_{(1,i)} = 0$ for all $i \in \{2,\ldots,n\}$. Continuing in a similar fashion row by row, it follows that $\hat{B}$ is diagonal. $\square$

Corollary 5.4.5. Let $A \in \mathbb{F}^{n \times n}$, assume that A is Hermitian, and define $r \triangleq \text{rank}\, A$. Then, there exist a unitary matrix $S \in \mathbb{F}^{n \times n}$ and a diagonal matrix $B \in \mathbb{R}^{r \times r}$ such that (5.4.4) is satisfied. In addition, A is positive semidefinite if and only if the diagonal entries of B are positive, and A is positive definite if and only if A is positive semidefinite and $r = n$.

Proof. Corollary 5.4.4 and $x)$, $xi)$ of Proposition 4.4.5 imply that there exist a unitary matrix $S \in \mathbb{F}^{n \times n}$ and a diagonal matrix $B \in \mathbb{R}^{r \times r}$ such that (5.4.4) is satisfied. If A is positive semidefinite, then $x^*Ax \geq 0$ for all $x \in \mathbb{F}^n$. Choosing $x = Se_i$, it follows that $B_{(i,i)} = e_i^{\mathrm{T}}S^*ASe_i \geq 0$ for all $i \in \{1,\ldots,r\}$. If A is positive definite, then $r = n$ and $B_{(i,i)} > 0$ for all $i = 1,\ldots,n$. $\square$

Proposition 5.4.6. Let $A \in \mathbb{F}^{n \times n}$ be Hermitian. Then, there exists a nonsingular matrix $S \in \mathbb{F}^{n \times n}$ such that

$$A = S \begin{bmatrix} -I_{\nu_-(A)} & 0 & 0 \\ 0 & 0_{\nu_0(A) \times \nu_0(A)} & 0 \\ 0 & 0 & I_{\nu_+(A)} \end{bmatrix} S^*. \tag{5.4.5}$$

Furthermore,

$$\text{rank}\, A = \nu_+(A) + \nu_-(A) \tag{5.4.6}$$

and

$$\text{def}\, A = \nu_0(A). \tag{5.4.7}$$

Proof. Since A is Hermitian, it follows from Corollary 5.4.5 that there exist a unitary matrix $\hat{S} \in \mathbb{F}^{n \times n}$ and a diagonal matrix $B \in \mathbb{R}^{n \times n}$ such that $A = \hat{S}B\hat{S}^*$. Choose S to order the diagonal entries of B such that $B = \text{diag}(B_1, 0, -B_2)$, where the diagonal matrices B_1, B_2 are both positive definite. Now, define $\hat{B} \triangleq \text{diag}(B_1, I, B_2)$. Then, $B = \hat{B}^{1/2}D\hat{B}^{1/2}$, where $D \triangleq \text{diag}(I_{\nu_-(A)}, 0_{\nu_0(A) \times \nu_0(A)}, -I_{\nu_+(A)})$. Hence, $A = \hat{S}\hat{B}^{1/2}D\hat{B}^{1/2}\hat{S}^*$. $\square$

The following result is *Sylvester's law of inertia*.

Corollary 5.4.7. Let $A, B \in \mathbb{F}^{n \times n}$ be Hermitian. Then, A and B are congruent if and only if $\operatorname{In} A = \operatorname{In} B$.

Proposition 4.5.4 shows that two or more eigenvectors associated with distinct eigenvalues of a normal matrix are mutually orthogonal. Thus, a normal matrix has at least as many mutually orthogonal eigenvectors as it has distinct eigenvalues. The next result, which is an immediate consequence of Corollary 5.4.4, shows that every $n \times n$ normal matrix has n mutually orthogonal eigenvectors. In fact, the converse is also true.

Corollary 5.4.8. Let $A \in \mathbb{C}^{n \times n}$. Then, A is normal if and only if A has n mutually orthogonal eigenvectors.

The following result concerns the *real normal form*.

Corollary 5.4.9. Let $A \in \mathbb{R}^{n \times n}$ be range symmetric. Then, there exist an orthogonal matrix $S \in \mathbb{R}^{n \times n}$ and a nonsingular matrix $B \in \mathbb{R}^{r \times r}$, where $r \triangleq \operatorname{rank} A$, such that

$$A = S \begin{bmatrix} B & 0 \\ 0 & 0 \end{bmatrix} S^{\mathrm{T}}. \tag{5.4.8}$$

In addition, assume that A is normal, and let $\operatorname{mspec}(A) = \{\lambda_1, \ldots, \lambda_r\}_{\mathrm{ms}} \cup \{\nu_1 + \jmath\omega_1, \nu_1 - \jmath\omega_1, \ldots, \nu_l + \jmath\omega_l, \nu_l - \jmath\omega_l\}_{\mathrm{ms}}$, where $\lambda_1, \ldots, \lambda_r \in \mathbb{R}$ and, for all $i \in \{1, \ldots, l,\}$ $\nu_i, \omega_i \in \mathbb{R}$ and $\omega_i \neq 0$. Then, there exists an orthogonal matrix $S \in \mathbb{R}^{n \times n}$ such that

$$A = SBS^{\mathrm{T}}, \tag{5.4.9}$$

where $B \triangleq \operatorname{diag}(B_1, \ldots, B_r, \hat{B}_1, \ldots, \hat{B}_l)$, $B_i \triangleq [\lambda_i]$ for all $i \in \{1, \ldots, r\}$, and $\hat{B}_i \triangleq \begin{bmatrix} \nu_i & \omega_i \\ -\omega_i & \nu_i \end{bmatrix}$ for all $i \in \{1, \ldots, l\}$.

5.5 Eigenstructure Properties

Definition 5.5.1. Let $A \in \mathbb{F}^{n \times n}$, and let $\lambda \in \mathbb{C}$. Then, the *index of λ with respect to A*, denoted by $\operatorname{ind}_A(\lambda)$, is the smallest nonnegative integer k such that

$$\mathcal{R}\left[(\lambda I - A)^k\right] = \mathcal{R}\left[(\lambda I - A)^{k+1}\right]. \tag{5.5.1}$$

That is,

$$\operatorname{ind}_A(\lambda) = \operatorname{ind}(\lambda I - A). \tag{5.5.2}$$

Note that $\lambda \notin \operatorname{spec}(A)$ if and only if $\operatorname{ind}_A(\lambda) = 0$. Hence, $0 \notin \operatorname{spec}(A)$ if and only if $\operatorname{ind} A = \operatorname{ind}_A(0) = 0$.

Proposition 5.5.2. Let $A \in \mathbb{F}^{n \times n}$, and let $\lambda \in \mathbb{C}$. Then, $\operatorname{ind}_A(\lambda)$ is the smallest nonnegative integer k such that

$$\operatorname{rank}\left[(\lambda I - A)^k\right] = \operatorname{rank}\left[(\lambda I - A)^{k+1}\right]. \tag{5.5.3}$$

Furthermore, $\operatorname{ind} A$ is the smallest nonnegative integer k such that

$$\operatorname{rank}(A^k) = \operatorname{rank}(A^{k+1}). \tag{5.5.4}$$

Proof. Corollary 2.4.2 implies that $\mathcal{R}\big[(\lambda I - A)^k\big] \subseteq \mathcal{R}\big[(\lambda I - A)^{k+1}\big]$. Consequently, Lemma 2.3.4 implies that $\mathcal{R}\big[(\lambda I - A)^k\big] = \mathcal{R}\big[(\lambda I - A)^{k+1}\big]$ if and only if $\operatorname{rank}\big[(\lambda I - A)^k\big] = \operatorname{rank}\big[(\lambda I - A)^{k+1}\big]$. $\qquad\square$

Proposition 5.5.3. Let $A \in \mathbb{F}^{n \times n}$, and let $\lambda \in \operatorname{spec}(A)$. Then, the following statements hold:

i) The order of the largest Jordan block of A associated with λ is $\operatorname{ind}_A(\lambda)$.

ii) The number of Jordan blocks of A associated with λ is $\operatorname{gmult}_A(\lambda)$.

iii) The number of linearly independent eigenvectors of A associated with λ is $\operatorname{gmult}_A(\lambda)$.

iv) $\operatorname{ind}_A(\lambda) \leq \operatorname{amult}_A(\lambda)$.

v) $\operatorname{ind}_A(\lambda) = \operatorname{amult}_A(\lambda)$ if and only if exactly one block is associated with λ.

vi) $\operatorname{gmult}_A(\lambda) \leq \operatorname{amult}_A(\lambda)$.

vii) $\operatorname{gmult}_A(\lambda) = \operatorname{amult}_A(\lambda)$ if and only if every block associated with λ is of order equal to 1.

viii) $\operatorname{ind}_A(\lambda) + \operatorname{gmult}_A(\lambda) \leq \operatorname{amult}_A(\lambda) + 1$.

ix) $\operatorname{ind}_A(\lambda) + \operatorname{gmult}_A(\lambda) = \operatorname{amult}_A(\lambda) + 1$ if and only if at most one block associated with λ is of order greater than 1.

Definition 5.5.4. Let $A \in \mathbb{F}^{n \times n}$, and let $\lambda \in \operatorname{spec}(A)$. Then, the following terminology is defined:

i) λ is *simple* if $\operatorname{amult}_A(\lambda) = 1$.

ii) A is *simple* if every eigenvalue of A is simple.

iii) λ is *cyclic* (or *nonderogatory*) if $\operatorname{gmult}_A(\lambda) = 1$.

iv) A is *cyclic* (or *nonderogatory*) if every eigenvalue of A is cyclic.

v) λ is *derogatory* if $\operatorname{gmult}_A(\lambda) > 1$.

vi) A is *derogatory* if A has at least one derogatory eigenvalue.

vii) λ is *semisimple* if $\operatorname{gmult}_A(\lambda) = \operatorname{amult}_A(\lambda)$.

viii) A is *semisimple* if every eigenvalue of A is semisimple.

ix) λ is *defective* if $\operatorname{gmult}_A(\lambda) < \operatorname{amult}_A(\lambda)$.

x) A is *defective* if A has at least one defective eigenvalue.

xi) A is *diagonalizable over* $\mathbb{C}$ if A is semisimple.

xii) $A \in \mathbb{R}^{n \times n}$ is *diagonalizable over* $\mathbb{R}$ if A is semisimple and every eigenvalue of A is real.

Proposition 5.5.5. Let $A \in \mathbb{F}^{n \times n}$, and let $\lambda \in \operatorname{spec}(A)$. Then, λ is simple if and only if λ is cyclic and semisimple.

Proposition 5.5.6. Let $A \in \mathbb{F}^{n \times n}$, and let $\lambda \in \mathrm{spec}(A)$. Then,

$$\mathrm{def}\left[(\lambda I - A)^{\mathrm{ind}_A(\lambda)}\right] = \mathrm{amult}_A(\lambda). \qquad (5.5.5)$$

Theorem 5.3.3 yields the following result, which shows that the subspaces $\mathcal{N}\left[(\lambda I - A)^k\right]$, where $\lambda \in \mathrm{spec}(A)$ and $k = \mathrm{ind}_A(\lambda)$, provide a decomposition of $\mathbb{F}^n$.

Proposition 5.5.7. Let $A \in \mathbb{F}^{n \times n}$, let $\mathrm{spec}(A) = \{\lambda_1, \ldots, \lambda_r\}$, and, for all $i \in \{1, \ldots, r\}$, let $k_i \triangleq \mathrm{ind}_A(\lambda_i)$. Then, the following statements hold:

i) $\mathcal{N}\left[(\lambda_i I - A)^{k_i}\right] \cap \mathcal{N}\left[(\lambda_j I - A)^{k_j}\right] = \{0\}$ for all $i, j \in \{1, \ldots, r\}$ such that $i \neq j$.

ii) $\sum_{i=1}^r \mathcal{N}\left[(\lambda_i I - A)^{k_i}\right] = \mathbb{F}^n$.

Proposition 5.5.8. Let $A \in \mathbb{F}^{n \times n}$, and let $\lambda \in \mathrm{spec}(A)$. Then, the following statements are equivalent:

i) λ is semisimple.

ii) $\mathrm{def}(\lambda I - A) = \mathrm{def}\left[(\lambda I - A)^2\right]$.

iii) $\mathcal{N}(\lambda I - A) = \mathcal{N}\left[(\lambda I - A)^2\right]$.

iv) $\mathrm{ind}_A(\lambda) = 1$.

Proof. To prove that i) implies ii), suppose that λ is semisimple so that $\mathrm{gmult}_A(\lambda) = \mathrm{amult}_A(\lambda)$, and thus $\mathrm{def}(\lambda I - A) = \mathrm{amult}_A(\lambda)$. Then, it follows from Proposition 5.5.6 that $\mathrm{def}\left[(\lambda I - A)^k\right] = \mathrm{amult}_A(\lambda)$, where $k \triangleq \mathrm{ind}_A(\lambda)$. Therefore, it follows from Corollary 2.5.7 that $\mathrm{amult}_A(\lambda) = \mathrm{def}(\lambda I - A) \leq \mathrm{def}\left[(\lambda I - A)^2\right] \leq \mathrm{def}\left[(\lambda I - A)^k\right] = \mathrm{amult}_A(\lambda)$, which implies that $\mathrm{def}(\lambda I - A) = \mathrm{def}\left[(\lambda I - A)^2\right]$.

To prove that ii) implies iii), note that it follows from Corollary 2.5.7 that $\mathcal{N}(\lambda I - A) \subseteq \mathcal{N}\left[(\lambda I - A)^2\right]$. Since, by ii), these subspaces have equal dimension, it follows from Lemma 2.3.4 that these subspaces are equal.

To prove that iii) implies iv), note that iii) implies ii), and thus $\mathrm{rank}(\lambda I - A) = n - \mathrm{def}(\lambda I - A) = n - \mathrm{def}\left[(\lambda I - A)^2\right] = \mathrm{rank}\left[(\lambda I - A)^2\right]$. Therefore, since $\mathcal{R}(\lambda I - A) \subseteq \mathcal{R}\left[(\lambda I - A)^2\right]$ it follows from Corollary 2.5.7 that $\mathcal{R}(\lambda I - A) = \mathcal{R}\left[(\lambda I - A)^2\right]$. Finally, since $\lambda \in \mathrm{spec}(A)$, it follows from Definition 5.5.1 that $\mathrm{ind}_A(\lambda) = 1$.

Finally, to prove that iv) implies i), note that iv) is equivalent to the fact that every Jordan block of A associated with λ has order 1, which is equivalent to the fact that the geometric multiplicity of λ is equal to the algebraic multiplicity of λ, that is, that λ is semisimple. $\square$

Corollary 5.5.9. Let $A \in \mathbb{F}^{n \times n}$. Then, A is group invertible if and only if $\mathrm{ind}\, A \leq 1$.

Proposition 5.5.10. Assume that $A, B \in \mathbb{F}^{n \times n}$ are similar. Then, the following statements hold:

i) $\operatorname{mspec}(A) = \operatorname{mspec}(B)$.

ii) For all $\lambda \in \operatorname{spec}(A)$, $\operatorname{gmult}_A(\lambda) = \operatorname{gmult}_B(\lambda)$.

Proposition 5.5.11. Let $A \in \mathbb{F}^{n \times n}$. Then, A is semisimple if and only if A is similar to a normal matrix.

The following result is an extension of Corollary 5.3.9.

Proposition 5.5.12. Let $A \in \mathbb{F}^{n \times n}$. Then, the following statements are equivalent:

i) A is semisimple, and $\operatorname{spec}(A) \subset \mathbb{R}$.

ii) There exists a positive-definite matrix $S \in \mathbb{F}^{n \times n}$ such that $A = SA^*S^{-1}$.

iii) There exist a Hermitian matrix $S_1 \in \mathbb{F}^{n \times n}$ and a positive-definite matrix $S_2 \in \mathbb{F}^{n \times n}$ such that $A = S_1 S_2$.

Proof. To prove that *i*) implies *ii*), let $\hat{S} \in \mathbb{F}^{n \times n}$ be a nonsingular matrix such that $A = \hat{S} B \hat{S}^{-1}$, where $B \in \mathbb{R}^{n \times n}$ is diagonal. Then, $B = \hat{S}^{-1} A \hat{S} = \hat{S}^* A^* \hat{S}^{-*}$. Hence, $A = \hat{S} B \hat{S}^{-1} = \hat{S}(\hat{S}^* A^* \hat{S}^{-*}) \hat{S}^{-1} = (\hat{S}\hat{S}^*) A^* (\hat{S}\hat{S}^*)^{-1} = SA^*S^{-1}$, where $S \triangleq \hat{S}\hat{S}^*$ is positive definite. To show that *ii*) implies *iii*), note that $A = SA^*S^{-1} = S_1 S_2$, where $S_1 \triangleq SA^*$ and $S_2 \triangleq S^{-1}$. Since $S_1^* = (SA^*)^* = AS^* = AS = SA^* = S_1$, it follows that S_1 is Hermitian. Furthermore, since S is positive definite, it follows that S^{-1}, and hence S_2, is also positive definite. Finally, to prove that *iii*) implies *i*), note that $A = S_1 S_2 = S_2^{-1/2}\left(S_2^{1/2} S_1 S_2^{1/2}\right) S_2^{1/2}$. Since $S_2^{1/2} S_1 S_2^{1/2}$ is Hermitian, it follows from Corollary 5.4.5 that $S_2^{1/2} S_1 S_2^{1/2}$ is unitarily similar to a real diagonal matrix. Consequently, A is semisimple and $\operatorname{spec}(A) \subset \mathbb{R}$. $\square$

If a matrix is block triangular, then the following result shows that its eigenvalues and their algebraic multiplicity are determined by the diagonally located blocks. If, in addition, the matrix is block diagonal, then the geometric multiplicities of its eigenvalues are determined by the diagonally located blocks.

Proposition 5.5.13. Let $A \in \mathbb{F}^{n \times n}$, assume that A is partitioned as $A = \begin{bmatrix} A_{11} & \cdots & A_{1k} \\ \vdots & \ddots & \vdots \\ A_{k1} & \cdots & A_{kk} \end{bmatrix}$, where, for all $i, j \in \{1, \ldots, k\}$, $A_{ij} \in \mathbb{F}^{n_i \times n_j}$, and let $\lambda \in \operatorname{spec}(A)$. Then, the following statements hold:

i) If A_{ii} is the only nonzero block in the ith column of blocks, then

$$\operatorname{amult}_{A_{ii}}(\lambda) \le \operatorname{amult}_A(\lambda). \tag{5.5.6}$$

ii) If A is upper block triangular or lower block triangular, then

$$\operatorname{amult}_A(\lambda) = \sum_{i=1}^{r} \operatorname{amult}_{A_{ii}}(\lambda) \tag{5.5.7}$$

and

$$\mathrm{mspec}(A) = \bigcup_{i=1}^{k} \mathrm{mspec}(A_{ii}). \qquad (5.5.8)$$

iii) If A_{ii} is the only nonzero block in the ith column of blocks, then

$$\mathrm{gmult}_{A_{ii}}(\lambda) \leq \mathrm{gmult}_A(\lambda). \qquad (5.5.9)$$

iv) If A is upper block triangular, then

$$\mathrm{gmult}_{A_{11}}(\lambda) \leq \mathrm{gmult}_A(\lambda). \qquad (5.5.10)$$

v) If A is lower block triangular, then

$$\mathrm{gmult}_{A_{kk}}(\lambda) \leq \mathrm{gmult}_A(\lambda). \qquad (5.5.11)$$

vi) If A is block diagonal, then

$$\mathrm{gmult}_A(\lambda) = \sum_{i=1}^{r} \mathrm{gmult}_{A_{ii}}(\lambda). \qquad (5.5.12)$$

Proposition 5.5.14. Let $A \in \mathbb{F}^{n \times n}$, let $\mathrm{spec}(A) = \{\lambda_1, \ldots, \lambda_r\}$, and let $k_i \triangleq \mathrm{ind}_A(\lambda_i)$ for all $i \in \{1, \ldots, r\}$. Then,

$$\mu_A(s) = \prod_{i=1}^{r} (s - \lambda_i)^{k_i} \qquad (5.5.13)$$

and

$$\deg \mu_A = \sum_{i=1}^{r} k_i. \qquad (5.5.14)$$

Furthermore, the following statements are equivalent:

i) $\mu_A = \chi_A$.

ii) A is cyclic.

iii) For all $\lambda \in \mathrm{spec}(A)$, the Jordan form of A contains exactly one block associated with λ.

iv) A is similar to $C(\chi_A)$.

Proof. Let $A = SBS^{-1}$, where $B = \mathrm{diag}(B_1, \ldots, B_{n_{\mathrm{h}}})$ denotes the Jordan form of A given by (5.3.4). Let $\lambda_i \in \mathrm{spec}(A)$, and let B_j be a Jordan block associated with λ_i. Then, the order of B_j is less than or equal to k_i. Consequently, $(B_j - \lambda_i I)^{k_i} = 0$.

Next, let $p(s)$ denote the right-hand side of (5.5.13). Thus,

$$p(A) = \prod_{i=1}^{r} (A - \lambda_i I)^{k_i} = S \left[\prod_{i=1}^{r} (B - \lambda_i I)^{k_i} \right] S^{-1}$$

$$= S \mathrm{diag}\left(\prod_{i=1}^{r} (B_1 - \lambda_i I)^{k_i}, \ldots, \prod_{i=1}^{r} (B_{n_{\mathrm{h}}} - \lambda_i I)^{k_i} \right) S^{-1} = 0.$$

Therefore, it follows from Theorem 4.6.1 that μ_A divides p. Furthermore, note that, if k_i is replaced by $\hat{k}_i < k_i$, then $p(A) \neq 0$. Hence, p is the minimal polynomial of A. The equivalence of *i)* and *ii)* is now immediate, while the equivalence of *ii)* and *iii)* follows from Theorem 5.3.5. The equivalence of *i)* and *iv)* is given by Corollary 5.2.4. $\qquad\qquad\square$

Example 5.5.15. The standard nilpotent matrix N_n is in companion form, and thus is cyclic. In fact, N_n consists of a single Jordan block, and $\chi_{N_n}(s) = \mu_{N_n}(s) = s^n$.

Example 5.5.16. The matrix $\left[\begin{smallmatrix} 1 & 1 \\ -1 & 1 \end{smallmatrix}\right]$ is normal but is neither symmetric nor skew symmetric, while the matrix $\left[\begin{smallmatrix} 0 & 1 \\ -1 & 0 \end{smallmatrix}\right]$ is normal but is neither symmetric nor semisimple with real eigenvalues.

Example 5.5.17. The matrices $\left[\begin{smallmatrix} 1 & 0 \\ 2 & -1 \end{smallmatrix}\right]$ and $\left[\begin{smallmatrix} 1 & 1 \\ 0 & 2 \end{smallmatrix}\right]$ are diagonalizable over $\mathbb{R}$ but not normal, while the matrix $\left[\begin{smallmatrix} -1 & 1 \\ -2 & 1 \end{smallmatrix}\right]$ is diagonalizable but is neither normal nor diagonalizable over $\mathbb{R}$.

Example 5.5.18. The product of the Hermitian matrices $\left[\begin{smallmatrix} 1 & 2 \\ 2 & 1 \end{smallmatrix}\right]$ and $\left[\begin{smallmatrix} 2 & 1 \\ 1 & -2 \end{smallmatrix}\right]$ has no real eigenvalues.

Example 5.5.19. The matrices $\left[\begin{smallmatrix} 1 & 0 \\ 0 & 2 \end{smallmatrix}\right]$ and $\left[\begin{smallmatrix} 0 & 1 \\ -2 & 3 \end{smallmatrix}\right]$ are similar, whereas $\left[\begin{smallmatrix} 1 & 0 \\ 0 & 1 \end{smallmatrix}\right]$ and $\left[\begin{smallmatrix} 0 & 1 \\ -1 & 2 \end{smallmatrix}\right]$ have the same spectrum but are not similar.

Proposition 5.5.20. Let $A \in \mathbb{F}^{n \times n}$. Then, the following statements hold:

i) A is singular if and only if $0 \in \operatorname{spec}(A)$.

ii) A is group invertible if and only if either A is nonsingular or $0 \in \operatorname{spec}(A)$ is semisimple.

iii) A is Hermitian if and only if A is normal and $\operatorname{spec}(A) \subset \mathbb{R}$.

iv) A is skew Hermitian if and only if A is normal and $\operatorname{spec}(A) \subset \jmath\mathbb{R}$.

v) A is positive semidefinite if and only if A is normal and $\operatorname{spec}(A) \subset [0, \infty)$.

vi) A is positive definite if and only if A is normal and $\operatorname{spec}(A) \subset (0, \infty)$.

vii) A is unitary if and only if A is normal and $\operatorname{spec}(A) \subset \{\lambda \in \mathbb{C}: |\lambda| = 1\}$.

viii) A is shifted unitary if and only if A is normal and

$$\operatorname{spec}(A) \subset \{\lambda \in \mathbb{C}: |\lambda - \tfrac{1}{2}| = \tfrac{1}{2}\}. \qquad (5.5.15)$$

ix) A is involutory if and only if A is semisimple and $\operatorname{spec}(A) \subseteq \{-1, 1\}$.

x) A is skew involutory if and only if A is semisimple and $\operatorname{spec}(A) \subseteq \{-\jmath, \jmath\}$.

xi) A is idempotent if and only if A is semisimple and $\operatorname{spec}(A) \subseteq \{0, 1\}$.

xii) A is skew idempotent if and only if A is semisimple and $\operatorname{spec}(A) \subseteq \{0, -1\}$.

xiii) A is tripotent if and only if A is semisimple and $\operatorname{spec}(A) \subseteq \{-1, 0, 1\}$.

xiv) A is nilpotent if and only if $\operatorname{spec}(A) = \{0\}$.

xv) A is unipotent if and only if $\mathrm{spec}(A) = \{1\}$.

xvi) A is a projector if and only if A is normal and $\mathrm{spec}(A) \subseteq \{0, 1\}$.

xvii) A is a reflector if and only if A is normal and $\mathrm{spec}(A) \subseteq \{-1, 1\}$.

xviii) A is a skew reflector if and only if A is normal and $\mathrm{spec}(A) \subseteq \{-\jmath, \jmath\}$.

xix) A is an elementary projector if and only if A is normal and $\mathrm{mspec}(A) = \{0, 1, \ldots, 1\}_{\mathrm{ms}}$.

xx) A is an elementary reflector if and only if A is normal and $\mathrm{mspec}(A) = \{-1, 1, \ldots, 1\}_{\mathrm{ms}}$.

If, furthermore, $A \in \mathbb{F}^{2n \times 2n}$, then the following statements hold:

xxi) If A is Hamiltonian, then $\mathrm{mspec}(A) = \mathrm{mspec}(-A)$.

xxii) If A is symplectic, then $\mathrm{mspec}(A) = \mathrm{mspec}(A^{-1})$.

The following result is a consequence of Proposition 5.5.12 and Proposition 5.5.20.

Corollary 5.5.21. Let $A \in \mathbb{F}^{n \times n}$, and assume that A is either involutory, idempotent, skew idempotent, tripotent, a projector, or a reflector. Then, the following statements hold:

i) There exists a positive-definite matrix $S \in \mathbb{F}^{n \times n}$ such that $A = SA^*S^{-1}$.

ii) There exist a Hermitian matrix $S_1 \in \mathbb{F}^{n \times n}$ and a positive-definite matrix $S_2 \in \mathbb{F}^{n \times n}$ such that $A = S_1 S_2$.

Proposition 5.5.22. Let $A, B \in \mathbb{F}^{n \times n}$. Then, the following statements hold:

i) Assume that A and B are normal. Then, A and B are unitarily similar if and only if $\mathrm{mspec}(A) = \mathrm{mspec}(B)$.

ii) Assume that A and B are projectors. Then, A and B are unitarily similar if and only if $\mathrm{rank}\, A = \mathrm{rank}\, B$.

iii) Assume that A and B are (projectors, reflectors). Then, A and B are unitarily similar if and only if $\mathrm{tr}\, A = \mathrm{tr}\, B$.

iv) Assume that A and B are semisimple. Then, A and B are similar if and only if $\mathrm{mspec}(A) = \mathrm{mspec}(B)$.

v) Assume that A and B are (involutory, skew involutory, idempotent). Then, A and B are similar if and only if $\mathrm{tr}\, A = \mathrm{tr}\, B$.

vi) Assume that A and B are idempotent. Then, A and B are similar if and only if $\mathrm{rank}\, A = \mathrm{rank}\, B$.

vii) Assume that A and B are tripotent. Then, A and B are similar if and only if $\mathrm{rank}\, A = \mathrm{rank}\, B$ and $\mathrm{tr}\, A = \mathrm{tr}\, B$.

5.6 Singular Value Decomposition

The third matrix decomposition that we consider is the *singular value decomposition*. Unlike the Jordan and Schur decompositions, the singular value decomposition applies to matrices that are not necessarily square. Let $A \in \mathbb{F}^{n \times m}$, where $A \neq 0$, and consider the positive-semidefinite matrices $AA^* \in \mathbb{F}^{n \times n}$ and $A^*A \in \mathbb{F}^{m \times m}$. It follows from Proposition 4.4.10 that AA^* and A^*A have the same nonzero eigenvalues with the same algebraic multiplicities. Since AA^* and A^*A are positive semidefinite, it follows that they have the same *positive* eigenvalues with the same algebraic multiplicities. Furthermore, since AA^* is Hermitian, it follows that the number of positive eigenvalues of AA^* (or A^*A), counting algebraic multiplicity, is equal to the rank of AA^* (or A^*A). Since $\operatorname{rank} A = \operatorname{rank} AA^* = \operatorname{rank} A^*A$, it thus follows that AA^* and A^*A both have r positive eigenvalues, where $r \triangleq \operatorname{rank} A$.

Definition 5.6.1. Let $A \in \mathbb{F}^{n \times m}$. Then, the *singular values* of A are the $\min\{n, m\}$ nonnegative numbers $\sigma_1(A), \ldots, \sigma_{\min\{n,m\}}(A)$, where, for all $i \in \{1, \ldots, \min\{n, m\}\}$,

$$\sigma_i(A) \triangleq \lambda_i^{1/2}(AA^*) = \lambda_i^{1/2}(A^*A). \tag{5.6.1}$$

Hence,

$$\sigma_1(A) \geq \cdots \geq \sigma_{\min\{n,m\}}(A) \geq 0. \tag{5.6.2}$$

Let $A \in \mathbb{F}^{n \times m}$, and define $r \triangleq \operatorname{rank} A$. If $1 \leq r < \min\{n, m\}$, then

$$\sigma_1(A) \geq \cdots \geq \sigma_r(A) > \sigma_{r+1}(A) = \cdots = \sigma_{\min\{n,m\}}(A) = 0, \tag{5.6.3}$$

whereas, if $r = \min\{m, n\}$, then

$$\sigma_1(A) \geq \cdots \geq \sigma_r(A) = \sigma_{\min\{n,m\}}(A) > 0. \tag{5.6.4}$$

Consequently, $\operatorname{rank} A$ is the number of positive singular values of A. For convenience, define

$$\sigma_{\max}(A) \triangleq \sigma_1(A) \tag{5.6.5}$$

and, if $n = m$,

$$\sigma_{\min}(A) \triangleq \sigma_n(A). \tag{5.6.6}$$

If $n \neq m$, then $\sigma_{\min}(A)$ is not defined. By convention, we define

$$\sigma_{\max}(0_{n \times m}) = \sigma_{\min}(0_{n \times n}) = 0, \tag{5.6.7}$$

and, for all $i \in \{1, \ldots, \min\{n, m\}\}$,

$$\sigma_i(A) = \sigma_i(A^*) = \sigma_i(\overline{A}) = \sigma_i(A^{\mathrm{T}}). \tag{5.6.8}$$

Now, suppose that $n = m$. If A is Hermitian, then, for all $i \in \{1, \ldots, n\}$,

$$\sigma_i(A) = |\lambda_i(A)|, \tag{5.6.9}$$

while, if A is positive semidefinite, then, for all $i \in \{1, \ldots, n\}$,

$$\sigma_i(A) = \lambda_i(A). \tag{5.6.10}$$

Proposition 5.6.2. Let $A \in \mathbb{F}^{n \times m}$. If $n \leq m$, then the following statements are equivalent:

i) $\operatorname{rank} A = n$.

ii) $\sigma_n(A) > 0$.

If $m \leq n$, then the following statements are equivalent:

iii) $\operatorname{rank} A = m$.

iv) $\sigma_m(A) > 0$.

If $n = m$, then the following statements are equivalent:

v) A is nonsingular.

vi) $\sigma_{\min}(A) > 0$.

We now state the singular value decomposition.

Theorem 5.6.3. Let $A \in \mathbb{F}^{n \times m}$, assume that A is nonzero, let $r \triangleq \operatorname{rank} A$, and define $B \triangleq \operatorname{diag}[\sigma_1(A), \ldots, \sigma_r(A)]$. Then, there exist unitary matrices $S_1 \in \mathbb{F}^{n \times n}$ and $S_2 \in \mathbb{F}^{m \times m}$ such that

$$A = S_1 \begin{bmatrix} B & 0_{r \times (m-r)} \\ 0_{(n-r) \times r} & 0_{(n-r) \times (m-r)} \end{bmatrix} S_2. \tag{5.6.11}$$

Furthermore, each column of S_1 is an eigenvector of AA^*, while each column of S_2^* is an eigenvector of A^*A.

Proof. For convenience, assume that $r < \min\{n, m\}$, since otherwise the zero matrices become empty matrices. By Corollary 5.4.5 there exists a unitary matrix $U \in \mathbb{F}^{n \times n}$ such that

$$AA^* = U \begin{bmatrix} B^2 & 0 \\ 0 & 0 \end{bmatrix} U^*.$$

Partition $U = \begin{bmatrix} U_1 & U_2 \end{bmatrix}$, where $U_1 \in \mathbb{F}^{n \times r}$ and $U_2 \in \mathbb{F}^{n \times (n-r)}$. Since $U^*U = I_n$, it follows that $U_1^*U_1 = I_r$ and $U_1^*U = \begin{bmatrix} I_r & 0_{r \times (n-r)} \end{bmatrix}$. Now, define $V_1 \triangleq A^*U_1 B^{-1} \in \mathbb{F}^{m \times r}$, and note that

$$V_1^*V_1 = B^{-1}U_1^*AA^*U_1 B^{-1} = B^{-1}U_1^*U \begin{bmatrix} B^2 & 0 \\ 0 & 0 \end{bmatrix} U^*U_1 B^{-1} = I_r.$$

Next, note that, since $U_2^*U = \begin{bmatrix} 0_{(n-r) \times r} & I_{n-r} \end{bmatrix}$, it follows that

$$U_2^*AA^* = \begin{bmatrix} 0 & I \end{bmatrix} \begin{bmatrix} B^2 & 0 \\ 0 & 0 \end{bmatrix} U^* = 0.$$

However, since $\mathcal{R}(A) = \mathcal{R}(AA^*)$, it follows that $U_2^*A = 0$. Finally, let $V_2 \in \mathbb{F}^{m \times (m-r)}$ be such that $V \triangleq \begin{bmatrix} V_1 & V_2 \end{bmatrix} \in \mathbb{F}^{m \times m}$ is unitary. Hence, we have

$$U \begin{bmatrix} B & 0 \\ 0 & 0 \end{bmatrix} V^* = \begin{bmatrix} U_1 & U_2 \end{bmatrix} \begin{bmatrix} B & 0 \\ 0 & 0 \end{bmatrix} \begin{bmatrix} V_1^* \\ V_2^* \end{bmatrix} = U_1 B V_1^* = U_1 B B^{-1} U_1^* A$$

$$= U_1 U_1^* A = (U_1 U_1^* + U_2 U_2^*)A = UU^*A = A,$$

which yields (5.6.11) with $S_1 = U$ and $S_2 = V^*$. $\qquad\square$

An immediate corollary of the singular value decomposition is the *polar decomposition*.

Corollary 5.6.4. Let $A \in \mathbb{F}^{n\times n}$. Then, there exists a positive-semidefinite matrix $M \in \mathbb{F}^{n\times n}$ and a unitary matrix $S \in \mathbb{F}^{n\times n}$ such that

$$A = MS. \tag{5.6.12}$$

Proof. It follows from the singular value decomposition that there exist unitary matrices $S_1, S_2 \in \mathbb{F}^{n\times n}$ and a diagonal positive-definite matrix $B \in \mathbb{F}^{r\times r}$, where $r \triangleq \text{rank}\,A$, such that $A = S_1 \left[\begin{smallmatrix} B & 0 \\ 0 & 0 \end{smallmatrix}\right] S_2$. Hence,

$$A = S_1 \begin{bmatrix} B & 0 \\ 0 & 0 \end{bmatrix} S_1^* S_1 S_2 = MS,$$

where $M \triangleq S_1 \left[\begin{smallmatrix} B & 0 \\ 0 & 0 \end{smallmatrix}\right] S_1^*$ is positive semidefinite and $S \triangleq S_1 S_2$ is unitary. $\qquad\square$

Proposition 5.6.5. Let $A \in \mathbb{F}^{n\times m}$, let $r \triangleq \text{rank}\,A$, and define the Hermitian matrix $\mathcal{A} \triangleq \left[\begin{smallmatrix} 0 & A \\ A^* & 0 \end{smallmatrix}\right] \in \mathbb{F}^{(n+m)\times(n+m)}$. Then, $\text{In}\,\mathcal{A} = \begin{bmatrix} r & 0 & r \end{bmatrix}^{\text{T}}$, and the $2r$ nonzero eigenvalues of $\mathcal{A}$ are the r positive singular values of A and their negatives.

Proof. Since $\chi_{\mathcal{A}}(s) = \det(s^2 I - A^*A)$, it follows that

$$\text{mspec}(\mathcal{A})\backslash\{0,\ldots,0\}_{\text{ms}} = \{\sigma_1(A), -\sigma_1(A), \ldots, \sigma_r(A), -\sigma_r(A)\}_{\text{ms}}. \qquad\square$$

5.7 Pencils and the Kronecker Canonical Form

Let $A, B \in \mathbb{F}^{n\times m}$, and define the polynomial matrix $P_{A,B} \in \mathbb{F}^{n\times m}[s]$, called a *pencil*, by

$$P_{A,B}(s) \triangleq sB - A.$$

The pencil $P_{A,B}$ is *regular* if $\text{rank}\,P_{A,B} = \min\{n, m\}$ (see Definition 4.2.4). Otherwise, $P_{A,B}$ is *singular*.

Let $A, B \in \mathbb{F}^{n\times m}$. Since $P_{A,B} \in \mathbb{F}^{n\times m}$ we define the *generalized spectrum* of $P_{A,B}$ by

$$\text{spec}(A, B) \triangleq \text{Szeros}(P_{A,B}) \tag{5.7.1}$$

and the *generalized multispectrum* of $P_{A,B}$ by

$$\text{mspec}(A, B) \triangleq \text{mSzeros}(P_{A,B}). \tag{5.7.2}$$

Furthermore, the elements of $\text{spec}(A, B)$ are the *generalized eigenvalues* of $P_{A,B}$.

The structure of a pencil is illuminated by the following result known as the *Kronecker canonical form*.

Theorem 5.7.1. Let $A, B \in \mathbb{C}^{n \times m}$. Then, there exist nonsingular matrices $S_1 \in \mathbb{C}^{n \times n}$ and $S_2 \in \mathbb{C}^{m \times m}$ such that, for all $s \in \mathbb{C}$,

$$P_{A,B}(s) = S_1 \mathrm{diag}(sI_{r_1} - A_1, sB_2 - I_{r_2}, [sI_{k_1} - N_{k_1} \ -e_{k_1}], \ldots, [sI_{k_p} - N_{k_p} \ -e_{k_p}],$$

$$[sI_{l_1} - N_{l_1} \ -e_{l_1}]^{\mathrm{T}}, \ldots, [sI_{l_q} - N_{l_q} \ -e_{l_q}]^{\mathrm{T}}, 0_{t \times u}) S_2, \qquad (5.7.3)$$

where $A_1 \in \mathbb{C}^{r_1 \times r_1}$ is in Jordan form, $B_2 \in \mathbb{R}^{r_2 \times r_2}$ is nilpotent and in Jordan form, $k_1, \ldots, k_p, l_1, \ldots, l_q$ are positive integers, and $[sI_l - N_l \ -e_l] \in \mathbb{C}^{l \times (l+1)}$. Furthermore,

$$\mathrm{rank}\, P_{A,B} = r_1 + r_2 + \sum_{i=1}^{p} k_i + \sum_{i=1}^{q} l_i. \qquad (5.7.4)$$

Proof. See [68, Chapter 2], [555, Chapter XII], [809, pp. 395–398], [891], [897, pp. 128, 129], and [1261, Chapter VI]. $\qquad\qquad\square$

In Theorem 5.7.1, note that

$$n = r_1 + r_2 + \sum_{i=1}^{p} k_i + \sum_{i=1}^{q} l_i + q + t \qquad (5.7.5)$$

and

$$m = r_1 + r_2 + \sum_{i=1}^{p} k_i + \sum_{i=1}^{q} l_i + p + u. \qquad (5.7.6)$$

Proposition 5.7.2. Let $A, B \in \mathbb{C}^{n \times m}$, and consider the notation of Theorem 5.7.1. Then, $P_{A,B}$ is regular if and only if $t = u = 0$ and either $p = 0$ or $q = 0$.

Let $A, B \in \mathbb{F}^{n \times m}$, and let $\lambda \in \mathbb{C}$. Then,

$$\mathrm{rank}\, P_{A,B}(\lambda) = \mathrm{rank}(\lambda I - A_1) + r_2 + \sum_{i=1}^{p} k_i + \sum_{i=1}^{q} l_i. \qquad (5.7.7)$$

Note that λ is a generalized eigenvalue of $P_{A,B}$ if and only if $\mathrm{rank}\, P_{A,B}(\lambda) < \mathrm{rank}\, P_{A,B}$. Consequently, λ is a generalized eigenvalue of $P_{A,B}$ if and only if λ is an eigenvalue of A_1, that is,

$$\mathrm{spec}(A, B) = \mathrm{spec}(A_1). \qquad (5.7.8)$$

Furthermore,

$$\mathrm{mspec}(A, B) = \mathrm{mspec}(A_1). \qquad (5.7.9)$$

The *generalized algebraic multiplicity* $\mathrm{amult}_{A,B}(\lambda)$ of $\lambda \in \mathrm{spec}(A, B)$ is defined by

$$\mathrm{amult}_{A,B}(\lambda) \triangleq \mathrm{amult}_{A_1}(\lambda). \qquad (5.7.10)$$

It can be seen that, for $\lambda \in \mathrm{spec}(A, B)$,

$$\mathrm{gmult}_{A_1}(\lambda) \triangleq \mathrm{rank}\, P_{A,B} - \mathrm{rank}\, P_{A,B}(\lambda).$$

The *generalized geometric multiplicity* $\mathrm{gmult}_{A,B}(\lambda)$ of $\lambda \in \mathrm{spec}(A, B)$ is defined by

$$\mathrm{gmult}_{A,B}(\lambda) \triangleq \mathrm{gmult}_{A_1}(\lambda). \qquad (5.7.11)$$

Now, assume that $A, B \in \mathbb{F}^{n \times n}$, that is, A and B are square, which, from (5.7.5) and (5.7.6), is equivalent to $q+t = p+u$. Then, the *characteristic polynomial* $\chi_{A,B} \in \mathbb{F}[s]$ of (A, B) is defined by

$$\chi_{A,B}(s) \triangleq \det P_{A,B}(s) = \det(sB - A).$$

Proposition 5.7.3. Let $A, B \in \mathbb{F}^{n \times n}$. Then, the following statements hold:

i) $P_{A,B}$ is singular if and only if $\chi_{A,B} = 0$.

ii) $P_{A,B}$ is singular if and only if $\deg \chi_{A,B} = -\infty$.

iii) $P_{A,B}$ is regular if and only if $\chi_{A,B}$ is not the zero polynomial.

iv) $P_{A,B}$ is regular if and only if $0 \leq \deg \chi_{A,B} \leq n$.

v) If $P_{A,B}$ is regular, then $\text{mult}_{\chi_{A,B}}(0) = n - \deg \chi_{B,A}$.

vi) $\deg \chi_{A,B} = n$ if and only if B is nonsingular.

vii) If B is nonsingular, then $\chi_{A,B} = \chi_{B^{-1}A}$, $\text{spec}(A, B) = \text{spec}(B^{-1}A)$, and $\text{mspec}(A, B) = \text{mspec}(B^{-1}A)$.

viii) $\text{roots}(\chi_{A,B}) = \text{spec}(A, B)$.

ix) $\text{mroots}(\chi_{A,B}) = \text{mspec}(A, B)$.

x) If A or B is nonsingular, then $P_{A,B}$ is regular.

xi) If all of the generalized eigenvalues of (A, B) are real, then $P_{A,B}$ is regular.

xii) If $P_{A,B}$ is regular, then $\mathcal{N}(A) \cap \mathcal{N}(B) = \{0\}$.

xiii) If $P_{A,B}$ is regular, then there exist nonsingular matrices $S_1, S_2 \in \mathbb{C}^{n \times n}$ such that, for all $s \in \mathbb{C}$,

$$P_{A,B}(s) = S_1\left(s\begin{bmatrix} I_r & 0 \\ 0 & B_2 \end{bmatrix} - \begin{bmatrix} A_1 & 0 \\ 0 & I_{n-r} \end{bmatrix}\right)S_2, \qquad (5.7.12)$$

where $r \triangleq \deg \chi_{A,B}$, $A_1 \in \mathbb{C}^{r \times r}$ is in Jordan form, and $B_2 \in \mathbb{R}^{(n-r) \times (n-r)}$ is nilpotent and in Jordan form. Furthermore,

$$\chi_{A,B} = \chi_{A_1}, \qquad (5.7.13)$$
$$\text{roots}(\chi_{A,B}) = \text{spec}(A_1), \qquad (5.7.14)$$
$$\text{mroots}(\chi_{A,B}) = \text{mspec}(A_1). \qquad (5.7.15)$$

Proof. See [897, p. 128] and [1261, Chapter VI]. □

Statement *xiii)* is the *Weierstrass canonical form* for a square, regular pencil.

Proposition 5.7.4. Let $A, B \in \mathbb{F}^{n \times n}$, assume that A is positive semidefinite, and assume that B is Hermitian. Then, the following statements hold:

i) $P_{A,B}$ is regular.

ii) There exists $\alpha \in \mathbb{F}$ such that $A + \alpha B$ is nonsingular.

iii) $\mathcal{N}(A) \cap \mathcal{N}(B) = \{0\}$.

iv) $\mathcal{N}(\left[\begin{smallmatrix} A \\ B \end{smallmatrix}\right]) = \{0\}$.

v) There exists nonzero $\alpha \in \mathbb{F}$ such that $\mathcal{N}(A) \cap \mathcal{N}(B + \alpha A) = \{0\}$.

vi) For all nonzero $\alpha \in \mathbb{F}$, $\mathcal{N}(A) \cap \mathcal{N}(B + \alpha A) = \{0\}$.

vii) All generalized eigenvalues of (A, B) are real.

If, in addition, B is positive semidefinite, then the following statement is equivalent to *i)–vii)*:

viii) There exists $\beta > 0$ such that $\beta B < A$.

Proof. The results *i)* $\implies$ *ii)* and *ii)* $\implies$ *iii)* are immediate. Next, Fact 2.10.10 and Fact 2.11.3 imply that *iii)*, *iv)*, *v)*, and *vi)* are equivalent. Next, to prove *iii)* $\implies$ *vii)*, let $\lambda \in \mathbb{C}$ be a generalized eigenvalue of (A, B). Since $\lambda = 0$ is real, suppose $\lambda \neq 0$. Since $\det(\lambda B - A) = 0$, let nonzero $\theta \in \mathbb{C}^n$ satisfy $(\lambda B - A)\theta = 0$, and thus it follows that $\theta^* A \theta = \lambda \theta^* B \theta$. Furthermore, note that $\theta^* A \theta$ and $\theta^* B \theta$ are real. Now, suppose $\theta \in \mathcal{N}(A)$. Then, it follows from $(\lambda B - A)\theta = 0$ that $\theta \in \mathcal{N}(B)$, which contradicts $\mathcal{N}(A) \cap \mathcal{N}(B) = \{0\}$. Hence, $\theta \notin \mathcal{N}(A)$, and thus $\theta^* A \theta > 0$ and, consequently, $\theta^* B \theta \neq 0$. Hence, it follows that $\lambda = \theta^* A \theta / \theta^* B \theta$, and thus λ is real. Hence, all generalized eigenvalues of (A, B) are real.

Next, to prove *vii)* $\implies$ *i)*, let $\lambda \in \mathbb{C} \backslash \mathbb{R}$ so that λ is not a generalized eigenvalue of (A, B). Consequently, $\chi_{A,B}(s)$ is not the zero polynomial, and thus (A, B) is regular.

Next, to prove *i)–vii)* $\implies$ *viii)*, let $\theta \in \mathbb{R}^n$ be nonzero, and note that $\mathcal{N}(A) \cap \mathcal{N}(B) = \{0\}$ implies that either $A\theta \neq 0$ or $B\theta \neq 0$. Hence, either $\theta^{\mathrm{T}} A \theta > 0$ or $\theta^{\mathrm{T}} B \theta > 0$. Thus, $\theta^{\mathrm{T}} (A + B) \theta > 0$, which implies $A + B > 0$ and hence $-B < A$.

Finally, to prove *viii)* $\implies$ *i)–vii)*, let $\beta \in \mathbb{R}$ be such that $\beta B < A$, so that $\beta \theta^{\mathrm{T}} B \theta < \theta^{\mathrm{T}} A \theta$ for all nonzero $\theta \in \mathbb{R}^n$. Next, suppose $\hat{\theta} \in \mathcal{N}(A) \cap \mathcal{N}(B)$ is nonzero. Hence, $A\hat{\theta} = 0$ and $B\hat{\theta} = 0$. Consequently, $\hat{\theta}^{\mathrm{T}} B \hat{\theta} = 0$ and $\hat{\theta}^{\mathrm{T}} A \hat{\theta} = 0$, which contradicts $\beta \hat{\theta}^{\mathrm{T}} B \hat{\theta} < \hat{\theta}^{\mathrm{T}} A \hat{\theta}$. Thus, $\mathcal{N}(A) \cap \mathcal{N}(B) = \{0\}$. $\square$

5.8 Facts on the Inertia

Fact 5.8.1. Let $A \in \mathbb{F}^{n \times n}$, and assume that A is idempotent. Then,

$$\operatorname{rank} A = \operatorname{sig} A = \operatorname{tr} A$$

and

$$\operatorname{In} A = \left[\begin{array}{c} 0 \\ n - \operatorname{tr} A \\ \operatorname{tr} A \end{array} \right].$$

Fact 5.8.2. Let $A \in \mathbb{F}^{n \times n}$, and assume that A is involutory. Then,

$$\operatorname{rank} A = n,$$
$$\operatorname{sig} A = \operatorname{tr} A,$$

and

$$\operatorname{In} A = \left[\begin{array}{c} \frac{1}{2}(n - \operatorname{tr} A) \\ 0 \\ \frac{1}{2}(n + \operatorname{tr} A) \end{array} \right].$$

Fact 5.8.3. Let $A \in \mathbb{F}^{n \times n}$, and assume that A is tripotent. Then,

$$\operatorname{rank} A = \operatorname{tr} A^2,$$
$$\operatorname{sig} A = \operatorname{tr} A,$$

and

$$\operatorname{In} A = \left[\begin{array}{c} \frac{1}{2}(\operatorname{tr} A^2 - \operatorname{tr} A) \\ n - \operatorname{tr} A^2 \\ \frac{1}{2}(\operatorname{tr} A^2 + \operatorname{tr} A) \end{array} \right].$$

Fact 5.8.4. Let $A \in \mathbb{F}^{n \times n}$, and assume that A is either skew Hermitian, skew involutory, or nilpotent. Then,

$$\operatorname{sig} A = \nu_-(A) = \nu_+(A) = 0$$

and

$$\operatorname{In} A = \left[\begin{array}{c} 0 \\ n \\ 0 \end{array} \right].$$

Fact 5.8.5. Let $A \in \mathbb{F}^{n \times n}$, assume that A is group invertible, and assume that $\operatorname{spec}(A) \cap \jmath\mathbb{R} \subseteq \{0\}$. Then,

$$\operatorname{rank} A = \nu_-(A) + \nu_+(A)$$

and

$$\operatorname{def} A = \nu_0(A) = \operatorname{amult}_A(0).$$

Fact 5.8.6. Let $A \in \mathbb{F}^{n \times n}$, and assume that A is Hermitian. Then,

$$\operatorname{rank} A = \nu_-(A) + \nu_+(A)$$

and

$$\text{In } A = \begin{bmatrix} \nu_-(A) \\ \nu_0(A) \\ \nu_+(A) \end{bmatrix} = \begin{bmatrix} \frac{1}{2}(\text{rank } A - \text{sig } A) \\ n - \text{rank } A \\ \frac{1}{2}(\text{rank } A + \text{sig } A) \end{bmatrix}.$$

Fact 5.8.7. Let $A, B \in \mathbb{F}^{n \times n}$, and assume that A and B are Hermitian. Then, $\text{In } A = \text{In } B$ if and only if $\text{rank } A = \text{rank } B$ and $\text{sig } A = \text{sig } B$.

Fact 5.8.8. Let $A \in \mathbb{F}^{n \times n}$, assume that A is Hermitian, and let A_0 be a principal submatrix of A. Then,

$$\nu_-(A_0) \leq \nu_-(A)$$

and

$$\nu_+(A_0) \leq \nu_+(A).$$

Proof: See [792].

Fact 5.8.9. Let $A \in \mathbb{F}^{n \times n}$, and assume that A is positive semidefinite. Then,

$$\text{rank } A = \text{sig } A = \nu_+(A)$$

and

$$\text{In } A = \begin{bmatrix} 0 \\ \text{def } A \\ \text{rank } A \end{bmatrix}.$$

Fact 5.8.10. Let $A \in \mathbb{F}^{n \times n}$, and assume that A is positive semidefinite. Then,

$$\text{In } A = \begin{bmatrix} 0 \\ \text{def } A \\ \text{rank } A \end{bmatrix}.$$

If, in addition, A is positive definite, then

$$\text{In } A = \begin{bmatrix} 0 \\ 0 \\ n \end{bmatrix}.$$

Fact 5.8.11. Let $A \in \mathbb{F}^{n \times n}$. Then, the following statements are equivalent:

i) A is an elementary projector.

ii) A is a projector, and $\text{tr } A = n - 1$.

iii) A is a projector, and $\text{In } A = \begin{bmatrix} 0 \\ 1 \\ n-1 \end{bmatrix}$.

Furthermore, the following statements are equivalent:

iv) A is an elementary reflector.

v) A is a reflector, and $\text{tr } A = n - 2$.

vi) A is a reflector, and $\text{In } A = \begin{bmatrix} 1 \\ 0 \\ n-1 \end{bmatrix}$.

Proof: See Proposition 5.5.20.

Fact 5.8.12. Let $A \in \mathbb{F}^{n \times n}$. Then, the following statements are equivalent:

i) $A + A^*$ is positive definite.

ii) For all Hermitian matrices $B \in \mathbb{F}^{n \times n}$, $\mathrm{In}\, B = \mathrm{In}\, AB$.

Proof: See [287].

Fact 5.8.13. Let $A, B \in \mathbb{F}^{n \times n}$, assume that AB and B are Hermitian, and assume that $\mathrm{spec}(A) \cap [0, \infty) = \varnothing$. Then,

$$\mathrm{In}(-AB) = \mathrm{In}\, B.$$

Proof: See [287].

Fact 5.8.14. Let $A, B \in \mathbb{F}^{n \times n}$, assume that A and B are Hermitian and nonsingular, and assume that $\mathrm{spec}(AB) \cap [0, \infty) = \varnothing$. Then,

$$\nu_+(A) + \nu_+(B) = n.$$

Proof: Use Fact 5.8.13. See [287].

Remark: Weaker versions of this result are given in [784, 1063].

Fact 5.8.15. Let $A \in \mathbb{F}^{n \times n}$, assume that A is Hermitian, and let $S \in \mathbb{F}^{m \times n}$. Then,

$$\nu_-(SAS^*) \leq \nu_-(A)$$

and

$$\nu_+(SAS^*) \leq \nu_+(A).$$

Furthermore, consider the following conditions:

i) $\mathrm{rank}\, S = n$.

ii) $\mathrm{rank}\, SAS^* = \mathrm{rank}\, A$.

iii) $\nu_-(SAS^*) = \nu_-(A)$ and $\nu_+(SAS^*) = \nu_+(A)$.

Then, $i) \implies ii) \iff iii)$.

Proof: See [459, pp. 430, 431] and [521, p. 194].

Fact 5.8.16. Let $A \in \mathbb{F}^{n \times n}$, assume that A is Hermitian, and let $S \in \mathbb{F}^{m \times n}$. Then,

$$\nu_-(SAS^*) + \nu_+(SAS^*) = \mathrm{rank}\, SAS^* \leq \min\{\mathrm{rank}\, A, \mathrm{rank}\, S\},$$

$$\nu_-(A) + \mathrm{rank}\, S - n \leq \nu_-(SAS^*) \leq \nu_-(A),$$

$$\nu_+(A) + \mathrm{rank}\, S - n \leq \nu_+(SAS^*) \leq \nu_+(A).$$

Proof: See [1087].

Fact 5.8.17. Let $A, S \in \mathbb{F}^{n \times n}$, assume that A is Hermitian, and assume that S is nonsingular. Then, there exist $\alpha_1, \ldots, \alpha_n \in [\lambda_{\min}(SS^*), \lambda_{\max}(SS^*)]$ such that, for all $i \in \{1, \ldots, n\}$,

$$\lambda_i(SAS^*) = \alpha_i \lambda_i(A).$$

Proof: See [1473].

Remark: This result, which is due to Ostrowski, is a quantitative version of Sylvester's law of inertia given by Corollary 5.4.7.

Fact 5.8.18. Let $A, S \in \mathbb{F}^{n \times n}$, assume that A is Hermitian, and assume that S is nonsingular. Then, the following statements are equivalent:

i) $\operatorname{In}(SAS^*) = \operatorname{In} A$.

ii) $\operatorname{rank}(SAS^*) = \operatorname{rank} A$.

iii) $\mathcal{R}(A) \cap \mathcal{N}(A) = \{0\}$.

Proof: See [112].

Fact 5.8.19. Let $A \in \mathbb{F}^{n \times n}$, $B \in \mathbb{F}^{n \times m}$, and $C \in \mathbb{F}^{m \times m}$, and assume that A is positive definite and C is negative definite. Then,

$$\operatorname{In} \begin{bmatrix} A & B & 0 \\ B^* & C & 0 \\ 0 & 0 & 0_{l \times l} \end{bmatrix} = \begin{bmatrix} n \\ m \\ l \end{bmatrix}.$$

Proof: This result follows from Fact 5.8.6. See [792].

Fact 5.8.20. Let $A \in \mathbb{R}^{n \times m}$. Then,

$$\operatorname{In} \begin{bmatrix} 0 & A \\ A^* & 0 \end{bmatrix} = \operatorname{In} \begin{bmatrix} AA^* & 0 \\ 0 & -A^*A \end{bmatrix}$$

$$= \operatorname{In} \begin{bmatrix} AA^+ & 0 \\ 0 & -A^+A \end{bmatrix}$$

$$= \begin{bmatrix} \operatorname{rank} A \\ n + m - 2\operatorname{rank} A \\ \operatorname{rank} A \end{bmatrix}.$$

Proof: See [459, pp. 432, 434].

Fact 5.8.21. Let $A \in \mathbb{C}^{n \times n}$, assume that A is Hermitian, and let $B \in \mathbb{C}^{n \times m}$. Then,

$$\operatorname{In} \begin{bmatrix} A & B \\ B^* & 0 \end{bmatrix} \geq\geq \begin{bmatrix} \operatorname{rank} B \\ n - \operatorname{rank} B \\ \operatorname{rank} B \end{bmatrix}.$$

Furthermore, if $\mathcal{R}(A) \subseteq \mathcal{R}(B)$, then

$$\operatorname{In} \begin{bmatrix} A & B \\ B^* & 0 \end{bmatrix} = \begin{bmatrix} \operatorname{rank} B \\ n + m - 2\operatorname{rank} B \\ \operatorname{rank} B \end{bmatrix}.$$

Finally, if $\operatorname{rank} B = n$, Then,

$$\operatorname{In} \begin{bmatrix} A & B \\ B^* & 0 \end{bmatrix} = \begin{bmatrix} n \\ m - n \\ n \end{bmatrix}.$$

Proof: See [459, pp. 433, 434] or [970].

Remark: Extensions are given in [970].

Remark: See Fact 8.15.28.

Fact 5.8.22. Let $A \in \mathbb{F}^{n \times n}$. Then, there exist a nonsingular matrix $S \in \mathbb{F}^{n \times n}$ and a skew-Hermitian matrix $B \in \mathbb{F}^{n \times n}$ such that

$$A = S \left(\begin{bmatrix} I_{\nu_-(A+A^*)} & 0 & 0 \\ 0 & 0_{\nu_0(A+A^*) \times \nu_0(A+A^*)} & 0 \\ 0 & 0 & -I_{\nu_+(A+A^*)} \end{bmatrix} + B \right) S^*.$$

Proof: Write $A = \frac{1}{2}(A+A^*) + \frac{1}{2}(A-A^*)$, and apply Proposition 5.4.6 to $\frac{1}{2}(A+A^*)$.

5.9 Facts on Matrix Transformations for One Matrix

Fact 5.9.1. Define $S \in \mathbb{C}^{3 \times 3}$ by

$$S \triangleq \frac{1}{\sqrt{2}} \begin{bmatrix} 0 & 1 & 1 \\ 0 & -\jmath & \jmath \\ \sqrt{2} & 0 & 0 \end{bmatrix}.$$

Then, S is unitary, and

$$K(e_3) = S \operatorname{diag}(0, \jmath, -\jmath) S^{-1}.$$

Remark: See Fact 5.9.2.

Fact 5.9.2. Let $x \in \mathbb{R}^3$, assume that either $x_{(1)} \neq 0$ or $x_{(2)} \neq 0$, and define

$$\begin{bmatrix} a \\ b \\ c \end{bmatrix} \triangleq \frac{1}{\|x\|_2} x,$$

$$\alpha \triangleq \sqrt{a^2 c^2 + b^2 c^2 + 2 - 3c^2 + c^4},$$

and

$$S \triangleq \begin{bmatrix} a & \frac{-b+\jmath ac}{\alpha} & \frac{-b-\jmath ac}{\alpha} \\ b & \frac{a+\jmath bc}{\alpha} & \frac{a-\jmath bc}{\alpha} \\ c & \frac{-\jmath(1-c^2)}{\alpha} & \frac{\jmath(1-c^2)}{\alpha} \end{bmatrix}.$$

Then, $\alpha \geq \min\{|a|, |b|\} > 0$, S is unitary, and

$$K(x) = S \operatorname{diag}(0, \jmath\|x\|_2, -\jmath\|x\|_2) S^{-1}.$$

Proof: See [855, p. 154].

Remark: If $x_{(1)} = x_{(2)} = 0$, then $a = b = 0$, $c = 1$, and $\alpha = 0$. This case is considered by Fact 5.9.1.

Problem: Find a decomposition of $K(x)$ that holds for all $x \in \mathbb{R}^3$.

Fact 5.9.3. Let $A \in \mathbb{F}^{n \times n}$, and assume that $\text{spec}(A) = \{1\}$. Then, A^k is similar to A for all $k \geq 1$.

Fact 5.9.4. Let $A \in \mathbb{F}^{n \times n}$, and assume there exists a nonsingular matrix $S \in \mathbb{F}^{n \times n}$ such that $S^{-1}AS$ is upper triangular. Then, for all $r \in \{1, \ldots, n\}$, $\mathcal{R}\left(S\begin{bmatrix} I_r \\ 0 \end{bmatrix}\right)$ is an invariant subspace of A.

Remark: Analogous results hold for lower triangular matrices and block-triangular matrices.

Fact 5.9.5. Let $A \in \mathbb{F}^{n \times n}$. Then, there exist unique matrices $B, C \in \mathbb{F}^{n \times n}$ such that the following properties are satisfied:

i) B is diagonalizable over $\mathbb{F}$.

ii) C is nilpotent.

iii) $A = B + C$.

iv) $BC = CB$.

Furthermore, $\text{mspec}(A) = \text{mspec}(B)$.

Proof: See [709, p. 112] or [746, p. 74]. Existence follows from the real Jordan form. The last statement follows from Fact 5.17.4.

Remark: This result is the *S-N decomposition* or the *Jordan-Chevalley decomposition*.

Fact 5.9.6. Let $A \in \mathbb{F}^{n \times n}$. Then, the following statements are equivalent:

i) A is similar to a skew-Hermitian matrix.

ii) A is semisimple, and $\text{spec}(A) \subset \jmath\mathbb{R}$.

Remark: See Fact 11.18.12.

Fact 5.9.7. Let $A \in \mathbb{F}^{n \times n}$, and let $r \triangleq \text{rank}\, A$. Then, A is group invertible if and only if there exist a nonsingular matrix $B \in \mathbb{F}^{r \times r}$ and a nonsingular matrix $S \in \mathbb{F}^{n \times n}$ such that
$$A = S\begin{bmatrix} B & 0 \\ 0 & 0 \end{bmatrix} S^{-1}.$$

Fact 5.9.8. Let $A \in \mathbb{F}^{n \times n}$, and let $r \triangleq \text{rank}\, A$. Then, A is range Hermitian if and only if there exist a nonsingular matrix $S \in \mathbb{F}^{n \times n}$ and a nonsingular matrix $B \in \mathbb{F}^{r \times r}$ such that
$$A = S\begin{bmatrix} B & 0 \\ 0 & 0 \end{bmatrix} S^*.$$

Remark: S need not be unitary for sufficiency. See Corollary 5.4.4.

Proof: Use the QR decomposition Fact 5.15.9 to let $S \triangleq \hat{S}R$, where $\hat{S}$ is unitary and R is upper triangular. See [1309].

Fact 5.9.9. Let $A \in \mathbb{F}^{n \times n}$. Then, there exists an involutory matrix $S \in \mathbb{F}^{n \times n}$ such that

$$A^{\mathrm{T}} = SAS^{\mathrm{T}}.$$

Proof: See [430] and [591].

Remark: Note A^{T} rather than A^*.

Fact 5.9.10. Let $A \in \mathbb{F}^{n \times n}$. Then, there exists a nonsingular matrix $S \in \mathbb{F}^{n \times n}$ such that $A = SA^*S^{-1}$ if and only if there exist Hermitian matrices $S_1, S_2 \in \mathbb{F}^{n \times n}$ such that $A = S_1 S_2$.

Proof: See [1526, pp. 215, 216].

Remark: See Proposition 5.5.12.

Remark: An analogous result in Hilbert space is given in [779] under the assumption that A is normal.

Fact 5.9.11. Let $A \in \mathbb{F}^{n \times n}$, and assume that A is normal. Then, there exists a symmetric, nonsingular matrix $S \in \mathbb{F}^{n \times n}$ such that

$$A^{\mathrm{T}} = SAS^{-1}$$

and such that $S^{-1} = \overline{S}$.

Proof: For $\mathbb{F} = \mathbb{C}$, let $A = UBU^*$, where U is unitary and B is diagonal. Then, $A^{\mathrm{T}} = SA\overline{S} = SAS^{-1}$, where $S \triangleq \overline{U}U^{-1}$. For $\mathbb{F} = \mathbb{R}$, use the real normal form and let $S \triangleq U\tilde{I}U^{\mathrm{T}}$, where U is orthogonal and $\tilde{I} \triangleq \mathrm{diag}(\hat{I}, \ldots, \hat{I})$.

Remark: See Corollary 5.3.8.

Fact 5.9.12. Let $A \in \mathbb{R}^{n \times n}$, and assume that A is normal. Then, there exists a reflector $S \in \mathbb{R}^{n \times n}$ such that

$$A^{\mathrm{T}} = SAS^{-1}.$$

Consequently, A and A^{T} are orthogonally similar. Finally, if A is skew symmetric, then A and $-A$ are orthogonally similar.

Proof: Specialize Fact 5.9.11 to the case $\mathbb{F} = \mathbb{R}$.

Fact 5.9.13. Let $A \in \mathbb{F}^{n \times n}$. Then, there exists a reverse-symmetric, nonsingular matrix $S \in \mathbb{F}^{n \times n}$ such that $A^{\hat{\mathrm{T}}} = SAS^{-1}$.

Proof: This result follows from Corollary 5.3.8. See [907].

Fact 5.9.14. Let $A \in \mathbb{F}^{n \times n}$. Then, there exist reverse-symmetric matrices $S_1, S_2 \in \mathbb{F}^{n \times n}$ such that S_2 is nonsingular and $A = S_1 S_2$.

Proof: This result follows from Corollary 5.3.9. See [907].

Fact 5.9.15. Let $A \in \mathbb{R}^{n \times n}$, and assume that A is not of the form aI, where $a \in \mathbb{R}$. Then, A is similar to a matrix with diagonal entries $0, \ldots, 0, \mathrm{tr}\, A$.

Proof: See [1125, p. 77].

Remark: This result is due to Gibson.

Fact 5.9.16. Let $A \in \mathbb{R}^{n \times n}$, and assume that A is not zero. Then, A is similar to a matrix whose diagonal entries are all nonzero.

Proof: See [1125, p. 79].

Remark: This result is due to Marcus and Purves.

Fact 5.9.17. Let $A \in \mathbb{R}^{n \times n}$, and assume that A is symmetric. Then, there exists an orthogonal matrix $S \in \mathbb{R}^{n \times n}$ such that $-1 \notin \text{spec}(S)$ and SAS^{T} is diagonal.

Proof: See [1125, p. 101].

Remark: This result is due to Hsu.

Fact 5.9.18. Let $A \in \mathbb{R}^{n \times n}$, and assume that A is symmetric. Then, there exist a diagonal matrix $B \in \mathbb{R}^{n \times n}$ and a skew-symmetric matrix $C \in \mathbb{R}^{n \times n}$ such that

$$A = [2(I + C)^{-1} - I]B[2(I + C)^{-1} - I]^{\mathrm{T}}.$$

Proof: Use Fact 5.9.17. See [1125, p. 101].

Fact 5.9.19. Let $A \in \mathbb{F}^{n \times n}$. Then, there exists a unitary matrix $S \in \mathbb{F}^{n \times n}$ such that S^*AS has equal diagonal entries.

Proof: See [501] or [1125, p. 78], or use Fact 5.9.20.

Remark: The diagonal entries are equal to $(\text{tr } A)/n$.

Remark: This result is due to Parker. See [549].

Fact 5.9.20. Let $A \in \mathbb{F}^{n \times n}$. Then, the following statements are equivalent:

i) $\text{tr } A = 0$.

ii) There exist matrices $B, C \in \mathbb{F}^{n \times n}$ such that $A = [B, C]$.

iii) A is unitarily similar to a matrix whose diagonal entries are zero.

Proof: See [14, 549, 822, 838] or [641, p. 146].

Remark: This result is *Shoda's theorem.*

Remark: See Fact 5.9.21.

Fact 5.9.21. Let $R \in \mathbb{F}^{n \times n}$, and assume that R is Hermitian. Then, the following statements are equivalent:

i) $\text{tr } R < 0$.

ii) R is unitarily similar to a matrix all of whose diagonal entries are negative.

iii) There exists an asymptotically stable matrix $A \in \mathbb{F}^{n \times n}$ such that $R = A + A^*$.

Proof: See [124].

Remark: See Fact 5.9.20.

Fact 5.9.22. Let $A \in \mathbb{F}^{n \times n}$. Then, AA^* and A^*A are unitarily similar.

Fact 5.9.23. Let $A \in \mathbb{F}^{n \times n}$, and assume that A is idempotent. Then, A and A^* are unitarily similar.

Proof: This result follows from Fact 5.9.29 and the fact that $\left[\begin{smallmatrix} 1 & a \\ 0 & 0 \end{smallmatrix}\right]$ and $\left[\begin{smallmatrix} 1 & 0 \\ a & 0 \end{smallmatrix}\right]$ are unitarily similar. See [429].

Fact 5.9.24. Let $A \in \mathbb{F}^{n \times n}$, and assume that A is symmetric. Then, there exists a unitary matrix $S \in \mathbb{F}^{n \times n}$ such that

$$A = SBS^{\mathrm{T}},$$

where

$$B \triangleq \operatorname{diag}[\sigma_1(A), \ldots, \sigma_n(A)].$$

Proof: See [728, p. 207].

Remark: A is symmetric, complex, and T-congruent to B.

Fact 5.9.25. Let $A \in \mathbb{F}^{n \times n}$. Then, $\left[\begin{smallmatrix} A & 0 \\ 0 & -A \end{smallmatrix}\right]$ and $\left[\begin{smallmatrix} 0 & A \\ A & 0 \end{smallmatrix}\right]$ are unitarily similar.

Proof: Use the unitary transformation $\frac{1}{\sqrt{2}}\left[\begin{smallmatrix} I & -I \\ I & I \end{smallmatrix}\right]$.

Fact 5.9.26. Let n be a positive integer. Then,

$$\hat{I}_n = \begin{cases} S\begin{bmatrix} -I_{n/2} & 0 \\ 0 & -I_{n/2} \end{bmatrix} S^{\mathrm{T}}, & n \text{ even}, \\[2em] S\begin{bmatrix} -I_{n/2} & 0 & 0 \\ 0 & 1 & 0 \\ 0 & 0 & I_{n/2} \end{bmatrix} S^{\mathrm{T}}, & n \text{ odd}, \end{cases}$$

where

$$S \triangleq \begin{cases} \dfrac{1}{\sqrt{2}}\begin{bmatrix} I_{n/2} & -\hat{I}_{n/2} \\ \hat{I}_{n/2} & I_{n/2} \end{bmatrix}, & n \text{ even}, \\[2em] \dfrac{1}{\sqrt{2}}\begin{bmatrix} I_{n/2} & 0 & -\hat{I}_{n/2} \\ 0 & \sqrt{2} & 0 \\ \hat{I}_{n/2} & 0 & I_{n/2} \end{bmatrix}, & n \text{ odd}. \end{cases}$$

Therefore,

$$\operatorname{mspec}(\hat{I}_n) = \begin{cases} \{-1, 1, \ldots, -1, 1\}_{\mathrm{ms}}, & n \text{ even}, \\ \{1, -1, 1, \ldots, -1, 1\}_{\mathrm{ms}}, & n \text{ odd}. \end{cases}$$

Remark: For even n, Fact 3.20.3 shows that $\hat{I}_n$ is Hamiltonian, and thus, by Fact 4.9.22, $\operatorname{mspec}(\hat{I}_n) = -\operatorname{mspec}(\hat{I}_n)$.

Remark: See [1444].

Fact 5.9.27. Let n be a positive integer. Then,

$$J_{2n} = S \begin{bmatrix} \jmath I_n & 0 \\ 0 & -\jmath I_n \end{bmatrix} S^*,$$

where

$$S \triangleq \frac{1}{\sqrt{2}} \begin{bmatrix} I & -I \\ \jmath I & -\jmath I \end{bmatrix}.$$

Hence,

$$\mathrm{mspec}(J_{2n}) = \{\jmath, -\jmath, \ldots, \jmath, -\jmath\}_{\mathrm{ms}}$$

and

$$\det J_{2n} = 1.$$

Proof: See Fact 2.19.3.

Remark: Fact 3.20.3 shows that J_{2n} is Hamiltonian, and thus, by Fact 4.9.22, $\mathrm{mspec}(J_{2n}) = -\mathrm{mspec}(J_{2n})$.

Fact 5.9.28. Let $A \in \mathbb{F}^{n \times n}$, assume that A is idempotent, and let $r \triangleq \mathrm{rank}\, A$. Then, there exists a matrix $B \in \mathbb{F}^{r \times (n-r)}$ and a unitary matrix $S \in \mathbb{F}^{n \times n}$ such that

$$A = S \begin{bmatrix} I_r & B \\ 0 & 0_{(n-r)\times(n-r)} \end{bmatrix} S^*.$$

Proof: See [550, p. 46].

Fact 5.9.29. Let $A \in \mathbb{F}^{n \times n}$, assume that A is idempotent, and let $r \triangleq \mathrm{rank}\, A$. Then, there exist a unitary matrix $S \in \mathbb{F}^{n \times n}$ and positive numbers $a_1, \ldots, a_k$ such that

$$A = S\mathrm{diag}\left(\begin{bmatrix} 1 & a_1 \\ 0 & 0 \end{bmatrix}, \ldots, \begin{bmatrix} 1 & a_k \\ 0 & 0 \end{bmatrix}, I_{r-k}, 0_{(n-r-k)\times(n-r-k)} \right) S^*.$$

Proof: See [429].

Remark: This result provides a canonical form for idempotent matrices under unitary similarity. See also [551].

Remark: See Fact 5.9.23.

Fact 5.9.30. Let $A \in \mathbb{F}^{n \times m}$, assume that A is nonzero, let $r \triangleq \mathrm{rank}\, A$, define $B \triangleq \mathrm{diag}[\sigma_1(A), \ldots, \sigma_r(A)]$, and let $S_1 \in \mathbb{F}^{n \times n}$ and $S_2 \in \mathbb{F}^{m \times m}$ be unitary matrices such that

$$A = S_1 \begin{bmatrix} B & 0_{r \times (m-r)} \\ 0_{(n-r)\times r} & 0_{(n-r)\times(m-r)} \end{bmatrix} S_2.$$

Then, there exist $K \in \mathbb{F}^{r \times r}$ and $L \in \mathbb{F}^{r \times (m-r)}$ such that

$$KK^* + LL^* = I_r$$

and

$$A = S_1 \begin{bmatrix} BK & BL \\ 0_{(n-r) \times r} & 0_{(n-r) \times (m-r)} \end{bmatrix} S_1^*.$$

Proof: See [119, 668].

Remark: See Fact 6.3.14 and Fact 6.6.16.

Fact 5.9.31. Let $A \in \mathbb{F}^{n \times n}$, assume that A is unitary, and partition A as

$$A = \begin{bmatrix} A_{11} & A_{12} \\ A_{21} & A_{22} \end{bmatrix},$$

where $A_{11} \in \mathbb{F}^{m \times k}$, $A_{12} \in \mathbb{F}^{m \times q}$, $A_{21} \in \mathbb{F}^{p \times k}$, $A_{22} \in \mathbb{F}^{p \times q}$, and $m + p = k + q = n$. Then, there exist unitary matrices $U, V \in \mathbb{F}^{n \times n}$ and nonnegative integers l, r such that

$$A = U \begin{bmatrix} I_r & 0 & 0 & 0 & 0 & 0 \\ 0 & \Gamma & 0 & 0 & \Sigma & 0 \\ 0 & 0 & 0 & 0 & 0 & I_{m-r-l} \\ 0 & 0 & 0 & I_{q-m+r} & 0 & 0 \\ 0 & \Sigma & 0 & 0 & -\Gamma & 0 \\ 0 & 0 & I_{k-r-l} & 0 & 0 & 0 \end{bmatrix} V,$$

where $\Gamma, \Sigma \in \mathbb{R}^{l \times l}$ are diagonal and satisfy

$$0 < \Gamma_{(l,l)} \le \cdots \le \Gamma_{(1,1)} < 1, \qquad (5.9.1)$$

$$0 < \Sigma_{(1,1)} \le \cdots \le \Sigma_{(l,l)} < 1, \qquad (5.9.2)$$

and

$$\Gamma^2 + \Sigma^2 = I_m.$$

Proof: See [550, p. 12] and [1261, p. 37].

Remark: This result is the *CS decomposition*. See [1086, 1088]. The entries $\Sigma_{(i,i)}$ and $\Gamma_{(i,i)}$ can be interpreted as sines and cosines, respectively, of the principal angles between a pair of subspaces $\mathcal{S}_1 = \mathcal{R}(X_1)$ and $\mathcal{S}_2 = \mathcal{R}(Y_1)$ such that $[X_1 \ X_2]$ and $[Y_1 \ Y_2]$ are unitary and $A = [X_1 \ X_2]^*[Y_1 \ Y_2]$; see [550, pp. 25–29], [1261, pp. 40–43], and Fact 2.9.19. Principal angles can also be defined recursively; see [550, p. 25] and [551]. See also [821].

Fact 5.9.32. Let $A \in \mathbb{F}^{n \times n}$, and let $r \triangleq \operatorname{rank} A$. Then, there exist $S_1 \in \mathbb{F}^{n \times r}$, $B \in \mathbb{R}^{r \times r}$, and $S_2 \in \mathbb{F}^{n \times r}$ such that S_1 is left inner, S_2 is right inner, B is upper triangular, $I \circ B = \alpha I$, where $\alpha \triangleq \prod_{i=1}^{r} \sigma_i(A)$, and

$$A = S_1 B S_2.$$

Proof: See [780].

Remark: Note that B is real.

Remark: This result is the *geometric mean decomposition*.

Fact 5.9.33. Let $A \in \mathbb{C}^{n \times n}$. Then, there exists a matrix $B \in \mathbb{R}^{n \times n}$ such that $A\overline{A}$ and B^2 are similar.

Proof: See [425].

5.10 Facts on Matrix Transformations for Two or More Matrices

Fact 5.10.1. Let $q(s) \triangleq s^2 - \beta_1 s - \beta_0 \in \mathbb{R}[s]$ be irreducible, and let $\lambda = \nu + \jmath \omega$ denote a root of q so that $\beta_1 = 2\nu$ and $\beta_0 = -(\nu^2 + \omega^2)$. Then,

$$\mathcal{H}_1(q) = \begin{bmatrix} 0 & 1 \\ \beta_0 & \beta_1 \end{bmatrix} = \begin{bmatrix} 1 & 0 \\ \nu & \omega \end{bmatrix} \begin{bmatrix} \nu & \omega \\ -\omega & \nu \end{bmatrix} \begin{bmatrix} 1 & 0 \\ -\nu/\omega & 1/\omega \end{bmatrix} = S\mathcal{J}_1(q)S^{-1}.$$

The transformation matrix $S = \begin{bmatrix} 1 & 0 \\ \nu & \omega \end{bmatrix}$ is not unique; an alternative choice is $S = \begin{bmatrix} \omega & \nu \\ 0 & \nu^2+\omega^2 \end{bmatrix}$. Similarly,

$$\mathcal{H}_2(q) = \begin{bmatrix} 0 & 1 & 0 & 0 \\ \beta_0 & \beta_1 & 1 & 0 \\ 0 & 0 & 0 & 1 \\ 0 & 0 & \beta_0 & \beta_1 \end{bmatrix} = S \begin{bmatrix} \nu & \omega & 1 & 0 \\ -\omega & \nu & 0 & 1 \\ 0 & 0 & \nu & \omega \\ 0 & 0 & -\omega & \nu \end{bmatrix} S^{-1} = S\mathcal{J}_2(q)S^{-1},$$

where

$$S \triangleq \begin{bmatrix} \omega & \nu & \omega & \nu \\ 0 & \nu^2 + \omega^2 & \omega & \nu^2 + \omega^2 + \nu \\ 0 & 0 & -2\omega\nu & 2\omega^2 \\ 0 & 0 & -2\omega(\nu^2 + \omega^2) & 0 \end{bmatrix}.$$

Fact 5.10.2. Let $q(s) \triangleq s^2 - 2\nu s + \nu^2 + \omega^2 \in \mathbb{R}[s]$ with roots $\lambda = \nu + \jmath \omega$ and $\overline{\lambda} = \nu - \jmath\omega$. Then,

$$\mathcal{H}_1(q) = \begin{bmatrix} \nu & \omega \\ -\omega & \nu \end{bmatrix} = \frac{1}{\sqrt{2}} \begin{bmatrix} 1 & 1 \\ \jmath & -\jmath \end{bmatrix} \begin{bmatrix} \lambda & 0 \\ 0 & \overline{\lambda} \end{bmatrix} \frac{1}{\sqrt{2}} \begin{bmatrix} 1 & -\jmath \\ 1 & \jmath \end{bmatrix}$$

and

$$\mathcal{H}_2(q) = \begin{bmatrix} \nu & \omega & 1 & 0 \\ -\omega & \nu & 0 & 1 \\ 0 & 0 & \nu & \omega \\ 0 & 0 & -\omega & \nu \end{bmatrix} = S \begin{bmatrix} \lambda & 1 & 0 & 0 \\ 0 & \lambda & 0 & 0 \\ 0 & 0 & \overline{\lambda} & 1 \\ 0 & 0 & 0 & \overline{\lambda} \end{bmatrix} S^{-1},$$

where

$$S \triangleq \frac{1}{\sqrt{2}} \begin{bmatrix} 1 & 0 & 1 & 0 \\ \jmath & 0 & -\jmath & 0 \\ 0 & 1 & 0 & 1 \\ 0 & \jmath & 0 & -\jmath \end{bmatrix}, \qquad S^{-1} = \frac{1}{\sqrt{2}} \begin{bmatrix} 1 & -\jmath & 0 & 0 \\ 0 & 0 & 1 & -\jmath \\ 1 & \jmath & 0 & 0 \\ 0 & 0 & 1 & \jmath \end{bmatrix}.$$

Fact 5.10.3. Left equivalence, right equivalence, biequivalence, unitary left equivalence, unitary right equivalence, and unitary biequivalence are equivalence relations on $\mathbb{F}^{n \times m}$. Similarity, congruence, and unitary similarity are equivalence relations on $\mathbb{F}^{n \times n}$.

Fact 5.10.4. Let $A, B \in \mathbb{F}^{n \times m}$. Then, A and B are in the same equivalence class of $\mathbb{F}^{n \times m}$ induced by biequivalent transformations if and only if A and B are biequivalent to $\left[\begin{smallmatrix} I & 0 \\ 0 & 0 \end{smallmatrix}\right]$. Now, let $n = m$. Then, A and B are in the same equivalence class of $\mathbb{F}^{n \times n}$ induced by similarity transformations if and only if A and B have the same Jordan form.

Fact 5.10.5. Let $A, B \in \mathbb{F}^{n \times n}$, and assume that A and B are similar. Then, A is semisimple if and only if B is.

Fact 5.10.6. Let $A \in \mathbb{F}^{n \times n}$, and assume that A is normal. Then, A is unitarily similar to its Jordan form.

Fact 5.10.7. Let $A, B \in \mathbb{F}^{n \times n}$, assume that A and B are normal, and assume that A and B are similar. Then, A and B are unitarily similar.

Proof: Since A and B are similar, it follows that $\mathrm{mspec}(A) = \mathrm{mspec}(B)$. Since A and B are normal, it follows that they are unitarily similar to the same diagonal matrix. See Fact 5.10.6. See [642, p. 104].

Remark: See [555, p. 8] for related results.

Fact 5.10.8. Let $A, B \in \mathbb{F}^{n \times n}$, and let $r \triangleq 2n^2$. Then, the following statements are equivalent:

i) A and B are unitarily similar.

ii) For all $k_1, \ldots, k_r, l_1, \ldots, l_r \in \mathbb{N}$ such that $\sum_{i,j=1}^{r}(k_i + l_j) \leq r$, it follows that
$$\mathrm{tr}\, A^{k_1} A^{l_1 *} \cdots A^{k_r} A^{l_r *} = \mathrm{tr}\, B^{k_1} B^{l_1 *} \cdots B^{k_r} B^{l_r *}.$$

Proof: See [1103].

Remark: See [812, pp. 71, 72] and [224, 1221].

Remark: The number of distinct tuples of positive integers whose sum is a positive integer k is 2^{k-1}. The number of expressions in *ii)* is thus $\sum_{k=1}^{2n^2} 2^{k-1} = 4^{n^2} - 1$. Because of properties of the trace function, the number of distinct expressions is less than this number. Furthermore, in special cases, the number of expressions that need to be checked is significantly less than the number of distinct expressions. In the case $n = 2$, it suffices to check three equalities, specifically, $\mathrm{tr}\, A = \mathrm{tr}\, B$, $\mathrm{tr}\, A^2 = \mathrm{tr}\, B^2$, and $\mathrm{tr}\, A^* A = \mathrm{tr}\, B^* B$. In the case $n = 3$, it suffices to check 7 equalities. See [224, 1221].

Fact 5.10.9. Let $A, B \in \mathbb{F}^{n \times n}$, assume that A and B are idempotent, assume that $\mathrm{sprad}(A - B) < 1$, and define
$$S \triangleq (AB + A_\perp B_\perp)\left[I - (A - B)^2\right]^{-1/2}.$$

Then, the following statements hold:

i) S is nonsingular.

ii) If $A = B$, then $S = I$.

iii) $S^{-1} = (BA + B_\perp A_\perp)\big[I - (B - A)^2\big]^{-1/2}$.

iv) A and B are similar. In fact, $A = SBS^{-1}$.

v) If A and B are projectors, then S is unitary and A and B are unitarily similar.

Proof: See [708, p. 412].

Remark: $[I - (A - B)^2]^{-1/2}$ is defined by *ix)* of Fact 10.13.1.

Fact 5.10.10. Let $A, B \in \mathbb{F}^{n \times n}$, and assume that A and B are idempotent. Then, the following statements are equivalent:

i) A and B are unitarily similar.

ii) $\operatorname{tr} A = \operatorname{tr} B$ and, for all $i \in \{1, \ldots, \lfloor n/2 \rfloor\}$, $\operatorname{tr}(AA^*)^i = \operatorname{tr}(BB^*)^i$.

iii) $\chi_{AA^*} = \chi_{BB^*}$.

Proof: This result follows from Fact 5.9.29. See [429].

Fact 5.10.11. Let $A, B \in \mathbb{F}^{n \times n}$, and assume that either A or B is nonsingular. Then, AB and BA are similar.

Proof: If A is nonsingular, then $AB = A(BA)A^{-1}$, whereas, if B is nonsingular, then $BA = B(AB)B^{-1}$.

Fact 5.10.12. Let $A, B \in \mathbb{F}^{n \times n}$, and assume that A and B are projectors. Then, AB and BA are unitarily similar.

Remark: This result is due to Dixmier. See [1141].

Fact 5.10.13. Let $A \in \mathbb{F}^{n \times n}$. Then, A is idempotent if and only if there exists an orthogonal matrix $B \in \mathbb{F}^{n \times n}$ such that A and B are similar.

Fact 5.10.14. Let $A, B \in \mathbb{F}^{n \times n}$, assume that A and B are idempotent, and assume that $A + B - I$ is nonsingular. Then, A and B are similar. In particular,

$$A = (A + B - I)^{-1} B (A + B - I).$$

Fact 5.10.15. Let $A_1, \ldots, A_r \in \mathbb{F}^{n \times n}$, and assume that $A_i A_j = A_j A_i$ for all $i, j \in \{1, \ldots, r\}$. Then,

$$\dim \operatorname{span}\left\{ \prod_{i=1}^{r} A_i^{n_i} \colon\ 0 \le n_i \le n - 1 \text{ for all } i \in \{1, \ldots, r\} \right\} \le \tfrac{1}{4}n^2 + 1.$$

Remark: This result gives a bound on the dimension of a commutative subalgebra.

Remark: This result is due to Schur. See [884].

Fact 5.10.16. Let $A, B \in \mathbb{F}^{n \times n}$, and assume that $AB = BA$. Then,

$$\dim \operatorname{span}\{ A^i B^j \colon\ 0 \le i \le n - 1,\ 0 \le j \le n - 1 \} \le n.$$

Remark: This result gives a bound on the dimension of a commutative subalgebra generated by two matrices.

Remark: This result is due to Gerstenhaber. See [154, 884].

Fact 5.10.17. Let $A, B \in \mathbb{F}^{n \times n}$, and assume that A and B are normal, nonsingular, and congruent. Then, $\text{In } A = \text{In } B$.

Remark: This result is due to Ando.

Fact 5.10.18. Let $A, B \in \mathbb{F}^{n \times m}$. Then, the following statements hold:

i) The matrices A and B are unitarily left equivalent if and only if $A^*A = B^*B$.

ii) The matrices A and B are unitarily right equivalent if and only if $AA^* = BB^*$.

iii) The matrices A and B are unitarily biequivalent if and only if A and B have the same singular values with the same multiplicity.

Proof: See [734] and [1157, pp. 372, 373].

Remark: In [734] A and B need not be the same size.

Remark: The singular value decomposition provides a canonical form under unitary biequivalence in analogy with the Smith form under biequivalence.

Remark: Note that $AA^* = BB^*$ implies that $\mathcal{R}(A) = \mathcal{R}(B)$, which implies right equivalence. This is an alternative proof of the immediate fact that unitary right equivalence implies right equivalence.

Fact 5.10.19. Let $A, B \in \mathbb{F}^{n \times n}$. Then, the following statements hold:

i) $A^*A = B^*B$ if and only if there exists a unitary matrix $S \in \mathbb{F}^{n \times n}$ such that $A = SB$.

ii) $A^*A \leq B^*B$ if and only if there exists a matrix $S \in \mathbb{F}^{n \times n}$ such that $A = SB$ and $S^*S \leq I$.

iii) $A^*B + B^*A = 0$ if and only if there exists a unitary matrix $S \in \mathbb{F}^{n \times n}$ such that $(I - S)A = (I + S)B$.

iv) $A^*B + B^*A \geq 0$ if and only if there exists a matrix $S \in \mathbb{F}^{n \times n}$ such that $(I - S)A = (I + S)B$ and $S^*S \leq I$.

Proof: See [728, p. 406] and [1144].

Remark: Statements iii) and iv) follow from i) and ii) by replacing A and B with $A - B$ and $A + B$, respectively.

Fact 5.10.20. Let $A \in \mathbb{F}^{n \times n}$, $B \in \mathbb{F}^{m \times m}$, and $C \in \mathbb{F}^{n \times m}$. Then, there exist matrices $X, Y \in \mathbb{F}^{n \times m}$ satisfying

$$AX + YB + C = 0$$

if and only if

$$\text{rank} \begin{bmatrix} A & 0 \\ 0 & -B \end{bmatrix} = \text{rank} \begin{bmatrix} A & C \\ 0 & -B \end{bmatrix}.$$

Proof: See [1125, pp. 194, 195] and [1437].

Remark: $AX + YB + C = 0$ is a generalization of Sylvester's equation. See Fact 5.10.21.

Remark: This result is due to Roth.

Remark: An explicit expression for all solutions is given by Fact 6.5.7, which applies to the case in which A and B are not necessarily square and thus X and Y are not necessarily the same size.

Fact 5.10.21. Let $A \in \mathbb{F}^{n \times n}$, $B \in \mathbb{F}^{m \times m}$, and $C \in \mathbb{F}^{n \times m}$. Then, there exists a matrix $X \in \mathbb{F}^{n \times m}$ satisfying

$$AX + XB + C = 0$$

if and only if the matrices

$$\begin{bmatrix} A & 0 \\ 0 & -B \end{bmatrix}, \quad \begin{bmatrix} A & C \\ 0 & -B \end{bmatrix}$$

are similar. In this case,

$$\begin{bmatrix} A & C \\ 0 & -B \end{bmatrix} = \begin{bmatrix} I & X \\ 0 & I \end{bmatrix} \begin{bmatrix} A & 0 \\ 0 & -B \end{bmatrix} \begin{bmatrix} I & -X \\ 0 & I \end{bmatrix}.$$

Proof: See [1437]. For sufficiency, see [892, pp. 422–424] or [1125, pp. 194, 195].

Remark: $AX + XB + C = 0$ is *Sylvester's equation*. See Proposition 7.2.4, Corollary 7.2.5, and Proposition 11.9.3.

Remark: This result is due to Roth. See [221].

Fact 5.10.22. Let $A, B \in \mathbb{F}^{n \times n}$, and assume that A and B are idempotent. Then, the matrices

$$\begin{bmatrix} A+B & A \\ 0 & -A-B \end{bmatrix}, \quad \begin{bmatrix} A+B & 0 \\ 0 & -A-B \end{bmatrix}$$

are similar. In fact,

$$\begin{bmatrix} A+B & A \\ 0 & -A-B \end{bmatrix} = \begin{bmatrix} I & X \\ 0 & I \end{bmatrix} \begin{bmatrix} A+B & 0 \\ 0 & -A-B \end{bmatrix} \begin{bmatrix} I & -X \\ 0 & I \end{bmatrix},$$

where $X \triangleq \frac{1}{4}(I + A - B)$.

Remark: This result is due to Tian.

Remark: See Fact 5.10.21.

Fact 5.10.23. Let $A \in \mathbb{F}^{n \times n}$, $B \in \mathbb{F}^{m \times m}$, and $C \in \mathbb{F}^{n \times m}$, and assume that A and B are nilpotent. Then, the matrices

$$\begin{bmatrix} A & C \\ 0 & B \end{bmatrix}, \quad \begin{bmatrix} A & 0 \\ 0 & B \end{bmatrix}$$

are similar if and only if

$$\text{rank} \begin{bmatrix} A & C \\ 0 & B \end{bmatrix} = \text{rank}\, A + \text{rank}\, B.$$

and

$$AC + CB = 0.$$

Proof: See [1326].

5.11 Facts on Eigenvalues and Singular Values for One Matrix

Fact 5.11.1. Let $A \in \mathbb{F}^{n \times n}$, and assume that A is singular. If A is either simple or cyclic, then $\operatorname{rank} A = n - 1$.

Fact 5.11.2. Let $A \in \mathbb{R}^{n \times n}$, and assume that $A \in \mathrm{SO}(n)$. Then, $\operatorname{amult}_A(-1)$ is even. Now, assume in addition that $n = 3$. Then, the following statements hold:

   i) $\operatorname{amult}_A(1)$ is either 1 or 3.

   ii) $\operatorname{tr} A \geq -1$.

   iii) $\operatorname{tr} A = -1$ if and only if $\operatorname{mspec}(A) = \{1, -1, -1\}_{\mathrm{ms}}$.

Fact 5.11.3. Let $A \in \mathbb{F}^{n \times n}$, let $\alpha \in \mathbb{F}$, and assume that $A^2 = \alpha A$. Then, $\operatorname{spec}(A) \subseteq \{0, \alpha\}$.

Fact 5.11.4. Let $A \in \mathbb{F}^{n \times n}$, assume that A is Hermitian, and let $\alpha \in \mathbb{R}$. Then, $A^2 = \alpha A$ if and only if $\operatorname{spec}(A) \subseteq \{0, \alpha\}$.

Remark: See Fact 3.7.22.

Fact 5.11.5. Let $A \in \mathbb{F}^{n \times n}$, and assume that A is Hermitian. Then,

$$\operatorname{spabs}(A) = \lambda_{\max}(A)$$

and

$$\operatorname{sprad}(A) = \sigma_{\max}(A) = \max\{|\lambda_{\min}(A)|, \lambda_{\max}(A)\}.$$

If, in addition, A is positive semidefinite, then

$$\operatorname{sprad}(A) = \sigma_{\max}(A) = \operatorname{spabs}(A) = \lambda_{\max}(A).$$

Remark: See Fact 5.12.2.

Fact 5.11.6. Let $A \in \mathbb{F}^{n \times n}$, and assume that A is skew Hermitian. Then, the eigenvalues of A are imaginary.

Proof: Let $\lambda \in \operatorname{spec}(A)$. Since $0 \leq AA^* = -A^2$, it follows that $-\lambda^2 \geq 0$, and thus $\lambda^2 \leq 0$.

Fact 5.11.7. Let $A, B \in \mathbb{F}^{n \times n}$, and assume that A and B are idempotent. Then, the following statements are equivalent:

   i) $\operatorname{mspec}(A) = \operatorname{mspec}(B)$.

   ii) $\operatorname{rank} A = \operatorname{rank} B$.

   iii) $\operatorname{tr} A = \operatorname{tr} B$.

Fact 5.11.8. Let $A \in \mathbb{F}^{n \times n}$. Then, the following statements are equivalent:

i) A is idempotent.

ii) $\operatorname{rank}(I - A) \leq \operatorname{tr}(I - A)$, A is group invertible, and every eigenvalue of A is nonnegative.

iii) A and $I - A$ are group invertible, and every eigenvalue of A is nonnegative.

Proof: See [666].

Fact 5.11.9. Let $A \in \mathbb{F}^{n \times n}$, and let $\operatorname{mspec}(A) = \{\lambda_1, \ldots, \lambda_k, 0, \ldots, 0\}_{\mathrm{ms}}$. Then,

$$|\operatorname{tr} A|^2 \leq \left(\sum_{i=1}^{k} |\lambda_i| \right)^2 \leq k \sum_{i=1}^{k} |\lambda_i|^2.$$

Proof: Use Fact 1.17.3.

Fact 5.11.10. Let $A \in \mathbb{F}^{n \times n}$, and assume that A has exactly k nonzero eigenvalues. Then,

$$\left. \begin{aligned} |\operatorname{tr} A|^2 \\ k|\operatorname{tr} A^2| \leq k \operatorname{tr} (A^{2*}A^2)^{1/2} \end{aligned} \right\} \leq k \operatorname{tr} A^*A \leq (\operatorname{rank} A) \operatorname{tr} A^*A.$$

Furthermore, the upper left-hand inequality is an equality if and only if A is normal and all of the nonzero eigenvalues of A have the same absolute value. Moreover, the right-hand inequality is an equality if and only if A is group invertible. If, in addition, all of the eigenvalues of A are real, then

$$(\operatorname{tr} A)^2 \leq k \operatorname{tr} A^2 \leq k \operatorname{tr} A^*A \leq (\operatorname{rank} A) \operatorname{tr} A^*A.$$

Proof: The upper left-hand inequality in the first string is given in [1483]. The lower left-hand inequality in the first string is given by Fact 9.11.3. When all of the eigenvalues of A are real, the inequality $(\operatorname{tr} A)^2 \leq k \operatorname{tr} A^2$ follows from Fact 5.11.9.

Remark: The inequality $|\operatorname{tr} A|^2 \leq k|\operatorname{tr} A^2|$ does not necessarily hold. Consider $\operatorname{mspec}(A) = \{1, 1, \jmath, -\jmath\}_{\mathrm{ms}}$.

Remark: See Fact 3.7.22, Fact 8.18.7, Fact 9.13.16, and Fact 9.13.17.

Fact 5.11.11. Let $A \in \mathbb{R}^{n \times n}$, and let $\operatorname{mspec}(A) = \{\lambda_1, \ldots, \lambda_n\}_{\mathrm{ms}}$. Then,

$$\sum_{i=1}^{n} (\operatorname{Re} \lambda_i)(\operatorname{Im} \lambda_i) = 0$$

and

$$\operatorname{tr} A^2 = \sum_{i=1}^{n} (\operatorname{Re} \lambda_i)^2 - \sum_{i=1}^{n} (\operatorname{Im} \lambda_i)^2.$$

Fact 5.11.12. Let $n \geq 2$, let $a_1, \ldots, a_n > 0$, and define the symmetric matrix $A \in \mathbb{R}^{n \times n}$ by $A_{(i,j)} \triangleq a_i + a_j$ for all $i, j \in \{1, \ldots, n\}$. Then,

$$\operatorname{rank} A \leq 2$$

and
$$\text{mspec}(A) = \{\lambda, \mu, 0, \ldots, 0\}_{\text{ms}},$$

where

$$\lambda \triangleq \sum_{i=1}^{n} a_i + \sqrt{n \sum_{i=1}^{n} a_i^2}, \quad \mu \triangleq \sum_{i=1}^{n} a_i - \sqrt{n \sum_{i=1}^{n} a_i^2}.$$

Furthermore, the following statements hold:

i) $\lambda > 0$.

ii) $\mu \le 0$.

Moreover, the following statements are equivalent:

iii) $\mu < 0$.

iv) At least two of the numbers $a_1, \ldots, a_n > 0$ are distinct.

v) $\text{rank}\, A = 2$.

In this case,

$$\lambda_{\min}(A) = \mu < 0 < \text{tr}\, A = 2 \sum_{i=1}^{n} a_i < \lambda_{\max}(A) = \lambda.$$

Proof: $A = a1_{1 \times n} + 1_{n \times 1} a^{\text{T}}$, where $a \triangleq \begin{bmatrix} a_1 & \cdots & a_n \end{bmatrix}^{\text{T}}$. Then, it follows from Fact 2.11.12 that $\text{rank}\, A \le \text{rank}(a1_{1 \times n}) + \text{rank}(1_{n \times 1} a^{\text{T}}) = 2$. Furthermore, $\text{mspec}(A)$ follows from Fact 5.11.13, while Fact 1.17.14 implies that $\mu \le 0$.

Remark: See Fact 8.8.7.

Fact 5.11.13. Let $x, y \in \mathbb{R}^n$. Then,

$$\text{mspec}(xy^{\text{T}} + yx^{\text{T}}) = \left\{ x^{\text{T}}y + \sqrt{x^{\text{T}}x y^{\text{T}}y}, x^{\text{T}}y - \sqrt{x^{\text{T}}x y^{\text{T}}y}, 0, \ldots, 0 \right\}_{\text{ms}},$$

$$\text{sprad}(xy^{\text{T}} + yx^{\text{T}}) = \begin{cases} x^{\text{T}}y + \sqrt{x^{\text{T}}x y^{\text{T}}y}, & x^{\text{T}}y \ge 0, \\ \left| x^{\text{T}}y - \sqrt{x^{\text{T}}x y^{\text{T}}y} \right|, & x^{\text{T}}y \le 0, \end{cases}$$

and
$$\text{spabs}(xy^{\text{T}} + yx^{\text{T}}) = x^{\text{T}}y + \sqrt{x^{\text{T}}x y^{\text{T}}y}.$$

If, in addition, x and y are nonzero, then $v_1, v_2 \in \mathbb{R}^n$ defined by

$$v_1 \triangleq \frac{1}{\|x\|}x + \frac{1}{\|y\|}y, \quad v_2 \triangleq \frac{1}{\|x\|}x - \frac{1}{\|y\|}y$$

are eigenvectors of $xy^{\text{T}} + yx^{\text{T}}$ corresponding to $x^{\text{T}}y + \sqrt{x^{\text{T}}x y^{\text{T}}y}$ and $x^{\text{T}}y - \sqrt{x^{\text{T}}x y^{\text{T}}y}$, respectively.

Proof: See [382, p. 539].

Example: The spectrum of $\begin{bmatrix} 0_{n \times n} & 1_{n \times 1} \\ 1 1_{1 \times n} & 0 \end{bmatrix}$ is $\{-\sqrt{n}, 0, \ldots, 0, \sqrt{n}\}_{\text{ms}}$.

Problem: Extend this result to $\mathbb{C}$ and $xy^{\text{T}} + zw^{\text{T}}$. See Fact 4.9.17.

Fact 5.11.14. Let $A \in \mathbb{F}^{n \times n}$, and let $\mathrm{mspec}(A) = \{\lambda_1, \ldots, \lambda_n\}_{\mathrm{ms}}$. Then,

$$\mathrm{mspec}\left[(I + A)^2\right] = \left\{(1 + \lambda_1)^2, \ldots, (1 + \lambda_n)^2\right\}_{\mathrm{ms}}.$$

If A is nonsingular, then

$$\mathrm{mspec}\left(A^{-1}\right) = \left\{\lambda_1^{-1}, \ldots, \lambda_n^{-1}\right\}_{\mathrm{ms}}.$$

Finally, if $I + A$ is nonsingular, then

$$\mathrm{mspec}\left[(I + A)^{-1}\right] = \left\{(1 + \lambda_1)^{-1}, \ldots, (1 + \lambda_n)^{-1}\right\}_{\mathrm{ms}}$$

and

$$\mathrm{mspec}\left[A(I + A)^{-1}\right] = \left\{\lambda_1(1 + \lambda_1)^{-1}, \ldots, \lambda_n(1 + \lambda_n)^{-1}\right\}_{\mathrm{ms}}.$$

Proof: Use Fact 5.11.15.

Fact 5.11.15. Let $p, q \in \mathbb{F}[s]$, assume that p and q are coprime, and define $g \triangleq p/q \in \mathbb{F}(s)$. Furthermore, let $A \in \mathbb{F}^{n \times n}$, let $\mathrm{mspec}(A) = \{\lambda_1, \ldots, \lambda_n\}_{\mathrm{ms}}$, assume that $\mathrm{roots}(q) \cap \mathrm{spec}(A) = \varnothing$, and define $g(A) \triangleq p(A)[q(A)]^{-1}$. Then,

$$\mathrm{mspec}[g(A)] = \{g(\lambda_1), \ldots, g(\lambda_n)\}_{\mathrm{ms}}.$$

Proof: Statement ii) of Fact 4.9.27 implies that $q(A)$ is nonsingular.

Fact 5.11.16. Let $x \in \mathbb{F}^n$ and $y \in \mathbb{F}^m$. Then,

$$\sigma_{\max}(xy^*) = \sqrt{x^* x y^* y}.$$

If, in addition, $m = n$, then

$$\mathrm{mspec}(xy^*) = \{x^* y, 0, \ldots, 0\}_{\mathrm{ms}},$$

$$\mathrm{mspec}(I + xy^*) = \{1 + x^* y, 1, \ldots, 1\}_{\mathrm{ms}},$$

$$\mathrm{sprad}(xy^*) = |x^* y|,$$

$$\mathrm{spabs}(xy^*) = \max\{0, \mathrm{Re}\, x^* y\}.$$

Remark: See Fact 9.7.26.

Fact 5.11.17. Let $A \in \mathbb{F}^{n \times n}$, and assume that $\mathrm{rank}\, A = 1$. Then,

$$\sigma_{\max}(A) = (\mathrm{tr}\, AA^*)^{1/2}.$$

Fact 5.11.18. Let $x, y \in \mathbb{F}^n$, and assume that $x^* y \neq 0$. Then,

$$\sigma_{\max}\left[(x^* y)^{-1} xy^*\right] \geq 1.$$

Fact 5.11.19. Let $A \in \mathbb{F}^{n \times m}$, and let $\alpha \in \mathbb{F}$. Then, for all $i \in \{1, \ldots, \min\{n, m\}\}$,

$$\sigma_i(\alpha A) = |\alpha| \sigma_i(A).$$

Fact 5.11.20. Let $A \in \mathbb{F}^{n \times m}$. Then, for all $i \in \{1, \ldots, \mathrm{rank}\, A\}$, it follows that

$$\sigma_i(A) = \sigma_i(A^*).$$

Fact 5.11.21. Let $A \in \mathbb{F}^{n \times n}$, and let $\lambda \in \mathrm{spec}(A)$. Then, the following inequalities hold:

i) $\sigma_{\min}(A) \leq |\lambda| \leq \sigma_{\max}(A)$.

ii) $\lambda_{\min}[\frac{1}{2}(A + A^*)] \leq \mathrm{Re}\,\lambda \leq \lambda_{\max}[\frac{1}{2}(A + A^*)]$.

iii) $\lambda_{\min}[\frac{1}{j2}(A - A^*)] \leq \mathrm{Im}\,\lambda \leq \lambda_{\max}[\frac{1}{j2}(A - A^*)]$.

Remark: i) is *Browne's theorem*, ii) is *Bendixson's theorem*, and iii) is *Hirsch's theorem*. See [319, p. 17] and [988, pp. 140–144].

Remark: See Fact 5.11.22, Fact 5.12.3, and Fact 9.11.8.

Fact 5.11.22. Let $A \in \mathbb{F}^{n \times n}$, and let $\mathrm{mspec}(A) = \{\lambda_1, \ldots, \lambda_n\}_{\mathrm{ms}}$. Then, for all $k \in \{1, \ldots, n\}$,

$$\sum_{i=1}^{k} \left[\sigma_{n-i+1}^2(A) - |\lambda_i|^2\right] \leq 2 \sum_{i=1}^{k} \left(\sigma_i^2[\tfrac{1}{j2}(A - A^*)] - |\mathrm{Im}\,\lambda_i|^2\right)$$

and

$$2 \sum_{i=1}^{k} \left(\sigma_{n-i+1}^2[\tfrac{1}{j2}(A - A^*)] - |\mathrm{Im}\,\lambda_i|^2\right) \leq \sum_{i=1}^{k} \left[\sigma_i^2(A) - |\lambda_i|^2\right].$$

Furthermore,

$$\sum_{i=1}^{n} \left[\sigma_i^2(A) - |\lambda_i|^2\right] = 2 \sum_{i=1}^{n} \left(\sigma_i^2[\tfrac{1}{j2}(A - A^*)] - |\mathrm{Im}\,\lambda_i|^2\right).$$

Finally, for all $i \in \{1, \ldots, n\}$,

$$\sigma_n(A) \leq |\mathrm{Re}\,\lambda_i| \leq \sigma_1(A)$$

and

$$\sigma_n[\tfrac{1}{j2}(A - A^*)] \leq |\mathrm{Im}\,\lambda_i| \leq \sigma_1[\tfrac{1}{j2}(A - A^*)].$$

Proof: See [566].

Remark: See Fact 9.11.7.

Fact 5.11.23. Let $A \in \mathbb{F}^{n \times n}$, let $\mathrm{mspec}(A) = \{\lambda_1, \ldots, \lambda_n\}_{\mathrm{ms}}$, and let r denote the number of Jordan blocks in the Jordan decomposition of A. Then, for all $k \in \{1, \ldots, r\}$,

$$\sum_{i=1}^{k} \sigma_{n-i+1}^2(A) \leq \sum_{i=1}^{k} |\lambda_i|^2 \leq \sum_{i=1}^{k} \sigma_i^2(A)$$

and

$$\sum_{i=1}^{k} \sigma_{n-i+1}^2[\tfrac{1}{j2}(A - A^*)] \leq \sum_{i=1}^{k} |\mathrm{Im}\,\lambda_i|^2 \leq \sum_{i=1}^{k} \sigma_i^2[\tfrac{1}{j2}(A - A^*)].$$

Proof: See [566].

Fact 5.11.24. Let $A \in \mathbb{F}^{n \times n}$, and let $\mathrm{mspec}(A) = \{\lambda_1(A), \ldots, \lambda_n(A)\}_{\mathrm{ms}}$, where $\lambda_1(A), \ldots, \lambda_n(A)$ are ordered such that $\mathrm{Re}\,\lambda_1(A) \geq \cdots \geq \mathrm{Re}\,\lambda_n(A)$. Then, for all $k \in \{1, \ldots, n\}$,

$$\sum_{i=1}^{k} \mathrm{Re}\,\lambda_i(A) \leq \sum_{i=1}^{k} \lambda_i\left[\tfrac{1}{2}(A + A^*)\right]$$

and

$$\sum_{i=1}^{n} \mathrm{Re}\,\lambda_i(A) = \mathrm{Re\,tr}\,A = \mathrm{Re\,tr}\,\tfrac{1}{2}(A + A^*) = \sum_{i=1}^{n} \lambda_i\left[\tfrac{1}{2}(A + A^*)\right].$$

In particular,

$$\lambda_{\min}\left[\tfrac{1}{2}(A + A^*)\right] \leq \mathrm{Re}\,\lambda_n(A) \leq \mathrm{spabs}(A) \leq \lambda_{\max}\left[\tfrac{1}{2}(A + A^*)\right].$$

Furthermore, the last right-hand inequality is an equality if and only if A is normal.

Proof: See [201, p. 74]. Also, see *xii)* and *xiv)* of Fact 11.15.7.

Remark: $\mathrm{spabs}(A) = \mathrm{Re}\,\lambda_1(A)$.

Remark: This result is due to Fan.

Fact 5.11.25. Let $A \in \mathbb{F}^{n \times n}$. Then, for all $i \in \{1, \ldots, n\}$,

$$-\sigma_i(A) \leq \lambda_i\left[\tfrac{1}{2}(A + A^*)\right] \leq \sigma_i(A).$$

In particular,

$$-\sigma_{\min}(A) \leq \lambda_{\min}\left[\tfrac{1}{2}(A + A^*)\right] \leq \sigma_{\min}(A)$$

and

$$-\sigma_{\max}(A) \leq \lambda_{\max}\left[\tfrac{1}{2}(A + A^*)\right] \leq \sigma_{\max}(A).$$

Proof: See [708, p. 447], [730, p. 151], or [996, p. 240].

Remark: This result generalizes $\mathrm{Re}\,z \leq |z|$ for $z \in \mathbb{C}$.

Remark: See Fact 5.11.27 and Fact 8.18.4.

Fact 5.11.26. Let $A \in \mathbb{F}^{n \times n}$. Then,

$$-\sigma_{\max}(A) \leq -\sigma_{\min}(A)$$
$$\leq \lambda_{\min}\left[\tfrac{1}{2}(A + A^*)\right]$$
$$\leq \mathrm{spabs}(A)$$
$$\leq \left\{ \begin{array}{c} |\,\mathrm{spabs}(A)| \leq \mathrm{sprad}(A) \\ \tfrac{1}{2}\lambda_{\max}(A + A^*) \end{array} \right\}$$
$$\leq \sigma_{\max}(A).$$

Proof: Combine Fact 5.11.24 and Fact 5.11.25.

Fact 5.11.27. Let $A \in \mathbb{F}^{n \times n}$, and let $\{\mu_1, \ldots, \mu_n\}_{\mathrm{ms}} = \{\tfrac{1}{2}|\lambda_1(A + A^*)|, \ldots, \tfrac{1}{2}|\lambda_n(A + A^*)|\}_{\mathrm{ms}}$, where $\mu_1 \geq \cdots \geq \mu_n \geq 0$. Then, $\begin{bmatrix} \sigma_1(A) & \cdots & \sigma_n(A) \end{bmatrix}^{\mathrm{T}}$ weakly majorizes $\begin{bmatrix} \mu_1 & \cdots & \mu_n \end{bmatrix}^{\mathrm{T}}$.

Proof: See [996, p. 240].

Remark: See Fact 5.11.25.

Fact 5.11.28. Let $A \in \mathbb{F}^{n \times n}$, and let $\mathrm{mspec}(A) = \{\lambda_1, \ldots, \lambda_n\}_{\mathrm{ms}}$, where $\lambda_1, \ldots, \lambda_n$ are ordered such that $|\lambda_1| \geq \cdots \geq |\lambda_n|$. Then, for all $k \in \{1, \ldots, n\}$,

$$\prod_{i=1}^{k} |\lambda_i| \leq \prod_{i=1}^{k} \sigma_i(A)$$

with equality for $k = n$, that is,

$$|\det A| = \prod_{i=1}^{n} |\lambda_i| = \prod_{i=1}^{n} \sigma_i(A).$$

Hence, for all $k \in \{1, \ldots, n\}$,

$$\prod_{i=k}^{n} \sigma_i(A) \leq \prod_{i=k}^{n} |\lambda_i|.$$

Proof: See [201, p. 43], [708, p. 445], [730, p. 171], or [1521, p. 19].

Remark: This result is due to Weyl.

Remark: See Fact 8.19.22 and Fact 9.13.18.

Fact 5.11.29. Let $A \in \mathbb{F}^{n \times n}$, and let $\mathrm{mspec}(A) = \{\lambda_1, \ldots, \lambda_n\}_{\mathrm{ms}}$, where $\lambda_1, \ldots, \lambda_n$ are ordered such that $|\lambda_1| \geq \cdots \geq |\lambda_n|$. Then,

$$\sigma_{\min}(A) \leq \sigma_{\max}^{1/n}(A)\sigma_{\min}^{(n-1)/n}(A) \leq |\lambda_n| \leq |\lambda_1| \leq \sigma_{\min}^{1/n}(A)\sigma_{\max}^{(n-1)/n}(A) \leq \sigma_{\max}(A)$$

and

$$\sigma_{\min}^n(A) \leq \sigma_{\max}(A)\sigma_{\min}^{n-1}(A) \leq |\det A| \leq \sigma_{\min}(A)\sigma_{\max}^{n-1}(A) \leq \sigma_{\max}^n(A).$$

Proof: Use Fact 5.11.28. See [708, p. 445].

Remark: See Fact 8.13.1 and Fact 11.20.13.

Fact 5.11.30. Let $\beta_0, \ldots, \beta_{n-1} \in \mathbb{F}$, define $A \in \mathbb{F}^{n \times n}$ by

$$A \triangleq \begin{bmatrix} 0 & 1 & 0 & \cdots & 0 & 0 \\ 0 & 0 & 1 & \ddots & 0 & 0 \\ 0 & 0 & 0 & \ddots & 0 & 0 \\ \vdots & \vdots & \vdots & \ddots & \ddots & \vdots \\ 0 & 0 & 0 & \cdots & 0 & 1 \\ -\beta_0 & -\beta_1 & -\beta_2 & \cdots & -\beta_{n-2} & -\beta_{n-1} \end{bmatrix},$$

and define $\alpha \triangleq 1 + \sum_{i=0}^{n-1} |\beta_i|^2$. Then,

$$\sigma_1(A) = \sqrt{\tfrac{1}{2}\left(\alpha + \sqrt{\alpha^2 - 4|\beta_0|^2}\right)},$$

$$\sigma_2(A) = \cdots = \sigma_{n-1}(A) = 1,$$

$$\sigma_n(A) = \sqrt{\tfrac{1}{2}\left(\alpha - \sqrt{\alpha^2 - 4|\beta_0|^2}\right)}.$$

In particular,

$$\sigma_1(N_n) = \cdots = \sigma_{n-1}(N_n) = 1$$

and

$$\sigma_{\min}(N_n) = 0.$$

Proof: See [699, p. 523] or [825, 841].

Remark: See Fact 6.3.27 and Fact 11.20.13.

Fact 5.11.31. Let $\beta \in \mathbb{C}$. Then,

$$\sigma_{\max}\left(\begin{bmatrix} 1 & 2\beta \\ 0 & 1 \end{bmatrix}\right) = |\beta| + \sqrt{1 + |\beta|^2}$$

and

$$\sigma_{\min}\left(\begin{bmatrix} 1 & 2\beta \\ 0 & 1 \end{bmatrix}\right) = \sqrt{1 + |\beta|^2} - |\beta|.$$

Proof: See [923].

Remark: Inequalities for the singular values of block-triangular matrices are given in [923].

Fact 5.11.32. Let $A \in \mathbb{F}^{n \times m}$. Then,

$$\sigma_{\max}\left(\begin{bmatrix} I & 2A \\ 0 & I \end{bmatrix}\right) = \sigma_{\max}(A) + \sqrt{1 + \sigma_{\max}^2(A)}.$$

Proof: See [699, p. 116].

Fact 5.11.33. For $i = 1, \ldots, l$, let $A_i \in \mathbb{F}^{n_i \times m_i}$. Then,

$$\sigma_{\max}[\mathrm{diag}(A_1, \ldots, A_l)] = \max\{\sigma_{\max}(A_1), \ldots, \sigma_{\max}(A_l)\}.$$

Fact 5.11.34. Let $A \in \mathbb{F}^{n \times m}$, and let $r \triangleq \mathrm{rank}\, A$. Then, for all $i \in \{1, \ldots, r\}$,

$$\lambda_i(AA^*) = \lambda_i(A^*A) = \sigma_i(AA^*) = \sigma_i(A^*A) = \sigma_i^2(A).$$

In particular,

$$\sigma_{\max}(AA^*) = \sigma_{\max}^2(A),$$

and, if $n = m$, then

$$\sigma_{\min}(AA^*) = \sigma_{\min}^2(A).$$

Furthermore, for all $i \in \{1, \ldots, r\}$,

$$\sigma_i(AA^*A) = \sigma_i^3(A).$$

Fact 5.11.35. Let $A \in \mathbb{F}^{n \times n}$. Then, $\sigma_{\max}(A) \leq 1$ if and only if $A^*A \leq I$.

Fact 5.11.36. Let $A \in \mathbb{F}^{n \times n}$. Then, for all $i \in \{1, \ldots, n\}$,

$$\sigma_i(A^{\mathrm{A}}) = \prod_{\substack{j=1 \\ j \neq n+1-i}}^{n} \sigma_j(A).$$

Proof: See Fact 4.10.9 and [1125, p. 149].

Fact 5.11.37. Let $A \in \mathbb{F}^{n \times n}$. Then, $\sigma_1(A) = \sigma_n(A)$ if and only if there exist $\lambda \in \mathbb{F}$ and a unitary matrix $B \in \mathbb{F}^{n \times n}$ such that $A = \lambda B$.

Proof: See [1125, pp. 149, 165].

Fact 5.11.38. Let $A \in \mathbb{F}^{n \times n}$, and assume that A is idempotent. Then, the following statements hold:

 i) If σ is a singular value of A, then either $\sigma = 0$ or $\sigma \geq 1$.

 ii) If $A \neq 0$, then $\sigma_{\max}(A) \geq 1$.

 iii) $\sigma_{\max}(A) = 1$ if and only if A is a projector.

 iv) If $1 \leq \operatorname{rank} A \leq n - 1$, then

$$\sigma_{\max}(A) = \sigma_{\max}(A_{\perp}).$$

 v) If $A \neq 0$, then

$$\sigma_{\max}(A) = \sigma_{\max}(A + A^* - I) = \sigma_{\max}(A + A^*) - 1$$

and

$$\sigma_{\max}(I - 2A) = \sigma_{\max}(A) + [\sigma_{\max}^2(A) - 1]^{1/2}.$$

Proof: See [551, 742, 765]. Statement *iv)* is given in [550, p. 61] and follows from Fact 5.11.39.

Problem: Use Fact 5.9.28 to prove *iv)*.

Fact 5.11.39. Let $A \in \mathbb{F}^{n \times n}$, assume that A is idempotent, and assume that $1 \leq \operatorname{rank} A \leq n - 1$. Then,

$$\sigma_{\max}(A) = \sigma_{\max}(A + A^* - I) = \frac{1}{\sin \theta},$$

where $\theta \in (0, \pi/2]$ is defined by

$$\cos \theta = \max\{|x^*y| \colon (x, y) \in \mathcal{R}(A) \times \mathcal{N}(A) \text{ and } x^*x = y^*y = 1\}.$$

Proof: See [551, 765].

Remark: θ is the minimal principal angle. See Fact 2.9.19 and Fact 5.12.17.

Remark: Note that $\mathcal{N}(A) = \mathcal{R}(A_{\perp})$. See Fact 3.12.3.

Remark: This result is due to Ljance.

Remark: This result yields statement *iii)* of Fact 5.11.38.

Remark: See Fact 10.9.19.

Fact 5.11.40. Let $A \in \mathbb{R}^{n \times n}$, where $n \geq 2$, be the tridiagonal matrix

$$
A \triangleq \begin{bmatrix}
b_1 & c_1 & 0 & \cdots & 0 & 0 \\
a_1 & b_2 & c_2 & \cdots & 0 & 0 \\
0 & a_2 & b_3 & \ddots & 0 & 0 \\
\vdots & \vdots & \ddots & \ddots & \ddots & \vdots \\
0 & 0 & 0 & \ddots & b_{n-1} & c_{n-1} \\
0 & 0 & 0 & \cdots & a_{n-1} & b_n
\end{bmatrix},
$$

and assume that, for all $i \in \{1, \ldots, n-1\}$, $a_i c_i > 0$ Then, A is simple, and every eigenvalue of A is real. Hence, rank $A \geq n - 1$.

Proof: SAS^{-1} is symmetric, where $S \triangleq \operatorname{diag}(d_1, \ldots, d_n)$, $d_1 \triangleq 1$, and $d_{i+1} \triangleq (c_i/a_i)^{1/2} d_i$ for all $i \in \{1, \ldots, n-1\}$. For a proof of the fact that A is simple, see [494, p. 198].

Remark: See Fact 5.11.41.

Fact 5.11.41. Let $A \in \mathbb{R}^{n \times n}$, where $n \geq 2$, be the tridiagonal matrix

$$
A \triangleq \begin{bmatrix}
b_1 & c_1 & 0 & \cdots & 0 & 0 \\
a_1 & b_2 & c_2 & \cdots & 0 & 0 \\
0 & a_2 & b_3 & \ddots & 0 & 0 \\
\vdots & \vdots & \ddots & \ddots & \ddots & \vdots \\
0 & 0 & 0 & \ddots & b_{n-1} & c_{n-1} \\
0 & 0 & 0 & \cdots & a_{n-1} & b_n
\end{bmatrix},
$$

and assume that, for all $i \in \{1, \ldots, n-1\}$, $a_i c_i \neq 0$. Then, A is reducible. Furthermore, let k_+ and k_- denote, respectively, the number of positive and negative numbers in the sequence

$$1, \quad a_1 c_1, \quad a_1 a_2 c_1 c_2, \quad \ldots, \quad a_1 a_2 \cdots a_{n-1} c_1 c_2 \cdots c_{n-1}.$$

Then, A has at least $|k_+ - k_-|$ distinct real eigenvalues, of which at least $\max\{0, n - 3\min\{k_+, k_-\}\}$ are simple.

Proof: See [1410].

Remark: Note that $k_+ + k_- = n$ and $|k_+ - k_-| = n - 2\min\{k_+, k_-\}$.

Remark: This result yields Fact 5.11.40 as a special case.

Fact 5.11.42. Let $A \in \mathbb{R}^{n \times n}$ be the tridiagonal matrix

$$A \triangleq \begin{bmatrix} 0 & 1 & 0 & & & & & \\ n-1 & 0 & 2 & & & 0 & & \\ 0 & n-2 & 0 & \ddots & & & & \\ & & \ddots & \ddots & \ddots & \ddots & & \\ & & & \ddots & \ddots & 0 & n-2 & 0 \\ & 0 & & & \ddots & 2 & 0 & n-1 \\ & & & & & 0 & 1 & 0 \end{bmatrix}.$$

Then,

$$\chi_A(s) = \prod_{i=1}^{n} [s - (n+1-2i)].$$

Hence,

$$\text{spec}(A) = \begin{cases} \{n-1, -(n-1), \ldots, 1, -1\}, & n \text{ even}, \\ \{n-1, -(n-1), \ldots, 2, -2, 0\}, & n \text{ odd}. \end{cases}$$

Proof: See [1291].

Fact 5.11.43. Let $A \in \mathbb{R}^{n \times n}$, where $n \geq 1$, be the tridiagonal, Toeplitz matrix

$$A \triangleq \begin{bmatrix} b & c & 0 & \cdots & 0 & 0 \\ a & b & c & \cdots & 0 & 0 \\ 0 & a & b & \ddots & 0 & 0 \\ \vdots & \vdots & \ddots & \ddots & \ddots & \vdots \\ 0 & 0 & 0 & \ddots & b & c \\ 0 & 0 & 0 & \cdots & a & b \end{bmatrix},$$

and assume that $ac > 0$. Then,

$$\text{spec}(A) = \left\{ b + 2\sqrt{ac} \cos \frac{i\pi}{n+1} : \ i \in \{1, \ldots, n\} \right\}.$$

Remark: See [699, p. 522].

Remark: See Fact 3.18.9.

Fact 5.11.44. Let $A \in \mathbb{R}^{n \times n}$, where $n \geq 1$, be the tridiagonal, Toeplitz matrix

$$A \triangleq \begin{bmatrix} 0 & 1/2 & 0 & \cdots & 0 & 0 \\ 1/2 & 0 & 1/2 & \cdots & 0 & 0 \\ 0 & 1/2 & 0 & \ddots & 0 & 0 \\ \vdots & \vdots & \ddots & \ddots & \ddots & \vdots \\ 0 & 0 & 0 & \ddots & 0 & 1/2 \\ 0 & 0 & 0 & \cdots & 1/2 & 0 \end{bmatrix}.$$

Then,

$$\text{spec}(A) = \left\{ \cos \frac{i\pi}{n+1} : \ i \in \{1, \ldots, n\} \right\}.$$

Furthermore, the associated eigenvectors $v_1, \ldots, v_n$ are given by

$$v_i = \sqrt{\frac{2}{n+1}} \begin{bmatrix} \sin \frac{i\pi}{n+1} \\ \sin \frac{2i\pi}{n+1} \\ \vdots \\ \sin \frac{ni\pi}{n+1} \end{bmatrix}$$

and satisfy $\|v_i\|_2 = 1$ for all $i \in \{1, \ldots, n\}$, and are mutually orthogonal.

Remark: See [846].

Fact 5.11.45. Let $A \in \mathbb{F}^{n \times n}$, and assume that A has real eigenvalues. Then,

$$\frac{1}{n}\text{tr}\, A - \sqrt{\frac{n-1}{n}\left[\text{tr}\, A^2 - \frac{1}{n}(\text{tr}\, A)^2\right]} \leq \lambda_{\min}(A)$$
$$\leq \frac{1}{n}\text{tr}\, A - \sqrt{\frac{1}{n^2-n}\left[\text{tr}\, A^2 - \frac{1}{n}(\text{tr}\, A)^2\right]}$$
$$\leq \frac{1}{n}\text{tr}\, A + \sqrt{\frac{1}{n^2-n}\left[\text{tr}\, A^2 - \frac{1}{n}(\text{tr}\, A)^2\right]}$$
$$\leq \lambda_{\max}(A)$$
$$\leq \frac{1}{n}\text{tr}\, A + \sqrt{\frac{n-1}{n}\left[\text{tr}\, A^2 - \frac{1}{n}(\text{tr}\, A)^2\right]}.$$

Furthermore, for all $i \in \{1, \ldots, n\}$,

$$\left|\lambda_i(A) - \frac{1}{n}\text{tr}\, A\right| \leq \sqrt{\frac{n-1}{n}\left[\text{tr}\, A^2 - \frac{1}{n}(\text{tr}\, A)^2\right]}.$$

Finally, if $n = 2$, then

$$\frac{1}{n}\text{tr}\, A - \sqrt{\frac{1}{n}\text{tr}\, A^2 - \frac{1}{n^2}(\text{tr}\, A)^2} = \lambda_{\min}(A) \leq \lambda_{\max}(A) = \frac{1}{n}\text{tr}\, A + \sqrt{\frac{1}{n}\text{tr}\, A^2 - \frac{1}{n^2}(\text{tr}\, A)^2}.$$

Proof: See [1483, 1484].

Remark: These inequalities are related to Fact 1.17.12.

Fact 5.11.46. Let $A \in \mathbb{F}^{n \times n}$, and let $\mu(A) \triangleq \min\{|\lambda| : \lambda \in \text{spec}(A)\}$. Then,

$$\frac{1}{n}|\text{tr}\, A| - \sqrt{\frac{n-1}{n}(\text{tr}\, AA^* - \frac{1}{n}|\text{tr}\, A|^2)} \leq \mu(A) \leq \sqrt{\frac{1}{n}\text{tr}\, AA^*}$$

and

$$\tfrac{1}{n}|\operatorname{tr} A| \leq \operatorname{sprad}(A) \leq \tfrac{1}{n}|\operatorname{tr} A| + \sqrt{\tfrac{n-1}{n}(\operatorname{tr} AA^* - \tfrac{1}{n}|\operatorname{tr} A|^2)}.$$

Proof: See Theorem 3.1 of [1483].

Fact 5.11.47. Let $A \in \mathbb{F}^{n \times n}$, where $n \geq 2$, be the bidiagonal matrix

$$A \triangleq \begin{bmatrix} a_1 & b_1 & 0 & \cdots & 0 & 0 \\ 0 & a_2 & b_2 & \cdots & 0 & 0 \\ 0 & 0 & a_3 & \ddots & 0 & 0 \\ \vdots & \vdots & \ddots & \ddots & \ddots & \vdots \\ 0 & 0 & 0 & \ddots & a_{n-1} & b_{n-1} \\ 0 & 0 & 0 & \cdots & 0 & a_n \end{bmatrix},$$

and assume that $a_1, \ldots, a_n, b_1, \ldots, b_{n-1}$ are nonzero. Then, the following statements hold:

  i) The singular values of A are distinct.

  ii) If $B \in \mathbb{F}^{n \times n}$ is bidiagonal and $|B| = |A|$, then A and B have the same singular values.

  iii) If $B \in \mathbb{F}^{n \times n}$ is bidiagonal, $|A| \leq |B|$, and $|A| \neq |B|$, then $\sigma_{\max}(A) < \sigma_{\max}(B)$.

  iv) If $B \in \mathbb{F}^{n \times n}$ is bidiagonal, $|I \circ A| \leq |I \circ B|$, and $|I \circ A| \neq |I \circ B|$, then $\sigma_{\min}(A) < \sigma_{\min}(B)$.

  v) If $B \in \mathbb{F}^{n \times n}$ is bidiagonal, $|I_{\sup} \circ A| \leq |I_{\sup} \circ B|$, and $|I_{\sup} \circ A| \neq |I_{\sup} \circ B|$, then $\sigma_{\min}(B) < \sigma_{\min}(A)$.

Proof: See [1006, p. 17-5].

Remark: $I_{\sup}$ denotes the matrix all of whose entries on the superdiagonal are 1 and are 0 otherwise.

5.12 Facts on Eigenvalues and Singular Values for Two or More Matrices

Fact 5.12.1. Let $A \in \mathbb{F}^{n \times n}$ and $B \in \mathbb{F}^{n \times m}$, let $r \triangleq \operatorname{rank} B$, and define $\mathcal{A} \triangleq \begin{bmatrix} A & B \\ B^* & 0 \end{bmatrix}$. Then, $\nu_-(\mathcal{A}) \geq r$, $\nu_0(\mathcal{A}) \geq 0$, and $\nu_+(\mathcal{A}) \geq r$. If, in addition, $n = m$ and B is nonsingular, then $\operatorname{In} \mathcal{A} = \begin{bmatrix} n & 0 & n \end{bmatrix}^{\mathrm{T}}$.

Proof: See [736].

Remark: See Proposition 5.6.5.

Fact 5.12.2. Let $A, B \in \mathbb{F}^{n \times n}$. Then,

$$\operatorname{sprad}(A + B) \leq \sigma_{\max}(A + B) \leq \sigma_{\max}(A) + \sigma_{\max}(B).$$

If, in addition, A and B are Hermitian, then

$$\text{sprad}(A+B) = \sigma_{\max}(A+B) \le \sigma_{\max}(A) + \sigma_{\max}(B) = \text{sprad}(A) + \text{sprad}(B)$$

and

$$\lambda_{\min}(A) + \lambda_{\min}(B) \le \lambda_{\min}(A+B) \le \lambda_{\max}(A+B) \le \lambda_{\max}(A) + \lambda_{\max}(B).$$

Proof: Use Lemma 8.4.3 for the last string of inequalities.

Remark: See Fact 5.11.5.

Fact 5.12.3. Let $A, B \in \mathbb{F}^{n \times n}$, and let λ be an eigenvalue of $A + B$. Then,

$$\tfrac{1}{2}\lambda_{\min}(A^* + A) + \tfrac{1}{2}\lambda_{\min}(B^* + B) \le \text{Re}\,\lambda \le \tfrac{1}{2}\lambda_{\max}(A^* + A) + \tfrac{1}{2}\lambda_{\max}(B^* + B).$$

Proof: See [319, p. 18].

Remark: See Fact 5.11.21.

Fact 5.12.4. Let $A, B \in \mathbb{F}^{n \times n}$ be normal, and let $\text{mspec}(A) = \{\lambda_1, \ldots, \lambda_n\}_{\text{ms}}$ and $\text{mspec}(B) = \{\mu_1, \ldots, \mu_n\}_{\text{ms}}$. Then,

$$\min \text{Re} \sum_{i=1}^{n} \lambda_i \mu_{\sigma(i)} \le \text{Re}\,\text{tr}\,AB \le \max \text{Re} \sum_{i=1}^{n} \lambda_i \mu_{\sigma(i)},$$

where "max" and "min" are taken over all permutations σ of the eigenvalues of B. Now, assume in addition that A and B are Hermitian. Then, $\text{tr}\,AB$ is real, and

$$\sum_{i=1}^{n} \lambda_i(A)\lambda_{n-i+1}(B) \le \text{tr}\,AB \le \sum_{i=1}^{n} \lambda_i(A)\lambda_i(B).$$

Furthermore, the last inequality is an equality if and only if there exists a unitary matrix $S \in \mathbb{F}^{n \times n}$ such that $A = S\text{diag}[\lambda_1(A), \ldots, \lambda_n(A)]S^*$ and $B = S\text{diag}[\lambda_1(B), \ldots, \lambda_n(B)]S^*$.

Proof: See [982]. For the second string of inequalities, use Fact 1.18.4. For the last statement, see [243, p. 10] or [916].

Remark: The upper bound for $\text{tr}\,AB$ is due to Fan.

Remark: See Fact 5.12.5, Fact 5.12.8, Proposition 8.4.13, Fact 8.12.29, and Fact 8.19.19.

Fact 5.12.5. Let $A, B \in \mathbb{F}^{n \times n}$, and assume that B is Hermitian. Then,

$$\sum_{i=1}^{n} \lambda_i[\tfrac{1}{2}(A + A^*)]\lambda_{n-i+1}(B) \le \text{Re}\,\text{tr}\,AB \le \sum_{i=1}^{n} \lambda_i[\tfrac{1}{2}(A + A^*)]\lambda_i(B).$$

Proof: Apply the second string of inequalities in Fact 5.12.4.

Remark: For A, B real, these inequalities are given in [861]. The complex case is given in [896].

Remark: See Proposition 8.4.13 for the case in which B is positive semidefinite.

Fact 5.12.6. Let $A \in \mathbb{F}^{n \times m}$ and $B \in \mathbb{F}^{m \times n}$, and let $r \triangleq \min\{\operatorname{rank} A, \operatorname{rank} B\}$. Then,

$$|\operatorname{tr} AB| \leq \sum_{i=1}^{r} \sigma_i(A)\sigma_i(B).$$

Proof: See [996, pp. 514, 515] or [1125, p. 148].

Remark: Applying Fact 5.12.4 to $\begin{bmatrix} 0 & A \\ A^* & 0 \end{bmatrix}$ and $\begin{bmatrix} 0 & B^* \\ B & 0 \end{bmatrix}$ and using Proposition 5.6.5 yields the weaker result

$$|\operatorname{Re} \operatorname{tr} AB| \leq \sum_{i=1}^{r} \sigma_i(A)\sigma_i(B).$$

Remark: See [243, p. 14].

Remark: This result is due to Mirsky.

Remark: See Fact 5.12.7.

Remark: A generalization of this result is given by Fact 9.14.3.

Fact 5.12.7. Let $A, B \in \mathbb{F}^{n \times n}$, and assume that B is positive semidefinite. Then,

$$|\operatorname{tr} AB| \leq \sigma_{\max}(A) \operatorname{tr} B.$$

Proof: Apply Fact 5.12.6.

Remark: A generalization of this result is given by Fact 9.14.4.

Fact 5.12.8. Let $A, B \in \mathbb{R}^{n \times n}$, assume that B is symmetric, and define $C \triangleq \frac{1}{2}(A + A^{\mathrm{T}})$. Then,

$$\lambda_{\min}(C) \operatorname{tr} B - \lambda_{\min}(B)[n\lambda_{\min}(C) - \operatorname{tr} A]$$
$$\leq \operatorname{tr} AB \leq \lambda_{\max}(C) \operatorname{tr} B - \lambda_{\max}(B)[n\lambda_{\max}(C) - \operatorname{tr} A].$$

Proof: See [481].

Remark: See Fact 5.12.4, Proposition 8.4.13, and Fact 8.12.29. Extensions are given in [1098].

Fact 5.12.9. Let $A, B, Q, S_1, S_2 \in \mathbb{R}^{n \times n}$, assume that A and B are symmetric, and assume that Q, S_1, and S_2 are orthogonal. Furthermore, assume that $S_1^{\mathrm{T}} A S_1$ and $S_2^{\mathrm{T}} B S_2$ are diagonal with the diagonal entries arranged in nonincreasing order, and define the orthogonal matrices $Q_1, Q_2 \in \mathbb{R}^{n \times n}$ by $Q_1 \triangleq S_1 \operatorname{revdiag}(\pm 1, \ldots, \pm 1) S_1^{\mathrm{T}}$ and $Q_2 \triangleq S_2 \operatorname{diag}(\pm 1, \ldots, \pm 1) S_2^{\mathrm{T}}$. Then,

$$\operatorname{tr} A Q_1 B Q_1^{\mathrm{T}} \leq \operatorname{tr} AQBQ^{\mathrm{T}} \leq \operatorname{tr} A Q_2 B Q_2^{\mathrm{T}}.$$

Proof: See [160, 916].

Remark: See Fact 5.12.8.

Fact 5.12.10. Let $A_1, \ldots, A_k, B_1, \ldots, B_k \in \mathbb{F}^{n \times n}$, and assume that $A_1, \ldots, A_k$ are unitary. Then,

$$|\operatorname{tr} A_1 B_1 \cdots A_k B_k| \leq \sum_{i=1}^n \sigma_i(B_1) \cdots \sigma_i(B_k).$$

Proof: See [996, p. 516].

Remark: This result is due to Fan.

Remark: See Fact 5.12.9.

Fact 5.12.11. Let $A, B \in \mathbb{R}^{n \times n}$, and assume that $AB = BA$. Then,

$$\operatorname{sprad}(AB) \leq \operatorname{sprad}(A) \operatorname{sprad}(B)$$

and

$$\operatorname{sprad}(A + B) \leq \operatorname{sprad}(A) + \operatorname{sprad}(B).$$

Proof: Use Fact 5.17.4.

Remark: If $AB \neq BA$, then both of these inequalities may be violated. Consider $A = \left[\begin{smallmatrix} 0 & 1 \\ 0 & 0 \end{smallmatrix}\right]$ and $B = \left[\begin{smallmatrix} 0 & 0 \\ 1 & 0 \end{smallmatrix}\right]$.

Fact 5.12.12. Let $A, B \in \mathbb{C}^{n \times n}$, assume that A and B are normal, and let $\operatorname{mspec}(A) = \{\lambda_1, \ldots, \lambda_n\}_{\mathrm{ms}}$ and $\operatorname{mspec}(B) = \{\mu_1, \ldots, \mu_n\}_{\mathrm{ms}}$. Then,

$$|\det(A + B)| \leq \min\left\{\prod_{i=1}^n \max_{j=1,\ldots,n} |\lambda_i + \mu_j|, \prod_{j=1}^n \max_{i=1,\ldots,n} |\lambda_i + \mu_j|\right\}.$$

Proof: See [1137].

Remark: Equality is discussed in [165].

Remark: See Fact 9.14.18.

Fact 5.12.13. Let $A \in \mathbb{F}^{n \times m}$ and $B \in \mathbb{F}^{n \times m}$. Then,

$$\det(ABB^*A^*) \leq \left[\prod_{i=1}^m \sigma_i(B)\right] \det(AA^*).$$

Proof: See [459, p. 218].

Fact 5.12.14. Let $A, B, C \in \mathbb{F}^{n \times n}$, assume that $\operatorname{spec}(A) \cap \operatorname{spec}(B) = \varnothing$, and assume that $[A + B, C] = 0$ and $[AB, C] = 0$. Then, $[A, C] = [B, C] = 0$.

Proof: This result follows from Corollary 7.2.5.

Remark: This result is due to Embry. See [221].

Fact 5.12.15. Let $A, B \in \mathbb{F}^{n \times n}$, and assume that A and B are projectors. Then,

$$\operatorname{spec}(AB) \subset [0, 1]$$

and

$$\operatorname{spec}(A - B) \subset [-1, 1].$$

Proof: See [40], [550, p. 53], or [1125, p. 147].

Remark: The first result is due to Afriat.

Fact 5.12.16. Let $A, B \in \mathbb{F}^{n \times n}$, and assume that A and B are projectors. Then, the following statements are equivalent:

i) AB is a projector.

ii) $\operatorname{spec}(A + B) \subset \{0\} \cup [1, \infty)$.

iii) $\operatorname{spec}(A - B) \subset \{-1, 0, 1\}$.

Proof: See [551, 612].

Remark: See Fact 3.13.20 and Fact 6.4.26.

Fact 5.12.17. Let $A, B \in \mathbb{F}^{n \times n}$, assume that A and B are nonzero projectors, and define the minimal principal angle $\theta \in [0, \pi/2]$ by

$$\cos \theta = \max\{|x^*y| : (x, y) \in \mathcal{R}(A) \times \mathcal{R}(B) \text{ and } x^*x = y^*y = 1\}.$$

Then, the following statements hold:

i) $\sigma_{\max}(AB) = \sigma_{\max}(BA) = \cos \theta$.

ii) $\sigma_{\max}(A + B) = 1 + \sigma_{\max}(AB) = 1 + \cos \theta$.

iii) $1 \leq \sigma_{\max}(AB) + \sigma_{\max}(A - B)$.

iv) If $\sigma_{\max}(A - B) < 1$, then $\operatorname{rank} A = \operatorname{rank} B$.

v) $\theta > 0$ if and only if $\mathcal{R}(A) \cap \mathcal{R}(B) = \{0\}$.

Furthermore, the following statements are equivalent:

vi) $A - B$ is nonsingular.

vii) $\mathcal{R}(A)$ and $\mathcal{R}(B)$ are complementary subspaces.

viii) $\sigma_{\max}(A + B - I) < 1$.

Now, assume in addition that $A - B$ is nonsingular. Then, the following statements hold:

ix) $\sigma_{\max}(AB) < 1$.

x) $\sigma_{\max}[(A - B)^{-1}] = \dfrac{1}{\sqrt{1 - \sigma_{\max}^2(AB)}} = 1/\sin \theta$.

xi) $\sigma_{\min}(A - B) = \sin \theta$.

xii) $\sigma_{\min}^2(A - B) + \sigma_{\max}^2(AB) = 1$.

xiii) $I - AB$ is nonsingular.

xiv) If $\operatorname{rank} A = \operatorname{rank} B$, then $\sigma_{\max}(A - B) = \sin \theta$.

Proof: Statement *i)* is given in [765]. Statement *ii)* is given in [551]. Statement *iii)* follows from the first inequality in Fact 8.19.11. For *iv)*, see [459, p. 195] or [574, p. 389]. Statement *v)* is given in [574, p. 393]. Fact 3.13.24 shows that *vi)* and *vii)* are equivalent. Statement *viii)* is given in [278]; see also [550, p. 236]. Statement *xiv*) follows from [1261, pp. 92, 93].

Remark: Additional conditions for the nonsingularity of $A - B$ are given in Fact 3.13.24.

Remark: See Fact 2.9.19, Fact 5.11.39, and Fact 5.12.18.

Fact 5.12.18. Let $A \in \mathbb{F}^{n \times n}$, and assume that A is idempotent. Furthermore, let $P, Q \in \mathbb{F}^{n \times n}$, where P is the projector onto $\mathcal{R}(A)$ and Q is the projector onto $\mathcal{N}(A)$. Then, the following statements hold:

  i) $P - Q$ is nonsingular.

  ii) $(P - Q)^{-1} = A + A^* - I = A - A_\perp^*$.

  iii) $\sigma_{\max}(A) = \dfrac{1}{\sqrt{1 - \sigma_{\max}^2(PQ)}} = \sigma_{\max}[(P - Q)^{-1}] = \sigma_{\max}(A + A^* - I)$.

  iv) $\sigma_{\max}(A) = 1/\sin\theta$, where θ is the minimal principal angle $\theta \in [0, \pi/2]$ defined by
$$\cos\theta = \max\{|x^*y| : (x, y) \in \mathcal{R}(P) \times \mathcal{R}(Q) \text{ and } x^*x = y^*y = 1\}.$$

  v) $\sigma_{\min}^2(P - Q) = 1 - \sigma_{\max}^2(PQ)$.

  vi) $\sigma_{\max}(PQ) = \sigma_{\max}(QP) = \sigma_{\max}(P + Q - I) < 1$.

Proof: See [1142] and Fact 5.12.17. The nonsingularity of $P - Q$ follows from Fact 3.13.24. Statement *ii*) is given by Fact 3.13.24 and Fact 6.3.24. The first equality in *iii*) is given in [278]. See also [551].

Remark: $A_\perp^*$ is the idempotent matrix onto $\mathcal{R}(A)^\perp$ along $\mathcal{N}(A)^\perp$. See Fact 3.12.3.

Remark: $P = AA^+$ and $Q = I - A^+A$.

Fact 5.12.19. Let $A, B \in \mathbb{F}^{n \times n}$, and assume that A and B are idempotent. Then, $A - B$ is idempotent if and only if $A - B$ is group invertible and every eigenvalue of $A - B$ is nonnegative.

Proof: See [666].

Remark: This result is due to Makelainen and Styan.

Remark: See Fact 3.12.29.

Remark: Conditions for a matrix to be expressible as a difference of idempotents are given in [666].

Fact 5.12.20. Let $A \in \mathbb{R}^{n \times n}$, $B \in \mathbb{R}^{n \times m}$, and $C \in \mathbb{R}^{m \times m}$, define $\mathcal{A} \triangleq \begin{bmatrix} A & B \\ B^{\mathrm{T}} & C \end{bmatrix} \in \mathbb{R}^{(n+m) \times (n+m)}$, and assume that $\mathcal{A}$ is symmetric. Then,
$$\lambda_{\min}(\mathcal{A}) + \lambda_{\max}(\mathcal{A}) \leq \lambda_{\max}(A) + \lambda_{\max}(C).$$

Proof: See [227, p. 56].

Fact 5.12.21. Let $M \in \mathbb{R}^{r \times r}$, assume that M is positive definite, let $C, K \in \mathbb{R}^{r \times r}$, assume that C and K are positive semidefinite, and consider the equation

$$M\ddot{q} + C\dot{q} + Kq = 0.$$

Then, $x(t) \triangleq \begin{bmatrix} q(t) \\ \dot{q}(t) \end{bmatrix}$ satisfies $\dot{x}(t) = Ax(t)$, where A is the $2r \times 2r$ matrix

$$A \triangleq \begin{bmatrix} 0 & I \\ -M^{-1}K & -M^{-1}C \end{bmatrix}.$$

Furthermore, the following statements hold:

i) A, K, and M satisfy

$$\det A = \frac{\det K}{\det M}.$$

ii) A and K satisfy

$$\operatorname{rank} A = r + \operatorname{rank} K.$$

iii) A is nonsingular if and only if K is positive definite. In this case,

$$A^{-1} = \begin{bmatrix} -K^{-1}C & -K^{-1}M \\ I & 0 \end{bmatrix}.$$

iv) Let $\lambda \in \mathbb{C}$. Then, $\lambda \in \operatorname{spec}(A)$ if and only if $\det(\lambda^2 M + \lambda C + K) = 0$.

v) If $\lambda \in \operatorname{spec}(A)$, $\operatorname{Re} \lambda = 0$, and $\operatorname{Im} \lambda \neq 0$, then λ is semisimple.

vi) $\operatorname{mspec}(A) \subset \operatorname{CLHP}$.

vii) If $C = 0$, then $\operatorname{spec}(A) \subset \jmath\mathbb{R}$.

viii) If C and K are positive definite, then $\operatorname{spec}(A) \subset \operatorname{OLHP}$.

ix) $\hat{x}(t) \triangleq \begin{bmatrix} \frac{1}{\sqrt{2}} K^{1/2} q(t) \\ \frac{1}{\sqrt{2}} M^{1/2} \dot{q}(t) \end{bmatrix}$ satisfies $\dot{x}(t) = \hat{A}x(t)$, where

$$\hat{A} \triangleq \begin{bmatrix} 0 & K^{1/2}M^{-1/2} \\ -M^{-1/2}K^{1/2} & -M^{-1/2}CM^{-1/2} \end{bmatrix}.$$

If, in addition, $C = 0$, then $\hat{A}$ is skew symmetric.

x) $\hat{x}(t) \triangleq \begin{bmatrix} M^{1/2} q(t) \\ M^{1/2} \dot{q}(t) \end{bmatrix}$ satisfies $\dot{x}(t) = \hat{A}x(t)$, where

$$\hat{A} \triangleq \begin{bmatrix} 0 & I \\ -M^{-1/2}KM^{-1/2} & -M^{-1/2}CM^{-1/2} \end{bmatrix}.$$

If, in addition, $C = 0$, then $\hat{A}$ is Hamiltonian.

Remark: M, C, and K are mass, damping, and stiffness matrices, respectively. See [190].

Remark: See Fact 5.14.34 and Fact 11.18.38.

Problem: Prove *v)*.

Fact 5.12.22. Let $A, B \in \mathbb{R}^{n \times n}$, and assume that A and B are positive semidefinite. Then, every eigenvalue λ of $\begin{bmatrix} 0 & B \\ -A & 0 \end{bmatrix}$ satisfies $\operatorname{Re} \lambda = 0$.

Proof: Square this matrix.

Problem: What happens if A and B have different sizes? In addition, let $C \in \mathbb{R}^{n \times n}$, and assume that C is (positive semidefinite, positive definite). Then, every eigenvalue of $\begin{bmatrix} 0 & A \\ -B & -C \end{bmatrix}$ satisfies ($\operatorname{Re} \lambda \leq 0$, $\operatorname{Re} \lambda < 0$).

Problem: Consider also $\begin{bmatrix} -C & A \\ -B & -C \end{bmatrix}$ and $\begin{bmatrix} -C & A \\ -A & -C \end{bmatrix}$.

5.13 Facts on Matrix Pencils

Fact 5.13.1. Let $A, B \in \mathbb{F}^{n \times n}$, and assume that $P_{A,B}$ is a regular pencil. Furthermore, let $\mathcal{S} \subseteq \mathbb{F}^n$, assume that $\mathcal{S}$ is a subspace, let $k \triangleq \dim \mathcal{S}$, let $S \in \mathbb{F}^{n \times k}$, and assume that $\mathcal{R}(S) = \mathcal{S}$. Then, the following statements are equivalent:

i) $\dim(A\mathcal{S} + B\mathcal{S}) = \dim \mathcal{S}$.

ii) There exists a matrix $M \in \mathbb{F}^{k \times k}$ such that $A S = B S M$.

Proof: See [897, p. 144].

Remark: $\mathcal{S}$ is a *deflating subspace* of $P_{A,B}$. This result generalizes Fact 2.9.25.

5.14 Facts on Matrix Eigenstructure

Fact 5.14.1. Let $A \in \mathbb{F}^{n \times n}$, let $\lambda \in \operatorname{spec}(A)$, assume that λ is cyclic, let $i \in \{1, \ldots, n\}$ be such that $\operatorname{rank}(A - \lambda I)_{(\{i\}^\sim, \{1, \ldots, n\})} = n - 1$, and define $x \in \mathbb{C}^n$ by

$$
x \triangleq \begin{bmatrix} \det(A - \lambda I)_{[i;1]} \\ -\det(A - \lambda I)_{[i;2]} \\ \vdots \\ (-1)^{n+1} \det(A - \lambda I)_{[i;n]} \end{bmatrix}.
$$

Then, x is an eigenvector of A associated with λ.

Proof: See [1372].

Fact 5.14.2. Let $A \in \mathbb{F}^{n \times n}$. Then, the following statements are equivalent:

i) A is group invertible.

ii) $\mathcal{R}(A) = \mathcal{R}(A^2)$.

iii) $\operatorname{ind} A \leq 1$.

iv) $\operatorname{rank} A = \sum_{i=1}^r \operatorname{amult}_A(\lambda_i)$, where $\lambda_1, \ldots, \lambda_r$ are the nonzero eigenvalues of A.

Fact 5.14.3. Let $n \geq 2$, $x, y \in \mathbb{F}^n$, define $A \triangleq xy^{\mathrm{T}}$, and assume that rank $A = 1$, that is, A is nonzero. Then, the following statements are equivalent:

i) A is semisimple.

ii) $y^{\mathrm{T}}x \neq 0$.

iii) tr $A \neq 0$.

iv) A is group invertible.

v) ind $A = 1$.

vi) $\mathrm{amult}_A(0) = n - 1$.

Furthermore, the following statements are equivalent:

vii) A is defective.

viii) $y^{\mathrm{T}}x = 0$.

ix) tr $A = 0$.

x) A is not group invertible.

xi) ind $A = 2$.

xii) A is nilpotent.

xiii) $\mathrm{amult}_A(0) = n$.

xiv) $\mathrm{spec}(A) = \{0\}$.

Remark: See Fact 2.10.19.

Fact 5.14.4. Let $A \in \mathbb{F}^{n \times n}$, and assume that A is diagonalizable over $\mathbb{F}$. Then, A^{T}, $\overline{A}$, A^*, and A^{A} are diagonalizable. If, in addition, A is nonsingular, then A^{-1} is diagonalizable.

Proof: See Fact 2.16.10 and Fact 3.7.10.

Fact 5.14.5. Let $A \in \mathbb{F}^{n \times n}$, assume that A is diagonalizable over $\mathbb{F}$ with eigenvalues $\lambda_1, \ldots, \lambda_n$, and let $B \triangleq \mathrm{diag}(\lambda_1, \ldots, \lambda_n)$. If, $x_1, \ldots, x_n \in \mathbb{F}^n$ are linearly independent eigenvectors of A associated with $\lambda_1, \ldots, \lambda_n$, respectively, then $A = SBS^{-1}$, where $S \triangleq \begin{bmatrix} x_1 & \cdots & x_n \end{bmatrix}$. Conversely, if $S \in \mathbb{F}^{n \times n}$ is nonsingular and $A = SBS^{-1}$, then, for all $i \in \{1, \ldots, n\}$, $\mathrm{col}_i(S)$ is an associated eigenvector.

Fact 5.14.6. Let $A, S \in \mathbb{F}^{n \times n}$, assume that S is nonsingular, let $\lambda \in \mathbb{C}$, and assume that $\mathrm{row}_1(S^{-1}AS) = \lambda e_1^{\mathrm{T}}$. Then, $\lambda \in \mathrm{spec}(A)$, and $\mathrm{col}_1(S)$ is an associated eigenvector.

Fact 5.14.7. Let $A \in \mathbb{F}^{n \times n}$. Then, A is cyclic if and only if there exists a vector $b \in \mathbb{F}^n$ such that $\begin{bmatrix} b & Ab & \cdots & A^{n-1}b \end{bmatrix}$ is nonsingular.

Proof: See Fact 12.20.13.

Remark: (A, b) is controllable. See Corollary 12.6.3.

Fact 5.14.8. Let $A \in \mathbb{F}^{n \times n}$, and define the positive integer m by

$$m \triangleq \max_{\lambda \in \mathrm{spec}(A)} \mathrm{gmult}_A(\lambda).$$

Then, m is the smallest integer such that there exists $B \in \mathbb{F}^{n \times m}$ such that rank $\begin{bmatrix} B & AB & \cdots & A^{n-1}B \end{bmatrix} = n$.

Proof: See Fact 12.20.13.

Remark: (A, B) is controllable. See Corollary 12.6.3.

Fact 5.14.9. Let $A \in \mathbb{C}^{n \times n}$. Then, there exist $v_1, \ldots, v_n \in \mathbb{C}^n$ such that the following statements hold:

i) $v_1, \ldots, v_n \in \mathbb{C}^n$ are linearly independent.

ii) If $\lambda \in \mathrm{spec}(A)$ and A has a $k \times k$ Jordan block associated with λ, then there exist distinct integers $i_1, \ldots, i_k$ such that

$$Av_{i_1} = \lambda v_{i_1},$$
$$Av_{i_2} = \lambda v_{i_2} + v_{i_1},$$
$$\vdots$$
$$Av_{i_k} = \lambda v_{i_k} + v_{i_{k-1}}.$$

iii) Let λ and $v_{i_1}, \ldots, v_{i_k}$ be given by ii). Then,

$$\mathrm{span}\,\{v_{i_1}, \ldots, v_{i_k}\} = \mathcal{N}[(\lambda I - A)^k].$$

Remark: $v_1, \ldots, v_n$ are *generalized eigenvectors* of A.

Remark: $(v_{i_1}, \ldots, v_{i_k})$ is a *Jordan chain* of A associated with λ. See [892, pp. 229–231].

Remark: See Fact 11.13.7.

Fact 5.14.10. Let $A \in \mathbb{R}^{n \times n}$. Then, A is cyclic and semisimple if and only if A is simple.

Fact 5.14.11. Let $A = \mathrm{revdiag}(a_1, \ldots, a_n) \in \mathbb{R}^{n \times n}$. Then, A is semisimple if and only if, for all $i \in \{1, \ldots, n\}$, a_i and a_{n+1-i} are either both zero or both nonzero.

Proof: See [641, p. 116], [827], or [1125, pp. 68, 86].

Fact 5.14.12. Let $A \in \mathbb{F}^{n \times n}$. Then, A has at least m real eigenvalues and m associated linearly independent eigenvectors if and only if there exists a positive-semidefinite matrix $S \in \mathbb{F}^{n \times n}$ such that rank $S = m$ and $AS = SA^*$.

Proof: See [1125, pp. 68, 86].

Remark: The case $m = n$ is given by Proposition 5.5.12.

Remark: This result is due to Drazin and Haynsworth.

Fact 5.14.13. Let $A \in \mathbb{F}^{n \times n}$, assume that A is normal, and let $\mathrm{mspec}(A) = \{\lambda_1, \ldots, \lambda_n\}_{\mathrm{ms}}$. Then, there exist vectors $x_1, \ldots, x_n \in \mathbb{C}^n$ such that $x_i^* x_j = \delta_{ij}$ for all $i, j \in \{1, \ldots, n\}$ and

$$A = \sum_{i=1}^{n} \lambda_i x_i x_i^*.$$

Furthermore, $x_1, \ldots, x_n$ are mutually orthogonal eigenvectors of A.

Remark: See Corollary 5.4.8.

Fact 5.14.14. Let $A \in \mathbb{F}^{n \times n}$, and let $\mathrm{mspec}(A) = \{\lambda_1, \ldots, \lambda_n\}_{\mathrm{ms}}$, where $|\lambda_1| \geq \cdots \geq |\lambda_n|$. Then, the following statements are equivalent:

i) A is normal.

ii) For all $i \in \{1, \ldots, n\}$, $|\lambda_i| = \sigma_i(A)$.

iii) $\sum_{i=1}^{n} |\lambda_i|^2 = \sum_{i=1}^{n} \sigma_i^2(A)$.

iv) There exists $p \in \mathbb{F}[s]$ such that $A = p(A^*)$.

v) Every eigenvector of A is also an eigenvector of A^*.

vi) $AA^* - A^*A$ is either positive semidefinite or negative semidefinite.

vii) For all $x \in \mathbb{F}^n$, $x^*A^*Ax = x^*AA^*x$.

viii) For all $x, y \in \mathbb{F}^n$, $x^*A^*Ay = x^*AA^*y$.

In this case,

$$\mathrm{sprad}(A) = \sigma_{\max}(A).$$

Proof: See [603] or [1125, p. 146].

Remark: See Fact 9.8.13 and Fact 9.11.2.

Fact 5.14.15. Let $A \in \mathbb{F}^{n \times n}$. Then, the following statements are equivalent:

i) A is (simple, cyclic, derogatory, semisimple, defective, diagonalizable over $\mathbb{F}$).

ii) There exists $\alpha \in \mathbb{F}$ such that $A + \alpha I$ is (simple, cyclic, derogatory, semisimple, defective, diagonalizable over $\mathbb{F}$).

iii) For all $\alpha \in \mathbb{F}$, $A + \alpha I$ is (simple, cyclic, derogatory, semisimple, defective, diagonalizable over $\mathbb{F}$).

Fact 5.14.16. Let $x, y \in \mathbb{F}^n$, assume that $x^\mathrm{T} y \neq 1$, and define the elementary matrix $A \triangleq I - xy^\mathrm{T}$. Then, A is semisimple if and only if either $xy^\mathrm{T} = 0$ or $x^\mathrm{T} y \neq 0$.

Remark: Use Fact 5.14.3 and Fact 5.14.15.

Fact 5.14.17. Let $A \in \mathbb{F}^{n \times n}$, and assume that A is nilpotent. Then, A is nonzero if and only if A is defective.

Fact 5.14.18. Let $A \in \mathbb{F}^{n \times n}$, and assume that A is either involutory or skew involutory. Then, A is semisimple.

Fact 5.14.19. Let $A \in \mathbb{R}^{n \times n}$, and assume that A is involutory. Then, A is diagonalizable over $\mathbb{R}$.

Fact 5.14.20. Let $A \in \mathbb{F}^{n \times n}$, assume that A is semisimple, and assume that $A^3 = A^2$. Then, A is idempotent.

Fact 5.14.21. Let $A \in \mathbb{F}^{n \times n}$. Then, A is cyclic if and only if every matrix $B \in \mathbb{F}^{n \times n}$ satisfying $AB = BA$ is a polynomial in A.

Proof: See [730, p. 275].

Remark: See Fact 2.18.9, Fact 5.14.22, Fact 5.14.23, and Fact 7.5.2.

Fact 5.14.22. Let $A \in \mathbb{F}^{n \times n}$, assume that A is simple, let $B \in \mathbb{F}^{n \times n}$, and assume that $AB = BA$. Then, B is a polynomial in A whose degree is not greater than $n - 1$.

Proof: See [1526, p. 59].

Remark: See Fact 5.14.21.

Fact 5.14.23. Let $A, B \in \mathbb{F}^{n \times n}$. Then, B is a polynomial in A if and only if B commutes with every matrix that commutes with A.

Proof: See [730, p. 276].

Remark: See Fact 4.8.13.

Remark: See Fact 2.18.9, Fact 5.14.21, Fact 5.14.22, and Fact 7.5.2.

Fact 5.14.24. Let $A, B \in \mathbb{C}^{n \times n}$, and assume that $AB = BA$. Furthermore, let $x \in \mathbb{C}^n$ be an eigenvector of A associated with the eigenvalue $\lambda \in \mathbb{C}$, and assume that $Bx \neq 0$. Then, Bx is an eigenvector of A associated with the eigenvalue $\lambda \in \mathbb{C}$.

Proof: $A(Bx) = BAx = B(\lambda x) = \lambda(Bx)$.

Fact 5.14.25. Let $A \in \mathbb{C}^{n \times n}$, and let $x \in \mathbb{C}^n$ be an eigenvector of A associated with the eigenvalue λ. If A is nonsingular, then x is an eigenvector of A^A associated with the eigenvalue $(\det A)/\lambda$. If $\operatorname{rank} A = n - 1$, then x is an eigenvector of A^A associated with the eigenvalue $\operatorname{tr} A^A$ or 0. Finally, if $\operatorname{rank} A \leq n - 2$, then x is an eigenvector of A^A associated with the eigenvalue 0.

Proof: Use Fact 5.14.24 and the fact that $A^A A = A A^A$. See [362].

Remark: See Fact 2.16.8 or Fact 6.3.6.

Fact 5.14.26. Let $A, B \in \mathbb{C}^{n \times n}$. Then, the following statements are equivalent:

i) $\cap_{k,l=1}^{n-1} \mathcal{N}([A^k, B^l]) \neq \{0\}$.

ii) $\sum_{k,l=1}^{n-1} [A^k, B^l]^* [A^k, B^l]$ is singular.

iii) A and B have a common eigenvector.

Proof: See [561].

Remark: This result is due to Shemesh.

Remark: See Fact 5.17.1.

Fact 5.14.27. Let $A, B \in \mathbb{C}^{n \times n}$, and assume that $AB = BA$. Then, there exists a nonzero vector $x \in \mathbb{C}^n$ that is an eigenvector of both A and B.

Proof: See [728, p. 51].

Fact 5.14.28. Let $A, B \in \mathbb{F}^{n \times n}$. Then, the following statements hold:

i) Assume that A and B are Hermitian. Then, AB is Hermitian if and only if $AB = BA$.

ii) A is normal if and only if, for all $C \in \mathbb{F}^{n \times n}$, $AC = CA$ implies that $A^*C = CA^*$.

iii) Assume that B is Hermitian and $AB = BA$. Then, $A^*B = BA^*$.

iv) Assume that A and B are normal and $AB = BA$. Then, AB is normal.

v) Assume that A, B, and AB are normal. Then, BA is normal.

vi) Assume that A and B are normal and either A or B has the property that distinct eigenvalues have unequal absolute values. Then, AB is normal if and only if $AB = BA$.

Proof: See [366, 1462], [645, p. 157], and [1125, p. 102].

Fact 5.14.29. Let $A, B, C \in \mathbb{F}^{n \times n}$, and assume that A and B are normal and $AC = CB$. Then, $A^*C = CB^*$.

Proof: Consider $\begin{bmatrix} A & 0 \\ 0 & B \end{bmatrix}$ and $\begin{bmatrix} 0 & C \\ 0 & 0 \end{bmatrix}$ in *ii)* of Fact 5.14.28. See [642, p. 104] or [645, p. 321].

Remark: This result is the *Putnam-Fuglede theorem*.

Fact 5.14.30. Let $A, B \in \mathbb{F}^{n \times n}$, and assume that A is dissipative and B is range Hermitian. Then,
$$\operatorname{ind} B = \operatorname{ind} AB.$$

Proof: See [193].

Fact 5.14.31. Let $A \in \mathbb{F}^{n \times n}$, $B \in \mathbb{F}^{n \times m}$, and $C \in \mathbb{F}^{m \times m}$. Then,
$$\max\{\operatorname{ind} A, \operatorname{ind} C\} \leq \operatorname{ind} \begin{bmatrix} A & B \\ 0 & C \end{bmatrix} \leq \operatorname{ind} A + \operatorname{ind} C.$$

If C is nonsingular, then
$$\operatorname{ind} \begin{bmatrix} A & B \\ 0 & C \end{bmatrix} = \operatorname{ind} A,$$

whereas, if A is nonsingular, then
$$\operatorname{ind} \begin{bmatrix} A & B \\ 0 & C \end{bmatrix} = \operatorname{ind} C.$$

Proof: See [269, 1024].

Remark: See Fact 6.6.14.

Remark: The eigenstructure of a partitioned Hamiltonian matrix is considered in Fact 12.23.1.

Fact 5.14.32. Let $A, B \in \mathbb{R}^{n \times n}$, and assume that A and B are skew symmetric. Then, there exists an orthogonal matrix $S \in \mathbb{R}^{n \times n}$ such that

$$A = S \begin{bmatrix} 0_{(n-l) \times (n-l)} & A_{12} \\ -A_{12}^{\mathrm{T}} & A_{22} \end{bmatrix} S^{\mathrm{T}}$$

and

$$B = S \begin{bmatrix} B_{11} & B_{12} \\ -B_{12}^{\mathrm{T}} & 0_{l \times l} \end{bmatrix} S^{\mathrm{T}},$$

where $l \triangleq \lfloor n/2 \rfloor$. Consequently,

$$\mathrm{mspec}(AB) = \mathrm{mspec}\left(-A_{12} B_{12}^{\mathrm{T}}\right) \cup \mathrm{mspec}\left(-A_{12}^{\mathrm{T}} B_{12}\right),$$

and thus every nonzero eigenvalue of AB has even algebraic multiplicity.

Proof: See [32].

Fact 5.14.33. Let $A, B \in \mathbb{R}^{n \times n}$, and assume that A and B are skew symmetric. If n is even, then there exists a monic polynomial p of degree $n/2$ such that $\chi_{AB}(s) = p^2(s)$ and $p(AB) = 0$. If n is odd, then there exists a monic polynomial $p(s)$ of degree $(n-1)/2$ such that $\chi_{AB}(s) = sp^2(s)$ and $ABp(AB) = 0$. Consequently, if n is (even, odd), then χ_{AB} is (even, odd) and (every, every nonzero) eigenvalue of AB has even algebraic multiplicity and geometric multiplicity of at least 2.

Proof: See [428, 592].

Fact 5.14.34. Let $q(t)$ denote the displacement of a mass $m > 0$ connected to a spring $k \geq 0$ and dashpot $c \geq 0$ and subject to a force $f(t)$. Then, $q(t)$ satisfies

$$m\ddot{q}(t) + c\dot{q}(t) + kq(t) = f(t)$$

or

$$\ddot{q}(t) + \frac{c}{m}\dot{q}(t) + \frac{k}{m}q(t) = \frac{1}{m}f(t).$$

Now, define the *natural frequency* $\omega_{\mathrm{n}} \triangleq \sqrt{k/m}$ and, if $k > 0$, the *damping ratio* $\zeta \triangleq c/2\sqrt{km}$ to obtain

$$\ddot{q}(t) + 2\zeta\omega_{\mathrm{n}}\dot{q}(t) + \omega_{\mathrm{n}}^2 q(t) = \frac{1}{m}f(t).$$

If $k = 0$, then set $\omega_{\mathrm{n}} = 0$ and $\zeta\omega_{\mathrm{n}} = c/2m$. Next, define $x_1(t) \triangleq q(t)$ and $x_2(t) \triangleq \dot{q}(t)$ so that this equation can be written as

$$\begin{bmatrix} \dot{x}_1(t) \\ \dot{x}_2(t) \end{bmatrix} = \begin{bmatrix} 0 & 1 \\ -\omega_{\mathrm{n}}^2 & -2\zeta\omega_{\mathrm{n}} \end{bmatrix} \begin{bmatrix} x_1(t) \\ x_2(t) \end{bmatrix} + \begin{bmatrix} 0 \\ 1/m \end{bmatrix} f(t).$$

The eigenvalues of the companion matrix $A_c \triangleq \begin{bmatrix} 0 & 1 \\ -\omega_n^2 & -2\zeta\omega_n \end{bmatrix}$ are given by

$$\text{mspec}(A_c) = \begin{cases} \{-\zeta\omega_n - \jmath\omega_d, -\zeta\omega_n + \jmath\omega_d\}_{ms}, & 0 \le \zeta \le 1, \\ \left\{(-\zeta - \sqrt{\zeta^2 - 1})\omega_n, (-\zeta + \sqrt{\zeta^2 - 1})\omega_n\right\}, & \zeta > 1, \end{cases}$$

where $\omega_d \triangleq \omega_n\sqrt{1 - \zeta^2}$ is the *damped natural frequency*. The matrix A_c has repeated eigenvalues in exactly two cases, namely,

$$\text{mspec}(A_c) = \begin{cases} \{0, 0\}_{ms}, & \omega_n = 0, \\ \{-\omega_n, -\omega_n\}_{ms}, & \zeta = 1. \end{cases}$$

In both of these cases the matrix A_c is defective. In the case $\omega_n = 0$, the matrix A_c is also in Jordan form. In particular, in the case $\zeta = 1$, it follows that $A_c = SA_JS^{-1}$, where $S \triangleq \begin{bmatrix} -1 & 0 \\ \omega_n & -1 \end{bmatrix}$ and A_J is the Jordan form matrix $A_J \triangleq \begin{bmatrix} -\omega_n & 1 \\ 0 & -\omega_n \end{bmatrix}$. If A_c is not defective, that is, if $\omega_n \ne 0$ and $\zeta \ne 1$, then the Jordan form A_J of A_c is given by

$$A_J \triangleq \begin{cases} \begin{bmatrix} -\zeta\omega_n + \jmath\omega_d & 0 \\ 0 & -\zeta\omega_n - \jmath\omega_d \end{bmatrix}, & 0 \le \zeta < 1, \omega_n \ne 0, \\ \begin{bmatrix} \left(-\zeta - \sqrt{\zeta^2 - 1}\right)\omega_n & 0 \\ 0 & \left(-\zeta + \sqrt{\zeta^2 - 1}\right)\omega_n \end{bmatrix}, & \zeta > 1, \omega_n \ne 0. \end{cases}$$

In the case $0 \le \zeta < 1$ and $\omega_n \ne 0$, define the real normal form

$$A_n \triangleq \begin{bmatrix} -\zeta\omega_n & \omega_d \\ -\omega_d & -\zeta\omega_n \end{bmatrix}.$$

The matrices A_c, A_J, and A_n are related by the similarity transformations

$$A_c = S_1 A_J S_1^{-1} = S_2 A_n S_2^{-1}, \quad A_J = S_3 A_n S_3^{-1},$$

where

$$S_1 \triangleq \begin{bmatrix} 1 & 1 \\ -\zeta\omega_n + \jmath\omega_d & -\zeta\omega_n - \jmath\omega_d \end{bmatrix}, \quad S_1^{-1} = \frac{\jmath}{2\omega_d}\begin{bmatrix} -\zeta\omega_n - \jmath\omega_d & -1 \\ \zeta\omega_n - \jmath\omega_d & 1 \end{bmatrix},$$

$$S_2 \triangleq \frac{1}{\omega_d}\begin{bmatrix} 1 & 0 \\ -\zeta\omega_n & \omega_d \end{bmatrix}, \quad S_2^{-1} = \begin{bmatrix} \omega_d & 0 \\ \zeta\omega_n & 1 \end{bmatrix},$$

$$S_3 \triangleq \frac{1}{2\omega_d}\begin{bmatrix} 1 & -\jmath \\ 1 & \jmath \end{bmatrix}, \quad S_3^{-1} = \omega_d\begin{bmatrix} 1 & 1 \\ \jmath & -\jmath \end{bmatrix}.$$

In the case $\zeta > 1$ and $\omega_n \ne 0$, the matrices A_c and A_J are related by

$$A_c = S_4 A_J S_4^{-1},$$

where

$$S_4 \triangleq \begin{bmatrix} 1 & 1 \\ -\zeta\omega_n + \jmath\omega_d & -\zeta\omega_n - \jmath\omega_d \end{bmatrix}, \quad S_4^{-1} = \frac{\jmath}{2\omega_d}\begin{bmatrix} -\zeta\omega_n - \jmath\omega_d & -1 \\ \zeta\omega_n - \jmath\omega_d & 1 \end{bmatrix}.$$

Finally, define the energy-coordinates matrix

$$A_e \triangleq \begin{bmatrix} 0 & \omega_n \\ -\omega_n & -2\zeta\omega_n \end{bmatrix}.$$

Then, $A_e = S_5 A_c S_5^{-1}$, where

$$S_5 \triangleq \sqrt{\tfrac{m}{2}} \begin{bmatrix} 1 & 0 \\ 0 & 1/\omega_n \end{bmatrix}.$$

Remark: See Fact 5.12.21.

5.15 Facts on Matrix Factorizations

Fact 5.15.1. Let $A \in \mathbb{F}^{n \times n}$. Then, A is normal if and only if there exists a unitary matrix $S \in \mathbb{F}^{n \times n}$ such that $A^* = AS$.

Proof: See [1125, pp. 102, 113].

Fact 5.15.2. Let $A \in \mathbb{C}^{n \times n}$. Then, there exists a nonsingular matrix $S \in \mathbb{C}^{n \times n}$ such that SAS^{-1} is symmetric.

Proof: See [728, p. 209].

Remark: The symmetric matrix is a *complex symmetric Jordan form*.

Remark: See Corollary 5.3.8.

Remark: The coefficient of the last matrix in [728, p. 209] should be $\jmath/2$.

Fact 5.15.3. Let $A \in \mathbb{C}^{n \times n}$, and assume that A^2 is normal. Then, the following statements hold:

   i) There exists a unitary matrix $S \in \mathbb{C}^{n \times n}$ such that SAS^{-1} is symmetric.

   ii) There exists a symmetric unitary matrix $S \in \mathbb{C}^{n \times n}$ such that $A^T = SAS^{-1}$.

Proof: See [1409].

Fact 5.15.4. Let $A \in \mathbb{F}^{n \times n}$, and assume that A is nonsingular. Then, A^{-1} and A^* are similar if and only if there exists a nonsingular matrix $B \in \mathbb{F}^{n \times n}$ such that $A = B^{-1}B^*$. Furthermore, A is unitary if and only if there exists a normal, nonsingular matrix $B \in \mathbb{F}^{n \times n}$ such that $A = B^{-1}B^*$.

Proof: See [407]. Sufficiency in the second statement follows from Fact 3.11.5.

Fact 5.15.5. Let $A \in \mathbb{F}^{n \times m}$, and assume that $\operatorname{rank} A = m$. Then, there exist a unique matrix $B \in \mathbb{F}^{n \times m}$ and a matrix $C \in \mathbb{F}^{m \times m}$ such that $B^*B = I_m$, C is upper triangular with positive diagonal entries, and $A = BC$.

Proof: See [728, p. 15] or [1157, p. 206].

Remark: $C \in \mathrm{UT}_+(n)$. See Fact 3.23.12.

Remark: This factorization is a consequence of *Gram-Schmidt orthonormalization*.

Fact 5.15.6. Let $A \in \mathbb{F}^{m \times m}$ and $B \in \mathbb{F}^{n \times n}$. Then, there exist matrices $C \in \mathbb{F}^{m \times n}$ and $D \in \mathbb{F}^{n \times m}$ such that $A = CD$ and $B = DC$ if and only if both of the following statements hold:

i) The Jordan blocks associated with nonzero eigenvalues are identical in A and B.

ii) Let $n_1 \geq n_2 \geq \cdots \geq n_r$ denote the orders of the Jordan blocks of A associated with $0 \in \mathrm{spec}(A)$, and let $m_1 \geq m_2 \geq \cdots \geq m_r$ denote the orders of the Jordan blocks of B associated with $0 \in \mathrm{spec}(B)$, where $n_i = 0$ or $m_i = 0$ as needed. Then, $|n_i - m_i| \leq 1$ for all $i \in \{1, \ldots, r\}$.

Proof: See [793].

Remark: See Fact 5.15.7.

Fact 5.15.7. Let $A, B \in \mathbb{F}^{n \times n}$, and assume that A and B are nonsingular. Then, A and B are similar if and only if there exist nonsingular matrices $C, D \in \mathbb{F}^{n \times n}$ such that $A = CD$ and $B = DC$.

Proof: Sufficiency follows from Fact 5.10.11. Necessity is a special case of Fact 5.15.6.

Fact 5.15.8. Let $A, B \in \mathbb{F}^{n \times n}$, and assume that A and B are nonsingular. Then, $\det A = \det B$ if and only if there exist nonsingular matrices $C, D, E \in \mathbb{R}^{n \times n}$ such that $A = CDE$ and $B = EDC$.

Remark: This result is due to Shoda and Taussky-Todd. See [262].

Fact 5.15.9. Let $A \in \mathbb{F}^{n \times n}$. Then, there exist matrices $B, C \in \mathbb{F}^{n \times n}$ such that B is unitary, C is upper triangular, and $A = BC$. If, in addition, A is nonsingular, then there exist unique matrices $B, C \in \mathbb{F}^{n \times n}$ such that B is unitary, C is upper triangular with positive diagonal entries, and $A = BC$.

Proof: See [728, p. 112] or [1157, p. 362].

Remark: This result is the *QR decomposition*. The orthogonal matrix B is constructed as a product of elementary reflectors.

Fact 5.15.10. Let $A \in \mathbb{F}^{n \times n}$, let $r \triangleq \mathrm{rank}\, A$, and assume that the first r leading principal subdeterminants of A are nonzero. Then, there exist matrices $B, C \in \mathbb{F}^{n \times n}$ such that B is lower triangular, C is upper triangular, and $A = BC$. Either B or C can be chosen to be nonsingular. Furthermore, both B and C are nonsingular if and only if A is nonsingular.

Proof: See [728, p. 160].

Remark: This result is the *LU decomposition*.

Remark: All LU factorizations of a singular matrix are characterized in [434].

Fact 5.15.11. Let $\theta \in (-\pi, \pi)$. Then,

$$\begin{bmatrix} \cos\theta & -\sin\theta \\ \sin\theta & \cos\theta \end{bmatrix} = \begin{bmatrix} 1 & -\tan(\theta/2) \\ 0 & 1 \end{bmatrix} \begin{bmatrix} 1 & 0 \\ \sin\theta & 1 \end{bmatrix} \begin{bmatrix} 1 & -\tan(\theta/2) \\ 0 & 1 \end{bmatrix}.$$

Remark: This result is a *ULU factorization* involving three *shear factors*. The matrix $-I_2$ requires four shear factors. In general, all shear factors may be different. See [1271, 1343].

Fact 5.15.12. Let $A \in \mathbb{F}^{n \times n}$. Then, A is nonsingular if and only if A is the product of elementary matrices.

Problem: How many factors are needed?

Fact 5.15.13. Let $A \in \mathbb{F}^{n \times n}$, assume that A is a projector, and let $r \triangleq \operatorname{rank} A$. Then, there exist nonzero vectors $x_1, \ldots, x_{n-r} \in \mathbb{F}^n$ such that $x_i^* x_j = 0$ for all $i \neq j$ and such that

$$A = \prod_{i=1}^{n-r} \left[I - (x_i^* x_i)^{-1} x_i x_i^* \right].$$

Proof: A is unitarily similar to $\operatorname{diag}(1, \ldots, 1, 0, \ldots, 0)$, which can be written as the product of elementary projectors.

Remark: Every projector is the product of mutually orthogonal elementary projectors.

Fact 5.15.14. Let $A \in \mathbb{F}^{n \times n}$. Then, A is a reflector if and only if there exist a positive integer $m \leq n$ and nonzero vectors $x_1, \ldots, x_m \in \mathbb{F}^n$ such that $x_i^* x_j = 0$ for all $i \neq j$ and such that

$$A = \prod_{i=1}^{m} \left[I - 2(x_i^* x_i)^{-1} x_i x_i^* \right].$$

In this case, m is the algebraic multiplicity of $-1 \in \operatorname{spec}(A)$.

Proof: A is unitarily similar to $\operatorname{diag}(\pm 1, \ldots, \pm 1)$, which can be written as the product of elementary reflectors.

Remark: Every reflector is the product of mutually orthogonal elementary reflectors.

Fact 5.15.15. Let $A \in \mathbb{R}^{n \times n}$. Then, A is orthogonal if and only if there exist a positive integer m and nonzero vectors $x_1, \ldots, x_m \in \mathbb{R}^n$ such that $\det A = (-1)^m$ and

$$A = \prod_{i=1}^{m} \left[I - 2(x_i^{\mathrm{T}} x_i)^{-1} x_i x_i^{\mathrm{T}} \right].$$

Remark: Every orthogonal matrix is the product of elementary reflectors. This factorization is a result of Cartan and Dieudonné. See [106, p. 24] and [1198, 1387]. The minimal number of factors is unsettled. See Fact 3.9.5 and Fact 3.14.4. The complex case is open.

Fact 5.15.16. Let $A \in \mathbb{R}^{n \times n}$, where $n \geq 2$. Then, A is orthogonal and $\det A = 1$ if and only if there exist a positive integer m such that $1 \leq m \leq n(n-1)/2$, $\theta_1, \ldots, \theta_m \in \mathbb{R}$, and $j_1, \ldots, j_m, k_1, \ldots, k_m \in \{1, \ldots, n\}$ such that

$$A = \prod_{i=1}^{m} P(\theta_i, j_i, k_i),$$

where

$$P(\theta, j, k) \triangleq I_n + [(\cos\theta) - 1](E_{j,j} + E_{k,k}) + (\sin\theta)(E_{j,k} - E_{k,j}).$$

Proof: See [484].

Remark: $P(\theta, j, k)$ is a *plane* or *Givens rotation*. See Fact 3.9.5.

Remark: Suppose that $\det A = -1$, and let $B \in \mathbb{R}^{n \times n}$ be an elementary reflector. Then, $AB \in \mathrm{SO}(n)$. Therefore, the factorization given above holds with an additional elementary reflector.

Remark: See [912].

Problem: Generalize this result to $\mathbb{C}^{n \times n}$.

Fact 5.15.17. Let $A \in \mathbb{F}^{n \times n}$. Then, $A^{2*}A = A^*A^2$ if and only if there exist a projector $B \in \mathbb{F}^{n \times n}$ and a Hermitian matrix $C \in \mathbb{F}^{n \times n}$ such that $A = BC$.

Proof: See [1141].

Fact 5.15.18. Let $A \in \mathbb{R}^{n \times n}$. Then, $|\det A| = 1$ if and only if A is the product of $n + 2$ or fewer involutory matrices that have exactly one negative eigenvalue. In addition, the following statements hold:

  i) If $n = 2$, then 3 or fewer factors are needed.

  ii) If $A \neq \alpha I$ for all $\alpha \in \mathbb{R}$ and $\det A = (-1)^n$, then n or fewer factors are needed.

  iii) If $\det A = (-1)^{n+1}$, then $n + 1$ or fewer factors are needed.

Proof: See [306, 1139].

Remark: The minimal number of factors for a unitary matrix A is given in [427].

Fact 5.15.19. Let $A \in \mathbb{C}^{n \times n}$, and define $r_0 \triangleq n$ and $r_k \triangleq \mathrm{rank}\, A^k$ for all $k \in \mathbb{P}$. Then, there exists a matrix $B \in \mathbb{C}^{n \times n}$ such that $A = B^2$ if and only if the sequence $(r_k - r_{k+1})_{k=0}^{\infty}$ does not contain two components that are the same odd integer and, if $r_0 - r_1$ is odd, then $r_0 + r_2 \geq 1 + 2r_1$. Now, assume in addition that $A \in \mathbb{R}^{n \times n}$. Then, there exists $B \in \mathbb{R}^{n \times n}$ such that $A = B^2$ if and only if the above condition holds and, for every negative eigenvalue λ of A and for every positive integer k, the Jordan form of A has an even number of $k \times k$ blocks associated with λ.

Proof: See [730, p. 472].

Remark: See Fact 11.18.36.

Remark: For all $l \geq 2$, $A \triangleq N_l$ does not have a square root.

Remark: Uniqueness is discussed in [791]. Square roots of A that are functions of A are defined in [696].

Remark: The principal square root is considered in Theorem 10.6.1.

Remark: mth roots are considered in [337, 701, 1128, 1294].

Fact 5.15.20. Let $A \in \mathbb{C}^{n \times n}$, and assume that A is group invertible. Then, there exists $B \in \mathbb{C}^{n \times n}$ such that $A = B^2$.

Fact 5.15.21. Let $A \in \mathbb{F}^{n \times n}$, and assume that A is nonsingular and has no negative eigenvalues. Furthermore, define $(P_k)_{k=0}^{\infty} \subset \mathbb{F}^{n \times n}$ and $(Q_k)_{k=0}^{\infty} \subset \mathbb{F}^{n \times n}$ by

$$P_0 \triangleq A, \qquad Q_0 \triangleq I,$$

and, for all $k \geq 1$,

$$P_{k+1} \triangleq \tfrac{1}{2}(P_k + Q_k^{-1}),$$

$$Q_{k+1} \triangleq \tfrac{1}{2}(Q_k + P_k^{-1}).$$

Then,

$$B \triangleq \lim_{k \to \infty} P_k$$

exists, satisfies $B^2 = A$, and is the unique square root of A satisfying $\operatorname{spec}(B) \subset$ ORHP. Furthermore,

$$\lim_{k \to \infty} Q_k = A^{-1}.$$

Proof: See [406, 695].

Remark: All indicated inverses exist.

Remark: This sequence is related to Newton's iteration for the matrix sign function. See Fact 10.10.2.

Remark: See Fact 8.9.33.

Fact 5.15.22. Let $A \in \mathbb{F}^{n \times n}$, assume that A is positive semidefinite, and let $r \triangleq \operatorname{rank} A$. Then, there exists $B \in \mathbb{F}^{n \times r}$ such that $A = BB^*$.

Fact 5.15.23. Let $A \in \mathbb{F}^{n \times n}$, and let $k \geq 1$. Then, there exists a unique matrix $B \in \mathbb{F}^{n \times n}$ such that

$$A = B(B^*B)^k.$$

Proof: See [1118].

Fact 5.15.24. Let $A \in \mathbb{F}^{n \times n}$. Then, there exist symmetric matrices $B, C \in \mathbb{F}^{n \times n}$, at least one of which is nonsingular, such that $A = BC$.

Proof: See [1125, p. 82].

Remark: Note that

$$\begin{bmatrix} \beta_1 & \beta_2 & 1 \\ \beta_2 & 1 & 0 \\ 1 & 0 & 0 \end{bmatrix} \begin{bmatrix} 0 & 1 & 0 \\ 0 & 0 & 1 \\ -\beta_0 & -\beta_1 & -\beta_2 \end{bmatrix} = \begin{bmatrix} -\beta_0 & 0 & 0 \\ 0 & \beta_2 & 1 \\ 0 & 1 & 0 \end{bmatrix}$$

and use Theorem 5.2.3.

Remark: This result is due to Frobenius. The equality is a *Bezout matrix factorization*; see Fact 4.8.6. See [244, 245, 643].

Remark: B and C are symmetric for $\mathbb{F} = \mathbb{C}$.

Fact 5.15.25. Let $A \in \mathbb{C}^{n \times n}$. Then, $\det A$ is real if and only if A is the product of four Hermitian matrices. Furthermore, four is the smallest number for which the previous statement is true.

Proof: See [1494].

Fact 5.15.26. Let $A \in \mathbb{R}^{n \times n}$. Then, the following statements hold:

i) A is the product of two positive-semidefinite matrices if and only if A is similar to a positive-semidefinite matrix.

ii) If A is nilpotent, then A is the product of three positive-semidefinite matrices.

iii) If A is singular, then A is the product of four positive-semidefinite matrices.

iv) $\det A > 0$ and $A \neq \alpha I$ for all $\alpha \leq 0$ if and only if A is the product of four positive-definite matrices.

v) $\det A > 0$ if and only if A is the product of five positive-definite matrices.

Proof: [121, 643, 1493, 1494].

Remark: See [1494] for factorizations of complex matrices and operators.

Example:

$$\begin{bmatrix} -1 & 0 \\ 0 & -1 \end{bmatrix} = \begin{bmatrix} 2 & 0 \\ 0 & 1/2 \end{bmatrix} \begin{bmatrix} 5 & 7 \\ 7 & 10 \end{bmatrix} \begin{bmatrix} 13/2 & -5 \\ -5 & 4 \end{bmatrix} \begin{bmatrix} 8 & 5 \\ 5 & 13/4 \end{bmatrix} \begin{bmatrix} 25/8 & -11/2 \\ -11/2 & 10 \end{bmatrix}.$$

Fact 5.15.27. Let $A \in \mathbb{R}^{n \times n}$. Then, the following statements hold:

i) $A = BC$, where $B \in \mathbb{R}^{n \times n}$ is symmetric and $C \in \mathbb{R}^{n \times n}$ is positive semidefinite, if and only if A^2 is diagonalizable over $\mathbb{R}$ and $\mathrm{spec}(A) \subset [0, \infty)$.

ii) $A = BC$, where $B \in \mathbb{R}^{n \times n}$ is symmetric and $C \in \mathbb{R}^{n \times n}$ is positive definite, if and only if A is diagonalizable over $\mathbb{R}$.

iii) $A = BC$, where $B, C \in \mathbb{R}^{n \times n}$ are positive semidefinite, if and only if $A = DE$, where $D \in \mathbb{R}^{n \times n}$ is positive semidefinite and $E \in \mathbb{R}^{n \times n}$ is positive definite.

iv) $A = BC$, where $B \in \mathbb{R}^{n \times n}$ is positive semidefinite and $C \in \mathbb{R}^{n \times n}$ is positive definite, if and only if A is diagonalizable over $\mathbb{R}$ and $\mathrm{spec}(A) \subset [0, \infty)$.

v) $A = BC$, where $B, C \in \mathbb{R}^{n \times n}$ are positive definite, if and only if A is diagonalizable over $\mathbb{R}$ and $\mathrm{spec}(A) \subset (0, \infty)$.

Proof: See [724, 1488, 1493].

Fact 5.15.28. Let $A \in \mathbb{F}^{n \times n}$. Then, A is either singular or the identity matrix if and only if A is the product of n or fewer idempotent matrices in $\mathbb{F}^{n \times n}$, each of whose rank is equal to rank A. Furthermore, $\operatorname{rank}(A - I) \leq k \operatorname{def} A$, where $k \geq 1$, if and only if A is the product of k idempotent matrices.

Proof: See [74, 129, 386, 473].

Example:

$$\begin{bmatrix} 0 & 1 \\ 0 & 0 \end{bmatrix} = \begin{bmatrix} 1 & 1/2 \\ 0 & 0 \end{bmatrix} \begin{bmatrix} 0 & 1/2 \\ 0 & 1 \end{bmatrix}$$

and

$$\begin{bmatrix} 2 & 0 \\ 0 & 0 \end{bmatrix} = \begin{bmatrix} 1 & 1 \\ 0 & 0 \end{bmatrix} \begin{bmatrix} 1 & 0 \\ 1 & 0 \end{bmatrix}.$$

Fact 5.15.29. Let $A \in \mathbb{R}^{n \times n}$, assume that A is singular, and assume that A is not a 2×2 nilpotent matrix. Then, there exist nilpotent matrices $B, C \in \mathbb{R}^{n \times n}$ such that $A = BC$ and rank $A = \operatorname{rank} B = \operatorname{rank} C$.

Proof: See [1246, 1492]. See also [1279].

Fact 5.15.30. Let $A \in \mathbb{F}^{n \times n}$, and assume that A is idempotent. Then, there exist $B, C \in \mathbb{F}^{n \times n}$ such that B is positive definite, C is positive semidefinite, and $A = BC$.

Proof: See [1356].

Fact 5.15.31. Let $A \in \mathbb{R}^{n \times n}$, and assume that A is nonsingular. Then, A is similar to A^{-1} if and only if A is the product of two involutory matrices. If, in addition, A is orthogonal, then A is the product of two reflectors.

Proof: See [127, 424, 1486, 1487] or [1125, p. 108].

Problem: Construct these reflectors for $A = \begin{bmatrix} \cos\theta & \sin\theta \\ -\sin\theta & \cos\theta \end{bmatrix}$.

Fact 5.15.32. Let $A \in \mathbb{R}^{n \times n}$. Then, $|\det A| = 1$ if and only if A is the product of four or fewer involutory matrices.

Proof: [128, 626, 1245].

Fact 5.15.33. Let $A \in \mathbb{R}^{n \times n}$, where $n \geq 2$. Then, A is the product of two commutators.

Proof: See [1494].

Fact 5.15.34. Let $A \in \mathbb{R}^{n \times n}$, and assume that $\det A = 1$. Then, there exist nonsingular matrices $B, C \in \mathbb{R}^{n \times n}$ such that $A = BCB^{-1}C^{-1}$.

Proof: See [1222].

Remark: The product is a *multiplicative commutator*. This result is due to Shoda.

Remark: For nonsingular matrices A, B, note that $[A, B] = 0$ if and only if $ABA^{-1}B = I$.

Remark: See Fact 5.15.35.

Fact 5.15.35. Let $A \in \mathbb{R}^{n \times n}$, assume that A is orthogonal, and assume that $\det A = 1$. Then, there exist reflectors $B, C \in \mathbb{R}^{n \times n}$ such that $A = BCB^{-1}C^{-1}$.

Proof: See [1299].

Remark: See Fact 5.15.34.

Fact 5.15.36. Let $A \in \mathbb{F}^{n \times n}$, and assume that A is nonsingular. Then, there exist an involutory matrix $B \in \mathbb{F}^{n \times n}$ and a symmetric matrix $C \in \mathbb{F}^{n \times n}$ such that $A = BC$.

Proof: See [591].

Fact 5.15.37. Let $A \in \mathbb{F}^{n \times n}$, and assume that n is even. Then, the following statements are equivalent:

i) A is the product of two skew-symmetric matrices.

ii) Every elementary divisor of A has even algebraic multiplicity.

iii) There exists a matrix $B \in \mathbb{F}^{n/2 \times n/2}$ such that A is similar to $\begin{bmatrix} B & 0 \\ 0 & B \end{bmatrix}$.

Remark: In *i)* the factors are skew symmetric even when A is complex.

Proof: See [592, 1494].

Fact 5.15.38. Let $A \in \mathbb{C}^{n \times n}$, and assume that $n \geq 4$ and n is even. Then, A is the product of five skew-symmetric matrices in $\mathbb{C}^{n \times n}$.

Proof: See [882, 883].

Fact 5.15.39. Let $A \in \mathbb{F}^{n \times n}$. Then, there exist a symmetric matrix $B \in \mathbb{F}^{n \times n}$ and a skew-symmetric matrix $C \in \mathbb{F}^{n \times n}$ such that $A = BC$ if and only if A is similar to $-A$.

Proof: See [1163].

Fact 5.15.40. Let $A \in \mathbb{F}^{n \times m}$, and let $r \triangleq \operatorname{rank} A$. Then, there exist matrices $B \in \mathbb{F}^{n \times r}$ and $C \in \mathbb{R}^{r \times m}$ such that $A = BC$ and $\operatorname{rank} B = \operatorname{rank} C = r$.

Fact 5.15.41. Let $A \in \mathbb{F}^{n \times n}$. Then, A is diagonalizable over $\mathbb{F}$ with (nonnegative, positive) eigenvalues if and only if there exist (positive-semidefinite, positive-definite) matrices $B, C \in \mathbb{F}^{n \times n}$ such that $A = BC$.

Proof: To prove sufficiency, use Theorem 8.3.6 and note that

$$A = S^{-1}(SBS^*)(S^{-*}CS^{-1})S.$$

5.16 Facts on Companion, Vandermonde, Circulant, and Hadamard Matrices

Fact 5.16.1. Let $p \in \mathbb{F}[s]$, where $p(s) = s^n + \beta_{n-1}s^{n-1} + \cdots + \beta_0$, and define $C_b(p), C_r(p), C_t(p), C_l(p) \in \mathbb{F}^{n \times n}$ by

$$
C_b(p) \triangleq \begin{bmatrix}
0 & 1 & 0 & \cdots & 0 & 0 \\
0 & 0 & 1 & \ddots & 0 & 0 \\
0 & 0 & 0 & \ddots & 0 & 0 \\
\vdots & \vdots & \vdots & \ddots & \ddots & \vdots \\
0 & 0 & 0 & \cdots & 0 & 1 \\
-\beta_0 & -\beta_1 & -\beta_2 & \cdots & -\beta_{n-2} & -\beta_{n-1}
\end{bmatrix},
$$

$$
C_r(p) \triangleq \begin{bmatrix}
0 & 0 & 0 & \cdots & 0 & -\beta_0 \\
1 & 0 & 0 & \cdots & 0 & -\beta_1 \\
0 & 1 & 0 & \cdots & 0 & -\beta_2 \\
\vdots & \ddots & \ddots & \ddots & \vdots & \vdots \\
0 & 0 & 0 & \ddots & 0 & -\beta_{n-2} \\
0 & 0 & 0 & \cdots & 1 & -\beta_{n-1}
\end{bmatrix},
$$

$$
C_t(p) \triangleq \begin{bmatrix}
-\beta_{n-1} & -\beta_{n-2} & \cdots & -\beta_2 & -\beta_1 & -\beta_0 \\
1 & 0 & \cdots & 0 & 0 & 0 \\
\vdots & \ddots & \ddots & \vdots & \vdots & \vdots \\
0 & 0 & \ddots & 0 & 0 & 0 \\
0 & 0 & \ddots & 1 & 0 & 0 \\
0 & 0 & \cdots & 0 & 1 & 0
\end{bmatrix},
$$

$$
C_l(p) \triangleq \begin{bmatrix}
-\beta_{n-1} & 1 & \cdots & 0 & 0 & 0 \\
-\beta_{n-2} & 0 & \ddots & 0 & 0 & 0 \\
\vdots & \vdots & \ddots & \ddots & \ddots & \vdots \\
-\beta_2 & 0 & \cdots & 0 & 1 & 0 \\
-\beta_1 & 0 & \cdots & 0 & 0 & 1 \\
-\beta_0 & 0 & \cdots & 0 & 0 & 0
\end{bmatrix}.
$$

Then,

$$
C_r(p) = C_b^{\mathrm{T}}(p), \quad C_l(p) = C_t^{\mathrm{T}}(p),
$$

$$
C_t(p) = \hat{I} C_b(p) \hat{I}, \quad C_l(p) = \hat{I} C_r(p) \hat{I},
$$

$$C_{\mathrm{l}}(p) = C_{\mathrm{b}}^{\hat{\mathrm{T}}}(p), \quad C_{\mathrm{t}}(p) = C_{\mathrm{r}}^{\hat{\mathrm{T}}}(p),$$

and

$$\chi_{C_{\mathrm{b}}(p)} = \chi_{C_{\mathrm{r}}(p)} = \chi_{C_{\mathrm{t}}(p)} = \chi_{C_{\mathrm{l}}(p)} = p.$$

Furthermore,

$$C_{\mathrm{r}}(p) = SC_{\mathrm{b}}(p)S^{-1}$$

and

$$C_{\mathrm{l}}(p) = \hat{S}C_{\mathrm{t}}(p)\hat{S}^{-1},$$

where $S, \hat{S} \in \mathbb{F}^{n \times n}$ are the Hankel matrices

$$S \triangleq \begin{bmatrix} \beta_1 & \beta_2 & \cdots & \beta_{n-1} & 1 \\ \beta_2 & \beta_3 & \cdot^{\cdot^{\cdot}} & 1 & 0 \\ \vdots & \cdot^{\cdot^{\cdot}} & \cdot^{\cdot^{\cdot}} & \cdot^{\cdot^{\cdot}} & \vdots \\ \beta_{n-1} & 1 & \cdot^{\cdot^{\cdot}} & 0 & 0 \\ 1 & 0 & \cdots & 0 & 0 \end{bmatrix}$$

and

$$\hat{S} \triangleq \hat{I}S\hat{I} = \begin{bmatrix} 0 & 0 & \cdots & 0 & 1 \\ 0 & 0 & \cdot^{\cdot^{\cdot}} & 1 & \beta_{n-1} \\ \vdots & \cdot^{\cdot^{\cdot}} & \cdot^{\cdot^{\cdot}} & \cdot^{\cdot^{\cdot}} & \vdots \\ 0 & 1 & \cdot^{\cdot^{\cdot}} & \beta_3 & \beta_2 \\ 1 & \beta_{n-1} & \cdots & \beta_2 & \beta_1 \end{bmatrix}.$$

Remark: $(C_{\mathrm{b}}(p), C_{\mathrm{r}}(p), C_{\mathrm{t}}(p), C_{\mathrm{l}}(p))$ are the (*bottom, right, top, left*) companion matrices. Note that $C_{\mathrm{b}}(p) = C(p)$. See [148, p. 282] and [809, p. 659].

Remark: $S = B(p, 1)$, where $B(p, 1)$ is a Bezout matrix. See Fact 4.8.6.

Fact 5.16.2. Let $p \in \mathbb{F}[s]$, where $p(s) = s^n + \beta_{n-1}s^{n-1} + \cdots + \beta_0$, assume that $\beta_0 \neq 0$, and let

$$C_{\mathrm{b}}(p) \triangleq \begin{bmatrix} 0 & 1 & 0 & \cdots & 0 & 0 \\ 0 & 0 & 1 & \ddots & 0 & 0 \\ 0 & 0 & 0 & \ddots & 0 & 0 \\ \vdots & \vdots & \vdots & \ddots & \ddots & \vdots \\ 0 & 0 & 0 & \cdots & 0 & 1 \\ -\beta_0 & -\beta_1 & -\beta_2 & \cdots & -\beta_{n-2} & -\beta_{n-1} \end{bmatrix}.$$

Then,

$$C_{\mathrm{b}}^{-1}(p) = C_{\mathrm{t}}(\hat{p}) = \begin{bmatrix} -\beta_1/\beta_0 & \cdots & -\beta_{n-2}/\beta_0 & -\beta_{n-1}/\beta_0 & -1/\beta_0 \\ 1 & \cdots & 0 & 0 & 0 \\ \vdots & \ddots & \vdots & \vdots & \vdots \\ 0 & \cdots & 1 & 0 & 0 \\ 0 & \cdots & 0 & 1 & 0 \end{bmatrix},$$

where $\hat{p}(s) \triangleq \beta_0^{-1} s^n p(1/s)$.

Remark: See Fact 4.9.9.

Fact 5.16.3. Let $\lambda_1, \ldots, \lambda_n \in \mathbb{F}$, and define the *Vandermonde matrix* $V(\lambda_1, \ldots, \lambda_n) \in \mathbb{F}^{n \times n}$ by

$$V(\lambda_1, \ldots, \lambda_n) \triangleq \begin{bmatrix} 1 & 1 & \cdots & 1 \\ \lambda_1 & \lambda_2 & \cdots & \lambda_n \\ \lambda_1^2 & \lambda_2^2 & \cdots & \lambda_n^2 \\ \lambda_1^3 & \lambda_2^3 & \cdots & \lambda_n^3 \\ \vdots & \vdots & \ddots & \vdots \\ \lambda_1^{n-1} & \lambda_2^{n-1} & \cdots & \lambda_n^{n-1} \end{bmatrix}.$$

Then,

$$\det V(\lambda_1, \ldots, \lambda_n) = \prod_{1 \leq i < j \leq n} (\lambda_i - \lambda_j).$$

Thus, $V(\lambda_1, \ldots, \lambda_n)$ is nonsingular if and only if $\lambda_1, \ldots, \lambda_n$ are distinct.

Remark: This result yields Proposition 4.5.4. Let $x_1, \ldots, x_k$ be eigenvectors of $V(\lambda_1, \ldots, \lambda_n)$ associated with distinct eigenvalues $\lambda_1, \ldots, \lambda_k$ of $V(\lambda_1, \ldots, \lambda_n)$. Suppose that $\alpha_1 x_1 + \cdots + \alpha_k x_k = 0$ so that $V^i(\lambda_1, \ldots, \lambda_n)(\alpha_1 x_1 + \cdots + \alpha_k x_k) = \alpha_1 \lambda_1^i x_i + \cdots + \alpha_k \lambda_k^i x_k = 0$ for all $i \in \{0, 1, \ldots, k-1\}$. Let $X \triangleq \begin{bmatrix} x_1 & \cdots & x_k \end{bmatrix} \in \mathbb{F}^{n \times k}$ and $D \triangleq \mathrm{diag}(\alpha_1, \ldots, \alpha_k)$. Then, $XDV^{\mathrm{T}}(\lambda_1, \ldots, \lambda_k) = 0$, which implies that $XD = 0$. Hence, $\alpha_i x_i = 0$ for all $i \in \{1, \ldots, k\}$, and thus $\alpha_1 = \cdots = \alpha_k = 0$.

Remark: Connections between the Vandermonde matrix and the Pascal matrix, *Stirling matrix*, *Bernoulli matrix*, *Bernstein matrix*, and companion matrices are discussed in [5]. See also Fact 11.11.4.

Fact 5.16.4. Let $p \in \mathbb{F}[s]$, where $p(s) = s^n + \beta_{n-1} s^{n-1} + \cdots + \beta_1 s + \beta_0$, and assume that p has distinct roots $\lambda_1, \ldots, \lambda_n \in \mathbb{C}$. Then,

$$C(p) = V(\lambda_1, \ldots, \lambda_n) \mathrm{diag}(\lambda_1, \ldots, \lambda_n) V^{-1}(\lambda_1, \ldots, \lambda_n).$$

Consequently, for all $i \in \{1, \ldots, n\}$, λ_i is an eigenvalue of $C(p)$ with associated eigenvector $\mathrm{col}_i(V)$. Finally,

$$(VV^{\mathrm{T}})^{-1} C V V^{\mathrm{T}} = C^{\mathrm{T}}.$$

Proof: See [143].

Remark: Case in which $C(p)$ has repeated eigenvalues is considered in [143].

Fact 5.16.5. Let $A \in \mathbb{F}^{n \times n}$. Then, A is cyclic if and only if A is similar to a companion matrix.

Proof: This result follows from Corollary 5.3.4. Alternatively, let $\mathrm{spec}(A) = \{\lambda_1, \ldots, \lambda_r\}$ and $A = SBS^{-1}$, where $S \in \mathbb{C}^{n \times n}$ is nonsingular and $B = \mathrm{diag}(B_1, \ldots, B_r)$ is the Jordan form of A, where, for all $i \in \{1, \ldots, r\}$, $B_i \in \mathbb{C}^{n_i \times n_i}$ and $\lambda_i, \ldots, \lambda_i$ are the diagonal entries of B_i. Now, define $R \in \mathbb{C}^{n \times n}$ by $R \triangleq \begin{bmatrix} R_1 & \cdots & R_r \end{bmatrix} \in \mathbb{C}^{n \times n}$, where, for all $i \in \{1, \ldots, r\}$, $R_i \in \mathbb{C}^{n \times n_i}$ is the matrix

$$R_i \triangleq \begin{bmatrix} 1 & 0 & \cdots & 0 \\ \lambda_i & 1 & \cdots & 0 \\ \vdots & \vdots & \ddots & \vdots \\ \lambda_i^{n-2} & \binom{n-2}{1}\lambda_i^{n-3} & \cdots & \binom{n-2}{n_i-1}\lambda_i^{n-n_i-1} \\ \lambda_i^{n-1} & \binom{n-1}{1}\lambda_i^{n-2} & \cdots & \binom{n-1}{n_i-1}\lambda_i^{n-n_i} \end{bmatrix}.$$

Then, since $\lambda_1, \ldots, \lambda_r$ are distinct, it follows that R is nonsingular. Furthermore, $C = RBR^{-1}$ is in companion form, and thus $A = SR^{-1}CRS$. If $n_i = 1$ for all $i \in \{1, \ldots, r\}$, then R is a Vandermonde matrix. See Fact 5.16.3 and Fact 5.16.4.

Fact 5.16.6. Let $\lambda_1, \ldots, \lambda_n \in \mathbb{F}$ and, for $i = 1, \ldots, n$, define

$$p_i(s) \triangleq \prod_{\substack{j=1 \\ j \neq i}}^{n} (s - \lambda_j).$$

Furthermore, define $A \in \mathbb{F}^{n \times n}$ by

$$A \triangleq \begin{bmatrix} p_1(0) & \frac{1}{1!}p_1'(0) & \cdots & \frac{1}{(n-1)!}p_1^{(n-1)}(0) \\ \vdots & \ddots & \ddots & \vdots \\ p_n(0) & \frac{1}{1!}p_n'(0) & \cdots & \frac{1}{(n-1)!}p_n^{(n-1)}(0) \end{bmatrix}.$$

Then,
$$\mathrm{diag}[p_1(\lambda_1), \ldots, p_n(\lambda_n)] = AV(\lambda_1, \ldots, \lambda_n).$$

Proof: See [494, p. 159].

Remark: p' is the derivative of p.

Fact 5.16.7. Let $n \geq 1$, let $a_0, \ldots, a_{n-1} \in \mathbb{F}$, and define $\mathrm{circ}(a_0, \ldots, a_{n-1}) \in \mathbb{F}^{n \times n}$ by

$$\mathrm{circ}(a_0, \ldots, a_{n-1}) \triangleq \begin{bmatrix} a_0 & a_1 & a_2 & \cdots & a_{n-2} & a_{n-1} \\ a_{n-1} & a_0 & a_1 & \cdots & a_{n-3} & a_{n-2} \\ a_{n-2} & a_{n-1} & a_0 & \ddots & a_{n-4} & a_{n-3} \\ \vdots & \vdots & \ddots & \ddots & \ddots & \vdots \\ a_2 & a_3 & a_4 & \ddots & a_0 & a_1 \\ a_1 & a_2 & a_3 & \cdots & a_{n-1} & a_0 \end{bmatrix}.$$

A matrix of this form is *circulant*. Furthermore, for $n \geq 2$, the $n \times n$ *primary circulant matrix* $\mathrm{circ}(0, 1, 0, \ldots, 0)$ is the cyclic permutation matrix P_n, that is,

$$\mathrm{circ}(0, 1, 0, \ldots, 0) = P_n = \begin{bmatrix} 0 & 1 & 0 & \cdots & 0 & 0 \\ 0 & 0 & 1 & \ddots & 0 & 0 \\ 0 & 0 & 0 & \ddots & 0 & 0 \\ \vdots & \ddots & \ddots & \ddots & \ddots & \vdots \\ 0 & 0 & 0 & \ddots & 0 & 1 \\ 1 & 0 & 0 & \cdots & 0 & 0 \end{bmatrix}.$$

Note that $\mathrm{circ}(1) = P_1 = 1$. Finally, define $p(s) \triangleq a_{n-1}s^{n-1} + \cdots + a_1 s + a_0 \in \mathbb{F}[s]$, and let $\theta \triangleq e^{j2\pi/n}$. Then, the following statements hold:

i) $p(P_n) = \mathrm{circ}(a_0, \ldots, a_{n-1})$.

ii) $P_n = C(q)$, where $q \in \mathbb{F}[s]$ is defined by $q(s) \triangleq s^n - 1$ and $C(q)$ is the companion matrix associated with q.

iii) $\mathrm{spec}(P_n) = \{1, \theta, \theta^2, \ldots, \theta^{n-1}\}$.

iv) $\det P_n = (-1)^{n-1}$.

v) $\mathrm{mspec}[\mathrm{circ}(a_0, \ldots, a_{n-1})] = \{p(1), p(\theta), p(\theta^2), \ldots, p(\theta^{n-1})\}_{\mathrm{ms}}$.

vi) If $A, B \in \mathbb{F}^{n \times n}$ are circulant and $\alpha, \beta \in \mathbb{F}$, then $\alpha A + \beta B$ is circulant.

vii) If $A, B \in \mathbb{F}^{n \times n}$ are circulant, then AB is circulant.

viii) If $A, B \in \mathbb{F}^{n \times n}$ are circulant, then $AB = BA$.

ix) If A is circulant, then $\overline{A}$, A^{T}, and A^* are circulant.

x) If A is circulant and k is a nonnegative integer, then A^k is circulant.

xi) If A is nonsingular and circulant, then A^{-1} is circulant.

xii) $A \in \mathbb{F}^{n \times n}$ is circulant if and only if $A = P_n A P_n^{\mathrm{T}}$.

xiii) P_n is an orthogonal matrix, and $P_n^n = I_n$.

xiv) If $A \in \mathbb{F}^{n \times n}$ is circulant, then A is reverse symmetric, Toeplitz, and normal.

xv) If $A \in \mathbb{F}^{n \times n}$ is circulant and nonzero, then A is irreducible.

xvi) $A \in \mathbb{F}^{n \times n}$ is normal if and only if A is unitarily similar to a circulant matrix.

Next, define the *Fourier matrix* $S \in \mathbb{C}^{n \times n}$ by

$$S \triangleq n^{-1/2} V(1, \theta, \ldots, \theta^{n-1}) = \frac{1}{\sqrt{n}} \begin{bmatrix} 1 & 1 & 1 & \cdots & 1 \\ 1 & \theta & \theta^2 & \cdots & \theta^{n-1} \\ 1 & \theta^2 & \theta^4 & \cdots & \theta^{n-2} \\ \vdots & \vdots & \vdots & \ddots & \vdots \\ 1 & \theta^{n-1} & \theta^{n-2} & \cdots & \theta \end{bmatrix}.$$

Then, the following statements hold:

xvii) S is symmetric and unitary, but not Hermitian.

xvii) $S^4 = I_n$.

xix) $\operatorname{spec}(S) \subseteq \{1, -1, \jmath, -\jmath\}$.

xx) $\operatorname{Re} S$ and $\operatorname{Im} S$ are symmetric, commute, and satisfy
$$(\operatorname{Re} S)^2 + (\operatorname{Im} S)^2 = I_n.$$

xxi) $S^{-1} P_n S = \operatorname{diag}(1, \theta, \ldots, \theta^{n-1})$.

xxii) $S^{-1} \operatorname{circ}(a_0, \ldots, a_{n-1}) S = \operatorname{diag}[p(1), p(\theta), \ldots, p(\theta^{n-1})]$.

Proof: See Fact 2.13.2, [18, pp. 81–98], [385, p. 81], and [1526, pp. 106–110].

Remark: Circulant matrices play a role in digital signal processing, specifically, in the efficient implementation of the *fast Fourier transform*. See [1022, pp. 356–380], [1171], and [1395, pp. 206, 207].

Remark: S is a *Fourier matrix* and a Vandermonde matrix.

Remark: If a real Toeplitz matrix is normal, then it must be either symmetric, skew symmetric, circulant, or skew circulant. See [75, 485].

Remark: A unified treatment of the solutions of quadratic, cubic, and quartic equations using circulant matrices is given in [810].

Remark: The set $\{I, P_k, P_k^2, \ldots, P_k^{k-1}\}$ is the group $C(k)$. See Fact 3.23.5.

Remark: Circulant matrices are generalized by *cycle matrices*, which correspond to visual geometric symmetries. See [562].

Fact 5.16.8. Let $A \in \mathbb{R}^{n \times n}$, and assume that A is a permutation matrix. Then, there exists a permutation matrix $S \in \mathbb{R}^{n \times n}$ such that
$$A = S \operatorname{diag}(P_{n_1}, \ldots, P_{n_r}) S^{-1},$$
and, for all $i \in \{1, \ldots, r\}$, P_{n_i} is the $n_i \times n_i$ cyclic permutation matrix. Furthermore, the cyclic permutation matrices $P_{n_1}, \ldots, P_{n_r}$ are unique up to a relabeling. Consequently,
$$\operatorname{mspec}(A) = \bigcup_{i=1}^{r} \{1, \theta_i, \ldots, \theta_i^{n_i-1}\}_{\mathrm{ms}},$$
where $\theta_i \triangleq e^{\jmath 2\pi/n_i}$. Hence,
$$\det A = (-1)^{n-r}.$$

Finally, the smallest positive integer m such that $A^m = I$ is given by the least common multiple of $n_1, \ldots, n_r$.

Proof: See [385, p. 29]. The last statement follows from [457, pp. 32, 33].

Remark: This result provides a canonical form for permutation matrices under unitary similarity with a permutation matrix.

Remark: It follows that A can be written as the product
$$A = S \begin{bmatrix} P_{n_1} & 0 \\ 0 & I \end{bmatrix} \cdots \begin{bmatrix} I & 0 & 0 \\ 0 & P_{n_i} & 0 \\ 0 & 0 & I \end{bmatrix} \cdots \begin{bmatrix} I & 0 \\ 0 & P_{n_r} \end{bmatrix} S^{-1},$$

where the factors represent disjoint cycles. This factorization reveals the *cycle decomposition* for an element of the permutation group S_n on a set having n elements, where S_n is represented by the group $P(n)$ of $n \times n$ permutation matrices. See [457, pp. 29–32], [1178, p. 18] and Fact 3.23.2.

Remark: The number of possible canonical forms is given by p_n, where p_n is the number of integral partitions of n. For example, $p_1 = 1$, $p_2 = 2$, $p_3 = 3$, $p_4 = 5$, and $p_5 = 7$. For all n, p_n is given by the expansion

$$1 + \sum_{n=1}^{\infty} p_n x^n = \frac{1}{(1-x)(1-x^2)(1-x^3)\cdots}.$$

See [63, p. 50] or [1540, pp. 210, 211].

Fact 5.16.9. Let $A \in \mathbb{R}^{n \times n}$, assume that every entry of A is either 1 or −1, and assume that $AA^T = nI$. A matrix with these properties is a *Hadamard matrix.* Then, the following statements hold:

i) Either $n = 1$, $n = 2$, or n is a multiple of 4.

ii) $n^{-1/2}A$ is orthogonal.

iii) $|\det A| = n^{n/2}$.

iv) $\begin{bmatrix} A & A \\ A & -A \end{bmatrix}$ is a Hadamard matrix.

v) If $B \in \mathbb{R}^{m \times m}$ is a Hadamard matrix, then $A \otimes B$ is a Hadamard matrix.

vi) For every positive integer k, there exists a Hadamard matrix of order 2^k.

Proof: See [726, p. 10] or [1208, pp. 333–335].

Remark: $\begin{bmatrix} 1 & 1 \\ 1 & -1 \end{bmatrix}$ is a Hadamard matrix.

Remark: It is not known whether there exists a Hadamard matrix for every integer n that is divisible by 4. See [726, p. 9].

5.17 Facts on Simultaneous Transformations

Fact 5.17.1. Let $A, B \in \mathbb{F}^{n \times n}$, and assume that there exists a nonsingular matrix $S \in \mathbb{F}^{n \times n}$ such that SAS^{-1} and SBS^{-1} are upper triangular. Then, A and B have a common eigenvector with corresponding eigenvalues $(SAS^{-1})_{(1,1)}$ and $(SAS^{-1})_{(1,1)}$.

Proof: See [561].

Remark: See Fact 5.14.26.

Fact 5.17.2. Let $A, B \in \mathbb{C}^{n \times n}$, and assume that $P_{A,B}$ is regular. Then, there exist unitary matrices $S_1, S_2 \in \mathbb{C}^{n \times n}$ such that $S_1 A S_2$ and $S_1 B S_2$ are upper triangular.

Proof: See [1261, p. 276].

Fact 5.17.3. Let $A, B \in \mathbb{R}^{n \times n}$, and assume that $P_{A,B}$ is regular. Then, there exist orthogonal matrices $S_1, S_2 \in \mathbb{R}^{n \times n}$ such that $S_1 A S_2$ is upper triangular and $S_1 B S_2$ is upper Hessenberg with 2×2 diagonally located blocks.

Proof: See [1261, p. 290].

Remark: This result is due to Moler and Stewart.

Fact 5.17.4. Let $\mathcal{S} \subset \mathbb{F}^{n \times n}$, and assume that $AB = BA$ for all $A, B \in \mathcal{S}$. Then, there exists a unitary matrix $S \in \mathbb{F}^{n \times n}$ such that, for all $A \in \mathcal{S}$, SAS^* is upper triangular.

Proof: See [728, p. 81] and [1140].

Remark: See Fact 5.17.9.

Fact 5.17.5. Let $A, B \in \mathbb{C}^{n \times n}$, and assume that either

$$[A, [A, B]] = [B, [A, B]] = 0$$

or

$$\text{rank } [A, B] \leq 1.$$

Then, there exists a nonsingular matrix $S \in \mathbb{C}^{n \times n}$ such that SAS^{-1} and SBS^{-1} are upper triangular.

Proof: The first result is due to McCoy, and the second result is due to Laffey. See [561, 1140].

Fact 5.17.6. Let $A, B \in \mathbb{C}^{n \times n}$, and assume that A and B are idempotent. Then, there exists a unitary matrix $S \in \mathbb{C}^{n \times n}$ such that SAS^* and SBS^* are upper triangular if and only if $[A, B]$ is nilpotent.

Proof: See [1282].

Remark: Necessity follows from Fact 3.17.11.

Remark: See Fact 5.17.4.

Fact 5.17.7. Let $\mathcal{S} \subset \mathbb{F}^{n \times n}$, and assume that every matrix $A \in \mathcal{S}$ is normal. Then, $AB = BA$ for all $A, B \in \mathcal{S}$ if and only if there exists a unitary matrix $S \in \mathbb{F}^{n \times n}$ such that, for all $A \in \mathcal{S}$, SAS^* is diagonal.

Remark: See Fact 8.17.1 and [728, pp. 103, 172].

Fact 5.17.8. Let $\mathcal{S} \subset \mathbb{F}^{n \times n}$, and assume that every matrix $A \in \mathcal{S}$ is diagonalizable over $\mathbb{F}$. Then, $AB = BA$ for all $A, B \in \mathcal{S}$ if and only if there exists a nonsingular matrix $S \in \mathbb{F}^{n \times n}$ such that, for all $A \in \mathcal{S}$, SAS^{-1} is diagonal.

Proof: See [728, p. 52].

Fact 5.17.9. Let $A, B \in \mathbb{F}^{n \times n}$, and assume that $\{x \in \mathbb{F}^n : x^*Ax = x^*Bx = 0\} = \{0\}$. Then, there exists a nonsingular matrix $S \in \mathbb{F}^{n \times n}$ such that SAS^* and SBS^* are upper triangular.

Proof: See [1125, p. 96].

Remark: A and B need not be Hermitian.

Remark: See Fact 5.17.4 and Fact 8.17.6.

Remark: Simultaneous triangularization by means of a unitary biequivalence transformation is given in Proposition 5.7.3.

5.18 Facts on the Polar Decomposition

Fact 5.18.1. Let $A \in \mathbb{F}^{n \times m}$. Then,

$$(AA^*)^{1/2}A = A(A^*A)^{1/2}.$$

Remark: See Fact 5.18.4.

Remark: The positive-semidefinite square root is defined in (8.5.4).

Fact 5.18.2. Let $A \in \mathbb{F}^{n \times m}$, where $n \le m$. Then, there exist $M \in \mathbb{F}^{n \times n}$ and $S \in \mathbb{F}^{n \times m}$ such that M is positive semidefinite, S satisfies $SS^* = I_n$, and $A = MS$. Furthermore, M is given uniquely by $M = (AA^*)^{1/2}$. If, in addition, $\operatorname{rank} A = n$, then S is given uniquely by

$$S = (AA^*)^{-1/2}A = \frac{2}{\pi}A^* \int_0^\infty (t^2 I + AA^*)^{-1} \, dt.$$

Proof: See [701, Chapter 8].

Fact 5.18.3. Let $A \in \mathbb{F}^{n \times m}$, where $m \le n$. Then, there exist $M \in \mathbb{F}^{m \times m}$ and $S \in \mathbb{F}^{n \times m}$ such that M is positive semidefinite, S satisfies $S^*S = I_m$, and $A = SM$. Furthermore, M is given uniquely by $M = (A^*A)^{1/2}$. If, in addition, $\operatorname{rank} A = m$, then M is positive definite and S is given uniquely by

$$S = A(A^*A)^{-1/2} = \frac{2}{\pi}A \int_0^\infty (t^2 I + A^*A)^{-1} \, dt.$$

Proof: See [701, Chapter 8].

Fact 5.18.4. Let $A \in \mathbb{F}^{n \times n}$, and assume that A is nonsingular. Then, there exist unique matrices $M, S \in \mathbb{F}^{n \times n}$ such that $A = MS$, M is positive definite, and S is unitary. In particular, $M = (AA^*)^{1/2}$ and $S = (AA^*)^{-1/2}A$.

Remark: See Fact 5.18.1.

Fact 5.18.5. Let $A \in \mathbb{F}^{n \times n}$, and assume that A is nonsingular. Then, there exist unique matrices $M, S \in \mathbb{F}^{n \times n}$ such that $A = SM$, M is positive definite, and S is unitary. In particular, $M = (A^*A)^{1/2}$ and $S = (AA^*)^{-1/2}A$.

Fact 5.18.6. Let $M_1, M_2 \in \mathbb{F}^{n \times n}$, and assume that M_1, M_2 are positive definite. Furthermore, let $S_1, S_2 \in \mathbb{F}^{n \times n}$, assume that S_1, S_2 are unitary, and assume that $M_1 S_1 = S_2 M_2$. Then, $S_1 = S_2$.

Proof: Let $A = M_1 S_1 = S_2 M_2$. Then, $S_1 = (S_2 M_2^2 S_2^*)^{-1/2} S_2 M_2 = S_2$.

Fact 5.18.7. Let $A \in \mathbb{F}^{n \times n}$, and assume that A is singular. Then, there exist a matrix $S \in \mathbb{F}^{n \times n}$ and unique matrices $M_1, M_2 \in \mathbb{F}^{n \times n}$ such that $A = M_1 S = SM_2$. In particular, $M_1 = (AA^*)^{1/2}$ and $M_2 = (A^*A)^{1/2}$.

Remark: S is not uniquely determined.

Fact 5.18.8. Let $A \in \mathbb{F}^{n \times n}$, assume that A is nonsingular, and let $M, S \in \mathbb{F}^{n \times n}$ be such that $A = MS$, M is positive semidefinite, and S is unitary. Then, A is normal if and only if $MS = SM$.

Proof: See [728, p. 414].

Fact 5.18.9. Let $A, B \in \mathbb{F}^{n \times n}$, assume that A and B are unitary, and assume that $A + B$ is nonsingular. Then, the unitary factor in the polar decomposition of $A + B$ is $A(A^*B)^{1/2}$.

Proof: See [701, p. 216] or [1038].

Remark: The principal square root of A^*B exists since $A + B$ is nonsingular.

5.19 Facts on Additive Decompositions

Fact 5.19.1. Let $A \in \mathbb{C}^{n \times n}$. Then, there exist unitary matrices $B, C \in \mathbb{C}^{n \times n}$ such that
$$A = \tfrac{1}{2}\sigma_{\max}(A)(B + C).$$

Proof: See [925, 1520].

Fact 5.19.2. Let $A \in \mathbb{R}^{n \times n}$. Then, there exist orthogonal matrices $B, C, D, E \in \mathbb{R}^{n \times n}$ such that
$$A = \tfrac{1}{2}\sigma_{\max}(A)(B + C + D - E).$$

Proof: See [925]. See also [1520].

Remark: $A/\sigma_{\max}(A)$ is expressed as an affine combination of B, C, D, E since the sum of the coefficients is 1.

Fact 5.19.3. Let $A \in \mathbb{R}^{n \times n}$, assume that $\sigma_{\max}(A) \leq 1$, and define $r \triangleq \operatorname{rank}(I - A^*A)$. Then, A is a convex combination of $h(r)$ or fewer orthogonal matrices, where
$$h(r) \triangleq \begin{cases} 1 + r, & r \leq 4, \\ 3 + \log_2 r, & r > 4. \end{cases}$$

Proof: See [925].

Fact 5.19.4. Let $A \in \mathbb{F}^{n \times n}$. Then, the following statements hold:

$i)$ A is positive semidefinite, $\operatorname{tr} A$ is an integer, and $\operatorname{rank} A \leq \operatorname{tr} A$.

$ii)$ There exist projectors $B_1, \ldots, B_l \in \mathbb{F}^{n \times n}$, where $l = \operatorname{tr} A$, such that $A = \sum_{i=1}^{l} B_i$.

Proof: See [502, 1495].

Remark: The minimal number of projectors needed in general is $\operatorname{tr} A$.

Remark: See Fact 5.19.7.

Fact 5.19.5. Let $A \in \mathbb{F}^{n \times n}$, assume that A is Hermitian, $0 \le A \le I$, and $\operatorname{tr} A$ is a rational number. Then, A is the average of a finite set of projectors in $\mathbb{F}^{n \times n}$.

Proof: See [335].

Remark: The required number of projectors can be arbitrarily large.

Fact 5.19.6. Let $A \in \mathbb{F}^{n \times n}$, assume that A is Hermitian, and assume that $0 \le A \le I$. Then, A is a convex combination of $\lfloor \log_2 n \rfloor + 2$ projectors in $\mathbb{F}^{n \times n}$.

Proof: See [335].

Fact 5.19.7. Let $A \in \mathbb{F}^{n \times n}$. Then, the following statements hold:

i) $\operatorname{tr} A$ is an integer, and $\operatorname{rank} A \le \operatorname{tr} A$.

ii) There exist idempotent matrices $B_1, \ldots, B_m \in \mathbb{F}^{n \times n}$ such that $A = \sum_{i=1}^{m} B_i$.

iii) There exist a positive integer m and idempotent matrices $B_1, \ldots, B_m \in \mathbb{F}^{n \times n}$ such that, for all $i \in \{1, \ldots, m\}$, $\operatorname{rank} B_i = 1$ and $\mathcal{R}(B_i) \subseteq A$, and such that $A = \sum_{i=1}^{m} B_i$.

iv) There exist idempotent matrices $B_1, \ldots, B_l \in \mathbb{F}^{n \times n}$, where $l \triangleq \operatorname{tr} A$, such that $A = \sum_{i=1}^{l} B_i$.

Proof: See [667, 1247, 1495].

Remark: The minimal number of idempotent matrices is discussed in [1431].

Remark: See Fact 5.19.8.

Fact 5.19.8. Let $A \in \mathbb{F}^{n \times n}$, and assume that $2(\operatorname{rank} A - 1) \le \operatorname{tr} A \le 2n$. Then, there exist idempotent matrices $B, C, D, E \in \mathbb{F}^{n \times n}$ such that $A = B + C + D + E$.

Proof: See [899].

Remark: See Fact 5.19.10.

Fact 5.19.9. Let $A \in \mathbb{F}^{n \times n}$. If $n = 2$ or $n = 3$, then there exist $b, c \in \mathbb{F}$ and idempotent matrices $B, C \in \mathbb{F}^{n \times n}$ such that $A = bB + cC$. Furthermore, if $n \ge 4$, then there exist $b, c, d \in \mathbb{F}$ and idempotent matrices $B, C, D \in \mathbb{F}^{n \times n}$ such that $A = bB + cC + dD$.

Proof: See [1138].

Fact 5.19.10. Let $A \in \mathbb{C}^{n \times n}$, and assume that A is Hermitian. If $n = 2$ or $n = 3$, then there exist $b, c \in \mathbb{C}$ and projectors $B, C \in \mathbb{C}^{n \times n}$ such that $A = bB + cC$. Furthermore, if $4 \le n \le 7$, then there exist $b, c, d \in \mathbb{F}$ and projectors $B, C, D \in \mathbb{F}^{n \times n}$ such that $A = bB + cC + dD$. If $n \ge 8$, then there exist $b, c, d, e \in \mathbb{C}$ and projectors $B, C, D, E \in \mathbb{C}^{n \times n}$ such that $A = bB + cC + dD + eE$.

Proof: See [1056].

Remark: See Fact 5.19.8.

5.20 Notes

The multicompanion form and the elementary multicompanion form are known as *rational canonical forms* [457, pp. 472–488], while the multicompanion form is traditionally called the *Frobenius canonical form* [150]. The derivation of the Jordan form by means of the elementary multicompanion form and the hypercompanion form follows [1108]. Corollary 5.3.8, Corollary 5.3.9, and Proposition 5.5.12 are given in [244, 245, 1288, 1289, 1292]. Corollary 5.3.9 is due to Frobenius. Canonical forms for congruence transformations are given in [909, 1306].

It is sometimes useful to define block-companion form matrices in which the scalars are replaced by matrix blocks [573, 574, 576]. The companion form provides only one of many connections between matrices and polynomials. Additional connections are given by the *Leslie, Schwarz*, and *Routh* forms [143]. Given a polynomial expressed in terms of an arbitrary polynomial basis, the corresponding matrix is in *confederate form*, which specializes to the *comrade form* when the basis polynomials are orthogonal. The comrade form specializes to the *colleague form* when Chebyshev polynomials are used. The companion, confederate, comrade, and colleague forms are called *congenial* matrices. See [143, 145, 148] and Fact 11.18.25 and Fact 11.18.27 for the Schwarz and Routh forms. The companion matrix is sometimes called a *Frobenius matrix* or the *Frobenius canonical form*, see [5].

Matrix pencils are discussed in [88, 167, 228, 866, 1373, 1385]. Computational algorithms for the Kronecker canonical form are given in [942, 1391]. Applications to linear system theory are discussed in [319, pp. 52–55] and [813].

Application of the polar decomposition to the elastic deformation of solids is discussed in [1099, pp. 140–142].

Chapter Six

Generalized Inverses

Generalized inverses provide a useful extension of the matrix inverse to singular matrices and to rectangular matrices that are neither left nor right invertible.

6.1 Moore-Penrose Generalized Inverse

Let $A \in \mathbb{F}^{n \times m}$. If A is nonzero, then, by the singular value decomposition Theorem 5.6.3, there exist orthogonal matrices $S_1 \in \mathbb{F}^{n \times n}$ and $S_2 \in \mathbb{F}^{m \times m}$ such that

$$A = S_1 \begin{bmatrix} B & 0_{r \times (m-r)} \\ 0_{(n-r) \times r} & 0_{(n-r) \times (m-r)} \end{bmatrix} S_2, \tag{6.1.1}$$

where $B \triangleq \mathrm{diag}[\sigma_1(A), \ldots, \sigma_r(A)]$, $r \triangleq \mathrm{rank}\, A$, and $\sigma_1(A) \geq \sigma_2(A) \geq \cdots \geq \sigma_r(A) > 0$ are the positive singular values of A. In (6.1.1), some of the bordering zero matrices may be empty. Then, the *(Moore-Penrose) generalized inverse* A^+ of A is the $m \times n$ matrix

$$A^+ \triangleq S_2^* \begin{bmatrix} B^{-1} & 0_{r \times (n-r)} \\ 0_{(m-r) \times r} & 0_{(m-r) \times (n-r)} \end{bmatrix} S_1^*. \tag{6.1.2}$$

If $A = 0_{n \times m}$, then $A^+ \triangleq 0_{m \times n}$, while, if $m = n$ and $\det A \neq 0$, then $A^+ = A^{-1}$. In general, it is helpful to remember that A^+ and A^* are the same size. It is easy to verify that A^+ satisfies

$$AA^+A = A, \tag{6.1.3}$$

$$A^+AA^+ = A^+, \tag{6.1.4}$$

$$(AA^+)^* = AA^+, \tag{6.1.5}$$

$$(A^+A)^* = A^+A. \tag{6.1.6}$$

Hence, for each $A \in \mathbb{F}^{n \times m}$ there exists a matrix $X \in \mathbb{F}^{m \times n}$ satisfying the four conditions

$$AXA = A, \tag{6.1.7}$$

$$XAX = X, \tag{6.1.8}$$

$$(AX)^* = AX, \tag{6.1.9}$$

$$(XA)^* = XA. \tag{6.1.10}$$

We now show that X is uniquely defined by (6.1.7)–(6.1.10).

Theorem 6.1.1. Let $A \in \mathbb{F}^{n \times m}$. Then, $X = A^+$ is the unique matrix $X \in \mathbb{F}^{m \times n}$ satisfying (6.1.7)–(6.1.10).

Proof. Suppose there exists a matrix $X \in \mathbb{F}^{m \times n}$ satisfying (6.1.7)–(6.1.10). Then,

$$\begin{aligned}
X &= XAX = X(AX)^* = XX^*A^* = XX^*(AA^+A)^* = XX^*A^*A^{+*}A^* \\
&= X(AX)^*(AA^+)^* = XAXAA^+ = XAA^+ = (XA)^*A^+ = A^*X^*A^+ \\
&= (AA^+A)^*X^*A^+ = A^*A^{+*}A^*X^*A^+ = (A^+A)^*(XA)^*A^+ \\
&= A^+AXAA^+ = A^+AA^+ = A^+.
\end{aligned}$$
$\hfill\square$

Given $A \in \mathbb{F}^{n \times m}$, $X \in \mathbb{F}^{m \times n}$ is a *(1)-inverse* of A if (6.1.7) holds, a *(1,2)-inverse* of A if (6.1.7) and (6.1.8) hold, and so forth.

Proposition 6.1.2. Let $A \in \mathbb{F}^{n \times m}$, and assume that A is right invertible. Then, $X \in \mathbb{F}^{m \times n}$ is a right inverse of A if and only if X is a (1)-inverse of A. Furthermore, every right inverse (or, equivalently, every (1)-inverse) of A is also a (2,3)-inverse of A.

Proof. Suppose that $AX = I_n$, that is, $X \in \mathbb{F}^{m \times n}$ is a right inverse of A. Then, $AXA = A$, which implies that X is a (1)-inverse of A. Conversely, let X be a (1)-inverse of A, that is, $AXA = A$. Then, letting $\hat{X} \in \mathbb{F}^{m \times n}$ denote a right inverse of A, it follows that $AX = AXA\hat{X} = A\hat{X} = I_n$. Hence, X is a right inverse of A. Finally, if X is a right inverse of A, then it is also a (2,3)-inverse of A. $\hfill\square$

Proposition 6.1.3. Let $A \in \mathbb{F}^{n \times m}$, and assume that A is left invertible. Then, $X \in \mathbb{F}^{m \times n}$ is a left inverse of A if and only if X is a (1)-inverse of A. Furthermore, every left inverse (or, equivalently, every (1)-inverse) of A is also a (2,4)-inverse of A.

It can now be seen that A^+ is a particular (right, left) inverse when A is (right, left) invertible.

Corollary 6.1.4. Let $A \in \mathbb{F}^{n \times m}$. If A is right invertible, then A^+ is a right inverse of A. Furthermore, if A is left invertible, then A^+ is a left inverse of A.

The following result provides an explicit expression for A^+ when A is either right invertible or left invertible. It is helpful to note that A is (right, left) invertible if and only if (AA^*, A^*A) is positive definite.

Proposition 6.1.5. Let $A \in \mathbb{F}^{n \times m}$. If A is right invertible, then

$$A^+ = A^*(AA^*)^{-1} \tag{6.1.11}$$

and A^+ is a right inverse of A. If A is left invertible, then

$$A^+ = (A^*A)^{-1}A^* \tag{6.1.12}$$

and A^+ is a left inverse of A.

Proof. It suffices to verify (6.1.7)–(6.1.10) with $X = A^+$. □

Proposition 6.1.6. Let $A \in \mathbb{F}^{n \times m}$. Then, the following statements hold:

i) $A = 0$ if and only if $A^+ = 0$.

ii) $(A^+)^+ = A$.

iii) $\overline{A}^+ = \overline{A^+}$.

iv) $A^{+\mathrm{T}} \triangleq (A^{\mathrm{T}})^+ = (A^+)^{\mathrm{T}}$.

v) $A^{+*} \triangleq (A^*)^+ = (A^+)^*$.

vi) $\mathcal{R}(A) = \mathcal{R}(AA^*) = \mathcal{R}(AA^+) = \mathcal{R}(A^{+*}) = \mathcal{N}(I - AA^+) = \mathcal{N}(A^*)^\perp$.

vii) $\mathcal{R}(A^*) = \mathcal{R}(A^*A) = \mathcal{R}(A^+A) = \mathcal{R}(A^+) = \mathcal{N}(I - A^+A) = \mathcal{N}(A)^\perp$.

viii) $\mathcal{N}(A) = \mathcal{N}(A^+A) = \mathcal{N}(A^*A) = \mathcal{N}(A^{+*}) = \mathcal{R}(I - A^+A) = \mathcal{R}(A^*)^\perp$.

ix) $\mathcal{N}(A^*) = \mathcal{N}(AA^+) = \mathcal{N}(AA^*) = \mathcal{N}(A^+) = \mathcal{R}(I - AA^+) = \mathcal{R}(A)^\perp$.

x) AA^+ and A^+A are positive semidefinite.

xi) $\mathrm{spec}(AA^+) \subseteq \{0, 1\}$ and $\mathrm{spec}(A^+A) \subseteq \{0, 1\}$.

xii) AA^+ is the projector onto $\mathcal{R}(A)$.

xiii) A^+A is the projector onto $\mathcal{R}(A^*)$.

xiv) $I_m - A^+A$ is the projector onto $\mathcal{N}(A)$.

xv) $I_n - AA^+$ is the projector onto $\mathcal{N}(A^*)$.

xvi) $x \in \mathcal{R}(A)$ if and only if $x = AA^+x$.

xvii) $\mathrm{rank}\, A = \mathrm{rank}\, A^+ = \mathrm{rank}\, AA^+ = \mathrm{rank}\, A^+A = \mathrm{tr}\, AA^+ = \mathrm{tr}\, A^+A$.

xviii) $\mathrm{rank}(I_m - A^+A) = m - \mathrm{rank}\, A$.

xix) $\mathrm{rank}(I_n - AA^+) = n - \mathrm{rank}\, A$.

xx) $(A^*A)^+ = A^+A^{+*}$.

xxi) $(AA^*)^+ = A^{+*}A^+$.

xxii) $AA^+ = A(A^*A)^+A^*$.

xxiii) $A^+A = A^*(AA^*)^+A$.

xxiv) $A = AA^*A^{*+} = A^{*+}A^*A$.

xxv) $A^* = A^*AA^+ = A^+AA^*$.

xxvi) $A^+ = A^*(AA^*)^+ = (A^*A)^+A^* = A^*(A^*AA^*)^+A^*$.

xxvii) $A^{+*} = (AA^*)^+A = A(A^*A)^+$.

xxviii) $A = A(A^*A)^+A^*A = AA^*A(A^*A)^+$.

xxix) $A = AA^*(AA^*)^+A = (AA^*)^+AA^*A$.

xxx) If $S_1 \in \mathbb{F}^{n \times n}$ and $S_2 \in \mathbb{F}^{m \times m}$ are orthogonal, then $(S_1AS_2)^+ = S_2^*A^+S_1^*$.

xxxi) A is (range Hermitian, normal, Hermitian, positive semidefinite, positive definite) if and only if A^+ is.

xxxii) If A is a projector, then $A^+ = A$.

xxxiii) $A^+ = A$ if and only if A is tripotent and A^2 is Hermitian.

xxxiv) If $B \in \mathbb{F}^{n \times l}$, then $\mathcal{R}(AA^+B)$ is the projection of $\mathcal{R}(B)$ onto $\mathcal{R}(A)$.

Proof. The last equality in *xxvi)* is given in [1539]. $\qquad\square$

Theorem 2.6.4 shows that the equation $Ax = b$, where $A \in \mathbb{F}^{n \times m}$ and $b \in \mathbb{F}^n$, has a solution $x \in \mathbb{F}^m$ if and only if $\operatorname{rank} A = \operatorname{rank} \begin{bmatrix} A & b \end{bmatrix}$. In particular, $Ax = b$ has a unique solution $x \in \mathbb{F}^m$ if and only if $\operatorname{rank} A = \operatorname{rank} \begin{bmatrix} A & b \end{bmatrix} = m$, while $Ax = b$ has infinitely many solutions if and only if $\operatorname{rank} A = \operatorname{rank} \begin{bmatrix} A & b \end{bmatrix} < m$. The following result characterizes these solutions in terms of the generalized inverse. Connections to least squares solutions are discussed in Fact 9.15.4.

Proposition 6.1.7. Let $A \in \mathbb{F}^{n \times m}$ and $b \in \mathbb{F}^n$. Then, the following statements are equivalent:

i) There exists a vector $x \in \mathbb{F}^m$ satisfying $Ax = b$.

ii) $\operatorname{rank} A = \operatorname{rank} \begin{bmatrix} A & b \end{bmatrix}$.

iii) $b \in \mathcal{R}(A)$.

iv) $AA^+b = b$.

Now, assume in addition that *i)–iv)* are satisfied. Then, the following statements hold:

v) $x \in \mathbb{F}^m$ satisfies $Ax = b$ if and only if

$$x = A^+b + (I - A^+A)x. \qquad (6.1.13)$$

vi) For all $y \in \mathbb{F}^m$, $x \in \mathbb{F}^m$ given by

$$x = A^+b + (I - A^+A)y \qquad (6.1.14)$$

satisfies $Ax = b$.

vii) Let $x \in \mathbb{F}^m$ be given by (6.1.14), where $y \in \mathbb{F}^m$. Then, $y = 0$ minimizes x^*x.

viii) Assume that $\operatorname{rank} A = m$. Then, there exists a unique vector $x \in \mathbb{F}^m$ satisfying $Ax = b$ given by $x = A^+b$. If, in addition, $A^{\mathrm{L}} \in \mathbb{F}^{m \times n}$ is a left inverse of A, then $A^{\mathrm{L}}b = A^+b$.

ix) Assume that $\operatorname{rank} A = n$, and let $A^{\mathrm{R}} \in \mathbb{F}^{m \times n}$ be a right inverse of A. Then, $x = A^{\mathrm{R}}b$ satisfies $Ax = b$.

Proof. The equivalence of *i)–iii)* is immediate. To prove the equivalence of *iv)*, note that, if there exists a vector $x \in \mathbb{F}^m$ satisfying $Ax = b$, then $b = Ax = AA^+Ax = AA^+b$. Conversely, if $b = AA^+b$, then $x = A^+b$ satisfies $Ax = b$.

Now, suppose that $i)$–$iv)$ hold. To prove $v)$, let $x \in \mathbb{F}^m$ satisfy $Ax = b$ so that $A^+Ax = A^+b$. Hence, $x = x + A^+b - A^+Ax = A^+b + (I - A^+A)x$. To prove $vi)$, let $y \in \mathbb{F}^m$, and let $x \in \mathbb{F}^m$ be given by (6.1.14). Then, $Ax = AA^+b = b$. To prove $vii)$, let $y \in \mathbb{F}^m$, and let $x \in \mathbb{F}^n$ be given by (6.1.14). Then, $x^*x = b^*A^{+*}A^+b + y^*(I - A^+A)y$. Therefore, x^*x is minimized by $y = 0$. See also Fact 9.15.4.

To prove $viii)$, suppose that $\operatorname{rank} A = m$. Then, A is left invertible, and it follows from Corollary 6.1.4 that A^+ is a left inverse of A. Hence, it follows from (6.1.13) that $x = A^+b$ is the unique solution of $Ax = b$. In addition, $x = A^{\mathrm{L}}b$. To prove $ix)$, let $x = A^{\mathrm{R}}b$, and note that $AA^{\mathrm{R}}b = b$. $\square$

Definition 6.1.8. Let $A \in \mathbb{F}^{n \times m}$, $B \in \mathbb{F}^{n \times l}$, $C \in \mathbb{F}^{k \times m}$, and $D \in \mathbb{F}^{k \times l}$, and define $\mathcal{A} \triangleq \begin{bmatrix} A & B \\ C & D \end{bmatrix} \in \mathbb{F}^{(n+k) \times (m+l)}$. Then, the *Schur complement* $A|\mathcal{A}$ of A with respect to $\mathcal{A}$ is defined by

$$A|\mathcal{A} \triangleq D - CA^+B. \tag{6.1.15}$$

Likewise, the *Schur complement* $D|\mathcal{A}$ of D with respect to $\mathcal{A}$ is defined by

$$D|\mathcal{A} \triangleq A - BD^+C. \tag{6.1.16}$$

6.2 Drazin Generalized Inverse

We now introduce a different type of generalized inverse, which applies only to square matrices yet is more useful in certain applications. Let $A \in \mathbb{F}^{n \times n}$. Then, A has a decomposition

$$A = S \begin{bmatrix} J_1 & 0 \\ 0 & J_2 \end{bmatrix} S^{-1}, \tag{6.2.1}$$

where $S \in \mathbb{F}^{n \times n}$ is nonsingular, $J_1 \in \mathbb{F}^{m \times m}$ is nonsingular, and $J_2 \in \mathbb{F}^{(n-m) \times (n-m)}$ is nilpotent. Then, the *Drazin generalized inverse* A^{D} of A is the matrix

$$A^{\mathrm{D}} \triangleq S \begin{bmatrix} J_1^{-1} & 0 \\ 0 & 0 \end{bmatrix} S^{-1}. \tag{6.2.2}$$

Let $A \in \mathbb{F}^{n \times n}$. Then, it follows from Definition 5.5.1 that $\operatorname{ind} A = \operatorname{ind}_A(0)$. Furthermore, A is nonsingular if and only if $\operatorname{ind} A = 0$, whereas $\operatorname{ind} A = 1$ if and only if A is singular and the zero eigenvalue of A is semisimple. In particular, $\operatorname{ind} 0_{n \times n} = 1$. Note that $\operatorname{ind} A$ is the order of the largest Jordan block of A associated with the zero eigenvalue of A.

It can be seen that A^{D} satisfies

$$A^{\mathrm{D}}AA^{\mathrm{D}} = A^{\mathrm{D}}, \tag{6.2.3}$$

$$AA^{\mathrm{D}} = A^{\mathrm{D}}A, \tag{6.2.4}$$

$$A^{k+1}A^{\mathrm{D}} = A^k, \tag{6.2.5}$$

where $k = \operatorname{ind} A$. Hence, for all $A \in \mathbb{F}^{n \times n}$ such that $\operatorname{ind} A = k$ there exists a matrix

$X \in \mathbb{F}^{n \times n}$ satisfying the three conditions

$$XAX = X, \qquad (6.2.6)$$

$$AX = XA, \qquad (6.2.7)$$

$$A^{k+1}X = A^k. \qquad (6.2.8)$$

We now show that X is uniquely defined by (6.2.6)–(6.2.8).

Theorem 6.2.1. Let $A \in \mathbb{F}^{n \times n}$, and let $k \triangleq \operatorname{ind} A$. Then, $X = A^D$ is the unique matrix $X \in \mathbb{F}^{n \times n}$ satisfying (6.2.6)–(6.2.8).

Proof. Let $X \in \mathbb{F}^{n \times n}$ satisfy (6.2.6)–(6.2.8). If $k = 0$, then it follows from (6.2.8) that $X = A^{-1}$. Hence, let $A = S \left[\begin{smallmatrix} J_1 & 0 \\ 0 & J_2 \end{smallmatrix} \right] S^{-1}$, where $k = \operatorname{ind} A \geq 1$, $S \in \mathbb{F}^{n \times n}$ is nonsingular, $J_1 \in \mathbb{F}^{m \times m}$ is nonsingular, and $J_2 \in \mathbb{F}^{(n-m) \times (n-m)}$ is nilpotent. Now, let $\hat{X} \triangleq S^{-1}XS = \left[\begin{smallmatrix} \hat{X}_1 & \hat{X}_{12} \\ \hat{X}_{21} & \hat{X}_2 \end{smallmatrix} \right]$ be partitioned conformably with $\hat{A} \triangleq S^{-1}AS = \left[\begin{smallmatrix} J_1 & 0 \\ 0 & J_2 \end{smallmatrix} \right]$. Since, by (6.2.7), $\hat{A}\hat{X} = \hat{X}\hat{A}$, it follows that $J_1\hat{X}_1 = \hat{X}_1 J_1$, $J_1\hat{X}_{12} = \hat{X}_{12}J_2$, $J_2\hat{X}_{21} = \hat{X}_{21}J_1$, and $J_2\hat{X}_2 = \hat{X}_2 J_2$. Since $J_2^k = 0$, it follows that $J_1\hat{X}_{12}J_2^{k-1} = 0$, and thus $\hat{X}_{12}J_2^{k-1} = 0$. By repeating this argument, it follows that $J_1\hat{X}_{12}J_2 = 0$, and thus $\hat{X}_{12}J_2 = 0$, which implies that $J_1\hat{X}_{12} = 0$, and thus $\hat{X}_{12} = 0$. Similarly, $\hat{X}_{21} = 0$, so that $\hat{X} = \left[\begin{smallmatrix} \hat{X}_1 & 0 \\ 0 & \hat{X}_2 \end{smallmatrix} \right]$. Now, (6.2.8) implies that $J_1^{k+1}\hat{X}_1 = J_1^k$, and hence $\hat{X}_1 = J_1^{-1}$. Next, (6.2.6) implies that $\hat{X}_2 J_2 \hat{X}_2 = \hat{X}_2$, which, together with $J_2\hat{X}_2 = \hat{X}_2 J_2$, yields $\hat{X}_2^2 J_2 = \hat{X}_2$. Consequently, $0 = \hat{X}_2^2 J_2^k = \hat{X}_2 J_2^{k-1}$, and thus, by repeating this argument, $\hat{X}_2 = 0$. Hence, $A^D = S \left[\begin{smallmatrix} J_1^{-1} & 0 \\ 0 & 0 \end{smallmatrix} \right] S^{-1} = S \left[\begin{smallmatrix} \hat{X}_1 & 0 \\ 0 & 0 \end{smallmatrix} \right] S^{-1} = S\hat{X}S^{-1} = X$. $\qquad \square$

Proposition 6.2.2. Let $A \in \mathbb{F}^{n \times n}$, and define $k \triangleq \operatorname{ind} A$. Then, the following statements hold:

i) $\overline{A}^D = \overline{A^D}$.

ii) $A^{DT} \triangleq A^{TD} \triangleq \left(A^T\right)^D = (A^D)^T$.

iii) $A^{D*} \triangleq A^{*D} \triangleq \left(A^*\right)^D = (A^D)^*$.

iv) If $r \in \mathbb{P}$, then $A^{Dr} \triangleq A^{rD} \triangleq \left(A^D\right)^r = (A^r)^D$.

v) $\mathcal{R}(A^k) = \mathcal{R}(A^D) = \mathcal{R}(AA^D) = \mathcal{N}(I - AA^D)$.

vi) $\mathcal{N}(A^k) = \mathcal{N}(A^D) = \mathcal{N}(AA^D) = \mathcal{R}(I - AA^D)$.

vii) $\operatorname{rank} A^k = \operatorname{rank} A^D = \operatorname{rank} AA^D = \operatorname{def}(I - AA^D)$.

viii) $\operatorname{def} A^k = \operatorname{def} A^D = \operatorname{def} AA^D = \operatorname{rank}(I - AA^D)$.

ix) AA^D is the idempotent matrix onto $\mathcal{R}(A^D)$ along $\mathcal{N}(A^D)$.

x) $A^D = 0$ if and only if A is nilpotent.

xi) A^D is group invertible.

xii) $\operatorname{ind} A^D = 0$ if and only if A is nonsingular.

xiii) $\operatorname{ind} A^D = 1$ if and only if A is singular.

xiv) $(A^{\mathrm{D}})^{\mathrm{D}} = (A^{\mathrm{D}})^{\#} = A^2 A^{\mathrm{D}}.$

xv) $(A^{\mathrm{D}})^{\mathrm{D}} = A$ if and only if A is group invertible.

xvi) If A is idempotent, then $k = 1$ and $A^{\mathrm{D}} = A.$

xvii) $A = A^{\mathrm{D}}$ if and only if A is tripotent.

Let $A \in \mathbb{F}^{n \times n}$, and assume that ind $A \leq 1$ so that, by Corollary 5.5.9, A is group invertible. In this case, the Drazin generalized inverse A^{D} is denoted by $A^{\#}$, which is the *group generalized inverse* of A. Therefore, $A^{\#}$ satisfies

$$A^{\#}AA^{\#} = A^{\#}, \tag{6.2.9}$$

$$AA^{\#} = A^{\#}A, \tag{6.2.10}$$

$$AA^{\#}A = A, \tag{6.2.11}$$

while $A^{\#}$ is the unique matrix $X \in \mathbb{F}^{n \times n}$ satisfying

$$XAX = X, \tag{6.2.12}$$

$$AX = XA, \tag{6.2.13}$$

$$AXA = A. \tag{6.2.14}$$

Proposition 6.2.3. Let $A \in \mathbb{F}^{n \times n}$, and assume that A is group invertible. Then, the following statements hold:

i) $\overline{A}^{\#} = \overline{A^{\#}}.$

ii) $A^{\#\mathrm{T}} \triangleq A^{\mathrm{T}\#} \triangleq (A^{\mathrm{T}})^{\#} = (A^{\#})^{\mathrm{T}}.$

iii) $A^{\#*} \triangleq A^{*\#} \triangleq (A^*)^{\#} = (A^{\#})^*.$

iv) If $r \in \mathbb{P}$, then $A^{\#r} \triangleq A^{r\#} \triangleq (A^{\#})^r = (A^r)^{\#}.$

v) $\mathcal{R}(A) = \mathcal{R}(AA^{\#}) = \mathcal{N}(I - AA^{\#}) = \mathcal{R}(AA^+) = \mathcal{N}(I - AA^+).$

vi) $\mathcal{N}(A) = \mathcal{N}(AA^{\#}) = \mathcal{R}(I - AA^{\#}) = \mathcal{N}(A^+A) = \mathcal{R}(I - A^+A).$

vii) rank $A = $ rank $A^{\#} = $ rank $AA^{\#} = $ rank $A^{\#}A.$

viii) def $A = $ def $A^{\#} = $ def $AA^{\#} = $ def $A^{\#}A.$

ix) $AA^{\#}$ is the idempotent matrix onto $\mathcal{R}(A)$ along $\mathcal{N}(A).$

x) $A^{\#} = 0$ if and only if $A = 0.$

xi) $A^{\#}$ is group invertible.

xii) $(A^{\#})^{\#} = A.$

xiii) If A is idempotent, then $A^{\#} = A.$

xiv) $A = A^{\#}$ if and only if A is tripotent.

An alternative expression for the idempotent matrix onto $\mathcal{R}(A)$ along $\mathcal{N}(A)$ is given by Proposition 3.5.9.

6.3 Facts on the Moore-Penrose Generalized Inverse for One Matrix

Fact 6.3.1. Let $A \in \mathbb{F}^{n \times m}$, $x \in \mathbb{F}^m$, $b \in \mathbb{F}^n$, and $y \in \mathbb{F}^m$, assume that A is right invertible, and assume that

$$x = A^+ b + (I - A^+ A)y,$$

which satisfies $Ax = b$. Then, there exists a right inverse $A^{\mathrm{R}} \in \mathbb{F}^{m \times n}$ of A such that $x = A^{\mathrm{R}} b$. Furthermore, if $S \in \mathbb{F}^{m \times n}$ is such that $z^{\mathrm{T}} S b \neq 0$, where $z \triangleq (I - A^+ A)y$, then one such right inverse is given by

$$A^{\mathrm{R}} = A^+ + \frac{1}{z^{\mathrm{T}} S b} z z^{\mathrm{T}} S.$$

Fact 6.3.2. Let $A \in \mathbb{F}^{n \times m}$, and assume that $\operatorname{rank} A = 1$. Then,

$$A^+ = (\operatorname{tr} AA^*)^{-1} A^*.$$

Consequently, if $x \in \mathbb{F}^n$ and $y \in \mathbb{F}^n$ are nonzero, then

$$(xy^*)^+ = (x^* x y^* y)^{-1} y x^* = \frac{1}{\|x\|_2^2 \|y\|_2^2} y x^*.$$

In particular,

$$1_{n \times m}^+ = \tfrac{1}{nm} 1_{m \times n}.$$

Fact 6.3.3. Let $x \in \mathbb{F}^n$, and assume that x is nonzero. Then, the projector $A \in \mathbb{F}^{n \times n}$ onto $\operatorname{span}\{x\}$ is given by

$$A = (x^* x)^{-1} x x^*.$$

Fact 6.3.4. Let $x, y \in \mathbb{F}^n$, assume that x, y are nonzero, and assume that $x^* y = 0$. Then, the projector $A \in \mathbb{F}^{n \times n}$ onto $\operatorname{span}\{x, y\}$ is given by

$$A = (x^* x)^{-1} x x^* + (y^* y)^{-1} y y^*.$$

Fact 6.3.5. Let $x, y \in \mathbb{F}^n$, and assume that x, y are linearly independent. Then, the projector $A \in \mathbb{F}^{n \times n}$ onto $\operatorname{span}\{x, y\}$ is given by

$$A = (x^* x y^* y - |x^* y|^2)^{-1} (y^* y x x^* - y^* x y x^* - x^* y x y^* + x^* x y y^*).$$

Furthermore, define $z \triangleq [I - (x^* x)^{-1} x x^*]y$. Then,

$$A = (x^* x)^{-1} x x^* + (z^* z)^{-1} z z^*.$$

Remark: For $\mathbb{F} = \mathbb{R}$, this result is given in [1237, p. 178].

Fact 6.3.6. Let $A \in \mathbb{F}^{n \times m}$, assume that $\operatorname{rank} A = n - 1$, let $x \in \mathcal{N}(A)$ be nonzero, let $y \in \mathcal{N}(A^*)$ be nonzero, let $\alpha = 1$ if $\operatorname{spec}(A) = \{0\}$ and the product of the nonzero eigenvalues of A otherwise, and define $k \triangleq \operatorname{amult}_A(0)$. Then,

$$A^{\mathrm{A}} = \frac{(-1)^{k+1} \alpha}{y^* (A^{k-1})^+ x} xy^*.$$

In particular,

$$N_n^{\mathrm{A}} = (-1)^{n+1} E_{1,n}.$$

If, in addition, $k = 1$, then

$$A^{\mathrm{A}} = \frac{\alpha}{y^* x} x y^*.$$

Proof: See [973, p. 41] and Fact 3.17.4.

Remark: This result provides an expression for $ii)$ of Fact 2.16.8.

Remark: If A is range Hermitian, then $\mathcal{N}(A) = \mathcal{N}(A^*)$ and $y^* x \neq 0$, and thus Fact 5.14.3 implies that A^{A} is semisimple.

Remark: See Fact 5.14.25.

Fact 6.3.7. Let $A \in \mathbb{F}^{n \times m}$, and assume that $\operatorname{rank} A = n - 1$. Then,

$$A^+ = \frac{1}{\det[AA^* + (AA^*)^{\mathrm{A}}]} A^* [AA^* + (AA^*)^{\mathrm{A}}]^{\mathrm{A}}.$$

Proof: See [353].

Remark: Extensions to matrices of arbitrary rank are given in [353].

Fact 6.3.8. Let $A \in \mathbb{F}^{n \times m}$, $B \in \mathbb{F}^{k \times n}$, and $C \in \mathbb{F}^{m \times l}$, and assume that B is left inner and C is right inner. Then,

$$(BAC)^+ = C^* A^+ B^*.$$

Proof: See [671, p. 506].

Fact 6.3.9. Let $A \in \mathbb{F}^{n \times n}$. Then,

$$\begin{aligned}
\operatorname{rank}[A, A^+] &= 2(\operatorname{rank}\begin{bmatrix} A & A^* \end{bmatrix} - \operatorname{rank} A) \\
&= \operatorname{rank}(A - A^2 A^+) \\
&= \operatorname{rank}(A - A^+ A^2).
\end{aligned}$$

Furthermore, the following statements are equivalent:

$i)$ A is range Hermitian.

$ii)$ $[A, A^+] = 0$.

$iii)$ $\operatorname{rank}\begin{bmatrix} A & A^* \end{bmatrix} = \operatorname{rank} A$.

$iv)$ $A = A^2 A^+$.

$v)$ $A = A^+ A^2$.

Proof: See [1338].

Remark: See Fact 3.6.3, Fact 6.3.10, and Fact 6.4.13.

Fact 6.3.10. Let $A \in \mathbb{F}^{n \times n}$. Then, the following statements are equivalent:

$i)$ A is range Hermitian.

$ii)$ $\mathcal{R}(A) = \mathcal{R}(A^+)$.

$iii)$ $A^+ A = AA^+$.

$iv)$ $(I - A^+ A)_\perp = AA^+$.

v) $A = A^2 A^+$.

vi) $A = A^+ A^2$.

vii) $AA^+ = A^2 (A^+)^2$.

viii) $(AA^+)^2 = A^2 (A^+)^2$.

ix) $(A^+A)^2 = (A^+)^2 A^2$.

x) ind $A \le 1$, and $(A^+)^2 = (A^2)^+$.

xi) ind $A \le 1$, and $AA^+ A^* A = A^* A^2 A^+$.

xii) $A^2 A^+ + A^* A^{+*} A = 2A$.

xiii) $A^2 A^+ + (A^2 A^+)^* = A + A^*$.

xiv) $\mathcal{R}(A - A^+) = \mathcal{R}(A - A^3)$.

xv) $\mathcal{R}(A + A^+) = \mathcal{R}(A + A^3)$.

Proof: See [331, 1313, 1328, 1363] and Fact 6.6.9.

Remark: See Fact 3.6.3, Fact 6.3.9, and Fact 6.4.13.

Fact 6.3.11. Let $A \in \mathbb{F}^{n \times n}$. Then, the following statements are equivalent:

i) $A + A^+ = 2AA^+$.

ii) $A + A^+ = 2A^+A$.

iii) $A + A^+ = AA^+ + A^+A$.

iv) A is range Hermitian, and $A^2 + AA^+ = 2A$.

v) A is range Hermitian, and $(I - A)^2 A = 0$.

Proof: See [1355, 1362].

Fact 6.3.12. Let $A \in \mathbb{F}^{n \times n}$. Then, the following statements are equivalent:

i) $A^+ A^* = A^* A^+$.

ii) $AA^+ A^* A = AA^* A^+ A$.

iii) $AA^* A^2 = A^2 A^* A$.

If these conditions hold, then A is *star-dagger*. If A is star-dagger, then $A^2 (A^+)^2$ and $(A^+)^2 A^2$ are positive semidefinite.

Proof: See [668, 1313].

Remark: See Fact 6.3.15.

Fact 6.3.13. Let $A \in \mathbb{F}^{n \times m}$, let $B, C \in \mathbb{F}^{m \times n}$, assume that B is a $(1, 3)$ inverse of A, and assume that C is a $(1, 4)$ inverse of A. Then,

$$A^+ = CAB.$$

Proof: See [178, p. 48].

Remark: This result is due to Urquhart.

Fact 6.3.14. Let $A \in \mathbb{F}^{n \times m}$, assume that A is nonzero, let $r \triangleq \operatorname{rank} A$, define $B \triangleq \operatorname{diag}[\sigma_1(A), \ldots, \sigma_r(A)]$, and let $S \in \mathbb{F}^{n \times n}$, $K \in \mathbb{F}^{r \times r}$, and $L \in \mathbb{F}^{r \times (m-r)}$ be such that S is unitary,

$$KK^* + LL^* = I_r,$$

and

$$A = S \left[\begin{array}{cc} BK & BL \\ 0_{(n-r) \times r} & 0_{(n-r) \times (m-r)} \end{array} \right] S^*.$$

Then,

$$A^+ = S \left[\begin{array}{cc} K^*B^{-1} & 0_{r \times (n-r)} \\ L^*B^{-1} & 0_{(m-r) \times (n-r)} \end{array} \right] S^*.$$

Proof: See [119, 668].

Remark: See Fact 5.9.30 and Fact 6.6.16.

Fact 6.3.15. Let $A \in \mathbb{F}^{n \times n}$. Then, the following statements are equivalent:

i) A is normal.

ii) $AA^*A^+ = A^+AA^*$.

iii) A is range Hermitian, and $A^+A^* = A^*A^+$.

iv) $A(AA^*A)^+ = (AA^*A)^+A$.

v) $AA^+A^*A^2A^+ = AA^*$.

vi) $A(A^* + A^+) = (A^* + A^+)A$.

vii) $A^*A(AA^*)^+A^*A = AA^*$.

viii) $2AA^*(AA^* + A^*A)^+AA^* = AA^*$.

ix) There exists a matrix $X \in \mathbb{F}^{n \times n}$ such that $AA^*X = A^*A$ and $A^*AX = AA^*$.

x) There exists a matrix $X \in \mathbb{F}^{n \times n}$ such that $AX = A^*$ and $A^{+*}X = A^+$.

Proof: See [331].

Remark: See Fact 3.7.12, Fact 3.11.5, Fact 5.15.4, Fact 6.3.12, and Fact 6.6.11.

Fact 6.3.16. Let $A \in \mathbb{F}^{n \times n}$. Then, the following statements are equivalent:

i) A is Hermitian.

ii) $AA^+ = A^*A^+$.

iii) $A^2A^+ = A^*$.

iv) $AA^*A^+ = A$.

Proof: See [119].

Fact 6.3.17. Let $A \in \mathbb{F}^{n \times m}$, and assume that $\operatorname{rank} A = m$. Then,

$$(AA^*)^+ = A(A^*A)^{-2}A^*.$$

Remark: See Fact 6.4.10.

Fact 6.3.18. Let $A \in \mathbb{F}^{n \times m}$. Then,

$$A^+ = \lim_{\alpha \downarrow 0} A^*(AA^* + \alpha I)^{-1} = \lim_{\alpha \downarrow 0} (A^*A + \alpha I)^{-1}A^*.$$

Fact 6.3.19. Let $A \in \mathbb{F}^{n \times m}$, let $\chi_{AA^*}(s) = s^n + \beta_{n-1}s^{n-1} + \cdots + \beta_1 s + \beta_0$, and let k denote the largest integer in $\{0, \ldots, n-1\}$ such that $\beta_{n-k} \neq 0$. Then,

$$A^+ = -\beta_{n-k}^{-1} A^* \big[(AA^*)^{k-1} + \beta_{n-1}(AA^*)^{k-2} + \cdots + \beta_{n-k+1}I \big].$$

Proof: See [402].

Fact 6.3.20. Let $A \in \mathbb{F}^{n \times n}$, and assume that A is Hermitian. Then,

$$\operatorname{In} A = \operatorname{In} A^+ = \operatorname{In} A^{\mathrm{D}}.$$

If, in addition, A is nonsingular, then

$$\operatorname{In} A = \operatorname{In} A^{-1}.$$

Fact 6.3.21. Let $A \in \mathbb{F}^{n \times n}$, and consider the following statements:

i) A is idempotent.

ii) $\operatorname{rank} A = \operatorname{tr} A$.

iii) $\operatorname{rank} A \leq \operatorname{tr} A^2 A^+ A^*$.

Then, *i)* $\implies$ *ii)* $\implies$ *iii)*. Furthermore, the following statements are equivalent:

iv) A is idempotent.

v) $\operatorname{rank} A = \operatorname{tr} A = \operatorname{tr} A^2 A^+ A^*$.

vi) There exist projectors $B, C \in \mathbb{F}^{n \times n}$ such that $A^+ = BC$.

vii) $A^*A^+ = A^+$.

viii) $A^+A^* = A^+$.

Proof: See [830] and [1215, p. 166].

Fact 6.3.22. Let $A \in \mathbb{F}^{n \times n}$, and assume that A is idempotent. Then,

$$A^*A^+A = A^+A$$

and

$$AA^+A^* = AA^+.$$

Proof: Note that A^*A^+A is a projector, and $\mathcal{R}(A^*A^+A) = \mathcal{R}(A^*) = \mathcal{R}(A^+A)$. Alternatively, use Fact 6.3.21.

Fact 6.3.23. Let $A \in \mathbb{F}^{n \times n}$, and assume that A is idempotent. Then,

$$A^+A + (I - A)(I - A)^+ = I$$

and

$$AA^+ + (I - A)^+(I - A) = I.$$

Proof: $\mathcal{N}(A) = \mathcal{R}(I - A^+A) = \mathcal{R}(I - A) = \mathcal{R}[(I - A)(I - A^+)]$.

Remark: The first equality states that the projector onto the null space of A is the same as the projector onto the range of $I - A$, while the second equality states that the projector onto the range of A is the same as the projector onto the null space of $I - A$.

Remark: See Fact 3.13.24 and Fact 5.12.18.

Fact 6.3.24. Let $A \in \mathbb{F}^{n \times n}$, and assume that A is idempotent. Then, $A + A^* - I$ is nonsingular, and

$$(A + A^* - I)^{-1} = AA^+ + A^+A - I.$$

Proof: Use Fact 6.3.22.

Remark: See Fact 3.13.24, Fact 5.12.18, or [1023, p. 457] for a geometric interpretation of this equality.

Fact 6.3.25. Let $A \in \mathbb{F}^{n \times n}$, and assume that A is idempotent. Then, $2A(A + A^*)^+A^*$ is the projector onto $\mathcal{R}(A) \cap \mathcal{R}(A^*)$.

Proof: See [1352].

Fact 6.3.26. Let $A \in \mathbb{F}^{n \times n}$. Then, the following statements are equivalent:

i) A^+ is idempotent.

ii) $AA^*A = A^2$.

If A is range Hermitian, then the following statements are equivalent:

iii) A^+ is idempotent.

iv) $AA^* = A^*A = A$.

The following statements are equivalent:

v) A^+ is a projector.

vi) A is a projector.

vii) A is idempotent, and A and A^+ are similar.

viii) A is idempotent, and $A = A^+$.

ix) A is idempotent, and $AA^+ = AA^*$.

x) $A^+ = A$, and $A^2 = A^*$.

xi) A and A^+ are idempotent.

xii) $A = AA^+$.

Proof: See [1215, pp. 167, 168] and [1313, 1358, 1457].

Remark: See Fact 3.13.1.

Fact 6.3.27. Let $A \in \mathbb{F}^{n \times m}$, and let $r \triangleq \operatorname{rank} A$. Then, the following statements are equivalent:

i) AA^* is a projector.

ii) A^*A is a projector.

iii) $AA^*A = A$.

iv) $A^*AA^* = A^*$.

v) $A^+ = A^*$.

vi) $\sigma_1(A) = \sigma_r(A) = 1$.

In particular, $N_n^+ = N_n^{\mathrm{T}}$.

Proof: See [178, pp. 219–220].

Remark: A is a *partial isometry*, which preserves lengths and distances with respect to the Euclidean norm on $\mathcal{R}(A^*)$. See [178, p. 219].

Remark: See Fact 5.11.30.

Fact 6.3.28. Let $A \in \mathbb{F}^{n \times m}$, assume that A is nonzero, and let $r \triangleq \operatorname{rank} A$. Then, for all $i \in \{1, \ldots, r\}$, the singular values of A^+ are given by

$$\sigma_i(A^+) = \sigma_{r+1-i}^{-1}(A).$$

In particular,

$$\sigma_r(A) = 1/\sigma_{\max}(A^+).$$

If, in addition, $A \in \mathbb{F}^{n \times n}$ and A is nonsingular, then

$$\sigma_{\min}(A) = 1/\sigma_{\max}(A^{-1}).$$

Fact 6.3.29. Let $A \in \mathbb{F}^{n \times m}$. Then, $X = A^+$ is the unique matrix satisfying

$$\operatorname{rank} \begin{bmatrix} A & AA^+ \\ A^+A & X \end{bmatrix} = \operatorname{rank} A.$$

Proof: See [496].

Remark: See Fact 2.17.10 and Fact 6.6.2.

Fact 6.3.30. Let $A \in \mathbb{F}^{n \times n}$, and assume that A is centrohermitian. Then, A^+ is centrohermitian.

Proof: See [908].

Fact 6.3.31. Let $A \in \mathbb{F}^{n \times n}$. Then, the following statements are equivalent:

i) $A^2 = AA^*A$.

ii) A is the product of two projectors.

iii) $A = A(A^+)^2A$.

Remark: This result is due to Crimmins. See [1141].

Fact 6.3.32. Let $A \in \mathbb{F}^{n \times m}$. Then,

$$A^+ = 4(I + A^+A)^+A^+(I + AA^+)^+.$$

Proof: Use Fact 6.4.41 with $B = A$.

Fact 6.3.33. Let $A \in \mathbb{F}^{n \times n}$, and assume that A is unitary. Then,

$$\lim_{k \to \infty} \tfrac{1}{k} \sum_{i=0}^{k-1} A^i = I - (A - I)(A - I)^+.$$

Proof: Use Fact 11.21.13 and Fact 11.21.15, and note that $(A - I)^* = (A - I)^+$. See [641, p. 185].

Remark: $I - (A - I)(A - I)^+$ is the projector onto $\{x: Ax = x\} = \mathcal{N}(A - I)$.

Remark: This result is the *ergodic theorem*.

Fact 6.3.34. Let $A \in \mathbb{F}^{n \times m}$, and define the sequence $(B_i)_{i=1}^\infty$ by

$$B_{i+1} \triangleq 2B_i - B_iAB_i,$$

where $B_0 \triangleq \alpha A^*$ and $\alpha \in (0, 2/\sigma_{\max}^2(A))$. Then,

$$\lim_{i \to \infty} B_i = A^+.$$

Proof: See [148, p. 259] or [291, p. 250]. This result is due to Ben-Israel.

Remark: This sequence is a Newton-Raphson algorithm.

Remark: B_0 satisfies $\mathrm{sprad}(I - B_0A) < 1$.

Remark: For the case in which A is square and nonsingular, see Fact 2.16.29.

Problem: Does convergence hold for all $B_0 \in \mathbb{F}^{n \times n}$ satisfying $\mathrm{sprad}(I - B_0A) < 1$?

Fact 6.3.35. Let $A \in \mathbb{F}^{n \times m}$, let $(A_i)_{i=1}^\infty \subset \mathbb{F}^{n \times m}$, and assume that $\lim_{i \to \infty} A_i = A$. Then, $\lim_{i \to \infty} A_i^+ = A^+$ if and only if there exists a positive integer k such that, for all $i > k$, $\mathrm{rank}\, A_i = \mathrm{rank}\, A$.

Proof: See [291, pp. 218, 219] or [1208, pp. 199, 200].

6.4 Facts on the Moore-Penrose Generalized Inverse for Two or More Matrices

Fact 6.4.1. Let $A \in \mathbb{F}^{n \times m}$ and $B \in \mathbb{F}^{m \times n}$. Then, the following statements are equivalent:

  i) $B = A^+$.

  ii) $A^*AB = A^*$ and $B^*BA = B^*$.

  iii) $BAA^* = A^*$ and $ABB^* = B^*$.

Remark: See [671, pp. 503, 513].

Fact 6.4.2. Let $A \in \mathbb{F}^{n \times m}$, and let $x \in \mathbb{F}^n$ and $y \in \mathbb{F}^m$ be nonzero. Furthermore, define

$$d \triangleq A^+x, \quad e \triangleq A^{+*}y, \quad f \triangleq (I - AA^+)x, \quad g \triangleq (I - A^+A)y,$$

$$\delta \triangleq d^*d, \quad \eta \triangleq e^*e, \quad \phi \triangleq f^*f, \quad \psi \triangleq g^*g,$$

$$\lambda \triangleq 1 + y^*A^+x, \quad \mu \triangleq |\lambda|^2 + \delta\psi, \quad \nu \triangleq |\lambda|^2 + \eta\phi.$$

Then,

$$\operatorname{rank}(A + xy^*) = \operatorname{rank} A - 1$$

if and only if

$$x \in \mathcal{R}(A), \quad y \in \mathcal{R}(A^*), \quad \lambda = 0.$$

In this case,

$$(A + xy^*)^+ = A^+ - \delta^{-1}dd^*A^+ - \eta^{-1}A^+ee^* + (\delta\eta)^{-1}d^*A^+ede^*.$$

Furthermore,

$$\operatorname{rank}(A + xy^*) = \operatorname{rank} A$$

if and only if one of the following conditions is satisfied:

$$\begin{cases} x \in \mathcal{R}(A), \quad y \in \mathcal{R}(A^*), \quad \lambda \neq 0, \\ x \in \mathcal{R}(A), \quad y \notin \mathcal{R}(A^*), \\ x \notin \mathcal{R}(A), \quad y \in \mathcal{R}(A^*). \end{cases}$$

In this case, respectively,

$$\begin{cases} (A + xy^*)^+ = A^+ - \lambda^{-1}de^*, \\ (A + xy^*)^+ = A^+ - \mu^{-1}(\psi dd^*A^+ + \delta ge^*) + \mu^{-1}(\lambda gd^*A^+ - \bar{\lambda}de^*), \\ (A + xy^*)^+ = A^+ - \nu^{-1}(\phi A^+ee^* + \eta df^*) + \nu^{-1}(\lambda A^+ef^* - \bar{\lambda}de^*). \end{cases}$$

Finally,

$$\operatorname{rank}(A + xy^*) = \operatorname{rank} A + 1$$

if and only if

$$x \notin \mathcal{R}(A), \quad y \notin \mathcal{R}(A^*).$$

In this case,

$$(A + xy^*)^+ = A^+ - \phi^{-1}df^* - \psi^{-1}ge^* + \lambda(\phi\psi)^{-1}gf^*.$$

Proof: See [111]. To prove sufficiency in the first alternative of the third statement, let $\hat{x} \in \mathbb{F}^m$ and $\hat{y} \in \mathbb{F}^n$ be such that $x = A\hat{x}$ and $y = A^*\hat{y}$. Then, $A + xy^* = A(I + \hat{x}y^*)$. Since $\alpha \neq 0$ it follows that $-1 \neq y^*A^+x = \hat{y}^*AA^+A\hat{x} = \hat{y}^*A\hat{x} = y^*\hat{x}$. It now follows that $I + \hat{x}y^*$ is an elementary matrix and thus, by Fact 3.7.19, is nonsingular.

Remark: An equivalent version of the first statement is given in [338] and [740, p. 33]. A detailed treatment of the generalized inverse of an outer-product perturbation is given in [1430, pp. 152–157].

Remark: See Fact 2.10.25.

Fact 6.4.3. Let $A \in \mathbb{F}^{n \times n}$, assume that A is Hermitian and nonsingular, and let $x \in \mathbb{F}^n$ and $y \in \mathbb{F}^n$ be nonzero. Then, $A + xy^*$ is singular if and only if $y^*A^{-1}x + 1 = 0$. In this case,

$$(A + xy^*)^+ = (I - aa^+)A^{-1}(I - bb^+),$$

where $a \triangleq A^{-1}x$ and $b \triangleq A^{-1}y$.

Proof: See [1208, pp. 197, 198].

Fact 6.4.4. Let $A \in \mathbb{F}^{n \times n}$, assume that A is Hermitian, let $b \in \mathbb{F}^n$, and define $S \triangleq I - A^+A$. Then,

$(A + bb^*)^+$

$$= \begin{cases} \left[I - (b^*(A^+)^2b)^{-1}A^+bb^*A^+\right]A^+\left[I - (b^*(A^+)^2b)^{-1}A^+bb^*A^+\right], & 1 + b^*A^+b = 0, \\ A^+ - (1 + b^*A^+b)^{-1}A^+bb^*A^+, & 1 + b^*A^+b \neq 0, \\ \left[I - (b^*Sb)^{-1}Sbb^*\right]A^+\left[I - (b^*Sb)^{-1}bb^*S\right] + (b^*Sb)^{-2}Sbb^*S, & b^*Sb \neq 0. \end{cases}$$

Proof: See [1031].

Fact 6.4.5. Let $A \in \mathbb{F}^{n \times n}$, assume that A is positive semidefinite, let $C \in \mathbb{F}^{m \times m}$, assume that C is positive definite, and let $B \in \mathbb{F}^{n \times m}$. Then,

$$(A + BCB^*)^+ = A^+ - A^+B(C^{-1} + B^*A^+B)^{-1}B^*A^+$$

if and only if

$$AA^+B = B.$$

Proof: See [1076].

Remark: $AA^+B = B$ is equivalent to $\mathcal{R}(B) \subseteq \mathcal{R}(A)$.

Remark: Extensions of the matrix inversion lemma are considered in [392, 500, 1031, 1154] and [671, pp. 426–428, 447, 448].

Fact 6.4.6. Let $A \in \mathbb{F}^{n \times m}$ and $B \in \mathbb{F}^{m \times l}$. Then, $AB = 0$ if and only if $B^+A^+ = 0$.

Proof: The result follows from $ix) \implies i)$ of Fact 6.4.16.

Fact 6.4.7. Let $A \in \mathbb{F}^{n \times m}$ and $B \in \mathbb{F}^{n \times l}$. Then, $A^+B = 0$ if and only if $A^*B = 0$.

Proof: The result follows from Proposition 6.1.6.

Fact 6.4.8. Let $A \in \mathbb{F}^{n \times m}$, let $B \in \mathbb{F}^{m \times p}$, and assume that $\operatorname{rank} B = m$. Then,

$$AB(AB)^+ = AA^+.$$

Proof: See [1208, p. 215].

Fact 6.4.9. Let $A \in \mathbb{F}^{n \times m}$, let $B \in \mathbb{F}^{m \times m}$, and assume that B is positive definite. Then,

$$ABA^*(ABA^*)^+A = A.$$

Proof: See [1208, p. 215].

Fact 6.4.10. Let $A \in \mathbb{F}^{n \times m}$, assume that $\operatorname{rank} A = m$, let $B \in \mathbb{F}^{n \times n}$, and assume that B is positive definite. Then,

$$(ABA^*)^+ = A(A^*A)^{-1}B^{-1}(A^*A)^{-1}A^*.$$

Proof: Use Fact 6.3.17.

Fact 6.4.11. Let $A \in \mathbb{F}^{n \times m}$, let $S \in \mathbb{F}^{m \times m}$, assume that S is nonsingular, and define $B \triangleq AS$. Then,

$$BB^+ = AA^+.$$

Proof: See [1215, p. 144].

Fact 6.4.12. Let $A \in \mathbb{F}^{n \times r}$ and $B \in \mathbb{F}^{r \times m}$, and assume that $\operatorname{rank} A = \operatorname{rank} B = r$. Then,

$$(AB)^+ = B^+A^+ = B^*(BB^*)^{-1}(A^*A)^{-1}A^*.$$

Remark: AB is a full-rank factorization.

Remark: See Fact 6.4.13.

Fact 6.4.13. Let $A \in \mathbb{F}^{n \times n}$, let $r \triangleq \operatorname{rank} A$, let $B \in \mathbb{F}^{n \times r}$ and $C \in \mathbb{F}^{r \times n}$, and assume that $A = BC$ and $\operatorname{rank} B = \operatorname{rank} C = r$. Then, the following statements are equivalent:

i) A is range Hermitian.

ii) $BB^+ = C^+C$.

iii) $\mathcal{N}(B^*) = \mathcal{N}(C)$.

iv) $B = C^+CB$ and $C = CBB^+$.

v) $B^+ = B^+C^+C$ and $C = CBB^+$.

vi) $B = C^+CB$ and $C^+ = BB^+C^+$.

vii) $B^+ = B^+C^+C$ and $C^+ = BB^+C^+$.

Proof: See [448].

Remark: See Fact 3.6.3, Fact 6.3.9, Fact 6.3.10, and Fact 6.4.12.

Fact 6.4.14. Let $A \in \mathbb{F}^{n \times m}$ and $B \in \mathbb{F}^{m \times l}$. Then,

$$(AB)^+ = (A^+AB)^+(ABB^+)^+.$$

If, in addition, $\mathcal{R}(B) = \mathcal{R}(A^*)$, then $A^+AB = B$, $ABB^+ = A$, and

$$(AB)^+ = B^+A^+.$$

Proof: See [1208, pp. 192] or [1333].

Remark: This result is due to Cline and Greville.

Fact 6.4.15. Let $A \in \mathbb{F}^{n \times m}$ and $B \in \mathbb{F}^{m \times l}$, and define $A_1 \triangleq AB_1 B_1^+$ and $B_1 \triangleq A^+ AB$. Then,
$$AB = A_1 B_1$$
and
$$(AB)^+ = B_1^+ A_1^+.$$

Proof: See [1208, pp. 191, 192].

Fact 6.4.16. Let $A \in \mathbb{F}^{n \times m}$ and $B \in \mathbb{F}^{m \times l}$. Then, the following statements are equivalent:

i) $(AB)^+ = B^+ A^+$.

ii) $\mathcal{R}(A^*AB) \subseteq \mathcal{R}(B)$ and $\mathcal{R}(BB^*A^*) \subseteq \mathcal{R}(A^*)$.

iii) $(AB)(AB)^+ = (AB)B^+A^+$ and $(AB)^+(AB) = B^+A^+AB$.

iv) $A^*AB = BB^+A^*AB$ and $ABB^* = ABB^*A^+A$.

v) $AB(AB)^+A = ABB^+$ and $A^+AB = B(AB)^+AB$.

vi) A^*ABB^+ and A^+ABB^* are Hermitian.

vii) $(ABB^+)^+ = BB^+A^+$ and $(A^+AB)^+ = B^+A^+A$.

viii) $B^+(ABB^+)^+ = B^+A^+$ and $(A^+AB)^+A = B^+A^+$.

ix) $A^*ABB^* = BB^+A^*ABB^*A^+A$.

Proof: See [16, p. 53], [1208, pp. 190, 191], and [601, 1323].

Remark: The equivalence of *i)* and *ii)* is due to Greville.

Remark: Conditions under which B^+A^+ is a (1)-inverse of AB are given in [1323].

Remark: See [1450].

Fact 6.4.17. Let $A \in \mathbb{F}^{n \times m}$ and $B \in \mathbb{F}^{m \times l}$. Then, the following statements are equivalent:

i) $(AB)^+ = B^+A^+ - B^+[(I - BB^+)(I - A^+A)]^+A^+$.

ii) $\mathcal{R}(AA^*AB) = \mathcal{R}(AB)$ and $\mathcal{R}[(ABB^*B)^*] = \mathcal{R}[(AB)^*]$.

Proof: See [1321].

Fact 6.4.18. Let $A, B \in \mathbb{F}^{n \times n}$, and assume that A and B are projectors. Then,
$$\mathcal{R}([A, B]) = \mathcal{R}[(A - B)^+ - (A - B)].$$
Consequently, $(A - B)^+ = (A - B)$ if and only if $AB = BA$.

Proof: See [1320].

Fact 6.4.19. Let $A, B \in \mathbb{F}^{n \times n}$, and assume that A and B are projectors. Then, the following statements hold:

i) $(AB)^+ = B(AB)^+$.

ii) $(AB)^+ = (AB)^+A$.

iii) $(AB)^+ = B(AB)^+A$.

iv) $(AB)^+ = BA - B(B_\perp A_\perp)^+A$.

v) $(AB)^+$, $B(AB)^+$, $(AB)^+A$, $B(AB)^+A$, and $BA - B(B_\perp A_\perp)^+A$ are idempotent.

vi) $AB = A(AB)^+B$.

vii) $(AB)^2 = AB + AB(B_\perp A_\perp)^+AB$.

Proof: To prove *i)* note that $\mathcal{R}[(AB)^+] = \mathcal{R}[(AB)^*] = \mathcal{R}(BA)$, and thus $\mathcal{R}[B(AB)^+] = \mathcal{R}[B(AB)^*] = \mathcal{R}(BA)$. Hence, $\mathcal{R}[(AB)^+] = \mathcal{R}[B(AB)^+]$. It now follows from Fact 3.13.14 that $(AB)^+ = B(AB)^+$. Statement *iv)* follows from Fact 6.4.17. Statements *v)* and *vi)* follow from *iii)*. Statement *vii)* follows from *iv)* and *vi)*.

Remark: The converse of the first result in *v)* is given by Fact 6.4.20.

Remark: See Fact 6.3.26, Fact 6.4.14, and Fact 6.4.24. See [1321, 1457].

Fact 6.4.20. Let $A \in \mathbb{F}^{n \times n}$, and assume that A is idempotent. Then, there exist projectors $B, C \in \mathbb{F}^{n \times n}$ such that $A = (BC)^+$.

Proof: See [330, 551].

Remark: The converse of this result is given by *v)* of Fact 6.4.19.

Remark: This result is due to Penrose.

Fact 6.4.21. Let $A, B \in \mathbb{F}^{n \times n}$, and assume that $\mathcal{R}(A)$ and $\mathcal{R}(B)$ are complementary subspaces. Furthermore, define $P \triangleq AA^+$ and $Q \triangleq BB^+$. Then, the matrix $(Q_\perp P)^+$ is the idempotent matrix onto $\mathcal{R}(B)$ along $\mathcal{R}(A)$.

Proof: See [602].

Remark: See Fact 3.12.33, Fact 3.13.24, and Fact 6.4.22.

Fact 6.4.22. Let $A, B \in \mathbb{F}^{n \times n}$, assume that A and B are projectors, and assume that $\mathcal{R}(A)$ and $\mathcal{R}(B)$ are complementary subspaces. Then, $(A_\perp B)^+$ is the idempotent matrix onto $\mathcal{R}(B)$ along $\mathcal{R}(A)$.

Proof: See Fact 6.4.21, [607], or [765].

Remark: It follows from Fact 6.4.19 that $(A_\perp B)^+$ is idempotent.

Remark: See Fact 3.12.33, Fact 3.13.24, and Fact 6.4.21.

Fact 6.4.23. Let $A, B \in \mathbb{F}^{n \times n}$, assume that A and B are projectors, and assume that $A - B$ is nonsingular. Then, $I - BA$ is nonsingular, and

$$(A_\perp B)^+ = (I - BA)^{-1}B(I - BA).$$

Proof: Combine Fact 3.13.24 and Fact 6.4.22.

Fact 6.4.24. Let $k \geq 1$, let $A_1, \ldots, A_k \in \mathbb{F}^{n \times n}$, assume that $A_1, \ldots, A_k$ are projectors, and define $B_1, \ldots, B_{k-1} \in \mathbb{F}^{n \times n}$ by

$$B_i \triangleq (A_1 \cdots A_{k-i+1})^+ A_1 \cdots A_{k-i}, \quad i = 1, \ldots, k-2,$$

and

$$B_{k-1} \triangleq A_2 \cdots A_k (A_1 \cdots A_k)^+.$$

Then, $B_1, \ldots, B_{k-1}$ are idempotent, and

$$(A_1 \cdots A_k)^+ = B_1 \cdots B_{k-1}.$$

Proof: See [1330].

Remark: When $k = 2$, the result that B_1 is idempotent is given by *vi*) of Fact 6.4.19.

Fact 6.4.25. Let $A \in \mathbb{F}^{n \times n}$ and $B \in \mathbb{F}^{m \times n}$, and assume that A is idempotent. Then,

$$A^*(BA)^+ = (BA)^+.$$

Proof: See [671, p. 514].

Fact 6.4.26. Let $A, B \in \mathbb{F}^{n \times n}$, and assume that A and B are projectors. Then, the following statements are equivalent:

i) AB is a projector.

ii) $[(AB)^+]^2 = [(AB)^2]^+$.

Proof: See [1353].

Remark: See Fact 3.13.20 and Fact 5.12.16.

Fact 6.4.27. Let $A \in \mathbb{F}^{n \times m}$. Then, $B \in \mathbb{F}^{m \times n}$ satisfies $BAB = B$ if and only if there exist projectors $C \in \mathbb{F}^{n \times n}$ and $D \in \mathbb{F}^{m \times m}$ such that $B = (CAD)^+$.

Proof: See [602].

Fact 6.4.28. Let $A \in \mathbb{F}^{n \times n}$. Then, A is idempotent if and only if there exist projectors $B, C \in \mathbb{F}^{n \times n}$ such that $A = (BC)^+$.

Proof: Let $A = I$ in Fact 6.4.27.

Remark: See [608].

Fact 6.4.29. Let $A, B \in \mathbb{F}^{n \times n}$, and assume that A is range Hermitian. Then, $AB = BA$ if and only if $A^+B = BA^+$.

Proof: See [1312].

Fact 6.4.30. Let $A, B \in \mathbb{F}^{n \times n}$, and assume that A and B are range Hermitian. Then, the following statements are equivalent:

i) $AB = BA$.

ii) $A^+B = BA^+$.

iii) $AB^+ = B^+A$.

iv) $A^+B^+ = B^+A^+$.

Proof: See [1312].

Fact 6.4.31. Let $A, B \in \mathbb{F}^{n \times n}$, assume that A and B are range Hermitian, and assume that $(AB)^+ = A^+B^+$. Then, AB is range Hermitian.

Proof: See [665].

Remark: See Fact 8.21.21.

Fact 6.4.32. Let $A, B \in \mathbb{F}^{n \times n}$, and assume that A and B are range Hermitian. Then, the following statements are equivalent:

i) AB is range Hermitian.

ii) $AB(I - A^+A) = 0$ and $(I - B^+B)AB = 0$.

iii) $\mathcal{N}(A) \subseteq \mathcal{N}(AB)$ and $\mathcal{R}(AB) \subseteq \mathcal{R}(B)$.

iv) $\mathcal{N}(AB) = \mathcal{N}(A) + \mathcal{N}(B)$ and $\mathcal{R}(AB) = \mathcal{R}(A) \cap \mathcal{R}(B)$.

Proof: See [665, 856].

Fact 6.4.33. Let $A \in \mathbb{F}^{n \times m}$ and $B \in \mathbb{F}^{m \times l}$, and assume that $\operatorname{rank} B = m$. Then,
$$AB(AB)^+ = AA^+.$$

Fact 6.4.34. Let $A \in \mathbb{F}^{n \times m}$, $B \in \mathbb{F}^{m \times n}$, and $C \in \mathbb{F}^{m \times n}$, and assume that $BAA^* = A^*$ and $A^*AC = A^*$. Then,
$$A^+ = BAC.$$

Proof: See [16, p. 36].

Remark: This result is due to Decell.

Fact 6.4.35. Let $A, B \in \mathbb{F}^{n \times n}$, and assume that $A + B$ is nonsingular. Then, the following statements are equivalent:

i) $\operatorname{rank} A + \operatorname{rank} B = n$.

ii) $A(A + B)^{-1}B = 0$.

iii) $B(A + B)^{-1}A = 0$.

iv) $A(A + B)^{-1}A = A$.

v) $B(A + B)^{-1}B = B$.

vi) $A(A + B)^{-1}B + B(A + B)^{-1}A = 0$.

vii) $A(A + B)^{-1}A + B(A + B)^{-1}B = A + B.$

viii) $(A + B)^{-1} = [(I - BB^+)A(I - B^+B)]^+ + [(I - AA^+)B(I - A^+A)]^+.$

Proof: See [1334].

Remark: See Fact 2.11.4 and Fact 8.21.23.

Fact 6.4.36. Let $A, B \in \mathbb{F}^{n \times n}$, and assume that A and B are projectors. Then, the following statements hold:

i) $A(A - B)^+B = B(A - B)^+A = 0.$

ii) $A - B = A(A - B)^+A - B(B - A)^+B.$

iii) $(A - B)^+ = (A - AB)^+ + (AB - B)^+.$

iv) $(A - B)^+ = (A - BA)^+ + (BA - B)^+.$

v) $(A - B)^+ = A - B + B(A - BA)^+ - (B - BA)^+A.$

vi) $(A - B)^+ = A - B + (A - AB)^+B - A(B - AB)^+.$

vii) $(I - A - B)^+ = (A_\perp B_\perp)^+ - (AB)^+.$

viii) $(I - A - B)^+ = (B_\perp A_\perp)^+ - (BA)^+.$

Furthermore, the following statements are equivalent:

ix) $AB = BA.$

x) $(A - B)^+ = A - B.$

xi) $B(A - BA)^+ = (B - BA)^+A.$

xii) $(A - B)^3 = A - B.$

xiii) $A - B$ is tripotent.

Proof: See [330].

Remark: See Fact 3.12.22.

Fact 6.4.37. Let $A \in \mathbb{F}^{n \times m}$, and let $B \in \mathbb{F}^{n \times p}$. Then,

$$(AA^* + BB^*)^+ = (I - C^{+*}B^*)A^{+*}EA^+(I - BC^+) + (CC^*)^+,$$

where

$$C \triangleq (I - AA^+)B$$

and

$$E \triangleq I - A^*B(I - C^+C)[I + (I - C^+C)B^*(AA^*)^+B(I - C^+C)]^{-1}(A^+B)^*.$$

Proof: See [1208, p. 196].

Fact 6.4.38. Let $A, B \in \mathbb{F}^{n \times m}$, and assume that $A^*B = 0$. Then,

$$(A + B)^+ = A^+ + (I - A^+B)(C^+ + D),$$

where

$$C \triangleq (I - AA^+)B$$

and

$$D \triangleq (I - C^+C)[I + (I - C^+C)B^*(AA^*)^+B(I - C^+C)]^{-1}B^*(AA^*)^+(I - BC^+).$$

Proof: See [1208, p. 196].

Remark: See Fact 6.5.18, from which it follows that $BA^* = 0$ implies that $(I - A^+B)(C^+ + D) = B^+$.

Fact 6.4.39. Let $A, B \in \mathbb{F}^{n \times m}$, and assume that $A^*B = 0$ and $BA^* = 0$. Then,

$$(A + B)^+ = A^+ + B^+.$$

Proof: Use Fact 2.10.29 and Fact 6.4.40. See [347], [671, p. 513], or [1208, p. 197].

Remark: This result is due to Penrose.

Fact 6.4.40. Let $A, B \in \mathbb{F}^{n \times m}$, and assume that $\operatorname{rank}(A + B) = \operatorname{rank} A + \operatorname{rank} B$. Then,

$$(A + B)^+ = (I - C^+B)A^+(I - BC^+) + C^+,$$

where $C \triangleq (I - AA^+)B(I - A^+A)$.

Proof: See [347].

Fact 6.4.41. Let $A, B \in \mathbb{F}^{n \times m}$. Then,

$$(A + B)^+ = (I + A^+B)^+(A^+ + A^+BA^+)(I + BA^+)^+$$

if and only if $AA^+B = B = BA^+A$. Furthermore, if $n = m$ and A is nonsingular, then

$$(A + B)^+ = (I + A^{-1}B)^+(A^{-1} + A^{-1}BA^{-1})(I + BA^{-1})^+.$$

Proof: See [347].

Remark: If A and $A+B$ are nonsingular, then the last statement yields $(A+B)^{-1} = (A + B)^{-1}(A + B)(A + B)^{-1}$ for which the assumption that A is nonsingular is superfluous.

Fact 6.4.42. Let $A, B \in \mathbb{F}^{n \times m}$. Then,

$A^+ - B^+$

$$= B^+(B - A)A^+ + (I - B^+B)(A^* - B^*)A^{+*}A^+ + B^+B^{+*}(A^* - B^*)(I - AA^+)$$

$$= A^+(B - A)B^+ + (I - A^+A)(A^* - B^*)B^{+*}B^+ + A^+A^{+*}(A^* - B^*)(I - BB^+).$$

Furthermore, if B is left invertible, then

$$A^+ - B^+ = B^+(B - A)A^+ + B^+B^{+*}(A^* - B^*)(I - AA^+),$$

while, if B is right invertible, then

$$A^+ - B^+ = A^+(B - A)B^+ + (I - A^+A)(A^* - B^*)B^{+*}B^+.$$

Proof: See [291, p. 224].

Fact 6.4.43. Let $A \in \mathbb{F}^{n \times m}$, $B \in \mathbb{F}^{l \times k}$, and $C \in \mathbb{F}^{n \times k}$. Then, there exists a matrix $X \in \mathbb{F}^{m \times l}$ satisfying $AXB = C$ if and only if $AA^{+}CB^{+}B = C$. Furthermore, X satisfies $AXB = C$ if and only if there exists a matrix $Y \in \mathbb{F}^{m \times l}$ such that

$$X = A^{+}CB^{+} + Y - A^{+}AYBB^{+}.$$

Finally, if $Y = 0$, then $\operatorname{tr} X^{*}X$ is minimized.

Proof: Use Proposition 6.1.7. See [973, p. 37] and, for Hermitian solutions, see [831].

Fact 6.4.44. Let $A \in \mathbb{F}^{n \times m}$, and assume that $\operatorname{rank} A = m$. Then, $A^{\mathrm{L}} \in \mathbb{F}^{m \times n}$ is a left inverse of A if and only if there exists a matrix $B \in \mathbb{F}^{m \times n}$ such that

$$A^{\mathrm{L}} = A^{+} + B(I - AA^{+}).$$

Proof: Use Fact 6.4.43 with $A = C = I_{n}$.

Fact 6.4.45. Let $A \in \mathbb{F}^{n \times m}$, and assume that $\operatorname{rank} A = n$. Then, $A^{\mathrm{R}} \in \mathbb{F}^{m \times n}$ is a right inverse of A if and only if there exists a matrix $B \in \mathbb{F}^{m \times n}$ such that

$$A^{\mathrm{R}} = A^{+} + (I - A^{+}A)B.$$

Proof: Use Fact 6.4.43 with $B = C = I_{n}$.

Fact 6.4.46. Let $A, B \in \mathbb{F}^{n \times n}$, and assume that A and B are projectors. Then,

$$\operatorname{glb}\{A, B\} = \lim_{k \to \infty} A(BA)^{k} = 2A(A + B)^{+}B.$$

Furthermore, $2A(A + B)^{+}B$ is the projector onto $\mathcal{R}(A) \cap \mathcal{R}(B)$.

Proof: See [41] and [642, pp. 64, 65, 121, 122].

Remark: See Fact 6.4.47 and Fact 8.21.18.

Fact 6.4.47. Let $A \in \mathbb{R}^{n \times m}$ and $B \in \mathbb{R}^{n \times l}$. Then,

$$\mathcal{R}(A) \cap \mathcal{R}(B) = \mathcal{R}[AA^{+}(AA^{+} + BB^{+})^{+}BB^{+}].$$

Remark: See Theorem 2.3.1 and Fact 8.21.18.

Fact 6.4.48. Let $A \in \mathbb{R}^{n \times m}$ and $B \in \mathbb{R}^{n \times l}$. Then, $\mathcal{R}(A) \subseteq \mathcal{R}(B)$ if and only if $BB^{+}A = A$.

Proof: See [16, p. 35].

Fact 6.4.49. Let $A \in \mathbb{R}^{n \times m}$ and $B \in \mathbb{R}^{n \times l}$. Then,

$$\dim[\mathcal{R}(A) \cap \mathcal{R}(B)] = \operatorname{rank} AA^{+}(AA^{+} + BB^{+})^{+}BB^{+}$$
$$= \operatorname{rank} A + \operatorname{rank} B - \operatorname{rank} \begin{bmatrix} A & B \end{bmatrix}.$$

Proof: Use Fact 2.11.1, Fact 2.11.12, and Fact 6.4.47.

Remark: See Fact 2.11.8.

Fact 6.4.50. Let $A, B \in \mathbb{F}^{n \times n}$, and assume that A and B are projectors. Then,
$$\mathrm{lub}\{A, B\} = (A + B)(A + B)^+.$$

Furthermore, $\mathrm{lub}\{A, B\}$ is the projector onto $\mathcal{R}(A) + \mathcal{R}(B) = \mathrm{span}[\mathcal{R}(A) \cup \mathcal{R}(B)]$.

Proof: Use Fact 2.9.13 and Fact 8.7.5.

Remark: See Fact 8.7.2.

Fact 6.4.51. Let $A, B \in \mathbb{F}^{n \times n}$, and assume that A and B are projectors. Then,
$$\mathrm{lub}\{A, B\} = I - \lim_{k \to \infty} A_\perp (B_\perp A_\perp)^k = I - 2A_\perp (A_\perp + B_\perp)^+ B_\perp.$$

Furthermore, $I - 2A_\perp (A_\perp + B_\perp)^+ B_\perp$ is the projector onto
$$\begin{aligned}
[\mathcal{R}(A_\perp) \cap \mathcal{R}(B_\perp)]^\perp &= [\mathcal{N}(A) \cap \mathcal{N}(B)]^\perp \\
&= [\mathcal{N}(A)]^\perp + [\mathcal{N}(B)]^\perp \\
&= \mathcal{R}(A) + \mathcal{R}(B) \\
&= \mathrm{span}[\mathcal{R}(A) \cup \mathcal{R}(B)].
\end{aligned}$$

Consequently,
$$I - 2A_\perp (A_\perp + B_\perp)^+ B_\perp = (A + B)(A + B)^+.$$

Proof: See [41] and [642, pp. 64, 65, 121, 122].

Remark: See Fact 6.4.47 and Fact 8.21.18.

Fact 6.4.52. Let $A, B \in \mathbb{F}^{n \times m}$. Then,
$$A \overset{*}{\leq} B$$

if and only if
$$A^+ A = A^+ B$$

and
$$AA^+ = BA^+.$$

Proof: See [669].

Remark: See Fact 2.10.35.

6.5 Facts on the Moore-Penrose Generalized Inverse for Partitioned Matrices

Fact 6.5.1. Let $A, B \in \mathbb{F}^{n \times m}$. Then,
$$(A + B)^+ = \tfrac{1}{2} \begin{bmatrix} I_m & I_m \end{bmatrix} \begin{bmatrix} A & B \\ B & A \end{bmatrix}^+ \begin{bmatrix} I_n \\ I_n \end{bmatrix}.$$

Proof: See [1310, 1314, 1334].

Remark: See Fact 2.17.5 and Fact 2.19.7.

Fact 6.5.2. Let $A_1, \ldots, A_k \in \mathbb{F}^{n \times m}$. Then,

$$(A_1 + \cdots + A_k)^+ = \tfrac{1}{k} \begin{bmatrix} I_m & \cdots & I_m \end{bmatrix} \begin{bmatrix} A_1 & A_2 & \cdots & A_k \\ A_k & A_1 & \cdots & A_{k-1} \\ \vdots & \vdots & \ddots & \vdots \\ A_2 & A_3 & \cdots & A_1 \end{bmatrix}^+ \begin{bmatrix} I_n \\ \vdots \\ I_n \end{bmatrix}.$$

Proof: See [1314].

Remark: The partitioned matrix is *block circulant*. See Fact 2.17.6 and Fact 6.6.1.

Fact 6.5.3. Let $A, B \in \mathbb{F}^{n \times m}$. Then, the following statements are equivalent:

i) $\mathcal{R}(\begin{bmatrix} A \\ A^*A \end{bmatrix}) = \mathcal{R}(\begin{bmatrix} B \\ B^*B \end{bmatrix})$.

ii) $\mathcal{R}(\begin{bmatrix} A \\ A^+A \end{bmatrix}) = \mathcal{R}(\begin{bmatrix} B \\ B^+B \end{bmatrix})$.

iii) $A = B$.

Remark: This result is due to Tian.

Fact 6.5.4. Let $A \in \mathbb{F}^{n \times m}$, $B \in \mathbb{F}^{n \times l}$, $C \in \mathbb{F}^{k \times m}$, and $D \in \mathbb{F}^{k \times l}$. Then,

$$\begin{bmatrix} A & B \\ C & D \end{bmatrix} = \begin{bmatrix} I & 0 \\ CA^+ & I \end{bmatrix} \begin{bmatrix} A & B - AA^+B \\ C - CA^+A & D - CA^+B \end{bmatrix} \begin{bmatrix} I & A^+B \\ 0 & I \end{bmatrix}.$$

Remark: See Fact 6.5.25.

Remark: See [1322].

Fact 6.5.5. Let $A \in \mathbb{F}^{n \times n}$, $B \in \mathbb{F}^{n \times m}$, and $C \in \mathbb{F}^{m \times m}$, define $\mathcal{A} \triangleq \begin{bmatrix} A & B \\ B^* & C \end{bmatrix}$, and assume that $B = AA^+B$. Then,

$$\text{In}\,\mathcal{A} = \text{In}\,A + \text{In}(A|\mathcal{A}).$$

Remark: This result is the *Haynsworth inertia additivity formula*. See [1130].

Remark: If $\mathcal{A}$ is positive semidefinite, then $B = AA^+B$. See Proposition 8.2.4.

Fact 6.5.6. Let $A \in \mathbb{F}^{n \times m}$, $B \in \mathbb{F}^{n \times l}$, $C \in \mathbb{F}^{k \times m}$, and $D \in \mathbb{F}^{k \times l}$. Then,

$$\begin{aligned} \text{rank} \begin{bmatrix} A & B \end{bmatrix} &= \text{rank}\,A + \text{rank}(B - AA^+B) \\ &= \text{rank}\,B + \text{rank}(A - BB^+A) \\ &= \text{rank}\,A + \text{rank}\,B - \dim[\mathcal{R}(A) \cap \mathcal{R}(B)], \end{aligned}$$

$$\begin{aligned} \text{rank} \begin{bmatrix} A \\ C \end{bmatrix} &= \text{rank}\,A + \text{rank}(C - CA^+A) \\ &= \text{rank}\,C + \text{rank}(A - AC^+C) \\ &= \text{rank}\,A + \text{rank}\,C - \dim[\mathcal{R}(A^*) \cap \mathcal{R}(C^*)], \end{aligned}$$

$$\text{rank} \begin{bmatrix} A & B \\ C & 0 \end{bmatrix} = \text{rank}\,B + \text{rank}\,C + \text{rank}[(I_n - BB^+)A(I_m - C^+C)],$$

and
$$\text{rank}\begin{bmatrix} A & B \\ C & D \end{bmatrix} = \text{rank}\,A + \text{rank}\,X + \text{rank}\,Y$$
$$+ \text{rank}\big[(I_k - YY^+)(D - CA^+B)(I_l - X^+X)\big],$$

where $X \triangleq B - AA^+B$ and $Y \triangleq C - CA^+A$. Consequently,

$$\text{rank}\,A + \text{rank}(D - CA^+B) \le \text{rank}\begin{bmatrix} A & B \\ C & D \end{bmatrix},$$

and, if $AA^+B = B$ and $CA^+A = C$, then

$$\text{rank}\,A + \text{rank}(D - CA^+B) = \text{rank}\begin{bmatrix} A & B \\ C & D \end{bmatrix}.$$

Finally, if $n = m$ and A is nonsingular, then

$$n + \text{rank}(D - CA^{-1}B) = \text{rank}\begin{bmatrix} A & B \\ C & D \end{bmatrix}.$$

Proof: See [298, 993], Fact 2.11.8, and Fact 2.11.11.

Remark: With certain restrictions the generalized inverses can be replaced by (1)-inverses.

Remark: See Proposition 2.8.3 and Proposition 8.2.3.

    Fact 6.5.7. Let $A \in \mathbb{F}^{n \times m}$, $B \in \mathbb{F}^{k \times l}$, and $C \in \mathbb{F}^{n \times l}$. Then,

$$\min_{X \in \mathbb{F}^{m \times l}, Y \in \mathbb{F}^{n \times k}} \text{rank}(AX + YB + C) = \text{rank}\begin{bmatrix} A & C \\ 0 & -B \end{bmatrix} - \text{rank}\,A - \text{rank}\,B.$$

Furthermore, X, Y is a minimizing solution if and only if there exist $U \in \mathbb{F}^{m \times k}$, $U_1 \in \mathbb{F}^{m \times l}$, and $U_2 \in \mathbb{F}^{n \times k}$, such that

$$X = -A^+C + UB + (I_m - A^+A)U_1,$$

$$Y = (AA^+ - I)CB^+ - AU + U_2(I_k - BB^+).$$

Finally, all such matrices $X \in \mathbb{F}^{m \times l}$ and $Y \in \mathbb{F}^{n \times k}$ satisfy

$$AX + YB + C = 0$$

if and only if

$$\text{rank}\begin{bmatrix} A & C \\ 0 & -B \end{bmatrix} = \text{rank}\,A + \text{rank}\,B.$$

Proof: See [1317, 1335].

Remark: See Fact 5.10.20. Note that A and B are square in Fact 5.10.20.

    Fact 6.5.8. Let $A \in \mathbb{F}^{n \times n}$, $B \in \mathbb{F}^{n \times m}$, and $C \in \mathbb{F}^{m \times m}$, and assume that $\begin{bmatrix} A & B \\ B^* & C \end{bmatrix}$ is a projector. Then,

$$\text{rank}(D - B^*A^+B) = \text{rank}\,C - \text{rank}\,B^*A^+B.$$

Proof: See [1327].

Remark: See [110].

Fact 6.5.9. Let $A \in \mathbb{F}^{n \times m}$ and $B \in \mathbb{F}^{n \times l}$. Then, the following statements are equivalent:

i) $\operatorname{rank} \begin{bmatrix} A & B \end{bmatrix} = \operatorname{rank} A + \operatorname{rank} B$.

ii) $\mathcal{R}(A) \cap \mathcal{R}(B) = \{0\}$.

iii) $\operatorname{rank}(AA^* + BB^*) = \operatorname{rank} A + \operatorname{rank} B$.

iv) $A^*(AA^* + BB^*)^+A$ is idempotent.

v) $A^*(AA^* + BB^*)^+A = A^+A$.

vi) $A^*(AA^* + BB^*)^+B = 0$.

Proof: See [973, pp. 56, 57].

Remark: See Fact 2.11.8.

Fact 6.5.10. Let $A \in \mathbb{F}^{n \times m}$ and $B \in \mathbb{F}^{n \times l}$, and define the projectors $P \triangleq AA^+$ and $Q \triangleq BB^+$. Then, the following statements are equivalent:

i) $\operatorname{rank} \begin{bmatrix} A & B \end{bmatrix} = \operatorname{rank} A + \operatorname{rank} B = n$.

ii) $P - Q$ is nonsingular.

In this case,

$$
\begin{aligned}
(P - Q)^{-1} &= (P - PQ)^+ + (PQ - Q)^+ \\
&= (P - QP)^+ + (QP - Q)^+ \\
&= P - Q + Q(P - QP)^+ - (Q - QP)^+P.
\end{aligned}
$$

Proof: See [330].

Fact 6.5.11. Let $A \in \mathbb{F}^{n \times m}$, $B \in \mathbb{F}^{n \times l}$, $C \in \mathbb{F}^{l \times n}$, $D \in \mathbb{F}^{l \times l}$, and assume that D is nonsingular. Then,

$$
\operatorname{rank} A = \operatorname{rank}(A - BD^{-1}C) + \operatorname{rank} BD^{-1}C
$$

if and only if there exist matrices $X \in \mathbb{F}^{m \times l}$ and $Y \in \mathbb{F}^{l \times n}$ such that $B = AX$, $C = YA$, and $D = YAX$.

Proof: See [338].

Fact 6.5.12. Let $A \in \mathbb{F}^{n \times m}$, $B \in \mathbb{F}^{n \times l}$, $C \in \mathbb{F}^{k \times m}$, and $D \in \mathbb{F}^{k \times l}$. Then,

$$
\operatorname{rank} A + \operatorname{rank}(D - CA^+B) = \operatorname{rank} \begin{bmatrix} A^*AA^* & A^*B \\ CA^* & D \end{bmatrix}.
$$

Proof: See [1318].

Fact 6.5.13. Let $A_{11} \in \mathbb{F}^{n \times m}$, $A_{12} \in \mathbb{F}^{n \times l}$, $A_{21} \in \mathbb{F}^{k \times m}$, and $A_{22} \in \mathbb{F}^{k \times l}$, and define $A \triangleq \left[\begin{smallmatrix} A_{11} & A_{12} \\ A_{21} & A_{22} \end{smallmatrix} \right] \in \mathbb{F}^{(n+k) \times (m+l)}$ and $B \triangleq AA^+ = \left[\begin{smallmatrix} B_{11} & B_{12} \\ B_{12}^{\mathrm{T}} & B_{22} \end{smallmatrix} \right]$, where $B_{11} \in \mathbb{F}^{n \times m}$, $B_{12} \in \mathbb{F}^{n \times l}$, $B_{21} \in \mathbb{F}^{k \times m}$, and $B_{22} \in \mathbb{F}^{k \times l}$. Then,

$$\operatorname{rank} B_{12} = \operatorname{rank} \begin{bmatrix} A_{11} & A_{12} \end{bmatrix} + \operatorname{rank} \begin{bmatrix} A_{21} & A_{22} \end{bmatrix} - \operatorname{rank} A.$$

Proof: See [1340].

Remark: See Fact 3.12.20 and Fact 3.13.12.

Fact 6.5.14. Let $A, B \in \mathbb{F}^{n \times n}$. Then,

$$\operatorname{rank} \begin{bmatrix} 0 & A \\ B & I \end{bmatrix} = \operatorname{rank} A + \operatorname{rank} \begin{bmatrix} B & I - A^+A \end{bmatrix}$$

$$= \operatorname{rank} \begin{bmatrix} A \\ I - BB^+ \end{bmatrix} + \operatorname{rank} B$$

$$= \operatorname{rank} A + \operatorname{rank} B + \operatorname{rank} \left[(I - BB^+)(I - A^+A) \right]$$

$$= n + \operatorname{rank} AB.$$

Hence, the following statements hold:

i) $\operatorname{rank} AB = \operatorname{rank} A + \operatorname{rank} B - n$ if and only if $(I - BB^+)(I - A^+A) = 0$.

ii) $\operatorname{rank} AB = \operatorname{rank} A$ if and only if $\begin{bmatrix} B & I - A^+A \end{bmatrix}$ is right invertible.

iii) $\operatorname{rank} AB = \operatorname{rank} B$ if and only if $\begin{bmatrix} A \\ I - BB^+ \end{bmatrix}$ is left invertible.

Proof: See [993].

Remark: The generalized inverses can be replaced by arbitrary (1)-inverses.

Fact 6.5.15. Let $A \in \mathbb{F}^{n \times m}$, $B \in \mathbb{F}^{m \times l}$, and $C \in \mathbb{F}^{l \times k}$. Then,

$$\operatorname{rank} \begin{bmatrix} 0 & AB \\ BC & B \end{bmatrix} = \operatorname{rank} B + \operatorname{rank} ABC$$

$$= \operatorname{rank} AB + \operatorname{rank} BC$$

$$\quad + \operatorname{rank} \left[(I - BC)(BC)^+ \right] B \left[(I - (AB)^+(AB) \right].$$

Furthermore, the following statements are equivalent:

i) $\operatorname{rank} \begin{bmatrix} 0 & AB \\ BC & B \end{bmatrix} = \operatorname{rank} AB + \operatorname{rank} BC$.

ii) $\operatorname{rank} ABC = \operatorname{rank} AB + \operatorname{rank} BC - \operatorname{rank} B$.

iii) There exist matrices $X \in \mathbb{F}^{k \times l}$ and $Y \in \mathbb{F}^{m \times n}$ such that

$$BCX + YAB = B.$$

Proof: See [993, 1340] and Fact 5.10.20.

Remark: This result is related to the Frobenius inequality. See Fact 2.11.14.

Fact 6.5.16. Let $x, y \in \mathbb{R}^3$, and assume that x and y are linearly independent. Then,

$$\begin{bmatrix} x & y \end{bmatrix}^+ = \begin{bmatrix} x^+(I_3 - y\phi^{\mathrm{T}}) \\ \phi^{\mathrm{T}} \end{bmatrix},$$

where $x^+ = (x^{\mathrm{T}}x)^{-1}x^{\mathrm{T}}$, $\alpha \triangleq y^{\mathrm{T}}(I - xx^+)y$, and $\phi \triangleq \alpha^{-1}(I - xx^+)y$. Now, let $x, y, z \in \mathbb{R}^3$, and assume that x and y are linearly independent. Then,

$$\begin{bmatrix} x & y & z \end{bmatrix}^+ = \begin{bmatrix} (I_2 - \beta ww^{\mathrm{T}}) \begin{bmatrix} x & y \end{bmatrix}^+ \\ \beta w^{\mathrm{T}} \begin{bmatrix} x & y \end{bmatrix}^+ \end{bmatrix},$$

where $w \triangleq \begin{bmatrix} x & y \end{bmatrix}^+ z$ and $\beta \triangleq 1/(1 + w^{\mathrm{T}}w)$.

Proof: See [1351].

Fact 6.5.17. Let $A \in \mathbb{F}^{n \times m}$ and $b \in \mathbb{F}^n$. Then,

$$\begin{bmatrix} A & b \end{bmatrix}^+ = \begin{bmatrix} A^+(I_n - b\phi^*) \\ \phi^* \end{bmatrix}$$

and

$$\begin{bmatrix} b & A \end{bmatrix}^+ = \begin{bmatrix} \phi^* \\ A^+(I_n - b\phi^*) \end{bmatrix},$$

where

$$\phi \triangleq \begin{cases} (b - AA^+b)^{+*}, & b \neq AA^+b, \\ \gamma^{-1}(AA^*)^+b, & b = AA^+b. \end{cases}$$

and $\gamma \triangleq 1 + b^*(AA^*)^+b$.

Proof: See [16, p. 44], [494, p. 270], or [1217, p. 148].

Remark: This result is due to Greville.

Fact 6.5.18. Let $A \in \mathbb{F}^{n \times m}$ and $B \in \mathbb{F}^{n \times l}$. Then,

$$\begin{bmatrix} A & B \end{bmatrix}^+ = \begin{bmatrix} A^+ - A^+B(C^+ + D) \\ C^+ + D \end{bmatrix},$$

where

$$C \triangleq (I - AA^+)B$$

and

$$D \triangleq (I - C^+C)[I + (I - C^+C)B^*(AA^*)^+B(I - C^+C)]^{-1}B^*(AA^*)^+(I - BC^+).$$

Furthermore,

$$
\begin{bmatrix} A & B \end{bmatrix}^+ = \begin{cases} \begin{bmatrix} A^*(AA^* + BB^*)^{-1} \\ B^*(AA^* + BB^*)^{-1} \end{bmatrix}, & \operatorname{rank}\begin{bmatrix} A & B \end{bmatrix} = n, \\[2em] \begin{bmatrix} A^*A & A^*B \\ B^*A & B^*B \end{bmatrix}^{-1} \begin{bmatrix} A^* \\ B^* \end{bmatrix}, & \operatorname{rank}\begin{bmatrix} A & B \end{bmatrix} = m + l, \\[2em] \begin{bmatrix} A^*(AA^*)^{-1}(I - BE) \\ E \end{bmatrix}, & \operatorname{rank} A = n, \end{cases}
$$

where

$$
E \triangleq \left[I + B^*(AA^*)^{-1}B \right]^{-1} B^*(AA^*)^{-1}.
$$

Proof: See [346], [972, p. 14], or [1208, pp. 193–195].

Remark: If $\begin{bmatrix} A & B \end{bmatrix}$ is square and nonsingular and $A^*B = 0$, then the second expression yields Fact 2.17.8.

Remark: See Fact 6.4.38.

Fact 6.5.19. Let $A \in \mathbb{F}^{n \times m}$ and $B \in \mathbb{F}^{n \times l}$. Then,

$$
\operatorname{rank}\left(\begin{bmatrix} A & B \end{bmatrix}^+ - \begin{bmatrix} A^+ \\ B^+ \end{bmatrix} \right) = \operatorname{rank}\begin{bmatrix} AA^*B & BB^*A \end{bmatrix}.
$$

Hence, if $A^*B = 0$, then

$$
\begin{bmatrix} A & B \end{bmatrix}^+ = \begin{bmatrix} A^+ \\ B^+ \end{bmatrix}.
$$

Proof: See [1321].

Fact 6.5.20. Let $A \in \mathbb{F}^{n \times m}$ and $B \in \mathbb{F}^{n \times l}$. Then, the following statements are equivalent:

i) $\begin{bmatrix} A & B \end{bmatrix}\begin{bmatrix} A & B \end{bmatrix}^+ = \frac{1}{2}(AA^+ + BB^+)$.

ii) $\mathcal{R}(A) = \mathcal{R}(B)$.

Furthermore, the following statements are equivalent:

iii) $\begin{bmatrix} A & B \end{bmatrix}^+ = \frac{1}{2}\begin{bmatrix} A^+ \\ B^+ \end{bmatrix}$.

iv) $AA^* = BB^*$.

Proof: See [1332].

Fact 6.5.21. Let $A \in \mathbb{F}^{n \times m}$ and $B \in \mathbb{F}^{k \times l}$. Then,

$$
\begin{bmatrix} A & 0 \\ 0 & B \end{bmatrix}^+ = \begin{bmatrix} A^+ & 0 \\ 0 & B^+ \end{bmatrix}.
$$

Fact 6.5.22. Let $A \in \mathbb{F}^{n \times m}$. Then,

$$
\begin{bmatrix} I_n & A \\ 0_{m \times n} & 0_{m \times m} \end{bmatrix}^+ = \begin{bmatrix} (I_n + AA^*)^{-1} & 0_{n \times m} \\ A^*(I_n + AA^*)^{-1} & 0_{m \times m} \end{bmatrix}.
$$

Proof: See [19, 1358].

Fact 6.5.23. Let $A \in \mathbb{F}^{n \times n}$, let $B \in \mathbb{F}^{n \times m}$, and assume that $BB^* = I$. Then,

$$\begin{bmatrix} A & B \\ B^* & 0 \end{bmatrix}^+ = \begin{bmatrix} 0 & B \\ B^* & -B^*AB \end{bmatrix}.$$

Proof: See [459, p. 237].

Fact 6.5.24. Let $A \in \mathbb{F}^{n \times n}$, assume that A is positive semidefinite, and let $B \in \mathbb{F}^{n \times m}$. Then,

$$\begin{bmatrix} A & B \\ B^* & 0 \end{bmatrix}^+ = \begin{bmatrix} C^+ - C^+BD^+B^*C^+ & C^+BD^+ \\ (C^+BD^+)^* & DD^+ - D^+ \end{bmatrix},$$

where

$$C \triangleq A + BB^*, \qquad D \triangleq B^*C^+B.$$

Proof: See [973, p. 58].

Remark: Representations for the generalized inverse of a partitioned matrix are given in [178, Chapter 5] and [108, 115, 138, 176, 284, 291, 304, 609, 660, 662, 757, 930, 1021, 1022, 1024, 1025, 1026, 1073, 1147, 1165, 1310, 1342, 1308, 1452].

Problem: Show that the generalized inverses in this result and in Fact 6.5.23 are identical when A is positive semidefinite and $BB^* = I$.

Fact 6.5.25. Let $A \in \mathbb{F}^{n \times n}$, $x, y \in \mathbb{F}^n$, and $a \in \mathbb{F}$, and assume that $x \in \mathcal{R}(A)$. Then,

$$\begin{bmatrix} A & x \\ y^\mathrm{T} & a \end{bmatrix} = \begin{bmatrix} I & 0 \\ y^\mathrm{T} & 1 \end{bmatrix} \begin{bmatrix} A & 0 \\ y^\mathrm{T} - y^\mathrm{T}A & a - y^\mathrm{T}A^+x \end{bmatrix} \begin{bmatrix} I & A^+x \\ 0 & 1 \end{bmatrix}.$$

Remark: This factorization holds for the case in which A is singular and $a = 0$. See Fact 2.14.9, Fact 2.16.2, and Fact 6.5.4, and note that $x = AA^+x$.

Problem: Obtain a factorization for the case in which $x \notin \mathcal{R}(A)$ (and thus x is nonzero and A is singular) and $a = 0$.

Fact 6.5.26. Let $A \in \mathbb{F}^{n \times m}$, assume that A is partitioned as

$$A = \begin{bmatrix} A_1 \\ \vdots \\ A_k \end{bmatrix},$$

and define

$$B \triangleq \begin{bmatrix} A_1^+ & \cdots & A_k^+ \end{bmatrix}.$$

Then, the following statements hold:

i) $\det AB = 0$ if and only if $\operatorname{rank} A < n$.

ii) $0 < \det AB \leq 1$ if and only if $\operatorname{rank} A = n$.

iii) If $\operatorname{rank} A = n$, then

$$\det AB = \frac{\det AA^*}{\prod_{i=1}^k \det A_iA_i^*},$$

and thus

$$\det AA^* \leq \prod_{i=1}^{k} \det A_i A_i^*.$$

iv) $\det AB = 1$ if and only if $AB = I$.

v) AB is group invertible.

vi) Every eigenvalue of AB is nonnegative.

vii) $\operatorname{rank} A = \operatorname{rank} B = \operatorname{rank} AB = \operatorname{rank} BA$.

Now, assume in addition that $\operatorname{rank} A = \sum_{i=1}^{k} \operatorname{rank} A_i$, and let β denote the product of the positive eigenvalues of AB. Then, the following statements hold:

viii) $0 < \beta \leq 1$.

ix) $\beta = 1$ if and only if $B = A^+$.

Proof: See [900, 1278].

Remark: Result *iii*) yields Hadamard's inequality given by Fact 8.13.35 in the case in which A is square and each A_i has a single row.

Fact 6.5.27. Let $A \in \mathbb{F}^{n \times m}$ and $B \in \mathbb{F}^{n \times l}$. Then,

$$\det \begin{bmatrix} A^*A & B^*A \\ B^*A & B^*B \end{bmatrix} = \det(A^*A)\det[B^*(I - AA^+)B]$$

$$= \det(B^*B)\det[A^*(I - BB^+)A].$$

Remark: See Fact 2.14.25.

Fact 6.5.28. Let $A \in \mathbb{F}^{n \times n}$, $B \in \mathbb{F}^{n \times m}$, $C \in \mathbb{F}^{m \times n}$, and $D \in \mathbb{F}^{m \times m}$, assume that either $\operatorname{rank} \begin{bmatrix} A & B \end{bmatrix} = \operatorname{rank} A$ or $\operatorname{rank} \begin{bmatrix} A \\ C \end{bmatrix} = \operatorname{rank} A$, and let $A^- \in \mathbb{F}^{n \times n}$ be a (1)-inverse of A. Then,

$$\det \begin{bmatrix} A & B \\ C & D \end{bmatrix} = (\det A)\det(D - CA^-B).$$

Proof: See [148, p. 266].

Fact 6.5.29. Let $A \triangleq \begin{bmatrix} A_{11} & A_{12} \\ A_{21} & A_{22} \end{bmatrix} \in \mathbb{F}^{(n+m) \times (n+m)}$, $B \in \mathbb{F}^{(n+m) \times l}$, $C \in \mathbb{F}^{l \times (n+m)}$, $D \in \mathbb{F}^{l \times l}$, and $\mathcal{A} \triangleq \begin{bmatrix} A & B \\ C & D \end{bmatrix}$, and assume that A and A_{11} are nonsingular. Then,

$$A|\mathcal{A} = (A_{11}|A)|(A_{11}|\mathcal{A}).$$

Proof: See [1125, pp. 18, 19].

Remark: This result is the *Crabtree-Haynsworth quotient formula*. See [736].

Remark: Extensions are given in [1531].

Problem: Extend this result to the case in which either A or A_{11} is singular.

Fact 6.5.30. Let $A, B \in \mathbb{F}^{n \times m}$. Then, the following statements are equivalent:

i) $A \overset{\mathrm{rs}}{\leq} B$.

ii) $AA^+B = BA^+A = BA^+B = B$.

iii) $\operatorname{rank} A = \operatorname{rank} \begin{bmatrix} A & B \end{bmatrix} = \operatorname{rank} \begin{bmatrix} A \\ B \end{bmatrix}$ and $BA^+B = B$.

Proof: See [1215, p. 45].

Remark: See Fact 8.21.7.

6.6 Facts on the Drazin and Group Generalized Inverses

Fact 6.6.1. Let $A_1, \ldots, A_k \in \mathbb{F}^{n \times m}$. Then,

$$(A_1 + \cdots + A_k)^{\mathrm{D}} = \tfrac{1}{k} \begin{bmatrix} I_n & \cdots & I_n \end{bmatrix} \begin{bmatrix} A_1 & A_2 & \cdots & A_k \\ A_k & A_1 & \cdots & A_{k-1} \\ \vdots & \vdots & \ddots & \vdots \\ A_2 & A_3 & \cdots & A_1 \end{bmatrix}^{\mathrm{D}} \begin{bmatrix} I_m \\ \vdots \\ I_m \end{bmatrix}.$$

Proof: See [1314].

Remark: See Fact 6.5.2.

Fact 6.6.2. Let $A \in \mathbb{F}^{n \times n}$. Then, $X = A^{\mathrm{D}}$ is the unique matrix satisfying

$$\operatorname{rank} \begin{bmatrix} A & AA^{\mathrm{D}} \\ A^{\mathrm{D}}A & X \end{bmatrix} = \operatorname{rank} A.$$

Proof: See [1451, 1532].

Remark: See Fact 2.17.10 and Fact 6.3.29.

Fact 6.6.3. Let $A, B \in \mathbb{F}^{n \times n}$, and assume that $AB = 0$. Then,

$$(AB)^{\mathrm{D}} = A(BA)^{2\mathrm{D}}B.$$

Remark: This result is *Cline's formula*.

Fact 6.6.4. Let $A, B \in \mathbb{F}^{n \times n}$, and assume that $AB = BA$. Then,

$$(AB)^{\mathrm{D}} = B^{\mathrm{D}}A^{\mathrm{D}},$$
$$A^{\mathrm{D}}B = BA^{\mathrm{D}},$$
$$AB^{\mathrm{D}} = B^{\mathrm{D}}A.$$

Fact 6.6.5. Let $A, B \in \mathbb{F}^{n \times n}$, and assume that $AB = BA = 0$. Then,

$$(A + B)^{\mathrm{D}} = A^{\mathrm{D}} + B^{\mathrm{D}}.$$

Proof: See [670].

Remark: This result is due to Drazin.

Fact 6.6.6. Let $A, B \in \mathbb{F}^{n \times n}$, and assume that A and B are idempotent. Then, the following statements hold:

i) If $AB = 0$, then

$$(A + B)^{\mathrm{D}} = A + B - 2BA$$

and

$$(A - B)^{\mathrm{D}} = A - B.$$

ii) If $BA = 0$, then

$$(A + B)^{\mathrm{D}} = A + B - 2AB$$

and

$$(A - B)^{\mathrm{D}} = A - B.$$

iii) If $AB = A$, then

$$(A + B)^{\mathrm{D}} = \tfrac{1}{4}A + B - \tfrac{3}{4}BA$$

and

$$(A - B)^{\mathrm{D}} = BA - B.$$

iv) If $AB = B$, then

$$(A + B)^{\mathrm{D}} = A + \tfrac{1}{4}B - \tfrac{3}{4}BA$$

and

$$(A - B)^{\mathrm{D}} = A - BA.$$

v) If $BA = A$, then

$$(A + B)^{\mathrm{D}} = \tfrac{1}{4}A + B - \tfrac{3}{4}AB$$

and

$$(A - B)^{\mathrm{D}} = AB - B.$$

vi) If $BA = B$, then

$$(A + B)^{\mathrm{D}} = A + \tfrac{1}{4}B - \tfrac{3}{4}AB$$

and

$$(A - B)^{\mathrm{D}} = A - AB.$$

vii) If $AB = BA$, then

$$(A + B)^{\mathrm{D}} = A + B - \tfrac{3}{2}AB$$

and

$$(A - B)^{\mathrm{D}} = A - B.$$

viii) If $ABA = 0$, then

$$(A + B)^{\mathrm{D}} = A + B - 2AB - 2BA + 3BAB$$

and

$$(A - B)^{\mathrm{D}} = A - B - BAB.$$

ix) If $BAB = 0$, then

$$(A + B)^{\mathrm{D}} = A + B - 2AB - 2BA + 3ABA$$

and

$$(A - B)^{\mathrm{D}} = A - B + ABA.$$

x) If $ABA = A$, then

$$(A + B)^{\mathrm{D}} = \tfrac{1}{8}(A + B)^2 + \tfrac{7}{8}BA_\perp B$$

and

$$(A - B)^{\mathrm{D}} = -BA_\perp B.$$

xi) If $BAB = B$, then

$$(A + B)^{\mathrm{D}} = \tfrac{1}{8}(A + B)^2 + \tfrac{7}{8}AB_\perp A$$

and

$$(A - B)^{\mathrm{D}} = AB_\perp A.$$

xii) If $ABA = B$, then

$$(A + B)^{\mathrm{D}} = A - \tfrac{1}{2}B$$

and

$$(A - B)^{\mathrm{D}} = A - B.$$

$xiii$) If $BAB = A$, then

$$(A + B)^{\mathrm{D}} = -\tfrac{1}{2}A + B$$

and

$$(A - B)^{\mathrm{D}} = A - B.$$

xiv) If $ABA = AB$, then

$$(A + B)^{\mathrm{D}} = A + B - 2BA - \tfrac{3}{4}AB + \tfrac{5}{4}BAB$$

and

$$(A - B)^{\mathrm{D}} = A - B - AB + BAB.$$

xv) If $ABA = BA$, then

$$(A + B)^{\mathrm{D}} = A + B - 2AB - \tfrac{3}{4}BA + \tfrac{5}{4}BAB$$

and

$$(A - B)^{\mathrm{D}} = A - B - BA + BAB.$$

Now, assume in addition that A and B are projectors. Then, the following statements hold:

xvi) If $AB = A$, then

$$(A + B)^{\mathrm{D}} = -\tfrac{1}{2}A + B$$

and

$$(A - B)^{\mathrm{D}} = A - B.$$

$xvii$) If $AB = B$, then

$$(A + B)^{\mathrm{D}} = A - \tfrac{1}{2}B$$

and

$$(A - B)^{\mathrm{D}} = A - B.$$

$xviii$) $AB = BA$ if and only if

$$(A - B)^{\mathrm{D}} = A - B.$$

Proof: See [405].

Fact 6.6.7. Let $A \in \mathbb{F}^{n \times n}$, and assume that $\operatorname{ind} A = \operatorname{rank} A = 1$. Then,
$$A^{\#} = \left(\operatorname{tr} A^2\right)^{-1} A.$$

Consequently, if $x, y \in \mathbb{F}^n$ satisfy $x^* y \neq 0$, then
$$(xy^*)^{\#} = (x^* y)^{-2} xy^*.$$

In particular,
$$1_{n \times n}^{\#} = n^{-2} 1_{n \times n}.$$

Fact 6.6.8. Let $A \in \mathbb{F}^{n \times n}$, and let $k \triangleq \operatorname{ind} A$. Then,
$$A^{\mathrm{D}} = A^k \left(A^{2k+1}\right)^{+} A^k.$$

If, in particular, $\operatorname{ind} A \leq 1$, then
$$A^{\#} = A\left(A^3\right)^{+} A.$$

Proof: See [178, pp. 165, 174].

Fact 6.6.9. Let $A \in \mathbb{F}^{n \times n}$. Then, the following statements are equivalent:

i) A is range Hermitian.

ii) $A^{+} = A^{\mathrm{D}}$.

iii) $\operatorname{ind} A \leq 1$, and $A^{+} = A^{\#}$.

iv) $\operatorname{ind} A \leq 1$, and $A^* A^{\#} A + A A^{\#} A^* = 2A^*$.

v) $\operatorname{ind} A \leq 1$, and $A^{+} A^{\#} A + A A^{\#} A^{+} = 2A^{+}$.

Proof: See [331].

Remark: See Fact 6.3.10.

Fact 6.6.10. Let $A \in \mathbb{F}^{n \times n}$, assume that A is group invertible, and let $S, B \in \mathbb{F}^{n \times n}$, where S is nonsingular, B is a Jordan canonical form of A, and $A = SBS^{-1}$. Then,
$$A^{\#} = SB^{\#}S^{-1} = SB^{+}S^{-1}.$$

Proof: Since B is range Hermitian, it follows from Fact 6.6.9 that $B^{\#} = B^{+}$. See [178, p. 158].

Fact 6.6.11. Let $A \in \mathbb{F}^{n \times n}$. Then, the following statements are equivalent:

i) A is normal.

ii) $\operatorname{ind} A \leq 1$, and $A^{\#} A^* = A^* A^{\#}$.

Proof: See [331].

Remark: See Fact 3.7.12, Fact 3.11.5, Fact 5.15.4, and Fact 6.3.15.

Fact 6.6.12. Let $A \in \mathbb{F}^{n \times n}$, and let $k \geq 1$. Then, the following statements are equivalent:

i) $k \geq \operatorname{ind} A$.

ii) $\lim_{\alpha \to 0} \alpha^k (A + \alpha I)^{-1}$ exists.

iii) $\lim_{\alpha \to 0} (A^{k+1} + \alpha I)^{-1} A^k$ exists.

In this case,
$$A^{\mathrm{D}} = \lim_{\alpha \to 0} (A^{k+1} + \alpha I)^{-1} A^k$$

and

$$\lim_{\alpha \to 0} \alpha^k (A + \alpha I)^{-1} = \begin{cases} (-1)^{k-1} (I - AA^{\mathrm{D}}) A^{k-1}, & k = \operatorname{ind} A > 0, \\ A^{-1}, & k = \operatorname{ind} A = 0, \\ 0, & k > \operatorname{ind} A. \end{cases}$$

Proof: See [1024].

Fact 6.6.13. Let $A \in \mathbb{F}^{n \times n}$, let $r \triangleq \operatorname{rank} A$, let $B \in \mathbb{R}^{n \times r}$ and $C \in \mathbb{R}^{r \times n}$, and assume that $A = BC$. Then, A is group invertible if and only if BA is nonsingular. In this case,
$$A^\# = B(CB)^{-2} C.$$

Proof: See [178, p. 157].

Remark: This result is due to Cline.

Fact 6.6.14. Let $A \in \mathbb{F}^{n \times n}$, $B \in \mathbb{F}^{n \times m}$, and $C \in \mathbb{F}^{m \times m}$. If A and C are singular, then $\operatorname{ind} \left[\begin{smallmatrix} A & B \\ 0 & C \end{smallmatrix} \right] = 1$ if and only if $\operatorname{ind} A = \operatorname{ind} C = 1$, and $(I - AA^{\mathrm{D}})B(I - CC^{\mathrm{D}}) = 0$.

Proof: See [1024].

Remark: See Fact 5.14.31.

Fact 6.6.15. Let $A \in \mathbb{F}^{n \times n}$. Then, A is group invertible if and only if $\lim_{\alpha \to 0} (A + \alpha I)^{-1} A$ exists. In this case,
$$\lim_{\alpha \to 0} (A + \alpha I)^{-1} A = AA^\#.$$

Proof: See [291, p. 138].

Fact 6.6.16. Let $A \in \mathbb{F}^{n \times n}$, assume that A is nonzero and group invertible, let $r \triangleq \operatorname{rank} A$, define $B \triangleq \operatorname{diag}[\sigma_1(A), \ldots, \sigma_r(A)]$, and let $S \in \mathbb{F}^{n \times n}$, $K \in \mathbb{F}^{r \times r}$, and $L \in \mathbb{F}^{r \times (n-r)}$ be such that S is unitary,
$$KK^* + LL^* = I_r,$$

and

$$A = S \begin{bmatrix} BK & BL \\ 0_{(n-r) \times r} & 0_{(n-r) \times (n-r)} \end{bmatrix} S^*.$$

Then,

$$A^{\#} = S \begin{bmatrix} K^{-1}B^{-1} & K^{-1}B^{-1}K^{-1}L \\ 0_{(n-r)\times r} & 0_{(n-r)\times(n-r)} \end{bmatrix} S^{*}.$$

Proof: See [119, 668].

Remark: See Fact 5.9.30 and Fact 6.3.14.

Fact 6.6.17. Let $A \in \mathbb{F}^{n \times n}$. Then, the following statements are equivalent:

  $i)$ A is range Hermitian.

 $ii)$ A is group invertible and $AA^{+}A^{+} = A^{\#}$.

 $iii)$ A is group invertible and $AA^{\#}A^{+} = A^{\#}$.

 $iv)$ A is group invertible and $A^{*}AA^{\#} = A^{*}$.

  $v)$ A is group invertible and $A^{+}AA^{\#} = A^{+}$.

 $vi)$ A is group invertible and $A^{\#}A^{+}A = A^{+}$.

 $vii)$ A is group invertible and $AA^{\#} = A^{+}A$.

 $viii)$ A is group invertible and $A^{*}A^{+} = A^{*}A^{\#}$.

 $ix)$ A is group invertible and $A^{+}A^{*} = A^{\#}A^{*}$.

  $x)$ A is group invertible and $A^{+}A^{+} = A^{+}A^{\#}$.

 $xi)$ A is group invertible and $A^{+}A^{+} = A^{\#}A^{+}$.

 $xii)$ A is group invertible and $A^{+}A^{+} = A^{\#}A^{\#}$.

 $xiii)$ A is group invertible and $A^{+}A^{\#} = A^{\#}A^{\#}$.

 $xiv)$ A is group invertible and $A^{\#}A^{+} = A^{\#}A^{\#}$.

 $xv)$ A is group invertible and $A^{+}A^{\#} = A^{\#}A^{+}$.

 $xvi)$ A is group invertible and $AA^{+}A^{*} = A^{*}AA^{+}$.

 $xvii)$ A is group invertible and $AA^{+}A^{\#} = A^{+}A^{\#}A$.

 $xviii)$ A is group invertible and $AA^{+}A^{\#} = A^{\#}AA^{+}$.

 $xix)$ A is group invertible and $AA^{\#}A^{*} = A^{*}AA^{\#}$.

 $xx)$ A is group invertible and $AA^{\#}A^{+} = A^{+}AA^{\#}$.

 $xxi)$ A is group invertible and $AA^{\#}A^{+} = A^{\#}A^{+}A$.

 $xxii)$ A is group invertible and $A^{*}A^{+}A = A^{+}AA^{*}$.

 $xxiii)$ A is group invertible and $A^{+}AA^{\#} = A^{\#}A^{+}A$.

 $xxiv)$ A is group invertible and $A^{+}A^{+}A^{\#} = A^{+}A^{\#}A^{+}$.

 $xxv)$ A is group invertible and $A^{+}A^{+}A^{\#} = A^{\#}A^{+}A^{+}$.

 $xxvi)$ A is group invertible and $A^{+}A^{\#}A^{+} = A^{\#}A^{+}A^{+}$.

 $xxvii)$ A is group invertible and $A^{+}A^{\#}A^{\#} = A^{\#}A^{+}A^{\#}$.

xxviii) A is group invertible and $A^+A^\#A^\# = A^\#A^\#A^+$.

xxix) A is group invertible and $A^\#A^\#A^+ = A^\#A^+A^\#$.

Proof: See [119].

Fact 6.6.18. Let $A \in \mathbb{F}^{n \times n}$. Then, the following statements are equivalent:

i) A is normal.

ii) A is group invertible and $A^*A^+ = A^\#A^*$.

iii) A is group invertible and $A^*A^\# = A^+A^*$.

iv) A is group invertible and $A^*A^\# = A^\#A^*$.

v) A is group invertible and $AA^*A^\# = A^*A^\#A$.

vi) A is group invertible and $AA^*A^\# = A^\#AA^*$.

vii) A is group invertible and $AA^\#A^* = A^\#A^*A$.

viii) A is group invertible and $A^*AA^\# = A^\#A^*A$.

ix) A is group invertible and $A^{*2}A^\# = A^*A^\#A^*$.

x) A is group invertible and $A^*A^+A^\# = A^\#A^*A^+$.

xi) A is group invertible and $A^*A^\#A^* = A^\#A^{2*}$.

xii) A is group invertible and $A^*A^\#A^+ = A^+A^*A^\#$.

xiii) A is group invertible and $A^*A^\#A^\# = A^\#A^*A^\#$.

xiv) A is group invertible and $A^+A^*A^\# = A^\#A^+A^*$.

xv) A is group invertible and $A^+A^\#A^* = A^\#A^*A^+$.

xvi) A is group invertible and $A^\#A^*A^\# = A^\#A^*A^*$.

Proof: See [119].

Fact 6.6.19. Let $A \in \mathbb{F}^{n \times n}$. Then, the following statements are equivalent:

i) A is Hermitian.

ii) A is group invertible and $AA^\# = A^*A^+$.

iii) A is group invertible and $AA^\# = A^*A^\#$.

iv) A is group invertible and $AA^\# = A^+A^*$.

v) A is group invertible and $A^+A = A^\#A^*$.

vi) A is group invertible and $A^*AA^\# = A$.

vii) A is group invertible and $A^{2*}A^\# = A^*$.

viii) A is group invertible and $A^*A^+A^+ = A^\#$.

ix) A is group invertible and $A^*A^+A^\# = A^+$.

x) A is group invertible and $A^*A^+A^\# = A^\#$.

xi) A is group invertible and $A^*A^\#A^\# = A^\#$.

xii) A is group invertible and $A^\#A^*A^\# = A^+$.

Proof: See [119].

Fact 6.6.20. Let $A, B \in \mathbb{F}^{n \times n}$, assume that A and B are group invertible, and consider the following conditions:

i) $ABA = B$.

ii) $BAB = A$.

iii) $A^2 = B^2$.

Then, if two of the above conditions are satisfied, then the third condition is satisfied. Furthermore, if *i*)–*iii*) are satisfied, then the following statements hold:

iv) A and B are group invertible.

v) $A^\# = A^3$ and $B^\# = B^3$.

vi) $A^5 = A$ and $B^5 = B$.

vii) $A^4 = B^4 = (AB)^4$.

viii) If A and B are nonsingular, then $A^4 = B^4 = (AB)^4 = I$.

Proof: See [482].

Fact 6.6.21. Let $A \in \mathbb{R}^{n \times n}$, where $n \geq 2$, assume that A is positive, define $B \triangleq \operatorname{sprad}(A)I - A$, let $x, y \in \mathbb{R}^n$ be positive, and assume that $Ax = \operatorname{sprad}(A)x$ and $A^\mathrm{T}y = \operatorname{sprad}(A)y$. Then, the following statements hold:

i) $B + \frac{1}{x^\mathrm{T}y}xy^\mathrm{T}$ is nonsingular.

ii) $B^\# = (B + \frac{1}{x^\mathrm{T}y}xy^\mathrm{T})^{-1}(I - \frac{1}{x^\mathrm{T}y}xy^\mathrm{T})$.

iii) $I - BB^\# = \frac{1}{x^\mathrm{T}y}xy^\mathrm{T}$.

iv) $B^\# = \lim_{k \to \infty} \left[\sum_{i=0}^{k-1} \frac{1}{[\operatorname{sprad}(A)]^i} A^i - \frac{k}{x^\mathrm{T}y}xy^\mathrm{T} \right]$.

Proof: See [1177, p. 9-4].

Remark: See Fact 4.11.4.

6.7 Notes

A brief history of the generalized inverse is given in [177] and [178, p. 4]. The proof of the uniqueness of A^+ is given in [973, p. 32]. Additional books on generalized inverses include [178, 249, 1145, 1430]. The terminology "range Hermitian" is used in [178]; the terminology "EP" is more common. Generalized inverses are widely used in least squares methods; see [241, 291, 901]. Applications to singular differential equations are considered in [290]. Applications to Markov chains are discussed in [758].

Chapter Seven

Kronecker and Schur Algebra

In this chapter we introduce Kronecker matrix algebra, which is useful for solving linear matrix equations.

7.1 Kronecker Product

For $A \in \mathbb{F}^{n \times m}$ define the *vec* operator as

$$\text{vec}\, A \triangleq \begin{bmatrix} \text{col}_1(A) \\ \vdots \\ \text{col}_m(A) \end{bmatrix} \in \mathbb{F}^{nm}, \tag{7.1.1}$$

which is the column vector of size $nm \times 1$ obtained by stacking the columns of A. We recover A from $\text{vec}\, A$ by writing

$$A = \text{vec}^{-1}(\text{vec}\, A). \tag{7.1.2}$$

Proposition 7.1.1. Let $A \in \mathbb{F}^{n \times m}$ and $B \in \mathbb{F}^{m \times n}$. Then,

$$\text{tr}\, AB = \left(\text{vec}\, A^{\mathrm{T}}\right)^{\mathrm{T}} \text{vec}\, B = \left(\text{vec}\, B^{\mathrm{T}}\right)^{\mathrm{T}} \text{vec}\, A. \tag{7.1.3}$$

Proof. Note that

$$\begin{aligned}
\text{tr}\, AB &= \sum_{i=1}^{n} \text{row}_i(A)\text{col}_i(B) \\
&= \sum_{i=1}^{n} \left[\text{col}_i\!\left(A^{\mathrm{T}}\right)\right]^{\mathrm{T}} \text{col}_i(B) \\
&= \begin{bmatrix} \text{col}_1^{\mathrm{T}}\!\left(A^{\mathrm{T}}\right) & \cdots & \text{col}_n^{\mathrm{T}}\!\left(A^{\mathrm{T}}\right) \end{bmatrix} \begin{bmatrix} \text{col}_1(B) \\ \vdots \\ \text{col}_n(B) \end{bmatrix} \\
&= \left(\text{vec}\, A^{\mathrm{T}}\right)^{\mathrm{T}} \text{vec}\, B. \qquad \square
\end{aligned}$$

Next, we introduce the Kronecker product.

Definition 7.1.2. Let $A \in \mathbb{F}^{n \times m}$ and $B \in \mathbb{F}^{l \times k}$. Then, the *Kronecker product* $A \otimes B \in \mathbb{F}^{nl \times mk}$ of A and B is the partitioned matrix

$$A \otimes B \triangleq \begin{bmatrix} A_{(1,1)}B & A_{(1,2)}B & \cdots & A_{(1,m)}B \\ \vdots & \vdots & \ddots & \vdots \\ A_{(n,1)}B & A_{(n,2)}B & \cdots & A_{(n,m)}B \end{bmatrix}. \tag{7.1.4}$$

Unlike matrix multiplication, the Kronecker product $A \otimes B$ does not entail a restriction on either the size of A or the size of B.

The following results are immediate consequences of the definition of the Kronecker product.

Proposition 7.1.3. Let $\alpha \in \mathbb{F}$, $A \in \mathbb{F}^{n \times m}$, and $B \in \mathbb{F}^{l \times k}$. Then,

$$A \otimes (\alpha B) = (\alpha A) \otimes B = \alpha(A \otimes B), \tag{7.1.5}$$

$$\overline{A \otimes B} = \overline{A} \otimes \overline{B}, \tag{7.1.6}$$

$$(A \otimes B)^{\mathrm{T}} = A^{\mathrm{T}} \otimes B^{\mathrm{T}}, \tag{7.1.7}$$

$$(A \otimes B)^{*} = A^{*} \otimes B^{*}. \tag{7.1.8}$$

Proposition 7.1.4. Let $A, B \in \mathbb{F}^{n \times m}$ and $C \in \mathbb{F}^{l \times k}$. Then,

$$(A + B) \otimes C = A \otimes C + B \otimes C \tag{7.1.9}$$

and

$$C \otimes (A + B) = C \otimes A + C \otimes B. \tag{7.1.10}$$

The next result shows that the Kronecker product is associative.

Proposition 7.1.5. Let $A \in \mathbb{F}^{n \times m}$, $B \in \mathbb{F}^{l \times k}$, and $C \in \mathbb{F}^{p \times q}$. Then,

$$A \otimes (B \otimes C) = (A \otimes B) \otimes C. \tag{7.1.11}$$

Hence, we write $A \otimes B \otimes C$ for $A \otimes (B \otimes C)$ and $(A \otimes B) \otimes C$.

The next result illustrates a useful form of compatibility between matrix multiplication and the Kronecker product.

Proposition 7.1.6. Let $A \in \mathbb{F}^{n \times m}$, $B \in \mathbb{F}^{l \times k}$, $C \in \mathbb{F}^{m \times q}$, and $D \in \mathbb{F}^{k \times p}$. Then,

$$(A \otimes B)(C \otimes D) = AC \otimes BD. \tag{7.1.12}$$

Proof. Note that the ij block of $(A \otimes B)(C \otimes D)$ is given by

$$[(A \otimes B)(C \otimes D)]_{ij} = \begin{bmatrix} A_{(i,1)}B & \cdots & A_{(i,m)}B \end{bmatrix} \begin{bmatrix} C_{(1,j)}D \\ \vdots \\ C_{(m,j)}D \end{bmatrix}$$

$$= \sum_{k=1}^{m} A_{(i,k)}C_{(k,j)}BD = (AC)_{(i,j)}BD$$

$$= (AC \otimes BD)_{ij}. \qquad \square$$

Next, we consider the inverse of a Kronecker product.

Proposition 7.1.7. Assume that $A \in \mathbb{F}^{n \times n}$ and $B \in \mathbb{F}^{m \times m}$ are nonsingular. Then, $A \otimes B$ is nonsingular, and

$$(A \otimes B)^{-1} = A^{-1} \otimes B^{-1}. \qquad (7.1.13)$$

Proof. Note that

$$(A \otimes B)\left(A^{-1} \otimes B^{-1}\right) = AA^{-1} \otimes BB^{-1} = I_n \otimes I_m = I_{nm}. \qquad \square$$

Proposition 7.1.8. Let $x \in \mathbb{F}^n$ and $y \in \mathbb{F}^m$. Then,

$$xy^{\mathrm{T}} = x \otimes y^{\mathrm{T}} = y^{\mathrm{T}} \otimes x \qquad (7.1.14)$$

and

$$\mathrm{vec}\, xy^{\mathrm{T}} = y \otimes x. \qquad (7.1.15)$$

The following result concerns the vec of the product of three matrices.

Proposition 7.1.9. Let $A \in \mathbb{F}^{n \times m}$, $B \in \mathbb{F}^{m \times l}$, and $C \in \mathbb{F}^{l \times k}$. Then,

$$\mathrm{vec}(ABC) = \left(C^{\mathrm{T}} \otimes A\right) \mathrm{vec}\, B. \qquad (7.1.16)$$

Proof. Using (7.1.12) and (7.1.15), it follows that

$$\mathrm{vec}\, ABC = \mathrm{vec} \sum_{i=1}^{l} A\mathrm{col}_i(B)e_i^{\mathrm{T}}C = \sum_{i=1}^{l} \mathrm{vec}\left[A\mathrm{col}_i(B)\left(C^{\mathrm{T}}e_i\right)^{\mathrm{T}}\right]$$

$$= \sum_{i=1}^{l} \left[C^{\mathrm{T}}e_i\right] \otimes \left[A\mathrm{col}_i(B)\right] = \left(C^{\mathrm{T}} \otimes A\right) \sum_{i=1}^{l} e_i \otimes \mathrm{col}_i(B)$$

$$= \left(C^{\mathrm{T}} \otimes A\right) \sum_{i=1}^{l} \mathrm{vec}\left[\mathrm{col}_i(B)e_i^{\mathrm{T}}\right] = \left(C^{\mathrm{T}} \otimes A\right) \mathrm{vec}\, B. \qquad \square$$

The following result concerns the eigenvalues and eigenvectors of the Kronecker product of two matrices.

Proposition 7.1.10. Let $A \in \mathbb{F}^{n \times n}$ and $B \in \mathbb{F}^{m \times m}$. Then,

$$\mathrm{mspec}(A \otimes B) = \{\lambda\mu \colon \lambda \in \mathrm{mspec}(A), \mu \in \mathrm{mspec}(B)\}_{\mathrm{ms}}. \qquad (7.1.17)$$

If, in addition, $x \in \mathbb{C}^n$ is an eigenvector of A associated with $\lambda \in \mathrm{spec}(A)$ and $y \in \mathbb{C}^m$ is an eigenvector of B associated with $\mu \in \mathrm{spec}(B)$, then $x \otimes y$ is an eigenvector of $A \otimes B$ associated with $\lambda\mu$.

Proof. Using (7.1.12), we have

$$(A \otimes B)(x \otimes y) = (Ax) \otimes (By) = (\lambda x) \otimes (\mu y) = \lambda\mu(x \otimes y). \qquad \square$$

Proposition 7.1.10 shows that $\mathrm{mspec}(A \otimes B) = \mathrm{mspec}(B \otimes A)$. Consequently, it follows that $\det(A \otimes B) = \det(B \otimes A)$ and $\mathrm{tr}(A \otimes B) = \mathrm{tr}(B \otimes A)$. The following results are generalizations of these equalities.

Proposition 7.1.11. Let $A \in \mathbb{F}^{n \times n}$ and $B \in \mathbb{F}^{m \times m}$. Then,

$$\det(A \otimes B) = \det(B \otimes A) = (\det A)^m (\det B)^n. \qquad (7.1.18)$$

Proof. Let $\mathrm{mspec}(A) = \{\lambda_1, \ldots, \lambda_n\}_{\mathrm{ms}}$ and $\mathrm{mspec}(B) = \{\mu_1, \ldots, \mu_m\}_{\mathrm{ms}}$. Then, Proposition 7.1.10 implies that

$$\det(A \otimes B) = \prod_{i,j=1}^{n,m} \lambda_i \mu_j = \left(\lambda_1^m \prod_{j=1}^m \mu_j\right) \cdots \left(\lambda_n^m \prod_{j=1}^m \mu_j\right)$$
$$= (\lambda_1 \cdots \lambda_n)^m (\mu_1 \cdots \mu_m)^n = (\det A)^m (\det B)^n. \qquad \square$$

Proposition 7.1.12. Let $A \in \mathbb{F}^{n \times n}$ and $B \in \mathbb{F}^{m \times m}$. Then,

$$\mathrm{tr}(A \otimes B) = \mathrm{tr}(B \otimes A) = (\mathrm{tr}\, A)(\mathrm{tr}\, B). \qquad (7.1.19)$$

Proof. Note that

$$\mathrm{tr}(A \otimes B) = \mathrm{tr}(A_{(1,1)}B) + \cdots + \mathrm{tr}(A_{(n,n)}B)$$
$$= [A_{(1,1)} + \cdots + A_{(n,n)}]\mathrm{tr}\, B$$
$$= (\mathrm{tr}\, A)(\mathrm{tr}\, B). \qquad \square$$

Next, define the *Kronecker permutation matrix* $P_{n,m} \in \mathbb{F}^{nm \times nm}$ by

$$P_{n,m} \triangleq \sum_{i,j=1}^{n,m} E_{i,j,n \times m} \otimes E_{j,i,m \times n}. \qquad (7.1.20)$$

Proposition 7.1.13. Let $A \in \mathbb{F}^{n \times m}$. Then,

$$\mathrm{vec}\, A^{\mathrm{T}} = P_{n,m} \mathrm{vec}\, A. \qquad (7.1.21)$$

7.2 Kronecker Sum and Linear Matrix Equations

Next, we define the Kronecker sum of two square matrices.

Definition 7.2.1. Let $A \in \mathbb{F}^{n \times n}$ and $B \in \mathbb{F}^{m \times m}$. Then, the *Kronecker sum* $A \oplus B \in \mathbb{F}^{nm \times nm}$ of A and B is

$$A \oplus B \triangleq A \otimes I_m + I_n \otimes B. \tag{7.2.1}$$

Proposition 7.2.2. Let $A \in \mathbb{F}^{n \times n}$, $B \in \mathbb{F}^{m \times m}$, and $C \in \mathbb{F}^{l \times l}$. Then,

$$A \oplus (B \oplus C) = (A \oplus B) \oplus C. \tag{7.2.2}$$

Hence, we write $A \oplus B \oplus C$ for $A \oplus (B \oplus C)$ and $(A \oplus B) \oplus C$.

Proposition 7.1.10 shows that, if $\lambda \in \mathrm{spec}(A)$ and $\mu \in \mathrm{spec}(B)$, then $\lambda\mu \in \mathrm{spec}(A \otimes B)$. Next, we present an analogous result involving Kronecker sums.

Proposition 7.2.3. Let $A \in \mathbb{F}^{n \times n}$ and $B \in \mathbb{F}^{m \times m}$. Then,

$$\mathrm{mspec}(A \oplus B) = \{\lambda + \mu \colon \ \lambda \in \mathrm{mspec}(A), \ \mu \in \mathrm{mspec}(B)\}_{\mathrm{ms}}. \tag{7.2.3}$$

Now, let $x \in \mathbb{C}^n$ be an eigenvector of A associated with $\lambda \in \mathrm{spec}(A)$, and let $y \in \mathbb{C}^m$ be an eigenvector of B associated with $\mu \in \mathrm{spec}(B)$. Then, $x \otimes y$ is an eigenvector of $A \oplus B$ associated with $\lambda + \mu$.

Proof. Using (7.1.12), we have

$$\begin{aligned}
(A \oplus B)(x \otimes y) &= (A \otimes I_m)(x \otimes y) + (I_n \otimes B)(x \otimes y) \\
&= (Ax \otimes y) + (x \otimes By) = (\lambda x \otimes y) + (x \otimes \mu y) \\
&= \lambda(x \otimes y) + \mu(x \otimes y) = (\lambda + \mu)(x \otimes y). \qquad \square
\end{aligned}$$

The next result concerns the existence and uniqueness of solutions to *Sylvester's equation*. See Fact 5.10.21 and Proposition 11.9.3.

Proposition 7.2.4. Let $A \in \mathbb{F}^{n \times n}$, $B \in \mathbb{F}^{m \times m}$, and $C \in \mathbb{F}^{n \times m}$. Then, $X \in \mathbb{F}^{n \times m}$ satisfies

$$AX + XB + C = 0 \tag{7.2.4}$$

if and only if X satisfies

$$\left(B^{\mathrm{T}} \oplus A\right) \mathrm{vec}\, X + \mathrm{vec}\, C = 0. \tag{7.2.5}$$

Consequently, $B^{\mathrm{T}} \oplus A$ is nonsingular if and only if there exists a unique matrix $X \in \mathbb{F}^{n \times m}$ satisfying (7.2.4). In this case, X is given by

$$X = -\mathrm{vec}^{-1}\left[\left(B^{\mathrm{T}} \oplus A\right)^{-1} \mathrm{vec}\, C\right]. \tag{7.2.6}$$

Furthermore, $B^{\mathrm{T}} \oplus A$ is singular and $\mathrm{rank}\, B^{\mathrm{T}} \oplus A = \mathrm{rank}\, [\ B^{\mathrm{T}} \oplus A \quad \mathrm{vec}\, C\ ]$ if and only if there exist infinitely many matrices $X \in \mathbb{F}^{n \times m}$ satisfying (7.5.8). In this case, the set of solutions of (7.2.4) is given by $X + \mathcal{N}(B^{\mathrm{T}} \oplus A)$.

Proof. Note that (7.2.4) is equivalent to

$$0 = \operatorname{vec}(AXI + IXB) + \operatorname{vec} C = (I \otimes A)\operatorname{vec} X + (B^{\mathrm{T}} \otimes I)\operatorname{vec} X + \operatorname{vec} C$$
$$= (B^{\mathrm{T}} \otimes I + I \otimes A)\operatorname{vec} X + \operatorname{vec} C = (B^{\mathrm{T}} \oplus A)\operatorname{vec} X + \operatorname{vec} C,$$

which yields (7.2.5). The remaining results follow from Corollary 2.6.7.   □

For the following corollary, note Fact 5.10.21.

Corollary 7.2.5. Let $A \in \mathbb{F}^{n \times n}$, $B \in \mathbb{F}^{m \times m}$, and $C \in \mathbb{F}^{n \times m}$, and assume that $\operatorname{spec}(A)$ and $\operatorname{spec}(-B)$ are disjoint. Then, there exists a unique matrix $X \in \mathbb{F}^{n \times m}$ satisfying (7.2.4). Furthermore, the matrices $\begin{bmatrix} A & 0 \\ 0 & -B \end{bmatrix}$ and $\begin{bmatrix} A & C \\ 0 & -B \end{bmatrix}$ are similar and satisfy

$$\begin{bmatrix} A & C \\ 0 & -B \end{bmatrix} = \begin{bmatrix} I & X \\ 0 & I \end{bmatrix} \begin{bmatrix} A & 0 \\ 0 & -B \end{bmatrix} \begin{bmatrix} I & -X \\ 0 & I \end{bmatrix}. \tag{7.2.7}$$

7.3 Schur Product

An alternative form of vector and matrix multiplication is given by the *Schur product*. If $A \in \mathbb{F}^{n \times m}$ and $B \in \mathbb{F}^{n \times m}$, then $A \circ B \in \mathbb{F}^{n \times m}$ is defined by

$$(A \circ B)_{(i,j)} \triangleq A_{(i,j)} B_{(i,j)}, \tag{7.3.1}$$

that is, $A \circ B$ is formed by means of entry-by-entry multiplication. For matrices $A, B, C \in \mathbb{F}^{n \times m}$, the commutative, associative, and distributive equalities

$$A \circ B = B \circ A, \tag{7.3.2}$$
$$A \circ (B \circ C) = (A \circ B) \circ C, \tag{7.3.3}$$
$$A \circ (B + C) = A \circ B + A \circ C \tag{7.3.4}$$

hold. For a real scalar $\alpha \geq 0$ and $A \in \mathbb{F}^{n \times m}$, the *Schur power* $A^{\circ \alpha}$ is defined by

$$(A^{\circ \alpha})_{(i,j)} \triangleq (A_{(i,j)})^{\alpha}. \tag{7.3.5}$$

Thus, $A^{\circ 2} = A \circ A$. Note that $A^{\circ 0} = 1_{n \times m}$. Furthermore, $\alpha < 0$ is allowed if A has no zero entries. In particular, $A^{\circ -1}$ is the matrix whose entries are the reciprocals of the entries of A. For all $A \in \mathbb{F}^{n \times m}$,

$$A \circ 1_{n \times m} = 1_{n \times m} \circ A = A. \tag{7.3.6}$$

Finally, if A is square, then $I \circ A$ is the diagonal part of A.

The following result shows that $A \circ B$ is a submatrix of $A \otimes B$.

Proposition 7.3.1. Let $A, B \in \mathbb{F}^{n \times m}$. Then,

$$A \circ B = (A \otimes B)_{(\{1, n+2, 2n+3, \ldots, n^2\}, \{1, m+2, 2m+3, \ldots, m^2\})}. \tag{7.3.7}$$

If, in addition, $n = m$, then

$$A \circ B = (A \otimes B)_{(\{1, n+2, 2n+3, \ldots, n^2\})}, \tag{7.3.8}$$

and thus $A \circ B$ is a principal submatrix of $A \otimes B$.

Proof. See [730, p. 304] or [987].   □

7.4 Facts on the Kronecker Product

Fact 7.4.1. Let $x, y \in \mathbb{F}^n$. Then,

$$x \otimes y = (x \otimes I_n)y = (I_n \otimes y)x.$$

Fact 7.4.2. Let $x, y, w, z \in \mathbb{F}^n$. Then,

$$x^{\mathrm{T}}wy^{\mathrm{T}}z = (x^{\mathrm{T}} \otimes y^{\mathrm{T}})(w \otimes z) = (x \otimes y)^{\mathrm{T}}(w \otimes z).$$

Fact 7.4.3. Let $A \in \mathbb{F}^{n \times m}$ and $B \in \mathbb{F}^{1 \times m}$. Then,

$$A(B \otimes I_m) = B \otimes A.$$

Fact 7.4.4. Let $A \in \mathbb{F}^{n \times n}$ and $B \in \mathbb{F}^{m \times m}$, and assume that A and B are (diagonal, upper triangular, lower triangular). Then, so is $A \otimes B$.

Fact 7.4.5. Let $A \in \mathbb{F}^{n \times n}$, $B \in \mathbb{F}^{m \times m}$, and $l \in \mathbb{P}$. Then,

$$(A \otimes B)^l = A^l \otimes B^l.$$

Fact 7.4.6. Let $A \in \mathbb{F}^{n \times m}$. Then,

$$\operatorname{vec} A = (I_m \otimes A)\operatorname{vec} I_m = (A^{\mathrm{T}} \otimes I_n)\operatorname{vec} I_n.$$

Fact 7.4.7. Let $A \in \mathbb{F}^{n \times m}$ and $B \in \mathbb{F}^{m \times l}$. Then,

$$\operatorname{vec} AB = (I_l \otimes A)\operatorname{vec} B = (B^{\mathrm{T}} \otimes A)\operatorname{vec} I_m = \sum_{i=1}^{m} \operatorname{col}_i(B^{\mathrm{T}}) \otimes \operatorname{col}_i(A).$$

Fact 7.4.8. Let $A \in \mathbb{F}^{n \times m}$, $B \in \mathbb{F}^{m \times l}$, and $C \in \mathbb{F}^{l \times n}$. Then,

$$\operatorname{tr} ABC = (\operatorname{vec} A)^{\mathrm{T}}(B \otimes I_n)\operatorname{vec} C^{\mathrm{T}}.$$

Fact 7.4.9. Let $A, B, C \in \mathbb{F}^{n \times n}$, and assume that C is symmetric. Then,

$$(\operatorname{vec} C)^{\mathrm{T}}(A \otimes B)\operatorname{vec} C = (\operatorname{vec} C)^{\mathrm{T}}(B \otimes A)\operatorname{vec} C.$$

Fact 7.4.10. Let $A \in \mathbb{F}^{n \times m}$, $B \in \mathbb{F}^{m \times l}$, $C \in \mathbb{F}^{l \times k}$, and $D \in \mathbb{F}^{k \times n}$. Then,

$$\operatorname{tr} ABCD = (\operatorname{vec} A)^{\mathrm{T}}(B \otimes D^{\mathrm{T}})\operatorname{vec} C^{\mathrm{T}}.$$

Fact 7.4.11. Let $A \in \mathbb{F}^{n \times m}$, $B \in \mathbb{F}^{m \times l}$, and $k \geq 1$. Then,

$$(AB)^{\otimes k} = A^{\otimes k}B^{\otimes k},$$

where $A^{\otimes k} \triangleq A \otimes A \otimes \cdots \otimes A$, with A appearing k times.

Fact 7.4.12. Let $A, C \in \mathbb{F}^{n \times m}$ and $B, D \in \mathbb{F}^{l \times k}$, assume that A is (left equivalent, right equivalent, biequivalent) to C, and assume that B is (left equivalent, right equivalent, biequivalent) to D. Then, $A \otimes B$ is (left equivalent, right equivalent, biequivalent) to $C \otimes D$.

Fact 7.4.13. Let $A, B, C, D \in \mathbb{F}^{n \times n}$, assume that A is (similar, congruent, unitarily similar) to C, and assume that B is (similar, congruent, unitarily similar) to D. Then, $A \otimes B$ is (similar, congruent, unitarily similar) to $C \otimes D$.

Fact 7.4.14. Let $A \in \mathbb{F}^{n \times n}$ and $B \in \mathbb{F}^{m \times m}$, and let $\gamma \in \operatorname{spec}(A \otimes B)$. Then,

$$\sum \operatorname{gmult}_A(\lambda)\operatorname{gmult}_B(\mu) \leq \operatorname{gmult}_{A \otimes B}(\gamma)$$
$$\leq \operatorname{amult}_{A \otimes B}(\gamma)$$
$$= \sum \operatorname{amult}_A(\lambda)\operatorname{amult}_B(\mu),$$

where both sums are taken over all $\lambda \in \operatorname{spec}(A)$ and $\mu \in \operatorname{spec}(B)$ such that $\lambda\mu = \gamma$.

Fact 7.4.15. Let $A \in \mathbb{F}^{n \times n}$. Then,

$$\operatorname{sprad}(A \otimes A) = [\operatorname{sprad}(A)]^2.$$

Fact 7.4.16. Let $A \in \mathbb{F}^{n \times n}$ and $B \in \mathbb{F}^{m \times m}$, and let $\gamma \in \operatorname{spec}(A \otimes B)$. Then, $\operatorname{ind}_{A \otimes B}(\gamma) = 1$ if and only if $\operatorname{ind}_A(\lambda) = 1$ and $\operatorname{ind}_B(\mu) = 1$ for all $\lambda \in \operatorname{spec}(A)$ and $\mu \in \operatorname{spec}(B)$ such that $\lambda\mu = \gamma$.

Fact 7.4.17. Let $A \in \mathbb{F}^{n \times n}$ and $B \in \mathbb{F}^{n \times n}$, and assume that A and B are (group invertible, range Hermitian, range symmetric, Hermitian, symmetric, normal, positive semidefinite, positive definite, unitary, orthogonal, projectors, reflectors, involutory, idempotent, tripotent, nilpotent, semisimple). Then, so is $A \otimes B$.

Remark: See Fact 7.4.33.

Fact 7.4.18. Let $A_1, \ldots, A_l \in \mathbb{F}^{n \times n}$, and assume that $A_1, \ldots, A_l$ are skew Hermitian. If l is (even, odd), then $A_1 \otimes \cdots \otimes A_l$ is (Hermitian, skew Hermitian).

Fact 7.4.19. Let $A_{i,j} \in \mathbb{F}^{n_i \times n_j}$ for all $i \in \{1, \ldots, k\}$ and $j \in \{1, \ldots, l\}$. Then,

$$\begin{bmatrix} A_{11} & A_{22} & \cdots & A_{1l} \\ A_{21} & A_{22} & \ddots & A_{2l} \\ \vdots & \ddots & \ddots & \vdots \\ A_{k1} & A_{k2} & \cdots & A_{kl} \end{bmatrix} \otimes B = \begin{bmatrix} A_{11} \otimes B & A_{22} \otimes B & \cdots & A_{1l} \otimes B \\ A_{21} \otimes B & A_{22} \otimes B & \ddots & A_{2l} \otimes B \\ \vdots & & \ddots & \vdots \\ A_{k1} \otimes B & A_{k2} \otimes B & \cdots & A_{kl} \otimes B \end{bmatrix}.$$

Fact 7.4.20. Let $x \in \mathbb{F}^k$, and let $A_i \in \mathbb{F}^{n \times n_i}$ for all $i \in \{1, \ldots, l\}$. Then,

$$x \otimes \begin{bmatrix} A_1 & \cdots & A_l \end{bmatrix} = \begin{bmatrix} x \otimes A_1 & \cdots & x \otimes A_l \end{bmatrix}.$$

Fact 7.4.21. Let $x \in \mathbb{F}^m$, let $A \in \mathbb{F}^{n \times m}$, and let $B \in \mathbb{F}^{m \times l}$. Then,

$$(A \otimes x)B = (A \otimes x)(B \otimes 1) = (AB) \otimes x.$$

Fact 7.4.22. Let $A \in \mathbb{F}^{n \times n}$ and $B \in \mathbb{F}^{m \times m}$. Then, the eigenvalues of $\sum_{i,j=1,1}^{k,l} \gamma_{ij} A^i \otimes B^j$ are of the form $\sum_{i,j=1,1}^{k,l} \gamma_{ij} \lambda^i \mu^j$, where $\lambda \in \operatorname{spec}(A)$ and $\mu \in \operatorname{spec}(B)$, and an associated eigenvector is given by $x \otimes y$, where $x \in \mathbb{F}^n$ is an eigenvector of A associated with $\lambda \in \operatorname{spec}(A)$ and $y \in \mathbb{F}^n$ is an eigenvector of B associated with $\mu \in \operatorname{spec}(B)$.

Proof: Let $Ax = \lambda x$ and $By = \mu y$. Then, $\gamma_{ij}(A^i \otimes B^j)(x \otimes y) = \gamma_{ij}\lambda^i\mu^j(x \otimes y)$. See [532], [892, p. 411], or [967, p. 83].

Remark: This result is due to Stephanos.

Fact 7.4.23. Let $A \in \mathbb{F}^{n \times m}$ and $B \in \mathbb{F}^{l \times k}$. Then,

$$\mathcal{R}(A \otimes B) = \mathcal{R}(A \otimes I_{l \times l}) \cap \mathcal{R}(I_{n \times n} \otimes B).$$

Proof: See [1325].

Fact 7.4.24. Let $A \in \mathbb{F}^{n \times m}$ and $B \in \mathbb{F}^{l \times k}$. Then,

$$\operatorname{rank}(A \otimes B) = (\operatorname{rank} A)(\operatorname{rank} B) = \operatorname{rank}(B \otimes A).$$

Consequently, $A \otimes B = 0$ if and only if either $A = 0$ or $B = 0$.

Proof: Use the singular value decomposition of $A \otimes B$.

Remark: See Fact 8.22.16.

Fact 7.4.25. Let $A \in \mathbb{F}^{n \times m}$, $B \in \mathbb{F}^{l \times k}$, $C \in \mathbb{F}^{n \times p}$, $D \in \mathbb{F}^{l \times q}$. Then,

$\operatorname{rank} \begin{bmatrix} A \otimes B & C \otimes D \end{bmatrix}$

$$\leq \begin{cases} (\operatorname{rank} A)\operatorname{rank} \begin{bmatrix} B & D \end{bmatrix} + (\operatorname{rank} D)\operatorname{rank} \begin{bmatrix} A & C \end{bmatrix} - (\operatorname{rank} A)\operatorname{rank} D \\ (\operatorname{rank} B)\operatorname{rank} \begin{bmatrix} A & C \end{bmatrix} + (\operatorname{rank} C)\operatorname{rank} \begin{bmatrix} B & D \end{bmatrix} - (\operatorname{rank} B)\operatorname{rank} C. \end{cases}$$

Proof: See [1329].

Fact 7.4.26. Let $A \in \mathbb{F}^{n \times n}$ and $B \in \mathbb{F}^{m \times m}$. Then,

$$\operatorname{rank}(I - A \otimes B) \leq nm - [n - \operatorname{rank}(I - A)][m - \operatorname{rank}(I - B)].$$

Proof: See [341].

Fact 7.4.27. Let $A \in \mathbb{F}^{n \times n}$ and $B \in \mathbb{F}^{m \times m}$. Then,

$$\operatorname{ind} A \otimes B = \max\{\operatorname{ind} A, \operatorname{ind} B\}.$$

Fact 7.4.28. Let $A \in \mathbb{F}^{n \times m}$ and $B \in \mathbb{F}^{m \times n}$. Then,

$$|n - m|\min\{n, m\} \leq \operatorname{amult}_{A \otimes B}(0).$$

Proof: See [730, p. 249].

Fact 7.4.29. Let $A \in \mathbb{F}^{n \times m}$ and $B \in \mathbb{F}^{l \times k}$, and assume that $nl = mk$ and $n \neq m$. Then, $A \otimes B$ and $B \otimes A$ are singular.

Proof: See [730, p. 250].

Fact 7.4.30. The Kronecker permutation matrix $P_{n,m} \in \mathbb{R}^{nm \times nm}$ has the following properties:

$i)$ $P_{n,m}$ is a permutation matrix.

$ii)$ $P_{n,m}^{\mathrm{T}} = P_{n,m}^{-1} = P_{m,n}.$

iii) $P_{n,m}$ is orthogonal.

iv) $P_{n,m}P_{m,n} = I_{nm}$.

v) $P_{n,n}$ is orthogonal, symmetric, and involutory.

vi) $P_{n,n}$ is a reflector.

vii) $P_{n,m} = \sum_{i=1}^{n} e_{i,n} \otimes I_m \otimes e_{i,n}^{\mathrm{T}}$.

viii) $P_{np,m} = (I_n \otimes P_{p,m})(P_{n,m} \otimes I_p) = (I_p \otimes P_{n,m})(P_{p,m} \otimes I_n)$.

ix) $\operatorname{sig} P_{n,n} = \operatorname{tr} P_{n,n} = n$.

x) The inertia of $P_{n,n}$ is given by

$$\operatorname{In} P_{n,n} = \begin{bmatrix} \frac{1}{2}(n^2 - n) \\ 0 \\ \frac{1}{2}(n^2 + n) \end{bmatrix}.$$

xi) $\det P_{n,n} = (-1)^{(n^2-n)/2}$.

xii) $P_{1,m} = I_m$ and $P_{n,1} = I_n$.

xiii) If $x \in \mathbb{F}^n$ and $y \in \mathbb{F}^m$, then

$$P_{n,m}(y \otimes x) = x \otimes y.$$

xiv) If $A \in \mathbb{F}^{n \times m}$ and $b \in \mathbb{F}^k$, then

$$P_{k,n}(A \otimes b) = b \otimes A$$

and

$$P_{n,k}(b \otimes A) = A \otimes b.$$

xv) If $A \in \mathbb{F}^{n \times m}$ and $B \in \mathbb{F}^{l \times k}$, then

$$P_{l,n}(A \otimes B)P_{m,k} = B \otimes A,$$

$$\operatorname{vec}(A \otimes B) = (I_m \otimes P_{k,n} \otimes I_l)[(\operatorname{vec} A) \otimes (\operatorname{vec} B)],$$

$$\operatorname{vec}(A^{\mathrm{T}} \otimes B) = (P_{nk,m} \otimes I_l)[(\operatorname{vec} A) \otimes (\operatorname{vec} B)],$$

$$\operatorname{vec}(A \otimes B^{\mathrm{T}}) = (I_m \otimes P_{l,nk})[(\operatorname{vec} A) \otimes (\operatorname{vec} B)].$$

xvi) If $A \in \mathbb{F}^{n \times m}$, $B \in \mathbb{F}^{l \times k}$, and $nl = mk$, then

$$\operatorname{tr}(A \otimes B) = [\operatorname{vec}(I_m) \otimes (I_k)]^{\mathrm{T}}[(\operatorname{vec} A) \otimes (\operatorname{vec} B^{\mathrm{T}})].$$

xvii) If $A \in \mathbb{F}^{n \times n}$ and $B \in \mathbb{F}^{l \times l}$, then

$$P_{l,n}(A \otimes B)P_{n,l} = P_{l,n}(A \otimes B)P_{l,n}^{-1} = B \otimes A.$$

Hence, $A \otimes B$ and $B \otimes A$ are similar.

xviii) If $A \in \mathbb{F}^{n \times m}$ and $B \in \mathbb{F}^{m \times n}$, then

$$\operatorname{tr} AB = \operatorname{tr}[P_{m,n}(A \otimes B)].$$

xix) $P_{np,m} = P_{n,pm}P_{p,nm} = P_{p,nm}P_{n,pm}$.

xx) $P_{np,m}P_{pm,n}P_{mn,p} = I$.

Now, let $A \in \mathbb{F}^{n \times m}$, let $r \triangleq \operatorname{rank} A$, and define $K \triangleq P_{n,m}(A^* \otimes A)$. Then, the following statements hold:

xxi) K is Hermitian.

xxii) $\operatorname{rank} K = r^2$.

xxiii) $\operatorname{tr} K = \operatorname{tr} A^*A$.

xxiv) $K^2 = (AA^*) \otimes (A^*A)$.

xxv) $\operatorname{mspec}(K) = \{\sigma_1^2(A), \ldots, \sigma_r^2(A)\} \cup \{\pm\sigma_i(A)\sigma_j(A) \colon i < j, i,j = 1, \ldots, r\}$.

Proof: See [1208, pp. 308–311, 342, 343].

Fact 7.4.31. Define $\Psi_n \in \mathbb{R}^{n^2 \times n^2}$ by

$$\Psi_n \triangleq \tfrac{1}{2}(I_{n^2} + P_{n,n}),$$

let $x, y \in \mathbb{F}^n$, and let $A, B \in \mathbb{F}^{n \times n}$. Then, the following statements hold:

i) Ψ_n is a projector.

ii) $\Psi_n = \Psi_n P_{n,n} = P_{n,n}\Psi_n$.

iii) $\Psi_n(x \otimes y) = \tfrac{1}{2}(x \otimes y + y \otimes x)$.

iv) $\Psi_n \operatorname{vec}(A) = \tfrac{1}{2}\operatorname{vec}(A + A^{\mathrm{T}})$.

v) $\Psi_n(A \otimes B)\Psi_n = \Psi_n(B \otimes A)\Psi_n$.

vi) $\Psi_n(A \otimes A)\Psi_n = \Psi_n(A \otimes A) = (A \otimes A)\Psi_n$.

vii) $\Psi_n(A \otimes B + B \otimes A)\Psi_n = \Psi_n(A \otimes B + B \otimes A) = (A \otimes B + B \otimes A)\Psi_n = 2\Psi_n(B \otimes A)\Psi_n$.

viii) $(A \otimes A)\Psi_n(A^{\mathrm{T}} \otimes A^{\mathrm{T}}) = \Psi_n(AA^{\mathrm{T}} \otimes AA^{\mathrm{T}})$.

Proof: See [1208, p. 312].

Fact 7.4.32. Let $A \in \mathbb{F}^{n \times m}$ and $B \in \mathbb{F}^{l \times k}$. Then,

$$(A \otimes B)^+ = A^+ \otimes B^+.$$

Fact 7.4.33. Let $A \in \mathbb{F}^{n \times n}$ and $B \in \mathbb{F}^{m \times m}$. Then,

$$(A \otimes B)^{\mathrm{D}} = A^{\mathrm{D}} \otimes B^{\mathrm{D}}.$$

Now, assume in addition that A and B are group invertible. Then, $A \otimes B$ is group invertible, and

$$(A \otimes B)^{\#} = A^{\#} \otimes B^{\#}.$$

Remark: See Fact 7.4.17.

Fact 7.4.34. For all $i \in \{1, \ldots, p\}$, let $A_i \in \mathbb{F}^{n_i \times n_i}$. Then,

$$\operatorname{mspec}(A_1 \otimes \cdots \otimes A_p)$$
$$= \{\lambda_1 \cdots \lambda_p \colon \lambda_i \in \operatorname{mspec}(A_i) \text{ for all } i \in \{1, \ldots, p\}\}_{\mathrm{ms}}.$$

If, in addition, for all $i \in \{1, \ldots, p\}$, $x_i \in \mathbb{C}^{n_i}$ is an eigenvector of A_i associated with $\lambda_i \in \text{spec}(A_i)$, then $x_1 \otimes \cdots \otimes x_p$ is an eigenvector of $A_1 \otimes \cdots \otimes A_p$ associated with the eigenvalue $\lambda_1 \cdots \lambda_p$.

7.5 Facts on the Kronecker Sum

Fact 7.5.1. Let $A \in \mathbb{F}^{n \times n}$. Then,

$$(A \oplus A)^2 = A^2 \oplus A^2 + 2A \otimes A.$$

Fact 7.5.2. Let $A \in \mathbb{F}^{n \times n}$. Then,

$$n \leq \text{def}(A^{\mathrm{T}} \oplus -A) = \dim \{X \in \mathbb{F}^{n \times n} : AX = XA\}$$

and

$$\text{rank}(A^{\mathrm{T}} \oplus -A) = \dim \{[A, X] : X \in \mathbb{F}^{n \times n}\} \leq n^2 - n.$$

Proof: See Fact 2.18.9.

Remark: $\text{rank}(A^{\mathrm{T}} \oplus -A)$ is the dimension of the commutant or centralizer of A. See Fact 2.18.9.

Remark: See Fact 5.14.21 and Fact 5.14.23.

Problem: Express $\text{rank}(A^{\mathrm{T}} \oplus -A)$ in terms of the eigenstructure of A.

Fact 7.5.3. Let $A \in \mathbb{F}^{n \times n}$, assume that A is nilpotent, and assume that $A^{\mathrm{T}} \oplus -A = 0$. Then, $A = 0$.

Proof: Note that $A^{\mathrm{T}} \otimes A^k = I \otimes A^{k+1}$, and use Fact 7.4.24.

Fact 7.5.4. Let $A \in \mathbb{F}^{n \times n}$, and assume that, for all $X \in \mathbb{F}^{n \times n}$, $AX = XA$. Then, there exists $\alpha \in \mathbb{F}$ such that $A = \alpha I$.

Proof: It follows from Proposition 7.2.3 that all of the eigenvalues of A are equal. Hence, there exists $\alpha \in \mathbb{F}$ such that $A = \alpha I + B$, where B is nilpotent. Now, Fact 7.5.3 implies that $B = 0$.

Fact 7.5.5. Let $A \in \mathbb{F}^{n \times n}$ and $B \in \mathbb{F}^{m \times m}$, and let $\gamma \in \text{spec}(A \oplus B)$. Then,

$$\sum \text{gmult}_A(\lambda)\text{gmult}_B(\mu) \leq \text{gmult}_{A \oplus B}(\gamma)$$
$$\leq \text{amult}_{A \oplus B}(\gamma)$$
$$= \sum \text{amult}_A(\lambda)\text{amult}_B(\mu),$$

where both sums are taken over all $\lambda \in \text{spec}(A)$ and $\mu \in \text{spec}(B)$ such that $\lambda + \mu = \gamma$.

Fact 7.5.6. Let $A \in \mathbb{F}^{n \times n}$. Then,

$$\text{spabs}(A \oplus A) = 2\,\text{spabs}(A).$$

Fact 7.5.7. Let $A \in \mathbb{F}^{n \times n}$ and $B \in \mathbb{F}^{m \times m}$, and let $\gamma \in \text{spec}(A \oplus B)$. Then, $\text{ind}_{A \oplus B}(\gamma) = 1$ if and only if $\text{ind}_A(\lambda) = 1$ and $\text{ind}_B(\mu) = 1$ for all $\lambda \in \text{spec}(A)$ and $\mu \in \text{spec}(B)$ such that $\lambda + \mu = \gamma$.

Fact 7.5.8. Let $A \in \mathbb{F}^{n \times n}$ and $B \in \mathbb{F}^{m \times m}$, and assume that A and B are (group invertible, range Hermitian, Hermitian, symmetric, skew Hermitian, skew symmetric, normal, positive semidefinite, positive definite, semidissipative, dissipative, nilpotent, semisimple). Then, so is $A \oplus B$.

Fact 7.5.9. Let $A \in \mathbb{F}^{n \times n}$ and $B \in \mathbb{F}^{m \times m}$. Then,

$$P_{m,n}(A \oplus B)P_{n,m} = P_{m,n}(A \oplus B)P_{m,n}^{-1} = B \oplus A.$$

Hence, $A \oplus B$ and $B \oplus A$ are similar, and thus

$$\mathrm{rank}(A \oplus B) = \mathrm{rank}(B \oplus A).$$

Proof: Use *xiii*) of Fact 7.4.30.

Fact 7.5.10. Let $A \in \mathbb{F}^{n \times n}$ and $B \in \mathbb{F}^{m \times m}$. Then,

$$n\,\mathrm{rank}\,B + m\,\mathrm{rank}\,A - 2(\mathrm{rank}\,A)(\mathrm{rank}\,B)$$
$$\leq \mathrm{rank}(A \oplus B)$$
$$\leq \begin{cases} nm - [n - \mathrm{rank}(I + A)][m - \mathrm{rank}(I - B)] \\ nm - [n - \mathrm{rank}(I - A)][m - \mathrm{rank}(I + B)]. \end{cases}$$

If, in addition, $-A$ and B are idempotent, then

$$\mathrm{rank}(A \oplus B) = n\,\mathrm{rank}\,B + m\,\mathrm{rank}\,A - 2(\mathrm{rank}\,A)(\mathrm{rank}\,B).$$

Equivalently,

$$\mathrm{rank}(A \oplus B) = (\mathrm{rank}\ (-A)_\perp)\,\mathrm{rank}\,B + (\mathrm{rank}\,B_\perp)\,\mathrm{rank}\,A.$$

Proof: See [341].

Remark: Equality may not hold for the upper bounds when $-A$ and B are idempotent.

Fact 7.5.11. Let $A \in \mathbb{F}^{n \times n}$, let $B \in \mathbb{F}^{m \times m}$, assume that A is positive definite, and define $p(s) \triangleq \det(I - sA)$, and let $\mathrm{mroots}(p) = \{\lambda_1, \ldots, \lambda_n\}_{\mathrm{ms}}$. Then,

$$\det(A \oplus B) = (\det A)^m \prod_{i=1}^{n} \det(\lambda_i B + I).$$

Proof: Specialize Fact 7.5.12.

Fact 7.5.12. Let $A, C \in \mathbb{F}^{n \times n}$, let $B, D \in \mathbb{F}^{m \times m}$, assume that A is positive definite, assume that C is positive semidefinite, define $p(s) \triangleq \det(C - sA)$, and let $\mathrm{mroots}(p) = \{\lambda_1, \ldots, \lambda_n\}_{\mathrm{ms}}$. Then,

$$\det(A \otimes B + C \otimes D) = (\det A)^m \prod_{i=1}^{n} \det(\lambda_i D + B).$$

Proof: See [1027, pp. 40, 41].

Remark: The Kronecker product definition in [1027] follows the convention of [967], where "$A \otimes B$" denotes $B \otimes A$.

Fact 7.5.13. Let $A, D \in \mathbb{F}^{n \times n}$, let $C, B \in \mathbb{F}^{m \times m}$, assume that $\operatorname{rank} C = 1$, and assume that A is nonsingular. Then,

$$\det(A \otimes B + C \otimes D) = (\det A)^m (\det B)^{n-1} \det\left[B + \left(\operatorname{tr} CA^{-1}\right)D\right].$$

Proof: See [1027, p. 41].

Fact 7.5.14. Let $A \in \mathbb{F}^{n \times n}$ and $B \in \mathbb{F}^{m \times m}$. Then, $\operatorname{spec}(A)$ and $\operatorname{spec}(-B)$ are disjoint if and only if, for all $C \in \mathbb{F}^{n \times m}$, the matrices $\left[\begin{smallmatrix} A & 0 \\ 0 & -B \end{smallmatrix}\right]$ and $\left[\begin{smallmatrix} A & C \\ 0 & -B \end{smallmatrix}\right]$ are similar.

Proof: Sufficiency follows from Fact 5.10.21, while necessity follows from Corollary 2.6.6 and Proposition 7.2.3.

Fact 7.5.15. Let $A \in \mathbb{F}^{n \times n}$, $B \in \mathbb{F}^{m \times m}$, and $C \in \mathbb{F}^{n \times m}$, and assume that $\det(B^{\mathrm{T}} \oplus A) \neq 0$. Then, $X \in \mathbb{F}^{n \times m}$ satisfies

$$A^2 X + 2AXB + XB^2 + C = 0$$

if and only if

$$X = -\operatorname{vec}^{-1}\left[\left(B^{\mathrm{T}} \oplus A\right)^{-2} \operatorname{vec} C\right].$$

Fact 7.5.16. For all $i \in \{1, \ldots, p\}$, let $A_i \in \mathbb{F}^{n_i \times n_i}$. Then,

$\operatorname{mspec}(A_1 \oplus \cdots \oplus A_p)$
$$= \{\lambda_1 + \cdots + \lambda_p \colon \ \lambda_i \in \operatorname{mspec}(A_i) \text{ for all } i \in \{1, \ldots, p\}\}_{\mathrm{ms}}.$$

If, in addition, for all $i \in \{1, \ldots, p\}$, $x_i \in \mathbb{C}^{n_i}$ is an eigenvector of A_i associated with $\lambda_i \in \operatorname{spec}(A_i)$, then $x_1 \oplus \cdots \oplus x_p$ is an eigenvector of $A_1 \oplus \cdots \oplus A_p$ associated with $\lambda_1 + \cdots + \lambda_p$.

Fact 7.5.17. Let $A \in \mathbb{F}^{n \times m}$, and let $k \in \mathbb{P}$ satisfy $1 \leq k \leq \min\{n, m\}$. Furthermore, define the kth *compound* $A^{(k)}$ to be the $\binom{n}{k} \times \binom{m}{k}$ matrix whose entries are $k \times k$ subdeterminants of A, ordered lexicographically. (Example: For $n = k = 3$, subsets of the rows and columns of A are chosen in the order $\{1, 1, 1\}, \{1, 1, 2\}, \{1, 1, 3\}, \{1, 2, 1\}, \{1, 2, 2\}, \ldots$.) Specifically, $\left(A^{(k)}\right)_{(i,j)}$ is the $k \times k$ subdeterminant of A corresponding to the ith selection of k rows of A and the jth selection of k columns of A. Then, the following statements hold:

i) $A^{(1)} = A$.

ii) $(\alpha A)^{(k)} = \alpha^k A^{(k)}$.

iii) $\left(A^{\mathrm{T}}\right)^{(k)} = \left(A^{(k)}\right)^{\mathrm{T}}$.

iv) $\overline{A}^{(k)} = \overline{A^{(k)}}$.

v) $(A^*)^{(k)} = \left(A^{(k)}\right)^*$.

vi) If $B \in \mathbb{F}^{m \times l}$ and $1 \leq k \leq \min\{n, m, l\}$, then $(AB)^{(k)} = A^{(k)} B^{(k)}$.

vii) If $B \in \mathbb{F}^{m \times n}$, then $\det AB = A^{(k)} B^{(k)}$.

Now, assume in addition that $m = n$, let $1 \leq k \leq n$, and let $\operatorname{mspec}(A) = \{\lambda_1, \ldots, \lambda_n\}_{\mathrm{ms}}$. Then, the following statements hold:

viii) If A is (diagonal, lower triangular, upper triangular, Hermitian, positive semidefinite, positive definite, unitary), then so is $A^{(k)}$.

ix) Assume that A is skew Hermitian. If k is odd, then $A^{(k)}$ is skew Hermitian. If k is even, then $A^{(k)}$ is Hermitian.

x) Assume that A is diagonal, upper triangular, or lower triangular, and let $1 \leq i_1 < \cdots < i_k \leq n$. Then, the $(i_1 + \cdots + i_k, i_1 + \cdots + i_k)$ entry of $A^{(k)}$ is $A_{(i_1,i_1)} \cdots A_{(i_k,i_k)}$. In particular, $I_n^{(k)} = I_{\binom{n}{k}}$.

xi) $\det A^{(k)} = (\det A)^{\binom{n-1}{k-1}}$.

xii) $A^{(n)} = \det A$.

xiii) $SA^{(n-1)\mathrm{T}}S = A^{\mathrm{A}}$, where $S \triangleq \mathrm{diag}(1, -1, 1, \ldots)$.

xiv) $\det A^{(n-1)} = \det A^{\mathrm{A}} = (\det A)^{n-1}$.

xv) $\mathrm{tr}\, A^{(n-1)} = \mathrm{tr}\, A^{\mathrm{A}}$.

xvi) If A is nonsingular, then $\left(A^{(k)}\right)^{-1} = \left(A^{-1}\right)^{(k)}$.

xvii) $\mathrm{mspec}\left(A^{(k)}\right) = \{\lambda_{i_1} \cdots \lambda_{i_k} : 1 \leq i_1 < \cdots < i_k \leq n\}_{\mathrm{ms}}$. In particular,

$$\mathrm{mspec}\left(A^{(2)}\right) = \{\lambda_i \lambda_j : i, j = 1, \ldots, n, i < j\}_{\mathrm{ms}}.$$

xviii) $\mathrm{tr}\, A^{(k)} = \sum_{1 \leq i_1 < \cdots < i_k \leq n} \lambda_{i_1} \cdots \lambda_{i_k}$.

xix) If A has exactly k nonzero eigenvalues, then $A^{(k)}$ has exactly one nonzero eigenvalue.

xx) If $k < n$ and A has exactly k nonzero eigenvalues, then $\mathrm{spec}\left(A^{(k+1)}\right) = \{0\}$, and thus $A^{(k+1)}$ is nilpotent.

xxi) If $B \in \mathbb{F}^{n \times n}$, then $\det(A + B) = \begin{bmatrix} A & I \end{bmatrix}^{(n)} \begin{bmatrix} I \\ B \end{bmatrix}^{(n)}$.

xxii) The characteristic polynomial of A is given by

$$\chi_A(s) = s^n + \sum_{i=1}^{n-1} (-1)^{n+i}[\mathrm{tr}\, A^{(n-i)}]s^i + (-1)^n \det A.$$

xxiii) $\det(I + A) = 1 + \det A + \sum_{i=1}^{n-1} \mathrm{tr}\, A^{(n-i)}$.

Now, for $i = 0, \ldots, k$, define $A^{(k,i)}$ by

$$(A + sI)^{(k)} = s^k A^{(k,0)} + s^{k-1} A^{(k,1)} + \cdots + sA^{(k,k-1)} + A^{(k,k)}.$$

Then, the following statements hold:

xxiv) $A^{(k,0)} = I$.

xxv) $A^{(k,k)} = A^{(k)}$.

xxvi) If $B \in \mathbb{F}^{n \times n}$ and $\alpha, \beta \in \mathbb{F}$, then

$$(\alpha A + \beta B)^{(k,1)} = \alpha A^{(k,1)} + \beta B^{(k,1)}.$$

xxvii) $\mathrm{mspec}\left(A^{(k,1)}\right) = \{\lambda_{i_1} + \cdots + \lambda_{i_k} : 1 \leq i_1 < \cdots < i_k \leq n\}_{\mathrm{ms}}$.

xxviii) $\operatorname{tr} A^{(k,1)} = \binom{n-1}{k-1} \operatorname{tr} A$.

xxix) $\operatorname{mspec}(A^{(2,1)}) = \{\lambda_i + \lambda_j : i, j = 1, \ldots, n, i < j\}_{\mathrm{ms}}$.

xxx) $\operatorname{mspec}\left[(A^{(2,1)})^2 - 4A^{(2)}\right] = \{(\lambda_i - \lambda_j)^2 : i, j = 1, \ldots, n, i < j\}_{\mathrm{ms}}$.

Proof: See [494, pp. 142–155], [728, p. 11], [983, pp. 116–130], [996, pp. 502–506], [1125, p. 124], and [1126].

Remark: Statement *vi*) is the *Binet-Cauchy theorem*. See [996, p. 503]. The special case given by statement *vii*) is also given by Fact 2.13.5. Another special case is given by statement *xxi*). Statement *xi*) is the *Sylvester-Franke theorem*. See [983, p. 130].

Remark: $A^{(k,1)}$ is the kth *additive compound* of A.

Remark: $(A^{(2,1)})^2 - 4A^{(2)}$ is the *discriminant* of A, which is singular if and only if A has a repeated eigenvalue.

Remark: Additional expressions for the determinant of a sum of matrices are given in [1126].

Remark: The compound operation is related to the *bialternate product* since $\operatorname{mspec}(2A \cdot I) = \operatorname{mspec}(A^{(2,1)})$ and $\operatorname{mspec}(A \cdot A) = \operatorname{mspec}(A^{(2)})$. See [532, 590], [804, pp. 313–320], and [967, pp. 84, 85].

Remark: Induced norms of compound matrices are considered in [464].

Remark: See Fact 4.9.2, Fact 8.13.43, and Fact 11.17.12.

Problem: Express $A \cdot B$ in terms of compounds.

7.6 Facts on the Schur Product

Fact 7.6.1. Let $x, y, z \in \mathbb{F}^n$. Then,
$$x^{\mathrm{T}}(y \circ z) = z^{\mathrm{T}}(x \circ y) = y^{\mathrm{T}}(x \circ z).$$

Fact 7.6.2. Let $w, y \in \mathbb{F}^n$ and $x, z \in \mathbb{F}^m$. Then,
$$(wx^{\mathrm{T}}) \circ (yz^{\mathrm{T}}) = (w \circ y)(x \circ z)^{\mathrm{T}}.$$

Fact 7.6.3. Let $A \in \mathbb{F}^{n \times n}$ and $d \in \mathbb{F}^n$. Then,
$$\operatorname{diag}(d)A = A \circ d1_{1 \times n}.$$

Fact 7.6.4. Let $A, B \in \mathbb{F}^{n \times m}$, $D_1 \in \mathbb{F}^{n \times n}$, and $D_2 \in \mathbb{F}^{m \times m}$, and assume that D_1 and D_2 are diagonal. Then,
$$(D_1 A) \circ (B D_2) = D_1(A \circ B)D_2.$$

Fact 7.6.5. Let $x, a \in \mathbb{F}^n$, $y, b \in \mathbb{F}^m$, and $A \in \mathbb{F}^{n \times m}$. Then,
$$x^{\mathrm{T}}(A \circ ab^{\mathrm{T}})y = (a \circ x)^{\mathrm{T}}A(b \circ y).$$

Fact 7.6.6. Let $A \in \mathbb{F}^{n \times m}$, $B \in \mathbb{F}^{m \times n}$, $x \in \mathbb{F}^m$, and $y \in \mathbb{F}^n$. Then,

$$x^{\mathrm{T}}(A^{\mathrm{T}} \circ B)y = \mathrm{tr}[\mathrm{diag}(x)A^{\mathrm{T}}\mathrm{diag}(y)B^{\mathrm{T}}] = \mathrm{tr}[A(B \circ xy^{\mathrm{T}})].$$

In particular,

$$\mathrm{tr}\, AB = 1_{1 \times m}(A^{\mathrm{T}} \circ B)1_{n \times 1}.$$

Fact 7.6.7. Let $A, B \in \mathbb{F}^{n \times m}$. Then,

$$\mathrm{tr}[(A \circ B)(A \circ B)^{\mathrm{T}}] = \mathrm{tr}[(A \circ A)(B \circ B)^{\mathrm{T}}].$$

Fact 7.6.8. Let $A, B \in \mathbb{F}^{n \times m}$ and $C \in \mathbb{F}^{m \times n}$. Then,

$$I_n \circ [A(B^{\mathrm{T}} \circ C)] = I_n \circ [(A \circ B)C] = I_n \circ [(A \circ C^{\mathrm{T}})B^{\mathrm{T}}].$$

Hence,

$$\mathrm{tr}[A(B^{\mathrm{T}} \circ C)] = \mathrm{tr}[(A \circ B)C] = \mathrm{tr}[(A \circ C^{\mathrm{T}})B^{\mathrm{T}}].$$

Fact 7.6.9. Let $A_1, \ldots, A_k \in \mathbb{F}^{n \times n}$. Then,

$$\mathcal{R}[(A_1 A_1^*) \circ \cdots \circ (A_k A_k^*)] = \mathrm{span}\,\{(A_1 x_1) \circ \cdots \circ (A_k x_k) \colon x_1, \cdots, x_k \in \mathbb{F}^n\}.$$

Furthermore, if $A_1, \ldots, A_k$ are positive semidefinite, then

$$\mathcal{R}(A_1 \circ \cdots \circ A_k) = \mathrm{span}\,\{(A_1 x_1) \circ \cdots \circ (A_k x_k) \colon x_1, \ldots, x_k \in \mathbb{F}^n\}$$
$$= \mathrm{span}\,\{(A_1 x) \circ \cdots \circ (A_k x) \colon x \in \mathbb{F}^n\}.$$

Proof: See [1136].

Fact 7.6.10. Let $A, B \in \mathbb{F}^{n \times m}$. Then,

$$\mathrm{rank}(A \circ B) \leq \mathrm{rank}(A \otimes B) = (\mathrm{rank}\, A)(\mathrm{rank}\, B).$$

Proof: Use Proposition 7.3.1.

Remark: See Fact 8.22.16.

Fact 7.6.11. Let $x \in \mathbb{R}^m$ and $A \in \mathbb{R}^{n \times m}$, and define $x^A \in \mathbb{R}^n$ by

$$x^A \triangleq \begin{bmatrix} \prod_{i=1}^m x_{(i)}^{A_{(1,i)}} \\ \vdots \\ \prod_{i=1}^m x_{(i)}^{A_{(n,i)}} \end{bmatrix},$$

where every component of x^A is assumed to exist. Then, the following statements hold:

i) If $a \in \mathbb{R}$, then $a^x = \begin{bmatrix} a^{x_{(1)}} \\ \vdots \\ a^{x_{(m)}} \end{bmatrix}$.

ii) $x^{-A} = (x^A)^{\circ -1}$.

iii) If $y \in \mathbb{R}^m$, then $(x \circ y)^A = x^A \circ y^A$.

iv) If $B \in \mathbb{R}^{n \times m}$, then $x^{A+B} = x^A \circ x^B$.

v) If $B \in \mathbb{R}^{l \times n}$, then $(x^A)^B = x^{BA}$.

vi) If $a \in \mathbb{R}$, then $(a^x)^A = a^{Ax}$.

vii) If $A^L \in \mathbb{R}^{m \times n}$ is a left inverse of A and $y = x^A$, then $x = y^{A^L}$.

viii) If $A \in \mathbb{R}^{n \times n}$ is nonsingular and $y = x^A$, then $x = y^{A^{-1}}$.

ix) Define $f(x) \triangleq x^A$. Then, $f'(x) = \operatorname{diag}(x^A) A \operatorname{diag}(x^{\circ -1})$.

x) Let $x_1, \ldots, x_n \in \mathbb{R}^n$, let $a \in \mathbb{R}^n$, and assume that $0 < x_1 < \cdots < x_n$ and $a_{(1)} < \cdots < a_{(n)}$. Then,

$$\det \begin{bmatrix} x_1^a & \cdots & x_n^a \end{bmatrix} > 0.$$

Remark: These operations arise in modeling chemical reaction kinetics. See [917].

Proof: Result *x)* is given in [1158].

Fact 7.6.12. Let $A \in \mathbb{R}^{n \times n}$, and assume that A is nonsingular. Then,

$$(A \circ A^{-\mathrm{T}}) 1_{n \times 1} = 1_{n \times 1}$$

and

$$1_{1 \times n} (A \circ A^{-\mathrm{T}}) = 1_{1 \times n}.$$

Proof: See [794].

Fact 7.6.13. Let $A \in \mathbb{R}^{n \times n}$, and assume that $A \geq\geq 0$. Then,

$$\operatorname{sprad}\left[(A \circ A^{\mathrm{T}})^{\circ 1/2}\right] \leq \operatorname{sprad}(A) \leq \operatorname{sprad}\left[\tfrac{1}{2}(A + A^{\mathrm{T}})\right].$$

Proof: See [1211].

Fact 7.6.14. Let $A_1, \ldots, A_r \in \mathbb{R}^{n \times n}$ and $\alpha_1, \ldots, \alpha_r \in \mathbb{R}$, and assume that $A_i \geq\geq 0$ for all $i \in \{1, \ldots, r\}$, $\alpha_i > 0$ for all $i \in \{1, \ldots, r\}$, and $\sum_{i=1}^{r} \alpha_i \geq 1$. Then,

$$\operatorname{sprad}(A_1^{\circ \alpha_1} \circ \cdots \circ A_r^{\circ \alpha_r}) \leq \prod_{i=1}^{r} [\operatorname{sprad}(A_i)]^{\alpha_i}.$$

In particular, let $A \in \mathbb{R}^{n \times n}$, and assume that $A \geq\geq 0$. Then, for all $\alpha \geq 1$,

$$\operatorname{sprad}(A^{\circ \alpha}) \leq [\operatorname{sprad}(A)]^{\alpha},$$

whereas, for all $\alpha \leq 1$,

$$[\operatorname{sprad}(A)]^{\alpha} \leq \operatorname{sprad}(A^{\circ \alpha}).$$

Furthermore,

$$\operatorname{sprad}\left(A^{\circ 1/2} \circ A^{\mathrm{T} \circ 1/2}\right) \leq \operatorname{sprad}(A)$$

and

$$[\operatorname{sprad}(A \circ A)]^{1/2} \leq \operatorname{sprad}(A) = [\operatorname{sprad}(A \otimes A)]^{1/2}.$$

If, in addition, $B \in \mathbb{R}^{n \times n}$ is such that $B \geq\geq 0$, then

$$\operatorname{sprad}(A \circ B) \leq [\operatorname{sprad}(A \circ A) \operatorname{sprad}(B \circ B)]^{1/2} \leq \operatorname{sprad}(A) \operatorname{sprad}(B),$$

$$\text{sprad}(A \circ B) \leq \text{sprad}(A) \text{sprad}(B)$$
$$+ \max_{i=1,\ldots,n} [2A_{(i,i)}B_{(i,i)} - \text{sprad}(A)B_{(i,i)} - \text{sprad}(B)A_{(i,i)}]$$
$$\leq \text{sprad}(A) \text{sprad}(B),$$

and

$$\text{sprad}\left(A^{\circ 1/2} \circ B^{\circ 1/2}\right) \leq \sqrt{\text{sprad}(A) \text{sprad}(B)}.$$

If, in addition, $A \gg 0$ and $B \gg 0$, then

$$\text{sprad}(A \circ B) < \text{sprad}(A) \text{sprad}(B).$$

Proof: See [466, 480, 814]. The equality $\text{sprad}(A) = [\text{sprad}(A \otimes A)]^{1/2}$ follows from Fact 7.4.15.

Remark: The inequality $\text{sprad}(A \circ A) \leq \text{sprad}(A \otimes A)$ follows from Fact 4.11.19 and Proposition 7.3.1.

Remark: Some extensions are given in [750].

Fact 7.6.15. Let $A, B \in \mathbb{R}^{n \times n}$, and assume that A and B are nonsingular M-matrices. Then, the following statements hold:

i) $A \circ B^{-1}$ is a nonsingular M-matrix.

ii) If $n = 2$, then $\tau(A \circ A^{-1}) = 1$.

iii) If $n \geq 3$, then $\frac{1}{n} < \tau(A \circ A^{-1}) \leq 1$.

iv) $\tau(A) \min_{i=1,\ldots,n}(B^{-1})_{(i,i)} \leq \tau(A \circ B^{-1})$.

v) $[\tau(A)\tau(B)]^n \leq |\det(A \circ B)|$.

vi) $|(A \circ B)^{-1}| \leq \leq A^{-1} \circ B^{-1}$.

Proof: See [730, pp. 359, 370, 375, 380].

Remark: The minimum eigenvalue $\tau(A)$ is defined in Fact 4.11.11.

Remark: Some extensions are given in [750].

Fact 7.6.16. Let $A, B \in \mathbb{F}^{n \times m}$. Then,

$$\text{sprad}(A \circ B) \leq \sqrt{\text{sprad}(A \circ \overline{A}) \text{sprad}(B \circ \overline{B})}.$$

Consequently,

$$\left.\begin{array}{c} \text{sprad}(A \circ A) \\ \text{sprad}(A \circ A^{\text{T}}) \\ \text{sprad}(A \circ A^*) \end{array}\right\} \leq \text{sprad}(A \circ \overline{A}).$$

Proof: See [1224].

Remark: See Fact 9.14.33.

Fact 7.6.17. Let $A, B \in \mathbb{R}^{n \times n}$, assume that A and B are nonnegative, and let $\alpha \in [0,1]$. Then,

$$\mathrm{sprad}(A^{\circ \alpha} \circ B^{\circ (1-\alpha)}) \le \mathrm{sprad}^\alpha(A) \, \mathrm{sprad}^{1-\alpha}(B).$$

In particular,

$$\mathrm{sprad}(A^{\circ 1/2} \circ B^{\circ 1/2}) \le \sqrt{\mathrm{sprad}(A) \, \mathrm{sprad}(B)}.$$

Finally,

$$\mathrm{sprad}(A^{\circ 1/2} \circ A^{\circ 1/2\mathrm{T}}) \le \mathrm{sprad}(A^{\circ \alpha} \circ A^{\circ (1-\alpha)\mathrm{T}}) \le \mathrm{sprad}(A).$$

Proof: See [1224].

Remark: See Fact 9.14.36.

7.7 Notes

A history of the Kronecker product is given in [683]. Kronecker matrix algebra is discussed in [263, 593, 685, 973, 1019, 1250, 1413]. Applications are discussed in [1148, 1149, 1396].

The fact that the Schur product is a principal submatrix of the Kronecker product is noted in [987]. A variation of Kronecker matrix algebra for symmetric matrices can be developed in terms of the half-vectorization operator "vech" and the associated elimination and duplication matrices [685, 972, 1377].

Generalizations of the Schur and Kronecker products, known as the block-Kronecker, strong Kronecker, Khatri-Rao, and Tracy-Singh products, are discussed in [393, 733, 760, 864, 948, 950, 951, 953] and [1146, pp. 216, 217]. A related operation is the *bialternate product*, which is a variation of the compound operation discussed in Fact 7.5.17. See [532, 590], [804, pp. 313–320], and [967, pp. 84, 85]. The Schur product is also called the Hadamard product.

The Kronecker product is associated with tensor analysis and multilinear algebra [431, 559, 599, 983, 984, 1019].

Chapter Eight

Positive-Semidefinite Matrices

In this chapter we focus on positive-semidefinite and positive-definite matrices. These matrices arise in a variety of applications, such as covariance analysis in signal processing and controllability analysis in linear system theory, and they have many special properties.

8.1 Positive-Semidefinite and Positive-Definite Orderings

Let $A \in \mathbb{F}^{n \times n}$ be a Hermitian matrix. As shown in Corollary 5.4.5, A is unitarily similar to a real diagonal matrix whose diagonal entries are the eigenvalues of A. We denote these eigenvalues by $\lambda_1, \ldots, \lambda_n$ or, for clarity, by $\lambda_1(A), \ldots, \lambda_n(A)$. As in Chapter 4, we employ the convention

$$\lambda_1 \geq \lambda_2 \geq \cdots \geq \lambda_n, \tag{8.1.1}$$

and, for convenience, we define

$$\lambda_{\max}(A) \triangleq \lambda_1, \quad \lambda_{\min}(A) \triangleq \lambda_n. \tag{8.1.2}$$

Then, A is positive semidefinite if and only if $\lambda_{\min}(A) \geq 0$, while A is positive definite if and only if $\lambda_{\min}(A) > 0$.

For convenience, let $\mathbf{H}^n, \mathbf{N}^n$, and $\mathbf{P}^n$ denote, respectively, the Hermitian, positive-semidefinite, and positive-definite matrices in $\mathbb{F}^{n \times n}$. Hence, $\mathbf{P}^n \subset \mathbf{N}^n \subset \mathbf{H}^n$. If $A \in \mathbf{N}^n$, then we write $A \geq 0$, while, if $A \in \mathbf{P}^n$, then we write $A > 0$. If $A, B \in \mathbf{H}^n$, then $A - B \in \mathbf{N}^n$ is possible even if neither A nor B is positive semidefinite. In this case, we write $A \geq B$ or $B \leq A$. Similarly, $A - B \in \mathbf{P}^n$ is denoted by $A > B$ or $B < A$. This notation is consistent with the case $n = 1$, where $\mathbf{H}^1 = \mathbb{R}$, $\mathbf{N}^1 = [0, \infty)$, and $\mathbf{P}^1 = (0, \infty)$.

Since $0 \in \mathbf{N}^n$, it follows that $\mathbf{N}^n$ is a pointed cone. Furthermore, if $A, -A \in \mathbf{N}^n$, then $x^*Ax = 0$ for all $x \in \mathbb{F}^n$, which implies that $A = 0$. Hence, $\mathbf{N}^n$ is a one-sided cone. Finally, $\mathbf{N}^n$ and $\mathbf{P}^n$ are convex cones since, if $A, B \in \mathbf{N}^n$, then $\alpha A + \beta B \in \mathbf{N}^n$ for all $\alpha, \beta > 0$, and likewise for $\mathbf{P}^n$. The following result shows that the relation "$\leq$" is a partial ordering on $\mathbf{H}^n$.

Proposition 8.1.1. The relation "$\leq$" is reflexive, antisymmetric, and transitive on $\mathbf{H}^n$, that is, if $A, B, C \in \mathbf{H}^n$, then the following statements hold:

  i) $A \leq A$.

  ii) If $A \leq B$ and $B \leq A$, then $A = B$.

  iii) If $A \leq B$ and $B \leq C$, then $A \leq C$.

Proof. Since $\mathbf{N}^n$ is a pointed, one-sided, convex cone, it follows from Proposition 2.3.6 that the relation "$\leq$" is reflexive, antisymmetric, and transitive. $\square$

Additional properties of "$\leq$" and "$<$" are given by the following result.

Proposition 8.1.2. Let $A, B, C, D \in \mathbf{H}^n$. Then, the following statements hold:

  i) If $A \geq 0$, then $\alpha A \geq 0$ for all $\alpha \geq 0$, and $\alpha A \leq 0$ for all $\alpha \leq 0$.

  ii) If $A > 0$, then $\alpha A > 0$ for all $\alpha > 0$, and $\alpha A < 0$ for all $\alpha < 0$.

  iii) $\alpha A + \beta B \in \mathbf{H}^n$ for all $\alpha, \beta \in \mathbb{R}$.

  iv) If $A \geq 0$ and $B \geq 0$, then $\alpha A + \beta B \geq 0$ for all $\alpha, \beta \geq 0$.

  v) If $A \geq 0$ and $B > 0$, then $A + B > 0$.

  vi) $A^2 \geq 0$.

  vii) $A^2 > 0$ if and only if $\det A \neq 0$.

  viii) If $A \leq B$ and $B < C$, then $A < C$.

  ix) If $A < B$ and $B \leq C$, then $A < C$.

  x) If $A \leq B$ and $C \leq D$, then $A + C \leq B + D$.

  xi) If $A \leq B$ and $C < D$, then $A + C < B + D$.

Furthermore, let $S \in \mathbb{F}^{m \times n}$. Then, the following statements hold:

  xii) If $A \leq B$, then $SAS^* \leq SBS^*$.

  xiii) If $A < B$ and $\operatorname{rank} S = m$, then $SAS^* < SBS^*$.

  xiv) If $SAS^* \leq SBS^*$ and $\operatorname{rank} S = n$, then $A \leq B$.

  xv) If $SAS^* < SBS^*$ and $\operatorname{rank} S = n$, then $m = n$ and $A < B$.

  xvi) If $A \leq B$, then $SAS^* < SBS^*$ if and only if $\operatorname{rank} S = m$ and $\mathcal{R}(S) \cap \mathcal{N}(B - A) = \{0\}$.

Proof. Results *i)*–*xi)* are immediate. To prove *xiii)*, note that $A < B$ implies that $(B - A)^{1/2}$ is positive definite. Thus, $\operatorname{rank} S(A - B)^{1/2} = m$, which implies that $S(A - B)S^*$ is positive definite. To prove *xiv)*, note that, since $\operatorname{rank} S = n$, it follows that S has a left inverse $S^{\mathrm{L}} \in \mathbb{F}^{n \times m}$. Thus, *xii)* implies that $A = S^{\mathrm{L}}SAS^*S^{\mathrm{L}*} \leq S^{\mathrm{L}}SBS^*S^{\mathrm{L}*} = B$. To prove *xv)*, note that, since $S(B - A)S^*$ is positive definite, it follows that $\operatorname{rank} S = m$. Hence, $m = n$ and S is nonsingular. Thus, *xiii)* implies that $A = S^{-1}SAS^*S^{-*} < S^{-1}SBS^*S^{-*} = B$. Statement *xvi)* is proved in [293]. $\square$

The following result is an immediate consequence of Corollary 5.4.7.

Corollary 8.1.3. Let $A, B \in \mathbf{H}^n$, and assume that A and B are congruent. Then, A is positive semidefinite if and only if B is positive semidefinite. Furthermore, A is positive definite if and only if B is positive definite.

8.2 Submatrices

We first consider some equalities for a partitioned positive-semidefinite matrix.

Lemma 8.2.1. Let $A = \begin{bmatrix} A_{11} & A_{12} \\ A_{12}^* & A_{22} \end{bmatrix} \in \mathbf{N}^{n+m}$. Then,

$$A_{12} = A_{11}A_{11}^+ A_{12}, \qquad (8.2.1)$$

$$A_{12} = A_{12}A_{22}A_{22}^+. \qquad (8.2.2)$$

Proof. Since $A \geq 0$, it follows from Corollary 5.4.5 that $A = BB^*$, where $B = \begin{bmatrix} B_1 \\ B_2 \end{bmatrix} \in \mathbb{F}^{(n+m)\times r}$ and $r \triangleq \operatorname{rank} A$. Thus, $A_{11} = B_1 B_1^*$, $A_{12} = B_1 B_2^*$, and $A_{22} = B_2 B_2^*$. Since A_{11} is Hermitian, it follows from *xxvii*) of Proposition 6.1.6 that A_{11}^+ is also Hermitian. Next, defining $S \triangleq B_1 - B_1 B_1^*(B_1 B_1^*)^+ B_1$, it follows that $SS^* = 0$, and thus $\operatorname{tr} SS^* = 0$. Hence, Lemma 2.2.3 implies that $S = 0$, and thus $B_1 = B_1 B_1^*(B_1 B_1^*)^+ B_1$. Consequently, $B_1 B_2^* = B_1 B_1^*(B_1 B_1^*)^+ B_1 B_2^*$, that is, $A_{12} = A_{11}A_{11}^+ A_{12}$. The second result is analogous. $\qquad \square$

Corollary 8.2.2. Let $A = \begin{bmatrix} A_{11} & A_{12} \\ A_{12}^* & A_{22} \end{bmatrix} \in \mathbf{N}^{n+m}$. Then, the following statements hold:

i) $\mathcal{R}(A_{12}) \subseteq \mathcal{R}(A_{11})$.

ii) $\mathcal{R}(A_{12}^*) \subseteq \mathcal{R}(A_{22})$.

iii) $\operatorname{rank} \begin{bmatrix} A_{11} & A_{12} \end{bmatrix} = \operatorname{rank} A_{11}$.

iv) $\operatorname{rank} \begin{bmatrix} A_{12}^* & A_{22} \end{bmatrix} = \operatorname{rank} A_{22}$.

Proof. Results *i)* and *ii)* follow from (8.2.1) and (8.2.2), while *iii)* and *iv)* are consequences of *i)* and *ii)*. $\qquad \square$

Next, if (8.2.1) holds, then the partitioned Hermitian matrix $A \triangleq \begin{bmatrix} A_{11} & A_{12} \\ A_{12}^* & A_{22} \end{bmatrix}$ can be factored as

$$\begin{bmatrix} A_{11} & A_{12} \\ A_{12}^* & A_{22} \end{bmatrix} = \begin{bmatrix} I & 0 \\ A_{12}^* A_{11}^+ & I \end{bmatrix} \begin{bmatrix} A_{11} & 0 \\ 0 & A_{11}|A \end{bmatrix} \begin{bmatrix} I & A_{11}^+ A_{12} \\ 0 & I \end{bmatrix}, \qquad (8.2.3)$$

while, if (8.2.2) holds, then

$$\begin{bmatrix} A_{11} & A_{12} \\ A_{12}^* & A_{22} \end{bmatrix} = \begin{bmatrix} I & A_{12}A_{22}^+ \\ 0 & I \end{bmatrix} \begin{bmatrix} A_{22}|A & 0 \\ 0 & A_{22} \end{bmatrix} \begin{bmatrix} I & 0 \\ A_{22}^+ A_{12}^* & I \end{bmatrix}, \qquad (8.2.4)$$

where

$$A_{11}|A = A_{22} - A_{12}^* A_{11}^+ A_{12} \qquad (8.2.5)$$

and

$$A_{22}|A = A_{11} - A_{12}A_{22}^+A_{12}^*. \qquad (8.2.6)$$

Hence, it follows from Lemma 8.2.1 that, if A is positive semidefinite, then (8.2.3) and (8.2.4) are valid, and, furthermore, the Schur complements (see Definition 6.1.8) $A_{11}|A$ and $A_{22}|A$ are both positive semidefinite. Consequently, we have the following results.

Proposition 8.2.3. Let $A \triangleq \begin{bmatrix} A_{11} & A_{12} \\ A_{12}^* & A_{22} \end{bmatrix} \in \mathbf{N}^{n+m}$. Then,

$$\operatorname{rank} A = \operatorname{rank} A_{11} + \operatorname{rank} A_{11}|A \qquad (8.2.7)$$

$$= \operatorname{rank} A_{22}|A + \operatorname{rank} A_{22} \qquad (8.2.8)$$

$$\leq \operatorname{rank} A_{11} + \operatorname{rank} A_{22}. \qquad (8.2.9)$$

Furthermore,

$$\det A = (\det A_{11})\det(A_{11}|A) \qquad (8.2.10)$$

and

$$\det A = (\det A_{22})\det(A_{22}|A). \qquad (8.2.11)$$

Proposition 8.2.4. Let $A \triangleq \begin{bmatrix} A_{11} & A_{12} \\ A_{12}^* & A_{22} \end{bmatrix} \in \mathbf{H}^{n+m}$. Then, the following statements are equivalent:

    i) $A \geq 0$.

    ii) $A_{11} \geq 0$, $A_{12} = A_{11}A_{11}^+A_{12}$, and $A_{12}^*A_{11}^+A_{12} \leq A_{22}$.

    iii) $A_{22} \geq 0$, $A_{12} = A_{12}A_{22}A_{22}^+$, and $A_{12}A_{22}^+A_{12}^* \leq A_{11}$.

The following statements are also equivalent:

    iv) $A > 0$.

    v) $A_{11} > 0$ and $A_{12}^*A_{11}^{-1}A_{12} < A_{22}$.

    vi) $A_{22} > 0$ and $A_{12}A_{22}^{-1}A_{12}^* < A_{11}$.

The following result follows from (2.8.16) and (2.8.17) or from (8.2.3) and (8.2.4).

Proposition 8.2.5. Let $A \triangleq \begin{bmatrix} A_{11} & A_{12} \\ A_{12}^* & A_{22} \end{bmatrix} \in \mathbf{P}^{n+m}$. Then,

$$A^{-1} = \begin{bmatrix} A_{11}^{-1} + A_{11}^{-1}A_{12}(A_{11}|A)^{-1}A_{12}^*A_{11}^{-1} & -A_{11}^{-1}A_{12}(A_{11}|A)^{-1} \\ -(A_{11}|A)^{-1}A_{12}^*A_{11}^{-1} & (A_{11}|A)^{-1} \end{bmatrix} \qquad (8.2.12)$$

and

$$A^{-1} = \begin{bmatrix} (A_{22}|A)^{-1} & -(A_{22}|A)^{-1}A_{12}A_{22}^{-1} \\ -A_{22}^{-1}A_{12}^*(A_{22}|A)^{-1} & A_{22}^{-1}A_{12}^*(A_{22}|A)^{-1}A_{12}A_{22}^{-1} + A_{22}^{-1} \end{bmatrix}, \qquad (8.2.13)$$

where

$$A_{11}|A = A_{22} - A_{12}^* A_{11}^{-1} A_{12} \tag{8.2.14}$$

and

$$A_{22}|A = A_{11} - A_{12} A_{22}^{-1} A_{12}^*. \tag{8.2.15}$$

Now, let $A^{-1} = \begin{bmatrix} B_{11} & B_{12} \\ B_{12}^* & B_{22} \end{bmatrix}$. Then,

$$B_{11}|A^{-1} = A_{22}^{-1} \tag{8.2.16}$$

and

$$B_{22}|A^{-1} = A_{11}^{-1}. \tag{8.2.17}$$

Lemma 8.2.6. Let $A \in \mathbb{F}^{n \times n}$, $b \in \mathbb{F}^n$, and $a \in \mathbb{R}$, and define $\mathcal{A} \triangleq \begin{bmatrix} A & b \\ b^* & a \end{bmatrix}$. Then, the following statements are equivalent:

i) $\mathcal{A}$ is positive semidefinite.

ii) A is positive semidefinite, $b = AA^+b$, and $b^*A^+b \leq a$.

iii) Either A is positive semidefinite, $a = 0$, and $b = 0$, or $a > 0$ and $bb^* \leq aA$.

Furthermore, the following statements are equivalent:

i) $\mathcal{A}$ is positive definite.

ii) A is positive definite, and $b^*A^{-1}b < a$.

iii) $a > 0$ and $bb^* < aA$.

In this case,

$$\det \mathcal{A} = (\det A)(a - b^*A^{-1}b). \tag{8.2.18}$$

For the following result note that a matrix is a principal submatrix of itself, while the determinant of a matrix is also a principal subdeterminant of the matrix.

Proposition 8.2.7. Let $A \in \mathbf{H}^n$. Then, the following statements are equivalent:

i) A is positive semidefinite.

ii) Every principal submatrix of A is positive semidefinite.

iii) Every principal subdeterminant of A is nonnegative.

iv) For all $i \in \{1, \ldots, n\}$, the sum of all $i \times i$ principal subdeterminants of A is nonnegative.

v) $\beta_0, \ldots, \beta_{n-1} \geq 0$, where $\chi_A(s) = s^n + \beta_{n-1}s^{n-1} + \cdots + \beta_1 s + \beta_0$.

Proof. To prove that $i) \implies ii)$, let $\hat{A} \in \mathbb{F}^{m \times m}$ be the principal submatrix of A obtained from A by retaining rows and columns $i_1, \ldots, i_m$. Then, $\hat{A} = S^{\mathrm{T}}AS$, where $S \triangleq \begin{bmatrix} e_{i_1} & \cdots & e_{i_m} \end{bmatrix} \in \mathbb{R}^{n \times m}$. Now, let $\hat{x} \in \mathbb{F}^m$. Since A is positive semidefinite, it follows that $\hat{x}^*\hat{A}\hat{x} = \hat{x}^*S^{\mathrm{T}}AS\hat{x} \geq 0$, and thus $\hat{A}$ is positive semidefinite.

Next, the statements $ii) \implies iii) \implies iv)$ are immediate. To prove that $iv)$ $\implies i)$, note that it follows from Proposition 4.4.6 that

$$\chi_A(s) = \sum_{i=0}^{n} \beta_i s^i = \sum_{i=0}^{n} (-1)^{n-i} \gamma_{n-i} s^i = (-1)^n \sum_{i=0}^{n} \gamma_{n-i}(-s)^i, \qquad (8.2.19)$$

where, for all $i \in \{1, \ldots, n\}$, γ_i is the sum of all $i \times i$ principal subdeterminants of A, and $\beta_n = \gamma_0 = 1$. By assumption, $\gamma_i \geq 0$ for all $i \in \{1, \ldots, n\}$. Now, suppose there exists $\lambda \in \mathrm{spec}(A)$ such that $\lambda < 0$. Then, $0 = (-1)^n \chi_A(\lambda) = \sum_{i=0}^{n} \gamma_{n-i}(-\lambda)^i > 0$, which is a contradiction. The equivalence of $iv)$ and $v)$ follows from Proposition 4.4.6. $\qquad \square$

Proposition 8.2.8. Let $A \in \mathbf{H}^n$. Then, the following statements are equivalent:

i) A is positive definite.

ii) Every principal submatrix of A is positive definite.

iii) Every principal subdeterminant of A is positive.

iv) Every leading principal submatrix of A is positive definite.

v) Every leading principal subdeterminant of A is positive.

Proof. To prove that $i) \implies ii)$, let $\hat{A} \in \mathbb{F}^{m \times m}$ and S be as in the proof of Proposition 8.2.7, and let $\hat{x}$ be nonzero so that $S\hat{x}$ is nonzero. Since A is positive definite, it follows that $\hat{x}^* \hat{A} \hat{x} = \hat{x}^* S^T A S \hat{x} > 0$, and hence $\hat{A}$ is positive definite.

Next, the implications $i) \implies ii) \implies iii) \implies v)$ and $ii) \implies iv) \implies v)$ are immediate. To prove that $v) \implies i)$, suppose that the leading principal submatrix $A_i \in \mathbb{F}^{i \times i}$ has positive determinant for all $i \in \{1, \ldots, n\}$. The result is true for $n = 1$. For $n \geq 2$, we show that, if A_i is positive definite, then so is A_{i+1}. Writing $A_{i+1} = \begin{bmatrix} A_i & b_i \\ b_i^* & a_i \end{bmatrix}$, it follows from Lemma 8.2.6 that $\det A_{i+1} = (\det A_i)(a_i - b_i^* A_i^{-1} b_i) > 0$, and hence $a_i - b_i^* A_i^{-1} b_i = \det A_{i+1}/\det A_i > 0$. Lemma 8.2.6 now implies that A_{i+1} is positive definite. Using this argument for all $i \in \{2, \ldots, n\}$ implies that A is positive definite. $\qquad \square$

The example $A = \begin{bmatrix} 0 & 0 \\ 0 & -1 \end{bmatrix}$ shows that every principal subdeterminant of A, rather than just the leading principal subdeterminants of A, must be checked to determine whether A is positive semidefinite. A less obvious example is $A = \begin{bmatrix} 1 & 1 & 1 \\ 1 & 1 & 1 \\ 1 & 1 & 0 \end{bmatrix}$, whose eigenvalues are 0, $1 + \sqrt{3}$, and $1 - \sqrt{3}$. In this case, the principal subdeterminant $\det A_{[1;1]} = \det \begin{bmatrix} 1 & 1 \\ 1 & 0 \end{bmatrix} < 0$.

Note that condition *iii)* of Proposition 8.2.8 includes $\det A > 0$ since the determinant of A is also a subdeterminant of A. The matrix $A = \begin{bmatrix} 3/2 & -1 & 1 \\ -1 & 2 & 1 \\ 1 & 1 & 2 \end{bmatrix}$ has the property that every 1×1 and 2×2 subdeterminant is positive but is not positive definite. This example shows that the result *iii)* $\implies$ *ii)* of Proposition 8.2.8 is false if the requirement that the determinant of A be positive is omitted.

8.3 Simultaneous Diagonalization

This section considers the simultaneous diagonalization of a pair of matrices $A, B \in \mathbf{H}^n$. There are two types of simultaneous diagonalization. *Cogredient diagonalization* involves a nonsingular matrix $S \in \mathbb{F}^{n \times n}$ such that SAS^* and SBS^* are both diagonal, whereas *contragredient diagonalization* involves finding a nonsingular matrix $S \in \mathbb{F}^{n \times n}$ such that SAS^* and $S^{-*}BS^{-1}$ are both diagonal. Both types of simultaneous transformation involve only congruence transformations. We begin by assuming that one of the matrices is positive definite, in which case the results are easy to prove. The first result involves cogredient diagonalization.

Theorem 8.3.1. Let $A, B \in \mathbf{H}^n$, and assume that A is positive definite. Then, there exists a nonsingular matrix $S \in \mathbb{F}^{n \times n}$ such that $SAS^* = I$ and SBS^* is diagonal.

Proof. Setting $S_1 = A^{-1/2}$, it follows that $S_1 A S_1^* = I$. Now, since $S_1 B S_1^*$ is Hermitian, it follows from Corollary 5.4.5 that there exists a unitary matrix $S_2 \in \mathbb{F}^{n \times n}$ such that $SBS^* = S_2 S_1 B S_1^* S_2^*$ is diagonal, where $S = S_2 S_1$. Finally, $SAS^* = S_2 S_1 A S_1^* S_2^* = S_2 I S_2^* = I$. $\square$

An analogous result holds for contragredient diagonalization.

Theorem 8.3.2. Let $A, B \in \mathbf{H}^n$, and assume that A is positive definite. Then, there exists a nonsingular matrix $S \in \mathbb{F}^{n \times n}$ such that $SAS^* = I$ and $S^{-*}BS^{-1}$ is diagonal.

Proof. Setting $S_1 = A^{-1/2}$, it follows that $S_1 A S_1^* = I$. Since $S_1^{-*}BS_1^{-1}$ is Hermitian, it follows that there exists a unitary matrix $S_2 \in \mathbb{F}^{n \times n}$ such that $S^{-*}BS^{-1} = S_2^{-*}S_1^{-*}BS_1^{-1}S_2^{-1} = S_2(S_1^{-*}BS_1^{-1})S_2^*$ is diagonal, where $S = S_2 S_1$. Finally, $SAS^* = S_2 S_1 A S_1^* S_2^* = S_2 I S_2^* = I$. $\square$

Corollary 8.3.3. Let $A, B \in \mathbf{H}^n$, and assume that A is positive definite. Then, AB is diagonalizable over $\mathbb{F}$, all of the eigenvalues of AB are real, and $\text{In}(AB) = \text{In}(B)$.

Corollary 8.3.4. Let $A, B \in \mathbf{P}^n$. Then, there exists a nonsingular matrix $S \in \mathbb{F}^{n \times n}$ such that SAS^* and $S^{-*}BS^{-1}$ are equal and diagonal.

Proof. By Theorem 8.3.2 there exists a nonsingular matrix $S_1 \in \mathbb{F}^{n \times n}$ such that $S_1 A S_1^* = I$ and $B_1 = S_1^{-*}BS_1^{-1}$ is diagonal. Defining $S \triangleq B_1^{1/4}S_1$ yields $SAS^* = S^{-*}BS^{-1} = B_1^{1/2}$. $\square$

The transformation S of Corollary 8.3.4 is a *balancing transformation*.

Next, we weaken the requirement in Theorem 8.3.1 and Theorem 8.3.2 that A be positive definite by assuming only that A is positive semidefinite. In this case, however, we assume that B is also positive semidefinite.

Theorem 8.3.5. Let $A, B \in \mathbf{N}^n$. Then, there exists a nonsingular matrix $S \in \mathbb{F}^{n \times n}$ such that $SAS^* = \begin{bmatrix} I & 0 \\ 0 & 0 \end{bmatrix}$ and SBS^* is diagonal.

Proof. Let the nonsingular matrix $S_1 \in \mathbb{F}^{n \times n}$ be such that $S_1 A S_1^* = \begin{bmatrix} I & 0 \\ 0 & 0 \end{bmatrix}$, and similarly partition $S_1 B S_1^* = \begin{bmatrix} B_{11} & B_{12} \\ B_{12}^* & B_{22} \end{bmatrix}$, which is positive semidefinite. Letting $S_2 \triangleq \begin{bmatrix} I & -B_{12}B_{22}^+ \\ 0 & I \end{bmatrix}$, it follows from Lemma 8.2.1 that

$$ S_2 S_1 B S_1^* S_2^* = \begin{bmatrix} B_{11} - B_{12}B_{22}^+ B_{12}^* & 0 \\ 0 & B_{22} \end{bmatrix}. $$

Next, let U_1 and U_2 be unitary matrices such that $U_1(B_{11} - B_{12}B_{22}^+ B_{12}^*)U_1^*$ and $U_2 B_{22} U_2^*$ are diagonal. Then, defining $S_3 \triangleq \begin{bmatrix} U_1 & 0 \\ 0 & U_2 \end{bmatrix}$ and $S \triangleq S_3 S_2 S_1$, it follows that $SAS^* = \begin{bmatrix} I & 0 \\ 0 & 0 \end{bmatrix}$ and $SBS^* = S_3 S_2 S_1 B S_1^* S_2^* S_3^*$ is diagonal. $\square$

Theorem 8.3.6. Let $A, B \in \mathbf{N}^n$. Then, there exists a nonsingular matrix $S \in \mathbb{F}^{n \times n}$ such that $SAS^* = \begin{bmatrix} I & 0 \\ 0 & 0 \end{bmatrix}$ and $S^{-*}BS^{-1}$ is diagonal.

Proof. Let $S_1 \in \mathbb{F}^{n \times n}$ be a nonsingular matrix such that $S_1 A S_1^* = \begin{bmatrix} I & 0 \\ 0 & 0 \end{bmatrix}$, and similarly partition $S_1^{-*}BS_1^{-1} = \begin{bmatrix} B_{11} & B_{12} \\ B_{12}^* & B_{22} \end{bmatrix}$, which is positive semidefinite. Letting $S_2 \triangleq \begin{bmatrix} I & B_{11}^+ B_{12} \\ 0 & I \end{bmatrix}$, it follows that

$$ S_2^{-*} S_1^{-*} B S_1^{-1} S_2^{-1} = \begin{bmatrix} B_{11} & 0 \\ 0 & B_{22} - B_{12}^* B_{11}^+ B_{12} \end{bmatrix}. $$

Now, let U_1 and U_2 be unitary matrices such that $U_1 B_{11} U_1^*$ and $U_2(B_{22} - B_{12}^* B_{11}^+ B_{12})U_2^*$ are diagonal. Then, defining $S_3 \triangleq \begin{bmatrix} U_1 & 0 \\ 0 & U_2 \end{bmatrix}$ and $S \triangleq S_3 S_2 S_1$, it follows that $SAS^* = \begin{bmatrix} I & 0 \\ 0 & 0 \end{bmatrix}$ and $S^{-*}BS^{-1} = S_3^{-*} S_2^{-*} S_1^{-*} B S_1^{-1} S_2^{-1} S_3^{-1}$ is diagonal. $\square$

Corollary 8.3.7. Let $A, B \in \mathbf{N}^n$. Then, AB is semisimple, and every eigenvalue of AB is nonnegative. If, in addition, A and B are positive definite, then every eigenvalue of AB is positive.

Proof. It follows from Theorem 8.3.6 that there exists a nonsingular matrix $S \in \mathbb{R}^{n \times n}$ such that $A_1 = SAS^*$ and $B_1 = S^{-*}BS^{-1}$ are diagonal with nonnegative diagonal entries. Hence, $AB = S^{-1}A_1 B_1 S$ is semisimple and has nonnegative eigenvalues. $\square$

A more direct approach to showing that AB has nonnegative eigenvalues is to use Corollary 4.4.11 and note that $\lambda_i(AB) = \lambda_i(B^{1/2}AB^{1/2}) \geq 0$.

Corollary 8.3.8. Let $A, B \in \mathbf{N}^n$, and assume that $\operatorname{rank} A = \operatorname{rank} B = \operatorname{rank} AB$. Then, there exists a nonsingular matrix $S \in \mathbb{F}^{n \times n}$ such that $SAS^* = S^{-*}BS^{-1}$ and such that SAS^* is diagonal.

Proof. By Theorem 8.3.6 there exists a nonsingular matrix $S_1 \in \mathbb{F}^{n \times n}$ such that $S_1 A S_1^* = \begin{bmatrix} I_r & 0 \\ 0 & 0 \end{bmatrix}$, where $r \triangleq \operatorname{rank} A$, and such that $B_1 = S_1^{-*}BS_1^{-1}$ is diagonal. Hence, $AB = S_1^{-1}\begin{bmatrix} I_r & 0 \\ 0 & 0 \end{bmatrix}B_1 S_1$. Since $\operatorname{rank} A = \operatorname{rank} B = \operatorname{rank} AB = r$, it follows that $B_1 = \begin{bmatrix} \hat{B}_1 & 0 \\ 0 & 0 \end{bmatrix}$, where $\hat{B}_1 \in \mathbb{F}^{r \times r}$ is diagonal with positive diagonal entries. Hence, $S_1^{-*}BS_1^{-1} = \begin{bmatrix} \hat{B}_1 & 0 \\ 0 & 0 \end{bmatrix}$. Now, define $S_2 \triangleq \begin{bmatrix} \hat{B}_1^{1/4} & 0 \\ 0 & I \end{bmatrix}$ and $S \triangleq S_2 S_1$. Then, $SAS^* = S_2 S_1 A S_1^* S_2^* = \begin{bmatrix} \hat{B}_1^{1/2} & 0 \\ 0 & 0 \end{bmatrix} = S_2^{-*}S_1^{-*}BS_1^{-1}S_2^{-1} = S^{-*}BS^{-1}$. $\square$

8.4 Eigenvalue Inequalities

Next, we turn our attention to inequalities for eigenvalues. We begin with a series of lemmas.

Lemma 8.4.1. Let $A \in \mathbf{H}^n$, and let $\beta \in \mathbb{R}$. Then, the following statements hold:

i) $\beta I \leq A$ if and only if $\beta \leq \lambda_{\min}(A)$.

ii) $\beta I < A$ if and only if $\beta < \lambda_{\min}(A)$.

iii) $A \leq \beta I$ if and only if $\lambda_{\max}(A) \leq \beta$.

iv) $A < \beta I$ if and only if $\lambda_{\max}(A) < \beta$.

Proof. To prove i), assume that $\beta I \leq A$, and let $S \in \mathbb{F}^{n \times n}$ be a unitary matrix such that $B = SAS^*$ is diagonal. Then, $\beta I \leq B$, which yields $\beta \leq \lambda_{\min}(B) = \lambda_{\min}(A)$. Conversely, let $S \in \mathbb{F}^{n \times n}$ be a unitary matrix such that $B = SAS^*$ is diagonal. Since the diagonal entries of B are the eigenvalues of A, it follows that $\lambda_{\min}(A)I \leq B$, which implies that $\beta I \leq \lambda_{\min}(A)I \leq S^*BS = A$. Results ii), iii), and iv) are proved in a similar manner. $\square$

Corollary 8.4.2. Let $A \in \mathbf{H}^n$. Then,

$$\lambda_{\min}(A)I \leq A \leq \lambda_{\max}(A)I. \tag{8.4.1}$$

Proof. This result follows from i) and iii) of Lemma 8.4.1 with $\beta = \lambda_{\min}(A)$ and $\beta = \lambda_{\max}(A)$, respectively. $\square$

The following result concerns the maximum and minimum values of the *Rayleigh quotient*.

Lemma 8.4.3. Let $A \in \mathbf{H}^n$. Then,

$$\lambda_{\min}(A) = \min_{x \in \mathbb{F}^n \setminus \{0\}} \frac{x^*Ax}{x^*x} \tag{8.4.2}$$

and

$$\lambda_{\max}(A) = \max_{x \in \mathbb{F}^n \setminus \{0\}} \frac{x^*Ax}{x^*x}. \tag{8.4.3}$$

Proof. It follows from (8.4.1) that $\lambda_{\min}(A) \leq x^*Ax/x^*x$ for all nonzero $x \in \mathbb{F}^n$. Letting $x \in \mathbb{F}^n$ be an eigenvector of A associated with $\lambda_{\min}(A)$, it follows that this lower bound is attained. This proves (8.4.2). An analogous argument yields (8.4.3). $\square$

The following result is the *Cauchy interlacing theorem*.

Lemma 8.4.4. Let $A \in \mathbf{H}^n$, and let A_0 be an $(n-1) \times (n-1)$ principal submatrix of A. Then, for all $i \in \{1, \ldots, n-1\}$,

$$\lambda_{i+1}(A) \leq \lambda_i(A_0) \leq \lambda_i(A). \tag{8.4.4}$$

Proof. Note that (8.4.4) is the chain of inequalities

$$\lambda_n(A) \leq \lambda_{n-1}(A_0) \leq \lambda_{n-1}(A) \leq \cdots \leq \lambda_2(A) \leq \lambda_1(A_0) \leq \lambda_1(A).$$

Suppose that this chain of inequalities does not hold. In particular, first suppose that the rightmost inequality that is not true is $\lambda_j(A_0) \leq \lambda_j(A)$, so that $\lambda_j(A) < \lambda_j(A_0)$. Choose δ such that $\lambda_j(A) < \delta < \lambda_j(A_0)$ and such that δ is not an eigenvalue of A_0. If $j = 1$, then $A - \delta I$ is negative definite, while, if $j \geq 2$, then $\lambda_j(A) < \delta < \lambda_j(A_0) \leq \lambda_{j-1}(A_0) \leq \lambda_{j-1}(A)$, so that $A - \delta I$ has $j - 1$ positive eigenvalues. Thus, $\nu_+(A - \delta I) = j - 1$. Furthermore, since $\delta < \lambda_j(A_0)$, it follows that $\nu_+(A_0 - \delta I) \geq j$.

Now, assume for convenience that the rows and columns of A are ordered so that A_0 is the $(n-1) \times (n-1)$ leading principal submatrix of A. Thus, $A = \begin{bmatrix} A_0 & \beta \\ \beta^* & \gamma \end{bmatrix}$, where $\beta \in \mathbb{F}^{n-1}$ and $\gamma \in \mathbb{F}$. Next, note the equality

$A - \delta I$

$$= \begin{bmatrix} I & 0 \\ \beta^*(A_0 - \delta I)^{-1} & 1 \end{bmatrix} \begin{bmatrix} A_0 - \delta I & 0 \\ 0 & \gamma - \delta - \beta^*(A_0 - \delta I)^{-1}\beta \end{bmatrix} \begin{bmatrix} I & (A_0 - \delta I)^{-1}\beta \\ 0 & 1 \end{bmatrix},$$

where $A_0 - \delta I$ is nonsingular since δ is chosen to not be an eigenvalue of A_0. Since the right-hand side of this equality involves a congruence transformation, and since $\nu_+(A_0 - \delta I) \geq j$, it follows from Corollary 5.4.7 that $\nu_+(A - \delta I) \geq j$. However, this inequality contradicts the fact that $\nu_+(A - \delta I) = j - 1$.

Finally, suppose that the rightmost inequality in (8.4.4) that is not true is $\lambda_{j+1}(A) \leq \lambda_j(A_0)$, so that $\lambda_j(A_0) < \lambda_{j+1}(A)$. Choose δ such that $\lambda_j(A_0) < \delta < \lambda_{j+1}(A)$ and such that δ is not an eigenvalue of A_0. Then, it follows that $\nu_+(A - \delta I) \geq j + 1$ and $\nu_+(A_0 - \delta I) = j - 1$. Using the congruence transformation as in the previous case, it follows that $\nu_+(A - \delta I) \leq j$, which contradicts the fact that $\nu_+(A - \delta I) \geq j + 1$. $\square$

The following result is the *inclusion principle*.

Theorem 8.4.5. Let $A \in \mathbf{H}^n$, and let $A_0 \in \mathbf{H}^k$ be a $k \times k$ principal submatrix of A. Then, for all $i \in \{1, \ldots, k\}$,

$$\lambda_{i+n-k}(A) \leq \lambda_i(A_0) \leq \lambda_i(A). \tag{8.4.5}$$

Proof. For $k = n-1$, the result is given by Lemma 8.4.4. Hence, let $k = n-2$, and let A_1 denote an $(n-1) \times (n-1)$ principal submatrix of A such that the $(n-2) \times (n-2)$ principal submatrix A_0 of A is also a principal submatrix of A_1. Therefore, Lemma 8.4.4 implies that $\lambda_n(A) \leq \lambda_{n-1}(A_1) \leq \cdots \leq \lambda_2(A) \leq \lambda_2(A) \leq \lambda_1(A_1) \leq \lambda_1(A)$ and $\lambda_{n-1}(A_1) \leq \lambda_{n-2}(A_0) \leq \cdots \leq \lambda_2(A_0) \leq \lambda_2(A_1) \leq \lambda_1(A_0) \leq \lambda_1(A_1)$. Combining these inequalities yields $\lambda_{i+2}(A) \leq \lambda_i(A_0) \leq \lambda_i(A)$

for all $i = 1, \ldots, n-2$, while proceeding in a similar manner with $k < n-2$ yields (8.4.5). $\qquad\square$

Corollary 8.4.6. Let $A \in \mathbf{H}^n$, and let $A_0 \in \mathbf{H}^k$ be a $k \times k$ principal submatrix of A. Then,

$$\lambda_{\min}(A) \le \lambda_{\min}(A_0) \le \lambda_{\max}(A_0) \le \lambda_{\max}(A) \qquad (8.4.6)$$

and

$$\lambda_{\min}(A_0) \le \lambda_k(A). \qquad (8.4.7)$$

The following result compares the maximum and minimum eigenvalues with the maximum and minimum diagonal entries.

Corollary 8.4.7. Let $A \in \mathbf{H}^n$. Then,

$$\lambda_{\min}(A) \le \mathrm{d}_{\min}(A) \le \mathrm{d}_{\max}(A) \le \lambda_{\max}(A). \qquad (8.4.8)$$

Lemma 8.4.8. Let $A, B \in \mathbf{H}^n$, and assume that $A \le B$ and $\mathrm{mspec}(A) = \mathrm{mspec}(B)$. Then, $A = B$.

Proof. Let $\alpha \ge 0$ be such that $0 < \hat{A} \le \hat{B}$, where $\hat{A} \triangleq A + \alpha I$ and $\hat{B} \triangleq B + \alpha I$. Note that $\mathrm{mspec}(\hat{A}) = \mathrm{mspec}(\hat{B})$, and thus $\det \hat{A} = \det \hat{B}$. Next, it follows that $I \le \hat{A}^{-1/2} \hat{B} \hat{A}^{-1/2}$. Hence, it follows from $i)$ of Lemma 8.4.1 that $\lambda_{\min}(\hat{A}^{-1/2} \hat{B} \hat{A}^{-1/2}) \ge 1$. Furthermore, $\det(\hat{A}^{-1/2} \hat{B} \hat{A}^{-1/2}) = \det \hat{B} / \det \hat{A} = 1$, which implies that $\lambda_i(\hat{A}^{-1/2} \hat{B} \hat{A}^{-1/2}) = 1$ for all $i \in \{1, \ldots, n\}$. Hence, $\hat{A}^{-1/2} \hat{B} \hat{A}^{-1/2} = I$, and thus $\hat{A} = \hat{B}$. Hence, $A = B$. $\qquad\square$

The following result is the *monotonicity theorem* or *Weyl's inequality*.

Theorem 8.4.9. Let $A, B \in \mathbf{H}^n$, and assume that $A \le B$. Then, for all $i \in \{1, \ldots, n\}$,

$$\lambda_i(A) \le \lambda_i(B). \qquad (8.4.9)$$

If $A \ne B$, then there exists $i \in \{1, \ldots, n\}$ such that

$$\lambda_i(A) < \lambda_i(B). \qquad (8.4.10)$$

If $A < B$, then (8.4.10) holds for all $i \in \{1, \ldots, n\}$.

Proof. Since $A \le B$, it follows from Corollary 8.4.2 that $\lambda_{\min}(A)I \le A \le B \le \lambda_{\max}(B)I$. Hence, it follows from $iii)$ and $i)$ of Lemma 8.4.1 that $\lambda_{\min}(A) \le \lambda_{\min}(B)$ and $\lambda_{\max}(A) \le \lambda_{\max}(B)$. Next, let $S \in \mathbb{F}^{n \times n}$ be a unitary matrix such that $SAS^* = \mathrm{diag}[\lambda_1(A), \ldots, \lambda_n(A)]$. Furthermore, for $2 \le i \le n-1$, let $A_0 = \mathrm{diag}[\lambda_1(A), \ldots, \lambda_i(A)]$, and let B_0 denote the $i \times i$ leading principal submatrices of SAS^* and SBS^*, respectively. Since $A \le B$, it follows that $A_0 \le B_0$, which implies that $\lambda_{\min}(A_0) \le \lambda_{\min}(B_0)$. It now follows from (8.4.7) that

$$\lambda_i(A) = \lambda_{\min}(A_0) \le \lambda_{\min}(B_0) \le \lambda_i(SBS^*) = \lambda_i(B),$$

which proves (8.4.9). If $A \ne B$, then it follows from Lemma 8.4.8 that $\mathrm{mspec}(A) \ne \mathrm{mspec}(B)$, and thus there exists $i \in \{1, \ldots, n\}$ such that (8.4.10) holds. If $A < B$, then $\lambda_{\min}(A_0) < \lambda_{\min}(B_0)$, which implies (8.4.10) for all $i \in \{1, \ldots, n\}$. $\qquad\square$

Corollary 8.4.10. Let $A, B \in \mathbf{H}^n$. Then, the following statements hold:

i) If $A \leq B$, then $\operatorname{tr} A \leq \operatorname{tr} B$.

ii) If $A \leq B$ and $\operatorname{tr} A = \operatorname{tr} B$, then $A = B$.

iii) If $A < B$, then $\operatorname{tr} A < \operatorname{tr} B$.

iv) If $0 \leq A \leq B$, then $0 \leq \det A \leq \det B$.

v) If $0 \leq A < B$, then $0 \leq \det A < \det B$.

vi) If $0 < A \leq B$ and $\det A = \det B$, then $A = B$.

Proof. Statements *i)*, *iii)*, *iv)*, and *v)* follow from Theorem 8.4.9. To prove *ii)*, note that, since $A \leq B$ and $\operatorname{tr} A = \operatorname{tr} B$, it follows from Theorem 8.4.9 that $\operatorname{mspec}(A) = \operatorname{mspec}(B)$. Now, Lemma 8.4.8 implies that $A = B$. A similar argument yields *vi)*. $\square$

The following result, which is a generalization of Theorem 8.4.9, is due to Weyl.

Theorem 8.4.11. Let $A, B \in \mathbf{H}^n$. If $i + j \geq n + 1$, then

$$\lambda_i(A) + \lambda_j(B) \leq \lambda_{i+j-n}(A + B). \tag{8.4.11}$$

If $i + j \leq n + 1$, then

$$\lambda_{i+j-1}(A + B) \leq \lambda_i(A) + \lambda_j(B). \tag{8.4.12}$$

In particular, for all $i \in \{1, \dots, n\}$,

$$\lambda_i(A) + \lambda_{\min}(B) \leq \lambda_i(A + B) \leq \lambda_i(A) + \lambda_{\max}(B), \tag{8.4.13}$$

$$\lambda_{\min}(A) + \lambda_{\min}(B) \leq \lambda_{\min}(A + B) \leq \lambda_{\min}(A) + \lambda_{\max}(B), \tag{8.4.14}$$

$$\lambda_{\max}(A) + \lambda_{\min}(B) \leq \lambda_{\max}(A + B) \leq \lambda_{\max}(A) + \lambda_{\max}(B). \tag{8.4.15}$$

Furthermore, if $\operatorname{rank} B \leq r$, then, for all $i = 1, \dots, n - r$,

$$\lambda_{i+r}(A) \leq \lambda_i(A + B), \tag{8.4.16}$$

$$\lambda_{i+r}(A + B) \leq \lambda_i(A). \tag{8.4.17}$$

Finally,

$$\nu_-(A + B) \leq \nu_-(A) + \nu_-(B), \tag{8.4.18}$$

$$\nu_+(A + B) \leq \nu_+(A) + \nu_+(B). \tag{8.4.19}$$

Proof. See [401], [728, p. 182], and [1208, pp. 112–115]. $\square$

Lemma 8.4.12. Let $A, B, C \in \mathbf{H}^n$. If $A \leq B$ and C is positive semidefinite, then

$$\operatorname{tr} AC \leq \operatorname{tr} BC. \qquad (8.4.20)$$

If $A < B$ and C is positive definite, then

$$\operatorname{tr} AC < \operatorname{tr} BC. \qquad (8.4.21)$$

Proof. Since $C^{1/2}AC^{1/2} \leq C^{1/2}BC^{1/2}$, it follows from $i)$ of Corollary 8.4.10 that

$$\operatorname{tr} AC = \operatorname{tr} C^{1/2}AC^{1/2} \leq \operatorname{tr} C^{1/2}BC^{1/2} = \operatorname{tr} BC.$$

Result (8.4.21) follows from $ii)$ of Corollary 8.4.10 in a similar fashion. $\qquad \square$

Proposition 8.4.13. Let $A, B \in \mathbb{F}^{n \times n}$, and assume that B is positive semidefinite. Then,

$$\tfrac{1}{2}\lambda_{\min}(A + A^*)\operatorname{tr} B \leq \operatorname{Re}\operatorname{tr} AB \leq \tfrac{1}{2}\lambda_{\max}(A + A^*)\operatorname{tr} B. \qquad (8.4.22)$$

If, in addition, A is Hermitian, then

$$\lambda_{\min}(A)\operatorname{tr} B \leq \operatorname{tr} AB \leq \lambda_{\max}(A)\operatorname{tr} B. \qquad (8.4.23)$$

Proof. It follows from Corollary 8.4.2 that $\tfrac{1}{2}\lambda_{\min}(A+A^*)I \leq \tfrac{1}{2}(A+A^*)$, while Lemma 8.4.12 implies that $\tfrac{1}{2}\lambda_{\min}(A + A^*)\operatorname{tr} B = \operatorname{tr} \tfrac{1}{2}\lambda_{\min}(A + A^*)IB \leq \operatorname{tr} \tfrac{1}{2}(A + A^*)B = \operatorname{Re}\operatorname{tr} AB$, which proves the left-hand inequality of (8.4.22). Similarly, the right-hand inequality holds. $\qquad \square$

For results relating to Proposition 8.4.13, see Fact 5.12.4, Fact 5.12.5, Fact 5.12.8, and Fact 8.19.19.

Proposition 8.4.14. Let $A, B \in \mathbf{P}^n$, and assume that $\det B = 1$. Then,

$$(\det A)^{1/n} \leq \tfrac{1}{n}\operatorname{tr} AB. \qquad (8.4.24)$$

Furthermore, equality holds if and only if $B = (\det A)^{1/n}A^{-1}$.

Proof. Using the arithmetic-mean–geometric-mean inequality given by Fact 1.17.14, it follows that

$$(\det A)^{1/n} = \left(\det B^{1/2}AB^{1/2}\right)^{1/n} = \left[\prod_{i=1}^{n} \lambda_i\left(B^{1/2}AB^{1/2}\right)\right]^{1/n}$$

$$\leq \tfrac{1}{n}\sum_{i=1}^{n} \lambda_i\left(B^{1/2}AB^{1/2}\right) = \tfrac{1}{n}\operatorname{tr} AB.$$

Equality holds if and only if there exists $\beta > 0$ such that $B^{1/2}AB^{1/2} = \beta I$. In this case, $\beta = (\det A)^{1/n}$ and $B = (\det A)^{1/n}A^{-1}$. $\qquad \square$

The following corollary of Proposition 8.4.14 is *Minkowski's determinant theorem*.

Corollary 8.4.15. Let $A, B \in \mathbf{N}^n$, and let $p \in [1, n]$. Then,

$$\det A + \det B \leq \left[(\det A)^{1/p} + (\det B)^{1/p}\right]^p \tag{8.4.25}$$

$$\leq \left[(\det A)^{1/n} + (\det B)^{1/n}\right]^n \tag{8.4.26}$$

$$\leq \det(A + B). \tag{8.4.27}$$

Furthermore, the following statements hold:

i) If $A = 0$ or $B = 0$ or $\det(A + B) = 0$, then (8.4.25)–(8.4.27) are equalities.

ii) If there exists $\alpha \geq 0$ such that $B = \alpha A$, then (8.4.27) is an equality.

iii) If $A + B$ is positive definite and (8.4.27) holds as an equality, then there exists $\alpha \geq 0$ such that either $B = \alpha A$ or $A = \alpha B$.

iv) If $n \geq 2$, $p > 1$, A is positive definite, and (8.4.25) holds as an equality, then $\det B = 0$.

v) If $n \geq 2$, $p < n$, A is positive definite, and (8.4.26) holds as an equality, then $\det B = 0$.

vi) If $n \geq 2$, A is positive definite, and $\det A + \det B = \det(A + B)$, then $B = 0$.

Proof. Inequalities (8.4.25) and (8.4.26) are consequences of the power-sum inequality Fact 1.17.35. Now, assume that $A + B$ is positive definite, since otherwise (8.4.25)–(8.4.27) are equalities. To prove (8.4.27), Proposition 8.4.14 implies that

$$(\det A)^{1/n} + (\det B)^{1/n} \leq \tfrac{1}{n}\operatorname{tr}\left[A[\det(A + B)]^{1/n}(A + B)^{-1}\right]$$

$$+ \tfrac{1}{n}\operatorname{tr}\left[B[\det(A + B)]^{1/n}(A + B)^{-1}\right]$$

$$= [\det(A + B)]^{1/n}.$$

Statements *i)* and *ii)* are immediate. To prove *iii)*, suppose that $A + B$ is positive definite and that (8.4.27) holds as an equality. Then, either A or B is positive definite. Hence, suppose that A is positive definite. Multiplying the equality $(\det A)^{1/n} + (\det B)^{1/n} = [\det(A + B)]^{1/n}$ by $(\det A)^{-1/n}$ yields

$$1 + \left(\det A^{-1/2}BA^{-1/2}\right)^{1/n} = \left[\det\left(I + A^{-1/2}BA^{-1/2}\right)\right]^{1/n}.$$

Letting $\lambda_1, \ldots, \lambda_n$ denote the eigenvalues of $A^{-1/2}BA^{-1/2}$, it follows that $1 + (\lambda_1 \cdots \lambda_n)^{1/n} = [(1 + \lambda_1) \cdots (1 + \lambda_n)]^{1/n}$. It now follows from Fact 1.17.34 that $\lambda_1 = \cdots = \lambda_n$.

To prove *iv)*, note that, since $1/p < 1$, $\det A > 0$, and equality holds in (8.4.25), it follows from Fact 1.17.35 that $\det B = 0$.

To prove *v)*, note that, since $1/n < 1/p$, $\det A > 0$, and equality holds in (8.4.26), it follows from Fact 1.17.35 that $\det B = 0$.

To prove *vi)*, note that (8.4.25) and (8.4.26) hold as equalities for all $p \in [1, n]$.

Therefore, $\det B = 0$. Consequently, $\det A = \det(A + B)$. Since $0 < A \le A + B$, it follows from *vi*) of Corollary 8.4.10 that $B = 0$. $\qquad\square$

8.5 Exponential, Square Root, and Logarithm of Hermitian Matrices

Let $B \in \mathbb{R}^{n \times n}$ be diagonal, let $\mathcal{D} \subseteq \mathbb{R}$, let $f \colon \mathcal{D} \mapsto \mathbb{R}$, and assume that, for all $i \in \{1, \ldots, n\}$, $B_{(i,i)} \in \mathcal{D}$. Then, we define

$$f(B) \triangleq \mathrm{diag}[f(B_{(1,1)}), \ldots, f(B_{(n,n)})]. \tag{8.5.1}$$

Furthermore, let $A = SBS^* \in \mathbb{F}^{n \times n}$ be Hermitian, where $S \in \mathbb{F}^{n \times n}$ is unitary, $B \in \mathbb{R}^{n \times n}$ is diagonal, and assume that $\mathrm{spec}(A) \subset \mathcal{D}$. Then, we define $f(A) \in \mathbf{H}^n$ by

$$f(A) \triangleq Sf(B)S^*. \tag{8.5.2}$$

Hence, with an obvious extension of notation, $f \colon \{X \in \mathbf{H}^n \colon \mathrm{spec}(X) \subset \mathcal{D}\} \mapsto \mathbf{H}^n$. If $f \colon \mathcal{D} \mapsto \mathbb{R}$ is one-to-one, then its inverse $f^{-1} \colon \{X \in \mathbf{H}^n \colon \mathrm{spec}(X) \subset f(\mathcal{D})\} \mapsto \mathbf{H}^n$ exists. It remains to be shown, however, that the definition of $f(A)$ given by (8.5.2) is independent of the matrices S and B in the decomposition $A = SBS^*$. The following lemma is needed.

Lemma 8.5.1. Let $S \in \mathbb{F}^{n \times n}$ be unitary, let $D, \hat{D} \in \mathbb{R}^{n \times n}$ denote the diagonal matrices $D = \mathrm{diag}(\lambda_1 I_{n_1}, \ldots, \lambda_{n_r} I_{n_r})$ and $\hat{D} = \mathrm{diag}(\mu_1 I_{n_1}, \ldots, \mu_{n_r} I_{n_r})$, where $\lambda_1, \ldots, \lambda_r, \mu_1, \ldots, \lambda_r \in \mathbb{R}$, and assume that $SD = DS$. Then, $S\hat{D} = \hat{D}S$.

Proof. Let $r = 2$, and partition $S = \left[\begin{smallmatrix} S_{11} & S_{12} \\ S_{21} & S_{22} \end{smallmatrix}\right]$. Then, it follows from $SD = DS$ that $\lambda_2 S_{12} = \lambda_1 S_{12}$ and $\lambda_1 S_{21} = \lambda_2 S_{21}$. Since $\lambda_1 \ne \lambda_2$, it follows that $S_{12} = 0$ and $S_{21} = 0$. Therefore, $S = \left[\begin{smallmatrix} S_{11} & 0 \\ 0 & S_{22} \end{smallmatrix}\right]$, and thus $S\hat{D} = \hat{D}S$. A similar argument holds for $r \ge 3$. $\qquad\square$

Proposition 8.5.2. Let $A = RBR^* = SCS^* \in \mathbb{F}^{n \times n}$ be Hermitian, where $R, S \in \mathbb{F}^{n \times n}$ are unitary and $B, C \in \mathbb{R}^{n \times n}$ are diagonal. Furthermore, let $\mathcal{D} \subseteq \mathbb{R}$, let $f \colon \mathcal{D} \mapsto \mathbb{R}$, and assume that all diagonal entries of B are contained in $\mathcal{D}$. Then, $Rf(B)R^* = Sf(C)S^*$.

Proof. Let $\mathrm{spec}(A) = \{\lambda_1, \ldots, \lambda_r\}$. Then, the columns of R and S can be rearranged to obtain unitary matrices $\tilde{R}, \tilde{S} \in \mathbb{F}^{n \times n}$ such that $A = \tilde{R}D\tilde{R}^* = \tilde{S}D\tilde{S}^*$, where $D \triangleq \mathrm{diag}(\lambda_1 I_{n_1}, \ldots, \lambda_{n_r} I_{n_r})$. Hence, $UD = DU$, where $U \triangleq \tilde{S}^*\tilde{R}$. It thus follows from Lemma 8.5.1 that $U\hat{D} = \hat{D}U$, where $\hat{D} \triangleq f(D) = \mathrm{diag}[f(\lambda_1)I_{n_1}, \ldots, f(\lambda_{n_r})I_{n_r}]$. Hence, $\tilde{R}\hat{D}\tilde{R}^* = \tilde{S}\hat{D}\tilde{S}^*$, while rearranging the columns of $\tilde{R}$ and $\tilde{S}$ as well as the diagonal entries of $\hat{D}$ yields $Rf(B)R^* = Sf(C)S^*$. $\qquad\square$

Let $A = SBS^* \in \mathbb{F}^{n \times n}$ be Hermitian, where $S \in \mathbb{F}^{n \times n}$ is unitary and $B \in \mathbb{R}^{n \times n}$ is diagonal. Then, the *matrix exponential* is defined by

$$e^A \triangleq Se^B S^* \in \mathbf{H}^n, \tag{8.5.3}$$

where, for all $i \in \{1, \ldots, n\}$, $(e^B)_{(i,i)} \triangleq e^{B_{(i,i)}}$.

Let $A = SBS^* \in \mathbb{F}^{n \times n}$ be positive semidefinite, where $S \in \mathbb{F}^{n \times n}$ is unitary and $B \in \mathbb{R}^{n \times n}$ is diagonal with nonnegative entries. Then, for all $r \geq 0$ (not necessarily an integer), $A^r = SB^r S^*$ is positive semidefinite, where, for all $i \in \{1, \ldots, n\}$, $(B^r)_{(i,i)} = \left[B_{(i,i)}\right]^r$. Note that $A^0 \triangleq I$. In particular, the positive-semidefinite matrix

$$A^{1/2} = SB^{1/2}S^* \tag{8.5.4}$$

is a square root of A since

$$A^{1/2}A^{1/2} = SB^{1/2}S^*SB^{1/2}S^* = SBS^* = A. \tag{8.5.5}$$

The uniqueness of the *positive-semidefinite square root* of A given by (8.5.4) follows from Theorem 10.6.1; see also [730, p. 410] or [902]. Uniqueness can also be shown directly; see [459, pp. 265, 266] or [728, p. 405]. Hence, if $C \in \mathbb{F}^{n \times m}$, then C^*C is positive semidefinite, and we define

$$\langle C \rangle \triangleq (C^*C)^{1/2}. \tag{8.5.6}$$

If A is positive definite, then A^r is positive definite for all $r \in \mathbb{R}$, and, if $r \neq 0$, then $(A^r)^{1/r} = A$.

Now, assume that $A \in \mathbb{F}^{n \times n}$ is positive definite. Then, the *matrix logarithm* is defined by

$$\log A \triangleq S(\log B)S^* \in \mathbf{H}^n, \tag{8.5.7}$$

where, for all $i \in \{1, \ldots, n\}$, $(\log B)_{(i,i)} \triangleq \log[B_{(i,i)}]$.

In chapters 10 and 11, the matrix exponential, square root, and logarithm are extended to matrices that are not necessarily Hermitian.

8.6 Matrix Inequalities

Lemma 8.6.1. Let $A, B \in \mathbb{F}^n$, assume that A and B are Hermitian, and assume that $0 \leq A \leq B$. Then, $\mathcal{R}(A) \subseteq \mathcal{R}(B)$.

Proof. Let $x \in \mathcal{N}(B)$. Then, $x^*Bx = 0$, and thus $x^*Ax = 0$, which implies that $Ax = 0$. Hence, $\mathcal{N}(B) \subseteq \mathcal{N}(A)$, and thus $\mathcal{N}(A)^\perp \subseteq \mathcal{N}(B)^\perp$. Since A and B are Hermitian, it follows from Theorem 2.4.3 that $\mathcal{R}(A) = \mathcal{N}(A)^\perp$ and $\mathcal{R}(B) = \mathcal{N}(B)^\perp$. Hence, $\mathcal{R}(A) \subseteq \mathcal{R}(B)$. $\square$

The following result is the *Douglas-Fillmore-Williams lemma* [437, 503].

Theorem 8.6.2. Let $A \in \mathbb{F}^{n \times m}$ and $B \in \mathbb{F}^{n \times l}$. Then, the following statements are equivalent:

i) There exists a matrix $C \in \mathbb{F}^{l \times m}$ such that $A = BC$.

ii) There exists $\alpha > 0$ such that $AA^* \leq \alpha BB^*$.

iii) $\mathcal{R}(A) \subseteq \mathcal{R}(B)$.

Proof. First we prove that *i*) implies *ii*). Since $A = BC$, it follows that $AA^* = BCC^*B^*$. Since $CC^* \leq \lambda_{\max}(CC^*)I$, it follows that $AA^* \leq \alpha BB^*$, where $\alpha \triangleq \lambda_{\max}(CC^*)$. To prove that *ii*) implies *iii*), first note that Lemma 8.6.1 implies that $\mathcal{R}(AA^*) \subseteq \mathcal{R}(\alpha BB^*) = \mathcal{R}(BB^*)$. Since, by Theorem 2.4.3, $\mathcal{R}(AA^*) = \mathcal{R}(A)$ and $\mathcal{R}(BB^*) = \mathcal{R}(B)$, it follows that $\mathcal{R}(A) \subseteq \mathcal{R}(B)$. Finally, to prove that *iii*) implies *i*), use Theorem 5.6.3 to write $B = S_1 \left[\begin{smallmatrix} D & 0 \\ 0 & 0 \end{smallmatrix}\right] S_2$, where $S_1 \in \mathbb{F}^{n \times n}$ and $S_2 \in \mathbb{F}^{l \times l}$ are unitary and $D \in \mathbb{R}^{r \times r}$ is diagonal with positive diagonal entries, where $r \triangleq \operatorname{rank} B$. Since $\mathcal{R}(S_1^*A) \subseteq \mathcal{R}(S_1^*B)$ and $S_1^*B = \left[\begin{smallmatrix} D & 0 \\ 0 & 0 \end{smallmatrix}\right] S_2$, it follows that $S_1^*A = \left[\begin{smallmatrix} A_1 \\ 0 \end{smallmatrix}\right]$, where $A_1 \in \mathbb{F}^{r \times m}$. Consequently,

$$A = S_1 \begin{bmatrix} A_1 \\ 0 \end{bmatrix} = S_1 \begin{bmatrix} D & 0 \\ 0 & 0 \end{bmatrix} S_2 S_2^* \begin{bmatrix} D^{-1} & 0 \\ 0 & 0 \end{bmatrix} \begin{bmatrix} A_1 \\ 0 \end{bmatrix} = BC,$$

where $C \triangleq S_2^* \left[\begin{smallmatrix} D^{-1} & 0 \\ 0 & 0 \end{smallmatrix}\right] \left[\begin{smallmatrix} A_1 \\ 0 \end{smallmatrix}\right] \in \mathbb{F}^{l \times m}$. $\square$

Proposition 8.6.3. Let $(A_i)_{i=1}^\infty \subset \mathbf{N}^n$ satisfy $0 \leq A_i \leq A_j$ for all $i \leq j$, and assume there exists $B \in \mathbf{N}^n$ satisfying $A_i \leq B$ for all $i \geq 1$. Then, $A \triangleq \lim_{i \to \infty} A_i$ exists and satisfies $0 \leq A \leq B$.

Proof. Let $k \in \{1, \dots, n\}$, and let $i < j$. Since $A_i \leq A_j \leq B$, it follows that the sequence $(A_{r(k,k)})_{r=1}^\infty$ is nondecreasing and bounded from above by $B_{(k,k)}$. Hence, $A_{(k,k)} \triangleq \lim_{r \to \infty} A_{r(k,k)}$ exists. Now, let $l \in \{1, \dots, n\}$, where $l \neq k$. Since $A_i \leq A_j$, it follows that $(e_k + e_l)^{\mathrm{T}} A_i (e_k + e_l) \leq (e_k + e_l)^{\mathrm{T}} A_j (e_k + e_l)$, which implies that $A_{i(k,l)} - A_{j(k,l)} \leq \frac{1}{2}[A_{j(k,k)} - A_{i(k,k)} + A_{j(l,l)} - A_{i(l,l)}]$. Likewise, $(e_k - e_l)^{\mathrm{T}} A_i (e_k - e_l) \leq (e_k - e_l)^{\mathrm{T}} A_j (e_k - e_l)$ implies that $A_{j(k,l)} - A_{i(k,l)} \leq \frac{1}{2}[A_{j(k,k)} - A_{i(k,k)} + A_{j(l,l)} - A_{i(l,l)}]$. Hence, $|A_{j(k,l)} - A_{i(k,l)}| \leq \frac{1}{2}[A_{j(k,k)} - A_{i(k,k)}] + \frac{1}{2}[A_{j(l,l)} - A_{i(l,l)}]$. Next, since $(A_{r(k,k)})_{r=1}^\infty$ and $(A_{r(l,l)})_{r=1}^\infty$ are convergent sequences and thus Cauchy sequences, it follows that $(A_{r(k,l)})_{r=1}^\infty$ is a Cauchy sequence. Consequently, $(A_{r(k,l)})_{r=1}^\infty$ is convergent, and thus $A_{(k,l)} \triangleq \lim_{i \to \infty} A_{i(k,l)}$ exists. Therefore, $(A_i)_{i=1}^\infty$ is convergent, and thus $A \triangleq \lim_{i \to \infty} A_i$ exists. Since $A_i \leq B$ for all $i \geq 1$, it follows that $A \leq B$. $\square$

Proposition 8.6.4. Let $A \in \mathbb{F}^{n \times n}$, assume that A is positive definite, and let $p > 0$. Then,

$$A^{-1}(A - I) \leq \log A \leq p^{-1}(A^p - I) \tag{8.6.1}$$

and

$$\log A = \lim_{p \downarrow 0} p^{-1}(A^p - I). \tag{8.6.2}$$

Proof. This result follows from Fact 1.11.26. $\square$

Lemma 8.6.5. Let $A \in \mathbf{P}^n$. If $A \leq I$, then $I \leq A^{-1}$. Furthermore, if $A < I$, then $I < A^{-1}$.

Proof. Since $A \leq I$, it follows from *xii*) of Proposition 8.1.2 that $I = A^{-1/2}AA^{-1/2} \leq A^{-1/2}IA^{-1/2} = A^{-1}$. Similarly, $A < I$ implies that $I = A^{-1/2}AA^{-1/2} < A^{-1/2}IA^{-1/2} = A^{-1}$. $\square$

Proposition 8.6.6. Let $A, B \in \mathbf{H}^n$, and assume that either A and B are positive definite or A and B are negative definite. If $A \leq B$, then $B^{-1} \leq A^{-1}$. If, in addition, $A < B$, then $B^{-1} < A^{-1}$.

Proof. Suppose that A and B are positive definite. Since $A \leq B$, it follows that $B^{-1/2}AB^{-1/2} \leq I$. Now, Lemma 8.6.5 implies that $I \leq B^{1/2}A^{-1}B^{1/2}$, which implies that $B^{-1} \leq A^{-1}$. If A and B are negative definite, then $A \leq B$ is equivalent to $-B \leq -A$. The case $A < B$ is proved in a similar manner. $\square$

The following result is the *Furuta inequality*.

Proposition 8.6.7. Let $A, B \in \mathbf{N}^n$, and assume that $0 \leq A \leq B$. Furthermore, let $p, q, r \in \mathbb{R}$ satisfy $p \geq 0$, $q \geq 1$, $r \geq 0$, and $p + 2r \leq (1 + 2r)q$. Then,

$$A^{(p+2r)/q} \leq (A^r B^p A^r)^{1/q} \tag{8.6.3}$$

and

$$(B^r A^p B^r)^{1/q} \leq B^{(p+2r)/q}. \tag{8.6.4}$$

Proof. See [536] or [544, pp. 129, 130]. $\square$

Corollary 8.6.8. Let $A, B \in \mathbf{N}^n$, and assume that $0 \leq A \leq B$. Then,

$$A^2 \leq \left(AB^2A\right)^{1/2} \tag{8.6.5}$$

and

$$\left(BA^2B\right)^{1/2} \leq B^2. \tag{8.6.6}$$

Proof. In Proposition 8.6.7 set $r = 1$, $p = 2$, and $q = 2$. $\square$

Corollary 8.6.9. Let $A, B, C \in \mathbf{N}^n$, and assume that $0 \leq A \leq C \leq B$. Then,

$$\left(CA^2C\right)^{1/2} \leq C^2 \leq \left(CB^2C\right)^{1/2}. \tag{8.6.7}$$

Proof. This result follows from Corollary 8.6.8. See also [1429]. $\square$

The following result provides representations for A^r, where $r \in (0, 1)$.

Proposition 8.6.10. Let $A \in \mathbf{P}^n$ and $r \in (0, 1)$. Then,

$$A^r = \left(\cos \frac{r\pi}{2}\right)I + \frac{\sin r\pi}{\pi} \int_0^\infty \left[\frac{x^{r+1}}{1+x^2}I - (A + xI)^{-1}x^r\right] dx \tag{8.6.8}$$

and

$$A^r = \frac{\sin r\pi}{\pi} \int_0^\infty (A + xI)^{-1}Ax^{r-1} dx. \tag{8.6.9}$$

Proof. Let $t \geq 0$. As shown in [197], [201, p. 143],

$$\int_0^\infty \left[\frac{x^{r+1}}{1+x^2} - \frac{x^r}{t+x}\right] dx = \frac{\pi}{\sin r\pi}\left(t^r - \cos \frac{r\pi}{2}\right).$$

Solving for t^r and replacing t by A yields (8.6.8). Likewise, replacing t by A in *xxxii)* of Fact 1.21.1 yields (8.6.9). □

The following result is the *Löwner-Heinz inequality.*

Corollary 8.6.11. Let $A, B \in \mathbf{N}^n$, assume that $0 \leq A \leq B$, and let $r \in [0,1]$. Then, $A^r \leq B^r$. If, in addition, $A < B$ and $r \in (0,1]$, then $A^r < B^r$.

Proof. Let $0 < A \leq B$, and let $r \in [0,1]$. In Proposition 8.6.7, replace p, q, r with $r, 1, 0$. The first result now follows from (8.6.3).

Now, assume that $A < B$. Then, it follows from (8.6.8) of Proposition 8.6.10 as well as Proposition 8.6.6 that, for all $r \in (0,1]$,

$$B^r - A^r = \frac{\sin r\pi}{\pi} \int_0^\infty [(A + xI)^{-1} - (B + xI)^{-1}]x^r \, \mathrm{d}x > 0.$$

Hence, $A^r < B^r$. By continuity, it follows that $A^r \leq B^r$ for all $A, B \in \mathbf{N}^n$ such that $0 \leq A \leq B$ and for all $r \in [0,1]$.

Alternatively, assume that $A < B$. Then, it follows from Proposition 8.6.6 that, for all $x \geq 0$,

$$(A + xI)^{-1}A = I - x(A + xI)^{-1} < I - x(B + xI)^{-1} = (B + xI)^{-1}B.$$

It thus follows from (8.6.9) of Proposition 8.6.10 that

$$B^r - A^r = \frac{\sin r\pi}{\pi} \int_0^\infty [(B + xI)^{-1}B - (A + xI)^{-1}A]x^{r-1} \, \mathrm{d}x > 0.$$

Hence, $A^r < B^r$. By continuity, it follows that $A^r \leq B^r$ for all $A, B \in \mathbf{N}^n$ such that $0 \leq A \leq B$ and for all $r \in [0,1]$.

Alternative proofs are given in [544, p. 127] and [1521, p. 2].

For the case $r = 1/2$, let $\lambda \in \mathbb{R}$ be an eigenvalue of $B^{1/2} - A^{1/2}$, and let $x \in \mathbb{F}^n$ be an associated eigenvector. Then,

$$\lambda x^* \left(B^{1/2} + A^{1/2}\right)x = x^* \left(B^{1/2} + A^{1/2}\right)\left(B^{1/2} - A^{1/2}\right)x$$
$$= x^* \left(B - B^{1/2}A^{1/2} + A^{1/2}B^{1/2} - A\right)x$$
$$= x^*(B - A)x \geq 0.$$

Since $B^{1/2} + A^{1/2}$ is positive semidefinite, it follows that either $\lambda \geq 0$ or $x^*(B^{1/2} + A^{1/2})x = 0$. In the latter case, $B^{1/2}x = A^{1/2}x = 0$, which implies that $\lambda = 0$. □

The Löwner-Heinz inequality does not extend to $r > 1$. In fact, $A \triangleq \left[\begin{smallmatrix} 2 & 1 \\ 1 & 1 \end{smallmatrix}\right]$ and $B \triangleq \left[\begin{smallmatrix} 1 & 0 \\ 0 & 0 \end{smallmatrix}\right]$ satisfy $A \geq B \geq 0$, whereas, for all $r > 1$, $A^r \not\geq B^r$. For details, see [544, pp. 127, 128].

Many of the results given so far involve functions that are nondecreasing or increasing on suitable sets of matrices.

Definition 8.6.12. Let $\mathcal{D} \subseteq \mathbf{H}^n$, and let $\phi \colon \mathcal{D} \mapsto \mathbf{H}^m$. Then, the following terminology is defined:

 i) ϕ is *nondecreasing* if, for all $A, B \in \mathcal{D}$ such that $A \leq B$, it follows that $\phi(A) \leq \phi(B)$.

 ii) ϕ is *increasing* if ϕ is nondecreasing and, for all $A, B \in \mathcal{D}$ such that $A < B$, it follows that $\phi(A) < \phi(B)$.

 iii) ϕ is *strongly increasing* if ϕ is nondecreasing and, for all $A, B \in \mathcal{D}$ such that $A \leq B$ and $A \neq B$, it follows that $\phi(A) < \phi(B)$.

 iv) ϕ is (*nonincreasing, decreasing, strongly decreasing*) if $-\phi$ is (nondecreasing, increasing, strongly increasing).

Proposition 8.6.13. The following functions are nondecreasing:

 i) $\phi \colon \mathbf{H}^n \mapsto \mathbf{H}^m$ defined by $\phi(A) \triangleq BAB^*$, where $B \in \mathbb{F}^{m \times n}$.

 ii) $\phi \colon \mathbf{H}^n \mapsto \mathbb{R}$ defined by $\phi(A) \triangleq \operatorname{tr} AB$, where $B \in \mathbf{N}^n$.

 iii) $\phi \colon \mathbf{N}^{n+m} \mapsto \mathbf{N}^n$ defined by $\phi(A) \triangleq A_{22} | A$, where $A \triangleq \begin{bmatrix} A_{11} & A_{12} \\ A_{12}^* & A_{22} \end{bmatrix}$.

 iv) $\phi \colon \mathbf{N}^n \times \mathbf{N}^m \mapsto \mathbf{N}^{nm}$ defined by $\phi(A, B) \triangleq A^{r_1} \otimes B^{r_2}$, where $r_1, r_2 \in [0, 1]$ satisfy $r_1 + r_2 \leq 1$.

 v) $\phi \colon \mathbf{N}^n \times \mathbf{N}^n \mapsto \mathbf{N}^n$ defined by $\phi(A, B) \triangleq A^{r_1} \circ B^{r_2}$, where $r_1, r_2 \in [0, 1]$ satisfy $r_1 + r_2 \leq 1$.

The following functions are increasing:

 vi) $\phi \colon \mathbf{H}^n \mapsto \mathbb{R}$ defined by $\phi(A) \triangleq \lambda_i(A)$, where $i \in \{1, \ldots, n\}$.

 vii) $\phi \colon \mathbf{N}^n \mapsto \mathbf{N}^n$ defined by $\phi(A) \triangleq A^r$, where $r \in [0, 1]$.

 viii) $\phi \colon \mathbf{N}^n \mapsto \mathbf{N}^n$ defined by $\phi(A) \triangleq A^{1/2}$.

 ix) $\phi \colon \mathbf{P}^n \mapsto -\mathbf{P}^n$ defined by $\phi(A) \triangleq -A^{-r}$, where $r \in [0, 1]$.

 x) $\phi \colon \mathbf{P}^n \mapsto -\mathbf{P}^n$ defined by $\phi(A) \triangleq -A^{-1}$.

 xi) $\phi \colon \mathbf{P}^n \mapsto -\mathbf{P}^n$ defined by $\phi(A) \triangleq -A^{-1/2}$.

 xii) $\phi \colon -\mathbf{P}^n \mapsto \mathbf{P}^n$ defined by $\phi(A) \triangleq (-A)^{-r}$, where $r \in [0, 1]$.

 xiii) $\phi \colon -\mathbf{P}^n \mapsto \mathbf{P}^n$ defined by $\phi(A) \triangleq -A^{-1}$.

 xiv) $\phi \colon -\mathbf{P}^n \mapsto \mathbf{P}^n$ defined by $\phi(A) \triangleq -A^{-1/2}$.

 xv) $\phi \colon \mathbf{H}^n \mapsto \mathbf{H}^m$ defined by $\phi(A) \triangleq BAB^*$, where $B \in \mathbb{F}^{m \times n}$ and $\operatorname{rank} B = m$.

 xvi) $\phi \colon \mathbf{P}^{n+m} \mapsto \mathbf{P}^n$ defined by $\phi(A) \triangleq A_{22} | A$, where $A \triangleq \begin{bmatrix} A_{11} & A_{12} \\ A_{12}^* & A_{22} \end{bmatrix}$.

xvii) ϕ: $\mathbf{P}^{n+m} \mapsto \mathbf{P}^n$ defined by $\phi(A) \triangleq -(A_{22}|A)^{-1}$, where $A \triangleq \begin{bmatrix} A_{11} & A_{12} \\ A_{12}^* & A_{22} \end{bmatrix}$.

xviii) ϕ: $\mathbf{P}^n \mapsto \mathbf{H}^n$ defined by $\phi(A) \triangleq \log A$.

The following functions are strongly increasing:

xix) ϕ: $\mathbf{H}^n \mapsto [0, \infty)$ defined by $\phi(A) \triangleq \operatorname{tr} BAB^*$, where $B \in \mathbb{F}^{m \times n}$ and $\operatorname{rank} B = m$.

xx) ϕ: $\mathbf{H}^n \mapsto \mathbb{R}$ defined by $\phi(A) \triangleq \operatorname{tr} AB$, where $B \in \mathbf{P}^n$.

xxi) ϕ: $\mathbf{N}^n \mapsto [0, \infty)$ defined by $\phi(A) \triangleq \operatorname{tr} A^r$, where $r > 0$.

xxii) ϕ: $\mathbf{N}^n \mapsto [0, \infty)$ defined by $\phi(A) \triangleq \det A$.

Proof. For the proof of *iii)*, see [922]. To prove *xviii)*, let $A, B \in \mathbf{P}^n$, and assume that $A \leq B$. Then, for all $r \in [0, 1]$, it follows from *vii)* that $r^{-1}(A^r - I) \leq r^{-1}(B^r - I)$. Letting $r \downarrow 0$ and using Proposition 8.6.4 yields $\log A \leq \log B$, which proves that log is nondecreasing. See [544, p. 139]. To prove that log is increasing, assume that $A < B$, and let $\varepsilon > 0$ be such that $A + \varepsilon I < B$. Then, it follows that $\log A < \log(A + \varepsilon I) \leq \log B$. $\qquad\square$

Finally, we consider convex functions defined with respect to matrix inequalities. The following definition generalizes Definition 1.4.3 in the case $n = m = p = 1$.

Definition 8.6.14. Let $\mathcal{D} \subseteq \mathbb{F}^{n \times m}$ be a convex set, and let ϕ: $\mathcal{D} \mapsto \mathbf{H}^p$. Then, the following terminology is defined:

i) ϕ is *convex* if, for all $\alpha \in [0, 1]$ and $A_1, A_2 \in \mathcal{D}$,

$$\phi[\alpha A_1 + (1 - \alpha)A_2] \leq \alpha\phi(A_1) + (1 - \alpha)\phi(A_2). \qquad (8.6.10)$$

ii) ϕ is *concave* if $-\phi$ is convex.

iii) ϕ is *strictly convex* if, for all $\alpha \in (0, 1)$ and distinct $A_1, A_2 \in \mathcal{D}$,

$$\phi[\alpha A_1 + (1 - \alpha)A_2] < \alpha\phi(A_1) + (1 - \alpha)\phi(A_2). \qquad (8.6.11)$$

iv) ϕ is *strictly concave* if $-\phi$ is strictly convex.

Theorem 8.6.15. Let $\mathcal{S} \subseteq \mathbb{R}$, let ϕ: $\mathcal{S}_1 \mapsto \mathcal{S}_2$, and assume that ϕ is continuous. Then, the following statements hold:

i) Assume that $\mathcal{S}_1 = \mathcal{S}_2 = (0, \infty)$ and ϕ: $\mathbf{P}^n \mapsto \mathbf{P}^n$ is increasing. Then, ψ: $\mathbf{P}^n \mapsto \mathbf{P}^n$ defined by $\psi(x) = 1/\phi(x)$ is convex.

ii) Assume that $\mathcal{S}_1 = \mathcal{S}_2 = [0, \infty)$. Then, ϕ: $\mathbf{N}^n \mapsto \mathbf{N}^n$ is increasing if and only if ϕ: $\mathbf{N}^n \mapsto \mathbf{N}^n$ is concave.

iii) Assume that $\mathcal{S}_1 = [0, \infty)$ and $\mathcal{S}_2 = \mathbb{R}$. Then, ϕ: $\mathbf{N}^n \mapsto \mathbf{H}^n$ is convex and $\phi(0) \leq 0$ if and only if ψ: $\mathbf{P}^n \mapsto \mathbf{H}^n$ defined by $\psi(x) = \phi(x)/x$ is increasing.

Proof. See [201, pp. 120–122]. $\qquad\square$

Lemma 8.6.16. Let $\mathcal{D} \subseteq \mathbb{F}^{n \times m}$ and $\mathcal{S} \subseteq \mathbf{H}^p$ be convex sets, and let $\phi_1\colon \mathcal{D} \mapsto \mathcal{S}$ and $\phi_2\colon \mathcal{S} \mapsto \mathbf{H}^q$. Then, the following statements hold:

i) If ϕ_1 is convex and ϕ_2 is nondecreasing and convex, then $\phi_2 \bullet \phi_1\colon \mathcal{D} \mapsto \mathbf{H}^q$ is convex.

ii) If ϕ_1 is concave and ϕ_2 is nonincreasing and convex, then $\phi_2 \bullet \phi_1\colon \mathcal{D} \mapsto \mathbf{H}^q$ is convex.

iii) If $\mathcal{S}$ is symmetric, $\phi_2(-A) = -\phi_2(A)$ for all $A \in \mathcal{S}$, ϕ_1 is concave, and ϕ_2 is nonincreasing and concave, then $\phi_2 \bullet \phi_1\colon \mathcal{D} \mapsto \mathbf{H}^q$ is convex.

iv) If $\mathcal{S}$ is symmetric, $\phi_2(-A) = -\phi_2(A)$ for all $A \in \mathcal{S}$, ϕ_1 is convex, and ϕ_2 is nondecreasing and concave, then $\phi_2 \bullet \phi_1\colon \mathcal{D} \mapsto \mathbf{H}^q$ is convex.

Proof. To prove i) and ii), let $\alpha \in [0,1]$ and $A_1, A_2 \in \mathcal{D}$. In both cases it follows that

$$\phi_2(\phi_1[\alpha A_1 + (1-\alpha)A_2]) \le \phi_2[\alpha \phi_1(A_1) + (1-\alpha)\phi_1(A_2)]$$
$$\le \alpha \phi_2[\phi_1(A_1)] + (1-\alpha)\phi_2[\phi_1(A_2)].$$

Statements iii) and iv) follow from i) and ii), respectively. $\square$

Proposition 8.6.17. The following functions are convex:

i) $\phi\colon \mathbf{N}^n \mapsto \mathbf{N}^n$ defined by $\phi(A) \triangleq A^r$, where $r \in [1,2]$.

ii) $\phi\colon \mathbf{N}^n \mapsto \mathbf{N}^n$ defined by $\phi(A) \triangleq A^2$.

iii) $\phi\colon \mathbf{P}^n \mapsto \mathbf{P}^n$ defined by $\phi(A) \triangleq A^{-r}$, where $r \in [0,1]$.

iv) $\phi\colon \mathbf{P}^n \mapsto \mathbf{P}^n$ defined by $\phi(A) \triangleq A^{-1}$.

v) $\phi\colon \mathbf{P}^n \mapsto \mathbf{P}^n$ defined by $\phi(A) \triangleq A^{-1/2}$.

vi) $\phi\colon \mathbf{N}^n \mapsto -\mathbf{N}^n$ defined by $\phi(A) \triangleq -A^r$, where $r \in [0,1]$.

vii) $\phi\colon \mathbf{N}^n \mapsto -\mathbf{N}^n$ defined by $\phi(A) \triangleq -A^{1/2}$.

viii) $\phi\colon \mathbf{N}^n \mapsto \mathbf{H}^m$ defined by $\phi(A) \triangleq \gamma BAB^*$, where $\gamma \in \mathbb{R}$ and $B \in \mathbb{F}^{m \times n}$.

ix) $\phi\colon \mathbf{N}^n \mapsto \mathbf{N}^m$ defined by $\phi(A) \triangleq BA^rB^*$, where $B \in \mathbb{F}^{m \times n}$ and $r \in [1,2]$.

x) $\phi\colon \mathbf{P}^n \mapsto \mathbf{N}^m$ defined by $\phi(A) \triangleq BA^{-r}B^*$, where $B \in \mathbb{F}^{m \times n}$ and $r \in [0,1]$.

xi) $\phi\colon \mathbf{N}^n \mapsto -\mathbf{N}^m$ defined by $\phi(A) \triangleq -BA^rB^*$, where $B \in \mathbb{F}^{m \times n}$ and $r \in [0,1]$.

xii) $\phi\colon \mathbf{P}^n \mapsto -\mathbf{P}^m$ defined by $\phi(A) \triangleq -(BA^{-r}B^*)^{-p}$, where $B \in \mathbb{F}^{m \times n}$ has rank m and $r, p \in [0,1]$.

xiii) $\phi\colon \mathbb{F}^{n \times m} \mapsto \mathbf{N}^n$ defined by $\phi(A) \triangleq ABA^*$, where $B \in \mathbf{N}^m$.

xiv) $\phi\colon \mathbf{P}^n \times \mathbb{F}^{m \times n} \mapsto \mathbf{N}^m$ defined by $\phi(A,B) \triangleq BA^{-1}B^*$.

xv) $\phi\colon \mathbf{P}^n \mapsto \mathbf{N}^m$ defined by $\phi(A) \triangleq \left(A^{-1} + A^{-*}\right)^{-1}$.

xvi) ϕ: $\mathbf{N}^n \times \mathbf{N}^n \mapsto \mathbf{N}^n$ defined by $\phi(A, B) \triangleq -A(A + B)^+ B$.

xvii) ϕ: $\mathbf{N}^{n+m} \mapsto \mathbf{N}^n$ defined by $\phi(A) \triangleq -A_{22}|A$, where $A \triangleq \begin{bmatrix} A_{11} & A_{12} \\ A_{12}^* & A_{22} \end{bmatrix}$.

xviii) ϕ: $\mathbf{P}^{n+m} \mapsto \mathbf{P}^n$ defined by $\phi(A) \triangleq (A_{22}|A)^{-1}$, where $A \triangleq \begin{bmatrix} A_{11} & A_{12} \\ A_{12}^* & A_{22} \end{bmatrix}$.

xix) ϕ: $\mathbf{H}^n \mapsto [0, \infty)$ defined by $\phi(A) \triangleq \operatorname{tr} A^k$, where k is a nonnegative even integer.

xx) ϕ: $\mathbf{P}^n \mapsto (0, \infty)$ defined by $\phi(A) \triangleq \operatorname{tr} A^{-r}$, where $r > 0$.

xxi) ϕ: $\mathbf{P}^n \mapsto (-\infty, 0)$ defined by $\phi(A) \triangleq -(\operatorname{tr} A^{-r})^{-p}$, where $r, p \in [0, 1]$.

xxii) ϕ: $\mathbf{N}^n \times \mathbf{N}^n \mapsto (-\infty, 0]$ defined by $\phi(A, B) \triangleq -\operatorname{tr} (A^r + B^r)^{1/r}$, where $r \in [0, 1]$.

xxiii) ϕ: $\mathbf{N}^n \times \mathbf{N}^n \mapsto [0, \infty)$ defined by $\phi(A, B) \triangleq \operatorname{tr} (A^2 + B^2)^{1/2}$.

xxiv) ϕ: $\mathbf{N}^n \times \mathbf{N}^m \mapsto \mathbb{R}$ defined by $\phi(A, B) \triangleq -\operatorname{tr} A^r X B^p X^*$, where $X \in \mathbb{F}^{n \times m}$, $r, p \geq 0$, and $r + p \leq 1$.

xxv) ϕ: $\mathbf{N}^n \mapsto (-\infty, 0)$ defined by $\phi(A) \triangleq -\operatorname{tr} A^r X A^p X^*$, where $X \in \mathbb{F}^{n \times n}$, $r, p \geq 0$, and $r + p \leq 1$.

xxvi) ϕ: $\mathbf{P}^n \times \mathbf{P}^m \times \mathbb{F}^{m \times n} \mapsto \mathbb{R}$ defined by $\phi(A, B, X) \triangleq (\operatorname{tr} A^{-p} X B^{-r} X^*)^q$, where $r, p \geq 0$, $r + p \leq 1$, and $q \geq (2 - r - p)^{-1}$.

xxvii) ϕ: $\mathbf{P}^n \times \mathbb{F}^{n \times n} \mapsto [0, \infty)$ defined by $\phi(A, X) \triangleq \operatorname{tr} A^{-p} X A^{-r} X^*$, where $r, p \geq 0$ and $r + p \leq 1$.

xxviii) ϕ: $\mathbf{P}^n \times \mathbb{F}^{n \times n} \mapsto [0, \infty)$ defined by $\phi(A) \triangleq \operatorname{tr} A^{-p} X A^{-r} X^*$, where $r, p \in [0, 1]$ and $X \in \mathbb{F}^{n \times n}$.

xxix) ϕ: $\mathbf{P}^n \mapsto \mathbb{R}$ defined by $\phi(A) \triangleq -\operatorname{tr}([A^r, X][A^{1-r}, X])$, where $r \in (0, 1)$ and $X \in \mathbf{H}^n$.

xxx) ϕ: $\mathbf{P}^n \mapsto \mathbf{H}^n$ defined by $\phi(A) \triangleq -\log A$.

xxxi) ϕ: $\mathbf{P}^n \mapsto \mathbf{H}^m$ defined by $\phi(A) \triangleq A \log A$.

xxxii) ϕ: $\mathbf{N}^n \backslash \{0\} \mapsto \mathbb{R}$ defined by $\phi(A) \triangleq -\log \operatorname{tr} A^r$, where $r \in [0, 1]$.

xxxiii) ϕ: $\mathbf{P}^n \mapsto \mathbb{R}$ defined by $\phi(A) \triangleq \log \operatorname{tr} A^{-1}$.

xxxiv) ϕ: $\mathbf{P}^n \times \mathbf{P}^n \mapsto (0, \infty)$ defined by $\phi(A, B) \triangleq \operatorname{tr}[A(\log A - \log B)]$.

xxxv) ϕ: $\mathbf{P}^n \times \mathbf{P}^n \to [0, \infty)$ defined by $\phi(A, B) \triangleq -e^{[1/(2n)]\operatorname{tr}(\log A + \log B)}$.

xxxvi) ϕ: $\mathbf{N}^n \mapsto (-\infty, 0]$ defined by $\phi(A) \triangleq -(\det A)^{1/n}$.

xxxvii) ϕ: $\mathbf{P}^n \mapsto (0, \infty)$ defined by $\phi(A) \triangleq \log \det BA^{-1}B^*$, where $B \in \mathbb{F}^{m \times n}$ and rank $B = m$.

xxxviii) ϕ: $\mathbf{P}^n \mapsto \mathbb{R}$ defined by $\phi(A) \triangleq -\log \det A$.

xxxix) ϕ: $\mathbf{P}^n \mapsto (0, \infty)$ defined by $\phi(A) \triangleq \det A^{-1}$.

xl) ϕ: $\mathbf{P}^n \mapsto \mathbb{R}$ defined by $\phi(A) \triangleq \log(\det A_k / \det A)$, where $k \in \{1, \ldots, n-1\}$ and A_k is the leading $k \times k$ principal submatrix of A.

xli) ϕ: $\mathbf{P}^n \mapsto \mathbb{R}$ defined by $\phi(A) \triangleq -\det A / \det A_{[n;n]}$.

xlii) ϕ: $\mathbf{N}^n \times \mathbf{N}^m \mapsto -\mathbf{N}^{nm}$ defined by $\phi(A, B) \triangleq -A^{r_1} \otimes B^{r_2}$, where $r_1, r_2 \in [0, 1]$ satisfy $r_1 + r_2 \leq 1$.

xliii) ϕ: $\mathbf{P}^n \times \mathbf{N}^m \mapsto \mathbf{N}^{nm}$ defined by $\phi(A, B) \triangleq A^{-r} \otimes B^{1+r}$, where $r \in [0, 1]$.

xliv) ϕ: $\mathbf{N}^n \times \mathbf{N}^n \mapsto -\mathbf{N}^n$ defined by $\phi(A, B) \triangleq -A^{r_1} \circ B^{r_2}$, where $r_1, r_2 \in [0, 1]$ satisfy $r_1 + r_2 \leq 1$.

xlv) ϕ: $\mathbf{H}^n \mapsto \mathbb{R}$ defined by $\phi(A) \triangleq \sum_{i=1}^{k} \lambda_i(A)$, where $k \in \{1, \ldots, n\}$.

xlvi) ϕ: $\mathbf{H}^n \mapsto \mathbb{R}$ defined by $\phi(A) \triangleq -\sum_{i=k}^{n} \lambda_i(A)$, where $k \in \{1, \ldots, n\}$.

Proof. Statements *i*) and *iii*) are proved in [45] and [201, p. 123].

Let $\alpha \in [0, 1]$ for the remainder of the proof.

To prove *ii*) directly, let $A_1, A_2 \in \mathbf{H}^n$. Since

$$\alpha(1 - \alpha) = \left(\alpha - \alpha^2\right)^{1/2} \left[(1 - \alpha) - (1 - \alpha)^2\right]^{1/2},$$

it follows that

$$0 \leq \left[\left(\alpha - \alpha^2\right)^{1/2} A_1 - \left[(1 - \alpha) - (1 - \alpha)^2\right]^{1/2} A_2\right]^2$$

$$= (\alpha - \alpha^2) A_1^2 + \left[(1 - \alpha) - (1 - \alpha)^2\right] A_2^2 - \alpha(1 - \alpha)(A_1 A_2 + A_2 A_1).$$

Hence,

$$[\alpha A_1 + (1 - \alpha) A_2]^2 \leq \alpha A_1^2 + (1 - \alpha) A_2^2,$$

which shows that $\phi(A) = A^2$ is convex.

To prove *iv*) directly, let $A_1, A_2 \in \mathbf{P}^n$. Then, $\begin{bmatrix} A_1^{-1} & I \\ I & A_1 \end{bmatrix}$ and $\begin{bmatrix} A_2^{-1} & I \\ I & A_2 \end{bmatrix}$ are positive semidefinite, and thus

$$\alpha \begin{bmatrix} A_1^{-1} & I \\ I & A_1 \end{bmatrix} + (1 - \alpha) \begin{bmatrix} A_2^{-1} & I \\ I & A_2 \end{bmatrix}$$

$$= \begin{bmatrix} \alpha A_1^{-1} + (1 - \alpha) A_2^{-1} & I \\ I & \alpha A_1 + (1 - \alpha) A_2 \end{bmatrix}$$

is positive semidefinite. It now follows from Proposition 8.2.4 that $[\alpha A_1 + (1 - \alpha) A_2]^{-1} \leq \alpha A_1^{-1} + (1 - \alpha) A_2^{-1}$, which shows that $\phi(A) = A^{-1}$ is convex.

To prove *v*) directly, note that $\phi(A) = A^{-1/2} = \phi_2[\phi_1(A)]$, where $\phi_1(A) \triangleq A^{1/2}$ and $\phi_2(B) \triangleq B^{-1}$. It follows from *vii*) that ϕ_1 is concave, while it follows from *iv*) that ϕ_2 is convex. Furthermore, *x*) of Proposition 8.6.13 implies that ϕ_2 is nonincreasing. It thus follows from *ii*) of Lemma 8.6.16 that $\phi(A) = A^{-1/2}$ is convex.

To prove $vi)$, let $A \in \mathbf{P}^n$, and note that $\phi(A) = -A^r = \phi_2[\phi_1(A)]$, where $\phi_1(A) \triangleq A^{-r}$ and $\phi_2(B) \triangleq -B^{-1}$. It follows from $iii)$ that ϕ_1 is convex, while it follows from $iv)$ that ϕ_2 is concave. Furthermore, $x)$ of Proposition 8.6.13 implies that ϕ_2 is nondecreasing. It thus follows from $iv)$ of Lemma 8.6.16 that $\phi(A) = A^r$ is convex on $\mathbf{P}^n$. Continuity implies that $\phi(A) = A^r$ is convex on $\mathbf{N}^n$.

To prove $vii)$ directly, let $A_1, A_2 \in \mathbf{N}^n$. Then,

$$0 \le \alpha(1 - \alpha)\left(A_1^{1/2} - A_2^{1/2}\right)^2,$$

which is equivalent to

$$\left[\alpha A_1^{1/2} + (1 - \alpha)A_2^{1/2}\right]^2 \le \alpha A_1 + (1 - \alpha)A_2.$$

Using $viii)$ of Proposition 8.6.13 yields

$$\alpha A_1^{1/2} + (1 - \alpha)A_2^{1/2} \le [\alpha A_1 + (1 - \alpha)A_2]^{1/2}.$$

Finally, multiplying by -1 shows that $\phi(A) = -A^{1/2}$ is convex.

The proof of $viii)$ is immediate. Statements $ix)$, $x)$, and $xi)$ follow from $i)$, $iii)$, and $vi)$, respectively.

To prove $xii)$, note that $\phi(A) = -(BA^{-r}B^*)^{-p} = \phi_2[\phi_1(A)]$, where $\phi_1(A) = -BA^{-r}B^*$ and $\phi_2(C) = C^{-p}$. Statement $x)$ implies that ϕ_1 is concave, while $iii)$ implies that ϕ_2 is convex. Furthermore, $ix)$ of Proposition 8.6.13 implies that ϕ_2 is nonincreasing. It thus follows from $ii)$ of Lemma 8.6.16 that $\phi(A) = -(BA^{-r}B^*)^{-p}$ is convex.

To prove $xiii)$, let $A_1, A_2 \in \mathbb{F}^{n \times m}$, and let $B \in \mathbf{N}^m$. Then,

$$0 \le \alpha(1 - \alpha)(A_1 - A_2)B(A_1 - A_2)^*$$
$$= \alpha A_1 B A_1^* + (1 - \alpha)A_2 B A_2^* - [\alpha A_1 + (1 - \alpha)A_2]B[\alpha A_1 + (1 - \alpha)A_2]^*.$$

Thus,

$$[\alpha A_1 + (1 - \alpha)A_2]B[\alpha A_1 + (1 - \alpha)A_2]^* \le \alpha A_1 B A_1^* + (1 - \alpha)A_2 B A_2^*,$$

which shows that $\phi(A) = ABA^*$ is convex.

To prove $xiv)$, let $A_1, A_2 \in \mathbf{P}^n$ and $B_1, B_2 \in \mathbb{F}^{m \times n}$. Then, it follows from Proposition 8.2.4 that $\begin{bmatrix} B_1 A_1^{-1} B_1^* & B_1 \\ B_1^* & A_1 \end{bmatrix}$ and $\begin{bmatrix} B_2 A_2^{-1} B_2^* & B_2 \\ B_2^* & A_2 \end{bmatrix}$ are positive semidefinite, and thus

$$\alpha \begin{bmatrix} B_1 A_1^{-1} B_1^* & B_1 \\ B_1^* & A_1 \end{bmatrix} + (1 - \alpha) \begin{bmatrix} B_2 A_2^{-1} B_2^* & B_2 \\ B_2^* & A_2 \end{bmatrix}$$

$$= \begin{bmatrix} \alpha B_1 A_1^{-1} B_1^* + (1 - \alpha)B_2 A_2^{-1} B_2^* & \alpha B_1 + (1 - \alpha)B_2 \\ \alpha B_1^* + (1 - \alpha)B_2^* & \alpha A_1 + (1 - \alpha)A_2 \end{bmatrix}$$

is positive semidefinite. It thus follows from Proposition 8.2.4 that

$$[\alpha B_1 + (1 - \alpha)B_2][\alpha A_1 + (1 - \alpha)A_2]^{-1}[\alpha B_1 + (1 - \alpha)B_2]^*$$

$$\leq \alpha B_1 A_1^{-1} B_1^* + (1 - \alpha)B_2 A_2^{-1} B_2^*,$$

which shows that $\phi(A, B) = BA^{-1}B^*$ is convex.

Result $xv)$ is given in [1003].

Result $xvi)$ follows from Fact 8.21.18.

To prove $xvii)$, let $A \triangleq \begin{bmatrix} A_{11} & A_{12} \\ A_{12}^* & A_{22} \end{bmatrix} \in \mathbf{P}^{n+m}$ and $B \triangleq \begin{bmatrix} B_{11} & B_{12} \\ B_{12}^* & B_{22} \end{bmatrix} \in \mathbf{P}^{n+m}$. Then, it follows from $xiv)$ with A_1, B_1, A_2, B_2 replaced by $A_{22}, A_{12}, B_{22}, B_{12}$, respectively, that

$$[\alpha A_{12} + (1 - \alpha)B_{12}][\alpha A_{22} + (1 - \alpha)B_{22}]^{-1}[\alpha A_{12} + (1 - \alpha)B_{12}]^*$$

$$\leq \alpha A_{12} A_{22}^{-1} A_{12}^* + (1 - \alpha)B_{12} B_{22}^{-1} B_{12}^*.$$

Hence,

$$-[\alpha A_{22} + (1 - \alpha)B_{22}]|[\alpha A + (1 - \alpha)B]$$

$$= [\alpha A_{12} + (1 - \alpha)B_{12}][\alpha A_{22} + (1 - \alpha)B_{22}]^{-1}[\alpha A_{12} + (1 - \alpha)B_{12}]^*$$

$$- [\alpha A_{11} + (1 - \alpha)B_{11}]$$

$$\leq \alpha \left(A_{12} A_{22}^{-1} A_{12}^* - A_{11} \right) + (1 - \alpha)(B_{12} B_{22}^{-1} B_{12}^* - B_{11})$$

$$= \alpha(-A_{22}|A) + (1 - \alpha)(-B_{22}|B),$$

which shows that $\phi(A) \triangleq -A_{22}|A$ is convex. By continuity, the result holds for $A \in \mathbf{N}^{n+m}$.

To prove $xviii)$, note that $\phi(A) = (A_{22}|A)^{-1} = \phi_2[\phi_1(A)]$, where $\phi_1(A) = A_{22}|A$ and $\phi_2(B) = B^{-1}$. It follows from $xv)$ that ϕ_1 is concave, while it follows from $iv)$ that ϕ_2 is convex. Furthermore, $x)$ of Proposition 8.6.13 implies that ϕ_2 is nonincreasing. It thus follows from Lemma 8.6.16 that $\phi(A) \triangleq (A_{22}|A)^{-1}$ is convex.

Result $xix)$ is given in [243, p. 106].

Result $xx)$ is given in by Theorem 9 of [931].

To prove $xxi)$, note that $\phi(A) = -(\operatorname{tr} A^{-r})^{-p} = \phi_2[\phi_1(A)]$, where $\phi_1(A) = \operatorname{tr} A^{-r}$ and $\phi_2(B) = -B^{-p}$. Statement $iii)$ implies that ϕ_1 is convex and that ϕ_2 is concave. Furthermore, $ix)$ of Proposition 8.6.13 implies that ϕ_2 is nondecreasing. It thus follows from $iv)$ of Lemma 8.6.16 that $\phi(A) = -(\operatorname{tr} A^{-r})^{-p}$ is convex.

Results $xxii)$ and $xxiii)$ are proved in [294].

Results $xxiv)$–$xxviii)$ are given by Corollary 1.1, Theorem 1, Corollary 2.1, Theorem 2, and Theorem 8, respectively, of [294]. A proof of $xxiv)$ in the case $p = 1 - r$ is given in [201, p. 273].

Result $xxix$) is proved in [201, p. 274] and [294].

Result xxx) is given in [205, p. 113].

Result $xxxi$) is given in [201, p. 123], [205, p. 113], and [543].

To prove $xxxii$), note that $\phi(A) = -\log \operatorname{tr} A^r = \phi_2[\phi_1(A)]$, where $\phi_1(A) = \operatorname{tr} A^r$ and $\phi_2(x) = -\log x$. Statement vi) implies that ϕ_1 is concave. Furthermore, ϕ_2 is convex and nonincreasing. It thus follows from ii) of Lemma 8.6.16 that $\phi(A) = -\log \operatorname{tr} A^r$ is convex.

Result $xxxiii$) is given in [1051].

Result $xxxiv$) is given in [201, p. 275].

Result $xxxv$) is given in [56].

To prove $xxxvi$), let $A_1, A_2 \in \mathbf{N}^n$. From Corollary 8.4.15 it follows that $(\det A_1)^{1/n} + (\det A_2)^{1/n} \leq [\det(A_1 + A_2)]^{1/n}$. Replacing A_1 and A_2 by αA_1 and $(1 - \alpha)A_2$, respectively, and multiplying by -1 shows that $\phi(A) = -(\det A)^{1/n}$ is convex.

Result $xxxvii$) is proved in [1051].

Result $xxxviii$) is a special case of result $xxxvii$). This result is due to Fan. See [360] or [361, p. 679]. To prove $xxxviii$), note that $\phi(A) = -n\log[(\det A)^{1/n}] = \phi_2[\phi_1(A)]$, where $\phi_1(A) = (\det A)^{1/n}$ and $\phi_2(x) = -n\log x$. It follows from xix) that ϕ_1 is concave. Since ϕ_2 is nonincreasing and convex, it follows from ii) of Lemma 8.6.16 that $\phi(A) = -\log \det A$ is convex.

To prove $xxxix$), note that $\phi(A) = \det A^{-1} = \phi_2[\phi_1(A)]$, where $\phi_1(A) = \log \det A^{-1}$ and $\phi_2(x) = e^x$. It follows from xx) that ϕ_1 is convex. Since ϕ_2 is nondecreasing and convex, it follows from i) of Lemma 8.6.16 that $\phi(A) = \det A^{-1}$ is convex.

Results xl) and xli) are given in [360] and [361, pp. 684, 685].

Next, $xlii$) is given in [201, p. 273], [205, p. 114], and [1521, p. 9]. Statement $xliii$) is given in [205, p. 114]. Statement $xliv$) is given in [1521, p. 9].

Finally, xlv) is given in [996, p. 478]. Statement $xlvi$) follows immediately from xlv). $\square$

The following result is a corollary of $xvii$) of Proposition 8.6.17 for the case $\alpha = 1/2$. Versions of this result appear in [298, 676, 922, 947] and [1125, p. 152].

Corollary 8.6.18. Let $A \triangleq \begin{bmatrix} A_{11} & A_{12} \\ A_{12}^* & A_{22} \end{bmatrix} \in \mathbb{F}^{n+m}$ and $B \triangleq \begin{bmatrix} B_{11} & B_{12} \\ B_{12}^* & B_{22} \end{bmatrix} \in \mathbb{F}^{n+m}$, and assume that A and B are positive semidefinite. Then,

$$A_{11}|A + B_{11}|B \le (A_{11} + B_{11})|(A + B). \tag{8.6.12}$$

The following corollary of *xlv)* and *xlvi)* of Proposition 8.6.17 gives a strong majorization condition for the eigenvalues of a pair of Hermitian matrices.

Corollary 8.6.19. Let $A, B \in \mathbf{H}^n$. Then, for all $k \in \{1, \ldots, n\}$,

$$\sum_{i=1}^{k} \lambda_i(A) + \sum_{i=n-k+1}^{n} \lambda_i(B) \le \sum_{i=1}^{k} \lambda_i(A + B) \le \sum_{i=1}^{k} [\lambda_i(A) + \lambda_i(B)] \tag{8.6.13}$$

with equality in both inequalities for $k = n$. Furthermore, for all $k \in \{1, \ldots, n\}$,

$$\sum_{i=k}^{n} [\lambda_i(A) + \lambda_i(B)] \le \sum_{i=k}^{n} \lambda_i(A + B) \tag{8.6.14}$$

with equality for $k = 1$.

Proof. The lower bound in (8.6.13) is given in [1208, p. 116]. See also [201, p. 69], [328], [730, p. 201], or [996, p. 478]. $\qquad\square$

Equality in Corollary 8.6.19 is discussed in [328].

8.7 Facts on Range and Rank

Fact 8.7.1. Let $A, B \in \mathbb{F}^{n \times n}$, and assume that A and B are positive semidefinite. Then, there exists $\alpha > 0$ such that $A \le \alpha B$ if and only if $\mathcal{R}(A) \subseteq \mathcal{R}(B)$. In this case, $\operatorname{rank} A \le \operatorname{rank} B$.

Proof: Use Theorem 8.6.2 and Corollary 8.6.11.

Fact 8.7.2. Let $A, B \in \mathbb{F}^{n \times n}$. Then,

$$\mathcal{R}(A) + \mathcal{R}(B) = \mathcal{R}[(AA^* + BB^*)^{1/2}].$$

Proof: This result follows from Fact 2.11.1 and Theorem 2.4.3.

Remark: See [42].

Fact 8.7.3. Let $A, B \in \mathbb{F}^{n \times n}$, and assume that A and B are positive semidefinite. Then, $(A + B)(A + B)^+$ is the projector onto the subspace $\mathcal{R}(A) + \mathcal{R}(B) = \operatorname{span}[\mathcal{R}(A) \cup \mathcal{R}(B)]$.

Proof: Use Fact 2.9.13 and Fact 8.7.5.

Remark: See Fact 6.4.50.

Fact 8.7.4. Let $A \in \mathbb{F}^{n \times n}$, and assume that $A + A^* \geq 0$. Then, the following statements hold:

i) $\mathcal{N}(A) = \mathcal{N}(A + A^*) \cap \mathcal{N}(A - A^*)$.

ii) $\mathcal{R}(A) = \mathcal{R}(A + A^*) + \mathcal{R}(A - A^*)$.

iii) $\operatorname{rank} A = \operatorname{rank} \begin{bmatrix} A + A^* & A - A^* \end{bmatrix}$.

Proof: Statements *i)* and *ii)* follow from Fact 8.7.5, while statement *iii)* follows from Fact 8.7.6.

Fact 8.7.5. Let $A, B \in \mathbb{F}^{n \times n}$, assume that A is positive semidefinite, and assume that B is either positive semidefinite or skew Hermitian. Then, the following statements hold:

i) $\mathcal{N}(A + B) = \mathcal{N}(A) \cap \mathcal{N}(B)$.

ii) $\mathcal{R}(A + B) = \mathcal{R}(A) + \mathcal{R}(B)$.

Proof: Use $[(\mathcal{N}(A) \cap \mathcal{N}(B)]^{\perp} = \mathcal{R}(A) + \mathcal{R}(B)$.

Fact 8.7.6. Let $A, B \in \mathbb{F}^{n \times n}$, assume that A is positive semidefinite, and assume that B is either positive semidefinite or skew Hermitian. Then,

$$\operatorname{rank}(A + B) = \operatorname{rank} \begin{bmatrix} A & B \end{bmatrix} = \operatorname{rank} \begin{bmatrix} A \\ B \end{bmatrix}$$

If, in addition, B is positive semidefinite, then

$$\operatorname{rank} \begin{bmatrix} A & B \\ 0 & A \end{bmatrix} = \operatorname{rank} \begin{bmatrix} A & A + B \\ 0 & A \end{bmatrix} = \operatorname{rank} A + \operatorname{rank}(A + B).$$

Proof: Using Fact 8.7.5,

$$\mathcal{R}\left(\begin{bmatrix} A & B \end{bmatrix} \right) = \mathcal{R}\left(\begin{bmatrix} A & B \end{bmatrix} \begin{bmatrix} A \\ B^* \end{bmatrix} \right) = \mathcal{R}(A^2 + BB^*)$$

$$= \mathcal{R}(A^2) + \mathcal{R}(BB^*) = \mathcal{R}(A) + \mathcal{R}(B) = \mathcal{R}(A + B).$$

Alternatively, for the case in which B is positive semidefinite, it follows from Fact 6.5.6 that

$$\operatorname{rank} \begin{bmatrix} A & B \end{bmatrix} = \operatorname{rank} \begin{bmatrix} A + B & B \end{bmatrix}$$
$$= \operatorname{rank}(A + B) + \operatorname{rank}[B - (A + B)(A + B)^+ B].$$

Next, note that

$$\operatorname{rank}[B - (A + B)(A + B)^+ B] = \operatorname{rank}\left(B^{1/2}[I - (A + B)(A + B)^+] B^{1/2} \right)$$

$$\leq \operatorname{rank}\left(B^{1/2}[I - BB^+] B^{1/2} \right) = 0.$$

Fact 8.7.7. Let $A, B \in \mathbb{F}^{n \times n}$, and assume that A and B are positive semidefinite. Then,

$$\operatorname{rank} \begin{bmatrix} A & B \\ 0 & A \end{bmatrix} = \operatorname{rank} \begin{bmatrix} A & A + B \\ 0 & A \end{bmatrix} = \operatorname{rank} A + \operatorname{rank}(A + B).$$

Proof: Use Theorem 8.3.5 to simultaneously diagonalize A and B.

Fact 8.7.8. Let $A \in \mathbb{F}^{n \times n}$, and let $S \subseteq \{1, \ldots, n\}$. If A is either positive semidefinite or an irreducible, singular M-matrix, then the following statements hold:

i) If $\alpha \subset \{1, \ldots, n\}$, then

$$\operatorname{rank} A \leq \operatorname{rank} A_{(\alpha)} + \operatorname{rank} A_{(\alpha\sim)}.$$

ii) If $\alpha, \beta \subseteq \{1, \ldots, n\}$, then

$$\operatorname{rank} A_{(\alpha \cup \beta)} \leq \operatorname{rank} A_{(\alpha)} + \operatorname{rank} A_{(\beta)} - \operatorname{rank} A_{(\alpha \cap \beta)}.$$

iii) If $1 \leq k \leq n - 1$, then

$$k \sum_{\{\alpha:\, \operatorname{card}(\alpha)=k+1\}} \det A_{(\alpha)} \leq (n - k) \sum_{\{\alpha:\, \operatorname{card}(\alpha)=k\}} \det A_{(\alpha)}.$$

If, in addition, A is either positive definite, a nonsingular M-matrix, or totally positive, then all three inclusions hold as equalities.

Proof: See [963].

Remark: See Fact 8.13.37.

Remark: Totally positive means that every subdeterminant of A is positive. See Fact 11.18.23.

8.8 Facts on Structured Positive-Semidefinite Matrices

Fact 8.8.1. Let $\phi \colon \mathbb{R} \mapsto \mathbb{C}$, and assume that, for all $x_1, \ldots, x_n \in \mathbb{R}$, the matrix $A \in \mathbb{C}^{n \times n}$, where $A_{(i,j)} \triangleq \phi(x_i - x_j)$, is positive semidefinite. (The function ϕ is *positive semidefinite*.) Then, the following statements hold:

i) For all $x_1, x_2 \in \mathbb{R}$, it follows that

$$|\phi(x_1) - \phi(x_2)|^2 \leq 2\phi(0)\operatorname{Re}[\phi(0) - \phi(x_1 - x_2)].$$

ii) The function $\psi \colon \mathbb{R} \mapsto \mathbb{C}$, where, for all $x \in \mathbb{R}$, $\psi(x) \triangleq \overline{\phi(x)}$, is positive semidefinite.

iii) For all $\alpha \in \mathbb{R}$, the function $\psi \colon \mathbb{R} \mapsto \mathbb{C}$, where, for all $x \in \mathbb{R}$, $\psi(x) \triangleq \phi(\alpha x)$, is positive semidefinite.

iv) The function $\psi \colon \mathbb{R} \mapsto \mathbb{C}$, where, for all $x \in \mathbb{R}$, $\psi(x) \triangleq |\phi(x)|$, is positive semidefinite.

v) The function $\psi \colon \mathbb{R} \mapsto \mathbb{C}$, where, for all $x \in \mathbb{R}$, $\psi(x) \triangleq \operatorname{Re} \phi(x)$, is positive semidefinite.

vi) If $\phi_1 \colon \mathbb{R} \mapsto \mathbb{C}$ and $\phi_2 \colon \mathbb{R} \mapsto \mathbb{C}$ are positive semidefinite, then $\phi_3 \colon \mathbb{R} \mapsto \mathbb{C}$, where, for all $x \in \mathbb{R}$, $\phi_3(x) \triangleq \phi_1(x)\phi_2(x)$, is positive semidefinite.

vii) If $\phi_1 \colon \mathbb{R} \mapsto \mathbb{C}$ and $\phi_2 \colon \mathbb{R} \mapsto \mathbb{C}$ are positive semidefinite and α_1, α_2 are positive numbers, then $\phi_3 \colon \mathbb{R} \mapsto \mathbb{C}$, where, for all $x \in \mathbb{R}$, $\phi_3(x) \triangleq \alpha_1\phi_1(x) + \alpha_2\phi_2(x)$, is positive semidefinite.

viii) Let $\phi\colon \mathbb{R} \mapsto \mathbb{C}$, and assume that ϕ is bounded and continuous. Furthermore, for all $x, y \in \mathbb{R}$, define $K\colon \mathbb{R} \times \mathbb{R} \mapsto \mathbb{C}$ by $K(x, y) \triangleq \phi(x - y)$. Then, ϕ is positive semidefinite if and only if, for every continuous integrable function $f\colon \mathbb{R} \mapsto \mathbb{C}$, it follows that

$$\int_{\mathbb{R}^2} K(x, y) f(x) \overline{f(y)} \, \mathrm{d}x \, \mathrm{d}y \geq 0.$$

Proof: See [205, pp. 141–144].

Remark: The function K is a *kernel function* associated with a reproducing kernel space. See [560] for extensions to vector arguments. For applications, see [1205] and Fact 8.8.4.

Fact 8.8.2. Let $a_1, \ldots, a_n$ be positive numbers, and define $A \in \mathbb{R}^{n \times n}$ by either of the following expressions:

i) $A_{(i,j)} \triangleq \min\{a_i, a_j\}$.

ii) $A_{(i,j)} \triangleq \dfrac{1}{\max\{a_i, a_j\}}$.

iii) $A_{(i,j)} \triangleq \dfrac{a_i}{a_j}$, where $a_1 \leq \cdots \leq a_n$.

iv) $A_{(i,j)} \triangleq \dfrac{a_i^p - a_j^p}{a_i - a_j}$, where $p \in [0, 1]$.

v) $A_{(i,j)} \triangleq \dfrac{a_i^p + a_j^p}{a_i + a_j}$, where $p \in [-1, 1]$.

vi) $A_{(i,j)} \triangleq \dfrac{\log a_i - \log a_j}{a_i - a_j}$.

Then, A is positive semidefinite. If, in addition, α is a positive number, then $A^{\circ \alpha}$ is positive semidefinite.

Proof: See [203], [205, pp. 153, 178, 189], and [432, p. 90].

Remark: The matrix A in *iii*) is the Schur product of the matrices defined in *i*) and *ii*).

Fact 8.8.3. Let $a_1 < \cdots < a_n$ be positive numbers, and define $A \in \mathbb{R}^{n \times n}$ by $A_{(i,j)} \triangleq \min\{a_i, a_j\}$. Then, A is positive definite,

$$\det A = \prod_{i=1}^{n} (a_i - a_{i-1}),$$

and, for all $x \in \mathbb{R}^n$,

$$x^{\mathrm{T}} A^{-1} x = \sum_{i=1}^{n} \frac{[x_{(i)} - x_{(i-1)}]^2}{a_i - a_{i-1}},$$

where $a_0 \triangleq 0$ and $x_0 \triangleq 0$.

Remark: The matrix A is a covariance matrix arising in the theory of Brownian motion. See [691, p. 132] and [1489, p. 50].

Fact 8.8.4. Let $a_1, \ldots, a_n \in \mathbb{R}$, and define $A \in \mathbb{C}^{n \times n}$ by either of the following expressions:

i) $A_{(i,j)} \triangleq \frac{1}{1+\jmath(a_i-a_j)}$.

ii) $A_{(i,j)} \triangleq \frac{1}{1-\jmath(a_i-a_j)}$.

iii) $A_{(i,j)} \triangleq \frac{1}{1+(a_i-a_j)^2}$.

iv) $A_{(i,j)} \triangleq \frac{1}{1+|a_i-a_j|}$.

v) $A_{(i,j)} \triangleq e^{\jmath(a_i-a_j)}$.

vi) $A_{(i,j)} \triangleq \cos(a_i - a_j)$.

vii) $A_{(i,j)} \triangleq \frac{\sin[(a_i-a_j)]}{a_i-a_j}$.

viii) $A_{(i,j)} \triangleq \frac{a_i-a_j}{\sinh[(a_i-a_j)]}$.

ix) $A_{(i,j)} \triangleq \frac{\sinh p(a_i-a_j)}{\sinh(a_i-a_j)}$, where $p \in (0,1)$.

x) $A_{(i,j)} \triangleq \frac{\tanh[(a_i-a_j)]}{a_i-a_j}$.

xi) $A_{(i,j)} \triangleq \frac{\sinh[(a_i-a_j)]}{(a_i-a_j)[\cosh(a_i-a_j)+p]}$, where $p \in (-1,1]$.

xii) $A_{(i,j)} \triangleq \frac{1}{\cosh(a_i-a_j)+p}$, where $p \in (-1,1]$.

xiii) $A_{(i,j)} \triangleq \frac{\cosh p(a_i-a_j)}{\cosh(a_i-a_j)}$, where $p \in [-1,1]$.

xiv) $A_{(i,j)} \triangleq e^{-(a_i-a_j)^2}$.

xv) $A_{(i,j)} \triangleq e^{-|a_i-a_j|^p}$, where $p \in [0,2]$.

xvi) $A_{(i,j)} \triangleq \frac{1}{1+|a_i-a_j|}$.

xvii) $A_{(i,j)} \triangleq \frac{1+p(a_i-a_j)^2}{1+q(a_i-a_j)^2}$, where $0 \le p \le q$.

xviii) $A_{(i,j)} \triangleq \operatorname{tr} e^{B+\jmath(a_i-a_j)C}$, where $B, C \in \mathbb{C}^{n \times n}$ are Hermitian and commute.

Then, A is positive semidefinite. Finally, if, α is a nonnegative number and A is defined by either ix), x), xi), $xiii$), xvi), or $xvii$), then $A^{\circ \alpha}$ is positive semidefinite.

Proof: See [205, pp. 141–144, 153, 177, 188], [220], [432, p. 90], and [728, pp. 400, 401, 456, 457, 462, 463].

Remark: In each case, A is associated with a positive-semidefinite function. See Fact 8.8.1.

Remark: xv) is related to the Bessis-Moussa-Villani conjecture. See Fact 8.12.31 and Fact 8.12.32.

Problem: In each case, determine rank A and determine when A is positive definite.

Fact 8.8.5. Define $A \in \mathbb{R}^{n \times n}$ by either of the following expressions:

i) $A_{(i,j)} \triangleq \binom{i+j}{i}$.

ii) $A_{(i,j)} \triangleq (i+j)!$.

iii) $A_{(i,j)} \triangleq \min\{i,j\}$.

iv) $A_{(i,j)} \triangleq \gcd\{i,j\}$.

v) $A_{(i,j)} \triangleq \frac{i}{j}$.

Then, A is positive semidefinite. If, in addition, α is a nonnegative number, then $A^{\circ \alpha}$ is positive semidefinite.

Remark: Fact 8.22.2 guarantees the weaker result that $A^{\circ \alpha}$ is positive semidefinite for all $\alpha \in [0, n-2]$.

Remark: i) is the *Pascal matrix*. See [5, 203, 460]. The fact that A is positive semidefinite follows from the equality

$$\binom{i+j}{i} = \sum_{k=0}^{\min\{i,j\}} \binom{i}{k}\binom{j}{k}.$$

Remark: The matrix defined in v), which is a special case of iii) of Fact 8.8.2, is the *Lehmer matrix*.

Remark: The determinant of A defined in iv) can be expressed in terms of the *Euler totient function*. See [69, 257].

Fact 8.8.6. Let $a_1, \ldots, a_n \geq 0$ and $p \in \mathbb{R}$, assume that either $a_1, \ldots, a_n$ are positive or p is positive, and, for all $i, j \in \{1, \ldots, n\}$, define $A \in \mathbb{R}^{n \times n}$ by

$$A_{(i,j)} \triangleq (a_i a_j)^p.$$

Then, A is positive semidefinite.

Proof: Let $a \triangleq \begin{bmatrix} a_1 & \cdots & a_n \end{bmatrix}^{\mathrm{T}}$ and $A \triangleq a^{\circ p} a^{\circ p\mathrm{T}}$.

Fact 8.8.7. Let $a_1, \ldots, a_n > 0$, let $\alpha > 0$, and, for all $i, j \in \{1, \ldots, n\}$, define $A \in \mathbb{R}^{n \times n}$ by

$$A_{(i,j)} \triangleq \frac{1}{(a_i + a_j)^\alpha}.$$

Then, A is positive semidefinite.

Proof: See [203], [205, pp. 24, 25], or [1119].

Remark: See Fact 5.11.12.

Remark: For $\alpha = 1$, A is a Cauchy matrix. See Fact 3.22.9.

Fact 8.8.8. Let $a_1, \ldots, a_n > 0$, let $r \in [-1, 1]$, and, for all $i, j \in \{1, \ldots, n\}$, define $A \in \mathbb{R}^{n \times n}$ by

$$A_{(i,j)} \triangleq \frac{a_i^r + a_j^r}{a_i + a_j}.$$

Then, A is positive semidefinite.

Proof: See [1521, p. 74].

Fact 8.8.9. Let $a_1, \ldots, a_n > 0$, let $q > 0$, let $p \in [-q, q]$, and, for all $i, j \in \{1, \ldots, n\}$, define $A \in \mathbb{R}^{n \times n}$ by

$$A_{(i,j)} \triangleq \frac{a_i^p + a_j^p}{a_i^q + a_j^q}.$$

Then, A is positive semidefinite.

Proof: Let $r = p/q$ and $b_i = a_i^q$. Then, $A_{(i,j)} = (b_i^r + b_j^r)/(b_i + b_j)$. Now, use Fact 8.8.8. See [1004] for the case $q \geq p \geq 0$.

Remark: The case $q = 1$ and $p = 0$ yields a Cauchy matrix. In the case $n = 2$, $A \geq 0$ yields Fact 1.12.33.

Problem: When is A positive definite?

Fact 8.8.10. Let $a_1, \ldots, a_n > 0$, let $p \in (-2, 2]$, and define $A \in \mathbb{R}^{n \times n}$ by

$$A_{(i,j)} \triangleq \frac{1}{a_i^2 + p a_i a_j + a_j^2}.$$

Then, A is positive semidefinite.

Proof: See [208].

Fact 8.8.11. Let $a_1, \ldots, a_n > 0$, let $p \in (-1, \infty)$, and define $A \in \mathbb{R}^{n \times n}$ by

$$A_{(i,j)} \triangleq \frac{1}{a_i^3 + p(a_i^2 a_j + a_i a_j^2) + a_j^3}.$$

Then, A is positive semidefinite.

Proof: See [208].

Fact 8.8.12. Let $a_1, \ldots, a_n > 0$, $p \in [-1, 1]$, $q \in (-2, 2]$, and, for all $i, j \in \{1, \ldots, n\}$, define $A \in \mathbb{R}^{n \times n}$ by

$$A_{(i,j)} \triangleq \frac{a_i^p + a_j^p}{a_i^2 + q a_i a_j + a_j^2}.$$

Then, A is positive semidefinite.

Proof: See [1518] or [1521, p. 76].

Fact 8.8.13. Let $A \in \mathbb{F}^{n \times n}$, and assume that A is Hermitian, $A_{(i,i)} > 0$ for all $i \in \{1, \ldots, n\}$, and, for all $i, j \in \{1, \ldots, n\}$,

$$|A_{(i,j)}| < \tfrac{1}{n-1}\sqrt{A_{(i,i)} A_{(j,j)}}.$$

Then, A is positive definite.

Proof: Note that

$$x^* A x = \sum_{i=1}^{n-1} \sum_{j=i+1}^{n} \begin{bmatrix} x_{(i)} \\ x_{(j)} \end{bmatrix}^* \begin{bmatrix} \frac{1}{n-1} A_{(i,i)} & A_{(i,j)} \\ \overline{A_{(i,j)}} & \frac{1}{n-1} A_{(j,j)} \end{bmatrix} \begin{bmatrix} x_{(i)} \\ x_{(j)} \end{bmatrix}.$$

Remark: This result is due to Roup.

Fact 8.8.14. Let $A \in \mathbb{R}^{n \times n}$, assume that A is positive semidefinite, assume that $A_{(i,i)} > 0$ for all $i \in \{1, \ldots, n\}$, and define $B \in \mathbb{R}^{n \times n}$ by

$$B_{(i,j)} \triangleq \frac{A_{(i,j)}}{\mu_\alpha(A_{(i,i)}, A_{(j,j)})},$$

where, for positive scalars α, x, y,

$$\mu_\alpha(x, y) \triangleq \left[\tfrac{1}{2}(x^\alpha + y^\alpha)\right]^{1/\alpha}.$$

Then, B is positive semidefinite. If, in addition, A is positive definite, then B is positive definite. In particular, letting $\alpha \downarrow 0$, $\alpha = 1$, and $\alpha \to \infty$, respectively, the matrices $C, D, E \in \mathbb{R}^{n \times n}$ defined by

$$C_{(i,j)} \triangleq \frac{A_{(i,j)}}{\sqrt{A_{(i,i)} A_{(j,j)}}},$$

$$D_{(i,j)} \triangleq \frac{2 A_{(i,j)}}{A_{(i,i)} + A_{(j,j)}},$$

$$E_{(i,j)} \triangleq \frac{A_{(i,j)}}{\max\{A_{(i,i)}, A_{(j,j)}\}}$$

are positive semidefinite. Finally, if A is positive definite, then C, D, and E are positive definite.

Proof: See [1180].

Remark: The assumption that all of the diagonal entries of A are positive can be weakened. See [1180].

Remark: See Fact 1.12.34.

Problem: Extend this result to Hermitian matrices.

Fact 8.8.15. Let $\alpha, \beta, \gamma \in [0, \pi]$, and define $A \in \mathbb{R}^{3 \times 3}$ by

$$A = \begin{bmatrix} 1 & \cos\alpha & \cos\gamma \\ \cos\alpha & 1 & \cos\beta \\ \cos\gamma & \cos\beta & 1 \end{bmatrix}.$$

Then, A is positive semidefinite if and only if the following conditions are satisfied:

i) $\alpha \leq \beta + \gamma$.

ii) $\beta \leq \alpha + \gamma$.

iii) $\gamma \leq \alpha + \beta$.

iv) $\alpha + \beta + \gamma \leq 2\pi$.

Furthermore, A is positive definite if and only if all of these inequalities are strict.

Proof: See [153].

Fact 8.8.16. Let $\lambda_1, \ldots, \lambda_n \in \mathbb{C}$, assume that, for all $i \in \{1, \ldots, n\}$, $\operatorname{Re} \lambda_i < 0$, and, for all $i, j \in \{1, \ldots, n\}$, define $A \in \mathbb{C}^{n \times n}$ by

$$A_{(i,j)} \triangleq \frac{-1}{\overline{\lambda_i} + \lambda_j}.$$

Then, A is positive definite.

Proof: Note that $A = 2B \circ (1_{n \times n} - C)^{\circ -1}$, where $B_{(i,j)} = \frac{1}{(\overline{\lambda_i} - 1)(\lambda_j - 1)}$ and $C_{(i,j)} = \frac{(\overline{\lambda_i} + 1)(\lambda_j + 1)}{(\overline{\lambda_i} - 1)(\lambda_j - 1)}$. Then, note that B is positive semidefinite and that $(1_{n \times n} - C)^{\circ -1} = 1_{n \times n} + C + C^{\circ 2} + C^{\circ 3} + \cdots$.)

Remark: A is the solution of a Lyapunov equation. See Fact 12.21.18 and Fact 12.21.19.

Remark: A is a Cauchy matrix. See Fact 3.18.4, Fact 3.22.9, and Fact 3.22.10.

Remark: A Cauchy matrix is also a Gram matrix defined in terms of the inner product of the functions $f_i(t) = e^{-\lambda_i t}$. See [205, p. 3].

Fact 8.8.17. Let $\lambda_1, \ldots, \lambda_n \in \mathrm{OUD}$, and let $w_1, \ldots, w_n \in \mathbb{C}$. Then, there exists a holomorphic function $\phi \colon \mathrm{OUD} \mapsto \mathrm{OUD}$ such that $\phi(\lambda_i) = w_i$ for all $i \in \{1, \ldots, n\}$ if and only if $A \in \mathbb{C}^{n \times n}$ is positive semidefinite, where, for all $i, j \in \{1, \ldots, n\}$,

$$A_{(i,j)} \triangleq \frac{1 - \overline{w_i} w_j}{1 - \overline{\lambda_i} \lambda_j}.$$

Proof: See [1010].

Remark: A is a *Pick matrix*.

Fact 8.8.18. Let $\alpha_0, \ldots, \alpha_n > 0$, and define the tridiagonal matrix $A \in \mathbb{R}^{n \times n}$ by

$$A \triangleq \begin{bmatrix} \alpha_0 + \alpha_1 & -\alpha_1 & 0 & 0 & \cdots & 0 \\ -\alpha_1 & \alpha_1 + \alpha_2 & -\alpha_2 & 0 & \cdots & 0 \\ 0 & -\alpha_2 & \alpha_2 + \alpha_3 & -\alpha_3 & \cdots & 0 \\ \vdots & \vdots & \vdots & \vdots & \ddots & \vdots \\ 0 & 0 & 0 & 0 & \cdots & \alpha_{n-1} + \alpha_n \end{bmatrix}.$$

Then, A is positive definite.

Proof: For $k = 2, \ldots, n$, the $k \times k$ leading principal subdeterminant of A is given by $\left[\sum_{i=0}^{k} \alpha_i^{-1} \right] \alpha_0 \alpha_1 \cdots \alpha_k$. See [150, p. 115].

Remark: A is a stiffness matrix arising in structural analysis.

Remark: See Fact 3.19.3.

8.9 Facts on Identities and Inequalities for One Matrix

Fact 8.9.1. Let $n \leq 3$, let $A \in \mathbb{F}^{n \times n}$, and assume that A is positive semidefinite. Then, $|A|$ is positive semidefinite.

Proof: See [989].

Remark: $|A|$ denotes the matrix whose entries are the absolute values of the entries of A.

Remark: This result does not hold for $n \geq 4$. Let

$$
A = \begin{bmatrix}
1 & \frac{1}{\sqrt{3}} & 0 & -\frac{1}{\sqrt{3}} \\
\frac{1}{\sqrt{3}} & 1 & \frac{1}{\sqrt{3}} & 0 \\
0 & \frac{1}{\sqrt{3}} & 0 & \frac{1}{\sqrt{3}} \\
-\frac{1}{\sqrt{3}} & 0 & \frac{1}{\sqrt{3}} & 1
\end{bmatrix}.
$$

Then, $\mathrm{mspec}(A) = \{1 - \sqrt{6}/3, 1 - \sqrt{6}/3, 1 + \sqrt{6}/3, 1 + \sqrt{6}/3\}_{\mathrm{ms}}$, whereas $\mathrm{mspec}(|A|)$
$= \{1, 1, 1 - \sqrt{12}/3, 1 + \sqrt{12}/3\}_{\mathrm{ms}}$.

Fact 8.9.2. Let $x \in \mathbb{F}^n$. Then,

$$
xx^* \leq x^*xI.
$$

Fact 8.9.3. Let $x \in \mathbb{F}^n$, assume that x is nonzero, and define $A \triangleq x^*xI - xx^*$. Then, A is positive semidefinite, $\mathrm{mspec}(A) = \{x^*x, \ldots, x^*x, 0\}_{\mathrm{ms}}$, and $\mathrm{rank}\, A = n - 1$.

Fact 8.9.4. Let $x, y \in \mathbb{F}^n$, assume that x and y are linearly independent, and define $A \triangleq (x^*x + y^*y)I - xx^* - yy^*$. Then, A is positive definite. Now, let $\mathbb{F} = \mathbb{R}$. Then,

$$
\begin{aligned}
\mathrm{mspec}(A) = \{ & x^\mathrm{T}x + y^\mathrm{T}y, \ldots, x^\mathrm{T}x + y^\mathrm{T}y, \\
& \tfrac{1}{2}(x^\mathrm{T}x + y^\mathrm{T}y) + \sqrt{\tfrac{1}{4}(x^\mathrm{T}x - y^\mathrm{T}y)^2 + (x^\mathrm{T}y)^2}, \\
& \tfrac{1}{2}(x^\mathrm{T}x + y^\mathrm{T}y) - \sqrt{\tfrac{1}{4}(x^\mathrm{T}x - y^\mathrm{T}y)^2 + (x^\mathrm{T}y)^2} \}_{\mathrm{ms}}.
\end{aligned}
$$

Proof: To show that A is positive definite, write $A = B + C$, where $B \triangleq x^*xI - xx^*$ and $C \triangleq y^*yI - yy^*$. Then, using Fact 8.9.3 it follows that $\mathcal{N}(B) = \mathrm{span}\,\{x\}$ and $\mathcal{N}(C) = \mathrm{span}\,\{y\}$. Now, it follows from Fact 8.7.5 that $\mathcal{N}(A) = \mathcal{N}(B) \cap \mathcal{N}(C) = \{0\}$. Therefore, A is nonsingular and thus positive definite. The expression for $\mathrm{mspec}(A)$ follows from Fact 4.9.17.

Fact 8.9.5. Let $x_1, \ldots, x_n \in \mathbb{R}^3$, assume that $\mathrm{span}\,\{x_1, \ldots, x_n\} = \mathbb{R}^3$, and define $A \triangleq \sum_{i=1}^n (x_i^\mathrm{T}x_i I - x_i x_i^\mathrm{T})$. Then, A is positive definite. Furthermore,

$$
\lambda_1(A) < \lambda_2(A) + \lambda_3(A)
$$

and

$$
\mathrm{d}_1(A) < \mathrm{d}_2(A) + \mathrm{d}_3(A).
$$

Proof: Suppose that $d_1(A) = A_{(1,1)}$. Then, $d_2(A) + d_3(A) - d_1(A) = 2\sum_{i=1}^n x_{i(3)}^2 > 0$. Now, let $S \in \mathbb{R}^{3\times3}$ be such that $SAS^T = \sum_{i=1}^n (\hat{x}_i^T \hat{x}_i I - \hat{x}_i \hat{x}_i^T)$ is diagonal, where, for $i = 1, \ldots, n$, $\hat{x}_i \triangleq Sx_i$. Then, for $i = 1, 2, 3$, $d_i(A) = \lambda_i(A)$.

Remark: A is the inertia matrix for a rigid body consisting of n discrete particles. For a homogeneous continuum body $\mathcal{B}$ whose density is ρ, the inertia matrix is given by

$$I = \rho \iiint_{\mathcal{B}} (r^T r I - r r^T)\, \mathrm{d}x\mathrm{d}y\mathrm{d}z,$$

where $r \triangleq \begin{bmatrix} x \\ y \\ z \end{bmatrix}$.

Remark: The eigenvalues and diagonal entries of A represent the lengths of the sides of triangles. See Fact 1.13.17 and [1096, p. 220].

Fact 8.9.6. Let $A \in \mathbb{F}^{2\times2}$, assume that A is positive semidefinite and nonzero, and define $B \in \mathbb{F}^{2\times2}$ by

$$B \triangleq \left(\operatorname{tr} A + 2\sqrt{\det A}\right)^{-1/2}\left(A + \sqrt{\det A}\, I\right).$$

Then, $B = A^{1/2}$.

Proof: See [644, pp. 84, 266, 267].

Fact 8.9.7. Let $A \in \mathbb{F}^{n\times n}$, and assume that A is Hermitian. Then,

$$\operatorname{rank} A = \nu_-(A) + \nu_+(A)$$

and

$$\operatorname{def} A = \nu_0(A).$$

Fact 8.9.8. Let $A \in \mathbb{F}^{n\times n}$, assume that A is positive semidefinite, and assume there exists $i \in \{1, \ldots, n\}$ such that $A_{(i,i)} = 0$. Then, $\operatorname{row}_i(A) = 0$ and $\operatorname{col}_i(A) = 0$.

Fact 8.9.9. Let $A \in \mathbb{F}^{n\times n}$, and assume that A is positive semidefinite. Then, $A_{(i,i)} \geq 0$ for all $i \in \{1, \ldots, n\}$, and $|A_{(i,j)}|^2 \leq A_{(i,i)}A_{(j,j)}$ for all $i, j \in \{1, \ldots, n\}$.

Fact 8.9.10. Let $A \in \mathbb{F}^{n\times n}$. Then, $A \geq 0$ if and only if $A \geq -A$.

Fact 8.9.11. Let $A \in \mathbb{F}^{n\times n}$, and assume that A is Hermitian. Then, $A^2 \geq 0$.

Fact 8.9.12. Let $A \in \mathbb{F}^{n\times n}$, and assume that A is skew Hermitian. Then, $A^2 \leq 0$.

Fact 8.9.13. Let $A \in \mathbb{F}^{n\times n}$, and let $\alpha > 0$. Then,

$$A^2 + A^{2*} \leq \alpha AA^* + \tfrac{1}{\alpha}A^*A.$$

Equality holds if and only if $\alpha A = A^*$.

Fact 8.9.14. Let $A \in \mathbb{F}^{n\times n}$. Then,

$$(A - A^*)^2 \leq 0 \leq (A + A^*)^2 \leq 2(AA^* + A^*A).$$

Fact 8.9.15. Let $A \in \mathbb{F}^{n \times n}$, and let $\alpha > 0$. Then,

$$A + A^* \leq \alpha I + \alpha^{-1} A A^*.$$

Equality holds if and only if $A = \alpha I$.

Fact 8.9.16. Let $A \in \mathbb{F}^{n \times n}$, and assume that A is positive definite. Then,

$$2I \leq A + A^{-1}.$$

Equality holds if and only if $A = I$. Furthermore,

$$2n \leq \operatorname{tr} A + \operatorname{tr} A^{-1}.$$

Fact 8.9.17. Let $A \in \mathbb{F}^{n \times n}$, and assume that A is positive definite. Then,

$$\left(1_{1 \times n} A^{-1} 1_{n \times 1}\right)^{-1} 1_{n \times n} \leq A.$$

Proof: Set $B = 1_{n \times n}$ in Fact 8.22.14. See [1528].

Fact 8.9.18. Let $A \in \mathbb{F}^{n \times n}$, and assume that A is positive definite. Then, $\left[\begin{smallmatrix} A & I \\ I & A^{-1} \end{smallmatrix}\right]$ is positive semidefinite.

Fact 8.9.19. Let $A \in \mathbb{F}^{n \times n}$, and assume that A is Hermitian. Then, $A^2 \leq A$ if and only if $0 \leq A \leq I$.

Fact 8.9.20. Let $A \in \mathbb{F}^{n \times n}$, and assume that A is Hermitian. Then, $\alpha I + A \geq 0$ if and only if $\alpha \geq -\lambda_{\min}(A)$. Furthermore,

$$A^2 + A + \tfrac{1}{4} I \geq 0.$$

Fact 8.9.21. Let $A \in \mathbb{F}^{n \times m}$. Then, $A A^* \leq I_n$ if and only if $A^* A \leq I_m$.

Fact 8.9.22. Let $A \in \mathbb{F}^{n \times n}$, and assume that either $A A^* \leq A^* A$ or $A^* A \leq A A^*$. Then, A is normal.

Proof: Use *ii*) of Corollary 8.4.10.

Fact 8.9.23. Let $A \in \mathbb{F}^{n \times n}$, and assume that A is a projector. Then,

$$0 \leq A \leq I.$$

Therefore,

$$0 \leq I \circ A \leq I.$$

Fact 8.9.24. Let $A \in \mathbb{F}^{n \times n}$, assume that A is (semisimple, Hermitian), and assume that there exists a nonnegative integer k such that $A^k = A^{k+1}$. Then, A is (idempotent, a projector).

Fact 8.9.25. Let $A \in \mathbb{F}^{n \times n}$, and assume that A is nonsingular. Then,

$$\langle A^{-1} \rangle = \langle A^* \rangle^{-1}.$$

Fact 8.9.26. Let $A \in \mathbb{F}^{n \times m}$, and assume that $A^* A$ is nonsingular. Then,

$$\langle A^* \rangle = A \langle A \rangle^{-1/2} A^*.$$

Fact 8.9.27. Let $A \in \mathbb{F}^{n \times n}$. Then, A is unitary if and only if there exists a nonsingular matrix $B \in \mathbb{F}^{n \times n}$ such that

$$A = \langle B^* \rangle^{-1/2} B.$$

If, in addition, A is real, then $\det B = \text{sign}(\det A)$.

Proof: For necessity, set $B = A$.

Remark: See Fact 3.11.31.

Fact 8.9.28. Let $A \in \mathbb{F}^{n \times n}$. Then, A is normal if and only if $\langle A \rangle = \langle A^* \rangle$.

Remark: See Fact 3.7.12.

Fact 8.9.29. Let $A \in \mathbb{F}^{n \times n}$. Then,

$$-\langle A \rangle - \langle A^* \rangle \leq A + A^* \leq \langle A \rangle + \langle A^* \rangle.$$

Proof: See [911].

Fact 8.9.30. Let $A \in \mathbb{F}^{n \times n}$, assume that A is normal, and let $\alpha, \beta \in (0, \infty)$. Then,
$$-\alpha \langle A \rangle - \beta \langle A^* \rangle \leq \langle \alpha A + \beta A^* \rangle \leq \alpha \langle A \rangle + \beta \langle A^* \rangle.$$

In particular,

$$-\langle A \rangle - \langle A^* \rangle \leq \langle A + A^* \rangle \leq \langle A \rangle + \langle A^* \rangle.$$

Proof: See [911, 1530].

Remark: See Fact 8.11.11.

Fact 8.9.31. Let $A \in \mathbb{F}^{n \times n}$. The following statements hold:

$i)$ If $A \in \mathbb{F}^{n \times n}$ is positive definite, then $I + A$ is nonsingular and the matrices $I - B$ and $I + B$ are positive definite, where $B \triangleq (I + A)^{-1}(I - A)$.

$ii)$ If $I + A$ is nonsingular and the matrices $I - B$ and $I + B$ are positive definite, where $B \triangleq (I + A)^{-1}(I - A)$, then A is positive definite.

Proof: See [476].

Remark: For additional results on the Cayley transform, see Fact 3.11.21, Fact 3.11.22, Fact 3.11.23, Fact 3.20.12, and Fact 11.21.9.

Fact 8.9.32. Let $A \in \mathbb{F}^{n \times n}$, and assume that $\frac{1}{\jmath 2}(A - A^*)$ is positive definite. Then,

$$B \triangleq \left[\tfrac{1}{2}(A + A^*) \right]^{1/2} A^{-1} A^* \left[\tfrac{1}{2}(A + A^*) \right]^{-1/2}$$

is unitary.

Proof: See [479].

Remark: A is *strictly dissipative* if $\frac{1}{\jmath 2}(A - A^*)$ is negative definite. A is strictly dissipative if and only if $-\jmath A$ is dissipative. See [477, 478].

Remark: $A^{-1} A^*$ is similar to a unitary matrix. See Fact 3.11.5.

Remark: See Fact 8.13.11 and Fact 8.18.12.

Fact 8.9.33. Let $A \in \mathbb{R}^{n \times n}$, assume that A is positive definite, assume that $A \leq I$, and define the sequence $(B_k)_{k=0}^{\infty}$ by $B_0 \triangleq 0$ and

$$B_{k+1} \triangleq B_k + \tfrac{1}{2}(A - B_k^2).$$

Then,
$$\lim_{k \to \infty} B_k = A^{1/2}.$$

Proof: See [174, p. 181].

Remark: See Fact 5.15.21.

Fact 8.9.34. Let $A \in \mathbb{R}^{n \times n}$, assume that A is nonsingular, and define the sequence $(B_k)_{k=0}^{\infty}$ by $B_0 \triangleq A$ and

$$B_{k+1} \triangleq \tfrac{1}{2}(B_k + B_k^{-T}).$$

Then,
$$\lim_{k \to \infty} B_k = (AA^T)^{-1/2}A.$$

Remark: The limit is a unitary matrix. See Fact 8.9.27. See [148, p. 224].

Fact 8.9.35. Let $a, b \in \mathbb{R}$, and define the symmetric, Toeplitz matrix $A \in \mathbb{R}^{n \times n}$ by

$$A \triangleq aI_n + b1_{n \times n}.$$

Then, A is positive definite if and only if $a + nb > 0$ and $a > 0$.

Remark: See Fact 2.13.12 and Fact 4.10.16.

Fact 8.9.36. Let $x_1, \ldots, x_n \in \mathbb{R}^m$, and define

$$\overline{x} \triangleq \tfrac{1}{n}\sum_{j=1}^n x_j, \qquad S \triangleq \tfrac{1}{n}\sum_{j=1}^n (x_j - \overline{x})(x_j - \overline{x})^T.$$

Then, for all $i \in \{1, \ldots, n\}$,
$$(x_i - \overline{x})(x_i - \overline{x})^T \leq (n-1)S.$$

Furthermore, equality holds if and only if all of the elements of $\{x_1, \ldots, x_n\} \backslash \{x_i\}$ are equal.

Proof: See [776, 1070, 1364].

Remark: This result is an extension of the Laguerre-Samuelson inequality. See Fact 1.17.12.

Fact 8.9.37. Let $x_1, \ldots, x_n \in \mathbb{F}^n$, and define $A \in \mathbb{F}^{n \times n}$ by $A_{(i,j)} \triangleq x_i^* x_j$ for all $i, j \in \{1, \ldots, n\}$, and $B \triangleq \begin{bmatrix} x_1 & \cdots & x_n \end{bmatrix}$. Then, $A = B^*B$. Consequently, A is positive semidefinite and $\operatorname{rank} A = \operatorname{rank} B$. Conversely, let $A \in \mathbb{F}^{n \times n}$, and assume that A is positive semidefinite. Then, there exist $x_1, \ldots, x_n \in \mathbb{F}^n$ such that $A = B^*B$, where $B = \begin{bmatrix} x_1 & \cdots & x_n \end{bmatrix}$.

Proof: The converse is an immediate consequence of Corollary 5.4.5.

Remark: A is the *Gram matrix* of $x_1, \ldots, x_n$.

Fact 8.9.38. Let $A \in \mathbb{F}^{n \times n}$, and assume that A is positive semidefinite. Then, there exists a matrix $B \in \mathbb{F}^{n \times n}$ such that B is lower triangular, B has nonnegative diagonal entries, and $A = BB^*$. If, in addition, A is positive definite, then B is unique and has positive diagonal entries.

Remark: This result is the *Cholesky decomposition*.

Fact 8.9.39. Let $A \in \mathbb{F}^{n \times m}$, and assume that rank $A = m$. Then,

$$0 \le A(A^*A)^{-1}A^* \le I.$$

Fact 8.9.40. Let $A \in \mathbb{F}^{n \times m}$. Then, $I - A^*A$ is positive definite if and only if $I - AA^*$ is positive definite. In this case,

$$(I - A^*A)^{-1} = I + A^*(I - AA^*)^{-1}A.$$

Fact 8.9.41. Let $A \in \mathbb{F}^{n \times m}$, let α be a positive number, and define $A_\alpha \triangleq (\alpha I + A^*A)^{-1}A^*$. Then, the following statements are equivalent:

  i) $AA_\alpha = A_\alpha A$.

 ii) $AA^* = A^*A$.

Furthermore, the following statements are equivalent:

 iii) $A_\alpha A^* = A^*A_\alpha$.

 iv) $AA^*A^2 = A^2A^*A$.

Proof: See [1331].

Remark: A_α is a *regularized Tikhonov inverse*.

Fact 8.9.42. Let $A \in \mathbb{F}^{n \times n}$, and assume that A is positive definite. Then,

$$A^{-1} \le \frac{\alpha + \beta}{\alpha\beta}I - \frac{1}{\alpha\beta}A \le \frac{(\alpha + \beta)^2}{4\alpha\beta}A^{-1},$$

where $\alpha \triangleq \lambda_{\max}(A)$ and $\beta \triangleq \lambda_{\min}(A)$.

Proof: See [997].

Fact 8.9.43. Let $A \in \mathbb{F}^{n \times n}$, and assume that A is positive semidefinite. Then, the following statements hold:

  i) If $\alpha \in [0, 1]$, then
$$A^\alpha \le \alpha A + (1 - \alpha)I.$$

 ii) If $\alpha \in [0, 1]$ and A is positive definite, then
$$[\alpha A^{-1} + (1 - \alpha)I]^{-1} \le A^\alpha \le \alpha A + (1 - \alpha)I.$$

 iii) If $\alpha \ge 1$, then
$$\alpha A + (1 - \alpha)I \le A^\alpha.$$

 iv) If A is positive definite and either $\alpha \ge 1$ or $\alpha \le 0$, then
$$\alpha A + (1 - \alpha)I \le A^\alpha \le [\alpha A^{-1} + (1 - \alpha)I]^{-1}.$$

Proof: See [544, pp. 122, 123].

Remark: This result is a special case of the Young inequality. See Fact 1.11.2 and Fact 8.10.46.

Remark: See Fact 8.12.27 and Fact 8.12.28.

Fact 8.9.44. Let $A \in \mathbb{F}^{n \times n}$, and assume that A is positive definite. Then,
$$I - A^{-1} \leq \log A \leq A - I.$$

Furthermore, if $A \geq I$, then $\log A$ is positive semidefinite, and, if $A > I$, then $\log A$ is positive definite.

Proof: See Fact 1.11.22.

8.10 Facts on Identities and Inequalities for Two or More Matrices

Fact 8.10.1. Let $(A_i)_{i=1}^{\infty} \subset \mathbf{H}^n$ and $(B_i)_{i=1}^{\infty} \subset \mathbf{H}^n$, assume that, for all $i \in \mathbb{P}$, $A_i \leq B_i$, and assume that $A \triangleq \lim_{i \to \infty} A_i$ and $B \triangleq \lim_{i \to \infty} B_i$ exist. Then, $A \leq B$.

Fact 8.10.2. Let $A, B \in \mathbb{F}^{n \times n}$, assume that A and B are positive semidefinite, and assume that $A \leq B$. Then, $\mathcal{R}(A) \subseteq \mathcal{R}(B)$ and $\operatorname{rank} A \leq \operatorname{rank} B$. Furthermore, $\mathcal{R}(A) = \mathcal{R}(B)$ if and only if $\operatorname{rank} A = \operatorname{rank} B$.

Fact 8.10.3. Let $A, B \in \mathbb{F}^{n \times n}$, and assume that A and B are Hermitian. Then, the following statements hold:

i) $\lambda_{\min}(A) \leq \lambda_{\min}(B)$ if and only if $\lambda_{\min}(A)I \leq B$.

ii) $\lambda_{\max}(A) \leq \lambda_{\max}(B)$ if and only if $A \leq \lambda_{\max}(B)I$.

Fact 8.10.4. Let $A, B \in \mathbb{F}^{n \times n}$, assume that A and B are Hermitian, and consider the following conditions:

i) $A \leq B$.

ii) For all $i \in \{1, \ldots, n\}$, $\lambda_i(A) \leq \lambda_i(B)$.

iii) There exists a unitary matrix $S \in \mathbb{F}^{n \times n}$ such that $A \leq SBS^*$.

Then, i) $\implies$ ii) $\iff$ iii).

Remark: i) $\implies$ ii) is the monotonicity theorem given by Theorem 8.4.9.

Fact 8.10.5. Let $A, B \in \mathbb{F}^{n \times n}$, and assume that A and B are positive semidefinite. Then, $0 \leq A < B$ if and only if $\operatorname{sprad}(AB^{-1}) < 1$.

Fact 8.10.6. Let $A, B \in \mathbb{F}^{n \times n}$, and assume that A and B are positive definite. Then,
$$\left(A^{-1} + B^{-1}\right)^{-1} = A(A + B)^{-1}B.$$

Fact 8.10.7. Let $A, B \in \mathbb{F}^{n \times n}$, and assume that A and B are positive definite. Then,
$$(A + B)^{-1} \leq \tfrac{1}{4}(A^{-1} + B^{-1}).$$
Equivalently,
$$A + B \leq AB^{-1}A + BA^{-1}B.$$
In both inequalities, equality holds if and only if $A = B$.

Proof: See [1526, p. 168].

Remark: See Fact 1.12.4.

Fact 8.10.8. Let $A, B \in \mathbb{F}^{n \times n}$, and assume that A is positive definite, B is Hermitian, and $A + B$ is nonsingular. Then,
$$(A + B)^{-1} + (A + B)^{-1}B(A + B)^{-1} \leq A^{-1}.$$
If, in addition, B is nonsingular, then the inequality is strict.

Proof: This inequality is equivalent to $BA^{-1}B \geq 0$. See [1077].

Fact 8.10.9. Let $A, B \in \mathbb{F}^{n \times n}$, assume that A and B are positive definite, and let $\alpha \in [0, 1]$. Then,
$$\beta[\alpha A^{-1} + (1 - \alpha)B^{-1}] \leq [\alpha A + (1 - \alpha)B]^{-1},$$
where
$$\beta \triangleq \min_{\mu \in \mathrm{mspec}(A^{-1}B)} \frac{4\mu}{(1 + \mu)^2}.$$

Proof: See [1042].

Remark: This result is a reverse form of an inequality based on convexity.

Fact 8.10.10. Let $A \in \mathbb{F}^{n \times m}$ and $B \in \mathbb{F}^{m \times m}$, and assume that B is positive semidefinite. Then, $ABA^* = 0$ if and only if $AB = 0$.

Fact 8.10.11. Let $A, B \in \mathbb{F}^{n \times n}$, and assume that A and B are positive semidefinite. Then, AB is positive semidefinite if and only if AB is normal.

Fact 8.10.12. Let $A, B \in \mathbb{F}^{n \times n}$, assume that A and B are Hermitian, and assume that either i) A and B are positive semidefinite or ii) either A or B is positive definite. Then, AB is group invertible.

Proof: Use Theorem 8.3.2 and Theorem 8.3.6.

Fact 8.10.13. Let $A, B \in \mathbb{F}^{n \times n}$, assume that A and B are Hermitian, and assume that A and $AB + BA$ are (positive semidefinite, positive definite). Then, B is (positive semidefinite, positive definite).

Proof: See [205, p. 8], [903, p. 120], or [1464]. Alternatively, the result follows from Corollary 11.9.4.

Fact 8.10.14. Let $A, B, C \in \mathbb{F}^{n \times n}$, assume that A, B, and C are positive semidefinite, and assume that $A = B + C$. Then, the following statements are equivalent:

i) $\operatorname{rank} A = \operatorname{rank} B + \operatorname{rank} C$.

ii) There exists $S \in \mathbb{F}^{m \times n}$ such that $\operatorname{rank} S = m$, $\mathcal{R}(S) \cap \mathcal{N}(A) = \{0\}$, and either $B = AS^*(SAS^*)^{-1}SA$ or $C = AS^*(SAS^*)^{-1}SA$.

Proof: See [293, 339].

Fact 8.10.15. Let $A, B \in \mathbb{F}^{n \times n}$, and assume that A and B are Hermitian and nonsingular. Then, the following statements hold:

i) If every eigenvalue of AB is positive, then $\operatorname{In} A = \operatorname{In} B$.

ii) $\operatorname{In} A - \operatorname{In} B = \operatorname{In}(A - B) + \operatorname{In}(A^{-1} - B^{-1})$.

iii) If $\operatorname{In} A = \operatorname{In} B$ and $A \leq B$, then $B^{-1} \leq A^{-1}$.

Proof: See [53, 112, 1074].

Remark: The equality *ii)* is due to Styan. See [1074].

Remark: An extension to singular A and B is given by Fact 8.21.14.

Fact 8.10.16. Let $A, B \in \mathbb{F}^{n \times n}$, assume that A and B are Hermitian, and assume that $A \leq B$. Then, $A_{(i,i)} \leq B_{(i,i)}$ for all $i \in \{1, \ldots, n\}$.

Fact 8.10.17. Let $A, B \in \mathbb{F}^{n \times n}$, assume that A and B are Hermitian, and assume that $A \leq B$. Then, $\operatorname{sig} A \leq \operatorname{sig} B$.

Proof: See [400, p. 148].

Fact 8.10.18. Let $A, B \in \mathbb{F}^{n \times n}$, assume that A and B are Hermitian, and assume that $\langle A \rangle \leq B$. Then, either $A \leq B$ or $-A \leq B$.

Proof: See [1529].

Fact 8.10.19. Let $A, B \in \mathbb{F}^{n \times n}$, and assume that A is positive semidefinite and B is positive definite. Then, $A \leq B$ if and only if $AB^{-1}A \leq A$.

Fact 8.10.20. Let $A, B \in \mathbb{F}^{n \times n}$, assume that A and B are positive semidefinite, and assume that $A \leq B$. Then, there exists a matrix $S \in \mathbb{F}^{n \times n}$ such that $A = S^*BS$ and $S^*S \leq I$.

Proof: See [459, p. 269].

Fact 8.10.21. Let $A, B, C, D \in \mathbb{F}^{n \times n}$, assume that A, B, C, D are positive semidefinite, and assume that $0 < D \leq C$ and $BCB \leq ADA$. Then, $B \leq A$.

Proof: See [87, 308].

Fact 8.10.22. Let $A, B \in \mathbb{F}^{n \times n}$, and assume that A and B are positive definite. Then, there exists a unitary matrix $S \in \mathbb{F}^{n \times n}$ such that

$$\langle AB \rangle \le \tfrac{1}{2} S(A^2 + B^2)S^*.$$

Proof: See [93, 213].

Fact 8.10.23. Let $A, B \in \mathbb{F}^{n \times n}$, and assume that A and B are projectors. Then, $ABA \le B$ if and only if $AB = BA$.

Proof: See [1357].

Fact 8.10.24. Let $A, B \in \mathbb{F}^{n \times n}$, and assume that A is positive definite, $0 \le A \le I$, and B is positive definite. Then,

$$ABA \le \frac{(\alpha + \beta)^2}{4\alpha\beta} B.$$

where $\alpha \triangleq \lambda_{\min}(B)$ and $\beta \triangleq \lambda_{\max}(B)$.

Proof: See [255].

Remark: This inequality is related to Fact 1.18.6.

Fact 8.10.25. Let $A, B \in \mathbb{F}^{n \times n}$, and assume that A and B are projectors. Then,
$$(A + B)^{1/2} \le A^{1/2} + B^{1/2}$$

if and only if $AB = BA$.

Proof: See [1349, p. 30].

Fact 8.10.26. Let $A, B \in \mathbb{F}^{n \times n}$, assume that A and B are positive semidefinite, and assume that $0 \le A \le B$. Then,

$$\left(A + \tfrac{1}{4}A^2\right)^{1/2} \le \left(B + \tfrac{1}{4}B^2\right)^{1/2}.$$

Proof: See [1037].

Fact 8.10.27. Let $A \in \mathbb{F}^{n \times n}$, assume that A is positive semidefinite, and let $B \in \mathbb{F}^{l \times n}$. Then, BAB^* is positive definite if and only if $B(A + A^2)B^*$ is positive definite.

Proof: Diagonalize A using a unitary transformation and note that $BA^{1/2}$ and $B(A + A^2)^{1/2}$ have the same rank.

Fact 8.10.28. Let $A, B, C \in \mathbb{F}^{n \times n}$, assume that A is positive definite, and assume that B and C are positive semidefinite. Then,

$$2\mathrm{tr}\,\langle B^{1/2}C^{1/2} \rangle \le \mathrm{tr}(AB + A^{-1}C).$$

Furthermore, there exists A such that equality holds if and only if $\mathrm{rank}\,B = \mathrm{rank}\,C = \mathrm{rank}\,B^{1/2}C^{1/2}$.

Proof: See [37, 507].

Remark: A matrix A for which equality holds is given in [37].

Remark: Applications to linear systems are given in [1476].

Fact 8.10.29. Let $A, B \in \mathbb{F}^{n \times n}$, assume that A and B are positive definite, let $S \in \mathbb{F}^{n \times n}$ be such that $SAS^* = \text{diag}(\alpha_1, \ldots, \alpha_n)$ and $SBS^* = \text{diag}(\beta_1, \ldots, \beta_n)$, and define

$$C_l \triangleq S^{-1} \text{diag}(\min\{\alpha_1, \beta_1\}, \ldots, \min\{\alpha_n, \beta_n\})S^{-*}$$

and

$$C_u \triangleq S^{-1} \text{diag}(\max\{\alpha_1, \beta_1\}, \ldots, \max\{\alpha_n, \beta_n\})S^{-*}.$$

Then, C_l and C_u are independent of the choice of S, and

$$C_l \leq A \leq C_u,$$
$$C_l \leq B \leq C_u.$$

Proof: See [926].

Fact 8.10.30. Let $A, B \in \mathbf{H}^{n \times n}$. Then, $\text{glb}\{A, B\}$ exists in $\mathbf{H}^n$ with respect to the ordering "$\leq$" if and only if either $A \leq B$ or $B \leq A$.

Proof: See [806].

Remark: Let $A = \begin{bmatrix} 1 & 0 \\ 0 & 0 \end{bmatrix}$ and $B = \begin{bmatrix} 0 & 0 \\ 0 & 1 \end{bmatrix}$. Then, $C = 0$ is a lower bound for $\{A, B\}$. Furthermore, $D = \begin{bmatrix} -1 & \sqrt{2} \\ \sqrt{2} & -1 \end{bmatrix}$, which has eigenvalues $-1 - \sqrt{2}$ and $-1 + \sqrt{2}$, is also a lower bound for $\{A, B\}$ but is not comparable with C.

Fact 8.10.31. Let $A, B \in \mathbf{H}^{n \times n}$, and assume that A and B are positive semidefinite. Then, the following statements hold:

i) $\{A, B\}$ does not necessarily have a least upper bound in $\mathbf{N}^n$.

ii) If A and B are positive definite, then $\{A, B\}$ has a greatest lower bound in $\mathbf{N}^n$ if and only if A and B are comparable.

iii) If A is a projector and $0 \leq B \leq I$, then $\{A, B\}$ has a greatest lower bound in $\mathbf{N}^n$.

iv) If $A, B \in \mathbf{N}^n$ are projectors, then the greatest lower bound of $\{A, B\}$ in $\mathbf{N}^n$ is given by

$$\text{glb}\{A, B\} = 2A(A + B)^+ B,$$

which is the projector onto $\mathcal{R}(A) \cap \mathcal{R}(B)$.

v) $\text{glb}\{A, B\}$ exists in $\mathbf{N}^n$ if and only if $\text{glb}\{A, \text{glb}\{AA^+, BB^+\}\}$ and $\text{glb}\{B, \text{glb}\{AA^+, BB^+\}\}$ are comparable. In this case,

$$\text{glb}\{A, B\} = \min\{\text{glb}\{A, \text{glb}\{AA^+, BB^+\}\}, \text{glb}\{B, \text{glb}\{AA^+, BB^+\}\}\}.$$

vi) $\text{glb}\{A, B\}$ exists if and only if $\text{sh}(A, B)$ and $\text{sh}(B, A)$ are comparable, where $\text{sh}(A, B) \triangleq \lim_{\alpha \to \infty}(\alpha B) : A$. In this case,

$$\text{glb}\{A, B\} = \min\{\text{sh}(A, B), \text{sh}(B, A)\}.$$

Proof: To prove *i)*, let $A = \begin{bmatrix} 1 & 0 \\ 0 & 0 \end{bmatrix}$ and $B = \begin{bmatrix} 0 & 0 \\ 0 & 1 \end{bmatrix}$, and suppose that Z is the least upper bound for A and B. Hence, $A \leq Z \leq I$ and $B \leq Z \leq I$, and thus $Z = I$. Next, note that $X \triangleq \begin{bmatrix} 4/3 & 2/3 \\ 2/3 & 4/3 \end{bmatrix}$ satisfies $A \leq X$ and $B \leq X$. However, it is not true that $Z \leq X$, which implies that $\{A, B\}$ does not have a least upper bound. See [243, p. 11]. Statement *ii*) is given in [451, 564, 1047]. Statements *iii*) and *v*) are given in [1047]. Statement *iv*) is given in [41]. The expression for the projector onto $\mathcal{R}(A) \cap \mathcal{R}(B)$ is given in Fact 6.4.46. Statement *vi*) is given in [52].

Remark: The partially ordered cones $\mathbf{H}^n$ and $\mathbf{N}^n$ with the ordering "$\leq$" are not lattices.

Remark: $\mathrm{sh}(A, B)$ is the shorted operator, see Fact 8.21.19. However, the usage here is more general since B need not be a projector. See [52].

Remark: An alternative approach to showing that $\mathbf{N}^n$ is not a lattice is given in [926].

Remark: The cone $\mathbf{N}^n$ is a partially ordered set under the spectral order, see Fact 8.10.35.

Fact 8.10.32. Let $A_1, \ldots, A_k \in \mathbb{F}^{n \times n}$, and assume that $A_1, \ldots, A_k$ are positive definite. Then,

$$n^2 \left(\sum_{i=1}^k A_i \right)^{-1} \leq \sum_{i=1}^k A_i^{-1}.$$

Remark: This result is an extension of Fact 1.17.38.

Fact 8.10.33. Let $A_1, \ldots, A_k \in \mathbb{F}^{n \times n}$, assume that $A_1, \ldots, A_k$ are positive semidefinite, and let $p, q \in \mathbb{R}$ satisfy $1 \leq p \leq q$. Then,

$$\left(\frac{1}{k} \sum_{i=1}^k A_i^p \right)^{1/p} \leq \left(\frac{1}{k} \sum_{i=1}^k A_i^q \right)^{1/q}.$$

Proof: See [197].

Fact 8.10.34. Let $A, B \in \mathbb{F}^{n \times n}$, assume that A and B are positive semidefinite, let p be a real number, and assume that either $p \in [1, 2]$ or A and B are positive definite and $p \in [-1, 0] \cup [1, 2]$. Then,

$$[\tfrac{1}{2}(A + B)]^p \leq \tfrac{1}{2}(A^p + B^p).$$

Proof: See [879].

Fact 8.10.35. Let $A, B \in \mathbb{F}^{n \times n}$, assume that A and B are positive semidefinite, and let $p, q \in \mathbb{R}$ satisfy $p \geq q \geq 1$. Then,

$$[\tfrac{1}{2}(A^q + B^q)]^{1/q} \leq [\tfrac{1}{2}(A^p + B^p)]^{1/p}.$$

Furthermore,

$$\mu(A, B) \triangleq \lim_{p \to \infty} [\tfrac{1}{2}(A^p + B^p)]^{1/p}$$

exists and satisfies

$$A \leq \mu(A, B), \quad B \leq \mu(A, B).$$

Proof: See [175].

Remark: $\mu(A, B)$ is the least upper bound of A and B with respect to the spectral order. See [56, 817] and Fact 8.20.3.

Remark: The result does not hold for $p = 1$ and $q = 1/3$. A counterexample is $A = \begin{bmatrix} 2 & 1 \\ 1 & 2 \end{bmatrix}^3 = \begin{bmatrix} 13 & 8 \\ 8 & 5 \end{bmatrix}$, $B = \begin{bmatrix} 1 & 0 \\ 0 & 0 \end{bmatrix}$.

Fact 8.10.36. Let $A, B \in \mathbb{F}^{n \times n}$, assume that A and B are positive semidefinite, let $p \in (1, \infty)$, and let $\alpha \in [0, 1]$. Then,

$$\alpha^{1-1/p}A + (1 - \alpha)^{1-1/p}B \leq (A^p + B^p)^{1/p}.$$

Proof: See [56].

Fact 8.10.37. Let $A, B, C \in \mathbb{F}^{n \times n}$. Then,

$$A^*A + B^*B = (B + CA)^*(I + CC^*)^{-1}(B + CA) + (A - C^*B)(I + C^*C)^{-1}(A - C^*B).$$

Proof: See [736].

Remark: See Fact 8.13.30.

Fact 8.10.38. Let $A \in \mathbb{F}^{n \times n}$, let $\alpha \in \mathbb{R}$, and assume that either A is nonsingular or $\alpha \geq 1$. Then,

$$(A^*A)^\alpha = A^*(AA^*)^{\alpha-1}A.$$

Proof: Use the singular value decomposition.

Remark: This result is given in [525, 540].

Fact 8.10.39. Let $A, B \in \mathbb{F}^{n \times n}$, let $\alpha \in \mathbb{R}$, assume that A and B are positive semidefinite, and assume that either A and B are positive definite or $\alpha \geq 1$. Then,

$$(AB^2A)^\alpha = AB(BA^2B)^{\alpha-1}BA.$$

Proof: Use Fact 8.10.38.

Fact 8.10.40. Let $A, B, C \in \mathbb{F}^{n \times n}$, assume that A is positive semidefinite, B is positive definite, and $B = C^*C$, and let $\alpha \in [0, 1]$. Then,

$$C^*(C^{-*}AC^{-1})^\alpha C \leq \alpha A + (1 - \alpha)B.$$

If, in addition, $\alpha \in (0, 1)$, then equality holds if and only if $A = B$.

Proof: See [1020].

Fact 8.10.41. Let $A, B \in \mathbb{F}^{n \times n}$, assume that A is positive semidefinite, and let $p \in \mathbb{R}$. Furthermore, assume that either A and B are nonsingular or $p \geq 1$. Then,

$$(BAB^*)^p = BA^{1/2}(A^{1/2}B^*BA^{1/2})^{p-1}A^{1/2}B^*.$$

Proof: See [540] or [544, p. 129].

Fact 8.10.42. Let $A, B \in \mathbb{F}^{n \times n}$, assume that A and B are positive definite, and let $p \in \mathbb{R}$. Then,

$$(BAB)^p = BA^{1/2}(A^{1/2}B^2A^{1/2})^{p-1}A^{1/2}B.$$

Proof: See [538, 692].

Fact 8.10.43. Let $A, B \in \mathbb{F}^{n \times n}$, and assume that A and B are positive semidefinite. Furthermore, if A is positive definite, then define

$$A \# B \triangleq A^{1/2} \left(A^{-1/2} B A^{-1/2} \right)^{1/2} A^{1/2},$$

whereas, if A is singular, then define

$$A \# B \triangleq \lim_{\varepsilon \downarrow 0} (A + \varepsilon I) \# B.$$

Then, the following statements hold:

i) $A \# B$ is positive semidefinite.

ii) $A \# A = A$.

iii) $A \# B = B \# A$.

iv) $\mathcal{R}(A \# B) = \mathcal{R}(A) \cap \mathcal{R}(B)$.

v) If $S \in \mathbb{F}^{m \times n}$ is right invertible, then $(SAS^*) \# (SBS^*) \le S(A \# B)S^*$.

vi) If $S \in \mathbb{F}^{n \times n}$ is nonsingular, then $(SAS^*) \# (SBS^*) = S(A \# B)S^*$.

vii) If $C, D \in \mathbf{P}^n$, $A \le C$, and $B \le D$, then $A \# B \le C \# D$.

viii) If $C, D \in \mathbf{P}^n$, then

$$(A \# C) + (B \# D) \le (A + B) \# (C + D).$$

ix) If $A \le B$, then

$$4A \# (B - A) = [A + A \# (4B - 3A)] \# [-A + A \# (4B - 3A)].$$

x) If $\alpha \in [0, 1]$, then

$$\sqrt{\alpha}(A \# B) \pm \tfrac{1}{2}\sqrt{1 - \alpha}(A - B) \le \tfrac{1}{2}(A + B).$$

xi) $A \# B = \max\{X \in \mathbf{H}: \left[\begin{smallmatrix} A & X \\ X & B \end{smallmatrix} \right]$ is positive semidefinite$\}$.

xii) Let $X \in \mathbb{F}^{n \times n}$, and assume that X is Hermitian and

$$\begin{bmatrix} A & X \\ X & B \end{bmatrix} \ge 0.$$

Then,

$$-A \# B \le X \le A \# B.$$

Furthermore, $\left[\begin{smallmatrix} A & A \# B \\ A \# B & B \end{smallmatrix} \right]$ and $\left[\begin{smallmatrix} A & -A \# B \\ -A \# B & B \end{smallmatrix} \right]$ are positive semidefinite.

xiii) If $S \in \mathbb{F}^{n \times n}$ is unitary and $A^{1/2}SB^{1/2}$ is positive semidefinite, then $A \# B = A^{1/2}SB^{1/2}$.

Now, assume in addition that A is positive definite. Then, the following statements hold:

xiv) $(A \# B)A^{-1}(A \# B) = B$.

xv) For all $\alpha \in \mathbb{R}$, $A \# B = A^{1-\alpha} \left(A^{\alpha-1} B A^{-\alpha} \right)^{1/2} A^{\alpha}$.

xvi) $A \# B = A \left(A^{-1} B \right)^{1/2} = (B A^{-1})^{1/2} A$.

xvii) $A\#B = (A + B)[(A + B)^{-1}A(A + B)^{-1}B]^{1/2}$.

Now, assume in addition that A and B are positive definite. Then, the following statements hold:

xviii) $A\#B$ is positive definite.

xix) $S \triangleq (A^{-1/2}BA^{-1/2})^{1/2}A^{1/2}B^{-1/2}$ is unitary, and $A\#B = A^{1/2}SB^{1/2}$.

xx) $\det A\#B = \sqrt{(\det A)\det B}$.

xxi) $\det (A\#B)^2 = \det AB$.

xxii) $(A\#B)^{-1} = A^{-1}\#B^{-1}$.

xxiii) Let $A_0 \triangleq A$ and $B_0 \triangleq B$, and, for all $k \in \mathbb{N}$, define $A_{k+1} \triangleq 2(A_k^{-1} + B_k^{-1})^{-1}$ and $B_{k+1} \triangleq \frac{1}{2}(A_k + B_k)$. Then, for all $k \in \mathbb{N}$,

$$A_k \leq A_{k+1} \leq A\#B \leq B_{k+1} \leq B_k$$

and

$$\lim_{k\to\infty} A_k = \lim_{k\to\infty} B_k = A\#B.$$

xxiv) For all $\alpha \in (-1,1)$, $\begin{bmatrix} A & \alpha A\#B \\ \alpha A\#B & B \end{bmatrix}$ is positive definite.

xxv) $\text{rank} \begin{bmatrix} A & A\#B \\ A\#B & B \end{bmatrix} = \text{rank} \begin{bmatrix} A & -A\#B \\ -A\#B & B \end{bmatrix} = n$.

Furthermore, the following statements hold:

xxvi) Assume that $n = 2$, and let $\alpha \triangleq \sqrt{\det A}$ and $\beta \triangleq \sqrt{\det B}$. Then,

$$A\#B = \frac{\sqrt{\alpha\beta}}{\sqrt{\det(\alpha^{-1}A + \beta^{-1}B)}}(\alpha^{-1}A + \beta^{-1}B).$$

xxvii) If $0 < A \leq B$, then $\phi: [0,\infty) \mapsto \mathbf{P}^n$ defined by $\phi(p) \triangleq A^{-p}\#B^p$ is nondecreasing.

xxviii) If B is positive definite and $A \leq B$, then

$$A^2\#B^{-2} \leq A\#B^{-1} \leq I.$$

xxix) If A and B are positive semidefinite and $A \leq B$, then

$$(BA^2B)^{1/2} \leq B^{1/2}(B^{1/2}AB^{1/2})^{1/2}B^{1/2} \leq B^2.$$

Finally, let $X \in \mathbf{H}^n$. Then, the following statements are equivalent:

xxx) $\begin{bmatrix} A & X \\ X & B \end{bmatrix}$ is positive semidefinite.

xxxi) $XA^{-1}X \leq B$.

xxxii) $XB^{-1}X \leq A$.

xxxiii) $-A\#B \leq X \leq A\#B$.

Proof: See [47, 499, 597, 902, 1346]. For *xiii)*, *xix)*, and *xxvi)*, see [205, pp. 108, 109, 111]. For *xxvii)*, see [48]. Statement *xxvii)* implies *xxviii)*, which, in turn, implies *xxix)*.

Remark: The square roots in $xvi)$ indicate a semisimple matrix with positive diagonal entries.

Remark: $A\#B$ is the *geometric mean* of A and B. A related mean is defined in [499]. Alternative means and their differences are considered in [22]. Geometric means for an arbitrary number of positive-definite matrices are discussed in [59, 832, 1039, 1111].

Remark: Inverse problems are considered in [43].

Remark: $xxix)$ interpolates (8.6.6).

Remark: Compare statements $xiii)$ and $xix)$ with Fact 8.11.6.

Remark: See Fact 10.10.4 and Fact 12.23.4.

Problem: For singular A and B, express $A\#B$ in terms of generalized inverses.

Fact 8.10.44. Let $A, B \in \mathbb{F}^{n \times n}$, and assume that A and B are Hermitian. Then, the following statements are equivalent:

i) $A \leq B$.

ii) For all $t \geq 0$, $I \leq e^{-tA}\#e^{tB}$.

iii) $\phi\colon [0, \infty) \mapsto \mathbf{P}^n$ defined by $\phi(t) \triangleq e^{-tA}\#e^{tB}$ is nondecreasing.

Proof: See [48].

Fact 8.10.45. Let $A, B \in \mathbb{F}^{n \times n}$, assume that A and B are positive semidefinite, and let $\alpha \in [0, 1]$. Furthermore, if A is positive definite, then define

$$A\#_\alpha B \triangleq A^{1/2}\Big(A^{-1/2}BA^{-1/2}\Big)^\alpha A^{1/2},$$

whereas, if A is singular, then define

$$A\#_\alpha B \triangleq \lim_{\varepsilon \downarrow 0}(A + \varepsilon I)\#_\alpha B.$$

Then, the following statements hold:

i) $A\#_\alpha B = B\#_{1-\alpha}A$.

ii) $(A\#_\alpha B)^{-1} = A^{-1}\#_\alpha B^{-1}$.

Fact 8.10.46. Let $A, B \in \mathbb{F}^{n \times n}$, assume that A and B are positive definite, and let $\alpha \in [0, 1]$. Then,

$$[\alpha A^{-1} + (1 - \alpha)B^{-1}]^{-1} \leq A^{1/2}\Big(A^{-1/2}BA^{-1/2}\Big)^{1-\alpha}A^{1/2} \leq \alpha A + (1 - \alpha)B,$$

or, equivalently,

$$[\alpha A^{-1} + (1 - \alpha)B^{-1}]^{-1} \leq A\#_{1-\alpha}B \leq \alpha A + (1 - \alpha)B,$$

or, equivalently,

$$[\alpha A + (1 - \alpha)B]^{-1} \leq A^{-1/2}\Big(A^{-1/2}BA^{-1/2}\Big)^{\alpha-1}A^{-1/2} \leq \alpha A^{-1} + (1 - \alpha)B^{-1}.$$

Consequently,

$$\operatorname{tr}\left[\alpha A + (1-\alpha)B\right]^{-1} \leq \operatorname{tr}\left[A^{-1}\left(A^{-1/2}BA^{-1/2}\right)^{\alpha-1}\right] \leq \operatorname{tr}\left[\alpha A^{-1} + (1-\alpha)B^{-1}\right].$$

Remark: The left-hand inequality in the first string of inequalities is the *Young inequality.* See [544, p. 122] and Fact 1.12.21. Setting $B = I$ yields Fact 8.9.43. The third string of inequalities interpolates the fact that $\phi(A) = A^{-1}$ is convex as shown by *iv)* of Proposition 8.6.17.

Remark: Related inequalities are given by Fact 8.12.27 and Fact 8.12.28. See also Fact 8.21.18.

Fact 8.10.47. Let $A, B \in \mathbb{F}^{n \times n}$, and assume that A and B are positive definite. Then,

$$\tfrac{2\alpha\beta}{(\alpha+\beta)^2}(A + B) \leq 2\left(A^{-1} + B^{-1}\right)^{-1} \leq A\#B \leq \tfrac{1}{2}(A + B) \leq \tfrac{(\alpha+\beta)^2}{2\alpha\beta}\left(A^{-1} + B^{-1}\right)^{-1},$$

where

$$\alpha \triangleq \min\{\lambda_{\min}(A), \lambda_{\min}(B)\}$$

and

$$\beta \triangleq \max\{\lambda_{\max}(A), \lambda_{\max}(B)\}.$$

Proof: Use the second string of inequalities of Fact 8.10.46 with $\alpha = 1/2$ along with results given in [1315] and [1526, p. 174].

Fact 8.10.48. Let $(x_i)_{i=1}^{\infty} \subset \mathbb{R}^n$, assume that $\sum_{i=1}^{\infty} x_i$ exists, and let $(A_i)_{i=1}^{\infty} \subset \mathbf{N}^n$ be such that $A_i \leq A_{i+1}$ for all $i \in \mathbb{P}$ and $\lim_{i \to \infty} \operatorname{tr} A_i = \infty$. Then,

$$\lim_{k \to \infty} (\operatorname{tr} A_k)^{-1} \sum_{i=1}^{k} A_i x_i = 0.$$

If, in addition A_i is positive definite for all $i \in \mathbb{P}$ and $\{\lambda_{\max}(A_i)/\lambda_{\min}(A_i)\}_{i=1}^{\infty}$ is bounded, then

$$\lim_{k \to \infty} A_k^{-1} \sum_{i=1}^{k} A_i x_i = 0.$$

Proof: See [35].

Remark: These equalities are matrix versions of the *Kronecker lemma.*

Remark: Extensions are given in [638].

Fact 8.10.49. Let $A, B \in \mathbb{F}^{n \times n}$, assume that A and B are positive definite, assume that $A \leq B$, and let $p \geq 1$. Then,

$$A^p \leq K(\lambda_{\min}(A), \lambda_{\min}(A), p)B^p \leq \left[\frac{\lambda_{\max}(A)}{\lambda_{\min}(A)}\right]^{p-1} B^p,$$

where

$$K(a, b, p) \triangleq \frac{a^p b - a b^p}{(p-1)(a-b)}\left[\frac{(p-1)(a^p - b^p)}{p(a^p b - a b^p)}\right]^p.$$

Proof: See [253, 542] and [544, pp. 193, 194].

Remark: $K(a, b, p)$ is the *Fan constant.*

Fact 8.10.50. Let $A, B \in \mathbb{F}^{n \times n}$, assume that A is positive definite and B is positive semidefinite, and let $p \geq 1$. Then, there exist unitary matrices $U, V \in \mathbb{F}^{n \times n}$ such that

$$\frac{1}{K(\lambda_{\min}(A), \lambda_{\min}(A), p)} U(BAB)^p U^* \leq B^p A^p B^p \leq K(\lambda_{\min}(A), \lambda_{\min}(A), p) V(BAB)^p V^*,$$

where $K(a, b, p)$ is the Fan constant defined in Fact 8.10.49.

Proof: See [253].

Remark: See Fact 8.12.22, Fact 8.19.27, and Fact 9.9.17.

Fact 8.10.51. Let $A, B \in \mathbb{F}^{n \times n}$, assume that A is positive definite, B is positive semidefinite, and $B \leq A$, and let $p \geq 1$ and $r \geq 1$. Then,

$$\left[A^{r/2} \left(A^{-1/2} B^p A^{-1/2} \right)^r A^{r/2} \right]^{1/p} \leq A^r.$$

In particular,

$$\left\langle A^{-1/2} B^p A^{1/2} \right\rangle^{2/p} \leq A^2.$$

Proof: See [55].

Fact 8.10.52. Let $A, B \in \mathbb{F}^{n \times n}$, and assume that A and B are positive definite. Then, the following statements are equivalent:

i) $B \leq A$.

ii) For all $r \in [0, \infty)$, $p \in [1, \infty)$, and $k \in \mathbb{N}$ such that $(k+1)(r+1) = p + r$,

$$B^{r+1} \leq \left(B^{r/2} A^p B^{r/2} \right)^{\frac{1}{k+1}}.$$

iii) For all $r \in [0, \infty)$, $p \in [1, \infty)$, and $k \in \mathbb{N}$ such that $(k+1)(r+1) = p + r$,

$$\left(A^{r/2} B^p A^{r/2} \right)^{\frac{1}{k+1}} \leq A^{r+1}.$$

Proof: See [940].

Remark: See Fact 8.20.1.

Fact 8.10.53. Let $A, B \in \mathbb{F}^{n \times n}$, and assume that A is positive definite and B is positive semidefinite. Then, the following statements are equivalent:

i) $B \leq A$.

ii) For all $p, q, r, t \in \mathbb{R}$ such that $p \geq 1$, $r \geq 0$, $t \geq 0$, and $q \in [1, 2]$,

$$\left[A^{r/2} \left(A^{t/2} B^p A^{t/2} \right)^q A^{r/2} \right]^{\frac{r+t+1}{r+qt+qp}} \leq A^{r+t+1}.$$

iii) For all $p, q, r, \tau \in \mathbb{R}$ such that $p \geq 1$, $r \geq \tau$, $q \geq 1$, and $\tau \in [0, 1]$,

$$\left[A^{r/2} \left(A^{-\tau/2} B^p A^{-\tau/2} \right)^q A^{r/2} \right]^{\frac{r-\tau}{r-q\tau+qp}} \leq A^{r-\tau}.$$

iv) For all $p, q, r, \tau \in \mathbb{R}$ be such that $p \geq 1$, $r \geq \tau$, $\tau \in [0, 1]$, and $q \geq 1$,

$$\left[A^{r/2} \left(A^{-\tau/2} B^p A^{-\tau/2} \right)^q A^{r/2} \right]^{\frac{r-\tau+1}{r-q\tau+qp}} \leq A^{r-\tau+1}.$$

In particular, if $B \leq A$, $p \geq 1$, and $r \geq 1$, then

$$\left[A^{r/2} \left(A^{-1/2} B^p A^{-1/2} \right)^r A^{r/2} \right]^{\frac{r-1}{pr}} \leq A^{r-1}.$$

Proof: Condition $ii)$ is given in [525], $iii)$ appears in [545], and $iv)$ appears in [540]. See also [525, 526] and [544, p. 133].

Remark: Setting $q = r$ and $\tau = 1$ in $iv)$ yields Fact 8.10.51.

Remark: Condition $iv)$, which is the *generalized Furuta inequality*, interpolates Proposition 8.6.7 and Fact 8.10.51.

Fact 8.10.54. Let $A, B \in \mathbb{F}^{n \times n}$, and assume that A is positive definite, B is positive semidefinite, and $B \leq A$. Furthermore, let $t \in [0, 1]$, $p \geq 1$, $r \geq t$, and $s \geq 1$, and define

$$F_{A,B}(r, s) \triangleq A^{-r/2} \left[A^{r/2} (A^{-t/2} B^p A^{-t/2})^s A^{r/2} \right]^{\frac{1-t+r}{(p-t)s+r}} A^{-r/2}.$$

Then,

$$F_{A,B}(r, s) \leq F_{A,A}(r, s),$$

that is,

$$\left[A^{r/2} (A^{-t/2} B^p A^{-t/2})^s A^{r/2} \right]^{\frac{1-t+r}{(p-t)s+r}} \leq A^{1-t+r}.$$

Furthermore, if $r' \geq r$ and $s' \geq s$, then

$$F_{A,B}(r', s') \leq F_{A,B}(r, s).$$

Proof: See [540] and [544, p. 143].

Remark: This result extends $iv)$ of Fact 8.10.53.

Fact 8.10.55. Each of the following functions $\phi \colon (0, \infty) \mapsto (0, \infty)$ yields an increasing function $\phi \colon \mathbf{P}^n \mapsto \mathbf{P}^n$:

$i)$ $\phi(x) = \frac{x^{p+1/2}}{x^{2p}+1}$, where $p \in [0, 1/2]$.

$ii)$ $\phi(x) = x(1+x) \log(1 + 1/x)$.

$iii)$ $\phi(x) = \frac{1}{(1+x) \log(1+1/x)}$.

$iv)$ $\phi(x) = \frac{x - 1 - \log x}{(\log x)^2}$.

$v)$ $\phi(x) = \frac{x(\log x)^2}{x - 1 - \log x}$.

$vi)$ $\phi(x) = \frac{x(x+2) \log(x+2)}{(x+1)^2}$.

$vii)$ $\phi(x) = \frac{x(x+1)}{(x+2) \log(x+2)}$.

$viii)$ $\phi(x) = \frac{(x^2 - 1) \log(1+x)}{x^2}$.

$ix)$ $\phi(x) = \frac{x(x-1)}{(x+1) \log(x+1)}$.

$x)$ $\phi(x) = \frac{(x-1)^2}{(x+1) \log x}$.

xi) $\phi(x) = \frac{p-1}{p}\left(\frac{x^p-1}{x^{p-1}-1}\right)$, where $p \in [-1, 2]$.

xii) $\phi(x) = \frac{x-1}{\log x}$.

xiii) $\phi(x) = \sqrt{x}$.

xiv) $\phi(x) = \frac{2x}{x+1}$.

xv) $\phi(x) = \frac{x-1}{x^p-1}$, where $p \in (0, 1]$.

Proof: See [548, 1111]. To obtain *xii*), *xiii*), and *xiv*), set $p = 1, 1/2, -1$, respectively, in *xi*).

Fact 8.10.56. Let $A, B \in \mathbb{F}^{n \times n}$, and assume that A and B are positive semidefinite, $A \le B$, and $AB = BA$. Then, $A^2 \le B^2$.

Proof: See [113].

8.11 Facts on Identities and Inequalities for Partitioned Matrices

Fact 8.11.1. Let $A \in \mathbb{F}^{n \times n}$, and assume that A is positive semidefinite. Then, the following statements hold:

i) $\begin{bmatrix} A & A \\ A & A \end{bmatrix}$ and $\begin{bmatrix} A & -A \\ -A & A \end{bmatrix}$ are positive semidefinite.

ii) If $\begin{bmatrix} \alpha & \beta \\ \bar\beta & \gamma \end{bmatrix} \in \mathbb{F}^{2 \times 2}$ is positive semidefinite, then $\begin{bmatrix} \alpha A & \beta A \\ \bar\beta A & \gamma A \end{bmatrix}$ is positive semidefinite.

iii) If A and $\begin{bmatrix} \alpha & \beta \\ \bar\beta & \gamma \end{bmatrix}$ are positive definite, then $\begin{bmatrix} \alpha A & \beta A \\ \bar\beta A & \gamma A \end{bmatrix}$ is positive definite.

Proof: Use Fact 7.4.17.

Fact 8.11.2. Let $A \in \mathbb{F}^{n \times n}$, $B \in \mathbb{F}^{n \times m}$, $C \in \mathbb{F}^{m \times m}$, assume that $\begin{bmatrix} A & B \\ B^* & C \end{bmatrix} \in \mathbb{F}^{(n+m) \times (n+m)}$ is positive semidefinite, and assume that $\begin{bmatrix} \alpha & \beta \\ \bar\beta & \gamma \end{bmatrix} \in \mathbb{F}^{2 \times 2}$ is positive semidefinite. Then, the following statements hold:

i) $\begin{bmatrix} \alpha 1_{n \times n} & \beta 1_{n \times m} \\ \bar\beta 1_{m \times n} & \gamma 1_{m \times m} \end{bmatrix}$ is positive semidefinite.

ii) $\begin{bmatrix} \alpha A & \beta B \\ \bar\beta B^* & \gamma C \end{bmatrix}$ is positive semidefinite.

iii) If $\begin{bmatrix} A & B \\ B^* & C \end{bmatrix}$ is positive definite and α and γ are positive, then $\begin{bmatrix} \alpha A & \beta B \\ \bar\beta B^* & \gamma C \end{bmatrix}$ is positive definite.

Proof: To prove *i*), use Proposition 8.2.4. Statements *ii*) and *iii*) follow from Fact 8.22.12.

Fact 8.11.3. Let $A, B \in \mathbb{F}^{n \times n}$, assume that A and B are positive semidefinite, and assume that A and B are partitioned identically as $A = \begin{bmatrix} A_{11} & A_{12} \\ A_{12}^* & A_{22} \end{bmatrix}$ and $B =$

$\begin{bmatrix} B_{11} & B_{12} \\ B_{12}^* & B_{22} \end{bmatrix}$. Then,

$$A_{22}|A + B_{22}|B \le (A_{22} + B_{22})|(A + B).$$

Now, assume in addition that A_{22} and B_{22} are positive definite. Then, equality holds if and only if $A_{12}A_{22}^{-1} = B_{12}B_{22}^{-1}$.

Proof: See [498, 1084].

Remark: The first inequality, which follows from *xvii*) of Proposition 8.6.17, is an extension of Bergstrom's inequality, which corresponds to the case in which A_{11} is a scalar. See Fact 8.15.19.

Fact 8.11.4. Let $A, B \in \mathbb{F}^{n \times n}$, assume that A and B are positive semidefinite, assume that A and B are partitioned identically as $A = \begin{bmatrix} A_{11} & A_{12} \\ A_{12}^* & A_{22} \end{bmatrix}$ and $B = \begin{bmatrix} B_{11} & B_{12} \\ B_{12}^* & B_{22} \end{bmatrix}$, and assume that A_{11} and B_{11} are positive definite. Then,

$$(A_{12} + B_{12})^*(A_{11} + B_{11})^{-1}(A_{12} + B_{12}) \le A_{12}^* A_{11}^{-1} A_{12} + B_{12}^* B_{11}^{-1} B_{12}$$

and

$$\text{rank}[A_{12}^* A_{11}^{-1} A_{12} + B_{12}^* B_{11}^{-1} B_{12} - (A_{12} + B_{12})^*(A_{11} + B_{11})^{-1}(A_{12} + B_{12})]$$

$$= \text{rank}(A_{12} - A_{11}B_{11}^{-1}B_{12}).$$

Furthermore,

$$\frac{\det A}{\det A_{11}} + \frac{\det B}{\det B_{11}} \le \frac{\det(A + B)}{\det(A_{11} + B_{11})} = \det[(A_{11} + B_{11})|(A + B)].$$

Remark: The last inequality generalizes Fact 8.13.18.

Fact 8.11.5. Let $A \in \mathbb{F}^{n \times n}$, $B \in \mathbb{F}^{n \times m}$, and $C \in \mathbb{F}^{m \times m}$, and define $\mathcal{A} \triangleq \begin{bmatrix} A & B \\ B^* & C \end{bmatrix}$. Then, the following statements hold:

i) If $\mathcal{A}$ is positive semidefinite, then

$$0 \le BC^+B^* \le A.$$

ii) If $\mathcal{A}$ is positive definite, then C is positive definite and

$$0 \le BC^{-1}B^* < A.$$

Now, assume in addition that $n = m$. Then, the following statements hold:

iii) If $\mathcal{A}$ is positive semidefinite, then

$$-A - C \le B + B^* \le A + C.$$

iv) If $\mathcal{A}$ is positive definite, then

$$-A - C < B + B^* < A + C.$$

Proof: The first two statements follow from Proposition 8.2.4. To prove the last two statements, consider SAS^T, where $S \triangleq \begin{bmatrix} I & I \end{bmatrix}$ and $S \triangleq \begin{bmatrix} I & -I \end{bmatrix}$.

Remark: See Fact 8.22.42.

Fact 8.11.6. Let $A \in \mathbb{F}^{n \times n}$, $B \in \mathbb{F}^{n \times m}$, and $C \in \mathbb{F}^{m \times m}$, and define $\mathcal{A} \triangleq \left[\begin{smallmatrix} A & B \\ B^* & C \end{smallmatrix} \right]$. Then, $\mathcal{A}$ is positive semidefinite if and only if A and C are positive semidefinite and there exists a semicontractive matrix $S \in \mathbb{F}^{n \times m}$ such that

$$B = A^{1/2} S C^{1/2}.$$

Proof: See [738].

Remark: Compare this result with statements *xiii*) and *xix*) of Fact 8.10.43.

Fact 8.11.7. Let $A, B, C \in \mathbb{F}^{n \times n}$, assume that $\left[\begin{smallmatrix} A & B \\ B^* & C \end{smallmatrix} \right] \in \mathbb{F}^{2n \times 2n}$ is positive semidefinite, and assume that $AB = BA$. Then,

$$B^* B \leq A^{1/2} C A^{1/2}.$$

Proof: See [1528].

Fact 8.11.8. Let $A, B \in \mathbb{F}^{n \times n}$, and assume that A and B are Hermitian. Then, $-A \leq B \leq A$ if and only if $\left[\begin{smallmatrix} A & B \\ B & A \end{smallmatrix} \right]$ is positive semidefinite. Furthermore, $-A < B < A$ if and only if $\left[\begin{smallmatrix} A & B \\ B & A \end{smallmatrix} \right]$ is positive definite.

Proof: Note that

$$\frac{1}{\sqrt{2}} \begin{bmatrix} I & -I \\ I & I \end{bmatrix} \begin{bmatrix} A & B \\ B & A \end{bmatrix} \frac{1}{\sqrt{2}} \begin{bmatrix} I & I \\ -I & I \end{bmatrix} = \begin{bmatrix} A - B & 0 \\ 0 & A + B \end{bmatrix}.$$

Fact 8.11.9. Let $A \in \mathbb{F}^{n \times n}$, $B \in \mathbb{F}^{n \times m}$, and $C \in \mathbb{F}^{m \times m}$, assume that $\left[\begin{smallmatrix} A & B \\ B^* & C \end{smallmatrix} \right]$ is positive semidefinite, and let $r \triangleq \operatorname{rank} B$. Then, for all $k \in \{1, \dots, r\}$,

$$\prod_{i=1}^{k} \sigma_i(B) \leq \prod_{i=1}^{k} \max\{\lambda_i(A), \lambda_i(C)\}.$$

Proof: See[1528].

Fact 8.11.10. Let $A \in \mathbb{F}^{n \times n}$, $B \in \mathbb{F}^{n \times m}$, and $C \in \mathbb{F}^{m \times m}$, define $\mathcal{A} \triangleq \left[\begin{smallmatrix} A & B \\ B^* & C \end{smallmatrix} \right]$, and assume that $\mathcal{A}$ is positive definite. Then,

$$\operatorname{tr} A^{-1} + \operatorname{tr} C^{-1} \leq \operatorname{tr} \mathcal{A}^{-1}.$$

Furthermore, B is nonzero if and only if

$$\operatorname{tr} A^{-1} + \operatorname{tr} C^{-1} < \operatorname{tr} \mathcal{A}^{-1}.$$

Proof: Use Proposition 8.2.5 or see [1020].

Fact 8.11.11. Let $A \in \mathbb{F}^{n \times m}$, and define

$$\mathcal{A} \triangleq \begin{bmatrix} \langle A^* \rangle & A \\ A^* & \langle A \rangle \end{bmatrix}.$$

Then, $\mathcal{A}$ is positive semidefinite. If, in addition, $n = m$, then

$$-\langle A^* \rangle - \langle A \rangle \leq A + A^* \leq \langle A^* \rangle + \langle A \rangle.$$

Proof: Use Fact 8.11.5.

Remark: See Fact 8.9.30 and Fact 8.21.4.

Fact 8.11.12. Let $A \in \mathbb{F}^{n \times n}$, assume that A is normal, and define

$$\mathcal{A} \triangleq \begin{bmatrix} \langle A \rangle & A \\ A^* & \langle A \rangle \end{bmatrix}.$$

Then, $\mathcal{A}$ is positive semidefinite.

Proof: See [730, p. 213].

Fact 8.11.13. Let $A \in \mathbb{F}^{n \times n}$, and define

$$\mathcal{A} \triangleq \begin{bmatrix} I & A \\ A^* & I \end{bmatrix}.$$

Then, $\mathcal{A}$ is (positive semidefinite, positive definite) if and only if A is (semicontractive, contractive).

Fact 8.11.14. Let $A \in \mathbb{F}^{n \times m}$ and $B \in \mathbb{F}^{n \times l}$, and define

$$\mathcal{A} \triangleq \begin{bmatrix} A^*A & A^*B \\ B^*A & B^*B \end{bmatrix}.$$

Then, $\mathcal{A}$ is positive semidefinite, and

$$0 \leq A^*B(B^*B)^+B^*A \leq A^*A.$$

If $m = l$, then

$$-A^*A - B^*B \leq A^*B + B^*A \leq A^*A + B^*B.$$

If, in addition, $m = l = 1$ and $B^*B \neq 0$, then

$$|A^*B|^2 \leq A^*AB^*B.$$

Remark: This result is the Cauchy-Schwarz inequality. See Fact 8.13.23.

Remark: See Fact 8.22.43.

Fact 8.11.15. Let $A, B \in \mathbb{F}^{n \times m}$, and define

$$\mathcal{A} \triangleq \begin{bmatrix} I + A^*A & I - A^*B \\ I - B^*A & I + B^*B \end{bmatrix}$$

and

$$\mathcal{B} \triangleq \begin{bmatrix} I + A^*A & I + A^*B \\ I + B^*A & I + B^*B \end{bmatrix}.$$

Then, $\mathcal{A}$ and $\mathcal{B}$ are positive semidefinite,

$$0 \leq (I - A^*B)(I + B^*B)^{-1}(I - B^*A) \leq I + A^*A,$$

and

$$0 \leq (I + A^*B)(I + B^*B)^{-1}(I + B^*A) \leq I + A^*A.$$

Remark: See Fact 8.13.26.

Fact 8.11.16. Let $A, B \in \mathbb{F}^{n \times m}$. Then,

$$I + AA^* = (A + B)(I + B^*B)^{-1}(A + B)^* + (I - AB^*)(I + BB^*)^{-1}(I - BA^*).$$

Therefore,

$$(A + B)(I + B^*B)^{-1}(A + B)^* \leq I + AA^*.$$

Proof: Set $C = A$ in Fact 2.16.23. See also [1526, p. 185].

Fact 8.11.17. Let $A \in \mathbb{F}^{n \times n}$ and $B \in \mathbb{F}^{n \times m}$, assume that A is positive semidefinite, and define

$$\mathcal{A} \triangleq \begin{bmatrix} A & AB \\ B^*A & B^*AB \end{bmatrix}.$$

Then,

$$\mathcal{A} = \begin{bmatrix} A^{1/2} \\ B^*A^{1/2} \end{bmatrix} \begin{bmatrix} A^{1/2} & A^{1/2}B \end{bmatrix},$$

and thus $\mathcal{A}$ is positive semidefinite. Furthermore,

$$0 \leq AB(B^*AB)^+B^*A \leq A.$$

Now, assume in addition that $n = m$. Then,

$$-A - B^*AB \leq AB + B^*A \leq A + B^*AB.$$

Fact 8.11.18. Let $A \in \mathbb{F}^{n \times n}$ and $B \in \mathbb{F}^{n \times m}$, assume that A is positive definite, and define

$$\mathcal{A} \triangleq \begin{bmatrix} A & B \\ B^* & B^*A^{-1}B \end{bmatrix}.$$

Then,

$$\mathcal{A} = \begin{bmatrix} A^{1/2} \\ B^*A^{-1/2} \end{bmatrix} \begin{bmatrix} A^{1/2} & A^{-1/2}B \end{bmatrix},$$

and thus $\mathcal{A}$ is positive semidefinite. Furthermore,

$$0 \leq B(B^*A^{-1}B)^+B^* \leq A.$$

Furthermore, if rank $B = m$, then

$$\mathrm{rank}\big[A - B(B^*A^{-1}B)^{-1}B^*\big] = n - m.$$

Now, assume in addition that $n = m$. Then,

$$-A - B^*A^{-1}B \leq B + B^* \leq A + B^*A^{-1}B.$$

Proof: Use Fact 8.11.5.

Remark: See Fact 8.22.44.

Remark: The matrix $I - A^{-1/2}B(B^*A^{-1}B)^+B^*A^{-1/2}$ is a projector.

Fact 8.11.19. Let $A \in \mathbb{F}^{n \times n}$ and $B \in \mathbb{F}^{n \times m}$, assume that A is positive definite, and define

$$\mathcal{A} \triangleq \begin{bmatrix} B^*AB & B^*B \\ B^*B & B^*A^{-1}B \end{bmatrix}.$$

Then,

$$\mathcal{A} = \begin{bmatrix} B^*A^{1/2} \\ B^*A^{-1/2} \end{bmatrix} \begin{bmatrix} A^{1/2}B & A^{-1/2}B \end{bmatrix},$$

and thus $\mathcal{A}$ is positive semidefinite. Furthermore,

$$0 \leq B^*B(B^*A^{-1}B)^+B^*B \leq B^*AB.$$

Now, assume in addition that $n = m$. Then,

$$-B^*AB - B^*A^{-1}B \leq 2B^*B \leq B^*AB + B^*A^{-1}B.$$

Proof: Use Fact 8.11.5.

Remark: See Fact 8.13.24 and Fact 8.22.44.

Fact 8.11.20. Let $A, B \in \mathbb{F}^{n \times m}$, let $\alpha, \beta \in (0, \infty)$, and define

$$\mathcal{A} \triangleq \begin{bmatrix} \beta^{-1}I + \alpha A^*A & (A+B)^* \\ A+B & \alpha^{-1}I + \beta BB^* \end{bmatrix}.$$

Then,

$$\mathcal{A} = \begin{bmatrix} \beta^{-1/2}I & \alpha^{1/2}A^* \\ \beta^{1/2}B & \alpha^{-1/2}I \end{bmatrix} \begin{bmatrix} \beta^{-1/2}I & \beta^{1/2}B^* \\ \alpha^{1/2}A & \alpha^{-1/2}I \end{bmatrix}$$

$$= \begin{bmatrix} \alpha A^*A & A^* \\ A & \alpha^{-1}I \end{bmatrix} + \begin{bmatrix} \beta^{-1}I & B^* \\ B & \beta BB^* \end{bmatrix},$$

and thus $\mathcal{A}$ is positive semidefinite. Furthermore,

$$(A+B)^*(\alpha^{-1}I + \beta BB^*)^{-1}(A+B) \leq \beta^{-1}I + \alpha A^*A.$$

Now, assume in addition that $n = m$. Then,

$$-\left(\beta^{-1/2} + \alpha^{-1/2}\right)I - \alpha A^*A - \beta BB^* \leq A + B + (A+B)^*$$

$$\leq \left(\beta^{-1/2} + \alpha^{-1/2}\right)I + \alpha A^*A + \beta BB^*.$$

Remark: See Fact 8.13.27 and Fact 8.22.45.

Fact 8.11.21. Let $A, B \in \mathbb{F}^{n \times m}$, and assume that $I - A^*A$ and thus $I - AA^*$ are nonsingular. Then,

$$I - B^*B - (I - B^*A)(I - A^*A)^{-1}(I - A^*B) = -(A - B)^*(I - AA^*)^{-1}(A - B).$$

Now, assume in addition that $I - A^*A$ is positive definite. Then,

$$I - B^*B \leq (I - B^*A)(I - A^*A)^{-1}(I - A^*B).$$

Now, assume in addition that $I - B^*B$ is positive definite. Then, $I - A^*B$ is nonsingular. Next, define

$$\mathcal{A} \triangleq \begin{bmatrix} (I - A^*A)^{-1} & (I - B^*A)^{-1} \\ (I - A^*B)^{-1} & (I - B^*B)^{-1} \end{bmatrix}.$$

Then, $\mathcal{A}$ is positive semidefinite. Finally,

$$-(I - A^*A)^{-1} - (I - B^*B)^{-1} \leq (I - B^*A)^{-1} + (I - A^*B)^{-1}$$

$$\leq (I - A^*A)^{-1} + (I - B^*B)^{-1}.$$

Proof: For the first equality, set $D = -B^*$ and $C = -A^*$, and replace B with $-B$ in Fact 2.16.22. See [49, 1087]. The last statement follows from Fact 8.11.5.

Remark: The equality is *Hua's matrix equality*. This result does not assume that either $I - A^*A$ or $I - B^*B$ is positive semidefinite. The inequality and Fact 8.13.26 constitute *Hua's inequalities*. See [1087, 1502].

Remark: Extensions to the case in which $I - A^*A$ is singular are considered in [1087].

Remark: See Fact 8.9.40 and Fact 8.13.26.

Fact 8.11.22. Let $A \in \mathbb{F}^{n \times n}$ be semicontractive, and define $B \in \mathbb{F}^{2n \times 2n}$ by

$$B \triangleq \begin{bmatrix} A & (I - AA^*)^{1/2} \\ (I - A^*A)^{1/2} & -A^* \end{bmatrix}.$$

Then, B is unitary.

Remark: See [521, p. 180].

Fact 8.11.23. Let $A \in \mathbb{F}^{n \times m}$, and define $B \in \mathbb{F}^{(n+m) \times (n+m)}$ by

$$B \triangleq \begin{bmatrix} (I + A^*A)^{-1/2} & -A^*(I + AA^*)^{-1/2} \\ (I + AA^*)^{-1/2}A & (I + AA^*)^{-1/2} \end{bmatrix}.$$

Then, B is unitary and satisfies $A^* = \tilde{I}A\tilde{I}$, where $\tilde{I} \triangleq \mathrm{diag}(I_m, -I_n)$. Furthermore, $\det B = 1$.

Remark: See [655].

Fact 8.11.24. Let $A \in \mathbb{F}^{n \times m}$, assume that A is contractive, and define $B \in \mathbb{F}^{(n+m) \times (n+m)}$ by

$$B \triangleq \begin{bmatrix} (I - A^*A)^{-1/2} & A^*(I - AA^*)^{-1/2} \\ (I - AA^*)^{-1/2}A & (I - AA^*)^{-1/2} \end{bmatrix}.$$

Then, B is Hermitian and satisfies $A^*\tilde{I}A = \tilde{I}$, where $\tilde{I} \triangleq \mathrm{diag}(I_m, -I_n)$. Furthermore, $\det B = 1$.

Remark: See [655].

Fact 8.11.25. Let $X \in \mathbb{F}^{n \times m}$, and define $U \in \mathbb{F}^{(n+m) \times (n+m)}$ by

$$U \triangleq \begin{bmatrix} (I + X^*X)^{-1/2} & -X^*(I + XX^*)^{-1/2} \\ (I + XX^*)^{-1/2}X & (I + XX^*)^{-1/2} \end{bmatrix}.$$

Furthermore, let $A \in \mathbb{F}^{n \times n}$, $B \in \mathbb{F}^{n \times m}$, $C \in \mathbb{F}^{m \times n}$, $D \in \mathbb{F}^{m \times m}$. Then, the following statements hold:

i) Assume that D is nonsingular, and let $X \triangleq D^{-1}C$. Then,

$$\begin{bmatrix} A & B \\ C & D \end{bmatrix} = \begin{bmatrix} (A - BX)(I + X^*X)^{-1/2} & (B + AX^*)(I + XX^*)^{-1/2} \\ 0 & D(I + XX^*)^{1/2} \end{bmatrix} U.$$

ii) Assume that A is nonsingular, and let $X \triangleq CA^{-1}$. Then,

$$\begin{bmatrix} A & B \\ C & D \end{bmatrix} = U \begin{bmatrix} (I + X^*X)^{1/2}A & (I + X^*X)^{-1/2}(B + X^*D) \\ 0 & (I + XX^*)^{-1/2}(D - XB) \end{bmatrix}.$$

Remark: See Proposition 2.8.3 and Proposition 2.8.4.

Proof: See [655].

Fact 8.11.26. Let $X \in \mathbb{F}^{n \times m}$, and define $U \in \mathbb{F}^{(n+m) \times (n+m)}$ by

$$U \triangleq \begin{bmatrix} (I - X^*X)^{-1/2} & X^*(I - XX^*)^{-1/2} \\ (I - XX^*)^{-1/2}X & (I - XX^*)^{-1/2} \end{bmatrix}.$$

Furthermore, let $A \in \mathbb{F}^{n \times n}$, $B \in \mathbb{F}^{n \times m}$, $C \in \mathbb{F}^{m \times n}$, $D \in \mathbb{F}^{m \times m}$. Then, the following statements hold:

i) Assume that D is nonsingular, let $X \triangleq D^{-1}C$, and assume that $X^*X < I$. Then,

$$\begin{bmatrix} A & B \\ C & D \end{bmatrix} = \begin{bmatrix} (A - BX)(I - X^*X)^{-1/2} & (B + AX^*)(I - XX^*)^{-1/2} \\ 0 & D(I - XX^*)^{1/2} \end{bmatrix} U.$$

ii) Assume that A is nonsingular, let $X \triangleq CA^{-1}$, and assume that $X^*X < I$. Then,

$$\begin{bmatrix} A & B \\ C & D \end{bmatrix} = U \begin{bmatrix} (I - X^*X)^{1/2}A & (I - X^*X)^{-1/2}(B - X^*D) \\ 0 & (I - XX^*)^{-1/2}(D - XB) \end{bmatrix}.$$

Proof: See [655].

Remark: See Proposition 2.8.3 and Proposition 2.8.4.

Fact 8.11.27. Let $A, B \in \mathbb{F}^{n \times m}$ and $C, D \in \mathbb{F}^{m \times m}$, assume that C and D are positive definite, and define

$$\mathcal{A} \triangleq \begin{bmatrix} AC^{-1}A^* + BD^{-1}B^* & A + B \\ (A + B)^* & C + D \end{bmatrix}.$$

Then, $\mathcal{A}$ is positive semidefinite, and

$$(A + B)(C + D)^{-1}(A + B)^* \leq AC^{-1}A^* + BD^{-1}B^*.$$

Now, assume in addition that $n = m$. Then,

$$-AC^{-1}A^* - BD^{-1}B^* - C - D \leq A + B + (A + B)^*$$

$$\leq AC^{-1}A^* + BD^{-1}B^* + C + D.$$

Proof: See [676, 933] or [1125, p. 151].

Remark: Replacing A, B, C, D by $\alpha B_1, (1 - \alpha)B_2, \alpha A_1, (1 - \alpha)A_2$ yields *xiv*) of Proposition 8.6.17.

Fact 8.11.28. Let $A \in \mathbb{R}^{n \times n}$, assume that A is positive definite, and let $\mathcal{S} \subseteq \{1, \ldots, n\}$. Then,

$$\left(A_{(\mathcal{S})} \right)^{-1} \leq \left(A^{-1} \right)_{(\mathcal{S})}.$$

Proof: See [728, p. 474].

Remark: Generalizations of this result are given in [336].

Fact 8.11.29. Let $A_{ij} \in \mathbb{F}^{n_i \times n_j}$ for all $i, j \in \{1, \ldots, k\}$, define

$$A \triangleq \begin{bmatrix} A_{11} & \cdots & A_{1k} \\ \vdots & \ddots & \vdots \\ A_{1k} & \cdots & A_{kk} \end{bmatrix},$$

and assume that A is square and positive definite. Furthermore, define

$$\hat{A} \triangleq \begin{bmatrix} \hat{A}_{11} & \cdots & \hat{A}_{1k} \\ \vdots & \ddots & \vdots \\ \hat{A}_{1k} & \cdots & \hat{A}_{kk} \end{bmatrix},$$

where $\hat{A}_{ij} = 1_{1 \times n_i} A_{ij} 1_{n_j \times 1}$ is the sum of the entries of A_{ij} for all $i, j \in \{1, \ldots, k\}$. Then, $\hat{A}$ is positive definite.

Proof: $\hat{A} = BAB^{\mathrm{T}}$, where the entries of $B \in \mathbb{R}^{k \times \sum_{i=1}^{k} n_i}$ are 0's and 1's. See [44].

Fact 8.11.30. Let $A, D \in \mathbb{F}^{n \times n}$, $B \in \mathbb{F}^{n \times m}$, and $C \in \mathbb{F}^{m \times m}$, and assume that $\begin{bmatrix} A & B \\ B^* & C \end{bmatrix} \in \mathbb{F}^{n \times n}$ is positive semidefinite, C is positive definite, and D is positive definite. Then, $\begin{bmatrix} A+D & B \\ B^* & C \end{bmatrix}$ is positive definite.

Fact 8.11.31. Let $A \in \mathbb{F}^{(n+m+l) \times (n+m+l)}$, assume that A is positive semidefinite, and assume that A is of the form

$$A = \begin{bmatrix} A_{11} & A_{12} & 0 \\ A_{12}^* & A_{22} & A_{23} \\ 0 & A_{32}^* & A_{33} \end{bmatrix}.$$

Then, there exist positive-semidefinite matrices $B, C \in \mathbb{F}^{(n+m+l) \times (n+m+l)}$ such that $A = B + C$ and such that B and C have the form

$$B = \begin{bmatrix} B_{11} & B_{12} & 0 \\ B_{12}^* & B_{22} & 0 \\ 0 & 0 & 0 \end{bmatrix}$$

and

$$C = \begin{bmatrix} 0 & 0 & 0 \\ 0 & C_{22} & C_{23} \\ 0 & C_{23}^* & C_{33} \end{bmatrix}.$$

Proof: See [687].

8.12 Facts on the Trace

Fact 8.12.1. Let $A \in \mathbb{F}^{n \times n}$, assume that A is positive definite, let p and q be real numbers, and assume that $p \leq q$. Then,

$$\left(\tfrac{1}{n} \operatorname{tr} A^p\right)^{1/p} \leq \left(\tfrac{1}{n} \operatorname{tr} A^q\right)^{1/q}.$$

Furthermore,

$$\lim_{p \downarrow 0} \left(\tfrac{1}{n} \operatorname{tr} A^p\right)^{1/p} = \det A^{1/n}.$$

Proof: Use Fact 1.17.30.

Fact 8.12.2. Let $A \in \mathbb{F}^{n \times n}$, and assume that A is positive definite. Then,

$$n^2 \leq (\operatorname{tr} A) \operatorname{tr} A^{-1}.$$

Finally, equality holds if and only if $A = I_n$.

Remark: Bounds on $\operatorname{tr} A^{-1}$ are given in [103, 315, 1079, 1160].

Fact 8.12.3. Let $A \in \mathbb{F}^{n \times n}$, and assume that A is positive semidefinite. Then, the following statements hold:

i) Let $r \in [0, 1]$. Then, for all $k \in \{1, \ldots, n\}$,

$$\sum_{i=k}^{n} \lambda_i^r(A) \leq \sum_{i=k}^{n} \mathrm{d}_i^r(A).$$

In particular,

$$\operatorname{tr} A^r \leq \sum_{i=1}^{n} A_{(i,i)}^r.$$

ii) Let $r \geq 1$. Then, for all $k \in \{1, \ldots, n\}$,

$$\sum_{i=1}^{k} \mathrm{d}_i^r(A) \leq \sum_{i=1}^{k} \lambda_i^r(A).$$

In particular,

$$\sum_{i=1}^{n} A_{(i,i)}^r \leq \operatorname{tr} A^r.$$

iii) If either $r = 0$ or $r = 1$, then

$$\operatorname{tr} A^r = \sum_{i=1}^{n} A_{(i,i)}^r.$$

iv) If $r \neq 0$ and $r \neq 1$, then

$$\operatorname{tr} A^r = \sum_{i=1}^{n} A_{(i,i)}^r$$

if and only if A is diagonal.

Proof: Use Fact 2.21.7 and Fact 8.18.8. See [971] and [973, p. 217].

Remark: See Fact 8.18.8.

Fact 8.12.4. Let $A \in \mathbb{F}^{n \times n}$, and let $p, q \in [0, \infty)$. Then,

$$\text{tr}\, (A^{*p}A^p)^q \leq \text{tr}\, (A^*A)^{pq}.$$

Furthermore, equality holds if and only if $\text{tr}\, A^{*p}A^p = \text{tr}\, (A^*A)^p$.

Proof: See [1239].

Fact 8.12.5. Let $A \in \mathbb{F}^{n \times n}$, $p \in [2, \infty)$, and $q \in [1, \infty)$. Then, A is normal if and only if

$$\text{tr}\, (A^{*p}A^p)^q = \text{tr}\, (A^*A)^{pq}.$$

Proof: See [1239].

Fact 8.12.6. Let $A, B \in \mathbb{F}^{n \times n}$, and assume that either A and B are Hermitian or A and B are skew Hermitian. Then, $\text{tr}\, AB$ is real.

Proof: $\text{tr}\, AB = \text{tr}\, A^*B^* = \text{tr}\, (BA)^* = \overline{\text{tr}\, BA} = \overline{\text{tr}\, AB}$.

Remark: See [1511] or [1526, p. 213].

Fact 8.12.7. Let $A, B \in \mathbb{F}^{n \times n}$, assume that A and B are Hermitian, and let $k \in \mathbb{N}$. Then, $\text{tr}\, (AB)^k$ is real.

Proof: See [57].

Fact 8.12.8. Let $A, B \in \mathbb{F}^{n \times n}$, and assume that A and B are Hermitian. Then,

$$\text{tr}\, AB \leq |\text{tr}\, AB| \leq \sqrt{(\text{tr}\, A^2)\, \text{tr}\, B^2} \leq \tfrac{1}{2}\text{tr}(A^2 + B^2).$$

The second inequality is an equality if and only if A and B are linearly dependent. The third inequality is an equality if and only if $\text{tr}\, A^2 = \text{tr}\, B^2$. All four terms are equal if and only if $A = B$.

Proof: Use the Cauchy-Schwarz inequality given by Corollary 9.3.9.

Remark: See Fact 8.12.20.

Fact 8.12.9. Let $A, B \in \mathbb{F}^{n \times n}$, assume that A and B are Hermitian, and assume that $-A \leq B \leq A$. Then,

$$\text{tr}\, B^2 \leq \text{tr}\, A^2.$$

Proof: $0 \leq \text{tr}[(A - B)(A + B)] = \text{tr}\, A^2 - \text{tr}\, B^2$. See [1350].

Remark: For $0 \leq B \leq A$, this result is a special case of *xxi*) of Proposition 8.6.13.

Fact 8.12.10. Let $A, B \in \mathbb{F}^{n \times n}$, and assume that A and B are positive semidefinite. Then, $AB = 0$ if and only if $\text{tr}\, AB = 0$.

Fact 8.12.11. Let $A, B \in \mathbb{F}^{n \times n}$, assume that A and B are positive semidefinite, and let $p, q \geq 1$ satisfy $1/p + 1/q = 1$. Then,

$$\operatorname{tr} AB \leq \operatorname{tr} \langle AB \rangle \leq (\operatorname{tr} A^p)^{1/p} (\operatorname{tr} B^q)^{1/q}.$$

Furthermore, equality holds for both inequalities if and only if A^{p-1} and B are linearly dependent.

Proof: See [971] and [973, pp. 219, 222].

Remark: This result is a matrix version of Hölder's inequality.

Remark: See Fact 8.12.12 and Fact 8.12.19.

Fact 8.12.12. Let $A_1, \ldots, A_m \in \mathbb{F}^{n \times n}$, assume that $A_1, \ldots, A_m$ are positive semidefinite, and let $p_1, \ldots, p_m \in [1, \infty)$ satisfy $\frac{1}{p_1} + \cdots + \frac{1}{p_1} = 1$. Then,

$$\operatorname{tr} \langle A_1 \cdots A_m \rangle \leq \prod_{i=1}^{m} (\operatorname{tr} A_i^{p_i})^{1/p_i} \leq \operatorname{tr} \sum_{i=1}^{m} \frac{1}{p_i} A_i^{p_i}.$$

Furthermore, the following statements are equivalent:

i) $\operatorname{tr} \langle A_1 \cdots A_m \rangle = \prod_{i=1}^{m} (\operatorname{tr} A_i^{p_i})^{1/p_i}$.

ii) $\operatorname{tr} \langle A_1 \cdots A_m \rangle = \operatorname{tr} \sum_{i=1}^{m} \frac{1}{p_i} A_i^{p_i}$.

iii) $A_1^{p_1} = \cdots = A_m^{p_m}$.

Proof: See [979].

Remark: The first inequality is a matrix version of Hölder's inequality. The first and third terms constitute a matrix version of Young's inequality. See Fact 1.12.32 and Fact 1.17.31.

Fact 8.12.13. Let $A_1, \ldots, A_m \in \mathbb{F}^{n \times n}$, assume that $A_1, \ldots, A_m$ are positive semidefinite, let $\alpha_1, \ldots, \alpha_m$ be nonnegative numbers, and assume that $\sum_{i=1}^{m} \alpha_i \geq 1$. Then,

$$\left| \operatorname{tr} \prod_{i=1}^{m} A_i^{\alpha_i} \right| \leq \prod_{i=1}^{m} (\operatorname{tr} A_i)^{\alpha_i}.$$

Furthermore, if $\sum_{i=1}^{m} \alpha_i = 1$, then equality holds if and only if $A_2, \ldots, A_m$ are scalar multiples of A_1, whereas, if $\sum_{i=1}^{m} \alpha_i > 1$, then equality holds if and only if $A_2, \ldots, A_m$ are scalar multiples of A_1 and rank $A_1 = 1$.

Proof: See [325].

Remark: See Fact 8.12.11.

Fact 8.12.14. Let $A, B \in \mathbb{F}^{n \times n}$. Then,

$$|\operatorname{tr} AB|^2 \leq (\operatorname{tr} A^*A) \operatorname{tr} BB^*.$$

Proof: See [1526, p. 25] or Corollary 9.3.9.

Remark: See Fact 8.12.15.

Fact 8.12.15. Let $A \in \mathbb{F}^{n \times m}$ and $B \in \mathbb{F}^{m \times n}$, and let $k \in \mathbb{N}$. Then,

$$|\mathrm{tr}\,(AB)^{2k}| \leq \mathrm{tr}\,(A^*ABB^*)^k \leq \mathrm{tr}\,(A^*A)^k(BB^*)^k \leq [\mathrm{tr}\,(A^*A)^k]\,\mathrm{tr}\,(BB^*)^k.$$

In particular,

$$|\mathrm{tr}\,(AB)^2| \leq \mathrm{tr}\,A^*ABB^* \leq (\mathrm{tr}\,A^*A)\,\mathrm{tr}\,BB^*.$$

Proof: See [1511] for the case $n = m$. If $n \neq m$, then A and B can be augmented with 0's.

Problem: Show that

$$\left.\begin{array}{c} |\mathrm{tr}\,AB|^2 \\ |\mathrm{tr}\,(AB)^2| \end{array}\right\} \leq \mathrm{tr}\,A^*ABB^* \leq (\mathrm{tr}\,A^*A)\,\mathrm{tr}\,BB^*.$$

See Fact 8.12.14.

Fact 8.12.16. Let $A, B \in \mathbb{F}^{n \times n}$, assume that A and B are Hermitian, and let $k \geq 1$. Then,

$$\mathrm{tr}\,\left(A^2B^2\right)^k \leq \left(\mathrm{tr}\,A^2B^2\right)^k$$

and

$$\mathrm{tr}\,(AB)^{2k} \leq |\mathrm{tr}\,(AB)^{2k}| \leq \left\{\begin{array}{c} \mathrm{tr}\,(A^2B^2)^k \\ \mathrm{tr}\,\langle(AB)^{2k}\rangle \end{array}\right\} \leq \mathrm{tr}\,A^{2k}B^{2k}.$$

Proof: Use Fact 8.12.15, and see [57, 1511].

Remark: It follows from Fact 8.12.7 that $\mathrm{tr}\,(AB)^{2k}$ and $\mathrm{tr}\,(A^2B^2)^k$ are real.

Fact 8.12.17. Let $A, B \in \mathbb{F}^{n \times n}$, assume that A and B are positive semidefinite, and let $k, m \in \mathbb{P}$, where $m \leq k$. Then,

$$\mathrm{tr}\,(A^m B^m)^k \leq \mathrm{tr}\,\left(A^k B^k\right)^m.$$

In particular,

$$\mathrm{tr}\,(AB)^k \leq \mathrm{tr}\,A^k B^k.$$

If, in addition, k is even, then

$$\mathrm{tr}\,(AB)^k \leq \mathrm{tr}\,\left(A^2 B^2\right)^{k/2} \leq \mathrm{tr}\,A^k B^k.$$

Proof: Use Fact 8.19.21 and Fact 8.19.28.

Remark: It follows from Fact 8.12.7 that $\mathrm{tr}\,(AB)^k$ is real.

Remark: The result $\mathrm{tr}\,(AB)^k \leq \mathrm{tr}\,A^k B^k$ is the *Lieb-Thirring inequality*. See [201, p. 279]. The inequality $\mathrm{tr}\,(AB)^k \leq \mathrm{tr}\,\left(A^2 B^2\right)^{k/2}$ follows from Fact 8.12.22. See [1501, 1511].

Fact 8.12.18. Let $A, B \in \mathbb{F}^{n \times n}$, assume that A is positive semidefinite, assume that B is Hermitian, and let $\alpha \in [0, 1]$. Then,

$$\mathrm{tr}\,(AB)^2 \leq \mathrm{tr}\,A^{2\alpha}BA^{2-2\alpha}B \leq \mathrm{tr}\,A^2 B^2.$$

Proof: See [535].

Fact 8.12.19. Let $A, B \in \mathbb{F}^{n \times n}$, and assume that A and B are positive semidefinite. Then,

$$\operatorname{tr} AB \le \operatorname{tr} \left(AB^2A \right)^{1/2} = \operatorname{tr} \langle AB \rangle \le \tfrac{1}{4} \operatorname{tr} (A+B)^2$$

and

$$\operatorname{tr} (AB)^2 \le \operatorname{tr} A^2 B^2 \le \tfrac{1}{16} \operatorname{tr} (A+B)^4.$$

Proof: See Fact 8.12.22 and Fact 9.9.18.

Fact 8.12.20. Let $A, B \in \mathbb{F}^{n \times n}$, and assume that A and B are positive semidefinite. Then,

$$\operatorname{tr} AB = \operatorname{tr} A^{1/2} B A^{1/2}$$

$$= \operatorname{tr} \left[\left(A^{1/2} B A^{1/2} \right)^{1/2} \left(A^{1/2} B A^{1/2} \right)^{1/2} \right]$$

$$\le \left[\operatorname{tr} \left(A^{1/2} B A^{1/2} \right)^{1/2} \right]^2$$

$$\le (\operatorname{tr} A)(\operatorname{tr} B)$$

$$\le \tfrac{1}{4} (\operatorname{tr} A + \operatorname{tr} B)^2$$

$$\le \tfrac{1}{2} \left[(\operatorname{tr} A)^2 + (\operatorname{tr} B)^2 \right]$$

and

$$\operatorname{tr} AB \le \sqrt{\operatorname{tr} A^2} \sqrt{\operatorname{tr} B^2}$$

$$\le \tfrac{1}{4} \left(\sqrt{\operatorname{tr} A^2} + \sqrt{\operatorname{tr} B^2} \right)^2$$

$$\le \tfrac{1}{2} \left(\operatorname{tr} A^2 + \operatorname{tr} B^2 \right)$$

$$\le \tfrac{1}{2} \left[(\operatorname{tr} A)^2 + (\operatorname{tr} B)^2 \right].$$

Remark: Use Fact 1.12.4.

Remark: Note that

$$\operatorname{tr} \left(A^{1/2} B A^{1/2} \right)^{1/2} = \sum_{i=1}^{n} \lambda_i^{1/2}(AB).$$

The second inequality follows from Proposition 9.3.6 with $p = q = 2$, $r = 1$, and A and B replaced by $A^{1/2}$ and $B^{1/2}$.

Remark: See Fact 2.12.16.

Fact 8.12.21. Let $A, B \in \mathbb{F}^{n \times n}$, assume that A and B are positive semidefinite, and let $p \ge 1$. Then,

$$\operatorname{tr} AB \le \operatorname{tr} (A^{p/2} B^p A^{p/2})^{1/p}.$$

Proof: See [534].

Fact 8.12.22. Let $A, B \in \mathbb{F}^{n \times n}$, assume that A and B are positive semidefinite, and let $p \geq 0$ and $r \geq 1$. Then,

$$\mathrm{tr}\left(A^{1/2}BA^{1/2}\right)^{pr} \leq \mathrm{tr}\left(A^{r/2}B^rA^{r/2}\right)^p.$$

In particular,

$$\mathrm{tr}\left(A^{1/2}BA^{1/2}\right)^{2p} \leq \mathrm{tr}\left(AB^2A\right)^p$$

and

$$\mathrm{tr}\,AB \leq \mathrm{tr}\left(AB^2A\right)^{1/2} = \mathrm{tr}\,\langle AB \rangle.$$

Proof: Use Fact 8.19.21 and Fact 8.19.28.

Remark: This result is the *Araki-Lieb-Thirring inequality*. See [72, 91] and [201, p. 258]. See Fact 8.10.50, Fact 8.19.27, and Fact 9.9.17.

Problem: Referring to Fact 8.12.20, compare the upper bounds

$$\mathrm{tr}\,AB \leq \begin{cases} \left[\mathrm{tr}\left(A^{1/2}BA^{1/2}\right)^{1/2}\right]^2 \\ \sqrt{\mathrm{tr}\,A^2}\sqrt{\mathrm{tr}\,B^2} \\ \mathrm{tr}\left(AB^2A\right)^{1/2}. \end{cases}$$

Fact 8.12.23. Let $A, B \in \mathbb{F}^{n \times n}$, assume that A and B are positive semidefinite, and let $q \geq 0$ and $t \in [0, 1]$. Then,

$$\sigma_{\max}^{2tq}(A)\,\mathrm{tr}\,B^{tq} \leq \mathrm{tr}(A^tB^tA^t)^q \leq \mathrm{tr}\,(ABA)^{tq}.$$

Proof: See [91].

Remark: The right-hand inequality is equivalent to the Araki-Lieb-Thirring inequality, where $t = 1/r$ and $q = pr$. See Fact 8.12.22.

Fact 8.12.24. Let $A, B \in \mathbb{F}^{n \times n}$, assume that A and B are positive semidefinite, and let $p \geq r \geq 0$. Then,

$$\left[\mathrm{tr}\left(A^{1/2}BA^{1/2}\right)^p\right]^{1/p} \leq \left[\mathrm{tr}\left(A^{1/2}BA^{1/2}\right)^r\right]^{1/r}.$$

In particular,

$$\left[\mathrm{tr}\left(A^{1/2}BA^{1/2}\right)^2\right]^{1/2} \leq \mathrm{tr}\,AB \leq \begin{cases} \mathrm{tr}\left(AB^2A\right)^{1/2} \\ \left[\mathrm{tr}\left(A^{1/2}BA^{1/2}\right)^{1/2}\right]^2. \end{cases}$$

Proof: This result follows from the power-sum inequality Fact 1.17.35. See [377].

Fact 8.12.25. Let $A, B \in \mathbb{F}^{n \times n}$, assume that A and B are positive semidefinite, assume that $A \leq B$, and let $p, q \geq 0$. Then,

$$\mathrm{tr}\,A^pB^q \leq \mathrm{tr}\,B^{p+q}.$$

If, in addition, A and B are positive definite, then this inequality holds for all $p, q \in \mathbb{R}$ satisfying $q \geq -1$ and $p + q \geq 0$.

Proof: See [250].

Fact 8.12.26. Let $A, B \in \mathbb{F}^{n \times n}$, assume that A and B are positive semidefinite, assume that $A \leq B$, let $f \colon [0, \infty) \mapsto [0, \infty)$, and assume that $f(0) = 0$, f is continuous, and f is increasing. Then,

$$\mathrm{tr}\, f(A) \leq \mathrm{tr}\, f(B).$$

Now, let $p > 1$ and $q \geq \max\{-1, -p/2\}$, and, if $q < 0$, assume that A is positive definite. Then,

$$\mathrm{tr}\, f(A^{q/2} B^p A^{q/2}) \leq \mathrm{tr}\, f(A^{p+q}).$$

Proof: See [541].

Fact 8.12.27. Let $A, B \in \mathbb{F}^{n \times n}$, assume that A and B are positive semidefinite, and let $\alpha \in [0, 1]$. Then,

$$\mathrm{tr}\, A^{\alpha} B^{1-\alpha} \leq (\mathrm{tr}\, A)^{\alpha} (\mathrm{tr}\, B)^{1-\alpha} \leq \mathrm{tr}[\alpha A + (1-\alpha)B].$$

Furthermore, the first inequality is an equality if and only if A and B are linearly dependent, while the second inequality is an equality if and only if $A = B$.

Proof: Use Fact 8.12.11 or Fact 8.12.13 for the left-hand inequality and Fact 1.12.21 for the right-hand inequality.

Fact 8.12.28. Let $A, B \in \mathbb{F}^{n \times n}$, assume that A and B are positive definite, and let $\alpha \in [0, 1]$. Then,

$$\left. \begin{array}{c} \mathrm{tr}\, A^{-\alpha} B^{\alpha-1} \\[4pt] \mathrm{tr}\,[\alpha A + (1-\alpha)B]^{-1} \end{array} \right\} \leq (\mathrm{tr}\, A^{-1})^{\alpha} (\mathrm{tr}\, B^{-1})^{1-\alpha} \leq \mathrm{tr}[\alpha A^{-1} + (1-\alpha)B^{-1}]$$

and

$$\mathrm{tr}\,[\alpha A + (1-\alpha)B]^{-1} \leq \left\{ \begin{array}{c} (\mathrm{tr}\, A^{-1})^{\alpha} (\mathrm{tr}\, B^{-1})^{1-\alpha} \\[6pt] \mathrm{tr}\left[A^{-1} (A^{-1/2} B A^{-1/2})^{\alpha-1} \right] \end{array} \right\} \leq \mathrm{tr}[\alpha A^{-1} + (1-\alpha)B^{-1}].$$

Remark: In the first string of inequalities, the upper left inequality and right-hand inequality are equivalent to Fact 8.12.27. The lower left inequality is given by *xxxiii)* of Proposition 8.6.17. The second string of inequalities combines the lower left inequality in the first string of inequalities with the third string of inequalities in Fact 8.10.46.

Remark: These inequalities interpolate the convexity of $\phi(A) = \mathrm{tr}\, A^{-1}$. See Fact 1.12.21.

Fact 8.12.29. Let $A, B \in \mathbb{F}^{n \times n}$, and assume that B is positive semidefinite. Then,

$$|\mathrm{tr}\, AB| \leq \sigma_{\max}(A) \mathrm{tr}\, B.$$

Proof: Use Proposition 8.4.13 and $\sigma_{\max}(A + A^*) \leq 2\sigma_{\max}(A)$.

Remark: See Fact 5.12.4.

Fact 8.12.30. Let $A, B \in \mathbb{F}^{n \times n}$, assume that A and B are positive semidefinite, and let $p \geq 1$. Then,

$$\mathrm{tr}(A^p + B^p) \leq \mathrm{tr}\, (A + B)^p \leq \left[(\mathrm{tr}\, A^p)^{1/p} + (\mathrm{tr}\, B^p)^{1/p} \right]^p.$$

Furthermore, the second inequality is an equality if and only if A and B are linearly independent.

Proof: See [250] and [971].

Remark: The first inequality is the *McCarthy inequality*. The second inequality is a special case of the triangle inequality for the norm $\|\cdot\|_{\sigma p}$ and a matrix version of Minkowski's inequality.

Fact 8.12.31. Let $A, B \in \mathbb{F}^{n \times n}$, assume that A and B are positive semidefinite, let m be a positive integer, and define $p \in \mathbb{F}[s]$ by

$$p(s) = \operatorname{tr}(A + sB)^m.$$

Then, all of the coefficients of p are nonnegative.

Remark: This result is the *Bessis-Moussa-Villani trace conjecture*. See [705, 934] and Fact 8.12.32.

Fact 8.12.32. Let $A, B \in \mathbb{F}^{n \times n}$, assume that A is Hermitian and B is positive semidefinite, and define

$$f(t) = e^{A + tB}.$$

Then, for all $k \in \mathbb{N}$ and $t \geq 0$,

$$(-1)^{k+1} f^{(k)}(t) \geq 0.$$

Remark: This result is a consequence of the Bessis-Moussa-Villani trace conjecture. See [705, 934] and Fact 8.12.31.

Remark: See Fact 8.14.18.

Fact 8.12.33. Let $A, B \in \mathbb{F}^{n \times n}$, assume that A and B are Hermitian, and let $f: \mathbb{R} \mapsto \mathbb{R}$. Then, the following statements hold:

i) If f is convex, then there exist unitary matrices $S_1, S_2 \in \mathbb{F}^{n \times n}$ such that

$$f[\tfrac{1}{2}(A + B)] \leq \tfrac{1}{2}[S_1(\tfrac{1}{2}[f(A) + f(B)])S_1^* + S_2(\tfrac{1}{2}[f(A) + f(B)])S_2^*].$$

ii) If f is convex and even, then there exist unitary matrices $S_1, S_2 \in \mathbb{F}^{n \times n}$ such that
$$f[\tfrac{1}{2}(A + B)] \leq \tfrac{1}{2}[S_1 f(A)S_1^* + S_2 f(B)S_2^*].$$

iii) If f is convex and increasing, then there exists a unitary matrix $S \in \mathbb{F}^{n \times n}$ such that
$$f[\tfrac{1}{2}(A + B)] \leq S(\tfrac{1}{2}[f(A) + f(B)])S^*.$$

iv) There exist unitary matrices $S_1, S_2 \in \mathbb{F}^{n \times n}$ such that
$$\langle A + B \rangle \leq S_1 \langle A \rangle S_1^* + S_2 \langle B \rangle S_2^*.$$

v) If f is convex, then
$$\operatorname{tr} f[\tfrac{1}{2}(A + B)] \leq \operatorname{tr} \tfrac{1}{2}[f(A) + f(B)].$$

Proof: See [251, 252].

Remark: Result *v)*, which is a consequence of *i)*, is *von Neumann's trace inequality*.

Remark: See Fact 8.12.34.

Fact 8.12.34. Let $f\colon \mathbb{R} \mapsto \mathbb{R}$, and assume that f is convex. Then, the following statements hold:

i) If $f(0) \leq 0$, $A \in \mathbb{F}^{n \times n}$ is Hermitian, and $S \in \mathbb{F}^{n \times m}$ is a contractive matrix, then

$$\operatorname{tr} f(S^*AS) \leq \operatorname{tr} S^* f(A) S.$$

ii) If $A_1, \ldots, A_k \in \mathbb{F}^{n \times n}$ are Hermitian and $S_1, \ldots, S_k \in \mathbb{F}^{n \times m}$ satisfy $\sum_{i=1}^{k} S_i^* S_i = I$, then

$$\operatorname{tr} f\left(\sum_{i=1}^{k} S_i^* A_i S_i\right) \leq \operatorname{tr} \sum_{i=1}^{k} S_i^* f(A_i) S_i.$$

iii) If $A \in \mathbb{F}^{n \times n}$ is Hermitian and $S \in \mathbb{F}^{n \times n}$ is a projector, then

$$\operatorname{tr} Sf(SAS)S \leq \operatorname{tr} Sf(A)S.$$

Proof: See [252] and [1066, p. 36].

Remark: Special cases are considered in [807].

Remark: The first result is due to Brown and Kosaki, the second result is due to Hansen and Pedersen, and the third result is due to Berezin.

Remark: The second result generalizes statement v) of Fact 8.12.33.

Fact 8.12.35. Let $A, B \in \mathbb{F}^{n \times n}$, assume that B is positive semidefinite, and assume that $A^*A \leq B$. Then,

$$|\operatorname{tr} A| \leq \operatorname{tr} B^{1/2}.$$

Proof: Corollary 8.6.11 with $r = 2$ implies that $(A^*A)^{1/2} \leq \operatorname{tr} B^{1/2}$. Letting $\operatorname{mspec}(A) = \{\lambda_1, \ldots, \lambda_n\}_{\mathrm{ms}}$, it follows from Fact 9.11.2 that $|\operatorname{tr} A| \leq \sum_{i=1}^{n} |\lambda_i| \leq \sum_{i=1}^{n} \sigma_i(A) = \operatorname{tr} (A^*A)^{1/2} \leq \operatorname{tr} B^{1/2}$. See [171].

Fact 8.12.36. Let $A, B \in \mathbb{F}^{n \times n}$, assume that A is positive definite and B is positive semidefinite, let $\alpha \in [0, 1]$, and let $\beta \geq 0$. Then,

$$\operatorname{tr}(-BA^{-1}B + \beta B^\alpha) \leq \beta(1 - \tfrac{\alpha}{2}) \operatorname{tr} \left(\tfrac{\alpha\beta}{2}A\right)^{\alpha/(2-\alpha)}.$$

If, in addition, either A and B commute or B is a multiple of a projector, then

$$-BA^{-1}B + \beta B^\alpha \leq \beta(1 - \tfrac{\alpha}{2})\left(\tfrac{\alpha\beta}{2}A\right)^{\alpha/(2-\alpha)}.$$

Proof: See [649, 650].

Fact 8.12.37. Let $A, P \in \mathbb{F}^{n \times n}$, $B, Q \in \mathbb{F}^{n \times m}$, and $C, R \in \mathbb{F}^{m \times m}$, and assume that $\begin{bmatrix} A & B \\ B^* & C \end{bmatrix}, \begin{bmatrix} P & Q \\ Q^* & R \end{bmatrix} \in \mathbb{F}^{(n+m) \times (n+m)}$ are positive semidefinite. Then,

$$|\operatorname{tr} BQ^*|^2 \leq (\operatorname{tr} AP)(\operatorname{tr} CR).$$

Proof: See [911, 1530].

Fact 8.12.38. Let $A, B \in \mathbb{F}^{n \times n}$, and assume that A and B are projectors. Then,

$$(\operatorname{tr} AB)^2 \leq (\operatorname{rank} AB)(\operatorname{tr} ABAB).$$

Furthermore, equality holds if and only if there exists $\alpha > 0$ such that αAB is idempotent.

Proof: See [116].

Fact 8.12.39. Let $A, B \in \mathbb{F}^{n \times m}$, let $X \in \mathbb{F}^{n \times n}$, and assume that X is positive definite. Then,

$$|\operatorname{tr} A^*B|^2 \leq (\operatorname{tr} A^*XA)(\operatorname{tr} B^*X^{-1}A).$$

Proof: Use Fact 8.12.37 with $\left[\begin{smallmatrix} X & I \\ I & X^{-1} \end{smallmatrix}\right]$ and $\left[\begin{smallmatrix} AA^* & AB^* \\ BA^* & BB^* \end{smallmatrix}\right]$. See [911, 1530].

Fact 8.12.40. Let $A, B, C \in \mathbb{F}^{n \times n}$, and assume that A and B are Hermitian and C is positive semidefinite. Then,

$$|\operatorname{tr} ABC^2 - \operatorname{tr} ACBC| \leq \tfrac{1}{4}[\lambda_1(A) - \lambda_n(A)][\lambda_1(B) - \lambda_n(B)] \operatorname{tr} C^2.$$

Proof: See [254].

Fact 8.12.41. Let $A_{11} \in \mathbb{R}^{n \times n}$, $A_{12} \in \mathbb{R}^{n \times m}$, and $A_{22} \in \mathbb{R}^{m \times m}$, define $A \triangleq \left[\begin{smallmatrix} A_{11} & A_{12} \\ A_{12}^{\mathrm{T}} & A_{22} \end{smallmatrix}\right] \in \mathbb{R}^{(n+m) \times (n+m)}$, and assume that A is symmetric. Then, A is positive semidefinite if and only if, for all $B \in \mathbb{R}^{n \times m}$,

$$\operatorname{tr} BA_{12}^{\mathrm{T}} \leq \operatorname{tr} \left(A_{11}^{1/2} BA_{22}B^{\mathrm{T}}A_{11}^{1/2} \right)^{1/2}.$$

Proof: See [171].

Fact 8.12.42. Let $A \in \mathbb{F}^{n \times n}$, $B \in \mathbb{F}^{n \times m}$, and $C \in \mathbb{F}^{m \times m}$, and assume that $\left[\begin{smallmatrix} A & B \\ B^* & C \end{smallmatrix}\right] \in \mathbb{F}^{(n+m) \times (n+m)}$ is positive semidefinite. Then,

$$\operatorname{tr} B^*B \leq \sqrt{(\operatorname{tr} A^2)(\operatorname{tr} C^2)} \leq (\operatorname{tr} A)(\operatorname{tr} C).$$

Proof: Use Fact 8.12.37 with $P = A$, $Q = B$, and $R = C$.

Remark: The inequality consisting of the first and third terms is given in [1102].

Remark: See Fact 8.12.43 for the case $n = m$.

Fact 8.12.43. Let $A, B, C \in \mathbb{F}^{n \times n}$, and assume that $\left[\begin{smallmatrix} A & B \\ B^* & C \end{smallmatrix}\right] \in \mathbb{F}^{2n \times 2n}$ is positive semidefinite. Then,

$$|\operatorname{tr} B|^2 \leq (\operatorname{tr} A)(\operatorname{tr} C)$$

and

$$|\operatorname{tr} B^2| \leq \operatorname{tr} B^*B \leq \sqrt{(\operatorname{tr} A^2)(\operatorname{tr} C^2)} \leq (\operatorname{tr} A)(\operatorname{tr} C).$$

Remark: The first result follows from Fact 8.12.44. In the second string, the first inequality is given by Fact 9.11.3, while the second inequality is given by Fact 8.12.42. The inequality $|\operatorname{tr} B^2| \leq \sqrt{(\operatorname{tr} A^2)(\operatorname{tr} C^2)}$ is given in [989].

Fact 8.12.44. Let $A_{ij} \in \mathbb{F}^{n \times n}$ for all $i, j \in \{1, \dots, k\}$, define $A \in \mathbb{F}^{kn \times kn}$ by

$$A \triangleq \begin{bmatrix} A_{11} & \cdots & A_{1k} \\ \vdots & \ddots & \vdots \\ A_{1k}^* & \cdots & A_{kk} \end{bmatrix},$$

and assume that A is positive semidefinite. Then,

$$\begin{bmatrix} \operatorname{tr} A_{11} & \cdots & \operatorname{tr} A_{1k} \\ \vdots & \ddots & \vdots \\ \operatorname{tr} A_{1k}^* & \cdots & \operatorname{tr} A_{kk} \end{bmatrix} \geq 0$$

and

$$\begin{bmatrix} \operatorname{tr} A_{11}^2 & \cdots & \operatorname{tr} A_{1k}^* A_{1k} \\ \vdots & \ddots & \vdots \\ \operatorname{tr} A_{1k}^* A_{1k} & \cdots & \operatorname{tr} A_{kk}^2 \end{bmatrix} \geq 0.$$

Proof: See [394, 989, 1102].

Remark: See Fact 8.13.43.

8.13 Facts on the Determinant

Fact 8.13.1. Let $A \in \mathbb{F}^{n \times n}$, assume that A is positive semidefinite, and let $\operatorname{mspec}(A) = \{\lambda_1, \dots, \lambda_n\}_{\mathrm{ms}}$. Then,

$$\lambda_{\min}(A) \leq \lambda_{\max}^{1/n}(A) \lambda_{\min}^{(n-1)/n}(A)$$

$$\leq \lambda_n$$

$$\leq \lambda_1$$

$$\leq \lambda_{\min}^{1/n}(A) \lambda_{\max}^{(n-1)/n}(A)$$

$$\leq \lambda_{\max}(A)$$

and

$$\lambda_{\min}^n(A) \leq \lambda_{\max}(A) \lambda_{\min}^{n-1}(A)$$

$$\leq \det A$$

$$\leq \lambda_{\min}(A) \lambda_{\max}^{n-1}(A)$$

$$\leq \lambda_{\max}^n(A).$$

Proof: Use Fact 5.11.29.

Fact 8.13.2. Let $A \in \mathbb{F}^{n \times n}$, and assume that $A + A^*$ is positive semidefinite. Then,

$$\det \tfrac{1}{2}(A + A^*) \leq |\det A|.$$

Furthermore, if $A + A^*$ is positive definite, then equality holds if and only if A is

Hermitian.

Proof: The inequality follows from Fact 5.11.25 and Fact 5.11.28.

Remark: This result is the *Ostrowski-Taussky inequality*.

Remark: See Fact 8.13.2.

Fact 8.13.3. Let $A \in \mathbb{F}^{n \times n}$, and assume that $A + A^*$ is positive semidefinite. Then,

$$[\det \tfrac{1}{2}(A + A^*)]^{2/n} + |\det \tfrac{1}{2}(A - A^*)|^{2/n} \leq |\det A|^{2/n}.$$

Furthermore, if $A + A^*$ is positive definite, then equality holds if and only if every eigenvalue of $(A + A^*)^{-1}(A - A^*)$ has the same absolute value. Finally, if $n \geq 2$, then

$$\det \tfrac{1}{2}(A + A^*) \leq \det \tfrac{1}{2}(A + A^*) + |\det \tfrac{1}{2}(A - A^*)| \leq |\det A|.$$

Proof: See [479, 783]. To prove the last result, use Fact 1.12.30.

Remark: Setting $A = 1 + \jmath$ shows that the last result can fail for $n = 1$.

Remark: $-A$ is semidissipative.

Remark: The last result interpolates Fact 8.13.2.

Remark: Extensions to the case in which $A + A^*$ is positive definite are considered in [1300].

Fact 8.13.4. Let $A, B \in \mathbb{F}^{n \times n}$, and assume that A is positive semidefinite. Then,

$$(\det A)^{2/n} + |\det(A + B)|^{2/n} \leq |\det(A + B)|^{2/n}.$$

Furthermore, if A is positive definite, then equality holds if and only if every eigenvalue of $A^{-1}B$ has the same absolute value. Finally, if $n \geq 2$, then

$$\det A \leq \det A + |\det B| \leq |\det(A + B)|.$$

Remark: This result is a restatement of Fact 8.13.2 in terms of the Cartesian decomposition.

Fact 8.13.5. Let $A, B \in \mathbb{F}^{n \times n}$, assume that A is positive semidefinite, assume that B is positive definite. Then,

$$\prod_{i=1}^{n} [\lambda_i^2(A) + \lambda_i^2(B)]^{1/2} \leq |\det(A + \jmath B)| \leq \prod_{i=1}^{n} [\lambda_i^2(A) + \lambda_{n-i+1}^2(B)]^{1/2}.$$

Proof: See [162].

Fact 8.13.6. Let $A, B \in \mathbb{F}^{n \times n}$, and assume that A is positive semidefinite and B is skew Hermitian. Then,

$$\det A \leq |\det(A + B)|.$$

Furthermore, if A and B are real, then

$$\det A \leq \det(A + B).$$

Finally, if A is positive definite, then equality holds if and only if $B = 0$.

Proof: See [671, p. 447] and [1125, pp. 146, 163]. Now, suppose that A and B are real. If A is positive definite, then $A^{-1/2}BA^{-1/2}$ is skew symmetric, and thus $\det(A+B) = (\det A)\det(I + A^{-1/2}BA^{-1/2})$ is positive. If A is positive semidefinite, then a continuity argument implies that $\det(A + B)$ is nonnegative.

Remark: Extensions of this result are given in [223].

Fact 8.13.7. Let $A, B \in \mathbb{F}^{n \times n}$, and assume that A is positive definite and B is Hermitian. Then,

$$\det(A + \jmath B) = (\det A)\prod_{i=1}^{n}\left[1 + \sigma_i^2\left(A^{-1/2}BA^{-1/2}\right)\right]^{1/2}.$$

Proof: See [328].

Fact 8.13.8. Let $A \in \mathbb{F}^{n \times n}$, and assume that A is positive definite. Then,

$$n + \operatorname{tr}\log A = n + \log\det A \le n(\det A)^{1/n} \le \operatorname{tr} A \le \left(n\operatorname{tr} A^2\right)^{1/2},$$

with equality if and only if $A = I$.

Remark: The inequality

$$(\det A)^{1/n} \le \tfrac{1}{n}\operatorname{tr} A$$

is a consequence of the arithmetic-mean–geometric-mean inequality.

Fact 8.13.9. Let $A, B \in \mathbb{F}^{n \times n}$, assume that A and B are positive semidefinite, and assume that $A \le B$. Then,

$$n\det A + \det B \le \det(A + B).$$

Proof: See [1125, pp. 154, 166].

Remark: Under weaker conditions, Corollary 8.4.15 implies that $\det A + \det B \le \det(A + B)$.

Fact 8.13.10. Let $A, B \in \mathbb{F}^{n \times n}$, and assume that A and B are positive semidefinite. Then,

$$\det A + \det B + (2^n - 2)\sqrt{\det AB} \le \det(A + B).$$

If, in addition, $B \le A$, then

$$\det A + (2^n - 1)\det B \le \det A + \det B + (2^n - 2)\sqrt{\det AB} \le \det(A + B).$$

Proof: See [1084] or [1215, p. 231].

Fact 8.13.11. Let $A \in \mathbb{R}^{n \times n}$, and assume that $A + A^{\mathrm{T}}$ is positive semidefinite. Then,

$$\left[\tfrac{1}{2}(A + A^{\mathrm{T}})\right]^{\mathrm{A}} \le \tfrac{1}{2}(A^{\mathrm{A}} + A^{\mathrm{AT}}).$$

Now, assume in addition that $A + A^{\mathrm{T}}$ is positive definite. Then,

$$\left[\det\tfrac{1}{2}(A + A^{\mathrm{T}})\right]\left[\tfrac{1}{2}(A + A^{\mathrm{T}})\right]^{-1} \le (\det A)\left[\tfrac{1}{2}(A^{-1} + A^{-\mathrm{T}})\right].$$

Furthermore,

$$[\det \tfrac{1}{2}(A + A^{\mathrm{T}})][\tfrac{1}{2}(A + A^{\mathrm{T}})]^{-1} < (\det A)[\tfrac{1}{2}(A^{-1} + A^{-\mathrm{T}})]$$

if and only if $\mathrm{rank}(A - A^{\mathrm{T}}) \geq 4$. Finally, if $n \geq 4$ and $A - A^{\mathrm{T}}$ is nonsingular, then

$$(\det A)[\tfrac{1}{2}(A^{-1} + A^{-\mathrm{T}})] < [\det A - \det \tfrac{1}{2}(A - A^{\mathrm{T}})][\tfrac{1}{2}(A + A^{\mathrm{T}})]^{-1}.$$

Proof: See [478, 782].

Remark: This result does not hold for complex matrices.

Remark: See Fact 8.9.32 and Fact 8.18.12.

Fact 8.13.12. Let $A \in \mathbb{F}^{n \times n}$, assume that A is Hermitian, and assume that, for all $i = 1, \ldots, n - 1$, $\det A_{(\{1,\ldots,i\})} > 0$. Then, $\mathrm{sign}[\lambda_{\min}(A)] = \mathrm{sign}(\det A)$. Furthermore, A is (positive semidefinite, positive definite) if and only if ($\det A \geq 0$, $\det A > 0$). Finally, if $\det A = 0$, then $\mathrm{rank}\, A = n - 1$.

Proof: Use Proposition 8.2.8 and Theorem 8.4.5. See [1208, p. 278].

Fact 8.13.13. Let $A \in \mathbb{R}^{n \times n}$, and assume that A is positive definite. Then,

$$\sum_{i=1}^{n} [\det A_{(\{1,\ldots,i\})}]^{1/i} \leq (1 + \tfrac{1}{n})^n \mathrm{tr}\, A < e\, \mathrm{tr}\, A.$$

Proof: See [31].

Fact 8.13.14. Let $A \in \mathbb{F}^{n \times n}$, assume that A is positive definite and Toeplitz, and, for all $i \in \{1, \ldots, n\}$, define $A_i \triangleq A_{(\{1,\ldots,i\})} \in \mathbb{F}^{i \times i}$. Then,

$$(\det A)^{1/n} \leq (\det A_{n-1})^{1/(n-1)} \leq \cdots \leq (\det A_2)^{1/2} \leq \det A_1.$$

Furthermore,

$$\frac{\det A}{\det A_{n-1}} \leq \frac{\det A_{n-1}}{\det A_{n-2}} \leq \cdots \leq \frac{\det A_3}{\det A_2} \leq \frac{\det A_2}{\det A_1}.$$

Proof: See [360] or [361, p. 682].

Fact 8.13.15. Let $A, B \in \mathbb{F}^{n \times n}$, assume that B is Hermitian, and assume that $A^*BA < A + A^*$. Then, $\det A \neq 0$.

Fact 8.13.16. Let $A, B \in \mathbb{F}^{n \times n}$, assume that A and B are positive definite, and let $\alpha \in [0, 1]$. Then,

$$(\det A)^{\alpha}(\det B)^{1-\alpha} \leq \det[\alpha A + (1 - \alpha)B].$$

Furthermore, equality holds if and only if $A = B$.

Proof: This inequality is a restatement of *xxxviii*) of Proposition 8.6.17.

Remark: This result is due to Bergstrom.

Remark: $\alpha = 2$ yields $\sqrt{(\det A)\det B} \leq \det[\tfrac{1}{2}(A + B)]$.

Fact 8.13.17. Let $A, B \in \mathbb{F}^{n \times n}$, assume that A and B are positive semidefinite, assume that either $A \leq B$ or $B \leq A$, and let $\alpha \in [0, 1]$. Then,

$$\det[\alpha A + (1 - \alpha)B] \leq \alpha \det A + (1 - \alpha) \det B.$$

Proof: See [1440].

Fact 8.13.18. Let $A, B \in \mathbb{F}^{n \times n}$, and assume that A and B are positive definite. Then,

$$\frac{\det A}{\det A_{[1;1]}} + \frac{\det B}{\det B_{[1;1]}} \leq \frac{\det(A + B)}{\det\left(A_{[1;1]} + B_{[1;1]}\right)}.$$

Proof: See [1125, p. 145].

Remark: This inequality is a special case of xli) of Proposition 8.6.17.

Remark: See Fact 8.11.4.

Fact 8.13.19. Let $A_1, \ldots, A_k \in \mathbb{F}^{n \times n}$, assume that $A_1, \ldots, A_k$ are positive semidefinite, and let $\lambda_1, \ldots, \lambda_k \in \mathbb{C}$. Then,

$$\det\left(\sum_{i=1}^{k} \lambda_i A_i\right) \leq \det\left(\sum_{i=1}^{k} |\lambda_i| A_i\right).$$

Proof: See [1125, p. 144].

Fact 8.13.20. Let $A, B, C \in \mathbb{R}^{n \times n}$, let $D \triangleq A + \jmath B$, and assume that $CB + B^{\mathrm{T}} C^{\mathrm{T}} < D + D^*$. Then, $\det A \neq 0$.

Fact 8.13.21. Let $A, B \in \mathbb{F}^{n \times n}$, assume that A and B are positive semidefinite, and let m be a positive integer. Then,

$$n^{1/m}(\det AB)^{1/n} \leq (\operatorname{tr} A^m B^m)^{1/m}.$$

Proof: See [377].

Remark: Assuming $\det B = 1$ and setting $m = 1$ yields Proposition 8.4.14.

Fact 8.13.22. Let $A, B, C \in \mathbb{F}^{n \times n}$, define

$$\mathcal{A} \triangleq \begin{bmatrix} A & B \\ B^* & C \end{bmatrix},$$

and assume that $\mathcal{A}$ is positive semidefinite. Then,

$$|\det(B + B^*)| \leq \det(A + C).$$

If, in addition, $\mathcal{A}$ is positive definite, then

$$|\det(B + B^*)| < \det(A + C).$$

Remark: Use Fact 8.11.5.

Fact 8.13.23. Let $A, B \in \mathbb{F}^{n \times m}$. Then,

$$|\det A^* B|^2 \leq (\det A^* A)(\det B^* B).$$

Proof: Use Fact 8.11.14 or apply Fact 8.13.43 to $\begin{bmatrix} A^*A & B^*A \\ A^*B & B^*B \end{bmatrix}$.

Remark: This result is a determinantal version of the Cauchy-Schwarz inequality.

Fact 8.13.24. Let $A \in \mathbb{F}^{n \times n}$, assume that A is positive definite, and let $B \in \mathbb{F}^{m \times n}$, where $\operatorname{rank} B = m$. Then,

$$(\det BB^*)^2 \leq (\det BAB^*) \det BA^{-1}B^*.$$

Proof: Use Fact 8.11.19.

Fact 8.13.25. Let $A, B \in \mathbb{F}^{n \times n}$. Then,

$$|\det(A+B)|^2 + |\det(I - AB^*)|^2 \leq \det(I + AA^*) \det(I + B^*B)$$

and

$$|\det(A-B)|^2 + |\det(I + AB^*)|^2 \leq \det(I + AA^*) \det(I + B^*B).$$

Furthermore, the first inequality is an equality if and only if either $n = 1$, $A + B = 0$, or $AB^* = I$.

Proof: This result follows from Fact 8.11.16. See [1526, p. 184].

Fact 8.13.26. Let $A, B \in \mathbb{F}^{n \times m}$, and assume that $I - A^*A$ and $I - B^*B$ are positive semidefinite. Then,

$$0 \leq \det(I - A^*A) \det(I - B^*B)$$

$$\leq \left\{ \begin{array}{l} |\det(I - A^*B)|^2 \\ |\det(I + A^*B)|^2 \end{array} \right\}$$

$$\leq \det(I + A^*A) \det(I + B^*B).$$

Now, assume in addition that $n = m$. Then,

$$0 \leq \det(I - A^*A) \det(I - B^*B)$$

$$\leq |\det(I - A^*B)|^2 - |\det(A - B)|^2$$

$$\leq |\det(I - A^*B)|^2$$

$$\leq |\det(I - A^*B)|^2 + |\det(A + B)|^2$$

$$\leq \det(I + A^*A) \det(I + B^*B)$$

and

$$0 \leq \det(I - A^*A) \det(I - B^*B)$$

$$\leq |\det(I + A^*B)|^2 - |\det(A + B)|^2$$

$$\leq |\det(I + A^*B)|^2$$

$$\leq |\det(I + A^*B)|^2 + |\det(A - B)|^2$$

$$\leq \det(I + A^*A) \det(I + B^*B).$$

Finally,

$$\left[\begin{array}{cc} \det[(I - A^*A)^{-1}] & \det[(I - A^*B)^{-1}] \\ \det[(I - B^*A)^{-1}] & \det[(I - B^*B)^{-1}] \end{array} \right] \geq 0.$$

Proof: The second inequality and Fact 8.11.21 are *Hua's inequalities*. See [49]. The third inequality follows from Fact 8.11.15. The first interpolation in the case $n = m$ is given in [1087].

Remark: Generalizations of the last result are given in [1502].

Remark: See Fact 8.11.21 and Fact 8.15.20.

Fact 8.13.27. Let $A, B \in \mathbb{F}^{n \times n}$, and let $\alpha, \beta \in (0, \infty)$. Then,

$$|\det(A + B)|^2 \leq \det(\beta^{-1}I + \alpha A^*A)\det(\alpha^{-1}I + \beta BB^*).$$

Proof: Use Fact 8.11.20. See [1527].

Fact 8.13.28. Let $A \in \mathbb{F}^{n \times m}$, $B \in \mathbb{F}^{n \times l}$, $C \in \mathbb{F}^{n \times m}$, and $D \in \mathbb{F}^{n \times l}$. Then,

$$|\det(AC^* + BD^*)|^2 \leq \det(AA^* + BB^*)\det(CC^* + DD^*).$$

Proof: Use Fact 8.13.39 and $\mathcal{A}\mathcal{A}^* \geq 0$, where $\mathcal{A} \triangleq \left[\begin{smallmatrix} A & B \\ C & D \end{smallmatrix}\right]$.

Remark: See Fact 2.14.22.

Fact 8.13.29. Let $A \in \mathbb{F}^{n \times m}$, $B \in \mathbb{F}^{n \times m}$, $C \in \mathbb{F}^{k \times m}$, and $D \in \mathbb{F}^{k \times m}$. Then,

$$|\det(A^*B + C^*D)|^2 \leq \det(A^*A + C^*C)\det(B^*B + D^*D).$$

Proof: Use Fact 8.13.39 and $\mathcal{A}^*\mathcal{A} \geq 0$, where $\mathcal{A} \triangleq \left[\begin{smallmatrix} A & B \\ C & D \end{smallmatrix}\right]$.

Remark: See Fact 2.14.18.

Fact 8.13.30. Let $A, B, C \in \mathbb{F}^{n \times n}$. Then,

$$|\det(B + CA)|^2 \leq \det(A^*A + B^*B)\det(I + CC^*).$$

Proof: See [736].

Remark: See Fact 8.10.37.

Fact 8.13.31. Let $A, B \in \mathbb{F}^{n \times m}$. Then, there exist unitary matrices $S_1, S_2 \in \mathbb{F}^{n \times n}$ such that

$$I + \langle A + B \rangle \leq S_1(I + \langle A \rangle)^{1/2}S_2(I + \langle B \rangle)S_2^*(I + \langle A \rangle)^{1/2}S_1^*.$$

Therefore,

$$\det(I + \langle A + B \rangle) \leq \det(I + \langle A \rangle)\det(I + \langle B \rangle).$$

Proof: See [49, 1301].

Remark: This result is due to Seiler and Simon.

Fact 8.13.32. Let $A, B \in \mathbb{F}^{n \times n}$, assume that $A + A^* > 0$ and $B + B^* \geq 0$, and let $\alpha > 0$. Then, $\alpha I + AB$ is nonsingular and has no negative eigenvalues. Hence,

$$\det(\alpha I + AB) > 0.$$

Proof: See [628].

Remark: Equivalently, $-A$ is dissipative and $-B$ is semidissipative.

Problem: Find a positive lower bound for $\det(\alpha I + AB)$ in terms of α, A, and B.

Fact 8.13.33. Let $A \in \mathbb{F}^{n \times n}$, assume that A is positive definite, and define

$$\alpha \triangleq \tfrac{1}{n} \operatorname{tr} A$$

and

$$\beta \triangleq \frac{1}{n(n-1)} \sum_{\substack{i,j=1 \\ i \neq j}}^{n} |A_{(i,j)}|.$$

Then,

$$|\det A| \leq (\alpha - \beta)^{n-1}[\alpha + (n-1)\beta].$$

Furthermore, if $A = aI_n + b1_{n \times n}$, where $a + nb > 0$ and $a > 0$, then $\alpha = a + b$, $\beta = b$, and equality holds.

Proof: See [1060].

Remark: See Fact 2.13.12 and Fact 8.9.35.

Fact 8.13.34. Let $A \in \mathbb{F}^{n \times n}$, assume that A is positive definite, and define

$$\beta \triangleq \frac{1}{n(n-1)} \sum_{\substack{i,j=1 \\ i \neq j}}^{n} \frac{|A_{(i,j)}|}{\sqrt{A_{(i,i)}A_{(j,j)}}}.$$

Then,

$$|\det A| \leq (1 - \beta)^{n-1}[1 + (n-1)\beta] \prod_{i=1}^{n} A_{(i,i)}.$$

Proof: See [1060].

Remark: This inequality strengthens Hadamard's inequality. See Fact 8.18.11. See also [422].

Fact 8.13.35. Let $A \in \mathbb{F}^{n \times n}$. Then,

$$|\det A| \leq \prod_{i=1}^{n} \left(\sum_{j=1}^{n} |A_{(i,j)}|^2 \right)^{1/2} = \prod_{i=1}^{n} \|\operatorname{row}_i(A)\|_2.$$

Furthermore, equality holds if and only if AA^* is diagonal. Now, let $\alpha > 0$ be such that, for all $i, j \in \{1, \ldots, n\}$, $|A_{(i,j)}| \leq \alpha$. Then,

$$|\det A| \leq \alpha^n n^{n/2}.$$

If, in addition, at least one entry of A has absolute value less than α, then

$$|\det A| < \alpha^n n^{n/2}.$$

Remark: Replace A with AA^* in Fact 8.18.11.

Remark: This result is a direct consequence of Hadamard's inequality. See Fact 8.18.11.

Remark: See Fact 2.13.14 and Fact 6.5.26.

Fact 8.13.36. Let $A \in \mathbb{F}^{n \times n}$, $B \in \mathbb{F}^{n \times m}$, and $C \in \mathbb{F}^{m \times m}$, define $\mathcal{A} \triangleq \left[\begin{smallmatrix} A & B \\ B^* & C \end{smallmatrix} \right] \in \mathbb{F}^{(n+m) \times (n+m)}$, and assume that $\mathcal{A}$ is positive definite. Then,

$$\det \mathcal{A} = (\det A) \det(C - B^* A^{-1} B) \leq (\det A) \det C \leq \prod_{i=1}^{n+m} \mathcal{A}_{(i,i)}.$$

Proof: The second inequality is obtained by successive application of the first inequality.

Remark: $\det \mathcal{A} \leq (\det A) \det C$ is *Fischer's inequality*.

Fact 8.13.37. Let $A \in \mathbb{F}^{n \times n}$, $B \in \mathbb{F}^{n \times m}$, and $C \in \mathbb{F}^{m \times m}$, define $\mathcal{A} \triangleq \left[\begin{smallmatrix} A & B \\ B^* & C \end{smallmatrix} \right] \in \mathbb{F}^{(n+m) \times (n+m)}$, assume that $\mathcal{A}$ is positive definite, let $k \triangleq \min\{m, n\}$, and, for $i = 1, \ldots, n$, let $\lambda_i \triangleq \lambda_i(\mathcal{A})$. Then,

$$\prod_{i=1}^{n+m} \lambda_i \leq (\det A) \det C \leq \left(\prod_{i=k+1}^{n+m-k} \lambda_i \right) \prod_{i=1}^{k} \left[\tfrac{1}{2}(\lambda_i + \lambda_{n+m-i+1}) \right]^2.$$

Proof: The left-hand inequality is given by Fact 8.13.36. The right-hand inequality is given in [1052].

Fact 8.13.38. Let $A \in \mathbb{F}^{n \times n}$, and let $\mathcal{S} \subseteq \{1, \ldots, n\}$. Then, the following statements hold:

i) If $\alpha \subset \{1, \ldots, n\}$, then

$$\det A \leq [\det A_{(\alpha)}] \det A_{(\alpha \sim)}.$$

ii) If $\alpha, \beta \subseteq \{1, \ldots, n\}$, then

$$\det A_{(\alpha \cup \beta)} \leq \frac{[\det A_{(\alpha)}] \det A_{(\beta)}}{\det A_{(\alpha \cap \beta)}}.$$

iii) If $1 \leq k \leq n - 1$, then

$$\left(\prod_{\{\alpha:\, \mathrm{card}(\alpha) = k+1\}} \det A_{(\alpha)} \right)^{\binom{n-1}{k-1}} \leq \left(\prod_{\{\alpha:\, \mathrm{card}(\alpha) = k\}} \det A_{(\alpha)} \right)^{\binom{n-1}{k}}.$$

Proof: See [963].

Remark: The first result is Fischer's inequality, see Fact 8.13.36. The second result is the *Hadamard-Fischer inequality*. The third result is *Szasz's inequality*. See [361, p. 680], [728, p. 479], and [963].

Remark: See Fact 8.13.37.

Fact 8.13.39. Let $A, B, C \in \mathbb{F}^{n \times n}$, define $\mathcal{A} \triangleq \left[\begin{smallmatrix} A & B \\ B^* & C \end{smallmatrix} \right] \in \mathbb{F}^{2n \times 2n}$, and assume that $\mathcal{A}$ is positive semidefinite. Then,

$$0 \leq (\det A) \det C - |\det B|^2 \leq \det \mathcal{A} \leq (\det A) \det C.$$

Hence,

$$|\det B|^2 \leq (\det A) \det C.$$

Furthermore, $\mathcal{A}$ is positive definite if and only if

$$|\det B|^2 < (\det A)\det C.$$

Proof: Assuming that A is positive definite, it follows that $0 \leq B^*A^{-1}B \leq C$, which implies that $|\det B|^2/\det A \leq \det C$. Then, use continuity for the case in which A is singular. For an alternative proof, see [1125, p. 142]. For the case in which $\mathcal{A}$ is positive definite, note that $0 \leq B^*A^{-1}B < C$, and thus $|\det B|^2/\det A < \det C$.

Remark: This result is due to Everitt.

Remark: See Fact 8.13.43.

Remark: When B is nonsquare, it is not necessarily true that $|\det(B^*B)|^2 < (\det A)\det C$. See [1528].

Fact 8.13.40. Let $A \in \mathbb{F}^{n \times n}$, $B \in \mathbb{F}^{n \times m}$, and $C \in \mathbb{F}^{m \times m}$, define $\mathcal{A} \triangleq \left[\begin{smallmatrix} A & B \\ B^* & C \end{smallmatrix}\right] \in \mathbb{F}^{(n+m) \times (n+m)}$, and assume that $\mathcal{A}$ is positive semidefinite and A is positive definite. Then,

$$B^*A^{-1}B \leq \left[\frac{\lambda_{\max}(\mathcal{A}) - \lambda_{\min}(\mathcal{A})}{\lambda_{\max}(\mathcal{A}) + \lambda_{\min}(\mathcal{A})}\right]^2 C.$$

Proof: See [911, 1530].

Fact 8.13.41. Let $A, B, C \in \mathbb{F}^{n \times n}$, define $\mathcal{A} \triangleq \left[\begin{smallmatrix} A & B \\ B^* & C \end{smallmatrix}\right] \in \mathbb{F}^{2n \times 2n}$, and assume that $\mathcal{A}$ is positive semidefinite. Then,

$$|\det B|^2 \leq \left[\frac{\lambda_{\max}(\mathcal{A}) - \lambda_{\min}(\mathcal{A})}{\lambda_{\max}(\mathcal{A}) + \lambda_{\min}(\mathcal{A})}\right]^{2n} (\det A)\det C.$$

Hence,

$$|\det B|^2 \leq \left[\frac{\lambda_{\max}(\mathcal{A}) - \lambda_{\min}(\mathcal{A})}{\lambda_{\max}(\mathcal{A}) + \lambda_{\min}(\mathcal{A})}\right]^2 (\det A)\det C.$$

Now, define $\hat{A} \triangleq \left[\begin{smallmatrix} \det A & \det B \\ \det B^* & \det C \end{smallmatrix}\right] \in \mathbb{F}^{2 \times 2}$. Then,

$$|\det B|^2 \leq \left[\frac{\lambda_{\max}(\hat{A}) - \lambda_{\min}(\hat{A})}{\lambda_{\max}(\hat{A}) + \lambda_{\min}(\hat{A})}\right]^2 (\det A)\det C.$$

Proof: See [911, 1530].

Remark: The second and third bounds are not comparable. See [911, 1530].

Fact 8.13.42. Let $A \in \mathbb{F}^{n \times n}$, $B \in \mathbb{F}^{n \times m}$, and $C \in \mathbb{F}^{m \times m}$, define $\mathcal{A} \triangleq \left[\begin{smallmatrix} A & B \\ B^* & C \end{smallmatrix}\right] \in \mathbb{F}^{(n+m) \times (n+m)}$, assume that $\mathcal{A}$ is positive semidefinite, and assume that A and C are positive definite. Then,

$$\det(A|\mathcal{A})\det(C|\mathcal{A}) \leq \det \mathcal{A}.$$

Proof: See [736].

Remark: This result is the *reverse Fischer inequality*.

Fact 8.13.43. Let $A_{ij} \in \mathbb{F}^{n \times n}$ for all $i, j \in \{1, \ldots, k\}$, define

$$
A \triangleq \begin{bmatrix} A_{11} & \cdots & A_{1k} \\ \vdots & \ddots & \vdots \\ A_{1k}^* & \cdots & A_{kk} \end{bmatrix},
$$

assume that A is positive semidefinite, let $1 \le k \le n$, and define

$$
\tilde{A}_k \triangleq \begin{bmatrix} A_{11}^{(k)} & \cdots & A_{1k}^{(k)} \\ \vdots & \ddots & \vdots \\ A_{1k}^{*(k)} & \cdots & A_{kk}^{(k)} \end{bmatrix}.
$$

Then, $\tilde{A}_k$ is positive semidefinite. In particular,

$$
\tilde{A}_n = \begin{bmatrix} \det A_{11} & \cdots & \det A_{1k} \\ \vdots & \ddots & \vdots \\ \det A_{1k}^* & \cdots & \det A_{kk} \end{bmatrix}
$$

is positive semidefinite. Furthermore,

$$
\det A \le \det \tilde{A}.
$$

Now, assume in addition that A is positive definite. Then, $\det A = \det \tilde{A}$ if and only if, for all distinct $i, j \in \{1, \ldots, k\}$, $A_{ij} = 0$.

Proof: The first statement is given in [394]. The inequality as well as the final statement are given in [1298].

Remark: $B^{(k)}$ is the kth compound of B. See Fact 7.5.17.

Remark: Note that every principal subdeterminant of $\tilde{A}_n$ is lower bounded by the determinant of a positive-semidefinite matrix. Hence, the inequality implies that $\tilde{A}_n$ is positive semidefinite.

Remark: A weaker result is given in [396] and quoted in [986] in terms of elementary symmetric functions of the eigenvalues of each block.

Remark: The example $A = \begin{bmatrix} 1 & 0 & 1 & 0 \\ 0 & 1 & 0 & 0 \\ 1 & 0 & 1 & 0 \\ 0 & 0 & 0 & 1 \end{bmatrix}$ shows that $\tilde{A}$ can be positive definite while A is singular.

Remark: The matrix whose (i, j) entry is $\det A_{ij}$ is a *determinantal compression* of A. See [395, 989, 1298].

Remark: See Fact 8.12.44.

8.14 Facts on Convex Sets and Convex Functions

Fact 8.14.1. Let $f: \mathbb{R}^n \mapsto \mathbb{R}^n$, and assume that f is convex. Then, for all $\alpha \in \mathbb{R}$, the sets $\{x \in \mathbb{R}^n: f(x) \le \alpha\}$ and $\{x \in \mathbb{R}^n: f(x) < \alpha\}$ are convex.

Proof: See [508, p. 108].

Remark: The converse is not true. Consider the function $f(x) = x^3$.

Fact 8.14.2. Let $A \in \mathbb{F}^{n \times n}$, assume that A is Hermitian, let $\alpha \geq 0$, and define the set $\mathcal{S} \triangleq \{x \in \mathbb{F}^n \colon x^*Ax < \alpha\}$. Then, the following statements hold:

i) $\mathcal{S}$ is open.

ii) $\mathcal{S}$ is a blunt cone if and only if $\alpha = 0$.

iii) $\mathcal{S}$ is nonempty if and only if either $\alpha > 0$ or $\lambda_{\min}(A) < 0$.

iv) $\mathcal{S}$ is convex if and only if $A \geq 0$.

v) $\mathcal{S}$ is convex and nonempty if and only if $\alpha > 0$ and $A \geq 0$.

vi) The following statements are equivalent:

    a) $\mathcal{S}$ is bounded.

    b) $\mathcal{S}$ is convex and bounded.

    c) $A > 0$.

vii) The following statements are equivalent:

    a) $\mathcal{S}$ is bounded and nonempty.

    b) $\mathcal{S}$ is convex, bounded, and nonempty.

    c) $\alpha > 0$ and $A > 0$.

Fact 8.14.3. Let $A \in \mathbb{F}^{n \times n}$, assume that A is Hermitian, let $\alpha \geq 0$, and define the set $\mathcal{S} \triangleq \{x \in \mathbb{F}^n \colon x^*Ax \leq \alpha\}$. Then, the following statements hold:

i) $\mathcal{S}$ is closed.

ii) $0 \in \mathcal{S}$, and thus $\mathcal{S}$ is nonempty.

iii) $\mathcal{S}$ is a pointed cone if and only if $\alpha = 0$ or $A \leq 0$.

iv) $\mathcal{S}$ is convex if and only if $A \geq 0$.

v) The following statements are equivalent:

    a) $\mathcal{S}$ is bounded.

    b) $\mathcal{S}$ is convex and bounded.

    c) $A > 0$.

Fact 8.14.4. Let $A \in \mathbb{F}^{n \times n}$, assume that A is Hermitian, let $\alpha \geq 0$, and define the set $\mathcal{S} \triangleq \{x \in \mathbb{F}^n \colon x^*Ax = \alpha\}$. Then, the following statements hold:

i) $\mathcal{S}$ is closed.

ii) $\mathcal{S}$ is nonempty if and only if either $\alpha = 0$ or $\lambda_{\max}(A) > 0$.

iii) The following statements are equivalent:

    a) $\mathcal{S}$ is a pointed cone.

    b) $0 \in \mathcal{S}$.

    c) $\alpha = 0$.

iv) $\mathcal{S} = \{0\}$ if and only if $\alpha = 0$ and either $A > 0$ or $A < 0$.

v) S is bounded if and only if either $A > 0$ or both $\alpha > 0$ and $A \leq 0$.

vi) S is bounded and nonempty if and only if $A > 0$.

vii) The following statements are equivalent:

   a) S is convex.

   b) S is convex and nonempty.

   c) $\alpha = 0$ and either $A > 0$ or $A < 0$.

$viii$) If $\alpha > 0$, then the following statements are equivalent:

   a) S is nonempty.

   b) S is not convex.

   c) $\lambda_{\max}(A) > 0$.

ix) The following statements are equivalent:

   a) S is convex and bounded.

   b) S is convex, bounded, and nonempty.

   c) $\alpha = 0$ and $A > 0$.

Fact 8.14.5. Let $A \in \mathbb{F}^{n \times n}$, assume that A is Hermitian, let $\alpha \geq 0$, and define the set $S \triangleq \{x \in \mathbb{F}^n : x^*Ax \geq \alpha\}$. Then, the following statements hold:

i) S is closed.

ii) S is a pointed cone if and only if $\alpha = 0$.

iii) S is nonempty if and only if either $\alpha = 0$ or $\lambda_{\max}(A) > 0$.

iv) S is bounded if and only if $S \subseteq \{0\}$.

v) The following statements are equivalent:

   a) S is bounded and nonempty.

   b) $S = \{0\}$.

   c) $\alpha = 0$ and $A < 0$.

vi) S is convex if and only if either S is empty or $S = \mathbb{F}^n$.

vii) S is convex and bounded if and only if S is empty.

$viii$) The following statements are equivalent:

   a) S is convex and nonempty.

   b) $S = \mathbb{F}^n$.

   c) $\alpha = 0$ and $A \geq 0$.

Fact 8.14.6. Let $A \in \mathbb{F}^{n \times n}$, assume that A is Hermitian, let $\alpha \geq 0$, and define the set $S \triangleq \{x \in \mathbb{F}^n : x^*Ax > \alpha\}$. Then, the following statements hold:

i) S is open.

ii) S is a blunt cone if and only if $\alpha = 0$.

iii) $\mathcal{S}$ is nonempty if and only if $\lambda_{\max}(A) > 0$.

iv) The following statements are equivalent:

   a) $\mathcal{S}$ is empty.

   b) $\lambda_{\max}(A) \leq 0$.

   c) $\mathcal{S}$ is bounded.

   d) $\mathcal{S}$ is convex.

Fact 8.14.7. Let $A \in \mathbb{C}^{n \times n}$, and define the *numerical range* of A by

$$\Theta_1(A) \triangleq \{x^*Ax: \ x \in \mathbb{C}^n \text{ and } x^*x = 1\}$$

and the set

$$\Theta(A) \triangleq \{x^*Ax: \ x \in \mathbb{C}^n\}.$$

Then, the following statements hold:

   i) $\Theta_1(A)$ is a closed, bounded, convex subset of $\mathbb{C}$.

   ii) $\Theta(A) = \{0\} \cup \text{cone}\,\Theta_1(A)$.

   iii) $\Theta(A)$ is a pointed, closed, convex cone contained in $\mathbb{C}$.

   iv) If A is Hermitian, then $\Theta_1(A)$ is a closed, bounded interval contained in $\mathbb{R}$.

   v) If A is Hermitian, then $\Theta(A)$ is either $(-\infty, 0]$, $[0, \infty)$, or $\mathbb{R}$.

   vi) $\Theta_1(A)$ satisfies

$$\text{co spec}(A) \subseteq \Theta_1(A) \subseteq \text{co}\{\nu_1 + \jmath\mu_1, \nu_1 + \jmath\mu_n, \nu_n + \jmath\mu_1, \nu_n + \jmath\mu_n\},$$

   where
$$\nu_1 \triangleq \lambda_{\max}\left[\tfrac{1}{2}(A + A^*)\right], \qquad \nu_n \triangleq \lambda_{\min}\left[\tfrac{1}{2}(A + A^*)\right],$$

$$\mu_1 \triangleq \lambda_{\max}\left[\tfrac{1}{\jmath 2}(A - A^*)\right], \qquad \mu_n \triangleq \lambda_{\min}\left[\tfrac{1}{\jmath 2}(A - A^*)\right].$$

   vii) If A is normal, then
$$\Theta_1(A) = \text{co spec}(A).$$

   viii) If $n \leq 4$ and $\Theta_1(A) = \text{co spec}(A)$, then A is normal.

   ix) $\Theta_1(A) = \text{co spec}(A)$ if and only if either A is normal or there exist matrices $A_1 \in \mathbb{F}^{n_1 \times n_1}$ and $A_2 \in \mathbb{F}^{n_2 \times n_2}$ such that $n_1 + n_2 = n$, $\Theta_1(A_1) \subseteq \Theta_1(A_2)$, and A is unitarily similar to $\left[\begin{smallmatrix} A_1 & 0 \\ 0 & A_2 \end{smallmatrix}\right]$.

Proof: See [625] or [730, pp. 11, 52].

Remark: $\Theta_1(A)$ is called the *field of values* in [730, p. 5].

Remark: See Fact 4.10.25 and Fact 8.14.7.

Remark: *viii*) is an example of the *quartic barrier*. See [359], Fact 8.15.33, and Fact 11.17.3.

Fact 8.14.8. Let $A \in \mathbb{R}^{n \times n}$, and define the *real numerical range* of A by

$$\Psi_1(A) \triangleq \{x^{\mathrm{T}}Ax: \ x \in \mathbb{R}^n \text{ and } x^{\mathrm{T}}x = 1\}$$

and the set

$$\Psi(A) \triangleq \{x^{\mathrm{T}}Ax: \ x \in \mathbb{R}^n\}.$$

Then, the following statements hold:

i) $\Psi_1(A) = \Psi_1[\frac{1}{2}(A + A^{\mathrm{T}})]$.

ii) $\Psi_1(A) = [\lambda_{\min}[\frac{1}{2}(A + A^{\mathrm{T}})], \lambda_{\min}[\frac{1}{2}(A + A^{\mathrm{T}})]]$.

iii) If A is symmetric, then $\Psi_1(A) = [\lambda_{\min}(A), \lambda_{\max}(A)]$.

iv) $\Psi(A) = \{0\} \cup \operatorname{cone} \Psi_1(A)$.

v) $\Psi(A)$ is either $(-\infty, 0]$, $[0, \infty)$, or $\mathbb{R}$.

vi) $\Psi_1(A) = \Theta_1(A)$ if and only if A is symmetric.

Proof: See [730, p. 83].

Remark: $\Theta_1(A)$ is defined in Fact 8.14.7.

Fact 8.14.9. Let $A, B \in \mathbb{C}^{n \times n}$, assume that A and B are Hermitian, and define

$$\Theta_1(A, B) \triangleq \left\{ \begin{bmatrix} x^*Ax \\ x^*Bx \end{bmatrix} : \ x \in \mathbb{C}^n \text{ and } x^*x = 1 \right\} \subseteq \mathbb{R}^2.$$

Then, $\Theta_1(A, B)$ is convex.

Proof: See [1117].

Remark: This result is an immediate consequence of Fact 8.14.7.

Fact 8.14.10. Let $A, B \in \mathbb{R}^{n \times n}$, assume that A and B are symmetric, and let α, β be real numbers. Then, the following statements are equivalent:

i) There exists $x \in \mathbb{R}^n$ such that $x^{\mathrm{T}}Ax = \alpha$ and $x^{\mathrm{T}}Bx = \beta$.

ii) There exists a positive-semidefinite matrix $X \in \mathbb{R}^{n \times n}$ such that $\operatorname{tr} AX = \alpha$ and $\operatorname{tr} BX = \beta$.

Proof: See [157, p. 84].

Fact 8.14.11. Let $A, B \in \mathbb{R}^{n \times n}$, assume that A and B are symmetric, and define

$$\Psi_1(A, B) \triangleq \left\{ \begin{bmatrix} x^{\mathrm{T}}Ax \\ x^{\mathrm{T}}Bx \end{bmatrix} : \ x \in \mathbb{R}^n \text{ and } x^{\mathrm{T}}x = 1 \right\} \subseteq \mathbb{R}^2$$

and

$$\Psi(A, B) \triangleq \left\{ \begin{bmatrix} x^{\mathrm{T}}Ax \\ x^{\mathrm{T}}Bx \end{bmatrix} : \ x \in \mathbb{R}^n \right\} \subseteq \mathbb{R}^2.$$

Then, $\Psi(A, B)$ is a pointed, convex cone. If, in addition, $n \geq 3$, then $\Psi_1(A, B)$ is convex.

Proof: See [157, pp. 84, 89] or [416, 1117].

Remark: $\Psi(A, B) = [\operatorname{cone} \Psi_1(A, B)] \cup \{[\begin{smallmatrix} 0 \\ 0 \end{smallmatrix}]\}$.

Remark: The set $\Psi(A, B)$ is not necessarily closed. See [416, 1090, 1091].

Fact 8.14.12. Let $A, B \in \mathbb{R}^{n \times n}$, where $n \geq 2$, assume that A and B are symmetric, let $a, b \in \mathbb{R}^n$, let $a_0, b_0 \in \mathbb{R}$, assume that there exist real numbers α, β such that $\alpha A + \beta B > 0$, and define

$$\Psi(A, a, a_0, B, b, b_0) \triangleq \left\{ \begin{bmatrix} x^{\mathrm{T}}Ax + a^{\mathrm{T}}x + a_0 \\ x^{\mathrm{T}}Bx + b^{\mathrm{T}}x + b_0 \end{bmatrix} : x \in \mathbb{R}^n \right\} \subseteq \mathbb{R}^2.$$

Then, $\Psi(A, a, a_0, B, b, b_0)$ is closed and convex.

Proof: See [1117].

Fact 8.14.13. Let $A, B, C \in \mathbb{R}^{n \times n}$, where $n \geq 3$, assume that A, B, and C are symmetric, and define

$$\Phi_1(A, B, C) \triangleq \left\{ \begin{bmatrix} x^{\mathrm{T}}Ax \\ x^{\mathrm{T}}Bx \\ x^{\mathrm{T}}Cx \end{bmatrix} : x \in \mathbb{R}^n \text{ and } x^{\mathrm{T}}x = 1 \right\} \subseteq \mathbb{R}^3$$

and

$$\Phi(A, B, C) \triangleq \left\{ \begin{bmatrix} x^{\mathrm{T}}Ax \\ x^{\mathrm{T}}Bx \\ x^{\mathrm{T}}Cx \end{bmatrix} : x \in \mathbb{R}^n \right\} \subseteq \mathbb{R}^3.$$

Then, $\Phi_1(A, B, C)$ is convex and $\Phi(A, B, C)$ is a pointed, convex cone.

Proof: See [264, 1114, 1117].

Fact 8.14.14. Let $A, B, C \in \mathbb{R}^{n \times n}$, where $n \geq 3$, assume that A, B, and C are symmetric, and define

$$\Phi(A, B, C) \triangleq \left\{ \begin{bmatrix} x^{\mathrm{T}}Ax \\ x^{\mathrm{T}}Bx \\ x^{\mathrm{T}}Cx \end{bmatrix} : x \in \mathbb{R}^n \right\} \subseteq \mathbb{R}^3.$$

Then, the following statements are equivalent:

i) There exist real numbers α, β, γ such that $\alpha A + \beta B + \gamma C$ is positive definite.

ii) $\Phi(A, B, C)$ is a pointed, one-sided, closed, convex cone, and, if $x \in \mathbb{R}^n$ satisfies $x^{\mathrm{T}}Ax = x^{\mathrm{T}}Bx = x^{\mathrm{T}}Cx = 0$, then $x = 0$.

Proof: See [1117].

Fact 8.14.15. Let $A \in \mathbb{F}^{n \times n}$, assume that A is Hermitian, let $b \in \mathbb{F}^n$ and $c \in \mathbb{R}$, and define $f \colon \mathbb{F}^n \mapsto \mathbb{R}$ by

$$f(x) \triangleq x^*Ax + \mathrm{Re}(b^*x) + c.$$

Then, the following statements hold:

i) f is convex if and only if A is positive semidefinite.

ii) f is strictly convex if and only if A is positive definite.

Now, assume in addition that A is positive semidefinite. Then, f has a minimizer if and only if $b \in \mathcal{R}(A)$. In this case, the following statements hold.

iii) The vector $x_0 \in \mathbb{F}^n$ is a minimizer of f if and only if x_0 satisfies $Ax_0 = -\frac{1}{2}b$.

iv) $x_0 \in \mathbb{F}^m$ minimizes f if and only if there exists a vector $y \in \mathbb{F}^m$ such that
$$x_0 = -\tfrac{1}{2}A^+b + (I - A^+A)y.$$

v) The minimum of f is given by
$$f(x_0) = c - x_0^*Ax_0 = c - \tfrac{1}{4}b^*A^+b.$$

vi) If A is positive definite, then $x_0 = -\frac{1}{2}A^{-1}b$ is the unique minimizer of f, and the minimum of f is given by
$$f(x_0) = c - x_0^*Ax_0 = c - \tfrac{1}{4}b^*A^{-1}b.$$

Proof: Use Proposition 6.1.7 and note that, for every x_0 satisfying $Ax_0 = -\frac{1}{2}b$, it follows that
$$\begin{aligned} f(x_0) &= (x - x_0)^*A(x - x_0) + c - x_0^*Ax_0 \\ &= (x - x_0)^*A(x - x_0) + c - \tfrac{1}{4}b^*A^+b. \end{aligned}$$

Remark: This result is the *quadratic minimization lemma*.

Remark: See Fact 9.15.4.

Fact 8.14.16. Let $A \in \mathbb{F}^{n \times n}$, assume that A is positive definite, and define $\phi \colon \mathbb{F}^{m \times n} \mapsto \mathbb{R}$ by $\phi(B) \triangleq \operatorname{tr} BAB^*$. Then, ϕ is strictly convex.

Proof: $\operatorname{tr}[\alpha(1 - \alpha)(B_1 - B_2)A(B_1 - B_2)^*] > 0$.

Fact 8.14.17. Let $p, q \in \mathbb{R}$, and define $\phi \colon \mathbf{P}^n \times \mathbf{P}^n \to (0, \infty)$ by
$$\phi(A, B) \triangleq \operatorname{tr} A^p B^q.$$
Then, the following statements hold:

i) If $p, q \in (0, 1)$ and $p + q \le 1$, then $-\phi$ is convex.

ii) If either $p, q \in [-1, 0)$ or $p \in [-1, 0)$, $q \in [1, 2]$, and $p + q \ge 1$, or $p \in [1, 2]$, $q \in [-1, 0]$, and $p + q \ge 1$, then ϕ is convex.

iii) If p, q do not satisfy the hypotheses of either *i)* or *ii)*, then neither ϕ nor $-\phi$ is convex.

Proof: See [170].

Fact 8.14.18. Let $B \in \mathbb{F}^{n \times n}$, assume that B is Hermitian, let $\alpha_1, \ldots, \alpha_k \in (0, \infty)$, define $r \triangleq \sum_{i=1}^{k} \alpha_i$, assume that $r \le 1$, let $q \in \mathbb{R}$, and define $\phi \colon \mathbf{P}^n \times \cdots \times \mathbf{P}^n \to [0, \infty)$ by
$$\phi(A_1, \ldots, A_k) \triangleq -\left[\operatorname{tr} e^{B + \sum_{i=1}^{k} \alpha_i \log A_i}\right]^q.$$
If $q \in (0, 1/r]$, then ϕ is convex. Furthermore, if $q < 0$, then $-\phi$ is convex.

Proof: See [931, 958].

Remark: See [1014] and Fact 8.12.32.

8.15 Facts on Quadratic Forms

Fact 8.15.1. Let $\mathcal{G} = (\mathcal{X}, \mathcal{R})$ be a symmetric graph, where $\mathcal{X} = \{x_1, \ldots, x_n\}$. Then, for all $z \in \mathbb{R}^n$, it follows that

$$z^\mathrm{T} L z = \tfrac{1}{2} \sum (z_{(i)} - z_{(j)})^2,$$

where the sum is over the set $\{(i, j) \colon (x_i, x_j) \in \mathcal{R}\}$.

Proof: See [275, pp. 29, 30] or [1018].

Fact 8.15.2. Let $A \in \mathbb{F}^{n \times n}$, and assume that A is Hermitian. Then,

$$\mathcal{N}(A) \subseteq \{x \in \mathbb{F}^n \colon x^* A x = 0\}.$$

Furthermore,

$$\mathcal{N}(A) = \{x \in \mathbb{F}^n \colon x^* A x = 0\}$$

if and only if either $A \geq 0$ or $A \leq 0$.

Fact 8.15.3. Let $x, y \in \mathbb{F}^n$. Then, $xx^* \leq yy^*$ if and only if there exists $\alpha \in \mathbb{F}$ such that $|\alpha| \in [0, 1]$ and $x = \alpha y$.

Fact 8.15.4. Let $x, y \in \mathbb{F}^n$. Then, $xy^* + yx^* \geq 0$ if and only if x and y are linearly dependent.

Proof: Evaluate the product of the nonzero eigenvalues of $xy^* + yx^*$, and use the Cauchy-Schwarz inequality $|x^* y|^2 \leq x^* x y^* y$.

Fact 8.15.5. Let $A \in \mathbb{F}^{n \times n}$, assume that A is positive definite, let $x \in \mathbb{F}^n$, and let $a \in [0, \infty)$. Then, the following statements are equivalent:

i) $xx^* \leq aA$.

ii) $x^* A^{-1} x \leq a$.

iii) $\begin{bmatrix} A & x \\ x^* & a \end{bmatrix} \geq 0$.

Proof: Use Fact 2.14.3 and Proposition 8.2.4. Note that, if $a = 0$, then $x = 0$.

Fact 8.15.6. Let $A, B \in \mathbb{F}^{n \times n}$, assume that A and B are Hermitian, assume that $A + B$ is nonsingular, let $x, a, b \in \mathbb{F}^n$, and define $c \triangleq (A + B)^{-1}(Aa + Bb)$. Then,

$$(x-a)^* A(x-a) + (x-b)^* B(x-b) = (x-c)^*(A+B)(x-c) = (a-b)^* A(A+B)^{-1} B(a-b).$$

Proof: See [1215, p. 278].

Fact 8.15.7. Let $A, B \in \mathbb{R}^{n \times n}$, assume that A is symmetric and B is skew symmetric, and let $x, y \in \mathbb{R}^n$. Then,

$$\begin{bmatrix} x \\ y \end{bmatrix}^\mathrm{T} \begin{bmatrix} A & B \\ B^\mathrm{T} & A \end{bmatrix} \begin{bmatrix} x \\ y \end{bmatrix} = (x + \jmath y)^*(A + \jmath B)(x + \jmath y).$$

Remark: See Fact 4.10.27.

Fact 8.15.8. Let $A \in \mathbb{F}^{n \times n}$, assume that A is positive definite, and let $x, y \in \mathbb{F}^n$. Then,

$$2 \operatorname{Re} x^* y \le x^* A x + y^* A^{-1} y.$$

Furthermore, if $y = Ax$, then equality holds. Therefore,

$$x^* A x = \max_{z \in \mathbb{F}^n} [2 \operatorname{Re} x^* z - z^* A z].$$

Proof: $\left(A^{1/2}x - A^{-1/2}y\right)^* \left(A^{1/2}x - A^{-1/2}y\right) \ge 0.$

Remark: This result is due to Bellman. See [911, 1530].

Fact 8.15.9. Let $A \in \mathbb{F}^{n \times n}$, assume that A is positive definite, and let $x, y \in \mathbb{F}^n$. Then,

$$|x^* y|^2 \le (x^* A x)(y^* A^{-1} y).$$

Proof: Use Fact 8.11.14 with A replaced by $A^{1/2}x$ and B replaced by $A^{-1/2}y$.

Fact 8.15.10. Let $A \in \mathbb{F}^{n \times n}$, assume that A is positive definite, and let $x \in \mathbb{F}^n$. Then,

$$(x^* x)^2 \le (x^* A x)(x^* A^{-1} x) \le \frac{(\alpha + \beta)^2}{4\alpha\beta} (x^* x)^2,$$

where $\alpha \triangleq \lambda_{\min}(A)$ and $\beta \triangleq \lambda_{\max}(A)$.

Remark: The second inequality is the *Kantorovich inequality*. See Fact 1.17.37 and [24]. See also [952].

Fact 8.15.11. Let $A \in \mathbb{F}^{n \times n}$, assume that A is positive definite, and let $x \in \mathbb{F}^n$. Then,

$$(x^* x)^{1/2}(x^* A x)^{1/2} - x^* A x \le \frac{(\alpha - \beta)^2}{4(\alpha + \beta)} x^* x$$

and

$$(x^* x)(x^* A^2 x) - (x^* A x)^2 \le \tfrac{1}{4}(\alpha - \beta)^2 (x^* x)^2,$$

where $\alpha \triangleq \lambda_{\min}(A)$ and $\beta \triangleq \lambda_{\max}(A)$.

Proof: See [1106].

Remark: Extensions of these results are given in [770, 1106].

Fact 8.15.12. Let $A \in \mathbb{F}^{n \times n}$, assume that A is positive semidefinite, let $r \triangleq \operatorname{rank} A$, let $x \in \mathbb{F}^n$, and assume that $x \notin \mathcal{N}(A)$. Then,

$$\frac{x^* A x}{x^* x} - \frac{x^* x}{x^* A^+ x} \le [\lambda_{\max}^{1/2}(A) - \lambda_r^{1/2}(A)]^2.$$

If, in addition, A is positive definite, then, for all nonzero $x \in \mathbb{F}^n$,

$$0 \le \frac{x^* A x}{x^* x} - \frac{x^* x}{x^* A^{-1} x} \le [\lambda_{\max}^{1/2}(A) - \lambda_{\min}^{1/2}(A)]^2.$$

Proof: See [1041, 1106]. The left-hand inequality in the last string of inequalities is given by Fact 8.15.10.

Fact 8.15.13. Let $A \in \mathbb{F}^{n \times n}$, assume that A is positive definite, let $y \in \mathbb{F}^n$, let $\alpha > 0$, and define $f \colon \mathbb{F}^n \mapsto \mathbb{R}$ by $f(x) \triangleq |x^*y|^2$. Then,

$$x_0 = \sqrt{\frac{\alpha}{y^*A^{-1}y}} A^{-1}y$$

minimizes $f(x)$ subject to $x^*Ax \leq \alpha$. Furthermore, $f(x_0) = \alpha y^*A^{-1}y$.

Proof: See [33].

Fact 8.15.14. Let $A \in \mathbb{F}^{n \times n}$, assume that A is positive semidefinite, and let $x \in \mathbb{F}^n$. Then,

$$\left(x^*A^2x\right)^2 \leq (x^*Ax)\left(x^*A^3x\right)$$

and

$$(x^*Ax)^2 \leq (x^*x)\left(x^*A^2x\right).$$

Proof: Apply the Cauchy-Schwarz inequality given by Corollary 9.1.7.

Fact 8.15.15. Let $A \in \mathbb{F}^{n \times n}$, assume that A is positive semidefinite, and let $x \in \mathbb{F}^n$. If $\alpha \in [0, 1]$, then

$$x^*A^\alpha x \leq (x^*x)^{1-\alpha}(x^*Ax)^\alpha.$$

Furthermore, if $\alpha > 1$, then

$$(x^*Ax)^\alpha \leq (x^*x)^{\alpha-1}x^*A^\alpha x.$$

Remark: The first inequality is the *Hölder-McCarthy inequality*, which is equivalent to the Young inequality. See Fact 8.9.43, Fact 8.10.43, [544, p. 125], and [546]. Matrix versions of the second inequality are given in [715].

Fact 8.15.16. Let $A \in \mathbb{F}^{n \times n}$, assume that A is positive semidefinite, let $x \in \mathbb{F}^n$, and let $\alpha, \beta \in [1, \infty)$, where $\alpha \leq \beta$. Then,

$$(x^*A^\alpha x)^{1/\alpha} \leq (x^*A^\beta x)^{1/\beta}.$$

Now, assume in addition that A is positive definite. Then,

$$x^*(\log A)x \leq \log x^*Ax \leq \tfrac{1}{\alpha}\log x^*A^\alpha x \leq \tfrac{1}{\beta}\log x^*A^\beta x.$$

Proof: See [522].

Fact 8.15.17. Let $A \in \mathbb{F}^{n \times n}$, $x, y \in \mathbb{F}^n$, and $\alpha \in (0, 1)$. Then,

$$|x^*Ay| \leq \|\langle A\rangle^\alpha x\|_2\|\langle A^*\rangle^{1-\alpha}y\|_2.$$

Consequently,

$$|x^*Ay| \leq [x^*\langle A\rangle x]^{1/2}[y^*\langle A^*\rangle y]^{1/2}.$$

Proof: See [797].

Fact 8.15.18. Let $A, B \in \mathbb{F}^{n \times n}$, assume that A is positive semidefinite, assume that AB is Hermitian, and let $x \in \mathbb{F}^n$. Then,

$$|x^*ABx| \leq \mathrm{sprad}(B)x^*Ax.$$

Proof: See [937].

Remark: This result is the sharpening by Halmos of Reid's inequality. Related results are given in [938].

Fact 8.15.19. Let $A, B \in \mathbb{F}^{n \times n}$, assume that A and B are positive definite, and let $x \in \mathbb{F}^n$. Then,

$$x^*(A + B)^{-1}x \le \frac{x^*A^{-1}xx^*B^{-1}x}{x^*A^{-1}x + x^*B^{-1}x} \le \tfrac{1}{4}\left(x^*A^{-1}x + x^*B^{-1}x\right).$$

In particular,

$$\frac{1}{(A^{-1})_{(i,i)}} + \frac{1}{(B^{-1})_{(i,i)}} \le \frac{1}{[(A + B)^{-1}]_{(i,i)}}.$$

Proof: See [973, p. 201]. The right-hand inequality follows from Fact 1.12.4.

Remark: This result is *Bergstrom's inequality*.

Remark: This result is a special case of Fact 8.11.3, which is a special case of *xvii)* of Proposition 8.6.17.

Fact 8.15.20. Let $A, B \in \mathbb{F}^{n \times m}$, assume that $I - A^*A$ and $I - B^*B$ are positive semidefinite, and let $x \in \mathbb{C}^n$. Then,

$$x^*(I - A^*A)xx^*(I - B^*B)x \le |x^*(I - A^*B)x|^2.$$

Remark: This result is due to Marcus. See [1087].

Remark: See Fact 8.13.26.

Fact 8.15.21. Let $A, B \in \mathbb{R}^n$, and assume that A is Hermitian and B is positive definite. Then,

$$\lambda_{\max}\left(AB^{-1}\right) = \max\{\lambda \in \mathbb{R}: \ \det(A - \lambda B) = 0\} = \min_{x \in \mathbb{F}^n \setminus \{0\}} \frac{x^*Ax}{x^*Bx}.$$

Proof: Use Lemma 8.4.3.

Fact 8.15.22. Let $A, B \in \mathbb{F}^{n \times n}$, and assume that A is positive definite and B is positive semidefinite. Then,

$$4(x^*x)(x^*Bx) < (x^*Ax)^2$$

for all nonzero $x \in \mathbb{F}^n$ if and only if there exists $\alpha > 0$ such that

$$\alpha I + \alpha^{-1}B < A.$$

In this case, $4B < A^2$, and hence $2B^{1/2} < A$.

Proof: Sufficiency follows from $\alpha x^*x + \alpha^{-1}x^*Bx < x^*Ax$. Necessity follows from Fact 8.15.23. The last result follows from $(A - 2\alpha I)^2 \ge 0$ or $2B^{1/2} \le \alpha I + \alpha^{-1}B$.

Fact 8.15.23. Let $A, B, C \in \mathbb{F}^{n \times n}$, assume that A, B, C are positive semidefinite, and assume that

$$4(x^*Cx)(x^*Bx) < (x^*Ax)^2$$

for all nonzero $x \in \mathbb{F}^n$. Then, there exists $\alpha > 0$ such that

$$\alpha C + \alpha^{-1}B < A.$$

Proof: See [1110].

Fact 8.15.24. Let $A, B \in \mathbb{F}^{n \times n}$, and assume that A is Hermitian and B is positive semidefinite. Then, $x^*Ax < 0$ for all $x \in \mathbb{F}^n$ such that $Bx = 0$ and $x \neq 0$ if and only if there exists $\alpha > 0$ such that $A < \alpha B$.

Proof: To prove necessity, suppose that, for every $\alpha > 0$, there exists a nonzero vector x such that $x^*Ax \geq \alpha x^*Bx$. Now, $Bx = 0$ implies that $x^*Ax \geq 0$. Sufficiency is immediate.

Fact 8.15.25. Let $A, B \in \mathbb{C}^{n \times n}$, and assume that A and B are Hermitian. Then, the following statements are equivalent:

i) There exist $\alpha, \beta \in \mathbb{R}$ such that $\alpha A + \beta B$ is positive definite.

ii) $\{x \in \mathbb{C}^n : x^*Ax = x^*Bx = 0\} = \{0\}$.

Remark: This result is *Finsler's lemma*. See [86, 167, 891, 1373, 1385].

Remark: See Fact 8.15.26, Fact 8.17.5, and Fact 8.17.6.

Fact 8.15.26. Let $A, B \in \mathbb{R}^{n \times n}$, and assume that A and B are symmetric. Then, the following statements are equivalent:

i) There exist $\alpha, \beta \in \mathbb{R}$ such that $\alpha A + \beta B$ is positive definite.

ii) Either $x^\mathrm{T}Ax > 0$ for all nonzero $x \in \{y \in \mathbb{F}^n : y^\mathrm{T}By = 0\}$ or $x^\mathrm{T}Ax < 0$ for all nonzero $x \in \{y \in \mathbb{F}^n : y^\mathrm{T}By = 0\}$.

Now, assume in addition that $n \geq 3$. Then, the following statement is equivalent to *i)* and *ii)*:

iii) $\{x \in \mathbb{R}^n : x^\mathrm{T}Ax = x^\mathrm{T}Bx = 0\} = \{0\}$.

Remark: This result is related to Finsler's lemma. See [86, 167, 1385].

Remark: See Fact 8.15.25, Fact 8.17.5, and Fact 8.17.6.

Fact 8.15.27. Let $A, B \in \mathbb{C}^{n \times n}$, assume that A and B are Hermitian, and assume that $x^*(A + \jmath B)x$ is nonzero for all nonzero $x \in \mathbb{C}^n$. Then, there exists $t \in [0, \pi)$ such that $(\sin t)A + (\cos t)B$ is positive definite.

Proof: See [363] or [1261, p. 282].

Fact 8.15.28. Let $A \in \mathbb{R}^{n \times n}$, assume that A is symmetric, and let $B \in \mathbb{R}^{n \times m}$. Then, the following statements are equivalent:

i) $x^\mathrm{T}Ax > 0$ for all nonzero $x \in \mathcal{N}(B^\mathrm{T})$.

ii) $\nu_+\left(\begin{bmatrix} A & B \\ B^\mathrm{T} & 0 \end{bmatrix}\right) = n$.

Furthermore, the following statements are equivalent:

iii) $x^\mathrm{T}Ax \geq 0$ for all $x \in \mathcal{N}(B^\mathrm{T})$.

iv) $\nu_-\left(\begin{bmatrix} A & B \\ B^\mathrm{T} & 0 \end{bmatrix}\right) = \operatorname{rank} B$.

Proof: See [307, 970].

Remark: See Fact 5.8.21 and Fact 8.15.29.

Fact 8.15.29. Let $A \in \mathbb{R}^{n \times n}$, assume that A is symmetric, let $B \in \mathbb{R}^{n \times m}$, where $m \leq n$, and assume that $\begin{bmatrix} I_m & 0 \end{bmatrix} B$ is nonsingular. Then, the following statements are equivalent:

i) $x^{\mathrm{T}} A x > 0$ for all nonzero $x \in \mathcal{N}(B^{\mathrm{T}})$.

ii) For all $i \in \{m+1, \ldots, n\}$, the sign of the $i \times i$ leading principal subdeterminant of the matrix $\begin{bmatrix} 0 & B^{\mathrm{T}} \\ B & A \end{bmatrix}$ is $(-1)^m$.

Proof: See [97, p. 20], [961, p. 312], or [980].

Remark: See Fact 8.15.28.

Fact 8.15.30. Let $A \in \mathbb{F}^{n \times n}$, assume that A is positive semidefinite and nonzero, let $x, y \in \mathbb{F}^n$, and assume that $x^* y = 0$. Then,

$$|x^* A y|^2 \leq \left[\frac{\lambda_{\max}(A) - \lambda_{\min}(A)}{\lambda_{\max}(A) + \lambda_{\min}(A)} \right]^2 (x^* A x)(y^* A y).$$

Furthermore, there exist vectors $x, y \in \mathbb{F}^n$ satisfying $x^* y = 0$ for which equality holds.

Proof: See [730, p. 443] or [911, 1530].

Remark: This result is the *Wielandt inequality*.

Fact 8.15.31. Let $A \in \mathbb{F}^{n \times n}$, $B \in \mathbb{F}^{n \times m}$, and $C \in \mathbb{F}^{m \times m}$, define $\mathcal{A} \triangleq \begin{bmatrix} A & B \\ B^* & C \end{bmatrix}$, and assume that A and C are positive semidefinite. Then, the following statements are equivalent:

i) $\mathcal{A}$ is positive semidefinite.

ii) $|x^* B y|^2 \leq (x^* A x)(y^* C y)$ for all $x \in \mathbb{F}^n$ and $y \in \mathbb{F}^m$.

iii) $2|x^* B y| \leq x^* A x + y^* C y$ for all $x \in \mathbb{F}^n$ and $y \in \mathbb{F}^m$.

If, in addition, A and C are positive definite, then the following statement is equivalent to i)–iii):

iv) $\operatorname{sprad}(B^* A^{-1} B C^{-1}) \leq 1$.

Finally, if $\mathcal{A}$ is positive semidefinite and nonzero, then, for all $x \in \mathbb{F}^n$ and $y \in \mathbb{F}^m$,

$$|x^* B y|^2 \leq \left[\frac{\lambda_{\max}(\mathcal{A}) - \lambda_{\min}(\mathcal{A})}{\lambda_{\max}(\mathcal{A}) + \lambda_{\min}(\mathcal{A})} \right]^2 (x^* A x)(y^* C y).$$

Proof: See [728, p. 473] and [911, 1530].

Fact 8.15.32. Let $A \in \mathbb{F}^{n \times n}$, assume that A is Hermitian, let $x, y \in \mathbb{F}^n$, and assume that $x^* x = y^* y = 1$ and $x^* y = 0$. Then,

$$2|x^* A y| \leq \lambda_{\max}(A) - \lambda_{\min}(A).$$

Furthermore, there exist vectors $x, y \in \mathbb{F}^n$ satisfying $x^* x = y^* y = 1$ and $x^* y = 0$ for which equality holds.

Proof: See [911, 1530].

Remark: $\lambda_{\max}(A) - \lambda_{\min}(A)$ is the *spread* of A. See Fact 9.9.30 and Fact 9.9.31.

Fact 8.15.33. Let $n \leq 4$, let $A \in \mathbb{R}^{n \times n}$, assume that A is symmetric, and assume that, for all nonnegative vectors $x \in \mathbb{R}^n$, $x^{\mathrm{T}} A x \geq 0$. Then, there exist $B, C \in \mathbb{R}^{n \times n}$ such that B is positive semidefinite, C is symmetric and nonnegative, and $A = B + C$.

Remark: This result does not hold for all $n > 5$. Hence, this result is an example of the *quartic barrier*. See [359], Fact 8.14.7, and Fact 11.17.3.

Remark: A is *copositive*.

8.16 Facts on the Gaussian Density

Fact 8.16.1. Let $A \in \mathbb{R}^{n \times n}$, and assume that A is positive definite. Then,

$$\int_{\mathbb{R}^n} e^{-x^{\mathrm{T}} A x} \, \mathrm{d}x = \frac{\pi^{n/2}}{\sqrt{\det A}}.$$

Remark: See Fact 11.13.16.

Fact 8.16.2. Let $A \in \mathbb{R}^{n \times n}$, assume that A is positive definite, and define $f \colon \mathbb{R}^n \mapsto \mathbb{R}$ by

$$f(x) = \frac{e^{-\frac{1}{2} x^{\mathrm{T}} A^{-1} x}}{(2\pi)^{n/2} \sqrt{\det A}}.$$

Then,

$$\int_{\mathbb{R}^n} f(x) \, \mathrm{d}x = 1,$$

$$\int_{\mathbb{R}^n} f(x) x x^{\mathrm{T}} \, \mathrm{d}x = A,$$

and

$$-\int_{\mathbb{R}^n} f(x) \log f(x) \, \mathrm{d}x = \tfrac{1}{2} \log[(2\pi e)^n \det A].$$

Proof: See [360] or use Fact 8.16.5.

Remark: f is the multivariate normal density. The last expression is the *entropy*.

Fact 8.16.3. Let $A, B \in \mathbb{R}^{n \times n}$, assume that A and B are positive definite, and, for $k = 0, 1, 2, 3$, define

$$\mathcal{I}_k \triangleq \frac{1}{(2\pi)^{n/2} \sqrt{\det A}} \int_{\mathbb{R}^n} \left(x^{\mathrm{T}} B x \right)^k e^{-\frac{1}{2} x^{\mathrm{T}} A^{-1} x} \, \mathrm{d}x.$$

Then,

$$\mathfrak{I}_0 = 1,$$
$$\mathfrak{I}_1 = \operatorname{tr} AB,$$
$$\mathfrak{I}_2 = (\operatorname{tr} AB)^2 + 2\operatorname{tr}(AB)^2,$$
$$\mathfrak{I}_3 = (\operatorname{tr} AB)^3 + 6(\operatorname{tr} AB)\left[\operatorname{tr}(AB)^2\right] + 8\operatorname{tr}(AB)^3.$$

Proof: See [1027, p. 80].

Remark: These equalities are *Lancaster's formulas*.

Fact 8.16.4. Let $A, B, C \in \mathbb{R}^{n \times n}$, assume that A is positive definite, assume that B and C are symmetric, and let $\mu \in \mathbb{R}^n$. Then,

$$\frac{1}{(2\pi)^{n/2}\sqrt{\det A}} \int_{\mathbb{R}^n} x^{\mathrm{T}}Bxx^{\mathrm{T}}Cx e^{-\frac{1}{2}(x-\mu)^{\mathrm{T}}A^{-1}(x-\mu)} \, dx = \operatorname{tr}(AB)\operatorname{tr}(AC) + 2\operatorname{tr}(ACAB)$$
$$+ \operatorname{tr}(AB)\mu^{\mathrm{T}}C\mu + 4\mu^{\mathrm{T}}BAC\mu + \mu^{\mathrm{T}}B\mu\operatorname{tr}(CA) + \mu^{\mathrm{T}}B\mu\mu^{\mathrm{T}}C\mu.$$

Proof: See [1208, p. 418, 419].

Remark: Setting $\mu = 0$ and $C = B$ yields $\mathfrak{I}_2$ of Fact 8.16.3.

Fact 8.16.5. Let $A \in \mathbb{R}^{n \times n}$, assume that A is positive definite, let $B \in \mathbb{R}^{n \times n}$, let $a, b \in \mathbb{R}^n$, and let $\alpha, \beta \in \mathbb{R}$. Then,

$$\int_{\mathbb{R}^n} \left(x^{\mathrm{T}}Bx + b^{\mathrm{T}}x + \beta\right) e^{-\left(x^{\mathrm{T}}Ax + a^{\mathrm{T}}x + \alpha\right)} \, dx$$
$$= \frac{\pi^{n/2}}{2\sqrt{\det A}}\left[2\beta + \operatorname{tr}\left(A^{-1}B\right) - b^{\mathrm{T}}A^{-1}a + \tfrac{1}{2}a^{\mathrm{T}}A^{-1}BA^{-1}a\right]e^{\frac{1}{4}a^{\mathrm{T}}A^{-1}a - \alpha}.$$

Proof: See [671, p. 322].

Fact 8.16.6. Let $A_1, A_2 \in \mathbb{R}^{n \times n}$, assume that A_1 and A_2 are positive definite, and let $\mu_1, \mu_2 \in \mathbb{R}^n$. Then,

$$e^{\frac{1}{2}(x-\mu_1)^{\mathrm{T}}A_1^{-1}(x-\mu_1)}e^{\frac{1}{2}(x-\mu_2)^{\mathrm{T}}A_2^{-1}(x-\mu_2)} = \alpha e^{\frac{1}{2}(x-\mu_3)^{\mathrm{T}}A_3^{-1}(x-\mu_3)},$$

where

$$A_3 \triangleq (A_1^{-1} + A_2^{-1})^{-1},$$
$$\mu_3 \triangleq A_3(A_1^{-1}\mu_1 + A_2^{-1}\mu_2),$$

and

$$\alpha \triangleq e^{\frac{1}{2}(\mu_1^{\mathrm{T}}A_1^{-1}\mu_1 + \mu_2^{\mathrm{T}}A_2^{-1}\mu_2 - \mu_3^{\mathrm{T}}A_3^{-1}\mu_3)}.$$

Remark: A product of Gaussian densities is a weighted Gaussian density.

8.17 Facts on Simultaneous Diagonalization

Fact 8.17.1. Let $A, B \in \mathbb{F}^{n \times n}$, assume that A and B are Hermitian. Then, the following statements are equivalent:

i) There exists a unitary matrix $S \in \mathbb{F}^{n \times n}$ such that SAS^* and SBS^* are diagonal.

ii) $AB = BA$.

iii) AB and BA are Hermitian.

If, in addition, A is nonsingular, then the following condition is equivalent to *i)–iii)*:

iv) $A^{-1}B$ is Hermitian.

Proof: See [178, p. 208], [459, pp. 188–190], or [728, p. 229].

Remark: The equivalence of *i)* and *ii)* is given by Fact 5.17.7.

Fact 8.17.2. Let $A, B \in \mathbb{F}^{n \times n}$, assume that A and B are Hermitian, and assume that A is nonsingular. Then, there exists a nonsingular matrix $S \in \mathbb{F}^{n \times n}$ such that SAS^* and SBS^* are diagonal if and only if $A^{-1}B$ is diagonalizable over $\mathbb{R}$.

Proof: See [728, p. 229] or [1125, p. 95].

Fact 8.17.3. Let $A, B \in \mathbb{F}^{n \times n}$, assume that A and B are symmetric, and assume that A is nonsingular. Then, there exists a nonsingular matrix $S \in \mathbb{F}^{n \times n}$ such that SAS^{T} and SBS^{T} are diagonal if and only if $A^{-1}B$ is diagonalizable.

Proof: See [728, p. 229] and [1385].

Remark: A and B are complex symmetric.

Fact 8.17.4. Let $A, B \in \mathbb{F}^{n \times n}$, and assume that A and B are Hermitian. Then, there exists a nonsingular matrix $S \in \mathbb{F}^{n \times n}$ such that SAS^* and SBS^* are diagonal if and only if there exists a positive-definite matrix $M \in \mathbb{F}^{n \times n}$ such that $AMB = BMA$.

Proof: See [86].

Fact 8.17.5. Let $A, B \in \mathbb{F}^{n \times n}$, assume that A and B are Hermitian, and assume there exist $\alpha, \beta \in \mathbb{R}$ such that $\alpha A + \beta B$ is positive definite. Then, there exists a nonsingular matrix $S \in \mathbb{F}^{n \times n}$ such that SAS^* and SBS^* are diagonal.

Proof: See [728, p. 465].

Remark: This result extends a result due to Weierstrass. See [1385].

Remark: Suppose that B is positive definite. Then, by necessity of Fact 8.17.2, it follows that $A^{-1}B$ is diagonalizable over $\mathbb{R}$, which proves *iii)* $\implies$ *i)* of Proposition 5.5.12.

Remark: See Fact 8.17.6.

Fact 8.17.6. Let $A, B \in \mathbb{F}^{n \times n}$, assume that A and B are Hermitian, assume that $\{x \in \mathbb{F}^n: x^*Ax = x^*Bx = 0\} = \{0\}$, and, if $\mathbb{F} = \mathbb{R}$, assume that $n \geq 3$. Then, there exists a nonsingular matrix $S \in \mathbb{F}^{n \times n}$ such that SAS^* and SBS^* are diagonal.

Proof: This result follows from Fact 5.17.9. See [975] or [1125, p. 96].

Remark: For $\mathbb{F} = \mathbb{R}$, this result is due to Pesonen and Milnor. See [1385].

Remark: See Fact 5.17.9, Fact 8.15.25, Fact 8.15.26, and Fact 8.17.5.

8.18 Facts on Eigenvalues and Singular Values for One Matrix

Fact 8.18.1. Let $A = \begin{bmatrix} a & b \\ b & c \end{bmatrix} \in \mathbb{F}^{2 \times 2}$, assume that A is Hermitian, and let $\mathrm{mspec}(A) = \{\lambda_1, \lambda_2\}_{\mathrm{ms}}$. Then,

$$2|b| \leq \lambda_1 - \lambda_2.$$

Now, assume in addition that A is positive semidefinite. Then,

$$\sqrt{2}|b| \leq \left(\sqrt{\lambda_1} - \sqrt{\lambda_2}\right)\sqrt{\lambda_1 + \lambda_2}.$$

If $c > 0$, then

$$\frac{|b|}{\sqrt{c}} \leq \sqrt{\lambda_1} - \sqrt{\lambda_2}.$$

If $a > 0$ and $c > 0$, then

$$\frac{|b|}{\sqrt{ac}} \leq \frac{\lambda_1 - \lambda_2}{\lambda_1 + \lambda_2}.$$

Finally, if A is positive definite, then

$$\frac{|b|}{a} \leq \frac{\lambda_1 - \lambda_2}{2\sqrt{\lambda_1 \lambda_2}}$$

and

$$4|b| \leq \frac{\lambda_1^2 - \lambda_2^2}{\sqrt{\lambda_1 \lambda_2}}.$$

Proof: See [911, 1530].

Remark: These inequalities are useful for deriving inequalities involving quadratic forms. See Fact 8.15.30 and Fact 8.15.31.

Fact 8.18.2. Let $A \in \mathbb{F}^{n \times m}$. Then, for all $i \in \{1, \ldots, \min\{n, m\}\}$,

$$\lambda_i(\langle A \rangle) = \sigma_i(A).$$

Hence,

$$\mathrm{tr}\,\langle A \rangle = \sum_{i=1}^{\min\{n,m\}} \sigma_i(A).$$

Fact 8.18.3. Let $A \in \mathbb{F}^{n \times n}$, and define

$$\mathcal{A} \triangleq \begin{bmatrix} \sigma_{\max}(A)I & A^* \\ A & \sigma_{\max}(A)I \end{bmatrix}.$$

Then, $\mathcal{A}$ is positive semidefinite. Furthermore,

$$\langle A + A^* \rangle \le \left\{ \begin{array}{c} \langle A \rangle + \langle A^* \rangle \le 2\sigma_{\max}(A)I \\ \\ A^*A + I \end{array} \right\} \le [\sigma_{\max}^2(A) + 1]I.$$

Proof: See [1528].

Fact 8.18.4. Let $A \in \mathbb{F}^{n \times n}$. Then, for all $i \in \{1, \ldots, n\}$,

$$-\sigma_i(A) \le \lambda_i[\tfrac{1}{2}(A + A^*)] \le \sigma_i(A).$$

Hence,

$$|\operatorname{tr} A| \le \operatorname{tr} \langle A \rangle.$$

Proof: See [1242].

Remark: See Fact 5.11.25.

Fact 8.18.5. Let $A \in \mathbb{F}^{n \times n}$, and let $\operatorname{mspec}(A) = \{\lambda_1, \ldots, \lambda_n\}_{\mathrm{ms}}$, where $\lambda_1, \ldots, \lambda_n$ are ordered such that $|\lambda_1| \ge \cdots \ge |\lambda_n|$. If $p > 0$, then, for all $k \in \{1, \ldots, n\}$,

$$\sum_{i=1}^{k} |\lambda_i|^p \le \sum_{i=1}^{k} \sigma_i^p(A).$$

In particular, for all $k \in \{1, \ldots, n\}$,

$$\sum_{i=1}^{k} |\lambda_i| \le \sum_{i=1}^{k} \sigma_i(A).$$

Hence,

$$|\operatorname{tr} A| \le \sum_{i=1}^{n} |\lambda_i| \le \sum_{i=1}^{n} \sigma_i(A) = \operatorname{tr} \langle A \rangle.$$

Furthermore, for all $k \in \{1, \ldots, n\}$,

$$\sum_{i=1}^{k} |\lambda_i|^2 \le \sum_{i=1}^{k} \sigma_i^2(A).$$

Hence,

$$\operatorname{Re} \operatorname{tr} A^2 \le |\operatorname{tr} A^2| \le \sum_{i=1}^{n} |\lambda_i|^2 \le \sum_{i=1}^{n} \sigma_i(A^2) = \operatorname{tr} \langle A^2 \rangle \le \sum_{i=1}^{n} \sigma_i^2(A) = \operatorname{tr} A^*A.$$

Furthermore,

$$\sum_{i=1}^{n} |\lambda_i|^2 = \operatorname{tr} A^*A$$

if and only if A is normal. Finally,

$$\sum_{i=1}^{n} \lambda_i^2 = \operatorname{tr} A^*A$$

if and only if A is Hermitian.

Proof: This result follows from Fact 2.21.12 and Fact 5.11.28. See [201, p. 42], [730, p. 176], or [1521, p. 19]. See Fact 9.13.16 for the inequality $\operatorname{tr} \langle A^2 \rangle =$

$\text{tr}\,\left(A^{2*}A^2\right)^{1/2} \le \text{tr}\,A^*A$. See Fact 3.7.13 and Fact 5.14.14.

Remark: The first result is *Weyl's inequalities*. The result $\sum_{i=1}^{n} |\lambda_i|^2 \le \text{tr}\,A^*A$ is *Schur's inequality*. See Fact 9.11.3.

Problem: Determine when equality holds for the remaining inequalities.

Fact 8.18.6. Let $A \in \mathbb{F}^{n \times n}$, let $\text{mspec}(A) = \{\lambda_1, \ldots, \lambda_n\}_{\text{ms}}$, where $\lambda_1, \ldots, \lambda_n$ are ordered such that $|\lambda_1| \ge \cdots \ge |\lambda_n|$, and let $r > 0$. Then, for all $k \in \{1, \ldots, n\}$,

$$\prod_{i=1}^{k} (1 + r|\lambda_i|) \le \prod_{i=1}^{k} [1 + \sigma_i(A)].$$

Proof: See [459, p. 222].

Fact 8.18.7. Let $A \in \mathbb{F}^{n \times n}$. Then,

$$|\text{tr}\,A^2| \le \begin{cases} \text{tr}\,\langle A \rangle \langle A^* \rangle \\ \text{tr}\,\langle A^2 \rangle \le \text{tr}\,\langle A \rangle^2 = \text{tr}\,A^*A. \end{cases}$$

Proof: For the upper inequality, see [911, 1530]. For the lower inequalities, use Fact 8.18.4 and Fact 9.11.3.

Remark: See Fact 5.11.10, Fact 9.13.16, and Fact 9.13.17.

Fact 8.18.8. Let $A \in \mathbb{F}^{n \times n}$, and assume that A is Hermitian. Then, for all $k \in \{1, \ldots, n\}$,

$$\sum_{i=1}^{k} \text{d}_i(A) \le \sum_{i=1}^{k} \lambda_i(A)$$

with equality for $k = n$, that is,

$$\text{tr}\,A = \sum_{i=1}^{n} \text{d}_i(A) = \sum_{i=1}^{n} \lambda_i(A).$$

That is, $\begin{bmatrix} \lambda_1(A) & \cdots & \lambda_n(A) \end{bmatrix}^{\text{T}}$ strongly majorizes $\begin{bmatrix} \text{d}_1(A) & \cdots & \text{d}_n(A) \end{bmatrix}^{\text{T}}$, and thus, for all $k \in \{1, \ldots, n\}$,

$$\sum_{i=k}^{n} \lambda_i(A) \le \sum_{i=k}^{n} \text{d}_i(A).$$

In particular,

$$\lambda_{\min}(A) \le \text{d}_{\min}(A) \le \text{d}_{\max}(A) \le \lambda_{\max}(A).$$

Furthermore, the vector $\begin{bmatrix} \text{d}_1(A) & \cdots & \text{d}_n(A) \end{bmatrix}^{\text{T}}$ is an element of the convex hull of the $n!$ vectors obtaining by permuting the components of $\begin{bmatrix} \lambda_1(A) & \cdots & \lambda_n(A) \end{bmatrix}^{\text{T}}$.

Proof: See [201, p. 35], [728, p. 193], [996, p. 218], or [1521, p. 18]. The last statement follows from Fact 3.9.6.

Remark: This result is *Schur's theorem*.

Remark: See Fact 8.12.3.

Fact 8.18.9. Let $A \in \mathbb{F}^{n \times n}$, assume that A is Hermitian, let k denote the number of positive diagonal entries of A, and let l denote the number of positive eigenvalues of A. Then,

$$\sum_{i=1}^{k} d_i^2(A) \leq \sum_{i=1}^{l} \lambda_i^2(A).$$

Proof: Write $A = B + C$, where B is positive semidefinite, C is negative semidefinite, and $\mathrm{mspec}(A) = \mathrm{mspec}(B) \cup \mathrm{mspec}(C)$. Furthermore, without loss of generality, assume that $A_{(1,1)}, \ldots, A_{(k,k)}$ are the positive diagonal entries of A. Then,

$$\sum_{i=1}^{k} d_i^2(A) = \sum_{i=1}^{k} A_{(i,i)}^2 \leq \sum_{i=1}^{k} (A_{(i,i)} - C_{(i,i)})^2$$

$$= \sum_{i=1}^{k} B_{(i,i)}^2 \leq \sum_{i=1}^{n} B_{(i,i)}^2 \leq \mathrm{tr}\, B^2 = \sum_{i=1}^{l} \lambda_i^2(A).$$

Remark: This inequality can be written as

$$\mathrm{tr}\, (A + |A|)^{\circ 2} \leq \mathrm{tr}\, (A + \langle A \rangle)^2.$$

Remark: This result is due to Y. Li.

Fact 8.18.10. Let $x, y \in \mathbb{R}^n$, where $n \geq 2$. Then, the following statements are equivalent:

i) y strongly majorizes by x.

ii) x is an element of the convex hull of the vectors $y_1, \ldots, y_{n!} \in \mathbb{R}^n$, where each of these $n!$ vectors is formed by permuting the components of y.

iii) There exists a Hermitian matrix $A \in \mathbb{C}^{n \times n}$ such that $\begin{bmatrix} A_{(1,1)} & \cdots & A_{(n,n)} \end{bmatrix}^{\mathrm{T}} = x$ and $\mathrm{mspec}(A) = \{y_{(1)}, \ldots, y_{(n)}\}_{\mathrm{ms}}$.

Remark: This result is the *Schur-Horn theorem*. Schur's theorem given by Fact 8.18.8 is *iii)* $\implies$ *i)*, while the result *i)* $\implies$ *iii)* is due to [727]. The equivalence of *ii)* is given by Fact 3.9.6. The significance of this result is discussed in [157, 202, 266].

Remark: An equivalent version is given by Fact 3.11.14.

Fact 8.18.11. Let $A \in \mathbb{F}^{n \times n}$, and assume that A is positive semidefinite. Then, for all $k \in \{1, \ldots, n\}$,

$$\prod_{i=k}^{n} \lambda_i(A) \leq \prod_{i=k}^{n} d_i(A).$$

In particular,

$$\det A \leq \prod_{i=1}^{n} A_{(i,i)}.$$

Now, assume in addition that A is positive definite. Then, equality holds if and only if A is diagonal.

Proof: See [544, pp. 21–24], [728, pp. 200, 477], or [1521, p. 18].

Remark: The case $k = 1$ is *Hadamard's inequality.*

Remark: See Fact 8.13.35 and Fact 9.11.1.

Remark: A strengthened version is given by Fact 8.13.34.

Remark: A geometric interpretation is discussed in [553].

Fact 8.18.12. Let $A \in \mathbb{F}^{n \times n}$, define $H \triangleq \frac{1}{2}(A + A^*)$ and $S \triangleq \frac{1}{2}(A - A^*)$, and assume that H is positive definite. Then, the following statements hold:

i) A is nonsingular.

ii) $\frac{1}{2}(A^{-1} + A^{-*}) = (H + S^*H^{-1}S)^{-1}$.

iii) $\sigma_{\max}(A^{-1}) \leq \sigma_{\max}(H^{-1})$.

iv) $\sigma_{\max}(A) \leq \sigma_{\max}(H + S^*H^{-1}S)$.

Proof: See [1003].

Remark: See Fact 8.9.32 and Fact 8.13.11.

Fact 8.18.13. Let $A \in \mathbb{F}^{n \times n}$, and assume that A is Hermitian. Then, $\{A_{(1,1)}, \ldots, A_{(n,n)}\}_{\mathrm{ms}} = \mathrm{mspec}(A)$ if and only if A is diagonal.

Proof: Apply Fact 8.18.11 with $A + \beta I > 0$.

Fact 8.18.14. Let $A \in \mathbb{F}^{n \times n}$. Then, $\left[\begin{smallmatrix} I & A \\ A^* & I \end{smallmatrix} \right]$ is positive semidefinite if and only if $\sigma_{\max}(A) \leq 1$. Furthermore, $\left[\begin{smallmatrix} I & A \\ A^* & I \end{smallmatrix} \right]$ is positive definite if and only if $\sigma_{\max}(A) < 1$.

Proof: Note that

$$\begin{bmatrix} I & A \\ A^* & I \end{bmatrix} = \begin{bmatrix} I & 0 \\ A^* & I \end{bmatrix} \begin{bmatrix} I & 0 \\ 0 & I - A^*A \end{bmatrix} \begin{bmatrix} I & A \\ 0 & I \end{bmatrix}.$$

Fact 8.18.15. Let $A \in \mathbb{F}^{n \times n}$, and assume that A is Hermitian. Then, for all $k \in \{1, \ldots, n\}$,

$$\sum_{i=1}^{k} \lambda_i = \max\{\mathrm{tr}\, S^*AS \colon \ S \in \mathbb{F}^{n \times k} \text{ and } S^*S = I_k\}$$

and

$$\sum_{i=n+1-k}^{n} \lambda_i = \min\{\mathrm{tr}\, S^*AS \colon \ S \in \mathbb{F}^{n \times k} \text{ and } S^*S = I_k\}.$$

Proof: See [728, p. 191].

Remark: This result is the *minimum principle.*

Fact 8.18.16. Let $A \in \mathbb{F}^{n \times n}$, assume that A is Hermitian, and let $S \in \mathbb{R}^{k \times n}$ satisfy $SS^* = I_k$. Then, for all $i \in \{1, \ldots, k\}$,

$$\lambda_{i+n-k}(A) \leq \lambda_i(SAS^*) \leq \lambda_i(A).$$

Consequently,

$$\sum_{i=1}^{k} \lambda_{i+n-k}(A) \leq \operatorname{tr} SAS^* \leq \sum_{i=1}^{k} \lambda_i(A)$$

and

$$\prod_{i=1}^{k} \lambda_{i+n-k}(A) \leq \det SAS^* \leq \prod_{i=1}^{k} \lambda_i(A).$$

Proof: See [728, p. 190] or [1208, p. 111].

Remark: This result is the *Poincaré separation theorem*.

8.19 Facts on Eigenvalues and Singular Values for Two or More Matrices

Fact 8.19.1. Let $A \in \mathbb{F}^{n \times n}$, $B \in \mathbb{F}^{n \times m}$, and $C \in \mathbb{F}^{m \times m}$, and assume that A and C are positive definite. Then, $\left[\begin{smallmatrix} A & B \\ B^* & C \end{smallmatrix} \right] \in \mathbb{F}^{(n+m) \times (n+m)}$ is positive semidefinite if and only if

$$\sigma_{\max}(A^{-1/2}BC^{-1/2}) \leq 1.$$

Furthermore, $\left[\begin{smallmatrix} A & B \\ B^* & C \end{smallmatrix} \right] \in \mathbb{F}^{(n+m) \times (n+m)}$ is positive definite if and only if

$$\sigma_{\max}(A^{-1/2}BC^{-1/2}) < 1.$$

Proof: See [989].

Fact 8.19.2. Let $A \in \mathbb{F}^{n \times n}$, $B \in \mathbb{F}^{n \times m}$, and $C \in \mathbb{F}^{m \times m}$, assume that A and C are positive definite, and assume that

$$\sigma_{\max}^2(B) \leq \sigma_{\min}(A)\sigma_{\min}(C).$$

Then, $\left[\begin{smallmatrix} A & B \\ B^* & C \end{smallmatrix} \right] \in \mathbb{F}^{(n+m) \times (n+m)}$ is positive semidefinite. If, in addition,

$$\sigma_{\max}^2(B) < \sigma_{\min}(A)\sigma_{\min}(C),$$

then $\left[\begin{smallmatrix} A & B \\ B^* & C \end{smallmatrix} \right] \in \mathbb{F}^{(n+m) \times (n+m)}$ is positive definite.

Proof: Note that

$$\sigma_{\max}^2(A^{-1/2}BC^{-1/2}) \leq \lambda_{\max}(A^{-1/2}BC^{-1}B^*A^{-1/2})$$

$$\leq \sigma_{\max}(C^{-1})\lambda_{\max}(A^{-1/2}BB^*A^{-1/2})$$

$$\leq \frac{1}{\sigma_{\min}(C)}\lambda_{\max}(B^*A^{-1}B)$$

$$\leq \frac{\sigma_{\max}(A^{-1})}{\sigma_{\min}(C)}\lambda_{\max}(B^*B)$$

$$= \frac{1}{\sigma_{\min}(A)\sigma_{\min}(C)}\sigma_{\max}^2(B)$$

$$\leq 1.$$

The result now follows from Fact 8.19.1.

Fact 8.19.3. Let $A, B \in \mathbb{F}^n$, assume that A and B are Hermitian, and define $\gamma \triangleq \begin{bmatrix} \gamma_1 \cdots \gamma_n \end{bmatrix}$, where the components of γ are the components of $\begin{bmatrix} \lambda_1(A) \cdots \lambda_n(A) \end{bmatrix} + \begin{bmatrix} \lambda_n(B) \cdots \lambda_1(B) \end{bmatrix}$ arranged in decreasing order. Then, for all $k \in \{1, \ldots, n\}$,

$$\sum_{i=1}^{k} \gamma_i \leq \sum_{i=1}^{k} \lambda_i(A + B).$$

Proof: This result follows from the Lidskii-Wielandt inequalities. See [201, p. 71] or [202, 388].

Remark: This result provides an alternative lower bound for (8.6.13).

Fact 8.19.4. Let $A, B \in \mathbf{H}^n$, let $k \in \{1, \ldots, n\}$, and let $1 \leq i_1 \leq \cdots \leq i_k \leq n$. Then,

$$\sum_{j=1}^{k} \lambda_{i_j}(A) + \sum_{j=n-k+1}^{n} \lambda_j(B)] \leq \sum_{j=1}^{k} \lambda_{i_j}(A + B) \leq \sum_{j=1}^{k} [\lambda_{i_j}(A) + \lambda_j(B)].$$

Proof: See [1208, pp. 115, 116].

Fact 8.19.5. Let $f \colon \mathbb{R} \mapsto \mathbb{R}$ be convex, define $f \colon \mathbf{H}^n \mapsto \mathbf{H}^n$ by (8.5.2), let $A, B \in \mathbb{F}^{n \times n}$, and assume that A and B are Hermitian. Then, for all $\alpha \in [0, 1]$,

$$\begin{bmatrix} \alpha \lambda_1[f(A)] + (1 - \alpha)\lambda_1[f(B)] & \cdots & \alpha \lambda_n[f(A)] + (1 - \alpha)\lambda_n[f(B)] \end{bmatrix}$$

weakly majorizes

$$\begin{bmatrix} \lambda_1[f(\alpha A + (1 - \alpha)B)] & \cdots & \lambda_n[f(\alpha A + (1 - \alpha)B)] \end{bmatrix}.$$

If, in addition, f is either nonincreasing or nondecreasing, then, for all $i \in \{1, \ldots, n\}$,

$$\lambda_i[f(\alpha A + (1 - \alpha)B)] \leq \alpha \lambda_i[f(A)] + (1 - \alpha)\lambda_i[f(B)].$$

Proof: See [94].

Remark: Convexity of $f \colon \mathbb{R} \mapsto \mathbb{R}$ does not imply convexity of $f \colon \mathbf{H}^n \mapsto \mathbf{H}^n$.

Fact 8.19.6. Let $A, B \in \mathbb{F}^{n \times n}$, and assume that A and B are positive semidefinite. If $r \in [0, 1]$, then

$$\begin{bmatrix} \lambda_1(A^r + B^r) & \cdots & \lambda_n(A^r + B^r) \end{bmatrix}^{\mathrm{T}}$$

weakly majorizes

$$\begin{bmatrix} \lambda_1[(A + B)^r] & \cdots & \lambda_n[(A + B)^r] \end{bmatrix}^{\mathrm{T}},$$

and, for all $i \in \{1, \ldots, n\}$,

$$2^{1-r}\lambda_i[(A + B)^r] \leq \lambda_i(A^r + B^r).$$

If $r \geq 1$, then

$$\begin{bmatrix} \lambda_1[(A + B)^r] & \cdots & \lambda_n[(A + B)^r] \end{bmatrix}^{\mathrm{T}}$$

weakly majorizes

$$\begin{bmatrix} \lambda_1(A^r + B^r) & \cdots & \lambda_n(A^r + B^r) \end{bmatrix}^{\mathrm{T}},$$

and, for all $i \in \{1, \ldots, n\}$,

$$\lambda_i(A^r + B^r) \leq 2^{r-1}\lambda_i[(A+B)^r].$$

Proof: This result follows from Fact 8.19.5. See [60, 92, 94].

Fact 8.19.7. Let $A, B \in \mathbb{F}^{n \times n}$, and assume that A and B are positive semidefinite. Then, for all $k \in \{1, \ldots, n\}$,

$$\sum_{i=1}^{k} \sigma_i^2(A + \jmath B) \leq \sum_{i=1}^{k} [\sigma_i^2(A) + \sigma_i^2(B)],$$

$$\sum_{i=1}^{n} \sigma_i^2(A + \jmath B) = \sum_{i=1}^{n} [\sigma_i^2(A) + \sigma_i^2(B)],$$

$$\sum_{i=1}^{k} [\sigma_i^2(A + \jmath B) + \sigma_{n-i}^2(A + \jmath B)] \leq \sum_{i=1}^{k} [\sigma_i^2(A) + \sigma_i^2(B)],$$

$$\sum_{i=1}^{n} [\sigma_i^2(A + \jmath B) + \sigma_{n-i}^2(A + \jmath B)] = \sum_{i=1}^{n} [\sigma_i^2(A) + \sigma_i^2(B)],$$

and

$$\sum_{i=1}^{k} [\sigma_i^2(A) + \sigma_{n-i}^2(B)] \leq \sum_{i=1}^{k} \sigma_i^2(A + \jmath B),$$

$$\sum_{i=1}^{n} [\sigma_i^2(A) + \sigma_{n-i}^2(B)] = \sum_{i=1}^{n} \sigma_i^2(A + \jmath B).$$

Proof: See [54, 328].

Remark: The first equality is given by Fact 9.9.40.

Fact 8.19.8. Let $A, B \in \mathbb{F}^{n \times n}$, and assume that A and B are positive semidefinite. Then, the following statements hold:

i) If $p \in [0, 1]$, then

$$\sigma_{\max}(A^p - B^p) \leq \sigma_{\max}^p(A - B).$$

ii) If $p \geq \sqrt{2}$, then

$$\sigma_{\max}(A^p - B^p) \leq p[\max\{\sigma_{\max}(A), \sigma_{\max}(B)\}]^{p-1}\sigma_{\max}(A - B).$$

iii) If a and b are positive numbers such that $aI \leq A \leq bI$ and $aI \leq B \leq bI$, then

$$\sigma_{\max}(A^p - B^p) \leq b[b^{p-2} + (p-1)a^{p-2}]\sigma_{\max}(A - B).$$

Proof: See [210, 840].

Fact 8.19.9. Let $A, B \in \mathbb{F}^{n \times n}$, and assume that A and B are positive semidefinite. Then, for all $i \in \{1, \ldots, n\}$,

$$\sigma_i(A - B) \leq \sigma_i\left(\begin{bmatrix} A & 0 \\ 0 & B \end{bmatrix}\right).$$

Proof: See [1286, 1519].

Fact 8.19.10. Let $A \in \mathbb{F}^{n \times n}$, $B \in \mathbb{F}^{n \times m}$, and $C \in \mathbb{F}^{m \times m}$, and assume that $\mathcal{A} \in \mathbb{F}^{(n+m) \times (n+m)}$ defined by

$$\mathcal{A} \triangleq \begin{bmatrix} A & B \\ B^* & C \end{bmatrix}$$

is positive semidefinite. Then, for all $i \in \{1, \ldots, \min\{n, m\}\}$,

$$2\sigma_i(B) \leq \sigma_i(\mathcal{A}).$$

Proof: See [219, 1286].

Fact 8.19.11. Let $A, B \in \mathbb{F}^{n \times n}$. Then,

$$\max\{\sigma_{\max}^2(A), \sigma_{\max}^2(B)\} - \sigma_{\max}(AB) \leq \sigma_{\max}(A^*A - BB^*)$$

and

$$\sigma_{\max}(A^*A - BB^*) \leq \max\{\sigma_{\max}^2(A), \sigma_{\max}^2(B)\} - \min\{\sigma_{\min}^2(A), \sigma_{\min}^2(B)\}.$$

Furthermore,

$$\max\{\sigma_{\max}^2(A), \sigma_{\max}^2(B)\} + \min\{\sigma_{\min}^2(A), \sigma_{\min}^2(B)\} \leq \sigma_{\max}(A^*A + BB^*)$$

and

$$\sigma_{\max}(A^*A + BB^*) \leq \max\{\sigma_{\max}^2(A), \sigma_{\max}^2(B)\} + \sigma_{\max}(AB).$$

Now, assume in addition that A and B are positive semidefinite. Then,

$$\max\{\lambda_{\max}(A), \lambda_{\max}(B)\} - \sigma_{\max}(A^{1/2}B^{1/2}) \leq \sigma_{\max}(A - B)$$

and

$$\sigma_{\max}(A - B) \leq \max\{\lambda_{\max}(A), \lambda_{\max}(B)\} - \min\{\lambda_{\min}(A), \lambda_{\min}(B)\}.$$

Furthermore,

$$\max\{\lambda_{\max}(A), \lambda_{\max}(B)\} + \min\{\lambda_{\min}(A), \lambda_{\min}(B)\} \leq \lambda_{\max}(A + B)$$

and

$$\lambda_{\max}(A + B) \leq \max\{\lambda_{\max}(A), \lambda_{\max}(B)\} + \sigma_{\max}(A^{1/2}B^{1/2}).$$

Proof: See [848, 1522].

Remark: See Fact 8.19.14 and Fact 9.13.8.

Fact 8.19.12. Let $A, B \in \mathbb{F}^{n \times n}$, and assume that A and B are positive semidefinite. Then,

$$\max\{\sigma_{\max}(A), \sigma_{\max}(B)\} - \sigma_{\max}(A^{1/2}B^{1/2})$$
$$\leq \sigma_{\max}(A - B)$$
$$\leq \max\{\sigma_{\max}(A), \sigma_{\max}(B)\}$$
$$\leq \sigma_{\max}(A + B)$$
$$\leq \left\{ \begin{array}{c} \max\{\sigma_{\max}(A), \sigma_{\max}(B)\} + \sigma_{\max}(A^{1/2}B^{1/2}) \\ \sigma_{\max}(A) + \sigma_{\max}(B) \end{array} \right\}$$
$$\leq 2\max\{\sigma_{\max}(A), \sigma_{\max}(B)\}.$$

Proof: See [842, 848], and use Fact 8.19.13.

Remark: See Fact 8.19.14.

Fact 8.19.13. Let $A, B \in \mathbb{F}^{n \times n}$, and assume that A and B are positive semidefinite, and let $k \geq 1$. Then, for all $i \in \{1, \ldots, n\}$,

$$2\sigma_i\left[A^{1/2}(A + B)^{k-1}B^{1/2}\right] \leq \lambda_i\left[(A + B)^k\right].$$

Hence,

$$2\sigma_{\max}(A^{1/2}B^{1/2}) \leq \lambda_{\max}(A + B)$$

and

$$\sigma_{\max}(A^{1/2}B^{1/2}) \leq \max\{\lambda_{\max}(A), \lambda_{\max}(B)\}.$$

Proof: See Fact 8.19.11 and Fact 9.9.18.

Fact 8.19.14. Let $A, B \in \mathbb{F}^{n \times n}$, and assume that A and B are positive semidefinite. Then,

$$\max\{\lambda_{\max}(A), \lambda_{\max}(B)\} - \sigma_{\max}(A^{1/2}B^{1/2}) \leq \sigma_{\max}(A - B)$$

and

$$\lambda_{\max}(A + B)$$
$$\leq \tfrac{1}{2}\left[\lambda_{\max}(A) + \lambda_{\max}(B) + \sqrt{[\lambda_{\max}(A) - \lambda_{\max}(B)]^2 + 4\sigma_{\max}^2(A^{1/2}B^{1/2})}\right]$$
$$\leq \left\{ \begin{array}{c} \max\{\lambda_{\max}(A), \lambda_{\max}(B)\} + \sigma_{\max}(A^{1/2}B^{1/2}) \\ \lambda_{\max}(A) + \lambda_{\max}(B). \end{array} \right.$$

Furthermore,

$$\lambda_{\max}(A + B) = \lambda_{\max}(A) + \lambda_{\max}(B)$$

if and only if

$$\sigma_{\max}(A^{1/2}B^{1/2}) = \lambda_{\max}^{1/2}(A)\lambda_{\max}^{1/2}(B).$$

Proof: See [842, 845, 848].

Remark: See Fact 8.19.11, Fact 8.19.12, Fact 9.14.15, and Fact 9.9.46.

Problem: Is $\sigma_{\max}(A - B) \leq \sigma_{\max}(A + B)$?

Fact 8.19.15. Let $A, B \in \mathbb{F}^{n \times n}$, and assume that A and B are positive semidefinite. Then,

$$\sigma_{\max}\left(A^{1/2}B^{1/2}\right) \leq \sigma_{\max}^{1/2}(AB).$$

Equivalently,

$$\lambda_{\max}\left(A^{1/2}BA^{1/2}\right) \leq \lambda_{\max}^{1/2}(AB^2A).$$

Furthermore, $AB = 0$ if and only if $A^{1/2}B^{1/2} = 0$.

Proof: See [842] and [848].

Fact 8.19.16. Let $A, B \in \mathbb{F}^{n \times n}$, and assume that A and B are positive semidefinite. Then,

$$\operatorname{tr} AB \leq \operatorname{tr}\left(AB^2A\right)^{1/2} \leq \tfrac{1}{4}\operatorname{tr}\left(A + B\right)^2,$$
$$\operatorname{tr}\left(AB\right)^2 \leq \operatorname{tr} A^2B^2 \leq \tfrac{1}{16}\operatorname{tr}\left(A + B\right)^4,$$

and

$$\sigma_{\max}(AB) \leq \tfrac{1}{4}\sigma_{\max}\left[(A + B)^2\right]$$

$$\leq \left\{ \begin{array}{c} \tfrac{1}{2}\sigma_{\max}(A^2 + B^2) \leq \tfrac{1}{2}\sigma_{\max}(A^2) + \tfrac{1}{2}\sigma_{\max}(B^2) \\ \tfrac{1}{4}\sigma_{\max}^2(A + B) \leq \tfrac{1}{4}[\sigma_{\max}(A) + \sigma_{\max}(B)]^2 \end{array} \right\}$$

$$\leq \tfrac{1}{2}\sigma_{\max}^2(A) + \tfrac{1}{2}\sigma_{\max}^2(B).$$

Proof: See Fact 9.9.18. The inequalities $\operatorname{tr} AB \leq \operatorname{tr}\left(AB^2A\right)^{1/2}$ and $\operatorname{tr}\left(AB\right)^2 \leq \operatorname{tr} A^2B^2$ follow from Fact 8.12.22.

Fact 8.19.17. Let $A, B \in \mathbb{F}^{n \times n}$, assume that A is positive semidefinite, and assume that B is positive definite. Then, for all $i, j, k \in \{1, \ldots, n\}$ such that $j + k \leq i + 1$,

$$\lambda_i(AB) \leq \lambda_j(A)\lambda_k(B)$$

and

$$\lambda_{n-j+1}(A)\lambda_{n-k+1}(B) \leq \lambda_{n-i+1}(AB).$$

In particular, for all $i \in \{1, \ldots, n\}$,

$$\lambda_i(A)\lambda_n(B) \leq \lambda_i(AB) \leq \lambda_i(A)\lambda_1(B).$$

Proof: See [1208, pp. 126, 127].

Fact 8.19.18. Let $A, B \in \mathbb{F}^{n \times n}$, and assume that A and B are positive definite. Then, for all $i = 1, \ldots, n$,

$$\frac{\lambda_i^2(AB)}{\lambda_1(A)\lambda_1(B)} \leq \lambda_i(A)\lambda_i(B) \leq \frac{\lambda_i^2(AB)}{\lambda_n(A)\lambda_n(B)}.$$

Proof: See [1208, p. 137].

Fact 8.19.19. Let $A, B \in \mathbb{F}^{n \times n}$, assume that A is positive semidefinite, and assume that B is Hermitian. Then, for all $k \in \{1, \ldots, n\}$,

$$\sum_{i=1}^{k} \lambda_i(A)\lambda_{n-i+1}(B) \le \sum_{i=1}^{k} \lambda_i(AB)$$

and

$$\sum_{i=1}^{k} \lambda_{n-i+1}(AB) \le \sum_{i=1}^{k} \lambda_i(A)\lambda_i(B).$$

In particular,

$$\sum_{i=1}^{n} \lambda_i(A)\lambda_{n-i+1}(B) \le \operatorname{tr} AB \le \sum_{i=1}^{n} \lambda_i(A)\lambda_i(B).$$

Proof: See [862] and [1208, p. 128].

Remark: See Fact 5.12.4, Fact 5.12.5, Fact 5.12.8, and Proposition 8.4.13.

Remark: The upper and lower bounds for $\operatorname{tr} AB$ are related to Fact 1.18.4. See [204, p. 140].

Fact 8.19.20. Let $A, B \in \mathbb{F}^{n \times n}$, assume that A and B are positive semidefinite, let $\lambda_1(AB) \ge \cdots \ge \lambda_n(AB) \ge 0$ denote the eigenvalues of AB, and let $1 \le i_1 < \cdots < i_k \le n$. Then,

$$\sum_{j=1}^{k} \lambda_{i_j}(A)\lambda_{n-j+1}(B) \le \sum_{j=1}^{k} \lambda_{i_j}(AB) \le \sum_{j=1}^{k} \lambda_{i_j}(A)\lambda_j(B).$$

Furthermore, for all $k = 1, \ldots, n$,

$$\sum_{j=1}^{k} \lambda_{i_j}(A)\lambda_{n-i_j+1}(B) \le \sum_{j=1}^{k} \lambda_j(AB).$$

In particular, for all $k = 1, \ldots, n$,

$$\sum_{i=1}^{k} \lambda_i(A)\lambda_{n-i+1}(B) \le \sum_{i=1}^{k} \lambda_i(AB) \le \sum_{i=1}^{k} \lambda_i(A)\lambda_i(B).$$

Proof: See [1208, p. 128] and [1422].

Remark: See Fact 8.19.23 and Fact 9.14.27.

Fact 8.19.21. Let $A, B \in \mathbb{F}^{n \times n}$, and assume that A and B are positive semidefinite. If $p \ge 1$, then

$$\sum_{i=1}^{n} \lambda_i^p(A)\lambda_{n-i+1}^p(B) \le \operatorname{tr}\left(B^{1/2}AB^{1/2}\right)^p \le \operatorname{tr} A^p B^p \le \sum_{i=1}^{n} \lambda_i^p(A)\lambda_i^p(B).$$

If $0 \le p \le 1$, then

$$\sum_{i=1}^{n} \lambda_i^p(A)\lambda_{n-i+1}^p(B) \le \operatorname{tr} A^p B^p \le \operatorname{tr}\left(B^{1/2}AB^{1/2}\right)^p \le \sum_{i=1}^{n} \lambda_i^p(A)\lambda_i^p(B).$$

Now, suppose that A and B are positive definite. If $p \leq -1$, then

$$\sum_{i=1}^{n} \lambda_i^p(A)\lambda_{n-i+1}^p(B) \leq \operatorname{tr}\left(B^{1/2}AB^{1/2}\right)^p \leq \operatorname{tr} A^pB^p \leq \sum_{i=1}^{n} \lambda_i^p(A)\lambda_i^p(B).$$

If $-1 \leq p \leq 0$, then

$$\sum_{i=1}^{n} \lambda_i^p(A)\lambda_{n-i+1}^p(B) \leq \operatorname{tr} A^pB^p \leq \operatorname{tr}\left(B^{1/2}AB^{1/2}\right)^p \leq \sum_{i=1}^{n} \lambda_i^p(A)\lambda_i^p(B).$$

Proof: See [1423]. See also [285, 906, 935, 1426].

Remark: See Fact 8.12.22. See Fact 8.12.15 for the indefinite case.

Fact 8.19.22. Let $A, B \in \mathbb{F}^{n \times n}$, and assume that A and B are positive semidefinite. Then, for all $k \in \{1, \ldots, n\}$,

$$\prod_{i=1}^{k} \lambda_i(AB) \leq \prod_{i=1}^{k} \sigma_i(AB) \leq \prod_{i=1}^{k} \lambda_i(A)\lambda_i(B)$$

with equality for $k = n$. Furthermore, for all $k \in \{1, \ldots, n\}$,

$$\prod_{i=k}^{n} \lambda_i(A)\lambda_i(B) \leq \prod_{i=k}^{n} \sigma_i(AB) \leq \prod_{i=k}^{n} \lambda_i(AB)$$

with equality for $k = 1$.

Proof: Use Fact 5.11.28 and Fact 9.13.18.

Fact 8.19.23. Let $A, B \in \mathbb{F}^{n \times n}$, assume that A and B are positive semidefinite, let $\lambda_1(AB) \geq \cdots \geq \lambda_n(AB) \geq 0$ denote the eigenvalues of AB, and let $1 \leq i_1 < \cdots < i_k \leq n$. Then,

$$\prod_{j=1}^{k} \lambda_{i_j}(AB) \leq \prod_{j=1}^{k} \lambda_{i_j}(A)\lambda_j(B)$$

with equality for $k = n$. Furthermore,

$$\prod_{j=1}^{k} \lambda_{i_j}(A)\lambda_{n-i_j+1}(B) \leq \prod_{j=1}^{k} \lambda_j(AB)$$

with equality for $k = n$. In particular,

$$\prod_{i=1}^{k} \lambda_i(A)\lambda_{n-i+1}(B) \leq \prod_{i=1}^{k} \lambda_i(AB) \leq \prod_{i=1}^{k} \lambda_i(A)\lambda_i(B)$$

with equality for $k = n$.

Proof: See [1208, p. 127] and [1422].

Remark: The first inequality is due to Lidskii.

Remark: See Fact 8.19.20 and Fact 9.14.27.

Fact 8.19.24. Let $A, B \in \mathbb{F}^{n \times n}$, assume that A and B are positive definite, and let $\lambda \in \operatorname{spec}(A)$. Then,

$$\frac{2}{n} \left[\frac{\lambda_{\min}^2(A)\lambda_{\min}^2(B)}{\lambda_{\min}^2(A) + \lambda_{\min}^2(B)} \right] < \lambda < \frac{n}{2} \left[\lambda_{\max}^2(A) + \lambda_{\max}^2(B) \right].$$

Proof: See [748].

Fact 8.19.25. Let $A, B \in \mathbb{F}^{n \times n}$, assume that A and B are positive definite, and define

$$k_A \triangleq \frac{\lambda_{\max}(A)}{\lambda_{\min}(A)}, \qquad k_B \triangleq \frac{\lambda_{\max}(B)}{\lambda_{\min}(B)},$$

and

$$\gamma \triangleq \frac{(\sqrt{k_A} + 1)^2}{\sqrt{k_A}} - \frac{k_B(\sqrt{k_A} - 1)^2}{\sqrt{k_A}}.$$

Then, if $\gamma < 0$, then

$$\tfrac{1}{2}\lambda_{\max}(A)\lambda_{\max}(B)\gamma \leq \lambda_{\min}(AB + BA) \leq \lambda_{\max}(AB + BA) \leq 2\lambda_{\max}(A)\lambda_{\max}(B),$$

whereas, if $\gamma > 0$, then

$$\tfrac{1}{2}\lambda_{\min}(A)\lambda_{\min}(B)\gamma \leq \lambda_{\min}(AB + BA) \leq \lambda_{\max}(AB + BA) \leq 2\lambda_{\max}(A)\lambda_{\max}(B).$$

Furthermore, if

$$\sqrt{k_A k_B} < 1 + \sqrt{k_A} + \sqrt{k_B},$$

then $AB + BA$ is positive definite.

Proof: See [1065].

Fact 8.19.26. Let $A, B \in \mathbb{F}^{n \times n}$, assume that A is positive definite, assume that B is positive semidefinite, and let $\alpha > 0$ and $\beta > 0$ be such that $\alpha I \leq A \leq \beta I$. Then,

$$\sigma_{\max}(AB) \leq \tfrac{\alpha + \beta}{2\sqrt{\alpha\beta}} \operatorname{sprad}(AB) \leq \tfrac{\alpha + \beta}{2\sqrt{\alpha\beta}} \sigma_{\max}(AB).$$

In particular,

$$\sigma_{\max}(A) \leq \tfrac{\alpha + \beta}{2\sqrt{\alpha\beta}} \operatorname{sprad}(A) \leq \tfrac{\alpha + \beta}{2\sqrt{\alpha\beta}} \sigma_{\max}(A).$$

Proof: See [1344].

Remark: The left-hand inequality is tightest for $\alpha = \lambda_{\min}(A)$ and $\beta = \lambda_{\max}(A)$.

Remark: This result is due to Bourin.

Fact 8.19.27. Let $A, B \in \mathbb{F}^{n \times n}$, and assume that A and B are positive semidefinite. Then, the following statements hold:

i) If $q \in [0, 1]$, then

$$\sigma_{\max}(A^q B^q) \leq \sigma_{\max}^q(AB)$$

and

$$\sigma_{\max}(B^q A^q B^q) \leq \sigma_{\max}^q(BAB).$$

ii) If $q \in [0, 1]$, then

$$\lambda_{\max}(A^q B^q) \leq \lambda_{\max}^q(AB).$$

iii) If $q \geq 1$, then
$$\sigma_{\max}^q(AB) \leq \sigma_{\max}(A^q B^q).$$

iv) If $q \geq 1$, then
$$\lambda_{\max}^q(AB) \leq \lambda_{\max}(A^q B^q).$$

v) If $p \geq q > 0$, then
$$\sigma_{\max}^{1/q}(A^q B^q) \leq \sigma_{\max}^{1/p}(A^p B^p).$$

vi) If $p \geq q > 0$, then
$$\lambda_{\max}^{1/q}(A^q B^q) \leq \lambda_{\max}^{1/p}(A^p B^p).$$

Proof: See [201, pp. 255–258] and [537].

Remark: See Fact 8.10.50, Fact 8.12.22, Fact 9.9.16, and Fact 9.9.17.

Remark: *ii)* is the *Cordes inequality*.

Fact 8.19.28. Let $A, B \in \mathbb{F}^{n \times n}$, assume that A and B are positive semidefinite, and let $p \geq r \geq 0$. Then,
$$\begin{bmatrix} \lambda_1^{1/p}(A^p B^p) & \cdots & \lambda_n^{1/p}(A^p B^p) \end{bmatrix}^{\mathrm{T}}$$
strongly log majorizes
$$\begin{bmatrix} \lambda_1^{1/r}(A^r B^r) & \cdots & \lambda_n^{1/r}(A^r B^r) \end{bmatrix}^{\mathrm{T}}.$$
In fact, for all $q > 0$,
$$\det(A^q B^q)^{1/q} = (\det A) \det B.$$

Proof: See [201, p. 257] or [1521, p. 20] and Fact 2.21.12.

Fact 8.19.29. Let $A \in \mathbb{F}^{n \times n}$, $B \in \mathbb{F}^{n \times m}$, and $C \in \mathbb{F}^{m \times m}$, and assume that
$$\mathcal{A} \triangleq \begin{bmatrix} A & B \\ B^* & C \end{bmatrix} \in \mathbb{F}^{(n+m) \times (n+m)}$$
is positive semidefinite. Then,
$$\max\{\sigma_{\max}(A), \sigma_{\max}(B)\}$$
$$\leq \sigma_{\max}(\mathcal{A})$$
$$\leq \tfrac{1}{2}\left[\sigma_{\max}(A) + \sigma_{\max}(B) + \sqrt{[\sigma_{\max}(A) - \sigma_{\max}(B)]^2 + 4\sigma_{\max}^2(C)} \right]$$
$$\leq \sigma_{\max}(A) + \sigma_{\max}(B)$$
and
$$\max\{\sigma_{\max}(A), \sigma_{\max}(B)\} \leq \sigma_{\max}(\mathcal{A}) \leq \max\{\sigma_{\max}(A), \sigma_{\max}(B)\} + \sigma_{\max}(C).$$

Proof: See [738].

Remark: See Fact 9.14.12.

Fact 8.19.30. Let $A, B \in \mathbb{F}^{n \times n}$, and assume that A and B are positive definite. Then,

$$\begin{bmatrix} \lambda_1(\log A + \log B) & \cdots & \lambda_n(\log A + \log B) \end{bmatrix}^{\mathrm{T}}$$

strongly log majorizes

$$\begin{bmatrix} \lambda_1(\log A^{1/2}BA^{1/2}) & \cdots & \lambda_n(\log A^{1/2}BA^{1/2}) \end{bmatrix}^{\mathrm{T}}.$$

Consequently,

$$\log \det AB = \mathrm{tr}(\log A + \log B) = \mathrm{tr} \log A^{1/2}BA^{1/2} = \log \det A^{1/2}BA^{1/2}.$$

Proof: See [93].

Fact 8.19.31. Let $A, B \in \mathbb{F}^{n \times n}$, and assume that A and B are positive semidefinite. Then, the following statements hold:

i) $\sigma_{\max}[\log(I + A)\log(I + B)] \leq \left(\log\left[1 + \sigma_{\max}^{1/2}(AB)\right]\right)^2.$

ii) $\sigma_{\max}[\log(I + B)\log(I + A)\log(I + B)] \leq \left(\log\left[1 + \sigma_{\max}^{1/3}(BAB)\right]\right)^3.$

iii) $\det[\log(I + A)\log(I + B)] \leq \det\left[\log(I + \langle AB \rangle^{1/2})\right]^2.$

iv) $\det[\log(I + B)\log(I + A)\log(I + B)] \leq \det\left(\log[I + (BAB)^{1/3}]\right)^3.$

Proof: See [1382].

Remark: See Fact 11.16.6.

Fact 8.19.32. Let $A, B \in \mathbb{F}^{n \times n}$, and assume that A and B are positive semidefinite. Then,

$$\sigma_{\max}\left[(I + A)^{-1}AB(I + B)^{-1}\right] \leq \frac{\sigma_{\max}(AB)}{\left[1 + \sigma_{\max}^{1/2}(AB)\right]^2}.$$

Proof: See [1382].

8.20 Facts on Alternative Partial Orderings

Fact 8.20.1. Let $A, B \in \mathbb{F}^{n \times n}$, and assume that A and B are positive definite. Then, the following statements are equivalent:

i) $\log B \leq \log A.$

ii) For all $r \in (0, \infty)$,

$$B^r \leq \left(B^{r/2}A^rB^{r/2}\right)^{1/2}.$$

iii) For all $r \in (0, \infty)$,

$$\left(A^{r/2}B^rA^{r/2}\right)^{1/2} \leq A^r.$$

iv) For all $p, r \in (0, \infty)$ and $k \in \mathbb{N}$ such that $(k + 1)r = p + r$,

$$B^r \leq \left(B^{r/2} A^p B^{r/2} \right)^{\frac{1}{k+1}}.$$

v) For all $p, r \in (0, \infty)$ and $k \in \mathbb{N}$ such that $(k+1)r = p + r$,

$$\left(A^{r/2} B^p A^{r/2} \right)^{\frac{1}{k+1}} \leq A^r.$$

vi) For all $p, r \in [0, \infty)$,

$$B^r \leq \left(B^{r/2} A^p B^{r/2} \right)^{\frac{r}{r+p}}.$$

vii) For all $p, r \in [0, \infty)$,

$$\left(A^{r/2} B^p A^{r/2} \right)^{\frac{r}{r+p}} \leq A^r.$$

$viii$) For all $p, q, r, t \in \mathbb{R}$ such that $p \geq 0$, $r \geq 0$, $t \geq 0$, and $q \in [1, 2]$,

$$\left[A^{r/2} \left(A^{t/2} B^p A^{t/2} \right)^q A^{r/2} \right]^{\frac{r+t}{r+qt+qp}} \leq A^{r+t}.$$

Proof: See [525, 940, 1506] and [544, pp. 139, 200].

Remark: $\log B \leq \log A$ is the *chaotic order*. This order is weaker than the Löwner order since $B \leq A$ implies that $\log B \leq \log A$, but not vice versa.

Remark: Additional conditions are given in [940].

Fact 8.20.2. Let $A, B \in \mathbb{F}^{n \times n}$, assume that A is positive definite and B is positive semidefinite, and let $\alpha > 0$. Then, the following statements are equivalent:

i) $B^\alpha \leq A^\alpha$.

ii) For all $p, q, r, \tau \in \mathbb{R}$ such that $p \geq \alpha$, $r \geq \tau$, $q \geq 1$, and $\tau \in [0, \alpha]$,

$$\left[A^{r/2} \left(A^{-\tau/2} B^p A^{-\tau/2} \right)^q A^{r/2} \right]^{\frac{r-\tau}{r-q\tau+qp}} \leq A^{r-\tau}.$$

Proof: See [525].

Fact 8.20.3. Let $A, B \in \mathbb{F}^{n \times n}$, and assume that A is positive definite and B is positive semidefinite. Then, the following statements are equivalent:

i) For all $k \in \mathbb{N}$, $B^k \leq A^k$.

ii) For all $\alpha > 0$, $B^\alpha \leq A^\alpha$.

iii) For all $p, r \in \mathbb{R}$ such that $p > r \geq 0$,

$$\left(A^{-r/2} B^p A^{-r/2} \right)^{\frac{2p-r}{p-r}} \leq A^{2p-r}.$$

iv) For all $p, q, r, \tau \in \mathbb{R}$ such that $p \geq \tau$, $r \geq \tau$, $q \geq 1$, and $\tau \geq 0$,

$$\left[A^{r/2} \left(A^{-\tau/2} B^p A^{-\tau/2} \right)^q A^{r/2} \right]^{\frac{r-\tau}{r-q\tau+qp}} \leq A^{r-\tau}.$$

Proof: See [545].

Remark: A and B are related by the *spectral order*.

Fact 8.20.4. Let $A, B \in \mathbb{F}^{n \times n}$, and assume that A and B are positive semidefinite. Then, if two of the following statements hold, then the remaining statement also holds:

i) $A \overset{\mathrm{rs}}{\leq} B$.

ii) $A^2 \overset{\mathrm{rs}}{\leq} B^2$.

iii) $AB = BA$.

Proof: See [113, 604, 605].

Remark: The rank subtractivity partial ordering is defined in Fact 2.10.32.

Fact 8.20.5. Let $A, B, C \in \mathbb{F}^{n \times n}$, and assume that A, B, and C are positive semidefinite. Then, the following statements hold:

i) If $A^2 = AB$ and $B^2 = BA$, then $A = B$.

ii) If $A^2 = AB$ and $B^2 = BC$, then $A^2 = AC$.

Proof: Use Fact 2.10.33 and Fact 2.10.34.

Fact 8.20.6. Let $A, B \in \mathbb{F}^{n \times n}$, and assume that A and B are positive semidefinite, and define
$$A \overset{*}{\leq} B$$
if and only if
$$A^2 = AB.$$
Then, "$\overset{*}{\leq}$" is a partial ordering on $\mathbf{N}^{n \times n}$.

Proof: Use Fact 2.10.35 or Fact 8.20.5.

Remark: The relation "$\overset{*}{\leq}$" is the *star partial ordering*.

Fact 8.20.7. Let $A, B \in \mathbb{F}^{n \times n}$, and assume that A and B are positive semidefinite. Then,
$$A \overset{*}{\leq} B$$
if and only if
$$B^+ \overset{*}{\leq} A^+.$$

Proof: See [663].

Remark: The star partial ordering is defined in Fact 8.20.6.

Fact 8.20.8. Let $A, B \in \mathbb{F}^{n \times n}$, and assume that A and B are positive semidefinite. Then, the following statements are equivalent:

i) $A \overset{*}{\leq} B$.

ii) $A \overset{\mathrm{rs}}{\leq} B$ and $A^2 \overset{\mathrm{rs}}{\leq} B^2$.

Remark: See [615].

Remark: The star partial ordering is defined in Fact 8.20.6.

Fact 8.20.9. Let $A, B \in \mathbb{F}^{n \times m}$, and define

$$A \overset{\text{GL}}{\leq} B$$

if and only if all of the following conditions hold:

i) $\langle A \rangle \leq \langle B \rangle$.

ii) $\mathcal{R}(A^*) \subseteq \mathcal{R}(B^*)$.

iii) $AB^* = \langle A \rangle \langle B \rangle$.

Then, "$\overset{\text{GL}}{\leq}$" is a partial ordering on $\mathbb{F}^{n \times m}$. Furthermore, the following statements are equivalent:

iv) $A \overset{\text{GL}}{\leq} B$.

v) $A^* \overset{\text{GL}}{\leq} B^*$.

vi) $\operatorname{sprad}(B^+ A) \leq 1$, $\mathcal{R}(A) \subseteq \mathcal{R}(B)$, $\mathcal{R}(A^*) \subseteq \mathcal{R}(B^*)$, and $AB^* = \langle A \rangle \langle B \rangle$.

Furthermore, if $A \overset{\text{rs}}{\leq} B$, then $A \overset{\text{GL}}{\leq} B$. Finally, if $A, B \in \mathbf{N}^n$, then $A \leq B$ if and only if $A \overset{\text{GL}}{\leq} B$.

Proof: See [672].

Remark: The relation "$\overset{\text{GL}}{\leq}$" is the *generalized Löwner partial ordering*. Remarkably, the Löwner, generalized Löwner, and star partial orderings are linked through the polar decomposition. See [672].

8.21 Facts on Generalized Inverses

Fact 8.21.1. Let $A \in \mathbb{F}^{n \times n}$. Then, the following statements are equivalent:

i) $A + A^* \geq 0$.

ii) $A^+ + A^{+*} \geq 0$.

If, in addition, A is group invertible, then the following statement is equivalent to *i)* and *ii)*:

iii) $A^\# + A^{\#*} \geq 0$.

Proof: See [1361].

Fact 8.21.2. Let $A \in \mathbb{F}^{n \times n}$, and assume that A is positive semidefinite. Then, the following statements hold:

i) $A^+ = A^{\text{D}} = A^\# \geq 0$.

ii) $\operatorname{rank} A = \operatorname{rank} A^+$.

iii) $A^{+1/2} \triangleq \left(A^{1/2} \right)^+ = (A^+)^{1/2}$.

iv) $A^{1/2} = A(A^+)^{1/2} = (A^+)^{1/2}A.$

v) $AA^+ = A^{1/2}(A^{1/2})^+.$

vi) $\begin{bmatrix} A & AA^+ \\ A^+A & A^+ \end{bmatrix}$ is positive semidefinite.

vii) $A^+A + AA^+ \le A + A^+.$

viii) $A^+A \circ AA^+ \le A \circ A^+.$

Proof: See [1528] or Fact 8.11.5 and Fact 8.22.42 for *vi)*–*viii)*.

Fact 8.21.3. Let $A \in \mathbb{F}^{n \times n}$, and assume that A is positive semidefinite. Then,

$$\operatorname{rank} A \le (\operatorname{tr} A) \operatorname{tr} A^+.$$

Furthermore, equality holds if and only if $\operatorname{rank} A \le 1$.

Proof: See [117].

Fact 8.21.4. Let $A \in \mathbb{F}^{n \times m}$. Then,

$$\langle A^* \rangle = A \langle A \rangle^{+1/2} A^*.$$

Remark: See Fact 8.11.11.

Fact 8.21.5. Let $A \in \mathbb{F}^{n \times m}$, and define $S \in \mathbb{F}^{n \times n}$ by

$$S \triangleq \langle A \rangle + I_n - AA^+.$$

Then, S is positive definite, and

$$SAA^+S = \langle A \rangle AA^+ \langle A \rangle = AA^*.$$

Proof: See [459, p. 432].

Remark: This result provides an explicit congruence transformation for AA^+ and AA^*.

Remark: See Fact 5.8.20.

Fact 8.21.6. Let $A, B \in \mathbb{F}^{n \times n}$, and assume that A and B are positive semidefinite. Then,

$$A = (A + B)(A + B)^+ A.$$

Fact 8.21.7. Let $A, B \in \mathbb{F}^{n \times n}$, and assume that A and B are Hermitian. Then, the following statements are equivalent:

i) $A \overset{\mathrm{rs}}{\le} B.$

ii) $\mathcal{R}(A) \subseteq \mathcal{R}(B)$ and $AB^+A = A.$

Proof: See [604, 605].

Remark: See Fact 6.5.30.

Fact 8.21.8. Let $A, B \in \mathbb{F}^{n \times n}$, assume that A and B are Hermitian, assume that $\nu_-(A) = \nu_-(B)$, and consider the following statements:

i) $A \overset{*}{\leq} B$.

ii) $A \overset{\mathrm{rs}}{\leq} B$.

iii) $A \leq B$.

iv) $\mathcal{R}(A) \subseteq \mathcal{R}(B)$ and $AB^+A \leq A$.

Then, $i) \Longrightarrow ii) \Longrightarrow iii) \Longleftrightarrow iv)$. If, in addition, A and B are positive semidefinite, then the following statement is equivalent to $iii)$ and $iv)$:

v) $\mathcal{R}(A) \subseteq \mathcal{R}(B)$ and $\mathrm{sprad}(B^+A) \leq 1$.

Proof: $i) \Longrightarrow ii)$ is given in [669]. See [113, 604, 615, 1254] and [1215, p. 229].

Remark: See Fact 8.21.7.

Fact 8.21.9. Let $A, B \in \mathbb{F}^{n \times n}$, and assume that A and B are positive semidefinite. Then, the following statements are equivalent:

i) $A^2 \leq B^2$.

ii) $\mathcal{R}(A) \subseteq \mathcal{R}(B)$ and $\sigma_{\max}(B^+A) \leq 1$.

Proof: See [615].

Fact 8.21.10. Let $A, B \in \mathbb{F}^{n \times n}$, assume that A and B are positive semidefinite, and assume that $A \leq B$. Then, the following statements are equivalent:

i) $B^+ \leq A^+$.

ii) $\mathrm{rank}\, A = \mathrm{rank}\, B$.

iii) $\mathcal{R}(A) = \mathcal{R}(B)$.

Furthermore, the following statements are equivalent:

iv) $A^+ \leq B^+$.

v) $A^2 = AB$.

vi) $A^+ \overset{*}{\leq} B^+$.

Proof: See [663, 1028].

Fact 8.21.11. Let $A, B \in \mathbb{F}^{n \times n}$, and assume that A and B are positive semidefinite. Then, if two of the following statements hold, then the remaining statement also holds:

i) $A \leq B$.

ii) $B^+ \leq A^+$.

iii) $\mathrm{rank}\, A = \mathrm{rank}\, B$.

Proof: See [114, 1028, 1456, 1491].

Fact 8.21.12. Let $A, B \in \mathbb{F}^{n \times n}$, and assume that A and B are Hermitian. Then, if two of the following statements hold, then the remaining statement also holds:

i) $A \leq B$.

ii) $B^+ \leq A^+$.

iii) $\operatorname{In} A = \operatorname{In} B$.

Proof: See [112].

Fact 8.21.13. Let $A, B \in \mathbb{F}^{n \times n}$, assume that A and B are positive semidefinite, and assume that $A \leq B$. Then,

$$0 \leq AA^+ \leq BB^+.$$

If, in addition, $\operatorname{rank} A = \operatorname{rank} B$, then

$$AA^+ = BB^+.$$

Fact 8.21.14. Let $A, B \in \mathbb{F}^{n \times n}$, assume that A and B are Hermitian, and assume that $\mathcal{R}(A) = \mathcal{R}(B)$. Then,

$$\operatorname{In} A - \operatorname{In} B = \operatorname{In}(A - B) + \operatorname{In}(A^+ - B^+).$$

Proof: See [1074].

Remark: See Fact 8.10.15.

Fact 8.21.15. Let $A, B \in \mathbb{F}^{n \times n}$, assume that A and B are positive semidefinite, and assume that $A \leq B$. Then,

$$0 \leq AB^+A \leq A \leq A + B\big[(I - AA^+)B(I - AA^+)\big]^+ B \leq B.$$

Proof: See [663].

Fact 8.21.16. Let $A, B \in \mathbb{F}^{n \times n}$, and assume that A and B are positive semidefinite. Then,

$$\operatorname{spec}\big[(A + B)^+A\big] \subset [0, 1].$$

Proof: Let C be positive definite and satisfy $B \leq C$. Then,

$$(A + C)^{-1/2}C(A + C)^{-1/2} \leq I.$$

The result now follows from Fact 8.21.17.

Fact 8.21.17. Let $A, B, C \in \mathbb{F}^{n \times n}$, assume that A, B, C are positive semidefinite, and assume that $B \leq C$. Then, for all $i \in \{1, \ldots, n\}$,

$$\lambda_i\big[(A + B)^+B\big] \leq \lambda_i\big[(A + C)^+C\big].$$

Consequently,

$$\operatorname{tr}\big[(A + B)^+B\big] \leq \operatorname{tr}\big[(A + C)^+C\big].$$

Proof: See [1424].

Remark: See Fact 8.21.16.

Fact 8.21.18. Let $A, B \in \mathbb{F}^{n \times n}$, assume that A and B are positive semidefinite, and define

$$A:B \triangleq A(A + B)^+ B.$$

Then, the following statements hold:

  i) $A:B$ is positive semidefinite.

  ii) $A:B = \lim_{\varepsilon \downarrow 0} (A + \varepsilon I):(B + \varepsilon I)$.

  iii) $A:A = \frac{1}{2} A$.

  iv) $A:B = B:A = B - B(A + B)^+ B = A - A(A + B)^+ A$.

  v) $A:B \le A$.

  vi) $A:B \le B$.

  vii) $A:B = -\begin{bmatrix} 0 & 0 & I \end{bmatrix} \begin{bmatrix} A & 0 & I \\ 0 & B & I \\ I & I & 0 \end{bmatrix}^+ \begin{bmatrix} 0 \\ 0 \\ I \end{bmatrix}$.

  viii) $A:B = (A^+ + B^+)^+$ if and only if $\mathcal{R}(A) = \mathcal{R}(B)$.

  ix) $A(A + B)^+ B = ACB$ for every (1)-inverse C of $A + B$.

  x) $\mathrm{tr}(A:B) \le (\mathrm{tr}\, B):(\mathrm{tr}\, A)$.

  xi) $\mathrm{tr}(A:B) = (\mathrm{tr}\, B):(\mathrm{tr}\, A)$ if and only if there exists $\alpha \in [0, \infty)$ such that either $A = \alpha B$ or $B = \alpha A$.

  xii) $\det(A:B) \le (\det B):(\det A)$.

  xiii) $\mathcal{R}(A:B) = \mathcal{R}(A) \cap \mathcal{R}(B)$.

  xiv) $\mathcal{N}(A:B) = \mathcal{N}(A) + \mathcal{N}(B)$.

  xv) $\mathrm{rank}(A:B) = \mathrm{rank}\, A + \mathrm{rank}\, B - \mathrm{rank}(A + B)$.

  xvi) Let $S \in \mathbb{F}^{p \times n}$, and assume that S is right invertible. Then,

$$S(A:B)S^* \le (SAS^*):(SBS^*).$$

  xvii) Let $S \in \mathbb{F}^{n \times n}$, and assume that S is nonsingular. Then,

$$S(A:B)S^* = (SAS^*):(SBS^*).$$

  xviii) For all positive numbers α, β,

$$(\alpha^{-1} A):(\beta^{-1} B) \le \alpha A + \beta B.$$

  xix) Let $X \in \mathbb{F}^{n \times n}$, and assume that X is Hermitian and

$$\begin{bmatrix} A + B & A \\ A & A - X \end{bmatrix} \ge 0.$$

Then,

$$X \le A:B.$$

Furthermore,

$$\begin{bmatrix} A + B & A \\ A & A - A:B \end{bmatrix} \ge 0.$$

xx) $\phi\colon \mathbf{N}^n \times \mathbf{N}^n \mapsto -\mathbf{N}^n$ defined by $\phi(A,B) \triangleq -A{:}B$ is convex.

xxi) If A and B are projectors, then $2(A{:}B)$ is the projector onto $\mathcal{R}(A) \cap \mathcal{R}(B)$.

$xxii$) If $A + B$ is positive definite, then

$$A{:}B = A(A+B)^{-1}B.$$

$xxiii$) $A\#B = [\tfrac{1}{2}(A+B)]\#[2(A{:}B)]$.

$xxiv$) If $C, D \in \mathbb{F}^{n \times n}$ are positive semidefinite, then

$$(A{:}B){:}C = A{:}(B{:}C)$$

and

$$A{:}C + B{:}D \le (A+B){:}(C+D).$$

xxv) If $C, D \in \mathbb{F}^{n \times n}$ are positive semidefinite, $A \le C$, and $B \le D$, then

$$A{:}B \le C{:}D.$$

$xxvi$) If A and B are positive definite, then

$$A{:}B = \left(A^{-1} + B^{-1}\right)^{-1} \le \tfrac{1}{2}(A\#B) \le \tfrac{1}{4}(A+B).$$

$xxvii$) Let $x, y \in \mathbb{F}^n$. Then,

$$(x+y)^*(A{:}B)(x+y) \le x^*Ax + y^*By.$$

$xxviii$) Let $x, y \in \mathbb{F}^n$. Then,

$$x^*(A{:}B)x \le y^*Ay + (x-y)^*B(x-y).$$

$xxix$) Let $x \in \mathbb{F}^n$. Then,

$$x^*(A{:}B)x = \inf_{y \in \mathbb{F}^n} [y^*Ay + (x-y)^*B(x-y)].$$

xxx) Let $x \in \mathbb{F}^n$. Then,

$$x^*(A{:}B)x \le (x^*Ax){:}(x^*Bx).$$

Proof: See [38, 39, 42, 597, 867, 1316], [1145, p. 189], and [1521, p. 9].

Remark: $A{:}B$ is the *parallel sum* of A and B.

Remark: See Fact 6.4.46 and Fact 6.4.47.

Remark: A symmetric expression for the parallel sum of three or more positive-semidefinite matrices is given in [1316].

Fact 8.21.19. Let $A, B \in \mathbb{F}^{n \times n}$, assume that A is positive semidefinite, and assume that B is a projector. Then,

$$\mathrm{sh}(A,B) \triangleq \min\{X \in \mathbf{N}^n\colon 0 \le X \le A \text{ and } \mathcal{R}(X) \subseteq \mathcal{R}(B)\}$$

exists. Furthermore,

$$\mathrm{sh}(A,B) = A - AB_\perp(B_\perp AB_\perp)^+ B_\perp A.$$

That is,

$$\mathrm{sh}(A,B) = A \left\| \begin{bmatrix} A & AB_\perp \\ B_\perp A & B_\perp AB_\perp \end{bmatrix} \right..$$

Finally,

$$\text{sh}(A, B) = \lim_{\alpha \to \infty} (\alpha B) : A.$$

Proof: Existence of the minimum is proved in [42]. The expression for $\text{sh}(A, B)$ is given in [582]; a related expression involving the Schur complement is given in [38]. The last equality is shown in [42]. See also [52].

Remark: $\text{sh}(A, B)$ is the *shorted operator*.

Fact 8.21.20. Let $B \in \mathbb{R}^{m \times n}$, define

$$\mathcal{S} \triangleq \{A \in \mathbb{R}^{n \times n} \colon A \geq 0 \text{ and } \mathcal{R}(B^{\mathrm{T}}BA) \subseteq \mathcal{R}(A)\},$$

and define $\phi \colon \mathcal{S} \mapsto -\mathbf{N}^m$ by $\phi(A) \triangleq -(BA^+B^{\mathrm{T}})^+$. Then, $\mathcal{S}$ is a convex cone, and ϕ is convex.

Proof: See [606].

Remark: This result generalizes $xii)$ of Proposition 8.6.17 in the case $r = p = 1$.

Fact 8.21.21. Let $A, B \in \mathbb{F}^{n \times n}$, and assume that A and B are positive semidefinite. If $(AB)^+ = B^+A^+$, then AB is range Hermitian. Furthermore, the following statements are equivalent:

$i)$ AB is range Hermitian.

$ii)$ $(AB)^\# = B^+A^+$.

$iii)$ $(AB)^+ = B^+A^+$.

Proof: See [1013].

Remark: See Fact 6.4.31.

Fact 8.21.22. Let $A \in \mathbb{F}^{n \times n}$ and $C \in \mathbb{F}^{m \times m}$, assume that A and C are positive semidefinite, let $B \in \mathbb{F}^{n \times m}$, and define $X \triangleq A^{+1/2}BC^{+1/2}$. Then, the following statements are equivalent:

$i)$ $\begin{bmatrix} A & B \\ B^* & C \end{bmatrix}$ is positive semidefinite.

$ii)$ $AA^+B = B$ and $X^*X \leq I_m$.

$iii)$ $BC^+C = B$ and $X^*X \leq I_m$.

$iv)$ $B = A^{1/2}XC^{1/2}$ and $X^*X \leq I_m$.

$v)$ There exists a matrix $Y \in \mathbb{F}^{n \times m}$ such that $B = A^{1/2}YC^{1/2}$ and $Y^*Y \leq I_m$.

Proof: See [1521, p. 15].

Fact 8.21.23. Let $A, B \in \mathbb{F}^{n \times n}$, and assume that A and B are positive semidefinite. Then, the following statements are equivalent:

$i)$ $A(A + B)^+B = 0$.

$ii)$ $B(A + B)^+A = 0$.

$iii)$ $A(A + B)^+A = A$.

iv) $B(A + B)^+B = B.$

v) $A(A + B)^+B + B(A + B)^+A = 0.$

vi) $A(A + B)^+A + B(A + B)^+B = A + B.$

vii) $\operatorname{rank} \begin{bmatrix} A & B \end{bmatrix} = \operatorname{rank} A + \operatorname{rank} B.$

viii) $\mathcal{R}(A) \cap \mathcal{R}(B) = \{0\}.$

ix) $(A + B)^+ = [(I - BB^+)A(I - B^+B]^+ + [(I - AA^+)B(I - A^+A]^+.$

Proof: See [1334].

Remark: See Fact 6.4.35.

8.22 Facts on the Kronecker and Schur Products

Fact 8.22.1. Let $A \in \mathbb{F}^{n \times n}$, assume that A is positive semidefinite, and assume that every entry of A is nonzero. Then, $A^{\circ -1}$ is positive semidefinite if and only if $\operatorname{rank} A = 1$.

Proof: See [914].

Fact 8.22.2. Let $A \in \mathbb{F}^{n \times n}$, assume that A is positive semidefinite, assume that every entry of A is nonnegative, and let $\alpha \in [0, n - 2]$. Then, $A^{\circ \alpha}$ is positive semidefinite.

Proof: See [203, 504].

Remark: In many cases, $A^{\circ \alpha}$ is positive semidefinite for all $\alpha \geq 0$. See Fact 8.8.5.

Fact 8.22.3. Let $A \in \mathbb{F}^{n \times n}$, assume that A is positive semidefinite, and let $k \geq 1$. If $r \in [0, 1]$, then
$$(A^r)^{\circ k} \leq (A^{\circ k})^r.$$

If $r \in [1, 2]$, then
$$(A^{\circ k})^r \leq (A^r)^{\circ k}.$$

If A is positive definite and $r \in [0, 1]$, then
$$(A^{\circ k})^{-r} \leq (A^{-r})^{\circ k}.$$

Proof: See [1521, p. 8].

Fact 8.22.4. Let $A \in \mathbb{F}^{n \times n}$, and assume that A is positive semidefinite. Then,
$$(I \circ A)^2 \leq \tfrac{1}{2}(I \circ A^2 + A \circ A) \leq I \circ A^2$$

and
$$A \circ A \leq I \circ A^2.$$

Hence,
$$\sum_{i=1}^{n} A_{(i,i)}^2 \leq \sum_{i=1}^{n} \lambda_i^2(A).$$

Now, assume in addition that A is positive definite. Then,

$$(A \circ A)^{-1} \le A^{-1} \circ A^{-1}$$

and

$$\left(A \circ A^{-1}\right)^{-1} \le I \le \left(A^{1/2} \circ A^{-1/2}\right)^2 \le \tfrac{1}{2}\left(I + A \circ A^{-1}\right) \le A \circ A^{-1}.$$

Furthermore,

$$\left(A \circ A^{-1}\right) 1_{n \times 1} = 1_{n \times 1}$$

and

$$1 \in \operatorname{spec}\left(A \circ A^{-1}\right).$$

Next, let $\alpha \triangleq \lambda_{\min}(A)$ and $\beta \triangleq \lambda_{\max}(A)$. Then,

$$\frac{2\alpha\beta}{\alpha^2 + \beta^2} I \le \frac{2\alpha\beta}{\alpha^2 + \beta^2}\left(A^2 \circ A^{-2}\right)^{1/2} \le \frac{\alpha\beta}{\alpha^2 + \beta^2}\left(I + A^2 \circ A^{-2}\right) \le A \circ A^{-1}.$$

Define $\Phi(A) \triangleq A \circ A^{-1}$, and, for all $k \ge 1$, define

$$\Phi^{(k+1)}(A) \triangleq \Phi\left[\Phi^{(k)}(A)\right],$$

where $\Phi^{(1)}(A) \triangleq \Phi(A)$. Then, for all $k \ge 1$,

$$\Phi^{(k)}(A) \ge I$$

and

$$\lim_{k \to \infty} \Phi^{(k)}(A) = I.$$

Proof: See [493, 794, 1417, 1418], [728, p. 475], [1208, pp. 304, 305], and set $B = A^{-1}$ in Fact 8.22.33.

Remark: The convergence result also holds if A is an *H-matrix* [794]. $A \circ A^{-1}$ is the *relative gain array*.

Remark: See Fact 8.22.40.

Fact 8.22.5. Let $A \in \mathbb{F}^{n \times n}$, and assume that A is positive definite. Then, for all $i \in \{1, \ldots, n\}$,

$$1 \le A_{(i,i)}(A^{-1})_{(i,i)}.$$

Furthermore,

$$\max_{i=1,\ldots,n} \sqrt{A_{(i,i)}(A^{-1})_{(i,i)} - 1} \le \sum_{i=1}^{n} \sqrt{A_{(i,i)}(A^{-1})_{(i,i)} - 1}$$

and

$$\max_{i=1,\ldots,n} \sqrt{A_{(i,i)}(A^{-1})_{(i,i)}} - 1 \le \sum_{i=1}^{n} \left[\sqrt{A_{(i,i)}(A^{-1})_{(i,i)}} - 1\right].$$

Proof: See [495, p. 66-6].

Fact 8.22.6. Let $\mathcal{A} \triangleq \left[\begin{smallmatrix} A & B \\ B^* & C \end{smallmatrix}\right] \in \mathbb{F}^{(n+m) \times (n+m)}$, assume that $\mathcal{A}$ is positive definite, and partition $\mathcal{A}^{-1} = \left[\begin{smallmatrix} X & Y \\ Y^* & Z \end{smallmatrix}\right]$ conformably with $\mathcal{A}$. Then,

$$I \le \begin{bmatrix} A \circ A^{-1} & 0 \\ 0 & Z \circ Z^{-1} \end{bmatrix} \le \mathcal{A} \circ \mathcal{A}^{-1}$$

and
$$I \leq \left[\begin{array}{cc} X \circ X^{-1} & 0 \\ 0 & C \circ C^{-1} \end{array} \right] \leq \mathcal{A} \circ \mathcal{A}^{-1}.$$

Proof: See [136].

Fact 8.22.7. Let $A \in \mathbb{F}^{n \times n}$, let $p, q \in \mathbb{R}$, assume that A is positive semidefinite, and assume that either p and q are nonnegative or A is positive definite. Then,
$$A^{(p+q)/2} \circ A^{(p+q)/2} \leq A^p \circ A^q.$$

In particular,
$$I \leq A \circ A^{-1}.$$

Proof: See [95].

Fact 8.22.8. Let $A \in \mathbb{F}^{n \times n}$, assume that A is positive semidefinite, and assume that $I_n \circ A = I_n$. Then,
$$\det A \leq \lambda_{\min}(A \circ \overline{A}).$$

Proof: See [1442].

Fact 8.22.9. Let $A \in \mathbb{F}^{n \times n}$. Then,
$$-A^*A \circ I \leq A^* \circ A \leq A^*A \circ I.$$

Proof: Use Fact 8.22.43 with $B = I$.

Fact 8.22.10. Let $A \in \mathbb{F}^{n \times n}$. Then,
$$\langle A \circ A^* \rangle \leq \left\{ \begin{array}{c} A^*A \circ I \\ \langle A \rangle \circ \langle A^* \rangle \end{array} \right\} \leq \sigma_{\max}^2(A)I.$$

Proof: See [1528] and Fact 8.22.23.

Fact 8.22.11. Let $A \triangleq \left[\begin{array}{cc} A_{11} & A_{12} \\ A_{12}^* & A_{22} \end{array} \right] \in \mathbb{F}^{(n+m) \times (n+m)}$ and $B \triangleq \left[\begin{array}{cc} B_{11} & B_{12} \\ B_{12}^* & B_{22} \end{array} \right] \in \mathbb{F}^{(n+m) \times (n+m)}$, and assume that A and B are positive semidefinite. Then,
$$(A_{11}|A) \circ (B_{11}|B) \leq (A_{11}|A) \circ B_{22} \leq (A_{11} \circ B_{11})|(A \circ B).$$

Proof: See [922].

Fact 8.22.12. Let $A, B \in \mathbb{F}^{n \times n}$, and assume that A and B are positive semidefinite. Then, $A \circ B$ is positive semidefinite. If, in addition, B is positive definite and $I \circ A$ is positive definite, then $A \circ B$ is positive definite.

Proof: By Fact 7.4.17, $A \otimes B$ is positive semidefinite, and the Schur product $A \circ B$ is a principal submatrix of the Kronecker product. If A is positive definite, use Fact 8.22.20 to obtain $\det(A \circ B) > 0$. See [1208, p. 300].

Remark: The first result is *Schur's theorem*. The second result is *Schott's theorem*. See [950] and Fact 8.22.20.

Fact 8.22.13. Let $A \in \mathbb{F}^{n \times n}$, and assume that A is positive definite. Then, there exist positive-definite matrices $B, C \in \mathbb{F}^{n \times n}$ such that $A = B \circ C$.

Remark: See [1125, pp. 154, 166].

Remark: This result is due to Djokovic.

Fact 8.22.14. Let $A, B \in \mathbb{F}^{n \times n}$, and assume that A is positive definite and B is positive semidefinite. Then,

$$\left(1_{1 \times n} A^{-1} 1_{n \times 1}\right)^{-1} B \le A \circ B.$$

Proof: See [497].

Remark: Setting $B = 1_{n \times n}$ yields Fact 8.9.17.

Fact 8.22.15. Let $A, B \in \mathbb{F}^{n \times n}$, and assume that A and B are positive definite. Then,
$$\left(1_{1 \times n} A^{-1} 1_{n \times 1} 1_{1 \times n} B^{-1} 1_{n \times 1}\right)^{-1} 1_{n \times n} \le A \circ B.$$

Proof: See [1528].

Fact 8.22.16. Let $A \in \mathbb{F}^{n \times n}$, assume that A is positive definite, let $B \in \mathbb{F}^{n \times n}$, and assume that B is positive semidefinite. Then,

$$\operatorname{rank} B \le \operatorname{rank}(A \circ B) \le \operatorname{rank}(A \otimes B) = (\operatorname{rank} A)(\operatorname{rank} B).$$

Remark: See Fact 7.4.24, Fact 7.6.10, and Fact 8.22.14.

Remark: The first inequality is due to Djokovic. See [1125, pp. 154, 166].

Fact 8.22.17. Let $A, B \in \mathbb{F}^{n \times n}$, and assume that A and B are positive semidefinite. If $p \ge 1$, then

$$\operatorname{tr}\,(A \circ B)^p \le \operatorname{tr} A^p \circ B^p.$$

If $0 \le p \le 1$, then
$$\operatorname{tr} A^p \circ B^p \le \operatorname{tr}\,(A \circ B)^p.$$

Now, assume in addition that A and B are positive definite. If $p \le 0$, then

$$\operatorname{tr}\,(A \circ B)^p \le \operatorname{tr} A^p \circ B^p.$$

Proof: See [1426].

Fact 8.22.18. Let $A, B \in \mathbb{F}^{n \times n}$, and assume that A and B are positive semidefinite. Then,
$$\lambda_{\min}(AB) \le \lambda_{\min}(A \circ B).$$

Hence,
$$\lambda_{\min}(AB)I \le \lambda_{\min}(A \circ B)I \le A \circ B.$$

Proof: See [787].

Remark: This result interpolates the penultimate inequality in Fact 8.22.21.

Fact 8.22.19. Let $A, B \in \mathbb{F}^{n \times n}$, and assume that A and B are positive semidefinite. Then, for all $i = 1, \dots, n$,

$$\lambda_{\min}(A)\mathrm{d}_{\min}(B) \le \lambda_i(A \circ B) \le \lambda_{\max}(A)\mathrm{d}_{\max}(B).$$

Proof: See [1208, pp. 303, 304].

Fact 8.22.20. Let $A, B \in \mathbb{F}^{n \times n}$, and assume that A and B are positive semidefinite. Then,

$$\det AB \leq \left(\prod_{i=1}^{n} A_{(i,i)} \right) \det B \leq \det(A \circ B) \leq \prod_{i=1}^{n} A_{(i,i)} B_{(i,i)}.$$

Equivalently,

$$\det AB \leq [\det(I \circ A)] \det B \leq \det(A \circ B) \leq \prod_{i=1}^{n} A_{(i,i)} B_{(i,i)}.$$

Furthermore,

$$2 \det AB \leq \left(\prod_{i=1}^{n} A_{(i,i)} \right) \det B + \left(\prod_{i=1}^{n} B_{(i,i)} \right) \det A \leq \det(A \circ B) + (\det A) \det B.$$

Finally, the following statements hold:

  i) If $I \circ A$ and B are positive definite, then $A \circ B$ is positive definite.

  ii) If $I \circ A$ and B are positive definite and $\operatorname{rank} A = 1$, then equality holds in the right-hand equality.

  iii) If A and B are positive definite, then equality holds in the right-hand equality if and only if B is diagonal.

Proof: See [992, 1512] and [1215, p. 253].

Remark: In the first string, the first and third inequalities follow from Hadamard's inequality Fact 8.18.11, while the second inequality is *Oppenheim's inequality*. See Fact 8.22.12.

Remark: The right-hand inequality in the third string of inequalities is valid when A and B are M-matrices. See [46, 326].

Problem: Compare the lower bounds $\det(A \# B)^2$ and $\left(\prod_{i=1}^{n} A_{(i,i)} \right) \det B$ for $\det(A \circ B)$. See Fact 8.22.21.

Fact 8.22.21. Let $A, B \in \mathbb{F}^{n \times n}$, assume that A and B are positive semidefinite, let $k \in \{1, \ldots, n\}$, and let $r \in (0, 1]$. Then,

$$\prod_{i=k}^{n} \lambda_i(A) \lambda_i(B) \leq \prod_{i=k}^{n} \sigma_i(AB) \leq \prod_{i=k}^{n} \lambda_i(AB) \leq \prod_{i=k}^{n} \lambda_i^2(A\#B) \leq \prod_{i=k}^{n} \lambda_i(A \circ B)$$

and

$$\prod_{i=k}^{n} \lambda_i(A) \lambda_i(B) \leq \prod_{i=k}^{n} \sigma_i(AB) \leq \prod_{i=k}^{n} \lambda_i(AB) \leq \prod_{i=k}^{n} \lambda_i^{1/r}(A^r B^r)$$

$$\leq \prod_{i=k}^{n} e^{\lambda_i (\log A + \log B)} \leq \prod_{i=k}^{n} e^{\lambda_i [I \circ (\log A + \log B)]}$$

$$\leq \prod_{i=k}^{n} \lambda_i^{1/r}(A^r \circ B^r) \leq \prod_{i=k}^{n} \lambda_i(A \circ B).$$

Consequently,
$$\lambda_{\min}(AB) \le \lambda_{\min}(A \circ B)$$
and
$$\det AB = \det(A\#B)^2 \le \det(A \circ B).$$

Proof: See [50, 493, 1416], [1208, p. 305], [1521, p. 21], Fact 8.10.43, and Fact 8.19.22.

Fact 8.22.22. Let $A, B \in \mathbb{F}^{n \times n}$, assume that A and B are positive definite, let $k \in \{1, \ldots, n\}$, and let $r > 0$. Then,
$$\prod_{i=k}^{n} \lambda_i^{-r}(A \circ B) \le \prod_{i=k}^{n} \lambda_i^{-r}(AB).$$

Proof: See [1415].

Fact 8.22.23. Let $A, B \in \mathbb{F}^{n \times n}$, let $C, D \in \mathbb{F}^{m \times m}$, assume that A, B, C, and D are Hermitian, $A \le B$, $C \le D$, and that either A and C are positive semidefinite, A and D are positive semidefinite, or B and D are positive semidefinite. Then,
$$A \otimes C \le B \otimes D.$$
If, in addition, $n = m$, then
$$A \circ C \le B \circ D.$$

Proof: See [45, 114].

Problem: Under which conditions are these inequalities strict?

Fact 8.22.24. Let $A, B, C, D \in \mathbb{F}^{n \times n}$, assume that A, B, C, D are positive semidefinite, and assume that $A \le B$ and $C \le D$. Then,
$$0 \le A \otimes C \le B \otimes D$$
and
$$0 \le A \circ C \le B \circ D.$$

Proof: See Fact 8.22.23.

Fact 8.22.25. Let $A, B \in \mathbb{F}^{n \times n}$, and assume that A and B are positive semidefinite. Then, $A \le B$ if and only if $A \otimes A \le B \otimes B$.

Proof: See [950].

Fact 8.22.26. Let $A, B \in \mathbb{F}^{n \times n}$, assume that A and B are positive semidefinite, assume that $0 \le A \le B$, and let $k \ge 1$. Then,
$$A^{\circ k} \le B^{\circ k}.$$

Proof: $0 \le (B - A) \circ (B + A)$ implies that $A \circ A \le B \circ B$, that is, $A^{\circ 2} \le B^{\circ 2}$.

Fact 8.22.27. Let $A_1, \ldots, A_k, B_1, \ldots, B_k \in \mathbb{F}^{n \times n}$, and assume that $A_1, \ldots, A_k, B_1, \ldots, B_k$ are positive semidefinite. Then,
$$(A_1 + B_1) \otimes \cdots \otimes (A_k + B_k) \le A_1 \otimes \cdots \otimes A_k + B_1 \otimes \cdots \otimes B_k.$$

Proof: See [1019, p. 143].

Fact 8.22.28. Let $A_1, A_2, B_1, B_2 \in \mathbb{F}^{n \times n}$, assume that A_1, A_2, B_1, B_2 are positive semidefinite, assume that $0 \leq A_1 \leq B_1$ and $0 \leq A_2 \leq B_2$, and let $\alpha \in [0,1]$. Then,

$$[\alpha A_1 + (1-\alpha)B_1] \otimes [\alpha A_2 + (1-\alpha)B_2] \leq \alpha(A_1 \otimes A_2) + (1-\alpha)(B_1 \otimes B_2).$$

Proof: See [1440].

Fact 8.22.29. Let $A, B \in \mathbb{F}^{n \times n}$, and assume that A and B are Hermitian. Then, for all $i \in \{1, \ldots, n\}$,

$$\lambda_n(A)\lambda_n(B) \leq \lambda_{i+n^2-n}(A \otimes B) \leq \lambda_i(A \circ B) \leq \lambda_i(A \otimes B) \leq \lambda_1(A)\lambda_1(B).$$

Proof: This result follows from Proposition 7.3.1 and Theorem 8.4.5. For A, B positive semidefinite, the result is given in [987].

Fact 8.22.30. Let $A \in \mathbb{F}^{n \times n}$ and $B \in \mathbb{F}^{m \times m}$, assume that A and B are positive semidefinite, let $r \in \mathbb{R}$, and assume that either A and B are positive definite or r is positive. Then,

$$(A \otimes B)^r = A^r \otimes B^r.$$

Proof: See [1044].

Fact 8.22.31. Let $A \in \mathbb{F}^{n \times m}$ and $B \in \mathbb{F}^{k \times l}$. Then,

$$\langle A \otimes B \rangle = \langle A \rangle \otimes \langle B \rangle.$$

Fact 8.22.32. Let $A, B \in \mathbb{F}^{n \times n}$, let $C, D \in \mathbb{F}^{m \times m}$, assume that A, B, C, D are positive semidefinite, let α and β be nonnegative numbers, and let $r \in [0,1]$. Then,

$$\alpha(A^r \otimes C^{1-r}) + \beta(B^r \otimes D^{1-r}) \leq (\alpha A + \beta B)^r \otimes (\alpha C + \beta D)^{1-r}.$$

Proof: See [918].

Fact 8.22.33. Let $A, B \in \mathbb{F}^{n \times n}$, and assume that A and B are positive semidefinite. If $r \in [0,1]$, then

$$A^r \circ B^r \leq (A \circ B)^r.$$

If $r \in [1,2]$, then

$$(A \circ B)^r \leq A^r \circ B^r.$$

If A and B are positive definite and $r \in [0,1]$, then

$$(A \circ B)^{-r} \leq A^{-r} \circ B^{-r}.$$

Therefore,

$$(A \circ B)^{-1} \leq A^{-1} \circ B^{-1},$$

$$(A \circ B)^2 \leq A^2 \circ B^2,$$

$$A \circ B \leq (A^2 \circ B^2)^{1/2},$$

$$A^{1/2} \circ B^{1/2} \leq (A \circ B)^{1/2}.$$

Furthermore,

$$A^2 \circ B^2 - \tfrac{1}{4}(\beta - \alpha)^2 I \leq (A \circ B)^2 \leq \tfrac{1}{2}[A^2 \circ B^2 + (AB)^{\circ 2}] \leq A^2 \circ B^2$$

and

$$A \circ B \leq \left(A^2 \circ B^2\right)^{1/2} \leq \frac{\alpha + \beta}{2\sqrt{\alpha\beta}} A \circ B,$$

where $\alpha \triangleq \lambda_{\min}(A \otimes B)$ and $\beta \triangleq \lambda_{\max}(A \otimes B)$. Hence,

$$\begin{aligned}
A \circ B - \tfrac{1}{4}\left(\sqrt{\beta} - \sqrt{\alpha}\right)^2 I &\leq \left(A^{1/2} \circ B^{1/2}\right)^2 \\
&\leq \tfrac{1}{2}\left[A \circ B + \left(A^{1/2}B^{1/2}\right)^{\circ 2}\right] \\
&\leq A \circ B \\
&\leq \left(A^2 \circ B^2\right)^{1/2} \\
&\leq \frac{\alpha + \beta}{2\sqrt{\alpha\beta}} A \circ B.
\end{aligned}$$

Proof: See [45, 1043, 1417], [728, p. 475], and [1521, p. 8].

Fact 8.22.34. Let $A, B \in \mathbb{F}^{n \times n}$, and assume that A and B are Hermitian. Then, there exist unitary matrices $S_1, S_2 \in \mathbb{F}^{n \times n}$ such that

$$\langle A \circ B \rangle \leq \tfrac{1}{2}[S_1(\langle A \rangle \circ \langle B \rangle)S_1^* + S_2(\langle A \rangle \circ \langle B \rangle)S_2^*].$$

Proof: See [93].

Fact 8.22.35. Let $A, B \in \mathbb{F}^{n \times n}$, assume that A and B are positive definite, and let k, l be nonzero integers such that $k \leq l$. Then,

$$\left(A^k \circ B^k\right)^{1/k} \leq \left(A^l \circ B^l\right)^{1/l}.$$

In particular,

$$\left(A^{-1} \circ B^{-1}\right)^{-1} \leq A \circ B$$

and

$$(A \circ B)^{-1} \leq A^{-1} \circ B^{-1}.$$

Furthermore, for all $k \geq 1$,

$$A \circ B \leq (A^k \circ B^k)^{1/k}$$

and

$$A^{1/k} \circ B^{1/k} \leq (A \circ B)^{1/k}.$$

Finally, $$(A \circ B)^{-1} \leq A^{-1} \circ B^{-1} \leq \frac{(\alpha + \beta)^2}{4\alpha\beta}(A \circ B)^{-1},$$

where $\alpha \triangleq \lambda_{\min}(A \otimes B)$ and $\beta \triangleq \lambda_{\max}(A \otimes B)$.

Proof: See [1043].

Fact 8.22.36. Let $A, B \in \mathbb{F}^{n \times n}$, and assume that A is positive definite, B is positive semidefinite, and $I \circ B$ is positive definite. Then, for all $i \in \{1, \ldots, n\}$,

$$[(A \circ B)^{-1}]_{(i,i)} \leq \frac{(A^{-1})_{(i,i)}}{B_{(i,i)}}.$$

Furthermore, if rank $B = 1$, then equality holds.

Proof: See [1512].

Fact 8.22.37. Let $A, B \in \mathbb{F}^{n \times n}$. Then, A is positive semidefinite if and only if, for every positive-semidefinite matrix $B \in \mathbb{F}^{n \times n}$,

$$1_{1 \times n}(A \circ B)1_{n \times 1} \geq 0.$$

Proof: See [728, p. 459].

Remark: This result is *Fejer's theorem*.

Fact 8.22.38. Let $A, B \in \mathbb{F}^{n \times n}$, and assume that A and B are positive definite. Then,

$$1_{1 \times n}[(A - B) \circ (A^{-1} - B^{-1})]1_{n \times 1} \leq 0.$$

Furthermore, equality holds if and only if $A = B$.

Proof: See [152, p. 8-8].

Fact 8.22.39. Let $A, B \in \mathbb{F}^{n \times n}$, assume that A and B are positive semidefinite, let $p, q \in \mathbb{R}$, and assume that one of the following conditions is satisfied:

 i) $p \leq q \leq -1$, and A and B are positive definite.

 ii) $p \leq -1 < 1 \leq q$, and A and B are positive definite.

 iii) $1 \leq p \leq q$.

 iv) $\frac{1}{2} \leq p \leq 1 \leq q$.

 v) $p \leq -1 \leq q \leq -\frac{1}{2}$, and A and B are positive definite.

Then,

$$(A^p \circ B^p)^{1/p} \leq (A^q \circ B^q)^{1/q}.$$

Proof: See [1044]. Consider case *iii)*. Since $p/q \leq 1$, it follows from Fact 8.22.33 that $A^p \circ B^p = (A^q)^{p/q} \circ (A^q)^{p/q} \leq (A^q \circ B^q)^{p/q}$. Then, use Corollary 8.6.11 with p replaced by $1/p$. See [1521, p. 8].

Remark: See [95].

Fact 8.22.40. Let $A, B \in \mathbb{F}^{n \times n}$, and assume that A and B are positive definite. Then,

$$2I \leq A \circ B^{-1} + B \circ A^{-1}.$$

Proof: See [1417, 1528].

Remark: Setting $B = A$ yields an inequality given by Fact 8.22.4.

Fact 8.22.41. Let $A, B \in \mathbb{F}^{n \times m}$, and define

$$\mathcal{A} \triangleq \left[\begin{array}{cc} A^*A \circ B^*B & (A \circ B)^* \\ A \circ B & I \end{array} \right].$$

Then, $\mathcal{A}$ is positive semidefinite. Furthermore,

$$(A \circ B)^*(A \circ B) \leq \tfrac{1}{2}(A^*A \circ B^*B + A^*B \circ B^*A) \leq A^*A \circ B^*B.$$

Proof: See [732, 1417, 1528].

Remark: The inequality $(A \circ B)^*(A \circ B) \leq A^*A \circ B^*B$ is *Amemiya's inequality*. See [950].

Fact 8.22.42. Let $A, B, C \in \mathbb{F}^{n \times n}$, define

$$\mathcal{A} \triangleq \left[\begin{array}{cc} A & B \\ B^* & C \end{array} \right],$$

and assume that $\mathcal{A}$ is positive semidefinite. Then,

$$-A \circ C \leq B \circ B^* \leq A \circ C$$

and

$$|\det(B \circ B^*)| \leq \det(A \circ C).$$

If, in addition, $\mathcal{A}$ is positive definite, then

$$-A \circ C < B \circ B^* < A \circ C$$

and

$$|\det(B \circ B^*)| < \det(A \circ C).$$

Proof: See [1528].

Remark: See Fact 8.11.5.

Fact 8.22.43. Let $A, B \in \mathbb{F}^{n \times m}$. Then,

$$-A^*A \circ B^*B \leq A^*B \circ B^*A \leq A^*A \circ B^*B$$

and

$$|\det(A^*B \circ B^*A)| \leq \det(A^*A \circ B^*B).$$

Proof: Apply Fact 8.22.42 to $\left[\begin{smallmatrix} A^*A & A^*B \\ B^*A & B^*B \end{smallmatrix} \right]$.

Remark: See Fact 8.11.14 and Fact 8.22.9.

Fact 8.22.44. Let $A, B \in \mathbb{F}^{n \times n}$, and assume that A is positive definite. Then,

$$-A \circ B^*A^{-1}B \leq B \circ B^* \leq A \circ B^*A^{-1}B$$

and

$$|\det(B \circ B^*)| \leq \det(A \circ B^*A^{-1}B).$$

Proof: Use Fact 8.11.19 and Fact 8.22.42.

Fact 8.22.45. Let $A, B \in \mathbb{F}^{n \times n}$, and let $\alpha, \beta \in (0, \infty)$.

$$-\left(\beta^{-1/2}I + \alpha A^*A\right) \circ \left(\alpha^{-1/2}I + \beta BB^*\right) \leq (A+B) \circ (A+B)^*$$
$$\leq \left(\beta^{-1/2}I + \alpha A^*A\right) \circ \left(\alpha^{-1/2}I + \beta BB^*\right).$$

Remark: See Fact 8.11.20.

Fact 8.22.46. Let $A, B \in \mathbb{F}^{n \times m}$, and define

$$\mathcal{A} \triangleq \left[\begin{array}{cc} A^*A \circ I & (A \circ B)^* \\ A \circ B & BB^* \circ I \end{array} \right].$$

Then, $\mathcal{A}$ is positive semidefinite. Now, assume in addition that $n = m$. Then,

$$-A^*A \circ I - BB^* \circ I \leq A \circ B + (A \circ B)^* \leq A^*A \circ I + BB^* \circ I$$

and

$$-A^*A \circ BB^* \circ I \leq A \circ A^* \circ B \circ B^* \leq A^*A \circ BB^* \circ I.$$

Remark: See Fact 8.22.42.

Fact 8.22.47. Let $A, B \in \mathbb{F}^{n \times n}$, and assume that A and B are positive semidefinite. Then,
$$A \circ B \leq \tfrac{1}{2}(A^2 + B^2) \circ I.$$

Proof: Use Fact 8.22.46.

Fact 8.22.48. Let $A \in \mathbb{F}^{n \times n}$, assume that A is positive semidefinite, and define $e^{\circ A} \in \mathbb{F}^{n \times n}$ by $[e^{\circ A}]_{(i,j)} \triangleq e^{A_{(i,j)}}$. Then, $e^{\circ A}$ is positive semidefinite.

Proof: Note that $e^{\circ A} = 1_{n \times n} + \tfrac{1}{2}A \circ A + \tfrac{1}{3!}A \circ A \circ A + \cdots$, and use Fact 8.22.12. See [432, p. 10].

Fact 8.22.49. Let $A, B \in \mathbb{F}^{n \times n}$, assume that A and B are positive definite, and let $p, q \in (0, \infty)$ satisfy $p \leq q$. Then,

$$I \circ (\log A + \log B) \leq \log \left(A^p \circ B^p\right)^{1/p} \leq \log \left(A^q \circ B^q\right)^{1/q}$$

and

$$I \circ (\log A + \log B) = \lim_{p \downarrow 0} \log \left(A^p \circ B^p\right)^{1/p}.$$

Proof: See [1416].

Remark: $\log \left(A^p \circ B^p\right)^{1/p} = \tfrac{1}{p}\log(A^p \circ B^p)$.

Fact 8.22.50. Let $A, B \in \mathbb{F}^{n \times n}$, and assume that A and B are positive definite. Then,
$$I \circ (\log A + \log B) \leq \log(A \circ B).$$

Proof: Set $p = 1$ in Fact 8.22.49. See [45] and [1521, p. 8].

Remark: See Fact 11.14.21.

Fact 8.22.51. Let $A, B \in \mathbb{F}^{n \times n}$, assume that A and B are positive definite, and let $C, D \in \mathbb{F}^{m \times n}$. Then,

$$(C \circ D)(A \circ B)^{-1}(C \circ D)^* \leq (CA^{-1}C^*) \circ (DB^{-1}D^*).$$

In particular,

$$(A \circ B)^{-1} \leq A^{-1} \circ B^{-1}$$

and

$$(C \circ D)(C \circ D)^* \leq (CC^*) \circ (DD^*).$$

Proof: Form the Schur complement of the lower right block of the Schur product of the positive-semidefinite matrices $\begin{bmatrix} A & C^* \\ C & CA^{-1}C^* \end{bmatrix}$ and $\begin{bmatrix} B & D^* \\ D & DB^{-1}D^* \end{bmatrix}$. See [991, 1427], [1521, p. 13], or [1526, p. 198].

Fact 8.22.52. Let $A, B \in \mathbb{F}^{n \times n}$, assume that A and B are positive semidefinite, and let $p, q \in (1, \infty)$ satisfy $1/p + 1/q = 1$. Then,

$$(A \circ B) + (C \circ D) \leq (A^p + C^p)^{1/p} \circ (B^q + D^q)^{1/q}.$$

Proof: Use *xxiv*) of Proposition 8.6.17 with $r = 1/p$. See [1521, p. 10].

Remark: Note the relationship between the *conjugate parameters* p, q and the *barycentric coordinates* $\alpha, 1 - \alpha$. See Fact 1.18.11.

Fact 8.22.53. Let $A, B, C, D \in \mathbb{F}^{n \times n}$, assume that A, B, C, and D are positive definite. Then,

$$(A \# C) \circ (B \# D) \leq (A \circ B) \# (C \circ D).$$

Furthermore,

$$(A \# B) \circ (A \# B) \leq (A \circ B).$$

Proof: See [95].

8.23 Notes

The ordering $A \leq B$ is traditionally called the *Löwner ordering*. Proposition 8.2.4 is given in [15] and [870] with extensions in [171]. The proof of Proposition 8.2.7 is based on [268, p. 120], as suggested in [1280]. The proof given in [554, p. 307] is incomplete.

Theorem 8.3.5 is due to Newcomb [1062]. Proposition 8.4.13 is given in [717, 1048]. Special cases such as Fact 8.12.29 appear in numerous papers. The proofs of Lemma 8.4.4 and Theorem 8.4.5 are based on [1261]. Theorem 8.4.9 can also be obtained as a corollary of the *Fischer minimax theorem* given in [728, 996], which provides a geometric characterization of the eigenvalues of a symmetric matrix. Theorem 8.3.6 appears in [1145, p. 121]. Theorem 8.6.2 is given in [42]. Additional inequalities appear in [1032].

Functions that are nondecreasing on $\mathbf{P}^n$ are characterized by the theory of *monotone matrix functions* [201, 432]. See [1037] for a summary of the principal results.

The literature on convex maps is extensive. Result *xiv*) of Proposition 8.6.17 is due to Lieb and Ruskai [933]. Result *xxiv*) is the *Lieb concavity theorem*. See [201, p. 271] or [931]. Result *xxxiv*) is due to Ando. Results *xlv*) and *xlvi*) are due to Fan. Some extensions to strict convexity are considered in [996]. See also [45, 1051].

Products of positive-definite matrices are studied in [121, 122, 123, 125, 1493].

Essays on the legacy of Issai Schur appear in [802]. Schur complements are discussed in [296, 298, 676, 922, 947, 1084]. Majorization and eigenvalue inequalities for sums and products of matrices are discussed in [202].

Chapter Nine

Norms

Norms are used to quantify vectors and matrices, and they play a basic role in convergence analysis. This chapter introduces vector and matrix norms and their properties.

9.1 Vector Norms

For many applications it is useful to have a scalar measure of the magnitude of a vector x or a matrix A. *Norms* provide such measures.

Definition 9.1.1. A *norm* $\| \cdot \|$ on $\mathbb{F}^n$ is a function $\| \cdot \| \colon \mathbb{F}^n \mapsto [0, \infty)$ that satisfies the following conditions:

  $i)$ $\|x\| \geq 0$ for all $x \in \mathbb{F}^n$.

  $ii)$ $\|x\| = 0$ if and only if $x = 0$.

  $iii)$ $\|\alpha x\| = |\alpha| \, \|x\|$ for all $\alpha \in \mathbb{F}$ and $x \in \mathbb{F}^n$.

  $iv)$ $\|x + y\| \leq \|x\| + \|y\|$ for all $x, y \in \mathbb{F}^n$.

Condition $iv)$ is the *triangle inequality*.

A norm $\| \cdot \|$ on $\mathbb{F}^n$ is *monotone* if $|x| \leq\leq |y|$ implies that $\|x\| \leq \|y\|$ for all $x, y \in \mathbb{F}^n$, while $\| \cdot \|$ is *absolute* if $\| \, |x| \, \| = \|x\|$ for all $x \in \mathbb{F}^n$.

Proposition 9.1.2. Let $\| \cdot \|$ be a norm on $\mathbb{F}^n$. Then, $\| \cdot \|$ is monotone if and only if $\| \cdot \|$ is absolute.

Proof. First, suppose that $\| \cdot \|$ is monotone. Let $x \in \mathbb{F}^n$, and define $y \triangleq |x|$. Then, $|y| = |x|$, and thus $|y| \leq\leq |x|$ and $|x| \leq\leq |y|$. Hence, $\|x\| \leq \|y\|$ and $\|y\| \leq \|x\|$, which implies that $\|x\| = \|y\|$. Thus, $\| \, |x| \, \| = \|y\| = \|x\|$, which proves that $\| \cdot \|$ is absolute.

Conversely, suppose that $\| \cdot \|$ is absolute and, for convenience, let $n = 2$. Now, let $x, y \in \mathbb{F}^2$ be such that $|x| \leq\leq |y|$. Then, there exist $\alpha_1, \alpha_2 \in [0, 1]$ and $\theta_1, \theta_2 \in \mathbb{R}$ such that $x_{(i)} = \alpha_i e^{j\theta_i} y_{(i)}$ for $i = 1, 2$. Since $\| \cdot \|$ is absolute, it follows that

$$\|x\| = \left\| \left[\begin{array}{c} \alpha_1 e^{j\theta_1} y_{(1)} \\ \alpha_2 e^{j\theta_2} y_{(2)} \end{array} \right] \right\|$$

$$= \left\| \left[\begin{array}{c} \alpha_1 |y_{(1)}| \\ \alpha_2 |y_{(2)}| \end{array} \right] \right\|$$

$$= \left\| \tfrac{1}{2}(1-\alpha_1) \left[\begin{array}{c} -|y_{(1)}| \\ \alpha_2 |y_{(2)}| \end{array} \right] + \tfrac{1}{2}(1-\alpha_1) \left[\begin{array}{c} |y_{(1)}| \\ \alpha_2 |y_{(2)}| \end{array} \right] + \alpha_1 \left[\begin{array}{c} |y_{(1)}| \\ \alpha_2 |y_{(2)}| \end{array} \right] \right\|$$

$$\leq \left[\tfrac{1}{2}(1-\alpha_1) + \tfrac{1}{2}(1-\alpha_1) + \alpha_1 \right] \left\| \left[\begin{array}{c} |y_{(1)}| \\ \alpha_2 |y_{(2)}| \end{array} \right] \right\|$$

$$= \left\| \left[\begin{array}{c} |y_{(1)}| \\ \alpha_2 |y_{(2)}| \end{array} \right] \right\|$$

$$= \left\| \tfrac{1}{2}(1-\alpha_2) \left[\begin{array}{c} |y_{(1)}| \\ -|y_{(2)}| \end{array} \right] + \tfrac{1}{2}(1-\alpha_2) \left[\begin{array}{c} |y_{(1)}| \\ |y_{(2)}| \end{array} \right] + \alpha_2 \left[\begin{array}{c} |y_{(1)}| \\ |y_{(2)}| \end{array} \right] \right\|$$

$$\leq \left\| \left[\begin{array}{c} |y_{(1)}| \\ |y_{(2)}| \end{array} \right] \right\|$$

$$= \|y\|.$$

Thus, $\|\cdot\|$ is monotone. $\square$

As we shall see, there are many different norms. For $x \in \mathbb{F}^n$, a useful class of norms consists of the *Hölder norms* defined by

$$\|x\|_p \triangleq \begin{cases} \left(\sum_{i=1}^n |x_{(i)}|^p \right)^{1/p}, & 1 \leq p < \infty, \\ \max_{i \in \{1,\ldots,n\}} |x_{(i)}|, & p = \infty. \end{cases} \tag{9.1.1}$$

Note that, for all $x \in \mathbb{C}^n$ and $p \in [1, \infty]$, $\|\bar{x}\|_p = \|x\|_p$. These norms depend on *Minkowski's inequality* given by the following result.

Lemma 9.1.3. Let $p \in [1, \infty]$, and let $x, y \in \mathbb{F}^n$. Then,

$$\|x + y\|_p \leq \|x\|_p + \|y\|_p. \tag{9.1.2}$$

If $p = 1$, then equality holds if and only if, for all $i \in \{1, \ldots, n\}$, there exists $\alpha_i \geq 0$ such that either $x_{(i)} = \alpha_i y_{(i)}$ or $y_{(i)} = \alpha_i x_{(i)}$. If $p \in (1, \infty)$, then equality holds if and only if there exists $\alpha \geq 0$ such that either $x = \alpha y$ or $y = \alpha x$.

Proof. See [166, 988] and Fact 1.18.25. $\square$

Proposition 9.1.4. Let $p \in [1, \infty]$. Then, $\|\cdot\|_p$ is a norm on $\mathbb{F}^n$.

For $p = 1$,

$$\|x\|_1 = \sum_{i=1}^n |x_{(i)}| \tag{9.1.3}$$

is the *absolute sum norm*; for $p = 2$,

$$\|x\|_2 = \left(\sum_{i=1}^{n} |x_{(i)}|^2 \right)^{1/2} = \sqrt{x^*x} \tag{9.1.4}$$

is the *Euclidean norm*; and, for $p = \infty$,

$$\|x\|_\infty = \max_{i \in \{1,\ldots,n\}} |x_{(i)}| \tag{9.1.5}$$

is the *infinity norm*.

The Hölder norms satisfy the following monotonicity property, which is related to the power-sum inequality given by Fact 1.17.35.

Proposition 9.1.5. Let $1 \le p \le q \le \infty$, and let $x \in \mathbb{F}^n$. Then,

$$\|x\|_\infty \le \|x\|_q \le \|x\|_p \le \|x\|_1. \tag{9.1.6}$$

Assume, in addition, that $1 < p < q < \infty$. Then, x has at least two nonzero components if and only if

$$\|x\|_\infty < \|x\|_q < \|x\|_p < \|x\|_1. \tag{9.1.7}$$

Proof. If either $p = q$ or $x = 0$ or x has exactly one nonzero component, then $\|x\|_q = \|x\|_p$. Hence, to prove both (9.1.6) and (9.1.7), it suffices to prove (9.1.7) for the case in which $1 < p < q < \infty$ and x has at least two nonzero components. Thus, let $n \ge 2$, let $x \in \mathbb{F}^n$ have at least two nonzero components, and define $f \colon [1, \infty) \to [0, \infty)$ by $f(\beta) \triangleq \|x\|_\beta$. Hence,

$$f'(\beta) = \tfrac{1}{\beta} \|x\|_\beta^{1-\beta} \sum_{i=1}^{n} \gamma_i,$$

where, for all $i \in \{1, \ldots, n\}$,

$$\gamma_i \triangleq \begin{cases} |x_i|^\beta \big(\log |x_{(i)}| - \log \|x\|_\beta \big), & x_{(i)} \ne 0, \\ 0, & x_{(i)} = 0. \end{cases}$$

If $x_{(i)} \ne 0$, then $\log |x_{(i)}| < \log \|x\|_\beta$. It thus follows that $f'(\beta) < 0$, which implies that f is decreasing on $[1, \infty)$. Hence, (9.1.7) holds. $\qquad \square$

The following result is *Hölder's inequality*. For this result we interpret $1/\infty = 0$. Note that, for all $x, y \in \mathbb{F}^n$, $|x^{\mathrm{T}}y| \le |x|^{\mathrm{T}}|y| = \|x \circ y\|_1$.

Proposition 9.1.6. Let $p, q \in [1, \infty]$ satisfy $1/p + 1/q = 1$, and let $x, y \in \mathbb{F}^n$. Then,

$$|x^{\mathrm{T}}y| \le \|x\|_p \|y\|_q. \tag{9.1.8}$$

Furthermore, equality holds if and only if $|x^{\mathrm{T}}y| = |x|^{\mathrm{T}}|y|$ and

$$\begin{cases} |x| \circ |y| = \|y\|_\infty |x|, & p = 1, \\ \|y\|_q^{1/p} |x|^{\circ 1/q} = \|x\|_p^{1/q} |y|^{\circ 1/p}, & 1 < p < \infty, \\ |x| \circ |y| = \|x\|_\infty |y|, & p = \infty. \end{cases} \tag{9.1.9}$$

Proof. See [279, p. 127], [728, p. 536], [823, p. 71], Fact 1.18.11, and Fact 1.18.12. □

The case $p = q = 2$ is the *Cauchy-Schwarz inequality*.

Corollary 9.1.7. Let $x, y \in \mathbb{F}^n$. Then,

$$|x^{\mathrm{T}}y| \leq \|x\|_2\|y\|_2. \tag{9.1.10}$$

Furthermore, equality holds if and only if x and y are linearly dependent.

Proof. Suppose that $y \neq 0$, and define $M \triangleq [\ \sqrt{y^*y}I \quad (y^*y)^{-1/2}y\ ]$. Since $M^*M = \begin{bmatrix} y^*yI & y \\ y^* & 1 \end{bmatrix}$ is positive semidefinite, it follows from iii) of Proposition 8.2.4 that $yy^* \leq y^*yI$. Therefore, $x^*yy^*x \leq x^*xy^*y$, which is equivalent to (9.1.10) with x replaced by $\bar{x}$.

Now, suppose that x and y are linearly dependent. Then, there exists $\beta \in \mathbb{F}$ such that either $x = \beta y$ or $y = \beta x$. In both cases it follows that $|x^*y| = \|x\|_2\|y\|_2$. Conversely, define $f\colon \mathbb{F}^n \times \mathbb{F}^n \to [0, \infty)$ by $f(\mu, \nu) \triangleq \mu^*\mu\nu^*\nu - |\mu^*\nu|^2$. Now, suppose that $f(x, y) = 0$ so that (x, y) minimizes f. Then, it follows that $f_\mu(x, y) = 0$, which implies that $y^*yx = y^*xy$. Hence, x and y are linearly dependent. □

Let $x, y \in \mathbb{F}^n$, assume that x and y are both nonzero, let $p, q \in [1, \infty]$, and assume that $1/p + 1/q = 1$. Since $\|\bar{x}\|_p = \|x\|_p$, it follows that

$$|x^*y| \leq \|x\|_p\|y\|_q, \tag{9.1.11}$$

and, in particular,

$$|x^*y| \leq \|x\|_2\|y\|_2. \tag{9.1.12}$$

The angle $\theta \in [0, \pi]$ between x and y, which is defined by (2.2.20), is thus given by

$$\theta \triangleq \cos^{-1} \frac{x^*y}{\|x\|_2\|y\|_2}. \tag{9.1.13}$$

The norms $\|\cdot\|$ and $\|\cdot\|'$ on $\mathbb{F}^n$ are *equivalent* if there exist $\alpha, \beta > 0$ such that

$$\alpha\|x\| \leq \|x\|' \leq \beta\|x\| \tag{9.1.14}$$

for all $x \in \mathbb{F}^n$. Note that these inequalities can be written as

$$\tfrac{1}{\beta}\|x\|' \leq \|x\| \leq \tfrac{1}{\alpha}\|x\|'. \tag{9.1.15}$$

Hence, the word "equivalent" is justified.

The following result shows that every pair of norms on $\mathbb{F}^n$ is equivalent.

Theorem 9.1.8. Let $\|\cdot\|$ and $\|\cdot\|'$ be norms on $\mathbb{F}^n$. Then, $\|\cdot\|$ and $\|\cdot\|'$ are equivalent.

Proof. See [728, p. 272]. □

9.2 Matrix Norms

One way to define norms for matrices is by viewing a matrix $A \in \mathbb{F}^{n \times m}$ as a vector in $\mathbb{F}^{nm}$, for example, as vec A.

Definition 9.2.1. A *norm* $\| \cdot \|$ on $\mathbb{F}^{n \times m}$ is a function $\| \cdot \| \colon \mathbb{F}^{n \times m} \mapsto [0, \infty)$ that satisfies the following conditions:

  i) $\|A\| \geq 0$ for all $A \in \mathbb{F}^{n \times m}$.

 ii) $\|A\| = 0$ if and only if $A = 0$.

iii) $\|\alpha A\| = |\alpha| \|A\|$ for all $\alpha \in \mathbb{F}$ and $A \in \mathbb{F}^{n \times m}$.

iv) $\|A + B\| \leq \|A\| + \|B\|$ for all $A, B \in \mathbb{F}^{n \times m}$.

If $\| \cdot \|$ is a norm on $\mathbb{F}^{nm}$, then $\| \cdot \|'$ defined by $\|A\|' \triangleq \|\text{vec } A\|$ is a norm on $\mathbb{F}^{n \times m}$. For example, Hölder norms can be defined for matrices by choosing $\| \cdot \| = \| \cdot \|_p$. Hence, for all $A \in \mathbb{F}^{n \times m}$, define

$$
\|A\|_p \triangleq \begin{cases} \left(\displaystyle\sum_{i=1}^{n} \sum_{j=1}^{m} |A_{(i,j)}|^p \right)^{1/p}, & 1 \leq p < \infty, \\[2ex] \displaystyle\max_{\substack{i \in \{1,\dots,n\} \\ j \in \{1,\dots,m\}}} |A_{(i,j)}|, & p = \infty. \end{cases}
\tag{9.2.1}
$$

Note that the same symbol $\| \cdot \|_p$ is used to denote the Hölder norm for both vectors and matrices. This notation is consistent since, if $A \in \mathbb{F}^{n \times 1}$, then $\|A\|_p$ coincides with the vector Hölder norm. Furthermore, if $A \in \mathbb{F}^{n \times m}$ and $1 \leq p \leq \infty$, then

$$
\|A\|_p = \|\text{vec } A\|_p.
\tag{9.2.2}
$$

It follows from (9.1.6) that, if $A \in \mathbb{F}^{n \times m}$ and $1 \leq p \leq q \leq \infty$, then

$$
\|A\|_\infty \leq \|A\|_q \leq \|A\|_p \leq \|A\|_1.
\tag{9.2.3}
$$

If, in addition, $1 < p < q < \infty$ and A has at least two nonzero entries, then

$$
\|A\|_\infty < \|A\|_q < \|A\|_p < \|A\|_1.
\tag{9.2.4}
$$

The Hölder norms in the cases $p = 1, 2, \infty$ are the most commonly used. Let $A \in \mathbb{F}^{n \times m}$. For $p = 2$ we define the *Frobenius norm* $\| \cdot \|_\mathrm{F}$ by

$$
\|A\|_\mathrm{F} \triangleq \|A\|_2.
\tag{9.2.5}
$$

Since $\|A\|_2 = \|\text{vec } A\|_2$, it follows that

$$
\|A\|_\mathrm{F} = \|A\|_2 = \|\text{vec } A\|_2 = \|\text{vec } A\|_\mathrm{F}.
\tag{9.2.6}
$$

It is easy to see that

$$
\|A\|_\mathrm{F} = \sqrt{\text{tr } A^* A}.
\tag{9.2.7}
$$

Let $\| \cdot \|$ be a norm on $\mathbb{F}^{n \times m}$. If $\|S_1 A S_2\| = \|A\|$ for all $A \in \mathbb{F}^{n \times m}$ and for all unitary matrices $S_1 \in \mathbb{F}^{n \times n}$ and $S_2 \in \mathbb{F}^{m \times m}$, then $\| \cdot \|$ is *unitarily invariant*. Now, let $m = n$. If $\|A\| = \|A^*\|$ for all $A \in \mathbb{F}^{n \times n}$, then $\| \cdot \|$ is *self-adjoint*. If $\|I_n\| = 1$, then $\| \cdot \|$ is *normalized*. Note that the Frobenius norm is not normalized since $\|I_n\|_{\mathrm{F}} = \sqrt{n}$. If $\|SAS^*\| = \|A\|$ for all $A \in \mathbb{F}^{n \times n}$ and for all unitary $S \in \mathbb{F}^{n \times n}$, then $\| \cdot \|$ is *weakly unitarily invariant*.

Matrix norms can be defined in terms of singular values. Let $\sigma_1(A) \geq \sigma_2(A) \geq \cdots$ denote the singular values of $A \in \mathbb{F}^{n \times m}$. The following result gives a weak majorization condition for singular values.

Proposition 9.2.2. Let $A, B \in \mathbb{F}^{n \times m}$. Then, for all $k \in \{1, \ldots, \min\{n, m\}\}$,

$$\sum_{i=1}^{k} [\sigma_i(A) - \sigma_i(B)] \leq \sum_{i=1}^{k} \sigma_i(A + B) \leq \sum_{i=1}^{k} [\sigma_i(A) + \sigma_i(B)]. \qquad (9.2.8)$$

In particular,

$$\sigma_{\max}(A) - \sigma_{\max}(B) \leq \sigma_{\max}(A + B) \leq \sigma_{\max}(A) + \sigma_{\max}(B) \qquad (9.2.9)$$

and

$$\operatorname{tr} \langle A \rangle - \operatorname{tr} \langle B \rangle \leq \operatorname{tr} \langle A + B \rangle \leq \operatorname{tr} \langle A \rangle + \operatorname{tr} \langle B \rangle. \qquad (9.2.10)$$

Proof. Define $\mathcal{A}, \mathcal{B} \in \mathbf{H}^{n+m}$ by $\mathcal{A} \triangleq \left[\begin{smallmatrix} 0 & A \\ A^* & 0 \end{smallmatrix} \right]$ and $\mathcal{B} \triangleq \left[\begin{smallmatrix} 0 & B \\ B^* & 0 \end{smallmatrix} \right]$. Then, Corollary 8.6.19 implies that, for all $k \in \{1, \ldots, n + m\}$,

$$\sum_{i=1}^{k} \lambda_i(\mathcal{A} + \mathcal{B}) \leq \sum_{i=1}^{k} [\lambda_i(\mathcal{A}) + \lambda_i(\mathcal{B})].$$

Now, consider $k \leq \min\{n, m\}$. Then, it follows from Proposition 5.6.5 that, for all $i \in \{1, \ldots, k\}$, $\lambda_i(\mathcal{A}) = \sigma_i(A)$. Setting $k = 1$ yields (9.2.9), while setting $k = \min\{n, m\}$ and using Fact 8.18.2 yields (9.2.10). $\qquad \square$

Proposition 9.2.3. Let $p \in [1, \infty]$, and let $A \in \mathbb{F}^{n \times m}$. Then, $\| \cdot \|_{\sigma p}$ defined by

$$\|A\|_{\sigma p} \triangleq \begin{cases} \left(\displaystyle\sum_{i=1}^{\min\{n,m\}} \sigma_i^p(A) \right)^{1/p}, & 1 \leq p < \infty, \\ \sigma_{\max}(A), & p = \infty, \end{cases} \qquad (9.2.11)$$

is a norm on $\mathbb{F}^{n \times m}$.

Proof. Let $p \in [1, \infty]$. Then, it follows from Proposition 9.2.2 and Minkowski's inequality Fact 1.18.25 that

$$\|A + B\|_{\sigma p} = \left(\sum_{i=1}^{\min\{n,m\}} \sigma_i^p(A + B) \right)^{1/p}$$

$$\leq \left(\sum_{i=1}^{\min\{n,m\}} [\sigma_i(A) + \sigma_i(B)]^p \right)^{1/p}$$

$$\leq \left(\sum_{i=1}^{\min\{n,m\}} \sigma_i^p(A) \right)^{1/p} + \left(\sum_{i=1}^{\min\{n,m\}} \sigma_i^p(B) \right)^{1/p}$$

$$= \|A\|_{\sigma p} + \|B\|_{\sigma p}. \qquad \square$$

The norm $\| \cdot \|_{\sigma p}$ is a *Schatten norm*. Let $A \in \mathbb{F}^{n \times m}$. Then, for all $p \in [1, \infty)$,

$$\|A\|_{\sigma p} = (\operatorname{tr} \langle A \rangle^p)^{1/p}. \tag{9.2.12}$$

Special cases are

$$\|A\|_{\sigma 1} = \sigma_1(A) + \cdots + \sigma_{\min\{n,m\}}(A) = \operatorname{tr} \langle A \rangle, \tag{9.2.13}$$

$$\|A\|_{\sigma 2} = \left[\sigma_1^2(A) + \cdots + \sigma_{\min\{n,m\}}^2(A) \right]^{1/2} = (\operatorname{tr} A^*A)^{1/2} = \|A\|_{\mathrm{F}}, \tag{9.2.14}$$

and

$$\|A\|_{\sigma\infty} = \sigma_1(A) = \sigma_{\max}(A), \tag{9.2.15}$$

which are the *trace norm*, Frobenius norm, and *spectral norm*, respectively.

By applying Proposition 9.1.5 to the vector $\begin{bmatrix} \sigma_1(A) & \cdots & \sigma_{\min\{n,m\}}(A) \end{bmatrix}^{\mathrm{T}}$, we obtain the following result.

Proposition 9.2.4. Let $p, q \in [1, \infty)$, where $p \leq q$, and let $A \in \mathbb{F}^{n \times m}$. Then,

$$\|A\|_{\sigma\infty} \leq \|A\|_{\sigma q} \leq \|A\|_{\sigma p} \leq \|A\|_{\sigma 1}. \tag{9.2.16}$$

Assume, in addition, that $1 < p < q < \infty$ and $\operatorname{rank} A \geq 2$. Then,

$$\|A\|_{\sigma\infty} < \|A\|_{\sigma q} < \|A\|_{\sigma p} < \|A\|_{\sigma 1}. \tag{9.2.17}$$

The norms $\| \cdot \|_{\sigma p}$ are not very interesting when applied to vectors. Let $x \in \mathbb{F}^n = \mathbb{F}^{n \times 1}$. Then, $\sigma_{\max}(x) = (x^*x)^{1/2} = \|x\|_2$, and, since $\operatorname{rank} x \leq 1$, it follows that, for all $p \in [1, \infty]$,

$$\|x\|_{\sigma p} = \|x\|_2. \tag{9.2.18}$$

Proposition 9.2.5. Let $A \in \mathbb{F}^{n \times m}$. If $p \in (0, 2]$, then

$$\|A\|_{\sigma p} \leq \|A\|_p. \tag{9.2.19}$$

If $p \geq 2$, then

$$\|A\|_p \leq \|A\|_{\sigma p}. \tag{9.2.20}$$

Proof. See [1521, p. 50]. $\square$

Proposition 9.2.6. Let $\| \cdot \|$ be a norm on $\mathbb{F}^{n \times n}$, and let $A \in \mathbb{F}^{n \times n}$. Then,

$$\mathrm{sprad}(A) = \lim_{k \to \infty} \|A^k\|^{1/k}. \tag{9.2.21}$$

Proof. See [728, p. 322]. □

9.3 Compatible Norms

The norms $\| \cdot \|$, $\| \cdot \|'$, and $\| \cdot \|''$ on $\mathbb{F}^{n \times l}$, $\mathbb{F}^{n \times m}$, and $\mathbb{F}^{m \times l}$, respectively, are *compatible* if, for all $A \in \mathbb{F}^{n \times m}$ and $B \in \mathbb{F}^{m \times l}$,

$$\|AB\| \leq \|A\|'\|B\|''. \tag{9.3.1}$$

For $l = 1$, the norms $\| \cdot \|$, $\| \cdot \|'$, and $\| \cdot \|''$ on $\mathbb{F}^n$, $\mathbb{F}^{n \times m}$, and $\mathbb{F}^m$, respectively, are compatible if, for all $A \in \mathbb{F}^{n \times m}$ and $x \in \mathbb{F}^m$,

$$\|Ax\| \leq \|A\|'\|x\|''. \tag{9.3.2}$$

Furthermore, the norm $\| \cdot \|$ on $\mathbb{F}^n$ is *compatible* with the norm $\| \cdot \|'$ on $\mathbb{F}^{n \times n}$ if, for all $A \in \mathbb{F}^{n \times n}$ and $x \in \mathbb{F}^n$,

$$\|Ax\| \leq \|A\|'\|x\|. \tag{9.3.3}$$

Note that $\|I_n\|' \geq 1$. The norm $\| \cdot \|$ on $\mathbb{F}^{n \times n}$ is *submultiplicative* if, for all $A, B \in \mathbb{F}^{n \times n}$,

$$\|AB\| \leq \|A\|\|B\|. \tag{9.3.4}$$

Hence, the norm $\| \cdot \|$ on $\mathbb{F}^{n \times n}$ is submultiplicative if and only if $\| \cdot \|$, $\| \cdot \|$, and $\| \cdot \|$ are compatible. In this case, $\|I_n\| \geq 1$, while $\| \cdot \|$ is normalized if and only if $\|I_n\| = 1$.

Proposition 9.3.1. Let $\| \cdot \|'$ be a submultiplicative norm on $\mathbb{F}^{n \times n}$, and let $y \in \mathbb{F}^n$ be nonzero. Then, $\|x\| \triangleq \|xy^*\|'$ is a norm on $\mathbb{F}^n$, and $\| \cdot \|$ is compatible with $\| \cdot \|'$.

Proof. Note that

$$\|Ax\| = \|Axy^*\|' \leq \|A\|'\|xy^*\|' = \|A\|'\|x\|. \qquad \square$$

Proposition 9.3.2. Let $\| \cdot \|$ be a submultiplicative norm on $\mathbb{F}^{n \times n}$, and let $A \in \mathbb{F}^{n \times n}$. Then,

$$\mathrm{sprad}(A) \leq \|A\|. \tag{9.3.5}$$

Proof. Use Proposition 9.3.1 to construct a norm $\| \cdot \|'$ on $\mathbb{F}^n$ that is compatible with $\| \cdot \|$. Furthermore, let $A \in \mathbb{F}^{n \times n}$, let $\lambda \in \mathrm{spec}(A)$, and let $x \in \mathbb{C}^n$ be an eigenvector of A associated with λ. Then, $Ax = \lambda x$ implies that $|\lambda|\|x\|' = \|Ax\|' \leq \|A\|\|x\|'$, and thus $|\lambda| \leq \|A\|$, which implies (9.3.5). Alternatively, under the additional assumption that $\| \cdot \|$ is submultiplicative, it follows from Proposition 9.2.6 that

$$\mathrm{sprad}(A) = \lim_{k \to \infty} \|A^k\|^{1/k} \leq \lim_{k \to \infty} \|A\|^{k/k} = \|A\|. \qquad \square$$

Proposition 9.3.3. Let $A \in \mathbb{F}^{n \times n}$, and let $\varepsilon > 0$. Then, there exists a submultiplicative norm $\| \cdot \|$ on $\mathbb{F}^{n \times n}$ such that

$$\mathrm{sprad}(A) \leq \|A\| \leq \mathrm{sprad}(A) + \varepsilon. \tag{9.3.6}$$

Proof. See [728, p. 297] or [1208, p. 167]. $\square$

Corollary 9.3.4. Let $A \in \mathbb{F}^{n \times n}$, and assume that $\mathrm{sprad}(A) < 1$. Then, there exists a submultiplicative norm $\| \cdot \|$ on $\mathbb{F}^{n \times n}$ such that $\|A\| < 1$.

We now identify some compatible norms. We begin with the Hölder norms.

Proposition 9.3.5. Let $A \in \mathbb{F}^{n \times m}$ and $B \in \mathbb{F}^{m \times l}$. If $p \in [1, 2]$, then

$$\|AB\|_p \leq \|A\|_p \|B\|_p. \tag{9.3.7}$$

If $p \in [2, \infty]$ and q satisfies $1/p + 1/q = 1$, then

$$\|AB\|_p \leq \|A\|_p \|B\|_q \tag{9.3.8}$$

and

$$\|AB\|_p \leq \|A\|_q \|B\|_p. \tag{9.3.9}$$

Proof. First let $1 \leq p \leq 2$ so that $q \triangleq p/(p-1) \geq 2$. Using Hölder's inequality (9.1.8) and (9.1.6) with $p \leq q$ yields

$$\|AB\|_p = \left(\sum_{i,j=1}^{n,l} |\mathrm{row}_i(A)\mathrm{col}_j(B)|^p \right)^{1/p}$$

$$\leq \left(\sum_{i,j=1}^{n,l} \|\mathrm{row}_i(A)\|_p^p \|\mathrm{col}_j(B)\|_q^p \right)^{1/p}$$

$$= \left(\sum_{i=1}^{n} \|\mathrm{row}_i(A)\|_p^p \right)^{1/p} \left(\sum_{j=1}^{l} \|\mathrm{col}_j(B)\|_q^p \right)^{1/p}$$

$$\leq \left(\sum_{i=1}^{n} \|\mathrm{row}_i(A)\|_p^p \right)^{1/p} \left(\sum_{j=1}^{l} \|\mathrm{col}_j(B)\|_p^p \right)^{1/p}$$

$$= \|A\|_p \|B\|_p.$$

Next, let $2 \leq p \leq \infty$ so that $q \triangleq p/(p-1) \leq 2$. Using Hölder's inequality (9.1.8) and (9.1.6) with $q \leq p$ yields

$$\|AB\|_p \leq \left(\sum_{i=1}^{n} \|\mathrm{row}_i(A)\|_p^p\right)^{1/p} \left(\sum_{j=1}^{l} \|\mathrm{col}_j(B)\|_q^p\right)^{1/p}$$

$$\leq \left(\sum_{i=1}^{n} \|\mathrm{row}_i(A)\|_p^p\right)^{1/p} \left(\sum_{j=1}^{l} \|\mathrm{col}_j(B)\|_q^q\right)^{1/q}$$

$$= \|A\|_p \|B\|_q.$$

Similarly, it can be shown that (9.3.9) holds. $\qquad\square$

Proposition 9.3.6. Let $A \in \mathbb{F}^{n \times m}$, $B \in \mathbb{F}^{m \times l}$, and $p, q \in [1, \infty]$, define

$$r \triangleq \frac{1}{\frac{1}{p} + \frac{1}{q}},$$

and assume that $r \geq 1$. Then,

$$\|AB\|_{\sigma r} \leq \|A\|_{\sigma p} \|B\|_{\sigma q}. \tag{9.3.10}$$

In particular,

$$\|AB\|_{\sigma r} \leq \|A\|_{\sigma 2r} \|B\|_{\sigma 2r}. \tag{9.3.11}$$

Proof. Using Proposition 9.6.2 and Hölder's inequality with $1/(p/r) + 1/(q/r) = 1$, it follows that

$$\|AB\|_{\sigma r} = \left(\sum_{i=1}^{\min\{n,m,l\}} \sigma_i^r(AB)\right)^{1/r}$$

$$\leq \left(\sum_{i=1}^{\min\{n,m,l\}} \sigma_i^r(A)\sigma_i^r(B)\right)^{1/r}$$

$$\leq \left[\left(\sum_{i=1}^{\min\{n,m,l\}} \sigma_i^p(A)\right)^{r/p} \left(\sum_{i=1}^{\min\{n,m,l\}} \sigma_i^q(B)\right)^{r/q}\right]^{1/r}$$

$$= \|A\|_{\sigma p} \|B\|_{\sigma q}. \qquad\square$$

Corollary 9.3.7. Let $A \in \mathbb{F}^{n \times m}$ and $B \in \mathbb{F}^{m \times l}$. Then,

$$\|AB\|_{\sigma\infty} \leq \|AB\|_{\sigma 2} \leq \left\{ \begin{array}{c} \|A\|_{\sigma\infty}\|B\|_{\sigma 2} \\ \|A\|_{\sigma 2}\|B\|_{\sigma\infty} \\ \|AB\|_{\sigma 1} \end{array} \right\} \leq \|A\|_{\sigma 2}\|B\|_{\sigma 2} \tag{9.3.12}$$

or, equivalently,

$$\sigma_{\max}(AB) \leq \|AB\|_{\mathrm{F}} \leq \left\{ \begin{array}{c} \sigma_{\max}(A)\|B\|_{\mathrm{F}} \\ \|A\|_{\mathrm{F}}\sigma_{\max}(B) \\ \mathrm{tr}\,\langle AB \rangle \end{array} \right\} \leq \|A\|_{\mathrm{F}}\|B\|_{\mathrm{F}}. \tag{9.3.13}$$

Furthermore, for all $r \in [1, \infty]$,

$$\|AB\|_{\sigma 2r} \le \|AB\|_{\sigma r} \le \left\{ \begin{array}{c} \|A\|_{\sigma r} \sigma_{\max}(B) \\ \sigma_{\max}(A)\|B\|_{\sigma r} \\ \|A\|_{\sigma 2r}\|B\|_{\sigma 2r} \end{array} \right\} \le \|A\|_{\sigma r}\|B\|_{\sigma r}. \qquad (9.3.14)$$

In particular, setting $r = \infty$ yields

$$\sigma_{\max}(AB) \le \sigma_{\max}(A)\sigma_{\max}(B). \qquad (9.3.15)$$

Corollary 9.3.8. Let $A \in \mathbb{F}^{n \times m}$ and $B \in \mathbb{F}^{m \times l}$. Then,

$$\|AB\|_{\sigma 1} \le \left\{ \begin{array}{l} \sigma_{\max}(A)\|B\|_{\sigma 1} \\ \|A\|_{\sigma 1}\sigma_{\max}(B). \end{array} \right. \qquad (9.3.16)$$

Note that the inequality $\|AB\|_{\mathrm{F}} \le \|A\|_{\mathrm{F}}\|B\|_{\mathrm{F}}$ in (9.3.13) is equivalent to (9.3.7) with $p = 2$ as well as (9.3.8) and (9.3.9) with $p = q = 2$.

The following result is the matrix version of the Cauchy-Schwarz inequality given by Corollary 9.1.7.

Corollary 9.3.9. Let $A \in \mathbb{F}^{n \times m}$ and $B \in \mathbb{F}^{n \times m}$. Then,

$$|\mathrm{tr}\, A^*B| \le \|A\|_{\mathrm{F}}\|B\|_{\mathrm{F}}. \qquad (9.3.17)$$

Equality holds if and only if A and B are linearly dependent.

9.4 Induced Norms

In this section we consider the case in which there exists a nonzero vector $x \in \mathbb{F}^m$ such that (9.3.3) holds as an equality. This condition characterizes a special class of norms on $\mathbb{F}^{n \times n}$, namely, the *induced norms*.

Definition 9.4.1. Let $\|\cdot\|''$ and $\|\cdot\|$ be norms on $\mathbb{F}^m$ and $\mathbb{F}^n$, respectively. Then, $\|\cdot\|': \mathbb{F}^{n \times m} \mapsto \mathbb{F}$ defined by

$$\|A\|' = \max_{x \in \mathbb{F}^m \setminus \{0\}} \frac{\|Ax\|}{\|x\|''} \qquad (9.4.1)$$

is an *induced norm* on $\mathbb{F}^{n \times m}$. In this case, $\|\cdot\|'$ is *induced by* $\|\cdot\|''$ and $\|\cdot\|$. If $m = n$ and $\|\cdot\|'' = \|\cdot\|$, then $\|\cdot\|'$ is *induced by* $\|\cdot\|$, and $\|\cdot\|'$ is an *equi-induced norm*.

The next result confirms that $\|\cdot\|'$ defined by (9.4.1) is a norm.

Theorem 9.4.2. Every induced norm is a norm. Furthermore, every equi-induced norm is normalized.

Proof. See [728, p. 293]. $\qquad\qquad\qquad\qquad\qquad\qquad\qquad\qquad\qquad\qquad\qquad\qquad\qquad\square$

Let $A \in \mathbb{F}^{n \times m}$. It can be seen that (9.4.1) is equivalent to

$$\|A\|' = \max_{x \in \{y \in \mathbb{F}^m: \, \|y\|''=1\}} \|Ax\|. \tag{9.4.2}$$

Theorem 10.3.8 implies that the maximum in (9.4.2) exists. Since, for all $x \neq 0$,

$$\|A\|' = \max_{x \in \mathbb{F}^m \setminus \{0\}} \frac{\|Ax\|}{\|x\|''} \geq \frac{\|Ax\|}{\|x\|''}, \tag{9.4.3}$$

it follows that, for all $x \in \mathbb{F}^m$,

$$\|Ax\| \leq \|A\|'\|x\|'' \tag{9.4.4}$$

so that $\|\cdot\|$, $\|\cdot\|'$, and $\|\cdot\|''$ are compatible. If $m = n$ and $\|\cdot\|'' = \|\cdot\|$, then the norm $\|\cdot\|$ is compatible with the induced norm $\|\cdot\|'$. The next result shows that compatible norms can be obtained from induced norms.

Proposition 9.4.3. Let $\|\cdot\|$, $\|\cdot\|'$, and $\|\cdot\|''$ be norms on $\mathbb{F}^l$, $\mathbb{F}^m$, and $\mathbb{F}^n$, respectively. Furthermore, let $\|\cdot\|'''$ be the norm on $\mathbb{F}^{m \times l}$ induced by $\|\cdot\|$ and $\|\cdot\|'$, let $\|\cdot\|''''$ be the norm on $\mathbb{F}^{n \times m}$ induced by $\|\cdot\|'$ and $\|\cdot\|''$, and let $\|\cdot\|'''''$ be the norm on $\mathbb{F}^{n \times l}$ induced by $\|\cdot\|$ and $\|\cdot\|''$. If $A \in \mathbb{F}^{n \times m}$ and $B \in \mathbb{F}^{m \times l}$, then

$$\|AB\|''''' \leq \|A\|''''\|B\|'''. \tag{9.4.5}$$

Proof. Note that, for all $x \in \mathbb{F}^l$, $\|Bx\|' \leq \|B\|'''\|x\|$, and, for all $y \in \mathbb{F}^m$, $\|Ay\|'' \leq \|A\|''''\|y\|'$. Hence, for all $x \in \mathbb{F}^l$, it follows that

$$\|ABx\|'' \leq \|A\|''''\|Bx\|' \leq \|A\|''''\|B\|'''\|x\|,$$

which implies that

$$\|AB\|''''' = \max_{x \in \mathbb{F}^l \setminus \{0\}} \frac{\|ABx\|''}{\|x\|} \leq \|A\|''''\|B\|'''. \qquad \square$$

Corollary 9.4.4. Every equi-induced norm is submultiplicative.

The following result is a consequence of Corollary 9.4.4 and Proposition 9.3.2.

Corollary 9.4.5. Let $\|\cdot\|$ be an equi-induced norm on $\mathbb{F}^{n \times n}$, and let $A \in \mathbb{F}^{n \times n}$. Then,

$$\mathrm{sprad}(A) \leq \|A\|. \tag{9.4.6}$$

By assigning $\|\cdot\|_p$ to $\mathbb{F}^m$ and $\|\cdot\|_q$ to $\mathbb{F}^n$, where $p \geq 1$ and $q \geq 1$, the *Hölder-induced norm* on $\mathbb{F}^{n \times m}$ is defined by

$$\|A\|_{q,p} \triangleq \max_{x \in \mathbb{F}^m \setminus \{0\}} \frac{\|Ax\|_q}{\|x\|_p}. \tag{9.4.7}$$

Proposition 9.4.6. Let $p, q, p', q' \in [1, \infty]$, where $p \leq p'$ and $q \leq q'$, and let $A \in \mathbb{F}^{n \times m}$. Then,

$$\|A\|_{q',p} \leq \|A\|_{q,p} \leq \|A\|_{q,p'}. \tag{9.4.8}$$

Proof. This result follows from Proposition 9.1.5. $\qquad \square$

A subtlety of induced norms is that the value of an induced norm may depend on the underlying field. In particular, the value of the induced norm of a real matrix A computed over the complex field may be different from the induced norm of A computed over the real field. Although the chosen field is usually not made explicit, we do so in special cases for clarity.

Proposition 9.4.7. Let $A \in \mathbb{R}^{n \times m}$, and let $\|A\|_{p,q,\mathbb{F}}$ denote the Hölder-induced norm of A evaluated over the field $\mathbb{F}$, where $p \geq 1$ and $q \geq 1$. Then,

$$\|A\|_{p,q,\mathbb{R}} \leq \|A\|_{p,q,\mathbb{C}}. \tag{9.4.9}$$

If $p \in [1, \infty]$, then

$$\|A\|_{p,1,\mathbb{R}} = \|A\|_{p,1,\mathbb{C}}. \tag{9.4.10}$$

Finally, if $p, q \in [1, \infty]$ satisfy $1/p + 1/q = 1$, then

$$\|A\|_{\infty,p,\mathbb{R}} = \|A\|_{\infty,p,\mathbb{C}}. \tag{9.4.11}$$

Proof. See [708, p. 716] and [1183]. $\square$

Example 9.4.8. Let $A = \begin{bmatrix} 1 & -1 \\ 1 & 1 \end{bmatrix}$ and $x = \begin{bmatrix} x_1 & x_2 \end{bmatrix}^{\mathrm{T}}$. Then, $\|Ax\|_1 = |x_1 - x_2| + |x_1 + x_2|$. Letting $x = \begin{bmatrix} 1 & \jmath \end{bmatrix}^{\mathrm{T}}$ so that $\|x\|_\infty = 1$, it follows that $\|A\|_{1,\infty,\mathbb{C}} \geq 2\sqrt{2}$. On the other hand, $\|A\|_{1,\infty,\mathbb{R}} = 2$. Hence, in this case, the inequality (9.4.9) is strict. See [708, p. 716].

The following result gives explicit expressions for several Hölder-induced norms.

Proposition 9.4.9. Let $A \in \mathbb{F}^{n \times m}$. Then,

$$\|A\|_{2,2} = \sigma_{\max}(A). \tag{9.4.12}$$

If $p \in [1, \infty]$, then

$$\|A\|_{p,1} = \max_{i \in \{1, \ldots, m\}} \|\mathrm{col}_i(A)\|_p. \tag{9.4.13}$$

Finally, if $p, q \in [1, \infty]$ satisfy $1/p + 1/q = 1$, then

$$\|A\|_{\infty,p} = \max_{i \in \{1, \ldots, n\}} \|\mathrm{row}_i(A)\|_q. \tag{9.4.14}$$

Proof. Since A^*A is Hermitian, it follows from Corollary 8.4.2 that, for all $x \in \mathbb{F}^m$,

$$x^*A^*Ax \leq \lambda_{\max}(A^*A)x^*x,$$

which implies that, for all $x \in \mathbb{F}^m$, $\|Ax\|_2 \leq \sigma_{\max}(A)\|x\|_2$, and thus $\|A\|_{2,2} \leq \sigma_{\max}(A)$. Now, let $x \in \mathbb{F}^{n \times n}$ be an eigenvector associated with $\lambda_{\max}(A^*A)$ so that $\|Ax\|_2 = \sigma_{\max}(A)\|x\|_2$, which implies that $\sigma_{\max}(A) \leq \|A\|_{2,2}$. Hence, (9.4.12) holds.

Next, note that, for all $x \in \mathbb{F}^m$,

$$\|Ax\|_p = \left\| \sum_{i=1}^m x_{(i)} \mathrm{col}_i(A) \right\|_p \leq \sum_{i=1}^m |x_{(i)}| \|\mathrm{col}_i(A)\|_p \leq \max_{i \in \{1,\ldots,m\}} \|\mathrm{col}_i(A)\|_p \|x\|_1,$$

and hence $\|A\|_{p,1} \leq \max_{i \in \{1,\ldots,m\}} \|\mathrm{col}_i(A)\|_p$. Next, let $j \in \{1,\ldots,m\}$ be such that $\|\mathrm{col}_j(A)\|_p = \max_{i \in \{1,\ldots,m\}} \|\mathrm{col}_i(A)\|_p$. Now, since $\|e_j\|_1 = 1$, it follows that $\|Ae_j\|_p = \|\mathrm{col}_j(A)\|_p \|e_j\|_1$, which implies that

$$\max_{i \in \{1,\ldots,n\}} \|\mathrm{col}_i(A)\|_p = \|\mathrm{col}_j(A)\|_p \leq \|A\|_{p,1},$$

and hence (9.4.13) holds.

Next, for all $x \in \mathbb{F}^m$, it follows from Hölder's inequality (9.1.8) that

$$\|Ax\|_\infty = \max_{i \in \{1,\ldots,n\}} |\mathrm{row}_i(A)x| \leq \max_{i \in \{1,\ldots,n\}} \|\mathrm{row}_i(A)\|_q \|x\|_p,$$

which implies that $\|A\|_{\infty,p} \leq \max_{i \in \{1,\ldots,n\}} \|\mathrm{row}_i(A)\|_q$. Next, let $j \in \{1,\ldots,n\}$ be such that $\|\mathrm{row}_j(A)\|_q = \max_{i \in \{1,\ldots,n\}} \|\mathrm{row}_i(A)\|_q$, and let nonzero $x \in \mathbb{F}^m$ be such that $|\mathrm{row}_j(A)x| = \|\mathrm{row}_j(A)\|_q \|x\|_p$. Hence,

$$\|Ax\|_\infty = \max_{i \in \{1,\ldots,n\}} |\mathrm{row}_i(A)x| \geq |\mathrm{row}_j(A)x| = \|\mathrm{row}_j(A)\|_q \|x\|_p,$$

which implies that

$$\max_{i \in \{1,\ldots,n\}} \|\mathrm{row}_i(A)\|_q = \|\mathrm{row}_j(A)\|_q \leq \|A\|_{\infty,p},$$

and thus (9.4.14) holds. $\qquad \square$

Let $A \in \mathbb{F}^{n \times m}$. Then,

$$\max_{i \in \{1,\ldots,m\}} \|\mathrm{col}_i(A)\|_2 = \mathrm{d}_{\max}^{1/2}(A^*A) \tag{9.4.15}$$

and

$$\max_{i \in \{1,\ldots,n\}} \|\mathrm{row}_i(A)\|_2 = \mathrm{d}_{\max}^{1/2}(AA^*). \tag{9.4.16}$$

Therefore, it follows from Proposition 9.4.9 that

$$\|A\|_{1,1} = \max_{i \in \{1,\ldots,m\}} \|\mathrm{col}_i(A)\|_1, \tag{9.4.17}$$

$$\|A\|_{2,1} = \max_{i \in \{1,\ldots,m\}} \|\mathrm{col}_i(A)\|_2 = \mathrm{d}_{\max}^{1/2}(A^*A), \tag{9.4.18}$$

$$\|A\|_{\infty,1} = \|A\|_\infty = \max_{\substack{i \in \{1,\ldots,n\} \\ j \in \{1,\ldots,m\}}} |A_{(i,j)}|, \tag{9.4.19}$$

$$\|A\|_{\infty,2} = \max_{i \in \{1,\ldots,n\}} \|\mathrm{row}_i(A)\|_2 = \mathrm{d}_{\max}^{1/2}(AA^*), \tag{9.4.20}$$

$$\|A\|_{\infty,\infty} = \max_{i \in \{1,\ldots,n\}} \|\mathrm{row}_i(A)\|_1. \tag{9.4.21}$$

For convenience, we define the *column norm*

$$\|A\|_{\mathrm{col}} \triangleq \|A\|_{1,1} \tag{9.4.22}$$

and the *row norm*

$$\|A\|_{\mathrm{row}} \triangleq \|A\|_{\infty,\infty}. \tag{9.4.23}$$

Note that

$$\|A^{\mathrm{T}}\|_{\mathrm{col}} = \|A\|_{\mathrm{row}}. \tag{9.4.24}$$

The following result follows from Corollary 9.4.5.

Corollary 9.4.10. Let $A \in \mathbb{F}^{n \times n}$. Then,

$$\mathrm{sprad}(A) \le \sigma_{\max}(A), \tag{9.4.25}$$

$$\mathrm{sprad}(A) \le \|A\|_{\mathrm{col}}, \tag{9.4.26}$$

$$\mathrm{sprad}(A) \le \|A\|_{\mathrm{row}}. \tag{9.4.27}$$

Proposition 9.4.11. Let $p, q \in [1, \infty]$ be such that $1/p + 1/q = 1$, and let $A \in \mathbb{F}^{n \times m}$. Then,

$$\|A\|_{q,p} \le \|A\|_q. \tag{9.4.28}$$

Proof. For $p = 1$ and $q = \infty$, (9.4.28) follows from (9.4.19). For $q < \infty$ and $x \in \mathbb{F}^n$, it follows from Hölder's inequality (9.1.8) that

$$\|Ax\|_q = \left(\sum_{i=1}^n |\mathrm{row}_i(A)x|^q \right)^{1/q} \le \left(\sum_{i=1}^n \|\mathrm{row}_i(A)\|_q^q \|x\|_p^q \right)^{1/q}$$

$$= \left(\sum_{i=1}^n \sum_{j=1}^m |A_{(i,j)}|^q \right)^{1/q} \|x\|_p = \|A\|_q \|x\|_p,$$

which implies (9.4.28). $\square$

Next, we specialize Proposition 9.4.3 to the Hölder-induced norms.

Corollary 9.4.12. Let $p, q, r \in [1, \infty]$, and let $A \in \mathbb{F}^{n \times m}$ and $A \in \mathbb{F}^{m \times l}$. Then,

$$\|AB\|_{r,p} \le \|A\|_{r,q} \|B\|_{q,p}. \tag{9.4.29}$$

In particular,

$$\|AB\|_{\mathrm{col}} \le \|A\|_{\mathrm{col}} \|B\|_{\mathrm{col}}, \tag{9.4.30}$$

$$\sigma_{\max}(AB) \le \sigma_{\max}(A)\sigma_{\max}(B), \tag{9.4.31}$$

$$\|AB\|_{\mathrm{row}} \le \|A\|_{\mathrm{row}} \|B\|_{\mathrm{row}}, \tag{9.4.32}$$

$$\|AB\|_{\infty} \le \|A\|_{\infty} \|B\|_{\mathrm{col}}, \tag{9.4.33}$$

$$\|AB\|_{\infty} \le \|A\|_{\mathrm{row}} \|B\|_{\infty}, \tag{9.4.34}$$

$$d_{max}^{1/2}(B^*A^*AB) \leq d_{max}^{1/2}(A^*A)\|B\|_{col}, \tag{9.4.35}$$

$$d_{max}^{1/2}(B^*A^*AB) \leq \sigma_{max}(A)d_{max}^{1/2}(B^*B), \tag{9.4.36}$$

$$d_{max}^{1/2}(ABB^*A^*) \leq d_{max}^{1/2}(AA^*)\sigma_{max}(B), \tag{9.4.37}$$

$$d_{max}^{1/2}(ABB^*A^*) \leq \|B\|_{row}d_{max}^{1/2}(BB^*). \tag{9.4.38}$$

The following result is often useful.

Proposition 9.4.13. Let $A \in \mathbb{F}^{n \times n}$, and assume that $\mathrm{sprad}(A) < 1$. Then, there exists a submultiplicative norm $\|\cdot\|$ on $\mathbb{F}^{n \times n}$ such that $\|A\| < 1$. Furthermore, the series $\sum_{k=0}^{\infty} A^k$ converges absolutely, and

$$(I - A)^{-1} = \sum_{k=0}^{\infty} A^k. \tag{9.4.39}$$

Finally,

$$\frac{1}{1 + \|A\|} \leq \|(I - A)^{-1}\| \leq \frac{1}{1 - \|A\|} + \|I\| - 1. \tag{9.4.40}$$

If, in addition, $\|\cdot\|$ is normalized, then

$$\frac{1}{1 + \|A\|} \leq \|(I - A)^{-1}\| \leq \frac{1}{1 - \|A\|}. \tag{9.4.41}$$

Proof. Corollary 9.3.4 implies that there exists a submultiplicative norm $\|\cdot\|$ on $\mathbb{F}^{n \times n}$ such that $\|A\| < 1$. It thus follows that

$$\left\|\sum_{k=0}^{\infty} A^k\right\| \leq \sum_{k=0}^{\infty} \|A^k\| \leq \|I\| - 1 + \sum_{k=0}^{\infty} \|A\|^k = \frac{1}{1 - \|A\|} + \|I\| - 1,$$

which proves that the series $\sum_{k=0}^{\infty} A^k$ converges absolutely.

Next, we show that $I - A$ is nonsingular. If $I - A$ is singular, then there exists a nonzero vector $x \in \mathbb{C}^n$ such that $Ax = x$. Hence, $1 \in \mathrm{spec}(A)$, which contradicts $\mathrm{sprad}(A) < 1$. Next, to verify (9.4.39), note that

$$(I - A)\sum_{k=0}^{\infty} A^k = \sum_{k=0}^{\infty} A^k - \sum_{k=1}^{\infty} A^k = I + \sum_{k=1}^{\infty} A^k - \sum_{k=1}^{\infty} A^k = I,$$

which implies (9.4.39) and thus the right-hand inequality in (9.4.40). Furthermore,

$$\begin{aligned}
1 &\leq \|I\| \\
&= \|(I - A)(I - A)^{-1}\| \\
&\leq \|I - A\| \|(I - A)^{-1}\| \\
&\leq (1 + \|A\|) \|(I - A)^{-1}\|,
\end{aligned}$$

which yields the left-hand inequality in (9.4.40). $\square$

9.5 Induced Lower Bound

We now consider a variation of the induced norm.

Definition 9.5.1. Let $\|\cdot\|$ and $\|\cdot\|'$ denote norms on $\mathbb{F}^m$ and $\mathbb{F}^n$, respectively, and let $A \in \mathbb{F}^{n\times m}$. Then, $\ell\colon \mathbb{F}^{n\times m} \mapsto \mathbb{R}$ defined by

$$\ell(A) \triangleq \begin{cases} \displaystyle\min_{y\in\mathcal{R}(A)\setminus\{0\}} \; \max_{x\in\{z\in\mathbb{F}^m\colon Az=y\}} \frac{\|y\|'}{\|x\|}, & A \neq 0, \\ 0, & A = 0, \end{cases} \tag{9.5.1}$$

is the *lower bound induced by* $\|\cdot\|$ and $\|\cdot\|'$. Equivalently,

$$\ell(A) \triangleq \begin{cases} \displaystyle\min_{x\in\mathbb{F}^m\setminus\mathcal{N}(A)} \; \max_{z\in\mathcal{N}(A)} \frac{\|Ax\|'}{\|x+z\|}, & A \neq 0, \\ 0, & A = 0. \end{cases} \tag{9.5.2}$$

Proposition 9.5.2. Let $\|\cdot\|$ and $\|\cdot\|'$ be norms on $\mathbb{F}^m$ and $\mathbb{F}^n$, respectively, let $\|\cdot\|''$ be the norm induced by $\|\cdot\|$ and $\|\cdot\|'$, let $\|\cdot\|'''$ be the norm induced by $\|\cdot\|'$ and $\|\cdot\|$, and let ℓ be the lower bound induced by $\|\cdot\|$ and $\|\cdot\|'$. Then, the following statements hold:

i) $\ell(A)$ exists for all $A \in \mathbb{F}^{n\times m}$, that is, the minimum in (9.5.1) is attained.

ii) If $A \in \mathbb{F}^{n\times m}$, then $\ell(A) = 0$ if and only if $A = 0$.

iii) For all $A \in \mathbb{F}^{n\times m}$ there exists a vector $x \in \mathbb{F}^m$ such that

$$\ell(A)\|x\| = \|Ax\|'. \tag{9.5.3}$$

iv) For all $A \in \mathbb{F}^{n\times m}$,

$$\ell(A) \leq \|A\|''. \tag{9.5.4}$$

v) If $A \neq 0$ and B is a (1)-inverse of A, then

$$1/\|B\|''' \leq \ell(A) \leq \|B\|'''. \tag{9.5.5}$$

vi) If $A, B \in \mathbb{F}^{n\times m}$ and either $\mathcal{R}(A) \subseteq \mathcal{R}(A+B)$ or $\mathcal{N}(A) \subseteq \mathcal{N}(A+B)$, then

$$\ell(A) - \|B\|''' \leq \ell(A+B). \tag{9.5.6}$$

vii) If $A, B \in \mathbb{F}^{n\times m}$ and either $\mathcal{R}(A+B) \subseteq \mathcal{R}(A)$ or $\mathcal{N}(A+B) \subseteq \mathcal{N}(A)$, then

$$\ell(A+B) \leq \ell(A) + \|B\|'''. \tag{9.5.7}$$

viii) If $n = m$ and $A \in \mathbb{F}^{n\times n}$ is nonsingular, then

$$\ell(A) = 1/\|A^{-1}\|'''. \tag{9.5.8}$$

Proof. See [596]. $\square$

Proposition 9.5.3. Let $\|\cdot\|$, $\|\cdot\|'$, and $\|\cdot\|''$ be norms on $\mathbb{F}^l$, $\mathbb{F}^m$, and $\mathbb{F}^n$, respectively, let $\|\cdot\|'''$ denote the norm on $\mathbb{F}^{m\times l}$ induced by $\|\cdot\|$ and $\|\cdot\|'$, let $\|\cdot\|''''$ denote the norm on $\mathbb{F}^{n\times m}$ induced by $\|\cdot\|'$ and $\|\cdot\|''$, and let $\|\cdot\|'''''$ denote the norm on $\mathbb{F}^{n \times l}$ induced by $\| \cdot \|$ and $\| \cdot \|''$. If $A \in \mathbb{F}^{n \times m}$ and $B \in \mathbb{F}^{m \times l}$, then

$$\ell(A)\ell'(B) \leq \ell''(AB). \tag{9.5.9}$$

In addition, the following statements hold:

i) If either $\operatorname{rank} B = \operatorname{rank} AB$ or $\operatorname{def} B = \operatorname{def} AB$, then

$$\ell''(AB) \leq \|A\|''\ell(B). \tag{9.5.10}$$

ii) If $\operatorname{rank} A = \operatorname{rank} AB$, then

$$\ell''(AB) \leq \ell(A)\|B\|''''. \tag{9.5.11}$$

iii) If $\operatorname{rank} B = m$, then

$$\|A\|''\ell(B) \leq \|AB\|''''. \tag{9.5.12}$$

iv) If $\operatorname{rank} A = m$, then

$$\ell(A)\|B\|'''' \leq \|AB\|''''''. \tag{9.5.13}$$

Proof. See [596]. $\qquad\square$

By assigning $\| \cdot \|_p$ to $\mathbb{F}^m$ and $\| \cdot \|_q$ to $\mathbb{F}^n$, where $p \geq 1$ and $q \geq 1$, the *Hölder-induced lower bound* on $\mathbb{F}^{n \times m}$ is defined by

$$\ell_{q,p}(A) \triangleq \begin{cases} \displaystyle\min_{y \in \mathcal{R}(A) \setminus \{0\}} \max_{x \in \{z \in \mathbb{F}^m : Az = y\}} \dfrac{\|y\|_q}{\|x\|_p}, & A \neq 0, \\[4mm] 0, & A = 0. \end{cases} \tag{9.5.14}$$

The following result shows that $\ell_{2,2}(A)$ is the smallest positive singular value of A.

Proposition 9.5.4. Let $A \in \mathbb{F}^{n \times m}$, assume that A is nonzero, and let $r \triangleq \operatorname{rank} A$. Then,

$$\ell_{2,2}(A) = \sigma_r(A). \tag{9.5.15}$$

Proof. This result follows from the singular value decomposition. $\qquad\square$

Corollary 9.5.5. Let $A \in \mathbb{F}^{n \times m}$. If $n \leq m$ and A is right invertible, then

$$\ell_{2,2}(A) = \sigma_{\min}(A) = \sigma_n(A). \tag{9.5.16}$$

If $m \leq n$ and A is left invertible, then

$$\ell_{2,2}(A) = \sigma_{\min}(A) = \sigma_m(A). \tag{9.5.17}$$

Finally, if $n = m$ and A is nonsingular, then

$$\ell_{2,2}(A^{-1}) = \sigma_{\min}(A^{-1}) = \frac{1}{\sigma_{\max}(A)}. \tag{9.5.18}$$

Proof. Use Proposition 5.6.2 and Fact 6.3.28. $\qquad\square$

In contrast to the submultiplicativity condition (9.4.4) satisfied by the induced norm, the induced lower bound satisfies a supermultiplicativity condition. The following result is analogous to Proposition 9.4.3.

Proposition 9.5.6. Let $\|\cdot\|$, $\|\cdot\|'$, and $\|\cdot\|''$ be norms on $\mathbb{F}^l$, $\mathbb{F}^m$, and $\mathbb{F}^n$, respectively. Let $\ell(\cdot)$ be the lower bound induced by $\|\cdot\|$ and $\|\cdot\|'$, let $\ell'(\cdot)$ be the lower bound induced by $\|\cdot\|'$ and $\|\cdot\|''$, let $\ell''(\cdot)$ be the lower bound induced by $\|\cdot\|$ and $\|\cdot\|''$, let $A \in \mathbb{F}^{n \times m}$ and $B \in \mathbb{F}^{m \times l}$, and assume that either A or B is right invertible. Then,

$$\ell'(A)\ell(B) \leq \ell''(AB). \tag{9.5.19}$$

Furthermore, if $1 \leq p, q, r \leq \infty$, then

$$\ell_{r,q}(A)\ell_{q,p}(B) \leq \ell_{r,p}(AB). \tag{9.5.20}$$

In particular,

$$\sigma_m(A)\sigma_l(B) \leq \sigma_l(AB). \tag{9.5.21}$$

Proof. See [596] and [892, pp. 369, 370]. $\qquad\square$

9.6 Singular Value Inequalities

Proposition 9.6.1. Let $A \in \mathbb{F}^{n \times m}$ and $B \in \mathbb{F}^{m \times l}$. Then, for all $i \in \{1, \ldots, \min\{n, m\}\}$ and $j \in \{1, \ldots, \min\{m, l\}\}$ such that $i + j \leq \min\{n, l\} + 1$,

$$\sigma_{i+j-1}(AB) \leq \sigma_i(A)\sigma_j(B). \tag{9.6.1}$$

In particular, for all $i \in \{1, \ldots, \min\{n, m, l\}\}$,

$$\sigma_i(AB) \leq \sigma_{\max}(A)\sigma_i(B) \tag{9.6.2}$$

and

$$\sigma_i(AB) \leq \sigma_i(A)\sigma_{\max}(B). \tag{9.6.3}$$

Proof. See [730, p. 178]. $\qquad\square$

Proposition 9.6.2. Let $A \in \mathbb{F}^{n \times m}$ and $B \in \mathbb{F}^{m \times l}$. If $r \geq 0$, then, for all $k \in \{1, \ldots, \min\{n, m, l\}\}$,

$$\sum_{i=1}^{k} \sigma_i^r(AB) \leq \sum_{i=1}^{k} \sigma_i^r(A)\sigma_i^r(B). \tag{9.6.4}$$

In particular, for all $k \in \{1, \ldots, \min\{n, m, l\}\}$,

$$\sum_{i=1}^{k} \sigma_i(AB) \leq \sum_{i=1}^{k} \sigma_i(A)\sigma_i(B). \tag{9.6.5}$$

If $r < 0$, $n = m = l$, and A and B are nonsingular, then

$$\sum_{i=1}^{n} \sigma_i^r(AB) \leq \sum_{i=1}^{n} \sigma_i^r(A)\sigma_i^r(B). \tag{9.6.6}$$

Proof. The first statement follows from Proposition 9.6.3 and Fact 2.21.8. For the case $r < 0$, use Fact 2.21.11. See [201, p. 94] or [730, p. 177]. $\qquad\square$

Proposition 9.6.3. Let $A \in \mathbb{F}^{n \times m}$ and $B \in \mathbb{F}^{m \times l}$. Then, for all $k \in \{1, \ldots, \min\{n, m, l\}\}$,

$$\prod_{i=1}^{k} \sigma_i(AB) \leq \prod_{i=1}^{k} \sigma_i(A)\sigma_i(B).$$

If, in addition, $n = m = l$, then

$$\prod_{i=1}^{n} \sigma_i(AB) = \prod_{i=1}^{n} \sigma_i(A)\sigma_i(B).$$

Proof. See [730, p. 172]. □

Proposition 9.6.4. Let $A \in \mathbb{F}^{n \times m}$ and $B \in \mathbb{F}^{m \times l}$. If $m \leq n$, then, for all $i \in \{1, \ldots, \min\{n, m, l\}\}$,

$$\sigma_{\min}(A)\sigma_i(B) = \sigma_m(A)\sigma_i(B) \leq \sigma_i(AB). \tag{9.6.7}$$

If $m \leq l$, then, for all $i \in \{1, \ldots, \min\{n, m, l\}\}$,

$$\sigma_i(A)\sigma_{\min}(B) = \sigma_i(A)\sigma_m(B) \leq \sigma_i(AB). \tag{9.6.8}$$

Proof. Corollary 8.4.2 implies that $\sigma_m^2(A)I_m = \lambda_{\min}(A^*A)I_m \leq A^*A$, which implies that $\sigma_m^2(A)B^*B \leq B^*A^*AB$. Hence, it follows from the monotonicity theorem Theorem 8.4.9 that, for all $i \in \{1, \ldots, \min\{n, m, l\}\}$,

$$\sigma_m(A)\sigma_i(B) = \lambda_i\big[\sigma_m^2(A)B^*B\big]^{1/2} \leq \lambda_i^{1/2}(B^*A^*AB) = \sigma_i(AB),$$

which proves the left-hand inequality in (9.6.7). Similarly, for all $i \in \{1, \ldots, \min\{n, m, l\}\}$,

$$\sigma_i(A)\sigma_m(B) = \lambda_i\big[\sigma_m^2(B)AA^*\big]^{1/2} \leq \lambda_i^{1/2}(ABB^*A^*) = \sigma_i(AB). \qquad \square$$

Corollary 9.6.5. Let $A \in \mathbb{F}^{n \times m}$ and $B \in \mathbb{F}^{m \times l}$. Then,

$$\sigma_m(A)\sigma_{\min\{n,m,l\}}(B) \leq \sigma_{\min\{n,m,l\}}(AB) \leq \sigma_{\max}(A)\sigma_{\min\{n,m,l\}}(B), \tag{9.6.9}$$

$$\sigma_m(A)\sigma_{\max}(B) \leq \sigma_{\max}(AB) \leq \sigma_{\max}(A)\sigma_{\max}(B), \tag{9.6.10}$$

$$\sigma_{\min\{n,m,l\}}(A)\sigma_m(B) \leq \sigma_{\min\{n,m,l\}}(AB) \leq \sigma_{\min\{n,m,l\}}(A)\sigma_{\max}(B), \tag{9.6.11}$$

$$\sigma_{\max}(A)\sigma_m(B) \leq \sigma_{\max}(AB) \leq \sigma_{\max}(A)\sigma_{\max}(B). \tag{9.6.12}$$

Specializing Corollary 9.6.5 to the case in which A or B is square yields the following result.

Corollary 9.6.6. Let $A \in \mathbb{F}^{n \times n}$ and $B \in \mathbb{F}^{n \times l}$. Then, for all $i \in \{1, \ldots, \min\{n, l\}\}$,

$$\sigma_{\min}(A)\sigma_i(B) \leq \sigma_i(AB) \leq \sigma_{\max}(A)\sigma_i(B). \tag{9.6.13}$$

In particular,

$$\sigma_{\min}(A)\sigma_{\max}(B) \leq \sigma_{\max}(AB) \leq \sigma_{\max}(A)\sigma_{\max}(B). \tag{9.6.14}$$

If $A \in \mathbb{F}^{n \times m}$ and $B \in \mathbb{F}^{m \times m}$, then, for all $i \in \{1, \ldots, \min\{n, m\}\}$,

$$\sigma_i(A)\sigma_{\min}(B) \leq \sigma_i(AB) \leq \sigma_i(A)\sigma_{\max}(B). \qquad (9.6.15)$$

In particular,

$$\sigma_{\max}(A)\sigma_{\min}(B) \leq \sigma_{\max}(AB) \leq \sigma_{\max}(A)\sigma_{\max}(B). \qquad (9.6.16)$$

Corollary 9.6.7. Let $A \in \mathbb{F}^{n \times m}$ and $B \in \mathbb{F}^{m \times l}$. If $m \leq n$, then

$$\sigma_{\min}(A)\|B\|_{\mathrm{F}} = \sigma_m(A)\|B\|_{\mathrm{F}} \leq \|AB\|_{\mathrm{F}}. \qquad (9.6.17)$$

If $m \leq l$, then

$$\|A\|_{\mathrm{F}}\sigma_{\min}(B) = \|A\|_{\mathrm{F}}\sigma_m(B) \leq \|AB\|_{\mathrm{F}}. \qquad (9.6.18)$$

Proposition 9.6.8. Let $A, B \in \mathbb{F}^{n \times m}$. Then, for all $i, j \in \{1, \ldots, \min\{n, m\}\}$ such that $i + j \leq \min\{n, m\} + 1$,

$$\sigma_{i+j-1}(A + B) \leq \sigma_i(A) + \sigma_j(B) \qquad (9.6.19)$$

and

$$\sigma_{i+j-1}(A) - \sigma_j(B) \leq \sigma_i(A + B). \qquad (9.6.20)$$

Proof. See [730, p. 178]. $\qquad \square$

Corollary 9.6.9. Let $A, B \in \mathbb{F}^{n \times m}$. Then,

$$\sigma_n(A) - \sigma_{\max}(B) \leq \sigma_n(A + B) \leq \sigma_n(A) + \sigma_{\max}(B). \qquad (9.6.21)$$

If, in addition, $n = m$, then

$$\sigma_{\min}(A) - \sigma_{\max}(B) \leq \sigma_{\min}(A + B) \leq \sigma_{\min}(A) + \sigma_{\max}(B). \qquad (9.6.22)$$

Proof. This result follows from Proposition 9.6.8. Alternatively, it follows from Lemma 8.4.3 and the Cauchy-Schwarz inequality given by Corollary 9.1.7 that, for all nonzero $x \in \mathbb{F}^n$,

$$\lambda_{\min}[(A + B)(A + B)^*] \leq \frac{x^*(AA^* + BB^* + AB^* + BA^*)x}{x^*x}$$

$$= \frac{x^*AA^*x}{\|x\|_2^2} + \frac{x^*BB^*x}{\|x\|_2^2} + \mathrm{Re}\,\frac{2x^*AB^*x}{\|x\|_2^2}$$

$$\leq \frac{x^*AA^*x}{\|x\|_2^2} + \sigma_{\max}^2(B) + 2\frac{(x^*AA^*x)^{1/2}}{\|x\|_2}\sigma_{\max}(B).$$

Minimizing with respect to x and using Lemma 8.4.3 yields

$$\sigma_n^2(A + B) = \lambda_{\min}[(A + B)(A + B)^*]$$

$$\leq \lambda_{\min}(AA^*) + \sigma_{\max}^2(B) + 2\lambda_{\min}^{1/2}(AA^*)\sigma_{\max}(B)$$

$$= [\sigma_n(A) + \sigma_{\max}(B)]^2,$$

which proves the right-hand inequality of (9.6.21). Finally, the left-hand inequality follows from the right-hand inequality with A and B replaced by $A + B$ and $-B$, respectively. $\qquad \square$

9.7 Facts on Vector Norms

Fact 9.7.1. Let $x, y \in \mathbb{F}^n$. Then, x and y are linearly dependent if and only if $|x|^{\circ 2}$ and $|y|^{\circ 2}$ are linearly dependent and $|x^*y| = |x|^{\mathrm{T}}|y|$.

Remark: This equivalence clarifies the relationship between (9.1.9) with $p = 2$ and Corollary 9.1.7.

Fact 9.7.2. Let $x, y \in \mathbb{F}^n$, and let $\|\cdot\|$ be a norm on $\mathbb{F}^n$. Then,

$$\big| \|x\| - \|y\| \big| \leq \begin{cases} \|x + y\| \\ \|x - y\|. \end{cases}$$

Fact 9.7.3. Let $x, y \in \mathbb{F}^n$, and let $\|\cdot\|$ be a norm on $\mathbb{F}^n$. Then, the following statements hold:

  i) If there exists $\beta \geq 0$ such that either $x = \beta y$ or $y = \beta x$, then $\|x + y\| = \|x\| + \|y\|$.

  ii) If $\|x + y\| = \|x\| + \|y\|$ and x and y are linearly dependent, then there exists $\beta \geq 0$ such that either $x = \beta y$ or $y = \beta x$.

  iii) If $\|x + y\|_2 = \|x\|_2 + \|y\|_2$, then there exists $\beta \geq 0$ such that either $x = \beta y$ or $y = \beta x$.

  iv) If x and y are linearly independent, then $\|x + y\|_2 < \|x\|_2 + \|y\|_2$.

Proof: For *iii)*, use *v)* of Fact 9.7.4.

Remark: Let $x = \begin{bmatrix} 1 & 0 \end{bmatrix}$ and $y = \begin{bmatrix} 1 & 1 \end{bmatrix}$, which are linearly independent. Then, $\|x + y\|_\infty = \|x\|_\infty + \|y\|_\infty = 2$.

Problem: If x and y are linearly independent and $p \in [1, \infty)$, then does it follow that $\|x + y\|_p < \|x\|_p + \|y\|_p$?

Fact 9.7.4. Let $x, y, z \in \mathbb{F}^n$. Then, the following statements hold:

  i) $\frac{1}{2}(\|x + y\|_2^2 + \|x - y\|_2^2) = \|x\|_2^2 + \|y\|_2^2$.

  ii) If x and y are nonzero, then

$$\frac{1}{2}(\|x\|_2 + \|y\|_2)\left\| \frac{x}{\|x\|_2} - \frac{y}{\|y\|_2} \right\|_2 \leq \|x - y\|_2.$$

  iii) If x and y are nonzero, then

$$\left\| \frac{1}{\|x\|_2}x - \|x\|_2 y \right\|_2 = \left\| \frac{1}{\|y\|_2}y - \|y\|_2 x \right\|_2.$$

  iv) If $\mathbb{F} = \mathbb{R}$, then

$$4x^{\mathrm{T}}y = \|x + y\|_2^2 - \|x - y\|_2^2.$$

  v) If $\mathbb{F} = \mathbb{C}$, then

$$4x^*y = \|x + y\|_2^2 - \|x - y\|_2^2 + \jmath(\|x + \jmath y\|_2^2 - \|x - \jmath y\|_2^2).$$

  vi) $\operatorname{Re} x^*y = \frac{1}{4}(\|x + y\|_2^2 - \|x - y\|_2^2) = \frac{1}{2}(\|x + y\|_2^2 - \|x\|_2^2 - \|y\|_2^2)$.

vii) If $\mathbb{F} = \mathbb{C}$, then $\operatorname{Im} x^*y = \frac{1}{4}(\|x + \jmath y\|_2^2 - \|x - \jmath y\|_2^2)$.

viii) $\|x + y\|_2 = \sqrt{\|x\|_2^2 + \|y\|_2^2 + 2\operatorname{Re} x^*y}$.

ix) $\|x - y\|_2 = \sqrt{\|x\|_2^2 + \|y\|_2^2 - 2\operatorname{Re} x^*y}$.

x) $\|x + y\|_2\|x - y\|_2 \leq \|x\|_2^2 + \|y\|_2^2$.

xi) If $\|x + y\|_2 = \|x\|_2 + \|y\|_2$, then $\operatorname{Im} x^*y = 0$ and $\operatorname{Re} x^*y \geq 0$.

xii) $|x^*y| \leq \|x\|_2\|y\|_2$.

xiii) If $\|x + y\|_2 \leq 2$, then

$$(1 - \|x\|_2^2)(1 - \|y\|_2^2) \leq |1 - \operatorname{Re} x^*y|^2.$$

xiv) For all nonzero $\alpha \in \mathbb{R}$,

$$\|x\|_2^2\|y\|_2^2 - |x^*y|^2 \leq \alpha^{-2}\|\alpha y - x\|_2^2\|x\|_2^2.$$

xv) If $\operatorname{Re} x^*y \neq 0$, then, for all nonzero $\alpha \in \mathbb{R}$,

$$\|x\|_2^2\|y\|_2^2 - |x^*y|^2 \leq \alpha_0^{-2}\|\alpha_0 y - x\|_2^2\|x\|_2^2 \leq \alpha^{-2}\|\alpha y - x\|_2^2\|x\|_2^2,$$

where $\alpha_0 \triangleq x^*x/(\operatorname{Re} x^*y)$.

xvi) x, y, z satisfy

$$\|x + y\|_2^2 + \|y + z\|_2^2 + \|z + x\|_2^2 = \|x\|_2^2 + \|y\|_2^2 + \|z\|_2^2 + \|x + y + z\|_2^2$$

and

$$\|x + y\|_2 + \|y + z\|_2 + \|z + x\|_2 \leq \|x\|_2 + \|y\|_2 + \|z\|_2 + \|x + y + z\|_2.$$

xvii) $|x^*zz^*y - \frac{1}{2}x^*y\|z\|_2^2| \leq \frac{1}{2}\|x\|_2\|y\|_2\|z\|_2^2$.

xviii) $|\operatorname{Re}(x^*zz^*y - \frac{1}{2}x^*y\|z\|_2^2)| \leq \frac{1}{2}\|z\|_2^2\sqrt{\|x\|_2^2\|y\|_2^2 - (\operatorname{Im} x^*y)^2}$.

xix) $|\operatorname{Im}(x^*zz^*y - \frac{1}{2}x^*y\|z\|_2^2)| \leq \frac{1}{2}\|z\|_2^2\sqrt{\|x\|_2^2\|y\|_2^2 - (\operatorname{Re} x^*y)^2}$.

Furthermore, the following statements are equivalent:

xx) $\|x - y\|_2 = \|x + y\|_2$.

xxi) $\|x + y\|_2^2 = \|x\|_2^2 + \|y\|_2^2$.

xxii) $\operatorname{Re} x^*y = 0$.

Now, let $x_1, \ldots, x_k \in \mathbb{F}^n$, and assume that, for all $i, j \in \{1, \ldots, n\}$, $x_i^*x_j = \delta_{ij}$. Then, the following statement holds:

xxiii) $\sum_{i=1}^{k} |y^*x_i|^2 \leq \|y\|_2^2$.

If, in addition, $k = n$, then the following statement holds:

xxiv) $\sum_{i=1}^{n} |y^*x_i|^2 = \|y\|_2^2$.

Remark: *i*) is the *parallelogram law*, which relates the diagonals and the sides of a parallelogram; *ii*) is the *Dunkl-Williams inequality*, which compares the distance between x and y with the distance between the projections of x and y onto the unit sphere (see [458], [1035, p. 515], and [1526, p. 28]); *iv*) and *v*) are the *polarization* *identity* (see [376, p. 54], [1057, p. 276], and Fact 1.20.2); *ix*) is the *cosine law* (see Fact 9.9.13 for a matrix version); *xiii*) is given in [1502] and implies Aczel's inequality given by Fact 1.18.19; *xv*) is given in [939]; *xvi*) is *Hlawka's identity* and *Hlawka's inequality* (see Fact 1.10.6, Fact 1.20.2, [1035, p. 521], and [1066, p. 100]); *xvii*) is *Buzano's inequality* (see [527] and Fact 1.19.2); *xviii*) and *xix*) are given in [1120]; the equivalence of *xxi*) and *xxii*) is the *Pythagorean theorem*; *xxiii*) is *Bessel's inequality*; and *xxiv*) is *Parseval's identity*. Note that *xxiv*) implies *xxiii*).

Remark: Hlawka's inequality is called the *quadrilateral inequality* in [1233], which gives a geometric interpretation. In addition, [1233] provides an extension and geometric interpretation to the *polygonal inequalities*. See Fact 9.7.7.

Remark: When $\mathbb{F} = \mathbb{R}$ and $n = 2$ the Euclidean norm of $\| \begin{bmatrix} x \\ y \end{bmatrix} \|_2$ is equivalent to the absolute value $|z| = |x + jy|$. See Fact 1.20.2.

Remark: δ_{ij} is the Kronecker delta.

Fact 9.7.5. Let $x, y \in \mathbb{R}^3$, and let $\mathcal{S} \subset \mathbb{R}^3$ be the parallelogram with vertices 0, x, y, and $x + y$. Then,
$$\mathrm{area}(\mathcal{S}) = \|x \times y\|_2.$$

Remark: See Fact 2.20.13, Fact 2.20.14 and Fact 3.10.1.

Remark: The parallelogram associated with the cross product can be interpreted as a bivector. See [436, pp. 86–88] or [620, 895].

Fact 9.7.6. Let $x, y \in \mathbb{R}^n$, and assume that x and y are nonzero. Then,
$$\frac{x^{\mathrm{T}} y}{\|x\|_2 \|y\|_2} (\|x\|_2 + \|y\|_2) \leq \|x + y\|_2 \leq \|x\|_2 + \|y\|_2.$$

Hence, if $x^{\mathrm{T}} y = \|x\|_2 \|y\|_2$, then $\|x\|_2 + \|y\|_2 = \|x + y\|_2$.

Proof: See [1035, p. 517].

Remark: This result is a *reverse triangle inequality*.

Problem: Extend this result to complex vectors.

Fact 9.7.7. Let $x_1, \ldots, x_n \in \mathbb{F}^n$, and let $\alpha_1, \ldots, \alpha_n$ be nonnegative numbers. Then,
$$\sum_{i=1}^{n} \alpha_i \left\| x_i - \sum_{j=1}^{n} \alpha_j x_j \right\|_2 \leq \sum_{i=1}^{n} \alpha_i \|x_i\|_2 + \left[\left(\sum_{i=1}^{n} \alpha_i \right) - 2 \right] \left\| \sum_{i=1}^{n} \alpha_i x_i \right\|_2.$$

In particular,
$$\sum_{i=1}^{n} \left\| \sum_{j=1, j \neq i}^{n} x_j \right\|_2 \leq \sum_{i=1}^{n} \|x_i\|_2 + (n - 2) \left\| \sum_{i=1}^{n} x_i \right\|_2.$$

Remark: The first inequality is the *generalized Hlawka inequality* or *polygonal inequalities*. The second inequality is the *Djokovic inequality*. See [1285] and Fact 9.7.4.

Fact 9.7.8. Let $x, y \in \mathbb{R}^n$, let α and δ, be positive numbers, and let $p, q \in (0, \infty)$ satisfy $1/p + 1/q = 1$. Then,

$$\left(\frac{\alpha}{\alpha + \|y\|_2^q}\right)^{p-1} \delta^p \leq |\delta - x^{\mathrm{T}}y|^p + \alpha^{p-1}\|x\|_2^p.$$

Equality holds if and only if $x = [\delta\|y\|_2^{q-2}/(\alpha + \|y\|_2^q)]y$. In particular,

$$\frac{\alpha\delta^2}{\alpha + \|y\|_2^2} \leq (\delta - x^{\mathrm{T}}y)^2 + \alpha\|x\|_2^2.$$

Equality holds if and only if $x = [\delta/(\alpha + \|y\|_2^2)]y$.

Proof: See [1284].

Remark: The first inequality is due to Pecaric. The case $p = q = 2$ is due to Dragomir and Yang. These results are generalizations of Hua's inequality. See Fact 1.17.13 and Fact 9.7.9.

Fact 9.7.9. Let $x_1, \ldots, x_n, y \in \mathbb{R}^n$, and let α be a positive number. Then,

$$\frac{\alpha}{\alpha + n}\|y\|_2^2 \leq \left\|y - \sum_{i=1}^n x_i\right\|_2^2 + \alpha\sum_{i=1}^n \|x_i\|_2^2.$$

Equality holds if and only if $x_1 = \cdots = x_n = [1/(\alpha + n)]y$.

Proof: See [1284].

Remark: This inequality, which is due to Dragomir and Yang, is a generalization of Hua's inequality. See Fact 1.17.13 and Fact 9.7.8.

Fact 9.7.10. Let $x, y \in \mathbb{F}^n$, and assume that x and y are nonzero. Then,

$$\frac{\|x - y\|_2 - \left|\|x\|_2 - \|y\|_2\right|}{\min\{\|x\|_2, \|y\|_2\}} \leq \left\|\frac{x}{\|x\|_2} - \frac{y}{\|y\|_2}\right\|_2$$

$$\leq \left\{\begin{array}{c} \dfrac{\|x - y\|_2 + \left|\|x\|_2 - \|y\|_2\right|}{\max\{\|x\|_2, \|y\|_2\}} \\[3mm] \dfrac{2\|x - y\|_2}{\|x\|_2 + \|y\|_2} \end{array}\right\}$$

$$\leq \left\{\begin{array}{c} \dfrac{2\|x - y\|_2}{\max\{\|x\|_2, \|y\|_2\}} \\[3mm] \dfrac{2(\|x - y\|_2 + \left|\|x\|_2 - \|y\|_2\right|)}{\|x\|_2 + \|y\|_2} \end{array}\right\}$$

$$\leq \frac{4\|x - y\|_2}{\|x\|_2 + \|y\|_2}.$$

Proof: See Fact 9.7.13 and [1016].

Remark: In the last string of inequalities, the first inequality is the *reverse Maligranda inequality*, the second and upper third terms constitute the *Maligranda inequality*, the second and lower third terms constitute the *Dunkl-Williams in-*

equality in an inner product space, the second and upper fourth terms constitute the *Massera-Schaffer inequality.*

Remark: See Fact 1.20.5.

Fact 9.7.11. Let $x, y \in \mathbb{F}^n$, and let $\| \cdot \|$ be a norm on $\mathbb{F}^n$. Then, there exists a unique number $\alpha \in [1, 2]$ such that, for all $x, y \in \mathbb{F}^n$, at least one of which is nonzero,

$$\frac{2}{\alpha} \leq \frac{\|x+y\|^2 + \|x-y\|^2}{\|x\|^2 + \|y\|^2} \leq 2\alpha.$$

Furthermore, if $\| \cdot \| = \| \cdot \|_p$, then

$$\alpha = \begin{cases} 2^{(2-p)/p}, & 1 \leq p \leq 2, \\ 2^{(p-2)/p}, & p \geq 2. \end{cases}$$

Proof: See [281, p. 258].

Remark: This result is the *von Neumann–Jordan* inequality.

Remark: When $p = 2$, it follows that $\alpha = 2$, and this result yields *i*) of Fact 9.7.4.

Fact 9.7.12. Let $x, y \in \mathbb{F}^n$, and let $\| \cdot \|$ be a norm on $\mathbb{F}^n$. Then,

$$\|x+y\| \leq \|x\| + \|y\| - \min\{\|x\|, \|y\|\}\left(2 - \left\|\frac{x}{\|x\|} + \frac{y}{\|y\|}\right\|\right) \leq \|x\| + \|y\|,$$

$$\|x-y\| \leq \|x\| + \|y\| - \min\{\|x\|, \|y\|\}\left(2 - \left\|\frac{x}{\|x\|} - \frac{y}{\|y\|}\right\|\right) \leq \|x\| + \|y\|,$$

$$\|x\| + \|y\| - \max\{\|x\|, \|y\|\}\left(2 - \left\|\frac{x}{\|x\|} + \frac{y}{\|y\|}\right\|\right) \leq \|x+y\| \leq \|x\| + \|y\|,$$

and

$$\|x\| + \|y\| - \max\{\|x\|, \|y\|\}\left(2 - \left\|\frac{x}{\|x\|} - \frac{y}{\|y\|}\right\|\right) \leq \|x-y\| \leq \|x\| + \|y\|.$$

Proof: See [976].

Fact 9.7.13. Let $x, y \in \mathbb{F}^n$, assume that x and y are nonzero, and let $\| \cdot \|$ be a norm on $\mathbb{F}^n$. Then,

$$\frac{(\|x\| + \|y\|)(\|x+y\| - |\|x\| - \|y\||)}{4\min\{\|x\|, \|y\|\}} \leq \tfrac{1}{4}(\|x\| + \|y\|)\left\|\frac{x}{\|x\|} + \frac{y}{\|y\|}\right\|$$

$$\leq \tfrac{1}{2}\max\{\|x\|, \|y\|\}\left\|\frac{x}{\|x\|} + \frac{y}{\|y\|}\right\|$$

$$\leq \tfrac{1}{2}(\|x+y\| + \max\{\|x\|, \|y\|\} - \|x\| - \|y\|)$$

$$\leq \tfrac{1}{2}(\|x+y\| + |\|x\| - \|y\||)$$

$$\leq \|x+y\|$$

and

$$\frac{(\|x\| + \|y\|)(\|x - y\| - |\|x\| - \|y\||)}{4\min\{\|x\|, \|y\|\}} \leq \tfrac{1}{4}(\|x\| + \|y\|)\left\|\frac{x}{\|x\|} - \frac{y}{\|y\|}\right\|$$

$$\leq \tfrac{1}{2}\max\{\|x\|, \|y\|\}\left\|\frac{x}{\|x\|} - \frac{y}{\|y\|}\right\|$$

$$\leq \tfrac{1}{2}(\|x - y\| + \max\{\|x\|, \|y\|\} - \|x\| - \|y\|)$$

$$\leq \tfrac{1}{2}(\|x - y\| + |\|x\| - \|y\||)$$

$$\leq \|x - y\|.$$

Furthermore,

$$\frac{\|x - y\| - |\|x\| - \|y\||}{\min\{\|x\|, \|y\|\}} \leq \left\|\frac{x}{\|x\|} - \frac{y}{\|y\|}\right\|$$

$$\leq \frac{\|x - y\| + |\|x\| - \|y\||}{\max\{\|x\|, \|y\|\}}$$

$$\leq \left\{ \begin{array}{c} \dfrac{2\|x - y\|}{\max\{\|x\|, \|y\|\}} \\[2mm] \dfrac{2(\|x - y\| + |\|x\| - \|y\||)}{\|x\| + \|y\|} \end{array} \right\}$$

$$\leq \frac{4\|x - y\|}{\|x\| + \|y\|}.$$

Proof: This result follows from Fact 9.7.12, [976, 1016] and [1035, p. 516].

Remark: In the last string of inequalities, the first inequality is the *reverse Maligranda inequality*, the second inequality is the *Maligranda inequality*, the second and upper fourth terms constitute the *Massera-Schaffer inequality*, and the second and fifth terms constitute the Dunkl-Williams inequality. See Fact 1.20.2 and Fact 9.7.4 for the case of the Euclidean norm.

Remark: Extensions to more than two vectors are given in [816, 1105].

Fact 9.7.14. Let $x, y \in \mathbb{F}^n$, and let $\| \cdot \|$ be a norm on $\mathbb{F}^n$. Then,

$$\left. \begin{array}{c} \|x\|^2 + \|y\|^2 \\ 2\|x\|^2 - 4\|x\|\|y\| + 2\|y\|^2 \end{array} \right\} \leq \|x + y\|^2 + \|x - y\|^2$$

$$\leq 2\|x\|^2 + 4\|x\|\|y\| + 2\|y\|^2$$

$$\leq 4(\|x\|^2 + \|y\|^2).$$

Proof: See [544, pp. 9, 10] and [1057, p. 278].

Fact 9.7.15. Let $x, y \in \mathbb{F}^n$, let $\alpha \in [0, 1]$, and let $\| \cdot \|$ be a norm on $\mathbb{F}^n$. Then,

$$\|x + y\| \leq \|\alpha x + (1 - \alpha)y\| + \|(1 - \alpha)x + \alpha y\| \leq \|x\| + \|y\|.$$

Fact 9.7.16. Let $x, y \in \mathbb{F}^n$, assume that x and y are nonzero, let $\| \cdot \|$ be a norm on $\mathbb{F}^n$, and let $p \in \mathbb{R}$. Then, the following statements hold:

i) If $p \leq 0$, then

$$\left\| \|x\|^{p-1} x - \|y\|^{p-1} y \right\| \leq (2 - p) \frac{\max\{\|x\|^p, \|y\|^p\}}{\max\{\|x\|, \|y\|\}} \|x - y\|.$$

ii) If $p \in [0, 1]$, then

$$\left\| \|x\|^{p-1} x - \|y\|^{p-1} y \right\| \leq (2 - p) \frac{\|x - y\|}{[\max\{\|x\|, \|y\|\}]^{1-p}}.$$

iii) If $p \geq 1$, then

$$\left\| \|x\|^{p-1} x - \|y\|^{p-1} y \right\| \leq p[\max\{\|x\|, \|y\|\}]^{p-1} \|x - y\|.$$

Proof: See [976].

Fact 9.7.17. Let $x, y \in \mathbb{F}^n$, let $\| \cdot \|$ be a norm on $\mathbb{F}^n$, assume that $\|x\| \neq \|y\|$, and let $p > 0$. Then,

$$\left| \|x\| - \|y\| \right| \leq \frac{\left\| \|x\|^p x - \|y\|^p y \right\|}{\left| \|x\|^{p+1} - \|y\|^{p+1} \right|} \left| \|x\| - \|y\| \right| \leq \|x - y\|.$$

Proof: See [1035, p. 516].

Fact 9.7.18. Let $x \in \mathbb{F}^n$, and let $p, q \in [1, \infty]$ satisfy $1/p + 1/q = 1$. Then,

$$\|x\|_2 \leq \sqrt{\|x\|_p \|x\|_q}.$$

Fact 9.7.19. Let $x, y \in \mathbb{R}^n$, assume that x and y are nonnegative, let $p \in (0, 1]$, and define

$$\|x\|_p \triangleq \left(\sum_{i=1}^{n} |x_{(i)}|^p \right)^{1/p}.$$

Then,

$$\|x\|_p + \|y\|_p \leq \|x + y\|_p.$$

Remark: The notation is for convenience only since $\| \cdot \|_p$ is not a norm for all $p \in (0, 1)$.

Remark: This result is the *reverse Minkowski inequality*.

Fact 9.7.20. Let $x, y \in \mathbb{F}^n$, let $\| \cdot \|$ be a norm on $\mathbb{F}^n$, let p and q be real numbers, and assume that $1 \leq p \leq q$. Then,

$$[\tfrac{1}{2}(\|x + \tfrac{1}{\sqrt{q-1}} y\|^q + \|x - \tfrac{1}{\sqrt{q-1}} y\|^q)]^{1/q} \leq [\tfrac{1}{2}(\|x + \tfrac{1}{\sqrt{p-1}} y\|^p + \|x - \tfrac{1}{\sqrt{p-1}} y\|^p)]^{1/p}.$$

Proof: See [556, p. 207].

Remark: This result is *Bonami's inequality*. See Fact 1.12.16.

Fact 9.7.21. Let $x, y \in \mathbb{F}^{n \times n}$. If $p \in [1, 2]$, then

$$(\|x\|_p + \|y\|_p)^p + |\|x\|_p - \|y\|_p|^p \le \|x + y\|_p^p + \|x - y\|_p^p$$

and

$$(\|x + y\|_p + \|x - y\|_p)^p + |\|x + y\|_p + \|x - y\|_p|^p \le 2^p(\|x\|_p^p + \|y\|_p^p).$$

If $p \in [2, \infty]$, then

$$\|x + y\|_p^p + \|x - y\|_p^p \le (\|x\|_p + \|y\|_{\sigma p})^p + |\|x\|_p - \|y\|_p|^p$$

and

$$2^p(\|x\|_p^p + \|y\|_p^p) \le (\|x + y\|_p + \|x - y\|_p)^p + |\|x + y\|_p + \|x - y\|_p|^p.$$

Proof: See [120, 932].

Remark: These inequalities are versions of *Hanner's inequality*. These vector versions follow from inequalities on L_p by appropriate choice of measure.

Remark: Matrix versions are given in Fact 9.9.36.

Fact 9.7.22. Let $y \in \mathbb{F}^n$, let $\| \cdot \|$ be a norm on $\mathbb{F}^n$, let $\| \cdot \|'$ be the norm on $\mathbb{F}^{n \times n}$ induced by $\| \cdot \|$, and define

$$\|y\|_{\mathrm{D}} \triangleq \max_{x \in \{z \in \mathbb{F}^n : \|z\| = 1\}} |y^* x|.$$

Then, $\| \cdot \|_{\mathrm{D}}$ is a norm on $\mathbb{F}^n$. Furthermore,

$$\|y\| = \max_{x \in \{z \in \mathbb{F}^n : \|z\|_{\mathrm{D}} = 1\}} |y^* x|.$$

Hence, for all $x \in \mathbb{F}^n$,

$$|x^* y| \le \|x\| \|y\|_{\mathrm{D}}.$$

In addition,

$$\|xy^*\|' = \|x\| \|y\|_{\mathrm{D}}.$$

Finally, let $p \in [1, \infty]$, and let $1/p + 1/q = 1$. Then,

$$\| \cdot \|_{p\mathrm{D}} = \| \cdot \|_q.$$

Hence, for all $x \in \mathbb{F}^n$,

$$|x^* y| \le \|x\|_p \|y\|_q$$

and

$$\|xy^*\|_{p,p} = \|x\|_p \|y\|_q.$$

Proof: See [1261, p. 57].

Remark: $\| \cdot \|_{\mathrm{D}}$ is the *dual norm* of $\| \cdot \|$.

Fact 9.7.23. Let $\| \cdot \|$ be a norm on $\mathbb{F}^n$, and define $f : \mathbb{F}^n \mapsto [0, \infty)$ by $f(x) = \|x\|$. Then, f is convex.

Fact 9.7.24. Let $x \in \mathbb{R}^n$, and let $\| \cdot \|$ be a norm on $\mathbb{R}^n$. Then, $x^{\mathrm{T}} y > 0$ for all $y \in \{z \in \mathbb{R}^n : \|z - x\| < \|x\|\}$.

Fact 9.7.25. Let $x, y \in \mathbb{R}^n$, assume that x and y are nonzero, assume that $x^{\mathrm{T}}y = 0$, and let $\|\cdot\|$ be a norm on $\mathbb{R}^n$. Then, $\|x\| \leq \|x + y\|$.

Proof: If $\|x + y\| < \|x\|$, then $x + y \in \mathbb{B}_{\|x\|}(0)$, and thus $y \in \mathbb{B}_{\|x\|}(-x)$. By Fact 9.7.24, $x^{\mathrm{T}}y < 0$.

Remark: See [222, 927] for related results concerning matrices.

Fact 9.7.26. Let $x \in \mathbb{F}^n$ and $y \in \mathbb{F}^m$. Then,

$$\sigma_{\max}(xy^*) = \|xy^*\|_{\mathrm{F}} = \|x\|_2 \|y\|_2$$

and

$$\sigma_{\max}(xx^*) = \|xx^*\|_{\mathrm{F}} = \|x\|_2^2.$$

Remark: See Fact 5.11.16.

Fact 9.7.27. Let $x \in \mathbb{F}^n$ and $y \in \mathbb{F}^m$. Then,

$$\|x \otimes y\|_2 = \left\|\mathrm{vec}\left(x \otimes y^{\mathrm{T}}\right)\right\|_2 = \left\|\mathrm{vec}\left(yx^{\mathrm{T}}\right)\right\|_2 = \left\|yx^{\mathrm{T}}\right\|_2 = \|x\|_2 \|y\|_2.$$

Fact 9.7.28. Let $x \in \mathbb{F}^n$, and let $1 \leq p, q \leq \infty$. Then,

$$\|x\|_p = \|x\|_{p,q}.$$

Fact 9.7.29. Let $x \in \mathbb{F}^n$, and let $p, q \in [1, \infty)$, where $p \leq q$. Then,

$$\|x\|_q \leq \|x\|_p \leq n^{1/p - 1/q} \|x\|_q.$$

Proof: See [698], [699, p. 107].

Remark: See Fact 1.17.5 and Fact 9.8.21.

Fact 9.7.30. Let $A \in \mathbb{F}^{n \times n}$, and assume that A is positive definite. Then,

$$\|x\|_A \triangleq (x^*Ax)^{1/2}$$

is a norm on $\mathbb{F}^n$.

Fact 9.7.31. Let $\|\cdot\|$ and $\|\cdot\|'$ be norms on $\mathbb{F}^n$, and let $\alpha, \beta > 0$. Then, $\alpha\|\cdot\| + \beta\|\cdot\|'$ is also a norm on $\mathbb{F}^n$. Furthermore, $\max\{\|\cdot\|, \|\cdot\|'\}$ is a norm on $\mathbb{F}^n$.

Remark: $\min\{\|\cdot\|, \|\cdot\|'\}$ is not necessarily a norm.

Fact 9.7.32. Let $A \in \mathbb{F}^{n \times n}$, assume that A is nonsingular, and let $\|\cdot\|$ be a norm on $\mathbb{F}^n$. Then, $\|x\|' \triangleq \|Ax\|$ is a norm on $\mathbb{F}^n$.

Fact 9.7.33. Let $x \in \mathbb{F}^n$, and let $p \in [1, \infty]$. Then,

$$\|\bar{x}\|_p = \|x\|_p.$$

Fact 9.7.34. Let $x_1, \ldots, x_k \in \mathbb{F}^n$, let $\alpha_1, \ldots, \alpha_k$ be positive numbers, and assume that $\sum_{i=1}^k \alpha_i = 1$. Then,

$$|1_{1 \times n}(x_1 \circ \cdots \circ x_k)| \leq \prod_{i=1}^k \|x_i\|_{1/\alpha_i}.$$

Remark: This result is the *generalized Hölder inequality*. See [279, p. 128].

9.8 Facts on Matrix Norms for One Matrix

Fact 9.8.1. Let $\mathcal{S} \subseteq \mathbb{F}^m$, assume that $\mathcal{S}$ is bounded, and let $A \in \mathbb{F}^{n \times m}$. Then, $A\mathcal{S}$ is bounded.

Remark: $A\mathcal{S} = \{Ax \colon x \in \mathcal{S}\}$.

Fact 9.8.2. Let $A \in \mathbb{F}^{n \times n}$, assume that A is a idempotent, and assume that, for all $x \in \mathbb{F}^n$,
$$\|Ax\|_2 \leq \|x\|_2.$$

Then, A is a projector.

Proof: See [550, p. 42].

Fact 9.8.3. Let $A, B \in \mathbb{F}^{n \times n}$, and assume that A and B are projectors. Then, the following statements are equivalent:

i) $A \leq B$.

ii) For all $x \in \mathbb{F}^n$, $\|Ax\|_2 \leq \|Bx\|_2$.

iii) $\mathcal{R}(A) \subseteq \mathcal{R}(B)$.

iv) $AB = A$.

v) $BA = A$.

vi) $B - A$ is a projector.

Proof: See [550, p. 43] and [1215, p. 24].

Remark: See Fact 3.13.14 and Fact 3.13.17.

Fact 9.8.4. Let $A \in \mathbb{F}^{n \times n}$, and assume that $\mathrm{sprad}(A) < 1$. Then, there exists a submultiplicative matrix norm $\|\cdot\|$ on $\mathbb{F}^{n \times n}$ such that $\|A\| < 1$. Furthermore,
$$\lim_{k \to \infty} A^k = 0.$$

Fact 9.8.5. Let $A \in \mathbb{F}^{n \times n}$, assume that A is nonsingular, and let $\|\cdot\|$ be a submultiplicative norm on $\mathbb{F}^{n \times n}$. Then,
$$\|A^{-1}\| \geq \|I_n\|/\|A\|.$$

Fact 9.8.6. Let $A \in \mathbb{F}^{n \times n}$, assume that A is nonzero and idempotent, and let $\|\cdot\|$ be a submultiplicative norm on $\mathbb{F}^{n \times n}$. Then,
$$\|A\| \geq 1.$$

Fact 9.8.7. Let $\|\cdot\|$ be a unitarily invariant norm on $\mathbb{F}^{n \times n}$. Then, $\|\cdot\|$ is self-adjoint.

Fact 9.8.8. Let $A \in \mathbb{F}^{n \times m}$, let $\|\cdot\|$ be a norm on $\mathbb{F}^{n \times m}$, and define $\|A\|' \triangleq \|A^*\|$. Then, $\|\cdot\|'$ is a norm on $\mathbb{F}^{m \times n}$. If, in addition, $n = m$ and $\|\cdot\|$ is induced by $\|\cdot\|''$, then $\|\cdot\|'$ is induced by $\|\cdot\|''_{\mathrm{D}}$.

Proof: See [728, p. 309] and Fact 9.8.10.

Remark: See Fact 9.7.22 for the definition of the dual norm. $\|\cdot\|'$ is the *adjoint norm* of $\|\cdot\|$.

Problem: Generalize this result to nonsquare matrices and norms that are not equi-induced.

Fact 9.8.9. Let $1 \le p \le \infty$. Then, $\|\cdot\|_{\sigma p}$ is unitarily invariant.

Fact 9.8.10. Let $A \in \mathbb{F}^{n \times m}$, and let $p, q \in [1, \infty]$ satisfy $1/p + 1/q = 1$. Then,

$$\|A^*\|_{p,p} = \|A\|_{q,q}.$$

In particular,

$$\|A^*\|_{\mathrm{col}} = \|A\|_{\mathrm{row}}.$$

Proof: See Fact 9.8.8.

Fact 9.8.11. Let $A \in \mathbb{F}^{n \times m}$, and let $p, q \in [1, \infty]$ satisfy $1/p + 1/q = 1$. Then,

$$\left\| \begin{bmatrix} 0 & A \\ A^* & 0 \end{bmatrix} \right\|_{p,p} = \max\{\|A\|_{p,p}, \|A\|_{q,q}\}.$$

In particular,

$$\left\| \begin{bmatrix} 0 & A \\ A^* & 0 \end{bmatrix} \right\|_{\mathrm{col}} = \left\| \begin{bmatrix} 0 & A \\ A^* & 0 \end{bmatrix} \right\|_{\mathrm{row}} = \max\{\|A\|_{\mathrm{col}}, \|A\|_{\mathrm{row}}\}.$$

Fact 9.8.12. Let $A \in \mathbb{F}^{n \times m}$. Then, the following inequalities hold:

i) $\|A\|_{\mathrm{F}} \le \|A\|_1 \le \sqrt{mn}\|A\|_{\mathrm{F}}$.

ii) $\|A\|_\infty \le \|A\|_1 \le mn\|A\|_\infty$.

iii) $\|A\|_{\mathrm{col}} \le \|A\|_1 \le m\|A\|_{\mathrm{col}}$.

iv) $\|A\|_{\mathrm{row}} \le \|A\|_1 \le n\|A\|_{\mathrm{row}}$.

v) $\sigma_{\max}(A) \le \|A\|_1 \le \sqrt{mn \, \mathrm{rank}\, A}\, \sigma_{\max}(A)$.

vi) $\|A\|_\infty \le \|A\|_{\mathrm{F}} \le \sqrt{mn}\|A\|_\infty$.

vii) $\frac{1}{\sqrt{n}}\|A\|_{\mathrm{col}} \le \|A\|_{\mathrm{F}} \le \sqrt{m}\|A\|_{\mathrm{col}}$.

viii) $\frac{1}{\sqrt{m}}\|A\|_{\mathrm{row}} \le \|A\|_{\mathrm{F}} \le \sqrt{n}\|A\|_{\mathrm{row}}$.

ix) $\sigma_{\max}(A) \le \|A\|_{\mathrm{F}} \le \sqrt{\mathrm{rank}\, A}\, \sigma_{\max}(A)$.

x) $\frac{1}{n}\|A\|_{\mathrm{col}} \le \|A\|_\infty \le \|A\|_{\mathrm{col}}$.

xi) $\frac{1}{m}\|A\|_{\mathrm{row}} \le \|A\|_\infty \le \|A\|_{\mathrm{row}}$.

xii) $\frac{1}{\sqrt{mn}}\sigma_{\max}(A) \le \|A\|_\infty \le \sigma_{\max}(A)$.

xiii) $\frac{1}{m}\|A\|_{\mathrm{row}} \le \|A\|_{\mathrm{col}} \le n\|A\|_{\mathrm{row}}$.

xiv) $\frac{1}{\sqrt{m}}\sigma_{\max}(A) \le \|A\|_{\mathrm{col}} \le \sqrt{n}\sigma_{\max}(A)$.

xv) $\frac{1}{\sqrt{n}}\sigma_{\max}(A) \le \|A\|_{\mathrm{row}} \le \sqrt{m}\sigma_{\max}(A)$.

Proof: See [728, p. 314] and [1538].

Remark: See [699, p. 115] for matrices that attain these bounds.

Fact 9.8.13. Let $A \in \mathbb{F}^{n \times m}$, and assume that A is normal. Then,

$$\frac{1}{\sqrt{mn}} \sigma_{\max}(A) \le \|A\|_\infty \le \mathrm{sprad}(A) = \sigma_{\max}(A).$$

Proof: Use Fact 5.14.14 and statement $xii)$ of Fact 9.8.12.

Fact 9.8.14. Let $A \in \mathbb{R}^{n \times n}$, assume that A is symmetric, and assume that every diagonal entry of A is zero. Then, the following conditions are equivalent:

$i)$ For all $x \in \mathbb{R}^n$ such that $1_{1 \times n} x = 0$, it follows that $x^{\mathrm{T}} A x \le 0$.

$ii)$ There exists a positive integer k and vectors $x_1, \ldots, x_n \in \mathbb{R}^k$ such that, for all $i, j \in \{1, \ldots, n\}$, $A_{(i,j)} = \|x_i - x_j\|_2^2$.

Proof: See [20].

Remark: This result is due to Schoenberg.

Remark: A is a *Euclidean distance matrix*.

Fact 9.8.15. Let $A \in \mathbb{F}^{n \times n}$. Then,

$$\|A^{\mathrm{A}}\|_{\mathrm{F}} \le n^{(2-n)/2} \|A\|_{\mathrm{F}}^{n-1}.$$

Proof: See [1125, pp. 151, 165].

Fact 9.8.16. Let $A \in \mathbb{F}^{n \times n}$, let $\|\cdot\|$ and $\|\cdot\|'$ be norms on $\mathbb{F}^n$, and define the induced norms

$$\|A\|'' \triangleq \max_{x \in \{y \in \mathbb{F}^m: \|y\|=1\}} \|Ax\|$$

and

$$\|A\|''' \triangleq \max_{x \in \{y \in \mathbb{F}^m: \|y\|'=1\}} \|Ax\|'.$$

Then,

$$\max_{A \in \{X \in \mathbb{F}^{n \times n}: X \ne 0\}} \frac{\|A\|''}{\|A\|'''} = \max_{A \in \{X \in \mathbb{F}^{n \times n}: X \ne 0\}} \frac{\|A\|'''}{\|A\|''}$$

$$= \max_{x \in \{y \in \mathbb{F}^n: y \ne 0\}} \frac{\|x\|}{\|x\|'} \max_{x \in \{y \in \mathbb{F}^n: y \ne 0\}} \frac{\|x\|'}{\|x\|}.$$

Proof: See [728, p. 303].

Remark: This symmetry property is evident in Fact 9.8.12.

Fact 9.8.17. Let $A \in \mathbb{F}^{n \times m}$, let $q, r \in [1, \infty]$, assume that $1 \le q \le r$, define

$$p \triangleq \frac{1}{\frac{1}{q} - \frac{1}{r}},$$

and assume that $p \ge 2$. Then,

$$\|A\|_p \le \|A\|_{q,r}.$$

In particular,

$$\|A\|_\infty \le \|A\|_{\infty,\infty}.$$

Proof: See [489].

Remark: This result is due to Hardy and Littlewood.

Fact 9.8.18. Let $A \in \mathbb{R}^{n \times m}$. Then,

$$\left\| \begin{bmatrix} \|\mathrm{row}_1(A)\|_2 \\ \vdots \\ \|\mathrm{row}_n(A)\|_2 \end{bmatrix} \right\|_1 \le \sqrt{2}\|A\|_{1,\infty},$$

$$\left\| \begin{bmatrix} \|\mathrm{row}_1(A)\|_1 \\ \vdots \\ \|\mathrm{row}_n(A)\|_1 \end{bmatrix} \right\|_2 \le \sqrt{2}\|A\|_{1,\infty},$$

$$\|A\|_{4/3}^{3/4} \le \sqrt{2}\|A\|_{1,\infty}.$$

Proof: See [556, p. 303].

Remark: The first and third results are due to Littlewood, while the second result is due to Orlicz.

Fact 9.8.19. Let $A \in \mathbb{F}^{n \times n}$, and assume that A is positive semidefinite. Then,

$$\|A\|_{1,\infty} = \max_{x \in \{z \in \mathbb{F}^n: \|z\|_\infty = 1\}} x^* A x.$$

Remark: This result is due to Tao. See [699, p. 116] and [1166].

Fact 9.8.20. Let $A \in \mathbb{F}^{n \times n}$. If $p \in [1, 2]$, then

$$\|A\|_{\mathrm{F}} \le \|A\|_{\sigma p} \le n^{1/p - 1/2}\|A\|_{\mathrm{F}}.$$

If $p \in [2, \infty]$, then

$$\|A\|_{\sigma p} \le \|A\|_{\mathrm{F}} \le n^{1/2 - 1/p}\|A\|_{\sigma p}.$$

Proof: See [204, p. 174].

Fact 9.8.21. Let $A \in \mathbb{F}^{n \times n}$, and let $p, q \in [1, \infty]$. Then,

$$\|A\|_{p,p} \le \begin{cases} n^{1/p - 1/q}\|A\|_{q,q}, & p \le q, \\ n^{1/q - 1/p}\|A\|_{q,q}, & q \le p. \end{cases}$$

Consequently,

$$n^{1/p - 1}\|A\|_{\mathrm{col}} \le \|A\|_{p,p} \le n^{1 - 1/p}\|A\|_{\mathrm{col}},$$

$$n^{-|1/p - 1/2|}\sigma_{\max}(A) \le \|A\|_{p,p} \le n^{|1/p - 1/2|}\sigma_{\max}(A),$$

$$n^{-1/p}\|A\|_{\mathrm{col}} \le \|A\|_{p,p} \le n^{1/p}\|A\|_{\mathrm{row}}.$$

Proof: See [698] and [699, p. 112].

Remark: See Fact 9.7.29.

Problem: Extend these inequalities to nonsquare matrices.

Fact 9.8.22. Let $A \in \mathbb{F}^{n \times m}$, $p, q \in [1, \infty]$, and $\alpha \in [0, 1]$, and let $r \triangleq pq/[(1 - \alpha)p + \alpha q]$. Then,
$$\|A\|_{r,r} \leq \|A\|_{p,p}^{\alpha} \|A\|_{q,q}^{1-\alpha}.$$

Proof: See [698] or [699, p. 113].

Fact 9.8.23. Let $A \in \mathbb{F}^{n \times m}$, and let $p \in [1, \infty]$. Then,
$$\|A\|_{p,p} \leq \|A\|_{\mathrm{col}}^{1/p} \|A\|_{\mathrm{row}}^{1-1/p}.$$

In particular,
$$\sigma_{\max}(A) \leq \sqrt{\|A\|_{\mathrm{col}} \|A\|_{\mathrm{row}}}.$$

Proof: Set $\alpha = 1/p$, $p = 1$, and $q = \infty$ in Fact 9.8.22. See [699, p. 113]. To prove the special case $p = 2$ directly, note that $\lambda_{\max}(A^*A) \leq \|A^*A\|_{\mathrm{col}} \leq \|A^*\|_{\mathrm{col}} \|A\|_{\mathrm{col}} = \|A\|_{\mathrm{row}} \|A\|_{\mathrm{col}}$.

Fact 9.8.24. Let $A \in \mathbb{F}^{n \times m}$. Then,
$$\left.\begin{array}{c} \|A\|_{2,1} \\ \|A\|_{\infty,2} \end{array}\right\} \leq \sigma_{\max}(A).$$

Proof: This result follows from Proposition 9.1.5.

Fact 9.8.25. Let $A \in \mathbb{F}^{n \times m}$, and let $p \in [1, 2]$. Then,
$$\|A\|_{p,p} \leq \|A\|_{\mathrm{col}}^{2/p-1} \sigma_{\max}^{2-2/p}(A).$$

Proof: Let $\alpha = 2/p - 1$, $p = 1$, and $q = 2$ in Fact 9.8.22. See [699, p. 113].

Fact 9.8.26. Let $A \in \mathbb{F}^{n \times n}$, and let $p \in [1, \infty]$. Then,
$$\|A\|_{p,p} \leq \|\,|A|\,\|_{p,p} \leq n^{\min\{1/p, 1-1/p\}} \|A\|_{p,p} \leq \sqrt{n} \|A\|_{p,p}.$$

Remark: See [699, p. 117].

Fact 9.8.27. Let $A \in \mathbb{F}^{n \times m}$, and let $p, q \in [1, \infty]$. Then,
$$\|\overline{A}\|_{q,p} = \|A\|_{q,p}.$$

Fact 9.8.28. Let $A \in \mathbb{F}^{n \times m}$, and let $p, q \in [1, \infty]$. Then,
$$\|A^*\|_{q,p} = \|A\|_{p/(p-1), q/(q-1)}.$$

Fact 9.8.29. Let $A \in \mathbb{F}^{n \times m}$, and let $p, q \in [1, \infty]$. Then,
$$\|A\|_{q,p} \leq \begin{cases} \|A\|_{p/(p-1)}, & 1/p + 1/q \leq 1, \\ \|A\|_q, & 1/p + 1/q \geq 1. \end{cases}$$

Fact 9.8.30. Let $A \in \mathbb{F}^{n \times n}$, and let $\| \cdot \|$ be a unitarily invariant norm on $\mathbb{F}^{n \times n}$. Then,

$$\|\langle A \rangle\| = \|A\|.$$

Fact 9.8.31. Let $A, S \in \mathbb{F}^{n \times n}$, assume that S is nonsingular, and let $\| \cdot \|$ be a unitarily invariant norm on $\mathbb{F}^{n \times n}$. Then,

$$\|A\| \leq \tfrac{1}{2}\|SAS^{-1} + S^{-*}AS^*\|.$$

Proof: See [64, 250].

Fact 9.8.32. Let $A \in \mathbb{F}^{n \times n}$, assume that A is positive semidefinite, and let $\| \cdot \|$ be a submultiplicative norm on $\mathbb{F}^{n \times n}$. Then,

$$\|A\|^{1/2} \leq \|A^{1/2}\|.$$

In particular,

$$\sigma_{\max}^{1/2}(A) = \sigma_{\max}(A^{1/2}).$$

Fact 9.8.33. Let $A_{11} \in \mathbb{F}^{n \times n}$, $A_{12} \in \mathbb{F}^{n \times m}$, and $A_{22} \in \mathbb{F}^{m \times m}$, assume that $\begin{bmatrix} A_{11} & A_{12} \\ A_{12}^* & A_{22} \end{bmatrix} \in \mathbb{F}^{(n+m) \times (n+m)}$ is positive semidefinite, let $\| \cdot \|$ and $\| \cdot \|'$ be unitarily invariant norms on $\mathbb{F}^{n \times n}$ and $\mathbb{F}^{m \times m}$, respectively, and let $p > 0$. Then,

$$\|\langle A_{12} \rangle^p\|'^2 \leq \|A_{11}^p\| \, \|A_{22}^p\|'.$$

Proof: See [732].

Fact 9.8.34. Let $A \in \mathbb{F}^{n \times n}$, let $\| \cdot \|$ be a norm on $\mathbb{F}^n$, let $\| \cdot \|_{\mathrm{D}}$ denote the dual norm on $\mathbb{F}^n$, and let $\| \cdot \|'$ denote the norm induced by $\| \cdot \|$ on $\mathbb{F}^{n \times n}$. Then,

$$\|A\|' = \max_{\substack{x,y \in \mathbb{F}^n \\ x,y \neq 0}} \frac{\mathrm{Re}\, y^* A x}{\|y\|_{\mathrm{D}} \|x\|}.$$

Proof: See [699, p. 115].

Remark: See Fact 9.7.22 for the definition of the dual norm.

Problem: Generalize this result to obtain Fact 9.8.35 as a special case.

Fact 9.8.35. Let $A \in \mathbb{F}^{n \times m}$, and let $p, q \in [1, \infty]$. Then,

$$\|A\|_{q,p} = \max_{\substack{x \in \mathbb{F}^m, y \in \mathbb{F}^n \\ x,y \neq 0}} \frac{|y^* A x|}{\|y\|_{q/(q-1)} \|x\|_p}.$$

Fact 9.8.36. Let $A \in \mathbb{F}^{n \times m}$, and let $p, q \in [1, \infty]$ satisfy $1/p + 1/q = 1$. Then,

$$\|A\|_{p,p} = \max_{\substack{x \in \mathbb{F}^m, y \in \mathbb{F}^n \\ x,y \neq 0}} \frac{|y^* A x|}{\|y\|_q \|x\|_p} = \max_{\substack{x \in \mathbb{F}^m, y \in \mathbb{F}^n \\ x,y \neq 0}} \frac{|y^* A x|}{\|y\|_{p/(p-1)} \|x\|_p}.$$

Remark: See Fact 9.13.2 for the case $p = 2$.

Fact 9.8.37. Let $A \in \mathbb{F}^{n \times n}$, and assume that A is positive definite. Then,

$$\min_{x \in \mathbb{F}^n \setminus \{0\}} \frac{x^* A x}{\|Ax\|_2 \|x\|_2} = \frac{2\sqrt{\alpha\beta}}{\alpha + \beta}$$

and

$$\min_{\alpha \geq 0} \sigma_{\max}(\alpha A - I) = \frac{\alpha - \beta}{\alpha + \beta},$$

where $\alpha \triangleq \lambda_{\max}(A)$ and $\beta \triangleq \lambda_{\min}(A)$.

Proof: See [624].

Remark: These quantities are *antieigenvalues*.

Fact 9.8.38. Let $A \in \mathbb{F}^{n \times n}$, and define

$$\mathrm{nrad}(A) \triangleq \max\{|x^*Ax|: \ x \in \mathbb{C}^n \text{ and } x^*x \leq 1\}.$$

Then, the following statements hold:

i) $\mathrm{nrad}(A) = \max\{|z|: \ z \in \Theta(A)\}$.

ii) $\mathrm{sprad}(A) \leq \mathrm{nrad}(A) \leq \mathrm{nrad}(|A|) = \frac{1}{2}\mathrm{sprad}(|A| + |A|^{\mathrm{T}})$.

iii) $\frac{1}{2}\sigma_{\max}(A) \leq \mathrm{nrad}(A) \leq \frac{1}{2}\left[\sigma_{\max}(A) + \sigma_{\max}^{1/2}(A^2)\right] \leq \sigma_{\max}(A)$.

iv) If $A^2 = 0$, then $\mathrm{nrad}(A) = \sigma_{\max}(A)$.

v) If $\mathrm{nrad}(A) = \sigma_{\max}(A)$, then $\sigma_{\max}(A^2) = \sigma_{\max}^2(A)$.

vi) If A is normal, then $\mathrm{nrad}(A) = \mathrm{sprad}(A)$.

vii) $\mathrm{nrad}(A^k) \leq [\mathrm{nrad}(A)]^k$ for all $k \in \mathbb{N}$.

viii) $\mathrm{nrad}(\cdot)$ is a weakly unitarily invariant norm on $\mathbb{F}^{n \times n}$.

ix) $\mathrm{nrad}(\cdot)$ is not a submultiplicative norm on $\mathbb{F}^{n \times n}$.

x) $\|\cdot\| \triangleq \alpha\mathrm{nrad}(\cdot)$ is a submultiplicative norm on $\mathbb{F}^{n \times n}$ if and only if $\alpha \geq 4$.

xi) $\mathrm{nrad}(AB) \leq \mathrm{nrad}(A)\mathrm{nrad}(B)$ for all $A, B \in \mathbb{F}^{n \times n}$ such that A and B are normal.

xii) $\mathrm{nrad}(A \circ B) \leq \alpha\mathrm{nrad}(A)\mathrm{nrad}(B)$ for all $A, B \in \mathbb{F}^{n \times n}$ if and only if $\alpha \geq 2$.

xiii) $\mathrm{nrad}(A \oplus B) = \max\{\mathrm{nrad}(A), \mathrm{nrad}(B)\}$ for all $A \in \mathbb{F}^{n \times n}$ and $B \in \mathbb{F}^{m \times m}$.

Proof: See [728, p. 331] and [730, pp. 43, 44]. For *iii*), see [847].

Remark: $\mathrm{nrad}(A)$ is the *numerical radius* of A. $\Theta(A)$ is the numerical range. See Fact 8.14.7.

Remark: $\mathrm{nrad}(\cdot)$ is not submultiplicative. The example $A = \begin{bmatrix} 0 & 1 \\ 0 & 0 \end{bmatrix}$, $B = \begin{bmatrix} 0 & 2 \\ 2 & 0 \end{bmatrix}$, where B is normal, $\mathrm{nrad}(A) = 1/2$, $\mathrm{nrad}(B) = 2$, and $\mathrm{nrad}(AB) = 2$, shows that *xi*) is not valid if only one of the matrices A and B is normal, which corrects [730, pp. 43, 73].

Remark: *vii*) is the *power inequality*.

Fact 9.8.39. Let $A \in \mathbb{F}^{n \times m}$, let $\gamma > \sigma_{\max}(A)$, and define $\beta \triangleq \sigma_{\max}(A)/\gamma$. Then,

$$\|A\|_{\mathrm{F}} \leq \sqrt{-[\gamma^2/(2\pi)]\log\det(I - \gamma^{-2}A^*A)} \leq \beta^{-1}\sqrt{-\log(1 - \beta^2)}\|A\|_{\mathrm{F}}.$$

Proof: See [258].

Fact 9.8.40. Let $\| \cdot \|$ be a unitarily invariant norm on $\mathbb{F}^{n \times n}$. Then, $\|A\| = 1$ for all $A \in \mathbb{F}^{n \times n}$ such that rank $A = 1$ if and only if $\|E_{1,1}\| = 1$.

Proof: $\|A\| = \|E_{1,1}\|\sigma_{\max}(A)$.

Remark: These equivalent normalizations are used in [1261, p. 74] and [201], respectively.

Fact 9.8.41. Let $\| \cdot \|$ be a unitarily invariant norm on $\mathbb{F}^{n \times n}$. Then, the following statements are equivalent:

i) $\sigma_{\max}(A) \leq \|A\|$ for all $A \in \mathbb{F}^{n \times n}$.

ii) $\| \cdot \|$ is submultiplicative.

iii) $\|A^2\| \leq \|A\|^2$ for all $A \in \mathbb{F}^{n \times n}$.

iv) $\|A^k\| \leq \|A\|^k$ for all $k \geq 1$ and $A \in \mathbb{F}^{n \times n}$.

v) $\|A \circ B\| \leq \|A\|\|B\|$ for all $A, B \in \mathbb{F}^{n \times n}$.

vi) $\operatorname{sprad}(A) \leq \|A\|$ for all $A \in \mathbb{F}^{n \times n}$.

vii) $\|Ax\|_2 \leq \|A\|\|x\|_2$ for all $A \in \mathbb{F}^{n \times n}$ and $x \in \mathbb{F}^n$.

viii) $\|A\|_\infty \leq \|A\|$ for all $A \in \mathbb{F}^{n \times n}$.

ix) $\|E_{1,1}\| \geq 1$.

x) $\sigma_{\max}(A) \leq \|A\|$ for all $A \in \mathbb{F}^{n \times n}$ such that rank $A = 1$.

Proof: The equivalence of *i)*–*vii)* is given in [729] and [730, p. 211]. Since $\|A\| = \|E_{1,1}\|\sigma_{\max}(A)$ for all $A \in \mathbb{F}^{n \times n}$ such that rank $A = 1$, it follows that *vii)* and *viii)* are equivalent. To prove that *ix)* $\implies$ *x)*, let $A \in \mathbb{F}^{n \times n}$ satisfy rank $A = 1$. Then, $\|A\| = \sigma_{\max}(A)\|E_{1,1}\| \geq \sigma_{\max}(A)$. To show *x)* $\implies$ *ii)*, define $\| \cdot \|' \triangleq \|E_{1,1}\|^{-1}\| \cdot \|$. Since $\|E_{1,1}\|' = 1$, it follows from [201, p. 94] that $\| \cdot \|'$ is submultiplicative. Since $\|E_{1,1}\|^{-1} \leq 1$, it follows that $\| \cdot \|$ is also submultiplicative. Alternatively, $\|A\|' = \sigma_{\max}(A)$ for all $A \in \mathbb{F}^{n \times n}$ having rank 1. Then, Corollary 3.10 of [1261, p. 80] implies that $\| \cdot \|'$, and thus $\| \cdot \|$, is submultiplicative.

Fact 9.8.42. Let $\Phi \colon \mathbb{F}^n \mapsto [0, \infty)$ satisfy the following conditions:

i) If $x \neq 0$, then $\Phi(x) > 0$.

ii) $\Phi(\alpha x) = |\alpha|\Phi(x)$ for all $\alpha \in \mathbb{R}$.

iii) $\Phi(x + y) \leq \Phi(x) + \Phi(y)$ for all $x, y \in \mathbb{F}^n$.

iv) If $A \in \mathbb{F}^{n \times n}$ is a permutation matrix, then $\Phi(Ax) = \Phi(x)$ for all $x \in \mathbb{F}^n$.

v) $\Phi(|x|) = \Phi(x)$ for all $x \in \mathbb{F}^n$.

Furthermore, for $A \in \mathbb{F}^{n \times m}$, where $n \leq m$, define

$$\|A\| \triangleq \Phi[\sigma_1(A), \ldots, \sigma_n(A)].$$

Then, $\| \cdot \|$ is a unitarily invariant norm on $\mathbb{F}^{n \times m}$. Conversely, if $\| \cdot \|$ is a unitarily invariant norm on $\mathbb{F}^{n \times m}$, where $n \leq m$, then Φ: $\mathbb{F}^n \mapsto [0, \infty)$ defined by

$$\Phi(x) \triangleq \left\| \begin{bmatrix} x_{(1)} & \cdots & 0 & 0_{n \times (m-n)} \\ \vdots & \ddots & \vdots & \vdots \\ 0 & \cdots & x_{(n)} & 0_{n \times (m-n)} \end{bmatrix} \right\|$$

satisfies *i*)–*v*).

Proof: See [1261, pp. 75, 76].

Remark: Φ is a *symmetric gauge function*. This result is due to von Neumann. See Fact 2.21.13.

Fact 9.8.43. Let $\| \cdot \|$ and $\| \cdot \|'$ denote norms on $\mathbb{F}^m$ and $\mathbb{F}^n$, respectively, and define $\hat{\ell}$: $\mathbb{F}^{n \times m} \mapsto \mathbb{R}$ by

$$\hat{\ell}(A) \triangleq \min_{x \in \mathbb{F}^m \setminus \{0\}} \frac{\|Ax\|'}{\|x\|},$$

or, equivalently,

$$\hat{\ell}(A) \triangleq \min_{x \in \{y \in \mathbb{F}^m : \|y\| = 1\}} \|Ax\|'.$$

Then, for $A \in \mathbb{F}^{n \times m}$, the following statements hold:

i) $\hat{\ell}(A) \geq 0$.

ii) $\hat{\ell}(A) > 0$ if and only if rank $A = m$.

iii) $\hat{\ell}(A) = \ell(A)$ if and only if either $A = 0$ or rank $A = m$.

Proof: See [892, pp. 369, 370].

Remark: $\hat{\ell}$ is a weaker version of ℓ.

Fact 9.8.44. Let $\| \cdot \|$ and $\| \cdot \|'$ denote norms on $\mathbb{F}^m$ and $\mathbb{F}^n$, respectively, let $\| \cdot \|'''$ denote the norm induced by $\| \cdot \|'$ and $\| \cdot \|$, and define $\hat{\ell}$: $\mathbb{F}^{n \times m} \mapsto \mathbb{R}$ by

$$\hat{\ell}(A) \triangleq \min_{x \in \mathcal{R}(A^*) \setminus \{0\}} \frac{\|Ax\|'}{\|x\|}.$$

If A is nonzero, then

$$\frac{1}{\|A^+\|'''} \leq \hat{\ell}(A).$$

If, in addition, rank $A = m$, then

$$\frac{1}{\|A^+\|'''} = \hat{\ell}(A) = \ell(A).$$

Proof: See [1371].

Fact 9.8.45. Let $A \in \mathbb{F}^{n \times n}$, let $\| \cdot \|$ be a normalized, submultiplicative norm on $\mathbb{F}^{n \times n}$, and assume that $\|I - A\| < 1$. Then, A is nonsingular.

Remark: See Fact 9.9.56.

Fact 9.8.46. Let $\|\cdot\|$ be a normalized, submultiplicative norm on $\mathbb{F}^{n \times n}$. Then, $\|\cdot\|$ is equi-induced if and only if $\|A\| \leq \|A\|'$ for all $A \in \mathbb{F}^{n \times n}$ and for all normalized submultiplicative norms $\|\cdot\|'$ on $\mathbb{F}^{n \times n}$.

Proof: See [1265].

Remark: As shown in [316, 391], not every normalized submultiplicative norm on $\mathbb{F}^{n \times n}$ is equi-induced or induced.

9.9 Facts on Matrix Norms for Two or More Matrices

Fact 9.9.1. $\|\cdot\|'_\infty \triangleq n\|\cdot\|_\infty$ is submultiplicative on $\mathbb{F}^{n \times n}$.

Remark: It is not necessarily true that $\|AB\|_\infty \leq \|A\|_\infty \|B\|_\infty$. For example, let $A = B = \left[\begin{smallmatrix} 1 & 1 \\ 1 & 1 \end{smallmatrix}\right]$.

Fact 9.9.2. Let $A \in \mathbb{F}^{n \times m}$ and $B \in \mathbb{F}^{m \times l}$. Then,

$$\|AB\|_\infty \leq m\|A\|_\infty \|B\|_\infty.$$

Furthermore, if $A = 1_{n \times m}$ and $B = 1_{m \times l}$, then $\|AB\|_\infty = m\|A\|_\infty \|B\|_\infty$.

Fact 9.9.3. Let $A, B \in \mathbb{F}^{n \times n}$, and let $\|\cdot\|$ be a submultiplicative norm on $\mathbb{F}^{n \times n}$. Then, $\|AB\| \leq \|A\|\|B\|$. Hence, if $\|A\| \leq 1$ and $\|B\| \leq 1$, then $\|AB\| \leq 1$. Finally, if either $\|A\| < 1$ or $\|B\| < 1$, then $\|AB\| < 1$.

Remark: $\operatorname{sprad}(A) < 1$ and $\operatorname{sprad}(B) < 1$ do not imply that $\operatorname{sprad}(AB) < 1$. Let $A = B^{\mathrm{T}} = \left[\begin{smallmatrix} 0 & 2 \\ 0 & 0 \end{smallmatrix}\right]$.

Fact 9.9.4. Let $\|\cdot\|$ be a norm on $\mathbb{F}^{m \times m}$, and let

$$\delta > \sup\left\{ \frac{\|AB\|}{\|A\|\|B\|}: \ A, B \in \mathbb{F}^{m \times m}, A, B \neq 0 \right\}.$$

Then, $\|\cdot\|' \triangleq \delta\|\cdot\|$ is a submultiplicative norm on $\mathbb{F}^{m \times m}$.

Proof: See [728, p. 323].

Fact 9.9.5. Let $A, B \in \mathbb{F}^{n \times n}$, let $\|\cdot\|$ be a unitarily invariant norm on $\mathbb{F}^{n \times n}$, assume that A and B are Hermitian, and assume that $A \leq B$. Then,

$$\|A\| \leq \|B\|.$$

Proof: See [219].

Fact 9.9.6. Let $A, B \in \mathbb{F}^{n \times n}$, let $\|\cdot\|$ be a unitarily invariant norm on $\mathbb{F}^{n \times n}$, and assume that AB is normal. Then,

$$\|AB\| \leq \|BA\|.$$

Proof: See [201, p. 253].

Fact 9.9.7. Let $A, B \in \mathbb{F}^{n \times n}$, assume that A and B are positive semidefinite and nonzero, and let $\| \cdot \|$ be a submultiplicative unitarily invariant norm on $\mathbb{F}^{n \times n}$. Then,

$$\frac{\|AB\|}{\|A\|\|B\|} \leq \frac{\|A+B\|}{\|A\|+\|B\|}$$

and

$$\frac{\|A \circ B\|}{\|A\|\|B\|} \leq \frac{\|A+B\|}{\|A\|+\|B\|}.$$

Proof: See [693].

Remark: See Fact 9.8.41.

Fact 9.9.8. Let $A, B \in \mathbb{F}^{n \times n}$, and let $\| \cdot \|$ be a submultiplicative norm on $\mathbb{F}^{n \times n}$. Then, $\| \cdot \|' \triangleq 2\| \cdot \|$ is a submultiplicative norm on $\mathbb{F}^{n \times n}$ and satisfies

$$\|[A, B]\|' \leq \|A\|'\|B\|'.$$

Fact 9.9.9. Let $A \in \mathbb{F}^{n \times n}$. Then, the following statements are equivalent:

i) There exist projectors $Q, P \in \mathbb{R}^{n \times n}$ such that $A = [P, Q]$.

ii) $\sigma_{\max}(A) \leq 1/2$, A and $-A$ are unitarily similar, and A is skew Hermitian.

Proof: See [929].

Remark: Extensions are discussed in [1009].

Remark: See Fact 3.12.16 for the case of idempotent matrices.

Remark: In the case $\mathbb{F} = \mathbb{R}$, the condition that A is skew symmetric implies that A and $-A$ are orthogonally similar. See Fact 5.9.12.

Fact 9.9.10. Let $A, B \in \mathbb{F}^{n \times n}$, and let $\| \cdot \|$ be a unitarily invariant norm on $\mathbb{F}^{n \times n}$. Then,

$$\|AB\| \leq \sigma_{\max}(A)\|B\|$$

and

$$\|AB\| \leq \|A\|\sigma_{\max}(B).$$

Consequently, if $C \in \mathbb{F}^{n \times n}$, then

$$\|ABC\| \leq \sigma_{\max}(A)\|B\|\sigma_{\max}(C).$$

Proof: See [844].

Fact 9.9.11. Let $A, B \in \mathbb{F}^{n \times m}$, and let $\| \cdot \|$ be a unitarily invariant norm on $\mathbb{F}^{m \times m}$. If $p > 0$, then

$$\|\langle A^*B \rangle^p\|^2 \leq \|(A^*A)^p\|\|(B^*B)^p\|.$$

In particular,

$$\|(A^*BB^*A)^{1/4}\|^2 \leq \|\langle A \rangle\|\|\langle B \rangle\|$$

and

$$\|\langle A^*B \rangle\| = \|A^*B\|^2 \leq \|A^*A\|\|B^*B\|.$$

Furthermore,

$$\operatorname{tr} \langle A^*B \rangle \leq \|A\|_{\mathrm{F}}\|B\|_{\mathrm{F}}$$

and
$$\left[\operatorname{tr}\left(A^*BB^*A\right)^{1/4}\right]^2 \le (\operatorname{tr}\langle A\rangle)(\operatorname{tr}\langle B\rangle).$$

Proof: See [732] and use Fact 9.8.30.

Problem: Noting Fact 9.12.1 and Fact 9.12.2, compare the lower bounds for $\|A\|_{\mathrm{F}}\|B\|_{\mathrm{F}}$ given by

$$\left.\begin{array}{r}\operatorname{tr}\langle A^*B\rangle \\ |\operatorname{tr} A^*B| \\ \sqrt{|\operatorname{tr}(A^*B)^2|} \le \sqrt{\operatorname{tr} AA^*BB^*}\end{array}\right\} \le \|A\|_{\mathrm{F}}\|B\|_{\mathrm{F}}.$$

Fact 9.9.12. Let $A, B \in \mathbb{F}^{n\times n}$, and assume that A and B are positive semidefinite. Then,

$$\begin{aligned}(2\|A\|_{\mathrm{F}}\|B\|_{\mathrm{F}})^{1/2} &\le \left(\|A\|_{\mathrm{F}}^2 + \|B\|_{\mathrm{F}}^2\right)^{1/2} \\ &= \|\left(A^2 + B^2\right)^{1/2}\|_{\mathrm{F}} \\ &\le \|A+B\|_{\mathrm{F}} \\ &\le \sqrt{2}\left(\|A\|_{\mathrm{F}}^2 + \|B\|_{\mathrm{F}}^2\right)^{1/2}.\end{aligned}$$

Fact 9.9.13. Let $A, B \in \mathbb{F}^{n\times m}$. Then,

$$\|A+B\|_{\mathrm{F}} = \sqrt{\|A\|_{\mathrm{F}}^2 + \|B\|_{\mathrm{F}}^2 + 2\operatorname{tr} AB^*} \le \|A\|_{\mathrm{F}} + \|B\|_{\mathrm{F}}.$$

In particular,

$$\|A-B\|_{\mathrm{F}} = \sqrt{\|A\|_{\mathrm{F}}^2 + \|B\|_{\mathrm{F}}^2 - 2\operatorname{tr} AB^*}.$$

If, in addition, A is Hermitian and B is skew Hermitian, then $\operatorname{tr} AB^* = 0$, and thus

$$\|A+B\|_{\mathrm{F}}^2 = \|A-B\|_{\mathrm{F}}^2 = \|A\|_{\mathrm{F}}^2 + \|B\|_{\mathrm{F}}^2.$$

Remark: The second equality is a matrix version of the cosine law given by $ix)$ of Fact 9.7.4.

Fact 9.9.14. Let $A, B \in \mathbb{F}^{n\times n}$, and let $\|\cdot\|$ be a unitarily invariant norm on $\mathbb{F}^{n\times n}$. Then,
$$\|AB\| \le \tfrac{1}{4}\|(\langle A\rangle + \langle B^*\rangle)^2\|.$$

Proof: See [216].

Fact 9.9.15. Let $A, B \in \mathbb{F}^{n\times n}$, assume that A and B are positive semidefinite, and let $\|\cdot\|$ be a unitarily invariant norm on $\mathbb{F}^{n\times n}$. Then,
$$\|AB\| \le \tfrac{1}{4}\|(A+B)^2\|.$$

Proof: See [216] or [1521, p. 77].

Problem: Noting Fact 9.9.12, compare the lower bounds for $\|A+B\|_{\mathrm{F}}$ given by

$$(2\|A\|_{\mathrm{F}}\|B\|_{\mathrm{F}})^{1/2} \le \|\left(A^2+B^2\right)^{1/2}\|_{\mathrm{F}} \le \|A+B\|_{\mathrm{F}}$$

and

$$2\|AB\|_{\mathrm{F}}^{1/2} \le \|(A+B)^2\|_{\mathrm{F}}^{1/2} \le \|A+B\|_{\mathrm{F}}.$$

Fact 9.9.16. Let $A, B \in \mathbb{F}^{n \times n}$, assume that A and B are positive semidefinite, let $\| \cdot \|$ be a unitarily invariant norm on $\mathbb{F}^{n \times n}$, and let $p \in [0, \infty)$. If $p \in [0, 1]$, then

$$\|A^p B^p\| \le \|AB\|^p.$$

If $p \in [1, \infty)$, then

$$\|AB\|^p \le \|A^p B^p\|.$$

Proof: See [207, 537].

Remark: See Fact 8.19.27.

Fact 9.9.17. Let $A, B \in \mathbb{F}^{n \times n}$, assume that A and B are positive semidefinite, and let $\| \cdot \|$ be a unitarily invariant norm on $\mathbb{F}^{n \times n}$. If $p \in [0, 1]$, then

$$\|B^p A^p B^p\| \le \|(BAB)^p\|.$$

Furthermore, if $p \ge 1$, then

$$\|(BAB)^p\| \le \|B^p A^p B^p\|.$$

Proof: See [72] and [201, p. 258].

Remark: Extensions and a reverse inequality are given in Fact 8.10.50.

Remark: See Fact 8.12.22 and Fact 8.19.27.

Fact 9.9.18. Let $A, B \in \mathbb{F}^{n \times n}$, assume that A and B are positive semidefinite, and let $\| \cdot \|$ be a unitarily invariant norm on $\mathbb{F}^{n \times n}$. Then,

$$\|A^{1/2} B^{1/2}\| \le \tfrac{1}{2}\|A + B\|.$$

Hence,

$$\|AB\| \le \tfrac{1}{2}\|A^2 + B^2\|,$$

and thus

$$\|(A + B)^2\| \le 2\|A^2 + B^2\|.$$

Consequently,

$$\|AB\| \le \tfrac{1}{4}\|(A + B)^2\| \le \tfrac{1}{2}\|A^2 + B^2\|.$$

Proof: Let $p = 1/2$ and $X = I$ in Fact 9.9.49. The last inequality follows from Fact 9.9.15.

Remark: See Fact 8.19.13.

Fact 9.9.19. Let $A, B \in \mathbb{F}^{n \times n}$, assume that A and B are positive semidefinite, and let either $p = 1$ or $p \in [2, \infty]$. Then,

$$\|\langle AB \rangle^{1/2}\|_{\sigma p} \le \tfrac{1}{2}\|A + B\|_{\sigma p}.$$

Proof: See [93, 216].

Remark: The inequality holds for all Q-norms. See [201].

Remark: See Fact 8.19.13.

Fact 9.9.20. Let $A \in \mathbb{F}^{n \times m}$, $B \in \mathbb{F}^{m \times l}$, and $p, q, q', r \in [1, \infty]$, and assume that $1/q + 1/q' = 1$. Then,

$$\|AB\|_p \le \varepsilon_{pq}(n) \varepsilon_{pr}(l) \varepsilon_{q'r}(m) \|A\|_q \|B\|_r,$$

where

$$\varepsilon_{pq}(n) \triangleq \begin{cases} 1, & p \ge q, \\ n^{1/p - 1/q}, & q \ge p. \end{cases}$$

Furthermore, there exist matrices $A \in \mathbb{F}^{n \times m}$ and $B \in \mathbb{F}^{m \times l}$ such that equality holds.

Proof: See [578].

Remark: Related results are given in [488, 489, 578, 579, 580, 852, 1345].

Fact 9.9.21. Let $A, B \in \mathbb{C}^{n \times m}$. Then, there exist unitary matrices $S_1, S_2 \in \mathbb{C}^{m \times m}$ such that
$$\langle A + B \rangle \le S_1 \langle A \rangle S_1^* + S_2 \langle B \rangle S_2^*.$$

Remark: This result is a matrix version of the triangle inequality. See [49, 1302].

Fact 9.9.22. Let $A, B \in \mathbb{F}^{n \times n}$, assume that A and B are positive semidefinite, and let $p \in [1, \infty]$. Then,
$$\|A - B\|_{\sigma 2p}^2 \le \|A^2 - B^2\|_{\sigma p}.$$

Proof: See [837].

Remark: The case $p = 1$ is due to Powers and Stormer.

Fact 9.9.23. Let $A, B \in \mathbb{F}^{n \times n}$, and let $p \in [1, \infty]$. Then,
$$\|\langle A \rangle - \langle B \rangle\|_{\sigma p}^2 \le \|A + B\|_{\sigma 2p} \|A - B\|_{\sigma 2p}.$$

Proof: See [851].

Fact 9.9.24. Let $A, B \in \mathbb{F}^{n \times n}$. Then,
$$\|\langle A \rangle - \langle B \rangle\|_{\sigma 1}^2 \le 2\|A + B\|_{\sigma 1} \|A - B\|_{\sigma 1}.$$

Proof: See [851].

Remark: This result is due to Borchers and Kosaki. See [851].

Fact 9.9.25. Let $A, B \in \mathbb{F}^{n \times n}$. Then,
$$\|\langle A \rangle - \langle B \rangle\|_{\mathrm{F}} \le \sqrt{2}\|A - B\|_{\mathrm{F}}$$

and

$$\|\langle A \rangle - \langle B \rangle\|_{\mathrm{F}}^2 + \|\langle A^* \rangle - \langle B^* \rangle\|_{\mathrm{F}}^2 \le 2\|A - B\|_{\mathrm{F}}^2.$$

If, in addition, A and B are normal, then
$$\|\langle A \rangle - \langle B \rangle\|_{\mathrm{F}} \le \|A - B\|_{\mathrm{F}}.$$

Proof: See [49, 73, 836, 851] and [701, pp. 217, 218].

Fact 9.9.26. Let $A, B \in \mathbb{R}^{n \times n}$. Then,

$$\|AB - BA\|_{\mathrm{F}} \le \sqrt{2}\|A\|_{\mathrm{F}}\|B\|_{\mathrm{F}}.$$

Proof: See [246, 1419].

Remark: The constant $\sqrt{2}$ holds for all n.

Remark: Extensions to complex matrices are given in [247].

Fact 9.9.27. Let $A, B \in \mathbb{F}^{n \times n}$, and assume that A and B are positive semidefinite. Then,

$$\|AB - BA\|_{\mathrm{F}}^2 + \|(A - B)^2\|_{\mathrm{F}}^2 \le \|A^2 - B^2\|_{\mathrm{F}}^2.$$

Proof: See [844].

Fact 9.9.28. Let $A, B \in \mathbb{F}^{n \times n}$, let p be a positive number, and assume that either A is normal and $p \in [2, \infty]$, or A is Hermitian and $p \ge 1$. Then,

$$\|\langle A \rangle B - B \langle A \rangle\|_{\sigma p} \le \|AB - BA\|_{\sigma p}.$$

Proof: See [1].

Fact 9.9.29. Let $\|\cdot\|$ be a unitarily invariant norm on $\mathbb{F}^{n \times n}$, and let $A, X, B \in \mathbb{F}^{n \times n}$. Then,

$$\|AX - XB\| \le [\sigma_{\max}(A) + \sigma_{\max}(B)]\|X\|.$$

In particular,

$$\sigma_{\max}(AX - XA) \le 2\sigma_{\max}(A)\sigma_{\max}(X).$$

Now, assume in addition that A and B are positive semidefinite. Then,

$$\|AX - XB\| \le \max\{\sigma_{\max}(A), \sigma_{\max}(B)\}\|X\|.$$

In particular,

$$\sigma_{\max}(AX - XA) \le \sigma_{\max}(A)\sigma_{\max}(X).$$

Finally, assume that A and X are positive semidefinite. Then,

$$\|AX - XA\| \le \tfrac{1}{2}\sigma_{\max}(A)\left\|\begin{bmatrix} X & 0 \\ 0 & X \end{bmatrix}\right\|.$$

In particular,

$$\sigma_{\max}(AX - XA) \le \tfrac{1}{2}\sigma_{\max}(A)\sigma_{\max}(X).$$

Proof: See [218].

Remark: The first inequality is sharp since equality holds for $A = B = \begin{bmatrix} 1 & 0 \\ 0 & -1 \end{bmatrix}$ and $X = \begin{bmatrix} 0 & 1 \\ -1 & 0 \end{bmatrix}$.

Remark: $\|\cdot\|$ can be extended to $\mathbb{F}^{2n \times 2n}$ by considering the n largest singular values of matrices in $\mathbb{F}^{2n \times 2n}$. For details, see [201, pp. 90, 98].

Fact 9.9.30. Let $\|\cdot\|$ be a unitarily invariant norm on $\mathbb{F}^{n \times n}$, let $A, X \in \mathbb{F}^{n \times n}$, and assume that A is Hermitian. Then,

$$\|AX - XA\| \le [\lambda_{\max}(A) - \lambda_{\min}(A)]\|X\|.$$

Proof: See [218].

Remark: $\lambda_{\max}(A) - \lambda_{\min}(A)$ is the spread of A. See Fact 8.15.32 and Fact 9.9.31.

Fact 9.9.31. Let $\|\cdot\|$ be a unitarily invariant norm on $\mathbb{F}^{n \times n}$, let $A, X \in \mathbb{F}^{n \times n}$, assume that A is normal, let $\mathrm{spec}(A) = \{\lambda_1, \ldots, \lambda_r\}$, and define

$$\mathrm{spd}(A) \triangleq \max\{|\lambda_i(A) - \lambda_j(A)| \colon i, j \in \{1, \ldots, r\}\}.$$

Then,

$$\|AX - XA\| \leq \sqrt{2}\,\mathrm{spd}(A)\|X\|.$$

Furthermore, let $p \in [1, \infty]$. Then,

$$\|AX - XA\|_{\sigma p} \leq 2^{|2-p|/(2p)}\mathrm{spd}(A)\|X\|_{\sigma p}.$$

In particular,

$$\|AX - XA\|_{\mathrm{F}} \leq \mathrm{spd}(A)\|X\|_{\mathrm{F}}$$

and

$$\sigma_{\max}(AX - XA) \leq \sqrt{2}\,\mathrm{spd}(A)\sigma_{\max}(X).$$

Proof: See [218].

Remark: $\mathrm{spd}(A)$ is the spread of A. See Fact 8.15.32 and Fact 9.9.30.

Fact 9.9.32. Let $A, B \in \mathbb{F}^{n \times n}$. Then,

$$\sigma_{\max}(\langle A \rangle - \langle B \rangle) \leq \tfrac{2}{\pi}\left[2 + \log \frac{\sigma_{\max}(A) + \sigma_{\max}(B)}{\sigma_{\max}(A - B)}\right]\sigma_{\max}(A - B).$$

Remark: This result is due to Kato. See [851].

Fact 9.9.33. Let $A \in \mathbb{F}^{n \times m}$ and $B \in \mathbb{F}^{m \times l}$, and let $r = 1$ or $r = 2$. Then,

$$\|AB\|_{\sigma r} = \|A\|_{\sigma 2r}\|B\|_{\sigma 2r}$$

if and only if there exists $\alpha \geq 0$ such that $AA^* = \alpha B^*B$. Furthermore,

$$\|AB\|_\infty = \|A\|_\infty\|B\|_\infty$$

if and only if AA^* and B^*B have a common eigenvector associated with $\lambda_1(AA^*)$ and $\lambda_1(B^*B)$.

Proof: See [1476].

Fact 9.9.34. Let $A, B \in \mathbb{F}^{n \times n}$. If $p \in (0, 2]$, then

$$2^{p-1}(\|A\|_{\sigma p}^p + \|B\|_{\sigma p}^p) \leq \|A + B\|_{\sigma p}^p + \|A - B\|_{\sigma p}^p \leq 2(\|A\|_{\sigma p}^p + \|B\|_{\sigma p}^p).$$

If $p \in [2, \infty)$, then

$$2(\|A\|_{\sigma p}^p + \|B\|_{\sigma p}^p) \leq \|A + B\|_{\sigma p}^p + \|A - B\|_{\sigma p}^p \leq 2^{p-1}(\|A\|_{\sigma p}^p + \|B\|_{\sigma p}^p).$$

If $p \in (1, 2]$ and $1/p + 1/q = 1$, then

$$\|A + B\|_{\sigma p}^q + \|A - B\|_{\sigma p}^q \leq 2(\|A\|_{\sigma p}^p + \|B\|_{\sigma p}^p)^{q/p}.$$

If $p \in [2, \infty)$ and $1/p + 1/q = 1$, then

$$2(\|A\|_{\sigma p}^p + \|B\|_{\sigma p}^p)^{q/p} \leq \|A + B\|_{\sigma p}^q + \|A - B\|_{\sigma p}^q.$$

Proof: See [714].

Remark: These inequalities are versions of the *Clarkson inequalities*. See Fact 1.20.2.

Remark: See [714] for extensions to unitarily invariant norms. See [217] for additional extensions.

Fact 9.9.35. Let $A, B \in \mathbb{C}^{n \times m}$. If $p \in [1, 2]$, then

$$[\|A\|^2 + (p-1)\|B\|^2]^{1/2} \le [\tfrac{1}{2}(\|A+B\|^p + \|A-B\|^p)]^{1/p}.$$

If $p \in [2, \infty]$, then

$$[\tfrac{1}{2}(\|A+B\|^p + \|A-B\|^p)]^{1/p} \le [\|A\|^2 + (p-1)\|B\|^2]^{1/2}.$$

Proof: See [120, 168].

Remark: This result is *Beckner's two-point inequality* or *optimal 2-uniform convexity*.

Fact 9.9.36. Let $A, B \in \mathbb{F}^{n \times n}$. If either $p \in [1, 4/3]$ or both $p \in (4/3, 2]$ and $A + B$ and $A - B$ are positive semidefinite, then

$$(\|A\|_{\sigma p} + \|B\|_{\sigma p})^p + |\|A\|_{\sigma p} - \|B\|_{\sigma p}|^p \le \|A+B\|_{\sigma p}^p + \|A-B\|_{\sigma p}^p.$$

Furthermore, if either $p \in [4, \infty]$ or both $p \in [2, 4)$ and A and B are positive semidefinite, then

$$\|A+B\|_{\sigma p}^p + \|A-B\|_{\sigma p}^p \le (\|A\|_{\sigma p} + \|B\|_{\sigma p})^p + |\|A\|_{\sigma p} - \|B\|_{\sigma p}|^p.$$

Proof: See [120, 834].

Remark: These inequalities are versions of *Hanner's inequality*.

Remark: Vector versions are given in Fact 9.7.21.

Fact 9.9.37. Let $A, B \in \mathbb{C}^{n \times n}$, and assume that A and B are Hermitian. If $p \in [1, 2]$, then

$$2^{1/2 - 1/p} \|(A^2 + B^2)^{1/2}\|_p \le \|A + \jmath B\|_{\sigma p} \le \|(A^2 + B^2)^{1/2}\|_p$$

and

$$2^{1-2/p}(\|A\|_{\sigma p}^2 + \|B\|_{\sigma p}^2) \le \|A + \jmath B\|_{\sigma p}^2 \le 2^{2/p-1}(\|A\|_{\sigma p}^2 + \|B\|_{\sigma p}^2).$$

Furthermore, if $p \in [2, \infty)$, then

$$\|(A^2 + B^2)^{1/2}\|_p \le \|A + \jmath B\|_{\sigma p} \le 2^{1/2 - 1/p} \|(A^2 + B^2)^{1/2}\|_p$$

and

$$2^{2/p-1}(\|A\|_{\sigma p}^2 + \|B\|_{\sigma p}^2) \le \|A + \jmath B\|_{\sigma p}^2 \le 2^{1-2/p}(\|A\|_{\sigma p}^2 + \|B\|_{\sigma p}^2).$$

Proof: See [215].

Fact 9.9.38. Let $A, B \in \mathbb{C}^{n \times n}$, and assume that A and B are Hermitian. If $p \in [1, 2]$, then

$$2^{1-2/p}(\|A\|_{\sigma p}^p + \|B\|_{\sigma p}^p) \le \|A + \jmath B\|_{\sigma p}^p.$$

If $p \in [2, \infty]$, then

$$\|A + \jmath B\|_{\sigma p}^p \le 2^{1-2/p}(\|A\|_{\sigma p}^p + \|B\|_{\sigma p}^p).$$

In particular,

$$\|A + _{J}B\|_{\mathrm{F}}^2 = \|A\|_{\mathrm{F}}^2 + \|B\|_{\mathrm{F}}^2 = \|(A^2 + B^2)^{1/2}\|_{\mathrm{F}}^2.$$

Proof: See [215, 223].

Fact 9.9.39. Let $A, B \in \mathbb{C}^{n \times n}$, and assume that A is positive semidefinite and B is Hermitian. If $p \in [1, 2]$, then

$$\|A\|_{\sigma p}^2 + 2^{1-2/p}\|B\|_{\sigma p}^2 \leq \|A + _{J}B\|_{\sigma p}^2.$$

If $p \in [2, \infty]$, then

$$\|A + _{J}B\|_{\sigma p}^2 \leq \|A\|_{\sigma p}^2 + 2^{1-2/p}\|B\|_{\sigma p}^2.$$

In particular,

$$\|A\|_{\sigma 1}^2 + \tfrac{1}{2}\|B\|_{\sigma 1}^2 \leq \|A + _{J}B\|_{\sigma 1}^2,$$

$$\|A + _{J}B\|_{\mathrm{F}}^2 = \|A\|_{\mathrm{F}}^2 + \|B\|_{\mathrm{F}}^2,$$

and

$$\sigma_{\max}^2(A + _{J}B) \leq \sigma_{\max}^2(A) + 2\sigma_{\max}^2(B).$$

In fact,

$$\|A\|_{\sigma 1}^2 + \|B\|_{\sigma 1}^2 \leq \|A + _{J}B\|_{\sigma 1}^2.$$

Proof: See [223].

Fact 9.9.40. Let $A, B \in \mathbb{C}^{n \times n}$, and assume that A and B are positive semidefinite. If $p \in [1, 2]$, then

$$\|A\|_{\sigma p}^2 + \|B\|_{\sigma p}^2 \leq \|A + _{J}B\|_{\sigma p}^2.$$

If $p \in [2, \infty]$, then

$$\|A + _{J}B\|_{\sigma p}^2 \leq \|A\|_{\sigma p}^2 + \|B\|_{\sigma p}^2.$$

Hence,

$$\|A\|_{\sigma 2}^2 + \|B\|_{\sigma 2}^2 = \|A + _{J}B\|_{\sigma 2}^2.$$

In particular,

$$(\operatorname{tr} \langle A \rangle)^2 + \langle B \rangle)^2 \leq (\operatorname{tr} \langle A + _{J}B \rangle)^2,$$

$$\sigma_{\max}^2(A + _{J}B) \leq \sigma_{\max}^2(A) + \sigma_{\max}^2(A),$$

$$\|A + _{J}B\|_{\mathrm{F}}^2 = \|A\|_{\mathrm{F}}^2 + \|B\|_{\mathrm{F}}^2.$$

Proof: See [223].

Remark: See Fact 8.19.7.

Fact 9.9.41. Let $A \in \mathbb{F}^{n \times n}$, let $B \in \mathbb{F}^{n \times n}$, assume that B is Hermitian, and let $\| \cdot \|$ be a unitarily invariant norm on $\mathbb{F}^{n \times n}$. Then,

$$\|A - \tfrac{1}{2}(A + A^*)\| \leq \|A - B\|.$$

In particular,

$$\|A - \tfrac{1}{2}(A + A^*)\|_{\mathrm{F}} \leq \|A - B\|_{\mathrm{F}}$$

and
$$\sigma_{\max}\left[A - \tfrac{1}{2}(A + A^*)\right] \leq \sigma_{\max}(A - B).$$

Proof: See [201, p. 275] and [1125, p. 150].

Fact 9.9.42. Let $A, M, S, B \in \mathbb{F}^{n \times n}$, assume that $A = MS$, M is positive semidefinite, and S and B are unitary, and let $\| \cdot \|$ be a unitarily invariant norm on $\mathbb{F}^{n \times n}$. Then,
$$\|A - S\| \leq \|A - B\|.$$

In particular,
$$\|A - S\|_{\mathrm{F}} \leq \|A - B\|_{\mathrm{F}}.$$

Proof: See [201, p. 276] and [1125, p. 150].

Remark: $A = MS$ is the polar decomposition of A. See Corollary 5.6.4.

Fact 9.9.43. Let $A, B \in \mathbb{F}^{n \times n}$, assume that A and B are Hermitian, let $\| \cdot \|$ be a unitarily invariant norm on $\mathbb{F}^{n \times n}$, and let $k \in \mathbb{N}$. Then,
$$\|(A - B)^{2k+1}\| \leq 2^{2k}\|A^{2k+1} - B^{2k+1}\|.$$

Proof: See [201, p. 294] or [781].

Fact 9.9.44. Let $A, B \in \mathbb{F}^{n \times n}$, and let $\| \cdot \|$ be a unitarily invariant norm on $\mathbb{F}^{n \times n}$. Then,
$$\|\langle A \rangle - \langle B \rangle\| \leq \sqrt{2\|A + B\|\|A - B\|}.$$

Proof: See [49].

Remark: This result is due to Kosaki and Bhatia.

Fact 9.9.45. Let $A, B \in \mathbb{F}^{n \times n}$, and let $p \geq 1$. Then,
$$\|\langle A \rangle - \langle B \rangle\|_{\sigma p} \leq \max\left\{2^{1/p - 1/2}, 1\right\}\sqrt{\|A + B\|_{\sigma p}\|A - B\|_{\sigma p}}.$$

Proof: See [49].

Remark: This result is due to Kittaneh, Kosaki, and Bhatia.

Fact 9.9.46. Let $A, B \in \mathbb{F}^{n \times n}$, assume that A and B are positive semidefinite, and let $\| \cdot \|$ be a unitarily invariant norm on $\mathbb{F}^{2n \times 2n}$. Then,
$$\left\| \begin{bmatrix} A + B & 0 \\ 0 & 0 \end{bmatrix} \right\| \leq \left\| \begin{bmatrix} A & 0 \\ 0 & B \end{bmatrix} \right\| + \left\| \begin{bmatrix} A^{1/2}B^{1/2} & 0 \\ 0 & A^{1/2}B^{1/2} \end{bmatrix} \right\|.$$

In particular,
$$\sigma_{\max}(A + B) \leq \max\{\sigma_{\max}(A), \sigma_{\max}(B)\} + \sigma_{\max}(A^{1/2}B^{1/2})$$

and, for all $p \in [1, \infty)$,
$$\|A + B\|_{\sigma p} \leq \left(\|A\|_{\sigma p}^p + \|B\|_{\sigma p}^p\right)^{1/p} + 2^{1/p}\|A^{1/2}B^{1/2}\|_{\sigma p}.$$

Proof: See [842, 845, 849].

Remark: See Fact 9.14.15 for a tighter upper bound for $\sigma_{\max}(A + B)$.

Fact 9.9.47. Let $A, X, B \in \mathbb{F}^{n \times n}$, and let $\| \cdot \|$ be a unitarily invariant norm on $\mathbb{F}^{n \times n}$. Then,

$$\|A^*XB\| \le \tfrac{1}{2}\|AA^*X + XBB^*\|.$$

In particular,

$$\|A^*B\| \le \tfrac{1}{2}\|AA^* + BB^*\|.$$

Proof: See [64, 206, 213, 539, 839].

Remark: The first result is *McIntosh's inequality*.

Remark: See Fact 9.14.23.

Fact 9.9.48. Let $A, X, B \in \mathbb{F}^{n \times n}$, assume that X is positive semidefinite, and let $\| \cdot \|$ be a unitarily invariant norm on $\mathbb{F}^{n \times n}$. Then,

$$\|A^*XB + B^*XA\| \le \|A^*XA + B^*XB\|.$$

In particular,

$$\|A^*B + B^*A\| \le \|A^*A + B^*B\|.$$

Proof: See [843].

Remark: See [843] for extensions to the case in which X is not necessarily positive semidefinite.

Fact 9.9.49. Let $A, X, B \in \mathbb{F}^{n \times n}$, assume that A and B are positive semidefinite, let $p \in [0, 1]$, and let $\| \cdot \|$ be a unitarily invariant norm on $\mathbb{F}^{n \times n}$. Then,

$$\|A^pXB^{1-p} + A^{1-p}XB^p\| \le \|AX + XB\|$$

and

$$\|A^pXB^{1-p} - A^{1-p}XB^p\| \le |2p - 1|\|AX - XB\|.$$

Proof: See [64, 207, 220, 523].

Remark: These results are the *Heinz inequalities*.

Fact 9.9.50. Let $A, B \in \mathbb{F}^{n \times n}$, assume that A is nonsingular and B is Hermitian, and let $\| \cdot \|$ be a unitarily invariant norm on $\mathbb{F}^{n \times n}$. Then,

$$\|B\| \le \tfrac{1}{2}\|ABA^{-1} + A^{-1}BA\|.$$

Proof: See [355, 530].

Fact 9.9.51. Let $A, B \in \mathbb{F}^{n \times n}$, assume that A and B are positive semidefinite, and let $\| \cdot \|$ be a unitarily invariant norm on $\mathbb{F}^{n \times n}$. If $r \in [0, 1]$, then

$$\|A^r - B^r\| \le \|\langle A - B\rangle^r\|.$$

Furthermore, if $r \in [1, \infty)$, then

$$\|\langle A - B\rangle^r\| \le \|A^r - B^r\|.$$

In particular,

$$\|(A - B)^2\| \le \|A^2 - B^2\|.$$

Proof: See [201, pp. 293, 294] and [844].

Fact 9.9.52. Let $A, B \in \mathbb{F}^{n \times n}$, assume that A and B are positive semidefinite, let $\| \cdot \|$ be a unitarily invariant norm on $\mathbb{F}^{n \times n}$, and let $z \in \mathbb{F}$. Then,

$$\|A - |z|B\| \leq \|A + zB\| \leq \|A + |z|B\|.$$

In particular,

$$\|A - B\| \leq \|A + B\|.$$

Proof: See [214].

Remark: Extensions to weak log majorization are given in [1519].

Remark: The special case $z = 1$ is given in [219].

Fact 9.9.53. Let $A, B \in \mathbb{F}^{n \times n}$, assume that A and B are positive semidefinite, and let $\| \cdot \|$ be a unitarily invariant norm on $\mathbb{F}^{n \times n}$. If $r \in [0, 1]$, then

$$\|(A + B)^r\| \leq \|A^r + B^r\|.$$

Furthermore, if $r \in [1, \infty)$, then

$$\|A^r + B^r\| \leq \|(A + B)^r\|.$$

In particular, if k is a positive integer, then

$$\|A^k + B^k\| \leq \|(A + B)^k\|.$$

Proof: See [60].

Fact 9.9.54. Let $A, B \in \mathbb{F}^{n \times n}$, assume that A and B are positive semidefinite, and let $\| \cdot \|$ be a unitarily invariant norm on $\mathbb{F}^{n \times n}$. Then,

$$\|\log(I + A) - \log(I + B)\| \leq \|\log(I + \langle A - B \rangle)\|$$

and

$$\|\log(I + A + B)\| \leq \|\log(I + A) + \log(I + B)\|.$$

Proof: See [60] and [201, p. 293].

Remark: See Fact 11.16.16.

Fact 9.9.55. Let $A, X, B \in \mathbb{F}^{n \times n}$, assume that A and B are positive definite, and let $\| \cdot \|$ be a unitarily invariant norm on $\mathbb{F}^{n \times n}$. Then,

$$\|(\log A)X - X(\log B)\| \leq \|A^{1/2}XB^{-1/2} - A^{-1/2}XB^{1/2}\|.$$

Proof: See [220].

Remark: See Fact 11.16.17.

Fact 9.9.56. Let $A, B \in \mathbb{F}^{n \times n}$, assume that A is nonsingular, let $\| \cdot \|$ be a normalized submultiplicative norm on $\mathbb{F}^{n \times n}$, and assume that $\|A - B\| < 1/\|A^{-1}\|$. Then, B is nonsingular.

Remark: See Fact 9.8.45.

Fact 9.9.57. Let $A, B \in \mathbb{F}^{n \times n}$, assume that A is nonsingular, let $\|\cdot\|$ be a normalized submultiplicative norm on $\mathbb{F}^{n \times n}$, let $\gamma > 0$, and assume that $\|A^{-1}\| < \gamma$ and $\|A - B\| < 1/\gamma$. Then, B is nonsingular,

$$\|B^{-1}\| \le \frac{\gamma}{1 - \gamma \|B - A\|},$$

and

$$\|A^{-1} - B^{-1}\| \le \gamma^2 \|A - B\|.$$

Proof: See [459, p. 148].

Remark: See Fact 9.8.45.

Fact 9.9.58. Let $A, B \in \mathbb{F}^{n \times n}$, let $\lambda \in \mathbb{C}$, assume that $\lambda I - A$ is nonsingular, let $\|\cdot\|$ be a normalized submultiplicative norm on $\mathbb{F}^{n \times n}$, let $\gamma > 0$, and assume that $\|(\lambda I - A)^{-1}\| < \gamma$ and $\|A - B\| < 1/\gamma$. Then, $\lambda I - B$ is nonsingular,

$$\|(\lambda I - B)^{-1}\| \le \frac{\gamma}{1 - \gamma \|B - A\|},$$

and

$$\|(\lambda I - A)^{-1} - (\lambda I - B)^{-1}\| \le \frac{\gamma^2 \|A - B\|}{1 - \gamma \|A - B\|}.$$

Proof: See [459, pp. 149, 150].

Remark: See Fact 9.9.57.

Fact 9.9.59. Let $A, B \in \mathbb{F}^{n \times n}$, assume that A and $A + B$ are nonsingular, and let $\|\cdot\|$ be a normalized submultiplicative norm on $\mathbb{F}^{n \times n}$. Then,

$$\|A^{-1} - (A + B)^{-1}\| \le \|A^{-1}\| \|(A + B)^{-1}\| \|B\|.$$

If, in addition, $\|A^{-1}B\| < 1$, then

$$\|A^{-1} + (A + B)^{-1}\| \le \frac{\|A^{-1}\| \|A^{-1}B\|}{1 - \|A^{-1}B\|}.$$

Furthermore, if $\|A^{-1}B\| < 1$ and $\|B\| < 1/\|A^{-1}\|$, then

$$\|A^{-1} - (A + B)^{-1}\| \le \frac{\|A^{-1}\|^2 \|B\|}{1 - \|A^{-1}\| \|B\|}.$$

Fact 9.9.60. Let $A \in \mathbb{F}^{n \times n}$, assume that A is nonsingular, let $E \in \mathbb{F}^{n \times n}$, and let $\|\cdot\|$ be a normalized norm on $\mathbb{F}^{n \times n}$. Then,

$$\begin{aligned}
(A + E)^{-1} &= A^{-1}(I + EA^{-1})^{-1} \\
&= A^{-1} - A^{-1}EA^{-1} + O(\|E\|^2).
\end{aligned}$$

Fact 9.9.61. Let $A \in \mathbb{F}^{n \times m}$ and $B \in \mathbb{F}^{l \times k}$. Then,

$$\|A \otimes B\|_{\mathrm{col}} = \|A\|_{\mathrm{col}} \|B\|_{\mathrm{col}},$$

$$\|A \otimes B\|_{\infty} = \|A\|_{\infty} \|B\|_{\infty},$$

$$\|A \otimes B\|_{\mathrm{row}} = \|A\|_{\mathrm{row}} \|B\|_{\mathrm{row}}.$$

Furthermore, if $p \in [1, \infty]$, then

$$\|A \otimes B\|_p = \|A\|_p \|B\|_p.$$

Fact 9.9.62. Let $A, B \in \mathbb{F}^{n \times n}$, and let $\| \cdot \|$ be a unitarily invariant norm on $\mathbb{F}^{n \times n}$. Then,

$$\|A \circ B\|^2 \leq \|A^*A\| \|B^*B\|.$$

Proof: See [731].

Fact 9.9.63. Let $A, B \in \mathbb{F}^{n \times n}$, assume that A and B are normal, and let $\| \cdot \|$ be a unitarily invariant norm on $\mathbb{F}^{n \times n}$. Then,

$$\|A + B\| \leq \|\langle A \rangle + \langle B \rangle\|$$

and

$$\|A \circ B\| \leq \|\langle A \rangle \circ \langle B \rangle\|.$$

Proof: See [93, 849] and [730, p. 213].

9.10 Facts on Matrix Norms for Partitioned Matrices

Fact 9.10.1. Let $A \in \mathbb{F}^{n \times m}$ be the partitioned matrix

$$A = \begin{bmatrix} A_{11} & A_{12} & \cdots & A_{1k} \\ A_{21} & A_{22} & \cdots & A_{2k} \\ \vdots & \vdots & \ddots & \vdots \\ A_{k1} & A_{k2} & \cdots & A_{kk} \end{bmatrix},$$

where $A_{ij} \in \mathbb{F}^{n_i \times n_j}$ for all $i, j \in \{1, \ldots, k\}$. Furthermore, define $\mu(A) \in \mathbb{R}^{k \times k}$ by

$$\mu(A) \triangleq \begin{bmatrix} \sigma_{\max}(A_{11}) & \sigma_{\max}(A_{12}) & \cdots & \sigma_{\max}(A_{1k}) \\ \sigma_{\max}(A_{21}) & \sigma_{\max}(A_{22}) & \cdots & \sigma_{\max}(A_{2k}) \\ \vdots & \vdots & \ddots & \vdots \\ \sigma_{\max}(A_{k1}) & \sigma_{\max}(A_{k2}) & \cdots & \sigma_{\max}(A_{kk}) \end{bmatrix}.$$

Finally, let $B \in \mathbb{F}^{n \times m}$ be partitioned conformally with A. Then, the following statements hold:

i) For all $\alpha \in \mathbb{F}$, $\mu(\alpha A) \leq |\alpha| \mu(A)$.

ii) $\mu(A + B) \leq \mu(A) + \mu(B)$.

iii) $\mu(AB) \leq \mu(A) \mu(B)$.

iv) $\text{sprad}(A) \leq \text{sprad}[\mu(A)]$.

v) $\sigma_{\max}(A) \leq \sigma_{\max}[\mu(A)]$.

Proof: See [410, 1082, 1236].

Remark: $\mu(A)$ is a *matricial norm*.

Remark: This result is a norm-compression inequality.

Fact 9.10.2. Let $A \in \mathbb{F}^{n \times m}$ be the partitioned matrix

$$A = \begin{bmatrix} A_{11} & A_{12} & \cdots & A_{1k} \\ A_{21} & A_{22} & \cdots & A_{2k} \\ \vdots & \vdots & \ddots & \vdots \\ A_{k1} & A_{k2} & \cdots & A_{kk} \end{bmatrix},$$

where $A_{ij} \in \mathbb{F}^{n_i \times n_j}$ for all $i,j \in \{1,\ldots,k\}$. Then, the following statements hold:

i) If $p \in [1,2]$, then

$$\sum_{i,j=1}^{k} \|A_{ij}\|_{\sigma p}^2 \leq \|A\|_{\sigma p}^2 \leq k^{4/p-2} \sum_{i,j=1}^{k} \|A_{ij}\|_{\sigma p}^2.$$

ii) If $p \in [2,\infty]$, then

$$k^{4/p-2} \sum_{i,j=1}^{k} \|A_{ij}\|_{\sigma p}^2 \leq \|A\|_{\sigma p}^2 \leq \sum_{i,j=1}^{k} \|A_{ij}\|_{\sigma p}^2.$$

iii) If $p \in [1,2]$, then

$$\|A\|_{\sigma p}^p \leq \sum_{i,j=1}^{k} \|A_{ij}\|_{\sigma p}^p \leq k^{2-p}\|A\|_{\sigma p}^p.$$

iv) If $p \in [2,\infty)$, then

$$k^{2-p}\|A\|_{\sigma p}^p \leq \sum_{i,j=1}^{k} \|A_{ij}\|_{\sigma p}^p \leq \|A\|_{\sigma p}^p.$$

v) $\|A\|_{\sigma 2}^2 = \sum_{i,j=1}^{k}\|A_{ij}\|_{\sigma 2}^2$.

vi) For all $p \in [1,\infty)$,

$$\left(\sum_{i=1}^{k} \|A_{ii}\|_{\sigma p}^p \right)^{1/p} \leq \|A\|_{\sigma p}.$$

vii) For all $i \in \{1,\ldots,k\}$,

$$\sigma_{\max}(A_{ii}) \leq \sigma_{\max}(A).$$

Proof: See [133, 212].

Fact 9.10.3. Let $A, B \in \mathbb{F}^{n \times n}$, and define $\mathcal{A} \in \mathbb{F}^{kn \times kn}$ by

$$\mathcal{A} \triangleq \begin{bmatrix} A & B & B & \cdots & B \\ B & A & B & \cdots & B \\ B & B & A & \ddots & B \\ \vdots & \vdots & \ddots & \ddots & \vdots \\ B & B & B & \cdots & A \end{bmatrix}.$$

Then,

$$\sigma_{\max}(\mathcal{A}) = \max\{\sigma_{\max}(A + (k-1)B), \sigma_{\max}(A - B)\}.$$

Now, let $p \in [1, \infty)$. Then,

$$\|\mathcal{A}\|_{\sigma p} = (\|A + (k-1)B\|_{\sigma p}^p + (k-1)\|A - B\|_{\sigma p}^p)^{1/p}.$$

Proof: See [133].

Fact 9.10.4. Let $A \in \mathbb{F}^{n \times n}$, and define $\mathcal{A} \in \mathbb{F}^{kn \times kn}$ by

$$\mathcal{A} \triangleq \begin{bmatrix} A & A & A & \cdots & A \\ -A & A & A & \cdots & A \\ -A & -A & A & \ddots & A \\ \vdots & \vdots & \ddots & \ddots & \vdots \\ -A & -A & -A & \cdots & A \end{bmatrix}.$$

Then,

$$\sigma_{\max}(\mathcal{A}) = \sqrt{\frac{2}{1 - \cos(\pi/k)}} \sigma_{\max}(A).$$

Furthermore, define $\mathcal{A}_0 \in \mathbb{F}^{kn \times kn}$ by

$$\mathcal{A}_0 \triangleq \begin{bmatrix} 0 & A & A & \cdots & A \\ -A & 0 & A & \cdots & A \\ -A & -A & 0 & \ddots & A \\ \vdots & \vdots & \ddots & \ddots & \vdots \\ -A & -A & -A & \cdots & 0 \end{bmatrix}.$$

Then,

$$\sigma_{\max}(\mathcal{A}_0) = \sqrt{\frac{1 + \cos(\pi/k)}{1 - \cos(\pi/k)}} \sigma_{\max}(A).$$

Proof: See [133].

Remark: Extensions to Schatten norms are given in [133].

Fact 9.10.5. Let $A, B, C, D \in \mathbb{F}^{n \times n}$. Then,

$$\tfrac{1}{2} \max\{\sigma_{\max}(A + B + C + D), \sigma_{\max}(A - B - C + D)\} \leq \sigma_{\max}\left(\begin{bmatrix} A & B \\ C & D \end{bmatrix}\right).$$

Now, let $p \in [1, \infty)$. Then,

$$\tfrac{1}{2}(\|A + B + C + D\|_{\sigma p}^p + \|A - B - C + D\|_{\sigma p}^p)^{1/p} \leq \left\|\begin{bmatrix} A & B \\ C & D \end{bmatrix}\right\|_{\sigma p}.$$

Proof: See [133].

Fact 9.10.6. Let $A, B, C \in \mathbb{F}^{n \times n}$, define

$$\mathcal{A} \triangleq \begin{bmatrix} A & B \\ B^* & C \end{bmatrix},$$

assume that $\mathcal{A}$ is positive semidefinite, let $p \in [1, \infty]$, and define

$$\mathcal{A}_0 \triangleq \begin{bmatrix} \|A\|_{\sigma p} & \|B\|_{\sigma p} \\ \|B\|_{\sigma p} & \|C\|_{\sigma p} \end{bmatrix}.$$

If $p \in [1, 2]$, then

$$\|\mathcal{A}_0\|_{\sigma p} \le \|\mathcal{A}\|_{\sigma p}.$$

Furthermore, if $p \in [2, \infty]$, then

$$\|\mathcal{A}\|_{\sigma p} \le \|\mathcal{A}_0\|_{\sigma p}.$$

Hence, if $p = 2$, then

$$\|\mathcal{A}_0\|_{\sigma p} = \|\mathcal{A}\|_{\sigma p}.$$

Finally, if $A = C$, B is Hermitian, and p is an integer, then

$$\|\mathcal{A}\|_{\sigma p}^p = \|A + B\|_{\sigma p}^p + \|A - B\|_{\sigma p}^p$$

and

$$\|\mathcal{A}_0\|_{\sigma p}^p = (\|A\|_{\sigma p} + \|B\|_{\sigma p})^p + |\|A\|_{\sigma p} - \|B\|_{\sigma p}|^p.$$

Proof: See [833].

Remark: This result is a norm-compression inequality.

Fact 9.10.7. Let $A \in \mathbb{F}^{n \times n}$, $B \in \mathbb{F}^{n \times m}$, and $C \in \mathbb{F}^{m \times m}$, define

$$\mathcal{A} \triangleq \begin{bmatrix} A & B \\ B^* & C \end{bmatrix},$$

assume that $\mathcal{A}$ is positive semidefinite, and let $p \ge 1$. If $p \in [1, 2]$, then

$$\|\mathcal{A}\|_{\sigma p}^p \le \|A\|_{\sigma p}^p + (2^p - 2)\|B\|_{\sigma p}^p + \|C\|_{\sigma p}^p.$$

Furthermore, if $p \ge 2$, then

$$\|A\|_{\sigma p}^p + (2^p - 2)\|B\|_{\sigma p}^p + \|C\|_{\sigma p}^p \le \|\mathcal{A}\|_{\sigma p}^p.$$

Finally, if $p = 2$, then

$$\|\mathcal{A}\|_{\sigma p}^p = \|A\|_{\sigma p}^p + (2^p - 2)\|B\|_{\sigma p}^p + \|C\|_{\sigma p}^p.$$

Proof: See [89].

Fact 9.10.8. Let $A \in \mathbb{F}^{n \times m}$ be the partitioned matrix

$$A = \begin{bmatrix} A_{11} & \cdots & A_{1k} \\ A_{21} & \cdots & A_{2k} \end{bmatrix},$$

where $A_{ij} \in \mathbb{F}^{n_i \times n_j}$ for all $i, j \in \{1, \ldots, k\}$. Then, the following statements are conjectured to hold:

i) If $p \in [1, 2]$, then

$$\left\| \left[\begin{array}{ccc} \|A_{11}\|_{\sigma p} & \cdots & \|A_{1k}\|_{\sigma p} \\ \|A_{21}\|_{\sigma p} & \cdots & \|A_{2k}\|_{\sigma p} \end{array} \right] \right\|_{\sigma p} \le \|A\|_{\sigma p}.$$

ii) If $p \ge 2$, then

$$\|A\|_{\sigma p} \le \left\| \left[\begin{array}{ccc} \|A_{11}\|_{\sigma p} & \cdots & \|A_{1k}\|_{\sigma p} \\ \|A_{21}\|_{\sigma p} & \cdots & \|A_{2k}\|_{\sigma p} \end{array} \right] \right\|_{\sigma p}.$$

Proof: See [90]. This result is true when all blocks have rank 1 or when $p \ge 4$.

Remark: This result is a norm-compression inequality.

9.11 Facts on Matrix Norms and Eigenvalues for One Matrix

Fact 9.11.1. Let $A \in \mathbb{F}^{n \times n}$. Then,

$$|\det A| \le \prod_{i=1}^{n} \|\mathrm{row}_i(A)\|_2$$

and

$$|\det A| \le \prod_{i=1}^{n} \|\mathrm{col}_i(A)\|_2.$$

Proof: This result follows from Hadamard's inequality. See Fact 8.18.11.

Fact 9.11.2. Let $A \in \mathbb{F}^{n \times n}$, and let $\mathrm{mspec}(A) = \{\lambda_1, \ldots, \lambda_n\}_{\mathrm{ms}}$. Then,

$$\mathrm{Re}\,\mathrm{tr}\,A \le |\mathrm{tr}\,A| \le \sum_{i=1}^{n} |\lambda_i| \le \|A\|_{\sigma 1} = \mathrm{tr}\,\langle A \rangle = \sum_{i=1}^{n} \sigma_i(A).$$

In addition, if A is normal, then

$$\|A\|_{\sigma 1} = \sum_{i=1}^{n} |\lambda_i|.$$

Finally, A is positive semidefinite if and only if

$$\|A\|_{\sigma 1} = \mathrm{tr}\,A.$$

Proof: See Fact 5.14.14 and Fact 9.13.18.

Remark: See Fact 5.11.9 and Fact 5.14.14.

Problem: Refine the second statement for necessity and sufficiency. See [763].

Fact 9.11.3. Let $A \in \mathbb{F}^{n \times n}$, and let $\mathrm{mspec}(A) = \{\lambda_1, \ldots, \lambda_n\}_{\mathrm{ms}}$. Then,

$$\mathrm{Re}\,\mathrm{tr}\,A^2 \leq |\mathrm{tr}\,A^2| \leq \sum_{i=1}^{n} |\lambda_i|^2 \leq \|A^2\|_{\sigma 1} = \mathrm{tr}\,\langle A^2 \rangle = \sum_{i=1}^{n} \sigma_i(A^2)$$

$$\leq \sum_{i=1}^{n} \sigma_i^2(A) = \mathrm{tr}\,A^*A = \mathrm{tr}\,\langle A \rangle^2 = \|A\|_{\sigma 2}^2 = \|A\|_{\mathrm{F}}^2$$

and

$$\|A\|_{\mathrm{F}}^2 - \sqrt{\tfrac{n^3 - n}{12}}\|[A, A^*]\|_{\mathrm{F}} \leq \sum_{i=1}^{n} |\lambda_i|^2 \leq \sqrt{\|A\|_{\mathrm{F}}^4 - \tfrac{1}{2}\|[A, A^*]\|_{\mathrm{F}}^2} \leq \|A\|_{\mathrm{F}}^2.$$

Consequently, A is normal if and only if

$$\|A\|_{\mathrm{F}}^2 = \sum_{i=1}^{n} |\lambda_i|^2.$$

Furthermore,

$$\sum_{i=1}^{n} |\lambda_i|^2 \leq \sqrt{\|A\|_{\mathrm{F}}^4 - \tfrac{1}{4}(\mathrm{tr}\,|[A, A^*]|)^2} \leq \|A\|_{\mathrm{F}}^2$$

and

$$\sum_{i=1}^{n} |\lambda_i|^2 \leq \sqrt{\|A\|_{\mathrm{F}}^4 - \tfrac{n^2}{4}|\det[A, A^*]|^{2/n}} \leq \|A\|_{\mathrm{F}}^2.$$

Finally, A is Hermitian if and only if

$$\|A\|_{\mathrm{F}}^2 = \mathrm{tr}\,A^2.$$

Proof: Use Fact 8.18.5 and Fact 9.11.2. The lower bound involving the commutator is due to Henrici; the corresponding upper bound is given in [871]. The bounds in the penultimate statement are given in [871]. The last statement follows from Fact 3.7.13.

Remark: $\mathrm{tr}\,(A + A^*)^2 \geq 0$ and $\mathrm{tr}\,(A - A^*)^2 \leq 0$ yield $|\mathrm{tr}\,A^2| \leq \|A\|_{\mathrm{F}}^2$.

Remark: The result $\sum_{i=1}^{n} |\lambda_i|^2 \leq \|A\|_{\mathrm{F}}^2$ is *Schur's inequality*. See Fact 8.18.5.

Remark: See Fact 5.11.10, Fact 9.11.5, Fact 9.13.16, and Fact 9.13.19.

Problem: Merge the first two strings.

Fact 9.11.4. Let $A \in \mathbb{F}^{n \times n}$. Then,

$$|\mathrm{tr}\,A^2| \leq (\mathrm{rank}\,A)\sqrt{\|A\|_{\mathrm{F}}^4 - \tfrac{1}{2}\|[A, A^*]\|_{\mathrm{F}}^2}.$$

Proof: See [323].

Fact 9.11.5. Let $A \in \mathbb{F}^{n \times n}$, let $\text{mspec}(A) = \{\lambda_1, \ldots, \lambda_n\}_{\text{ms}}$, and define

$$\alpha \triangleq \sqrt{\left(\|A\|_{\text{F}}^2 - \tfrac{1}{n}|\text{tr}\, A|^2\right)^2 - \tfrac{1}{2}\|[A, A^*]\|_{\text{F}}^2} + \tfrac{1}{n}|\text{tr}\, A|^2.$$

Then,

$$\sum_{i=1}^n |\lambda_i|^2 \leq \alpha \leq \sqrt{\|A\|_{\text{F}}^4 - \tfrac{1}{2}\|[A, A^*]\|_{\text{F}}^2} \leq \|A\|_{\text{F}}^2,$$

$$\sum_{i=1}^n (\text{Re}\, \lambda_i)^2 \leq \tfrac{1}{2}(\alpha + \text{Re}\,\text{tr}\, A^2),$$

$$\sum_{i=1}^n (\text{Im}\, \lambda_i)^2 \leq \tfrac{1}{2}(\alpha - \text{Re}\,\text{tr}\, A^2).$$

Proof: See [751].

Remark: The first string of inequalities interpolates the upper bound for $\sum_{i=1}^n |\lambda_i|^2$ in the second string of inequalities in Fact 9.11.3.

Fact 9.11.6. Let $A \in \mathbb{F}^{n \times n}$, let $\text{mspec}(A) = \{\lambda_1, \ldots, \lambda_n\}_{\text{ms}}$, and let $p \in (0, 2]$. Then,

$$\sum_{i=1}^n |\lambda_i|^p \leq \sum_{i=1}^n \sigma_i^p(A) = \|A\|_{\sigma p}^p \leq \|A\|_p^p.$$

Proof: The left-hand inequality, which holds for all $p > 0$, follows from Weyl's inequality in Fact 8.18.5. The right-hand inequality is given by Proposition 9.2.5.

Remark: This result is the *generalized Schur inequality*.

Remark: The case of equality is discussed in [763] for $p \in [1, 2)$.

Fact 9.11.7. Let $A \in \mathbb{F}^{n \times n}$, and let $\text{mspec}(A) = \{\lambda_1, \ldots, \lambda_n\}_{\text{ms}}$. Then,

$$\|A\|_{\text{F}}^2 - \sum_{i=1}^n |\lambda_i|^2 = 2\left(\|\tfrac{1}{\jmath 2}(A - A^*)\|_{\text{F}}^2 - \sum_{i=1}^n |\text{Im}\, \lambda_i|^2\right).$$

Proof: See Fact 5.11.22.

Remark: This result is an extension of Browne's theorem.

Fact 9.11.8. Let $A \in \mathbb{R}^{n \times n}$, and let $\lambda \in \text{spec}(A)$. Then, the following inequalities hold:

i) $|\lambda| \leq n\|A\|_\infty.$

ii) $|\text{Re}\, \lambda| \leq \tfrac{n}{2}\|A + A^{\text{T}}\|_\infty.$

iii) $|\text{Im}\, \lambda| \leq \frac{\sqrt{n^2 - n}}{2\sqrt{2}}\|A - A^{\text{T}}\|_\infty.$

Proof: See [988, p. 140].

Remark: *i)* and *ii)* are *Hirsch's theorems*, while *iii)* is *Bendixson's theorem*. See Fact 5.11.21.

Fact 9.11.9. Let $A \in \mathbb{R}^{n \times n}$, and assume that A is nonnegative. Then,

$$\min_{i=1,\ldots,n} \|\mathrm{col}_i(A)\|_1 \leq \mathrm{sprad}(A) \leq \|A\|_{\mathrm{col}}$$

and

$$\min_{i=1,\ldots,n} \|\mathrm{col}_i(A^{\mathrm{T}})\|_1 \leq \mathrm{sprad}(A) \leq \|A\|_{\mathrm{row}}.$$

Proof: See [1208, pp. 318, 319].

Remark: The upper bounds are given by (9.4.26) and (9.4.27).

9.12 Facts on Matrix Norms and Eigenvalues for Two or More Matrices

Fact 9.12.1. Let $A, B \in \mathbb{F}^{n \times m}$, let $\mathrm{mspec}(A^*B) = \{\lambda_1, \ldots, \lambda_m\}_{\mathrm{ms}}$, let $p, q \in [1, \infty]$ satisfy $1/p + 1/q = 1$, and define $r \triangleq \min\{m, n\}$. Then,

$$|\mathrm{tr}\, A^*B| \leq \sum_{i=1}^{m} |\lambda_i| \leq \|A^*B\|_{\sigma 1} = \sum_{i=1}^{m} \sigma_i(A^*B) \leq \sum_{i=1}^{r} \sigma_i(A)\sigma_i(B) \leq \|A\|_{\sigma p}\|B\|_{\sigma q}.$$

In particular,

$$|\mathrm{tr}\, A^*B| \leq \|A\|_{\mathrm{F}}\|B\|_{\mathrm{F}}.$$

Proof: Use Proposition 9.6.2 and Fact 9.11.2. The last inequality in the string of inequalities is Hölder's inequality.

Remark: See Fact 9.9.11.

Remark: The result

$$|\mathrm{tr}\, A^*B| \leq \sum_{i=1}^{r} \sigma_i(A)\sigma_i(B)$$

is *von Neumann's trace inequality*. See [254].

Fact 9.12.2. Let $A, B \in \mathbb{F}^{n \times m}$, and let $\mathrm{mspec}(A^*B) = \{\lambda_1, \ldots, \lambda_m\}_{\mathrm{ms}}$. Then,

$$|\mathrm{tr}\, (A^*B)^2| \leq \sum_{i=1}^{m} |\lambda_i|^2 \leq \sum_{i=1}^{m} \sigma_i^2(A^*B) = \mathrm{tr}\, AA^*BB^* = \|A^*B\|_{\mathrm{F}}^2 \leq \|A\|_{\mathrm{F}}^2\|B\|_{\mathrm{F}}^2.$$

Proof: Use Fact 8.18.5.

Fact 9.12.3. Let $A, B \in \mathbb{F}^{n \times n}$, assume that A and B are Hermitian, and let $\mathrm{mspec}(A + \jmath B) = \{\lambda_1, \ldots, \lambda_n\}_{\mathrm{ms}}$. Then,

$$\sum_{i=1}^{n} |\mathrm{Re}\, \lambda_i|^2 \leq \|A\|_{\mathrm{F}}^2$$

and

$$\sum_{i=1}^{n} |\mathrm{Im}\, \lambda_i|^2 \leq \|B\|_{\mathrm{F}}^2.$$

Proof: See [1125, p. 146].

Fact 9.12.4. Let $A, B \in \mathbb{F}^{n \times n}$, assume that A and B are Hermitian, and let $\| \cdot \|$ be a weakly unitarily invariant norm on $\mathbb{F}^{n \times n}$. Then,

$$\left\| \begin{bmatrix} \lambda_1(A) & & 0 \\ & \ddots & \\ 0 & & \lambda_n(A) \end{bmatrix} - \begin{bmatrix} \lambda_1(B) & & 0 \\ & \ddots & \\ 0 & & \lambda_n(B) \end{bmatrix} \right\| \leq \|A - B\|$$

$$\leq \left\| \begin{bmatrix} \lambda_1(A) & & 0 \\ & \ddots & \\ 0 & & \lambda_n(A) \end{bmatrix} - \begin{bmatrix} \lambda_n(B) & & 0 \\ & \ddots & \\ 0 & & \lambda_1(B) \end{bmatrix} \right\|.$$

In particular,

$$\max_{i \in \{1, \ldots, n\}} |\lambda_i(A) - \lambda_i(B)| \leq \sigma_{\max}(A - B) \leq \max_{i \in \{1, \ldots, n\}} |\lambda_i(A) - \lambda_{n-i+1}(B)|$$

and

$$\sum_{i=1}^{n} [\lambda_i(A) - \lambda_i(B)]^2 \leq \|A - B\|_{\mathrm{F}}^2 \leq \sum_{i=1}^{n} [\lambda_i(A) - \lambda_{n-i+1}(B)]^2.$$

Proof: See [49], [200, p. 38], [201, pp. 63, 69], [204, p. 38], [818, p. 126], [903, p. 134], [921], or [1261, p. 202].

Remark: The first inequality is the *Lidskii-Mirsky-Wielandt theorem*. This result can be stated without norms using Fact 9.8.42. See [921].

Remark: See Fact 9.14.29.

Remark: The case in which A and B are normal is considered in Fact 9.12.8.

Fact 9.12.5. Let $A, B \in \mathbb{F}^{n \times n}$, let $\mathrm{mspec}(A) = \{\lambda_1, \ldots, \lambda_n\}_{\mathrm{ms}}$ and $\mathrm{mspec}(B) = \{\mu_1, \ldots, \mu_n\}_{\mathrm{ms}}$, and assume that A and B satisfy at least one of the following conditions:

i) A and B are Hermitian.

ii) A is Hermitian, and B is skew Hermitian.

iii) A is skew Hermitian, and B is Hermitian.

iv) A and B are unitary.

v) There exist nonzero $\alpha, \beta \in \mathbb{C}$ such that αA and βB are unitary.

vi) A, B, and $A - B$ are normal.

Then,

$$\min \sigma_{\max} \left(\begin{bmatrix} \lambda_1 & & 0 \\ & \ddots & \\ 0 & & \lambda_n \end{bmatrix} - \begin{bmatrix} \mu_{\sigma(1)} & & 0 \\ & \ddots & \\ 0 & & \mu_{\sigma(n)} \end{bmatrix} \right) \leq \sigma_{\max}(A - B),$$

where the minimum is taken over all permutations σ of $\{1, \ldots, n\}$.

Proof: See [204, pp. 52, 152].

Fact 9.12.6. Let $A, B \in \mathbb{F}^{n \times n}$, let $\mathrm{mspec}(A) = \{\lambda_1, \ldots, \lambda_n\}_{\mathrm{ms}}$ and $\mathrm{mspec}(B) = \{\mu_1, \ldots, \mu_n\}_{\mathrm{ms}}$, and assume that A is normal. Then,

$$\min \left\| \begin{bmatrix} \lambda_1 & & 0 \\ & \ddots & \\ 0 & & \lambda_n \end{bmatrix} - \begin{bmatrix} \mu_{\sigma(1)} & & 0 \\ & \ddots & \\ 0 & & \mu_{\sigma(n)} \end{bmatrix} \right\|_{\mathrm{F}} \leq \sqrt{n} \|A - B\|_{\mathrm{F}},$$

where the minimum is taken over all permutations σ of $\{1, \ldots, n\}$. If, in addition, B is normal, then there exists $c \in (0, 2.9039)$ such that

$$\min \sigma_{\max} \left(\begin{bmatrix} \lambda_1 & & 0 \\ & \ddots & \\ 0 & & \lambda_n \end{bmatrix} - \begin{bmatrix} \mu_{\sigma(1)} & & 0 \\ & \ddots & \\ 0 & & \mu_{\sigma(n)} \end{bmatrix} \right) \leq c\sigma_{\max}(A - B).$$

Proof: See [204, pp. 152, 153, 173].

Remark: Constants c for alternative Schatten norms are given in [204, p. 159].

Remark: If, in addition, $A - B$ is normal, then it follows from Fact 9.12.5 that the last inequality holds with $c = 1$.

Fact 9.12.7. Let $A, B \in \mathbb{F}^{n \times n}$, let $\mathrm{mspec}(A) = \{\lambda_1, \ldots, \lambda_n\}_{\mathrm{ms}}$ and $\mathrm{mspec}(B) = \{\mu_1, \ldots, \mu_n\}_{\mathrm{ms}}$, and assume that A is Hermitian. Then,

$$\min \left\| \begin{bmatrix} \lambda_1 & & 0 \\ & \ddots & \\ 0 & & \lambda_n \end{bmatrix} - \begin{bmatrix} \mu_{\sigma(1)} & & 0 \\ & \ddots & \\ 0 & & \mu_{\sigma(n)} \end{bmatrix} \right\|_{\mathrm{F}} \leq \sqrt{2} \|A - B\|_{\mathrm{F}},$$

where the minimum is taken over all permutations σ of $\{1, \ldots, n\}$.

Proof: See [204, p. 174].

Fact 9.12.8. Let $A, B \in \mathbb{F}^{n \times n}$, assume that A and B are normal, and let $\mathrm{spec}(A) = \{\lambda_1, \ldots, \lambda_q\}$ and $\mathrm{spec}(B) = \{\mu_1, \ldots, \mu_r\}$. Then,

$$\sigma_{\max}(A - B) \leq \max\{|\lambda_i - \lambda_j| : i = 1, \ldots, q, \ j = 1, \ldots, r\}.$$

Proof: See [201, p. 164].

Remark: The case in which A and B are Hermitian is considered in Fact 9.12.4.

Fact 9.12.9. Let $A, B \in \mathbb{F}^{n \times n}$, and assume that A and B are normal. Then, there exists a permutation σ of $1, \ldots, n$ such that

$$\sum_{i=1}^{n} |\lambda_{\sigma(i)}(A) - \lambda_i(B)|^2 \leq \|A - B\|_{\mathrm{F}}^2.$$

Proof: See [728, p. 368] or [1125, pp. 160, 161].

Remark: This inequality is the *Hoffman-Wielandt theorem*.

Remark: The case in which A and B are Hermitian is considered in Fact 9.12.4.

Fact 9.12.10. Let $A, B \in \mathbb{F}^{n \times n}$, and assume that A is Hermitian and B is normal. Furthermore, let $\mathrm{mspec}(B) = \{\lambda_1(B), \ldots, \lambda_n(B)\}_{\mathrm{ms}}$, where $\mathrm{Re}\,\lambda_n(B) \leq \cdots \leq \mathrm{Re}\,\lambda_1(B)$. Then,

$$\sum_{i=1}^{n} |\lambda_i(A) - \lambda_i(B)|^2 \leq \|A - B\|_{\mathrm{F}}^2.$$

Proof: See [728, p. 370].

Remark: This result is a special case of Fact 9.12.9.

Remark: The left-hand side has the same value for all orderings that satisfy $\mathrm{Re}\,\lambda_n(B) \leq \cdots \leq \mathrm{Re}\,\lambda_1(B)$.

Fact 9.12.11. Let $A, B \in \mathbb{F}^{n \times n}$, and let $\| \cdot \|$ be an induced norm on $\mathbb{F}^{n \times n}$. Then,

$$|\det A - \det B| \leq \begin{cases} \|A - B\| \frac{\|A\|^n - \|B\|^n}{\|A\| - \|B\|}, & \|A\| \neq \|B\|, \\ n\|A - B\|\|A\|^{n-1}, & \|A\| = \|B\|. \end{cases}$$

Proof: See [518].

Remark: See Fact 1.20.2.

9.13 Facts on Matrix Norms and Singular Values for One Matrix

Fact 9.13.1. Let $A \in \mathbb{F}^{n \times m}$. Then,

$$\sigma_{\max}(A) = \max_{x \in \mathbb{F}^m \setminus \{0\}} \left(\frac{x^* A^* A x}{x^* x} \right)^{1/2},$$

and thus

$$\|Ax\|_2 \leq \sigma_{\max}(A)\|x\|_2.$$

Furthermore,

$$\lambda_{\min}^{1/2}(A^*A) = \min_{x \in \mathbb{F}^n \setminus \{0\}} \left(\frac{x^* A^* A x}{x^* x} \right)^{1/2},$$

and thus

$$\lambda_{\min}^{1/2}(A^*A)\|x\|_2 \leq \|Ax\|_2.$$

If, in addition, $m \leq n$, then

$$\sigma_m(A) = \min_{x \in \mathbb{F}^n \setminus \{0\}} \left(\frac{x^* A^* A x}{x^* x} \right)^{1/2},$$

and thus

$$\sigma_m(A)\|x\|_2 \leq \|Ax\|_2.$$

Finally, if $m = n$, then

$$\sigma_{\min}(A) = \min_{x \in \mathbb{F}^n \setminus \{0\}} \left(\frac{x^* A^* A x}{x^* x} \right)^{1/2},$$

and thus

$$\sigma_{\min}(A)\|x\|_2 \leq \|Ax\|_2.$$

Proof: See Lemma 8.4.3.

Fact 9.13.2. Let $A \in \mathbb{F}^{n \times m}$. Then,

$$\sigma_{\max}(A) = \max\{|y^*Ax|: \ x \in \mathbb{F}^m, \ y \in \mathbb{F}^n, \ \|x\|_2 = \|y\|_2 = 1\}$$
$$= \max\{|y^*Ax|: \ x \in \mathbb{F}^m, \ y \in \mathbb{F}^n, \ \|x\|_2 \leq 1, \ \|y\|_2 \leq 1\}.$$

Remark: See Fact 9.8.36.

Fact 9.13.3. Let $x \in \mathbb{F}^n$ and $y \in \mathbb{F}^m$, and define $\mathcal{S} \triangleq \{A \in \mathbb{F}^{n \times m}: \ \sigma_{\max}(A) \leq 1\}$. Then,

$$\max_{A \in \mathcal{S}} x^*Ay = \sqrt{x^*xy^*y}.$$

Fact 9.13.4. Let $\| \cdot \|$ be an equi-induced unitarily invariant norm on $\mathbb{F}^{n \times n}$. Then, $\| \cdot \| = \sigma_{\max}(\cdot)$.

Fact 9.13.5. Let $\| \cdot \|$ be an equi-induced self-adjoint norm on $\mathbb{F}^{n \times n}$. Then, $\| \cdot \| = \sigma_{\max}(\cdot)$.

Fact 9.13.6. Let $A \in \mathbb{F}^{n \times n}$. Then,

$$\sigma_{\min}(A) - 1 \leq \sigma_{\min}(A + I) \leq \sigma_{\min}(A) + 1.$$

Proof: Use Proposition 9.6.8.

Fact 9.13.7. Let $A \in \mathbb{F}^{n \times n}$, assume that A is normal, and let $r \in \mathbb{N}$. Then,

$$\sigma_{\max}(A^r) = \sigma_{\max}^r(A).$$

Remark: Matrices that are not normal might also satisfy these conditions. Consider $\begin{bmatrix} 1 & 0 & 0 \\ 0 & 0 & 0 \\ 0 & 1 & 0 \end{bmatrix}$.

Fact 9.13.8. Let $A \in \mathbb{F}^{n \times n}$. Then,

$$\sigma_{\max}^2(A) - \sigma_{\max}(A^2) \leq \sigma_{\max}(A^*A - AA^*) \leq \sigma_{\max}^2(A) - \sigma_{\min}^2(A)$$

and

$$\sigma_{\max}^2(A) + \sigma_{\min}^2(A) \leq \sigma_{\max}(A^*A + AA^*) \leq \sigma_{\max}^2(A) + \sigma_{\max}(A^2).$$

If $A^2 = 0$, then

$$\sigma_{\max}(A^*A - AA^*) = \sigma_{\max}^2(A).$$

Proof: See [844, 848].

Remark: See Fact 8.19.11.

Remark: If A is normal, then it follows that $\sigma_{\max}^2(A) \leq \sigma_{\max}(A^2)$, although Fact 9.13.7 implies that equality holds.

Fact 9.13.9. Let $A \in \mathbb{F}^{n \times n}$. Then, the following statements are equivalent:

i) $\mathrm{sprad}(A) = \sigma_{\max}(A)$.

ii) $\sigma_{\max}(A^i) = \sigma_{\max}^i(A)$ for all $i \in \mathbb{P}$.

iii) $\sigma_{\max}(A^n) = \sigma_{\max}^n(A)$.

Proof: See [506] and [730, p. 44].

Remark: The implication $iii) \implies i)$ is due to Ptak.

Remark: Additional conditions are given in [581].

Fact 9.13.10. Let $A \in \mathbb{F}^{n \times n}$. Then,

$$\sigma_{\max}(A) \le \sigma_{\max}(|A|) \le \sqrt{\operatorname{rank} A}\, \sigma_{\max}(A).$$

Proof: See [699, p. 111].

Fact 9.13.11. Let $A \in \mathbb{F}^{n \times n}$, and let $p \in [2, \infty)$ be an even integer. Then,

$$\|A\|_{\sigma p} \le \| |A| \|_{\sigma p}.$$

In particular,

$$\|A\|_{\mathrm{F}} \le \| |A| \|_{\mathrm{F}}$$

and

$$\sigma_{\max}(A) \le \sigma_{\max}(|A|).$$

Finally, let $\| \cdot \|$ be a unitarily invariant norm on $\mathbb{C}^{n \times m}$. Then, $\|A\|_{\mathrm{F}} = \| |A| \|_{\mathrm{F}}$ for all $A \in \mathbb{C}^{n \times m}$ if and only if $\| \cdot \|$ is a constant multiple of $\| \cdot \|_{\mathrm{F}}$.

Proof: See [731] and [749].

Fact 9.13.12. Let $A \in \mathbb{R}^{n \times n}$, and assume that $r \triangleq \operatorname{rank} A \ge 2$. If $r \operatorname{tr} A^2 \le (\operatorname{tr} A)^2$, then

$$\sqrt{\frac{(\operatorname{tr} A)^2 - \operatorname{tr} A^2}{r(r-1)}} \le \operatorname{sprad}(A).$$

If $(\operatorname{tr} A)^2 \le r \operatorname{tr} A^2$, then

$$\frac{|\operatorname{tr} A|}{r} + \sqrt{\frac{r \operatorname{tr} A^2 - (\operatorname{tr} A)^2}{r^2(r-1)}} \le \operatorname{sprad}(A).$$

If $\operatorname{rank} A = 2$, then equality holds in both cases. Finally, if A is skew symmetric, then

$$\sqrt{\frac{3}{r(r-1)}}\, \|A\|_{\mathrm{F}} \le \operatorname{sprad}(A).$$

Proof: See [737].

Fact 9.13.13. Let $A \in \mathbb{R}^{n \times n}$. Then,

$$\sqrt{\tfrac{1}{2(n^2-n)}(\|A\|_{\mathrm{F}}^2 + \operatorname{tr} A^2)} \le \sigma_{\max}(A).$$

Furthermore, if $\|A\|_{\mathrm{F}} \le \operatorname{tr} A$, then

$$\sigma_{\max}(A) \le \tfrac{1}{n}\operatorname{tr} A + \sqrt{\tfrac{n-1}{n}\left[\|A\|_{\mathrm{F}}^2 - \tfrac{1}{n}(\operatorname{tr} A)^2\right]}.$$

Proof: See [1017], which considers the complex case.

Fact 9.13.14. Let $A \in \mathbb{F}^{n \times n}$. Then, the polynomial $p \in \mathbb{R}[s]$ defined by

$$p(s) \triangleq s^n - \|A\|_{\mathrm{F}}^2 s + (n-1)|\det A|^{2/(n-1)}$$

has either exactly one or exactly two positive roots $0 < \alpha \le \beta$. Furthermore, α and β satisfy

$$\alpha^{(n-1)/2} \le \sigma_{\min}(A) \le \sigma_{\max}(A) \le \beta^{(n-1)/2}.$$

Proof: See [1167].

Fact 9.13.15. Let $A \in \mathbb{F}^{n \times n}$. Then,

$$\mathrm{tr}\,\langle A \rangle = \mathrm{tr}\,\langle A^* \rangle.$$

Fact 9.13.16. Let $A \in \mathbb{F}^{n \times n}$. Then, for all $k \in \{1, \ldots, n\}$,

$$\sum_{i=1}^{k} \sigma_i(A^2) \le \sum_{i=1}^{k} \sigma_i^2(A).$$

Hence,

$$\mathrm{tr}\,\left(A^{2*}A^2\right)^{1/2} \le \mathrm{tr}\,A^*A,$$

that is,

$$\mathrm{tr}\,\langle A^2 \rangle \le \mathrm{tr}\,\langle A \rangle^2.$$

Proof: Let $B = A$ and $r = 1$ in Proposition 9.6.2. See also Fact 9.11.3.

Fact 9.13.17. Let $A \in \mathbb{F}^{n \times n}$, and let k denote the number of nonzero eigenvalues of A. Then,

$$\left.\begin{array}{c} |\mathrm{tr}\,A^2| \le \mathrm{tr}\,\langle A^2 \rangle \\[1mm] \mathrm{tr}\,\langle A \rangle\langle A^* \rangle \\[1mm] \frac{1}{k}|\mathrm{tr}\,A|^2 \end{array}\right\} \le \mathrm{tr}\,\langle A \rangle^2.$$

Proof: The upper bound for $|\mathrm{tr}\,A^2|$ is given by Fact 9.11.3. The upper bound for $\mathrm{tr}\,\langle A^2 \rangle$ is given by Fact 9.13.16. To prove the center inequality, let $A = S_1 D S_2$ denote the singular value decomposition of A. Then, $\mathrm{tr}\,\langle A \rangle\langle A^* \rangle = \mathrm{tr}\,S_3^* D S_3 D$, where $S_3 \triangleq S_1 S_2$, and $\mathrm{tr}\,A^*A = \mathrm{tr}\,D^2$. The result now follows using Fact 5.12.4. The remaining inequality is given by Fact 5.11.10.

Remark: See Fact 5.11.10 and Fact 9.11.3.

Fact 9.13.18. Let $A \in \mathbb{F}^{n \times n}$, and let $\mathrm{mspec}(A) = \{\lambda_1, \ldots, \lambda_n\}_{\mathrm{ms}}$, where $\lambda_1, \ldots, \lambda_n$ are ordered such that $|\lambda_1| \ge \cdots \ge |\lambda_n|$. Then, for all $k \in \{1, \ldots, n\}$,

$$\prod_{i=1}^{k} |\lambda_i|^2 \le \prod_{i=1}^{k} \sigma_i(A^2) \le \prod_{i=1}^{k} \sigma_i^2(A)$$

and

$$\prod_{i=1}^{n} |\lambda_i|^2 = \prod_{i=1}^{n} \sigma_i(A^2) = \prod_{i=1}^{n} \sigma_i^2(A) = |\det A|^2.$$

Furthermore, for all $k \in \{1, \ldots, n\}$,

$$\left| \sum_{i=1}^{k} \lambda_i \right| \leq \sum_{i=1}^{k} |\lambda_i| \leq \sum_{i=1}^{k} \sigma_i(A),$$

and thus

$$|\mathrm{tr}\, A| \leq \sum_{i=1}^{k} |\lambda_i| \leq \mathrm{tr}\, \langle A \rangle.$$

Proof: See [730, p. 172], and use Fact 5.11.28. For the last statement, use Fact 2.21.12.

Remark: See Fact 5.11.28, Fact 8.19.22, and Fact 9.11.2.

Remark: This result is due to Weyl.

Fact 9.13.19. Let $A \in \mathbb{F}^{n \times n}$, and let $\mathrm{mspec}(A) = \{\lambda_1, \ldots, \lambda_n\}_{\mathrm{ms}}$, where $\lambda_1, \ldots, \lambda_n$ are ordered such that $|\lambda_n| \leq \cdots \leq |\lambda_1|$, and let $p \geq 0$. Then, for all $k \in \{1, \ldots, n\}$,

$$\left| \sum_{i=1}^{k} \lambda_i^p \right| \leq \sum_{i=1}^{k} |\lambda_i|^p \leq \sum_{i=1}^{k} \sigma_i^p(A).$$

Proof: See [201, p. 42].

Remark: This result is *Weyl's majorant theorem*.

Remark: See Fact 9.11.3.

Fact 9.13.20. Let $A \in \mathbb{F}^{n \times n}$, and define

$$r_i \triangleq \sum_{j=1}^{n} |A_{(i,j)}|, \qquad c_i \triangleq \sum_{j=1}^{n} |A_{(j,i)}|,$$

$$r_{\min} \triangleq \min_{i=1,\ldots,n} \|r_i\|_2, \qquad c_{\min} \triangleq \min_{i=1,\ldots,n} \|c_i\|_2,$$

$$\hat{r}_i \triangleq \sum_{\substack{j=1 \\ j \neq i}}^{n} |A_{(i,j)}|, \qquad \hat{c}_i \triangleq \sum_{\substack{j=1 \\ j \neq i}}^{n} |A_{(j,i)}|,$$

and

$$\alpha \triangleq \min_{i=1,\ldots,n} \left(|A_{(i,i)}| - \hat{r}_i \right), \qquad \beta \triangleq \min_{i=1,\ldots,n} \left(|A_{(i,i)}| - \hat{c}_i \right).$$

Then, the following statements hold:

i) If $\alpha > 0$, then A is nonsingular and

$$\|A^{-1}\|_{\mathrm{row}} < 1/\alpha.$$

ii) If $\beta > 0$, then A is nonsingular and

$$\|A^{-1}\|_{\mathrm{col}} < 1/\beta.$$

iii) If $\alpha > 0$ and $\beta > 0$, then A is nonsingular, and

$$\sqrt{\alpha\beta} \leq \sigma_{\min}(A).$$

iv) $\sigma_{\min}(A)$ satisfies

$$\min_{i=1,\ldots,n} \tfrac{1}{2}\left[2|A_{(i,i)}| - \hat{r}_i - \hat{c}_i\right] \le \sigma_{\min}(A).$$

v) $\sigma_{\min}(A)$ satisfies

$$\min_{i=1,\ldots,n} \tfrac{1}{2}\left[\left(4|A_{(i,i)}|^2 + [\hat{r}_i - \hat{c}_i]^2\right)^{1/2} - \hat{r}_i - \hat{c}_i\right] \le \sigma_{\min}(A).$$

vi) $\sigma_{\min}(A)$ satisfies

$$\left(\tfrac{n-1}{n}\right)^{(n-1)/2}|\det A| \max\left\{\frac{c_{\min}}{\prod_{i=1}^n c_i}, \frac{r_{\min}}{\prod_{i=1}^n r_i}\right\} \le \sigma_{\min}(A).$$

Proof: See Fact 9.8.23, [730, pp. 227, 231], and [725, 786, 796, 1401].

Fact 9.13.21. Let $A \in \mathbb{F}^{n \times n}$, and let $\mathrm{mspec}(A) = \{\lambda_1, \cdots, \lambda_n\}_{\mathrm{ms}}$, where $\lambda_1, \ldots, \lambda_n$ are ordered such that $|\lambda_n| \le \cdots \le |\lambda_1|$. Then, for all $i \in \{1, \ldots, n\}$,

$$\lim_{k\to\infty} \sigma_i^{1/k}(A^k) = |\lambda_i|.$$

In particular,

$$\lim_{k\to\infty} \sigma_{\max}^{1/k}(A^k) = \mathrm{sprad}(A).$$

Proof: See [730, p. 180].

Remark: This equality is due to Yamamoto.

Remark: The expression for $\mathrm{sprad}(A)$ is a special case of Proposition 9.2.6.

Fact 9.13.22. Let $A \in \mathbb{F}^{n \times n}$, and assume that A is nonzero. Then,

$$\frac{1}{\sigma_{\max}(A)} = \min_{B \in \{X \in \mathbb{F}^{n \times n}:\ \det(I-AX)=0\}} \sigma_{\max}(B).$$

Furthermore, there exists $B_0 \in \mathbb{F}^{n \times n}$ such that $\mathrm{rank}\, B_0 = 1$, $\det(I - AB_0) = 0$, and

$$\frac{1}{\sigma_{\max}(A)} = \sigma_{\max}(B_0).$$

Proof: If $\sigma_{\max}(B) < 1/\sigma_{\max}(A)$, then $\mathrm{sprad}(AB) \le \sigma_{\max}(AB) < 1$, and thus $I - AB$ is nonsingular. Hence,

$$\begin{aligned}
\frac{1}{\sigma_{\max}(A)} &= \min_{B \in \{X \in \mathbb{F}^{n \times n}:\ \sigma_{\max}(X) \ge 1/\sigma_{\max}(A)\}} \sigma_{\max}(B) \\
&= \min_{B \in \{X \in \mathbb{F}^{n \times n}:\ \sigma_{\max}(X) < 1/\sigma_{\max}(A)\}^{\sim}} \sigma_{\max}(B) \\
&\le \min_{B \in \{X \in \mathbb{F}^{n \times n}:\ \det(I-AX)=0\}} \sigma_{\max}(B).
\end{aligned}$$

Using the singular value decomposition, equality holds by constructing B_0 to have rank 1 and singular value $1/\sigma_{\max}(A)$.

Remark: This result is related to the *small-gain theorem*. See [1535, pp. 276, 277].

9.14 Facts on Matrix Norms and Singular Values for Two or More Matrices

Fact 9.14.1. Let $a_1, \ldots, a_n \in \mathbb{F}^n$ be linearly independent, and, for all $i \in \{1, \ldots, n\}$, define

$$A_i \triangleq I - (a_i^* a_i)^{-1} a_i a_i^*.$$

Then,

$$\sigma_{\max}(A_n A_{n-1} \cdots A_1) < 1.$$

Proof: Define $A \triangleq A_n A_{n-1} \cdots A_1$. Since $\sigma_{\max}(A_i) \le 1$ for all $i \in \{1, \ldots, n\}$, it follows that $\sigma_{\max}(A) \le 1$. Suppose that $\sigma_{\max}(A) = 1$, and let $x \in \mathbb{F}^n$ satisfy $x^* x = 1$ and $\|Ax\|_2 = 1$. Then, for all $i \in \{1, \ldots, n\}$, $\|A_i A_{i-1} \cdots A_1 x\|_2 = 1$. Consequently, $\|A_1 x\|_2 = 1$, which implies that $a_1^* x = 0$, and thus $A_1 x = x$. Hence, $\|A_i A_{i-1} \cdots A_2 x\|_2 = 1$. Repeating this argument implies that, for all $i \in \{1, \ldots, n\}$, $a_i^* x = 0$. Since $a_1, \ldots, a_n$ are linearly independent, it follows that $x = 0$, which is a contradiction.

Remark: This result is due to Akers and Djokovic.

Fact 9.14.2. Let $A_1, \ldots, A_n \in \mathbb{F}^{n \times n}$, assume that, for all $i, j \in \{1, \ldots, n\}$, $[A_i, A_j] = 0$, and assume that, for all $i \in \{1, \ldots, n\}$, $\sigma_{\max}(A_i) = 1$ and $\mathrm{sprad}(A_i) = 1$. Then,

$$\sigma_{\max}(A_n A_{n-1} \cdots A_1) < 1.$$

Proof: See [1514].

Fact 9.14.3. Let $A \in \mathbb{F}^{n \times m}$ and $B \in \mathbb{F}^{m \times n}$. Then,

$$|\mathrm{tr}\, AB| \le \|AB\|_{\sigma 1} = \sum_{i=1}^r \sigma_i(AB) \le \sum_{i=1}^r \sigma_i(A) \sigma_i(B).$$

Proof: Use Proposition 9.6.2 and Fact 9.11.2.

Remark: This result generalizes Fact 5.12.6.

Remark: Sufficient conditions for equality are given in [1215, p. 107].

Fact 9.14.4. Let $A \in \mathbb{F}^{n \times m}$ and $B \in \mathbb{F}^{m \times n}$. Then,

$$|\mathrm{tr}\, AB| \le \|AB\|_{\sigma 1} \le \sigma_{\max}(A) \|B\|_{\sigma 1}.$$

Proof: Use Corollary 9.3.8 and Fact 9.11.2.

Remark: This result generalizes Fact 5.12.7.

Fact 9.14.5. Let $A \in \mathbb{F}^{n \times m}$, $B \in \mathbb{F}^{m \times n}$, and $p \in [1, \infty)$, and assume that AB is normal. Then,

$$\|AB\|_{\sigma p} \le \|BA\|_{\sigma p}.$$

In particular,

$$\operatorname{tr} \langle AB \rangle \le \operatorname{tr} \langle BA \rangle,$$

$$\|AB\|_{\mathrm{F}} \le \|BA\|_{\mathrm{F}},$$

$$\sigma_{\max}(AB) \le \sigma_{\max}(BA).$$

Proof: This result is due to Simon. See [250].

Fact 9.14.6. Let $A, B \in \mathbb{R}^{n \times n}$, assume that A is nonsingular, and assume that B is singular. Then,

$$\sigma_{\min}(A) \le \sigma_{\max}(A - B).$$

Furthermore, if $\sigma_{\max}(A^{-1}) = \operatorname{sprad}(A^{-1})$, then there exists a singular matrix $C \in \mathbb{R}^{n \times n}$ such that $\sigma_{\max}(A - C) = \sigma_{\min}(A)$.

Proof: See [1125, p. 151].

Remark: This result is due to Franck.

Fact 9.14.7. Let $A \in \mathbb{C}^{n \times n}$, assume that A is nonsingular, let $\| \cdot \|$ and $\| \cdot \|'$ be norms on $\mathbb{C}^n$, let $\| \cdot \|''$ be the norm on $\mathbb{C}^{n \times n}$ induced by $\| \cdot \|$ and $\| \cdot \|'$, and let $\| \cdot \|'''$ be the norm on $\mathbb{C}^{n \times n}$ induced by $\| \cdot \|'$ and $\| \cdot \|$. Then,

$$\min\{\|B\|'' : \ B \in \mathbb{C}^{n \times n} \text{ and } A + B \text{ is nonsingular}\} = 1/\|A^{-1}\|'''.$$

In particular,

$$\min\{\|B\|_{\mathrm{col}} : \ B \in \mathbb{C}^{n \times n} \text{ and } A + B \text{ is singular}\} = 1/\|A^{-1}\|_{\mathrm{col}},$$

$$\min\{\sigma_{\max}(B) : \ B \in \mathbb{C}^{n \times n} \text{ and } A + B \text{ is singular}\} = \sigma_{\min}(A),$$

$$\min\{\|B\|_{\mathrm{row}} : \ B \in \mathbb{C}^{n \times n} \text{ and } A + B \text{ is singular}\} = 1/\|A^{-1}\|_{\mathrm{row}}.$$

Proof: See [697] and [699, p. 111].

Remark: This result is due to Gastinel. See [697].

Remark: The result involving $\sigma_{\max}(B)$ is equivalent to the inequality in Fact 9.14.6.

Fact 9.14.8. Let $A, B \in \mathbb{F}^{n \times m}$, and assume that $\operatorname{rank} A = \operatorname{rank} B$ and $\alpha \triangleq \sigma_{\max}(A^+)\sigma_{\max}(A - B) < 1$. Then,

$$\sigma_{\max}(B^+) < \frac{1}{1 - \alpha}\sigma_{\max}(A^+).$$

If, in addition, $n = m$, A and B are nonsingular, and $\sigma_{\max}(A - B) < \sigma_{\min}(A)$, then

$$\sigma_{\max}(B^{-1}) < \frac{\sigma_{\min}(A)}{\sigma_{\min}(A) - \sigma_{\max}(A - B)}\sigma_{\max}(A^{-1}).$$

Proof: See [699, p. 400].

Fact 9.14.9. Let $A, B \in \mathbb{F}^{n \times n}$. Then,

$$\sigma_{\max}(I - [A, B]) \geq 1.$$

Proof: Since $\operatorname{tr}[A, B] = 0$, it follows that there exists $\lambda \in \operatorname{spec}(I - [A, B])$ such that $\operatorname{Re} \lambda \geq 1$, and thus $|\lambda| \geq 1$. Hence, Corollary 9.4.5 implies that $\sigma_{\max}(I - [A, B]) \geq \operatorname{sprad}(I - [A, B]) \geq |\lambda| \geq 1$.

Fact 9.14.10. Let $A \in \mathbb{F}^{n \times m}$, and let $B \in \mathbb{F}^{k \times l}$ be a submatrix of A. Then, for all $i \in \{1, \ldots, \min\{k, l\}\}$,

$$\sigma_i(B) \leq \sigma_i(A).$$

Proof: Use Proposition 9.6.1.

Remark: Sufficient conditions for singular value interlacing are given in [728, p. 419].

Fact 9.14.11. Let

$$\mathcal{A} \triangleq \begin{bmatrix} A & B \\ C & D \end{bmatrix} \in \mathbb{F}^{(n+m) \times (n+m)},$$

assume that $\mathcal{A}$ is nonsingular, and define $\begin{bmatrix} E & F \\ G & H \end{bmatrix} \in \mathbb{F}^{(n+m) \times (n+m)}$ by

$$\begin{bmatrix} E & F \\ G & H \end{bmatrix} \triangleq \mathcal{A}^{-1}.$$

Then, the following statements hold:

i) For all $i \in \{1, \ldots, \min\{n, m\} - 1\}$,

$$\frac{\sigma_{n-i}(A)}{\sigma_{\max}^2(\mathcal{A})} \leq \sigma_{m-i}(H) \leq \frac{\sigma_{n-i}(A)}{\sigma_{\min}^2(\mathcal{A})}.$$

ii) Assume that $n < m$. Then, for all $i \in \{1, \ldots, m - n\}$,

$$\frac{1}{\sigma_{\max}(\mathcal{A})} \leq \sigma_i(H) \leq \frac{1}{\sigma_{\min}(\mathcal{A})}.$$

iii) Assume that $m < n$. Then, for all $i \in \{1, \ldots, n - m\}$,

$$\sigma_{\min}(\mathcal{A}) \leq \sigma_i(H) \leq \sigma_{\max}(\mathcal{A}).$$

iv) Assume that $n = m$. Then, for all $i \in \{1, \ldots, n\}$,

$$\frac{\sigma_i(A)}{\sigma_{\max}^2(\mathcal{A})} \leq \sigma_i(H) \leq \frac{\sigma_i(A)}{\sigma_{\min}^2(\mathcal{A})}.$$

v) Assume that $m < n$. Then,

$$\sigma_{\max}(H) \leq \frac{\sigma_{n-m+1}(A)}{\sigma_{\min}^2(\mathcal{A})}.$$

vi) Assume that $m < n$. Then, $H = 0$ if and only if $\operatorname{def} A = m$.

Now, assume in addition that $\mathcal{A}$ is unitary. Then, the following statements hold:

vii) If $n < m$, then

$$\sigma_i(D) = \begin{cases} 1, & 1 \le i \le m - n, \\ \sigma_{i-m+n}(A), & m - n < i \le m. \end{cases}$$

viii) If $n = m$, then, for all $i \in \{1, \dots, n\}$,

$$\sigma_i(D) = \sigma_i(A).$$

ix) If $n \le m$, then

$$|\det D| = \prod_{i=1}^{m} \sigma_i(D) = \prod_{i=1}^{n} \sigma_i(A) = |\det A|.$$

Proof: See [589].

Remark: Statement *vi*) is a special case of the nullity theorem given by Fact 2.11.20.

Remark: Statement *ix*) follows from Fact 3.11.19 using Fact 5.11.28.

Fact 9.14.12. Let $A \in \mathbb{F}^{n \times m}$, $B \in \mathbb{F}^{n \times l}$, $C \in \mathbb{F}^{k \times m}$, and $D \in \mathbb{F}^{k \times l}$. Then,

$$\sigma_{\max}\left(\begin{bmatrix} A & B \\ C & D \end{bmatrix}\right) \le \sigma_{\max}\left(\begin{bmatrix} \sigma_{\max}(A) & \sigma_{\max}(B) \\ \sigma_{\max}(C) & \sigma_{\max}(D) \end{bmatrix}\right).$$

Proof: See [738, 845].

Remark: This result is due to Tomiyama.

Remark: See Fact 8.19.29.

Fact 9.14.13. Let $A \in \mathbb{F}^{n \times m}$, $B \in \mathbb{F}^{n \times l}$, and $C \in \mathbb{F}^{k \times m}$. Then, for all $X \in \mathbb{F}^{k \times l}$,

$$\max\left\{\sigma_{\max}([\begin{array}{cc} A & B \end{array}]), \sigma_{\max}\left(\begin{bmatrix} A \\ C \end{bmatrix}\right)\right\} \le \sigma_{\max}\left(\begin{bmatrix} A & B \\ C & X \end{bmatrix}\right).$$

Furthermore, there exists a matrix $X \in \mathbb{F}^{k \times l}$ such that equality holds.

Remark: This result is *Parrott's theorem*. See [374], [459, pp. 271, 272], and [1535, pp. 40–42].

Fact 9.14.14. Let $A \in \mathbb{F}^{n \times m}$ and $B \in \mathbb{F}^{n \times l}$. Then,

$$\max\{\sigma_{\max}(A), \sigma_{\max}(B)\} \le \sigma_{\max}([\begin{array}{cc} A & B \end{array}])$$

$$\le [\sigma_{\max}^2(A) + \sigma_{\max}^2(B)]^{1/2}$$

$$\le \sqrt{2}\max\{\sigma_{\max}(A), \sigma_{\max}(B)\}$$

and, if $n \leq \min\{m, l\}$,

$$[\sigma_n^2(A) + \sigma_n^2(B)]^{1/2} \leq \sigma_n([\begin{array}{cc} A & B \end{array}]) \leq \begin{cases} [\sigma_n^2(A) + \sigma_{\max}^2(B)]^{1/2} \\[2mm] [\sigma_{\max}^2(A) + \sigma_n^2(B)]^{1/2}. \end{cases}$$

Problem: Obtain analogous bounds for $\sigma_i([\begin{array}{cc} A & B \end{array}])$.

Fact 9.14.15. Let $A, B \in \mathbb{F}^{n \times n}$. Then,

$\sigma_{\max}(A + B)$

$\leq \frac{1}{2}\Big[\sigma_{\max}(A) + \sigma_{\max}(B)$

$\qquad + \sqrt{[\sigma_{\max}(A) - \sigma_{\max}(B)]^2 + 4\max\{\sigma_{\max}^2(\langle A\rangle^{1/2}\langle B\rangle^{1/2}), \sigma_{\max}^2(\langle A^*\rangle^{1/2}\langle B^*\rangle^{1/2})\}}\Big]$

$\leq \sigma_{\max}(A) + \sigma_{\max}(B)$.

Proof: See [845].

Remark: See Fact 8.19.14.

Remark: This result interpolates the triangle inequality for the maximum singular value.

Fact 9.14.16. Let $A, B \in \mathbb{F}^{n \times n}$, and let $\alpha > 0$. Then,

$$\sigma_{\max}(A + B) \leq \left[(1 + \alpha)\sigma_{\max}^2(A) + (1 + \alpha^{-1})\sigma_{\max}^2(B)\right]^{1/2}$$

and

$$\sigma_{\min}(A + B) \leq \left[(1 + \alpha)\sigma_{\min}^2(A) + (1 + \alpha^{-1})\sigma_{\max}^2(B)\right]^{1/2}.$$

Fact 9.14.17. Let $A, B \in \mathbb{F}^{n \times n}$. Then,

$$\sigma_{\min}(A) - \sigma_{\max}(B) \leq |\det(A + B)|^{1/n}$$

$$\leq \prod_{i=1}^{n} |\sigma_i(A) + \sigma_{n-i+1}(B)|^{1/n}$$

$$\leq \sigma_{\max}(A) + \sigma_{\max}(B).$$

Proof: See [740, p. 63] and [920].

Fact 9.14.18. Let $A, B \in \mathbb{F}^{n \times n}$, and assume that $\sigma_{\max}(B) \leq \sigma_{\min}(A)$. Then,

$$0 \leq [\sigma_{\min}(A) - \sigma_{\max}(B)]^n$$

$$\leq \prod_{i=1}^{n} |\sigma_i(A) - \sigma_{n-i+1}(B)|$$

$$\leq |\det(A + B)|$$

$$\leq \prod_{i=1}^{n} |\sigma_i(A) + \sigma_{n-i+1}(B)|$$

$$\leq [\sigma_{\max}(A) + \sigma_{\max}(B)]^n.$$

Hence, if $\sigma_{\max}(B) < \sigma_{\min}(A)$, then A is nonsingular and $A + \alpha B$ is nonsingular for all $-1 \leq \alpha \leq 1$.

Proof: See [920].

Remark: See Fact 5.12.12 and Fact 11.18.16.

Fact 9.14.19. Let $A, B \in \mathbb{F}^{n \times m}$. Then, the following statements are equivalent:

i) For all $k \in \{1, \ldots, \min\{n, m\}\}$,

$$\sum_{i=1}^{k} \sigma_i(A) \leq \sum_{i=1}^{k} \sigma_i(B).$$

ii) For all unitarily invariant norms $\|\cdot\|$ on $\mathbb{F}^{n \times m}$, $\|A\| \leq \|B\|$.

Proof: See [730, pp. 205, 206].

Remark: This result is the *Fan dominance theorem*.

Fact 9.14.20. Let $A, B \in \mathbb{F}^{n \times m}$. Then, for all $k \in \{1, \ldots, \min\{n, m\}\}$,

$$\sum_{i=1}^{k} [\sigma_i(A) + \sigma_{\min\{n,m\}+1-i}(B)] \leq \sum_{i=1}^{k} \sigma_i(A + B) \leq \sum_{i=1}^{k} [\sigma_i(A) + \sigma_i(B)].$$

Furthermore, if either $\sigma_{\max}(A) < \sigma_{\min}(B)$ or $\sigma_{\max}(B) < \sigma_{\min}(A)$, then, for all $k \in \{1, \ldots, \min\{n, m\}\}$,

$$\sum_{i=1}^{k} \sigma_i(A + B) \leq \sum_{i=1}^{k} |\sigma_i(A) - \sigma_{\min\{n,m\}+1-i}(B)|.$$

Proof: See Proposition 9.2.2, [730, pp. 196, 197] and [920].

Fact 9.14.21. Let $A, B \in \mathbb{F}^{n \times m}$, and let $\alpha \in [0, 1]$. Then, for all $i \in \{1, \ldots, \min\{n, m\}\}$,

$$\sigma_i[\alpha A + (1 - \alpha)B] \leq \begin{cases} \sigma_i\left(\begin{bmatrix} A & 0 \\ 0 & B \end{bmatrix}\right) \\ \sigma_i\left(\begin{bmatrix} \sqrt{2\alpha}A & 0 \\ 0 & \sqrt{2(1-\alpha)}B \end{bmatrix}\right), \end{cases}$$

and

$$2\sigma_i(AB^*) \leq \sigma_i(\langle A \rangle^2 + \langle B \rangle^2).$$

Furthermore,

$$\langle \alpha A + (1 - \alpha)B \rangle^2 \leq \alpha \langle A \rangle^2 + (1 - \alpha) \langle B \rangle^2.$$

If, in addition, $n = m$, then, for all $i \in \{1, \ldots, n\}$,

$$\tfrac{1}{2}\sigma_i(A + A^*) \leq \sigma_i\left(\begin{bmatrix} A & 0 \\ 0 & A \end{bmatrix}\right).$$

Proof: See [716].

Remark: See Fact 9.14.23.

Fact 9.14.22. Let $A \in \mathbb{F}^{n \times m}$ and $B \in \mathbb{F}^{l \times m}$, and let $p, q > 1$ satisfy $1/p + 1/q = 1$. Then, for all $i \in \{1, \ldots, \min\{n, m, l\}\}$,

$$\sigma_i(AB^*) \leq \sigma_i\left(\tfrac{1}{p}\langle A \rangle^p + \tfrac{1}{q}\langle B \rangle^q\right).$$

Equivalently, there exists a unitary matrix $S \in \mathbb{F}^{m \times m}$ such that

$$\langle AB^* \rangle^{1/2} \leq S^*\left(\tfrac{1}{p}\langle A \rangle^p + \tfrac{1}{q}\langle B \rangle^q\right)S.$$

Proof: See [49, 51, 712] or [1521, p. 28].

Remark: This result is a matrix version of Young's inequality. See Fact 1.12.32.

Fact 9.14.23. Let $A \in \mathbb{F}^{n \times m}$ and $B \in \mathbb{F}^{l \times m}$. Then, for all $i \in \{1, \ldots, \min\{n, m, l\}\}$,

$$\sigma_i(AB^*) \leq \tfrac{1}{2}\sigma_i(A^*A + B^*B).$$

Proof: Set $p = q = 2$ in Fact 9.14.22. See [213].

Remark: See Fact 9.9.47 and Fact 9.14.21.

Fact 9.14.24. Let $A, B, C, D \in \mathbb{F}^{n \times m}$. Then, for all $i \in \{1, \ldots, \min\{n, m\}\}$,

$$\sqrt{2}\sigma_i(\langle AB^* + CD^* \rangle) \leq \sigma_i\left(\begin{bmatrix} A & B \\ C & D \end{bmatrix}\right).$$

Proof: See [711].

Fact 9.14.25. Let $A, B, C, D, X \in \mathbb{F}^{n \times n}$, assume that A, B, C, D are positive semidefinite, and assume that $0 \leq A \leq C$ and $0 \leq B \leq D$. Then, for all $i \in \{1, \ldots, n\}$,

$$\sigma_i(A^{1/2}XB^{1/2}) \leq \sigma_i(C^{1/2}XD^{1/2}).$$

Proof: See [716, 840].

Fact 9.14.26. Let $A_1, \ldots, A_k \in \mathbb{F}^{n \times n}$. Then, for all $l \in \{1, \ldots, n\}$,

$$\sum_{i=1}^{l} \sigma_i\left(\prod_{j=1}^{k} A_j\right) \leq \sum_{i=1}^{l} \prod_{j=1}^{k} \sigma_i(A_j).$$

Proof: See [325].

Remark: This result is a weak majorization relation.

Fact 9.14.27. Let $A, B \in \mathbb{F}^{n \times m}$, and let $1 \leq l_1 < \cdots < l_k \leq \min\{n, m\}$. Then,

$$\sum_{i=1}^{k} \sigma_{l_i}(A)\sigma_{n-i+1}(B) \leq \sum_{i=1}^{k} \sigma_{l_i}(AB) \leq \sum_{i=1}^{k} \sigma_{l_i}(A)\sigma_i(B)$$

and

$$\sum_{i=1}^{k} \sigma_{l_i}(A)\sigma_{n-l_i+1}(B) \leq \sum_{i=1}^{k} \sigma_i(AB).$$

In particular,

$$\sum_{i=1}^{k}\sigma_i(A)\sigma_{n-i+1}(B) \le \sum_{i=1}^{k}\sigma_i(AB) \le \sum_{i=1}^{k}\sigma_i(A)\sigma_i(B).$$

Furthermore,

$$\prod_{i=1}^{k}\sigma_{l_i}(AB) \le \prod_{i=1}^{k}\sigma_{l_i}(A)\sigma_i(B)$$

with equality for $k = n$. Furthermore,

$$\prod_{i=1}^{k}\sigma_{l_i}(A)\sigma_{n-l_i+1}(B) \le \prod_{i=1}^{k}\sigma_i(AB)$$

with equality for $k = n$. In particular,

$$\prod_{i=1}^{k}\sigma_i(A)\sigma_{n-i+1}(B) \le \prod_{i=1}^{k}\sigma_i(AB) \le \prod_{i=1}^{k}\sigma_i(A)\sigma_i(B)$$

with equality for $k = n$.

Proof: See [1422].

Remark: See Fact 8.19.20 and Fact 8.19.23.

Remark: The left-hand inequalities in the first and third strings are conjectures. See [1422].

Fact 9.14.28. Let $A \in \mathbb{F}^{n \times m}$, let $k \ge 1$ satisfy $k < \operatorname{rank} A$, and let $\|\cdot\|$ be a unitarily invariant norm on $\mathbb{F}^{n \times m}$. Then,

$$\min_{B \in \{X \in \mathbb{F}^{n \times m}: \ \operatorname{rank} X \le k\}} \|A - B\| = \|A - B_0\|,$$

where B_0 is formed by replacing $(\operatorname{rank} A) - k$ smallest positive singular values in the singular value decomposition of A by 0's. Furthermore,

$$\sigma_{\max}(A - B_0) = \sigma_{k+1}(A)$$

and

$$\|A - B_0\|_F = \sqrt{\sum_{i=k+1}^{r}\sigma_i^2(A)}.$$

Furthermore, B_0 is the unique solution if and only if $\sigma_{k+1}(A) < \sigma_k(A)$.

Proof: This result follows from Fact 9.14.29 with $B_\sigma \triangleq \operatorname{diag}[\sigma_1(A), \dots, \sigma_k(A),$ $0_{(n-k)\times(m-k)}]$, $S_1 = I_n$, and $S_2 = I_m$. See [583] and [1261, p. 208].

Remark: This result is known as the *Schmidt-Mirsky theorem*. For the case of the Frobenius norm, the result is known as the *Eckart-Young theorem*. See [520] and [1261, p. 210].

Remark: See Fact 9.15.8.

Fact 9.14.29. Let $A, B \in \mathbb{F}^{n \times m}$, define $A_\sigma, B_\sigma \in \mathbb{F}^{n \times m}$ by

$$
A_\sigma \triangleq \begin{bmatrix} \sigma_1(A) & & & \\ & \ddots & & \\ & & \sigma_r(A) & \\ & & & 0_{(n-r) \times (m-r)} \end{bmatrix},
$$

where $r \triangleq \operatorname{rank} A$, and

$$
B_\sigma \triangleq \begin{bmatrix} \sigma_1(B) & & & \\ & \ddots & & \\ & & \sigma_l(B) & \\ & & & 0_{(n-l) \times (m-l)} \end{bmatrix},
$$

where $l \triangleq \operatorname{rank} B$, let $S_1 \in \mathbb{F}^{n \times n}$ and $S_2 \in \mathbb{F}^{m \times m}$ be unitary matrices, and let $\| \cdot \|$ be a unitarily invariant norm on $\mathbb{F}^{n \times m}$. Then,

$$
\|A_\sigma - B_\sigma\| \leq \|A - S_1 B S_2\| \leq \|A_\sigma + B_\sigma\|.
$$

In particular,

$$
\max_{i \in \{1, \ldots, \max\{r,l\}\}} |\sigma_i(A) - \sigma_i(B)| \leq \sigma_{\max}(A - B) \leq \sigma_{\max}(A) + \sigma_{\max}(B).
$$

Proof: See [1424].

Remark: In the case $S_1 = I_n$ and $S_2 = I_m$, the left-hand inequality is *Mirsky's theorem*. See [1261, p. 204].

Remark: See Fact 9.12.4.

Fact 9.14.30. Let $A, B \in \mathbb{F}^{n \times m}$, and assume that $\operatorname{rank} A = \operatorname{rank} B$. Then,

$$
\sigma_{\max}[AA^+(I - BB^+)] = \sigma_{\max}[BB^+(I - AA^+)]
$$
$$
\leq \min\{\sigma_{\max}(A^+), \sigma_{\max}(B^+)\}\sigma_{\max}(A - B).
$$

Proof: See [699, p. 400] and [1261, p. 141].

Fact 9.14.31. Let $A, B \in \mathbb{F}^{n \times m}$. Then, for all $k \in \{1, \ldots, \min\{n, m\}\}$,

$$
\sum_{i=1}^{k} \sigma_i(A \circ B) \leq \sum_{i=1}^{k} d_i^{1/2}(A^*A) d_i^{1/2}(BB^*)
$$

$$
\leq \left\{ \begin{array}{c} \sum_{i=1}^{k} d_i^{1/2}(A^*A)\sigma_i(B) \\ \sum_{i=1}^{k} \sigma_i(A) d_i^{1/2}(BB^*) \end{array} \right\}
$$

$$
\leq \sum_{i=1}^{k} \sigma_i(A)\sigma_i(B)
$$

and

$$\sum_{i=1}^{k} \sigma_i(A \circ B) \leq \sum_{i=1}^{k} d_i^{1/2}(AA^*) d_i^{1/2}(B^*B)$$

$$\leq \left\{ \begin{array}{c} \sum_{i=1}^{k} d_i^{1/2}(AA^*)\sigma_i(B) \\ \sum_{i=1}^{k} \sigma_i(A) d_i^{1/2}(B^*B) \end{array} \right\}$$

$$\leq \sum_{i=1}^{k} \sigma_i(A)\sigma_i(B).$$

In particular,

$$\sigma_{\max}(A \circ B) \leq \|A\|_{2,1}\|B\|_{\infty,2} \leq \left\{ \begin{array}{c} \|A\|_{2,1}\sigma_{\max}(B) \\ \sigma_{\max}(A)\|B\|_{\infty,2} \end{array} \right\} \leq \sigma_{\max}(A)\sigma_{\max}(B)$$

and

$$\sigma_{\max}(A \circ B) \leq \|A\|_{\infty,2}\|B\|_{2,1} \leq \left\{ \begin{array}{c} \|A\|_{\infty,2}\sigma_{\max}(B) \\ \sigma_{\max}(A)\|B\|_{2,1} \end{array} \right\} \leq \sigma_{\max}(A)\sigma_{\max}(B).$$

Proof: See [58, 1001, 1517] and [730, pp. 332, 334], and use Fact 2.21.2, Fact 8.18.8, and Fact 9.8.24.

Remark: $d_i^{1/2}(A^*A)$ and $d_i^{1/2}(AA^*)$ are the ith largest Euclidean norms of the columns and rows of A, respectively.

Remark: For related results, see [1378].

Remark: The case of equality is discussed in [327].

Fact 9.14.32. Let $A, B \in \mathbb{F}^{n \times m}$. Then,

$$\sigma_{\max}(A \circ B) \leq \sqrt{n}\|A\|_{\infty}\sigma_{\max}(B).$$

Now, assume in addition that $n = m$ and that either A is positive semidefinite and B is Hermitian or A and B are nonnegative and symmetric. Then,

$$\sigma_{\max}(A \circ B) \leq \|A\|_{\infty}\sigma_{\max}(B).$$

Next, assume that A and B are real, let β denote the smallest positive entry of $|B|$, and assume that B is symmetric and positive semidefinite. Then,

$$\mathrm{sprad}(A \circ B) \leq \frac{\|A\|_{\infty}\|B\|_{\infty}}{\beta}\sigma_{\max}(B)$$

and

$$\mathrm{sprad}(B) \leq \mathrm{sprad}(|B|) \leq \frac{\|B\|_{\infty}}{\beta}\,\mathrm{sprad}(B).$$

Proof: See [1107].

Fact 9.14.33. Let $A, B \in \mathbb{F}^{n \times m}$, and let $p \in [2, \infty)$ be an even integer. Then,

$$\|A \circ B\|_{\sigma p}^2 \le \|A \circ \overline{A}\|_{\sigma p} \|B \circ \overline{B}\|_{\sigma p}.$$

In particular,

$$\|A \circ B\|_{\mathrm{F}}^2 \le \|A \circ \overline{A}\|_{\mathrm{F}} \|B \circ \overline{B}\|_{\mathrm{F}}$$

and

$$\sigma_{\max}^2(A \circ B) \le \sigma_{\max}(A \circ \overline{A})\sigma_{\max}(B \circ \overline{B}) \le \sigma_{\max}^2(A)\sigma_{\max}^2(B).$$

Equality holds if $B = \overline{A}$. Furthermore,

$$\|A \circ A\|_{\sigma p} \le \|A \circ \overline{A}\|_{\sigma p}.$$

In particular,

$$\|A \circ A\|_{\mathrm{F}} \le \|A \circ \overline{A}\|_{\mathrm{F}}$$

and

$$\sigma_{\max}(A \circ A) \le \sigma_{\max}(A \circ \overline{A}).$$

Now, assume in addition that $n = m$. Then,

$$\|A \circ A^{\mathrm{T}}\|_{\sigma p} \le \|A \circ \overline{A}\|_{\sigma p}.$$

In particular,

$$\|A \circ A^{\mathrm{T}}\|_{\mathrm{F}} \le \|A \circ \overline{A}\|_{\mathrm{F}}$$

and

$$\sigma_{\max}(A \circ A^{\mathrm{T}}) \le \sigma_{\max}(A \circ \overline{A}).$$

Finally,

$$\|A \circ A^*\|_{\sigma p} \le \|A \circ \overline{A}\|_{\sigma p}.$$

In particular,

$$\|A \circ A^*\|_{\mathrm{F}} \le \|A \circ \overline{A}\|_{\mathrm{F}}$$

and

$$\sigma_{\max}(A \circ A^*) \le \sigma_{\max}(A \circ \overline{A}).$$

Proof: See [730, p. 340] and [731, 1224].

Remark: See Fact 7.6.16 and Fact 9.14.31.

Fact 9.14.34. Let $A, B \in \mathbb{C}^{n \times m}$. Then,

$$\begin{aligned}
\|A \circ B\|_{\mathrm{F}}^2 &= \sum_{i=1}^{n} \sigma_i^2(A \circ B) \\
&= \operatorname{tr}(A \circ B)(\overline{A} \circ \overline{B})^{\mathrm{T}} \\
&= \operatorname{tr}(A \circ \overline{A})(B \circ \overline{B})^{\mathrm{T}} \\
&\le \sum_{i=1}^{n} \sigma_i[(A \circ \overline{A})(B \circ \overline{B})^{\mathrm{T}}] \\
&\le \sum_{i=1}^{n} \sigma_i(A \circ \overline{A})\sigma_i(B \circ \overline{B}).
\end{aligned}$$

Proof: See [749].

Fact 9.14.35. Let $A, B \in \mathbb{F}^{n \times n}$. Then,

$$\|(A \circ B)(\overline{A} \circ \overline{B})^{\mathrm{T}}\|_{\sigma 1} \le \|(A \circ \overline{A})(B \circ \overline{B})^{\mathrm{T}}\|_{\sigma 1},$$

$$\|(A \circ B)(\overline{A} \circ \overline{B})^{\mathrm{T}}\|_{\mathrm{F}} \le \|(A \circ \overline{A})(B \circ \overline{B})^{\mathrm{T}}\|_{\mathrm{F}},$$

$$\sigma_{\max}[(A \circ B)(\overline{A} \circ \overline{B})^{\mathrm{T}}] \le \sigma_{\max}[(A \circ \overline{A})(B \circ \overline{B})^{\mathrm{T}}].$$

Proof: See [452].

Fact 9.14.36. Let $A, B \in \mathbb{R}^{n \times n}$, assume that A and B are nonnegative, and let $\alpha \in [0, 1]$. Then,

$$\sigma_{\max}(A^{\circ \alpha} \circ B^{\circ(1-\alpha)}) \le \sigma_{\max}^{\alpha}(A)\sigma_{\max}^{1-\alpha}(B).$$

In particular,

$$\sigma_{\max}(A^{\circ 1/2} \circ B^{\circ 1/2}) \le \sqrt{\sigma_{\max}(A)\sigma_{\max}(B)}.$$

Finally,

$$\sigma_{\max}(A^{\circ 1/2} \circ A^{\circ 1/2\mathrm{T}}) \le \sigma_{\max}(A^{\circ \alpha} \circ A^{\circ(1-\alpha)\mathrm{T}}) \le \sigma_{\max}(A).$$

Proof: See [1224].

Remark: See Fact 7.6.17.

Fact 9.14.37. Let $\|\cdot\|$ be a unitarily invariant norm on $\mathbb{C}^{n \times n}$, and let $A, X, B \in \mathbb{C}^{n \times n}$. Then,

$$\|A \circ X \circ B\| \le \tfrac{1}{2}\sqrt{n}\|A \circ X \circ \overline{A} + B \circ X \circ \overline{B}\|$$

and

$$\|A \circ X \circ B\|^2 \le n\|A \circ X \circ \overline{A}\|\|B \circ X \circ \overline{B}\|.$$

Furthermore,

$$\|A \circ X \circ B\|_{\mathrm{F}} \le \tfrac{1}{2}\|A \circ X \circ \overline{A} + B \circ X \circ \overline{B}\|_{\mathrm{F}}.$$

Proof: See [749].

Fact 9.14.38. Let $A \in \mathbb{F}^{n \times m}$, $B \in \mathbb{F}^{l \times k}$, and $p \in [1, \infty]$. Then,

$$\|A \otimes B\|_{\sigma p} = \|A\|_{\sigma p}\|B\|_{\sigma p}.$$

In particular,

$$\sigma_{\max}(A \otimes B) = \sigma_{\max}(A)\sigma_{\max}(B)$$

and

$$\|A \otimes B\|_{\mathrm{F}} = \|A\|_{\mathrm{F}}\|B\|_{\mathrm{F}}.$$

Proof: See [708, p. 722].

9.15 Facts on Linear Equations and Least Squares

Fact 9.15.1. Let $A \in \mathbb{R}^{n \times n}$, assume that A is nonsingular, let $b \in \mathbb{R}^n$, and let $\hat{x} \in \mathbb{R}^n$. Then,

$$\frac{1}{\kappa(A)}\frac{\|A\hat{x} - b\|}{\|b\|} \le \frac{\|\hat{x} - A^{-1}b\|}{\|A^{-1}b\|} \le \kappa(A)\frac{\|A\hat{x} - b\|}{\|b\|},$$

where $\kappa(A) \triangleq \|A\| \|A^{-1}\|$ and the vector and matrix norms are compatible. Equivalently, letting $\hat{b} \triangleq A\hat{x}$ and $x \triangleq A^{-1}b$, it follows that

$$\frac{1}{\kappa(A)} \frac{\|\hat{b} - b\|}{\|b\|} \le \frac{\|\hat{x} - x\|}{\|x\|} \le \kappa(A) \frac{\|\hat{b} - b\|}{\|b\|}.$$

Remark: This result estimates the accuracy of an approximate solution $\hat{x}$ to $Ax = b$. $\kappa(A)$ is the *condition number* of A.

Remark: For $\|A\| = \sigma_{\max}(A)$, note that $\kappa(A) = \kappa(A^{-1})$.

Remark: See [1538].

Fact 9.15.2. Let $A \in \mathbb{R}^{n \times n}$, assume that A is nonsingular, let $\tilde{A} \in \mathbb{R}^{n \times n}$, assume that $\|A^{-1}\tilde{A}\| < 1$, and let $b, \tilde{b} \in \mathbb{R}^n$. Furthermore, let $x \in \mathbb{R}^n$ satisfy $Ax = b$, and let $\hat{x} \in \mathbb{R}^n$ satisfy $(A + \tilde{A})\hat{x} = b + \tilde{b}$. Then,

$$\frac{\|\hat{x} - x\|}{\|x\|} \le \frac{\kappa(A)}{1 - \|A^{-1}\tilde{A}\|} \left(\frac{\|\tilde{b}\|}{\|b\|} + \frac{\|\tilde{A}\|}{\|A\|} \right),$$

where $\kappa(A) \triangleq \|A\| \|A^{-1}\|$ and the vector and matrix norms are compatible. If, in addition, $\|A^{-1}\| \|\tilde{A}\| < 1$, then

$$\frac{1}{\kappa(A) + 1} \frac{\|\tilde{b} - \tilde{A}x\|}{\|b\|} \le \frac{\|\hat{x} - x\|}{\|x\|} \le \frac{\kappa(A)}{1 - \|A^{-1}\tilde{A}\|} \frac{\|\tilde{b} - \tilde{A}x\|}{\|b\|}.$$

Proof: See [417, 418].

Fact 9.15.3. Let $A, \hat{A} \in \mathbb{R}^{n \times n}$ satisfy $\|A^+\hat{A}\| < 1$, let $b \in \mathcal{R}(A)$, let $\hat{b} \in \mathbb{R}^n$, and assume that $b + \hat{b} \in \mathcal{R}(A + \hat{A})$. Furthermore, let $\hat{x} \in \mathbb{R}^n$ satisfy $(A + \hat{A})\hat{x} = b + \hat{b}$. Then, $x \triangleq A^+b + (I - A^+A)\hat{x}$ satisfies $Ax = b$ and

$$\frac{\|\hat{x} - x\|}{\|x\|} \le \frac{\kappa(A)}{1 - \|A^+\hat{A}\|} \left(\frac{\|\hat{b}\|}{\|b\|} + \frac{\|\hat{A}\|}{\|A\|} \right),$$

where $\kappa(A) \triangleq \|A\| \|A^{-1}\|$ and the vector and matrix norms are compatible.

Proof: See [417].

Remark: See [418] for a lower bound.

Fact 9.15.4. Let $A \in \mathbb{F}^{n \times m}$ and $b \in \mathbb{F}^n$, and define

$$f(x) \triangleq (Ax - b)^*(Ax - b) = \|Ax - b\|_2^2,$$

where $x \in \mathbb{F}^m$. Then, f has a minimizer. Furthermore, $x \in \mathbb{F}^m$ minimizes f if and only if there exists a vector $y \in \mathbb{F}^m$ such that

$$x = A^+b + (I - A^+A)y.$$

In this case,

$$f(x) = b^*(I - AA^+)b.$$

Furthermore, if $y \in \mathbb{F}^m$ is such that $(I - A^+A)y$ is nonzero, then

$$\|A^+b\|_2 < \|A^+b + (I - A^+A)y\|_2 = \sqrt{\|A^+b\|_2^2 + \|(I - A^+A)y\|_2^2}.$$

Finally, A^+b is the unique minimizer of f if and only if A is left invertible.

Remark: The minimization of f is the *least squares problem*. See [16, 230, 1257]. Note that the expression for x is identical to the expression (6.1.13) for solutions of $Ax = b$. Therefore, x satisfies $Ax = b$ if and only if x is optimal in the least-squares sense. However, unlike Proposition 6.1.7, consistency is not assumed, that is, there need not exist a solution to $Ax = b$.

Remark: This result is a special case of Fact 8.14.15.

Fact 9.15.5. Let $A \in \mathbb{F}^{n \times m}$ and $b \in \mathbb{F}^n$, define

$$f(x) \triangleq (Ax - b)^*(Ax - b) = \|Ax - b\|_2^2,$$

where $x \in \mathbb{F}^m$, and let $B \in \mathbb{F}^{m \times n}$ be a (1,3)-inverse of A. Then, $\hat{x} = Bb$ minimizes f. Furthermore, let $z \in \mathbb{F}^m$. Then, the following statements are equivalent:

i) z minimizes f.

ii) $f(z) = b^*(I - AB)b$.

iii) $Az = ABb$.

iv) There exists $y \in \mathbb{R}^m$ such that

$$z = Bb + (I - BA)y.$$

Proof: See [1208, pp. 233–236].

Fact 9.15.6. Let $A \in \mathbb{F}^{n \times m}$, $B \in \mathbb{F}^{n \times l}$, and define

$$f(X) \triangleq \operatorname{tr}[(AX - B)^*(AX - B)] = \|AX - B\|_F^2,$$

where $X \in \mathbb{F}^{m \times l}$. Then, $X = A^+B$ minimizes f.

Problem: Determine all minimizers.

Problem: Consider $f(X) = \operatorname{tr}[(AX - B)^*C(AX - B)]$, where $C \in \mathbb{F}^{n \times n}$ is positive definite.

Fact 9.15.7. Let $A \in \mathbb{F}^{n \times m}$ and $B \in \mathbb{F}^{l \times m}$, and define

$$f(X) \triangleq \operatorname{tr}[(XA - B)^*(XA - B)] = \|XA - B\|_F^2,$$

where $X \in \mathbb{F}^{l \times n}$. Then, $X = BA^+$ minimizes f.

Fact 9.15.8. Let $A \in \mathbb{F}^{n \times m}$, $B \in \mathbb{F}^{n \times p}$, and $C \in \mathbb{F}^{q \times m}$, and let $k \geq 1$ satisfy $k < \operatorname{rank} A$. Then,

$$\min_{X \in \{Y \in \mathbb{F}^{p \times q}: \ \operatorname{rank} Y \leq k\}} \|A - BXC\|_F = \|A - BX_0C\|_F,$$

where $X_0 = B^+SC^+$ and S is formed by replacing all but the k largest singular values in the singular value decomposition of BB^+AC^+C by 0's. Furthermore, X_0 is a solution that minimizes $\|X\|_F$. Finally, X_0 is the unique solution if and only if either $\operatorname{rank} BB^+AC^+C \leq k$ or both $k \leq BB^+AC^+C$ and $\sigma_{k+1}(BB^+AC^+C) < \sigma_k(BB^+AC^+C)$.

Proof: See [520].

Remark: This result generalizes Fact 9.14.28.

Fact 9.15.9. Let $A, B \in \mathbb{F}^{n \times m}$, and define

$$f(X) \triangleq \mathrm{tr}[(AX - B)^*(AX - B)] = \|AX - B\|_\mathrm{F}^2,$$

where $X \in \mathbb{F}^{m \times m}$ is unitary. Then, $X = S_1 S_2$ minimizes f, where $S_1\left[\begin{smallmatrix} \hat{B} & 0 \\ 0 & 0 \end{smallmatrix}\right]S_2$ is the singular value decomposition of A^*B.

Proof: See [148, p. 224]. See also [996, pp. 269, 270].

Fact 9.15.10. Let $A, B \in \mathbb{R}^{n \times n}$, and define

$$f(X_1, X_2) \triangleq \mathrm{tr}\left[(X_1 A X_2 - B)^\mathrm{T}(X_1 A X_2 - B)\right] = \|X_1 A X_2 - B\|_\mathrm{F}^2,$$

where $X_1, X_2 \in \mathbb{R}^{n \times n}$ are orthogonal. Then, $(X_1, X_2) = (V_2^\mathrm{T} U_1^\mathrm{T}, V_1^\mathrm{T} U_2^\mathrm{T})$ minimizes f, where $U_1\left[\begin{smallmatrix} \hat{A} & 0 \\ 0 & 0 \end{smallmatrix}\right]V_1$ is the singular value decomposition of A and $U_2\left[\begin{smallmatrix} \hat{B} & 0 \\ 0 & 0 \end{smallmatrix}\right]V_2$ is the singular value decomposition of B.

Proof: See [996, p. 270].

Remark: This result is due to Kristof.

Remark: See Fact 3.9.5.

Problem: Extend this result to $\mathbb{C}$ and nonsquare matrices.

Fact 9.15.11. Let $A \in \mathbb{R}^{n \times m}$, let $b \in \mathbb{R}^n$, and assume that $\mathrm{rank}\left[\begin{array}{cc} A & b \end{array}\right] = m + 1$. Furthermore, consider the singular value decomposition of $\left[\begin{array}{cc} A & b \end{array}\right]$ given by

$$\left[\begin{array}{cc} A & b \end{array}\right] = U\left[\begin{array}{c} \Sigma \\ 0_{(n-m-1)\times(m+1)} \end{array}\right]V,$$

where $U \in \mathbb{R}^{n \times n}$ and $V \in \mathbb{R}^{(m+1)\times(m+1)}$ are orthogonal and

$$\Sigma \triangleq \mathrm{diag}[\sigma_1(A), \ldots, \sigma_{m+1}(A)].$$

Furthermore, define $\hat{A} \in \mathbb{R}^{n \times m}$ and $\hat{b} \in \mathbb{R}^n$ by

$$\left[\begin{array}{cc} \hat{A} & \hat{b} \end{array}\right] \triangleq U\left[\begin{array}{c} \Sigma_0 \\ 0_{(n-m-1)\times(m+1)} \end{array}\right]V,$$

where $\Sigma_0 \triangleq \mathrm{diag}[\sigma_1(A), \ldots, \sigma_m(A), 0]$. Finally, assume that $V_{(m+1,m+1)} \neq 0$, and define

$$\hat{x} \triangleq -\frac{1}{V_{(m+1,m+1)}}\left[\begin{array}{c} V_{(m+1,1)} \\ \vdots \\ V_{(m+1,m)} \end{array}\right].$$

Then, $\hat{A}\hat{x} = \hat{b}$.

Remark: $\hat{x}$ is the *total least squares solution*. See [1392].

Remark: The construction of $\left[\begin{array}{cc} \hat{A} & \hat{b} \end{array}\right]$ is based on Fact 9.14.28.

9.16 Notes

The equivalence of absolute and monotone norms given by Proposition 9.1.2 is due to [159]. More general monotonicity conditions are considered in [790]. Induced lower bounds are treated in [892, pp. 369, 370]. See also [1261, pp. 33, 80]. The induced norms (9.4.13) and (9.4.14) are given in [318] and [699, p. 116]. Alternative norms for the convolution operator are given in [318, 1469]. Proposition 9.3.6 is given in [1155, p. 97]. Norm-related topics are discussed in [173]. Spectral perturbation theory in finite and infinite dimensions is treated in [818], where the emphasis is on the regularity of the spectrum as a function of the perturbation rather than on bounds for finite perturbations.

Chapter Ten

Functions of Matrices and Their Derivatives

The norms discussed in Chapter 9 provide the foundation for the development in this chapter of some basic results in topology and analysis.

10.1 Open Sets and Closed Sets

Let $\|\cdot\|$ be a norm on $\mathbb{F}^n$, let $x \in \mathbb{F}^n$, and let $\varepsilon > 0$. Then, define the *open ball of radius ε centered at x* by

$$\mathbb{B}_\varepsilon(x) \triangleq \{y \in \mathbb{F}^n\colon \ \|x - y\| < \varepsilon\} \tag{10.1.1}$$

and the *sphere of radius ε centered at x* by

$$\mathbb{S}_\varepsilon(x) \triangleq \{y \in \mathbb{F}^n\colon \ \|x - y\| = \varepsilon\}. \tag{10.1.2}$$

Definition 10.1.1. Let $\mathcal{S} \subseteq \mathbb{F}^n$. The vector $x \in \mathcal{S}$ is an *interior point* of $\mathcal{S}$ if there exists $\varepsilon > 0$ such that $\mathbb{B}_\varepsilon(x) \subseteq \mathcal{S}$. The *interior* of $\mathcal{S}$ is the set

$$\operatorname{int} \mathcal{S} \triangleq \{x \in \mathcal{S}\colon \ x \text{ is an interior point of } \mathcal{S}\}. \tag{10.1.3}$$

Finally, $\mathcal{S}$ is *open* if every element of $\mathcal{S}$ is an interior point, that is, if $\mathcal{S} = \operatorname{int} \mathcal{S}$.

Definition 10.1.2. Let $\mathcal{S} \subseteq \mathcal{S}' \subseteq \mathbb{F}^n$. The vector $x \in \mathcal{S}$ is an *interior point* of $\mathcal{S}$ *relative* to $\mathcal{S}'$ if there exists $\varepsilon > 0$ such that $\mathbb{B}_\varepsilon(x) \cap \mathcal{S}' \subseteq \mathcal{S}$ or, equivalently, $\mathbb{B}_\varepsilon(x) \cap \mathcal{S} = \mathbb{B}_\varepsilon(x) \cap \mathcal{S}'$. The *interior* of $\mathcal{S}$ *relative* to $\mathcal{S}'$ is the set

$$\operatorname{int}_{\mathcal{S}'} \mathcal{S} \triangleq \{x \in \mathcal{S}\colon \ x \text{ is an interior point of } \mathcal{S} \text{ relative to } \mathcal{S}'\}. \tag{10.1.4}$$

Finally, $\mathcal{S}$ is *open relative* to $\mathcal{S}'$ if $\mathcal{S} = \operatorname{int}_{\mathcal{S}'} \mathcal{S}$.

As an example, the interval $[0, 1)$ is open relative to the interval $[0, 2]$.

Definition 10.1.3. Let $\mathcal{S} \subseteq \mathbb{F}^n$. The vector $x \in \mathbb{F}^n$ is a *closure point* of $\mathcal{S}$ if, for all $\varepsilon > 0$, the set $\mathcal{S} \cap \mathbb{B}_\varepsilon(x)$ is not empty. The *closure* of $\mathcal{S}$ is the set

$$\operatorname{cl} \mathcal{S} \triangleq \{x \in \mathbb{F}^n\colon \ x \text{ is a closure point of } \mathcal{S}\}. \tag{10.1.5}$$

Finally, the set $\mathcal{S}$ is *closed* if every closure point of $\mathcal{S}$ is an element of $\mathcal{S}$, that is, if $\mathcal{S} = \operatorname{cl} \mathcal{S}$.

Definition 10.1.4. Let $\mathcal{S} \subseteq \mathcal{S}' \subseteq \mathbb{F}^n$. The vector $x \in \mathcal{S}'$ is a *closure point* of $\mathcal{S}$ relative to $\mathcal{S}'$ if, for all $\varepsilon > 0$, the set $\mathcal{S} \cap \mathbb{B}_\varepsilon(x)$ is not empty. The *closure* of $\mathcal{S}$ *relative* to $\mathcal{S}'$ is the set

$$\mathrm{cl}_{\mathcal{S}'}\, \mathcal{S} \triangleq \{x \in \mathbb{F}^n\colon \ x \text{ is a closure point of } \mathcal{S} \text{ relative to } \mathcal{S}'\}. \tag{10.1.6}$$

Finally, $\mathcal{S}$ is *closed relative* to $\mathcal{S}'$ if $\mathcal{S} = \mathrm{cl}_{\mathcal{S}'}\, \mathcal{S}$.

As an example, the interval $(0, 1]$ is closed relative to the interval $(0, 2]$.

It follows from Theorem 9.1.8 on the equivalence of norms on $\mathbb{F}^n$ that these definitions are independent of the norm assigned to $\mathbb{F}^n$.

Let $\mathcal{S} \subseteq \mathcal{S}' \subseteq \mathbb{F}^n$. Then,

$$\mathrm{cl}_{\mathcal{S}'}\, \mathcal{S} = (\mathrm{cl}\,\mathcal{S}) \cap \mathcal{S}', \tag{10.1.7}$$

$$\mathrm{int}_{\mathcal{S}'}\, \mathcal{S} = \mathcal{S}' \backslash \mathrm{cl}(\mathcal{S}' \backslash \mathcal{S}), \tag{10.1.8}$$

and

$$\mathrm{int}\,\mathcal{S} \subseteq \mathrm{int}_{\mathcal{S}'}\, \mathcal{S} \subseteq \mathcal{S} \subseteq \mathrm{cl}_{\mathcal{S}'}\, \mathcal{S} \subseteq \mathrm{cl}\,\mathcal{S}. \tag{10.1.9}$$

In particular,

$$\mathrm{int}\,\mathcal{S} = [\mathrm{cl}(\mathcal{S}^\sim)]^\sim \tag{10.1.10}$$

and

$$\mathrm{int}\,\mathcal{S} \subseteq \mathcal{S} \subseteq \mathrm{cl}\,\mathcal{S}. \tag{10.1.11}$$

The set $\mathcal{S}$ is *solid* if $\mathrm{int}\,\mathcal{S}$ is not empty, while $\mathcal{S}$ is *completely solid* if $\mathrm{cl}(\mathrm{int}\,\mathcal{S}) = \mathrm{cl}\,\mathcal{S}$. If $\mathcal{S}$ is completely solid, then $\mathcal{S}$ is solid. The *boundary* of $\mathcal{S}$ is the set

$$\mathrm{bd}\,\mathcal{S} \triangleq \mathrm{cl}\,\mathcal{S} \backslash \mathrm{int}\,\mathcal{S}, \tag{10.1.12}$$

while the *boundary* of $\mathcal{S}$ *relative to* $\mathcal{S}'$ is the set

$$\mathrm{bd}_{\mathcal{S}'}\, \mathcal{S} \triangleq \mathrm{cl}_{\mathcal{S}'}\, \mathcal{S} \backslash \mathrm{int}_{\mathcal{S}'}\, \mathcal{S}. \tag{10.1.13}$$

Note that the empty set is both open and closed, although it is not solid.

The set $\mathcal{S} \subset \mathbb{F}^n$ is *bounded* if there exists $\delta > 0$ such that, for all $x, y \in \mathcal{S}$,

$$\|x - y\| < \delta. \tag{10.1.14}$$

The set $\mathcal{S} \subset \mathbb{F}^n$ is *compact* if it is both closed and bounded.

10.2 Limits

Definition 10.2.1. The *sequence* $(x_1, x_2, \ldots)$ is a tuple with a countably infinite number of components. We write $(x_i)_{i=1}^\infty$ for $(x_1, x_2, \ldots)$.

Definition 10.2.2. The sequence $(\alpha_i)_{i=1}^\infty \subset \mathbb{F}$ *converges* to $\alpha \in \mathbb{F}$ if, for all $\varepsilon > 0$, there exists a positive integer p such that $|\alpha_i - \alpha| < \varepsilon$ for all $i > p$. In this case, we write $\alpha = \lim_{i \to \infty} \alpha_i$ or $\alpha_i \to \alpha$ as $i \to \infty$, where $i \in \mathbb{P}$. Finally, the sequence $(\alpha_i)_{i=1}^\infty \subset \mathbb{F}$ *converges* if there exists $\alpha \in \mathbb{F}$ such that $(\alpha_i)_{i=1}^\infty$ converges to α.

Definition 10.2.3. The sequence $(x_i)_{i=1}^\infty \subset \mathbb{F}^n$ *converges* to $x \in \mathbb{F}^n$ if $\lim_{i\to\infty} \|x-x_i\| = 0$, where $\|\cdot\|$ is a norm on $\mathbb{F}^n$. In this case, we write $x = \lim_{i\to\infty} x_i$ or $x_i \to x$ as $i \to \infty$, where $i \in \mathbb{P}$. The sequence $(x_i)_{i=1}^\infty \subset \mathbb{F}^n$ *converges* if there exists $x \in \mathbb{F}^n$ such that $(x_i)_{i=1}^\infty$ converges to x. Similarly, $(A_i)_{i=1}^\infty \subset \mathbb{F}^{n\times m}$ *converges* to $A \in \mathbb{F}^{n\times m}$ if $\lim_{i\to\infty} \|A - A_i\| = 0$, where $\|\cdot\|$ is a norm on $\mathbb{F}^{n\times m}$. In this case, we write $A = \lim_{i\to\infty} A_i$ or $A_i \to A$ as $i \to \infty$, where $i \in \mathbb{P}$. Finally, the sequence $(A_i)_{i=1}^\infty \subset \mathbb{F}^{n\times m}$ *converges* if there exists $A \in \mathbb{F}^{n\times m}$ such that $(A_i)_{i=1}^\infty$ converges to A.

It follows from Theorem 9.1.8 that convergence of a sequence is independent of the choice of norm.

Proposition 10.2.4. Let $\mathcal{S} \subseteq \mathbb{F}^n$. The vector $x \in \mathbb{F}^n$ is a closure point of $\mathcal{S}$ if and only if there exists a sequence $(x_i)_{i=1}^\infty \subseteq \mathcal{S}$ that converges to x.

Proof. Suppose that $x \in \mathbb{F}^n$ is a closure point of $\mathcal{S}$. Then, for all $i \in \mathbb{P}$, there exists a vector $x_i \in \mathcal{S}$ such that $\|x - x_i\| < 1/i$. Hence, $x - x_i \to 0$ as $i \to \infty$. Conversely, suppose that $(x_i)_{i=1}^\infty \subseteq \mathcal{S}$ is such that $x_i \to x$ as $i \to \infty$, and let $\varepsilon > 0$. Then, there exists a positive integer p such that $\|x - x_i\| < \varepsilon$ for all $i > p$. Therefore, $x_{p+1} \in \mathcal{S} \cap \mathbb{B}_\varepsilon(x)$, and thus $\mathcal{S} \cap \mathbb{B}_\varepsilon(x)$ is not empty. Hence, x is a closure point of $\mathcal{S}$. $\square$

Theorem 10.2.5. Let $\mathcal{S} \subset \mathbb{F}^n$ be compact, and let $(x_i)_{i=1}^\infty \subseteq \mathcal{S}$. Then, there exists a subsequence $(x_{i_j})_{j=1}^\infty$ of $(x_i)_{i=1}^\infty$ such that $(x_{i_j})_{j=1}^\infty$ converges and $\lim_{j\to\infty} x_{i_j} \in \mathcal{S}$.

Proof. See [1057, p. 145]. $\square$

Next, we define convergence for the *series* $\sum_{i=1}^\infty x_i$ in terms of the *partial sums* $\sum_{i=1}^k x_i$.

Definition 10.2.6. Let $(x_i)_{i=1}^\infty \subset \mathbb{F}^n$, and let $\|\cdot\|$ be a norm on $\mathbb{F}^n$. Then, the series $\sum_{i=1}^\infty x_i$ *converges* if $(\sum_{i=1}^k x_i)_{k=1}^\infty$ converges. Furthermore, $\sum_{i=1}^\infty x_i$ *converges absolutely* if the series $\sum_{i=1}^\infty \|x_i\|$ converges.

Proposition 10.2.7. Let $(x_i)_{i=1}^\infty \subset \mathbb{F}^n$, and assume that the series $\sum_{i=1}^\infty x_i$ converges absolutely. Then, the series $\sum_{i=1}^\infty x_i$ converges.

Definition 10.2.8. Let $(A_i)_{i=1}^\infty \subset \mathbb{F}^{n\times m}$, and let $\|\cdot\|$ be a norm on $\mathbb{F}^{n\times m}$. Then, the series $\sum_{i=1}^\infty A_i$ *converges* if $(\sum_{i=1}^k A_i)_{k=1}^\infty$ converges. Furthermore, $\sum_{i=1}^\infty A_i$ *converges absolutely* if the series $\sum_{i=1}^\infty \|A_i\|$ converges.

Proposition 10.2.9. Let $(A_i)_{i=1}^\infty \subset \mathbb{F}^{n\times m}$, and assume that the series $\sum_{i=1}^\infty A_i$ converges absolutely. Then, the series $\sum_{i=1}^\infty A_i$ converges.

10.3 Continuity

Definition 10.3.1. Let $\mathcal{D} \subseteq \mathbb{F}^m$, $f\colon \mathcal{D} \mapsto \mathbb{F}^n$, and $x \in \mathcal{D}$. Then, f is *continuous* at x if, for every convergent sequence $(x_i)_{i=1}^\infty \subseteq \mathcal{D}$ such that $\lim_{i\to\infty} x_i = x$, it follows that $\lim_{i\to\infty} f(x_i) = f(x)$. Furthermore, let $\mathcal{D}_0 \subseteq \mathcal{D}$. Then, f is *continuous* on $\mathcal{D}_0$ if f is continuous at x for all $x \in \mathcal{D}_0$. Finally, f is *continuous* if it is continuous on $\mathcal{D}$.

Theorem 10.3.2. Let $\mathcal{D} \subseteq \mathbb{F}^n$ be convex, and let $f\colon \mathcal{D} \to \mathbb{F}$ be convex. Then, f is continuous on $\mathrm{int}_{\mathrm{aff}\,\mathcal{D}}\,\mathcal{D}$.

Proof. See [161, p. 81] and [1161, p. 82]. $\qquad\square$

Corollary 10.3.3. Let $A \in \mathbb{F}^{n\times m}$, and define $f\colon \mathbb{F}^m \to \mathbb{F}^n$ by $f(x) \triangleq Ax$. Then, f is continuous.

Proof. This result is a consequence of Theorem 10.3.2. Alternatively, let $x \in \mathbb{F}^m$, and let $(x_i)_{i=1}^\infty \subset \mathbb{F}^m$ be such that $x_i \to x$ as $i \to \infty$. Furthermore, let $\|\cdot\|$, $\|\cdot\|'$, and $\|\cdot\|''$ be compatible norms on $\mathbb{F}^n$, $\mathbb{F}^{n\times m}$, and $\mathbb{F}^m$, respectively. Since $\|Ax - Ax_i\| \le \|A\|'\|x - x_i\|''$, it follows that $Ax_i \to Ax$ as $i \to \infty$. $\qquad\square$

Theorem 10.3.4. Let $\mathcal{D} \subseteq \mathbb{F}^m$, and let $f\colon \mathcal{D} \mapsto \mathbb{F}^n$. Then, the following statements are equivalent:

i) f is continuous.

ii) For all open $\mathcal{S} \subseteq \mathbb{F}^n$, the set $f^{-1}(\mathcal{S})$ is open relative to $\mathcal{D}$.

iii) For all closed $\mathcal{S} \subseteq \mathbb{F}^n$, the set $f^{-1}(\mathcal{S})$ is closed relative to $\mathcal{D}$.

Proof. See [1057, pp. 87, 110]. $\qquad\square$

Corollary 10.3.5. Let $A \in \mathbb{F}^{n\times m}$ and $\mathcal{S} \subseteq \mathbb{F}^n$, and define $\mathcal{S}' \triangleq \{x \in \mathbb{F}^m\colon Ax \in \mathcal{S}\}$. If $\mathcal{S}$ is open, then $\mathcal{S}'$ is open. If $\mathcal{S}$ is closed, then $\mathcal{S}'$ is closed.

The following result is the *open mapping theorem*.

Theorem 10.3.6. Let $\mathcal{D} \subseteq \mathbb{F}^m$, let $A \in \mathbb{F}^{n\times m}$, assume that $\mathcal{D}$ is open, and assume that A is right invertible. Then, $A\mathcal{D}$ is open.

The following result is the *invariance of domain*.

Theorem 10.3.7. Let $\mathcal{D} \subseteq \mathbb{F}^n$, let $f\colon \mathcal{D} \mapsto \mathbb{F}^n$, assume that $\mathcal{D}$ is open, and assume that f is continuous and one-to-one. Then, $f(\mathcal{D})$ is open.

Proof. See [1248, p. 3]. $\qquad\square$

Theorem 10.3.8. Let $\mathcal{D} \subset \mathbb{F}^m$ be compact, and let $f\colon \mathcal{D} \mapsto \mathbb{F}^n$ be continuous. Then, $f(\mathcal{D})$ is compact.

Proof. See [1057, p. 146]. □

The following corollary of Theorem 10.3.8 shows that a continuous real-valued function defined on a compact set has a minimizer and a maximizer.

Corollary 10.3.9. Let $\mathcal{D} \subset \mathbb{F}^m$ be compact, and let $f\colon \mathcal{D} \mapsto \mathbb{R}$ be continuous. Then, there exist $x_0, x_1 \in \mathcal{D}$ such that, for all $x \in \mathcal{D}$, $f(x_0) \le f(x) \le f(x_1)$.

The following result is the *Schauder fixed-point theorem*.

Theorem 10.3.10. Let $\mathcal{D} \subseteq \mathbb{F}^m$, assume that $\mathcal{D}$ is nonempty, closed, and convex, let $f\colon \mathcal{D} \to \mathcal{D}$, assume that f is continuous, and assume that $f(\mathcal{D})$ is bounded. Then, there exists $x \in \mathcal{D}$ such that $f(x) = x$.

Proof. See [1438, p. 167]. □

The following corollary for the case of a bounded domain is the *Brouwer fixed-point theorem*.

Corollary 10.3.11. Let $\mathcal{D} \subseteq \mathbb{F}^m$, assume that $\mathcal{D}$ is nonempty, compact, and convex, let $f\colon \mathcal{D} \to \mathcal{D}$, and assume that f is continuous. Then, there exists $x \in \mathcal{D}$ such that $f(x) = x$.

Proof. See [1438, p. 163]. □

Definition 10.3.12. Let $\mathcal{S} \subseteq \mathbb{F}^{n \times n}$. Then, $\mathcal{S}$ is *pathwise connected* if, for all $B_1, B_2 \in \mathcal{S}$, there exists a continuous function $f\colon [0,1] \mapsto \mathcal{S}$ such that $f(0) = B_1$ and $f(1) = B_2$.

10.4 Derivatives

Let $\mathcal{D} \subseteq \mathbb{F}^m$, and let $x_0 \in \mathcal{D}$. Then, the *variational cone of $\mathcal{D}$ with respect to x_0* is the set

$$\mathrm{vcone}(\mathcal{D}, x_0) \triangleq \{\xi \in \mathbb{F}^m\colon \quad \text{there exists } \alpha_0 > 0 \text{ such that,}$$
$$\text{for all } \alpha \in [0, \alpha_0), x_0 + \alpha\xi \in \mathcal{D}\}. \qquad (10.4.1)$$

Note that $\mathrm{vcone}(\mathcal{D}, x_0)$ is a pointed cone, although it may consist of only $\xi = 0$ as can be seen from the example $x_0 = 0$ and

$$\mathcal{D} = \left\{ x \in \mathbb{R}^2\colon \ 0 \le x_{(1)} \le 1, \ x_{(1)}^3 \le x_{(2)} \le x_{(1)}^2 \right\}.$$

Now, let $\mathcal{D} \subseteq \mathbb{F}^m$ and $f\colon \mathcal{D} \to \mathbb{F}^n$. If $\xi \in \mathrm{vcone}(\mathcal{D}, x_0)$, then the *one-sided directional differential of f at x_0 in the direction ξ* is defined by

$$\mathrm{D}_+ f(x_0; \xi) \triangleq \lim_{\alpha \downarrow 0} \tfrac{1}{\alpha} [f(x_0 + \alpha\xi) - f(x_0)] \qquad (10.4.2)$$

if the limit exists. Similarly, if $\xi \in \mathrm{vcone}(\mathcal{D}, x_0)$ and $-\xi \in \mathrm{vcone}(\mathcal{D}, x_0)$, then the *two-sided directional differential* $\mathrm{D}f(x_0; \xi)$ *of* f *at* x_0 *in the direction* ξ *is defined by*

$$\mathrm{D}f(x_0; \xi) \triangleq \lim_{\alpha \to 0} \tfrac{1}{\alpha}[f(x_0 + \alpha\xi) - f(x_0)] \tag{10.4.3}$$

if the limit exists. If $\xi = e_i$ so that the direction ξ is one of the coordinate axes, then the *partial derivative* of f *with respect to* $x_{(i)}$ *at* x_0, denoted by $\frac{\partial f(x_0)}{\partial x_{(i)}}$, is given by

$$\frac{\partial f(x_0)}{\partial x_{(i)}} \triangleq \lim_{\alpha \to 0} \tfrac{1}{\alpha}[f(x_0 + \alpha e_i) - f(x_0)], \tag{10.4.4}$$

that is,

$$\frac{\partial f(x_0)}{\partial x_{(i)}} = \mathrm{D}f(x_0; e_i), \tag{10.4.5}$$

when the two-sided directional differential $\mathrm{D}f(x_0; e_i)$ exists. Note that $\frac{\partial f(x_0)}{\partial x_{(i)}} \in \mathbb{F}^{n \times 1}$.

Proposition 10.4.1. Let $\mathcal{D} \subseteq \mathbb{F}^m$ be a convex set, let $f \colon \mathcal{D} \to \mathbb{F}^n$ be convex, and let $x_0 \in \mathrm{int}\, \mathcal{D}$. Then, $\mathrm{D}_+f(x_0; \xi)$ exists for all $\xi \in \mathbb{F}^m$.

Proof. See [161, p. 83]. $\square$

Note that $\mathrm{D}_+f(x_0; \xi) = \pm\infty$ is possible if x_0 is an element of the boundary of $\mathcal{D}$. For example, consider the convex, continuous function $f \colon [0, \infty) \mapsto \mathbb{R}$ given by $f(x) = 1 - \sqrt{x}$. In this case, $\mathrm{D}_+f(0; 1) = -\infty$ and thus does not exist.

Next, we consider a stronger form of differentiation.

Proposition 10.4.2. Let $\mathcal{D} \subseteq \mathbb{F}^m$ be solid and convex, let $f \colon \mathcal{D} \to \mathbb{F}^n$, and let $x_0 \in \mathcal{D}$. Then, there exists at most one matrix $F \in \mathbb{F}^{n \times m}$ satisfying

$$\lim_{\substack{x \to x_0 \\ x \in \mathcal{D}\backslash\{x_0\}}} \|x - x_0\|^{-1}[f(x) - f(x_0) - F(x - x_0)] = 0. \tag{10.4.6}$$

Proof. See [1438, p. 170]. $\square$

In (10.4.6) the limit is taken over all sequences that are contained in $\mathcal{D}$, do not include x_0, and converge to x_0. Note that $\mathcal{D}$ is not necessarily open, and x_0 may be an element of the boundary of $\mathcal{D}$.

Definition 10.4.3. Let $\mathcal{D} \subseteq \mathbb{F}^m$ be solid and convex, let $f \colon \mathcal{D} \to \mathbb{F}^n$, let $x_0 \in \mathcal{D}$, and assume there exists a matrix $F \in \mathbb{F}^{n \times m}$ satisfying (10.4.6). Then, f is *differentiable at* x_0, and the matrix F is the *derivative of* f *at* x_0. In this case, we write $f'(x_0) = F$ and

$$\lim_{\substack{x \to x_0 \\ x \in \mathcal{D}\backslash\{x_0\}}} \|x - x_0\|^{-1}[f(x) - f(x_0) - f'(x_0)(x - x_0)] = 0. \tag{10.4.7}$$

Note that Proposition 10.4.2 and Definition 10.4.3 do not require that x_0 lie in the interior of $\mathcal{D}$. We alternatively write $\frac{\mathrm{d}f(x_0)}{\mathrm{d}x}$ for $f'(x_0)$.

Proposition 10.4.4. Let $\mathcal{D} \subseteq \mathbb{F}^m$ be solid and convex, let $f\colon \mathcal{D} \to \mathbb{F}^n$, let $x \in \mathcal{D}$, and assume that f is differentiable at x_0. Then, f is continuous at x_0.

Let $\mathcal{D} \subseteq \mathbb{F}^m$ be solid and convex, and let $f\colon \mathcal{D} \mapsto \mathbb{F}^n$. In terms of its scalar components, f can be written as $f = [\ f_1 \ \cdots \ f_n\ ]^{\mathrm{T}}$, where $f_i\colon \mathcal{D} \mapsto \mathbb{F}$ for all $i \in \{1,\dots,n\}$ and $f(x) = [\ f_1(x) \ \cdots \ f_n(x)\ ]^{\mathrm{T}}$ for all $x \in \mathcal{D}$. With this notation, if $f'(x_0)$ exists, then it can be written as

$$f'(x_0) = \begin{bmatrix} f_1'(x_0) \\ \vdots \\ f_n'(x_0) \end{bmatrix}, \tag{10.4.8}$$

where $f_i'(x_0) \in \mathbb{F}^{1\times m}$ is the *gradient of* f_i *at* x_0 and $f'(x_0)$ is the *Jacobian of* f *at* x_0. Furthermore, if $x \in \operatorname{int} \mathcal{D}$, then $f'(x_0)$ is related to the partial derivatives of f by

$$f'(x_0) = \begin{bmatrix} \dfrac{\partial f(x_0)}{\partial x_{(1)}} & \cdots & \dfrac{\partial f(x_0)}{\partial x_{(m)}} \end{bmatrix}, \tag{10.4.9}$$

where $\frac{\partial f(x_0)}{\partial x_{(i)}} \in \mathbb{F}^{n\times 1}$ for all $i \in \{1,\dots,m\}$. Finally, note that the (i,j) entry of the $n \times m$ matrix $f'(x_0)$ is $\frac{\partial f_i(x_0)}{\partial x_{(j)}}$. For example, if $x \in \mathbb{F}^n$ and $A \in \mathbb{F}^{n\times n}$, then

$$\frac{\mathrm{d}}{\mathrm{d}x} Ax = A. \tag{10.4.10}$$

Note that the existence of the partial derivatives of f at x_0 does not imply that $f'(x_0)$ exists. That is, f may not be differentiable at x_0 since $f'(x_0)$ given by (10.4.9) may not satisfy (10.4.7).

Let $\mathcal{D} \subseteq \mathbb{F}^m$ and $f\colon \mathcal{D} \mapsto \mathbb{F}^n$, and assume that $f'(x)$ exists for all $x \in \mathcal{D}$ and $f'\colon \mathcal{D} \mapsto \mathbb{F}^{n\times m}$ is continuous. Then, f is *continuously differentiable*, or C^1. Note that, for all $x_0 \in \mathcal{D}$, $f'(x_0) \in \mathbb{F}^{n\times m}$, and thus $f'(x_0)\colon \mathbb{F}^m \mapsto \mathbb{F}^n$. The *second derivative* of f at $x_0 \in \mathcal{D}$, denoted by $f''(x_0)$, is the derivative of $f'\colon \mathcal{D} \mapsto \mathbb{F}^{n\times m}$ at $x_0 \in \mathcal{D}$. By analogy with the first derivative, it follows that $f''(x_0)\colon \mathbb{F}^m \mapsto \mathbb{F}^{n\times m}$ is linear. Therefore, for all $\eta \in \mathbb{F}^m$, it follows that $f''(x_0)\eta \in \mathbb{F}^{n\times m}$, and, thus, for all $\eta, \hat{\eta} \in \mathbb{F}^m$, it follows that $[f''(x_0)\eta]\hat{\eta} \in \mathbb{F}^n$. Defining $f''(x_0)(\eta,\hat{\eta}) \triangleq [f''(x_0)\eta]\hat{\eta}$, it follows that $f''(x_0)\colon \mathbb{F}^m \times \mathbb{F}^m \mapsto \mathbb{F}^n$ is *bilinear*, that is, for all $\hat{\eta} \in \mathbb{F}^m$, the mapping $\eta \mapsto f''(x_0)(\eta,\hat{\eta})$ is linear, and, for all $\eta \in \mathbb{F}^m$, the mapping $\hat{\eta} \mapsto f''(x_0)(\eta,\hat{\eta})$ is linear. Letting $f = [\ f_1 \ \cdots \ f_n\ ]^{\mathrm{T}}$, it follows that

$$f''(x_0)\eta = \begin{bmatrix} \eta^{\mathrm{T}} f_1''(x_0) \\ \vdots \\ \eta^{\mathrm{T}} f_n''(x_0) \end{bmatrix}, \tag{10.4.11}$$

and

$$f''(x_0)(\eta,\hat{\eta}) = \begin{bmatrix} \eta^{\mathrm{T}} f_1''(x_0)\hat{\eta} \\ \vdots \\ \eta^{\mathrm{T}} f_n''(x_0)\hat{\eta} \end{bmatrix}, \tag{10.4.12}$$

where, for all $i \in \{1, \ldots, n\}$, the matrix $f_i''(x_0)$ is the $m \times m$ *Hessian* of f_i at x_0. We write $f^{(2)}(x_0)$ for $f''(x_0)$ and $f^{(k)}(x_0)$ for the kth derivative of f at x_0. The function f is C^k if $f^{(k)}(x)$ exists for all $x \in \mathcal{D}$ and $f^{(k)}$ is continuous on $\mathcal{D}$.

The following result is the *inverse function theorem* [461, p. 185].

Theorem 10.4.5. Let $\mathcal{D} \subseteq \mathbb{F}^n$ be open, let $f\colon \mathcal{D} \mapsto \mathbb{F}^n$, and assume that f is C^k. Furthermore, let $x_0 \in \mathcal{D}$ be such that $\det f'(x_0) \neq 0$. Then, there exist open sets $\mathcal{N} \subset \mathbb{F}^n$ containing x_0 and $\mathcal{M} \subset \mathbb{F}^n$ containing $f(x_0)$ and a C^k function $g\colon \mathcal{M} \mapsto \mathcal{N}$ such that, for all $x \in \mathcal{N}$, $g[f(x)] = x$, and, for all $y \in \mathcal{M}$, $f[g(y)] = y$.

Let $S\colon [t_0, t_1] \mapsto \mathbb{F}^{n \times m}$, and assume that every entry of $S(t)$ is differentiable. Then, define $\dot{S}(t) \triangleq \frac{\mathrm{d}S(t)}{\mathrm{d}t} \in \mathbb{F}^{n \times m}$ for all $t \in [t_0, t_1]$ entrywise, that is, for all $i \in \{1, \ldots, n\}$ and $j \in \{1, \ldots, m\}$,

$$[\dot{S}(t)]_{(i,j)} \triangleq \frac{\mathrm{d}}{\mathrm{d}t} S_{(i,j)}(t). \tag{10.4.13}$$

If $t = t_0$ or $t = t_1$, then $\mathrm{d}^+/\mathrm{d}t$ or $\mathrm{d}^-/\mathrm{d}t$ (or just $\mathrm{d}/\mathrm{d}t$) denotes the right and left one-sided derivatives, respectively. Finally, define $\int_{t_0}^{t_1} S(t)\,\mathrm{d}t$ entrywise, that is, for all $i \in \{1, \ldots, n\}$ and $j \in \{1, \ldots, m\}$,

$$\left[\int_{t_0}^{t_1} S(t)\,\mathrm{d}t\right]_{(i,j)} \triangleq \int_{t_0}^{t_1} [S(t)]_{(i,j)}\,\mathrm{d}t. \tag{10.4.14}$$

10.5 Functions of a Matrix

Let $\mathcal{D} \subseteq \mathbb{C}$, let $f\colon \mathcal{D} \mapsto \mathbb{C}$, and assume that there exists $\gamma > 0$ such that $\mathbb{B}_\gamma(0) \subseteq \mathcal{D}$. Furthermore, assume that there exist complex numbers $\beta_0, \beta_1, \ldots$ such that, for all $s \in \mathbb{B}_\gamma(0)$, $f(s)$ is given by the convergent series

$$f(s) = \sum_{i=0}^{\infty} \beta_i s^i. \tag{10.5.1}$$

Next, let $A \in \mathbb{C}^{n \times n}$ be such that $\mathrm{sprad}(A) < \gamma$, and define $f(A)$ by the convergent series

$$f(A) \triangleq \sum_{i=0}^{\infty} \beta_i A^i. \tag{10.5.2}$$

Expressing A as $A = SBS^{-1}$, where $S \in \mathbb{C}^{n \times n}$ is nonsingular and $B \in \mathbb{C}^{n \times n}$, it follows that

$$f(A) = Sf(B)S^{-1}. \tag{10.5.3}$$

If, in addition, $B = \mathrm{diag}(J_1, \ldots, J_r)$ is the Jordan form of A, then

$$f(A) = S\,\mathrm{diag}[f(J_1), \ldots, f(J_r)]S^{-1}. \tag{10.5.4}$$

Letting $J = \lambda I_k + N_k$ denote a $k \times k$ Jordan block, expanding and rearranging the infinite series $\sum_{i=1}^{\infty} \beta_i J^i$ shows that $f(J)$ is the $k \times k$ upper triangular Toeplitz matrix

$$f(J) = f(\lambda)I_k + f'(\lambda)N_k + \tfrac{1}{2}f''(\lambda)N_k^2 + \cdots + \frac{1}{(k-1)!}f^{(k-1)}(\lambda)N_k^{k-1}$$

$$= \begin{bmatrix} f(\lambda) & f'(\lambda) & \tfrac{1}{2}f''(\lambda) & \cdots & \frac{1}{(k-1)!}f^{(k-1)}(\lambda) \\ 0 & f(\lambda) & f'(\lambda) & \cdots & \frac{1}{(k-2)!}f^{(k-2)}(\lambda) \\ 0 & 0 & f(\lambda) & \cdots & \frac{1}{(k-3)!}f^{(k-3)}(\lambda) \\ \vdots & \vdots & \ddots & \ddots & \vdots \\ 0 & 0 & 0 & \cdots & f(\lambda) \end{bmatrix}. \qquad (10.5.5)$$

Next, we extend the definition $f(A)$ to functions $f\colon \mathcal{D} \subseteq \mathbb{C} \mapsto \mathbb{C}$ that are not necessarily of the form (10.5.1).

Definition 10.5.1. Let $f\colon \mathcal{D} \subseteq \mathbb{C} \mapsto \mathbb{C}$, let $A \in \mathbb{C}^{n\times n}$, where $\mathrm{spec}(A) \subset \mathcal{D}$, and assume that, for all $\lambda_i \in \mathrm{spec}(A)$, f is $k_i - 1$ times differentiable at λ_i, where $k_i \triangleq \mathrm{ind}_A(\lambda_i)$ is the order of the largest Jordan block associated with λ_i as given by Theorem 5.3.3. Then, f is *defined* at A, and $f(A)$ is given by (10.5.3) and (10.5.4), where $f(J_i)$ is defined by (10.5.5) with $k = k_i$ and $\lambda = \lambda_i$.

The following result shows that the definition of $f(A)$ in Definition 10.5.1 is uniquely defined in the sense that $f(A)$ is independent of the decomposition $A = SBS^{-1}$ used to define $f(A)$ in (10.5.3).

Theorem 10.5.2. Let $A \in \mathbb{F}^{n\times n}$, let $\mathrm{spec}(A) = \{\lambda_1, \ldots, \lambda_r\}$, and, for all $i \in \{1, \ldots, r\}$, let $k_i \triangleq \mathrm{ind}_A(\lambda_i)$. Furthermore, suppose that $f\colon \mathcal{D} \subseteq \mathbb{C} \mapsto \mathbb{C}$ is defined at A. Then, there exists a polynomial $p \in \mathbb{F}[s]$ such that $f(A) = p(A)$. Furthermore, there exists a unique polynomial p of degree less than $\sum_{i=1}^{r} k_i$ satisfying $f(A) = p(A)$ and such that, for all $i \in \{1, \ldots, r\}$ and $j \in \{0, 1, \ldots, k_i - 1\}$,

$$f^{(j)}(\lambda_i) = p^{(j)}(\lambda_i). \qquad (10.5.6)$$

This polynomial is given by

$$p(s) = \sum_{i=1}^{r} \left(\left[\prod_{\substack{j=1 \\ j\neq i}}^{r} (s - \lambda_j)^{n_j} \right] \sum_{k=0}^{k_i-1} \frac{1}{k!} \frac{\mathrm{d}^k}{\mathrm{d}s^k} \left. \frac{f(s)}{\prod_{\substack{l=1 \\ l\neq i}}^{r} (s - \lambda_l)^{k_l}} \right|_{s=\lambda_i} (s - \lambda_i)^k \right). \qquad (10.5.7)$$

If, in addition, A is simple, then p is given by

$$p(s) = \sum_{i=1}^{r} f(\lambda_i) \prod_{\substack{j=1 \\ j\neq i}}^{r} \frac{s - \lambda_j}{\lambda_i - \lambda_j}. \qquad (10.5.8)$$

Proof. See [367, pp. 263, 264]. $\qquad\square$

The polynomial (10.5.7) is the *Lagrange-Hermite interpolation polynomial* for f.

The following result, which is known as the *identity theorem*, is a special case of Theorem 10.5.2.

Theorem 10.5.3. Let $A \in \mathbb{F}^{n \times n}$, let $\mathrm{spec}(A) = \{\lambda_1, \ldots, \lambda_r\}$, and, for all $i \in \{1, \ldots, r\}$, let $k_i \triangleq \mathrm{ind}_A(\lambda_i)$. Furthermore, let $f \colon \mathcal{D} \subseteq \mathbb{C} \mapsto \mathbb{C}$ and $g \colon \mathcal{D} \subseteq \mathbb{C} \mapsto \mathbb{C}$ be analytic on a neighborhood of $\mathrm{spec}(A)$. Then, $f(A) = g(A)$ if and only if, for all $i \in \{1, \ldots, r\}$ and $j \in \{0, 1, \ldots, k_i - 1\}$,

$$f^{(j)}(\lambda_i) = g^{(j)}(\lambda_i). \tag{10.5.9}$$

Corollary 10.5.4. Let $A \in \mathbb{F}^{n \times n}$, and let $f \colon \mathcal{D} \subset \mathbb{C} \to \mathbb{C}$ be analytic on a neighborhood of $\mathrm{mspec}(A)$. Then,

$$\mathrm{mspec}[f(A)] = f[\mathrm{mspec}(A)]. \tag{10.5.10}$$

10.6 Matrix Square Root and Matrix Sign Functions

Theorem 10.6.1. Let $A \in \mathbb{C}^{n \times n}$, and assume that A is group invertible and has no eigenvalues in $(-\infty, 0)$. Then, there exists a unique matrix $B \in \mathbb{C}^{n \times n}$ such that $\mathrm{spec}(B) \subset \mathrm{ORHP} \cup \{0\}$ and such that $B^2 = A$. If, in addition, A is real, then B is real.

Proof. See [701, pp. 20, 31]. $\qquad\square$

The matrix B given by Theorem 10.6.1 is the *principal square root* of A. This matrix is denoted by $A^{1/2}$. The existence of a square root that is not necessarily the principal square root is discussed in Fact 5.15.19.

The following result defines the *matrix sign function*.

Definition 10.6.2. Let $A \in \mathbb{C}^{n \times n}$, assume that A has no eigenvalues on the imaginary axis, and let

$$A = S \begin{bmatrix} J_1 & 0 \\ 0 & J_2 \end{bmatrix} S^{-1},$$

where $S \in \mathbb{C}^{n \times n}$ is nonsingular, $J_1 \in \mathbb{C}^{p \times p}$ and $J_2 \in \mathbb{C}^{q \times q}$ are in Jordan canonical form, and $\mathrm{spec}(J_1) \subset \mathrm{OLHP}$ and $\mathrm{spec}(J_2) \subset \mathrm{ORHP}$. Then, the *matrix sign* of A is defined by

$$\mathrm{Sign}(A) \triangleq S \begin{bmatrix} -I_p & 0 \\ 0 & I_q \end{bmatrix} S^{-1}.$$

10.7 Matrix Derivatives

In this section we consider derivatives of differentiable scalar-valued functions with matrix arguments. Consider the linear function $f \colon \mathbb{F}^{m \times n} \mapsto \mathbb{F}$ given by $f(X) = \mathrm{tr}\, AX$, where $A \in \mathbb{F}^{n \times m}$ and $X \in \mathbb{F}^{m \times n}$. In terms of vectors $x = \mathrm{vec}\, X \in \mathbb{F}^{mn}$, we can define the linear function $\hat{f}(x) \triangleq f(X) = (\mathrm{vec}\, A^{\mathrm{T}})^{\mathrm{T}} x$. Consequently, for all $Y \in \mathbb{F}^{m \times n}$, $\frac{d}{dX} f(X_0) \colon \mathbb{F}^{m \times n} \mapsto \mathbb{F}$ can be represented by

$$\frac{d}{dX} f(X_0)Y = \hat{f}'(\operatorname{vec} X_0) \operatorname{vec} Y = (\operatorname{vec} A^{\mathrm{T}})^{\mathrm{T}} \operatorname{vec} Y = \operatorname{tr} AY. \tag{10.7.1}$$

Noting that $\hat{f}'(\operatorname{vec} X_0) = (\operatorname{vec} A^{\mathrm{T}})^{\mathrm{T}}$ and identifying $\frac{d}{dX} f(X_0)$ with the matrix A, we define the *matrix derivative* of $f \colon \mathcal{D} \subseteq \mathbb{F}^{m \times n} \mapsto \mathbb{F}$ by

$$\frac{d}{dX} f(X) \triangleq \left(\operatorname{vec}^{-1} [\hat{f}'(\operatorname{vec} X)]^{\mathrm{T}} \right)^{\mathrm{T}}, \tag{10.7.2}$$

which is the $n \times m$ matrix A whose (i,j) entry is $\frac{\partial f(X)}{\partial X_{(j,i)}}$. Note the order of indices. The matrix derivative is a representation of the derivative in the sense that

$$\lim_{\substack{X \to X_0 \\ X \in \mathcal{D} \backslash \{X_0\}}} \frac{f(X) - f(X_0) - \operatorname{tr}[F(X - X_0)]}{\|X - X_0\|} = 0, \tag{10.7.3}$$

where F denotes $\frac{d}{dX} f(X_0)$ and $\| \cdot \|$ is a norm on $\mathbb{F}^{m \times n}$.

Proposition 10.7.1. Let $x \in \mathbb{F}^n$. Then, the following statements hold:

i) If $A \in \mathbb{F}^{n \times n}$, then

$$\frac{d}{dx} x^{\mathrm{T}} A x = x^{\mathrm{T}} (A + A^{\mathrm{T}}). \tag{10.7.4}$$

ii) If $A \in \mathbb{F}^{n \times n}$ is symmetric, then

$$\frac{d}{dx} x^{\mathrm{T}} A x = 2 x^{\mathrm{T}} A. \tag{10.7.5}$$

iii) If $A \in \mathbb{F}^{n \times n}$ is Hermitian, then

$$\frac{d}{dx} x^* A x = 2 x^* A. \tag{10.7.6}$$

Proposition 10.7.2. Let $A \in \mathbb{F}^{n \times m}$ and $B \in \mathbb{F}^{l \times n}$. Then, the following statements hold:

i) For all $X \in \mathbb{F}^{m \times n}$,

$$\frac{d}{dX} \operatorname{tr} AX = A. \tag{10.7.7}$$

ii) For all $X \in \mathbb{F}^{m \times l}$,

$$\frac{d}{dX} \operatorname{tr} AXB = BA. \tag{10.7.8}$$

iii) For all $X \in \mathbb{F}^{l \times m}$,

$$\frac{d}{dX} \operatorname{tr} AX^{\mathrm{T}} B = A^{\mathrm{T}} B^{\mathrm{T}}. \tag{10.7.9}$$

iv) For all $X \in \mathbb{F}^{m \times l}$ and $k \geq 1$,

$$\frac{d}{dX} \operatorname{tr} (AXB)^k = kB(AXB)^{k-1}A. \tag{10.7.10}$$

v) For all $X \in \mathbb{F}^{m \times l}$,

$$\frac{d}{dX} \det AXB = B(AXB)^{\mathrm{A}} A. \tag{10.7.11}$$

vi) For all $X \in \mathbb{F}^{m \times l}$ such that AXB is nonsingular,

$$\frac{\mathrm{d}}{\mathrm{d}X} \log \det AXB = B(AXB)^{-1}A. \tag{10.7.12}$$

Proposition 10.7.3. Let $A \in \mathbb{F}^{n \times m}$ and $B \in \mathbb{F}^{m \times n}$. Then, the following statements hold:

i) For all $X \in \mathbb{F}^{m \times m}$ and $k \geq 1$,

$$\frac{\mathrm{d}}{\mathrm{d}X} \operatorname{tr} AX^k B = \sum_{i=0}^{k-1} X^{k-1-i} BAX^i. \tag{10.7.13}$$

ii) For all nonsingular $X \in \mathbb{F}^{m \times m}$,

$$\frac{\mathrm{d}}{\mathrm{d}X} \operatorname{tr} AX^{-1}B = -X^{-1}BAX^{-1}. \tag{10.7.14}$$

iii) For all nonsingular $X \in \mathbb{F}^{m \times m}$,

$$\frac{\mathrm{d}}{\mathrm{d}X} \det AX^{-1}B = -X^{-1}B(AX^{-1}B)^{\mathrm{A}}AX^{-1}. \tag{10.7.15}$$

iv) For all nonsingular $X \in \mathbb{F}^{m \times m}$,

$$\frac{\mathrm{d}}{\mathrm{d}X} \log \det AX^{-1}B = -X^{-1}B(AX^{-1}B)^{-1}AX^{-1}. \tag{10.7.16}$$

Proposition 10.7.4. The following statements hold:

i) Let $A, B \in \mathbb{F}^{n \times m}$. Then, for all $X \in \mathbb{F}^{m \times n}$,

$$\frac{\mathrm{d}}{\mathrm{d}X} \operatorname{tr} AXBX = AXB + BXA. \tag{10.7.17}$$

ii) Let $A \in \mathbb{F}^{n \times n}$ and $B \in \mathbb{F}^{m \times m}$. Then, for all $X \in \mathbb{F}^{n \times m}$,

$$\frac{\mathrm{d}}{\mathrm{d}X} \operatorname{tr} AXBX^{\mathrm{T}} = BX^{\mathrm{T}}A + B^{\mathrm{T}}X^{\mathrm{T}}A^{\mathrm{T}}. \tag{10.7.18}$$

iii) Let $A \in \mathbb{F}^{n \times n}$. Then, for all $X \in \mathbb{F}^{n \times m}$,

$$\frac{\mathrm{d}}{\mathrm{d}X} \operatorname{tr} X^{\mathrm{T}}AX = X^{\mathrm{T}}(A + A^{\mathrm{T}}). \tag{10.7.19}$$

iv) Let $A \in \mathbb{F}^{k \times l}$, $B \in \mathbb{F}^{l \times m}$, $C \in \mathbb{F}^{n \times l}$, $D \in \mathbb{F}^{l \times l}$, and $E \in \mathbb{F}^{l \times k}$. Then, for all $X \in \mathbb{F}^{m \times n}$,

$$\frac{\mathrm{d}}{\mathrm{d}X} \operatorname{tr} A(D + BXC)^{-1}E = -C(D + BXC)^{-1}EA(D + BXC)^{-1}B. \tag{10.7.20}$$

v) Let $A \in \mathbb{F}^{k \times l}$, $B \in \mathbb{F}^{l \times m}$, $C \in \mathbb{F}^{n \times l}$, $D \in \mathbb{F}^{l \times l}$, and $E \in \mathbb{F}^{l \times k}$. Then, for all $X \in \mathbb{F}^{n \times m}$,

$$\frac{\mathrm{d}}{\mathrm{d}X} \operatorname{tr} A(D + BX^{\mathrm{T}}C)^{-1}E$$
$$= -B^{\mathrm{T}}(D + BX^{\mathrm{T}}C)^{-\mathrm{T}}A^{\mathrm{T}}E^{\mathrm{T}}(D + BX^{\mathrm{T}}C)^{-\mathrm{T}}C^{\mathrm{T}}. \tag{10.7.21}$$

10.8 Facts on One Set

Fact 10.8.1. Let $x \in \mathbb{F}^n$, and let $\varepsilon > 0$. Then, $\mathbb{B}_\varepsilon(x)$ is completely solid and convex.

Fact 10.8.2. Let $\mathcal{S} \subset \mathbb{F}^n$, let $\| \cdot \|$ be a norm on $\mathbb{F}^n$, assume that there exists $\delta > 0$ such that, for all $x, y \in \mathcal{S}$, $\|x - y\| < \delta$, and let $x_0 \in \mathcal{S}$. Then, $\mathcal{S} \subseteq \mathbb{B}_\delta(x_0)$.

Fact 10.8.3. Let $\mathcal{S} \subseteq \mathbb{F}^n$. Then, $\mathrm{cl}\,\mathcal{S}$ is the smallest closed set containing $\mathcal{S}$, and $\mathrm{int}\,\mathcal{S}$ is the largest open set contained in $\mathcal{S}$.

Fact 10.8.4. Let $\mathcal{S} \subseteq \mathbb{F}^n$. If $\mathcal{S}$ is (open, closed), then $\mathcal{S}^\sim$ is (closed, open).

Fact 10.8.5. Let $\mathcal{S} \subseteq \mathcal{S}' \subseteq \mathbb{F}^n$. If $\mathcal{S}$ is (open relative to $\mathcal{S}'$, closed relative to $\mathcal{S}'$), then $\mathcal{S}' \backslash \mathcal{S}$ is (closed relative to $\mathcal{S}'$, open relative to $\mathcal{S}'$).

Fact 10.8.6. Let $\mathcal{S} \subseteq \mathbb{F}^n$. Then,

$$(\mathrm{int}\,\mathcal{S})^\sim = \mathrm{cl}(\mathcal{S}^\sim)$$

and

$$\mathrm{bd}\,\mathcal{S} = \mathrm{bd}\,\mathcal{S}^\sim = (\mathrm{cl}\,\mathcal{S}) \cap (\mathrm{cl}\,\mathcal{S}^\sim) = [(\mathrm{int}\,\mathcal{S}) \cup \mathrm{int}(\mathcal{S}^\sim)]^\sim.$$

Hence, $\mathrm{bd}\,\mathcal{S}$ is closed.

Fact 10.8.7. Let $\mathcal{S} \subseteq \mathbb{F}^n$, and assume that $\mathcal{S}$ is either open or closed. Then, $\mathrm{int}\,\mathrm{bd}\,\mathcal{S}$ is empty.

Proof: See [71, p. 68].

Fact 10.8.8. Let $\mathcal{S} \subseteq \mathbb{F}^n$, and assume that $\mathcal{S}$ is convex. Then, $\mathrm{cl}\,\mathcal{S}$, $\mathrm{int}\,\mathcal{S}$, and $\mathrm{int}_{\mathrm{aff}\,\mathcal{S}}\,\mathcal{S}$ are convex.

Proof: See [1161, p. 45] and [1162, p. 64].

Fact 10.8.9. Let $\mathcal{S} \subseteq \mathbb{F}^n$, and assume that $\mathcal{S}$ is convex. Then, the following statements are equivalent:

  $i)$ $\mathcal{S}$ is solid.

  $ii)$ $\mathcal{S}$ is completely solid.

  $iii)$ $\dim \mathcal{S} = n$.

  $iv)$ $\mathrm{aff}\,\mathcal{S} = \mathbb{F}^n$.

Fact 10.8.10. Let $\mathcal{S} \subseteq \mathbb{F}^n$, and assume that $\mathcal{S}$ is solid. Then, $\mathrm{co}\,\mathcal{S}$ is completely solid.

Fact 10.8.11. Let $\mathcal{S} \subseteq \mathbb{F}^n$. Then,

$$\mathrm{cl}\,\mathcal{S} \subseteq \mathrm{aff}\,\mathrm{cl}\,\mathcal{S} = \mathrm{aff}\,\mathcal{S}.$$

Proof: See [243, p. 7].

Fact 10.8.12. Let $k \leq n$, and let $x_1, \ldots, x_k \in \mathbb{F}^n$. Then,

$$\text{int aff } \{x_1, \ldots, x_k\} = \varnothing.$$

Remark: See Fact 2.9.7.

Fact 10.8.13. Let $\mathcal{S} \subseteq \mathbb{F}^n$. Then,

$$\text{co cl } \mathcal{S} \subseteq \text{cl co } \mathcal{S}.$$

Now, assume in addition that $\mathcal{S}$ is either bounded or convex. Then,

$$\text{co cl } \mathcal{S} = \text{cl co } \mathcal{S}.$$

Proof: Use Fact 10.8.8 and Fact 10.8.13.

Remark: Although

$$\mathcal{S} = \left\{ x \in \mathbb{R}^2 \colon \ x_{(1)}^2 x_{(2)}^2 = 1 \text{ for all } x_{(1)} > 0 \right\}$$

is closed, $\text{co } \mathcal{S}$ is not closed. Hence, $\text{co cl } \mathcal{S} \subset \text{cl co } \mathcal{S}$.

Fact 10.8.14. Let $\mathcal{S} \subseteq \mathbb{F}^n$, and assume that $\mathcal{S}$ is open. Then, $\text{co } \mathcal{S}$ is open.

Fact 10.8.15. Let $\mathcal{S} \subseteq \mathbb{F}^n$, and assume that $\mathcal{S}$ is compact. Then, $\text{co } \mathcal{S}$ is compact.

Fact 10.8.16. Let $\mathcal{S} \subseteq \mathbb{F}^n$, and assume that $\mathcal{S}$ is solid. Then, $\dim \mathcal{S} = n$.

Fact 10.8.17. Let $\mathcal{S} \subseteq \mathbb{F}^m$, assume that $\mathcal{S}$ is solid, let $A \in \mathbb{F}^{n \times m}$, and assume that A is right invertible. Then, $A\mathcal{S}$ is solid.

Proof: Use Theorem 10.3.6.

Remark: See Fact 2.10.4.

Fact 10.8.18. $\mathbf{N}^n$ is a closed and completely solid subset of $\mathbb{F}^{n(n+1)/2}$. Furthermore,

$$\text{int } \mathbf{N}^n = \mathbf{P}^n.$$

Fact 10.8.19. Let $\mathcal{S} \subseteq \mathbb{F}^n$, and assume that $\mathcal{S}$ is convex. Then,

$$\text{int cl } \mathcal{S} = \text{int } \mathcal{S}.$$

Remark: The result follows from Fact 10.9.4.

Fact 10.8.20. Let $\mathcal{D} \subseteq \mathbb{F}^n$, and let x_0 belong to a solid, convex subset of $\mathcal{D}$. Then,

$$\dim \text{vcone}(\mathcal{D}, x_0) = n.$$

Fact 10.8.21. Let $\mathcal{S} \subseteq \mathbb{F}^n$, and assume that $\mathcal{S}$ is a subspace. Then, $\mathcal{S}$ is closed.

Fact 10.8.22. Let $\mathcal{S} \subset \mathbb{F}^n$, assume that $\mathcal{S}$ is symmetric, convex, bounded, solid, and closed, and, for all $x \in \mathbb{F}^n$, define

$$\|x\| \triangleq \min\{\alpha \geq 0 \colon \ x \in \alpha\mathcal{S}\} = \max\{\alpha \geq 0 \colon \ \alpha x \in \mathcal{S}\}.$$

Then, $\| \cdot \|$ is a norm on $\mathbb{F}^n$, and $\mathbb{B}_1(0) = \text{int } \mathcal{S}$. Conversely, let $\| \cdot \|$ be a norm on $\mathbb{F}^n$. Then, $\mathbb{B}_1(0)$ is symmetric, convex, bounded, and solid.

Proof: See [740, pp. 38, 39].

Remark: In all cases, $\mathbb{B}_1(0)$ is defined with respect to $\| \cdot \|$. This result is due to Minkowski.

Remark: See Fact 9.7.23.

Fact 10.8.23. Let $\mathcal{S} \subseteq \mathbb{R}^n$, assume that $\mathcal{S}$ is nonempty, closed, and convex, and let $\mathcal{E} \subseteq \mathcal{S}$ denote the set of elements of $\mathcal{S}$ that cannot be represented as nontrivial convex combinations of two distinct elements of $\mathcal{S}$. Then, $\mathcal{E}$ is nonempty and satisfies

$$\mathcal{S} = \text{co } \mathcal{E}.$$

If, in addition, $n = 2$, then $\mathcal{E}$ is closed.

Proof: See [459, pp. 482–484].

Remark: $\mathcal{E}$ is the set of *extreme points* of $\mathcal{S}$.

Remark: $\mathcal{E}$ is not necessarily closed for $n > 2$. See [459, p. 483].

Remark: The last result is the *Krein-Milman theorem*.

10.9 Facts on Two or More Sets

Fact 10.9.1. Let $\mathcal{S}_1 \subseteq \mathcal{S}_2 \subseteq \mathbb{F}^n$. Then,

$$\text{cl } \mathcal{S}_1 \subseteq \text{cl } \mathcal{S}_2$$

and

$$\text{int } \mathcal{S}_1 \subseteq \text{int } \mathcal{S}_2.$$

Fact 10.9.2. Let $\mathcal{S}_1, \mathcal{S}_2 \subseteq \mathbb{F}^n$. Then, the following statements hold:

i) $(\text{int } \mathcal{S}_1) \cap (\text{int } \mathcal{S}_2) = \text{int}(\mathcal{S}_1 \cap \mathcal{S}_2)$.

ii) $(\text{int } \mathcal{S}_1) \cup (\text{int } \mathcal{S}_2) \subseteq \text{int}(\mathcal{S}_1 \cup \mathcal{S}_2)$.

iii) $(\text{cl } \mathcal{S}_1) \cup (\text{cl } \mathcal{S}_2) = \text{cl}(\mathcal{S}_1 \cup \mathcal{S}_2)$.

iv) $\text{bd}(\mathcal{S}_1 \cup \mathcal{S}_2) \subseteq (\text{bd } \mathcal{S}_1) \cup (\text{bd } \mathcal{S}_2)$.

v) If $(\text{cl } \mathcal{S}_1) \cap (\text{cl } \mathcal{S}_2) = \varnothing$, then $\text{bd}(\mathcal{S}_1 \cup \mathcal{S}_2) = (\text{bd } \mathcal{S}_1) \cup (\text{bd } \mathcal{S}_2)$.

Proof: See [71, p. 65].

Fact 10.9.3. Let $\mathcal{S}_1, \mathcal{S}_2 \subseteq \mathbb{F}^n$, assume that either $\mathcal{S}_1$ or $\mathcal{S}_2$ is closed, and assume that $\text{int } \mathcal{S}_1 = \text{int } \mathcal{S}_2 = \varnothing$. Then, $\text{int}(\mathcal{S}_1 \cup \mathcal{S}_2)$ is empty.

Proof: See [71, p. 69].

Remark: The set $\text{int}(\mathcal{S}_1 \cup \mathcal{S}_2)$ is not necessarily empty if neither $\mathcal{S}_1$ nor $\mathcal{S}_2$ is closed. Consider the sets of rational and irrational numbers.

Fact 10.9.4. Let $\mathcal{S}_1, \mathcal{S}_2 \subseteq \mathbb{F}^n$, and assume that $\mathcal{S}_1$ is open, $\mathcal{S}_2$ is convex, and $\mathcal{S}_1 \subseteq \mathrm{cl}\,\mathcal{S}_2$. Then, $\mathcal{S}_1 \subseteq \mathcal{S}_2$.

Proof: See [1208, pp. 72, 73].

Remark: See Fact 10.8.19.

Fact 10.9.5. Let $\mathcal{S}_1, \mathcal{S}_2 \subseteq \mathbb{F}^n$, and assume that $\mathcal{S}_1$ is closed and $\mathcal{S}_2$ is compact. Then, $\mathcal{S}_1 + \mathcal{S}_2$ is closed.

Proof: See [454, p. 209].

Fact 10.9.6. Let $\mathcal{S}_1, \mathcal{S}_2 \subseteq \mathbb{F}^n$, and assume that $\mathcal{S}_1$ and $\mathcal{S}_2$ are closed and compact. Then, $\mathcal{S}_1 + \mathcal{S}_2$ is closed and compact.

Proof: See [157, p. 34].

Fact 10.9.7. Let $\mathcal{S}_1, \mathcal{S}_2, \mathcal{S}_3 \subseteq \mathbb{F}^n$, assume that $\mathcal{S}_1$, $\mathcal{S}_2$, and $\mathcal{S}_3$ are closed and convex, assume that $\mathcal{S}_1 \cap \mathcal{S}_2 \neq \varnothing$, $\mathcal{S}_2 \cap \mathcal{S}_3 \neq \varnothing$, and $\mathcal{S}_3 \cap \mathcal{S}_1 \neq \varnothing$, and assume that $\mathcal{S}_1 \cup \mathcal{S}_2 \cup \mathcal{S}_3$ is convex. Then, $\mathcal{S}_1 \cap \mathcal{S}_2 \cap \mathcal{S}_3 \neq \varnothing$.

Proof: See [157, p. 32].

Fact 10.9.8. Let $\mathcal{S}_1, \mathcal{S}_2, \mathcal{S}_3 \subseteq \mathbb{F}^n$, assume that $\mathcal{S}_1$ and $\mathcal{S}_2$ are convex, $\mathcal{S}_2$ is closed, and $\mathcal{S}_3$ is bounded, and assume that $\mathcal{S}_1 + \mathcal{S}_3 \subseteq \mathcal{S}_2 + \mathcal{S}_3$. Then, $\mathcal{S}_1 \subseteq \mathcal{S}_2$.

Proof: See [243, p. 5].

Remark: This result is due to Radstrom.

Fact 10.9.9. Let $\mathcal{S} \subseteq \mathbb{F}^m$, assume that $\mathcal{S}$ is closed, let $A \in \mathbb{F}^{n \times m}$, and assume that A has full row rank. Then, $A\mathcal{S}$ is not necessarily closed.

Remark: See Theorem 10.3.6.

Fact 10.9.10. Let $\mathcal{A}$ be a collection of open subsets of $\mathbb{R}^n$. Then, the union of all elements of $\mathcal{A}$ is open. If, in addition, $\mathcal{A}$ is finite, then the intersection of all elements of $\mathcal{A}$ is open.

Proof: See [71, p. 50].

Fact 10.9.11. Let $\mathcal{A}$ be a collection of closed subsets of $\mathbb{R}^n$. Then, the intersection of all elements of $\mathcal{A}$ is closed. If, in addition, $\mathcal{A}$ is finite, then the union of all elements of $\mathcal{A}$ is closed.

Proof: See [71, p. 50].

Fact 10.9.12. Let $\mathcal{A} = \{A_1, A_2, \ldots\}$ be a collection of nonempty, closed subsets of $\mathbb{R}^n$ such that A_1 is bounded and such that, for all $i \in \mathbb{P}$, $A_{i+1} \subseteq A_i$. Then, $\cap_{i=1}^{\infty} A_i$ is closed and nonempty.

Proof: See [71, p. 56].

Remark: This result is the *Cantor intersection theorem*.

Fact 10.9.13. Let $\| \cdot \|$ be a norm on $\mathbb{F}^n$, let $\mathcal{S} \subset \mathbb{F}^n$, assume that $\mathcal{S}$ is a subspace, let $y \in \mathbb{F}^n$, and define

$$\mu \triangleq \max_{x \in \{z \in \mathcal{S}: \ \|z\|=1\}} |y^*x|.$$

Then, there exists a vector $w \in \mathcal{S}^\perp$ such that

$$\max_{x \in \{z \in \mathbb{F}^n: \ \|z\|=1\}} |(y+w)^*x| = \mu.$$

Proof: See [1261, p. 57].

Remark: This result is a version of the *Hahn-Banach theorem*.

Problem: Find a simple interpretation in $\mathbb{R}^2$.

Fact 10.9.14. Let $\mathcal{S} \subset \mathbb{R}^n$, assume that $\mathcal{S}$ is a convex cone, let $x \in \mathbb{R}^n$, and assume that $x \notin \text{int}\,\mathcal{S}$. Then, there exists a nonzero vector $\lambda \in \mathbb{R}^n$ such that $\lambda^T x \leq 0$ and $\lambda^T z \geq 0$ for all $z \in \mathcal{S}$.

Remark: This result is a *separation theorem*. See [904, p. 37], [1123, p. 443], [1161, pp. 95–101], and [1266, pp. 96–100].

Fact 10.9.15. Let $\mathcal{S}_1, \mathcal{S}_2 \subset \mathbb{R}^n$, and assume that $\mathcal{S}_1$ and $\mathcal{S}_2$ are convex. Then, the following statements are equivalent:

i) There exist a nonzero vector $\lambda \in \mathbb{R}^n$ and $\alpha \in \mathbb{R}$ such that $\lambda^T x \leq \alpha$ for all $x \in \mathcal{S}_1$, $\lambda^T y \geq \alpha$ for all $y \in \mathcal{S}_2$, and either $\mathcal{S}_1$ or $\mathcal{S}_2$ is not contained in the affine hyperplane $\{x \in \mathbb{R}^n: \ \lambda^T x = \alpha\}$.

ii) $\text{int}_{\text{aff}\,\mathcal{S}_1}\,\mathcal{S}_1$ and $\text{int}_{\text{aff}\,\mathcal{S}_2}\,\mathcal{S}_2$ are disjoint.

Proof: See [184, p. 82].

Remark: This result is a *proper separation theorem*.

Fact 10.9.16. Let $\| \cdot \|$ be a norm on $\mathbb{F}^n$, let $y \in \mathbb{F}^n$, let $\mathcal{S} \subseteq \mathbb{F}^n$, and assume that $\mathcal{S}$ is nonempty and closed. Then, there exists a vector $x_0 \in \mathcal{S}$ such that

$$\|y - x_0\| = \min_{x \in \mathcal{S}} \|y - x\|.$$

Now, assume in addition that $\mathcal{S}$ is convex. Then, there exists a unique vector $x_0 \in \mathcal{S}$ such that

$$\|y - x_0\| = \min_{x \in \mathcal{S}} \|y - x\|.$$

In other words, there exists a vector $x_0 \in \mathcal{S}$ such that, for all $x \in \mathcal{S} \backslash \{x_0\}$,

$$\|y - x_0\| < \|y - x\|.$$

Proof: See [459, pp. 470, 471].

Remark: See Fact 10.9.18.

Fact 10.9.17. Let $\|\cdot\|$ be a norm on $\mathbb{F}^n$, let $y_1, y_2 \in \mathbb{F}^n$, let $\mathcal{S} \subseteq \mathbb{F}^n$, assume that $\mathcal{S}$ is nonempty, closed, and convex, and let x_1 and x_2 denote the unique elements of $\mathcal{S}$ that are closest to y_1 and y_2, respectively. Then,

$$\|x_1 - x_2\| \le \|y_1 - y_2\|.$$

Proof: See [459, pp. 474, 475].

Fact 10.9.18. Let $\mathcal{S} \subseteq \mathbb{R}^n$, assume that $\mathcal{S}$ is a subspace, let $A \in \mathbb{F}^{n \times n}$ be the projector onto $\mathcal{S}$, and let $x \in \mathbb{F}^n$. Then,

$$\min_{y \in \mathcal{S}} \|x - y\|_2 = \|A_\perp x\|_2.$$

Proof: See [550, p. 41] or [1261, p. 91].

Remark: See Fact 10.9.16.

Fact 10.9.19. Let $\mathcal{S}_1, \mathcal{S}_2 \subseteq \mathbb{R}^n$, assume that $\mathcal{S}_1$ and $\mathcal{S}_2$ are subspaces, let A_1 and A_2 be the projectors onto $\mathcal{S}_1$ and $\mathcal{S}_2$, respectively, and define

$$\operatorname{dist}(\mathcal{S}_1, \mathcal{S}_2) \triangleq \max \left\{ \max_{\substack{x \in \mathcal{S}_1 \\ \|x\|=1}} \min_{y \in \mathcal{S}_2} \|x - y\|_2, \ \max_{\substack{y \in \mathcal{S}_2 \\ \|y\|_2=1}} \min_{x \in \mathcal{S}_1} \|x - y\|_2 \right\}.$$

Then,

$$\operatorname{dist}(\mathcal{S}_1, \mathcal{S}_2) = \sigma_{\max}(A_1 - A_2).$$

If, in addition, $\dim \mathcal{S}_1 = \dim \mathcal{S}_2$, then

$$\operatorname{dist}(\mathcal{S}_1, \mathcal{S}_2) = \sin \theta,$$

where θ is the minimal principal angle defined in Fact 5.11.39.

Proof: See [574, Chapter 13] and [1261, pp. 92, 93].

Remark: If $\|\cdot\|$ is a norm on $\mathbb{F}^{n \times n}$, then

$$\operatorname{dist}(\mathcal{S}_1, \mathcal{S}_2) \triangleq \|A_1 - A_2\|_2$$

defines a metric on the set of all subspaces of $\mathbb{F}^n$, yielding the *gap topology*.

Remark: See Fact 5.12.17.

10.10 Facts on Matrix Functions

Fact 10.10.1. Let $A \in \mathbb{C}^{n \times n}$, and assume that A is group invertible and has no eigenvalues in $(-\infty, 0)$. Then,

$$A^{1/2} = \tfrac{2}{\pi} A \int_0^\infty (t^2 I + A)^{-1} \, \mathrm{d}t.$$

Proof: See [701, p. 133].

Fact 10.10.2. Let $A \in \mathbb{C}^{n \times n}$, and assume that A has no eigenvalues on the imaginary axis. Then, the following statements hold:

i) $\mathrm{Sign}(A)$ is involutory.

ii) $A = \mathrm{Sign}(A)$ if and only if A is involutory.

iii) $[A, \mathrm{Sign}(A)] = 0$.

iv) $\mathrm{Sign}(A) = \mathrm{Sign}(A^{-1})$.

v) If A is real, then $\mathrm{Sign}(A)$ is real.

vi) $\mathrm{Sign}(A) = A(A^2)^{-1/2} = A^{-1}(A^2)^{1/2}$.

vii) $\mathrm{Sign}(A)$ is given by

$$\mathrm{Sign}(A) = \tfrac{2}{\pi} A \int_0^\infty (t^2 I + A^2)^{-1} \, dt.$$

Proof: See [701, pp. 39, 40 and Chapter 5] and [826].

Remark: The square root in *vi)* is the principal square root.

Fact 10.10.3. Let $A, B \in \mathbb{C}^{n \times n}$, assume that AB has no eigenvalues on the imaginary axis, and define $C \triangleq A(BA)^{-1/2}$. Then,

$$\mathrm{Sign}\left(\begin{bmatrix} 0 & A \\ B & 0 \end{bmatrix}\right) = \begin{bmatrix} 0 & C \\ C^{-1} & 0 \end{bmatrix}.$$

If, in addition, A has no eigenvalues on the imaginary axis, then

$$\mathrm{Sign}\left(\begin{bmatrix} 0 & A \\ I & 0 \end{bmatrix}\right) = \begin{bmatrix} 0 & A^{1/2} \\ A^{-1/2} & 0 \end{bmatrix}.$$

Proof: Use *vi)* of Fact 10.10.2. See [701, p. 108].

Remark: The square root is the principal square root.

Fact 10.10.4. Let $A, B \in \mathbb{C}^{n \times n}$, and assume that A and B are positive definite. Then,

$$\mathrm{Sign}\left(\begin{bmatrix} 0 & B \\ A^{-1} & 0 \end{bmatrix}\right) = \begin{bmatrix} 0 & A\#B \\ (A\#B)^{-1} & 0 \end{bmatrix}.$$

Proof: See [701, p. 131].

Remark: The geometric mean is defined in Fact 8.10.43.

10.11 Facts on Functions

Fact 10.11.1. Let $(x_i)_{i=1}^\infty \subset \mathbb{F}^n$. Then, $\lim_{i \to \infty} x_i = x$ if and only if, for all $j \in \{1, \ldots, n\}$, $\lim_{i \to \infty} x_{i(j)} = x_{(j)}$.

Fact 10.11.2. Let $p \in \mathbb{C}[s]$, where $p(s) = s^n + a_{n-1}s^{n-1} + \cdots + a_0$, let roots$(p) = \{\lambda_1, \ldots, \lambda_r\}$, and, for all $i \in \{1, \ldots, r\}$, let $\alpha_i \in \mathbb{R}$ satisfy $0 < \alpha_i < \min_{j \neq i} |\lambda_i - \lambda_j|$. Furthermore, for all $\varepsilon_0, \ldots, \varepsilon_{n-1} \in \mathbb{R}$, define

$$p_{\varepsilon_0, \ldots, \varepsilon_{n-1}}(s) \triangleq s^n + (a_{n-1} + \varepsilon_{n-1})s^{n-1} + \cdots + (a_1 + \varepsilon_1)s + a_0 + \varepsilon_0.$$

Then, there exists $\varepsilon > 0$ such that, for all $\varepsilon_0, \ldots, \varepsilon_{n-1}$ satisfying $|\varepsilon_i| < \varepsilon$ for all $i \in \{1, \ldots, n-1\}$, it follows that, for all $i \in \{1, \ldots, r\}$, the polynomial $p_{\varepsilon_0, \ldots, \varepsilon_{n-1}}$ has exactly $\text{mult}_p(\lambda_i)$ roots in the open disk $\{s \in \mathbb{C}: |s - \lambda_i| < \alpha_i\}$.

Proof: See [1030].

Remark: This result shows that the roots of a polynomial are continuous functions of the coefficients.

Remark: $\lambda_1, \ldots, \lambda_r$ are the distinct roots of p.

Fact 10.11.3. Let $\mathcal{S}_1 \subseteq \mathbb{F}^n$, assume that $\mathcal{S}_1$ is compact, let $\mathcal{S}_2 \subset \mathbb{F}^m$, let $f: \mathcal{S}_1 \times \mathcal{S}_2 \to \mathbb{R}$, and assume that f is continuous. Then, $g: \mathcal{S}_2 \to \mathbb{R}$ defined by $g(y) \triangleq \max_{x \in \mathcal{S}_1} f(x, y)$ is continuous.

Remark: A related result is given in [454, p. 208].

Fact 10.11.4. Define $f: \mathbb{R}^2 \mapsto \mathbb{R}$ by

$$f(x, y) \triangleq \begin{cases} \min\{x/y, y/x\}, & x > 0 \text{ and } y > 0, \\ 0, & \text{otherwise.} \end{cases}$$

Furthermore, for all $x \in \mathbb{R}$, define $g_x: \mathbb{R} \mapsto \mathbb{R}$ by $g_x(y) \triangleq f(x, y)$, and, for all $y \in \mathbb{R}$, define $h_y: \mathbb{R} \mapsto \mathbb{R}$ by $h_y(x) \triangleq f(x, y)$. Then, the following statements hold:

i) For all $x \in \mathbb{R}$, g_x is continuous.

ii) For all $y \in \mathbb{R}$, h_y is continuous.

iii) f is not continuous at $(0, 0)$.

Fact 10.11.5. Let $\mathcal{S} \subseteq \mathbb{F}^n$, assume that $\mathcal{S}$ is pathwise connected, let $f: \mathcal{S} \mapsto \mathbb{F}^n$, and assume that f is continuous. Then, $f(\mathcal{S})$ is pathwise connected.

Proof: See [1287, p. 65].

Fact 10.11.6. Let $f: [0, \infty) \to \mathbb{R}$, assume that f is continuous, and assume that $\lim_{t \to \infty} f(t)$ exists. Then,

$$\lim_{t \to \infty} \frac{1}{t} \int_0^t f(\tau) \, d\tau = \lim_{t \to \infty} f(t).$$

Remark: The assumption that f is continuous can be weakened.

Fact 10.11.7. Let $\mathcal{J} \subseteq \mathbb{R}$ be a finite or infinite interval, let $f: \mathcal{J} \to \mathbb{R}$, assume that f is continuous, and assume that, for all $x, y \in \mathcal{J}$, it follows that $f[\frac{1}{2}(x + y)] \leq \frac{1}{2}f(x + y)$. Then, f is convex.

Proof: See [1066, p. 10].

Remark: This result is due to Jensen.

Remark: See Fact 1.10.4.

Fact 10.11.8. Let $A_0 \in \mathbb{F}^{n \times n}$, let $\| \cdot \|$ be a norm on $\mathbb{F}^{n \times n}$, and let $\varepsilon > 0$. Then, there exists $\delta > 0$ such that, if $A \in \mathbb{F}^{n \times n}$ and $\|A - A_0\| < \delta$, then

$$\mathrm{dist}[\mathrm{mspec}(A) - \mathrm{mspec}(A_0)] < \varepsilon,$$

where

$$\mathrm{dist}[\mathrm{mspec}(A) - \mathrm{mspec}(A_0)] \triangleq \min_{\sigma} \max_{i=1,\ldots,n} |\lambda_{\sigma(i)}(A) - \lambda_i(A_0)|$$

and the minimum is taken over all permutations σ of $\{1, \ldots, n\}$.

Proof: See [708, p. 399].

Fact 10.11.9. Let $\mathbb{I} \subseteq \mathbb{R}$ be an interval, let $A \colon \mathbb{I} \mapsto \mathbb{F}^{n \times n}$, and assume that A is continuous. Then, for $i = 1, \ldots, n$, there exist continuous functions $\lambda_i \colon \mathbb{I} \mapsto \mathbb{C}$ such that, for all $t \in \mathbb{I}$, $\mathrm{mspec}(A(t)) = \{\lambda_1(t), \ldots, \lambda_n(t)\}_{\mathrm{ms}}$.

Proof: See [708, p. 399].

Remark: The spectrum cannot always be continuously parameterized by more than one variable. See [708, p. 399].

Fact 10.11.10. Let $\mathcal{D} \subseteq \mathbb{R}^m$, assume that $\mathcal{D}$ is a convex set, and let $f \colon \mathcal{D} \to \mathbb{R}$. Then, f is convex if and only if the set $\{(x, y) \in \mathbb{R}^n \times \mathbb{R} \colon y \geq f(x)\}$ is convex.

Fact 10.11.11. Let $\mathcal{D} \subseteq \mathbb{R}^m$, assume that $\mathcal{D}$ is a convex set, let $f \colon \mathcal{D} \to \mathbb{R}$, and assume that f is convex. Then, f is continuous on $\mathrm{int}_{\mathrm{aff}\,\mathcal{D}}\,\mathcal{D}$.

Fact 10.11.12. Let $\mathcal{D} \subseteq \mathbb{R}^m$, assume that $\mathcal{D}$ is a convex set, let $f \colon \mathcal{D} \to \mathbb{R}$, and assume that f is convex. Then, $f^{-1}((-\infty, \alpha]) = \{x \in \mathcal{D} \colon f(x) \leq \alpha\}$ is convex.

10.12 Facts on Derivatives

Fact 10.12.1. Let $p \in \mathbb{C}[s]$. Then,

$$\mathrm{roots}(p') \subseteq \mathrm{co\,roots}(p).$$

Proof: See [459, p. 488].

Remark: p' is the derivative of p.

Fact 10.12.2. Let $f \colon \mathbb{R}^2 \to \mathbb{R}$, $g \colon \mathbb{R} \to \mathbb{R}$, and $h \colon \mathbb{R} \to \mathbb{R}$. Then, assuming each of the following integrals exists,

$$\frac{\mathrm{d}}{\mathrm{d}\alpha} \int_{g(\alpha)}^{h(\alpha)} f(t, \alpha)\,\mathrm{d}t = f(h(\alpha), \alpha)h'(\alpha) - f(g(\alpha), \alpha)g'(\alpha) + \int_{g(\alpha)}^{h(\alpha)} \frac{\partial}{\partial \alpha} f(t, \alpha)\,\mathrm{d}t.$$

Remark: This equality is *Leibniz's rule*.

Fact 10.12.3. Let $\mathcal{D} \subseteq \mathbb{R}^m$, assume that $\mathcal{D}$ is open and convex, let $f \colon \mathcal{D} \to \mathbb{R}$, and assume that f is C^1 on $\mathcal{D}$. Then, the following statements hold:

i) f is convex if and only if, for all $x, y \in \mathcal{D}$,

$$f(x) + (y - x)^{\mathrm{T}} f'(x) \leq f(y).$$

ii) f is strictly convex if and only if, for all distinct $x, y \in \mathcal{D}$,

$$f(x) + (y - x)^{\mathrm{T}} f'(x) < f(y).$$

Remark: If f is not differentiable, then these inequalities can be stated in terms of directional differentials of f or the *subdifferential* of f. See [1066, pp. 29–31, 128–145].

Fact 10.12.4. Let $f \colon \mathcal{D} \subseteq \mathbb{F}^m \mapsto \mathbb{F}^n$, and assume that $\mathrm{D}_+ f(0; \xi)$ exists. Then, for all $\beta > 0$,

$$\mathrm{D}_+ f(0; \beta\xi) = \beta \mathrm{D}_+ f(0; \xi).$$

Fact 10.12.5. Define $f \colon \mathbb{R} \to \mathbb{R}$ by $f(x) \triangleq |x|$. Then, for all $\xi \in \mathbb{R}$,

$$\mathrm{D}_+ f(0; \xi) = |\xi|.$$

Now, define $f \colon \mathbb{R}^n \to \mathbb{R}^n$ by $f(x) \triangleq \sqrt{x^{\mathrm{T}} x}$. Then, for all $\xi \in \mathbb{R}^n$,

$$\mathrm{D}_+ f(0; \xi) = \sqrt{\xi^{\mathrm{T}} \xi}.$$

Fact 10.12.6. Let $A, B \in \mathbb{F}^{n \times n}$. Then, for all $s \in \mathbb{F}$,

$$\frac{\mathrm{d}}{\mathrm{d}s}(A + sB)^2 = AB + BA + 2sB.$$

Hence,

$$\left.\frac{\mathrm{d}}{\mathrm{d}s}(A + sB)^2\right|_{s=0} = AB + BA.$$

Furthermore, for all $k \geq 1$,

$$\left.\frac{\mathrm{d}}{\mathrm{d}s}(A + sB)^k\right|_{s=0} = \sum_{i=0}^{k-1} A^i B A^{k-1-i}.$$

Fact 10.12.7. Let $A, B \in \mathbb{F}^{n \times n}$, and let $\mathcal{D} \triangleq \{s \in \mathbb{F} \colon \det(A + sB) \neq 0\}$. Then, for all $s \in \mathcal{D}$,

$$\frac{\mathrm{d}}{\mathrm{d}s}(A + sB)^{-1} = -(A + sB)^{-1} B (A + sB)^{-1}.$$

Hence, if A is nonsingular, then

$$\left.\frac{\mathrm{d}}{\mathrm{d}s}(A + sB)^{-1}\right|_{s=0} = -A^{-1} B A^{-1}.$$

Fact 10.12.8. Let $\mathcal{D} \subseteq \mathbb{F}$, let $A: \mathcal{D} \longrightarrow \mathbb{F}^{n \times n}$, and assume that A is differentiable. Then,

$$\frac{\mathrm{d}}{\mathrm{d}s} \det A(s) = \mathrm{tr}\left[A^{\mathrm{A}}(s)\frac{\mathrm{d}}{\mathrm{d}s}A(s)\right] = \frac{1}{n-1}\,\mathrm{tr}\left[A(s)\frac{\mathrm{d}}{\mathrm{d}s}A^{\mathrm{A}}(s)\right] = \sum_{i=1}^{n} \det A_i(s),$$

where $A_i(s)$ is obtained by differentiating the entries of the ith row of $A(s)$. If, in addition, $A(s)$ is nonsingular for all $s \in \mathcal{D}$, then

$$\frac{\mathrm{d}}{\mathrm{d}s} \log \det A(s) = \mathrm{tr}\left[A^{-1}(s)\frac{\mathrm{d}}{\mathrm{d}s}A(s)\right].$$

If $A(s)$ is positive definite for all $s \in \mathcal{D}$, then

$$\frac{\mathrm{d}}{\mathrm{d}s} \det A^{1/n}(s) = \frac{1}{n}[\det A^{1/n}(s)]\,\mathrm{tr}\left[A^{-1}(s)\frac{\mathrm{d}}{\mathrm{d}s}A(s)\right].$$

Finally, if $A(s)$ is nonsingular and has no negative eigenvalues for all $s \in \mathcal{D}$, then

$$\frac{\mathrm{d}}{\mathrm{d}s} \log^2 A(s) = 2\,\mathrm{tr}\left[[\log A(s)]A^{-1}(s)\frac{\mathrm{d}}{\mathrm{d}s}A(s)\right]$$

and

$$\frac{\mathrm{d}}{\mathrm{d}s} \log A(s) = \int_0^1 [(A(s) - I)t + I]^{-1}\frac{\mathrm{d}}{\mathrm{d}s}A(s)[(A(s) - I)t + I]^{-1}\,\mathrm{d}t.$$

Proof: See [367, p. 267], [577], [1039], [1125, pp. 199, 212], [1157, p. 430], and [1214].

Remark: See Fact 11.13.4.

Fact 10.12.9. Let $\mathcal{D} \subseteq \mathbb{F}$, let $A: \mathcal{D} \longrightarrow \mathbb{F}^{n \times n}$, assume that A is differentiable, and assume that $A(s)$ is nonsingular for all $x \in \mathcal{D}$. Then,

$$\frac{\mathrm{d}}{\mathrm{d}s}A^{-1}(s) = -A^{-1}(s)\left[\frac{\mathrm{d}}{\mathrm{d}s}A(s)\right]A^{-1}(s)$$

and

$$\mathrm{tr}\left[A^{-1}(s)\frac{\mathrm{d}}{\mathrm{d}s}A(s)\right] = -\mathrm{tr}\left[A(s)\frac{\mathrm{d}}{\mathrm{d}s}A^{-1}(s)\right].$$

Proof: See [730, p. 491] and [1125, pp. 198, 212].

Fact 10.12.10. Let $A, B \in \mathbb{F}^{n \times n}$. Then, for all $s \in \mathbb{F}$,

$$\frac{\mathrm{d}}{\mathrm{d}s} \det(A + sB) = \mathrm{tr}[B(A + sB)^{\mathrm{A}}].$$

Hence,

$$\frac{\mathrm{d}}{\mathrm{d}s} \det(A + sB)\bigg|_{s=0} = \mathrm{tr}\,BA^{\mathrm{A}} = \sum_{i=1}^{n} \det\left[A \overset{i}{\leftarrow} \mathrm{col}_i(B)\right].$$

Proof: Use Fact 2.16.9 and Fact 10.12.8.

Remark: This result generalizes Lemma 4.4.8.

Fact 10.12.11. Let $A \in \mathbb{F}^{n \times n}$, $r \in \mathbb{R}$, and $k \geq 1$. Then, for all $s \in \mathbb{C}$,

$$\frac{\mathrm{d}^k}{\mathrm{d}s^k}[\det(I + sA)]^r = (r \operatorname{tr} A)^k [\det(I + sA)]^r.$$

Hence,

$$\frac{\mathrm{d}^k}{\mathrm{d}s^k}[\det(I + sA)]^r \bigg|_{s=0} = (r \operatorname{tr} A)^k.$$

Fact 10.12.12. Let $A \in \mathbb{R}^{n \times n}$, assume that A is symmetric, let $X \in \mathbb{R}^{m \times n}$, and assume that XAX^{T} is nonsingular. Then,

$$\left(\frac{\mathrm{d}}{\mathrm{d}X} \det XAX^{\mathrm{T}}\right) = 2(\det XAX^{\mathrm{T}})A^{\mathrm{T}}X^{\mathrm{T}}(XAX^{\mathrm{T}})^{-1}.$$

Proof: See [358].

10.13 Facts on Infinite Series

Fact 10.13.1. The following infinite series converge for $A \in \mathbb{F}^{n \times n}$ with the given bounds on $\operatorname{sprad}(A)$:

i) For all $A \in \mathbb{F}^{n \times n}$,
$$\sin A = A - \tfrac{1}{3!}A^3 + \tfrac{1}{5!}A^5 - \tfrac{1}{7!}A^7 + \cdots.$$

ii) For all $A \in \mathbb{F}^{n \times n}$,
$$\cos A = I - \tfrac{1}{2!}A^2 + \tfrac{1}{4!}A^4 - \tfrac{1}{6!}A^6 + \cdots.$$

iii) For all $A \in \mathbb{F}^{n \times n}$ such that $\operatorname{sprad}(A) < \pi/2$,
$$\tan A = A + \tfrac{1}{3}A^3 + \tfrac{2}{15}A^5 + \tfrac{17}{315}A^7 + \tfrac{62}{2835}A^9 + \cdots.$$

iv) For all $A \in \mathbb{F}^{n \times n}$ such that $\operatorname{sprad}(A) < 1$,
$$e^A = I + A + \tfrac{1}{2!}A^2 + \tfrac{1}{3!}A^3 + \tfrac{1}{4!}A^4 + \cdots.$$

v) For all $A \in \mathbb{F}^{n \times n}$ such that $\operatorname{sprad}(A - I) < 1$,
$$\log A = -\left[I - A + \tfrac{1}{2}(I - A)^2 + \tfrac{1}{3}(I - A)^3 + \tfrac{1}{4}(I - A)^4 + \cdots\right].$$

vi) For all $A \in \mathbb{F}^{n \times n}$ such that $\operatorname{sprad}(A) < 1$,
$$\log(I - A) = -\left(A + \tfrac{1}{2}A^2 + \tfrac{1}{3}A^3 + \tfrac{1}{4}A^4 + \cdots\right).$$

vii) For all $A \in \mathbb{F}^{n \times n}$ such that $\operatorname{sprad}(A) < 1$,
$$\log(I + A) = A - \tfrac{1}{2}A^2 + \tfrac{1}{3}A^3 - \tfrac{1}{4}A^4 + \cdots.$$

viii) For all $A \in \mathbb{F}^{n \times n}$ such that $\operatorname{spec}(A) \subset \mathrm{ORHP}$,
$$\log A = \sum_{i=0}^{\infty} \frac{2}{2i + 1}\left[(A - I)(A + I)^{-1}\right]^{2i+1}.$$

ix) For all $A \in \mathbb{F}^{n \times n}$,
$$\sinh A = \sin \jmath A = A + \tfrac{1}{3!}A^3 + \tfrac{1}{5!}A^5 + \tfrac{1}{7!}A^7 + \cdots.$$

x) For all $A \in \mathbb{F}^{n \times n}$,

$$\cosh A = \cos \jmath A = I + \tfrac{1}{2!}A^2 + \tfrac{1}{4!}A^4 + \tfrac{1}{6!}A^6 + \cdots.$$

xi) For all $A \in \mathbb{F}^{n \times n}$ such that $\mathrm{sprad}(A) < \pi/2$,

$$\tanh A = \tan \jmath A = A - \tfrac{1}{3}A^3 + \tfrac{2}{15}A^5 - \tfrac{17}{315}A^7 + \tfrac{62}{2835}A^9 - \cdots.$$

xii) Let $\alpha \in \mathbb{R}$. For all $A \in \mathbb{F}^{n \times n}$ such that $\mathrm{sprad}(A) < 1$,

$$(I + A)^\alpha = I + \alpha A + \tfrac{\alpha(\alpha-1)}{2!}A^2 + \tfrac{\alpha(\alpha-1)(\alpha-2)}{3!}A^3 + \tfrac{1}{4}A^4 + \cdots$$

$$= I + \binom{\alpha}{1}A + \binom{\alpha}{2}A^2 + \binom{\alpha}{3}A^3 + \binom{\alpha}{4}A^4 + \cdots.$$

xiii) For all $A \in \mathbb{F}^{n \times n}$ such that $\mathrm{sprad}(A) < 1$,

$$(I - A)^{-1} = I + A + A^2 + A^3 + A^4 + \cdots.$$

Proof: See Fact 1.20.8.

Remark: The coefficients in *iii*) can be expressed in terms of Bernoulli numbers. See [772, p. 129].

10.14 Notes

An introductory treatment of limits and continuity is given in [1057]. The derivative and the directional differential are typically called the Fréchet derivative and the Gâteaux differential, respectively [509]. Differentiation of matrix functions is considered in [671, 973, 1000, 1116, 1164, 1213]. In [1161, 1162] the set $\mathrm{int}_{\mathrm{aff}\,\mathcal{S}}\,\mathcal{S}$ is called the relative interior of $\mathcal{S}$. An extensive treatment of matrix functions is given in Chapter 6 of [730]; see also [735]. The identity theorem is discussed in [762]. A chain rule for matrix functions is considered in [973, 1005]. Differentiation with respect to complex matrices is discussed in [798]. Extensive tables of derivatives of matrix functions are given in [382, pp. 586–593].

Chapter Eleven

The Matrix Exponential and Stability Theory

The matrix exponential function is fundamental to the study of linear ordinary differential equations. This chapter focuses on the properties of the matrix exponential as well as on stability theory.

11.1 Definition of the Matrix Exponential

The scalar initial value problem

$$\dot{x}(t) = ax(t), \tag{11.1.1}$$

$$x(0) = x_0, \tag{11.1.2}$$

where $t \in [0, \infty)$ and $a, x(t) \in \mathbb{R}$, has the solution

$$x(t) = e^{at}x_0, \tag{11.1.3}$$

where $t \in [0, \infty)$. We are interested in systems of linear differential equations of the form

$$\dot{x}(t) = Ax(t), \tag{11.1.4}$$

$$x(0) = x_0, \tag{11.1.5}$$

where $t \in [0, \infty)$, $x(t) \in \mathbb{R}^n$, and $A \in \mathbb{R}^{n \times n}$. Here $\dot{x}(t)$ denotes $\frac{dx(t)}{dt}$, where the derivative is one sided for $t = 0$ and two sided for $t > 0$. The solution of (11.1.4), (11.1.5) is given by

$$x(t) = e^{tA}x_0, \tag{11.1.6}$$

where $t \in [0, \infty)$ and e^{tA} is the *matrix exponential*. The following definition is based on (10.5.2).

Definition 11.1.1. Let $A \in \mathbb{F}^{n \times n}$. Then, the *matrix exponential* $e^A \in \mathbb{F}^{n \times n}$ or $\exp(A) \in \mathbb{F}^{n \times n}$ is the matrix

$$e^A \triangleq \sum_{k=0}^{\infty} \tfrac{1}{k!}A^k. \tag{11.1.7}$$

Note that $0! \triangleq 1$ and $e^{0_{n \times n}} = I_n$.

Proposition 11.1.2. Let $A \in \mathbb{F}^{n \times n}$. Then, the following statements hold:

i) The series (11.1.7) converges absolutely.

ii) The series (11.1.7) converges to e^A.

iii) Let $\| \cdot \|$ be a normalized submultiplicative norm on $\mathbb{F}^{n \times n}$. Then,

$$e^{-\|A\|} \leq \|e^A\| \leq e^{\|A\|}. \tag{11.1.8}$$

Proof. To prove *i)*, let $\| \cdot \|$ be a normalized submultiplicative norm on $\mathbb{F}^{n \times n}$. Then, for all $k \geq 1$,

$$\sum_{i=0}^{k} \tfrac{1}{i!} \|A^i\| \leq \sum_{i=0}^{k} \tfrac{1}{i!} \|A\|^i \leq e^{\|A\|}.$$

Since the sequence $\left(\sum_{i=0}^{k} \tfrac{1}{i!} \|A^i\| \right)_{k=0}^{\infty}$ of partial sums is increasing and bounded, there exists $\alpha > 0$ such that the series $\sum_{i=0}^{\infty} \tfrac{1}{i!} \|A^i\|$ converges to α. Hence, the series $\sum_{i=0}^{\infty} \tfrac{1}{i!} A^i$ converges absolutely.

Next, *ii)* follows from *i)* using Proposition 10.2.9.

Next, we have

$$\|e^A\| = \left\| \sum_{i=0}^{\infty} \tfrac{1}{i!} A^i \right\| \leq \sum_{i=0}^{\infty} \tfrac{1}{i!} \|A^i\| \leq \sum_{i=0}^{\infty} \tfrac{1}{i!} \|A\|^i = e^{\|A\|},$$

which proves the second inequality in (11.1.8). Finally, note that

$$1 \leq \|e^A\| \|e^{-A}\| \leq \|e^A\| e^{\|A\|},$$

and thus

$$e^{-\|A\|} \leq \|e^A\|. \qquad \square$$

The following result generalizes the well-known corresponding scalar result.

Proposition 11.1.3. Let $A \in \mathbb{F}^{n \times n}$. Then,

$$e^A = \lim_{k \to \infty} \left(I + \tfrac{1}{k} A \right)^k. \tag{11.1.9}$$

Proof. It follows from the binomial theorem that

$$\left(I + \tfrac{1}{k} A \right)^k = \sum_{i=0}^{k} \alpha_i(k) A^i,$$

where

$$\alpha_i(k) \triangleq \frac{1}{k^i} \binom{k}{i} = \frac{1}{k^i} \frac{k!}{i!(k-i)!}.$$

For all $i \in \mathbb{P}$, it follows that $\alpha_i(k) \to 1/i!$ as $k \to \infty$. Hence,

$$\lim_{k \to \infty} \left(I + \tfrac{1}{k} A \right)^k = \lim_{k \to \infty} \sum_{i=0}^{k} \alpha_i(k) A^i = \sum_{i=0}^{\infty} \tfrac{1}{i!} A^i = e^A. \qquad \square$$

Proposition 11.1.4. Let $A \in \mathbb{F}^{n \times n}$. Then, for all $t \in \mathbb{R}$,

$$e^{tA} - I = \int_0^t A e^{\tau A} \, d\tau \tag{11.1.10}$$

and

$$\frac{d}{dt} e^{tA} = A e^{tA}. \tag{11.1.11}$$

Proof. Note that

$$\int_0^t A e^{\tau A} \, d\tau = \int_0^t \sum_{k=0}^{\infty} \frac{1}{k!} \tau^k A^{k+1} \, d\tau = \sum_{k=0}^{\infty} \frac{1}{k!} \frac{t^{k+1}}{k+1} A^{k+1} = e^{tA} - I,$$

which yields (11.1.10), while differentiating (11.1.10) with respect to t yields (11.1.11). $\square$

Proposition 11.1.5. Let $A, B \in \mathbb{F}^{n \times n}$. Then, $AB = BA$ if and only if, for all $t \in [0, \infty)$,
$$e^{tA} e^{tB} = e^{t(A+B)}. \tag{11.1.12}$$

Proof. Suppose that $AB = BA$. By expanding e^{tA}, e^{tB}, and $e^{t(A+B)}$, it can be seen that the expansions of $e^{tA} e^{tB}$ and $e^{t(A+B)}$ are identical. Conversely, differentiating (11.1.12) twice with respect to t and setting $t = 0$ yields $AB = BA$. $\square$

Corollary 11.1.6. Let $A, B \in \mathbb{F}^{n \times n}$, and assume that $AB = BA$. Then,
$$e^A e^B = e^B e^A = e^{A+B}. \tag{11.1.13}$$

The converse of Corollary 11.1.6 is not true. For example, if $A \triangleq \begin{bmatrix} 0 & \pi \\ -\pi & 0 \end{bmatrix}$ and $B \triangleq \begin{bmatrix} 0 & (7+4\sqrt{3})\pi \\ (-7+4\sqrt{3})\pi & 0 \end{bmatrix}$, then $e^A = e^B = -I$ and $e^{A+B} = I$, although $AB \neq BA$. A partial converse is given by Fact 11.14.1.

Proposition 11.1.7. Let $A \in \mathbb{F}^{n \times n}$ and $B \in \mathbb{F}^{m \times m}$. Then,
$$e^{A \otimes I_m} = e^A \otimes I_m, \tag{11.1.14}$$

$$e^{I_n \otimes B} = I_n \otimes e^B, \tag{11.1.15}$$

$$e^{A \oplus B} = e^A \otimes e^B. \tag{11.1.16}$$

Proof. Note that

$$\begin{aligned} e^{A \otimes I_m} &= I_{nm} + A \otimes I_m + \tfrac{1}{2!}(A \otimes I_m)^2 + \cdots \\ &= I_n \otimes I_m + A \otimes I_m + \tfrac{1}{2!}(A^2 \otimes I_m) + \cdots \\ &= (I_n + A + \tfrac{1}{2!}A^2 + \cdots) \otimes I_m \\ &= e^A \otimes I_m \end{aligned}$$

and similarly for (11.1.15). To prove (11.1.16), note that $(A \otimes I_m)(I_n \otimes B) = A \otimes B$ and $(I_n \otimes B)(A \otimes I_m) = A \otimes B$, which shows that $A \otimes I_m$ and $I_n \otimes B$ commute. Thus, by Corollary 11.1.6,

$$e^{A \oplus B} = e^{A \otimes I_m + I_n \otimes B} = e^{A \otimes I_m} e^{I_n \otimes B} = \left(e^A \otimes I_m\right)\left(I_n \otimes e^B\right) = e^A \otimes e^B. \qquad \square$$

11.2 Structure of the Matrix Exponential

To elucidate the structure of the matrix exponential, recall that, by Theorem 4.6.1, every term A^k in (11.1.7) for $k > r \triangleq \deg \mu_A$ can be expressed as a linear combination of $I, A, \ldots, A^{r-1}$. The following result provides an expression for e^{tA} in terms of $I, A, \ldots, A^{r-1}$.

Proposition 11.2.1. Let $A \in \mathbb{F}^{n \times n}$. Then, for all $t \in \mathbb{R}$,

$$e^{tA} = \frac{1}{j2\pi} \oint_{\mathcal{C}} (zI - A)^{-1} e^{tz} \, \mathrm{d}z = \sum_{i=0}^{n-1} \psi_i(t) A^i, \qquad (11.2.1)$$

where, for all $i \in \{0, \ldots, n-1\}$, $\psi_i(t)$ is given by

$$\psi_i(t) \triangleq \frac{1}{j2\pi} \oint_{\mathcal{C}} \frac{\chi_A^{[i+1]}(z)}{\chi_A(z)} e^{tz} \, \mathrm{d}z, \qquad (11.2.2)$$

where $\mathcal{C}$ is a simple, closed contour in the complex plane enclosing $\operatorname{spec}(A)$,

$$\chi_A(s) = s^n + \beta_{n-1} s^{n-1} + \cdots + \beta_1 s + \beta_0, \qquad (11.2.3)$$

and the polynomials $\chi_A^{[1]}, \ldots, \chi_A^{[n]}$ are defined by the recursion

$$s \chi_A^{[i+1]}(s) = \chi_A^{[i]}(s) - \beta_i, \quad i = 0, \ldots, n-1,$$

where $\chi_A^{[0]} \triangleq \chi_A$ and $\chi_A^{[n]}(s) = 1$. Furthermore, for all $i \in \{0, \ldots, n-1\}$ and $t \geq 0$, $\psi_i(t)$ satisfies

$$\psi_i^{(n)}(t) + \beta_{n-1} \psi_i^{(n-1)}(t) + \cdots + \beta_1 \psi_i'(t) + \beta_0 \psi_i(t) = 0, \qquad (11.2.4)$$

where, for all $i, j \in \{0, \ldots, n-1\}$,

$$\psi_i^{(j)}(0) = \delta_{ij}. \qquad (11.2.5)$$

Proof. See [583, p. 381], [913, 954], [1490, p. 31], and Fact 4.9.11. $\qquad \square$

The coefficient $\psi_i(t)$ of A^i in (11.2.1) can be further characterized in terms of the Laplace transform. Define

$$\hat{x}(s) \triangleq \mathcal{L}\{x(t)\} \triangleq \int_0^\infty e^{-st} x(t) \, \mathrm{d}t. \qquad (11.2.6)$$

Note that

$$\mathcal{L}\{\dot{x}(t)\} = s\hat{x}(s) - x(0) \qquad (11.2.7)$$

and

$$\mathcal{L}\{\ddot{x}(t)\} = s^2 \hat{x}(s) - sx(0) - \dot{x}(0). \qquad (11.2.8)$$

The following result shows that the resolvent of A is the Laplace transform of the exponential of A. See (4.4.23).

Proposition 11.2.2. Let $A \in \mathbb{F}^{n \times n}$, and define $\psi_0, \ldots, \psi_{n-1}$ as in Proposition 11.2.1. Then, for all $s \in \mathbb{C}\backslash\mathrm{spec}(A)$,

$$\mathcal{L}\{e^{tA}\} = \int_0^\infty e^{-st}e^{tA}\,dt = (sI - A)^{-1}. \tag{11.2.9}$$

Furthermore, for all $i \in \{0, \ldots, n-1\}$, the Laplace transform $\hat{\psi}_i(s)$ of $\psi_i(t)$ is given by

$$\hat{\psi}_i(s) = \frac{\chi_A^{[i+1]}(s)}{\chi_A(s)} \tag{11.2.10}$$

and

$$(sI - A)^{-1} = \sum_{i=0}^{n-1} \hat{\psi}_i(s)A^i. \tag{11.2.11}$$

Proof. Let $s \in \mathbb{C}$ satisfy $\mathrm{Re}\,s > \mathrm{spabs}(A)$ so that $A - sI$ is asymptotically stable. Thus, it follows from Lemma 11.9.2 that

$$\mathcal{L}\{e^{tA}\} = \int_0^\infty e^{-st}e^{tA}\,dt = \int_0^\infty e^{t(A-sI)}\,dt = (sI - A)^{-1}.$$

By analytic continuation, the expression $\mathcal{L}\{e^{tA}\}$ is given by (11.2.9) for all $s \in \mathbb{C}\backslash\mathrm{spec}(A)$. $\qquad\square$

Comparing (11.2.11) with the expression for $(sI - A)^{-1}$ given by (4.4.23) shows that there exist $B_0, \ldots, B_{n-2} \in \mathbb{F}^{n \times n}$ such that

$$\sum_{i=0}^{n-1} \hat{\psi}_i(s)A^i = \frac{s^{n-1}}{\chi_A(s)}I + \frac{s^{n-2}}{\chi_A(s)}B_{n-2} + \cdots + \frac{s}{\chi_A(s)}B_1 + \frac{1}{\chi_A(s)}B_0. \tag{11.2.12}$$

To further illustrate the structure of e^{tA}, where $A \in \mathbb{F}^{n \times n}$, let $A = SBS^{-1}$, where $B = \mathrm{diag}(B_1, \ldots, B_k)$ is the Jordan form of A. Hence, by Proposition 11.2.8,

$$e^{tA} = Se^{tB}S^{-1}, \tag{11.2.13}$$

where

$$e^{tB} = \mathrm{diag}(e^{tB_1}, \ldots, e^{tB_k}). \tag{11.2.14}$$

The structure of e^{tB} can thus be determined by considering the block $B_i \in \mathbb{F}^{\alpha_i \times \alpha_i}$, which, for all $i \in \{1, \ldots, k\}$ has the form

$$B_i = \lambda_i I_{\alpha_i} + N_{\alpha_i}. \tag{11.2.15}$$

Since $\lambda_i I_{\alpha_i}$ and N_{α_i} commute, it follows from Proposition 11.1.5 that

$$e^{tB_i} = e^{t(\lambda_i I_{\alpha_i} + N_{\alpha_i})} = e^{\lambda_i t I_{\alpha_i}}e^{tN_{\alpha_i}} = e^{\lambda_i t}e^{tN_{\alpha_i}}. \tag{11.2.16}$$

Since $N_{\alpha_i}^{\alpha_i} = 0$, it follows that $e^{tN_{\alpha_i}}$ is a finite sum of powers of tN_{α_i}. Specifically,

$$e^{tN_{\alpha_i}} = I_{\alpha_i} + tN_{\alpha_i} + \tfrac{1}{2}t^2N_{\alpha_i}^2 + \cdots + \frac{1}{(\alpha_i-1)!}t^{\alpha_i-1}N_{\alpha_i}^{\alpha_i-1}, \tag{11.2.17}$$

and thus

$$e^{tN_{\alpha_i}} = \begin{bmatrix} 1 & t & \frac{t^2}{2} & \cdots & \frac{t^{\alpha_i-2}}{(\alpha_i-2)!} & \frac{t^{\alpha_i-1}}{(\alpha_i-1)!} \\ 0 & 1 & t & \ddots & \frac{t^{\alpha_i-3}}{(\alpha_i-3)!} & \frac{t^{\alpha_i-2}}{(\alpha_i-2)!} \\ 0 & 0 & 1 & \ddots & \frac{t^{\alpha_i-4}}{(\alpha_i-4)!} & \frac{t^{\alpha_i-3}}{(\alpha_i-3)!} \\ \vdots & \vdots & \ddots & \ddots & \ddots & \vdots \\ 0 & 0 & 0 & \ddots & 1 & t \\ 0 & 0 & 0 & \cdots & 0 & 1 \end{bmatrix}, \tag{11.2.18}$$

which is upper triangular and Toeplitz (see Fact 11.13.1). Alternatively, (11.2.18) follows from (10.5.5) with $f(s) = e^{st}$.

Note that (11.2.16) follows from (10.5.5) with $f(\lambda) = e^{\lambda t}$. Furthermore, every entry of e^{tB_i} is of the form $\frac{1}{r!}t^r e^{\lambda_i t}$, where $r \in \{0, \alpha_i - 1\}$ and λ_i is an eigenvalue of A. Reconstructing A by means of $A = SBS^{-1}$ shows that every entry of A is a linear combination of the entries of the blocks e^{tB_i}. If A is real, then e^{tA} is also real. Thus, the term $e^{\lambda_i t}$ for complex $\lambda_i = \nu_i + \jmath\omega_i \in \mathrm{spec}(A)$, where ν_i and ω_i are real, yields terms of the form $e^{\nu_i t}\cos\omega_i t$ and $e^{\nu_i t}\sin\omega_i t$.

The following result follows from (11.2.18) or Corollary 10.5.4.

Proposition 11.2.3. Let $A \in \mathbb{F}^{n\times n}$. Then,

$$\mathrm{mspec}(e^A) = \{e^\lambda\colon \lambda \in \mathrm{mspec}(A)\}_{\mathrm{ms}}. \tag{11.2.19}$$

Proof. It can be seen that every diagonal entry of the Jordan form of e^A is of the form e^λ, where $\lambda \in \mathrm{spec}(A)$. $\square$

Corollary 11.2.4. Let $A \in \mathbb{F}^{n\times n}$. Then,

$$\det e^A = e^{\mathrm{tr}\,A}. \tag{11.2.20}$$

Corollary 11.2.5. Let $A \in \mathbb{F}^{n\times n}$, and assume that $\mathrm{tr}\,A = 0$. Then, $\det e^A = 1$.

Corollary 11.2.6. Let $A \in \mathbb{F}^{n\times n}$. Then, the following statements hold:

i) If e^A is unitary, then, $\mathrm{spec}(A) \subset \jmath\mathbb{R}$.

ii) $\mathrm{spec}(e^A)$ is real if and only if $\mathrm{Im}\,\mathrm{spec}(A) \subset \pi\mathbb{Z}$.

Proposition 11.2.7. Let $A \in \mathbb{F}^{n\times n}$. Then, the following statements hold:

i) A and e^A have the same number of Jordan blocks of corresponding sizes.

ii) e^A is semisimple if and only if A is semisimple.

iii) If $\mu \in \mathrm{spec}(e^A)$, then

$$\mathrm{amult}_{\exp(A)}(\mu) = \sum_{\{\lambda \in \mathrm{spec}(A):\ e^\lambda = \mu\}} \mathrm{amult}_A(\lambda) \tag{11.2.21}$$

and

$$\mathrm{gmult}_{\exp(A)}(\mu) = \sum_{\{\lambda \in \mathrm{spec}(A):\ e^\lambda = \mu\}} \mathrm{gmult}_A(\lambda). \tag{11.2.22}$$

iv) If e^A is simple, then A is simple.

v) If e^A is cyclic, then A is cyclic.

vi) e^A is a scalar multiple of the identity matrix if and only if A is semisimple and every pair of eigenvalues of A differs by an integer multiple of $j2\pi$.

vii) e^A is a real scalar multiple of the identity matrix if and only if A is semisimple, every pair of eigenvalues of A differs by an integer multiple of $j2\pi$, and the imaginary part of every eigenvalue of A is an integer multiple of $j\pi$.

Proof. To prove *i)*, note that, for all $t \neq 0$, $\mathrm{def}(e^{tN_{\alpha_i}} - I_{\alpha_i}) = 1$, and thus the geometric multiplicity of (11.2.18) is 1. Since (11.2.18) has one distinct eigenvalue, it follows that (11.2.18) is cyclic. Hence, by Proposition 5.5.14, (11.2.18) is similar to a single Jordan block. Now, *i)* follows by setting $t = 1$ and applying this argument to each Jordan block of A. Statements *ii)–v)* follow by similar arguments.

To prove *vi)*, note that, for all $\lambda_i, \lambda_j \in \mathrm{spec}(A)$, it follows that $e^{\lambda_i} = e^{\lambda_j}$. Furthermore, since A is semisimple, it follows from *ii)* that e^A is also semisimple. Since all of the eigenvalues of e^A are equal, it follows that e^A is a scalar multiple of the identity matrix. Finally, *vii)* is an immediate consequence of *vii)*. $\qquad\square$

Proposition 11.2.8. Let $A \in \mathbb{F}^{n \times n}$. Then, the following statements hold:

i) $\left(e^A\right)^{\mathrm{T}} = e^{A^{\mathrm{T}}}$.

ii) $\left(e^{\overline{A}}\right) = \overline{e^A}$.

iii) $\left(e^A\right)^* = e^{A^*}$.

iv) e^A is nonsingular, and $\left(e^A\right)^{-1} = e^{-A}$.

v) If $S \in \mathbb{F}^{n \times n}$ is nonsingular, then $e^{SAS^{-1}} = Se^A S^{-1}$.

vi) If $A = \mathrm{diag}(A_1, \ldots, A_k)$, where $A_i \in \mathbb{F}^{n_i \times n_i}$ for all $i \in \{1, \ldots, k\}$, then $e^A = \mathrm{diag}\left(e^{A_1}, \ldots, e^{A_k}\right)$.

vii) If A is Hermitian, then e^A is positive definite.

Furthermore, the following statements are equivalent:

viii) A is normal.

ix) $\mathrm{tr}\, e^{A^*}e^A = \mathrm{tr}\, e^{A^*+A}$.

x) $e^{A^*}e^A = e^{A^*+A}$.

xi) $e^A e^{A^*} = e^{A^*} e^A = e^{A^*+A}$.

Finally, the following statements hold:

xii) If A is normal, then e^A is normal.

xiii) e^A is normal if and only if A is unitarily similar to a block-diagonal matrix $\mathrm{diag}(A_1, \ldots, A_k)$ such that, for all $i \in \{1, \ldots, k\}$, A_i is semisimple and each pair of eigenvalues of A_i differ by an integer multiple of $\jmath 2\pi$, and, for all distinct $i, j \in \{1, \ldots, k\}$, $\mathrm{spec}(e^{A_i}) \neq \mathrm{spec}(e^{A_j})$.

xiv) If e^A is normal and no pair of eigenvalues of A differ by an integer multiple of $\jmath 2\pi$, then A is normal.

xv) A is skew Hermitian if and only if A is normal and e^A is unitary.

xvi) If $\mathbb{F} = \mathbb{R}$ and A is skew symmetric, then e^A is orthogonal and $\det e^A = 1$.

xvii) If e^A is unitary, then either A is skew Hermitian or at least two eigenvalues of A differ by a nonzero integer multiple of $\jmath 2\pi$.

xviii) e^A is unitary if and only if A is unitarily similar to a block-diagonal matrix $\mathrm{diag}(A_1, \ldots, A_k)$ such that, for all $i \in \{1, \ldots, k\}$, A_i is semisimple, every eigenvalue of A_i is imaginary, and each pair of eigenvalues of A_i differ by an integer multiple of $\jmath 2\pi$, and, for all distinct $i, j \in \{1, \ldots, k\}$, $\mathrm{spec}(e^{A_i}) \neq \mathrm{spec}(e^{A_j})$.

xix) e^A is Hermitian if and only if A is unitarily similar to a block-diagonal matrix $\mathrm{diag}(A_1, \ldots, A_k)$ such that, for all $i \in \{1, \ldots, k\}$, A_i is semisimple, the imaginary part of every eigenvalue of A_i is an integer multiple of $\pi\jmath$, and each pair of eigenvalues of A_i differ by an integer multiple of $\jmath 2\pi$, and, for all distinct $i, j \in \{1, \ldots, k\}$, $\mathrm{spec}(e^{A_i}) \neq \mathrm{spec}(e^{A_j})$.

xx) e^A is positive definite if and only if A is unitarily similar to a block-diagonal matrix $\mathrm{diag}(A_1, \ldots, A_k)$ such that, for all $i \in \{1, \ldots, k\}$, A_i is semisimple, the imaginary part of every eigenvalue of A_i is an integer multiple of $2\pi\jmath$, and each pair of eigenvalues of A_i differ by an integer multiple of $\jmath 2\pi$, and, for all distinct $i, j \in \{1, \ldots, k\}$, $\mathrm{spec}(e^{A_i}) \neq \mathrm{spec}(e^{A_j})$.

Proof. The equivalence of *viii)* and *ix)* is given in [465, 1239], while the equivalence of *viii)* and *xi)* is given in [1202]. Note that *xi)* $\implies$ *x)* $\implies$ *ix)*. Statement *xii)* follows from the fact that *viii)* $\implies$ *xi)*. Statement *xiii)* is given in [1503]. Statement *xiv)* is a consequence of *xiii)*. To prove sufficiency in *xv)*, note that $e^{A+A^*} = e^A e^{A^*} = e^A (e^A)^* = I = e^0$. Since $A + A^*$ is Hermitian, it follows from *iii)* of Proposition 11.2.9 that $A + A^* = 0$. The converse is immediate. To prove *xvi)*, note that $e^A (e^A)^{\mathrm{T}} = e^A e^{A^{\mathrm{T}}} = e^A e^{-A} = e^A (e^A)^{-1} = I$, and, using Corollary 11.2.5, $\det e^A = e^{\mathrm{tr}\, A} = e^0 = 1$. To prove *xvii)*, note that it follows from *xiii)* that, if every block A_i is scalar, then A is skew Hermitian, while, if at least one block A_i is not scalar, then A has at least two eigenvalues that differ by an integer multiple of $\jmath 2\pi$. Finally, *xviii)–xx)* are analogous to *xiii)*. $\qquad\square$

The converse of *xii)* is false. For example, the matrix $A \triangleq \begin{bmatrix} -2\pi & 4\pi \\ -2\pi & 2\pi \end{bmatrix}$ satisfies $e^A = I$ but is not normal. Likewise, $A = \begin{bmatrix} \jmath\pi & 1 \\ 0 & -\jmath\pi \end{bmatrix}$ satisfies $e^A = -I$ but is not normal. For both matrices, $e^{A^*} e^A = e^A e^{A^*} = I$, but $e^{A^*} e^A \neq e^{A^*+A}$, which confirms that xi) does not hold. Both matrices have eigenvalues $\pm j\pi$.

Proposition 11.2.9. The following statements hold:

i) If $A, B \in \mathbb{F}^{n \times n}$ are similar, then e^A and e^B are similar.

ii) If $A, B \in \mathbb{F}^{n \times n}$ are unitarily similar, then e^A and e^B are unitarily similar.

iii) If $B \in \mathbb{F}^{n \times n}$ is positive definite, then there exists a unique Hermitian matrix $A \in \mathbb{F}^{n \times n}$ such that $e^A = B$.

iv) $B \in \mathbb{F}^{n \times n}$ is Hermitian and nonsingular if and only if there exists a normal matrix $A \in \mathbb{C}^{n \times n}$ such that, for all $\lambda \in \operatorname{spec}(A)$, $\operatorname{Im} \lambda$ is an integer multiple of $j\pi$ and $e^A = B$.

v) $B \in \mathbb{F}^{n \times n}$ is normal and nonsingular if and only if there exists a normal matrix $A \in \mathbb{F}^{n \times n}$ such that $e^A = B$.

vi) $B \in \mathbb{F}^{n \times n}$ is unitary if and only if there exists a normal matrix $A \in \mathbb{C}^{n \times n}$ such that $\operatorname{mspec}(A) \subset j\mathbb{R}$ and $e^A = B$.

vii) $B \in \mathbb{F}^{n \times n}$ is unitary if and only if there exists a skew-Hermitian matrix $A \in \mathbb{C}^{n \times n}$ such that $e^A = B$.

$viii$) $B \in \mathbb{F}^{n \times n}$ is unitary if and only if there exists a Hermitian matrix $A \in \mathbb{F}^{n \times n}$ such that $e^{jA} = B$.

ix) $B \in \mathbb{R}^{n \times n}$ is orthogonal and $\det B = 1$ if and only if there exists a skew-symmetric matrix $A \in \mathbb{R}^{n \times n}$ such that $e^A = B$.

x) If A and B are normal and $e^A = e^B$, then $A + A^* = B + B^*$.

Proof. To prove iii), let $B = S\operatorname{diag}(b_1, \ldots, b_n)S^{-1}$, where $S \in \mathbb{F}^{n \times n}$ is unitary and $b_1, \ldots, b_n$ are positive. Then, define $A \triangleq S\operatorname{diag}(\log b_1, \ldots, \log b_n)S^{-1}$. Next, to prove uniqueness, let A and $\hat{A}$ be Hermitian matrices such that $B = e^A = e^{\hat{A}}$. Then, for all $t \geq 0$, it follows that $e^{tA} = (e^A)^t = (e^{\hat{A}})^t = e^{t\hat{A}}$. Differentiating yields $Ae^{tA} = \hat{A}e^{t\hat{A}}$, while setting $t = 0$ implies that $A = \hat{A}$. As another proof, it follows from x) of Fact 11.14.1 that A and $\hat{A}$ commute. Therefore, $I = e^A e^{-\hat{A}} = e^{A - \hat{A}}$, which, since $A - \hat{A}$ is Hermitian, implies that $A - \hat{A} = 0$. As yet another proof, the result follows directly from $xiii$) of Fact 11.14.1. Finally, iii) is given by Theorem 4.4 of [1202]. Statement vii) is given by v) of Proposition 11.6.7. To prove x), note that $e^{A+A^*} = e^{B+B^*}$, which, by vii) of Proposition 11.2.8, is positive definite. The result now follows from iii). $\square$

The converse of i) is false. For example, $A \triangleq \left[\begin{smallmatrix} 0 & 0 \\ 0 & 0 \end{smallmatrix}\right]$ and $B \triangleq \left[\begin{smallmatrix} 0 & 2\pi \\ -2\pi & 0 \end{smallmatrix}\right]$ satisfy $e^A = e^B = I$, although A and B are not similar.

11.3 Explicit Expressions

In this section we present explicit expressions for the exponential of a general 2×2 real matrix A. Expressions are given in terms of both the entries of A and the eigenvalues of A.

Lemma 11.3.1. Let $A \triangleq \begin{bmatrix} a & b \\ 0 & d \end{bmatrix} \in \mathbb{C}^{2 \times 2}$. Then,

$$e^A = \begin{cases} e^a \begin{bmatrix} 1 & b \\ 0 & 1 \end{bmatrix}, & a = d, \\[4mm] \begin{bmatrix} e^a & b\frac{e^a - e^d}{a-d} \\ 0 & e^d \end{bmatrix}, & a \neq d. \end{cases} \tag{11.3.1}$$

The following result gives an expression for e^A in terms of the eigenvalues of A.

Proposition 11.3.2. Let $A \in \mathbb{C}^{2 \times 2}$, and let $\mathrm{mspec}(A) = \{\lambda, \mu\}_{\mathrm{ms}}$. Then,

$$e^A = \begin{cases} e^\lambda[(1 - \lambda)I + A], & \lambda = \mu, \\[3mm] \frac{\mu e^\lambda - \lambda e^\mu}{\mu - \lambda}I + \frac{e^\mu - e^\lambda}{\mu - \lambda}A, & \lambda \neq \mu. \end{cases} \tag{11.3.2}$$

Proof. This result follows from Theorem 10.5.2. Alternatively, suppose that $\lambda = \mu$. Then, there exists a nonsingular matrix $S \in \mathbb{C}^{2 \times 2}$ such that $A = S\begin{bmatrix} \lambda & \alpha \\ 0 & \lambda \end{bmatrix}S^{-1}$, where $\alpha \in \mathbb{C}$. Hence, $e^A = e^\lambda S\begin{bmatrix} 1 & \alpha \\ 0 & 1 \end{bmatrix}S^{-1} = e^\lambda[(1 - \lambda)I + A]$. Now, suppose that $\lambda \neq \mu$. Then, there exists a nonsingular matrix $S \in \mathbb{C}^{2 \times 2}$ such that $A = S\begin{bmatrix} \lambda & 0 \\ 0 & \mu \end{bmatrix}S^{-1}$. Hence, $e^A = S\begin{bmatrix} e^\lambda & 0 \\ 0 & e^\mu \end{bmatrix}S^{-1}$. Then, the equality $\begin{bmatrix} e^\lambda & 0 \\ 0 & e^\mu \end{bmatrix} = \frac{\mu e^\lambda - \lambda e^\mu}{\mu - \lambda}I + \frac{e^\mu - e^\lambda}{\mu - \lambda}\begin{bmatrix} \lambda & 0 \\ 0 & \mu \end{bmatrix}$ yields the desired result. $\square$

Next, we give an expression for e^A in terms of the entries of $A \in \mathbb{R}^{2 \times 2}$.

Corollary 11.3.3. Let $A \triangleq \begin{bmatrix} a & b \\ c & d \end{bmatrix} \in \mathbb{R}^{2 \times 2}$, and define $\gamma \triangleq (a - d)^2 + 4bc$ and $\delta \triangleq \frac{1}{2}|\gamma|^{1/2}$. Then,

$$e^A = \begin{cases} e^{\frac{a+d}{2}} \begin{bmatrix} \cos\delta + \frac{a-d}{2\delta}\sin\delta & \frac{b}{\delta}\sin\delta \\[2mm] \frac{c}{\delta}\sin\delta & \cos\delta - \frac{a-d}{2\delta}\sin\delta \end{bmatrix}, & \gamma < 0, \\[6mm] e^{\frac{a+d}{2}} \begin{bmatrix} 1 + \frac{a-d}{2} & b \\[2mm] c & 1 - \frac{a-d}{2} \end{bmatrix}, & \gamma = 0, \\[6mm] e^{\frac{a+d}{2}} \begin{bmatrix} \cosh\delta + \frac{a-d}{2\delta}\sinh\delta & \frac{b}{\delta}\sinh\delta \\[2mm] \frac{c}{\delta}\sinh\delta & \cosh\delta - \frac{a-d}{2\delta}\sinh\delta \end{bmatrix}, & \gamma > 0. \end{cases} \tag{11.3.3}$$

Proof. The eigenvalues of A are $\lambda \triangleq \frac{1}{2}(a + d - \sqrt{\gamma})$ and $\mu \triangleq \frac{1}{2}(a + d + \sqrt{\gamma})$. Hence, $\lambda = \mu$ if and only if $\gamma = 0$. The result now follows from Proposition 11.3.2. $\square$

Example 11.3.4. Let $A \triangleq \begin{bmatrix} \nu & \omega \\ -\omega & \nu \end{bmatrix} \in \mathbb{R}^{2 \times 2}$. Then,

$$e^{tA} = e^{\nu t} \begin{bmatrix} \cos\omega t & \sin\omega t \\ -\sin\omega t & \cos\omega t \end{bmatrix}. \tag{11.3.4}$$

On the other hand, if $A \triangleq \left[\begin{smallmatrix} \nu & \omega \\ \omega & -\nu \end{smallmatrix} \right]$, then

$$e^{tA} = \left[\begin{array}{cc} \cosh \delta t + \frac{\nu}{\delta} \sinh \delta t & \frac{\omega}{\delta} \sinh \delta t \\ \frac{\omega}{\delta} \sinh \delta t & \cosh \delta t - \frac{\nu}{\delta} \sinh \delta t \end{array} \right], \tag{11.3.5}$$

where $\delta \triangleq \sqrt{\omega^2 + \nu^2}$.

Example 11.3.5. Let $\alpha \in \mathbb{F}$, and define $A \triangleq \left[\begin{smallmatrix} 0 & 1 \\ 0 & \alpha \end{smallmatrix} \right]$. Then,

$$e^{tA} = \begin{cases} \left[\begin{array}{cc} 1 & \alpha^{-1}(e^{\alpha t} - 1) \\ 0 & e^{\alpha t} \end{array} \right], & \alpha \neq 0, \\[2em] \left[\begin{array}{cc} 1 & t \\ 0 & 1 \end{array} \right], & \alpha = 0. \end{cases}$$

Example 11.3.6. Let $\theta \in \mathbb{R}$, and define $A \triangleq \left[\begin{smallmatrix} 0 & \theta \\ -\theta & 0 \end{smallmatrix} \right]$. Then,

$$e^A = \left[\begin{array}{cc} \cos \theta & \sin \theta \\ -\sin \theta & \cos \theta \end{array} \right].$$

Furthermore, define $B \triangleq \left[\begin{smallmatrix} 0 & \frac{\pi}{2} - \theta \\ -\frac{\pi}{2} + \theta & 0 \end{smallmatrix} \right]$. Then,

$$e^B = \left[\begin{array}{cc} \sin \theta & \cos \theta \\ -\cos \theta & \sin \theta \end{array} \right].$$

Example 11.3.7. Consider the second-order mechanical vibration equation

$$m\ddot{q} + c\dot{q} + kq = 0, \tag{11.3.6}$$

where m is positive and c and k are nonnegative. Here m, c, and k denote mass, damping, and stiffness parameters, respectively. Equation (11.3.6) can be written in companion form as the system

$$\dot{x} = Ax, \tag{11.3.7}$$

where

$$x \triangleq \left[\begin{array}{c} q \\ \dot{q} \end{array} \right], \qquad A \triangleq \left[\begin{array}{cc} 0 & 1 \\ -k/m & -c/m \end{array} \right]. \tag{11.3.8}$$

The inelastic case $k = 0$ is the simplest one since A is upper triangular. In this case,

$$e^{tA} = \begin{cases} \left[\begin{array}{cc} 1 & t \\ 0 & 1 \end{array} \right], & k = c = 0, \\[2em] \left[\begin{array}{cc} 1 & \frac{m}{c}(1 - e^{-ct/m}) \\ 0 & e^{-ct/m} \end{array} \right], & k = 0, \ c > 0, \end{cases} \tag{11.3.9}$$

where $c = 0$ and $c > 0$ correspond to a rigid body and a damped rigid body, respectively.

Next, we consider the elastic case $c \geq 0$ and $k > 0$. In this case, we define

$$\omega_{\mathrm{n}} \triangleq \sqrt{\frac{k}{m}}, \qquad \zeta \triangleq \frac{c}{2\sqrt{mk}}, \qquad (11.3.10)$$

where $\omega_{\mathrm{n}} > 0$ denotes the (undamped) *natural frequency* of vibration and $\zeta \geq 0$ denotes the *damping ratio*. Now, A can be written as

$$A = \begin{bmatrix} 0 & 1 \\ -\omega_{\mathrm{n}}^2 & -2\zeta\omega_{\mathrm{n}} \end{bmatrix}, \qquad (11.3.11)$$

and Corollary 11.3.3 yields

$$e^{tA} \qquad (11.3.12)$$

$$= \begin{cases} \begin{bmatrix} \cos\omega_{\mathrm{n}}t & \frac{1}{\omega_{\mathrm{n}}}\sin\omega_{\mathrm{n}}t \\ -\omega_{\mathrm{n}}\sin\omega_{\mathrm{n}}t & \cos\omega_{\mathrm{n}}t \end{bmatrix}, & \zeta = 0, \\[20pt] e^{-\zeta\omega_{\mathrm{n}}t}\begin{bmatrix} \cos\omega_{\mathrm{d}}t + \frac{\zeta}{\sqrt{1-\zeta^2}}\sin\omega_{\mathrm{d}}t & \frac{1}{\omega_{\mathrm{d}}}\sin\omega_{\mathrm{d}}t \\ \frac{-\omega_{\mathrm{d}}}{1-\zeta^2}\sin\omega_{\mathrm{d}}t & \cos\omega_{\mathrm{d}}t - \frac{\zeta}{\sqrt{1-\zeta^2}}\sin\omega_{\mathrm{d}}t \end{bmatrix}, & 0 < \zeta < 1, \\[20pt] e^{-\omega_{\mathrm{n}}t}\begin{bmatrix} 1 + \omega_{\mathrm{n}}t & t \\ -\omega_{\mathrm{n}}^2 t & 1 - \omega_{\mathrm{n}}t \end{bmatrix}, & \zeta = 1, \\[20pt] e^{-\zeta\omega_{\mathrm{n}}t}\begin{bmatrix} \cosh\omega_{\mathrm{d}}t + \frac{\zeta}{\sqrt{\zeta^2-1}}\sinh\omega_{\mathrm{d}}t & \frac{1}{\omega_{\mathrm{d}}}\sinh\omega_{\mathrm{d}}t \\ \frac{-\omega_{\mathrm{d}}}{\zeta^2-1}\sinh\omega_{\mathrm{d}}t & \cosh\omega_{\mathrm{d}}t - \frac{\zeta}{\sqrt{\zeta^2-1}}\sinh\omega_{\mathrm{d}}t \end{bmatrix}, & \zeta > 1, \end{cases}$$

where $\zeta = 0$, $0 < \zeta < 1$, $\zeta = 1$, and $\zeta > 1$ correspond to *undamped, underdamped, critically damped,* and *overdamped oscillators*, respectively, and where the *damped natural frequency* ω_{d} is the positive number

$$\omega_{\mathrm{d}} \triangleq \begin{cases} \omega_{\mathrm{n}}\sqrt{1-\zeta^2}, & 0 < \zeta < 1, \\ \omega_{\mathrm{n}}\sqrt{\zeta^2-1}, & \zeta > 1. \end{cases} \qquad (11.3.13)$$

Note that m and k are not integers here.

11.4 Matrix Logarithms

Definition 11.4.1. Let $A \in \mathbb{C}^{n \times n}$. Then, $B \in \mathbb{C}^{n \times n}$ is a *logarithm* of A if $e^B = A$.

The following result shows that every complex, nonsingular matrix has a complex logarithm.

Proposition 11.4.2. Let $A \in \mathbb{C}^{n \times n}$. Then, there exists a matrix $B \in \mathbb{C}^{n \times n}$ such that $A = e^B$ if and only if A is nonsingular.

Proof. See [639, pp. 35, 60] or [730, p. 474]. $\qquad\qquad\qquad\qquad\qquad \square$

Although the real number -1 does not have a real logarithm, the real matrix $B = \begin{bmatrix} 0 & \pi \\ -\pi & 0 \end{bmatrix}$ satisfies $e^B = \begin{bmatrix} -1 & 0 \\ 0 & -1 \end{bmatrix}$. These examples suggest that not all real matrices have a real logarithm.

Proposition 11.4.3. Let $A \in \mathbb{R}^{n \times n}$. Then, there exists a matrix $B \in \mathbb{R}^{n \times n}$ such that $A = e^B$ if and only if A is nonsingular and, for every negative eigenvalue λ of A and for every positive integer k, the Jordan form of A has an even number of $k \times k$ blocks associated with λ.

Proof. See [730, p. 475]. $\qquad\square$

Replacing A and B in Proposition 11.4.3 by e^A and A, respectively, yields the following result.

Corollary 11.4.4. Let $A \in \mathbb{R}^{n \times n}$. Then, for every negative eigenvalue λ of e^A and for every positive integer k, the Jordan form of e^A has an even number of $k \times k$ blocks associated with λ.

Since the matrix $A \triangleq \begin{bmatrix} -2\pi & 4\pi \\ -2\pi & 2\pi \end{bmatrix}$ satisfies $e^A = I$, it follows that a positive-definite matrix can have a logarithm that is not normal. However, the following result shows that every positive-definite matrix has exactly one Hermitian logarithm.

Proposition 11.4.5. The function exp: $\mathbf{H}^n \mapsto \mathbf{P}^n$ is one-to-one and onto.

Proof. The result follows from *vii*) of Proposition 11.2.8 and *iii*) of Proposition 11.2.9. $\qquad\square$

Let $A \in \mathbb{R}^{n \times n}$. If there exists a matrix $B \in \mathbb{R}^{n \times n}$ such that $A = e^B$, then Corollary 11.2.4 implies that $\det A = \det e^B = e^{\operatorname{tr} B} > 0$. However, the converse is not true. Consider, for example, $A \triangleq \begin{bmatrix} -1 & 0 \\ 0 & -2 \end{bmatrix}$, which satisfies $\det A > 0$. However, Proposition 11.4.3 implies that there does not exist a matrix $B \in \mathbb{R}^{2 \times 2}$ such that $A = e^B$. On the other hand, note that $A = e^B e^C$, where $B \triangleq \begin{bmatrix} 0 & \pi \\ -\pi & 0 \end{bmatrix}$ and $C \triangleq \begin{bmatrix} 0 & 0 \\ 0 & \log 2 \end{bmatrix}$. While the product of two exponentials of real matrices has positive determinant, the following result shows that the converse is also true.

Proposition 11.4.6. Let $A \in \mathbb{R}^{n \times n}$. Then, there exist matrices $B, C \in \mathbb{R}^{n \times n}$ such that $A = e^B e^C$ if and only if $\det A > 0$.

Proof. Suppose that there exist $B, C \in \mathbb{R}^{n \times n}$ such that $A = e^B e^C$. Then, $\det A = (\det e^B)(\det e^C) > 0$. Conversely, suppose that $\det A > 0$. If A has no negative eigenvalues, then it follows from Proposition 11.4.3 that there exists $B \in \mathbb{R}^{n \times n}$ such that $A = e^B$. Hence, $A = e^B e^{0_{n \times n}}$. Now, suppose that A has at least one negative eigenvalue. Then, Theorem 5.3.5 on the real Jordan form implies that there exist a nonsingular matrix $S \in \mathbb{R}^{n \times n}$ and matrices $A_1 \in \mathbb{R}^{n_1 \times n_1}$ and $A_2 \in \mathbb{R}^{n_2 \times n_2}$ such that $A = S \begin{bmatrix} A_1 & 0 \\ 0 & A_2 \end{bmatrix} S^{-1}$, where every eigenvalue of A_1 is negative and where none of the eigenvalues of A_2 are negative. Since $\det A$ and $\det A_2$ are positive, it follows that n_1 is even. Now, write $A = S \begin{bmatrix} -I_{n_1} & 0 \\ 0 & I_{n_2} \end{bmatrix} \begin{bmatrix} -A_1 & 0 \\ 0 & A_2 \end{bmatrix} S^{-1}$.

Since the eigenvalue -1 of $\begin{bmatrix} -I_{n_1} & 0 \\ 0 & I_{n_2} \end{bmatrix}$ appears in an even number of 1×1 Jordan blocks, it follows from Proposition 11.4.3 that there exists a matrix $\hat{B} \in \mathbb{R}^{n \times n}$ such that $\begin{bmatrix} -I_{n_1} & 0 \\ 0 & I_{n_2} \end{bmatrix} = e^{\hat{B}}$. Furthermore, since $\begin{bmatrix} -A_1 & 0 \\ 0 & A_2 \end{bmatrix}$ has no negative eigenvalues, it follows that there exists a matrix $\hat{C} \in \mathbb{R}^{n \times n}$ such that $\begin{bmatrix} -A_1 & 0 \\ 0 & A_2 \end{bmatrix} = e^{\hat{C}}$. Hence, $e^A = Se^{\hat{B}}e^{\hat{C}}S^{-1} = e^{S\hat{B}S^{-1}}e^{S\hat{C}S^{-1}}$. $\qquad \square$

Although $e^A e^B$ may be different from e^{A+B}, the following result, known as the *Baker-Campbell-Hausdorff* series, provides an expansion for a matrix function $C(t)$ that satisfies $e^{C(t)} = e^{tA}e^{tB}$.

Proposition 11.4.7. Let $A_1, \ldots, A_l \in \mathbb{F}^{n \times n}$. Then, there exists $\varepsilon > 0$ such that, for all $t \in (-\varepsilon, \varepsilon)$,

$$e^{tA_1} \cdots e^{tA_l} = e^{C(t)}, \tag{11.4.1}$$

where

$$C(t) \triangleq \sum_{i=1}^{l} tA_i + \sum_{1 \le i < j \le l} \tfrac{1}{2} t^2 [A_i, A_j] + O(t^3). \tag{11.4.2}$$

Proof. See [639, Chapter 3], [1192, p. 35], or [1400, p. 97]. $\qquad \square$

To illustrate (11.4.1), let $l = 2$, $A = A_1$, and $B = A_2$. Then, the first few terms of the series are given by

$$e^{tA}e^{tB} = e^{tA+tB+(t^2/2)[A,B]+(t^3/12)[[B,A],A+B]+\cdots}. \tag{11.4.3}$$

The radius of convergence of this series is discussed in [387, 1064].

The following result is the *Lie-Trotter product formula*.

Corollary 11.4.8. Let $A, B \in \mathbb{F}^{n \times n}$. Then,

$$e^{A+B} = \lim_{p \to \infty} \left[e^{\frac{1}{p}A} e^{\frac{1}{p}B} \right]^p. \tag{11.4.4}$$

Proof. Setting $l = 2$ and $t = 1/p$ in (11.4.1) yields, as $p \to \infty$,

$$\left[e^{\frac{1}{p}A} e^{\frac{1}{p}B} \right]^p = \left[e^{\frac{1}{p}(A+B)+O(1/p^2)} \right]^p = e^{A+B+O(1/p)} \to e^{A+B}. \qquad \square$$

11.5 Principal Logarithm

Let $A \in \mathbb{F}^{n \times n}$ be positive definite so that $A = SBS^* \in \mathbb{F}^{n \times n}$, where $S \in \mathbb{F}^{n \times n}$ is unitary and $B \in \mathbb{R}^{n \times n}$ is diagonal with positive diagonal entries. In Section 8.5, $\log A$ is defined as $\log A = S(\log B)S^* \in \mathbf{H}^n$, where $(\log B)_{(i,i)} \triangleq \log B_{(i,i)}$. Since $\log A$ satisfies $A = e^{\log A}$, it follows that $\log A$ is a logarithm of A. The following result extends the definition of $\log A$ to a larger class of matrices $A \in \mathbb{C}^{n \times n}$. A logarithm function defined on the set of nonsingular matrices can be based on the principal branch of the log function given by Fact 1.20.7 along with Definition 10.5.1; however, this function is not continuous, and thus an alternative approach to defining the principal logarithm is taken.

Definition 11.5.1. Let $A \in \mathbb{C}^{n \times n}$, and assume that A has no eigenvalues in $(-\infty, 0]$. Then, the *principal logarithm* of A is the unique logarithm of A whose eigenvalues are elements of $\{z \in \mathbb{C} : \pi < \operatorname{Im} z < \pi\}$. The notation $\log A$ denotes the principal logarithm of A.

Theorem 11.5.2. Let $A \in \mathbb{C}^{n \times n}$. Then, the following statements hold:

i) If A is nonsingular, then $\log A$ is a logarithm of A, that is, $e^{\log A} = A$.

ii) $\log e^A = A$ if and only if, for all $\lambda \in \operatorname{spec}(A)$, it follows that $|\operatorname{Im} \lambda| < \pi$.

iii) If A is nonsingular and $\operatorname{sprad}(A - I) \leq 1$, then $\log A$ is given by the series

$$\log A = \sum_{i=1}^{\infty} \frac{(-1)^{i+1}}{i} (A - I)^i, \qquad (11.5.1)$$

which converges absolutely with respect to every submultiplicative norm $\|\cdot\|$ such that $\|A - I\| < 1$.

iv) If $\operatorname{spec}(A) \subset \operatorname{ORHP}$, then $\log A$ is given by the series

$$\log A = \sum_{i=0}^{\infty} \frac{2}{2i+1} \left[(A - I)(A + I)^{-1} \right]^{2i+1}.$$

v) If A has no eigenvalues in $(-\infty, 0]$, then

$$\log A = \int_0^1 (A - I)[t(A - I) + I]^{-1} \, dt.$$

vi) If A has no eigenvalues in $(-\infty, 0]$ and $\alpha \in [-1, 1]$, then

$$\log A^\alpha = \alpha \log A.$$

In particular,
$$\log A^{-1} = -\log A$$

and
$$\log A^{1/2} = \tfrac{1}{2} \log A.$$

vii) If A is real and $\operatorname{spec}(A) \subset \operatorname{ORHP}$, then $\log A$ is real.

viii) If A is real and nonsingular, then A has a real logarithm if and only if A is nonsingular and, for every negative eigenvalue λ of A and for every positive integer k, the Jordan form of A has an even number of $k \times k$ blocks associated with λ.

Now, let $\|\cdot\|$ be a submultiplicative norm on $\mathbb{C}^{n \times n}$. Then, the following statements hold:

ix) The function log is continuous on $\{X \in \mathbb{C}^{n \times n} : \|X - I\| < 1\}$.

x) If $B \in \mathbb{C}^{n \times n}$ and $\|B\| < \log 2$, then $\|e^B - I\| < 1$ and $\log e^B = B$.

xi) $\exp \colon \mathbb{B}_{\log 2}(0) \mapsto \mathbb{F}^{n \times n}$ is one-to-one.

xii) If $\|A - I\| < 1$, then

$$\|\log A\| \leq -\log(1 - \|A - I\|) \leq \frac{\|A - I\|}{1 - \|A - I\|}.$$

xiii) If $\|A - I\| < 2/3$, then

$$\|A - I\| \left[1 - \frac{\|A - I\|}{2(1 - \|A - I\|)} \right] \leq \|\log A\|.$$

xiv) Assume that A is nonsingular, and let $\mathrm{mspec}(A) = \{\lambda_1, \ldots, \lambda_n\}_{\mathrm{ms}}$. Then,

$$\mathrm{mspec}(\log A) = \{\log \lambda_1, \ldots, \log \lambda_n\}_{\mathrm{ms}}.$$

Proof. Statement *i*) follows from the discussion in [730, p. 420]. Statement *ii*) is given in [701, p. 32]. Statements *iii*) and *iv*) are given by Fact 10.13.1. See [639, pp. 34–35] and [701, p. 273]. Statement *v*) is given in [701, p. 269]. Statement *vi*) is given in [701, p. 270]. The proof of *vii*) is immediate. Statement *viii*) follows from Proposition 11.4.3 and the discussion in [730, pp. 474–475].

Statements *ix*) and *x*) are proved in [639, pp. 34–35]. To prove the inequality in *x*), let $\|B\| < 2$, so that $e^{\|B\|} < 2$, and thus

$$\|e^B - I\| \leq \sum_{i=1}^{\infty} (i!)^{-1} \|B\|^i = e^{\|B\|} - 1 < 1.$$

To prove *xi*), let $B_1, B_2 \in \mathbb{B}_{\log 2}(0)$, and assume that $e^{B_1} = e^{B_2}$. Then, it follows from *ii*) that $B_1 = \log e^{B_1} = \log e^{B_2} = B_2$.

Finally, to prove *xii*), let $\alpha \triangleq \|A - I\| < 1$. Then, it follows from (11.5.1) and *iv*) of Fact 1.20.7 that $\|\log A\| \leq \sum_{i=1}^{\infty} \alpha^i/i = -\log(1 - \alpha)$. For *xiii*), see [701, p. 647]. $\qquad\qquad\square$

11.6 Lie Groups

Definition 11.6.1. Let $\mathcal{S} \subset \mathbb{F}^{n \times n}$, and assume that $\mathcal{S}$ is a group. Then, $\mathcal{S}$ is a *Lie group* if $\mathcal{S}$ is closed relative to $\mathrm{GL}_{\mathbb{F}}(n)$.

Proposition 11.6.2. Let $\mathcal{S} \subset \mathbb{F}^{n \times n}$, and assume that $\mathcal{S}$ is a group. Then, $\mathcal{S}$ is a Lie group if and only if the limit of every convergent sequence in $\mathcal{S}$ is either an element of $\mathcal{S}$ or is singular.

The groups $\mathrm{SL}_{\mathbb{F}}(n)$, $\mathrm{U}(n)$, $\mathrm{O}(n)$, $\mathrm{SU}(n)$, $\mathrm{SO}(n)$, $\mathrm{U}(n, m)$, $\mathrm{O}(n, m)$, $\mathrm{SU}(n, m)$, $\mathrm{SO}(n, m)$, $\mathrm{Symp}_{\mathbb{F}}(2n)$, $\mathrm{Aff}_{\mathbb{F}}(n)$, $\mathrm{SE}_{\mathbb{F}}(n)$, and $\mathrm{Trans}_{\mathbb{F}}(n)$ defined in Proposition 3.3.6 are closed sets, and thus are Lie groups. Although the groups $\mathrm{GL}_{\mathbb{F}}(n)$, $\mathrm{PL}_{\mathbb{F}}(n)$, and $\mathrm{UT}(n)$ (see Fact 3.23.12) are not closed sets, they are closed relative to $\mathrm{GL}_{\mathbb{F}}(n)$, and thus they are Lie groups. Finally, the group $\mathcal{S} \subset \mathbb{C}^{2 \times 2}$ defined by

$$\mathcal{S} \triangleq \left\{ \begin{bmatrix} e^{\jmath t} & 0 \\ 0 & e^{\jmath \pi t} \end{bmatrix} : t \in \mathbb{R} \right\} \tag{11.6.1}$$

is not closed relative to $\mathrm{GL}_{\mathbb{C}}(2)$, and thus is not a Lie group. For details, see [639, p. 4].

Proposition 11.6.3. Let $\mathcal{S} \subset \mathbb{F}^{n \times n}$, and assume that $\mathcal{S}$ is a Lie group. Furthermore, define

$$\mathcal{S}_0 \triangleq \{ A \in \mathbb{F}^{n \times n} \colon e^{tA} \in \mathcal{S} \text{ for all } t \in \mathbb{R} \}. \tag{11.6.2}$$

Then, $\mathcal{S}_0$ is a Lie algebra.

Proof. See [639, pp. 39, 43, 44]. $\qquad\square$

The Lie algebra $\mathcal{S}_0$ defined by (11.6.2) is *the Lie algebra of* $\mathcal{S}$.

Proposition 11.6.4. Let $\mathcal{S} \subset \mathbb{F}^{n \times n}$, assume that $\mathcal{S}$ is a Lie group, and let $\mathcal{S}_0 \subseteq \mathbb{F}^{n \times n}$ be the Lie algebra of $\mathcal{S}$. Furthermore, let $S \in \mathcal{S}$ and $A \in \mathcal{S}_0$. Then, $SAS^{-1} \in \mathcal{S}_0$.

Proof. For all $t \in \mathbb{R}$, $e^{tA} \in \mathcal{S}$, and thus $e^{tSAS^{-1}} = Se^{tA}S^{-1} \in \mathcal{S}$. Hence, $SAS^{-1} \in \mathcal{S}_0$. $\qquad\square$

Proposition 11.6.5. The following statements hold:

i) $\mathrm{gl}_{\mathbb{F}}(n)$ is the Lie algebra of $\mathrm{GL}_{\mathbb{F}}(n)$.

ii) $\mathrm{gl}_{\mathbb{R}}(n) = \mathrm{pl}_{\mathbb{R}}(n)$ is the Lie algebra of $\mathrm{PL}_{\mathbb{R}}(n)$.

iii) $\mathrm{pl}_{\mathbb{C}}(n)$ is the Lie algebra of $\mathrm{PL}_{\mathbb{C}}(n)$.

iv) $\mathrm{sl}_{\mathbb{F}}(n)$ is the Lie algebra of $\mathrm{SL}_{\mathbb{F}}(n)$.

v) $\mathrm{u}(n)$ is the Lie algebra of $\mathrm{U}(n)$.

vi) $\mathrm{so}(n)$ is the Lie algebra of $\mathrm{O}(n)$.

vii) $\mathrm{su}(n)$ is the Lie algebra of $\mathrm{SU}(n)$.

viii) $\mathrm{so}(n)$ is the Lie algebra of $\mathrm{SO}(n)$.

ix) $\mathrm{su}(n, m)$ is the Lie algebra of $\mathrm{U}(n, m)$.

x) $\mathrm{so}(n, m)$ is the Lie algebra of $\mathrm{O}(n, m)$.

xi) $\mathrm{su}(n, m)$ is the Lie algebra of $\mathrm{SU}(n, m)$.

xii) $\mathrm{so}(n, m)$ is the Lie algebra of $\mathrm{SO}(n, m)$.

xiii) $\mathrm{symp}_{\mathbb{F}}(2n)$ is the Lie algebra of $\mathrm{Symp}_{\mathbb{F}}(2n)$.

xiv) $\mathrm{osymp}_{\mathbb{F}}(2n)$ is the Lie algebra of $\mathrm{OSymp}_{\mathbb{F}}(2n)$.

xv) $\mathrm{aff}_{\mathbb{F}}(n)$ is the Lie algebra of $\mathrm{Aff}_{\mathbb{F}}(n)$.

xvi) $\mathrm{se}_{\mathbb{C}}(n)$ is the Lie algebra of $\mathrm{SE}_{\mathbb{C}}(n)$.

xvii) $\mathrm{se}_{\mathbb{R}}(n)$ is the Lie algebra of $\mathrm{SE}_{\mathbb{R}}(n)$.

xviii) $\mathrm{trans}_{\mathbb{F}}(n)$ is the Lie algebra of $\mathrm{Trans}_{\mathbb{F}}(n)$.

Proof. See [639, pp. 38–41]. $\qquad\square$

Proposition 11.6.6. Let $S \subset \mathbb{F}^{n \times n}$, assume that S is a Lie group, and let $S_0 \subseteq \mathbb{F}^{n \times n}$ be the Lie algebra of S. Then, exp: $S_0 \mapsto S$. Furthermore, if exp is onto, then S is pathwise connected.

Proof. Let $A \in S_0$ so that $e^{tA} \in S$ for all $t \in \mathbb{R}$. Hence, setting $t = 1$ implies that exp: $S_0 \mapsto S$. Now, suppose that exp is onto, let $B \in S$, and let $A \in S_0$ be such that $e^A = B$. Then, $f(t) \triangleq e^{tA}$ satisfies $f(0) = I$ and $f(1) = B$, which implies that S is pathwise connected. $\qquad \square$

A Lie group can consist of multiple pathwise-connected components.

Proposition 11.6.7. Let $n \geq 1$. Then, the following functions are onto:

i) exp: $\mathrm{gl}_{\mathbb{C}}(n) \mapsto \mathrm{GL}_{\mathbb{C}}(n)$.

ii) exp: $\mathrm{gl}_{\mathbb{R}}(1) \mapsto \mathrm{PL}_{\mathbb{R}}(1)$.

iii) exp: $\mathrm{pl}_{\mathbb{C}}(n) \mapsto \mathrm{PL}_{\mathbb{C}}(n)$.

iv) exp: $\mathrm{sl}_{\mathbb{C}}(n) \mapsto \mathrm{SL}_{\mathbb{C}}(n)$.

v) exp: $\mathrm{u}(n) \mapsto \mathrm{U}(n)$.

vi) exp: $\mathrm{su}(n) \mapsto \mathrm{SU}(n)$.

vii) exp: $\mathrm{so}(n) \mapsto \mathrm{SO}(n)$.

Furthermore, the following functions are not onto:

viii) exp: $\mathrm{gl}_{\mathbb{R}}(n) \mapsto \mathrm{PL}_{\mathbb{R}}(n)$, where $n \geq 2$.

ix) exp: $\mathrm{sl}_{\mathbb{R}}(n) \mapsto \mathrm{SL}_{\mathbb{R}}(n)$.

x) exp: $\mathrm{so}(n) \mapsto \mathrm{O}(n)$.

xi) exp: $\mathrm{symp}_{\mathbb{R}}(2n) \mapsto \mathrm{Symp}_{\mathbb{R}}(2n)$.

Proof. Statement *i)* follows from Proposition 11.4.2, while *ii)* is immediate. Statements *iii)–vii)* can be verified by construction; see [1125, pp. 199, 212] for the proof of *v)* and *vii)*. The example $A \triangleq \left[\begin{smallmatrix} -1 & 0 \\ 0 & -2 \end{smallmatrix} \right]$ and Proposition 11.4.3 show that *viii)* is not onto. For $\lambda < 0$, $\lambda \neq -1$, Proposition 11.4.3 and the example $\left[\begin{smallmatrix} \lambda & 0 \\ 0 & 1/\lambda \end{smallmatrix} \right]$ given in [1192, p. 39] show that *ix)* is not onto. See also [106, pp. 84, 85]. Statement *viii)* shows that *x)* is not onto. For *xi)*, see [414]. $\qquad \square$

Proposition 11.6.8. The Lie groups $\mathrm{GL}_{\mathbb{C}}(n), \mathrm{SL}_{\mathbb{F}}(n), \mathrm{U}(n), \mathrm{SU}(n)$, and $\mathrm{SO}(n)$ are pathwise connected. The Lie groups $\mathrm{GL}_{\mathbb{R}}(n), \mathrm{O}(n), \mathrm{O}(n,1)$, and $\mathrm{SO}(n,1)$ are not pathwise connected.

Proof. See [639, p. 15]. $\qquad \square$

Proposition 11.6.8 and *ix)* of Proposition 11.6.7 show that the converse of Proposition 11.6.6 does not hold, that is, pathwise connectedness does not imply that exp is onto. See [1192, p. 39].

11.7 Lyapunov Stability Theory

Consider the dynamical system

$$\dot{x}(t) = f[x(t)], \tag{11.7.1}$$

where $t \geq 0$, $x(t) \in \mathcal{D} \subseteq \mathbb{R}^n$, and $f: \mathcal{D} \to \mathbb{R}^n$ is continuous. We assume that, for all $x_0 \in \mathcal{D}$ and for all $T > 0$, there exists a unique C^1 solution $x: [0, T) \mapsto \mathcal{D}$ satisfying (11.7.1). If $x_e \in \mathcal{D}$ satisfies $f(x_e) = 0$, then $x(t) \equiv x_e$ is an *equilibrium* of (11.7.1). The following definition concerns the stability of an equilibrium of (11.7.1). Throughout this section, $\|\cdot\|$ denotes a norm on $\mathbb{R}^n$.

Definition 11.7.1. Let $x_e \in \mathcal{D}$ be an equilibrium of (11.7.1). Then, x_e is *Lyapunov stable* if, for all $\varepsilon > 0$, there exists $\delta > 0$ such that, if $\|x(0) - x_e\| < \delta$, then $\|x(t) - x_e\| < \varepsilon$ for all $t \geq 0$. Furthermore, x_e is *asymptotically stable* if it is Lyapunov stable and there exists $\varepsilon > 0$ such that, if $\|x(0) - x_e\| < \varepsilon$, then $\lim_{t\to\infty} x(t) = x_e$. In addition, x_e is *globally asymptotically stable* if it is Lyapunov stable, $\mathcal{D} = \mathbb{R}^n$, and, for all $x(0) \in \mathbb{R}^n$, $\lim_{t\to\infty} x(t) = x_e$. Finally, x_e is *unstable* if it is not Lyapunov stable.

Note that, if $x_e \in \mathbb{R}^n$ is a globally asymptotically stable equilibrium, then x_e is the only equilibrium of (11.7.1).

The following result, known as *Lyapunov's direct method*, gives sufficient conditions for Lyapunov stability and asymptotic stability of an equilibrium of (11.7.1).

Theorem 11.7.2. Let $x_e \in \mathcal{D}$ be an equilibrium of the dynamical system (11.7.1), and assume there exists a C^1 function $V: \mathcal{D} \mapsto \mathbb{R}$ such that

$$V(x_e) = 0, \tag{11.7.2}$$

such that, for all $x \in \mathcal{D}\backslash\{x_e\}$,

$$V(x) > 0, \tag{11.7.3}$$

and such that, for all $x \in \mathcal{D}$,

$$V'(x)f(x) \leq 0. \tag{11.7.4}$$

Then, x_e is Lyapunov stable. If, in addition, for all $x \in \mathcal{D}\backslash\{x_e\}$,

$$V'(x)f(x) < 0, \tag{11.7.5}$$

then x_e is asymptotically stable. Finally, if $\mathcal{D} = \mathbb{R}^n$ and

$$\lim_{\|x\|\to\infty} V(x) = \infty, \tag{11.7.6}$$

then x_e is globally asymptotically stable.

Proof. To prove Lyapunov stability, let $\varepsilon > 0$ be such that $\mathbb{B}_\varepsilon(x_e) \subseteq \mathcal{D}$. Since $\mathbb{S}_\varepsilon(x_e)$ is compact and $V(x)$ is continuous, it follows from Theorem 10.3.8 that $V[\mathbb{S}_\varepsilon(x_e)]$ is compact. Since $0 \notin \mathbb{S}_\varepsilon(x_e)$, $V(x) > 0$ for all $x \in \mathcal{D}\backslash\{x_e\}$, and $V[\mathbb{S}_\varepsilon(x_e)]$ is compact, it follows that $\alpha \triangleq \min V[\mathbb{S}_\varepsilon(x_e)]$ is positive. Next, since V is continuous, it follows that there exists $\delta \in (0, \varepsilon)$ such that, for all $x \in \mathbb{B}_\delta(x_e)$, $V(x) < \alpha$. Now, let $x: [0, \infty) \mapsto \mathbb{R}^n$ satisfy (11.7.1), where $\|x(0)\| < \delta$. Hence,

$V[x(0)] < \alpha$. It thus follows from (11.7.4) that, for all $t \geq 0$,

$$V[x(t)] - V[x(0)] = \int_0^t V'[x(s)]f[x(s)]\,\mathrm{d}s \leq 0,$$

and hence, for all $t \geq 0$,

$$V[x(t)] \leq V[x(0)] < \alpha.$$

Now, since $V(x) \geq \alpha$ for all $x \in \mathbb{S}_\varepsilon(0)$, it follows that, for all $t \geq 0$, $x(t) \notin \mathbb{S}_\varepsilon(x_e)$. Hence, for all $t \geq 0$, $\|x(t)\| < \varepsilon$, which proves that $x_e = 0$ is Lyapunov stable.

To prove that x_e is asymptotically stable, let $\varepsilon > 0$ be such that $\mathbb{B}_\varepsilon(x_e) \subseteq \mathcal{D}$. Since (11.7.5) implies (11.7.4), it follows that there exists $\delta > 0$ such that, if $\|x(0)\| < \delta$, then $\|x(t)\| < \varepsilon$ for all $t \geq 0$. Furthermore, for all $t \geq 0$, $\frac{\mathrm{d}}{\mathrm{d}t}V[x(t)] = V'[x(t)]f[x(t)] < 0$, and thus $V[x(t)]$ is decreasing and bounded from below by zero. Now, suppose that $V[x(t)]$ does not converge to zero. Therefore, there exists $L > 0$ such that, for all $t \geq 0$, $V[x(t)] \geq L$. Now, let $\delta' > 0$ be such that, for all $x \in \mathbb{B}_{\delta'}(x_e)$, $V(x) < L$. Therefore, for all $t \geq 0$, $\|x(t)\| \geq \delta'$. Next, define $\gamma < 0$ by $\gamma \triangleq \max_{\delta' \leq \|x\| \leq \varepsilon} V'(x)f(x)$. Therefore, since $\|x(t)\| < \varepsilon$ for all $t \geq 0$, it follows that

$$V[x(t)] - V[x(0)] = \int_0^t V'[x(\tau)]f[x(\tau)]\,\mathrm{d}\tau \leq \gamma t,$$

and hence, for all $t \geq 0$,

$$V(x(t)) \leq V[x(0)] + \gamma t.$$

However, $t > -V[x(0)]/\gamma$ implies that $V[x(t)] < 0$, which is a contradiction.

To prove that x_e is globally asymptotically stable, let $x(0) \in \mathbb{R}^n$, and let $\beta \triangleq V[x(0)]$. It follows from (11.7.6) that there exists $\varepsilon > 0$ such that, for all $x \in \mathbb{R}^n$ such that $\|x\| > \varepsilon$, $V(x) > \beta$. Therefore, $\|x(0)\| \leq \varepsilon$, and, since $V[x(t)]$ is decreasing, it follows that, for all $t > 0$, $\|x(t)\| < \varepsilon$. The remainder of the proof is identical to the proof of asymptotic stability. $\square$

11.8 Linear Stability Theory

We now specialize Definition 11.7.1 to the linear system

$$\dot{x}(t) = Ax(t), \tag{11.8.1}$$

where $t \geq 0$, $x(t) \in \mathbb{R}^n$, and $A \in \mathbb{R}^{n \times n}$. Note that $x_e = 0$ is an equilibrium of (11.8.1), and that $x_e \in \mathbb{R}^n$ is an equilibrium of (11.8.1) if and only if $x_e \in \mathcal{N}(A)$. Hence, if x_e is the globally asymptotically stable equilibrium of (11.8.1), then A is nonsingular and $x_e = 0$.

We consider three types of stability for the linear system (11.8.1). Unlike Definition 11.7.1, these definitions are stated in terms of the dynamics matrix rather than the equilibrium.

Definition 11.8.1. For $A \in \mathbb{C}^{n \times n}$, define the following classes of matrices:

i) A is *Lyapunov stable* if spec$(A) \subset$ CLHP and, if $\lambda \in$ spec(A) and Re $\lambda = 0$, then λ is semisimple.

ii) A is *semistable* if spec$(A) \subset$ OLHP $\cup \{0\}$ and, if $0 \in$ spec(A), then 0 is semisimple.

iii) A is *asymptotically stable* if spec$(A) \subset$ OLHP.

The following result concerns Lyapunov stability, semistability, and asymptotic stability for (11.8.1).

Proposition 11.8.2. Let $A \in \mathbb{R}^{n \times n}$. Then, the following statements are equivalent:

i) $x_{\mathrm{e}} = 0$ is a Lyapunov-stable equilibrium of (11.8.1).

ii) At least one equilibrium of (11.8.1) is Lyapunov stable.

iii) Every equilibrium of (11.8.1) is Lyapunov stable.

iv) A is Lyapunov stable.

v) For every initial condition $x(0) \in \mathbb{R}^n$, $x(t)$ is bounded for all $t \geq 0$.

vi) $\|e^{tA}\|$ is bounded for all $t \geq 0$, where $\| \cdot \|$ is a norm on $\mathbb{R}^{n \times n}$.

vii) For every initial condition $x(0) \in \mathbb{R}^n$, $e^{tA}x(0)$ is bounded for all $t \geq 0$.

The following statements are equivalent:

viii) A is semistable.

ix) $\lim_{t \to \infty} e^{tA}$ exists.

x) For every initial condition $x(0)$, $\lim_{t \to \infty} x(t)$ exists.

In this case,

$$\lim_{t \to \infty} e^{tA} = I - AA^{\#}. \tag{11.8.2}$$

The following statements are equivalent:

xi) $x_{\mathrm{e}} = 0$ is an asymptotically stable equilibrium of (11.8.1).

xii) A is asymptotically stable.

xiii) spabs$(A) < 0$.

xiv) For every initial condition $x(0) \in \mathbb{R}^n$, $\lim_{t \to \infty} x(t) = 0$.

xv) For every initial condition $x(0) \in \mathbb{R}^n$, $e^{tA}x(0) \to 0$ as $t \to \infty$.

xvi) $e^{tA} \to 0$ as $t \to \infty$.

The following definition concerns the stability of a polynomial.

Definition 11.8.3. Let $p \in \mathbb{R}[s]$. Then, define the following terminology:

i) p is *Lyapunov stable* if roots$(p) \subset$ CLHP and, if λ is an imaginary root of p, then $m_p(\lambda) = 1$.

ii) p is *semistable* if roots$(p) \subset$ OLHP $\cup \{0\}$ and, if $0 \in$ roots(p), then $m_p(0) = 1$.

iii) p is *asymptotically stable* if roots$(p) \subset$ OLHP.

For the following result, recall Definition 11.8.1.

Proposition 11.8.4. Let $A \in \mathbb{R}^{n \times n}$. Then, the following statements hold:

i) A is Lyapunov stable if and only if μ_A is Lyapunov stable.

ii) A is semistable if and only if μ_A is semistable.

Furthermore, the following statements are equivalent:

iii) A is asymptotically stable

iv) μ_A is asymptotically stable.

v) χ_A is asymptotically stable.

Next, consider the factorization of the minimal polynomial μ_A of A given by

$$\mu_A = \mu_A^{\mathrm{s}} \mu_A^{\mathrm{u}}, \tag{11.8.3}$$

where μ_A^{s} and μ_A^{u} are monic polynomials such that

$$\mathrm{roots}(\mu_A^{\mathrm{s}}) \subset \mathrm{OLHP} \tag{11.8.4}$$

and

$$\mathrm{roots}(\mu_A^{\mathrm{u}}) \subset \mathrm{CRHP}. \tag{11.8.5}$$

Proposition 11.8.5. Let $A \in \mathbb{R}^{n \times n}$, and let $S \in \mathbb{R}^{n \times n}$ be a nonsingular matrix such that

$$A = S \begin{bmatrix} A_1 & A_{12} \\ 0 & A_2 \end{bmatrix} S^{-1}, \tag{11.8.6}$$

where $A_1 \in \mathbb{R}^{r \times r}$ is asymptotically stable, $A_{12} \in \mathbb{R}^{r \times (n-r)}$, and $A_2 \in \mathbb{R}^{(n-r) \times (n-r)}$ satisfies spec$(A_2) \subset$ CRHP. Then,

$$\mu_A^{\mathrm{s}}(A) = S \begin{bmatrix} 0 & C_{12\mathrm{s}} \\ 0 & \mu_A^{\mathrm{s}}(A_2) \end{bmatrix} S^{-1}, \tag{11.8.7}$$

where $C_{12\mathrm{s}} \in \mathbb{R}^{r \times (n-r)}$ and $\mu_A^{\mathrm{s}}(A_2)$ is nonsingular, and

$$\mu_A^{\mathrm{u}}(A) = S \begin{bmatrix} \mu_A^{\mathrm{u}}(A_1) & C_{12\mathrm{u}} \\ 0 & 0 \end{bmatrix} S^{-1}, \tag{11.8.8}$$

where $C_{12\mathrm{u}} \in \mathbb{R}^{r \times (n-r)}$ and $\mu_A^{\mathrm{u}}(A_1)$ is nonsingular. Consequently,

$$\mathcal{N}[\mu_A^{\mathrm{s}}(A)] = \mathcal{R}[\mu_A^{\mathrm{u}}(A)] = \mathcal{R}\left(S \begin{bmatrix} I_r \\ 0 \end{bmatrix}\right). \tag{11.8.9}$$

If, in addition, $A_{12} = 0$, then

$$\mu_A^s(A) = S \begin{bmatrix} 0 & 0 \\ 0 & \mu_A^s(A_2) \end{bmatrix} S^{-1} \tag{11.8.10}$$

and

$$\mu_A^u(A) = S \begin{bmatrix} \mu_A^u(A_1) & 0 \\ 0 & 0 \end{bmatrix} S^{-1}. \tag{11.8.11}$$

Consequently, $\quad \mathcal{R}[\mu_A^s(A)] = \mathcal{N}[\mu_A^u(A)] = \mathcal{R}\left(S \begin{bmatrix} 0 \\ I_{n-r} \end{bmatrix} \right). \tag{11.8.12}$

Corollary 11.8.6. Let $A \in \mathbb{R}^{n \times n}$. Then,

$$\mathcal{N}[\mu_A^s(A)] = \mathcal{R}[\mu_A^u(A)] \tag{11.8.13}$$

and

$$\mathcal{N}[\mu_A^u(A)] = \mathcal{R}[\mu_A^s(A)]. \tag{11.8.14}$$

Proof. It follows from Theorem 5.3.5 that there exists a nonsingular matrix $S \in \mathbb{R}^{n \times n}$ such that (11.8.6) is satisfied, where $A_1 \in \mathbb{R}^{r \times r}$ is asymptotically stable, $A_{12} = 0$, and $A_2 \in \mathbb{R}^{(n-r) \times (n-r)}$ satisfies $\mathrm{spec}(A_2) \subset \mathrm{CRHP}$. The result now follows from Proposition 11.8.5. $\qquad \square$

In view of Corollary 11.8.6, we define the *asymptotically stable subspace* $\mathcal{S}_s(A)$ of A by

$$\mathcal{S}_s(A) \triangleq \mathcal{N}[\mu_A^s(A)] = \mathcal{R}[\mu_A^u(A)] \tag{11.8.15}$$

and the *unstable subspace* $\mathcal{S}_u(A)$ of A by

$$\mathcal{S}_u(A) \triangleq \mathcal{N}[\mu_A^u(A)] = \mathcal{R}[\mu_A^s(A)]. \tag{11.8.16}$$

Note that

$$\dim \mathcal{S}_s(A) = \mathrm{def}\, \mu_A^s(A) = \mathrm{rank}\, \mu_A^u(A) = \sum_{\substack{\lambda \in \mathrm{spec}(A) \\ \mathrm{Re}\,\lambda < 0}} \mathrm{amult}_A(\lambda) \tag{11.8.17}$$

and

$$\dim \mathcal{S}_u(A) = \mathrm{def}\, \mu_A^u(A) = \mathrm{rank}\, \mu_A^s(A) = \sum_{\substack{\lambda \in \mathrm{spec}(A) \\ \mathrm{Re}\,\lambda \geq 0}} \mathrm{amult}_A(\lambda). \tag{11.8.18}$$

Lemma 11.8.7. Let $A \in \mathbb{R}^{n \times n}$, assume that $\mathrm{spec}(A) \subset \mathrm{CRHP}$, let $x \in \mathbb{R}^n$, and assume that $\lim_{t \to \infty} e^{tA} x = 0$. Then, $x = 0$.

For the following result, note Proposition 3.5.3, Fact 3.12.3, Proposition 6.1.7, Proposition 11.8.2, and Fact 11.18.3.

Proposition 11.8.8. Let $A \in \mathbb{R}^{n \times n}$. Then, the following statements hold:

i) $\mathcal{S}_s(A) = \{x \in \mathbb{R}^n: \lim_{t \to \infty} e^{tA} x = 0\}$.

ii) $\mu_A^s(A)$ and $\mu_A^u(A)$ are group invertible.

iii) $P_s \triangleq I - \mu_A^s(A)[\mu_A^s(A)]^\#$ and $P_u \triangleq I - \mu_A^u(A)[\mu_A^u(A)]^\#$ are idempotent.

iv) $P_s + P_u = I$.

v) $P_{s\perp} = P_u$ and $P_{u\perp} = P_s$.

vi) $\mathcal{S}_s(A) = \mathcal{R}(P_s) = \mathcal{N}(P_u)$.

vii) $\mathcal{S}_u(A) = \mathcal{R}(P_u) = \mathcal{N}(P_s)$.

viii) $\mathcal{S}_s(A)$ and $\mathcal{S}_u(A)$ are invariant subspaces of A.

ix) $\mathcal{S}_s(A)$ and $\mathcal{S}_u(A)$ are complementary subspaces.

x) P_s is the idempotent matrix onto $\mathcal{S}_s(A)$ along $\mathcal{S}_u(A)$.

xi) P_u is the idempotent matrix onto $\mathcal{S}_u(A)$ along $\mathcal{S}_s(A)$.

Proof. To prove *i*), let $S \in \mathbb{R}^{n \times n}$ be a nonsingular matrix such that

$$A = S \begin{bmatrix} A_1 & 0 \\ 0 & A_2 \end{bmatrix} S^{-1},$$

where $A_1 \in \mathbb{R}^{r \times r}$ is asymptotically stable and $\mathrm{spec}(A_2) \subset \mathrm{CRHP}$. It then follows from Proposition 11.8.5 that

$$\mathcal{S}_s(A) = \mathcal{N}[\mu_A^s(A)] = \mathcal{R}\left(S \begin{bmatrix} I_r \\ 0 \end{bmatrix}\right).$$

Furthermore,

$$e^{tA} = S \begin{bmatrix} e^{tA_1} & 0 \\ 0 & e^{tA_2} \end{bmatrix} S^{-1}.$$

To prove $\mathcal{S}_s(A) \subseteq \{z \in \mathbb{R}^n: \ \lim_{t \to \infty} e^{tA}z = 0\}$, let $x \triangleq S\begin{bmatrix} x_1 \\ 0 \end{bmatrix} \in \mathcal{S}_s(A)$, where $x_1 \in \mathbb{R}^r$. Then, $e^{tA}x = S\begin{bmatrix} e^{tA_1}x_1 \\ 0 \end{bmatrix} \to 0$ as $t \to \infty$. Hence, $x \in \{z \in \mathbb{R}^n: \ \lim_{t \to \infty} e^{tA}z = 0\}$. Conversely, to prove $\{z \in \mathbb{R}^n: \ \lim_{t \to \infty} e^{tA}z = 0\} \subseteq \mathcal{S}_s(A)$, let $x \triangleq S\begin{bmatrix} x_1 \\ x_2 \end{bmatrix} \in \mathbb{R}^n$ satisfy $\lim_{t \to \infty} e^{tA}x = 0$. Hence, $e^{tA_2}x_2 \to 0$ as $t \to \infty$. Since $\mathrm{spec}(A_2) \subset \mathrm{CRHP}$, it follows from Lemma 11.8.7 that $x_2 = 0$. Hence, $x \in \mathcal{R}(S\begin{bmatrix} I_r \\ 0 \end{bmatrix}) = \mathcal{S}_s(A)$.

The remaining statements follow directly from Proposition 11.8.5. $\square$

11.9 The Lyapunov Equation

In this section we specialize Theorem 11.7.2 to the linear system (11.8.1).

Corollary 11.9.1. Let $A \in \mathbb{R}^{n \times n}$, and assume there exist a positive-semidefinite matrix $R \in \mathbb{R}^{n \times n}$ and a positive-definite matrix $P \in \mathbb{R}^{n \times n}$ satisfying

$$A^{\mathrm{T}}P + PA + R = 0. \tag{11.9.1}$$

Then, A is Lyapunov stable. If, in addition, for all nonzero $\omega \in \mathbb{R}$,

$$\mathrm{rank} \begin{bmatrix} \jmath\omega I - A \\ R \end{bmatrix} = n, \tag{11.9.2}$$

then A is semistable. Finally, if R is positive definite, then A is asymptotically stable.

Proof. Define $V(x) \triangleq x^\mathrm{T}Px$, which satisfies (11.7.2) with $x_\mathrm{e} = 0$ and satisfies (11.7.3) for all nonzero $x \in \mathcal{D} = \mathbb{R}^n$. Furthermore, Theorem 11.7.2 implies that $V'(x)f(x) = 2x^\mathrm{T}PAx = x^\mathrm{T}(A^\mathrm{T}P + PA)x = -x^\mathrm{T}Rx$, which satisfies (11.7.4) for all $x \in \mathbb{R}^n$. Thus, Theorem 11.7.2 implies that A is Lyapunov stable. If, in addition, R is positive definite, then (11.7.5) is satisfied for all $x \neq 0$, and thus A is asymptotically stable.

Alternatively, we now prove the first and third statements without using Theorem 11.7.2. Letting $\lambda \in \mathrm{spec}(A)$, and letting $x \in \mathbb{C}^n$ be an associated eigenvector, it follows that $0 \geq -x^*Rx = x^*(A^\mathrm{T}P + PA)x = (\overline{\lambda} + \lambda)x^*Px$. Therefore, $\mathrm{spec}(A) \subset \mathrm{CLHP}$. Now, suppose that $\jmath\omega \in \mathrm{spec}(A)$, where $\omega \in \mathbb{R}$, and let $x \in \mathcal{N}[(\jmath\omega I - A)^2]$. Defining $y \triangleq (\jmath\omega I - A)x$, it follows that $(\jmath\omega I - A)y = 0$, and thus $Ay = \jmath\omega y$. Therefore, $-y^*Ry = y^*(A^\mathrm{T}P + PA)y = -\jmath\omega y^*Py + \jmath\omega y^*Py = 0$, and thus $Ry = 0$. Hence, $0 = x^*Ry = -x^*(A^\mathrm{T}P + PA)y = -x^*(A^\mathrm{T} + \jmath\omega I)Py = y^*Py$. Since P is positive definite, it follows that $y = 0$, that is, $(\jmath\omega I - A)x = 0$. Therefore, $x \in \mathcal{N}(\jmath\omega I - A)$. Now, Proposition 5.5.8 implies that $\jmath\omega$ is semisimple. Therefore, A is Lyapunov stable.

Next, to prove that A is asymptotically stable, let $\lambda \in \mathrm{spec}(A)$, and let $x \in \mathbb{C}^n$ be an associated eigenvector. Thus, $0 > -x^*Rx = (\overline{\lambda} + \lambda)x^*Px$, which implies that A is asymptotically stable.

Finally, to prove that A is semistable, let $\jmath\omega \in \mathrm{spec}(A)$, where $\omega \in \mathbb{R}$ is nonzero, and let $x \in \mathbb{C}^n$ be an associated eigenvector. Then,

$$-x^*Rx = x^*(A^\mathrm{T}P + PA)x = x^*[(\jmath\omega I - A)^*P + P(\jmath\omega I - A]x = 0.$$

Therefore, $Rx = 0$, and thus

$$\begin{bmatrix} \jmath\omega I - A \\ R \end{bmatrix} x = 0,$$

which implies that $x = 0$, which contradicts $x \neq 0$. Consequently, $\jmath\omega \notin \mathrm{spec}(A)$ for all nonzero $\omega \in \mathbb{R}$, and thus A is semistable. $\square$

Equation (11.9.1) is a *Lyapunov equation*. Converse results for Corollary 11.9.1 are given by Corollary 11.9.4, Proposition 11.9.6, Proposition 11.9.5, and Proposition 11.9.6. The following lemma is useful for analyzing (11.9.1).

Lemma 11.9.2. Assume that $A \in \mathbb{F}^{n \times n}$ is asymptotically stable. Then,

$$\int_0^\infty e^{tA}\,\mathrm{d}t = -A^{-1}. \tag{11.9.3}$$

Proof. Proposition 11.1.4 implies that $\int_0^t e^{\tau A}\,\mathrm{d}\tau = A^{-1}(e^{tA} - I)$. Letting $t \to \infty$ yields (11.9.3). $\square$

The following result concerns Sylvester's equation. See Fact 5.10.21 and Proposition 7.2.4.

Proposition 11.9.3. Let $A, B, C \in \mathbb{R}^{n \times n}$. Then, there exists a unique matrix $X \in \mathbb{R}^{n \times n}$ satisfying

$$AX + XB + C = 0 \tag{11.9.4}$$

if and only if $B^{\mathrm{T}} \oplus A$ is nonsingular. In this case, X is given by

$$X = -\operatorname{vec}^{-1}\left[\left(B^{\mathrm{T}} \oplus A\right)^{-1} \operatorname{vec} C\right]. \tag{11.9.5}$$

If, in addition, $B^{\mathrm{T}} \oplus A$ is asymptotically stable, then X is given by

$$X = \int_0^\infty e^{tA} C e^{tB} \, dt. \tag{11.9.6}$$

Proof. The first two statements follow from Proposition 7.2.4. If $B^{\mathrm{T}} \oplus A$ is asymptotically stable, then it follows from (11.9.5) using Lemma 11.9.2 and Proposition 11.1.7 that

$$X = \int_0^\infty \operatorname{vec}^{-1}\left(e^{t(B^{\mathrm{T}} \oplus A)} \operatorname{vec} C\right) dt = \int_0^\infty \operatorname{vec}^{-1}\left(e^{tB^{\mathrm{T}}} \otimes e^{tA}\right) \operatorname{vec} C \, dt$$

$$= \int_0^\infty \operatorname{vec}^{-1} \operatorname{vec}\left(e^{tA} C e^{tB}\right) dt = \int_0^\infty e^{tA} C e^{tB} \, dt. \qquad \square$$

The following result provides a converse to Corollary 11.9.1 for the case of asymptotic stability.

Corollary 11.9.4. Let $A \in \mathbb{R}^{n \times n}$, and let $R \in \mathbb{R}^{n \times n}$. Then, there exists a unique matrix $P \in \mathbb{R}^{n \times n}$ satisfying (11.9.1) if and only if $A \oplus A$ is nonsingular. In this case, if R is symmetric, then P is symmetric. Now, assume in addition that A is asymptotically stable. Then, $P \in \mathbf{S}^n$ is given by

$$P = \int_0^\infty e^{tA^{\mathrm{T}}} R e^{tA} \, dt. \tag{11.9.7}$$

Finally, if R is (positive semidefinite, positive definite), then P is (positive semidefinite, positive definite).

Proof. First note that $A \oplus A$ is nonsingular if and only if $(A \oplus A)^{\mathrm{T}} = A^{\mathrm{T}} \oplus A^{\mathrm{T}}$ is nonsingular. Now, the first statement follows from Proposition 11.9.3. To prove the second statement, note that $A^{\mathrm{T}}(P - P^{\mathrm{T}}) + (P - P^{\mathrm{T}})A = 0$, which implies that P is symmetric. Now, suppose that A is asymptotically stable. Then, Fact 11.18.33 implies that $A \oplus A$ is asymptotically stable. Consequently, (11.9.7) follows from (11.9.6). $\qquad \square$

The following results include converse statements. We first consider asymptotic stability.

Proposition 11.9.5. Let $A \in \mathbb{R}^{n \times n}$. Then, the following statements are equivalent:

i) A is asymptotically stable.

ii) For every positive-definite matrix $R \in \mathbb{R}^{n \times n}$ there exists a positive-definite matrix $P \in \mathbb{R}^{n \times n}$ such that (11.9.1) is satisfied.

iii) There exist a positive-definite matrix $R \in \mathbb{R}^{n \times n}$ and a positive-definite matrix $P \in \mathbb{R}^{n \times n}$ such that (11.9.1) is satisfied.

Proof. The result $i) \implies ii)$ follows from Corollary 11.9.1. The statement $ii) \implies iii)$ is immediate. To prove that $iii) \implies i)$, note that, since there exists a positive-semidefinite matrix P satisfying (11.9.1), it follows from Proposition 12.4.3 that (A, C) is observably asymptotically stable. Thus, there exists a nonsingular matrix $S \in \mathbb{R}^{n \times n}$ such that $A = S\left[\begin{smallmatrix} A_1 & 0 \\ A_{21} & A_2 \end{smallmatrix}\right]S^{-1}$ and $C = \left[\begin{smallmatrix} C_1 & 0 \end{smallmatrix}\right]S^{-1}$, where (C_1, A_1) is observable and A_1 is asymptotically stable. Furthermore, since $(S^{-1}AS, CS)$ is detectable, it follows that A_2 is also asymptotically stable. Consequently, A is asymptotically stable. $\qquad\square$

Next, we consider the case of Lyapunov stability.

Proposition 11.9.6. Let $A \in \mathbb{R}^{n \times n}$. Then, the following statements hold:

i) If A is Lyapunov stable, then there exist a positive-definite matrix $P \in \mathbb{R}^{n \times n}$ and a positive-semidefinite matrix $R \in \mathbb{R}^{n \times n}$ such that $\operatorname{rank} R = \nu_{-}(A)$ and such that (11.9.1) is satisfied.

ii) If there exist a positive-definite matrix $P \in \mathbb{R}^{n \times n}$ and a positive-semidefinite matrix $R \in \mathbb{R}^{n \times n}$ such that (11.9.1) is satisfied, then A is Lyapunov stable.

Proof. To prove $i)$, suppose that A is Lyapunov stable. Then, it follows from Theorem 5.3.5 and Definition 11.8.1 that there exists a nonsingular matrix $S \in \mathbb{R}^{n \times n}$ such that $A = S\left[\begin{smallmatrix} A_1 & 0 \\ 0 & A_2 \end{smallmatrix}\right]S^{-1}$ is in real Jordan form, where $A_1 \in \mathbb{R}^{n_1 \times n_1}$, $A_2 \in \mathbb{R}^{n_2 \times n_2}$, $\operatorname{spec}(A_1) \subset \jmath\mathbb{R}$, A_1 is semisimple, and $\operatorname{spec}(A_2) \subset \text{OLHP}$. Next, it follows from Fact 5.9.6 that there exists a nonsingular matrix $S_1 \in \mathbb{R}^{n_1 \times n_1}$ such that $A_1 = S_1 J_1 S_1^{-1}$, where $J_1 \in \mathbb{R}^{n_1 \times n_1}$ is skew symmetric. Then, it follows that $A_1^{\mathrm{T}}P_1 + P_1 A_1 = S_1^{-\mathrm{T}}(J_1 + J_1^{\mathrm{T}})S_1^{-1} = 0$, where $P_1 \triangleq S_1^{-\mathrm{T}}S_1^{-1}$ is positive definite. Next, let $R_2 \in \mathbb{R}^{n_2 \times n_2}$ be positive definite, and let $P_2 \in \mathbb{R}^{n_2 \times n_2}$ be the positive-definite solution of $A_2^{\mathrm{T}}P_2 + P_2 A_2 + R_2 = 0$. Hence, (11.9.1) is satisfied with $P \triangleq S^{-\mathrm{T}}\left[\begin{smallmatrix} P_1 & 0 \\ 0 & P_2 \end{smallmatrix}\right]S^{-1}$ and $R \triangleq S^{-\mathrm{T}}\left[\begin{smallmatrix} 0 & 0 \\ 0 & R_2 \end{smallmatrix}\right]S^{-1}$.

To prove $ii)$, suppose there exist a positive-semidefinite matrix $R \in \mathbb{R}^{n \times n}$ and a positive-definite matrix $P \in \mathbb{R}^{n \times n}$ such that (11.9.1) is satisfied. Let $\lambda \in \operatorname{spec}(A)$, and let $x \in \mathbb{C}^n$ be an eigenvector of A associated with λ. It thus follows from (11.9.1) that $0 = x^*A^{\mathrm{T}}Px + x^*PAx + x^*Rx = (\lambda + \bar{\lambda})x^*Px + x^*Rx$. Therefore, $\operatorname{Re}\lambda = -x^*Rx/(2x^*Px)$, which shows that $\operatorname{spec}(A) \subset \text{CLHP}$. Now, let $\jmath\omega \in \operatorname{spec}(A)$, and suppose that $x \in \mathbb{R}^n$ satisfies $(\jmath\omega I - A)^2 x = 0$. Then, $(\jmath\omega I - A)y = 0$, where $y = (\jmath\omega I - A)x$. Computing $0 = y^*\left(A^{\mathrm{T}}P + PA\right)y + y^*Ry$ yields $y^*Ry = 0$ and thus $Ry = 0$. Therefore, $\left(A^{\mathrm{T}}P + PA\right)y = 0$, and thus $y^*Py = (A^{\mathrm{T}} + \jmath\omega I)Py = 0$. Since

P is positive definite, it follows that $y = (\jmath\omega I - A)x = 0$. Therefore, $\mathcal{N}(\jmath\omega I - A) = \mathcal{N}\left[(\jmath\omega I - A)^2\right]$. Hence, it follows from Proposition 5.5.8 that $\jmath\omega$ is semisimple. $\square$

Corollary 11.9.7. Let $A \in \mathbb{R}^{n \times n}$. Then, the following statements hold:

i) A is Lyapunov stable if and only if there exists a positive-definite matrix $P \in \mathbb{R}^{n \times n}$ such that $A^{\mathrm{T}}P + PA$ is negative semidefinite.

ii) A is asymptotically stable if and only if there exists a positive-definite matrix $P \in \mathbb{R}^{n \times n}$ such that $A^{\mathrm{T}}P + PA$ is negative definite.

11.10 Discrete-Time Stability Theory

The theory of difference equations is concerned with solutions of discrete-time dynamical systems of the form

$$x(k + 1) = f[x(k)], \tag{11.10.1}$$

where $f\colon \mathbb{R}^n \to \mathbb{R}^n$, $k \in \mathbb{N}$, $x(k) \in \mathbb{R}^n$, and $x(0) = x_0$ is the initial condition. The solution $x(k) \equiv x_{\mathrm{e}}$ is an equilibrium of (11.10.1) if $x_{\mathrm{e}} = f(x_{\mathrm{e}})$.

A linear discrete-time system has the form

$$x(k + 1) = Ax(k), \tag{11.10.2}$$

where $A \in \mathbb{R}^{n \times n}$. For $k \in \mathbb{N}$, $x(k)$ is given by

$$x(k) = A^k x_0. \tag{11.10.3}$$

The behavior of the sequence $(x(k))_{k=0}^{\infty}$ is determined by the stability of A.

Definition 11.10.1. For $A \in \mathbb{C}^{n \times n}$, define the following classes of matrices:

i) A is *discrete-time Lyapunov stable* if $\mathrm{spec}(A) \subset \mathrm{CUD}$ and, if $\lambda \in \mathrm{spec}(A)$ and $|\lambda| = 1$, then λ is semisimple.

ii) A is *discrete-time semistable* if $\mathrm{spec}(A) \subset \mathrm{OUD} \cup \{1\}$ and, if $1 \in \mathrm{spec}(A)$, then 1 is semisimple.

iii) A is *discrete-time asymptotically stable* if $\mathrm{spec}(A) \subset \mathrm{OUD}$.

Proposition 11.10.2. Let $A \in \mathbb{R}^{n \times n}$ and consider the linear discrete-time system (11.10.2). Then, the following statements are equivalent:

i) A is discrete-time Lyapunov stable.

ii) For every initial condition $x_0 \in \mathbb{R}^n$, the sequence $(\|x(k)\|)_{k=1}^{\infty}$ is bounded, where $\|\cdot\|$ is a norm on $\mathbb{R}^n$.

iii) For every initial condition $x_0 \in \mathbb{R}^n$, the sequence $(\|A^k x_0\|)_{k=1}^{\infty}$ is bounded, where $\|\cdot\|$ is a norm on $\mathbb{R}^n$.

iv) The sequence $(\|A^k\|)_{k=1}^{\infty}$ is bounded, where $\|\cdot\|$ is a norm on $\mathbb{R}^{n \times n}$.

The following statements are equivalent:

v) A is discrete-time semistable.

vi) $\lim_{k \to \infty} A^k$ exists. In fact, $\lim_{k \to \infty} A^k = I - (I - A)(I - A)^{\#}$.

vii) For every initial condition $x_0 \in \mathbb{R}^n$, $\lim_{k \to \infty} x(k)$ exists.

The following statements are equivalent:

viii) A is discrete-time asymptotically stable.

ix) $\operatorname{sprad}(A) < 1$.

x) For every initial condition $x_0 \in \mathbb{R}^n$, $\lim_{k \to \infty} x(k) = 0$.

xi) For every initial condition $x_0 \in \mathbb{R}^n$, $A^k x_0 \to 0$ as $k \to \infty$.

xii) $A^k \to 0$ as $k \to \infty$.

The following definition concerns the discrete-time stability of a polynomial.

Definition 11.10.3. For $p \in \mathbb{R}[s]$, define the following terminology:

i) p is *discrete-time Lyapunov stable* if $\operatorname{roots}(p) \subset \text{CUD}$ and, if λ is an imaginary root of p, then $\mathrm{m}_p(\lambda) = 1$.

ii) p is *discrete-time semistable* if $\operatorname{roots}(p) \subset \text{OUD} \cup \{1\}$ and, if $1 \in \operatorname{roots}(p)$, then $\mathrm{m}_p(1) = 1$.

iii) p is *discrete-time asymptotically stable* if $\operatorname{roots}(p) \subset \text{OUD}$.

Proposition 11.10.4. Let $A \in \mathbb{R}^{n \times n}$. Then, the following statements hold:

i) A is discrete-time Lyapunov stable if and only if μ_A is discrete-time Lyapunov stable.

ii) A is discrete-time semistable if and only if μ_A is discrete-time semistable.

Furthermore, the following statements are equivalent:

iii) A is discrete-time asymptotically stable.

iv) μ_A is discrete-time asymptotically stable.

v) χ_A is discrete-time asymptotically stable.

We now consider the *discrete-time Lyapunov equation*

$$P = A^{\mathrm{T}} P A + R. \qquad (11.10.4)$$

Proposition 11.10.5. Let $A \in \mathbb{R}^{n \times n}$. Then, the following statements are equivalent:

i) A is discrete-time asymptotically stable.

ii) For every positive-definite matrix $R \in \mathbb{R}^{n \times n}$ there exists a positive-definite matrix $P \in \mathbb{R}^{n \times n}$ such that (11.10.4) is satisfied.

iii) There exist a positive-definite matrix $R \in \mathbb{R}^{n \times n}$ and a positive-definite matrix $P \in \mathbb{R}^{n \times n}$ such that (11.10.4) is satisfied.

Proposition 11.10.6. Let $A \in \mathbb{R}^{n \times n}$. Then, A is discrete-time Lyapunov-stable if and only if there exist a positive-definite matrix $P \in \mathbb{R}^{n \times n}$ and a positive-semidefinite matrix $R \in \mathbb{R}^{n \times n}$ such that (11.10.4) is satisfied.

11.11 Facts on Matrix Exponential Formulas

Fact 11.11.1. Let $A \in \mathbb{R}^{n \times n}$. Then, the following statements hold:

i) If $A^2 = 0$, then $e^{tA} = I + tA$.

ii) If $A^2 = I$, then $e^{tA} = (\cosh t)I + (\sinh t)A$.

iii) If $A^2 = -I$, then $e^{tA} = (\cos t)I + (\sin t)A$.

iv) If $A^2 = A$, then $e^{tA} = I + (e^t - 1)A$.

v) If $A^2 = -A$, then $e^{tA} = I + (1 - e^{-t})A$.

vi) If $\operatorname{rank} A = 1$ and $\operatorname{tr} A = 0$, then $e^{tA} = I + tA$.

vii) If $\operatorname{rank} A = 1$ and $\operatorname{tr} A \neq 0$, then $e^{tA} = I + \frac{e^{(\operatorname{tr} A)t} - 1}{\operatorname{tr} A} A$.

Remark: See [1112].

Fact 11.11.2. Let $A \triangleq \begin{bmatrix} 0 & I_n \\ I_n & 0 \end{bmatrix}$. Then,

$$e^{tA} = (\cosh t)I_{2n} + (\sinh t)A.$$

Furthermore,

$$e^{tJ_{2n}} = (\cos t)I_{2n} + (\sin t)J_{2n}.$$

Fact 11.11.3. Let $A \in \mathbb{R}^{n \times n}$, and assume that A is skew symmetric. Then, $\{e^{\theta A}: \ \theta \in \mathbb{R}\} \subseteq \operatorname{SO}(n)$ is a group. If, in addition, $n = 2$, then

$$\{e^{\theta J_2}: \ \theta \in \mathbb{R}\} = \operatorname{SO}(2).$$

Remark: Note that $e^{\theta J_2} = \begin{bmatrix} \cos\theta & \sin\theta \\ -\sin\theta & \cos\theta \end{bmatrix}$. See Fact 3.11.27.

Fact 11.11.4. Let $A \in \mathbb{R}^{n \times n}$, where

$$A \triangleq \begin{bmatrix} 0 & 1 & 0 & 0 & \cdots & 0 \\ 0 & 0 & 2 & 0 & \cdots & 0 \\ 0 & 0 & 0 & 3 & \cdots & 0 \\ \vdots & \vdots & \vdots & \ddots & \ddots & \vdots \\ 0 & 0 & 0 & 0 & \ddots & n-1 \\ 0 & 0 & 0 & 0 & \cdots & 0 \end{bmatrix}.$$

Then,

$$e^A = \begin{bmatrix} \binom{0}{0} & \binom{1}{0} & \binom{2}{0} & \binom{3}{0} & \cdots & \binom{n-1}{0} \\ 0 & \binom{1}{1} & \binom{2}{1} & \binom{3}{1} & \cdots & \binom{n-1}{1} \\ 0 & 0 & \binom{2}{2} & \binom{3}{2} & \cdots & \binom{n-1}{2} \\ \vdots & \vdots & \vdots & \ddots & \ddots & \vdots \\ 0 & 0 & 0 & 0 & \ddots & \binom{n-1}{n-2} \\ 0 & 0 & 0 & 0 & \cdots & \binom{n-1}{n-1} \end{bmatrix}.$$

Furthermore, if $k \geq n$, then

$$\sum_{i=1}^{k} i^{n-1} = \begin{bmatrix} 1^{n-1} & 2^{n-1} & \cdots & n^{n-1} \end{bmatrix} e^{-A} \begin{bmatrix} \binom{k}{1} \\ \vdots \\ \binom{k}{n} \end{bmatrix}.$$

Proof: See [76].

Remark: For related results, see [5], where A is called the *creation matrix*. See Fact 5.16.3.

Fact 11.11.5. Let $A \in \mathbb{F}^{3\times 3}$. If $\mathrm{spec}(A) = \{\lambda\}$, then

$$e^{tA} = e^{\lambda t}\left[I + t(A - \lambda I) + \tfrac{1}{2}t^2(A - \lambda I)^2 \right].$$

If $\mathrm{mspec}(A) = \{\lambda, \lambda, \mu\}_{\mathrm{ms}}$, where $\mu \neq \lambda$, then

$$e^{tA} = e^{\lambda t}[I + t(A - \lambda I)] + \left[\frac{e^{\mu t} - e^{\lambda t}}{(\mu - \lambda)^2} - \frac{te^{\lambda t}}{\mu - \lambda} \right](A - \lambda I)^2.$$

If $\mathrm{spec}(A) = \{\lambda, \mu, \nu\}$, then

$$e^{tA} = \frac{e^{\lambda t}}{(\lambda - \mu)(\lambda - \nu)}(A - \mu I)(A - \nu I) + \frac{e^{\mu t}}{(\mu - \lambda)(\mu - \nu)}(A - \lambda I)(A - \nu I)$$
$$+ \frac{e^{\nu t}}{(\nu - \lambda)(\nu - \mu)}(A - \lambda I)(A - \mu I).$$

Proof: See [70].

Remark: Additional expressions are given in [2, 179, 195, 329, 657, 1112, 1115].

Fact 11.11.6. Let $x \in \mathbb{R}^3$, assume that x is nonzero, and define $\theta \triangleq \|x\|_2$. Then,

$$e^{K(x)} = I + \frac{\sin\theta}{\theta}K(x) + \frac{1 - \cos\theta}{\theta^2}K^2(x)$$

$$= I + \frac{\sin\theta}{\theta}K(x) + \tfrac{1}{2}\left[\frac{\sin(\frac{1}{2}\theta)}{\frac{1}{2}\theta} \right]^2 K^2(x)$$

$$= (\cos\theta)I + \frac{\sin\theta}{\theta}K(x) + \frac{1 - \cos\theta}{\theta^2}xx^{\mathrm{T}}.$$

Furthermore,

$$e^{K(x)}x = x,$$

$$\mathrm{spec}[e^{K(x)}] = \{1, e^{\jmath\|x\|_2}, e^{-\jmath\|x\|_2}\},$$

and

$$\mathrm{tr}\, e^{K(x)} = 1 + 2\cos\theta = 1 + 2\cos\|x\|_2.$$

Proof: The Cayley-Hamilton theorem or Fact 3.10.1 implies that $K^3(x) + \theta^2 K(x) = 0$. Then, every term $K^k(x)$ in the expansion of $e^{K(x)}$ can be expressed in terms of $K(x)$ or $K^2(x)$. Finally, Fact 3.10.1 implies that $\theta^2 I + K^2(x) = xx^{\mathrm{T}}$.

Remark: Fact 11.11.7 shows that, for all $z \in \mathbb{R}^3$, $e^{K(x)}z$ is the counterclockwise (right-hand-rule) rotation of z about the vector x through the angle θ, which is given by the Euclidean norm of x. In Fact 3.11.29, the cross product is used to construct the pivot vector x for a given pair of vectors having the same length.

Fact 11.11.7. Let $x, y \in \mathbb{R}^3$, and assume that x and y are nonzero. Then, there exists a skew-symmetric matrix $A \in \mathbb{R}^{3\times 3}$ such that $y = e^A x$ if and only if $x^{\mathrm{T}}x = y^{\mathrm{T}}y$. If $x \neq -y$, then one such matrix is $A = \theta K(z)$, where

$$z \triangleq \frac{1}{\|x \times y\|_2} x \times y$$

and

$$\theta \triangleq \cos^{-1}\left(\frac{x^{\mathrm{T}}y}{\|x\|_2\|y\|_2}\right).$$

If $x = -y$, then one such matrix is $A = \pi K(z)$, where $z \triangleq \|y\|_2^{-1}\nu \times y$ and $\nu \in \{y\}^{\perp}$ satisfies $\nu^{\mathrm{T}}\nu = 1$.

Proof: This result follows from Fact 3.11.29 and Fact 11.11.6, which provide equivalent expressions for an orthogonal matrix that transforms a given vector into another given vector having the same length. This result thus provides a geometric interpretation for Fact 11.11.6.

Remark: Note that z is the unit vector perpendicular to the plane containing x and y, where the direction of z is determined by the right-hand rule. An intuitive proof is to let x be the initial condition to the differential equation $\dot{w}(t) = K(z)w(t)$, that is, $w(0) = x$, where $t \in [0, \theta]$. Then, the derivative $\dot{w}(t)$ lies in the x, y plane and is perpendicular to $w(t)$ for all $t \in [0, \theta]$. Hence, $y = w(\theta)$.

Remark: Since $\det e^A = e^{\mathrm{tr}\, A} = 1$, it follows that every pair of vectors in $\mathbb{R}^3$ having the same Euclidean length are related by a *proper rotation*. See Fact 3.9.5 and Fact 3.14.4. This is a linear interpolation problem. See Fact 3.9.5, Fact 3.11.29, and [795].

Remark: See Fact 3.11.29.

Remark: Parameterizations of SO(3) are considered in [1226, 1277].

Problem: Extend this result to $\mathbb{R}^n$. See [139, 1194].

Fact 11.11.8. Let $A \in \mathrm{SO}(3)$, let $z \in \mathbb{R}^3$ be an eigenvector of A corresponding to the eigenvalue 1 of A, assume that $\|z\|_2 = 1$, assume that $\mathrm{tr}\, A > -1$, and let $\theta \in (-\pi, \pi)$ satisfy $\mathrm{tr}\, A = 1 + 2\cos\theta$. Then,

$$A = e^{\theta K(z)}.$$

Remark: See Fact 5.11.2.

Fact 11.11.9. Let $x, y \in \mathbb{R}^3$, and assume that x and y are nonzero. Then, $x^T x = y^T y$ if and only if

$$y = e^{\frac{\theta}{\|x \times y\|_2}(yx^T - xy^T)}x,$$

where

$$\theta \triangleq \cos^{-1}\left(\frac{x^T y}{\|x\|_2 \|y\|_2}\right).$$

Proof: Use Fact 11.11.7.

Remark: Note that $K(x \times y) = yx^T - xy^T$.

Fact 11.11.10. Let $A \in \mathbb{R}^{3 \times 3}$, assume that $A \in \text{SO}(3)$ and $\text{tr}\, A > -1$, and let $\theta \in (-\pi, \pi)$ satisfy $\text{tr}\, A = 1 + 2\cos\theta$. Then,

$$\log A = \begin{cases} 0, & \theta = 0, \\ \frac{\theta}{2\sin\theta}(A - A^T), & \theta \neq 0. \end{cases}$$

Proof: See [767, p. 364] and [1038].

Remark: See Fact 11.15.10.

Fact 11.11.11. Let $x \in \mathbb{R}^3$, assume that x is nonzero, and define $\theta \triangleq \|x\|_2$. Then,

$$K(x) = \frac{\theta}{2\sin\theta}\left[e^{K(x)} - e^{-K(x)}\right].$$

Proof: Use Fact 11.11.10.

Remark: See Fact 3.10.1.

Fact 11.11.12. Let $A \in \text{SO}(3)$, let $x, y \in \mathbb{R}^3$, and assume that $x^T x = y^T y$. Then, $Ax = y$ if and only if, for all $t \in \mathbb{R}$,

$$Ae^{tK(x)}A^{-1} = e^{tK(y)}.$$

Proof: See [912].

Fact 11.11.13. Let $x, y, z \in \mathbb{R}^3$. Then, the following statements are equivalent:

i) For every $A \in \text{SO}(3)$, there exist $\alpha, \beta, \gamma \in \mathbb{R}$ such that

$$A = e^{\alpha K(x)}e^{\beta K(y)}e^{\gamma K(z)}.$$

ii) $y^T x = 0$ and $y^T z = 0$.

Proof: See [912].

Remark: This result is due to Davenport.

Problem: Given $A \in \text{SO}(3)$, determine α, β, γ.

Fact 11.11.14. Let $x, y, z \in \mathbb{R}^3$, assume that $x^{\mathrm{T}}(y \times z) \neq 0$, assume that $\|x\|_2 = \|y\|_2 = \|z\|_2 = 1$, and let $\alpha, \beta, \gamma \in [0, 2\pi)$ satisfy

$$\tan \frac{\alpha}{2} = \frac{z^{\mathrm{T}}(x \times y)}{(x \times y)^{\mathrm{T}}(x \times z)},$$

$$\tan \frac{\beta}{2} = \frac{x^{\mathrm{T}}(y \times z)}{(y \times z)^{\mathrm{T}}(y \times x)},$$

$$\tan \frac{\gamma}{2} = \frac{y^{\mathrm{T}}(z \times x)}{(z \times x)^{\mathrm{T}}(z \times y)}.$$

Then,

$$e^{\alpha K(x)} e^{\beta K(y)} e^{\gamma K(z)} = I.$$

Proof: See [1046, p. 5].

Remark: This result is the *Rodrigues-Hamilton theorem*. An equivalent result is given by *Donkin's theorem* [1046, p. 6].

Remark: See [1170, 1482].

Fact 11.11.15. Let $A \in \mathbb{R}^{4 \times 4}$, and assume that A is skew symmetric with $\mathrm{mspec}(A) = \{\jmath\omega, -\jmath\omega, \jmath\mu, -\jmath\mu\}_{\mathrm{ms}}$. If $\omega \neq \mu$, then

$$e^A = a_3 A^3 + a_2 A^2 + a_1 A + a_0 I,$$

where

$$a_3 = \left(\omega^2 - \mu^2\right)^{-1}\left(\tfrac{1}{\mu}\sin\mu - \tfrac{1}{\omega}\sin\omega\right),$$

$$a_2 = \left(\omega^2 - \mu^2\right)^{-1}(\cos\mu - \cos\omega),$$

$$a_1 = \left(\omega^2 - \mu^2\right)^{-1}\left(\tfrac{\omega^2}{\mu}\sin\mu - \tfrac{\mu^2}{\omega}\sin\omega\right),$$

$$a_0 = \left(\omega^2 - \mu^2\right)^{-1}\left(\omega^2\cos\mu - \mu^2\cos\omega\right).$$

If $\omega = \mu$, then

$$e^A = (\cos\omega)I + \frac{\sin\omega}{\omega}A.$$

Proof: See [622, p. 18] and [1115].

Remark: There are errors in [622, p. 18] and [1115].

Remark: See Fact 4.9.21 and Fact 4.10.4.

Fact 11.11.16. Let $a, b, c \in \mathbb{R}$, define the skew-symmetric matrix $A \in \mathbb{R}^{4 \times 4}$, by either

$$A \triangleq \begin{bmatrix} 0 & a & b & c \\ -a & 0 & -c & b \\ -b & c & 0 & -a \\ -c & -b & a & 0 \end{bmatrix}$$

or

$$A \triangleq \begin{bmatrix} 0 & a & b & c \\ -a & 0 & c & -b \\ -b & -c & 0 & a \\ -c & b & -a & 0 \end{bmatrix},$$

and define $\theta \triangleq \sqrt{a^2 + b^2 + c^2}$. Then,

$$\text{mspec}(A) = \{\jmath\theta, -\jmath\theta, \jmath\theta, -\jmath\theta\}_{\text{ms}}.$$

Furthermore,

$$A^k = \begin{cases} (-1)^{k/2}\theta^k I, & k \text{ even}, \\ (-1)^{(k-1)/2}\theta^{k-1}A, & k \text{ odd}, \end{cases}$$

and

$$e^A = (\cos\theta)I + \frac{\sin\theta}{\theta}A.$$

Proof: See [1390].

Remark: $(\sin 0)/0 = 1$.

Remark: The skew-symmetric matrix A arises in the kinematic relationship between the angular velocity vector and quaternion (Euler-parameter) rates. See [156, p. 385].

Remark: The two matrices A are similar. To show this, note that Fact 5.9.11 implies that A and $-A$ are similar. Then, apply the similarity transformation $S = \text{diag}(-1, 1, 1, 1)$.

Remark: See Fact 4.9.21 and Fact 4.10.4.

Fact 11.11.17. Let $x \in \mathbb{R}^3$, and define the skew-symmetric matrix $A \in \mathbb{R}^{4 \times 4}$ by

$$A = \begin{bmatrix} 0 & -x^{\text{T}} \\ x & -K(x) \end{bmatrix}.$$

Then, for all $t \in \mathbb{R}$,

$$e^{\frac{1}{2}tA} = \cos(\tfrac{1}{2}\|x\|t)I_4 + \frac{\sin(\tfrac{1}{2}\|x\|t)}{\|x\|}A.$$

Proof: See [753, p. 34].

Remark: The matrix $\frac{1}{2}A$ expresses quaternion rates in terms of the angular velocity vector.

Fact 11.11.18. Let $a, b \in \mathbb{R}^3$, define the skew-symmetric matrix $A \in \mathbb{R}^{4 \times 4}$ by

$$A = \begin{bmatrix} K(a) & b \\ -b^{\text{T}} & 0 \end{bmatrix},$$

assume that $a^{\text{T}}b = 0$, and define $\alpha \triangleq \sqrt{a^{\text{T}}a + b^{\text{T}}b}$. Then,

$$e^A = I_4 + \frac{\sin\alpha}{\alpha}A + \frac{1 - \cos\alpha}{\alpha^2}A^2.$$

Proof: See [1366].

Remark: See Fact 4.9.21 and Fact 4.10.4.

Fact 11.11.19. Let $a, b \in \mathbb{R}^{n-1}$, define $A \in \mathbb{R}^{n \times n}$ by

$$A \triangleq \begin{bmatrix} 0 & a^{\mathrm{T}} \\ b & 0_{(n-1) \times (n-1)} \end{bmatrix},$$

and define $\alpha \triangleq \sqrt{|a^{\mathrm{T}}b|}$. Then, the following statements hold:

i) If $a^{\mathrm{T}}b < 0$, then

$$e^{tA} = I + \frac{\sin \alpha}{\alpha} A + \frac{1}{2} \left[\frac{\sin(\alpha/2)}{\alpha/2} \right]^2 A^2.$$

ii) If $a^{\mathrm{T}}b = 0$, then

$$e^{tA} = I + A + \tfrac{1}{2} A^2.$$

iii) If $a^{\mathrm{T}}b > 0$, then

$$e^{tA} = I + \frac{\sinh \alpha}{\alpha} A + \frac{1}{2} \left[\frac{\sinh(\alpha/2)}{\alpha/2} \right]^2 A^2.$$

Proof: See [1516].

11.12 Facts on the Matrix Sine and Cosine

Fact 11.12.1. Let $A \in \mathbb{C}^{n \times n}$, and define

$$\sin A \triangleq A - \tfrac{1}{3!} A^3 + \tfrac{1}{5!} A^5 - \tfrac{1}{7!} A^7 + \cdots$$

and

$$\cos A \triangleq I - \tfrac{1}{2!} A^2 + \tfrac{1}{4!} A^4 - \tfrac{1}{6!} A^6 + \cdots.$$

Then, the following statements hold:

i) $\sin A = \frac{1}{\jmath 2} (e^{\jmath A} - e^{-\jmath A})$.

ii) $\cos A = \frac{1}{2} (e^{\jmath A} + e^{-\jmath A})$.

iii) $\sin^2 A + \cos^2 A = I$.

iv) $\sin(2A) = 2(\sin A) \cos A$.

v) $\cos(2A) = 2(\cos^2 A) - I$.

vi) If A is real, then $\sin A = \operatorname{Re} e^{\jmath A}$ and $\cos A = \operatorname{Re} e^{\jmath A}$.

vii) $\sin(A \oplus B) = (\sin A) \otimes \cos B - (\cos A) \otimes \sin B$.

viii) $\cos(A \oplus B) = (\cos A) \otimes \cos B - (\sin A) \otimes \sin B$.

ix) If A is involutory and k is an integer, then $\cos(k\pi A) = (-1)^k I$.

Furthermore, the following statements are equivalent:

x) For all $t \in \mathbb{R}$, $\sin[t(A + B)] = [\sin(tA)][\cos(tB)] + [\cos(tA)][\sin(tB)]$.

xi) For all $t \in \mathbb{R}$, $\cos[t(A + B)] = [\cos(tA)][\cos(tB)] - [\sin(tA)][\sin(tB)]$.

xii) $AB = BA$.

Proof: See [701, pp. 287, 288, 300].

11.13 Facts on the Matrix Exponential for One Matrix

Fact 11.13.1. Let $A \in \mathbb{F}^{n \times n}$, and assume that A is (lower triangular, upper triangular). Then, so is e^A. If, in addition, A is Toeplitz, then so is e^A.

Remark: See Fact 3.18.7.

Fact 11.13.2. Let $A \in \mathbb{F}^{n \times n}$. Then,

$$\mathrm{sprad}(e^A) = e^{\mathrm{spabs}(A)}.$$

Fact 11.13.3. Let $A \in \mathbb{R}^{n \times n}$, and let $X_0 \in \mathbb{R}^{n \times n}$. Then, the matrix differential equation

$$\dot{X}(t) = AX(t),$$
$$X(0) = X_0,$$

where $t \geq 0$, has the unique solution

$$X(t) = e^{tA}X_0.$$

Fact 11.13.4. Let $A \colon [0, T] \mapsto \mathbb{R}^{n \times n}$, assume that A is continuous, and let $X_0 \in \mathbb{R}^{n \times n}$. Then, the matrix differential equation

$$\dot{X}(t) = A(t)X(t),$$
$$X(0) = X_0$$

has a unique solution $X \colon [0, T] \mapsto \mathbb{R}^{n \times n}$. Furthermore, for all $t \in [0, T]$,

$$\det X(t) = e^{\int_0^t \mathrm{tr}\, A(\tau)\, d\tau} \det X_0.$$

Therefore, if X_0 is nonsingular, then $X(t)$ is nonsingular for all $t \in [0, T]$. If, in addition, for all $t_1, t_2 \in [0, T]$,

$$A(t_2) \int_{t_1}^{t_2} A(\tau)\, d\tau = \int_{t_1}^{t_2} A(\tau)\, d\tau A(t_2),$$

then, for all $t \in [0, T]$,

$$X(t) = e^{\int_0^t A(\tau)\, d\tau} X_0.$$

Proof: It follows from Fact 10.12.8 that $(d/dt) \det X = \mathrm{tr}(X^{\mathrm{A}}\dot{X}) = \mathrm{tr}(X^{\mathrm{A}}AX) = \mathrm{tr}(XX^{\mathrm{A}}A) = (\det X)\mathrm{tr}\, A$. This proof is given in [577]. See also [730, pp. 507, 508] and [1179, pp. 64–66].

Remark: See Fact 11.13.4.

Remark: The first result is *Jacobi's identity*.

Remark: If the commutativity assumption does not hold, then the solution is given by the *Peano-Baker series*. See [1179, Chapter 3]. Alternative expressions for $X(t)$ are given by the Magnus, Fer, Baker-Campbell-Hausdorff-Dynkin, Wei-Norman, Goldberg, and Zassenhaus expansions. See [232, 455, 766, 767, 854, 974, 1083, 1275, 1305, 1448, 1449, 1453] and [636, pp. 118–120].

Fact 11.13.5. Let $A\colon [0,T] \mapsto \mathbb{R}^{n \times n}$, assume that A is continuous, let $B\colon [0,T] \mapsto \mathbb{R}^{n \times m}$, assume that B is continuous, let $X\colon [0,T] \mapsto \mathbb{R}^{n \times n}$ satisfy the matrix differential equation

$$\dot{X}(t) = A(t)X(t),$$
$$X(0) = I,$$

define

$$\Phi(t,\tau) \triangleq X(t)X^{-1}(\tau),$$

let $u\colon [0,T] \mapsto \mathbb{R}^m$, and assume that u is continuous. Then, the vector differential equation

$$\dot{x}(t) = A(t)x(t) + B(t)u(t),$$
$$x(0) = x_0$$

has the unique solution

$$x(t) = X(t)x_0 + \int_0^t \Phi(t,\tau)B(\tau)u(\tau)\mathrm{d}\tau.$$

Remark: $\Phi(t,\tau)$ is the *state transition matrix*.

Fact 11.13.6. Let $A \in \mathbb{R}^{n \times n}$, let $\lambda \in \mathrm{spec}(A)$, and let $v \in \mathbb{C}^n$ be an eigenvector of A associated with λ. Then, for all $t \geq 0$,

$$x(t) \triangleq \mathrm{Re}\left(e^{\lambda t}v\right)$$

satisfies $\dot{x}(t) = Ax(t)$.

Remark: $x(t)$ is an *eigensolution*.

Fact 11.13.7. Let $A \in \mathbb{R}^{n \times n}$, let $\lambda \in \mathrm{spec}(A)$, and let $(v_1, \ldots, v_k) \in (\mathbb{C}^n)^k$ be a Jordan chain of A associated with λ. Then, for all i such that $1 \leq i \leq k$,

$$x(t) \triangleq \mathrm{Re}\left(e^{\lambda t}\left[\tfrac{1}{(i-1)!}t^{i-1}v_1 + \cdots + tv_{i-1} + v_i\right]\right)$$

satisfies $\dot{x}(t) = Ax(t)$ for all $t \geq 0$.

Remark: See Fact 5.14.9 for the definition of a Jordan chain.

Remark: $x(t)$ is a *generalized eigensolution*.

Example: Let $A = \left[\begin{smallmatrix} 0 & 1 \\ 0 & 0 \end{smallmatrix}\right]$, $\lambda = 0$, $\hat{k} = 2$, $v_1 = \left[\begin{smallmatrix} \beta \\ 0 \end{smallmatrix}\right]$, and $v_2 = \left[\begin{smallmatrix} 0 \\ \beta \end{smallmatrix}\right]$. Then, $x(t) = tv_1 + v_2 = \left[\begin{smallmatrix} \beta t \\ \beta \end{smallmatrix}\right]$ is a generalized eigensolution. Alternatively, choosing $\hat{k} = 1$ yields the eigensolution $x(t) = v_1 = \left[\begin{smallmatrix} \beta \\ 0 \end{smallmatrix}\right]$. Note that β represents velocity for the generalized eigensolution and position for the eigensolution. See [1089].

Fact 11.13.8. Let $S\colon [t_0,t_1] \to \mathbb{R}^{n \times n}$ be differentiable. Then, for all $t \in [t_0,t_1]$,

$$\frac{\mathrm{d}}{\mathrm{d}t}S^2(t) = \dot{S}(t)S(t) + S(t)\dot{S}(t).$$

Let $S_1\colon [t_0,t_1] \to \mathbb{R}^{n \times m}$ and $S_2\colon [t_0,t_1] \to \mathbb{R}^{m \times l}$ be differentiable. Then, for all $t \in [t_0,t_1]$,

$$\frac{\mathrm{d}}{\mathrm{d}t}S_1(t)S_2(t) = \dot{S}_1(t)S_2(t) + S_1(t)\dot{S}_2(t).$$

Fact 11.13.9. Let $A \in \mathbb{F}^{n \times n}$, and define $A_1 \triangleq \frac{1}{2}(A + A^*)$ and $A_2 \triangleq \frac{1}{2}(A - A^*)$. Then, $A_1 A_2 = A_2 A_1$ if and only if A is normal. In this case, $e^{A_1} e^{A_2}$ is the polar decomposition of e^A.

Remark: See Fact 3.7.28.

Problem: Obtain the polar decomposition of e^A for the case in which A is not normal.

Fact 11.13.10. Let $A \in \mathbb{F}^{n \times m}$, and assume that $\operatorname{rank} A = m$. Then,

$$A^+ = \int_0^\infty e^{-tA^*A} A^* \, dt.$$

Fact 11.13.11. Let $A \in \mathbb{F}^{n \times n}$, and assume that A is nonsingular. Then,

$$A^{-1} = \int_0^\infty e^{-tA^*A} \, dt A^*.$$

Fact 11.13.12. Let $A \in \mathbb{F}^{n \times n}$, and let $k \triangleq \operatorname{ind} A$. Then,

$$A^{\mathrm{D}} = \int_0^\infty e^{-tA^k A^{(2k+1)*} A^{k+1}} \, dt A^k A^{(2k+1)*} A^k.$$

Proof: See [584].

Fact 11.13.13. Let $A \in \mathbb{F}^{n \times n}$, and assume that $\operatorname{ind} A = 1$. Then,

$$A^{\#} = \int_0^\infty e^{-tAA^{3*}A^2} \, dt A A^{3*} A.$$

Proof: See Fact 11.13.12.

Fact 11.13.14. Let $A \in \mathbb{F}^{n \times n}$, and let $k \triangleq \operatorname{ind} A$. Then,

$$\int_0^t e^{\tau A} \, d\tau = A^{\mathrm{D}}(e^{tA} - I) + (I - AA^{\mathrm{D}})(tI + \frac{1}{2!}t^2 A + \cdots + \frac{1}{k!}t^k A^{k-1}).$$

If, in particular, A is group invertible, then

$$\int_0^t e^{\tau A} \, d\tau = A^{\#}(e^{tA} - I) + (I - AA^{\#})t.$$

Fact 11.13.15. Let $A \in \mathbb{F}^{n \times n}$, let $\operatorname{mspec}(A) = \{\lambda_1, \ldots, \lambda_r, 0, \ldots, 0\}_{\mathrm{ms}}$, where $\lambda_1, \ldots, \lambda_r$ are nonzero, and let $t > 0$. Then,

$$\det \int_0^t e^{\tau A} \, d\tau = t^{n-r} \prod_{i=1}^r \lambda_i^{-1}(e^{\lambda_i t} - 1).$$

Hence, $\det \int_0^t e^{\tau A} \, \mathrm{d}\tau \neq 0$ if and only if, for every nonzero integer k, $j2k\pi/t \notin \mathrm{spec}(A)$. Finally, $\det(e^{tA} - I) \neq 0$ if and only if $\det A \neq 0$ and $\det \int_0^t e^{\tau A} \, \mathrm{d}\tau \neq 0$.

Fact 11.13.16. Let $A \in \mathbb{R}^{n \times n}$, and let σ be a positive number. Then,

$$\frac{1}{\sqrt{2\pi\sigma}} \int\limits_{-\infty}^{\infty} e^{xA} e^{-x^2/(2\sigma)} \, \mathrm{d}x = e^{(\sigma/2)A^2}.$$

Remark: This result is due to Bhat.

Remark: See Fact 8.16.1.

Fact 11.13.17. Let $A \in \mathbb{F}^{n \times n}$, and assume that there exists $\alpha \in \mathbb{R}$ such that $\mathrm{spec}(A) \subset \{z \in \mathbb{C} : \alpha \leq \mathrm{Im}\, z < 2\pi + \alpha\}$. Then, e^A is (diagonal, upper triangular, lower triangular) if and only if A is.

Proof: See [957].

Fact 11.13.18. Let $A \in \mathbb{F}^{n \times n}$. Then, the following statements hold:

i) If A is unipotent, then the series (11.5.1) is finite, $\log A$ exists and is nilpotent, and $e^{\log A} = A$.

ii) If A is nilpotent, then e^A is unipotent and $\log e^A = A$.

Proof: See [639, p. 60].

Fact 11.13.19. Let $B \in \mathbb{R}^{n \times n}$. Then, there exists a normal matrix $A \in \mathbb{R}^{n \times n}$ such that $B = e^A$ if and only if B is normal, nonsingular, and every negative eigenvalue of B has even algebraic multiplicity.

Fact 11.13.20. Let $C \in \mathbb{R}^{n \times n}$, assume that C is nonsingular, and let $k \geq 1$. Then, there exists a matrix $B \in \mathbb{R}^{n \times n}$ such that $C^{2k} = e^B$.

Proof: Use Proposition 11.4.3 with $A = C^2$, and note that every negative eigenvalue $-\alpha < 0$ of C^2 arises as the square of complex conjugate eigenvalues $\pm j\sqrt{\alpha}$ of C.

11.14 Facts on the Matrix Exponential for Two or More Matrices

Fact 11.14.1. Let $A, B \in \mathbb{F}^{n \times n}$, and consider the following conditions:

i) $A = B$.

ii) $e^A = e^B$.

iii) $AB = BA$.

iv) $Ae^B = e^B A$.

v) $e^A e^B = e^B e^A$.

vi) $e^A e^B = e^{A+B}$.

vii) $e^A e^B = e^B e^A = e^{A+B}$.

Then, the following statements hold:

viii) *iii*) $\Longrightarrow$ *iv*) $\Longrightarrow$ *v*).

ix) *iii*) $\Longrightarrow$ *vii*) $\Longrightarrow$ *vi*).

x) If spec(A) is $\jmath 2\pi$ congruence free, then *ii*) $\Longrightarrow$ *iii*) $\Longrightarrow$ *iv*) $\Longleftrightarrow$ *v*).

xi) If spec(A) and spec(B) are $\jmath 2\pi$ congruence free, then *ii*) $\Longrightarrow$ *iii*) $\Longleftrightarrow$ *iv*) $\Longleftrightarrow$ *v*).

xii) If spec($A + B$) is $\jmath 2\pi$ congruence free, then *iii*) $\Longleftrightarrow$ *vii*).

xiii) If, for all $\lambda \in$ spec(A) and all $\mu \in$ spec(B), it follows that $(\lambda - \mu)/(\jmath 2\pi)$ is not a nonzero integer, then *ii*) $\Longrightarrow$ *i*).

xiv) If A and B are Hermitian, then *i*) $\Longleftrightarrow$ *ii*) $\Longrightarrow$ *iii*) $\Longleftrightarrow$ *iv*) $\Longleftrightarrow$ *v*) $\Longleftrightarrow$ *vi*).

Remark: The set $\mathcal{S} \subset \mathbb{C}$ is $\jmath 2\pi$ *congruence free* if no two elements of $\mathcal{S}$ differ by a nonzero integer multiple of $\jmath 2\pi$.

Proof: See [644, pp. 88, 89, 270–272] and [1092, 1199, 1200, 1201, 1239, 1454, 1455]. The assumption of normality in operator versions of some of these statements in [1092, 1201] is not needed in the matrix case. Statement *xiii*) and the first implication in *x*) are given in [701, p. 32].

Remark: The matrices

$$A = \begin{bmatrix} 0 & 1 \\ 0 & \jmath 2\pi \end{bmatrix}, \qquad B = \begin{bmatrix} \jmath 2\pi & 0 \\ 0 & -\jmath 2\pi \end{bmatrix}$$

satisfy $e^A = e^B = e^{A+B} = I$, but $AB \neq BA$. Therefore, *vii*) $\Longrightarrow$ *iii*) does not hold. Furthermore, since *vii*) holds and *iii*) does not hold, it follows from *xii*) that spec($A + B$) is not $\jmath 2\pi$ congruence free. In fact, spec($A + B$) = $\{0, \jmath 2\pi\}$. The same observation holds for the real matrices

$$A = 2\pi \begin{bmatrix} 0 & 0 & \sqrt{3}/2 \\ 0 & 0 & -1/2 \\ -\sqrt{3}/2 & 1/2 & 0 \end{bmatrix}, \qquad B = 2\pi \begin{bmatrix} 0 & 0 & -\sqrt{3}/2 \\ 0 & 0 & -1/2 \\ \sqrt{3}/2 & 1/2 & 0 \end{bmatrix}.$$

Remark: The matrices

$$A = \begin{bmatrix} (\log 2) + \jmath\pi & 0 \\ 0 & 0 \end{bmatrix}, \qquad B = \begin{bmatrix} 2(\log 2) + \jmath 2\pi & 1 \\ 0 & 0 \end{bmatrix}$$

satisfy $e^A e^B = e^{A+B} \neq e^B e^A$. Therefore, *vi*) $\Longrightarrow$ *vii*) does not hold. Furthermore, spec($A + B$) = $\{3(\log 2) + \jmath 3\pi, 0\}$, which is $\jmath 2\pi$ congruence free. Therefore, since *vii*) does not hold, it follows from *xii*) that *iii*) does not hold, that is, $AB \neq BA$. In fact,

$$AB - BA = \begin{bmatrix} 0 & (\log 2) + \jmath\pi \\ 0 & 0 \end{bmatrix}.$$

Consequently, under the additional condition that spec($A + B$) is $\jmath 2\pi$ congruence free, *vi*) $\Longrightarrow$ *iii*) does not hold. However, *xiv*) implies that *vi*) $\Longrightarrow$ *iii*) holds under the additional assumption that A and B are Hermitian, in which case spec($A + B$) is $\jmath 2\pi$ congruence free. This example is due to Bourgeois.

Fact 11.14.2. Let $A \in \mathbb{F}^{n \times n}$, $B \in \mathbb{F}^{n \times m}$, and $C \in \mathbb{F}^{m \times m}$. Then,

$$e^{t\left[\begin{smallmatrix} A & B \\ 0 & C \end{smallmatrix}\right]} = \begin{bmatrix} e^{tA} & \int_0^t e^{(t-\tau)A} B e^{\tau C} \, d\tau \\ 0 & e^{tC} \end{bmatrix}.$$

Furthermore,

$$\int_0^t e^{\tau A} \, d\tau = \begin{bmatrix} I & 0 \end{bmatrix} e^{t\left[\begin{smallmatrix} A & I \\ 0 & 0 \end{smallmatrix}\right]} \begin{bmatrix} 0 \\ I \end{bmatrix}.$$

Remark: The result can be extended to arbitrary upper block-triangular matrices. See [1393]. For an application to sampled-data control, see [1080].

Fact 11.14.3. Let $A, B \in \mathbb{R}^{n \times n}$. Then,

$$\frac{d}{dt} e^{A+tB} = \int_0^1 e^{\tau(A+tB)} B e^{(1-\tau)(A+tB)} \, d\tau.$$

Hence,

$$\mathrm{Dexp}(A;B) = \frac{d}{dt} e^{A+tB} \bigg|_{t=0} = \int_0^1 e^{\tau A} B e^{(1-\tau)A} \, d\tau.$$

Furthermore,

$$\frac{d}{dt} \mathrm{tr}\, e^{A+tB} = \mathrm{tr}\left(e^{A+tB} B \right).$$

Hence,

$$\frac{d}{dt} \mathrm{tr}\, e^{A+tB} \bigg|_{t=0} = \mathrm{tr}\left(e^A B \right).$$

Proof: See [174, p. 175], [454, p. 371], or [906, 1002, 1054].

Fact 11.14.4. Let $A, B \in \mathbb{F}^{n \times n}$. Then,

$$\frac{d}{dt} e^{A+tB} \bigg|_{t=0} = \left(\frac{e^{\mathrm{ad}_A} - I}{\mathrm{ad}_A} \right)(B) e^A$$

$$= e^A \left(\frac{I - e^{-\mathrm{ad}_A}}{\mathrm{ad}_A} \right)(B)$$

$$= \sum_{k=0}^{\infty} \frac{1}{(k+1)!} \mathrm{ad}_A^k(B) e^A.$$

Proof: The second and fourth expressions are given in [106, p. 49] and [767, p. 248], while the third expression appears in [1380]. See also [1400, pp. 107–110].

Remark: See Fact 2.18.6.

Fact 11.14.5. Let $A, B \in \mathbb{F}^{n \times n}$, and assume that $e^A = e^B$. Then, the following statements hold:

i) If $|\lambda| < \pi$ for all $\lambda \in \mathrm{spec}(A) \cup \mathrm{spec}(B)$, then $A = B$.

ii) If $\lambda - \mu \neq j2k\pi$ for all $\lambda \in \mathrm{spec}(A)$, $\mu \in \mathrm{spec}(B)$, and $k \in \mathbb{Z}$, then $[A, B] = 0$.

iii) If A is normal and $\sigma_{\max}(A) < \pi$, then $[A, B] = 0$.

iv) If A is normal and $\sigma_{\max}(A) = \pi$, then $[A^2, B] = 0$.

Proof: See [1203, 1239] and [1400, p. 111].

Remark: If $[A, B] = 0$, then $[A^2, B] = 0$.

Fact 11.14.6. Let $A, B \in \mathbb{F}^{n \times n}$, and assume that A and B are skew Hermitian. Then, $e^{tA}e^{tB}$ is unitary, and there exists a skew-Hermitian matrix $C(t)$ such that $e^{tA}e^{tB} = e^{C(t)}$.

Problem: Does (11.4.1) converge in this case? See [231, 471, 1150].

Fact 11.14.7. Let $A, B \in \mathbb{F}^{n \times n}$, and assume that A and B are Hermitian. Then,

$$\lim_{p \to 0} \left(e^{\frac{p}{2}A} e^{pB} e^{\frac{p}{2}A} \right)^{1/p} = e^{A+B}.$$

Proof: See [55].

Remark: This result is related to the Lie-Trotter formula given by Corollary 11.4.8. For extensions, see [10, 547].

Fact 11.14.8. Let $A, B \in \mathbb{F}^{n \times n}$, and assume that A and B are Hermitian. Then,

$$\lim_{p \to \infty} \left[\tfrac{1}{2}\left(e^{pA} + e^{pB} \right) \right]^{1/p} = e^{\frac{1}{2}(A+B)}.$$

Proof: See [197].

Fact 11.14.9. Let $A, B \in \mathbb{F}^{n \times n}$. Then,

$$\lim_{k \to \infty} \left[e^{\frac{1}{k}A} e^{\frac{1}{k}B} e^{-\frac{1}{k}A} e^{-\frac{1}{k}B} \right]^{k^2} = e^{[A,B]}.$$

Fact 11.14.10. Let $A \in \mathbb{F}^{n \times m}$, $X \in \mathbb{F}^{m \times l}$, and $B \in \mathbb{F}^{l \times n}$. Then,

$$\frac{\mathrm{d}}{\mathrm{d}X} \operatorname{tr} e^{AXB} = Be^{AXB}A.$$

Fact 11.14.11. Let $A, B \in \mathbb{F}^{n \times n}$. Then,

$$\frac{\mathrm{d}}{\mathrm{d}t} e^{tA} e^{tB} e^{-tA} e^{-tB} \bigg|_{t=0} = 0$$

and

$$\frac{\mathrm{d}}{\mathrm{d}t} e^{\sqrt{t}A} e^{\sqrt{t}B} e^{-\sqrt{t}A} e^{-\sqrt{t}B} \bigg|_{t=0} = AB - BA.$$

Fact 11.14.12. Let $A, B, C \in \mathbb{F}^{n \times n}$, assume there exists $\beta \in \mathbb{F}$ such that $[A, B] = \beta B + C$, and assume that $[A, C] = [B, C] = 0$. Then,

$$e^{A+B} = e^A e^{\phi(\beta)B} e^{\psi(\beta)C},$$

where

$$\phi(\beta) \triangleq \begin{cases} \frac{1}{\beta}(1 - e^{-\beta}), & \beta \neq 0, \\ 1, & \beta = 0, \end{cases}$$

and

$$\psi(\beta) \triangleq \begin{cases} \frac{1}{\beta^2}(1 - \beta - e^{-\beta}), & \beta \neq 0, \\ -\frac{1}{2}, & \beta = 0. \end{cases}$$

Proof: See [570, 1295].

Fact 11.14.13. Let $A, B \in \mathbb{F}^{n \times n}$, and assume there exist $\alpha, \beta \in \mathbb{F}$ such that $[A, B] = \alpha A + \beta B$. Then,

$$e^{t(A+B)} = e^{\phi(t)A} e^{\psi(t)B},$$

where

$$\phi(t) \triangleq \begin{cases} t, & \alpha = \beta = 0, \\ \alpha^{-1} \log(1 + \alpha t), & \alpha = \beta \neq 0, \ 1 + \alpha t > 0, \\ \int_0^t \frac{\alpha - \beta}{\alpha e^{(\alpha-\beta)\tau} - \beta} \, d\tau, & \alpha \neq \beta, \end{cases}$$

and

$$\psi(t) \triangleq \int_0^t e^{-\beta \phi(\tau)} \, d\tau.$$

Proof: See [1296].

Fact 11.14.14. Let $A, B \in \mathbb{F}^{n \times n}$, and assume there exists nonzero $\beta \in \mathbb{F}$ such that $[A, B] = \alpha B$. Then, for all $t > 0$,

$$e^{t(A+B)} = e^{tA} e^{[(1-e^{-\alpha t})/\alpha]B}.$$

Proof: Apply Fact 11.14.12 with $[tA, tB] = \alpha t(tB)$ and $\beta = \alpha t$.

Fact 11.14.15. Let $A, B \in \mathbb{F}^{n \times n}$, and assume that $[[A, B], A] = 0$ and $[[A, B], B] = 0$. Then, for all $t \in \mathbb{R}$,

$$e^{tA} e^{tB} = e^{tA + tB + (t^2/2)[A, B]}.$$

In particular,

$$e^A e^B = e^{A + B + \frac{1}{2}[A, B]} = e^{A+B} e^{\frac{1}{2}[A, B]} = e^{\frac{1}{2}[A, B]} e^{A+B}$$

and

$$e^B e^{2A} e^B = e^{2A + 2B}.$$

Proof: See [639, pp. 64–66] and [1465].

Fact 11.14.16. Let $A, B \in \mathbb{F}^{n \times n}$, and assume that $[A, B] = B^2$. Then,

$$e^{A+B} = e^A(I + B).$$

Fact 11.14.17. Let $A, B \in \mathbb{F}^{n \times n}$. Then, for all $t \in [0, \infty)$,

$$e^{t(A+B)} = e^{tA}e^{tB} + \sum_{k=2}^{\infty} C_k t^k,$$

where, for all $k \in \mathbb{N}$,

$$C_{k+1} \triangleq \tfrac{1}{k+1}\left([A+B]C_k + [B, D_k]\right), \quad C_0 \triangleq 0,$$

and

$$D_{k+1} \triangleq \tfrac{1}{k+1}\left(AD_k + D_k B\right), \quad D_0 \triangleq I.$$

Proof: See [1153].

Fact 11.14.18. Let $A, B \in \mathbb{F}^{n \times n}$. Then, for all $t \in [0, \infty)$,

$$e^{t(A+B)} = e^{tA}e^{tB}e^{tC_2}e^{tC_3}\cdots,$$

where

$$C_2 \triangleq -\tfrac{1}{2}[A, B], \quad C_3 \triangleq \tfrac{1}{3}[B, [A, B]] + \tfrac{1}{6}[A, [A, B]].$$

Remark: This result is the *Zassenhaus product formula*. See [701, p. 236] and [1206].

Remark: Higher order terms are given in [1206].

Remark: Conditions for convergence do not seem to be available.

Fact 11.14.19. Let $A \in \mathbb{R}^{2n \times 2n}$, and assume that A is symplectic and discrete-time Lyapunov stable. Then, $\operatorname{spec}(A) \subset \{s \in \mathbb{C}: |s| = 1\}$, $\operatorname{amult}_A(1)$ and $\operatorname{amult}_A(-1)$ are even, A is semisimple, and there exists a Hamiltonian matrix $B \in \mathbb{R}^{2n \times 2n}$ such that $A = e^B$.

Proof: Since A is symplectic and discrete-time Lyapunov stable, it follows that the spectrum of A is a subset of the unit circle and A is semisimple. Therefore, the only negative eigenvalue that A can have is -1. Since all nonreal eigenvalues appear in complex conjugate pairs and A has even order, and since, by Fact 3.20.10, $\det A = 1$, it follows that the eigenvalues -1 and 1 (if present) have even algebraic multiplicity. The fact that A has a Hamiltonian logarithm now follows from Theorem 2.6 of [414].

Remark: See *xiii*) of Proposition 11.6.5.

Fact 11.14.20. Let $A, B \in \mathbb{F}^{n \times n}$, assume that A is positive definite, and assume that B is positive semidefinite. Then,

$$A + B \leq A^{1/2}e^{A^{-1/2}BA^{-1/2}}A^{1/2}.$$

Hence,

$$\frac{\det(A+B)}{\det A} \leq e^{\operatorname{tr} A^{-1}B}.$$

Furthermore, for each inequality, equality holds if and only if $B = 0$.

Proof: For positive-semidefinite A it follows that $I + A \leq e^A$.

Fact 11.14.21. Let $A, B \in \mathbb{F}^{n \times n}$, and assume that A and B are Hermitian. Then,

$$I \circ (A + B) \le \log(e^A \circ e^B).$$

Proof: See [45, 1521].

Remark: See Fact 8.22.50.

Fact 11.14.22. Let $A, B \in \mathbb{F}^{n \times n}$, assume that A and B are Hermitian, assume that $A \le B$, let $\alpha, \beta \in \mathbb{R}$, assume that either $\alpha I \le A \le \beta I$ or $\alpha I \le B \le \beta I$, and let $t > 0$. Then,

$$e^{tA} \le S(t, e^{\beta - \alpha}) e^{tB},$$

where, for $t > 0$ and $h > 0$,

$$S(t, h) \triangleq \begin{cases} \dfrac{(h^t - 1) h^{t/(h^t - 1)}}{et \log h}, & h \ne 1, \\ 1, & h = 1. \end{cases}$$

Proof: See [531].

Remark: $S(t, h)$ is Specht's ratio. See Fact 1.12.22 and Fact 1.17.19.

Fact 11.14.23. Let $A, B \in \mathbb{F}^{n \times n}$, assume that A and B are Hermitian, let $\alpha, \beta \in \mathbb{R}$, assume that $\alpha I \le A \le \beta I$ and $\alpha I \le B \le \beta I$, and let $t > 0$. Then,

$$\frac{1}{S(1, e^{\beta - \alpha}) S^{1/t}(t, e^{\beta - \alpha})} \left[\alpha e^{tA} + (1 - \alpha) e^{tB} \right]^{1/t}$$

$$\le e^{\alpha A + (1 - \alpha) B}$$

$$\le S(1, e^{\beta - \alpha}) \left[\alpha e^{tA} + (1 - \alpha) e^{tB} \right]^{1/t},$$

where $S(t, h)$ is defined in Fact 11.14.22.

Proof: See [531].

Fact 11.14.24. Let $A, B \in \mathbb{F}^{n \times n}$, and assume that A and B are positive definite. Then,

$$\log \det A = \operatorname{tr} \log A$$

and

$$\log \det AB = \operatorname{tr}(\log A + \log B).$$

Fact 11.14.25. Let $A, B \in \mathbb{F}^{n \times n}$, and assume that A and B are positive definite. Then,

$$\operatorname{tr}(A - B) \le \operatorname{tr}[A(\log A - \log B)]$$

and

$$(\log \operatorname{tr} A - \log \operatorname{tr} B) \operatorname{tr} A \le \operatorname{tr}[A(\log A - \log B)].$$

Proof: See [163] and [201, p. 281].

Remark: The first inequality is *Klein's inequality*. See [205, p. 118].

Remark: The second inequality is equivalent to the thermodynamic inequality. See Fact 11.14.31.

Remark: $\text{tr}[A(\log A - \log B)]$ is the *relative entropy of Umegaki*.

Fact 11.14.26. Let $A, B \in \mathbb{F}^{n \times n}$, assume that A and B are positive definite, and define

$$\mu(A, B) \triangleq e^{\frac{1}{2}(\log A + \log B)}.$$

Then, the following statements hold:

  i) $\mu(A, A^{-1}) = I$.

  ii) $\mu(A, B) = \mu(B, A)$.

  iii) If $AB = BA$, then $\mu(A, B) = AB$.

Proof: See [77].

Remark: With multiplication defined by μ, the set of $n \times n$ positive-definite matrices is a commutative Lie group. See [77].

Fact 11.14.27. Let $A, B \in \mathbb{F}^{n \times n}$, assume that A and B are positive definite, and let $p > 0$. Then,

$$\tfrac{1}{p}\text{tr}[A\log(B^{p/2}A^pB^{p/2})] \leq \text{tr}[A(\log A + \log B)] \leq \tfrac{1}{p}\text{tr}[A\log(A^{p/2}B^pA^{p/2})].$$

Furthermore,

$$\lim_{p\downarrow 0} \tfrac{1}{p}\text{tr}[A\log(B^{p/2}A^pB^{p/2})] = \text{tr}[A(\log A + \log B)] = \lim_{p\downarrow 0} \tfrac{1}{p}\text{tr}[A\log(A^{p/2}B^pA^{p/2})].$$

Proof: See [55, 164, 547, 692].

Remark: This inequality has applications to quantum information theory.

Fact 11.14.28. Let $A, B \in \mathbb{F}^{n \times n}$, assume that A and B are Hermitian, let $q \geq p > 0$, let $h \triangleq \lambda_{\max}(e^A)/\lambda_{\min}(e^B)$, and define

$$S(1, h) \triangleq \frac{(h-1)h^{1/(h-1)}}{e \log h}.$$

Then, there exist unitary matrices $U, V \in \mathbb{F}^{n \times n}$ such that

$$\tfrac{1}{S(1,h)}Ue^{A+B}U^* \leq e^{\frac{1}{2}A}e^Be^{\frac{1}{2}A} \leq S(1,h)Ve^{A+B}V^*.$$

Furthermore,

$$\text{tr}\,e^{A+B} \leq \text{tr}\,e^Ae^B \leq S(1,h)\,\text{tr}\,e^{A+B},$$

$$\text{tr}\,(e^{pA}\#e^{pB})^{2/p} \leq \text{tr}\,e^{A+B} \leq \text{tr}\,(e^{\frac{p}{2}B}e^{pA}e^{\frac{p}{2}B})^{1/p} \leq \text{tr}\,(e^{\frac{q}{2}B}e^{qA}e^{\frac{q}{2}B})^{1/q},$$

$$\text{tr}\,e^{A+B} = \lim_{p\downarrow 0}\text{tr}\,(e^{\frac{p}{2}B}e^{pA}e^{\frac{p}{2}B})^{1/p},$$

$$e^{A+B} = \lim_{p\downarrow 0}(e^{pA}\#e^{pB})^{2/p}.$$

Moreover, $\text{tr}\,e^{A+B} = \text{tr}\,e^Ae^B$ if and only if $AB = BA$. Furthermore, for all $i \in \{1, \ldots, n\}$,

$$\tfrac{1}{S(1,h)}\lambda_i(e^{A+B}) \leq \lambda_i(e^Ae^B) \leq S(1,h)\lambda_i(e^{A+B}).$$

Finally, let $\alpha \in [0, 1]$. Then,

$$\lim_{p\downarrow 0} (e^{pA} \#_\alpha e^{pB})^{1/p} = e^{(1-\alpha)A + \alpha B}$$

and

$$\mathrm{tr}\, (e^{pA} \#_\alpha e^{pB})^{1/p} \le \mathrm{tr}\, e^{(1-\alpha)A + \alpha B}.$$

Proof: See [256].

Remark: The left-hand inequality in the second string of inequalities is the *Golden-Thompson inequality*. See Fact 11.16.4.

Remark: Since $S(1, h) > 1$ for all $h > 1$, the left-hand inequality in the first string of inequalities does not imply the Golden-Thompson inequality.

Remark: For $i = 1$, the stronger eigenvalue inequality $\lambda_{\max}(e^{A+B}) \le \lambda_{\max}(e^A e^B)$ holds. See Fact 11.16.4.

Remark: $S(1, h)$ is Specht's ratio given by Fact 11.14.22.

Remark: The generalized geometric mean is defined in Fact 8.10.45.

Fact 11.14.29. Let $A, B \in \mathbb{F}^{n\times n}$, and assume that A and B are Hermitian. Then,

$$(\mathrm{tr}\, e^A) e^{\mathrm{tr}(e^A B)/\mathrm{tr}\, e^A} \le \mathrm{tr}\, e^{A+B}.$$

Proof: See [163].

Remark: This result is the *Peierls-Bogoliubov inequality*.

Remark: This inequality is equivalent to the thermodynamic inequality. See Fact 11.14.31.

Fact 11.14.30. Let $A, B, C \in \mathbb{F}^{n\times n}$, and assume that $A, B,$ and C are positive definite. Then,

$$\mathrm{tr}\, e^{\log A - \log B + \log C} \le \mathrm{tr} \int_0^\infty A(B + xI)^{-1} C(B + xI)^{-1}\, \mathrm{d}x.$$

Proof: See [931, 958].

Remark: $-\log B$ is correct.

Remark: $\mathrm{tr}\, e^{A+B+C} \le |\mathrm{tr}\, e^A e^B e^C|$ is not necessarily true.

Fact 11.14.31. Let $A, B \in \mathbb{F}^{n\times n}$, and assume that A is positive definite, $\mathrm{tr}\, A = 1$, and B is Hermitian. Then,

$$\mathrm{tr}\, AB \le \mathrm{tr}(A \log A) + \log \mathrm{tr}\, e^B.$$

Furthermore, equality holds if and only if

$$A = (\mathrm{tr}\, e^B)^{-1} e^B.$$

Proof: See [163].

Remark: This result is the *thermodynamic inequality*. Equivalent forms are given by Fact 11.14.25 and Fact 11.14.29.

Fact 11.14.32. Let $A, B \in \mathbb{F}^{n \times n}$, and assume that A and B are Hermitian. Then,

$$\|A - B\|_{\mathrm{F}} \leq \|\log(e^{-\frac{1}{2}A} e^B e^{\frac{1}{2}A})\|_{\mathrm{F}}.$$

Proof: See [205, p. 203].

Remark: This result has a distance interpretation in terms of geodesics. See [205, p. 203] and [211, 1038, 1039].

Fact 11.14.33. Let $A, B \in \mathbb{F}^{n \times n}$, and assume that A and B are skew Hermitian. Then, there exist unitary matrices $S_1, S_2 \in \mathbb{F}^{n \times n}$ such that

$$e^A e^B = e^{S_1 A S_1^{-1} + S_2 B S_2^{-1}}.$$

Proof: See [1241, 1303, 1304].

Fact 11.14.34. Let $A, B \in \mathbb{F}^{n \times n}$, and assume that A and B are Hermitian. Then, there exist unitary matrices $S_1, S_2 \in \mathbb{F}^{n \times n}$ such that

$$e^{\frac{1}{2}A} e^B e^{\frac{1}{2}A} = e^{S_1 A S_1^{-1} + S_2 B S_2^{-1}}.$$

Proof: See [1240, 1241, 1303, 1304].

Problem: Determine the relationship between this result and Fact 11.14.33.

Fact 11.14.35. Let $A, B \in \mathbb{F}^{n \times n}$, assume that A and B are positive semidefinite, and assume that $B \leq A$. Furthermore, let $p, q, r, t \in \mathbb{R}$, and assume that $r \geq t \geq 0$, $p \geq 0$, $p + q \geq 0$, and $p + q + r > 0$. Then,

$$\left[e^{\frac{r}{2}A} e^{qA + pB} e^{\frac{r}{2}A}\right]^{t/(p+q+r)} \leq e^{tA}.$$

Proof: See [1383].

Fact 11.14.36. Let $A \in \mathbb{F}^{n \times n}$ and $B \in \mathbb{F}^{m \times m}$. Then,

$$\operatorname{tr} e^{A \oplus B} = \left(\operatorname{tr} e^A\right)\left(\operatorname{tr} e^B\right).$$

Fact 11.14.37. Let $A \in \mathbb{F}^{n \times n}$, $B \in \mathbb{F}^{m \times m}$, and $C \in \mathbb{F}^{l \times l}$. Then,

$$e^{A \oplus B \oplus C} = e^A \otimes e^B \otimes e^C.$$

Fact 11.14.38. Let $A \in \mathbb{F}^{n \times n}$, $B \in \mathbb{F}^{m \times m}$, $C \in \mathbb{F}^{k \times k}$, and $D \in \mathbb{F}^{l \times l}$. Then,

$$\operatorname{tr} e^{A \otimes I \otimes B \otimes I + I \otimes C \otimes I \otimes D} = \operatorname{tr} e^{A \otimes B} \operatorname{tr} e^{C \otimes D}.$$

Proof: By Fact 7.4.30, a similarity transformation involving the Kronecker permutation matrix can be used to reorder the inner two terms. See [1251].

Fact 11.14.39. Let $A, B \in \mathbb{R}^{n \times n}$, and assume that A and B are positive definite. Then, $A \# B$ is the unique positive-definite solution X of the matrix equation

$$\log(A^{-1}X) + \log(B^{-1}X) = 0.$$

Proof: See [1039].

11.15 Facts on the Matrix Exponential and Eigenvalues, Singular Values, and Norms for One Matrix

Fact 11.15.1. Let $A \in \mathbb{F}^{n \times n}$, assume that e^A is positive definite, and assume that $\sigma_{\max}(A) < 2\pi$. Then, A is Hermitian.

Proof: See [876, 1202].

Fact 11.15.2. Let $A \in \mathbb{F}^{n \times n}$, and define $f \colon [0, \infty) \mapsto (0, \infty)$ by $f(t) \triangleq \sigma_{\max}(e^{At})$. Then,

$$f'(0) = \tfrac{1}{2}\lambda_{\max}(A + A^*).$$

Hence, there exists $\varepsilon > 0$ such that $f(t) \triangleq \sigma_{\max}(e^{tA})$ is decreasing on $[0, \varepsilon)$ if and only if A is dissipative.

Proof: This result follows from *iii*) of Fact 11.15.7. See [1436].

Remark: The derivative is one sided.

Fact 11.15.3. Let $A \in \mathbb{F}^{n \times n}$. Then, for all $t \geq 0$,

$$\frac{\mathrm{d}}{\mathrm{d}t} \|e^{tA}\|_{\mathrm{F}}^2 = \operatorname{tr} e^{tA}(A + A^*)e^{tA^*}.$$

Hence, if A is dissipative, then $f(t) \triangleq \|e^{tA}\|_{\mathrm{F}}$ is decreasing on $[0, \infty)$.

Proof: See [1436].

Fact 11.15.4. Let $A \in \mathbb{F}^{n \times n}$. Then,

$$\left|\operatorname{tr} e^{2A}\right| \leq \operatorname{tr} e^A e^{A^*} \leq \operatorname{tr} e^{A+A^*} \leq \left[n\operatorname{tr} e^{2(A+A^*)}\right]^{1/2} \leq \tfrac{n}{2} + \tfrac{1}{2}\operatorname{tr} e^{2(A+A^*)}.$$

In addition, $\operatorname{tr} e^A e^{A^*} = \operatorname{tr} e^{A+A^*}$ if and only if A is normal.

Proof: See [188], [730, p. 515], and [1239].

Remark: $\operatorname{tr} e^A e^{A^*} \leq \operatorname{tr} e^{A+A^*}$ is *Bernstein's inequality*. See [49].

Remark: See Fact 3.7.12.

Fact 11.15.5. Let $A \in \mathbb{F}^{n \times n}$. Then, for all $k \in \{1, \ldots, n\}$,

$$\prod_{i=1}^{k} \sigma_i(e^A) \leq \prod_{i=1}^{k} \lambda_i\!\left[e^{\frac{1}{2}(A+A^*)}\right] = \prod_{i=1}^{k} e^{\lambda_i\left[\frac{1}{2}(A+A^*)\right]} \leq \prod_{i=1}^{k} e^{\sigma_i(A)}.$$

Furthermore, for all $k \in \{1, \ldots, n\}$,

$$\sum_{i=1}^{k} \sigma_i(e^A) \leq \sum_{i=1}^{k} \lambda_i\!\left[e^{\frac{1}{2}(A+A^*)}\right] = \sum_{i=1}^{k} e^{\lambda_i\left[\frac{1}{2}(A+A^*)\right]} \leq \sum_{i=1}^{k} e^{\sigma_i(A)}.$$

In particular,

$$\sigma_{\max}(e^A) \leq \lambda_{\max}\!\left[e^{\frac{1}{2}(A+A^*)}\right] = e^{\frac{1}{2}\lambda_{\max}(A+A^*)} \leq e^{\sigma_{\max}(A)}$$

or, equivalently,

$$\lambda_{\max}(e^A e^{A^*}) \leq \lambda_{\max}(e^{A+A^*}) = e^{\lambda_{\max}(A+A^*)} \leq e^{2\sigma_{\max}(A)}.$$

Furthermore,

$$\left|\det e^A\right| = \left|e^{\operatorname{tr} A}\right| \leq e^{|\operatorname{tr} A|} \leq e^{\operatorname{tr}\langle A\rangle}$$

and

$$\operatorname{tr}\langle e^A\rangle \leq \sum_{i=1}^{n} e^{\sigma_i(A)}.$$

Proof: See [1242], Fact 2.21.12, Fact 8.18.4, and Fact 8.18.5.

Fact 11.15.6. Let $A \in \mathbb{F}^{n\times n}$, and let $\|\cdot\|$ be a unitarily invariant norm on $\mathbb{F}^{n\times n}$. Then,

$$\|e^A e^{A^*}\| \leq \|e^{A+A^*}\|.$$

In particular,

$$\lambda_{\max}(e^A e^{A^*}) \leq \lambda_{\max}(e^{A+A^*})$$

and

$$\operatorname{tr} e^A e^{A^*} \leq \operatorname{tr} e^{A+A^*}.$$

Proof: See [350].

Fact 11.15.7. Let $A, B \in \mathbb{F}^{n\times n}$, let $\|\cdot\|$ be the norm on $\mathbb{F}^{n\times n}$ induced by the norm $\|\cdot\|'$ on $\mathbb{F}^n$, let $\operatorname{mspec}(A) = \{\lambda_1, \ldots, \lambda_n\}_{\text{ms}}$, and define

$$\mu(A) \triangleq \lim_{\varepsilon\downarrow 0} \frac{\|I + \varepsilon A\| - 1}{\varepsilon}.$$

Then, the following statements hold:

 i) $\mu(A) = \mathrm{D}_+ f(A; I)$, where $f\colon \mathbb{F}^{n\times n} \mapsto \mathbb{R}$ is defined by $f(A) \triangleq \|A\|$.

 ii) $\mu(A) = \lim_{t\downarrow 0} t^{-1}\log\|e^{tA}\| = \sup_{t>0} t^{-1}\log\|e^{tA}\|$.

 iii) $\mu(A) = \left.\frac{\mathrm{d}^+}{\mathrm{d}t}\|e^{tA}\|\right|_{t=0} = \left.\frac{\mathrm{d}^+}{\mathrm{d}t}\log\|e^{tA}\|\right|_{t=0}$.

 iv) $\mu(I) = 1$, $\mu(-I) = -1$, and $\mu(0) = 0$.

 v) $\operatorname{spabs}(A) = \lim_{t\to\infty} t^{-1}\log\|e^{tA}\| = \inf_{t>0} t^{-1}\log\|e^{tA}\|$.

 vi) For all $i \in \{1, \ldots, n\}$,

$$-\|A\| \leq -\mu(-A) \leq \operatorname{Re}\lambda_i \leq \operatorname{spabs}(A) \leq \mu(A) \leq \|A\|.$$

 vii) For all $\alpha \in \mathbb{R}$, $\mu(\alpha A) = |\alpha|\mu[(\operatorname{sign}\alpha)A]$.

 viii) For all $\alpha \in \mathbb{F}$, $\mu(A + \alpha I) = \mu(A) + \operatorname{Re}\alpha$.

 ix) $\max\{\mu(A) - \mu(-B), -\mu(-A) + \mu(B)\} \leq \mu(A + B) \leq \mu(A) + \mu(B)$.

 x) $\mu\colon \mathbb{F}^{n\times n} \mapsto \mathbb{R}$ is convex.

 xi) $|\mu(A) - \mu(B)| \leq \max\{|\mu(A - B)|, |\mu(B - A)|\} \leq \|A - B\|$.

 xii) For all $x \in \mathbb{F}^n$, $\max\{-\mu(-A), -\mu(A)\}\|x\|' \leq \|Ax\|'$.

 xiii) If A is nonsingular, then $\max\{-\mu(-A), -\mu(A)\} \leq 1/\|A^{-1}\|$.

xiv) For all $t \geq 0$ and all $i = 1, \ldots, n$,

$$e^{-\|A\|t} \leq e^{-\mu(-A)t} \leq e^{(\operatorname{Re}\lambda_i)t} \leq e^{\operatorname{spabs}(A)t} \leq \|e^{tA}\| \leq e^{\mu(A)t} \leq e^{\|A\|t}.$$

xv) $\mu(A) = \min\{\beta \in \mathbb{R} : \|e^{tA}\| \leq e^{\beta t} \text{ for all } t \geq 0\}$.

xvi) If $\|\cdot\|' = \|\cdot\|_1$, and thus $\|\cdot\| = \|\cdot\|_{\operatorname{col}}$, then

$$\mu(A) = \max_{j \in \{1,\ldots,n\}} \left(\operatorname{Re} A_{(j,j)} + \sum_{\substack{i=1 \\ i \neq j}}^{n} |A_{(i,j)}| \right).$$

xvii) If $\|\cdot\|' = \|\cdot\|_2$ and thus $\|\cdot\| = \sigma_{\max}(\cdot)$, then

$$\mu(A) = \lambda_{\max}[\tfrac{1}{2}(A + A^*)].$$

xviii) If $\|\cdot\|' = \|\cdot\|_\infty$, and thus $\|\cdot\| = \|\cdot\|_{\operatorname{row}}$, then

$$\mu(A) = \max_{i \in \{1,\ldots,n\}} \left(\operatorname{Re} A_{(i,i)} + \sum_{\substack{j=1 \\ j \neq i}}^{n} |A_{(i,j)}| \right).$$

Proof: See [408, 412, 1094, 1276], [708, pp. 653–655], and [1348, p. 150].

Remark: $\mu(\cdot)$ is the *matrix measure, logarithmic derivative,* or *initial growth rate.* For applications, see [708] and [1414]. See Fact 11.18.11 for the logarithmic derivative of an asymptotically stable matrix.

Remark: The directional differential $\mathrm{D}_+ f(A; I)$ is defined in (10.4.2).

Remark: *vi)* and *xvii)* yield Fact 5.11.24.

Remark: Higher order logarithmic derivatives are studied in [209].

Fact 11.15.8. Let $A \in \mathbb{F}^{n \times n}$, let $\beta > \operatorname{spabs}(A)$, let $\gamma \geq 1$, and let $\|\cdot\|$ be a normalized, submultiplicative norm on $\mathbb{F}^{n \times n}$. Then, for all $t \geq 0$,

$$\|e^{tA}\| \leq \gamma e^{\beta t}$$

if and only if, for all $k \geq 1$ and $\alpha > \beta$,

$$\|(\alpha I - A)^{-k}\| \leq \frac{\gamma}{(\alpha - \beta)^k}.$$

Remark: This result is a consequence of the *Hille-Yosida theorem.* See [369, pp. 26] and [708, p. 672].

Fact 11.15.9. Let $A \in \mathbb{R}^{n \times n}$, let $\beta \in \mathbb{R}$, and assume there exists a positive-definite matrix $P \in \mathbb{R}^{n \times n}$ such that

$$A^{\mathrm{T}} P + PA \leq 2\beta P.$$

Then, for all $t \geq 0$,

$$\sigma_{\max}(e^{tA}) \leq \sqrt{\sigma_{\max}(P)/\sigma_{\min}(P)} \, e^{\beta t}.$$

Remark: See [708, p. 665].

Remark: See Fact 11.18.9.

Fact 11.15.10. Let $A \in \mathrm{SO}(3)$. Then,

$$\theta \triangleq 2\cos^{-1}(\tfrac{1}{2}\sqrt{1 + \mathrm{tr}\, A}).$$

Then,

$$\theta = \sigma_{\max}(\log A) = \frac{1}{\sqrt{2}}\|\log A\|_{\mathrm{F}}.$$

Remark: See Fact 3.11.31 and Fact 11.11.10.

Remark: θ is a Riemannian metric giving the length of the shortest geodesic curve on $\mathrm{SO}(3)$ between A and I. See [1038].

11.16 Facts on the Matrix Exponential and Eigenvalues, Singular Values, and Norms for Two or More Matrices

Fact 11.16.1. Let $A, B \in \mathbb{F}^{n \times n}$. Then,

$$\left|\mathrm{tr}\, e^{A+B}\right| \le \mathrm{tr}\, e^{\frac{1}{2}(A+B)}e^{\frac{1}{2}(A+B)^*}$$

$$\le \mathrm{tr}\, e^{\frac{1}{2}(A+A^*+B+B^*)}$$

$$\le \mathrm{tr}\, e^{\frac{1}{2}(A+A^*)}e^{\frac{1}{2}(B+B^*)}$$

$$\le \left(\mathrm{tr}\, e^{A+A^*}\right)^{1/2}\left(\mathrm{tr}\, e^{B+B^*}\right)^{1/2}$$

$$\le \tfrac{1}{2}\mathrm{tr}\left(e^{A+A^*} + e^{B+B^*}\right)$$

and

$$\left.\begin{array}{r}\mathrm{tr}\, e^A e^B \\[4pt] \tfrac{1}{2}\mathrm{tr}\left(e^{2A} + e^{2B}\right)\end{array}\right\} \le \tfrac{1}{2}\mathrm{tr}\left(e^A e^{A^*} + e^B e^{B^*}\right) \le \tfrac{1}{2}\mathrm{tr}\left(e^{A+A^*} + e^{B+B^*}\right).$$

Proof: See [188, 351, 1102] and [730, p. 514].

Fact 11.16.2. Let $A, B \in \mathbb{F}^{n \times n}$. Then, for all $p > 0$,

$$\sigma_{\max}\left[e^{A+B} - \left(e^{\frac{1}{p}A}e^{\frac{1}{p}B}\right)^p\right] \le \tfrac{1}{2p}\sigma_{\max}([A, B])e^{\sigma_{\max}(A) + \sigma_{\max}(B)}.$$

Proof: See [701, p. 237] and [1040].

Remark: See Corollary 10.8.8 and Fact 11.16.3.

Fact 11.16.3. Let $A \in \mathbb{F}^{n \times n}$, and define $A_{\mathrm{H}} \triangleq \tfrac{1}{2}(A+A^*)$ and $A_{\mathrm{S}} \triangleq \tfrac{1}{2}(A-A^*)$. Then, for all $p > 0$,

$$\sigma_{\max}\left[e^A - \left(e^{\frac{1}{p}A_{\mathrm{H}}}e^{\frac{1}{p}A_{\mathrm{S}}}\right)^p\right] \le \tfrac{1}{4p}\sigma_{\max}([A^*, A])e^{\frac{1}{2}\lambda_{\max}(A+A^*)}.$$

Proof: See [1040].

Remark: See Fact 10.8.8.

Fact 11.16.4. Let $A, B \in \mathbb{F}^{n \times n}$, assume that A and B are Hermitian, and let $\| \cdot \|$ be a unitarily invariant norm on $\mathbb{F}^{n \times n}$. Then,

$$\left\| e^{A+B} \right\| \leq \left\| e^{\frac{1}{2}A} e^B e^{\frac{1}{2}A} \right\| \leq \left\| e^A e^B \right\|.$$

If, in addition, $p > 0$, then

$$\left\| e^{A+B} \right\| \leq \left\| e^{\frac{p}{2}A} e^B e^{\frac{p}{2}A} \right\|^{1/p}$$

and

$$\left\| e^{A+B} \right\| = \lim_{p \downarrow 0} \left\| e^{\frac{p}{2}A} e^B e^{\frac{p}{2}A} \right\|^{1/p}.$$

Furthermore, for all $k \in \{1, \ldots, n\}$,

$$\prod_{i=1}^{k} \lambda_i \left(e^{A+B} \right) \leq \prod_{i=1}^{k} \lambda_i \left(e^A e^B \right) \leq \prod_{i=1}^{k} \sigma_i \left(e^A e^B \right)$$

with equality for $k = n$, that is,

$$\prod_{i=1}^{n} \lambda_i \left(e^{A+B} \right) = \prod_{i=1}^{n} \lambda_i \left(e^A e^B \right) = \prod_{i=1}^{n} \sigma_i \left(e^A e^B \right) = \det \left(e^A e^B \right).$$

In fact,

$$\begin{aligned}
\det(e^{A+B}) &= \prod_{i=1}^{n} \lambda_i \left(e^{A+B} \right) \\
&= \prod_{i=1}^{n} e^{\lambda_i (A+B)} \\
&= e^{\operatorname{tr}(A+B)} \\
&= e^{(\operatorname{tr} A) + (\operatorname{tr} B)} \\
&= e^{\operatorname{tr} A} e^{\operatorname{tr} B} \\
&= \det(e^A) \det(e^B) \\
&= \det(e^A e^B) \\
&= \prod_{i=1}^{n} \sigma_i \left(e^A e^B \right).
\end{aligned}$$

Furthermore, for all $k \in \{1, \ldots, n\}$,

$$\sum_{i=1}^{k} \lambda_i \left(e^{A+B} \right) \leq \sum_{i=1}^{k} \lambda_i \left(e^A e^B \right) \leq \sum_{i=1}^{k} \sigma_i \left(e^A e^B \right).$$

In particular,

$$\lambda_{\max} \left(e^{A+B} \right) \leq \lambda_{\max} \left(e^A e^B \right) \leq \sigma_{\max} \left(e^A e^B \right),$$

$$\operatorname{tr} e^{A+B} \leq \operatorname{tr} e^A e^B \leq \operatorname{tr} \left\langle e^A e^B \right\rangle,$$

and, for all $p > 0$,

$$\operatorname{tr} e^{A+B} \leq \operatorname{tr}(e^{\frac{p}{2}A} e^B e^{\frac{p}{2}A}).$$

Finally, $\operatorname{tr} e^{A+B} = \operatorname{tr} e^A e^B$ if and only if A and B commute.

Proof: See [55], [201, p. 261], Fact 5.11.28, Fact 2.21.12, and Fact 9.11.2. For the last statement, see [1239].

Remark: Note that $\det(e^{A+B}) = \det(e^A)\det(e^B)$ even though e^{A+B} and $e^A e^B$ may not be equal. See [701, p. 265] or [730, p. 442].

Remark: $\operatorname{tr} e^{A+B} \leq \operatorname{tr} e^A e^B$ is the Golden-Thompson inequality. See Fact 11.14.28.

Remark: $\|e^{A+B}\| \leq \|e^{\frac{1}{2}A}e^B e^{\frac{1}{2}A}\|$ is *Segal's inequality*. See [49].

Problem: Compare the upper bound $\operatorname{tr}\langle e^A e^B \rangle$ for $\operatorname{tr} e^A e^B$ with the upper bound $S(1,h)\operatorname{tr} e^{A+B}$ given by Fact 11.14.28.

Fact 11.16.5. Let $A, B \in \mathbb{F}^{n\times n}$, assume that A and B are Hermitian, let $q, p > 0$, where $q \leq p$, and let $\|\cdot\|$ be a unitarily invariant norm on $\mathbb{F}^{n\times n}$. Then,

$$\left\|\left(e^{\frac{q}{2}A}e^{qB}e^{\frac{q}{2}A}\right)^{1/q}\right\| \leq \left\|\left(e^{\frac{p}{2}A}e^{pB}e^{\frac{p}{2}A}\right)^{1/p}\right\|.$$

Proof: See [55].

Fact 11.16.6. Let $A, B \in \mathbb{F}^{n\times n}$, and assume that A and B are positive semidefinite. Then,

$$e^{\sigma_{\max}^{1/2}(AB)} - 1 \leq \sigma_{\max}^{1/2}\left[(e^A - I)(e^B - I)\right]$$

and

$$e^{\sigma_{\max}^{1/3}(BAB)} - 1 \leq \sigma_{\max}^{1/3}\left[(e^B - I)(e^A - I)(e^B - I)\right].$$

Proof: See [1382].

Remark: See Fact 8.19.31.

Fact 11.16.7. Let $A, B \in \mathbb{F}^{n\times n}$, and let $t \geq 0$. Then,

$$e^{t(A+B)} = e^{tA} + \int_0^t e^{(t-\tau)A}Be^{\tau(A+B)}\, \mathrm{d}\tau.$$

Proof: See [701, p. 238].

Fact 11.16.8. Let $A, B \in \mathbb{F}^{n\times n}$, and let $\|\cdot\|$ be a submultiplicative norm on $\mathbb{F}^{n\times n}$. Then, for all $t \geq 0$,

$$\left\|e^{tA} - e^{tB}\right\| \leq e^{\|A\|t}\left(e^{\|A-B\|t} - 1\right).$$

Fact 11.16.9. Let $A, B \in \mathbb{F}^{n\times n}$, let $\|\cdot\|$ be a normalized submultiplicative norm on $\mathbb{F}^{n\times n}$, and let $t \geq 0$. Then,

$$\left\|e^{tA} - e^{tB}\right\| \leq t\|A - B\|e^{t\max\{\|A\|,\|B\|\}}.$$

Proof: See [701, p. 265].

Fact 11.16.10. Let $A, B \in \mathbb{R}^{n \times n}$, and assume that A is normal. Then, for all $t \geq 0$,

$$\sigma_{\max}(e^{tA} - e^{tB}) \leq \sigma_{\max}(e^{tA})\left[e^{\sigma_{\max}(A-B)t} - 1\right].$$

Proof: See [1454].

Fact 11.16.11. Let $A \in \mathbb{F}^{n \times n}$, let $\|\cdot\|$ be an induced norm on $\mathbb{F}^{n \times n}$, and let $\alpha > 0$ and $\beta \in \mathbb{R}$ be such that, for all $t \geq 0$,

$$\|e^{tA}\| \leq \alpha e^{\beta t}.$$

Then, for all $B \in \mathbb{F}^{n \times n}$ and $t \geq 0$,

$$\|e^{t(A+B)}\| \leq \alpha e^{(\beta + \alpha\|B\|)t}.$$

Proof: See [708, p. 406].

Fact 11.16.12. Let $A, B \in \mathbb{C}^{n \times n}$, assume that A and B are idempotent, assume that $A \neq B$, and let $\|\cdot\|$ be a norm on $\mathbb{C}^{n \times n}$. Then,

$$\|e^{\jmath A} - e^{\jmath B}\| = |e^{\jmath} - 1|\|A - B\| < \|A - B\|.$$

Proof: See [1055].

Remark: $|e^{\jmath} - 1| \approx 0.96$.

Fact 11.16.13. Let $A, B \in \mathbb{C}^{n \times n}$, assume that A and B are Hermitian, let $X \in \mathbb{C}^{n \times n}$, and let $\|\cdot\|$ be a unitarily invariant norm on $\mathbb{C}^{n \times n}$. Then,

$$\|e^{\jmath A}X - Xe^{\jmath B}\| \leq \|AX - XB\|.$$

Proof: See [1055].

Remark: This result is a matrix version of x) of Fact 1.20.6.

Fact 11.16.14. Let $A \in \mathbb{F}^{n \times n}$, and, for all $i \in \{1, \ldots, n\}$, define $f_i \colon [0, \infty) \mapsto \mathbb{R}$ by $f_i(t) \triangleq \log \sigma_i(e^{tA})$. Then, A is normal if and only if, for all $i \in \{1, \ldots, n\}$, f_i is convex.

Proof: See [96] and [465].

Remark: The statement in [96] that convexity holds on $\mathbb{R}$ is erroneous. A counterexample is $A \triangleq \begin{bmatrix} 1 & 0 \\ 0 & -1 \end{bmatrix}$ for which $\log \sigma_1(e^{tA}) = |t|$ and $\log \sigma_2(e^{tA}) = -|t|$.

Fact 11.16.15. Let $A \in \mathbb{F}^{n \times n}$, and, for nonzero $x \in \mathbb{F}^n$, define $f_x \colon \mathbb{R} \mapsto \mathbb{R}$ by $f_x(t) \triangleq \log \sigma_{\max}(e^{tA}x)$. Then, A is normal if and only if, for all nonzero $x \in \mathbb{F}^n$, f_x is convex.

Proof: See [96].

Remark: This result is due to Friedland.

Fact 11.16.16. Let $A, B \in \mathbb{F}^{n \times n}$, assume that A and B are positive semidefinite, and let $\|\cdot\|$ be a unitarily invariant norm on $\mathbb{F}^{n \times n}$. Then,

$$\|e^{\langle A - B \rangle} - I\| \leq \|e^A - e^B\|$$

and

$$\|e^A + e^B\| \le \|e^{A+B} + I\|.$$

Proof: See [60] and [201, p. 294].

Remark: See Fact 9.9.54.

Fact 11.16.17. Let $A, X, B \in \mathbb{F}^{n \times n}$, assume that A and B are Hermitian, and let $\| \cdot \|$ be a unitarily invariant norm on $\mathbb{F}^{n \times n}$. Then,

$$\|AX - XB\| \le \|e^{\frac{1}{2}A}Xe^{-\frac{1}{2}B} - e^{-\frac{1}{2}B}Xe^{\frac{1}{2}A}\|.$$

Proof: See [220].

Remark: See Fact 9.9.55.

11.17 Facts on Stable Polynomials

Fact 11.17.1. Let $a_1, \ldots, a_n$ be nonzero real numbers, let

$$\Delta \triangleq \{i \in \{1, \ldots, n-1\} \colon \tfrac{a_{i+1}}{a_i} < 0\},$$

let $b_1, \ldots, b_n$ be real numbers satisfying $b_1 < \cdots < b_n$, define $f \colon (0, \infty) \mapsto \mathbb{R}$ by

$$f(x) = a_n x^{b_n} + \cdots + a_1 x^{b_1},$$

and define

$$\mathcal{S} \triangleq \{x \in (0, \infty) \colon f(x) = 0\}.$$

Furthermore, for all $x \in \mathcal{S}$, define the multiplicity of x to be the positive integer m such that $f(x) = f'(x) = \cdots = f^{(m-1)} = 0$ and $f^{(m)}(x) \ne 0$, and let $\mathcal{S}'$ denote the multiset consisting of all elements of $\mathcal{S}$ counting multiplicity. Then,

$$\mathrm{card}(\mathcal{S}') \le \mathrm{card}(\Delta).$$

If, in addition, $b_1, \ldots, b_n$ are nonnegative integers, then $\mathrm{card}(\Delta) - \mathrm{card}(\mathcal{S}')$ is even.

Proof: See [863, 1434].

Remark: This result is the *Descartes rule of signs*.

Fact 11.17.2. Let $p \in \mathbb{R}[s]$, where $p(s) = s^n + a_{n-1}s^{n-1} + \cdots + a_0$. If p is asymptotically stable, then $a_0, \ldots, a_{n-1}$ are positive. Now, assume in addition that $a_0, \ldots, a_{n-1}$ are positive. Then, the following statements hold:

i) If $n = 1$ or $n = 2$, then p is asymptotically stable.

ii) If $n = 3$, then p is asymptotically stable if and only if

$$a_0 < a_1 a_2.$$

iii) If $n = 4$, then p is asymptotically stable if and only if

$$a_1^2 + a_0 a_3^2 < a_1 a_2 a_3.$$

iv) If $n = 5$, then p is asymptotically stable if and only if

$$a_2 < a_3 a_4,$$
$$a_2^2 + a_1 a_4^2 < a_0 a_4 + a_2 a_3 a_4,$$
$$a_0^2 + a_1 a_2^2 + a_1^2 a_4^2 + a_0 a_3^2 a_4 < a_0 a_2 a_3 + 2 a_0 a_1 a_4 + a_1 a_2 a_3 a_4.$$

Remark: These results are special cases of the *Routh criterion*, which provides stability criteria for polynomials of arbitrary degree n. See [309].

Remark: The Jury criterion for stability of continuous-time systems is given by Fact 11.20.1.

Fact 11.17.3. Let $\varepsilon \in [0,1]$, let $n \in \{2,3,4\}$, let $p_\varepsilon \in \mathbb{R}[s]$, where $p_\varepsilon(s) = s^n + a_{n-1} s^{n-1} + \cdots + a_1 s + \varepsilon a_0$, and assume that p_1 is asymptotically stable. Then, for all $\varepsilon \in (0,1]$, p_ε is asymptotically stable. Furthermore, $p_0(s)/s$ is asymptotically stable.

Remark: This result does not hold for $n = 5$. A counterexample is $p(s) = s^5 + 2s^4 + 3s^3 + 5s^2 + 2s + 2.5\varepsilon$, which is asymptotically stable if and only if $\varepsilon \in (4/5, 1]$. This result is another instance of the quartic barrier. See [359], Fact 8.14.7, and Fact 8.15.33.

Fact 11.17.4. Let $p \in \mathbb{R}[s]$ be monic, and define $q(s) \triangleq s^n p(1/s)$, where $n \triangleq \deg p$. Then, p is asymptotically stable if and only if q is asymptotically stable.

Remark: See Fact 4.8.1 and Fact 11.17.5.

Fact 11.17.5. Let $p \in \mathbb{R}[s]$ be monic, and assume that p is semistable. Then, $q(s) \triangleq p(s)/s$ and $\hat{q}(s) \triangleq s^n p(1/s)$ are asymptotically stable.

Remark: See Fact 4.8.1 and Fact 11.17.4.

Fact 11.17.6. Let $p, q \in \mathbb{R}[s]$, assume that p is even, assume that q is odd, and assume that every coefficient of $p + q$ is positive. Then, $p + q$ is asymptotically stable if and only if every root of p and every root of q is imaginary, and the roots of p and the roots of q are interlaced on the imaginary axis.

Proof: See [225, 309, 723].

Remark: This result is the *Hermite-Biehler* or *interlacing theorem*.

Example: $s^2 + 2s + 5 = (s^2 + 5) + 2s$.

Fact 11.17.7. Let $p \in \mathbb{R}[s]$ be asymptotically stable, and let $p(s) = \beta_n s^n + \beta_{n-1} s^{n-1} + \cdots + \beta_1 s + \beta_0$, where $\beta_n > 0$. Then, for all $i \in \{1, \ldots, n-2\}$,

$$\beta_{i-1} \beta_{i+2} < \beta_i \beta_{i+1}.$$

Remark: This result is a necessary condition for asymptotic stability, which can be used to show that a given polynomial with positive coefficients is unstable.

Remark: This result is due to Xie. See [1509]. For alternative conditions, see [225, p. 68].

Fact 11.17.8. Let $n \in \mathbb{P}$ be even, let $m \triangleq n/2$, let $p \in \mathbb{R}[s]$, where $p(s) = \beta_n s^n + \beta_{n-1} s^{n-1} + \cdots + \beta_1 s + \beta_0$ and $\beta_n > 0$, and assume that p is asymptotically stable. Then, for all $i \in \{1, \ldots, m-1\}$,

$$\binom{m}{i} \beta_0^{(m-i)/m} \beta_n^{i/m} \leq \beta_{2i}.$$

Remark: This result is a necessary condition for asymptotic stability, which can be used to show that a given polynomial with positive coefficients is unstable.

Remark: This result is due to Borobia and Dormido. See [1509, 1510] for extensions to polynomials of odd degree.

Fact 11.17.9. Let $p, q \in \mathbb{R}[s]$, where $p(s) = \alpha_n s^n + \alpha_{n-1} s^{n-1} + \cdots + \alpha_1 s + \alpha_0$ and $q(s) = \beta_m s^m + \beta_{m-1} s^{m-1} + \cdots + \beta_1 s + \beta_0$. If p and q are (Lyapunov, asymptotically) stable, then $r(s) \triangleq \alpha_l \beta_l s^l + \alpha_{l-1} \beta_{l-1} s^{l-1} + \cdots + \alpha_1 \beta_1 s + \alpha_0 \beta_0$, where $l \triangleq \min\{m, n\}$, is (Lyapunov, asymptotically) stable.

Proof: See [557].

Remark: The polynomial r is the *Schur product* of p and q. See [85, 785].

Fact 11.17.10. Let $A \in \mathbb{R}^{n \times n}$, and assume that A is diagonalizable over $\mathbb{R}$. Then, χ_A has all positive coefficients if and only if A is asymptotically stable.

Proof: Sufficiency follows from Fact 11.17.2. For necessity, note that all of the roots of χ_A are real and that $\chi_A(\lambda) > 0$ for all $\lambda \geq 0$. Hence, roots(χ_A) $\subset (-\infty, 0)$.

Fact 11.17.11. Let $A \in \mathbb{R}^{n \times n}$. Then, the following statements are equivalent:

i) $\chi_{A \oplus A}$ has all positive coefficients.

ii) $\chi_{A \oplus A}$ is asymptotically stable.

iii) $A \oplus A$ is asymptotically stable.

iv) A is asymptotically stable.

Proof: If A is not asymptotically stable, then Fact 11.18.32 implies that $A \oplus A$ has a nonnegative eigenvalue λ. Since $\chi_{A \oplus A}(\lambda) = 0$, it follows that $\chi_{A \oplus A}$ cannot have all positive coefficients. See [532, Theorem 5].

Remark: A similar method of proof is used in Proposition 8.2.7.

Fact 11.17.12. Let $A \in \mathbb{R}^{n \times n}$. Then, the following statements are equivalent:

i) χ_A and $\chi_{A^{(2,1)}}$ have all positive coefficients.

ii) A is asymptotically stable.

Proof: See [1274].

Remark: The additive compound $A^{(2,1)}$ is defined in Fact 7.5.17.

Fact 11.17.13. For $i = 1, \ldots, n-1$, let $a_i, b_i \in \mathbb{R}$ satisfy $0 < a_i \leq b_i$, define $\phi_1, \phi_2, \psi_1, \psi_2 \in \mathbb{R}[s]$ by

$$\phi_1(s) = b_n s^n + a_{n-2} s^{n-2} + b_{n-4} s^{n-4} + \cdots,$$
$$\phi_2(s) = a_n s^n + b_{n-2} s^{n-2} + a_{n-4} s^{n-4} + \cdots,$$
$$\psi_1(s) = b_{n-1} s^{n-1} + a_{n-3} s^{n-3} + b_{n-5} s^{n-5} + \cdots,$$
$$\psi_2(s) = a_{n-1} s^{n-1} + b_{n-3} s^{n-3} + a_{n-5} s^{n-5} + \cdots,$$

assume that $\phi_1 + \psi_1$, $\phi_1 + \psi_2$, $\phi_2 + \psi_1$, and $\phi_2 + \psi_2$ are asymptotically stable, let $p \in \mathbb{R}[s]$, where $p(s) = \beta_n s^n + \beta_{n-1} s^{n-1} + \cdots + \beta_1 s + \beta_0$, and assume that, for all $i \in \{1, \ldots, n\}$, $a_i \leq \beta_i \leq b_i$. Then, p is asymptotically stable.

Proof: See [459, pp. 466, 467].

Remark: This result is *Kharitonov's theorem*.

11.18 Facts on Stable Matrices

Fact 11.18.1. Let $A \in \mathbb{F}^{n \times n}$, and assume that A is semistable. Then, A is Lyapunov stable.

Fact 11.18.2. Let $A \in \mathbb{F}^{n \times n}$, and assume that A is Lyapunov stable. Then, A is group invertible.

Fact 11.18.3. Let $A \in \mathbb{F}^{n \times n}$, and assume that A is semistable. Then, A is group invertible.

Fact 11.18.4. Let $A, B \in \mathbb{F}^{n \times n}$, and assume that A and B are similar. Then, A is (Lyapunov stable, semistable, asymptotically stable, discrete-time Lyapunov stable, discrete-time semistable, discrete-time asymptotically stable) if and only if B is.

Fact 11.18.5. Let $A \in \mathbb{R}^{n \times n}$, and assume that A is Lyapunov stable. Then,

$$\lim_{t \to \infty} \frac{1}{t} \int_0^t e^{\tau A} \, \mathrm{d}\tau = I - AA^{\#}.$$

Remark: See Fact 11.18.2.

Fact 11.18.6. Let $A \in \mathbb{F}^{n \times n}$, and assume that A is semistable. Then,

$$\lim_{t \to \infty} e^{tA} = I - AA^{\#},$$

and thus

$$\lim_{t \to \infty} \frac{1}{t} \int_0^t e^{\tau A} \, \mathrm{d}\tau = I - AA^{\#}.$$

Remark: See Fact 10.11.6, Fact 11.18.1, and Fact 11.18.2.

Fact 11.18.7. Let $A, B \in \mathbb{F}^{n \times n}$. Then, $\lim_{\alpha \to \infty} e^{A + \alpha B}$ exists if and only if B is semistable. In this case,

$$\lim_{\alpha \to \infty} e^{A + \alpha B} = e^{(I - BB^\#)A}(I - BB^\#) = (I - BB^\#)e^{A(I - BB^\#)}.$$

Proof: See [292].

Fact 11.18.8. Let $A \in \mathbb{F}^{n \times n}$, assume that A is asymptotically stable, let $\beta > \mathrm{spabs}(A)$, and let $\| \cdot \|$ be a submultiplicative norm on $\mathbb{F}^{n \times n}$. Then, there exists $\gamma > 0$ such that, for all $t \geq 0$,

$$\|e^{tA}\| \leq \gamma e^{\beta t}.$$

Remark: See [572, pp. 201–206] and [808].

Fact 11.18.9. Let $A \in \mathbb{R}^{n \times n}$, assume that A is asymptotically stable, let $\beta \in (\mathrm{spabs}(A), 0)$, let $P \in \mathbb{R}^{n \times n}$ be positive definite and satisfy

$$A^{\mathrm{T}}P + PA \leq 2\beta P,$$

and let $\| \cdot \|$ be a normalized, submultiplicative norm on $\mathbb{R}^{n \times n}$. Then, for all $t \geq 0$,

$$\|e^{tA}\| \leq \sqrt{\|P\|\|P^{-1}\|}e^{\beta t}.$$

Remark: See [707].

Remark: See Fact 11.15.9.

Fact 11.18.10. Let $A \in \mathbb{F}^{n \times n}$, assume that A is asymptotically stable, let $R \in \mathbb{F}^{n \times n}$, assume that R is positive definite, and let $P \in \mathbb{F}^{n \times n}$ be the positive-definite solution of $A^*P + PA + R = 0$. Then,

$$\sigma_{\max}(e^{tA}) \leq \sqrt{\frac{\sigma_{\max}(P)}{\sigma_{\min}(P)}}e^{-t\lambda_{\min}(RP^{-1})/2}$$

and

$$\|e^{tA}\|_{\mathrm{F}} \leq \sqrt{\|P\|_{\mathrm{F}}\|P^{-1}\|_{\mathrm{F}}}e^{-t\lambda_{\min}(RP^{-1})/2}.$$

If, in addition, $A + A^*$ is negative definite, then

$$\|e^{tA}\|_{\mathrm{F}} \leq e^{-t\lambda_{\min}(-A-A^*)/2}.$$

Proof: See [977].

Fact 11.18.11. Let $A \in \mathbb{R}^{n \times n}$, assume that A is asymptotically stable, let $R \in \mathbb{R}^{n \times n}$, assume that R is positive definite, and let $P \in \mathbb{R}^{n \times n}$ be the positive-definite solution of $A^{\mathrm{T}}P + PA + R = 0$. Furthermore, define the vector norm $\|x\|' \triangleq \sqrt{x^{\mathrm{T}}Px}$ on $\mathbb{R}^n$, let $\| \cdot \|$ denote the induced norm on $\mathbb{R}^{n \times n}$, and let $\mu(\cdot)$ denote the corresponding logarithmic derivative. Then,

$$\mu(A) = -\lambda_{\min}(RP^{-1})/2.$$

Consequently,

$$\|e^{tA}\| \leq e^{-t\lambda_{\min}(RP^{-1})/2}.$$

Proof: See [747] and use *xiv)* of Fact 11.15.7.

Remark: See Fact 11.15.7 for the definition and properties of the logarithmic derivative.

Fact 11.18.12. Let $A \in \mathbb{F}^{n \times n}$. Then, A is similar to a skew-Hermitian matrix if and only if there exists a positive-definite matrix $P \in \mathbb{F}^{n \times n}$ such that $A^*P + PA = 0$.

Remark: See Fact 5.9.6.

Fact 11.18.13. Let $A \in \mathbb{R}^{n \times n}$. Then, A and A^2 are asymptotically stable if and only if, for all $\lambda \in \operatorname{spec}(A)$, there exist $r > 0$ and $\theta \in \left(\frac{\pi}{2}, \frac{3\pi}{4}\right) \cup \left(\frac{5\pi}{4}, \frac{3\pi}{2}\right)$ such that $\lambda = re^{\jmath\theta}$.

Fact 11.18.14. Let $A \in \mathbb{R}^{n \times n}$. Then, A is group invertible and $\jmath 2k\pi \notin \operatorname{spec}(A)$ for all $k \geq 1$ if and only if

$$AA^{\#} = (e^A - I)(e^A - I)^{\#}.$$

In particular, if A is semistable, then this equality holds.

Proof: Use ii) of Fact 11.21.12 and ix) of Proposition 11.8.2.

Fact 11.18.15. Let $A \in \mathbb{F}^{n \times n}$. Then, A is asymptotically stable if and only if A^{-1} is asymptotically stable. Hence, $e^{tA} \to 0$ as $t \to \infty$ if and only if $e^{tA^{-1}} \to 0$ as $t \to \infty$.

Fact 11.18.16. Let $A, B \in \mathbb{R}^{n \times n}$, assume that A is asymptotically stable, and assume that $\sigma_{\max}(B \oplus B) < \sigma_{\min}(A \oplus A)$. Then, $A + B$ is asymptotically stable.

Proof: Since $A \oplus A$ is nonsingular, Fact 9.14.18 implies that $A \oplus A + \alpha(B \oplus B) = (A + \alpha B) \oplus (A + \alpha B)$ is nonsingular for all $0 \leq \alpha \leq 1$. Now, suppose that $A + B$ is not asymptotically stable. Then, there exists $\alpha_0 \in (0, 1]$ such that $A + \alpha_0 B$ has an imaginary eigenvalue, and thus $(A + \alpha_0 B) \oplus (A + \alpha_0 B) = A \oplus A + \alpha_0(B \oplus B)$ is singular, which is a contradiction.

Remark: This result provides a suboptimal solution of a nearness problem. See [697, Section 7] and Fact 9.14.18.

Fact 11.18.17. Let $A \in \mathbb{C}^{n \times n}$, assume that A is asymptotically stable, let $\| \cdot \|$ denote either $\sigma_{\max}(\cdot)$ or $\| \cdot \|_{\mathrm{F}}$, and define

$$\beta(A) \triangleq \{\|B\|: \ B \in \mathbb{C}^{n \times n} \text{ and } A + B \text{ is not asymptotically stable}\}.$$

Then,

$$\begin{aligned}
\tfrac{1}{2}\sigma_{\min}(A \otimes A) &\leq \beta(A) \\
&= \min_{\gamma \in \mathbb{R}} \sigma_{\min}(A + \gamma \jmath I) \\
&\leq \min\{\operatorname{spabs}(A), \sigma_{\min}(A), \tfrac{1}{2}\sigma_{\max}(A + A^*)\}.
\end{aligned}$$

Furthermore, let $R \in \mathbb{F}^{n \times n}$, assume that R is positive definite, and let $P \in \mathbb{F}^{n \times n}$ be the positive-definite solution of $A^*P + PA + R = 0$. Then,

$$\tfrac{1}{2}\sigma_{\min}(R)/\|P\| \leq \beta(A).$$

If, in addition, $A + A^*$ is negative definite, then

$$-\tfrac{1}{2}\lambda_{\min}(A + A^*) \leq \beta(A).$$

Proof: See [697, 1394].

Remark: The analogous problem for real matrices and real perturbations is discussed in [1135].

    Fact 11.18.18. Let $A \in \mathbb{F}^{n \times n}$, assume that A is asymptotically stable, let $V \in \mathbb{F}^{n \times n}$, assume that V is positive definite, and let $Q \in \mathbb{F}^{n \times n}$ be the positive-definite solution of $AQ + QA^* + V = 0$. Then, for all $t \geq 0$,

$$\|e^{tA}\|_{\mathrm{F}}^2 = \operatorname{tr} e^{tA} e^{tA^*} \leq \kappa(Q) \operatorname{tr} e^{-tS^{-1}VS^{-*}} \leq \kappa(Q) \operatorname{tr} e^{-[t/\sigma_{\max}(Q)]V},$$

where $S \in \mathbb{F}^{n \times n}$ satisfies $Q = SS^*$ and $\kappa(Q) \triangleq \sigma_{\max}(Q)/\sigma_{\min}(Q)$. If, in particular, A satisfies $AQ + QA^* + I = 0$, then

$$\|e^{tA}\|_{\mathrm{F}}^2 \leq n\kappa(Q)e^{-t/\sigma_{\max}(Q)}.$$

Proof: See [1503].

Remark: Fact 11.15.4 yields $e^{tA}e^{tA^*} \leq e^{t(A+A^*)}$. However, this bound is poor in the case in which $A + A^*$ is not asymptotically stable. See [189].

Remark: See Fact 11.18.19.

    Fact 11.18.19. Let $A \in \mathbb{F}^{n \times n}$, assume that A is asymptotically stable, and let $Q \in \mathbb{F}^{n \times n}$ be the positive-definite solution of $AQ + QA^* + I = 0$. Then, for all $t \geq 0$,

$$\sigma_{\max}^2(e^{tA}) \leq \kappa(Q)e^{-t/\sigma_{\max}(Q)},$$

where $\kappa(Q) \triangleq \sigma_{\max}(Q)/\sigma_{\min}(Q)$.

Proof: See references in [1411, 1412].

Remark: Since $\|e^{tA}\|_{\mathrm{F}} \leq \sqrt{n}\sigma_{\max}(e^{tA})$, it follows that this inequality implies the last inequality in Fact 11.18.18.

    Fact 11.18.20. Let $A \in \mathbb{R}^{n \times n}$, and assume that every entry of $A \in \mathbb{R}^{n \times n}$ is positive. Then, A is unstable.

Proof: See Fact 4.11.4.

    Fact 11.18.21. Let $A \in \mathbb{R}^{n \times n}$. Then, A is asymptotically stable if and only if there exist matrices $B, C \in \mathbb{R}^{n \times n}$ such that B is positive definite, C is dissipative, and $A = BC$.

Proof: $A = P^{-1}(-A^{\mathrm{T}}P - R)$.

Remark: To reverse the order of factors, consider A^{T}.

Fact 11.18.22. Let $A \in \mathbb{F}^{n \times n}$. Then, the following statements hold:

i) All of the real eigenvalues of A are positive if and only if A is the product of two dissipative matrices.

ii) A is nonsingular and $A \neq \alpha I$ for all $\alpha < 0$ if and only if A is the product of two asymptotically stable matrices.

iii) A is nonsingular if and only if A is the product of three or fewer asymptotically stable matrices.

Proof: See [130, 1494].

Fact 11.18.23. Let $p \in \mathbb{R}[s]$, where $p(s) = s^n + \beta_{n-1} s^{n-1} + \cdots + \beta_1 s + \beta_0$ and $\beta_0, \ldots, \beta_n > 0$. Furthermore, define $A \in \mathbb{R}^{n \times n}$ by

$$A \triangleq \begin{bmatrix} \beta_{n-1} & \beta_{n-3} & \beta_{n-5} & \beta_{n-7} & \cdots & \cdots & 0 \\ 1 & \beta_{n-2} & \beta_{n-4} & \beta_{n-6} & \cdots & \cdots & 0 \\ 0 & \beta_{n-1} & \beta_{n-3} & \beta_{n-5} & \cdots & \cdots & 0 \\ 0 & 1 & \beta_{n-2} & \beta_{n-4} & \cdots & \cdots & 0 \\ \vdots & \vdots & \vdots & \vdots & \ddots & \vdots & \vdots \\ 0 & 0 & 0 & \cdots & \cdots & \beta_1 & 0 \\ 0 & 0 & 0 & \cdots & \cdots & \beta_2 & \beta_0 \end{bmatrix}.$$

If p is Lyapunov stable, then every subdeterminant of A is nonnegative.

Proof: See [85].

Remark: A is *totally nonnegative*. Furthermore, p is asymptotically stable if and only if every leading principal subdeterminant of A is positive.

Remark: The second statement is due to Hurwitz.

Remark: The diagonal entries of A are $\beta_{n-1}, \ldots, \beta_0$.

Problem: Show that this condition for stability is equivalent to the condition given in [494, p. 183] in terms of an alternative matrix $\hat{A}$.

Fact 11.18.24. Let $A \in \mathbb{R}^{n \times n}$, assume that A is tridiagonal, and assume that $A_{(i,i)} > 0$ for all $i \in \{1, \ldots, n\}$ and $A_{(i,i+1)} A_{(i+1,i)} > 0$ for all $i = 1, \ldots, n-1$. Then, A is asymptotically stable.

Proof: See [295].

Remark: This result is due to Barnett and Storey.

Fact 11.18.25. Let $A \in \mathbb{R}^{n \times n}$, and assume that A is cyclic. Then, there exists a nonsingular matrix $S \in \mathbb{R}^{n \times n}$ such that $A_{\mathrm{S}} \triangleq SAS^{-1}$ is given by the tridiagonal matrix

$$
A_{\mathrm{S}} = \begin{bmatrix}
0 & 1 & 0 & \cdots & 0 & 0 \\
-\alpha_n & 0 & 1 & \cdots & 0 & 0 \\
0 & -\alpha_{n-1} & 0 & \ddots & 0 & 0 \\
\vdots & \vdots & \ddots & \ddots & \ddots & \vdots \\
0 & 0 & 0 & \ddots & 0 & 1 \\
0 & 0 & 0 & \cdots & -\alpha_2 & -\alpha_1
\end{bmatrix},
$$

where $\alpha_1, \ldots, \alpha_n$ are real numbers. If $\alpha_1\alpha_2\cdots\alpha_n \neq 0$, then the number of eigenvalues of A in the OLHP is equal to the number of positive elements in $\{\alpha_1, \alpha_1\alpha_2, \ldots, \alpha_1\alpha_2\cdots\alpha_n\}_{\mathrm{ms}}$. Furthermore, $A_{\mathrm{S}}^{\mathrm{T}}P + PA_{\mathrm{S}} + R = 0$, where

$$
P \triangleq \mathrm{diag}(\alpha_1\alpha_2\cdots\alpha_n, \alpha_1\alpha_2\cdots\alpha_{n-1}, \ldots, \alpha_1\alpha_2, \alpha_1)
$$

and

$$
R \triangleq \mathrm{diag}\left(0, \ldots, 0, 2\alpha_1^2\right).
$$

Finally, A_{S} is asymptotically stable if and only if $\alpha_1, \ldots, \alpha_n$ are positive.

Proof: See [150, pp. 52, 95].

Remark: A_{S} is in *Schwarz form*. See Fact 11.18.26 and Fact 11.18.27.

Fact 11.18.26. Let $\alpha_1, \ldots, \alpha_n$ be real numbers, and define $A \in \mathbb{R}^{n \times n}$ by

$$
A = \begin{bmatrix}
0 & 1 & 0 & \cdots & 0 & 0 \\
-\alpha_n & 0 & 1 & \cdots & 0 & 0 \\
0 & -\alpha_{n-1} & 0 & \ddots & 0 & 0 \\
\vdots & \vdots & \ddots & \ddots & \ddots & \vdots \\
0 & 0 & 0 & \ddots & 0 & 1 \\
0 & 0 & 0 & \cdots & -\alpha_2 & \alpha_1
\end{bmatrix}.
$$

Then, $\mathrm{spec}(A) \subset \mathrm{ORHP}$ if and only if $\alpha_1, \ldots, \alpha_n$ are positive.

Proof: See [730, p. 111].

Remark: Note the absence of the minus sign in the (n, n) entry compared to the matrix in Fact 11.18.25. This minus sign changes the sign of all eigenvalues of A.

Fact 11.18.27. Let $\alpha_1, \alpha_2, \alpha_3 > 0$, and define $A_{\mathrm{R}}, P, R \in \mathbb{R}^{3 \times 3}$ by the tridiagonal matrix

$$
A_{\mathrm{R}} \triangleq \begin{bmatrix}
-\alpha_1 & \alpha_2^{1/2} & 0 \\
-\alpha_2^{1/2} & 0 & \alpha_3^{1/2} \\
0 & -\alpha_3^{1/2} & 0
\end{bmatrix}
$$

and the diagonal matrices

$$
P \triangleq I, \quad R \triangleq \mathrm{diag}(2\alpha_1, 0, 0).
$$

Then, $A_{\mathrm{R}}^{\mathrm{T}}P + PA_{\mathrm{R}} + R = 0$.

Remark: The matrix A_R is in *Routh form*. The Routh form A_R and the Schwarz form A_S are related by $A_R = S_{RS} A_S S_{RS}^{-1}$, where

$$S_{RS} \triangleq \begin{bmatrix} 0 & 0 & \alpha_1^{1/2} \\ 0 & -(\alpha_1\alpha_2)^{1/2} & 0 \\ (\alpha_1\alpha_2\alpha_3)^{1/2} & 0 & 0 \end{bmatrix}.$$

Remark: See Fact 11.18.25.

Fact 11.18.28. Let $\alpha_1, \alpha_2, \alpha_3 > 0$, and define $A_C, P, R \in \mathbb{R}^{3\times 3}$ by the tridiagonal matrix

$$A_C \triangleq \begin{bmatrix} 0 & 1/a_3 & 0 \\ -1/a_2 & 0 & 1/a_2 \\ 0 & -1/a_1 & -1/a_1 \end{bmatrix}$$

and the diagonal matrices

$$P \triangleq \mathrm{diag}(a_3, a_2, a_1), \quad R \triangleq \mathrm{diag}(0, 0, 2),$$

where $a_1 \triangleq 1/\alpha_1$, $a_2 \triangleq \alpha_1/\alpha_2$, and $a_3 \triangleq \alpha_2/(\alpha_1\alpha_3)$. Then, $A_C^T P + P A_C + R = 0$.

Proof: See [321, p. 346].

Remark: The matrix A_C is in *Chen form*. The Schwarz form A_S and the Chen form A_C are related by $A_S = S_{SC} A_C S_{SC}^{-1}$, where

$$S_{SC} \triangleq \begin{bmatrix} 1/(\alpha_1\alpha_3) & 0 & 0 \\ 0 & 1/\alpha_2 & 0 \\ 0 & 0 & 1/\alpha_1 \end{bmatrix}.$$

Remark: The Schwarz, Routh, and Chen forms provide the basis for the Routh criterion. See [34, 274, 321, 1100].

Remark: A circuit interpretation of the Chen form is given in [990].

Fact 11.18.29. Let $\alpha_1, \ldots, \alpha_n > 0$ and $\beta_1, \ldots, \beta_n > 0$, and define $A \in \mathbb{R}^{n\times n}$ by

$$A = \begin{bmatrix} -\alpha_1 & 0 & \cdots & 0 & -\beta_1 \\ \beta_2 & -\alpha_2 & \cdots & 0 & 0 \\ \vdots & \ddots & \ddots & \vdots & \vdots \\ 0 & 0 & \ddots & -\alpha_{n-1} & 0 \\ 0 & 0 & \cdots & \beta_n & -\alpha_n \end{bmatrix}.$$

Then,

$$\chi_A(s) = (s+\alpha_1)(s+\alpha_2)\cdots(s+\alpha_n) + \beta_1\beta_2\cdots\beta_n.$$

Furthermore, if

$$(\cos \pi/n)^n < \frac{\alpha_1 \cdots \alpha_n}{\beta_1 \cdots \beta_n},$$

then A is asymptotically stable.

Proof: See [1244].

Remark: If $n = 2$, then A is asymptotically stable for all positive $\alpha_1, \beta_1, \alpha_2, \beta_2$.

Remark: This result is the *secant condition*.

Fact 11.18.30. Let $A \in \mathbb{F}^{n \times n}$. Then, the following statements are equivalent:

i) A is asymptotically stable.

ii) There exist a negative-definite matrix $B \in \mathbb{F}^{n \times n}$, a skew-Hermitian matrix $C \in \mathbb{F}^{n \times n}$, and a nonsingular matrix $S \in \mathbb{F}^{n \times n}$ such that $A = B + SCS^{-1}$.

iii) There exist a negative-definite matrix $B \in \mathbb{F}^{n \times n}$, a skew-Hermitian matrix $C \in \mathbb{F}^{n \times n}$, and a nonsingular matrix $S \in \mathbb{F}^{n \times n}$ such that $A = S(B+C)S^{-1}$.

Proof: See [378].

Fact 11.18.31. Let $A \in \mathbb{R}^{n \times n}$, and let $k \geq 2$. Then, there exist asymptotically stable matrices $A_1, \ldots, A_k \in \mathbb{R}^{n \times n}$ such that $A = \sum_{i=1}^{k} A_i$ if and only if $\operatorname{tr} A < 0$.

Proof: See [769].

Fact 11.18.32. Let $A \in \mathbb{C}^{n \times n}$. Then, A is (Lyapunov stable, semistable, asymptotically stable) if and only if $A \oplus A$ is.

Proof: Use Fact 7.5.7 and the fact that $\operatorname{vec}(e^{tA} V e^{tA^*}) = e^{t(A \oplus \overline{A})} \operatorname{vec} V$.

Fact 11.18.33. Let $A \in \mathbb{R}^{n \times n}$ and $B \in \mathbb{R}^{m \times m}$. Then, the following statements hold:

i) If A and B are (Lyapunov stable, semistable, asymptotically stable), then so is $A \oplus B$.

ii) If $A \oplus B$ is (Lyapunov stable, semistable, asymptotically stable), then so is either A or B.

Proof: Use Fact 7.5.7.

Fact 11.18.34. Let $A \in \mathbb{R}^{n \times n}$, and assume that A is asymptotically stable. Then,

$$(A \oplus A)^{-1} = \int_{-\infty}^{\infty} (\jmath \omega I - A)^{-1} \otimes (\jmath \omega I - A)^{-1} \, d\omega$$

and

$$\int_{-\infty}^{\infty} (\omega^2 I + A^2) \, d\omega = -\pi A^{-1}.$$

Proof: Use $(\jmath \omega I - A)^{-1} + (-\jmath \omega I - A)^{-1} = -2A(\omega^2 I + A^2)^{-1}$.

Fact 11.18.35. Let $A \in \mathbb{R}^{2 \times 2}$. Then, A is asymptotically stable if and only if $\operatorname{tr} A < 0$ and $\det A > 0$.

Fact 11.18.36. Let $A \in \mathbb{C}^{n \times n}$. Then, there exists a unique asymptotically stable matrix $B \in \mathbb{C}^{n \times n}$ such that $B^2 = -A$.

Remark: This result is stated in [1262]. The uniqueness of the square root for complex matrices that have no eigenvalues in $(-\infty, 0]$ is implicitly assumed in [1263].

Remark: See Fact 5.15.19.

Fact 11.18.37. Let $A \in \mathbb{R}^{n \times n}$. Then, the following statements hold:

i) If A is semidissipative, then A is Lyapunov stable.

ii) If A is dissipative, then A is asymptotically stable.

iii) If A is Lyapunov stable and normal, then A is semidissipative.

iv) If A is asymptotically stable and normal, then A is dissipative.

v) If A is discrete-time Lyapunov stable and normal, then A is semicontractive.

Fact 11.18.38. Let $M \in \mathbb{R}^{r \times r}$, assume that M is positive definite, let $C, K \in \mathbb{R}^{r \times r}$, assume that C and K are positive semidefinite, and consider the equation

$$M\ddot{q} + C\dot{q} + Kq = 0.$$

Furthermore, define

$$A \triangleq \begin{bmatrix} 0 & I \\ -M^{-1}K & -M^{-1}C \end{bmatrix}.$$

Then, the following statements hold:

i) A is Lyapunov stable if and only if $C + K$ is positive definite.

ii) A is Lyapunov stable if and only if $\operatorname{rank} \begin{bmatrix} C \\ K \end{bmatrix} = r$.

iii) A is semistable if and only if $(M^{-1}K, C)$ is observable.

iv) A is asymptotically stable if and only if A is semistable and K is positive definite.

Proof: See [190].

Remark: See Fact 5.12.21.

11.19 Facts on Almost Nonnegative Matrices

Fact 11.19.1. Let $A \in \mathbb{R}^{n \times n}$. Then, e^{tA} is nonnegative for all $t \geq 0$ if and only if A is almost nonnegative.

Proof: Let $\alpha > 0$ be such that $\alpha I + A$ is nonnegative, and consider $e^{t(\alpha I + A)}$. See [185, p. 74], [186, p. 146], [194, 373], or [1228, p. 37].

Fact 11.19.2. Let $A \in \mathbb{R}^{n \times n}$, and assume that A is almost nonnegative. Then, e^{tA} is positive for all $t > 0$ if and only if A is irreducible.

Proof: See [1215, p. 208].

Fact 11.19.3. Let $A \in \mathbb{R}^{n \times n}$, where $n \geq 2$, and assume that A is almost nonnegative. Then, the following statements are equivalent:

i) There exist $\alpha \in (0, \infty)$ and $B \in \mathbb{R}^{n \times n}$ such that $A = B - \alpha I$, B is nonnegative, and $\text{sprad}(B) \leq \alpha$.

ii) $\text{spec}(A) \subset \text{OLHP} \cup \{0\}$.

iii) $\text{spec}(A) \subset \text{CLHP}$.

iv) If $\lambda \in \text{spec}(A)$ is real, then $\lambda \leq 0$.

v) Every principal subdeterminant of $-A$ is nonnegative.

vi) For every diagonal, positive-definite matrix $B \in \mathbb{R}^{n \times n}$, it follows that $A - B$ is nonsingular.

Remark: A is an *N-matrix* if A is almost nonnegative and *i)–vi)* hold.

Remark: This result follows from Fact 4.11.8.

Example: $A = \begin{bmatrix} 0 & 1 \\ 0 & 0 \end{bmatrix}$.

Fact 11.19.4. Let $A \in \mathbb{R}^{n \times n}$, where $n \geq 2$, and assume that A is almost nonnegative. Then, the following conditions are equivalent:

i) A is a group-invertible N-matrix.

ii) A is a Lyapunov-stable N-matrix.

iii) A is a semistable N-matrix.

iv) A is Lyapunov stable.

v) A is semistable.

vi) A is an N-matrix, and there exist $\alpha \in (0, \infty)$ and a nonnegative matrix $B \in \mathbb{R}^{n \times n}$ such that $A = B - \alpha I$ and $\alpha^{-1} B$ is discrete-time semistable.

vii) There exists a positive-definite matrix $P \in \mathbb{R}^{n \times n}$ such that $A^{\mathrm{T}} P + PA$ is negative semidefinite.

Furthermore, consider the following statements:

viii) There exists a positive vector $p \in \mathbb{R}^n$ such that $-Ap$ is nonnegative.

ix) There exists a nonzero nonnegative vector $p \in \mathbb{R}^n$ such that $-Ap$ is nonnegative.

Then, *viii)* $\implies$ [*i)–vii)*] $\implies$ *ix)*.

Proof: See [186, pp. 152–155] and [187]. The statement [*i)–vii)*] $\implies$ *ix)* is given by Fact 4.11.12.

Remark: The converse of *viii)* $\implies$ [*i)–vii)*] does not hold. For example, $A = \begin{bmatrix} 0 & 1 \\ 0 & -1 \end{bmatrix}$ is almost negative and semistable, but there does not exist a positive vector $p \in \mathbb{R}^2$ such that $-Ap$ is nonnegative. However, note that *viii)* holds for A^{T}, but not for $\text{diag}(A, A^{\mathrm{T}})$ or its transpose.

Remark: A discrete-time semistable matrix is called *semiconvergent* in [186, p. 152].

Remark: The last statement follows from the fact that the function $V(x) = p^T x$ is a Lyapunov function for the system $\dot{x} = -Ax$ for $x \in [0, \infty)^n$ with Lyapunov derivative $\dot{V}(x) = -A^T p$. See [191, 630].

Fact 11.19.5. Let $A \in \mathbb{R}^{n \times n}$, where $n \geq 2$, and assume that A is almost nonnegative. Then, the following conditions are equivalent:

i) A is a nonsingular N-matrix.

ii) A is asymptotically stable.

iii) A is an asymptotically stable N-matrix.

iv) There exist $\alpha \in (0, \infty)$ and a nonnegative matrix $B \in \mathbb{R}^{n \times n}$ such that $A = B - \alpha I$ and $\text{sprad}(B) < \alpha$.

v) If $\lambda \in \text{spec}(A)$ is real, then $\lambda < 0$.

vi) If $B \in \mathbb{R}^{n \times n}$ is nonnegative and diagonal, then $A - B$ is nonsingular.

vii) Every principal subdeterminant of $-A$ is positive.

viii) Every leading principal subdeterminant of $-A$ is positive.

ix) For all $i \in \{1, \ldots, n\}$, the sign of the ith leading principal subdeterminant of A is $(-1)^i$.

x) For all $k \in \{1, \ldots, n\}$, the sum of all $k \times k$ principal subdeterminants of $-A$ is positive.

xi) There exists a positive-definite matrix $P \in \mathbb{R}^{n \times n}$ such that $A^T P + PA$ is negative definite.

xii) There exists a positive vector $p \in \mathbb{R}^n$ such that $-Ap$ is positive.

xiii) There exists a nonnegative vector $p \in \mathbb{R}^n$ such that $-Ap$ is positive.

xiv) If $p \in \mathbb{R}^n$ and $-Ap$ is nonnegative, then $p \geq\geq 0$ is nonnegative.

xv) For every nonnegative vector $y \in \mathbb{R}^n$, there exists a unique nonnegative vector $x \in \mathbb{R}^n$ such that $Ax = -y$.

xvi) A is nonsingular and $-A^{-1}$ is nonnegative.

Proof: See [185, pp. 134–140] or [730, pp. 114–116].

Remark: $-A$ is a nonsingular M-matrix. See Fact 4.11.8.

Fact 11.19.6. For $i, j = 1, \ldots, n$, let $\sigma_{ij} \in [0, \infty)$, and define $A \in \mathbb{R}^{n \times n}$ by $A_{(i,j)} \triangleq \sigma_{ij}$ for all $i \neq j$ and $A_{(i,i)} \triangleq -\sum_{j=1}^{n} \sigma_{ij}$. Then, the following statements hold:

i) A is almost nonnegative.

ii) $-A 1_{n \times 1} = \begin{bmatrix} \sigma_{11} & \cdots & \sigma_{nn} \end{bmatrix}^T$ is nonnegative.

iii) $\text{spec}(A) \subset \text{OLHP} \cup \{0\}$.

iv) A is an N-matrix.

v) A is a group-invertible N-matrix.

vi) A is a Lyapunov-stable N-matrix.

vii) A is a semistable N-matrix.

If, in addition, $\sigma_{11}, \ldots, \sigma_{nn}$ are positive, then A is a nonsingular N-matrix.

Proof: It follows from the Gershgorin circle theorem given by Fact 4.10.17 that every eigenvalue λ of A is an element of a disk in $\mathbb{C}$ centered at $-\sum_{j=1}^{n} \sigma_{ij} \leq 0$ and with radius $\sum_{j=1, j\neq i}^{n} \sigma_{ij}$. Hence, if $\sigma_{ii} = 0$, then either $\lambda = 0$ or $\operatorname{Re}\lambda < 0$, whereas, if $\sigma_{ii} > 0$, then $\operatorname{Re}\lambda \leq \sigma_{ii} < 0$. Thus, *iii)* holds. Statements *iv)*–*vii)* follow from *ii)* and Fact 11.19.4. The last statement follows from the Gershgorin circle theorem.

Remark: A^{T} is a *compartmental matrix*. See [194, 632, 1421].

Problem: Determine necessary and sufficient conditions on the parameters σ_{ij} such that A is a nonsingular N-matrix.

Fact 11.19.7. Let $\mathcal{G} = (\mathcal{X}, \mathcal{R})$ be a graph, where $\mathcal{X} = \{x_1, \ldots, x_n\}$, and let $L \in \mathbb{R}^{n \times n}$ denote either the inbound Laplacian or the outbound Laplacian of $\mathcal{G}$. Then, the following statements hold:

i) $-L$ is semistable.

ii) $\lim_{t \to \infty} e^{-Lt}$ exists.

Remark: Use Fact 11.19.6.

Remark: The spectrum of the Laplacian is discussed in [7].

Fact 11.19.8. Let $A \in \mathbb{R}^{n \times n}$, and assume that A is asymptotically stable. Then, at least one of the following statements holds:

i) All of the diagonal entries of A are negative.

ii) At least one diagonal entry of A is negative and at least one off-diagonal entry of A is negative.

Proof: See [519].

Remark: *sign stability* is discussed in [773].

11.20 Facts on Discrete-Time-Stable Polynomials

Fact 11.20.1. Let $p \in \mathbb{R}[s]$, where $p(s) = s^n + a_{n-1}s^{n-1} + \cdots + a_0$. Then, the following statements hold:

i) If $n = 1$, then p is discrete-time asymptotically stable if and only if $|a_0| < 1$.

ii) If $n = 2$, then p is discrete-time asymptotically stable if and only if $|a_0| < 1$ and $|a_1| < 1 + a_0$.

iii) If $n = 3$, then p is discrete-time asymptotically stable if and only if $|a_0| < 1$, $|a_0 + a_2| < |1 + a_1|$, and $|a_1 - a_0 a_2| < 1 - a_0^2$.

Proof: See [140, p. 185], [289, p. 6], [708, p. 355], or [804, pp. 34, 35].

Remark: These results are the *Schur-Cohn criterion*. Conditions for polynomials of arbitrary degree n follow from the *Jury criterion*.

Remark: The Routh criterion for stability of continuous-time systems is given by Fact 11.17.2.

Problem: For $n = 3$, the conditions given in [289, p. 6] are $|a_0 + a_2| < 1 + a_1$, $|3a_0 - a_2| < 3 - a_1$, and $a_0^2 + a_1 - a_0 a_2 < 1$. Show that these conditions are equivalent to *iii*).

Fact 11.20.2. Let $p \in \mathbb{C}[s]$, where $p(s) = s^n + a_{n-1}s^{n-1} + \cdots + a_0$, and define $\hat{p} \in \mathbb{C}[s]$ by

$$\hat{p}(s) \triangleq s^{n-1} + \frac{a_{n-1} - a_0\bar{a}_1}{1 - |a_0|^2}s^{n-1} + \frac{a_{n-2} - a_0\bar{a}_2}{1 - |a_0|^2}s^{n-2} + \cdots + \frac{a_1 - a_0\bar{a}_{n-1}}{1 - |a_0|^2}.$$

Then, p is discrete-time asymptotically stable if and only if $|a_0| < 1$ and $\hat{p}$ is discrete-time asymptotically stable.

Proof: See [708, p. 354].

Fact 11.20.3. Let $p \in \mathbb{R}[s]$, where $p(s) = s^n + a_{n-1}s^{n-1} + \cdots + a_0$. Then, the following statements hold:

i) If $a_0 \leq \cdots \leq a_{n-1} \leq 1$, then $\text{roots}(p) \subset \{z \in \mathbb{C} : |z| \leq 1 + |a_0| - a_0\}$.

ii) If $0 < a_0 \leq \cdots \leq a_{n-1} \leq 1$, then $\text{roots}(p) \subset \text{CUD}$.

iii) If $0 < a_0 < \cdots < a_{n-1} < 1$, then p is discrete-time asymptotically stable.

Proof: For *i*), see [1220]. For *ii*), see [1029, p. 272]. For *iii*), use Fact 11.20.2. See [708, p. 355].

Remark: If there exists $r > 0$ such that $0 < ra_0 < \cdots < r^{n-1}a_{n-1} < r^n$, then $\text{roots}(p) \subset \{z \in \mathbb{C} : |z| \leq r\}$.

Remark: Statement *ii*) is the *Enestrom-Kakeya theorem*.

Fact 11.20.4. Let $p \in \mathbb{C}[s]$, where $p(s) = s^n + a_{n-1}s^{n-1} + \cdots + a_0$, assume that $a_0, \ldots, a_{n-1}$ are nonzero, and let $\lambda \in \text{roots}(p)$. Then,

$$|\lambda| \leq \max\{2|a_{n-1}|, 2|a_{n-2}/a_{n-1}|, \ldots, 2|a_1/a_2|, |a_0/a_1|\}.$$

Remark: This result is due to Bourbaki. See [1030].

Fact 11.20.5. Let $p \in \mathbb{C}[s]$, where $p(s) = s^n + a_{n-1}s^{n-1} + \cdots + a_0$, assume that $a_0, \ldots, a_{n-1}$ are nonzero, and let $\lambda \in \text{roots}(p)$. Then,

$$|\lambda| \leq \sum_{i=1}^{n-1} |a_i|^{1/(n-i)}$$

and

$$|\lambda + \tfrac{1}{2}a_{n-1}| \leq \tfrac{1}{2}|a_{n-1}| + \sum_{i=0}^{n-2} |a_i|^{1/(n-i)}.$$

Remark: These results are due to Walsh. See [1030].

Fact 11.20.6. Let $p \in \mathbb{C}[s]$, where $p(s) = s^n + a_{n-1}s^{n-1} + \cdots + a_0$, and let $\lambda \in \text{roots}(p)$. Then,

$$\frac{|a_0|}{|a_0| + \max\{|a_1|,\ldots,|a_{n-1}|,1\}} < |\lambda| \le \max\{|a_0|, 1 + |a_1|,\ldots,1 + |a_{n-1}|\}.$$

Proof: The lower bound is proved in [1030], while the upper bound is proved in [411].

Remark: The upper bound is *Cauchy's estimate*.

Remark: The weaker upper bound

$$|\lambda| < 1 + \max_{i=0,\ldots,n-1} |a_i|$$

is given in [140, p. 184] and [1030].

Fact 11.20.7. Let $p \in \mathbb{C}[s]$, where $p(s) = s^n + a_{n-1}s^{n-1} + \cdots + a_0$, and let $\lambda \in \text{roots}(p)$. Then,

$$|\lambda| \le \tfrac{1}{2}(1 + |a_{n-1}|) + \sqrt{\max_{i=0,\ldots,n-2} |a_i| + \tfrac{1}{4}(1 - |a_{n-1}|)^2},$$

$$|\lambda| \le \max\{2, |a_0| + |a_{n-1}|, |a_1| + |a_{n-1}|,\ldots,|a_{n-2}| + |a_{n-1}|\},$$

$$|\lambda| \le \sqrt{2 + \max_{i=0,\ldots,n-2} |a_i|^2 + |a_{n-1}|^2}.$$

Proof: See [411].

Remark: The first inequality is due to Joyal, Labelle, and Rahman. See [1030].

Fact 11.20.8. Let $p \in \mathbb{C}[s]$, where $p(s) = s^n + a_{n-1}s^{n-1} + \cdots + a_0$, assume that $a_0,\ldots,a_{n-1}$ are nonzero, define

$$\alpha \triangleq \max\left\{\left|\frac{a_0}{a_1}\right|, \left|\frac{a_1}{a_2}\right|,\ldots,\left|\frac{a_{n-2}}{a_{n-1}}\right|\right\}$$

and

$$\beta \triangleq \max\left\{\left|\frac{a_1}{a_2}\right|, \left|\frac{a_2}{a_3}\right|,\ldots,\left|\frac{a_{n-2}}{a_{n-1}}\right|\right\},$$

and let $\lambda \in \text{roots}(p)$. Then,

$$|\lambda| \le \tfrac{1}{2}(\beta + |a_{n-1}|) + \sqrt{\alpha|a_{n-1}| + \tfrac{1}{4}(\beta - |a_{n-1}|)^2},$$

$$|\lambda| \le |a_{n-1}| + \alpha,$$

$$|\lambda| \le \max\left\{\left|\frac{a_0}{a_1}\right|, 2\beta, 2|a_{n-1}|\right\},$$

$$|\lambda| \le 2 \max_{i=1,\ldots,n-1} |a_i|^{1/(n-i)},$$

$$|\lambda| \le \sqrt{2|a_{n-1}|^2 + \alpha^2 + \beta^2}.$$

Proof: See [411, 943].

Remark: The third inequality is *Kojima's bound*, while the fourth inequality is *Fujiwara's bound*.

Fact 11.20.9. Let $p \in \mathbb{C}[s]$, where $p(s) = s^n + a_{n-1}s^{n-1} + \cdots + a_0$, define $\alpha \triangleq 1 + \sum_{i=0}^{n-1} |a_i|^2$, and let $\lambda \in \mathrm{roots}(p)$. Then,

$$|\lambda| \leq \tfrac{1}{n}|a_{n-1}| + \sqrt{\tfrac{n}{n-1}\left(n - 1 + \sum_{i=0}^{n-1} |a_i|^2 - \tfrac{1}{n}|a_{n-1}|^2\right)},$$

$$|\lambda| \leq \tfrac{1}{2}\left(|a_{n-1}| + 1 + \sqrt{(|a_{n-1}| - 1)^2 + 4\sqrt{\sum_{i=0}^{n-2} |a_i|^2}}\right),$$

$$|\lambda| \leq \tfrac{1}{2}\left(|a_{n-1}| + \cos\tfrac{\pi}{n} + \sqrt{\left(|a_{n-1}| - \cos\tfrac{\pi}{n}\right)^2 + (|a_{n-2}| + 1)^2 + \sum_{i=0}^{n-3} |a_i|^2}\right),$$

$$|\lambda| \leq \cos\tfrac{\pi}{n+1} + \tfrac{1}{2}\left(|a_{n-1}| + \sqrt{\sum_{i=0}^{n-1} |a_i|^2}\right),$$

and

$$\sqrt{\tfrac{1}{2}\left(\alpha - \sqrt{\alpha^2 - 4|a_0|^2}\right)} \leq |\lambda| \leq \sqrt{\tfrac{1}{2}\left(\alpha + \sqrt{\alpha^2 - 4|a_0|^2}\right)}.$$

Furthermore,

$$|\mathrm{Re}\,\lambda| \leq \tfrac{1}{2}\left(|\mathrm{Re}\,a_{n-1}| + \cos\tfrac{\pi}{n} + \sqrt{\left(|\mathrm{Re}\,a_{n-1}| - \cos\tfrac{\pi}{n}\right)^2 + (|a_{n-2}| - 1)^2 + \sum_{i=0}^{n-3} |a_i|^2}\right)$$

and

$$|\mathrm{Im}\,\lambda| \leq \tfrac{1}{2}\left(|\mathrm{Im}\,a_{n-1}| + \cos\tfrac{\pi}{n} + \sqrt{\left(|\mathrm{Im}\,a_{n-1}| - \cos\tfrac{\pi}{n}\right)^2 + (|a_{n-2}| + 1)^2 + \sum_{i=0}^{n-3} |a_i|^2}\right).$$

Proof: See [527, 846, 850, 943].

Remark: The first bound is due to Linden (see [850]), the fourth bound is due to Fujii and Kubo, and the upper bound in the fifth result, which follows from Fact 5.11.21 and Fact 5.11.30, is due to Parodi, see also [825, 841].

Remark: The Parodi bound is a refinement of the Carmichael-Mason bound. See Fact 11.20.10.

Fact 11.20.10. Let $p \in \mathbb{C}[s]$, where $p(s) = s^n + a_{n-1}s^{n-1} + \cdots + a_0$, let $r, q \in (1, \infty)$, assume that $1/r + 1/q = 1$, define $\alpha \triangleq (\sum_{i=0}^{n-1} |a_i|^r)^{1/r}$, and let $\lambda \in \mathrm{roots}(p)$. Then,

$$|\lambda| \leq (1 + \alpha^q)^{1/q}.$$

In particular, if $r = q = 2$, then

$$|\lambda| \leq \sqrt{1 + |a_{n-1}|^2 + \cdots + |a_0|^2}.$$

Proof: See [943, 1030].

Remark: Letting $r \to \infty$ yields the upper bound in Fact 11.20.6.

Remark: The result for $r = q = 2$ is due to Carmichael and Mason.

Fact 11.20.11. Let $p \in \mathbb{C}[s]$, where $p(s) = s^n + a_{n-1}s^{n-1} + \cdots + a_0$, and let $\lambda \in \mathrm{roots}(p)$. Then,

$$|\lambda| \leq \sqrt{1 + |1 - a_{n-1}|^2 + |a_{n-1} - a_{n-2}|^2 + \cdots + |a_1 - a_0|^2 + |a_0|^2}.$$

Proof: See [1515].

Remark: This result is due to Williams.

Fact 11.20.12. Let $p \in \mathbb{C}[s]$, where $p(s) = s^n + a_{n-1}s^{n-1} + \cdots + a_0$, let $\mathrm{mroots}(p) = \{\lambda_1, \ldots, \lambda_n\}_{\mathrm{ms}}$, and let $r > 0$ be the unique positive root of $\hat{p}(s) \triangleq s^n - |a_{n-1}|s^{n-1} - \cdots - |a_0|$. Then,

$$(\sqrt[n]{2} - 1)r \leq \max_{i=1,\ldots,n} |\lambda_i| \leq r.$$

Furthermore,

$$(\sqrt[n]{2} - 1)r \leq \frac{1}{n}\sum_{i=1}^{n} |\lambda_i| < r.$$

Finally, the third inequality is an equality if and only if $\lambda_1 = \cdots = \lambda_n$.

Remark: The first inequality is due to Cohn, the second inequality is due to Cauchy, and the third and fourth inequalities are due to Berwald. See [1030] and [1029, p. 245].

Fact 11.20.13. Let $p \in \mathbb{C}[s]$, where $p(s) = s^n + a_{n-1}s^{n-1} + \cdots + a_0$, define $\alpha \triangleq 1 + \sum_{i=0}^{n-1} |a_i|^2$, and let $\lambda \in \mathrm{roots}(p)$. Then,

$$\sqrt{\frac{1}{2}\left(\alpha - \sqrt{\alpha^2 - 4|a_0|^2}\right)} \leq |\lambda| \leq \sqrt{\frac{1}{2}\left(\alpha + \sqrt{\alpha^2 - 4|a_0|^2}\right)}.$$

Proof: See [847]. This result follows from Fact 5.11.29 and Fact 5.11.30.

Fact 11.20.14. Let $p \in \mathbb{R}[s]$, where $p(s) = s^n + a_{n-1}s^{n-1} + \cdots + a_0$, assume that $a_0, \ldots, a_{n-1}$ are nonnegative, and let $x_1, \ldots, x_m \in [0, \infty)$. Then,

$$p(\sqrt[m]{x_1 \cdots x_m}) \leq \sqrt[m]{p(x_1) \cdots p(x_m)}.$$

Proof: See [1067].

Remark: This result, which is due to Mihet, extends a result of Huygens for the case $p(x) = x + 1$.

11.21 Facts on Discrete-Time-Stable Matrices

Fact 11.21.1. Let $A \in \mathbb{R}^{2\times 2}$. Then, A is discrete-time asymptotically stable if and only if $|\operatorname{tr} A| < 1 + \det A$ and $|\det A| < 1$.

Fact 11.21.2. Let $A \in \mathbb{F}^{n\times n}$. Then, A is discrete-time (Lyapunov stable, semistable, asymptotically stable) if and only if A^2 is.

Fact 11.21.3. Let $A \in \mathbb{R}^{n\times n}$, and let $\chi_A(s) = s^n + a_{n-1}s^{n-1} + \cdots + a_1 s + a_0$. Then, for all $k \geq 0$,

$$A^k = x_1(k)I + x_2(k)A + \cdots + x_n(k)A^{n-1},$$

where, for all $i \in \{1, \ldots, n\}$ and all $k \geq 0$, $x_i \colon \mathbb{N} \mapsto \mathbb{R}$ satisfies

$$x_i(k+n) + a_{n-1}x_i(k+n-1) + \cdots + a_1 x_i(k+1) + a_0 x_i(k) = 0,$$

with, for all $i, j \in \{1, \ldots, n\}$, the initial conditions

$$x_i(j-1) = \delta_{ij}.$$

Proof: See [878].

Fact 11.21.4. Let $A \in \mathbb{R}^{n\times n}$. Then, the following statements hold:

i) If A is semicontractive, then A is discrete-time Lyapunov stable.

ii) If A is contractive, then A is discrete-time asymptotically stable.

iii) If A is discrete-time Lyapunov stable and normal, then A is semicontractive.

iv) If A is discrete-time asymptotically stable and normal, then A is contractive.

Problem: Prove these results by using Fact 11.15.6.

Fact 11.21.5. Let $x \in \mathbb{F}^n$, let $A \in \mathbb{F}^{n\times n}$, and assume that A is discrete-time semistable. Then, $\sum_{i=0}^{\infty} A^i x$ exists if and only if $x \in \mathcal{R}(A - I)$. In this case,

$$\sum_{i=0}^{\infty} A^i x = -(A - I)^{\#}x.$$

Proof: See [755].

Fact 11.21.6. Let $A \in \mathbb{F}^{n\times n}$. Then, A is discrete-time (Lyapunov stable, semistable, asymptotically stable) if and only if $A \otimes A$ is.

Proof: Use Fact 7.4.16.

Remark: See [755].

Fact 11.21.7. Let $A \in \mathbb{R}^{n \times n}$ and $B \in \mathbb{R}^{m \times m}$. Then, the following statements hold:

i) If A and B are discrete-time (Lyapunov stable, semistable, asymptotically stable), then $A \otimes B$ is discrete-time (Lyapunov stable, semistable, asymptotically stable).

ii) If $A \otimes B$ is discrete-time (Lyapunov stable, semistable, asymptotically stable), then either A or B is discrete-time (Lyapunov stable, semistable, asymptotically stable).

Proof: Use Fact 7.4.16.

Fact 11.21.8. Let $A \in \mathbb{R}^{n \times n}$, and assume that A is (Lyapunov stable, semistable, asymptotically stable). Then, e^A is discrete-time (Lyapunov stable, semistable, asymptotically stable).

Problem: If $B \in \mathbb{R}^{n \times n}$ is discrete-time (Lyapunov stable, semistable, asymptotically stable), when does there exist a (Lyapunov-stable, semistable, asymptotically stable) matrix $A \in \mathbb{R}^{n \times n}$ such that $B = e^A$? See Proposition 11.4.3.

Fact 11.21.9. The following statements hold:

i) If $A \in \mathbb{R}^{n \times n}$ is discrete-time asymptotically stable, then $B \triangleq (A+I)^{-1}(A-I)$ is asymptotically stable.

ii) If $B \in \mathbb{R}^{n \times n}$ is asymptotically stable, then $A \triangleq (I+B)(I-B)^{-1}$ is discrete-time asymptotically stable.

iii) If $A \in \mathbb{R}^{n \times n}$ is discrete-time asymptotically stable, then there exists a unique asymptotically stable matrix $B \in \mathbb{R}^{n \times n}$ such that $A = (I+B)(I-B)^{-1}$. In fact, $B = (A+I)^{-1}(A-I)$.

iv) If $B \in \mathbb{R}^{n \times n}$ is asymptotically stable, then there exists a unique discrete-time asymptotically stable matrix $A \in \mathbb{R}^{n \times n}$ such that $B = (A+I)^{-1}(A-I)$. In fact, $A = (I+B)(I-B)^{-1}$.

Proof: See [675].

Remark: For additional results on the Cayley transform, see Fact 3.11.21, Fact 3.11.22, Fact 3.11.23, Fact 3.20.12, and Fact 8.9.31.

Problem: Obtain analogous results for Lyapunov-stable and semistable matrices.

Fact 11.21.10. Let $\begin{bmatrix} P_1 & P_{12} \\ P_{12}^T & P_2 \end{bmatrix} \in \mathbb{R}^{2n \times 2n}$ be positive definite, where $P_1, P_{12}, P_2 \in \mathbb{R}^{n \times n}$. If $P_1 \geq P_2$, then $A \triangleq P_1^{-1}P_{12}^T$ is discrete-time asymptotically stable, while, if $P_2 \geq P_1$, then $A \triangleq P_2^{-1}P_{12}$ is discrete-time asymptotically stable.

Proof: If $P_1 \geq P_2$, then $P_1 - P_{12}P_1^{-1}P_1P_1^{-1}P_{12}^T \geq P_1 - P_{12}P_2^{-2}P_{12}^T > 0$. See [342].

Fact 11.21.11. Let $A \in \mathbb{R}^{n \times n}$, where $n \geq 2$, and assume that A is row stochastic. Then, the following statements hold:

i) A is discrete-time Lyapunov stable.

ii) If A is primitive, then A is discrete-time semistable.

Proof: For all $k \geq 1$, it follows that $A^k 1_{n \times 1} = 1_{n \times 1}$. Since A^k is nonnegative, it follows that every entry of A is bounded. If A is primitive, then the result follows from Fact 4.11.6, which implies that $\text{sprad}(A) = 1$, and *viii)* and *xv)* of Fact 4.11.4, which imply that 1 is a simple eigenvalue of A as well as the only eigenvalue of A on the unit circle.

Fact 11.21.12. Let $A \in \mathbb{R}^{n \times n}$, and let $\| \cdot \|$ be a norm on $\mathbb{R}^{n \times n}$. Then, the following statements hold:

i) A is discrete-time Lyapunov stable if and only if $\left\{ \|A^k\| \right\}_{k=0}^{\infty}$ is bounded.

ii) A is discrete-time semistable if and only if $A_\infty \triangleq \lim_{k \to \infty} A^k$ exists.

iii) Assume that A is discrete-time semistable. Then, $A_\infty \triangleq I - (A-I)(A-I)^{\#}$ is idempotent and $\text{rank } A_\infty = \text{amult}_A(1)$. If, in addition, $\text{rank } A = 1$, then, for every eigenvector x of A associated with the eigenvalue 1, there exists $y \in \mathbb{F}^n$ such that $y^* x = 1$ and $A_\infty = xy^*$.

iv) A is discrete-time asymptotically stable if and only if $\lim_{k \to \infty} A^k = 0$.

Remark: A proof of *ii)* is given in [1023, p. 640]. See Fact 11.21.16.

Fact 11.21.13. Let $A \in \mathbb{F}^{n \times n}$. Then, A is discrete-time Lyapunov stable if and only if

$$A_\infty \triangleq \lim_{k \to \infty} \tfrac{1}{k} \sum_{i=0}^{k-1} A^i$$

exists. In this case,

$$A_\infty = I - (A-I)(A-I)^{\#}.$$

Proof: See [1023, p. 633].

Remark: A is *Cesaro summable*.

Remark: See Fact 6.3.33.

Fact 11.21.14. Let $A \in \mathbb{F}^{n \times n}$. Then, A is discrete-time asymptotically stable if and only if

$$\lim_{k \to \infty} A^k = 0.$$

In this case,

$$(I - A)^{-1} = \sum_{i=1}^{\infty} A^i,$$

where the series converges absolutely.

Fact 11.21.15. Let $A \in \mathbb{F}^{n \times n}$, and assume that A is unitary. Then, A is discrete-time Lyapunov stable.

Fact 11.21.16. Let $A, B \in \mathbb{R}^{n \times n}$, assume that A is discrete-time semistable, and let $A_\infty \triangleq \lim_{k \to \infty} A^k$. Then,

$$\lim_{k \to \infty} \left(A + \tfrac{1}{k}B\right)^k = A_\infty e^{A_\infty B A_\infty}.$$

Proof: See [237, 1463].

Remark: If A is idempotent, then $A_\infty = A$. The existence of A_∞ is guaranteed by Fact 11.21.12, which also implies that A_∞ is idempotent.

Fact 11.21.17. Let $A \in \mathbb{R}^{n \times n}$. Then, the following statements hold:

i) A is discrete-time Lyapunov stable if and only if there exists a positive-definite matrix $P \in \mathbb{R}^{n \times n}$ such that $P - A^{\mathrm{T}}PA$ is positive semidefinite.

ii) A is discrete-time asymptotically stable if and only if there exists a positive-definite matrix $P \in \mathbb{R}^{n \times n}$ such that $P - A^{\mathrm{T}}PA$ is positive definite.

Remark: The *discrete-time Lyapunov equation* or the *Stein equation* is $P = A^{\mathrm{T}}PA + R$.

Fact 11.21.18. Let $A \in \mathbb{R}^{n \times n}$, assume that A is discrete-time asymptotically stable, let $P \in \mathbb{R}^{n \times n}$ be positive definite, and assume that P satisfies $P = A^{\mathrm{T}}PA + I$. Then, P is given by

$$P = \tfrac{1}{2\pi} \int_{-\pi}^{\pi} (A^{\mathrm{T}} - e^{-\jmath\theta}I)^{-1}(A - e^{\jmath\theta}I)\, \mathrm{d}\theta.$$

Furthermore,

$$1 \le \lambda_n(P) \le \lambda_1(P) \le \frac{[\sigma_{\max}(A) + 1]^{2n-2}}{[1 - \mathrm{sprad}(A)]^{2n}}.$$

Proof: See [659, pp. 167–169].

Fact 11.21.19. Let $(A_k)_{k=0}^{\infty} \subset \mathbb{R}^{n \times n}$ and, for $k \in \mathbb{N}$, consider the discrete-time, time-varying system

$$x(k + 1) = A_k x(k).$$

Furthermore, assume there exist real numbers $\beta \in (0, 1)$, $\gamma > 0$, and $\varepsilon > 0$ such that

$$\rho(\beta^2 + \rho\varepsilon^2)^2 < 1,$$

where

$$\rho \triangleq \frac{(\gamma + 1)^{2n-2}}{(1 - \beta)^{2n}},$$

and such that, for all $k \in \mathbb{N}$,

$$\mathrm{sprad}(A_k) < \beta,$$

$$\|A_k\| < \gamma,$$

$$\|A_{k+1} - A_k\| < \varepsilon,$$

where $\|\cdot\|$ is an induced norm on $\mathbb{R}^{n \times n}$. Then, $x(k) \to 0$ as $k \to \infty$.

Proof: See [659, pp. 170–173].

Remark: This result arises from the theory of *infinite matrix products*. See [79, 234, 235, 383, 623, 722, 886].

Fact 11.21.20. Let $A \in \mathbb{F}^{n \times n}$, and define

$$r(A) \triangleq \sup_{\{z \in \mathbb{C}: \ |z|>1\}} \frac{|z| - 1}{\sigma_{\min}(zI - A)}.$$

Then,

$$r(A) \leq \sup_{k \geq 0} \sigma_{\max}(A^k) \leq ner(A).$$

Hence, if A is discrete-time Lyapunov stable, then $r(A)$ is finite.

Proof: See [1447].

Remark: This result is the *Kreiss matrix theorem*.

Remark: The constant ne is the best possible. See [1447].

Fact 11.21.21. Let $p \in \mathbb{R}[s]$, and assume that p is discrete-time semistable. Then, $C(p)$ is discrete-time semistable, and there exists $v \in \mathbb{R}^n$ such that

$$\lim_{k \to \infty} C^k(p) = 1_{n \times 1} v^{\mathrm{T}}.$$

Proof: Since $C(p)$ is a companion form matrix, it follows from Proposition 11.10.4 that its minimal polynomial is p. Hence, $C(p)$ is discrete-time semistable. Now, it follows from Proposition 11.10.2 that $\lim_{k \to \infty} C^k(p)$ exists, and thus the state $x(k)$ of the difference equation $x(k+1) = C(p)x(k)$ converges for all initial conditions $x(0) = x_0$. The structure of $C(p)$ shows that all components of $x(k)$ converge to the same value as $k \to \infty$. Hence, all rows of $\lim_{k \to \infty} C^k(p)$ are equal.

11.22 Facts on Lie Groups

Fact 11.22.1. The groups $\mathrm{UT}(n), \mathrm{UT}_+(n), \mathrm{UT}_{\pm 1}(n), \mathrm{SUT}(n)$, and $\{I_n\}$ are Lie groups. Furthermore, $\mathrm{ut}(n)$ is the Lie algebra of $\mathrm{UT}(n)$, $\mathrm{sut}(n)$ is the Lie algebra of $\mathrm{SUT}(n)$, and $\{0_{n \times n}\}$ is the Lie algebra of $\{I_n\}$.

Remark: See Fact 3.23.11 and Fact 3.23.12.

Problem: Determine the Lie algebras of $\mathrm{UT}_+(n)$ and $\mathrm{UT}_{\pm 1}(n)$.

11.23 Facts on Subspace Decomposition

Fact 11.23.1. Let $A \in \mathbb{R}^{n \times n}$, and let $S \in \mathbb{R}^{n \times n}$ be a nonsingular matrix such that

$$A = S \begin{bmatrix} A_1 & A_{12} \\ 0 & A_2 \end{bmatrix} S^{-1},$$

where $A_1 \in \mathbb{R}^{r \times r}$ is asymptotically stable, $A_{12} \in \mathbb{R}^{r \times (n-r)}$, and $A_2 \in \mathbb{R}^{(n-r) \times (n-r)}$. Then,

$$\mu_A^{\mathrm{s}}(A) = S \begin{bmatrix} 0 & B_{12\mathrm{s}} \\ 0 & \mu_A^{\mathrm{s}}(A_2) \end{bmatrix} S^{-1},$$

where $B_{12s} \in \mathbb{R}^{r \times (n-r)}$, and

$$\mu_A^u(A) = S \begin{bmatrix} \mu_A^u(A_1) & B_{12u} \\ 0 & \mu_A^u(A_2) \end{bmatrix} S^{-1},$$

where $B_{12u} \in \mathbb{R}^{r \times (n-r)}$ and $\mu_A^u(A_1)$ is nonsingular. Consequently,

$$\mathcal{R}\left(S \begin{bmatrix} I_r \\ 0 \end{bmatrix}\right) \subseteq \mathcal{S}_s(A).$$

If, in addition, $A_{12} = 0$, then

$$\mu_A^s(A) = S \begin{bmatrix} 0 & 0 \\ 0 & \mu_A^s(A_2) \end{bmatrix} S^{-1},$$

$$\mu_A^u(A) = S \begin{bmatrix} \mu_A^u(A_1) & 0 \\ 0 & \mu_A^u(A_2) \end{bmatrix} S^{-1},$$

$$\mathcal{S}_u(A) \subseteq \mathcal{R}\left(S \begin{bmatrix} 0 \\ I_{n-r} \end{bmatrix}\right).$$

Proof: This result follows from Fact 4.10.13.

Fact 11.23.2. Let $A \in \mathbb{R}^{n \times n}$, and let $S \in \mathbb{R}^{n \times n}$ be a nonsingular matrix such that

$$A = S \begin{bmatrix} A_1 & A_{12} \\ 0 & A_2 \end{bmatrix} S^{-1},$$

where $A_1 \in \mathbb{R}^{r \times r}$, $A_{12} \in \mathbb{R}^{r \times (n-r)}$, and $A_2 \in \mathbb{R}^{(n-r) \times (n-r)}$ satisfies $\mathrm{spec}(A_2) \subset$ CRHP. Then,

$$\mu_A^s(A) = S \begin{bmatrix} \mu_A^s(A_1) & C_{12s} \\ 0 & \mu_A^s(A_2) \end{bmatrix} S^{-1},$$

where $C_{12s} \in \mathbb{R}^{r \times (n-r)}$ and $\mu_A^s(A_2)$ is nonsingular, and

$$\mu_A^u(A) = S \begin{bmatrix} \mu_A^u(A_1) & C_{12u} \\ 0 & 0 \end{bmatrix} S^{-1},$$

where $C_{12u} \in \mathbb{R}^{r \times (n-r)}$. Consequently,

$$\mathcal{S}_s(A) \subseteq \mathcal{R}\left(S \begin{bmatrix} I_r \\ 0 \end{bmatrix}\right).$$

If, in addition, $A_{12} = 0$, then

$$\mu_A^s(A) = S \begin{bmatrix} \mu_A^s(A_1) & 0 \\ 0 & \mu_A^s(A_2) \end{bmatrix} S^{-1},$$

$$\mu_A^u(A) = S \begin{bmatrix} \mu_A^u(A_1) & 0 \\ 0 & 0 \end{bmatrix} S^{-1},$$

$$\mathcal{R}\left(S \begin{bmatrix} 0 \\ I_{n-r} \end{bmatrix}\right) \subseteq \mathcal{S}_u(A).$$

Fact 11.23.3. Let $A \in \mathbb{R}^{n \times n}$, and let $S \in \mathbb{R}^{n \times n}$ be a nonsingular matrix such that

$$A = S \begin{bmatrix} A_1 & A_{12} \\ 0 & A_2 \end{bmatrix} S^{-1},$$

where $A_1 \in \mathbb{R}^{r \times r}$ satisfies $\mathrm{spec}(A_1) \subset \mathrm{CRHP}$, $A_{12} \in \mathbb{R}^{r \times (n-r)}$, and $A_2 \in \mathbb{R}^{(n-r) \times (n-r)}$. Then,

$$\mu_A^{\mathrm{s}}(A) = S \begin{bmatrix} \mu_A^{\mathrm{s}}(A_1) & B_{12\mathrm{s}} \\ 0 & \mu_A^{\mathrm{s}}(A_2) \end{bmatrix} S^{-1},$$

where $\mu_A^{\mathrm{s}}(A_1)$ is nonsingular and $B_{12\mathrm{s}} \in \mathbb{R}^{r \times (n-r)}$, and

$$\mu_A^{\mathrm{u}}(A) = S \begin{bmatrix} 0 & B_{12\mathrm{u}} \\ 0 & \mu_A^{\mathrm{u}}(A_2) \end{bmatrix} S^{-1},$$

where $B_{12\mathrm{u}} \in \mathbb{R}^{r \times (n-r)}$. Consequently,

$$\mathcal{R}\left(S \begin{bmatrix} I_r \\ 0 \end{bmatrix} \right) \subseteq \mathcal{S}_{\mathrm{u}}(A).$$

If, in addition, $A_{12} = 0$, then

$$\mu_A^{\mathrm{s}}(A) = S \begin{bmatrix} \mu_A^{\mathrm{s}}(A_1) & 0 \\ 0 & \mu_A^{\mathrm{s}}(A_2) \end{bmatrix} S^{-1},$$

$$\mu_A^{\mathrm{u}}(A) = S \begin{bmatrix} 0 & 0 \\ 0 & \mu_A^{\mathrm{u}}(A_2) \end{bmatrix} S^{-1},$$

$$\mathcal{S}_{\mathrm{s}}(A) \subseteq \mathcal{R}\left(S \begin{bmatrix} 0 \\ I_{n-r} \end{bmatrix} \right).$$

Fact 11.23.4. Let $A \in \mathbb{R}^{n \times n}$, and let $S \in \mathbb{R}^{n \times n}$ be a nonsingular matrix such that

$$A = S \begin{bmatrix} A_1 & A_{12} \\ 0 & A_2 \end{bmatrix} S^{-1},$$

where $A_1 \in \mathbb{R}^{r \times r}$, $A_{12} \in \mathbb{R}^{r \times (n-r)}$, and $A_2 \in \mathbb{R}^{(n-r) \times (n-r)}$ is asymptotically stable. Then,

$$\mu_A^{\mathrm{s}}(A) = S \begin{bmatrix} \mu_A^{\mathrm{s}}(A_1) & C_{12\mathrm{s}} \\ 0 & 0 \end{bmatrix} S^{-1},$$

where $C_{12\mathrm{s}} \in \mathbb{R}^{r \times (n-r)}$, and

$$\mu_A^{\mathrm{u}}(A) = S \begin{bmatrix} \mu_A^{\mathrm{u}}(A_1) & C_{12\mathrm{u}} \\ 0 & \mu_A^{\mathrm{u}}(A_2) \end{bmatrix} S^{-1},$$

where $\mu_A^{\mathrm{u}}(A_2)$ is nonsingular and $C_{12\mathrm{u}} \in \mathbb{R}^{r \times (n-r)}$. Consequently,

$$\mathcal{S}_{\mathrm{u}}(A) \subseteq \mathcal{R}\left(S \begin{bmatrix} I_r \\ 0 \end{bmatrix} \right).$$

If, in addition, $A_{12} = 0$, then

$$\mu_A^{\mathrm{s}}(A) = S \begin{bmatrix} \mu_A^{\mathrm{s}}(A_1) & 0 \\ 0 & 0 \end{bmatrix} S^{-1},$$

$$\mu_A^{\mathrm{u}}(A) = S\begin{bmatrix} \mu_A^{\mathrm{u}}(A_1) & 0 \\ 0 & \mu_A^{\mathrm{u}}(A_2) \end{bmatrix}S^{-1},$$

$$\mathcal{R}\left(S\begin{bmatrix} 0 \\ I_{n-r} \end{bmatrix}\right) \subseteq \mathcal{S}_{\mathrm{s}}(A).$$

Fact 11.23.5. Let $A \in \mathbb{R}^{n\times n}$, and let $S \in \mathbb{R}^{n\times n}$ be a nonsingular matrix such that

$$A = S\begin{bmatrix} A_1 & A_{12} \\ 0 & A_2 \end{bmatrix}S^{-1},$$

where $A_1 \in \mathbb{R}^{r\times r}$ satisfies $\mathrm{spec}(A_1) \subset \mathrm{CRHP}$, $A_{12} \in \mathbb{R}^{r\times(n-r)}$, and $A_2 \in \mathbb{R}^{(n-r)\times(n-r)}$ is asymptotically stable. Then,

$$\mu_A^{\mathrm{s}}(A) = S\begin{bmatrix} \mu_A^{\mathrm{s}}(A_1) & C_{12\mathrm{s}} \\ 0 & 0 \end{bmatrix}S^{-1},$$

where $C_{12\mathrm{s}} \in \mathbb{R}^{r\times(n-r)}$ and $\mu_A^{\mathrm{s}}(A_1)$ is nonsingular, and

$$\mu_A^{\mathrm{u}}(A) = S\begin{bmatrix} 0 & C_{12\mathrm{u}} \\ 0 & \mu_A^{\mathrm{u}}(A_2) \end{bmatrix}S^{-1},$$

where $C_{12\mathrm{u}} \in \mathbb{R}^{r\times(n-r)}$ and $\mu_A^{\mathrm{u}}(A_2)$ is nonsingular. Consequently,

$$\mathcal{S}_{\mathrm{u}}(A) = \mathcal{R}\left(S\begin{bmatrix} I_r \\ 0 \end{bmatrix}\right).$$

If, in addition, $A_{12} = 0$, then

$$\mu_A^{\mathrm{s}}(A) = S\begin{bmatrix} \mu_A^{\mathrm{s}}(A_1) & 0 \\ 0 & 0 \end{bmatrix}S^{-1}$$

$$\mu_A^{\mathrm{u}}(A) = S\begin{bmatrix} 0 & 0 \\ 0 & \mu_A^{\mathrm{u}}(A_2) \end{bmatrix}S^{-1},$$

$$\mathcal{S}_{\mathrm{s}}(A) = \mathcal{R}\left(S\begin{bmatrix} 0 \\ I_{n-r} \end{bmatrix}\right).$$

Fact 11.23.6. Let $A \in \mathbb{R}^{n\times n}$, and let $S \in \mathbb{R}^{n\times n}$ be a nonsingular matrix such that

$$A = S\begin{bmatrix} A_1 & 0 \\ A_{21} & A_2 \end{bmatrix}S^{-1},$$

where $A_1 \in \mathbb{R}^{r\times r}$ is asymptotically stable, $A_{21} \in \mathbb{R}^{(n-r)\times r}$, and $A_2 \in \mathbb{R}^{(n-r)\times(n-r)}$. Then,

$$\mu_A^{\mathrm{s}}(A) = S\begin{bmatrix} 0 & 0 \\ B_{21\mathrm{s}} & \mu_A^{\mathrm{s}}(A_2) \end{bmatrix}S^{-1},$$

where $B_{21\mathrm{s}} \in \mathbb{R}^{(n-r)\times r}$, and

$$\mu_A^{\mathrm{u}}(A) = S\begin{bmatrix} \mu_A^{\mathrm{u}}(A_1) & 0 \\ B_{21\mathrm{u}} & \mu_A^{\mathrm{u}}(A_2) \end{bmatrix}S^{-1},$$

where $B_{21\mathrm{u}} \in \mathbb{R}^{(n-r) \times r}$ and $\mu_A^{\mathrm{u}}(A_1)$ is nonsingular. Consequently,

$$\mathcal{S}_{\mathrm{u}}(A) \subseteq \mathcal{R}\left(S \begin{bmatrix} 0 \\ I_{n-r} \end{bmatrix}\right).$$

If, in addition, $A_{21} = 0$, then

$$\mu_A^{\mathrm{s}}(A) = S \begin{bmatrix} 0 & 0 \\ 0 & \mu_A^{\mathrm{s}}(A_2) \end{bmatrix} S^{-1},$$

$$\mu_A^{\mathrm{u}}(A) = S \begin{bmatrix} \mu_A^{\mathrm{u}}(A_1) & 0 \\ 0 & \mu_A^{\mathrm{u}}(A_2) \end{bmatrix} S^{-1},$$

$$\mathcal{R}\left(S \begin{bmatrix} I_r \\ 0 \end{bmatrix}\right) \subseteq \mathcal{S}_{\mathrm{s}}(A).$$

Fact 11.23.7. Let $A \in \mathbb{R}^{n \times n}$, and let $S \in \mathbb{R}^{n \times n}$ be a nonsingular matrix such that

$$A = S \begin{bmatrix} A_1 & 0 \\ A_{21} & A_2 \end{bmatrix} S^{-1},$$

where $A_1 \in \mathbb{R}^{r \times r}$, $A_{21} \in \mathbb{R}^{(n-r) \times r}$, and $A_2 \in \mathbb{R}^{(n-r) \times (n-r)}$ satisfies $\operatorname{spec}(A_2) \subset$ CRHP. Then,

$$\mu_A^{\mathrm{s}}(A) = S \begin{bmatrix} \mu_A^{\mathrm{s}}(A_1) & 0 \\ C_{21\mathrm{s}} & \mu_A^{\mathrm{s}}(A_2) \end{bmatrix} S^{-1},$$

where $C_{21\mathrm{s}} \in \mathbb{R}^{(n-r) \times r}$ and $\mu_A^{\mathrm{s}}(A_2)$ is nonsingular, and

$$\mu_A^{\mathrm{u}}(A) = S \begin{bmatrix} \mu_A^{\mathrm{u}}(A_1) & 0 \\ C_{21\mathrm{u}} & 0 \end{bmatrix} S^{-1},$$

where $C_{21\mathrm{u}} \in \mathbb{R}^{(n-r) \times r}$. Consequently,

$$\mathcal{R}\left(S \begin{bmatrix} 0 \\ I_{n-r} \end{bmatrix}\right) \subseteq \mathcal{S}_{\mathrm{u}}(A).$$

If, in addition, $A_{21} = 0$, then

$$\mu_A^{\mathrm{s}}(A) = S \begin{bmatrix} \mu_A^{\mathrm{s}}(A_1) & 0 \\ 0 & \mu_A^{\mathrm{s}}(A_2) \end{bmatrix} S^{-1},$$

$$\mu_A^{\mathrm{u}}(A) = S \begin{bmatrix} \mu_A^{\mathrm{u}}(A_1) & 0 \\ 0 & 0 \end{bmatrix} S^{-1},$$

$$\mathcal{S}_{\mathrm{s}}(A) \subseteq \mathcal{R}\left(S \begin{bmatrix} I_r \\ 0 \end{bmatrix}\right).$$

Fact 11.23.8. Let $A \in \mathbb{R}^{n \times n}$, and let $S \in \mathbb{R}^{n \times n}$ be a nonsingular matrix such that

$$A = S \begin{bmatrix} A_1 & 0 \\ A_{21} & A_2 \end{bmatrix} S^{-1},$$

where $A_1 \in \mathbb{R}^{r \times r}$ is asymptotically stable, $A_{21} \in \mathbb{R}^{(n-r) \times r}$, and $A_2 \in \mathbb{R}^{(n-r) \times (n-r)}$ satisfies $\mathrm{spec}(A_2) \subset \mathrm{CRHP}$. Then,

$$\mu_A^s(A) = S \begin{bmatrix} 0 & 0 \\ C_{21s} & \mu_A^s(A_2) \end{bmatrix} S^{-1},$$

where $C_{21s} \in \mathbb{R}^{(n-r) \times r}$ and $\mu_A^s(A_2)$ is nonsingular, and

$$\mu_A^u(A) = S \begin{bmatrix} \mu_A^u(A_1) & 0 \\ C_{21u} & 0 \end{bmatrix} S^{-1},$$

where $C_{21u} \in \mathbb{R}^{(n-r) \times r}$ and $\mu_A^u(A_1)$ is nonsingular. Consequently,

$$\mathcal{S}_u(A) = \mathcal{R}\left(S \begin{bmatrix} 0 \\ I_{n-r} \end{bmatrix} \right).$$

If, in addition, $A_{21} = 0$, then

$$\mu_A^s(A) = S \begin{bmatrix} 0 & 0 \\ 0 & \mu_A^s(A_2) \end{bmatrix} S^{-1}$$

$$\mu_A^u(A) = S \begin{bmatrix} \mu_A^u(A_1) & 0 \\ 0 & 0 \end{bmatrix} S^{-1},$$

$$\mathcal{S}_s(A) = \mathcal{R}\left(S \begin{bmatrix} I_r \\ 0 \end{bmatrix} \right).$$

Fact 11.23.9. Let $A \in \mathbb{R}^{n \times n}$, and let $S \in \mathbb{R}^{n \times n}$ be a nonsingular matrix such that

$$A = S \begin{bmatrix} A_1 & 0 \\ A_{21} & A_2 \end{bmatrix} S^{-1},$$

where $A_1 \in \mathbb{R}^{r \times r}$, $A_{21} \in \mathbb{R}^{(n-r) \times r}$, and $A_2 \in \mathbb{R}^{(n-r) \times (n-r)}$ is asymptotically stable. Then,

$$\mu_A^s(A) = S \begin{bmatrix} \mu_A^s(A_1) & 0 \\ B_{21s} & 0 \end{bmatrix} S^{-1},$$

where $B_{21s} \in \mathbb{R}^{(n-r) \times r}$, and

$$\mu_A^u(A) = S \begin{bmatrix} \mu_A^u(A_1) & 0 \\ B_{21u} & \mu_A^u(A_2) \end{bmatrix} S^{-1},$$

where $B_{21u} \in \mathbb{R}^{(n-r) \times r}$ and $\mu_A^u(A_2)$ is nonsingular. Consequently,

$$\mathcal{R}\left(S \begin{bmatrix} 0 \\ I_{n-r} \end{bmatrix} \right) \subseteq \mathcal{S}(A).$$

If, in addition, $A_{21} = 0$, then

$$\mu_A^s(A) = S \begin{bmatrix} \mu_A^s(A_1) & 0 \\ 0 & 0 \end{bmatrix} S^{-1}$$

$$\mu_A^u(A) = S \begin{bmatrix} \mu_A^u(A_1) & 0 \\ 0 & \mu_A^u(A_2) \end{bmatrix} S^{-1},$$

$$\mathcal{S}_{\mathrm{u}}(A) \subseteq \mathcal{R}\left(S\begin{bmatrix} I_r \\ 0 \end{bmatrix}\right).$$

Fact 11.23.10. Let $A \in \mathbb{R}^{n \times n}$, and let $S \in \mathbb{R}^{n \times n}$ be a nonsingular matrix such that

$$A = S\begin{bmatrix} A_1 & 0 \\ A_{21} & A_2 \end{bmatrix} S^{-1},$$

where $A_1 \in \mathbb{R}^{r \times r}$ satisfies $\mathrm{spec}(A_1) \subset \mathrm{CRHP}$, $A_{21} \in \mathbb{R}^{(n-r) \times r}$, and $A_2 \in \mathbb{R}^{(n-r) \times (n-r)}$. Then,

$$\mu_A^{\mathrm{s}}(A) = S\begin{bmatrix} \mu_A^{\mathrm{s}}(A_1) & 0 \\ C_{12\mathrm{s}} & \mu_A^{\mathrm{s}}(A_2) \end{bmatrix} S^{-1},$$

where $C_{21\mathrm{s}} \in \mathbb{R}^{(n-r) \times r}$ and $\mu_A^{\mathrm{s}}(A_1)$ is nonsingular, and

$$\mu_A^{\mathrm{u}}(A) = S\begin{bmatrix} 0 & 0 \\ C_{21\mathrm{u}} & \mu_A^{\mathrm{u}}(A_2) \end{bmatrix} S^{-1},$$

where $C_{21\mathrm{u}} \in \mathbb{R}^{(n-r) \times r}$. Consequently,

$$\mathcal{S}_{\mathrm{s}}(A) \subseteq \mathcal{R}\left(S\begin{bmatrix} 0 \\ I_{n-r} \end{bmatrix}\right).$$

If, in addition, $A_{21} = 0$, then

$$\mu_A^{\mathrm{s}}(A) = S\begin{bmatrix} \mu_A^{\mathrm{s}}(A_1) & 0 \\ 0 & \mu_A^{\mathrm{s}}(A_2) \end{bmatrix} S^{-1},$$

$$\mu_A^{\mathrm{u}}(A) = S\begin{bmatrix} 0 & 0 \\ 0 & \mu_A^{\mathrm{u}}(A_2) \end{bmatrix} S^{-1},$$

$$\mathcal{R}\left(S\begin{bmatrix} I_r \\ 0 \end{bmatrix}\right) \subseteq \mathcal{S}_{\mathrm{u}}(A).$$

Fact 11.23.11. Let $A \in \mathbb{R}^{n \times n}$, and let $S \in \mathbb{R}^{n \times n}$ be a nonsingular matrix such that

$$A = S\begin{bmatrix} A_1 & 0 \\ A_{21} & A_2 \end{bmatrix} S^{-1},$$

where $A_1 \in \mathbb{R}^{r \times r}$ satisfies $\mathrm{spec}(A_1) \subset \mathrm{CRHP}$, $A_{21} \in \mathbb{R}^{(n-r) \times r}$, and $A_2 \in \mathbb{R}^{(n-r) \times (n-r)}$ is asymptotically stable. Then,

$$\mu_A^{\mathrm{s}}(A) = S\begin{bmatrix} \mu_A^{\mathrm{s}}(A_1) & 0 \\ C_{21\mathrm{s}} & 0 \end{bmatrix} S^{-1},$$

where $C_{21\mathrm{s}} \in \mathbb{R}^{(n-r) \times r}$ and $\mu_A^{\mathrm{s}}(A_1)$ is nonsingular, and

$$\mu_A^{\mathrm{u}}(A) = S\begin{bmatrix} 0 & 0 \\ C_{21\mathrm{u}} & \mu_A^{\mathrm{u}}(A_2) \end{bmatrix} S^{-1},$$

where $C_{21\mathrm{u}} \in \mathbb{R}^{(n-r) \times r}$ and $\mu_A^{\mathrm{u}}(A_2)$ is nonsingular. Consequently,

$$\mathcal{S}_{\mathrm{s}}(A) = \mathcal{R}\left(S\begin{bmatrix} 0 \\ I_{n-r} \end{bmatrix}\right).$$

If, in addition, $A_{21} = 0$, then

$$\mu_A^s(A) = S \begin{bmatrix} \mu_A^s(A_1) & 0 \\ 0 & 0 \end{bmatrix} S^{-1}$$

and

$$\mu_A^u(A) = S \begin{bmatrix} 0 & 0 \\ 0 & \mu_A^u(A_2) \end{bmatrix} S^{-1},$$

$$S_u(A) = \mathcal{R}\left(S \begin{bmatrix} I_r \\ 0 \end{bmatrix}\right).$$

11.24 Notes

The Laplace transform (11.2.10) is given in [1232, p. 34]. Computational methods are discussed in [701, 1040]. An arithmetic-mean–geometric-mean iteration for computing the matrix exponential and matrix logarithm is given in [1263].

The exponential function plays a central role in the theory of Lie groups, see [172, 303, 639, 743, 761, 1192, 1400]. Applications to robotics and kinematics are given in [1011, 1053, 1097]. Additional applications are discussed in [302].

The real logarithm is discussed in [368, 682, 1075, 1129]. The multiplicity and properties of logarithms are discussed in [475].

An asymptotically stable polynomial is traditionally called *Hurwitz*. Semistability is defined in [291] and developed in [190, 199]. Stability theory is treated in [635, 910, 1121] and [555, Chapter XV]. Solutions of the Lyapunov equation under weak conditions are considered in [1238]. Structured solutions of the Lyapunov equation are discussed in [815].

Linear and nonlinear difference equations are studied in [8, 289, 875].

Chapter Twelve

Linear Systems and Control Theory

This chapter considers linear state space systems with inputs and outputs. These systems are considered in both the time domain and frequency (Laplace) domain. Some basic results in control theory are also presented.

12.1 State Space and Transfer Function Models

Let $A \in \mathbb{R}^{n \times n}$ and $B \in \mathbb{R}^{n \times m}$, and, for $t \geq t_0$, consider the *state equation*

$$\dot{x}(t) = Ax(t) + Bu(t), \tag{12.1.1}$$

with the *initial condition*

$$x(t_0) = x_0. \tag{12.1.2}$$

In (12.1.1), $x \colon [0, \infty) \mapsto \mathbb{R}^n$ is the *state*, and $u \colon [0, \infty) \mapsto \mathbb{R}^m$ is the *input*. The function x is called the *solution* of (12.1.1).

The following result give the solution of (12.1.1) known as the *variation of constants formula*.

Proposition 12.1.1. For $t \geq t_0$ the state $x(t)$ of the dynamical equation (12.1.1) with initial condition (12.1.2) is given by

$$x(t) = e^{(t-t_0)A} x_0 + \int_{t_0}^{t} e^{(t-\tau)A} Bu(\tau) \, d\tau. \tag{12.1.3}$$

Proof. Multiplying (12.1.1) by e^{-tA} yields

$$e^{-tA}[\dot{x}(t) - Ax(t)] = e^{-tA} Bu(t),$$

which is equivalent to

$$\frac{d}{dt} \left[e^{-tA} x(t) \right] = e^{-tA} Bu(t).$$

Integrating over $[t_0, t]$ yields

$$e^{-tA} x(t) = e^{-t_0 A} x(t_0) + \int_{t_0}^{t} e^{-\tau A} Bu(\tau) \, d\tau.$$

Now, multiplying by e^{tA} yields (12.1.3).

Alternatively, let $x(t)$ be given by (12.1.3). Then, it follows from Leibniz's rule Fact 10.12.2 that

$$\dot{x}(t) = \frac{\mathrm{d}}{\mathrm{d}t}e^{(t-t_0)A}x_0 + \frac{\mathrm{d}}{\mathrm{d}t}\int_{t_0}^{t}e^{(t-\tau)A}Bu(\tau)\,\mathrm{d}\tau$$

$$= Ae^{(t-t_0)A}x_0 + \int_{t_0}^{t}Ae^{(t-\tau)A}Bu(\tau)\,\mathrm{d}\tau + Bu(t)$$

$$= Ax(t) + Bu(t). \qquad \square$$

For convenience, we can reset the clock by replacing t_0 by 0, and therefore assume without loss of generality that $t_0 = 0$. In this case, $x(t)$ for all $t \geq 0$ is given by

$$x(t) = e^{tA}x_0 + \int_{0}^{t}e^{(t-\tau)A}Bu(\tau)\,\mathrm{d}\tau. \tag{12.1.4}$$

If $u(t) = 0$ for all $t \geq 0$, then, for all $t \geq 0$, $x(t)$ is given by

$$x(t) = e^{tA}x_0. \tag{12.1.5}$$

Now, let $u(t) = \delta(t)v$, where $\delta(t)$ is the *unit impulse* at $t = 0$ and $v \in \mathbb{R}^m$. Loosely speaking, the unit impulse at $t = 0$ is zero for all $t \neq 0$ and is infinite at $t = 0$. More precisely, let $a < b$. Then,

$$\int_{a}^{b}\delta(\tau)\,\mathrm{d}\tau = \begin{cases} 0, & a > 0 \text{ or } b \leq 0, \\ 1, & a \leq 0 < b. \end{cases} \tag{12.1.6}$$

More generally, if $g\colon \mathcal{D} \to \mathbb{R}^n$, where $[a,b] \subseteq \mathcal{D} \subseteq \mathbb{R}$, $t_0 \in \mathcal{D}$, and g is continuous at t_0, then

$$\int_{a}^{b}\delta(\tau - t_0)g(\tau)\,\mathrm{d}\tau = \begin{cases} 0, & a > t_0 \text{ or } b \leq t_0, \\ g(t_0), & a \leq t_0 < b. \end{cases} \tag{12.1.7}$$

Consequently, for all $t \geq 0$, $x(t)$ is given by

$$x(t) = e^{tA}x_0 + e^{tA}Bv. \tag{12.1.8}$$

The unit impulse has the physical dimensions of 1/time. This convention makes the integral in (12.1.6) dimensionless.

Alternatively, let the input $u(t)$ be a *step function*, that is, $u(t) = 0$ for all $t < 0$ and $u(t) = v$ for all $t \geq 0$, where $v \in \mathbb{R}^m$. Then, by replacing $t - \tau$ by τ in the integral in (12.1.4), it follows that, for all $t \geq 0$,

$$x(t) = e^{tA}x_0 + \int_{0}^{t}e^{\tau A}\,\mathrm{d}\tau Bv. \tag{12.1.9}$$

Using Fact 11.13.14, (12.1.9) can be written for all $t \geq 0$ as

$$x(t) = e^{tA}x_0 + \left[A^{\mathrm{D}}(e^{tA} - I) + (I - AA^{\mathrm{D}}) \sum_{i=1}^{\mathrm{ind}\, A} (i!)^{-1} t^i A^{i-1} \right] Bv. \qquad (12.1.10)$$

If A is group invertible, then, for all $t \geq 0$, (12.1.10) becomes

$$x(t) = e^{tA}x_0 + \left[A^{\#}(e^{tA} - I) + t(I - AA^{\#}) \right] Bv. \qquad (12.1.11)$$

If, in addition, A is nonsingular, then, for all $t \geq 0$, (12.1.11) becomes

$$x(t) = e^{tA}x_0 + A^{-1}(e^{tA} - I)Bv. \qquad (12.1.12)$$

Next, consider the *output equation*

$$y(t) = Cx(t) + Du(t), \qquad (12.1.13)$$

where $t \geq 0$, $y(t) \in \mathbb{R}^l$ is the *output*, $C \in \mathbb{R}^{l \times n}$, and $D \in \mathbb{R}^{l \times m}$. Then, for all $t \geq 0$, the *total response* of (12.1.1), (12.1.13) is

$$y(t) = Ce^{tA}x_0 + \int_0^t Ce^{(t-\tau)A}Bu(\tau)\,\mathrm{d}\tau + Du(t). \qquad (12.1.14)$$

If $u(t) = 0$ for all $t \geq 0$, then the *free response* is given by

$$y(t) = Ce^{tA}x_0, \qquad (12.1.15)$$

while, if $x_0 = 0$, then the *forced response* is given by

$$y(t) = \int_0^t Ce^{(t-\tau)A}Bu(\tau)\,\mathrm{d}\tau + Du(t). \qquad (12.1.16)$$

Setting $u(t) = \delta(t)v$, where $v \in \mathbb{R}^m$, yields, for all $t > 0$, the total response

$$y(t) = Ce^{tA}x_0 + H(t)v, \qquad (12.1.17)$$

where, for all $t \geq 0$, the *impulse response function* $H(t)$ is defined by

$$H(t) \triangleq Ce^{tA}B + \delta(t)D. \qquad (12.1.18)$$

The corresponding forced response is the *impulse response*

$$y(t) = H(t)v = Ce^{tA}Bv + \delta(t)Dv. \qquad (12.1.19)$$

Alternatively, if $u(t) = v \in \mathbb{R}^m$ for all $t \geq 0$, then the total response is

$$y(t) = Ce^{tA}x_0 + \int_0^t Ce^{\tau A}\,\mathrm{d}\tau Bv + Dv, \qquad (12.1.20)$$

and the forced response is the *step response*

$$y(t) = \int_0^t H(\tau)\,\mathrm{d}\tau v = \int_0^t Ce^{\tau A}\,\mathrm{d}\tau Bv + Dv. \qquad (12.1.21)$$

In general, the forced response can be written as the *convolution integral*

$$y(t) = \int_0^t H(t - \tau)u(\tau)\,d\tau. \tag{12.1.22}$$

Setting $u(t) = \delta(t)v$ in (12.1.22) yields (12.1.20) by noting that

$$\int_0^t \delta(t - \tau)\delta(\tau)\,d\tau = \delta(t). \tag{12.1.23}$$

Proposition 12.1.2. Let $D = 0$ and $m = 1$, and assume that $x_0 = Bv$. Then, the free response and the impulse response are equal and given by

$$y(t) = Ce^{tA}x_0 = Ce^{tA}Bv. \tag{12.1.24}$$

12.2 Laplace Transform Analysis

Now, consider the linear system

$$\dot{x}(t) = Ax(t) + Bu(t), \tag{12.2.1}$$

$$y(t) = Cx(t) + Du(t), \tag{12.2.2}$$

with state $x(t) \in \mathbb{R}^n$, input $u(t) \in \mathbb{R}^m$, and *output* $y(t) \in \mathbb{R}^l$, where $t \geq 0$ and $x(0) = x_0$. Taking Laplace transforms yields

$$s\hat{x}(s) - x_0 = A\hat{x}(s) + B\hat{u}(s), \tag{12.2.3}$$

$$\hat{y}(s) = C\hat{x}(s) + D\hat{u}(s), \tag{12.2.4}$$

where

$$\hat{x}(s) \triangleq \mathcal{L}\{x(t)\} = \int_0^\infty e^{-st}x(t)\,dt, \tag{12.2.5}$$

$$\hat{u}(s) \triangleq \mathcal{L}\{u(t)\}, \tag{12.2.6}$$

and

$$\hat{y}(s) \triangleq \mathcal{L}\{y(t)\}. \tag{12.2.7}$$

Hence,

$$\hat{x}(s) = (sI - A)^{-1}x_0 + (sI - A)^{-1}B\hat{u}(s), \tag{12.2.8}$$

and thus

$$\hat{y}(s) = C(sI - A)^{-1}x_0 + \left[C(sI - A)^{-1}B + D\right]\hat{u}(s). \tag{12.2.9}$$

We can also obtain (12.2.9) from the time-domain expression for $y(t)$ given by (12.1.14). Using Proposition 11.2.2, it follows from (12.1.14) that

$$\hat{y}(s) = \mathcal{L}\{Ce^{tA}x_0\} + \mathcal{L}\left\{\int_0^t Ce^{(t-\tau)A}Bu(\tau)\,d\tau\right\} + D\hat{u}(s)$$

$$= C\mathcal{L}\{e^{tA}\}x_0 + C\mathcal{L}\{e^{tA}\}B\hat{u}(s) + D\hat{u}(s)$$

$$= C(sI - A)^{-1}x_0 + [C(sI - A)^{-1}B + D]\hat{u}(s), \qquad (12.2.10)$$

which coincides with (12.2.9). We define

$$G(s) \triangleq C(sI - A)^{-1}B + D. \qquad (12.2.11)$$

Note that $G \in \mathbb{R}^{l\times m}(s)$, that is, by Definition 4.7.2, G is a rational transfer function. Since $\mathcal{L}\{\delta(t)\} = 1$, it follows that

$$G(s) = \mathcal{L}\{H(t)\}. \qquad (12.2.12)$$

Using (4.7.2), G can be written as

$$G(s) = \frac{1}{\chi_A(s)}C(sI - A)^A B + D. \qquad (12.2.13)$$

It follows from (4.7.3) that G is a proper rational transfer function. Furthermore, G is a strictly proper rational transfer function if and only if $D = 0$, whereas G is an exactly proper rational transfer function if and only if $D \neq 0$. Finally, if A is nonsingular, then

$$G(0) = -CA^{-1}B + D. \qquad (12.2.14)$$

Let $A \in \mathbb{R}^{n\times n}$. If $|s| > \text{sprad}(A)$, then Proposition 9.4.13 implies that

$$(sI - A)^{-1} = \tfrac{1}{s}\left(I - \tfrac{1}{s}A\right)^{-1} = \sum_{k=0}^{\infty} \tfrac{1}{s^{k+1}}A^k, \qquad (12.2.15)$$

where the series is absolutely convergent, and thus

$$G(s) = D + \tfrac{1}{s}CB + \tfrac{1}{s^2}CAB + \cdots$$

$$= \sum_{k=0}^{\infty} \tfrac{1}{s^k}H_k, \qquad (12.2.16)$$

where, for $k \geq 0$, the *Markov parameter* $H_k \in \mathbb{R}^{l\times m}$ is defined by

$$H_k \triangleq \begin{cases} D, & k = 0, \\ CA^{k-1}B, & k \geq 1. \end{cases} \qquad (12.2.17)$$

It follows from (12.2.15) that $\lim_{s\to\infty}(sI - A)^{-1} = 0$, and thus

$$\lim_{s\to\infty} G(s) = D. \qquad (12.2.18)$$

Finally, it follows from Definition 4.7.3 that

$$\text{reldeg}\, G = \min\{k \geq 0\colon H_k \neq 0\}. \qquad (12.2.19)$$

12.3 The Unobservable Subspace and Observability

Let $A \in \mathbb{R}^{n \times n}$ and $C \in \mathbb{R}^{l \times n}$, and, for $t \geq 0$, consider the linear system

$$\dot{x}(t) = Ax(t), \tag{12.3.1}$$
$$x(0) = x_0, \tag{12.3.2}$$
$$y(t) = Cx(t). \tag{12.3.3}$$

Definition 12.3.1. The *unobservable subspace* $\mathcal{U}_{t_f}(A, C)$ of (A, C) at time $t_f > 0$ is the subspace

$$\mathcal{U}_{t_f}(A, C) \triangleq \{x_0 \in \mathbb{R}^n : \ y(t) = 0 \text{ for all } t \in [0, t_f]\}. \tag{12.3.4}$$

Let $t_f > 0$. Then, Definition 12.3.1 states that $x_0 \in \mathcal{U}_{t_f}(A, C)$ if and only if $y(t) = 0$ for all $t \in [0, t_f]$. Since $y(t) = 0$ for all $t \in [0, t_f]$ is the free response corresponding to $x_0 = 0$, it follows that $0 \in \mathcal{U}_{t_f}(A, C)$. Now, suppose there exists a nonzero vector $x_0 \in \mathcal{U}_{t_f}(A, C)$. Then, with $x(0) = x_0$, the free response is given by $y(t) = 0$ for all $t \in [0, t_f]$, and thus x_0 cannot be determined from knowledge of $y(t)$ for all $t \in [0, t_f]$.

The following result provides explicit expressions for $\mathcal{U}_{t_f}(A, C)$.

Lemma 12.3.2. Let $t_f > 0$. Then, the following subspaces are equal:

i) $\mathcal{U}_{t_f}(A, C)$.

ii) $\bigcap_{t \in [0, t_f]} \mathcal{N}(Ce^{tA})$.

iii) $\bigcap_{i=0}^{n-1} \mathcal{N}(CA^i)$.

iv) $\mathcal{N}\left(\begin{bmatrix} C \\ CA \\ \vdots \\ CA^{n-1} \end{bmatrix}\right)$.

v) $\mathcal{N}\left(\int_0^{t_f} e^{tA^{\mathrm{T}}} C^{\mathrm{T}} C e^{tA} \, \mathrm{d}t\right)$.

If, in addition, $\lim_{t_f \to \infty} \int_0^{t_f} e^{tA^{\mathrm{T}}} C^{\mathrm{T}} C e^{tA} \mathrm{d}t$ exists, then the following subspace is equal to *i)*–*v)*:

vi) $\mathcal{N}\left(\int_0^{\infty} e^{tA^{\mathrm{T}}} C^{\mathrm{T}} C e^{tA} \mathrm{d}t\right)$.

Proof. The proof is dual to the proof of Lemma 12.6.2. $\square$

Lemma 12.3.2 shows that $\mathcal{U}_{t_f}(A, C)$ is independent of t_f. We thus write $\mathcal{U}(A, C)$ for $\mathcal{U}_{t_f}(A, C)$, and call $\mathcal{U}(A, C)$ the *unobservable subspace* of (A, C). (A, C) is *observable* if $\mathcal{U}(A, C) = \{0\}$. For convenience, define the $nl \times n$ *observability matrix*

$$\mathcal{O}(A, C) \triangleq \begin{bmatrix} C \\ CA \\ \vdots \\ CA^{n-1} \end{bmatrix} \tag{12.3.5}$$

so that
$$\mathcal{U}(A,C) = \mathcal{N}[\mathcal{O}(A,C)]. \tag{12.3.6}$$

Define
$$p \triangleq n - \dim \mathcal{U}(A,C) = n - \operatorname{def} \mathcal{O}(A,C). \tag{12.3.7}$$

Corollary 12.3.3. For all $t_{\mathrm{f}} > 0$,
$$p = \dim \mathcal{U}(A,C)^{\perp} = \operatorname{rank} \mathcal{O}(A,C) = \operatorname{rank} \int\limits_{0}^{t_{\mathrm{f}}} e^{tA^{\mathrm{T}}} C^{\mathrm{T}} C e^{tA} dt. \tag{12.3.8}$$

If, in addition, $\lim_{t_{\mathrm{f}} \to \infty} \int_{0}^{t_{\mathrm{f}}} e^{tA^{\mathrm{T}}} C^{\mathrm{T}} C e^{tA} dt$ exists, then
$$p = \operatorname{rank} \int\limits_{0}^{\infty} e^{tA^{\mathrm{T}}} C^{\mathrm{T}} C e^{tA} dt. \tag{12.3.9}$$

Corollary 12.3.4. $\mathcal{U}(A,C)$ is an invariant subspace of A.

The following result shows that the unobservable subspace $\mathcal{U}(A,C)$ is unchanged by output injection
$$\dot{x}(t) = Ax(t) + Fy(t). \tag{12.3.10}$$

Proposition 12.3.5. Let $F \in \mathbb{R}^{n \times l}$. Then,
$$\mathcal{U}(A + FC, C) = \mathcal{U}(A, C). \tag{12.3.11}$$
In particular, (A, C) is observable if and only if $(A + FC, C)$ is observable.

Proof. The proof is dual to the proof of Proposition 12.6.5. $\qquad\square$

Let $\tilde{\mathcal{U}}(A,C) \subseteq \mathbb{R}^n$ be a subspace that is complementary to $\mathcal{U}(A,C)$. Then, $\tilde{\mathcal{U}}(A,C)$ is an *observable subspace* in the sense that, if $x_0 = x_0' + x_0''$, where $x_0' \in \tilde{\mathcal{U}}(A,C)$ is nonzero and $x_0'' \in \mathcal{U}(A,C)$, then it is possible to determine x_0' from knowledge of $y(t)$ for $t \in [0, t_{\mathrm{f}}]$. Using Proposition 3.5.3, let $\mathcal{P} \in \mathbb{R}^{n \times n}$ be the unique idempotent matrix such that $\mathcal{R}(\mathcal{P}) = \tilde{\mathcal{U}}(A,C)$ and $\mathcal{N}(\mathcal{P}) = \mathcal{U}(A,C)$. Then, $x_0' = \mathcal{P}x_0$. The following result constructs $\mathcal{P}$ and provides an expression for x_0' in terms of $y(t)$ for $\tilde{\mathcal{U}}(A,C) = \mathcal{U}(A,C)^{\perp}$. In this case, $\mathcal{P}$ is a projector.

Lemma 12.3.6. Let $t_{\mathrm{f}} > 0$, and define $\mathcal{P} \in \mathbb{R}^{n \times n}$ by
$$\mathcal{P} \triangleq \left(\int\limits_{0}^{t_{\mathrm{f}}} e^{tA^{\mathrm{T}}} C^{\mathrm{T}} C e^{tA} \, dt \right)^{+} \int\limits_{0}^{t_{\mathrm{f}}} e^{tA^{\mathrm{T}}} C^{\mathrm{T}} C e^{tA} \, dt. \tag{12.3.12}$$

Then, $\mathcal{P}$ is the projector onto $\mathcal{U}(A,C)^{\perp}$, and $\mathcal{P}_{\perp}$ is the projector onto $\mathcal{U}(A,C)$. Hence,
$$\mathcal{R}(\mathcal{P}) = \mathcal{N}(\mathcal{P}_{\perp}) = \mathcal{U}(A,C)^{\perp}, \tag{12.3.13}$$
$$\mathcal{N}(\mathcal{P}) = \mathcal{R}(\mathcal{P}_{\perp}) = \mathcal{U}(A,C), \tag{12.3.14}$$
$$\operatorname{rank} \mathcal{P} = \operatorname{def} \mathcal{P}_{\perp} = \dim \mathcal{U}(A,C)^{\perp} = p, \tag{12.3.15}$$
$$\operatorname{def} \mathcal{P} = \operatorname{rank} \mathcal{P}_{\perp} = \dim \mathcal{U}(A,C) = n - p. \tag{12.3.16}$$

If $x_0 = x_0' + x_0''$, where $x_0' \in \mathcal{U}(A, C)^\perp$ and $x_0'' \in \mathcal{U}(A, C)$, then

$$x_0' = \mathcal{P}x_0 = \left(\int_0^{t_f} e^{tA^{\mathrm{T}}} C^{\mathrm{T}} C e^{tA} \, \mathrm{d}t \right)^+ \int_0^{t_f} e^{tA^{\mathrm{T}}} C^{\mathrm{T}} y(t) \, \mathrm{d}t. \tag{12.3.17}$$

Finally, (A, C) is observable if and only if $\mathcal{P} = I_n$. In this case, for all $x_0 \in \mathbb{R}^n$,

$$x_0 = \left(\int_0^{t_f} e^{tA^{\mathrm{T}}} C^{\mathrm{T}} C e^{tA} \, \mathrm{d}t \right)^{-1} \int_0^{t_f} e^{tA^{\mathrm{T}}} C^{\mathrm{T}} y(t) \, \mathrm{d}t. \tag{12.3.18}$$

Lemma 12.3.7. Let $\alpha \in \mathbb{R}$. Then,

$$\mathcal{U}(A + \alpha I, C) = \mathcal{U}(A, C). \tag{12.3.19}$$

The following result uses a coordinate transformation to characterize the observable dynamics of a system.

Theorem 12.3.8. There exists an orthogonal matrix $S \in \mathbb{R}^{n \times n}$ such that

$$A = S \begin{bmatrix} A_1 & 0 \\ A_{21} & A_2 \end{bmatrix} S^{-1}, \qquad C = \begin{bmatrix} C_1 & 0 \end{bmatrix} S^{-1}, \tag{12.3.20}$$

where $A_1 \in \mathbb{R}^{p \times p}$, $C_1 \in \mathbb{R}^{l \times p}$, and (A_1, C_1) is observable.

Proof. The proof is dual to the proof of Theorem 12.6.8. $\qquad \square$

Proposition 12.3.9. Let $S \in \mathbb{R}^{n \times n}$, and assume that S is orthogonal. Then, the following conditions are equivalent:

i) A and C have the form (12.3.20), where $A_1 \in \mathbb{R}^{p \times p}$, $C_1 \in \mathbb{R}^{l \times p}$, and (A_1, C_1) is observable.

ii) $\mathcal{U}(A, C) = \mathcal{R}\left(S \begin{bmatrix} 0 \\ I_{n-p} \end{bmatrix} \right)$.

iii) $\mathcal{U}(A, C)^\perp = \mathcal{R}\left(S \begin{bmatrix} I_p \\ 0 \end{bmatrix} \right)$.

iv) $\mathcal{P} = S \begin{bmatrix} I_p & 0 \\ 0 & 0 \end{bmatrix} S^{\mathrm{T}}$.

Proposition 12.3.10. Let $S \in \mathbb{R}^{n \times n}$, and assume that S is nonsingular. Then, the following conditions are equivalent:

i) A and C have the form (12.3.20), where $A_1 \in \mathbb{R}^{p \times p}$, $C_1 \in \mathbb{R}^{l \times p}$, and (A_1, C_1) is observable.

ii) $\mathcal{U}(A, C) = \mathcal{R}\left(S \begin{bmatrix} 0 \\ I_{n-p} \end{bmatrix} \right)$ and $\tilde{\mathcal{U}}(A, C) = \mathcal{R}\left(S \begin{bmatrix} I_p \\ 0 \end{bmatrix} \right)$.

iii) $\mathcal{P} = S \begin{bmatrix} I_p & 0 \\ 0 & 0 \end{bmatrix} S^{-1}$.

Definition 12.3.11. Let $S \in \mathbb{R}^{n \times n}$, assume that S is nonsingular, and let A and C have the form (12.3.20), where $A_1 \in \mathbb{R}^{p \times p}$, $C_1 \in \mathbb{R}^{l \times p}$, and (A_1, C_1) is observable. Then, the *unobservable spectrum* of (A, C) is $\mathrm{spec}(A_2)$, while the *unobservable multispectrum* of (A, C) is $\mathrm{mspec}(A_2)$. Furthermore, $\lambda \in \mathbb{C}$ is an *unobservable eigenvalue* of (A, C) if $\lambda \in \mathrm{spec}(A_2)$.

Definition 12.3.12. The *observability pencil* $\mathcal{O}_{A,C}(s)$ is the pencil

$$\mathcal{O}_{A,C} = P_{\left[\begin{smallmatrix} A \\ -C \end{smallmatrix}\right], \left[\begin{smallmatrix} I \\ 0 \end{smallmatrix}\right]}, \tag{12.3.21}$$

that is,

$$\mathcal{O}_{A,C}(s) = \left[\begin{array}{c} sI - A \\ C \end{array} \right]. \tag{12.3.22}$$

Proposition 12.3.13. Let $\lambda \in \mathrm{spec}(A)$. Then, λ is an unobservable eigenvalue of (A, C) if and only if

$$\mathrm{rank} \left[\begin{array}{c} \lambda I - A \\ C \end{array} \right] < n. \tag{12.3.23}$$

Proof. The proof is dual to the proof of Proposition 12.6.13. $\square$

Proposition 12.3.14. Let $\lambda \in \mathrm{mspec}(A)$ and $F \in \mathbb{R}^{n \times m}$. Then, λ is an unobservable eigenvalue of (A, C) if and only if λ is an unobservable eigenvalue of $(A + FC, C)$.

Proof. The proof is dual to the proof of Proposition 12.6.14. $\square$

Proposition 12.3.15. Assume that (A, C) is observable. Then, the Smith form of $\mathcal{O}_{A,C}$ is $\left[\begin{array}{c} I_n \\ 0_{l \times n} \end{array} \right]$.

Proof. The proof is dual to the proof of Proposition 12.6.15. $\square$

Proposition 12.3.16. $S \in \mathbb{R}^{n \times n}$, assume that S is nonsingular, and let A and C have the form (12.3.20), where $A_1 \in \mathbb{R}^{p \times p}$, $C_1 \in \mathbb{R}^{l \times p}$, and (A_1, C_1) is observable. Furthermore, let $p_1, \ldots, p_{n-p}$ be the similarity invariants of A_2, where, for all $i \in \{1, \ldots, n-p-1\}$, p_i divides p_{i+1}. Then, there exist unimodular matrices $S_1 \in \mathbb{R}^{(n+l) \times (n+l)}[s]$ and $S_2 \in \mathbb{R}^{n \times n}[s]$ such that, for all $s \in \mathbb{C}$,

$$\left[\begin{array}{c} sI - A \\ C \end{array} \right] = S_1(s) \left[\begin{array}{ccccc} I_p & & & & \\ & p_1(s) & & & \\ & & \ddots & & \\ & & & p_{n-p}(s) & \\ & & 0_{l \times n} & & \end{array} \right] S_2(s). \tag{12.3.24}$$

Consequently,

$$\mathrm{Szeros}(\mathcal{O}_{A,C}) = \bigcup_{i=1}^{n-p} \mathrm{roots}(p_i) = \mathrm{roots}(\chi_{A_2}) = \mathrm{spec}(A_2) \tag{12.3.25}$$

and

$$\text{mSzeros}(\mathcal{O}_{A,C}) = \bigcup_{i=1}^{n-p} \text{mroots}(p_i) = \text{mroots}(\chi_{A_2}) = \text{mspec}(A_2). \qquad (12.3.26)$$

Proof. The proof is dual to the proof of Proposition 12.6.16. □

Proposition 12.3.17. Let $s \in \mathbb{C}$. Then,

$$\mathcal{O}(A,C) \subseteq \text{Re}\,\mathcal{R}\left(\begin{bmatrix} sI - A \\ C \end{bmatrix}\right). \qquad (12.3.27)$$

Proof. The proof is dual to the proof of Proposition 12.6.17. □

The next result characterizes observability in several equivalent ways.

Theorem 12.3.18. The following statements are equivalent:

i) (A,C) is observable.

ii) There exists $t > 0$ such that $\int_0^t e^{\tau A^{\text{T}}} C^{\text{T}} C e^{\tau A} \, d\tau$ is positive definite.

iii) $\int_0^t e^{\tau A^{\text{T}}} C^{\text{T}} C e^{\tau A} \, d\tau$ is positive definite for all $t > 0$.

iv) $\text{rank}\,\mathcal{O}(A,C) = n$.

v) Every eigenvalue of (A,C) is observable.

If, in addition, $\lim_{t \to \infty} \int_0^t e^{\tau A^{\text{T}}} C^{\text{T}} C e^{\tau A} \, d\tau$ exists, then the following condition is equivalent to *i)*–*v)*:

vi) $\int_0^\infty e^{t A^{\text{T}}} C^{\text{T}} C e^{tA} \, dt$ is positive definite.

Proof. The proof is dual to the proof of Theorem 12.6.18. □

The following result, which is a restatement of the equivalence of *i)* and *v)* of Theorem 12.3.18, is the *PBH test for observability*.

Corollary 12.3.19. The following statements are equivalent:

i) (A,C) is observable.

ii) For all $s \in \mathbb{C}$,

$$\text{rank}\begin{bmatrix} sI - A \\ C \end{bmatrix} = n. \qquad (12.3.28)$$

The following result implies that arbitrary eigenvalue placement is possible for (12.3.10) if (A,C) is observable.

Proposition 12.3.20. The pair (A,C) is observable if and only if, for every polynomial $p \in \mathbb{R}[s]$ such that $\deg p = n$, there exists a matrix $F \in \mathbb{R}^{m \times n}$ such that $\text{mspec}(A + FC) = \text{mroots}(p)$.

Proof. The proof is dual to the proof of Proposition 12.6.20. □

12.4 Observable Asymptotic Stability

Let $A \in \mathbb{R}^{n \times n}$ and $C \in \mathbb{R}^{l \times n}$, and define $p \triangleq n - \dim \mathcal{U}(A, C)$.

Definition 12.4.1. (A, C) is *observably asymptotically stable* if

$$\mathcal{S}_{\mathrm{u}}(A) \subseteq \mathcal{U}(A, C). \tag{12.4.1}$$

Proposition 12.4.2. Let $F \in \mathbb{R}^{n \times l}$. Then, (A, C) is observably asymptotically stable if and only if $(A + FC, C)$ is observably asymptotically stable.

Proposition 12.4.3. The following statements are equivalent:

i) (A, C) is observably asymptotically stable.

ii) There exists an orthogonal matrix $S \in \mathbb{R}^{n \times n}$ such that (12.3.20) holds, where $A_1 \in \mathbb{R}^{p \times p}$ is asymptotically stable and $C_1 \in \mathbb{R}^{l \times p}$.

iii) There exists a nonsingular matrix $S \in \mathbb{R}^{n \times n}$ such that (12.3.20) holds, where $A_1 \in \mathbb{R}^{p \times p}$ is asymptotically stable and $C_1 \in \mathbb{R}^{l \times p}$.

iv) $\lim_{t \to \infty} C e^{tA} = 0$.

v) The positive-semidefinite matrix $P \in \mathbb{R}^{n \times n}$ defined by

$$P \triangleq \int_0^\infty e^{tA^{\mathrm{T}}} C^{\mathrm{T}} C e^{tA} \, \mathrm{d}t \tag{12.4.2}$$

exists.

vi) There exists a positive-semidefinite matrix $P \in \mathbb{R}^{n \times n}$ satisfying

$$A^{\mathrm{T}} P + PA + C^{\mathrm{T}} C = 0. \tag{12.4.3}$$

In this case, the positive-semidefinite matrix $P \in \mathbb{R}^{n \times n}$ defined by (12.4.2) satisfies (12.4.3).

Proof. The proof is dual to the proof of Proposition 12.7.3. $\square$

The matrix P defined by (12.4.2) is the *observability Gramian*, and (12.4.3) is the *observation Lyapunov equation*.

Proposition 12.4.4. Assume that (A, C) is observably asymptotically stable, let $P \in \mathbb{R}^{n \times n}$ be the positive-semidefinite matrix defined by (12.4.2), and define $\mathcal{P} \in \mathbb{R}^{n \times n}$ by (12.3.12). Then, the following statements hold:

i) $PP^+ = \mathcal{P}$.

ii) $\mathcal{R}(P) = \mathcal{R}(\mathcal{P}) = \mathcal{U}(A, C)^\perp$.

iii) $\mathcal{N}(P) = \mathcal{N}(\mathcal{P}) = \mathcal{U}(A, C)$.

iv) $\operatorname{rank} P = \operatorname{rank} \mathcal{P} = p$.

v) P is the only positive-semidefinite solution of (12.4.3) whose rank is p.

Proof. The proof is dual to the proof of Proposition 12.7.4. $\square$

Proposition 12.4.5. Assume that (A, C) is observably asymptotically stable, let $P \in \mathbb{R}^{n \times n}$ be the positive-semidefinite matrix defined by (12.4.2), and let $\hat{P} \in \mathbb{R}^{n \times n}$. Then, the following statements are equivalent:

i) $\hat{P}$ is positive semidefinite and satisfies (12.4.3).

ii) There exists a positive-semidefinite matrix $P_0 \in \mathbb{R}^{n \times n}$ such that $\hat{P} = P+P_0$ and $A^{\mathrm{T}}P_0 + P_0 A = 0$.

In this case,

$$\operatorname{rank} \hat{P} = p + \operatorname{rank} P_0 \tag{12.4.4}$$

and

$$\operatorname{rank} P_0 \leq \sum_{\substack{\lambda \in \operatorname{spec}(A) \\ \lambda \in j\mathbb{R}}} \operatorname{gmult}_A(\lambda). \tag{12.4.5}$$

Proof. The proof is dual to the proof of Proposition 12.7.5. $\square$

Proposition 12.4.6. The following statements are equivalent:

i) (A, C) is observably asymptotically stable, every imaginary eigenvalue of A is semisimple, and A has no ORHP eigenvalues.

ii) (12.4.3) has a positive-definite solution $P \in \mathbb{R}^{n \times n}$.

Proof. The proof is dual to the proof of Proposition 12.7.6. $\square$

Proposition 12.4.7. The following statements are equivalent:

i) (A, C) is observably asymptotically stable, and A has no imaginary eigenvalues.

ii) (12.4.3) has exactly one positive-semidefinite solution $P \in \mathbb{R}^{n \times n}$.

In this case, $P \in \mathbb{R}^{n \times n}$ is given by (12.4.2) and satisfies $\operatorname{rank} P = p$.

Proof. The proof is dual to the proof of Proposition 12.7.7. $\square$

Corollary 12.4.8. Assume that A is asymptotically stable. Then, the positive-semidefinite matrix $P \in \mathbb{R}^{n \times n}$ defined by (12.4.2) is the unique solution of (12.4.3) and satisfies $\operatorname{rank} P = p$.

Proof. The proof is dual to the proof of Corollary 12.7.8. $\square$

Proposition 12.4.9. The following statements are equivalent:

i) (A, C) is observable, and A is asymptotically stable.

ii) (12.4.3) has exactly one positive-semidefinite solution $P \in \mathbb{R}^{n \times n}$, and P is positive definite.

In this case, $P \in \mathbb{R}^{n \times n}$ is given by (12.4.2).

Proof. The proof is dual to the proof of Proposition 12.7.9. $\square$

Corollary 12.4.10. Assume that A is asymptotically stable. Then, the positive-semidefinite matrix $P \in \mathbb{R}^{n \times n}$ defined by (12.4.2) exists. Furthermore, P is positive definite if and only if (A, C) is observable.

12.5 Detectability

Let $A \in \mathbb{R}^{n \times n}$ and $C \in \mathbb{R}^{l \times n}$, and define $p \triangleq n - \dim \mathcal{U}(A, C)$.

Definition 12.5.1. (A, C) is *detectable* if

$$\mathcal{U}(A, C) \subseteq \mathcal{S}_{\mathrm{s}}(A). \tag{12.5.1}$$

Proposition 12.5.2. Let $F \in \mathbb{R}^{n \times l}$. Then, (A, C) is detectable if and only if $(A + FC, C)$ is detectable.

Proposition 12.5.3. The following statements are equivalent:

i) (A, C) is detectable.

ii) There exists an orthogonal matrix $S \in \mathbb{R}^{n \times n}$ such that (12.3.20) holds, where $A_1 \in \mathbb{R}^{p \times p}$, $C_1 \in \mathbb{R}^{l \times p}$, (A_1, C_1) is observable, and $A_2 \in \mathbb{R}^{(n-p) \times (n-p)}$ is asymptotically stable.

iii) There exists a nonsingular matrix $S \in \mathbb{R}^{n \times n}$ such that (12.3.20) holds, where $A_1 \in \mathbb{R}^{p \times p}$, $C_1 \in \mathbb{R}^{l \times p}$, (A_1, C_1) is observable, and $A_2 \in \mathbb{R}^{(n-p) \times (n-p)}$ is asymptotically stable.

iv) Every CRHP eigenvalue of (A, C) is observable.

Proof. The proof is dual to the proof of Proposition 12.8.3. $\square$

The following result, which is a restatement of the equivalence of *i)* and *iv)* of Proposition 12.5.3, is the *PBH test for detectability*.

Corollary 12.5.4. The following statements are equivalent:

i) (A, C) is detectable.

ii) For all $s \in \mathrm{CRHP}$,

$$\mathrm{rank} \begin{bmatrix} sI - A \\ C \end{bmatrix} = n. \tag{12.5.2}$$

Proposition 12.5.5. The following statements are equivalent:

i) A is asymptotically stable.

ii) (A, C) is observably asymptotically stable and detectable.

Proof. The proof is dual to the proof of Proposition 12.8.5. $\square$

Corollary 12.5.6. The following statements are equivalent:

i) There exists a positive-semidefinite matrix $P \in \mathbb{R}^{n \times n}$ satisfying (12.4.3), and (A, C) is detectable.

ii) A is asymptotically stable.

Proof. The proof is dual to the proof of Proposition 12.8.6. $\square$

12.6 The Controllable Subspace and Controllability

Let $A \in \mathbb{R}^{n \times n}$ and $B \in \mathbb{R}^{n \times m}$, and, for $t \geq 0$, consider the linear system

$$\dot{x}(t) = Ax(t) + Bu(t), \tag{12.6.1}$$
$$x(0) = 0. \tag{12.6.2}$$

Definition 12.6.1. The *controllable subspace* $\mathcal{C}_{t_f}(A, B)$ of (A, B) at time $t_f > 0$ is the subspace

$$\mathcal{C}_{t_f}(A, B) \triangleq \{x_f \in \mathbb{R}^n: \quad \text{there exists a continuous control } u: \ [0, t_f] \mapsto \mathbb{R}^m$$

such that the solution $x(\cdot)$ of (12.6.1), (12.6.2) satisfies $x(t_f) = x_f\}$. (12.6.3)

Let $t_f > 0$. Then, Definition 12.6.1 states that $x_f \in \mathcal{C}_{t_f}(A, B)$ if and only if there exists a continuous control $u: \ [0, t_f] \mapsto \mathbb{R}^m$ such that

$$x_f = \int_0^{t_f} e^{(t_f - t)A} Bu(t) \, \mathrm{d}t. \tag{12.6.4}$$

The following result provides explicit expressions for $\mathcal{C}_{t_f}(A, B)$.

Lemma 12.6.2. Let $t_f > 0$. Then, the following subspaces are equal:

i) $\mathcal{C}_{t_f}(A, B)$.

ii) $\left[\bigcap_{t \in [0, t_f]} \mathcal{N}\left(B^T e^{tA^T} \right) \right]^{\perp}$.

iii) $\left[\bigcap_{i=0}^{n-1} \mathcal{N}(B^T A^{iT}) \right]^{\perp}$.

iv) $\mathcal{R}([\ B \quad AB \quad \cdots \quad A^{n-1}B \ ])$.

v) $\mathcal{R}\left(\int_0^{t_f} e^{tA} BB^T e^{tA^T} \, \mathrm{d}t \right)$.

If, in addition, $\lim_{t_f \to \infty} \int_0^{t_f} e^{tA} BB^T e^{tA^T} \, \mathrm{d}t$ exists, then the following subspace is equal to *i)–v)*:

vi) $\mathcal{R}\left(\int_0^{\infty} e^{tA} BB^T e^{tA^T} \, \mathrm{d}t \right)$.

Proof. To prove that $i) \subseteq ii)$, let $\eta \in \bigcap_{t\in[0,t_\mathrm{f}]}\mathcal{N}\left(B^\mathrm{T}e^{tA^\mathrm{T}}\right)$ so that $\eta^\mathrm{T}e^{tA}B = 0$ for all $t \in [0,t_\mathrm{f}]$. Now, let $u\colon [0,t_\mathrm{f}] \mapsto \mathbb{R}^m$ be continuous. Then, $\eta^\mathrm{T}\int_0^{t_\mathrm{f}} e^{(t_\mathrm{f}-t)A}Bu(t)\,\mathrm{d}t = 0$, which implies that $\eta \in \mathcal{C}_{t_\mathrm{f}}(A,B)^\perp$.

To prove that $ii) \subseteq iii)$, let $\eta \in \bigcap_{i=0}^{n-1}\mathcal{N}(B^\mathrm{T}A^{i\mathrm{T}})$ so that $\eta^\mathrm{T}A^iB = 0$ for all $i \in \{0,1,\ldots,n-1\}$. It follows from the Cayley-Hamilton theorem given by Theorem 4.4.7 that $\eta^\mathrm{T}A^iB = 0$ for all $i \geq 0$. Now, let $t \in [0,t_\mathrm{f}]$. Then, $\eta^\mathrm{T}e^{tA}B = \sum_{i=0}^\infty t^i(i!)^{-1}\eta^\mathrm{T}A^iB = 0$, and thus $\eta \in \mathcal{N}\left(B^\mathrm{T}e^{tA^\mathrm{T}}\right)$.

To show that $iii) \subseteq iv)$, let $\eta \in \mathcal{R}\left(\begin{bmatrix} B & AB & \cdots & A^{n-1}B \end{bmatrix}\right)^\perp$. Then, $\eta \in \mathcal{N}\left(\begin{bmatrix} B & AB & \cdots & A^{n-1}B \end{bmatrix}^\mathrm{T}\right)$, which implies that $\eta^\mathrm{T}A^iB = 0$ for all $i \in \{0,1,\ldots, n-1\}$.

To prove that $iv) \subseteq v)$, let $\eta \in \mathcal{N}\left(\int_0^{t_\mathrm{f}} e^{tA}BB^\mathrm{T}e^{tA^\mathrm{T}}\,\mathrm{d}t\right)$. Then,

$$\eta^\mathrm{T}\int_0^{t_\mathrm{f}} e^{tA}BB^\mathrm{T}e^{tA^\mathrm{T}}\,\mathrm{d}t\,\eta = 0,$$

which implies that $\eta^\mathrm{T}e^{tA}B = 0$ for all $t \in [0,t_\mathrm{f}]$. Differentiating with respect to t and setting $t = 0$ implies that $\eta^\mathrm{T}A^iB = 0$ for all $i \in \{0,1,\ldots,n-1\}$. Hence, $\eta \in \mathcal{R}\left(\begin{bmatrix} B & AB & \cdots & A^{n-1}B \end{bmatrix}\right)^\perp$.

To prove that $v) \subseteq i)$, let $\eta \in \mathcal{C}_{t_\mathrm{f}}(A,B)^\perp$. Then, $\eta^\mathrm{T}\int_0^{t_\mathrm{f}} e^{(t_\mathrm{f}-t)A}Bu(t)\,\mathrm{d}t = 0$ for all continuous $u\colon [0,t_\mathrm{f}] \mapsto \mathbb{R}^m$. Letting $u(t) = B^\mathrm{T}e^{(t_\mathrm{f}-t)A^\mathrm{T}}\eta^\mathrm{T}$, implies that $\eta^\mathrm{T}\int_0^{t_\mathrm{f}} e^{tA}BB^\mathrm{T}e^{tA^\mathrm{T}}\,\mathrm{d}t\,\eta = 0$, and thus $\eta \in \mathcal{N}\left(\int_0^{t_\mathrm{f}} e^{tA}BB^\mathrm{T}e^{tA^\mathrm{T}}\,\mathrm{d}t\right)$. $\square$

Lemma 12.6.2 shows that $\mathcal{C}_{t_\mathrm{f}}(A,B)$ is independent of t_f. We thus write $\mathcal{C}(A,B)$ for $\mathcal{C}_{t_\mathrm{f}}(A,B)$, and call $\mathcal{C}(A,B)$ the *controllable subspace* of (A,B). (A,B) is *controllable* if $\mathcal{C}(A,B) = \mathbb{R}^n$. For convenience, define the $m \times nm$ *controllability matrix*

$$\mathcal{K}(A,B) \triangleq \begin{bmatrix} B & AB & \cdots & A^{n-1}B \end{bmatrix} \tag{12.6.5}$$

so that

$$\mathcal{C}(A,B) = \mathcal{R}[\mathcal{K}(A,B)]. \tag{12.6.6}$$

Define

$$q \triangleq \dim \mathcal{C}(A,B) = \operatorname{rank}\mathcal{K}(A,B). \tag{12.6.7}$$

Corollary 12.6.3. For all $t_\mathrm{f} > 0$,

$$q = \dim \mathcal{C}(A,B) = \operatorname{rank}\mathcal{K}(A,B) = \operatorname{rank}\int_0^{t_\mathrm{f}} e^{tA}BB^\mathrm{T}e^{tA^\mathrm{T}}\,\mathrm{d}t. \tag{12.6.8}$$

If, in addition, $\lim_{t_f \to \infty} \int_0^{t_f} e^{tA} BB^T e^{tA^T} dt$ exists, then

$$q = \text{rank} \int_0^\infty e^{tA} BB^T e^{tA^T} dt. \tag{12.6.9}$$

Corollary 12.6.4. $\mathcal{C}(A, B)$ is an invariant subspace of A.

The following result shows that the controllable subspace $\mathcal{C}(A, B)$ is unchanged by full-state feedback $u(t) = Kx(t) + v(t)$.

Proposition 12.6.5. Let $K \in \mathbb{R}^{m \times n}$. Then,

$$\mathcal{C}(A + BK, B) = \mathcal{C}(A, B). \tag{12.6.10}$$

In particular, (A, B) is controllable if and only if $(A + BK, B)$ is controllable.

Proof. Note that

$$
\begin{aligned}
\mathcal{C}(A &+ BK, B) \\
&= \mathcal{R}[\mathcal{K}(A + BK, B)] \\
&= \mathcal{R}([\ B \quad AB + BKB \quad A^2B + ABKB + BKAB + BKBKB \quad \cdots\ ]) \\
&= \mathcal{R}[\mathcal{K}(A, B)] = \mathcal{C}(A, B). \qquad \square
\end{aligned}
$$

Let $\tilde{\mathcal{C}}(A, B) \subseteq \mathbb{R}^n$ be a subspace that is complementary to $\mathcal{C}(A, B)$. Then, $\tilde{\mathcal{C}}(A, B)$ is an *uncontrollable subspace* in the sense that, if $x_f = x_f' + x_f'' \in \mathbb{R}^n$, where $x_f' \in \mathcal{C}(A, B)$ and $x_f'' \in \tilde{\mathcal{C}}(A, B)$ is nonzero, then there exists a continuous control $u : [0, t_f] \to \mathbb{R}^m$ such that $x(t_f) = x_f'$, but there exists no continuous control such that $x(t_f) = x_f$. Using Proposition 3.5.3, let $\mathcal{Q} \in \mathbb{R}^{n \times n}$ be the unique idempotent matrix such that $\mathcal{R}(\mathcal{Q}) = \mathcal{C}(A, B)$ and $\mathcal{N}(\mathcal{Q}) = \tilde{\mathcal{C}}(A, B)$. Then, $x_f' = \mathcal{Q}x_f$. The following result constructs $\mathcal{Q}$ and a continuous control $u(\cdot)$ that yields $x(t_f) = x_f'$ for $\tilde{\mathcal{C}}(A, B) \triangleq \mathcal{C}(A, B)^\perp$. In this case, $\mathcal{Q}$ is a projector.

Lemma 12.6.6. Let $t_f > 0$, and define $\mathcal{Q} \in \mathbb{R}^{n \times n}$ by

$$\mathcal{Q} \triangleq \left(\int_0^{t_f} e^{tA} BB^T e^{tA^T} dt \right)^+ \int_0^{t_f} e^{tA} BB^T e^{tA^T} dt. \tag{12.6.11}$$

Then, $\mathcal{Q}$ is the projector onto $\mathcal{C}(A, B)$, and $\mathcal{Q}_\perp$ is the projector onto $\mathcal{C}(A, B)^\perp$. Hence,

$$\mathcal{R}(\mathcal{Q}) = \mathcal{N}(\mathcal{Q}_\perp) = \mathcal{C}(A, B), \tag{12.6.12}$$

$$\mathcal{N}(\mathcal{Q}) = \mathcal{R}(\mathcal{Q}) = \mathcal{C}(A, B)^\perp, \tag{12.6.13}$$

$$\text{rank}\, \mathcal{Q} = \text{def}\, \mathcal{Q}_\perp = \dim \mathcal{C}(A, B) = q, \tag{12.6.14}$$

$$\text{def}\, \mathcal{Q} = \text{rank}\, \mathcal{Q}_\perp = \dim \mathcal{C}(A, B)^\perp = n - q. \tag{12.6.15}$$

Now, define $u\colon\ [0,t_{\mathrm{f}}]\mapsto\mathbb{R}^m$ by

$$u(t)\triangleq B^{\mathrm{T}}e^{(t_{\mathrm{f}}-t)A^{\mathrm{T}}}\left(\int_0^{t_{\mathrm{f}}}e^{\tau A}BB^{\mathrm{T}}e^{\tau A^{\mathrm{T}}}\,\mathrm{d}\tau\right)^+x_{\mathrm{f}}.\qquad(12.6.16)$$

If $x_{\mathrm{f}}=x_{\mathrm{f}}'+x_{\mathrm{f}}''$, where $x_{\mathrm{f}}'\in\mathcal{C}(A,B)$ and $x_{\mathrm{f}}''\in\mathcal{C}(A,B)^{\perp}$, then

$$x_{\mathrm{f}}'=\mathcal{Q}x_{\mathrm{f}}=\int_0^{t_{\mathrm{f}}}e^{(t_{\mathrm{f}}-t)A}Bu(t)\,\mathrm{d}t.\qquad(12.6.17)$$

Finally, (A,B) is controllable if and only if $\mathcal{Q}=I_n$. In this case, for all $x_{\mathrm{f}}\in\mathbb{R}^n$,

$$x_{\mathrm{f}}=\int_0^{t_{\mathrm{f}}}e^{(t_{\mathrm{f}}-t)A}Bu(t)\,\mathrm{d}t,\qquad(12.6.18)$$

where $u\colon\ [0,t_{\mathrm{f}}]\mapsto\mathbb{R}^m$ is given by

$$u(t)=B^{\mathrm{T}}e^{(t_{\mathrm{f}}-t)A^{\mathrm{T}}}\left(\int_0^{t_{\mathrm{f}}}e^{\tau A}BB^{\mathrm{T}}e^{\tau A^{\mathrm{T}}}\,\mathrm{d}\tau\right)^{-1}x_{\mathrm{f}}.\qquad(12.6.19)$$

Lemma 12.6.7. Let $\alpha\in\mathbb{R}$. Then,

$$\mathcal{C}(A+\alpha I,B)=\mathcal{C}(A,B).\qquad(12.6.20)$$

The following result uses a coordinate transformation to characterize the controllable dynamics of (12.6.1).

Theorem 12.6.8. There exists an orthogonal matrix $S\in\mathbb{R}^{n\times n}$ such that

$$A=S\begin{bmatrix}A_1 & A_{12}\\ 0 & A_2\end{bmatrix}S^{-1},\qquad B=S\begin{bmatrix}B_1\\ 0\end{bmatrix},\qquad(12.6.21)$$

where $A_1\in\mathbb{R}^{q\times q}$, $B_1\in\mathbb{R}^{q\times m}$, and (A_1,B_1) is controllable.

Proof. Let $\alpha<0$ be such that $A_\alpha\triangleq A+\alpha I$ is asymptotically stable, and let $Q\in\mathbb{R}^{n\times n}$ be the positive-semidefinite solution of

$$A_\alpha Q+QA_\alpha^{\mathrm{T}}+BB^{\mathrm{T}}=0\qquad(12.6.22)$$

given by

$$Q=\int_0^{\infty}e^{tA_\alpha}BB^{\mathrm{T}}e^{tA_\alpha^{\mathrm{T}}}\,\mathrm{d}t.$$

It now follows from Lemma 12.6.2 and Lemma 12.6.7 that

$$\mathcal{R}(Q)=\mathcal{R}[\mathcal{C}(A_\alpha,B)]=\mathcal{R}[\mathcal{C}(A,B)].$$

Hence,

$$\operatorname{rank}Q=\dim\mathcal{C}(A_\alpha,B)=\dim\mathcal{C}(A,B)=q.$$

Next, let $S \in \mathbb{R}^{n \times n}$ be an orthogonal matrix such that $Q = S \begin{bmatrix} Q_1 & 0 \\ 0 & 0 \end{bmatrix} S^{\mathrm{T}}$, where $Q_1 \in \mathbb{R}^{q \times q}$ is positive definite. Writing $A_\alpha = S \begin{bmatrix} \hat{A}_1 & \hat{A}_{12} \\ \hat{A}_{21} & \hat{A}_2 \end{bmatrix} S^{-1}$ and $B = S \begin{bmatrix} B_1 \\ B_2 \end{bmatrix}$, where $\hat{A}_1 \in \mathbb{R}^{q \times q}$ and $B_1 \in \mathbb{R}^{q \times m}$, it follows from (12.6.22) that

$$\hat{A}_1 Q_1 + Q_1 \hat{A}_1^{\mathrm{T}} + B_1 B_1^{\mathrm{T}} = 0,$$
$$\hat{A}_{21} Q_1 + B_2 B_1^{\mathrm{T}} = 0,$$
$$B_2 B_2^{\mathrm{T}} = 0.$$

Therefore, $B_2 = 0$ and $\hat{A}_{21} = 0$, and thus

$$A_\alpha = S \begin{bmatrix} \hat{A}_1 & \hat{A}_{12} \\ 0 & \hat{A}_2 \end{bmatrix} S^{-1}, \quad B = S \begin{bmatrix} B_1 \\ 0 \end{bmatrix}.$$

Furthermore,

$$A = S \begin{bmatrix} \hat{A}_1 & \hat{A}_{12} \\ 0 & \hat{A}_2 \end{bmatrix} S^{-1} - \alpha I = S \left(\begin{bmatrix} \hat{A}_1 & \hat{A}_{12} \\ 0 & \hat{A}_2 \end{bmatrix} - \alpha I \right) S^{-1}.$$

Hence,

$$A = S \begin{bmatrix} A_1 & A_{12} \\ 0 & A_2 \end{bmatrix} S^{-1},$$

where $A_1 \triangleq \hat{A}_1 - \alpha I_q$, $A_{12} \triangleq \hat{A}_{12}$, and $A_2 \triangleq \hat{A}_2 - \alpha I_{n-q}$. $\qquad\square$

Proposition 12.6.9. Let $S \in \mathbb{R}^{n \times n}$, and assume that S is orthogonal. Then, the following conditions are equivalent:

i) A and B have the form (12.6.21), where $A_1 \in \mathbb{R}^{q \times q}$, $B_1 \in \mathbb{R}^{q \times m}$, and (A_1, B_1) is controllable.

ii) $\mathcal{C}(A, B) = \mathcal{R}\left(S \begin{bmatrix} I_q \\ 0 \end{bmatrix}\right)$.

iii) $\mathcal{C}(A, B)^\perp = \mathcal{R}\left(S \begin{bmatrix} 0 \\ I_{n-q} \end{bmatrix}\right)$.

iv) $\mathcal{Q} = S \begin{bmatrix} I_q & 0 \\ 0 & 0 \end{bmatrix} S^{\mathrm{T}}$.

Proposition 12.6.10. Let $S \in \mathbb{R}^{n \times n}$, and assume that S is nonsingular. Then, the following conditions are equivalent:

i) A and B have the form (12.6.21), where $A_1 \in \mathbb{R}^{q \times q}$, $B_1 \in \mathbb{R}^{q \times m}$, and (A_1, B_1) is controllable.

ii) $\mathcal{C}(A, B) = \mathcal{R}\left(S \begin{bmatrix} I_q \\ 0 \end{bmatrix}\right)$ and $\tilde{\mathcal{C}}(A, B) = \mathcal{R}\left(S \begin{bmatrix} 0 \\ I_{n-q} \end{bmatrix}\right)$.

iii) $\mathcal{Q} = S \begin{bmatrix} I_q & 0 \\ 0 & 0 \end{bmatrix} S^{-1}$.

Definition 12.6.11. Let $S \in \mathbb{R}^{n \times n}$, assume that S is nonsingular, and let A and B have the form (12.6.21), where $A_1 \in \mathbb{R}^{q \times q}$, $B_1 \in \mathbb{R}^{q \times m}$, and (A_1, B_1) is controllable. Then, the *uncontrollable spectrum* of (A, B) is $\mathrm{spec}(A_2)$, while the *uncontrollable multispectrum* of (A, B) is $\mathrm{mspec}(A_2)$. Furthermore, $\lambda \in \mathbb{C}$ is an *uncontrollable eigenvalue* of (A, B) if $\lambda \in \mathrm{spec}(A_2)$.

Definition 12.6.12. The *controllability pencil* $\mathcal{C}_{A,B}(s)$ is the pencil

$$\mathcal{C}_{A,B} = P_{[A \ -B],[I \ 0]}, \tag{12.6.23}$$

that is,

$$\mathcal{C}_{A,B}(s) = \left[\ sI - A \quad B\ \right]. \tag{12.6.24}$$

Proposition 12.6.13. Let $\lambda \in \operatorname{spec}(A)$. Then, λ is an uncontrollable eigenvalue of (A, B) if and only if

$$\operatorname{rank}\left[\ \lambda I - A \quad B\ \right] < n. \tag{12.6.25}$$

Proof. Since (A_1, B_1) is controllable, it follows from (12.6.21) that

$$\operatorname{rank}\left[\ \lambda I - A \quad B\ \right] = \operatorname{rank}\begin{bmatrix} \lambda I - A_1 & A_{12} & B_1 \\ 0 & \lambda I - A_2 & 0 \end{bmatrix}$$

$$= \operatorname{rank}\left[\ \lambda I - A_1 \quad B_1\ \right] + \operatorname{rank}(\lambda I - A_2)$$

$$= q + \operatorname{rank}(\lambda I - A_2).$$

Hence, $\operatorname{rank}\left[\ \lambda I - A \quad B\ \right] < n$ if and only if $\operatorname{rank}(\lambda I - A_2) < n - q$, that is, if and only if $\lambda \in \operatorname{spec}(A_2)$. $\square$

Proposition 12.6.14. Let $\lambda \in \operatorname{mspec}(A)$ and $K \in \mathbb{R}^{n \times m}$. Then, λ is an uncontrollable eigenvalue of (A, B) if and only if λ is an uncontrollable eigenvalue of $(A + BK, B)$.

Proof. In the notation of Theorem 12.6.8, partition $B_1 = \left[\ B_{11} \quad B_{12}\ \right]$, where $B_{11} \in \mathbb{F}^{q \times m}$, and partition $K = \begin{bmatrix} K_1 \\ K_2 \end{bmatrix}$, where $K_1 \in \mathbb{R}^{q \times m}$. Then,

$$A + BK = \begin{bmatrix} A_1 + B_{11}K_1 & A_{12} + B_{12}K_2 \\ 0 & A_2 \end{bmatrix}.$$

Consequently, the uncontrollable spectrum of $A + BK$ is $\operatorname{spec}(A_2)$. $\square$

Proposition 12.6.15. Assume that (A, B) is controllable. Then, the Smith form of $\mathcal{C}_{A,B}$ is $\left[\ I_n \quad 0_{n \times m}\ \right]$.

Proof. First, note that, if $\lambda \in \mathbb{C}$ is not an eigenvalue of A, then $n = \operatorname{rank}(\lambda I - A) = \operatorname{rank}\left[\ \lambda I - A \quad B\ \right] = \operatorname{rank}\mathcal{C}_{A,B}(\lambda)$. Therefore, $\operatorname{rank}\mathcal{C}_{A,B} = n$, and thus $\mathcal{C}_{A,B}$ has n Smith polynomials. Furthermore, since (A, B) is controllable, it follows that (A, B) has no uncontrollable eigenvalues. Therefore, it follows from Proposition 12.6.13 that, for all $\lambda \in \operatorname{spec}(A)$, $\operatorname{rank}\left[\ \lambda I - A \quad B\ \right] = n$. Consequently, $\operatorname{rank}\mathcal{C}_{A,B}(\lambda) = n$ for all $\lambda \in \mathbb{C}$. Thus, every Smith polynomial $\mathcal{C}_{A,B}$ is 1. $\square$

Proposition 12.6.16. Let $S \in \mathbb{R}^{n \times n}$, assume that S is nonsingular, and let A and B have the form (12.6.21), where $A_1 \in \mathbb{R}^{q \times q}$, $B_1 \in \mathbb{R}^{q \times m}$, and (A_1, B_1) is controllable. Furthermore, let $p_1, \ldots, p_{n-q}$ be the similarity invariants of A_2, where, for all $i \in \{1, \ldots, n-q-1\}$, p_i divides p_{i+1}. Then, there exist unimodular matrices $S_1 \in \mathbb{R}^{n \times n}[s]$ and $S_2 \in \mathbb{R}^{(n+m) \times (n+m)}[s]$ such that, for all $s \in \mathbb{C}$,

$$
\begin{bmatrix} sI - A & B \end{bmatrix} = S_1(s) \begin{bmatrix} I_q & & & & \\ & p_1(s) & & & 0_{n \times m} \\ & & \ddots & & \\ & & & p_{n-q}(s) & \end{bmatrix} S_2(s). \quad (12.6.26)
$$

Consequently,

$$
\mathrm{Szeros}(\mathcal{C}_{A,B}) = \bigcup_{i=1}^{n-q} \mathrm{roots}(p_i) = \mathrm{roots}(\chi_{A_2}) = \mathrm{spec}(A_2) \quad (12.6.27)
$$

and

$$
\mathrm{mSzeros}(\mathcal{C}_{A,B}) = \bigcup_{i=1}^{n-q} \mathrm{mroots}(p_i) = \mathrm{mroots}(\chi_{A_2}) = \mathrm{mspec}(A_2). \quad (12.6.28)
$$

Proof. Let $S \in \mathbb{R}^{n \times n}$ be as in Theorem 12.6.8, let $\hat{S}_1 \in \mathbb{R}^{q \times q}[s]$ and $\hat{S}_2 \in \mathbb{R}^{(q+m) \times (q+m)}[s]$ be unimodular matrices such that

$$
\hat{S}_1(s) \begin{bmatrix} sI_q - A_1 & B_1 \end{bmatrix} \hat{S}_2(s) = \begin{bmatrix} I_q & 0_{q \times m} \end{bmatrix},
$$

and let $\hat{S}_3, \hat{S}_4 \in \mathbb{R}^{(n-q) \times (n-q)}$ be unimodular matrices such that

$$
\hat{S}_3(s)(sI - A_2)\hat{S}_4(s) = \hat{P}(s),
$$

where $\hat{P} \triangleq \mathrm{diag}(p_1, \ldots, p_{n-q})$. Then,

$$
\begin{bmatrix} sI - A & B \end{bmatrix} = S \begin{bmatrix} \hat{S}_1^{-1}(s) & 0 \\ 0 & \hat{S}_3^{-1}(s) \end{bmatrix} \begin{bmatrix} I_q & 0 & 0_{q \times m} \\ 0 & \hat{P}(s) & 0 \end{bmatrix}
$$

$$
\times \begin{bmatrix} I_q & 0 & -\hat{S}_1(s)A_{12} \\ 0 & 0 & \hat{S}_4^{-1}(s) \\ 0 & I_m & 0 \end{bmatrix} \begin{bmatrix} \hat{S}_2^{-1}(s) & 0 \\ 0 & I_{n-q} \end{bmatrix} \begin{bmatrix} I_q & 0 & 0_{q \times m} \\ 0 & 0 & I_m \\ 0 & I_{n-q} & 0 \end{bmatrix} \begin{bmatrix} S^{-1} & 0 \\ 0 & I_m \end{bmatrix}. \quad \square
$$

Proposition 12.6.17. Let $s \in \mathbb{C}$. Then,

$$
\mathcal{C}(A, B) \subseteq \mathrm{Re}\,\mathcal{R}(\begin{bmatrix} sI - A & B \end{bmatrix}). \quad (12.6.29)
$$

Proof. Using Proposition 12.6.9 and the notation in the proof of Proposition 12.6.16, it follows that, for all $s \in \mathbb{C}$,

$$
\mathcal{C}(A, B) = \mathcal{R}\left(S \begin{bmatrix} I_q \\ 0 \end{bmatrix}\right) \subseteq \mathcal{R}\left(S \begin{bmatrix} \hat{S}_1^{-1}(s) & 0 \\ 0 & \hat{S}_3^{-1}(s)\hat{P}(s) \end{bmatrix}\right) = \mathcal{R}(\begin{bmatrix} sI - A & B \end{bmatrix}).
$$

Finally, (12.6.29) follows from the fact that $\mathcal{C}(A, B)$ is a real subspace. $\quad \square$

The next result characterizes controllability in several equivalent ways.

Theorem 12.6.18. The following statements are equivalent:

i) (A, B) is controllable.

ii) There exists $t > 0$ such that $\int_0^t e^{\tau A} BB^{\mathrm{T}} e^{\tau A^{\mathrm{T}}} \, d\tau$ is positive definite.

iii) $\int_0^t e^{\tau A} BB^{\mathrm{T}} e^{\tau A^{\mathrm{T}}} \, d\tau$ is positive definite for all $t > 0$.

iv) $\operatorname{rank} \mathcal{K}(A, B) = n$.

v) Every eigenvalue of (A, B) is controllable.

If, in addition, $\lim_{t\to\infty} \int_0^t e^{\tau A} BB^{\mathrm{T}} e^{\tau A^{\mathrm{T}}} \, d\tau$ exists, then the following condition is equivalent to *i)–v)*:

vi) $\int_0^\infty e^{tA} BB^{\mathrm{T}} e^{tA^{\mathrm{T}}} \, dt$ is positive definite.

Proof. The equivalence of *i)–iv)* follows from Lemma 12.6.2.

To prove that *iv)* $\implies$ *v)*, suppose that *v)* does not hold, that is, there exist $\lambda \in \operatorname{spec}(A)$ and a nonzero vector $x \in \mathbb{C}^n$ such that $x^*A = \lambda x^*$ and $x^*B = 0$. It thus follows that $x^*AB = \lambda x^*B = 0$. Similarly, $x^*A^iB = 0$ for all $i \in \{0, 1, \ldots, n-1\}$. Hence, $(\operatorname{Re} x)^{\mathrm{T}} \mathcal{K}(A, B) = 0$ and $(\operatorname{Im} x)^{\mathrm{T}} \mathcal{K}(A, B) = 0$. Since $\operatorname{Re} x$ and $\operatorname{Im} x$ are not both zero, it follows that $\dim \mathcal{C}(A, B) < n$.

Conversely, to show that *v)* implies *iv)*, suppose that $\operatorname{rank} \mathcal{K}(A, B) < n$. Then, there exists a nonzero vector $x \in \mathbb{R}^n$ such that $x^{\mathrm{T}}A^iB = 0$ for all $i \in \{0, \ldots, n-1\}$. Now, let $p \in \mathbb{R}[s]$ be a nonzero polynomial of minimal degree such that $x^{\mathrm{T}}p(A) = 0$. Note that p is not a constant polynomial and that $x^{\mathrm{T}}\mu_A(A) = 0$. Thus, $1 \leq \deg p \leq \deg \mu_A$. Now, let $\lambda \in \mathbb{C}$ be such that $p(\lambda) = 0$, and let $q \in \mathbb{R}[s]$ be such that $p(s) = q(s)(s - \lambda)$ for all $s \in \mathbb{C}$. Since $\deg q < \deg p$, it follows that $x^{\mathrm{T}}q(A) \neq 0$. Therefore, $\eta \triangleq q(A)x$ is nonzero. Furthermore, $\eta^{\mathrm{T}}(A - \lambda I) = x^{\mathrm{T}}p(A) = 0$. Since $x^{\mathrm{T}}A^iB = 0$ for all $i \in \{0, \ldots, n-1\}$, it follows that $\eta^{\mathrm{T}}B = x^{\mathrm{T}}q(A)B = 0$. Consequently, *v)* does not hold. $\square$

The following result, which is a restatement of the equivalence of *i)* and *v)* of Theorem 12.6.18, is the *PBH test for controllability*.

Corollary 12.6.19. The following statements are equivalent:

i) (A, B) is controllable.

ii) For all $s \in \mathbb{C}$,

$$\operatorname{rank} \begin{bmatrix} sI - A & B \end{bmatrix} = n. \tag{12.6.30}$$

The following result implies that arbitrary eigenvalue placement is possible for (12.6.1) when (A, B) is controllable.

Proposition 12.6.20. The pair (A, B) is controllable if and only if, for every polynomial $p \in \mathbb{R}[s]$ such that $\deg p = n$, there exists a matrix $K \in \mathbb{R}^{m \times n}$ such that $\operatorname{mspec}(A + BK) = \operatorname{mroots}(p)$.

Proof. For the case $m = 1$ let $A_c \triangleq C(\chi_A)$ and $B_c \triangleq e_n$ as in (12.9.5). Then, Proposition 12.9.3 implies that $\mathcal{K}(A_c, B_c)$ is nonsingular, while Corollary 12.9.9 implies that $A_c = S^{-1}AS$ and $B_c = S^{-1}B$. Now, let $\mathrm{mroots}(p) = \{\lambda_1, \ldots, \lambda_n\}_{\mathrm{ms}} \subset \mathbb{C}$. Letting $K \triangleq e_n^{\mathrm{T}}[C(p) - A_c]S^{-1}$ it follows that

$$\begin{aligned}
A + BK &= S(A_c + B_c KS)S^{-1} \\
&= S(A_c + E_{n,n}[C(p) - A_c])S^{-1} \\
&= SC(p)S^{-1}.
\end{aligned}$$

The case $m > 1$ requires the multivariable controllable canonical form. See [1179, p. 248]. $\qquad\square$

12.7 Controllable Asymptotic Stability

Let $A \in \mathbb{R}^{n \times n}$ and $B \in \mathbb{R}^{n \times m}$, and define $q \triangleq \dim \mathcal{C}(A, C)$.

Definition 12.7.1. (A, B) is *controllably asymptotically stable* if

$$\mathcal{C}(A, B) \subseteq \mathcal{S}_{\mathrm{s}}(A). \tag{12.7.1}$$

Proposition 12.7.2. Let $K \in \mathbb{R}^{m \times n}$. Then, (A, B) is controllably asymptotically stable if and only if $(A + BK, B)$ is controllably asymptotically stable.

Proposition 12.7.3. The following statements are equivalent:

i) (A, B) is controllably asymptotically stable.

ii) There exists an orthogonal matrix $S \in \mathbb{R}^{n \times n}$ such that (12.6.21) holds, where $A_1 \in \mathbb{R}^{q \times q}$ is asymptotically stable and $B_1 \in \mathbb{R}^{q \times m}$.

iii) There exists a nonsingular matrix $S \in \mathbb{R}^{n \times n}$ such that (12.6.21) holds, where $A_1 \in \mathbb{R}^{q \times q}$ is asymptotically stable and $B_1 \in \mathbb{R}^{q \times m}$.

iv) $\lim_{t \to \infty} e^{tA}B = 0$.

v) The positive-semidefinite matrix $Q \in \mathbb{R}^{n \times n}$ defined by

$$Q \triangleq \int_0^\infty e^{tA}BB^{\mathrm{T}}e^{tA^{\mathrm{T}}}\,\mathrm{d}t \tag{12.7.2}$$

exists.

vi) There exists a positive-semidefinite matrix $Q \in \mathbb{R}^{n \times n}$ satisfying

$$AQ + QA^{\mathrm{T}} + BB^{\mathrm{T}} = 0. \tag{12.7.3}$$

In this case, the positive-semidefinite matrix $Q \in \mathbb{R}^{n \times n}$ defined by (12.7.2) satisfies (12.7.3).

Proof. To prove that $i) \implies ii)$, assume that (A, B) is controllably asymptotically stable so that $\mathcal{C}(A, B) \subseteq \mathcal{S}_s(A) = \mathcal{N}[\mu_A^s(A)] = \mathcal{R}[\mu_A^u(A)]$. Using Theorem 12.6.8, it follows that there exists an orthogonal matrix $S \in \mathbb{R}^{n \times n}$ such that (12.6.21) is satisfied, where $A_1 \in \mathbb{R}^{q \times q}$ and (A_1, B_1) is controllable. Thus, $\mathcal{R}(S[\begin{smallmatrix} I_q \\ 0 \end{smallmatrix}]) = \mathcal{C}(A, B) \subseteq \mathcal{R}[\mu_A^s(A)]$.

Next, note that

$$\mu_A^s(A) = S \begin{bmatrix} \mu_A^s(A_1) & B_{12s} \\ 0 & \mu_A^s(A_2) \end{bmatrix} S^{-1},$$

where $B_{12s} \in \mathbb{R}^{q \times (n-q)}$, and suppose that A_1 is not asymptotically stable with CRHP eigenvalue λ. Then, $\lambda \notin \text{roots}(\mu_A^s)$, and thus $\mu_A^s(A_1) \neq 0$. Let $x_1 \in \mathbb{R}^{n-q}$ satisfy $\mu_A^s(A_1)x_1 \neq 0$. Then,

$$\begin{bmatrix} x_1 \\ 0 \end{bmatrix} \in \mathcal{R}\left(S \begin{bmatrix} I_q \\ 0 \end{bmatrix} \right) = \mathcal{C}(A, B)$$

and

$$\mu_A^s(A)S \begin{bmatrix} x_1 \\ 0 \end{bmatrix} = S \begin{bmatrix} \mu_A^s(A_1)x_1 \\ 0 \end{bmatrix},$$

and thus $[\begin{smallmatrix} x_1 \\ 0 \end{smallmatrix}] \notin \mathcal{N}[\mu_A^s(A)] = \mathcal{S}_s(A)$, which implies that $\mathcal{C}(A, B)$ is not contained in $\mathcal{S}_s(A)$. Hence, A_1 is asymptotically stable.

To prove that $iii) \implies iv)$, assume there exists a nonsingular matrix $S \in \mathbb{R}^{n \times n}$ such that (12.6.21) holds, where $A_1 \in \mathbb{R}^{k \times k}$ is asymptotically stable and $B_1 \in \mathbb{R}^{k \times m}$. Thus, $e^{tA}B = \begin{bmatrix} e^{tA_1}B_1 \\ 0 \end{bmatrix} S \to 0$ as $t \to \infty$.

Next, to prove that $iv)$ implies $v)$, assume that $e^{tA}B \to 0$ as $t \to \infty$. Then, every entry of $e^{tA}B$ involves exponentials of t, where the coefficients of t have negative real part. Hence, so does every entry of $e^{tA}BB^Te^{tA^T}$, which implies that $\int_0^\infty e^{tA}BB^Te^{tA^T} \, dt$ exists.

To prove that $v) \implies vi)$, note that, since $Q = \int_0^\infty e^{tA}BB^Te^{tA^T} \, dt$ exists, it follows that $e^{tA}BB^Te^{tA^T} \to 0$ as $t \to \infty$. Thus,

$$AQ + QA^T = \int\limits_0^\infty \left[Ae^{tA}BB^Te^{tA^T} + e^{tA}BB^Te^{tA^T}A \right] dt$$

$$= \int\limits_0^\infty \frac{\mathrm{d}}{\mathrm{d}t} e^{tA}BB^Te^{tA^T} \, dt$$

$$= \lim_{t \to \infty} e^{tA}BB^Te^{tA^T} - BB^T = -BB^T,$$

which shows that Q satisfies (12.4.3).

To prove that $vi) \implies i)$, suppose there exists a positive-semidefinite matrix $Q \in \mathbb{R}^{n \times n}$ satisfying (12.7.3). Then,

$$\int_0^t e^{tA} BB^{\mathrm{T}} e^{tA^{\mathrm{T}}} \, d\tau = -\int_0^t e^{\tau A} (AQ + QA^{\mathrm{T}}) e^{tA^{\mathrm{T}}} \, d\tau = -\int_0^t \frac{\mathrm{d}}{\mathrm{d}\tau} e^{\tau A} QA^{\mathrm{T}} \, d\tau$$

$$= Q - e^{tA} Q e^{tA^{\mathrm{T}}} \le Q.$$

Next, it follows from Theorem 12.6.8 that there exists an orthogonal matrix $S \in \mathbb{R}^{n \times n}$ such that (12.6.21) is satisfied, where $A_1 \in \mathbb{R}^{q \times q}$, $B_1 \in \mathbb{R}^{q \times m}$, and (A_1, B_1) is controllable. Consequently, we have

$$\int_0^t e^{\tau A_1} B_1 B_1^{\mathrm{T}} e^{\tau A_1^{\mathrm{T}}} \, d\tau = \begin{bmatrix} I & 0 \end{bmatrix} S \int_0^t e^{\tau A} BB^{\mathrm{T}} e^{\tau A^{\mathrm{T}}} \, d\tau S^{\mathrm{T}} \begin{bmatrix} I \\ 0 \end{bmatrix}$$

$$\le \begin{bmatrix} I & 0 \end{bmatrix} SQS^{\mathrm{T}} \begin{bmatrix} I \\ 0 \end{bmatrix}.$$

Thus, it follows from Proposition 8.6.3 that $Q_1 \triangleq \int_0^\infty e^{tA_1} B_1 B_1^{\mathrm{T}} e^{tA_1^{\mathrm{T}}} \, dt$ exists. Since (A_1, B_1) is controllable, it follows from $vii)$ of Theorem 12.6.18 that Q_1 is positive definite.

Now, let λ be an eigenvalue of A_1^{T}, and let $x_1 \in \mathbb{C}^n$ be an associated eigenvector. Consequently, $\alpha \triangleq x_1^* Q_1 x_1$ is positive, and

$$\alpha = x_1^* \int_0^\infty e^{\bar{\lambda} t} BB_1^{\mathrm{T}} e^{\lambda t} \, dt x_1 = x_1^* B_1 B_1^{\mathrm{T}} x_1 \int_0^\infty e^{2(\mathrm{Re}\,\lambda) t} \, dt.$$

Hence, $\int_0^\infty e^{2(\mathrm{Re}\,\lambda) t} \, dt = \alpha / x_1^* B_1 B_1^{\mathrm{T}} x_1$ exists, and thus $\mathrm{Re}\,\lambda < 0$. Consequently, A_1 is asymptotically stable, and thus $\mathcal{C}(A, B) \subseteq \mathcal{S}_{\mathrm{s}}(A)$, that is, (A, B) is controllably asymptotically stable. $\square$

The matrix $Q \in \mathbb{R}^{n \times n}$ defined by (12.7.2) is the *controllability Gramian*, and (12.7.3) is the *control Lyapunov equation*.

Proposition 12.7.4. Assume that (A, B) is controllably asymptotically stable, let $Q \in \mathbb{R}^{n \times n}$ be the positive-semidefinite matrix defined by (12.7.2), and define $\mathcal{Q} \in \mathbb{R}^{n \times n}$ by (12.6.11). Then, the following statements hold:

i) $QQ^+ = \mathcal{Q}$.

ii) $\mathcal{R}(Q) = \mathcal{R}(\mathcal{Q}) = \mathcal{C}(A, B)$.

iii) $\mathcal{N}(Q) = \mathcal{N}(\mathcal{Q}) = \mathcal{C}(A, B)^{\perp}$.

iv) $\mathrm{rank}\, Q = \mathrm{rank}\, \mathcal{Q} = q$.

v) Q is the only positive-semidefinite solution of (12.7.3) whose rank is q.

Proof. See [1238] for the proof of $v)$. $\square$

Proposition 12.7.5. Assume that (A, B) is controllably asymptotically stable, let $Q \in \mathbb{R}^{n \times n}$ be the positive-semidefinite matrix defined by (12.7.2), and let $\hat{Q} \in \mathbb{R}^{n \times n}$. Then, the following statements are equivalent:

i) $\hat{Q}$ is positive semidefinite and satisfies (12.7.3).

ii) There exists a positive-semidefinite matrix $Q_0 \in \mathbb{R}^{n \times n}$ such that $\hat{Q} = Q + Q_0$ and $AQ_0 + Q_0 A^{\mathrm{T}} = 0$.

In this case,
$$\mathrm{rank}\,\hat{Q} = q + \mathrm{rank}\,Q_0 \tag{12.7.4}$$
and
$$\mathrm{rank}\,Q_0 \leq \sum_{\substack{\lambda \in \mathrm{spec}(A) \\ \lambda \in \jmath \mathbb{R}}} \mathrm{gmult}_A(\lambda). \tag{12.7.5}$$

Proof. See [1238]. $\qquad\square$

Proposition 12.7.6. The following statements are equivalent:

i) (A, B) is controllably asymptotically stable, every imaginary eigenvalue of A is semisimple, and A has no ORHP eigenvalues.

ii) (12.7.3) has a positive-definite solution $Q \in \mathbb{R}^{n \times n}$.

Proof. See [1238]. $\qquad\square$

Proposition 12.7.7. The following statements are equivalent:

i) (A, B) is controllably asymptotically stable, and A has no imaginary eigenvalues.

ii) (12.7.3) has exactly one positive-semidefinite solution $Q \in \mathbb{R}^{n \times n}$.

In this case, $Q \in \mathbb{R}^{n \times n}$ is given by (12.7.2) and satisfies $\mathrm{rank}\,Q = q$.

Proof. See [1238]. $\qquad\square$

Corollary 12.7.8. Assume that A is asymptotically stable. Then, the positive-semidefinite matrix $Q \in \mathbb{R}^{n \times n}$ defined by (12.7.2) is the unique solution of (12.7.3) and satisfies $\mathrm{rank}\,Q = q$.

Proof. See [1238]. $\qquad\square$

Proposition 12.7.9. The following statements are equivalent:

i) (A, B) is controllable, and A is asymptotically stable.

ii) (12.7.3) has exactly one positive-semidefinite solution $Q \in \mathbb{R}^{n \times n}$, and Q is positive definite.

In this case, $Q \in \mathbb{R}^{n \times n}$ is given by (12.7.2).

Proof. See [1238]. $\qquad\square$

Corollary 12.7.10. Assume that A is asymptotically stable. Then, the positive-semidefinite matrix $Q \in \mathbb{R}^{n \times n}$ defined by (12.7.2) exists. Furthermore, Q is positive definite if and only if (A, B) is controllable.

12.8 Stabilizability

Let $A \in \mathbb{R}^{n \times n}$ and $B \in \mathbb{R}^{n \times m}$, and define $q \triangleq \dim \mathcal{C}(A, B)$.

Definition 12.8.1. (A, B) is *stabilizable* if

$$\mathcal{S}_{\mathrm{u}}(A) \subseteq \mathcal{C}(A, B). \tag{12.8.1}$$

Proposition 12.8.2. Let $K \in \mathbb{R}^{m \times n}$. Then, (A, B) is stabilizable if and only if $(A + BK, B)$ is stabilizable.

Proposition 12.8.3. The following statements are equivalent:

i) (A, B) is stabilizable.

ii) There exists an orthogonal matrix $S \in \mathbb{R}^{n \times n}$ such that (12.6.21) holds, where $A_1 \in \mathbb{R}^{q \times q}$, $B_1 \in \mathbb{R}^{q \times m}$, (A_1, B_1) is controllable, and $A_2 \in \mathbb{R}^{(n-q) \times (n-q)}$ is asymptotically stable.

iii) There exists a nonsingular matrix $S \in \mathbb{R}^{n \times n}$ such that (12.6.21) holds, where $A_1 \in \mathbb{R}^{q \times q}$, $B_1 \in \mathbb{R}^{q \times m}$, (A_1, B_1) is controllable, and $A_2 \in \mathbb{R}^{(n-q) \times (n-q)}$ is asymptotically stable.

iv) Every CRHP eigenvalue of (A, B) is controllable.

Proof. To prove that *i)* $\implies$ *ii)*, assume that (A, B) is stabilizable so that $\mathcal{S}_{\mathrm{u}}(A) = \mathcal{N}[\mu_A^{\mathrm{u}}(A)] = \mathcal{R}[\mu_A^{\mathrm{s}}(A)] \subseteq \mathcal{C}(A, B)$. Using Theorem 12.6.8, it follows that there exists an orthogonal matrix $S \in \mathbb{R}^{n \times n}$ such that (12.6.21) is satisfied, where $A_1 \in \mathbb{R}^{q \times q}$ and (A_1, B_1) is controllable. Thus, $\mathcal{R}[\mu_A^{\mathrm{s}}(A)] \subseteq \mathcal{C}(A, B) = \mathcal{R}(S[\begin{smallmatrix} I_q \\ 0 \end{smallmatrix}])$.

Next, note that

$$\mu_A^{\mathrm{s}}(A) = S \begin{bmatrix} \mu_A^{\mathrm{s}}(A_1) & B_{12\mathrm{s}} \\ 0 & \mu_A^{\mathrm{s}}(A_2) \end{bmatrix} S^{-1},$$

where $B_{12\mathrm{s}} \in \mathbb{R}^{q \times (n-q)}$, and suppose that A_2 is not asymptotically stable with CRHP eigenvalue λ. Then, $\lambda \notin \mathrm{roots}(\mu_A^{\mathrm{s}})$, and thus $\mu_A^{\mathrm{s}}(A_2) \neq 0$. Let $x_2 \in \mathbb{R}^{n-q}$ satisfy $\mu_A^{\mathrm{s}}(A_2)x_2 \neq 0$. Then,

$$\mu_A^{\mathrm{s}}(A)S \begin{bmatrix} 0 \\ x_2 \end{bmatrix} = S \begin{bmatrix} B_{12\mathrm{s}}x_2 \\ \mu_A^{\mathrm{s}}(A_2)x_2 \end{bmatrix} \notin \mathcal{R}\left(S \begin{bmatrix} I_q \\ 0 \end{bmatrix}\right) = \mathcal{C}(A, B),$$

which implies that $\mathcal{S}_{\mathrm{u}}(A)$ is not contained in $\mathcal{C}(A, B)$. Hence, A_2 is asymptotically stable.

The statement *ii)* $\implies$ *iii)* is immediate. To prove that *iii)* $\implies$ *iv)*, let $\lambda \in \mathrm{spec}(A)$ be a CRHP eigenvalue of A. Since A_2 is asymptotically stable, it follows that $\lambda \notin \mathrm{spec}(A_2)$. Consequently, Proposition 12.6.13 implies that λ is not an uncontrollable eigenvalue of (A, B), and thus λ is a controllable eigenvalue of (A, B).

To prove that $iv) \implies i)$, let $S \in \mathbb{R}^{n \times n}$ be nonsingular and such that A and B have the form (12.6.21), where $A_1 \in \mathbb{R}^{q \times q}$, $B_1 \in \mathbb{R}^{q \times m}$, and (A_1, B_1) is controllable. Since every CRHP eigenvalue of (A, B) is controllable, it follows from Proposition 12.6.13 that A_2 is asymptotically stable. From Fact 11.23.4 it follows that $S_u(A) \subseteq \mathcal{R}\left(S\left[\begin{smallmatrix} I_q \\ 0 \end{smallmatrix}\right]\right) = \mathcal{C}(A, B)$, which implies that (A, B) is stabilizable. $\qquad \square$

The following result, which is a restatement of the equivalence of $i)$ and $iv)$ of Proposition 12.8.3, is the *PBH test for stabilizability.*

Corollary 12.8.4. The following statements are equivalent:

i) (A, B) is stabilizable.

ii) For all $s \in$ CRHP,

$$\text{rank} \begin{bmatrix} sI - A & B \end{bmatrix} = n. \tag{12.8.2}$$

Proposition 12.8.5. The following statements are equivalent:

i) A is asymptotically stable.

ii) (A, B) is controllably asymptotically stable and stabilizable.

Proof. To prove that $i) \implies ii)$, assume that A is asymptotically stable. Then, $S_u(A) = \{0\}$, and $S_s(A) = \mathbb{R}^n$. Thus, $S_u(A) \subseteq \mathcal{C}(A, B)$, and $\mathcal{C}(A, B) \subseteq S_s(A)$.

To prove that $ii) \implies i)$, assume that (A, B) is stabilizable and controllably asymptotically stable. Then, $S_u(A) \subseteq \mathcal{C}(A, B) \subseteq S_s(A)$, and thus $S_u(A) = \{0\}$.

As an alternative proof that $ii) \implies i)$, note that, since (A, B) is stabilizable, it follows from Proposition 12.5.3 that there exists a nonsingular matrix $S \in \mathbb{R}^{n \times n}$ such that (12.6.21) holds, where $A_1 \in \mathbb{R}^{q \times q}$, $B_1 \in \mathbb{R}^{q \times m}$, (A_1, B_1) is controllable, and $A_2 \in \mathbb{R}^{(n-q) \times (n-q)}$ is asymptotically stable. Then,

$$\int_0^\infty e^{tA} BB^\mathrm{T} e^{tA^\mathrm{T}} \, dt = S \begin{bmatrix} \int_0^\infty e^{tA_1} B_1 B_1^\mathrm{T} e^{tA_1^\mathrm{T}} \, dt & 0 \\ 0 & 0 \end{bmatrix} S^{-1}.$$

Since the integral on the left-hand side exists by assumption, the integral on the right-hand side also exists. Since (A_1, B_1) is controllable, it follows from $vii)$ of Theorem 12.6.18 that $Q_1 \triangleq \int_0^\infty e^{tA_1} B_1 B_1^\mathrm{T} e^{tA_1^\mathrm{T}} \, dt$ is positive definite.

Now, let λ be an eigenvalue of A_1^T, and let $x_1 \in \mathbb{C}^q$ be an associated eigenvector. Consequently, $\alpha \triangleq x_1^* Q_1 x_1$ is positive, and

$$\alpha = x_1^* \int_0^\infty e^{\bar{\lambda}t} B_1 B_1^\mathrm{T} e^{\lambda t} \, dt x_1 = x_1^* B_1 B_1^\mathrm{T} x_1 \int_0^\infty e^{2(\mathrm{Re}\,\lambda)t} \, dt.$$

Hence, $\int_0^\infty e^{2(\mathrm{Re}\,\lambda)t} \, dt$ exists, and thus $\mathrm{Re}\,\lambda < 0$. Consequently, A_1 is asymptotically stable, and thus A is asymptotically stable. $\qquad \square$

Corollary 12.8.6. The following statements are equivalent:

i) There exists a positive-semidefinite matrix $Q \in \mathbb{R}^{n \times n}$ satisfying (12.7.3), and (A, B) is stabilizable.

ii) A is asymptotically stable.

Proof. This result follows from Proposition 12.7.3 and Proposition 12.8.5. $\square$

12.9 Realization Theory

Given a proper rational transfer function G, we wish to determine (A, B, C, D) such that (12.2.11) holds. The following terminology is convenient.

Definition 12.9.1. Let $G \in \mathbb{R}^{l \times m}(s)$. If $l = m = 1$, then G is a *single-input/single-output (SISO)* rational transfer function; if $l = 1$ and $m > 1$, then G is a *multiple-input/single-output (MISO)* rational transfer function; if $l > 1$ and $m = 1$, then G is a *single-input/multiple-output (SIMO)* rational transfer function; and, if $l > 1$ or $m > 1$, then G is a *multiple-input/multiple output (MIMO)* rational transfer function.

Definition 12.9.2. Let $G \in \mathbb{R}^{l \times m}_{\mathrm{prop}}(s)$, and assume that $A \in \mathbb{R}^{n \times n}$, $B \in \mathbb{R}^{n \times m}$, $C \in \mathbb{R}^{l \times n}$, and $D \in \mathbb{R}^{l \times m}$ satisfy $G(s) = C(sI - A)^{-1}B + D$. Then, $\left[\begin{array}{c|c} A & B \\ \hline C & D \end{array} \right]$ is a *realization* of G, which is written as

$$G \sim \left[\begin{array}{c|c} A & B \\ \hline C & D \end{array} \right]. \tag{12.9.1}$$

The *order* of the realization (12.9.1) is the order of A. Finally, (A, B, C) is *controllable and observable* if (A, B) is controllable and (A, C) is observable.

Suppose that $n = 0$. Then, A, B, and C are empty matrices, and $G \in \mathbb{R}^{l \times m}_{\mathrm{prop}}(s)$ is given by

$$G(s) = 0_{l \times 0}(sI_{0 \times 0} - 0_{0 \times 0})^{-1}0_{0 \times m} + D = 0_{l \times m} + D = D. \tag{12.9.2}$$

Therefore, the order of the realization $\left[\begin{array}{c|c} 0_{0 \times 0} & 0_{0 \times m} \\ \hline 0_{l \times 0} & D \end{array} \right]$ is zero.

Although the realization (12.9.1) is not unique, the matrix D is unique and is given by

$$D = G(\infty). \tag{12.9.3}$$

Furthermore, note that $G \sim \left[\begin{array}{c|c} A & B \\ \hline C & D \end{array} \right]$ if and only if $G - D \sim \left[\begin{array}{c|c} A & B \\ \hline C & 0 \end{array} \right]$. Therefore, it suffices to construct realizations for strictly proper transfer functions.

The following result shows that every strictly proper, SISO rational transfer function G has a realization. In fact, two realizations are the *controllable canonical form* $G \sim \left[\begin{array}{c|c} A_{\mathrm{c}} & B_{\mathrm{c}} \\ \hline C_{\mathrm{c}} & 0 \end{array} \right]$ and the *observable canonical form* $G \sim \left[\begin{array}{c|c} A_{\mathrm{o}} & B_{\mathrm{o}} \\ \hline C_{\mathrm{o}} & 0 \end{array} \right]$. If G is exactly proper, then a realization can be obtained for $G - G(\infty)$.

Proposition 12.9.3. Let $G \in \mathbb{R}_{\mathrm{prop}}(s)$ be the SISO strictly proper rational transfer function

$$G(s) = \frac{\alpha_{n-1}s^{n-1} + \alpha_{n-2}s^{n-2} + \cdots + \alpha_1 s + \alpha_0}{s^n + \beta_{n-1}s^{n-1} + \cdots + \beta_1 s + \beta_0}. \tag{12.9.4}$$

Then, $G \sim \left[\begin{array}{c|c} A_{\mathrm{c}} & B_{\mathrm{c}} \\ \hline C_{\mathrm{c}} & 0 \end{array} \right]$, where $A_{\mathrm{c}}, B_{\mathrm{c}}, C_{\mathrm{c}}$ are defined by

$$A_{\mathrm{c}} \triangleq \begin{bmatrix} 0 & 1 & 0 & \cdots & 0 \\ 0 & 0 & 1 & \cdots & 0 \\ \vdots & \vdots & \vdots & \ddots & \vdots \\ 0 & 0 & 0 & \cdots & 1 \\ -\beta_0 & -\beta_1 & -\beta_2 & \cdots & -\beta_{n-1} \end{bmatrix}, \quad B_{\mathrm{c}} \triangleq \begin{bmatrix} 0 \\ 0 \\ \vdots \\ 0 \\ 1 \end{bmatrix}, \tag{12.9.5}$$

$$C_{\mathrm{c}} \triangleq \begin{bmatrix} \alpha_0 & \alpha_1 & \alpha_2 & \cdots & \alpha_{n-1} \end{bmatrix}, \tag{12.9.6}$$

and $G \sim \left[\begin{array}{c|c} A_{\mathrm{o}} & B_{\mathrm{o}} \\ \hline C_{\mathrm{o}} & 0 \end{array} \right]$, where $A_{\mathrm{o}}, B_{\mathrm{o}}, C_{\mathrm{o}}$ are defined by

$$A_{\mathrm{o}} \triangleq \begin{bmatrix} 0 & 0 & \cdots & 0 & -\beta_0 \\ 1 & 0 & \cdots & 0 & -\beta_1 \\ 0 & 1 & \cdots & 0 & -\beta_2 \\ \vdots & \vdots & \ddots & & \vdots \\ 0 & 0 & \cdots & 1 & -\beta_{n-1} \end{bmatrix}, \quad B_{\mathrm{o}} \triangleq \begin{bmatrix} \alpha_0 \\ \alpha_1 \\ \alpha_2 \\ \vdots \\ \alpha_{n-1} \end{bmatrix}, \tag{12.9.7}$$

$$C_{\mathrm{o}} \triangleq \begin{bmatrix} 0 & 0 & \cdots & 0 & 1 \end{bmatrix}. \tag{12.9.8}$$

Furthermore, $(A_{\mathrm{c}}, B_{\mathrm{c}})$ is controllable, and $(A_{\mathrm{o}}, C_{\mathrm{o}})$ is observable. Finally, the following statements are equivalent:

i) The numerator and denominator of G given in (12.9.4) are coprime.

ii) $(A_{\mathrm{c}}, C_{\mathrm{c}})$ is observable.

iii) $(A_{\mathrm{c}}, B_{\mathrm{c}}, C_{\mathrm{c}})$ is controllable and observable.

iv) $(A_{\mathrm{o}}, B_{\mathrm{o}})$ is controllable.

v) $(A_{\mathrm{o}}, B_{\mathrm{o}}, C_{\mathrm{o}})$ is controllable and observable.

Proof. The realizations can be verified directly. Furthermore, note that

$$\mathcal{K}(A_{\mathrm{c}}, B_{\mathrm{c}}) = \mathcal{O}(A_{\mathrm{o}}, C_{\mathrm{o}}) = \begin{bmatrix} 0 & 0 & 0 & \cdots & 0 & 1 \\ 0 & 0 & 0 & \cdot^{\cdot^{\cdot}} & 1 & -\beta_{n-1} \\ \vdots & \vdots & & \cdot^{\cdot^{\cdot}} & \cdot^{\cdot^{\cdot}} & \vdots \\ 0 & 0 & 1 & \cdot^{\cdot^{\cdot}} & -\beta_3 & -\beta_2 \\ 0 & 1 & -\beta_{n-1} & \cdots & -\beta_2 & -\beta_1 \\ 1 & -\beta_{n-1} & -\beta_{n-2} & \cdots & -\beta_1 & -\beta_0 \end{bmatrix}.$$

It follows from Fact 2.13.9 that $\det \mathcal{K}(A_{\mathrm{c}}, B_{\mathrm{c}}) = \det \mathcal{O}(A_{\mathrm{o}}, C_{\mathrm{o}}) = (-1)^{\lfloor n/2 \rfloor}$, which implies that $(A_{\mathrm{c}}, B_{\mathrm{c}})$ is controllable and $(A_{\mathrm{o}}, C_{\mathrm{o}})$ is observable.

To prove the last statement, let $p, q \in \mathbb{R}[s]$ denote the numerator and denominator, respectively, of G in (12.9.4). Then, for $n = 2$,

$$\mathcal{K}(A_\mathrm{o}, B_\mathrm{o}) = \mathcal{O}^\mathrm{T}(A_\mathrm{c}, C_\mathrm{c}) = B(p,q)\hat{I} \begin{bmatrix} 1 & -\beta_1 \\ 0 & 1 \end{bmatrix},$$

where $B(p,q)$ is the Bezout matrix of p and q. It follows from $ix)$ of Fact 4.8.6 that $B(p,q)$ is nonsingular if and only if p and q are coprime. $\square$

The following result shows that every proper rational transfer function has a realization.

Theorem 12.9.4. Let $G \in \mathbb{R}^{l \times m}_\mathrm{prop}(s)$. Then, there exist $A \in \mathbb{R}^{n \times n}$, $B \in \mathbb{R}^{n \times m}$, $C \in \mathbb{R}^{l \times n}$, and $D \in \mathbb{R}^{l \times m}$ such that $G \sim \left[\begin{array}{c|c} A & B \\ \hline C & D \end{array} \right]$.

Proof. By Proposition 12.9.3, every entry $G_{(i,j)}$ of G has a realization $G_{(i,j)} \sim \left[\begin{array}{c|c} A_{ij} & B_{ij} \\ \hline C_{ij} & D_{ij} \end{array} \right]$. Combining these realizations yields a realization of G. $\square$

Proposition 12.9.5. Let $G \in \mathbb{R}^{l \times m}_\mathrm{prop}(s)$ have the nth-order realization $\left[\begin{array}{c|c} A & B \\ \hline C & D \end{array} \right]$, let $S \in \mathbb{R}^{n \times n}$, and assume that S is nonsingular. Then,

$$G \sim \left[\begin{array}{c|c} SAS^{-1} & SB \\ \hline CS^{-1} & D \end{array} \right]. \tag{12.9.9}$$

If, in addition, $\left[\begin{array}{c|c} A & B \\ \hline C & D \end{array} \right]$ is controllable and observable, then so is $\left[\begin{array}{c|c} SAS^{-1} & SB \\ \hline CS^{-1} & D \end{array} \right]$.

Definition 12.9.6. Let $G \in \mathbb{R}^{l \times m}_\mathrm{prop}(s)$, and let $\left[\begin{array}{c|c} A & B \\ \hline C & D \end{array} \right]$ and $\left[\begin{array}{c|c} \hat{A} & \hat{B} \\ \hline \hat{C} & D \end{array} \right]$ be nth-order realizations of G. Then, $\left[\begin{array}{c|c} A & B \\ \hline C & D \end{array} \right]$ and $\left[\begin{array}{c|c} \hat{A} & \hat{B} \\ \hline \hat{C} & D \end{array} \right]$ are *equivalent* if there exists a nonsingular matrix $S \in \mathbb{R}^{n \times n}$ such that $\hat{A} = SAS^{-1}$, $\hat{B} = SB$, and $\hat{C} = CS^{-1}$.

The following result shows that the Markov parameters of a rational transfer function are independent of the realization.

Proposition 12.9.7. Let $G \in \mathbb{R}^{l \times m}_\mathrm{prop}(s)$, and assume that $G \sim \left[\begin{array}{c|c} A & B \\ \hline C & D \end{array} \right]$, where $A \in \mathbb{R}^{n \times n}$, and $G \sim \left[\begin{array}{c|c} \hat{A} & \hat{B} \\ \hline \hat{C} & \hat{D} \end{array} \right]$, where $A \in \mathbb{R}^{\hat{n} \times \hat{n}}$. Then, $D = \hat{D}$, and, for all $k \geq 0$, $CA^k B = \hat{C}\hat{A}^k\hat{B}$.

Proposition 12.9.8. Let $G \in \mathbb{R}^{l \times m}_\mathrm{prop}(s)$, assume that G has the nth-order realizations $\left[\begin{array}{c|c} A_1 & B_1 \\ \hline C_1 & D \end{array} \right]$ and $\left[\begin{array}{c|c} A_2 & B_2 \\ \hline C_2 & D \end{array} \right]$, and assume that both of these realizations are controllable and observable. Then, these realizations are equivalent. Furthermore, there exists a unique matrix $S \in \mathbb{R}^{n \times n}$ such that

$$\left[\begin{array}{c|c} A_2 & B_2 \\ \hline C_2 & D \end{array} \right] = \left[\begin{array}{c|c} SA_1S^{-1} & SB_1 \\ \hline C_1S^{-1} & D \end{array} \right]. \tag{12.9.10}$$

In fact,

$$S = \left(O_2^T O_2\right)^{-1} O_2^T O_1, \qquad S^{-1} = \mathcal{K}_1 \mathcal{K}_2^T \left(\mathcal{K}_2 \mathcal{K}_2^T\right)^{-1}, \qquad (12.9.11)$$

where, for $i = 1, 2$, $\mathcal{K}_i \triangleq \mathcal{K}(A_i, B_i)$ and $O_i \triangleq O(A_i, C_i)$.

Proof. By Proposition 12.9.7, the realizations $\left[\begin{array}{c|c} A_1 & B_1 \\ \hline C_1 & D \end{array}\right]$ and $\left[\begin{array}{c|c} A_2 & B_2 \\ \hline C_2 & D \end{array}\right]$ generate the same Markov parameters. Hence, $O_1 A_1 \mathcal{K}_1 = O_2 A_2 \mathcal{K}_2$, $O_1 B_1 = O_2 B_2$, and $C_1 \mathcal{K}_1 = C_2 \mathcal{K}_2$. Since $\left[\begin{array}{c|c} A_2 & B_2 \\ \hline C_2 & D \end{array}\right]$ is controllable and observable, it follows that the $n \times n$ matrices $\mathcal{K}_2 \mathcal{K}_2^T$ and $O_2^T O_2$ are nonsingular. Consequently, $A_2 = S A_1 S^{-1}$, $B_2 = S B_1$, and $C_2 = C_1 S^{-1}$.

To prove uniqueness, assume there exists a matrix $\hat{S} \in \mathbb{R}^{n \times n}$ such that $A_2 = \hat{S} A_1 \hat{S}^{-1}$, $B_2 = \hat{S} B_1$, and $C_2 = C_1 \hat{S}^{-1}$. Then, it follows that $O_1 \hat{S} = O_2$. Since $O_1 S = O_2$, it follows that $O_1(S - \hat{S}) = 0$. Consequently, $S = \hat{S}$. □

Corollary 12.9.9. Let $G \in \mathbb{R}_{\text{prop}}(s)$ be given by (12.9.4), assume that G has the nth-order controllable and observable realization $\left[\begin{array}{c|c} A & B \\ \hline C & 0 \end{array}\right]$, and define A_c, B_c, C_c by (12.9.5), (12.9.6) and A_o, B_o, C_o by (12.9.7), (12.9.8). Furthermore, define $S_c \triangleq [O(A, B)]^{-1} O(A_c, B_c)$. Then,

$$S_c^{-1} = \mathcal{K}(A, B)[\mathcal{K}(A_c, B_c)]^{-1} \qquad (12.9.12)$$

and

$$\left[\begin{array}{c|c} S_c A S_c^{-1} & S_c B \\ \hline C S_c^{-1} & 0 \end{array}\right] = \left[\begin{array}{c|c} A_c & B_c \\ \hline C_c & 0 \end{array}\right]. \qquad (12.9.13)$$

Furthermore, define $S_o \triangleq [O(A, B)]^{-1} O(A_o, B_o)$. Then,

$$S_o^{-1} = \mathcal{K}(A, B)[\mathcal{K}(A_o, B_o)]^{-1} \qquad (12.9.14)$$

and

$$\left[\begin{array}{c|c} S_o A S_o^{-1} & S_o B \\ \hline C S_o^{-1} & 0 \end{array}\right] = \left[\begin{array}{c|c} A_o & B_o \\ \hline C_o & 0 \end{array}\right]. \qquad (12.9.15)$$

The following result, known as the *Kalman decomposition*, is useful for constructing controllable and observable realizations.

Proposition 12.9.10. Let $G \in \mathbb{R}_{\text{prop}}^{l \times m}(s)$, where $G \sim \left[\begin{array}{c|c} A & B \\ \hline C & D \end{array}\right]$. Then, there exists a nonsingular matrix $S \in \mathbb{R}^{n \times n}$ such that

$$A = S \begin{bmatrix} A_1 & 0 & A_{13} & 0 \\ A_{21} & A_2 & A_{23} & A_{24} \\ 0 & 0 & A_3 & 0 \\ 0 & 0 & A_{43} & A_4 \end{bmatrix} S^{-1}, \qquad B = S \begin{bmatrix} B_1 \\ B_2 \\ 0 \\ 0 \end{bmatrix}, \qquad (12.9.16)$$

$$C = \begin{bmatrix} C_1 & 0 & C_3 & 0 \end{bmatrix} S^{-1}, \qquad (12.9.17)$$

where, for $i \in \{1, \ldots, 4\}$, $A_i \in \mathbb{R}^{n_i \times n_i}$, $\left(\left[\begin{smallmatrix} A_1 & 0 \\ A_{21} & A_2 \end{smallmatrix} \right], \left[\begin{smallmatrix} B_1 \\ B_2 \end{smallmatrix} \right] \right)$ is controllable, and $\left(\left[\begin{smallmatrix} A_1 & A_{13} \\ 0 & A_3 \end{smallmatrix} \right], \left[\begin{smallmatrix} C_1 & C_3 \end{smallmatrix} \right] \right)$ is observable. Furthermore, the following statements hold:

i) (A, B) is stabilizable if and only if A_3 and A_4 are asymptotically stable.

ii) (A, B) is controllable if and only if A_3 and A_4 are empty.

iii) (A, C) is detectable if and only if A_2 and A_4 are asymptotically stable.

iv) (A, C) is observable if and only if A_2 and A_4 are empty.

v) $G \sim \left[\begin{array}{c|c} A_1 & B_1 \\ \hline C_1 & D \end{array} \right]$.

vi) The realization $\left[\begin{array}{c|c} A_1 & B_1 \\ \hline C_1 & D \end{array} \right]$ is controllable and observable.

Proof. Let $\alpha \leq 0$ be such that $A + \alpha I$ is asymptotically stable, and let $Q \in \mathbb{R}^{n \times n}$ and $P \in \mathbb{R}^{n \times n}$ denote the controllability and observability Gramians of the system $(A + \alpha I, B, C)$. Then, Theorem 8.3.5 implies that there exists a nonsingular matrix $S \in \mathbb{R}^{n \times n}$ such that

$$Q = S \begin{bmatrix} Q_1 & & & 0 \\ & Q_2 & & \\ & & 0 & \\ 0 & & & 0 \end{bmatrix} S^{\mathrm{T}}, \quad P = S^{-\mathrm{T}} \begin{bmatrix} P_1 & & & 0 \\ & 0 & & \\ & & P_2 & \\ 0 & & & 0 \end{bmatrix} S^{-1},$$

where Q_1 and P_1 are the same order, and where $Q_1, Q_2, P_1,$ and P_2 are positive definite and diagonal. The form of $SAS^{-1}, SB,$ and CS^{-1} given by (12.9.17) now follows from (12.7.3) and (12.4.3) with A replaced by $A + \alpha I$, where, as in the proof of Theorem 12.6.8, $SAS^{-1} = S(A + \alpha I)S^{-1} - \alpha I$. Finally, statements *i*)–*v*) are immediate, while it can be verified directly that $\left[\begin{array}{c|c} A_1 & B_1 \\ \hline C_1 & D_1 \end{array} \right]$ is a realization of G. $\square$

Note that the uncontrollable multispectrum of (A, B) is given by $\mathrm{mspec}(A_3) \cup \mathrm{mspec}(A_4)$, while the unobservable multispectrum of (A, C) is given by $\mathrm{mspec}(A_2) \cup \mathrm{mspec}(A_4)$. Likewise, the *uncontrollable-unobservable multispectrum* of (A, B, C) is given by $\mathrm{mspec}(A_4)$.

Let $G \sim \left[\begin{array}{c|c} A & B \\ \hline C & 0 \end{array} \right]$. Then, define the *i-step observability matrix* $\mathcal{O}_i(A, C) \in \mathbb{R}^{il \times n}$ by

$$\mathcal{O}_i(A, C) \triangleq \begin{bmatrix} C \\ CA \\ \vdots \\ CA^{i-1} \end{bmatrix} \tag{12.9.18}$$

and the *j-step controllability matrix* $\mathcal{K}_j(A, B) \in \mathbb{R}^{n \times jm}$ by

$$\mathcal{K}_j(A, B) \triangleq \begin{bmatrix} B & AB & \cdots & A^{j-1}B \end{bmatrix}. \tag{12.9.19}$$

Note that $\mathcal{O}(A, C) = \mathcal{O}_n(A, C)$ and $\mathcal{K}(A, B) = \mathcal{K}_n(A, B)$. Furthermore, define the *Markov block-Hankel matrix* $\mathcal{H}_{i,j,k}(G) \in \mathbb{R}^{il \times jm}$ of G by

$$\mathcal{H}_{i,j,k}(G) \triangleq \mathcal{O}_i(A, C) A^k \mathcal{K}_j(A, B). \tag{12.9.20}$$

Note that $\mathcal{H}_{i,j,k}(G)$ is the block-Hankel matrix of Markov parameters given by

$$\mathcal{H}_{i,j,k}(G) = \begin{bmatrix} CA^kB & CA^{k+1}B & CA^{k+2}B & \cdots & CA^{k+j-1}B \\ CA^{k+1}B & CA^{k+2}B & \reflectbox{$\ddots$} & \reflectbox{$\ddots$} & \reflectbox{$\ddots$} \\ CA^{k+2}B & \reflectbox{$\ddots$} & \reflectbox{$\ddots$} & \reflectbox{$\ddots$} & \reflectbox{$\ddots$} \\ \vdots & \reflectbox{$\ddots$} & \reflectbox{$\ddots$} & \reflectbox{$\ddots$} & \reflectbox{$\ddots$} \\ CA^{k+i-1}B & \reflectbox{$\ddots$} & & \reflectbox{$\ddots$} & CA^{k+j+i-2}B \end{bmatrix}$$

$$= \begin{bmatrix} H_{k+1} & H_{k+2} & H_{k+3} & \cdots & H_{k+j} \\ H_{k+2} & H_{k+3} & \reflectbox{$\ddots$} & \reflectbox{$\ddots$} & \reflectbox{$\ddots$} \\ H_{k+3} & \reflectbox{$\ddots$} & \reflectbox{$\ddots$} & \reflectbox{$\ddots$} & \reflectbox{$\ddots$} \\ \vdots & \reflectbox{$\ddots$} & \reflectbox{$\ddots$} & \reflectbox{$\ddots$} & \reflectbox{$\ddots$} \\ H_{k+i} & \reflectbox{$\ddots$} & \reflectbox{$\ddots$} & \reflectbox{$\ddots$} & H_{k+j+i-1} \end{bmatrix}. \qquad (12.9.21)$$

Note that

$$\mathcal{H}_{i,j,0}(G) = \mathcal{O}_i(A,C)\mathcal{K}_j(A,B) \qquad (12.9.22)$$

and

$$\mathcal{H}_{i,j,1}(G) = \mathcal{O}_i(A,C)A\mathcal{K}_j(A,B). \qquad (12.9.23)$$

Furthermore, define

$$\mathcal{H}(G) \triangleq \mathcal{H}_{n,n,0}(G) = \mathcal{O}(A,C)\mathcal{K}(A,B). \qquad (12.9.24)$$

The following result provides a MIMO extension of Fact 4.8.8.

Proposition 12.9.11. Let $G \sim \left[\begin{array}{c|c} A & B \\ \hline C & 0 \end{array}\right]$, where $A \in \mathbb{R}^{n \times n}$. Then, the following statements are equivalent:

  i) The realization $\left[\begin{array}{c|c} A & B \\ \hline C & 0 \end{array}\right]$ is controllable and observable.

  ii) $\operatorname{rank} \mathcal{H}(G) = n$.

  iii) For all $i, j \geq n$, $\operatorname{rank} \mathcal{H}_{i,j,0}(G) = n$.

  iv) There exist $i, j \geq n$ such that $\operatorname{rank} \mathcal{H}_{i,j,0}(G) = n$.

Proof. The equivalence of *ii)*, *iii)*, and *iv)* follows from Fact 2.11.7. To prove that *i)* $\implies$ *ii)*, note that, since the $n \times n$ matrices $\mathcal{O}^{\mathrm{T}}(A,C)\mathcal{O}(A,C)$ and $\mathcal{K}(A,B)\mathcal{K}^{\mathrm{T}}(A,B)$ are positive definite, it follows that

$$n = \operatorname{rank} \mathcal{O}^{\mathrm{T}}(A,C)\mathcal{O}(A,C)\mathcal{K}(A,B)\mathcal{K}^{\mathrm{T}}(A,B) \leq \operatorname{rank} \mathcal{H}(G) \leq n.$$

Conversely, $n = \operatorname{rank} \mathcal{H}(G) \leq \min\{\operatorname{rank} \mathcal{O}(A,C), \operatorname{rank} \mathcal{K}(A,B)\} \leq n$. $\qquad \square$

Proposition 12.9.12. Let $G \sim \left[\begin{array}{c|c} A & B \\ \hline C & 0 \end{array}\right]$, where $A \in \mathbb{R}^{n \times n}$, assume that $\left[\begin{array}{c|c} A & B \\ \hline C & 0 \end{array}\right]$ is controllable and observable, and let $i, j \geq 1$ be such that rank $\mathcal{O}_i(A, C)$ $= \text{rank } \mathcal{K}_j(A, B) = n$. Then, $G \sim \left[\begin{array}{c|c} \hat{A} & \hat{B} \\ \hline \hat{C} & 0 \end{array}\right]$, where

$$\hat{A} \triangleq \mathcal{O}_i^+(A, C)\mathcal{H}_{i,j,1}(G)\mathcal{K}_j^+(A, B), \tag{12.9.25}$$

$$\hat{B} \triangleq \mathcal{K}_j(A, B)\left[\begin{array}{c} I_m \\ 0_{(j-1)n \times m} \end{array}\right], \tag{12.9.26}$$

$$\hat{C} \triangleq \left[\begin{array}{cc} I_l & 0_{l \times (i-1)l} \end{array}\right]\mathcal{O}_i(A, C). \tag{12.9.27}$$

Proposition 12.9.13. Let $G \in \mathbb{R}_{\text{prop}}^{l \times m}(s)$, let $i, j \geq 1$, define $n \triangleq$ rank $\mathcal{H}_{i,j,0}(G)$, and let $L \in \mathbb{R}^{il \times n}$ and $R \in \mathbb{R}^{n \times jm}$ be such that $\mathcal{H}_{i,j,0}(G) = LR$. Then, the realization

$$G \sim \left[\begin{array}{c|c} L^+\mathcal{H}_{i,j,1}(G)R^+ & R\left[\begin{array}{c} I_m \\ 0_{(j-1)n \times m} \end{array}\right] \\ \hline \left[\begin{array}{cc} I_l & 0_{l \times (i-1)l} \end{array}\right]L & 0 \end{array}\right] \tag{12.9.28}$$

is controllable and observable.

A rational transfer function $G \in \mathbb{R}_{\text{prop}}^{l \times m}(s)$ can have realizations of different orders. For example, letting

$$A = 1, \qquad B = 1, \qquad C = 1, \qquad D = 0$$

and

$$\hat{A} = \left[\begin{array}{cc} 1 & 0 \\ 0 & 1 \end{array}\right], \qquad \hat{B} = \left[\begin{array}{c} 1 \\ 0 \end{array}\right], \qquad \hat{C} = \left[\begin{array}{cc} 1 & 0 \end{array}\right], \qquad \hat{D} = 0,$$

it follows that

$$G(s) = C(sI - A)^{-1}B + D = \hat{C}(sI - \hat{A})^{-1}\hat{B} + \hat{D} = \frac{1}{s-1}.$$

It is usually desirable to find realizations whose order is as small as possible.

Definition 12.9.14. Let $G \in \mathbb{R}_{\text{prop}}^{l \times m}(s)$, and assume that $G \sim \left[\begin{array}{c|c} A & B \\ \hline C & D \end{array}\right]$. Then, $\left[\begin{array}{c|c} A & B \\ \hline C & D \end{array}\right]$ is a *minimal realization* of G if its order is less than or equal to the order of every realization of G. In this case, we write

$$G \stackrel{\min}{\sim} \left[\begin{array}{c|c} A & B \\ \hline C & D \end{array}\right]. \tag{12.9.29}$$

Note that the minimality of a realization is independent of D.

The following result shows that the controllable and observable realization $\left[\begin{array}{c|c} A_1 & B_1 \\ \hline C_1 & D_1 \end{array}\right]$ of G in Proposition 12.9.10 is, in fact, minimal.

Corollary 12.9.15. Let $G \in \mathbb{R}^{l \times m}(s)$, and assume that $G \sim \left[\begin{array}{c|c} A & B \\ \hline C & D \end{array}\right]$. Then, $\left[\begin{array}{c|c} A & B \\ \hline C & D \end{array}\right]$ is minimal if and only if it is controllable and observable.

Proof. To prove necessity, suppose that $\left[\begin{array}{c|c} A & B \\ \hline C & D \end{array}\right]$ is either not controllable or not observable. Then, Proposition 12.9.10 can be used to construct a realization of G of order less than n. Hence, $\left[\begin{array}{c|c} A & B \\ \hline C & D \end{array}\right]$ is not minimal.

To prove sufficiency, assume that $A \in \mathbb{R}^{n \times n}$, and assume that $\left[\begin{array}{c|c} A & B \\ \hline C & D \end{array}\right]$ is not minimal. Hence, G has a minimal realization $\left[\begin{array}{c|c} \hat{A} & \hat{B} \\ \hline \hat{C} & D \end{array}\right]$ of order $\hat{n} < n$. Since the Markov parameters of G are independent of the realization, it follows from Proposition 12.9.11 that $\operatorname{rank} \mathcal{H}(G) = \hat{n} < n$. However, since $\left[\begin{array}{c|c} A & B \\ \hline C & D \end{array}\right]$ is observable and controllable, it follows from Proposition 12.9.11 that $\operatorname{rank} \mathcal{H}(G) = n$, which is a contradiction. $\square$

Theorem 12.9.16. Let $G \in \mathbb{R}^{l \times m}_{\text{prop}}(s)$, where $G \sim \left[\begin{array}{c|c} A & B \\ \hline C & D \end{array}\right]$ and $A \in \mathbb{R}^{n \times n}$. Then,

$$\operatorname{poles}(G) \subseteq \operatorname{spec}(A) \tag{12.9.30}$$

and

$$\operatorname{mpoles}(G) \subseteq \operatorname{mspec}(A). \tag{12.9.31}$$

Furthermore, the following statements are equivalent:

i) $G \overset{\min}{\sim} \left[\begin{array}{c|c} A & B \\ \hline C & D \end{array}\right]$.

ii) $\operatorname{Mcdeg}(G) = n$.

iii) $\operatorname{mpoles}(G) = \operatorname{mspec}(A)$.

Proof. See [1179, p. 319]. $\square$

Definition 12.9.17. Let $G \in \mathbb{R}^{l \times m}_{\text{prop}}(s)$, where $G \overset{\min}{\sim} \left[\begin{array}{c|c} A & B \\ \hline C & D \end{array}\right]$. Then, G is (asymptotically stable, semistable, Lyapunov stable) if A is.

Proposition 12.9.18. Let $G = p/q \in \mathbb{R}_{\text{prop}}(s)$, where $p, q \in \mathbb{R}[s]$, and assume that p and q are coprime. Then, G is (asymptotically stable, semistable, Lyapunov stable) if and only if q is.

Proposition 12.9.19. Let $G \in \mathbb{R}^{l \times m}_{\text{prop}}(s)$. Then, G is (asymptotically stable, semistable, Lyapunov stable) if and only if every entry of G is.

Definition 12.9.20. Let $G \in \mathbb{R}^{l \times m}_{\text{prop}}(s)$, where $G \overset{\min}{\sim} \left[\begin{array}{c|c} A & B \\ \hline C & D \end{array}\right]$ and A is asymptotically stable. Then, the realization $\left[\begin{array}{c|c} A & B \\ \hline C & D \end{array}\right]$ is *balanced* if the controllability and observability Gramians (12.7.2) and (12.4.2) are diagonal and equal.

Proposition 12.9.21. Let $G \in \mathbb{R}_{\mathrm{prop}}^{l \times m}(s)$, where $G \overset{\mathrm{min}}{\sim} \left[\begin{array}{c|c} A & B \\ \hline C & D \end{array}\right]$ and A is asymptotically stable. Then, there exists a nonsingular matrix $S \in \mathbb{R}^{n \times n}$ such that the realization $G \sim \left[\begin{array}{c|c} SAS^{-1} & SB \\ \hline CS^{-1} & D \end{array}\right]$ is balanced.

Proof. It follows from Corollary 8.3.8 that there exists a nonsingular matrix $S \in \mathbb{R}^{n \times n}$ such that SQS^{T} and $S^{-\mathrm{T}}PS^{-1}$ are diagonal, where Q and P are the controllability and observability Gramians (12.7.2) and (12.4.2). Hence, the realization $\left[\begin{array}{c|c} SAS^{-1} & SB \\ \hline CS^{-1} & D \end{array}\right]$ is balanced. $\square$

12.10 Zeros

In Section 4.7 the Smith-McMillan decomposition is used to define transmission zeros and blocking zeros of a transfer function $G(s)$. We now define the invariant zeros of a realization of $G(s)$ and relate these zeros to the transmission zeros. These zeros are related to the Smith zeros of a polynomial matrix as well as the spectrum of a pencil.

Definition 12.10.1. Let $G \in \mathbb{R}_{\mathrm{prop}}^{l \times m}(s)$, where $G \sim \left[\begin{array}{c|c} A & B \\ \hline C & D \end{array}\right]$. Then, the *Rosenbrock system matrix* $\mathcal{Z} \in \mathbb{R}^{(n+l) \times (n+m)}[s]$ is the polynomial matrix

$$\mathcal{Z}(s) \triangleq \left[\begin{array}{cc} sI - A & B \\ C & -D \end{array}\right]. \tag{12.10.1}$$

Furthermore, $z \in \mathbb{C}$ is an *invariant zero* of the realization $\left[\begin{array}{c|c} A & B \\ \hline C & D \end{array}\right]$ if

$$\mathrm{rank}\,\mathcal{Z}(z) < \mathrm{rank}\,\mathcal{Z}. \tag{12.10.2}$$

Let $G \in \mathbb{R}_{\mathrm{prop}}^{l \times m}(s)$, where $G \sim \left[\begin{array}{c|c} A & B \\ \hline C & D \end{array}\right]$ and $A \in \mathbb{R}^{n \times n}$, and note that $\mathcal{Z}$ is the pencil

$$\mathcal{Z}(s) = P_{\left[\begin{smallmatrix} A & -B \\ -C & D \end{smallmatrix}\right], \left[\begin{smallmatrix} I_n & 0 \\ 0 & 0 \end{smallmatrix}\right]}(s) \tag{12.10.3}$$

$$= s \left[\begin{array}{cc} I_n & 0 \\ 0 & 0 \end{array}\right] - \left[\begin{array}{cc} A & -B \\ -C & D \end{array}\right]. \tag{12.10.4}$$

Thus,

$$\mathrm{Szeros}(\mathcal{Z}) = \mathrm{spec}\left(\left[\begin{array}{cc} A & -B \\ -C & D \end{array}\right], \left[\begin{array}{cc} I_n & 0 \\ 0 & 0 \end{array}\right]\right) \tag{12.10.5}$$

and

$$\mathrm{mSzeros}(\mathcal{Z}) = \mathrm{mspec}\left(\left[\begin{array}{cc} A & -B \\ -C & D \end{array}\right], \left[\begin{array}{cc} I_n & 0 \\ 0 & 0 \end{array}\right]\right). \tag{12.10.6}$$

Hence, we define the set of invariant zeros of $\left[\begin{array}{c|c} A & B \\ \hline C & D \end{array}\right]$ by

$$\mathrm{izeros}\left(\left[\begin{array}{c|c} A & B \\ \hline C & D \end{array}\right]\right) \triangleq \mathrm{Szeros}(\mathcal{Z})$$

and the multiset of invariant zeros of $\left[\begin{array}{c|c} A & B \\ \hline C & D \end{array}\right]$ by

$$\mathrm{mizeros}\left(\left[\begin{array}{c|c} A & B \\ \hline C & D \end{array}\right]\right) \triangleq \mathrm{Szeros}(\mathcal{Z}).$$

Note that $P_{\left[\begin{smallmatrix} A & -B \\ -C & D \end{smallmatrix}\right],\left[\begin{smallmatrix} I_n & 0 \\ 0 & 0 \end{smallmatrix}\right]}$ is regular if and only if $\mathrm{rank}\,\mathcal{Z} = n + \min\{l, m\}$.

The following result shows that a strictly proper transfer function with full-state observation or full-state actuation has no invariant zeros.

Proposition 12.10.2. Let $G \in \mathbb{R}^{l\times m}_{\mathrm{prop}}(s)$, where $G \sim \left[\begin{array}{c|c} A & B \\ \hline C & 0 \end{array}\right]$ and $A \in \mathbb{R}^{n\times n}$. Then, the following statements hold:

 i) If $m = n$ and B is nonsingular, then $\mathrm{rank}\,\mathcal{Z} = n + \mathrm{rank}\,C$ and $\left[\begin{array}{c|c} A & B \\ \hline C & 0 \end{array}\right]$ has no invariant zeros.

 ii) If $l = n$ and C is nonsingular, then $\mathrm{rank}\,\mathcal{Z} = n + \mathrm{rank}\,B$ and $\left[\begin{array}{c|c} A & B \\ \hline C & 0 \end{array}\right]$ has no invariant zeros.

 iii) If $m = n$ and B is nonsingular, then $P_{\left[\begin{smallmatrix} I_n & 0 \\ 0 & 0 \end{smallmatrix}\right],\left[\begin{smallmatrix} A & -B \\ -C & 0 \end{smallmatrix}\right]}$ is regular if and only if $\mathrm{rank}\,C = \min\{l, n\}$.

 iv) If $l = n$ and C is nonsingular, then $P_{\left[\begin{smallmatrix} I_n & 0 \\ 0 & 0 \end{smallmatrix}\right],\left[\begin{smallmatrix} A & -B \\ -C & 0 \end{smallmatrix}\right]}$ is regular if and only if $\mathrm{rank}\,B = \min\{m, n\}$.

It is useful to note that, for all $s \notin \mathrm{spec}(A)$,

$$\mathcal{Z}(s) = \left[\begin{array}{cc} I & 0 \\ C(sI-A)^{-1} & I \end{array}\right]\left[\begin{array}{cc} sI-A & B \\ 0 & -G(s) \end{array}\right] \tag{12.10.7}$$

$$= \left[\begin{array}{cc} sI-A & 0 \\ C & -G(s) \end{array}\right]\left[\begin{array}{cc} I & (sI-A)^{-1}B \\ 0 & I \end{array}\right]. \tag{12.10.8}$$

Proposition 12.10.3. Let $G \in \mathbb{R}^{l\times m}_{\mathrm{prop}}(s)$, where $G \sim \left[\begin{array}{c|c} A & B \\ \hline C & D \end{array}\right]$. If $s \notin \mathrm{spec}(A)$, then

$$\mathrm{rank}\,\mathcal{Z}(s) = n + \mathrm{rank}\,G(s). \tag{12.10.9}$$

Furthermore,

$$\mathrm{rank}\,\mathcal{Z} = n + \mathrm{rank}\,G. \tag{12.10.10}$$

Proof. For $s \notin \mathrm{spec}(A)$, (12.10.9) follows from (12.10.7). Therefore, it follows from Proposition 4.3.6 and Proposition 4.7.8 that

$$\mathrm{rank}\,\mathcal{Z} = \max_{s\in\mathbb{C}} \mathrm{rank}\,\mathcal{Z}(s) = \max_{s\in\mathbb{C}\setminus\mathrm{spec}(A)} \mathrm{rank}\,\mathcal{Z}(s)$$

$$= n + \max_{s\in\mathbb{C}\setminus\mathrm{spec}(A)} \mathrm{rank}\,G(s) = n + \mathrm{rank}\,G. \qquad \square$$

Note that the realization in Proposition 12.10.3 is not assumed to be minimal. Therefore, $P_{\left[\begin{smallmatrix} A & -B \\ -C & D \end{smallmatrix}\right],\left[\begin{smallmatrix} I_n & 0 \\ 0 & 0 \end{smallmatrix}\right]}$ is (regular, singular) for one realization of G if and only if it is (regular, singular) for every realization of G. In fact, the following result shows that $P_{\left[\begin{smallmatrix} A & -B \\ -C & D \end{smallmatrix}\right], \left[\begin{smallmatrix} I_n & 0 \\ 0 & 0 \end{smallmatrix}\right]}$ is regular if and only if G has full rank.

Corollary 12.10.4. Let $G \in \mathbb{R}^{l \times m}_{\text{prop}}(s)$, where $G \sim \left[\begin{array}{c|c} A & B \\ \hline C & D \end{array}\right]$. Then, $P_{\left[\begin{smallmatrix} A & -B \\ -C & D \end{smallmatrix}\right], \left[\begin{smallmatrix} I_n & 0 \\ 0 & 0 \end{smallmatrix}\right]}$ is regular if and only if $\operatorname{rank} G = \min\{l, m\}$.

In the SISO case, it follows from (12.10.7) and (12.10.8) that, for all $s \in \mathbb{C} \backslash \operatorname{spec}(A)$,

$$\det \mathcal{Z}(s) = -[\det(sI - A)]G(s). \tag{12.10.11}$$

Consequently, for all $s \in \mathbb{C}$,

$$\det \mathcal{Z}(s) = -C(sI - A)^{\mathrm{A}}B - [\det(sI - A)]D. \tag{12.10.12}$$

The equality (12.10.12) also follows from Fact 2.14.2.

In particular, if $s \in \operatorname{spec}(A)$, then

$$\det \mathcal{Z}(s) = -C(sI - A)^{\mathrm{A}}B. \tag{12.10.13}$$

If, in addition, $n \geq 2$ and $\operatorname{rank}(sI - A) \leq n - 2$, then it follows from Fact 2.16.8 that $(sI - A)^{\mathrm{A}} = 0$, and thus

$$\det \mathcal{Z}(s) = 0. \tag{12.10.14}$$

Alternatively, in the case $n = 1$, it follows that, for all $s \in \mathbb{C}$, $(sI - A)^{\mathrm{A}} = 1$, and thus, for all $s \in \mathbb{C}$,

$$\det \mathcal{Z}(s) = -CB - (sI - A)D. \tag{12.10.15}$$

Next, it follows from (12.10.11) and (12.10.12) that

$$G(s) = \frac{C(sI - A)^{\mathrm{A}}B + [\det(sI - A)]D}{\det(sI - A)} = \frac{-\det \mathcal{Z}(s)}{\det(sI - A)}. \tag{12.10.16}$$

Consequently, $G \neq 0$ if and only if $\det \mathcal{Z} \neq 0$.

We now have the following result for scalar transfer functions.

Corollary 12.10.5. Let $G \in \mathbb{R}_{\text{prop}}(s)$, where $G \sim \left[\begin{array}{c|c} A & B \\ \hline C & D \end{array}\right]$. Then, the following statements are equivalent:

i) $P_{\left[\begin{smallmatrix} A & -B \\ -C & D \end{smallmatrix}\right], \left[\begin{smallmatrix} I_n & 0 \\ 0 & 0 \end{smallmatrix}\right]}$ is regular.

ii) $G \neq 0$.

iii) $\operatorname{rank} G = 1$.

iv) $\det \mathcal{Z} \neq 0$.

v) $\operatorname{rank} \mathcal{Z} = n + 1$.

vi) $C(sI - A)^{\mathrm{A}}B + [\det(sI - A)]D$ is not the zero polynomial.

In this case,

$$\text{mizeros}\left(\left[\begin{array}{c|c} A & B \\ \hline C & D \end{array}\right]\right) = \text{mroots}(\det \mathcal{Z}) \qquad (12.10.17)$$

and

$$\text{mizeros}\left(\left[\begin{array}{c|c} A & B \\ \hline C & D \end{array}\right]\right) = \text{mtzeros}(G) \cup [\text{mspec}(A)\backslash\text{mpoles}(G)]. \qquad (12.10.18)$$

If, in addition, $G \overset{\text{min}}{\sim} \left[\begin{array}{c|c} A & B \\ \hline C & D \end{array}\right]$, then

$$\text{mizeros}\left(\left[\begin{array}{c|c} A & B \\ \hline C & D \end{array}\right]\right) = \text{mtzeros}(G). \qquad (12.10.19)$$

Now, suppose that G is square, that is, $l = m$. Then, it follows from (12.10.7) and (12.10.8) that, for all $s \in \mathbb{C}\backslash\text{spec}(A)$,

$$\det \mathcal{Z}(s) = (-1)^l [\det(sI - A)] \det G(s), \qquad (12.10.20)$$

and thus

$$\det G(s) = \frac{(-1)^l \det \mathcal{Z}(s)}{\det(sI - A)}. \qquad (12.10.21)$$

Furthermore, for all $s \in \mathbb{C}$,

$$[\det(sI - A)]^{l-1} \det \mathcal{Z}(s) = (-1)^l \det\left(C(sI - A)^{\text{A}}B + [\det(sI - A)]D\right). \qquad (12.10.22)$$

Hence, for all $s \in \text{spec}(A)$, it follows that

$$\det\left[C(sI - A)^{\text{A}}B\right] = 0. \qquad (12.10.23)$$

We thus have the following result for square transfer functions G that satisfy $\det G \neq 0$.

Corollary 12.10.6. Let $G \in \mathbb{R}^{l\times l}_{\text{prop}}(s)$, where $G \sim \left[\begin{array}{c|c} A & B \\ \hline C & D \end{array}\right]$. Then, the following statements are equivalent:

i) $P_{\left[\begin{smallmatrix} A & -B \\ -C & D \end{smallmatrix}\right], \left[\begin{smallmatrix} I_n & 0 \\ 0 & 0 \end{smallmatrix}\right]}$ is regular.

ii) $\det G \neq 0$.

iii) $\text{rank}\, G = l$.

iv) $\det \mathcal{Z} \neq 0$.

v) $\text{rank}\, \mathcal{Z} = n + l$.

vi) $\det(C(sI - A)^{\text{A}}B + [\det(sI - A)]D)$ is not the zero polynomial.

In this case,

$$\text{mizeros}\left(\left[\begin{array}{c|c} A & B \\ \hline C & D \end{array}\right]\right) = \text{mroots}(\det \mathcal{Z}), \qquad (12.10.24)$$

$$\text{mizeros}\left(\left[\begin{array}{c|c} A & B \\ \hline C & D \end{array}\right]\right) = \text{mtzeros}(G) \cup [\text{mspec}(A)\backslash\text{mpoles}(G)], \qquad (12.10.25)$$

and

$$\text{izeros}\left(\left[\begin{array}{c|c} A & B \\ \hline C & D \end{array}\right]\right) = \text{tzeros}(G) \cup [\text{spec}(A)\backslash\text{poles}(G)]. \tag{12.10.26}$$

If, in addition, $G \overset{\min}{\sim} \left[\begin{array}{c|c} A & B \\ \hline C & D \end{array}\right]$, then

$$\text{mizeros}\left(\left[\begin{array}{c|c} A & B \\ \hline C & D \end{array}\right]\right) = \text{mtzeros}(G). \tag{12.10.27}$$

Example 12.10.7. Consider $G \in \mathbb{R}^{2\times 2}(s)$ defined by

$$G(s) \triangleq \left[\begin{array}{cc} \frac{s-1}{s+1} & 0 \\ 0 & \frac{s+1}{s-1} \end{array}\right]. \tag{12.10.28}$$

Then, the Smith-McMillan form of G is given by

$$G(s) \triangleq S_1(s) \left[\begin{array}{cc} \frac{1}{s^2-1} & 0 \\ 0 & s^2-1 \end{array}\right] S_2(s), \tag{12.10.29}$$

where $S_1, S_2 \in \mathbb{R}^{2\times 2}[s]$ are the unimodular matrices

$$S_1(s) \triangleq \left[\begin{array}{cc} (s-1)^2 & -1 \\ -\frac{1}{4}(s+1)^2(s-2) & \frac{1}{4}(s+2) \end{array}\right] \tag{12.10.30}$$

and

$$S_2(s) \triangleq \left[\begin{array}{cc} \frac{1}{4}(s-1)^2(s+2) & (s+1)^2 \\ \frac{1}{4}(s-2) & 1 \end{array}\right]. \tag{12.10.31}$$

Thus, $\text{mpoles}(G) = \text{mtzeros}(G) = \{1, -1\}$. Furthermore, a minimal realization of G is given by

$$G \overset{\min}{\sim} \left[\begin{array}{cc|cc} -1 & 0 & 1 & 0 \\ 0 & 1 & 0 & 1 \\ \hline -2 & 0 & 1 & 0 \\ 0 & 2 & 0 & 1 \end{array}\right]. \tag{12.10.32}$$

Finally, note that $\det \mathcal{Z}(s) = (-1)^2 [\det(sI - A)] \det G = s^2 - 1$, which confirms (12.10.27).

Theorem 12.10.8. Let $G \in \mathbb{R}^{l\times m}_{\text{prop}}(s)$, where $G \sim \left[\begin{array}{c|c} A & B \\ \hline C & D \end{array}\right]$. Then,

$$\text{izeros}\left(\left[\begin{array}{c|c} A & B \\ \hline C & D \end{array}\right]\right)\backslash\text{spec}(A) \subseteq \text{tzeros}(G) \tag{12.10.33}$$

and

$$\text{tzeros}(G)\backslash\text{poles}(G) \subseteq \text{izeros}\left(\left[\begin{array}{c|c} A & B \\ \hline C & D \end{array}\right]\right). \tag{12.10.34}$$

If, in addition, $G \overset{\min}{\sim} \left[\begin{array}{c|c} A & B \\ \hline C & D \end{array}\right]$, then

$$\text{izeros}\left(\left[\begin{array}{c|c} A & B \\ \hline C & D \end{array}\right]\right)\backslash\text{poles}(G) = \text{tzeros}(G)\backslash\text{poles}(G). \tag{12.10.35}$$

Proof. To prove (12.10.33), let $z \in \text{izeros}\left(\left[\begin{array}{c|c} A & B \\ \hline C & D \end{array}\right]\right) \backslash \text{spec}(A)$. Since $z \notin$ $\text{spec}(A)$ it follows from Theorem 12.9.16 that $z \notin \text{poles}(G)$. It now follows from Proposition 12.10.3 that $n + \text{rank}\, G(z) = \text{rank}\, \mathcal{Z}(z) < \text{rank}\, \mathcal{Z} = n + \text{rank}\, G$, which implies that $\text{rank}\, G(z) < \text{rank}\, G$. Thus, $z \in \text{tzeros}(G)$.

To prove (12.10.34), let $z \in \text{tzeros}(G) \backslash \text{poles}(G)$. Then, it follows from Proposition 12.10.3 that $\text{rank}\, \mathcal{Z}(z) = n + \text{rank}\, G(z) < n + \text{rank}\, G = \text{rank}\, \mathcal{Z}$, which implies that $z \in \text{izeros}\left(\left[\begin{array}{c|c} A & B \\ \hline C & D \end{array}\right]\right)$. The last statement follows from (12.10.33), (12.10.34), and Theorem 12.9.16. $\qquad\square$

The following result is a stronger form of Theorem 12.10.8.

Theorem 12.10.9. Let $G \in \mathbb{R}^{l \times m}_{\text{prop}}(s)$, where $G \sim \left[\begin{array}{c|c} A & B \\ \hline C & D \end{array}\right]$, let $S \in \mathbb{R}^{n \times n}$, assume that S is nonsingular, and let A, B, and C have the form (12.9.16), (12.9.17), where $\left(\left[\begin{smallmatrix} A_1 & 0 \\ A_{21} & A_2 \end{smallmatrix}\right], \left[\begin{smallmatrix} B_1 \\ B_2 \end{smallmatrix}\right]\right)$ is controllable and $\left(\left[\begin{smallmatrix} A_1 & A_{13} \\ 0 & A_3 \end{smallmatrix}\right], [\,C_1\ C_3\,]\right)$ is observable. Then,

$$\text{mtzeros}(G) = \text{mizeros}\left(\left[\begin{array}{c|c} A_1 & B_1 \\ \hline C_1 & D \end{array}\right]\right) \tag{12.10.36}$$

and

$$\text{mizeros}\left(\left[\begin{array}{c|c} A & B \\ \hline C & D \end{array}\right]\right) = \text{mspec}(A_2) \cup \text{mspec}(A_3) \cup \text{mspec}(A_4) \cup \text{mtzeros}(G). \tag{12.10.37}$$

Proof. Defining $\mathcal{Z}$ by (12.10.1), note that, in the notation of Proposition 12.9.10, $\mathcal{Z}$ has the same Smith form as

$$\tilde{\mathcal{Z}} \triangleq \begin{bmatrix} sI - A_4 & -A_{43} & 0 & 0 & 0 \\ 0 & sI - A_3 & 0 & 0 & 0 \\ -A_{24} & -A_{23} & sI - A_2 & -A_{21} & B_2 \\ 0 & -A_{13} & 0 & sI - A_1 & B_1 \\ 0 & C_3 & 0 & C_1 & -D \end{bmatrix}.$$

Hence, it follows from Proposition 12.10.3 that $\text{rank}\, \mathcal{Z} = \text{rank}\, \tilde{\mathcal{Z}} = n + r$, where $r \triangleq \text{rank}\, G$. Let $\tilde{p}_1, \ldots, \tilde{p}_{n+r}$ be the Smith polynomials of $\tilde{\mathcal{Z}}$. Then, since $\tilde{p}_{n+r}$ is the monic greatest common divisor of all $(n+r) \times (n+r)$ subdeterminants of $\tilde{\mathcal{Z}}$, it follows that $\tilde{p}_{n+r} = \chi_{A_1}\chi_{A_2}\chi_{A_3}p_r$, where p_r is the rth Smith polynomial of $\left[\begin{smallmatrix} sI - A_1 & B_1 \\ C_1 & -D \end{smallmatrix}\right]$. Therefore,

$$\text{mSzeros}(\mathcal{Z}) = \text{mspec}(A_2) \cup \text{mspec}(A_3) \cup \text{mspec}(A_4) \cup \text{mSzeros}\left(\left[\begin{smallmatrix} sI - A_1 & B_1 \\ C_1 & -D \end{smallmatrix}\right]\right).$$

Next, using the Smith-McMillan decomposition Theorem 4.7.5, it follows that there exist unimodular matrices $S_1 \in \mathbb{R}^{l \times l}[s]$ and $S_2 \in \mathbb{R}^{m \times m}[s]$ such that $G = S_1 D_0^{-1} N_0 S_2$, where

$$D_0 \triangleq \begin{bmatrix} q_1 & & & 0 \\ & \ddots & & \\ & & q_r & \\ 0 & & & I_{l-r} \end{bmatrix}, \quad N_0 \triangleq \begin{bmatrix} p_1 & & & 0 \\ & \ddots & & \\ & & p_r & \\ 0 & & & 0_{(l-r)\times(m-r)} \end{bmatrix}.$$

Now, define the polynomial matrix $\hat{\mathcal{Z}} \in \mathbb{R}^{(n+l)\times(n+m)}[s]$ by

$$\hat{\mathcal{Z}} \triangleq \begin{bmatrix} I_{n-l} & 0_{(n-l)\times l} & 0_{(n-l)\times m} \\ 0_{l\times(n-l)} & D_0 & N_0 S_2 \\ 0_{l\times(n-l)} & S_1 & 0_{l\times m} \end{bmatrix}.$$

Since S_1 is unimodular, it follows that the Smith form $\mathcal{S}$ of $\hat{\mathcal{Z}}$ is given by

$$\mathcal{S} = \begin{bmatrix} I_n & 0_{n\times m} \\ 0_{l\times n} & N_0 \end{bmatrix}.$$

Consequently, $\mathrm{mSzeros}(\hat{\mathcal{Z}}) = \mathrm{mSzeros}(\mathcal{S}) = \mathrm{mtzeros}(G)$.

Next, note that

$$\mathrm{rank}\begin{bmatrix} I_{n-l} & 0_{(n-l)\times l} & 0_{(n-l)\times m} \\ 0_{l\times(n-l)} & D_0 & N_0 S_2 \end{bmatrix} = \mathrm{rank}\begin{bmatrix} I_{n-l} & 0_{(n-l)\times l} \\ 0_{l\times(n-l)} & D_0 \\ 0_{l\times(n-l)} & S_1 \end{bmatrix} = n$$

and that

$$G = \begin{bmatrix} 0_{l\times(n-l)} & S_1 & 0_{l\times m} \end{bmatrix} \begin{bmatrix} I_{n-l} & 0_{(n-l)\times l} \\ 0_{l\times(n-l)} & D_0 \end{bmatrix}^{-1} \begin{bmatrix} 0_{(n-l)\times m} \\ N_0 S_2 \end{bmatrix}.$$

Furthermore, $G \overset{\min}{\sim} \left[\begin{array}{c|c} A_1 & B_1 \\ \hline C_1 & D \end{array}\right]$, Consequently, $\hat{\mathcal{Z}}$ and $\left[\begin{smallmatrix} sI-A_1 & B_1 \\ C_1 & D \end{smallmatrix}\right]$ have no decoupling zeros [1173, pp. 64–70], and it thus follows from Theorem 3.1 of [1173, p. 106] that $\hat{\mathcal{Z}}$ and $\left[\begin{smallmatrix} sI-A_1 & B_1 \\ C_1 & D \end{smallmatrix}\right]$ have the same Smith form. Thus,

$$\mathrm{mSzeros}\left(\begin{bmatrix} sI-A_1 & B_1 \\ C_1 & -D \end{bmatrix}\right) = \mathrm{mSzeros}(\hat{\mathcal{Z}}) = \mathrm{mtzeros}(G).$$

Consequently,

$$\mathrm{mizeros}\left(\left[\begin{array}{c|c} A_1 & B_1 \\ \hline C_1 & D \end{array}\right]\right) = \mathrm{mSzeros}\left(\left[\begin{smallmatrix} sI-A_1 & B_1 \\ C_1 & -D \end{smallmatrix}\right]\right) = \mathrm{mtzeros}(G),$$

which proves (12.10.36).

Finally, to prove (12.10.33) note that

$$\begin{aligned} \mathrm{mizeros}\left(\left[\begin{array}{c|c} A & B \\ \hline C & D \end{array}\right]\right) \\ = \mathrm{mSzeros}(\mathcal{Z}) \\ = \mathrm{mspec}(A_2) \cup \mathrm{mspec}(A_3) \cup \mathrm{mspec}(A_4) \cup \mathrm{mSzeros}\left(\left[\begin{smallmatrix} sI-A_1 & B_1 \\ -C_1 & -D \end{smallmatrix}\right]\right) \\ = \mathrm{mspec}(A_2) \cup \mathrm{mspec}(A_3) \cup \mathrm{mspec}(A_4) \cup \mathrm{mtzeros}(G). \qquad \square \end{aligned}$$

Proposition 12.10.10. Equivalent realizations have the same invariant zeros. Furthermore, invariant zeros are not changed by full-state feedback.

Proof. Let $u = Kx + v$, which leads to the rational transfer function

$$G_K \sim \left[\begin{array}{c|c} A + BK & B \\ \hline C + DK & D \end{array}\right]. \qquad (12.10.38)$$

Since

$$\left[\begin{array}{cc} zI - (A + BK) & B \\ C + DK & -D \end{array}\right] = \left[\begin{array}{cc} zI - A & B \\ C & -D \end{array}\right]\left[\begin{array}{cc} I & 0 \\ -K & I \end{array}\right], \qquad (12.10.39)$$

it follows that $\left[\begin{array}{c|c} A & B \\ \hline C & D \end{array}\right]$ and $\left[\begin{array}{c|c} A + BK & B \\ \hline C + DK & D \end{array}\right]$ have the same invariant zeros. $\qquad \square$

The following result provides an interpretation of condition i) of Theorem 12.17.9.

Proposition 12.10.11. Let $G \in \mathbb{R}^{l \times m}_{\mathrm{prop}}(s)$, where $G \sim \left[\begin{array}{c|c} A & B \\ \hline C & D \end{array}\right]$, and assume that $R \triangleq D^{\mathrm{T}}D$ is positive definite. Then, the following statements hold:

i) $\operatorname{rank} \mathcal{Z} = n + m$.

ii) $z \in \mathbb{C}$ is an invariant zero of $\left[\begin{array}{c|c} A & B \\ \hline C & D \end{array}\right]$ if and only if z is an unobservable eigenvalue of $(A - BR^{-1}D^{\mathrm{T}}C, [I - DR^{-1}D^{\mathrm{T}}]C)$.

Proof. To prove i), assume that $\operatorname{rank} \mathcal{Z} < n + m$. Then, for every $s \in \mathbb{C}$, there exists a nonzero vector $\left[\begin{smallmatrix} x \\ y \end{smallmatrix}\right] \in \mathcal{N}[\mathcal{Z}(s)]$, that is,

$$\left[\begin{array}{cc} sI - A & B \\ C & -D \end{array}\right]\left[\begin{array}{c} x \\ y \end{array}\right] = 0.$$

Consequently, $Cx - Dy = 0$, which implies that $D^{\mathrm{T}}Cx - Ry = 0$, and thus $y = R^{-1}D^{\mathrm{T}}Cx$. Furthermore, since $(sI - A + BR^{-1}D^{\mathrm{T}}C)x = 0$, choosing $s \notin \operatorname{spec}(A - BR^{-1}D^{\mathrm{T}}C)$ yields $x = 0$, and thus $y = 0$, which is a contradiction.

To prove ii), note that z is an invariant zero of $\left[\begin{array}{c|c} A & B \\ \hline C & D \end{array}\right]$ if and only if $\operatorname{rank} \mathcal{Z}(z) < n + m$, which holds if and only if there exists a nonzero vector $\left[\begin{smallmatrix} x \\ y \end{smallmatrix}\right] \in \mathcal{N}[\mathcal{Z}(z)]$. This condition is equivalent to

$$\left[\begin{array}{c} sI - A + BR^{-1}D^{\mathrm{T}}C \\ (I - DR^{-1}D^{\mathrm{T}})C \end{array}\right] x = 0,$$

where $x \neq 0$. This last condition is equivalent to the fact that z is an unobservable eigenvalue of $(A - BR^{-1}D^{\mathrm{T}}C, [I - DR^{-1}D^{\mathrm{T}}]C)$. $\qquad \square$

Corollary 12.10.12. Assume that $R \triangleq D^{\mathrm{T}}D$ is positive definite, and assume that $(A - BR^{-1}D^{\mathrm{T}}C, [I - DR^{-1}D^{\mathrm{T}}]C)$ is observable. Then, $\left[\begin{array}{c|c} A & B \\ \hline C & D \end{array}\right]$ has no invariant zeros.

12.11 H_2 System Norm

Consider the system

$$\dot{x}(t) = Ax(t) + Bu(t), \tag{12.11.1}$$
$$y(t) = Cx(t), \tag{12.11.2}$$

where $A \in \mathbb{R}^{n \times n}$ is asymptotically stable, $B \in \mathbb{R}^{n \times m}$, and $C \in \mathbb{R}^{l \times n}$. Then, for all $t \geq 0$, the impulse response function defined by (12.1.18) is given by

$$H(t) = Ce^{tA}B. \tag{12.11.3}$$

The L_2 *norm* of $H(\cdot)$ is given by

$$\|H\|_{L_2} \triangleq \left(\int_0^\infty \|H(t)\|_F^2 \, dt \right)^{1/2}. \tag{12.11.4}$$

The following result provides expressions for $\|H(\cdot)\|_{L_2}$ in terms of the controllability and observability Gramians.

Theorem 12.11.1. Assume that A is asymptotically stable. Then, the L_2 norm of H is given by

$$\|H\|_{L_2}^2 = \operatorname{tr} CQC^{\mathrm{T}} = \operatorname{tr} B^{\mathrm{T}}PB, \tag{12.11.5}$$

where $Q, P \in \mathbb{R}^{n \times n}$ satisfy

$$AQ + QA^{\mathrm{T}} + BB^{\mathrm{T}} = 0, \tag{12.11.6}$$
$$A^{\mathrm{T}}P + PA + C^{\mathrm{T}}C = 0. \tag{12.11.7}$$

Proof. Note that

$$\|H\|_{L_2}^2 = \int_0^\infty \operatorname{tr} Ce^{tA}BB^{\mathrm{T}}e^{tA^{\mathrm{T}}}C^{\mathrm{T}} dt = \operatorname{tr} CQC^{\mathrm{T}},$$

where Q satisfies (12.11.6). The dual expression (12.11.7) follows in a similar manner or by noting that

$$\operatorname{tr} CQC^{\mathrm{T}} = \operatorname{tr} C^{\mathrm{T}}CQ = -\operatorname{tr}\left(A^{\mathrm{T}}P + PA\right)Q$$
$$= -\operatorname{tr}\left(AQ + QA^{\mathrm{T}}\right)P = \operatorname{tr} BB^{\mathrm{T}}P = \operatorname{tr} B^{\mathrm{T}}PB. \qquad \square$$

For the following definition, note that

$$\|G(s)\|_F = [\operatorname{tr} G(s)G^*(s)]^{1/2}. \tag{12.11.8}$$

Definition 12.11.2. The H_2 *norm* of $G \in \mathbb{R}^{l \times m}(s)$ is the nonnegative number

$$\|G\|_{H_2} \triangleq \left(\frac{1}{2\pi} \int_{-\infty}^\infty \|G(\jmath\omega)\|_F^2 \, d\omega \right)^{1/2}. \tag{12.11.9}$$

The following result is *Parseval's theorem*, which relates the L_2 norm of the impulse response function to the H_2 norm of its transform.

Theorem 12.11.3. Let $G \in \mathbb{R}^{l \times m}_{\text{prop}}(s)$, where $G \sim \left[\begin{array}{c|c} A & B \\ \hline C & 0 \end{array} \right]$, define H by (12.11.3), and assume that $A \in \mathbb{R}^{n \times n}$ is asymptotically stable. Then,

$$\int_0^\infty H(t) H^{\mathrm{T}}(t) \, dt = \frac{1}{2\pi} \int_{-\infty}^\infty G(\jmath\omega) G^*(\jmath\omega) \, d\omega. \tag{12.11.10}$$

Therefore,

$$\|H\|_{\mathrm{L}_2} = \|G\|_{\mathrm{H}_2}. \tag{12.11.11}$$

Proof. First note that

$$G(s) = \mathcal{L}\{H(t)\} = \int_0^\infty H(t) e^{-st} \, dt$$

and that

$$H(t) = \frac{1}{2\pi} \int_{-\infty}^\infty G(\jmath\omega) e^{\jmath\omega t} \, d\omega.$$

Hence,

$$\int_0^\infty H(t) H^{\mathrm{T}}(t) e^{-st} \, dt = \int_0^\infty \left(\frac{1}{2\pi} \int_{-\infty}^\infty G(\jmath\omega) e^{\jmath\omega t} \, d\omega \right) H^{\mathrm{T}}(t) e^{-st} \, dt$$

$$= \frac{1}{2\pi} \int_{-\infty}^\infty G(\jmath\omega) \left(\int_0^\infty H^{\mathrm{T}}(t) e^{-(s-\jmath\omega)t} \, dt \right) d\omega$$

$$= \frac{1}{2\pi} \int_{-\infty}^\infty G(\jmath\omega) G^{\mathrm{T}}(s - \jmath\omega) \, d\omega.$$

Setting $s = 0$ yields (12.11.7), while taking the trace of (12.11.10) yields (12.11.11). $\square$

Corollary 12.11.4. Let $G \in \mathbb{R}^{l \times m}_{\text{prop}}(s)$, where $G \sim \left[\begin{array}{c|c} A & B \\ \hline C & 0 \end{array} \right]$, and assume that $A \in \mathbb{R}^{n \times n}$ is asymptotically stable. Then,

$$\|G\|^2_{\mathrm{H}_2} = \|H\|^2_{\mathrm{L}_2} = \operatorname{tr} CQC^{\mathrm{T}} = \operatorname{tr} B^{\mathrm{T}} PB, \tag{12.11.12}$$

where $Q, P \in \mathbb{R}^{n \times n}$ satisfy (12.11.6) and (12.11.7), respectively.

The following corollary of Theorem 12.11.3 provides a frequency-domain expression for the solution of the Lyapunov equation.

Corollary 12.11.5. Let $A \in \mathbb{R}^{n \times n}$, assume that A is asymptotically stable, let $B \in \mathbb{R}^{n \times m}$, and define $Q \in \mathbb{R}^{n \times n}$ by

$$Q = \tfrac{1}{2\pi} \int_{-\infty}^{\infty} (\jmath\omega I - A)^{-1} BB^{\mathrm{T}} (\jmath\omega I - A)^{-*} \, \mathrm{d}\omega. \tag{12.11.13}$$

Then, Q satisfies

$$AQ + QA^{\mathrm{T}} + BB^{\mathrm{T}} = 0. \tag{12.11.14}$$

Proof. This result follows directly from Theorem 12.11.3 with $H(t) = e^{tA}B$ and $G(s) = (sI - A)^{-1}B$. Alternatively, it follows from (12.11.14) that

$$\int_{-\infty}^{\infty} (\jmath\omega I - A)^{-1} \, \mathrm{d}\omega Q + Q \int_{-\infty}^{\infty} (\jmath\omega I - A)^{-*} \, \mathrm{d}\omega = \int_{-\infty}^{\infty} (\jmath\omega I - A)^{-1} BB^{\mathrm{T}} (\jmath\omega I - A)^{-*} \, \mathrm{d}\omega.$$

Assuming that A is diagonalizable with eigenvalues $\lambda_i = -\sigma_i + \jmath\omega_i$, it follows that

$$\int_{-\infty}^{\infty} \frac{\mathrm{d}\omega}{\jmath\omega - \lambda_i} = \int_{-\infty}^{\infty} \frac{\sigma_i - \jmath\omega}{\sigma_i^2 + \omega^2} \, \mathrm{d}\omega = \frac{\sigma_i \pi}{|\sigma_i|} - \jmath \lim_{r\to\infty} \int_{-r}^{r} \frac{\omega}{\sigma_i^2 + \omega^2} \, \mathrm{d}\omega = \pi,$$

which implies that

$$\int_{-\infty}^{\infty} (\jmath\omega I - A)^{-1} \, \mathrm{d}\omega = \pi I_n,$$

which yields (12.11.13). See [317] for a proof of the general case. $\square$

Proposition 12.11.6. Let $G_1, G_2 \in \mathbb{R}_{\mathrm{prop}}^{l \times m}(s)$ be asymptotically stable rational transfer functions. Then,

$$\|G_1 + G_2\|_{\mathrm{H}_2} \le \|G_1\|_{\mathrm{H}_2} + \|G_2\|_{\mathrm{H}_2}. \tag{12.11.15}$$

Proof. Let $G_1 \overset{\min}{\sim} \left[\begin{array}{c|c} A_1 & B_1 \\ \hline C_1 & 0 \end{array}\right]$ and $G_2 \overset{\min}{\sim} \left[\begin{array}{c|c} A_2 & B_2 \\ \hline C_2 & 0 \end{array}\right]$, where $A_1 \in \mathbb{R}^{n_1 \times n_1}$ and $A_2 \in \mathbb{R}^{n_2 \times n_2}$. It follows from Proposition 12.13.2 that

$$G_1 + G_2 \sim \left[\begin{array}{cc|c} A_1 & 0 & B_1 \\ 0 & A_2 & B_2 \\ \hline C_1 & C_2 & 0 \end{array}\right].$$

Now, Theorem 12.11.3 implies that $\|G_1\|_{\mathrm{H}_2} = \sqrt{\operatorname{tr} C_1 Q_1 C_1^{\mathrm{T}}}$ and $\|G_2\|_{\mathrm{H}_2} = \sqrt{\operatorname{tr} C_2 Q_2 C_2^{\mathrm{T}}}$, where $Q_1 \in \mathbb{R}^{n_1 \times n_1}$ and $Q_2 \in \mathbb{R}^{n_2 \times n_2}$ are the unique positive-definite matrices satisfying $A_1 Q_1 + Q_1 A_1^{\mathrm{T}} + B_1 B_1^{\mathrm{T}} = 0$ and $A_2 Q_2 + Q_2 A_2^{\mathrm{T}} + B_2 B_2^{\mathrm{T}} = 0$. Furthermore,

$$\|G_2 + G_2\|_{\mathrm{H}_2}^2 = \operatorname{tr} \left[\begin{array}{cc} C_1 & C_2 \end{array}\right] Q \left[\begin{array}{c} C_1^{\mathrm{T}} \\ C_2^{\mathrm{T}} \end{array}\right],$$

where $Q \in \mathbb{R}^{(n_1+n_2) \times (n_1+n_2)}$ is the unique, positive-semidefinite matrix satisfying

$$\left[\begin{array}{cc} A_1 & 0 \\ 0 & A_2 \end{array}\right] Q + Q \left[\begin{array}{cc} A_1 & 0 \\ 0 & A_2 \end{array}\right]^{\mathrm{T}} + \left[\begin{array}{c} B_1 \\ B_2 \end{array}\right] \left[\begin{array}{c} B_1 \\ B_2 \end{array}\right]^{\mathrm{T}} = 0.$$

It can be seen that $Q = \begin{bmatrix} Q_1 & Q_{12} \\ Q_{12}^T & Q_2 \end{bmatrix}$, where Q_1 and Q_2 are as given above and where Q_{12} satisfies $A_1 Q_{12} + Q_{12} A_2^T + B_1 B_2^T = 0$. Now, using the Cauchy-Schwarz inequality (9.3.17) and $iii)$ of Proposition 8.2.4, it follows that

$$
\begin{aligned}
\|G_1 + G_2\|_{H_2}^2 &= \operatorname{tr}(C_1 Q_1 C_1^T + C_2 Q_2 C_2^T + C_2 Q_{12}^T C_1^T + C_1 Q_{12} C_2^T) \\
&= \|G_1\|_{H_2}^2 + \|G_2\|_{H_2}^2 + 2\operatorname{tr} C_1 Q_{12} Q_2^{-1/2} Q_2^{1/2} C_2^T \\
&\leq \|G_1\|_{H_2}^2 + \|G_2\|_{H_2}^2 + 2\operatorname{tr}(C_1 Q_{12} Q_2^{-1} Q_{12}^T C_1^T)\operatorname{tr}(C_2 Q_2 C_2^T) \\
&\leq \|G_1\|_{H_2}^2 + \|G_2\|_{H_2}^2 + 2\operatorname{tr}(C_1 Q_1 C_1^T)\operatorname{tr}(C_2 Q_2 C_2^T) \\
&= (\|G_1\|_{H_2} + \|G_2\|_{H_2})^2. \qquad \square
\end{aligned}
$$

12.12 Harmonic Steady-State Response

The following result concerns the response of a linear system to a harmonic input.

Theorem 12.12.1. For $t \geq 0$, consider the linear system

$$\dot{x}(t) = Ax(t) + Bu(t), \qquad (12.12.1)$$

with harmonic input

$$u(t) = \operatorname{Re} u_0 e^{\jmath \omega_0 t}, \qquad (12.12.2)$$

where $u_0 \in \mathbb{C}^m$ and $\omega_0 \in \mathbb{R}$ is such that $\jmath \omega_0 \notin \operatorname{spec}(A)$. Then, $x(t)$ is given by

$$x(t) = e^{tA}\big(x(0) - \operatorname{Re}[(\jmath\omega_0 I - A)^{-1}Bu_0]\big) + \operatorname{Re}[(\jmath\omega_0 I - A)^{-1}Bu_0 e^{\jmath\omega_0 t}]. \quad (12.12.3)$$

Proof. We have

$$
\begin{aligned}
x(t) &= e^{tA}x(0) + \int_0^t e^{(t-\tau)A} B\operatorname{Re}(u_0 e^{\jmath\omega_0\tau})\, d\tau \\
&= e^{tA}x(0) + e^{tA}\operatorname{Re}\left[\int_0^t e^{-\tau A} e^{\jmath\omega_0\tau}\, d\tau B u_0\right] \\
&= e^{tA}x(0) + e^{tA}\operatorname{Re}\left[\int_0^t e^{\tau(\jmath\omega_0 I - A)}\, d\tau B u_0\right] \\
&= e^{tA}x(0) + e^{tA}\operatorname{Re}\left[(\jmath\omega_0 I - A)^{-1}\big(e^{t(\jmath\omega_0 I - A)} - I\big)B u_0\right] \\
&= e^{tA}x(0) + \operatorname{Re}\left[(\jmath\omega_0 I - A)^{-1}\big(e^{\jmath\omega_0 t I} - e^{tA}\big)B u_0\right] \\
&= e^{tA}x(0) + \operatorname{Re}\left[(\jmath\omega_0 I - A)^{-1}(-e^{tA})B u_0\right] + \operatorname{Re}\left[(\jmath\omega_0 I - A)^{-1}e^{\jmath\omega_0 t}B u_0\right] \\
&= e^{tA}\big(x(0) - \operatorname{Re}[(\jmath\omega_0 I - A)^{-1}B u_0]\big) + \operatorname{Re}[(\jmath\omega_0 I - A)^{-1}B u_0 e^{\jmath\omega_0 t}]. \qquad \square
\end{aligned}
$$

Theorem 12.12.1 shows that the total response $y(t)$ of the linear system $G \sim \left[\begin{array}{c|c} A & B \\ \hline C & 0 \end{array} \right]$ to a harmonic input can be written as $y(t) = y_{\text{trans}}(t) + y_{\text{hss}}(t)$, where the transient component

$$y_{\text{trans}}(t) \triangleq Ce^{tA}\big(x(0) - \text{Re}\big[(\jmath\omega_0 I - A)^{-1}Bu_0\big]\big) \qquad (12.12.4)$$

depends on the initial condition and the input, and the harmonic steady-state component

$$y_{\text{hss}}(t) = \text{Re}\big[G(\jmath\omega_0)u_0 e^{\jmath\omega_0 t}\big] \qquad (12.12.5)$$

depends only on the input.

If A is asymptotically stable, then $\lim_{t\to\infty} y_{\text{trans}}(t) = 0$, and thus $y(t)$ approaches its harmonic steady-state component $y_{\text{hss}}(t)$ for large t. Since the harmonic steady-state component is sinusoidal, it follows that $y(t)$ does not converge in the usual sense.

Finally, if A is semistable, then it follows from Proposition 11.8.2 that

$$\lim_{t\to\infty} y_{\text{trans}}(t) = C(I - AA^{\#})\big(x(0) - \text{Re}\big[(\jmath\omega_0 I - A)^{-1}Bu_0\big]\big), \qquad (12.12.6)$$

which represents a constant offset to the harmonic steady-state component.

In the SISO case, let $u_0 \triangleq a_0(\sin\phi_0 + \jmath\cos\phi_0)$, and consider the input

$$u(t) = a_0 \sin(\omega_0 t + \phi_0) = \text{Re}\, u_0 e^{\jmath\omega_0 t}. \qquad (12.12.7)$$

Then, writing $G(\jmath\omega_0) = \text{Re}\, Me^{\jmath\theta}$, it follows that

$$y_{\text{hss}}(t) = a_0 M \sin(\omega_0 t + \phi_0 + \theta). \qquad (12.12.8)$$

12.13 System Interconnections

Let $G \in \mathbb{R}^{l\times m}_{\text{prop}}(s)$. We define the *parahermitian conjugate* $G^{\sim}$ of G by

$$G^{\sim} \triangleq G^{\text{T}}(-s). \qquad (12.13.1)$$

The following result provides realizations for G^{T}, $G^{\sim}$, and G^{-1}.

Proposition 12.13.1. Let $G^{l\times m}_{\text{prop}}(s)$, and assume that $G \sim \left[\begin{array}{c|c} A & B \\ \hline C & D \end{array} \right]$. Then,

$$G^{\text{T}} \sim \left[\begin{array}{c|c} A^{\text{T}} & C^{\text{T}} \\ \hline B^{\text{T}} & D^{\text{T}} \end{array} \right] \qquad (12.13.2)$$

and

$$G^{\sim} \sim \left[\begin{array}{c|c} -A^{\text{T}} & -C^{\text{T}} \\ \hline B^{\text{T}} & D^{\text{T}} \end{array} \right]. \qquad (12.13.3)$$

Furthermore, if G is square and D is nonsingular, then

$$G^{-1} \sim \left[\begin{array}{c|c} A - BD^{-1}C & BD^{-1} \\ \hline -D^{-1}C & D^{-1} \end{array} \right]. \qquad (12.13.4)$$

Proof. Since $y = Gu$, it follows that G^{-1} satisfies $u = G^{-1}y$. Since $\dot{x} = Ax + Bu$ and $y = Cx + Du$, it follows that $u = -D^{-1}Cx + D^{-1}y$, and thus $\dot{x} = Ax + B(-D^{-1}Cx + D^{-1}y) = (A - BD^{-1}C)x + BD^{-1}y$. $\qquad\square$

Note that, if $G \in \mathbb{R}_{\mathrm{prop}}(s)$ and $G \sim \left[\begin{array}{c|c} A & B \\ \hline C & D \end{array}\right]$, then $G \sim \left[\begin{array}{c|c} A^{\mathrm{T}} & B^{\mathrm{T}} \\ \hline C^{\mathrm{T}} & D \end{array}\right]$.

Let $G_1 \in \mathbb{R}_{\mathrm{prop}}^{l_1 \times m_1}(s)$ and $G_2 \in \mathbb{R}_{\mathrm{prop}}^{l_2 \times m_2}(s)$. If $m_2 = l_2$, then the *cascade interconnection* of G_1 and G_2 shown in Figure 12.13.1 is the product $G_2 G_1$, while the *parallel interconnection* shown in Figure 12.13.2 is the sum $G_1 + G_2$. Note that $G_2 G_1$ is defined only if $m_2 = l_1$, whereas $G_1 + G_2$ requires that $m_1 = m_2$ and $l_1 = l_2$.

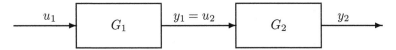

Figure 12.13.1
Cascade Interconnection of Linear Systems

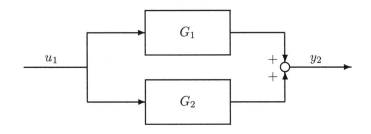

Figure 12.13.2
Parallel Interconnection of Linear Systems

Proposition 12.13.2. Let $G_1 \sim \left[\begin{array}{c|c} A_1 & B_1 \\ \hline C_1 & D_1 \end{array}\right]$ and $G_2 \sim \left[\begin{array}{c|c} A_2 & B_2 \\ \hline C_2 & D_2 \end{array}\right]$. Then,

$$G_2 G_1 \sim \left[\begin{array}{cc|c} A_1 & 0 & B_1 \\ B_2 C_1 & A_2 & B_2 D_1 \\ \hline D_2 C_1 & C_2 & D_2 D_1 \end{array}\right] \tag{12.13.5}$$

and

$$G_1 + G_2 \sim \left[\begin{array}{cc|c} A_1 & 0 & B_1 \\ 0 & A_2 & B_2 \\ \hline C_1 & C_2 & D_1 + D_2 \end{array}\right]. \tag{12.13.6}$$

Proof. Consider the state space equations

$$\dot{x}_1 = A_1 x_1 + B_1 u_1, \quad \dot{x}_2 = A_2 x_2 + B_2 u_2,$$
$$y_1 = C_1 x_1 + D_1 u_1, \quad y_2 = C_2 x_2 + D_2 u_2.$$

Since $u_2 = y_1$, it follows that

$$\dot{x}_2 = A_2 x_2 + B_2 C_1 x_1 + B_2 D_1 u_1,$$
$$y_2 = C_2 x_2 + D_2 C_1 x_1 + D_2 D_1 u_1,$$

and thus

$$\begin{bmatrix} \dot{x}_1 \\ \dot{x}_2 \end{bmatrix} = \begin{bmatrix} A_1 & 0 \\ B_2 C_1 & A_2 \end{bmatrix} \begin{bmatrix} x_1 \\ x_2 \end{bmatrix} + \begin{bmatrix} B_1 \\ B_2 D_1 \end{bmatrix} u_1,$$

$$y_2 = \begin{bmatrix} D_2 C_1 & C_2 \end{bmatrix} \begin{bmatrix} x_1 \\ x_2 \end{bmatrix} + D_2 D_1 u_1,$$

which yields the realization (12.13.5) of $G_2 G_1$. The realization (12.13.6) for $G_1 + G_2$ can be obtained by similar techniques. $\square$

It is sometimes useful to combine transfer functions by concatenating them into row, column, or block-diagonal transfer functions.

Proposition 12.13.3. Let $G_1 \sim \left[\begin{array}{c|c} A_1 & B_1 \\ \hline C_1 & D_1 \end{array} \right]$ and $G_2 \sim \left[\begin{array}{c|c} A_2 & B_2 \\ \hline C_2 & D_2 \end{array} \right]$. Then,

$$\begin{bmatrix} G_1 & G_2 \end{bmatrix} \sim \left[\begin{array}{cc|cc} A_1 & 0 & B_1 & 0 \\ 0 & A_2 & 0 & B_2 \\ \hline C_1 & C_2 & D_1 & D_2 \end{array} \right], \tag{12.13.7}$$

$$\begin{bmatrix} G_1 \\ G_2 \end{bmatrix} \sim \left[\begin{array}{cc|c} A_1 & 0 & B_1 \\ 0 & A_2 & B_2 \\ \hline C_1 & 0 & D_1 \\ 0 & C_2 & D_2 \end{array} \right], \tag{12.13.8}$$

$$\begin{bmatrix} G_1 & 0 \\ 0 & G_2 \end{bmatrix} \sim \left[\begin{array}{cc|cc} A_1 & 0 & B_1 & 0 \\ 0 & A_2 & 0 & B_2 \\ \hline C_1 & 0 & D_1 & 0 \\ 0 & C_2 & 0 & D_2 \end{array} \right]. \tag{12.13.9}$$

Next, we interconnect a pair of systems G_1, G_2 by means of feedback as shown in Figure 12.13.3. It can be seen that u and y are related by

$$\hat{y} = (I + G_1 G_2)^{-1} G_1 \hat{u} \tag{12.13.10}$$

or

$$\hat{y} = G_1 (I + G_2 G_1)^{-1} \hat{u}. \tag{12.13.11}$$

The equivalence of (12.13.10) and (12.13.11) follows from the push-through identity given by Fact 2.16.16,

$$(I + G_1 G_2)^{-1} G_1 = G_1 (I + G_2 G_1)^{-1}. \tag{12.13.12}$$

A realization of this rational transfer function is given by the following result.

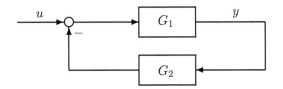

Figure 12.13.3
Feedback Interconnection of Linear Systems

Proposition 12.13.4. Let $G_1 \sim \left[\begin{array}{c|c} A_1 & B_1 \\ \hline C_1 & D_1 \end{array}\right]$ and $G_2 \sim \left[\begin{array}{c|c} A_2 & B_2 \\ \hline C_2 & D_2 \end{array}\right]$, and assume that $\det(I + D_1 D_2) \neq 0$. Then,

$$[I + G_1 G_2]^{-1} G_1$$

$$\sim \left[\begin{array}{cc|c} A_1 - B_1(I + D_2 D_1)^{-1} D_2 C_1 & -B_1(I + D_2 D_1)^{-1} C_2 & B_1(I + D_2 D_1)^{-1} \\ B_2(I + D_1 D_2)^{-1} C_1 & A_2 - B_2(I + D_1 D_2)^{-1} D_1 C_2 & B_2(I + D_1 D_2)^{-1} D_1 \\ \hline (I + D_1 D_2)^{-1} C_1 & -(I + D_1 D_2)^{-1} D_1 C_2 & (I + D_1 D_2)^{-1} D_1 \end{array}\right].$$
$$(12.13.13)$$

12.14 Standard Control Problem

The standard control problem shown in Figure 12.14.1 involves four distinct signals, namely, an *exogenous input* w, a *control input* u, a *performance variable* z, and a *feedback signal* y. This system can be written as

$$\left[\begin{array}{c} \hat{z}(s) \\ \hat{y}(s) \end{array}\right] = \mathcal{G}(s) \left[\begin{array}{c} \hat{w}(s) \\ \hat{u}(s) \end{array}\right], \tag{12.14.1}$$

where $\mathcal{G}(s)$ is partitioned as

$$\mathcal{G} \triangleq \left[\begin{array}{cc} G_{11} & G_{12} \\ G_{21} & G_{22} \end{array}\right] \tag{12.14.2}$$

with the realization

$$\mathcal{G} \sim \left[\begin{array}{c|cc} A & D_1 & B \\ \hline E_1 & E_0 & E_2 \\ C & D_2 & D \end{array}\right], \tag{12.14.3}$$

which represents the equations

$$\dot{x} = Ax + D_1 w + Bu, \tag{12.14.4}$$

$$z = E_1 x + E_0 w + E_2 u, \tag{12.14.5}$$

$$y = Cx + D_2 w + Du. \tag{12.14.6}$$

Consequently,

$$\mathcal{G}(s) = \left[\begin{array}{cc} E_1(sI-A)^{-1}D_1 + E_0 & E_1(sI-A)^{-1}B + E_2 \\ C(sI-A)^{-1}D_1 + D_2 & C(sI-A)^{-1}B + D \end{array} \right], \tag{12.14.7}$$

which shows that G_{11}, G_{12}, G_{21}, and G_{22} have the realizations

$$G_{11} \sim \left[\begin{array}{c|c} A & D_1 \\ \hline E_1 & E_0 \end{array} \right], \qquad G_{12} \sim \left[\begin{array}{c|c} A & B \\ \hline E_1 & E_2 \end{array} \right], \tag{12.14.8}$$

$$G_{21} \sim \left[\begin{array}{c|c} A & D_1 \\ \hline C & D_2 \end{array} \right], \qquad G_{22} \sim \left[\begin{array}{c|c} A & B \\ \hline C & D \end{array} \right]. \tag{12.14.9}$$

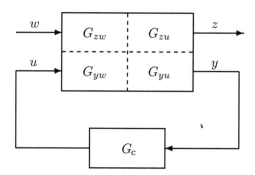

Figure 12.14.1
Standard Control Problem

Letting G_c denote a feedback controller with realization

$$G_c \sim \left[\begin{array}{c|c} A_c & B_c \\ \hline C_c & D_c \end{array} \right], \tag{12.14.10}$$

we interconnect G and G_c according to

$$\hat{u}(s) = G_c(s)\hat{y}(s). \tag{12.14.11}$$

The resulting rational transfer function $\tilde{\mathcal{G}}$ satisfying $\hat{z}(s) = \tilde{\mathcal{G}}(s)\hat{w}(s)$ is thus given by

$$\tilde{\mathcal{G}} = G_{11} + G_{12}G_c(I - G_{22}G_c)^{-1}G_{21} \tag{12.14.12}$$

or

$$\tilde{\mathcal{G}} = G_{11} + G_{12}(I - G_cG_{22})^{-1}G_cG_{21}. \tag{12.14.13}$$

A realization of $\tilde{\mathcal{G}}$ is given by the following result.

Proposition 12.14.1. Let $\mathcal{G}$ and G_c have the realizations (12.14.3) and (12.14.10), and assume that $\det(I - DD_c) \neq 0$. Then,

$$
\tilde{\mathcal{G}} \sim \left[
\begin{array}{cc|c}
A + BD_c(I - DD_c)^{-1}C & B(I - D_cD)^{-1}C_c & D_1 + BD_c(I + DD_c)^{-1}D_2 \\
B_c(I - DD_c)^{-1}C & A_c + B_c(I - DD_c)^{-1}DC_c & B_c(I - DD_c)^{-1}D_2 \\
\hline
E_1 + E_2D_c(I - DD_c)^{-1}C & E_2(I - D_cD)^{-1}C_c & E_0 + E_2D_c(I - DD_c)^{-1}D_2
\end{array}
\right].
$$
(12.14.14)

The realization (12.14.14) can be simplified when $DD_c = 0$. For example, if $D = 0$, then

$$
\tilde{\mathcal{G}} \sim \left[
\begin{array}{cc|c}
A + BD_cC & BC_c & D_1 + BD_cD_2 \\
B_cC & A_c & B_cD_2 \\
\hline
E_1 + E_2D_cC & E_2C_c & E_0 + E_2D_cD_2
\end{array}
\right],
$$
(12.14.15)

whereas, if $D_c = 0$, then

$$
\tilde{\mathcal{G}} \sim \left[
\begin{array}{cc|c}
A & BC_c & D_1 \\
B_cC & A_c + B_cDC_c & B_cD_2 \\
\hline
E_1 & E_2C_c & E_0
\end{array}
\right].
$$
(12.14.16)

Finally, if both $D = 0$ and $D_c = 0$, then

$$
\tilde{\mathcal{G}} \sim \left[
\begin{array}{cc|c}
A & BC_c & D_1 \\
B_cC & A_c & B_cD_2 \\
\hline
E_1 & E_2C_c & E_0
\end{array}
\right].
$$
(12.14.17)

The feedback interconnection shown in Figure 12.14.1 forms the basis for the *standard control problem* in feedback control. For this problem the signal w is an exogenous signal representing a command or a disturbance, while the signal z is the *performance variable*, that is, the variable whose behavior reflects the performance of the closed-loop system. The performance variable may or may not be physically measured. The *controlled input* or the *control* u is the output of the feedback controller G_c, while the *measurement* signal y serves as the input to the *feedback controller* G_c. The standard control problem is the following: Given knowledge of w, determine G_c to minimize a performance criterion $J(G_c)$.

12.15 Linear-Quadratic Control

Let $A \in \mathbb{R}^{n \times n}$ and $B \in \mathbb{R}^{n \times m}$, and consider the system

$$\dot{x}(t) = Ax(t) + Bu(t), \tag{12.15.1}$$

$$x(0) = x_0, \tag{12.15.2}$$

where $t \geq 0$. Furthermore, let $K \in \mathbb{R}^{m \times n}$, and consider the full-state-feedback control law

$$u(t) = Kx(t). \tag{12.15.3}$$

The objective of the *linear-quadratic control problem* is to minimize the *quadratic performance measure*

$$J(K, x_0) = \int_0^\infty \left[x^{\mathrm{T}}(t) R_1 x(t) + 2 x^{\mathrm{T}}(t) R_{12} u(t) + u^{\mathrm{T}}(t) R_2 u(t) \right] \mathrm{d}t, \qquad (12.15.4)$$

where $R_1 \in \mathbb{R}^{n \times n}$, $R_{12} \in \mathbb{R}^{n \times m}$, and $R_2 \in \mathbb{R}^{m \times m}$. We assume that $\begin{bmatrix} R_1 & R_{12} \\ R_{12}^{\mathrm{T}} & R_2 \end{bmatrix}$ is positive semidefinite and R_2 is positive definite.

The performance measure (12.15.4) indicates the desire to maintain the state vector $x(t)$ close to the zero equilibrium without an excessive expenditure of control effort. Specifically, the term $x^{\mathrm{T}}(t) R_1 x(t)$ is a measure of the deviation of the state $x(t)$ from the zero state, where the $n \times n$ positive-semidefinite matrix R_1 determines how much weighting is associated with each component of the state. Likewise, the $m \times m$ positive-definite matrix R_2 weights the magnitude of the control input. Finally, the cross-weighting term R_{12} arises naturally when additional filters are used to shape the system response or in specialized applications.

Using (12.15.1) and (12.15.3), the closed-loop dynamic system can be written as

$$\dot{x}(t) = (A + BK)x(t) \qquad (12.15.5)$$

so that

$$x(t) = e^{t\tilde{A}} x_0, \qquad (12.15.6)$$

where $\tilde{A} \triangleq A + BK$. Thus, the performance measure (12.15.4) becomes

$$J(K, x_0) = \int_0^\infty x^{\mathrm{T}}(t) \tilde{R} x(t) \, \mathrm{d}t = \int_0^\infty x_0^{\mathrm{T}} e^{t\tilde{A}^{\mathrm{T}}} \tilde{R} e^{t\tilde{A}} x_0 \, \mathrm{d}t$$

$$= \operatorname{tr} x_0^{\mathrm{T}} \int_0^\infty e^{t\tilde{A}^{\mathrm{T}}} \tilde{R} e^{t\tilde{A}} \, \mathrm{d}t x_0 = \operatorname{tr} \int_0^\infty e^{t\tilde{A}^{\mathrm{T}}} \tilde{R} e^{t\tilde{A}} \, \mathrm{d}t x_0 x_0^{\mathrm{T}}, \qquad (12.15.7)$$

where

$$\tilde{R} \triangleq R_1 + R_{12}K + K^{\mathrm{T}} R_{12}^{\mathrm{T}} + K^{\mathrm{T}} R_2 K. \qquad (12.15.8)$$

Now, consider the standard control problem with plant

$$\mathcal{G} \sim \left[\begin{array}{c|cc} A & D_1 & B \\ \hline E_1 & 0 & E_2 \\ I_n & 0 & 0 \end{array} \right] \qquad (12.15.9)$$

and full-state feedback $u = Kx$. Then, the closed-loop transfer function is given by

$$\tilde{\mathcal{G}} \sim \left[\begin{array}{c|c} A + BK & D_1 \\ \hline E_1 + E_2 K & 0 \end{array} \right]. \qquad (12.15.10)$$

The following result shows that the quadratic performance measure (12.15.4) is equal to the H$_2$ norm of a transfer function.

Proposition 12.15.1. Assume that $D_1 = x_0$ and

$$\begin{bmatrix} R_1 & R_{12} \\ R_{12}^{\mathrm{T}} & R_2 \end{bmatrix} = \begin{bmatrix} E_1^{\mathrm{T}} \\ E_2^{\mathrm{T}} \end{bmatrix} \begin{bmatrix} E_1 & E_2 \end{bmatrix}, \tag{12.15.11}$$

and let $\tilde{\mathcal{G}}$ be given by (12.15.10). Then,

$$J(K, x_0') = \|\tilde{\mathcal{G}}\|_{\mathrm{H}_2}^2. \tag{12.15.12}$$

Proof. This result follows from Proposition 12.1.2. $\qquad\square$

For the following development, we assume that (12.15.11) holds so that R_1, R_{12}, and R_2 are given by

$$R_1 = E_1^{\mathrm{T}}E_1, \quad R_{12} = E_1^{\mathrm{T}}E_2, \quad R_2 = E_2^{\mathrm{T}}E_2. \tag{12.15.13}$$

To develop necessary conditions for the linear-quadratic control problem, we restrict K to the set of stabilizing gains

$$\mathcal{S} \triangleq \{K \in \mathbb{R}^{m \times n}: \; A + BK \text{ is asymptotically stable}\}. \tag{12.15.14}$$

Obviously, $\mathcal{S}$ is nonempty if and only if (A, B) is stabilizable. The following result gives necessary conditions that characterize a stabilizing solution K of the linear-quadratic control problem.

Theorem 12.15.2. Assume that (A, B) is stabilizable, assume that $K \in \mathcal{S}$ solves the linear-quadratic control problem, and assume that $(A + BK, D_1)$ is controllable. Then, there exists an $n \times n$ positive-semidefinite matrix P such that K is given by

$$K = -R_2^{-1}(B^{\mathrm{T}}P + R_{12}^{\mathrm{T}}) \tag{12.15.15}$$

and such that P satisfies

$$\hat{A}_{\mathrm{R}}^{\mathrm{T}}P + P\hat{A}_{\mathrm{R}} + \hat{R}_1 - PBR_2^{-1}B^{\mathrm{T}}P = 0, \tag{12.15.16}$$

where

$$\hat{A}_{\mathrm{R}} \triangleq A - BR_2^{-1}R_{12}^{\mathrm{T}} \tag{12.15.17}$$

and

$$\hat{R}_1 \triangleq R_1 - R_{12}R_2^{-1}R_{12}^{\mathrm{T}}. \tag{12.15.18}$$

Furthermore, the minimal cost is given by

$$J(K) = \operatorname{tr} PV, \tag{12.15.19}$$

where $V \triangleq D_1 D_1^{\mathrm{T}}$.

Proof. Since $K \in \mathcal{S}$, it follows that $\tilde{A}$ is asymptotically stable. It then follows that $J(K)$ is given by (12.15.19), where $P \triangleq \int_0^\infty e^{t\tilde{A}^{\mathrm{T}}}\tilde{R}e^{t\tilde{A}}\,dt$ is positive semidefinite and satisfies the Lyapunov equation

$$\tilde{A}^{\mathrm{T}}P + P\tilde{A} + \tilde{R} = 0. \tag{12.15.20}$$

Note that (12.15.20) can be written as

$$(A + BK)^{\mathrm{T}}P + P(A + BK) + R_1 + R_{12}K + K^{\mathrm{T}}R_{12}^{\mathrm{T}} + K^{\mathrm{T}}R_2 K = 0. \quad (12.15.21)$$

To optimize (12.15.19) subject to the constraint (12.15.20) over the open set $\mathcal{S}$, form the Lagrangian

$$\mathcal{L}(K, P, Q, \lambda_0) \triangleq \mathrm{tr}\left[\lambda_0 PV + Q\left(\tilde{A}^{\mathrm{T}}P + P\tilde{A} + \tilde{R}\right)\right], \quad (12.15.22)$$

where the Lagrange multipliers $\lambda_0 \geq 0$ and $Q \in \mathbb{R}^{n \times n}$ are not both zero. Note that the $n \times n$ Lagrange multiplier Q accounts for the $n \times n$ constraint equation (12.15.20).

The necessary condition $\partial \mathcal{L}/\partial P = 0$ implies

$$\tilde{A}Q + Q\tilde{A}^{\mathrm{T}} + \lambda_0 V = 0. \quad (12.15.23)$$

Since $\tilde{A}$ is asymptotically stable, it follows from Proposition 11.9.3 that, for all $\lambda_0 \geq 0$, (12.15.23) has a unique solution Q and, furthermore, Q is positive semidefinite. In particular, if $\lambda_0 = 0$, then $Q = 0$. Since λ_0 and Q are not both zero, we can set $\lambda_0 = 1$ so that (12.15.23) becomes

$$\tilde{A}Q + Q\tilde{A}^{\mathrm{T}} + V = 0. \quad (12.15.24)$$

Since $(\tilde{A}, D_1)$ is controllable, it follows from Corollary 12.7.10 that Q is positive definite.

Next, evaluating $\partial \mathcal{L}/\partial K = 0$ yields

$$R_2 KQ + \left(B^{\mathrm{T}}P + R_{12}^{\mathrm{T}}\right)Q = 0. \quad (12.15.25)$$

Since Q is positive definite, it follows from (12.15.25) that (12.15.15) is satisfied. Furthermore, using (12.15.15), it follows that (12.15.20) is equivalent to (12.15.16). $\square$

With K given by (12.15.15), the closed-loop dynamics matrix $\tilde{A} = A + BK$ is given by

$$\tilde{A} = A - BR_2^{-1}\left(B^{\mathrm{T}}P + R_{12}^{\mathrm{T}}\right), \quad (12.15.26)$$

where P is the solution of the *Riccati equation* (12.15.16).

12.16 Solutions of the Riccati Equation

For convenience in the following development, we assume that $R_{12} = 0$. With this assumption, the gain K given by (12.15.15) becomes

$$K = -R_2^{-1}B^{\mathrm{T}}P. \quad (12.16.1)$$

Defining

$$\Sigma \triangleq BR_2^{-1}B^{\mathrm{T}}, \quad (12.16.2)$$

(12.15.26) becomes

$$\tilde{A} = A - \Sigma P, \quad (12.16.3)$$

while the Riccati equation (12.15.16) can be written as

$$A^\mathrm{T}P + PA + R_1 - P\Sigma P = 0. \tag{12.16.4}$$

Note that (12.16.4) has the alternative representation

$$(A - \Sigma P)^\mathrm{T}P + P(A - \Sigma P) + R_1 + P\Sigma P = 0, \tag{12.16.5}$$

which is equivalent to the Lyapunov equation

$$\tilde{A}^\mathrm{T}P + P\tilde{A} + \tilde{R} = 0, \tag{12.16.6}$$

where

$$\tilde{R} \triangleq R_1 + P\Sigma P. \tag{12.16.7}$$

By comparing (12.15.16) and (12.16.4), it can be seen that the linear-quadratic control problems with (A, B, R_1, R_{12}, R_2) and $(\hat{A}_\mathrm{R}, B, \hat{R}_1, 0, R_2)$ are equivalent. Hence, there is no loss of generality in assuming that $R_{12} = 0$ in the following development, where A and R_1 take the place of $\hat{A}_\mathrm{R}$ and $\hat{R}_1$, respectively.

To motivate the subsequent development, the following examples demonstrate the existence of solutions under various assumptions on (A, B, E_1). In the following four examples, (A, B) is not stabilizable.

Example 12.16.1. Let $n = 1$, $A = 1$, $B = 0$, $E_1 = 0$, and $R_2 > 0$. Hence, (A, B, E_1) has an ORHP eigenvalue that is uncontrollable and unobservable. In this case, (12.16.4) has the unique solution $P = 0$. Furthermore, since $B = 0$, it follows that $\tilde{A} = A$.

Example 12.16.2. Let $n = 1$, $A = 1$, $B = 0$, $E_1 = 1$, and $R_2 > 0$. Hence, (A, B, E_1) has an ORHP eigenvalue that is uncontrollable and observable. In this case, (12.16.4) has the unique solution $P = -1/2 < 0$. Furthermore, since $B = 0$, it follows that $\tilde{A} = A$.

Example 12.16.3. Let $n = 1$, $A = 0$, $B = 0$, $E_1 = 0$, and $R_2 > 0$. Hence, (A, B, E_1) has an imaginary eigenvalue that is uncontrollable and unobservable. In this case, (12.16.4) has infinitely many solutions $P \in \mathbb{R}$. Hence, (12.16.4) has no maximal solution. Furthermore, since $B = 0$, it follows that, for every solution P, $\tilde{A} = A$.

Example 12.16.4. Let $n = 1$, $A = 0$, $B = 0$, $E_1 = 1$, and $R_2 > 0$. Hence, (A, B, E_1) has an imaginary eigenvalue that is uncontrollable and observable. In this case, (12.16.4) becomes $R_1 = 0$. Thus, (12.16.4) has no solution.

In the remaining examples, (A, B) is controllable.

Example 12.16.5. Let $n = 1$, $A = 1$, $B = 1$, $E_1 = 0$, and $R_2 > 0$. Hence, (A, B, E_1) has an ORHP eigenvalue that is controllable and unobservable. In this case, (12.16.4) has the solutions $P = 0$ and $P = 2R_2 > 0$. The corresponding closed-loop dynamics matrices are $\tilde{A} = 1 > 0$ and $\tilde{A} = -1 < 0$. Hence, the solution $P = 2R_2$ is stabilizing, and the closed-loop eigenvalue 1, which does not depend on R_2, is the reflection of the open-loop eigenvalue -1 across the imaginary axis.

Example 12.16.6. Let $n = 1$, $A = 1$, $B = 1$, $E_1 = 1$, and $R_2 > 0$. Hence, (A, B, E_1) has an ORHP eigenvalue that is controllable and observable. In this case, (12.16.4) has the solutions $P = R_2 - \sqrt{R_2^2 + R_2} < 0$ and $P = R_2 + \sqrt{R_2^2 + R_2} > 0$. The corresponding closed-loop dynamics matrices are $\tilde{A} = \sqrt{1 + 1/R_2} > 0$ and $\tilde{A} = -\sqrt{1 + 1/R_2} < 0$. Hence, the positive-definite solution $P = R_2 + \sqrt{R_2^2 + R_2}$ is stabilizing.

Example 12.16.7. Let $n = 1$, $A = 0$, $B = 1$, $E_1 = 0$, and $R_2 > 0$. Hence, (A, B, E_1) has an imaginary eigenvalue that is controllable and unobservable. In this case, (12.16.4) has the unique solution $P = 0$, which is not stabilizing.

Example 12.16.8. Let $n = 1$, $A = 0$, $B = 1$, $E_1 = 1$, and $R_2 > 0$. Hence, (A, B, E_1) has an imaginary eigenvalue that is controllable and observable. In this case, (12.16.4) has the solutions $P = -\sqrt{R_2} < 0$ and $P = \sqrt{R_2} > 0$. The corresponding closed-loop dynamics matrices are $\tilde{A} = \sqrt{R_2} > 0$ and $\tilde{A} = -\sqrt{R_2} < 0$. Hence, the positive-definite solution $P = \sqrt{R_2}$ is stabilizing.

Example 12.16.9. Let $n = 2$, $A = \begin{bmatrix} 0 & 1 \\ -1 & 0 \end{bmatrix}$, $B = I_2$, $E_1 = 0$, and $R_2 = 1$. Hence, as in Example 12.16.7, both eigenvalues of (A, B, E_1) are imaginary, controllable, and unobservable. Taking the trace of (12.16.4) yields $\operatorname{tr} P^2 = 0$. Thus, the only symmetric matrix P satisfying (12.16.4) is $P = 0$, which implies that $\tilde{A} = A$. Consequently, the open-loop eigenvalues $\pm j$ are unaffected by the feedback gain (12.15.15) even though (A, B) is controllable.

Example 12.16.10. Let $n = 2$, $A = 0$, $B = I_2$, $E_1 = I_2$, and $R_2 = I$. Hence, as in Example 12.16.8, both eigenvalues of (A, B, E_1) are imaginary, controllable, and observable. Furthermore, (12.16.4) becomes $P^2 = I$. Requiring that P be symmetric, it follows that P is a reflector. Hence, $P = I$ is the only positive-semidefinite solution. In fact, P is positive definite and stabilizing since $\tilde{A} = -I$.

Example 12.16.11. Let $A = \begin{bmatrix} 1 & 0 \\ 0 & 2 \end{bmatrix}$, $B = \begin{bmatrix} 1 \\ 1 \end{bmatrix}$, $E_1 = 0$, and $R_2 = 1$ so that (A, B) is controllable, although neither of the states is weighted. In this case, (12.16.4) has four positive-semidefinite solutions, which are given by

$$P_1 = \begin{bmatrix} 18 & -24 \\ -24 & 36 \end{bmatrix}, \quad P_2 = \begin{bmatrix} 2 & 0 \\ 0 & 0 \end{bmatrix}, \quad P_3 = \begin{bmatrix} 0 & 0 \\ 0 & 4 \end{bmatrix}, \quad P_4 = \begin{bmatrix} 0 & 0 \\ 0 & 0 \end{bmatrix}.$$

The corresponding feedback matrices are given by $K_1 = \begin{bmatrix} 6 & -12 \end{bmatrix}$, $K_2 = \begin{bmatrix} -2 & 0 \end{bmatrix}$, $K_3 = \begin{bmatrix} 0 & -4 \end{bmatrix}$, and $K_4 = \begin{bmatrix} 0 & 0 \end{bmatrix}$. Letting $\tilde{A}_i = A - \Sigma P_i$, it follows that $\operatorname{spec}(\tilde{A}_1) = \{-1, -2\}$, $\operatorname{spec}(\tilde{A}_2) = \{-1, 2\}$, $\operatorname{spec}(\tilde{A}_3) = \{1, -2\}$, and $\operatorname{spec}(\tilde{A}_4) = \{1, 2\}$. Thus, P_1 is the only solution that stabilizes the closed-loop system, while the solutions P_2 and P_3 partially stabilize the closed-loop system. Note also that the closed-loop poles that differ from those of the open-loop system are mirror images of the open-loop poles as reflected across the imaginary axis. Finally, note that these solutions satisfy the partial ordering $P_1 \geq P_2 \geq P_4$ and $P_1 \geq P_3 \geq P_4$, and that "larger" solutions are more stabilizing than "smaller" solutions. Moreover, letting $J(K_i) = \operatorname{tr} P_i V$, it can be seen that larger solutions incur a greater closed-loop cost, with the greatest cost incurred by the stabilizing solution P_4. However, the cost expression $J(K) = \operatorname{tr} PV$ does not follow from (12.15.4) when $A + BK$ is not asymptotically stable.

The following definition concerns solutions of the Riccati equation.

Definition 12.16.12. A matrix $P \in \mathbb{R}^{n \times n}$ is a *solution* of the Riccati equation (12.16.4) if P is symmetric and satisfies (12.16.4). Furthermore, P is the *stabilizing solution* of (12.16.4) if $\tilde{A} = A - \Sigma P$ is asymptotically stable. Finally, a solution $P_{\max}$ of (12.16.4) is the *maximal solution* to (12.16.4) if $P \le P_{\max}$ for every solution P to (12.16.4).

Since the ordering "$\le$" is antisymmetric, it follows that (12.16.4) has at most one maximal solution. The uniqueness of the stabilizing solution is shown in the following section.

Next, define the $2n \times 2n$ *Hamiltonian*

$$\mathcal{H} \triangleq \begin{bmatrix} A & \Sigma \\ R_1 & -A^{\mathrm{T}} \end{bmatrix}. \tag{12.16.8}$$

Proposition 12.16.13. The following statements hold:

i) The Hamiltonian $\mathcal{H}$ is a Hamiltonian matrix.

ii) $\chi_{\mathcal{H}}$ has a spectral factorization, that is, there exists a monic polynomial $p \in \mathbb{R}[s]$ such that, for all $s \in \mathbb{C}$, $\chi_{\mathcal{H}}(s) = p(s)p(-s)$.

iii) $\chi_{\mathcal{H}}(\jmath\omega) \ge 0$ for all $\omega \in \mathbb{R}$.

iv) If either $R_1 = 0$ or $\Sigma = 0$, then $\mathrm{mspec}(\mathcal{H}) = \mathrm{mspec}(A) \cup \mathrm{mspec}(-A)$.

v) $\chi_{\mathcal{H}}$ is even.

vi) $\lambda \in \mathrm{spec}(\mathcal{H})$ if and only if $-\lambda \in \mathrm{spec}(\mathcal{H})$.

vii) If $\lambda \in \mathrm{spec}(\mathcal{H})$, then $\mathrm{amult}_{\mathcal{H}}(\lambda) = \mathrm{amult}_{\mathcal{H}}(-\lambda)$.

viii) Every imaginary root of $\chi_{\mathcal{H}}$ has even multiplicity.

ix) Every imaginary eigenvalue of $\mathcal{H}$ has even algebraic multiplicity.

Proof. This result follows from Proposition 4.1.1 and Fact 4.9.24. $\qquad\square$

It is helpful to keep in mind that spectral factorizations are not unique. For example, if $\chi_{\mathcal{H}}(s) = (s+1)(s+2)(-s+1)(-s+2)$, then $\chi_{\mathcal{H}}(s) = p(s)p(-s) = \hat{p}(s)\hat{p}(-s)$, where $p(s) = (s+1)(s+2)$ and $\hat{p}(s) = (s+1)(s-2)$. Thus, the spectral factors $p(s)$ and $p(-s)$ can "trade" roots. These roots are the eigenvalues of $\mathcal{H}$.

The following result shows that the Hamiltonian $\mathcal{H}$ is closely linked to the Riccati equation (12.16.4).

Proposition 12.16.14. Let $P \in \mathbb{R}^{n \times n}$ be symmetric. Then, the following statements are equivalent:

i) P is a solution of (12.16.4).

ii) P satisfies

$$\begin{bmatrix} P & I \end{bmatrix} \mathcal{H} \begin{bmatrix} I \\ -P \end{bmatrix} = 0. \tag{12.16.9}$$

iii) P satisfies

$$\mathcal{H}\begin{bmatrix} I \\ -P \end{bmatrix} = \begin{bmatrix} I \\ -P \end{bmatrix}(A - \Sigma P). \tag{12.16.10}$$

iv) P satisfies

$$\mathcal{H} = \begin{bmatrix} I & 0 \\ -P & I \end{bmatrix}\begin{bmatrix} A - \Sigma P & \Sigma \\ 0 & -(A - \Sigma P)^{\mathrm{T}} \end{bmatrix}\begin{bmatrix} I & 0 \\ P & I \end{bmatrix}. \tag{12.16.11}$$

In this case, the following statements hold:

v) $\mathrm{mspec}(\mathcal{H}) = \mathrm{mspec}(A - \Sigma P) \cup \mathrm{mspec}[-(A - \Sigma P)]$.

vi) $\chi_{\mathcal{H}}(s) = (-1)^n \chi_{A-\Sigma P}(s)\chi_{A-\Sigma P}(-s)$.

vii) $\mathcal{R}\left(\begin{bmatrix} I \\ -P \end{bmatrix}\right)$ is an invariant subspace of $\mathcal{H}$.

Corollary 12.16.15. Assume that (12.16.4) has a stabilizing solution. Then, $\mathcal{H}$ has no imaginary eigenvalues.

For the next two results, P is not necessarily a solution of (12.16.4).

Lemma 12.16.16. Assume that $\lambda \in \mathrm{spec}(A)$ is an observable eigenvalue of (A, R_1), and let $P \in \mathbb{R}^{n \times n}$ be symmetric. Then, $\lambda \in \mathrm{spec}(A)$ is an observable eigenvalue of $(\tilde{A}, \tilde{R})$.

Proof. Suppose that $\mathrm{rank}\begin{bmatrix} \lambda I - \tilde{A} \\ \tilde{R} \end{bmatrix} < n$. Then, there exists a nonzero vector $v \in \mathbb{C}^n$ such that $\tilde{A}v = \lambda v$ and $\tilde{R}v = 0$. Hence, $v^*R_1v = -v^*P\Sigma Pv \le 0$, which implies that $R_1v = 0$ and $P\Sigma Pv = 0$. Hence, $\Sigma Pv = 0$, and thus $Av = \lambda v$. Therefore, $\mathrm{rank}\begin{bmatrix} \lambda I - A \\ R_1 \end{bmatrix} < n$. $\square$

Lemma 12.16.17. Assume that (A, R_1) is (observable, detectable), and let $P \in \mathbb{R}^{n \times n}$ be symmetric. Then, $(\tilde{A}, \tilde{R})$ is (observable, detectable).

Lemma 12.16.18. Assume that (A, E_1) is observable, and assume that (12.16.4) has a solution P. Then, the following statements hold:

i) $\nu_-(\tilde{A}) = \nu_+(P)$.

ii) $\nu_0(\tilde{A}) = \nu_0(P) = 0$.

iii) $\nu_+(\tilde{A}) = \nu_-(P)$.

Proof. Since (A, R_1) is observable, it follows from Lemma 12.16.17 that $(\tilde{A}, \tilde{R})$ is observable. By writing (12.16.4) as the Lyapunov equation (12.16.6), the result now follows from Fact 12.21.1. $\square$

12.17 The Stabilizing Solution of the Riccati Equation

Proposition 12.17.1. The following statements hold:

i) (12.16.4) has at most one stabilizing solution.

ii) If P is the stabilizing solution of (12.16.4), then P is positive semidefinite.

iii) If P is the stabilizing solution of (12.16.4), then

$$\operatorname{rank} P = \operatorname{rank} \mathcal{O}(\tilde{A}, \tilde{R}). \tag{12.17.1}$$

Proof. To prove *i*), suppose that (12.16.4) has stabilizing solutions P_1 and P_2. Then,

$$A^{\mathrm{T}}P_1 + P_1 A + R_1 - P_1 \Sigma P_1 = 0,$$
$$A^{\mathrm{T}}P_2 + P_2 A + R_1 - P_2 \Sigma P_2 = 0.$$

Subtracting these equations and rearranging yields

$$(A - \Sigma P_1)^{\mathrm{T}}(P_1 - P_2) + (P_1 - P_2)(A - \Sigma P_2) = 0.$$

Since $A - \Sigma P_1$ and $A - \Sigma P_2$ are asymptotically stable, it follows from Proposition 11.9.3 and Fact 11.18.33 that $P_1 - P_2 = 0$. Hence, (12.16.4) has at most one stabilizing solution.

Next, to prove *ii*), suppose that P is a stabilizing solution of (12.16.4). Then, it follows from (12.16.4) that

$$P = \int_0^\infty e^{t(A-\Sigma P)^{\mathrm{T}}}(R_1 + P\Sigma P)e^{t(A-\Sigma P)}\,\mathrm{d}t,$$

which shows that P is positive semidefinite.

Finally, *iii*) follows from Corollary 12.3.3. $\qquad\square$

Theorem 12.17.2. Assume that (12.16.4) has a positive-semidefinite solution P, and assume that (A, E_1) is detectable. Then, P is the stabilizing solution of (12.16.4), and thus P is the only positive-semidefinite solution of (12.16.4). If, in addition, (A, E_1) is observable, then P is positive definite.

Proof. Since (A, R_1) is detectable, it follows from Lemma 12.16.17 that $(\tilde{A}, \tilde{R})$ is detectable. Next, since (12.16.4) has a positive-semidefinite solution P, it follows from Corollary 12.8.6 that $\tilde{A}$ is asymptotically stable. Hence, P is the stabilizing solution of (12.16.4). The last statement follows from Lemma 12.16.18. $\qquad\square$

Corollary 12.17.3. Assume that (A, E_1) is detectable. Then, (12.16.4) has at most one positive-semidefinite solution.

Lemma 12.17.4. Let $\lambda \in \mathbb{C}$, and assume that λ is either an uncontrollable eigenvalue of (A, B) or an unobservable eigenvalue of (A, E_1). Then, $\lambda \in \operatorname{spec}(\mathcal{H})$.

Proof. Note that

$$\lambda I - \mathcal{H} = \begin{bmatrix} \lambda I - A & -\Sigma \\ -R_1 & \lambda I + A^{\mathrm{T}} \end{bmatrix}.$$

If λ is an uncontrollable eigenvalue of (A, B), then the first n rows of $\lambda I - \mathcal{H}$ are linearly dependent, and thus $\lambda \in \mathrm{spec}(\mathcal{H})$. On the other hand, if λ is an unobservable eigenvalue of (A, E_1), then the first n columns of $\lambda I - \mathcal{H}$ are linearly dependent, and thus $\lambda \in \mathrm{spec}(\mathcal{H})$. $\square$

The following result is a consequence of Lemma 12.17.4.

Proposition 12.17.5. Let $S \in \mathbb{R}^{n \times n}$ be a nonsingular matrix such that

$$A = S \begin{bmatrix} A_1 & 0 & A_{13} & 0 \\ A_{21} & A_2 & A_{23} & A_{24} \\ 0 & 0 & A_3 & 0 \\ 0 & 0 & A_{43} & A_4 \end{bmatrix} S^{-1}, \quad B = S \begin{bmatrix} B_1 \\ B_2 \\ 0 \\ 0 \end{bmatrix}, \quad (12.17.2)$$

$$E_1 = \begin{bmatrix} E_{11} & 0 & E_{13} & 0 \end{bmatrix} S^{-1}, \quad (12.17.3)$$

where $\left(\begin{bmatrix} A_1 & 0 \\ A_{21} & A_2 \end{bmatrix}, \begin{bmatrix} B_1 \\ B_2 \end{bmatrix} \right)$ is controllable and $\left(\begin{bmatrix} A_1 & A_{13} \\ 0 & A_3 \end{bmatrix}, \begin{bmatrix} E_{11} & E_{13} \end{bmatrix} \right)$ is observable. Then,

$$\mathrm{mspec}(A_2) \cup \mathrm{mspec}(-A_2) \subseteq \mathrm{mspec}(\mathcal{H}), \quad (12.17.4)$$

$$\mathrm{mspec}(A_3) \cup \mathrm{mspec}(-A_3) \subseteq \mathrm{mspec}(\mathcal{H}), \quad (12.17.5)$$

$$\mathrm{mspec}(A_4) \cup \mathrm{mspec}(-A_4) \subseteq \mathrm{mspec}(\mathcal{H}). \quad (12.17.6)$$

Next, we present a partial converse of Lemma 12.17.4.

Lemma 12.17.6. Let $\lambda \in \mathrm{spec}(\mathcal{H})$, and assume that $\mathrm{Re}\,\lambda = 0$. Then, λ is either an uncontrollable eigenvalue of (A, B) or an unobservable eigenvalue of (A, E_1).

Proof. Suppose that $\lambda = \jmath\omega$ is an eigenvalue of $\mathcal{H}$, where $\omega \in \mathbb{R}$. Then, there exist $x, y \in \mathbb{C}^n$ such that $\begin{bmatrix} x \\ y \end{bmatrix} \neq 0$ and $\mathcal{H}\begin{bmatrix} x \\ y \end{bmatrix} = \jmath\omega\begin{bmatrix} x \\ y \end{bmatrix}$. Consequently,

$$Ax + \Sigma y = \jmath\omega x, \quad R_1 x - A^{\mathrm{T}} y = \jmath\omega y.$$

Rewriting these equalities as

$$(A - \jmath\omega I)x = -\Sigma y, \quad (A - \jmath\omega I)^* y = R_1 x$$

yields

$$y^*(A - \jmath\omega I)x = -y^*\Sigma y, \quad x^*(A - \jmath\omega I)^* y = x^* R_1 x.$$

Since $x^*(A - \jmath\omega I)^* y$ is real, it follows that $-y^*\Sigma y = x^* R_1 x$, and thus $y^*\Sigma y = x^* R_1 x = 0$, which implies that $B^{\mathrm{T}} y = 0$ and $E_1 x = 0$. Therefore,

$$(A - \jmath\omega I)x = 0, \quad (A - \jmath\omega I)^* y = 0,$$

and hence

$$\begin{bmatrix} A - \jmath\omega I \\ E_1 \end{bmatrix} x = 0, \quad y^* \begin{bmatrix} A - \jmath\omega I & B \end{bmatrix} = 0.$$

Since $\left[\begin{smallmatrix} x \\ y \end{smallmatrix}\right] \neq 0$, it follows that either $x \neq 0$ or $y \neq 0$, and thus either $\operatorname{rank} \left[\begin{smallmatrix} A - \jmath\omega I \\ E_1 \end{smallmatrix}\right] < n$ or $\operatorname{rank} \left[\begin{smallmatrix} A - \jmath\omega I & B \end{smallmatrix} \right] < n$. $\qquad\square$

The following result is a restatement of Lemma 12.17.6.

Proposition 12.17.7. Let $S \in \mathbb{R}^{n \times n}$ be a nonsingular matrix such that (12.17.2) and (12.17.3) are satisfied, where $\left(\left[\begin{smallmatrix} A_1 & 0 \\ A_{21} & A_2 \end{smallmatrix} \right], \left[\begin{smallmatrix} B_1 \\ B_2 \end{smallmatrix} \right] \right)$ is controllable and $\left(\left[\begin{smallmatrix} A_1 & A_{13} \\ 0 & A_3 \end{smallmatrix} \right], \left[\begin{smallmatrix} E_{11} & E_{13} \end{smallmatrix} \right] \right)$ is observable. Then,

$$\operatorname{mspec}(\mathcal{H}) \cap \jmath\mathbb{R} \subseteq \operatorname{mspec}(A_2) \cup \operatorname{mspec}(-A_2) \cup \operatorname{mspec}(A_3)$$
$$\cup \operatorname{mspec}(-A_3) \cup \operatorname{mspec}(A_4) \cup \operatorname{mspec}(-A_4). \qquad (12.17.7)$$

Combining Lemma 12.17.4 and Lemma 12.17.6 yields the following result.

Proposition 12.17.8. Let $\lambda \in \mathbb{C}$, assume that $\operatorname{Re} \lambda = 0$, and let $S \in \mathbb{R}^{n \times n}$ be a nonsingular matrix such that (12.17.2) and (12.17.3) are satisfied, where (A_1, B_1, E_{11}) is controllable and observable, (A_2, B_2) is controllable, and (A_3, E_{13}) is observable. Then, the following statements are equivalent:

i) λ is either an uncontrollable eigenvalue of (A, B) or an unobservable eigenvalue of (A, E_1).

ii) $\lambda \in \operatorname{mspec}(A_2) \cup \operatorname{mspec}(A_3) \cup \operatorname{mspec}(A_4)$.

iii) λ is an eigenvalue of $\mathcal{H}$.

The next result gives necessary and sufficient conditions under which (12.16.4) has a stabilizing solution. This result also provides a constructive characterization of the stabilizing solution. Result *ii)* of Proposition 12.10.11 shows that the condition in *i)* that every imaginary eigenvalue of (A, E_1) is observable is equivalent to the condition that $\left[\begin{array}{c|c} A & B \\ \hline E_1 & E_2 \end{array} \right]$ has no imaginary invariant zeros.

Theorem 12.17.9. The following statements are equivalent:

i) (A, B) is stabilizable, and every imaginary eigenvalue of (A, E_1) is observable.

ii) There exists a nonsingular matrix $S \in \mathbb{R}^{n \times n}$ such that (12.17.2) and (12.17.3) are satisfied, where $\left(\left[\begin{smallmatrix} A_1 & 0 \\ A_{21} & A_2 \end{smallmatrix} \right], \left[\begin{smallmatrix} B_1 \\ B_2 \end{smallmatrix} \right] \right)$ is controllable, $\left(\left[\begin{smallmatrix} A_1 & A_{13} \\ 0 & A_3 \end{smallmatrix} \right], \left[\begin{smallmatrix} E_{11} & E_{13} \end{smallmatrix} \right] \right)$ is observable, $\nu_0(A_2) = 0$, and A_3 and A_4 are asymptotically stable.

iii) (12.16.4) has a stabilizing solution.

In this case, let

$$M = \left[\begin{array}{cc} M_1 & M_{12} \\ M_{21} & M_2 \end{array} \right] \in \mathbb{R}^{2n \times 2n} \qquad (12.17.8)$$

be a nonsingular matrix such that $\mathcal{H} = MZM^{-1}$, where

$$Z = \left[\begin{array}{cc} Z_1 & Z_{12} \\ 0 & Z_2 \end{array} \right] \in \mathbb{R}^{2n \times 2n} \qquad (12.17.9)$$

and $Z_1 \in \mathbb{R}^{n \times n}$ is asymptotically stable. Then, M_1 is nonsingular, and

$$P \triangleq -M_{21} M_1^{-1} \qquad (12.17.10)$$

is the stabilizing solution of (12.16.4).

Proof. The equivalence of i) and ii) is immediate.

To prove that i) $\implies$ iii), first note that Lemma 12.17.6 implies that $\mathcal{H}$ has no imaginary eigenvalues. Hence, since $\mathcal{H}$ is Hamiltonian, it follows that there exists a matrix $M \in \mathbb{R}^{2n \times 2n}$ of the form (12.17.8) such that M is nonsingular and $\mathcal{H} = MZM^{-1}$, where $Z \in \mathbb{R}^{n \times n}$ is of the form (12.17.9) and $Z_1 \in \mathbb{R}^{n \times n}$ is asymptotically stable.

Next, note that $\mathcal{H}M = MZ$ implies that

$$\mathcal{H} \begin{bmatrix} M_1 \\ M_{21} \end{bmatrix} = M \begin{bmatrix} Z_1 \\ 0 \end{bmatrix} = \begin{bmatrix} M_1 \\ M_{21} \end{bmatrix} Z_1.$$

Therefore,

$$\begin{bmatrix} M_1 \\ M_{21} \end{bmatrix}^{\mathrm{T}} J_n \mathcal{H} \begin{bmatrix} M_1 \\ M_{21} \end{bmatrix} = \begin{bmatrix} M_1 \\ M_{21} \end{bmatrix}^{\mathrm{T}} J_n \begin{bmatrix} M_1 \\ M_{21} \end{bmatrix} Z_1$$

$$= \begin{bmatrix} M_1^{\mathrm{T}} & M_{21}^{\mathrm{T}} \end{bmatrix} \begin{bmatrix} M_{21} \\ -M_1 \end{bmatrix} Z_1$$

$$= L Z_1,$$

where $L \triangleq M_1^{\mathrm{T}} M_{21} - M_{21}^{\mathrm{T}} M_1$. Since $J_n \mathcal{H} = (J_n \mathcal{H})^{\mathrm{T}}$, it follows that $L Z_1$ is symmetric, that is, $L Z_1 = Z_1^{\mathrm{T}} L^{\mathrm{T}}$. Since, in addition, L is skew symmetric, it follows that $0 = Z_1^{\mathrm{T}} L + L Z_1$. Now, since Z_1 is asymptotically stable, it follows that $L = 0$. Hence, $M_1^{\mathrm{T}} M_{21} = M_{21}^{\mathrm{T}} M_1$, which shows that $M_{21}^{\mathrm{T}} M_1$ is symmetric.

To show that M_1 is nonsingular, note that it follows from the equality

$$\begin{bmatrix} I & 0 \end{bmatrix} \mathcal{H} \begin{bmatrix} M_1 \\ M_{21} \end{bmatrix} = \begin{bmatrix} I & 0 \end{bmatrix} \begin{bmatrix} M_1 \\ M_{21} \end{bmatrix} Z_1$$

that

$$A M_1 + \Sigma M_{21} = M_1 Z_1.$$

Now, let $x \in \mathbb{R}^n$ satisfy $M_1 x = 0$. We thus have

$$x^{\mathrm{T}} M_{21} \Sigma M_{21} x = x^{\mathrm{T}} M_{21}^{\mathrm{T}} (A M_1 + \Sigma M_{21}) x$$

$$= x^{\mathrm{T}} M_{21}^{\mathrm{T}} M_1 Z_1 x$$

$$= x^{\mathrm{T}} M_1^{\mathrm{T}} M_{21} Z_1 x$$

$$= 0,$$

which implies that $B^{\mathrm{T}} M_{21} x = 0$. Hence, $M_1 Z_1 x = (A M_1 + \Sigma M_{21}) x = 0$. Thus, $Z_1 \mathcal{N}(M_1) \subseteq \mathcal{N}(M_1)$.

Now, suppose that M_1 is singular. Since $Z_1 \mathcal{N}(M_1) \subseteq \mathcal{N}(M_1)$, it follows that there exist $\lambda \in \operatorname{spec}(Z_1)$ and $x \in \mathbb{C}^n$ such that $Z_1 x = \lambda x$ and $M_1 x = 0$. Forming

$$[\, 0 \ \ I\,]\mathcal{H}\begin{bmatrix} M_1 \\ M_{21} \end{bmatrix} x = [\, 0 \ \ I\,]\begin{bmatrix} M_1 \\ M_{21} \end{bmatrix} Z_1 x$$

yields $-A^{\mathrm{T}} M_{21} x = M_{21} \lambda Z$, and thus $(\lambda I + A^{\mathrm{T}}) M_{21} x = 0$. Since, in addition, as shown above, $B^{\mathrm{T}} M_{21} x = 0$, it follows that $x^* M_{21}^{\mathrm{T}}[\, -\overline{\lambda} I - A \ \ B\,] = 0$. Since $\lambda \in \operatorname{spec}(Z_1)$, it follows that $\operatorname{Re}(-\overline{\lambda}) > 0$. Furthermore, since, by assumption, (A, B) is stabilizable, it follows that $\operatorname{rank}[\, \overline{\lambda} I - A \ \ B\,] = n$. Therefore, $M_{21} x = 0$. Combining this fact with $M_1 x = 0$ yields $\begin{bmatrix} M_1 \\ M_{21} \end{bmatrix} x = 0$. Since x is nonzero, it follows that M is singular, which is a contradiction. Consequently, M_1 is nonsingular. Next, define $P \triangleq -M_{21} M_1^{-1}$ and note that, since $M_1^{\mathrm{T}} M_{21}$ is symmetric, it follows that $P = -M_1^{-\mathrm{T}}(M_1^{\mathrm{T}} M_{21}) M_1^{-1}$ is also symmetric.

Since $\mathcal{H}\begin{bmatrix} M_1 \\ M_{21} \end{bmatrix} = \begin{bmatrix} M_1 \\ M_{21} \end{bmatrix} Z_1$, it follows that

$$\mathcal{H}\begin{bmatrix} I \\ M_{21} M_1^{-1} \end{bmatrix} = \begin{bmatrix} I \\ M_{21} M_1^{-1} \end{bmatrix} M_1 Z_1 M_1^{-1},$$

and thus

$$\mathcal{H}\begin{bmatrix} I \\ -P \end{bmatrix} = \begin{bmatrix} I \\ -P \end{bmatrix} M_1 Z_1 M_1^{-1}.$$

Multiplying on the left by $[\, P \ \ I\,]$ yields

$$0 = [\, P \ \ I\,]\mathcal{H}\begin{bmatrix} I \\ -P \end{bmatrix} = A^{\mathrm{T}} P + PA + R_1 - P\Sigma P,$$

which shows that P is a solution of (12.16.4). Similarly, multiplying on the left by $[\, I \ \ 0\,]$ yields $A - \Sigma P = M_1 Z_1 M_1^{-1}$. Since Z_1 is asymptotically stable, it follows that $A - \Sigma P$ is also asymptotically stable.

To prove that $iii) \implies i)$, note that the existence of a stabilizing solution P implies that (A, B) is stabilizable, and that (12.16.11) implies that $\mathcal{H}$ has no imaginary eigenvalues. $\qquad\square$

Corollary 12.17.10. Assume that (A, B) is stabilizable and (A, E_1) is detectable. Then, (12.16.4) has a stabilizing solution.

12.18 The Maximal Solution of the Riccati Equation

In this section we consider the existence of the maximal solution of (12.16.4). Example 12.16.3 shows that the assumptions of Proposition 12.19.1 are not sufficient to guarantee that (12.16.4) has a maximal solution.

Theorem 12.18.1. The following statements are equivalent:

i) (A, B) is stabilizable.

ii) (12.16.4) has a solution $P_{\max}$ that is positive semidefinite, maximal, and satisfies $\operatorname{spec}(A - \Sigma P_{\max}) \subset \mathrm{CLHP}$.

Proof. The result $i) \Longrightarrow ii)$ is given by Theorem 2.1 and Theorem 2.2 of [575]. See also (i) of Theorem 13.11 of [1535]. The converse result follows from Corollary 3 of [1196]. $\square$

Proposition 12.18.2. Assume that (12.16.4) has a maximal solution P_{max}, let P be a solution of (12.16.4), and assume that $\mathrm{spec}(A - \Sigma P_{\mathrm{max}}) \subset \mathrm{CLHP}$ and $\mathrm{spec}(A - \Sigma P) \subset \mathrm{CLHP}$. Then, $P = P_{\mathrm{max}}$.

Proof. It follows from $i)$ of Proposition 12.16.14 that $\mathrm{spec}(A - \Sigma P) = \mathrm{spec}(A - \Sigma P_{\mathrm{max}})$. Since P_{max} is the maximal solution of (12.16.4), it follows that $P \leq P_{\mathrm{max}}$. Consequently, it follows from the contrapositive form of the second statement of Theorem 8.4.9 that $P = P_{\mathrm{max}}$. $\square$

Proposition 12.18.3. Assume that (12.16.4) has a solution P such that $\mathrm{spec}(A - \Sigma P) \subset \mathrm{CLHP}$. Then, P is stabilizing if and only if $\mathcal{H}$ has no imaginary eigenvalues.

It follows from Proposition 12.18.2 that (12.16.4) has at most one positive-semidefinite solution P such that $\mathrm{spec}(A - \Sigma P) \subset \mathrm{CLHP}$. Consequently, (12.16.4) has at most one positive-semidefinite stabilizing solution.

Theorem 12.18.4. The following statements hold:

$i)$ (12.16.4) has at most one stabilizing solution.

$ii)$ If P is the stabilizing solution of (12.16.4), then P is positive semidefinite.

$iii)$ If P is the stabilizing solution of (12.16.4), then P is maximal.

Proof. To prove $i)$, assume that (12.16.4) has stabilizing solutions P_1 and P_2. Then, (A, B) is stabilizable, and Theorem 12.18.1 implies that (12.16.4) has a maximal solution P_{max} such that $\mathrm{spec}(A - \Sigma P_{\mathrm{max}}) \subset \mathrm{CLHP}$. Now, Proposition 12.18.2 implies that $P_1 = P_{\mathrm{max}}$ and $P_2 = P_{\mathrm{max}}$. Hence, $P_1 = P_2$.

Alternatively, suppose that (12.16.4) has the stabilizing solutions P_1 and P_2. Then,

$$A^{\mathrm{T}}P_1 + P_1 A + R_1 - P_1 \Sigma P_1 = 0,$$
$$A^{\mathrm{T}}P_2 + P_2 A + R_1 - P_2 \Sigma P_2 = 0.$$

Subtracting these equations and rearranging yields

$$(A - \Sigma P_1)^{\mathrm{T}}(P_1 - P_2) + (P_1 - P_2)(A - \Sigma P_2) = 0.$$

Since $A - \Sigma P_1$ and $A - \Sigma P_2$ are asymptotically stable, it follows from Proposition 11.9.3 and Fact 11.18.33 that $P_1 - P_2 = 0$. Hence, (12.16.4) has at most one stabilizing solution.

Next, to prove *ii*), suppose that P is a stabilizing solution of (12.16.4). Then, it follows from (12.16.4) that

$$P = \int_0^\infty e^{t(A-\Sigma P)^{\mathrm{T}}}(R_1 + P\Sigma P)e^{t(A-\Sigma P)}\,\mathrm{d}t,$$

which shows that P is positive semidefinite.

To prove *iii*), let P' be a solution of (12.16.4). Then, it follows that

$$(A - \Sigma P)^{\mathrm{T}}(P - P') + (P - P')(A - \Sigma P) + (P - P')\Sigma(P - P') = 0,$$

which implies that $P' \leq P$. Thus, P is also the maximal solution of (12.16.4). $\square$

The following result concerns the monotonicity of solutions of the Riccati equation (12.16.4).

Proposition 12.18.5. Assume that (A, B) is stabilizable, and let $P_{\max}$ denote the maximal solution of (12.16.4). Furthermore, let $\hat{R}_1 \in \mathbb{R}^{n \times n}$ be positive semidefinite, let $\hat{R}_2 \in \mathbb{R}^{m \times m}$ be positive definite, let $\hat{A} \in \mathbb{R}^{n \times n}$, let $\hat{B} \in \mathbb{R}^{n \times m}$, define $\hat{\Sigma} \triangleq \hat{B}\hat{R}_2^{-1}B^{\mathrm{T}}$, assume that

$$\begin{bmatrix} \hat{R}_1 & \hat{A}^{\mathrm{T}} \\ \hat{A} & -\hat{\Sigma} \end{bmatrix} \leq \begin{bmatrix} R_1 & A^{\mathrm{T}} \\ A & -\Sigma \end{bmatrix},$$

and let $\hat{P}$ be a solution of

$$\hat{A}^{\mathrm{T}}\hat{P} + \hat{P}\hat{A} + \hat{R}_1 - \hat{P}\hat{\Sigma}\hat{P} = 0. \tag{12.18.1}$$

Then,

$$\hat{P} \leq P_{\max}. \tag{12.18.2}$$

Proof. This result is given by Theorem 1 of [1475]. $\square$

Corollary 12.18.6. Assume that (A, B) is stabilizable, let $\hat{R}_1 \in \mathbb{R}^{n \times n}$ be positive semidefinite, assume that $\hat{R}_1 \leq R_1$, and let $P_{\max}$ and $\hat{P}_{\max}$ denote, respectively, the maximal solutions of (12.16.4) and

$$A^{\mathrm{T}}P + PA + \hat{R}_1 - P\Sigma P = 0. \tag{12.18.3}$$

Then,

$$\hat{P}_{\max} \leq P_{\max}. \tag{12.18.4}$$

Proof. This result follows from Proposition 12.18.5 or Theorem 2.3 of [575].
$\square$

The following result shows that, if $R_1 = 0$, then the closed-loop eigenvalues of the closed-loop dynamics obtained from the maximal solution consist of the CLHP open-loop eigenvalues and reflections of the ORHP open-loop eigenvalues.

Proposition 12.18.7. Assume that (A, B) is stabilizable, assume that $R_1 = 0$, and let $P \in \mathbb{R}^{n \times n}$ be a positive-semidefinite solution of (12.16.4). Then, P is the maximal solution of (12.16.4) if and only if

$$\mathrm{mspec}(A - \Sigma P) = [\mathrm{mspec}(A) \cap \mathrm{CLHP}] \cup [\mathrm{mspec}(-A) \cap \mathrm{OLHP}]. \qquad (12.18.5)$$

Proof. Sufficiency follows from Proposition 12.18.2. To prove necessity, note that it follows from the definition (12.16.8) of $\mathcal{H}$ with $R_1 = 0$ and from iv) of Proposition 12.16.14 that

$$\mathrm{mspec}(A) \cup \mathrm{mspec}(-A) = \mathrm{mspec}(A - \Sigma P) \cup \mathrm{mspec}[-(A - \Sigma P)].$$

Now, Theorem 12.18.1 implies that $\mathrm{mspec}(A - \Sigma P) \subseteq \mathrm{CLHP}$, which implies that (12.18.5) is satisfied. $\qquad \square$

Corollary 12.18.8. Let $R_1 = 0$, and assume that $\mathrm{spec}(A) \subset \mathrm{CLHP}$. Then, $P = 0$ is the only positive-semidefinite solution of (12.16.4).

12.19 Positive-Semidefinite and Positive-Definite Solutions of the Riccati Equation

The following result gives sufficient conditions under which (12.16.4) has a positive-semidefinite solution.

Proposition 12.19.1. Assume that there exists a nonsingular matrix $S \in \mathbb{R}^{n \times n}$ such that (12.17.2) and (12.17.3) are satisfied, where $\left(\left[\begin{smallmatrix} A_1 & 0 \\ A_{21} & A_2 \end{smallmatrix} \right], \left[\begin{smallmatrix} B_1 \\ B_2 \end{smallmatrix} \right] \right)$ is controllable, $\left(\left[\begin{smallmatrix} A_1 & A_{13} \\ 0 & A_3 \end{smallmatrix} \right], \left[\begin{smallmatrix} E_{11} & E_{13} \end{smallmatrix} \right] \right)$ is observable, and A_3 is asymptotically stable. Then, (12.16.4) has a positive-semidefinite solution.

Proof. First, rewrite (12.17.2) and (12.17.3) as

$$A = S \begin{bmatrix} A_1 & A_{13} & 0 & 0 \\ 0 & A_3 & 0 & 0 \\ A_{21} & A_{23} & A_2 & A_{24} \\ 0 & A_{43} & 0 & A_4 \end{bmatrix} S^{-1}, \qquad B = S \begin{bmatrix} B_1 \\ 0 \\ B_2 \\ 0 \end{bmatrix},$$

$$E_1 = \begin{bmatrix} E_{11} & E_{13} & 0 & 0 \end{bmatrix} S^{-1},$$

where $\left(\left[\begin{smallmatrix} A_1 & 0 \\ A_{21} & A_2 \end{smallmatrix} \right], \left[\begin{smallmatrix} B_1 \\ B_2 \end{smallmatrix} \right] \right)$ is controllable, $\left(\left[\begin{smallmatrix} A_1 & A_{13} \\ 0 & A_3 \end{smallmatrix} \right], \left[\begin{smallmatrix} E_{11} & E_{13} \end{smallmatrix} \right] \right)$ is observable, and A_3 is asymptotically stable. Since $\left(\left[\begin{smallmatrix} A_1 & A_{13} \\ 0 & A_3 \end{smallmatrix} \right], \left[\begin{smallmatrix} B_1 \\ 0 \end{smallmatrix} \right] \right)$ is stabilizable, it follows from Theorem 12.18.1 that there exists a positive-semidefinite matrix $\hat{P}_1$ that satisfies

$$\begin{bmatrix} A_1 & A_{13} \\ 0 & A_3 \end{bmatrix}^{\mathrm{T}} \hat{P}_1 + \hat{P}_1 \begin{bmatrix} A_1 & A_{13} \\ 0 & A_3 \end{bmatrix} + \begin{bmatrix} E_{11}^{\mathrm{T}} E_{11} & E_{11}^{\mathrm{T}} E_{13} \\ E_{13}^{\mathrm{T}} E_{11} & E_{13}^{\mathrm{T}} E_{13} \end{bmatrix} - \hat{P}_1 \begin{bmatrix} B_1 R_2^{-1} B_1^{\mathrm{T}} & 0 \\ 0 & 0 \end{bmatrix} \hat{P}_1 = 0.$$

Consequently, $P \triangleq S^{\mathrm{T}} \mathrm{diag}(\hat{P}_1, 0, 0) S$ is a positive-semidefinite solution of (12.16.4). $\qquad \square$

Corollary 12.19.2. Assume that (A, B) is stabilizable. Then, (12.16.4) has a positive-semidefinite solution P. If, in addition, (A, E_1) is detectable, then P is the stabilizing solution of (12.16.4), and thus P is the only positive-semidefinite solution of (12.16.4). Finally, if (A, E_1) is observable, then P is positive definite.

Proof. The first statement is given by Theorem 12.18.1. Next, assume that (A, E_1) is detectable. Then, Theorem 12.17.2 implies that P is a stabilizing solution of (12.16.4), which is the only positive-semidefinite solution of (12.16.4). Finally, using Theorem 12.17.2, (A, E_1) observable implies that P is positive definite. $\square$

The next result gives necessary and sufficient conditions under which (12.16.4) has a positive-definite solution.

Proposition 12.19.3. The following statements are equivalent:

i) (12.16.4) has a positive-definite solution.

ii) There exists a nonsingular matrix $S \in \mathbb{R}^{n \times n}$ such that (12.17.2) and (12.17.3) are satisfied, where $\left(\left[\begin{smallmatrix} A_1 & 0 \\ A_{21} & A_2 \end{smallmatrix} \right], \left[\begin{smallmatrix} B_1 \\ B_2 \end{smallmatrix} \right] \right)$ is controllable, $\left(\left[\begin{smallmatrix} A_1 & A_{13} \\ 0 & A_3 \end{smallmatrix} \right], \left[E_{11} \; E_{13} \right] \right)$ is observable, A_3 is asymptotically stable, $-A_2$ is asymptotically stable, $\operatorname{spec}(A_4) \subset j\mathbb{R}$, and A_4 is semisimple.

In this case, (12.16.4) has exactly one positive-definite solution if and only if A_4 is empty, and infinitely many positive-definite solutions if and only if A_4 is not empty.

Proof. See [1151]. $\square$

Proposition 12.19.4. Assume that (12.16.4) has a stabilizing solution P, and let $S \in \mathbb{R}^{n \times n}$ be a nonsingular matrix such that (12.17.2) and (12.17.3) are satisfied, where (A_1, B_1, E_{11}) is controllable and observable, (A_2, B_2) is controllable, (A_3, E_{13}) is observable, $\nu_0(A_2) = 0$, and A_3 and A_4 are asymptotically stable. Then,

$$\operatorname{def} P = \nu_-(A_2). \tag{12.19.1}$$

Hence, P is positive definite if and only if $\operatorname{spec}(A_2) \subset \text{ORHP}$.

12.20 Facts on Stability, Observability, and Controllability

Fact 12.20.1. Let $A \in \mathbb{R}^{n \times n}$, $B \in \mathbb{R}^{n \times m}$, and $C \in \mathbb{R}^{p \times n}$, and assume that (A, B) is controllable and (A, C) is observable. Then, for all $v \in \mathbb{R}^m$, the step response

$$y(t) = \int_0^t C e^{tA} \, \mathrm{d}\tau B v + D v$$

is bounded on $[0, \infty)$ if and only if A is Lyapunov stable and nonsingular.

Fact 12.20.2. Let $A \in \mathbb{R}^{n \times n}$ and $C \in \mathbb{R}^{p \times n}$, assume that (A, C) is detectable, and let $x(t)$ and $y(t)$ satisfy $\dot{x}(t) = Ax(t)$ and $y(t) = Cx(t)$ for $t \in [0, \infty)$. Then, the following statements hold:

i) y is bounded if and only if x is bounded.

ii) $\lim_{t \to \infty} y(t)$ exists if and only if $\lim_{t \to \infty} x(t)$ exists.

iii) $y(t) \to 0$ as $t \to \infty$ if and only if $x(t) \to 0$ as $t \to \infty$.

Fact 12.20.3. Let $x(0) = x_0$, and let $x_f - e^{t_f A} x_0 \in \mathcal{C}(A, B)$. Then, for all $t \in [0, t_f]$, the control $u: [0, t_f] \mapsto \mathbb{R}^m$ defined by

$$u(t) \triangleq B^T e^{(t_f - t) A^T} \left(\int_0^{t_f} e^{\tau A} B B^T e^{\tau A^T} d\tau \right)^+ \left(x_f - e^{t_f A} x_0 \right)$$

yields $x(t_f) = x_f$.

Fact 12.20.4. Let $x(0) = x_0$, let $x_f \in \mathbb{R}^n$, and assume that (A, B) is controllable. Then, for all $t \in [0, t_f]$, the control $u: [0, t_f] \mapsto \mathbb{R}^m$ defined by

$$u(t) \triangleq B^T e^{(t_f - t) A^T} \left(\int_0^{t_f} e^{\tau A} B B^T e^{\tau A^T} d\tau \right)^{-1} \left(x_f - e^{t_f A} x_0 \right)$$

yields $x(t_f) = x_f$.

Fact 12.20.5. Let $A \in \mathbb{R}^{n \times n}$, let $B \in \mathbb{R}^{n \times m}$, assume that A is skew symmetric, and assume that (A, B) is controllable. Then, for all $\alpha > 0$, $A - \alpha B B^T$ is asymptotically stable.

Fact 12.20.6. Let $A \in \mathbb{R}^{n \times n}$ and $B \in \mathbb{R}^{n \times m}$. Then, (A, B) is (controllable, stabilizable) if and only if $(A, B B^T)$ is (controllable, stabilizable). Now, assume in addition that B is positive semidefinite. Then, (A, B) is (controllable, stabilizable) if and only if $(A, B^{1/2})$ is (controllable, stabilizable).

Fact 12.20.7. Let $A \in \mathbb{R}^{n \times n}$, $B \in \mathbb{R}^{n \times m}$, and $\hat{B} \in \mathbb{R}^{n \times \hat{m}}$, and assume that (A, B) is (controllable, stabilizable) and $\mathcal{R}(B) \subseteq \mathcal{R}(\hat{B})$. Then, $(A, \hat{B})$ is also (controllable, stabilizable).

Fact 12.20.8. Let $A \in \mathbb{R}^{n \times n}$, $B \in \mathbb{R}^{n \times m}$, and $\hat{B} \in \mathbb{R}^{n \times \hat{m}}$, and assume that (A, B) is (controllable, stabilizable) and $B B^T \leq \hat{B} \hat{B}^T$. Then, $(A, \hat{B})$ is also (controllable, stabilizable).

Proof: Use Lemma 8.6.1 and Fact 12.20.7.

Fact 12.20.9. Let $A \in \mathbb{R}^{n \times n}$, $B \in \mathbb{R}^{n \times m}$, $\hat{B} \in \mathbb{R}^{n \times \hat{m}}$, and $\hat{C} \in \mathbb{R}^{\hat{m} \times n}$, and assume that (A, B) is (controllable, stabilizable). Then,

$$\left(A + \hat{B} \hat{C}, [B B^T + \hat{B} \hat{B}^T]^{1/2} \right)$$

is also (controllable, stabilizable).

Proof: See [1490, p. 79].

Fact 12.20.10. Let $A \in \mathbb{R}^{n \times n}$ and $B \in \mathbb{R}^{n \times m}$. Then, the following statements are equivalent:

i) (A, B) is controllable.

ii) There exists $\alpha \in \mathbb{R}$ such that $(A + \alpha I, B)$ is controllable.

iii) $(A + \alpha I, B)$ is controllable for all $\alpha \in \mathbb{R}$.

Fact 12.20.11. Let $A \in \mathbb{R}^{n \times n}$ and $B \in \mathbb{R}^{n \times m}$. Then, the following statements are equivalent:

i) (A, B) is stabilizable.

ii) There exists $\alpha \leq \max\{0, -\operatorname{spabs}(A)\}$ such that $(A + \alpha I, B)$ is stabilizable.

iii) $(A + \alpha I, B)$ is stabilizable for all $\alpha \leq \max\{0, -\operatorname{spabs}(A)\}$.

Fact 12.20.12. Let $A \in \mathbb{R}^{n \times n}$, assume that A is diagonal, and let $B \in \mathbb{R}^{n \times 1}$. Then, (A, B) is controllable if and only if the diagonal entries of A are distinct and every entry of B is nonzero.

Proof: Note that

$$
\det \mathcal{K}(A, B) = \det \begin{bmatrix} b_1 & & 0 \\ & \ddots & \\ 0 & & b_n \end{bmatrix} \begin{bmatrix} 1 & a_1 & \cdots & a_1^{n-1} \\ \vdots & \vdots & \ddots & \vdots \\ 1 & a_n & \cdots & a_n^{n-1} \end{bmatrix}
$$

$$
= \left(\prod_{i=1}^{n} b_i \right) \prod_{i<j} (a_i - a_j).
$$

Fact 12.20.13. Let $A \in \mathbb{R}^{n \times n}$ and $B \in \mathbb{R}^{n \times 1}$, and assume that (A, B) is controllable. Then, A is cyclic.

Proof: See Fact 5.14.7.

Fact 12.20.14. Let $A \in \mathbb{R}^{n \times n}$ and $B \in \mathbb{R}^{n \times m}$, and assume that (A, B) is controllable. Then,

$$
\max_{\lambda \in \operatorname{spec}(A)} \operatorname{gmult}_A(\lambda) \leq m.
$$

Fact 12.20.15. Let $A \in \mathbb{R}^{n \times n}$ and $B \in \mathbb{R}^{n \times m}$. Then, the following conditions are equivalent:

i) (A, B) is (controllable, stabilizable) and A is nonsingular.

ii) (A, AB) is (controllable, stabilizable).

Fact 12.20.16. Let $A \in \mathbb{R}^{n \times n}$ and $B \in \mathbb{R}^{n \times m}$, and assume that (A, B) is controllable. Then, $(A, B^{\mathrm{T}} S^{-\mathrm{T}})$ is observable, where $S \in \mathbb{R}^{n \times n}$ is a nonsingular matrix satisfying $A^{\mathrm{T}} = S^{-1} A S$.

Fact 12.20.17. Let (A, B) be controllable, let $t_1 > 0$, and define

$$P = \left(\int_0^{t_1} e^{-tA} B B^T e^{-tA^T} \, dt \right)^{-1}.$$

Then, $A - BB^T P$ is asymptotically stable.

Proof: P satisfies

$$(A - BB^T P)^T P + P(A - BB^T P) + P\left(BB^T + e^{t_1 A} BB^T e^{t_1 A^T} \right) P = 0.$$

Since $(A - BB^T P, BB^T + e^{t_1 A} BB^T e^{t_1 A^T})$ is observable and P is positive definite, Proposition 11.9.5 implies that $A - BB^T P$ is asymptotically stable.

Remark: This result is due to Lukes and Kleinman. See [1181, pp. 113, 114].

Fact 12.20.18. Let $A \in \mathbb{R}^{n \times n}$ and $B \in \mathbb{R}^{n \times m}$, assume that A is asymptotically stable, and, for $t \geq 0$, consider the linear system $\dot{x} = Ax + Bu$. Then, if u is bounded, then x is bounded. Furthermore, if $u(t) \to 0$ as $t \to \infty$, then $x(t) \to 0$ as $t \to \infty$.

Proof: See [1243, p. 330].

Remark: These results are consequences of *input-to-state stability*.

Fact 12.20.19. Let $A \in \mathbb{R}^{n \times n}$ and $C \in \mathbb{R}^{p \times n}$, assume that (A, C) is observable, and let $k \geq n$. Then,

$$A = \left[\begin{array}{c} 0_{p \times n} \\ \mathcal{O}_k(A, C) \end{array} \right]^+ \mathcal{O}_{k+1}(A, C).$$

Remark: This result is due to Palanthandalam-Madapusi.

12.21 Facts on the Lyapunov Equation and Inertia

Fact 12.21.1. Let $A, P \in \mathbb{F}^{n \times n}$, assume that P is Hermitian, let $C \in \mathbb{F}^{l \times n}$, and assume that $A^* P + PA + C^* C = 0$. Then, the following statements hold:

i) $|\nu_-(A) - \nu_+(P)| \leq n - \operatorname{rank} \mathcal{O}(A, C)$.

ii) $|\nu_+(A) - \nu_-(P)| \leq n - \operatorname{rank} \mathcal{O}(A, C)$.

iii) If $\nu_0(A) = 0$, then

$$|\nu_-(A) - \nu_+(P)| + |\nu_+(A) - \nu_-(P)| \leq n - \operatorname{rank} \mathcal{O}(A, C).$$

If, in addition, (A, C) is observable, then the following statements hold:

iv) $\nu_-(A) = \nu_+(P)$.

v) $\nu_0(A) = \nu_0(P) = 0$.

vi) $\nu_+(A) = \nu_-(P)$.

vii) If P is positive definite, then A is asymptotically stable.

Proof: See [67, 320, 955, 1471] and [892, p. 448].

Remark: v) does not follow from i)–iii).

Remark: For related results, see [1081] and references given in [955]. See also [297, 380].

Fact 12.21.2. Let $A, P \in \mathbb{F}^{n \times n}$, assume that P is nonsingular and Hermitian, and assume that $A^*P + PA$ is negative semidefinite. Then, the following statements hold:

   i) $\nu_-(A) \leq \nu_+(P)$.

   ii) $\nu_+(A) \leq \nu_-(P)$.

   iii) If P is positive definite, then $\mathrm{spec}(A) \subset \mathrm{CLHP}$.

Proof: See [892, p. 447].

Remark: If P is positive definite, then A is Lyapunov stable, although this result does not follow from i) and ii).

Fact 12.21.3. Let $A, P \in \mathbb{F}^{n \times n}$, and assume that $\nu_0(A) = 0$, P is Hermitian, and $A^*P + PA$ is negative semidefinite. Then, the following statements hold:

   i) $\nu_-(P) \leq \nu_+(A)$.

   ii) $\nu_+(P) \leq \nu_-(A)$.

   iii) If P is nonsingular, then $\nu_-(P) = \nu_+(A)$ and $\nu_+(P) = \nu_-(A)$.

   iv) If P is positive definite, then A is asymptotically stable.

Proof: See [892, p. 447].

Fact 12.21.4. Let $A, P \in \mathbb{F}^{n \times n}$, and assume that $\nu_0(A) = 0$, P is nonsingular and Hermitian, and $A^*P + PA$ is negative semidefinite. Then, the following statements hold:

   i) $\nu_-(A) = \nu_+(P)$.

   ii) $\nu_+(A) = \nu_-(P)$.

Proof: Combine Fact 12.21.2 and Fact 12.21.3. See [892, p. 448].

Remark: This result is due to Carlson and Schneider.

Fact 12.21.5. Let $A, P \in \mathbb{F}^{n \times n}$, assume that P is Hermitian, and assume that $A^*P + PA$ is negative definite. Then, the following statements hold:

   i) $\nu_-(A) = \nu_+(P)$.

   ii) $\nu_0(A) = 0$.

   iii) $\nu_+(A) = \nu_-(P)$.

   iv) P is nonsingular.

   v) If P is positive definite, then A is asymptotically stable.

Proof: See [459, pp. 441, 442], [892, p. 445], or [1081]. This result follows from Fact 12.21.1 with positive-definite $C = -(A^*P + PA)^{1/2}$.

Remark: This result is due to Krein, Ostrowski, and Schneider.

Remark: These conditions are the *classical constraints*. An analogous result holds for the discrete-time Lyapunov equation, where the analogous definition of inertia counts the numbers of eigenvalues inside the open unit disk, outside the open unit disk, and on the unit circle. See [287, 401].

Fact 12.21.6. Let $A \in \mathbb{F}^{n \times n}$. Then, the following statements are equivalent:

i) $\nu_0(A) = 0$.

ii) There exists a nonsingular Hermitian matrix $P \in \mathbb{F}^{n \times n}$ such that $A^*P + PA$ is negative definite.

iii) There exists a Hermitian matrix $P \in \mathbb{F}^{n \times n}$ such that $A^*P + PA$ is negative definite.

In this case, the following statements hold for P given by *ii)* and *iii)*:

iv) $\nu_-(A) = \nu_+(P)$.

v) $\nu_0(A) = \nu_0(P) = 0$.

vi) $\nu_+(A) = \nu_-(P)$.

vii) P is nonsingular.

viii) If P is positive definite, then A is asymptotically stable.

Proof: For the result *i)* $\implies$ *ii)*, see [892, p. 445]. The result *iii)* $\implies$ *i)* follows from Fact 12.21.5. See [53, 287, 299].

Fact 12.21.7. Let $A \in \mathbb{F}^{n \times n}$. Then, the following statements are equivalent:

i) A is Lyapunov stable.

ii) There exists a positive-definite matrix $P \in \mathbb{F}^{n \times n}$ such that $A^*P + PA$ is negative semidefinite.

Furthermore, the following statements are equivalent:

iii) A is asymptotically stable.

iv) There exists a positive-definite matrix $P \in \mathbb{F}^{n \times n}$ such that $A^*P + PA$ is negative definite.

v) For every positive-definite matrix $R \in \mathbb{F}^{n \times n}$, there exists a positive-definite matrix $P \in \mathbb{F}^{n \times n}$ such that $A^*P + PA$ is negative definite.

Remark: See Proposition 11.9.5 and Proposition 11.9.6.

Fact 12.21.8. Let $A, P \in \mathbb{F}^{n \times n}$, and assume P is Hermitian. Then, the following statements hold:

i) $\nu_+(A^*P + PA) \leq \operatorname{rank} P$.

ii) $\nu_-(A^*P + PA) \leq \operatorname{rank} P$.

If, in addition, A is asymptotically stable, then the following statement holds:

iii) $1 \leq \nu_-(A^*P + PA) \leq \operatorname{rank} P$.

Proof: See [124, 401].

Fact 12.21.9. Let $A, P \in \mathbb{R}^{n \times n}$, assume that $\nu_0(A) = n$, and assume that P is positive semidefinite. Then, exactly one of the following statements holds:

i) $A^{\mathrm{T}}P + PA = 0$.

ii) $\nu_-(A^{\mathrm{T}}P + PA) \geq 1$ and $\nu_+(A^{\mathrm{T}}P + PA) \geq 1$.

Proof: See [1381].

Fact 12.21.10. Let $R \in \mathbb{F}^{n \times n}$, and assume that R is Hermitian and $\nu_+(R) \geq 1$. Then, there exist an asymptotically stable matrix $A \in \mathbb{F}^{n \times n}$ and a positive-definite matrix $P \in \mathbb{F}^{n \times n}$ such that $A^*P + PA + R = 0$.

Proof: See [124].

Fact 12.21.11. Let $A \in \mathbb{F}^{n \times n}$, assume that A is cyclic, and let a, b, c, d, e be nonnegative integers such that $a + b = c + d + e = n$, $c \geq 1$, and $e \geq 1$. Then, there exists a nonsingular, Hermitian matrix $P \in \mathbb{F}^{n \times n}$ such that

$$\operatorname{In} P = \begin{bmatrix} a \\ 0 \\ b \end{bmatrix}$$

and

$$\operatorname{In}(A^*P + PA) = \begin{bmatrix} c \\ d \\ e \end{bmatrix}.$$

Proof: See [1230].

Remark: See also [1229].

Fact 12.21.12. Let $P, R \in \mathbb{F}^{n \times n}$, and assume that P is positive and R is Hermitian. Then, the following statements are equivalent:

i) $\operatorname{tr} RP^{-1} > 0$.

ii) There exists an asymptotically stable matrix $A \in \mathbb{F}^{n \times n}$ such that $A^*P + PA + R = 0$.

Proof: See [124].

Fact 12.21.13. Let $A_1 \in \mathbb{R}^{n_1 \times n_1}$, $A_2 \in \mathbb{R}^{n_2 \times n_2}$, $B \in \mathbb{R}^{n_1 \times m}$, and $C \in \mathbb{R}^{m \times n_2}$, assume that $A_1 \oplus A_2$ is nonsingular, and assume that $\operatorname{rank} B = \operatorname{rank} C = m$. Furthermore, let $X \in \mathbb{R}^{n_1 \times n_2}$ be the unique solution of

$$A_1 X + X A_2 + BC = 0.$$

Then,

$$\operatorname{rank} X \leq \min\{\operatorname{rank} \mathcal{K}(A_1, B), \operatorname{rank} \mathcal{O}(A_2, C)\}.$$

Furthermore, equality holds if $m = 1$.

Proof: See [398].

Remark: Related results are given in [1471, 1477].

Fact 12.21.14. Let $A_1, A_2 \in \mathbb{R}^{n \times n}$, $B \in \mathbb{R}^n$, $C \in \mathbb{R}^{1 \times n}$, assume that $A_1 \oplus A_2$ is nonsingular, let $X \in \mathbb{R}^{n \times n}$ satisfy

$$A_1 X + X A_2 + BC = 0,$$

and assume that (A_1, B) is controllable and (A_2, C) is observable. Then, X is nonsingular.

Proof: See Fact 12.21.13 and [1477].

Fact 12.21.15. Let $A, P, R \in \mathbb{R}^{n \times n}$, and assume that P and R are positive semidefinite, $A^{\mathrm{T}}P + PA + R = 0$, and $\mathcal{N}[\mathcal{O}(A, R)] = \mathcal{N}(A)$. Then, A is semistable.

Proof: See [199].

Fact 12.21.16. Let $A, V \in \mathbb{R}^{n \times n}$, assume that A is asymptotically stable, assume that V is positive semidefinite, and let $Q \in \mathbb{R}^{n \times n}$ be the unique, positive-definite solution to $AQ + QA^{\mathrm{T}} + V = 0$. Furthermore, let $C \in \mathbb{R}^{l \times n}$, and assume that CVC^{T} is positive definite. Then, CQC^{T} is positive definite.

Fact 12.21.17. Let $A, R \in \mathbb{R}^{n \times n}$, assume that A is asymptotically stable, assume that $R \in \mathbb{R}^{n \times n}$ is positive semidefinite, and let $P \in \mathbb{R}^{n \times n}$ satisfy $A^{\mathrm{T}}P + PA + R = 0$. Then, for all $i, j \in \{1, \ldots, n\}$, there exist $\alpha_{ij} \in \mathbb{R}$ such that

$$P = \sum_{i,j=1}^{n} \alpha_{ij} A^{(i-1)\mathrm{T}} R A^{j-1}.$$

In particular, for all $i, j \in \{1, \ldots, n\}$, $\alpha_{ij} = \hat{P}_{(i,j)}$, where $\hat{P} \in \mathbb{R}^{n \times n}$ satisfies $\hat{A}^{\mathrm{T}}\hat{P} + \hat{P}\hat{A} + \hat{R} = 0$, where $\hat{A} = C(\chi_A)$ and $\hat{R} = E_{1,1}$.

Proof: See [1235].

Remark: This equality is *Smith's method*. See [399, 423, 661, 965] for finite-sum solutions of linear matrix equations.

Fact 12.21.18. Let $\lambda_1, \ldots, \lambda_n \in \mathbb{C}$, assume that, for all $i \in \{1, \ldots, n\}$, $\mathrm{Re}\,\lambda_i < 0$, define $\Lambda \triangleq \mathrm{diag}(\lambda_1, \ldots, \lambda_n)$, let k be a nonnegative integer, and, for all $i, j \in \{1, \ldots, n\}$, define $P \in \mathbb{C}^{n \times n}$ by

$$P \triangleq \frac{1}{k!} \int_0^\infty t^k e^{\overline{\Lambda}t} e^{\Lambda t} \, dt.$$

Then, P is positive definite, P satisfies the Lyapunov equation

$$\overline{\Lambda}P + P\Lambda + I = 0,$$

and, for all $i, j \in \{1, \ldots, n\}$,

$$P_{(i,j)} = \left(\frac{-1}{\overline{\lambda_i} + \lambda_j} \right)^{k+1}.$$

Proof: For all nonzero $x \in \mathbb{C}^n$, it follows that

$$x^*Px = \int_0^\infty t^k \|e^{\Lambda t}x\|_2^2 \, dt$$

is positive. Hence, P is positive definite. Furthermore, note that

$$P_{(i,j)} = \int_0^\infty t^k e^{\overline{\lambda_i}t} e^{\lambda_j t} \, dt = \frac{(-1)^{k+1} k!}{(\overline{\lambda_i} + \lambda_j)^{k+1}}.$$

Remark: See [266] and [730, p. 348].

Remark: See Fact 8.8.16 and Fact 12.21.19.

Fact 12.21.19. Let $\lambda_1, \ldots, \lambda_n \in \mathbb{C}$, assume that, for all $i \in \{1, \ldots, n\}$, $\mathrm{Re}\, \lambda_i < 0$, define $\Lambda \triangleq \mathrm{diag}(\lambda_1, \ldots, \lambda_n)$, let k be a nonnegative integer, let $R \in \mathbb{C}^{n \times n}$, assume that R is positive semidefinite, and, for all $i, j \in \{1, \ldots, n\}$, define $P \in \mathbb{C}^{n \times n}$ by

$$P \triangleq \frac{1}{k!} \int_0^\infty t^k e^{\overline{\Lambda}t} R e^{\Lambda t} \, dt.$$

Then, P is positive semidefinite, P satisfies the Lyapunov equation

$$\overline{\Lambda}P + P\Lambda + R = 0,$$

and, for all $i, j \in \{1, \ldots, n\}$,

$$P_{(i,j)} = R_{(i,j)} \left(\frac{-1}{\overline{\lambda_i} + \lambda_j} \right)^{k+1}.$$

If, in addition, $I \circ R$ is positive definite, then P is positive definite.

Proof: Use Fact 8.22.12 and Fact 12.21.18.

Remark: See Fact 8.8.16 and Fact 12.21.18. Note that $P = \hat{P} \circ R$, where $\hat{P}$ is the solution to the Lyapunov equation with $R = I$.

Fact 12.21.20. Let $A, R \in \mathbb{R}^{n \times n}$, assume that $R \in \mathbb{R}^{n \times n}$ is positive semidefinite, let $q, r \in \mathbb{R}$, where $r > 0$, and assume that there exists a positive-definite matrix $P \in \mathbb{R}^{n \times n}$ satisfying

$$[A - (q + r)I]^{\mathrm{T}}P + P[A - (q + r)I] + \tfrac{1}{r}A^{\mathrm{T}}PA + R = 0.$$

Then, the spectrum of A is contained in a disk centered at $q + \jmath 0$ with radius r.

Remark: The disk is an *eigenvalue inclusion region*. See [145, 629, 1435] for related results concerning elliptical, parabolic, hyperbolic, sector, and vertical strip regions.

12.22 Facts on Realizations and the H_2 System Norm

Fact 12.22.1. Let $x\colon [0,\infty) \mapsto \mathbb{R}^n$ and $y\colon [0,\infty) \mapsto \mathbb{R}^n$, assume that $\int_0^\infty x^{\mathrm{T}}(t)x(t)\,\mathrm{d}t$ and $\int_0^\infty y^{\mathrm{T}}(t)y(t)\,\mathrm{d}t$ exist, and let $\hat{x}\colon \jmath\mathbb{R} \mapsto \mathbb{C}^n$ and $\hat{y}\colon \jmath\mathbb{R} \mapsto \mathbb{C}^n$ denote the Fourier transforms of x and y, respectively. Then,

$$\int_0^\infty x^{\mathrm{T}}(t)x(t)\,\mathrm{d}t = \int_{-\infty}^\infty \hat{x}^*(\jmath\omega)\hat{x}(\jmath\omega)\,\mathrm{d}\omega$$

and

$$\int_0^\infty x^{\mathrm{T}}(t)y(t)\,\mathrm{d}t = \mathrm{Re}\int_{-\infty}^\infty \hat{x}^*(\jmath\omega)\hat{y}(\jmath\omega)\,\mathrm{d}\omega.$$

Remark: These equalities are equivalent versions of Parseval's theorem. The second equality follows from the first equality by replacing x with $x + y$.

Fact 12.22.2. Let $G \in \mathbb{R}^{l\times m}_{\mathrm{prop}}(s)$, where $G \overset{\mathrm{min}}{\sim} \left[\begin{array}{c|c} A & B \\ \hline C & D \end{array}\right]$, and assume that, for all $i \in \{1,\dots,l\}$ and $j \in \{1,\dots,m\}$, $G_{(i,j)} = p_{i,j}/q_{i,j}$, where $p_{i,j}, q_{i,j} \in \mathbb{R}[s]$ are coprime. Then,

$$\mathrm{spec}(A) = \bigcup_{i,j=1}^{l,m} \mathrm{roots}(p_{i,j}).$$

Fact 12.22.3. Let $G \sim \left[\begin{array}{c|c} A & B \\ \hline C & D \end{array}\right]$, let $a,b \in \mathbb{R}$, where $a \neq 0$, and define $H(s) \triangleq G(as + b)$. Then,

$$H \sim \left[\begin{array}{c|c} a^{-1}(A - bI) & B \\ \hline a^{-1}C & D \end{array}\right].$$

Fact 12.22.4. Let $G \sim \left[\begin{array}{c|c} A & B \\ \hline C & D \end{array}\right]$, where A is nonsingular, and define $H(s) \triangleq G(1/s)$. Then,

$$H \sim \left[\begin{array}{c|c} A^{-1} & -A^{-1}B \\ \hline CA^{-1} & D - CA^{-1}B \end{array}\right].$$

Fact 12.22.5. Let $G(s) = C(sI - A)^{-1}B$. Then,

$$G(\jmath\omega) = -CA(\omega^2 I + A^2)^{-1}B - \jmath\omega C(\omega^2 I + A^2)^{-1}B.$$

Fact 12.22.6. Let $G \sim \left[\begin{array}{c|c} A & B \\ \hline C & 0 \end{array}\right]$ and $H(s) = sG(s)$. Then,

$$H \sim \left[\begin{array}{c|c} A & B \\ \hline CA & CB \end{array}\right].$$

Consequently,

$$sC(sI - A)^{-1}B = CA(sI - A)^{-1}B + CB.$$

Fact 12.22.7. Let $G = \begin{bmatrix} G_{11} & G_{12} \\ G_{21} & G_{22} \end{bmatrix}$, where $G_{ij} \sim \left[\begin{array}{c|c} A_{ij} & B_{ij} \\ \hline C_{ij} & D_{ij} \end{array} \right]$ for all $i, j = 1, 2$. Then,

$$\begin{bmatrix} G_{11} & G_{12} \\ G_{21} & G_{22} \end{bmatrix} \sim \left[\begin{array}{cccc|cc} A_{11} & 0 & 0 & 0 & B_{11} & 0 \\ 0 & A_{12} & 0 & 0 & 0 & B_{12} \\ 0 & 0 & A_{21} & 0 & B_{21} & 0 \\ 0 & 0 & 0 & A_{22} & 0 & B_{22} \\ \hline C_{11} & C_{12} & 0 & 0 & D_{11} & D_{12} \\ 0 & 0 & C_{21} & C_{22} & D_{21} & D_{22} \end{array} \right].$$

Fact 12.22.8. Let $G \sim \left[\begin{array}{c|c} A & B \\ \hline C & 0 \end{array} \right]$, where $G \in \mathbb{R}^{l \times m}(s)$, and let $M \in \mathbb{R}^{m \times l}$. Then,

$$[I + GM]^{-1} \sim \left[\begin{array}{c|c} A - BMC & B \\ \hline -C & I \end{array} \right]$$

and

$$[I + GM]^{-1}G \sim \left[\begin{array}{c|c} A - BMC & B \\ \hline C & 0 \end{array} \right].$$

Fact 12.22.9. Let $G \sim \left[\begin{array}{c|c} A & B \\ \hline C & D \end{array} \right]$, where $G \in \mathbb{R}^{l \times m}(s)$. If D has a left inverse $D^{\mathrm{L}} \in \mathbb{R}^{m \times l}$, then

$$G^{\mathrm{L}} \sim \left[\begin{array}{c|c} A - BD^{\mathrm{L}}C & BD^{\mathrm{L}} \\ \hline -D^{\mathrm{L}}C & D^{\mathrm{L}} \end{array} \right]$$

satisfies $G^{\mathrm{L}}G = I$. If D has a right inverse $D^{\mathrm{R}} \in \mathbb{R}^{m \times l}$, then

$$G^{\mathrm{R}} \sim \left[\begin{array}{c|c} A - BD^{\mathrm{R}}C & BD^{\mathrm{R}} \\ \hline -D^{\mathrm{R}}C & D^{\mathrm{R}} \end{array} \right]$$

satisfies $GG^{\mathrm{R}} = I$.

Fact 12.22.10. Let $G \sim \left[\begin{array}{c|c} A & B \\ \hline C & 0 \end{array} \right]$ be a SISO rational transfer function, and let $\lambda \in \mathbb{C}$. Then, there exists a rational function H such that

$$G(s) = \frac{1}{(s + \lambda)^r} H(s)$$

and such that λ is neither a pole nor a zero of H if and only if the Jordan form of A has exactly one block associated with λ, which is of order r.

Fact 12.22.11. Let $G \sim \left[\begin{array}{c|c} A & B \\ \hline C & D \end{array} \right]$. Then, $G(s)$ is given by

$$G(s) = (A - sI) \left\| \begin{bmatrix} A - sI & B \\ C & D \end{bmatrix} \right.$$

Remark: See [155].

Remark: The vertical bar denotes the Schur complement.

Fact 12.22.12. Let $G \in \mathbb{F}^{n \times m}(s)$, where $G \overset{\min}{\sim} \left[\begin{array}{c|c} A & B \\ \hline C & D \end{array} \right]$, and, for all $i \in \{1, \ldots, n\}$ and $j \in \{1, \ldots, m\}$, let $G_{(i,j)} = p_{ij}/q_{ij}$, where $p_{ij}, q_{ij} \in \mathbb{F}[s]$ are coprime. Then,

$$\bigcup_{i,j=1}^{n,m} \mathrm{roots}(q_{ij}) = \mathrm{spec}(A).$$

Fact 12.22.13. Let $A \in \mathbb{R}^{n \times n}$, $B \in \mathbb{R}^{n \times m}$, and $C \in \mathbb{R}^{m \times n}$. Then,

$$\det[sI - (A + BC)] = \det\left[I - C(sI - A)^{-1}B\right]\det(sI - A).$$

If, in addition, $n = m = 1$, then

$$\det[sI - (A + BC)] = \det(sI - A) - C(sI - A)^{\mathrm{A}}B.$$

Remark: The last expression is used in [1034] to compute the frequency response of a transfer function.

Proof: Note that

$$\det\left[I - C(sI - A)^{-1}B\right]\det(sI - A) = \det\left[\begin{array}{cc} sI - A & B \\ C & I \end{array} \right]$$

$$= \det\left[\begin{array}{cc} sI - A & B \\ C & I \end{array} \right]\left[\begin{array}{cc} I & 0 \\ -C & I \end{array} \right]$$

$$= \det\left[\begin{array}{cc} sI - A - BC & B \\ 0 & I \end{array} \right]$$

$$= \det(sI - A - BC).$$

Fact 12.22.14. Let $A \in \mathbb{R}^{n \times n}$, $B \in \mathbb{R}^{n \times m}$, $C \in \mathbb{R}^{m \times n}$, and $K \in \mathbb{R}^{m \times n}$, and assume that $A + BK$ is nonsingular. Then,

$$\det\left[\begin{array}{cc} A & B \\ C & 0 \end{array} \right] = (-1)^m \det(A + BK)\det\left[C(A + BK)^{-1}B\right].$$

Hence, $\left[\begin{smallmatrix} A & B \\ C & 0 \end{smallmatrix} \right]$ is nonsingular if and only if $C(A + BK)^{-1}B$ is nonsingular.

Proof: Note that

$$\det\left[\begin{array}{cc} A & B \\ C & 0 \end{array} \right] = \det\left[\begin{array}{cc} A & B \\ C & 0 \end{array} \right]\left[\begin{array}{cc} I & 0 \\ K & I \end{array} \right]$$

$$= \det\left[\begin{array}{cc} A + BK & B \\ C & 0 \end{array} \right]$$

$$= \det(A + BK)\det\left[-C(A + BK)^{-1}B\right].$$

Fact 12.22.15. Let $A_1 \in \mathbb{R}^{n \times n}$, $C_1 \in \mathbb{R}^{1 \times n}$, $A_2 \in \mathbb{R}^{m \times m}$, and $B_2 \in \mathbb{R}^{m \times 1}$, let $\lambda \in \mathbb{C}$, assume that λ is an observable eigenvalue of (A_1, C_1) and a controllable eigenvalue of (A_2, B_2), and define the dynamics matrix $\mathcal{A}$ of the cascaded system by

$$\mathcal{A} \triangleq \left[\begin{array}{cc} A_1 & 0 \\ B_2 C_1 & A_2 \end{array} \right].$$

Then,

$$\mathrm{amult}_{\mathcal{A}}(\lambda) = \mathrm{amult}_{A_1}(\lambda) + \mathrm{amult}_{A_2}(\lambda)$$

and

$$\text{gmult}_A(\lambda) = 1.$$

Remark: The eigenvalue λ is a cyclic eigenvalue of both subsystems as well as the cascaded system. In other words, λ, which occurs in a single Jordan block of each subsystem, occurs in a single Jordan block in the cascaded system. Effectively, the Jordan blocks of the subsystems corresponding to λ are merged.

Fact 12.22.16. Let $G_1 \in \mathbb{R}^{l_1 \times m}(s)$ and $G_2 \in \mathbb{R}^{l_2 \times m}(s)$ be strictly proper. Then,

$$\left\| \begin{bmatrix} G_1 \\ G_2 \end{bmatrix} \right\|_{\mathrm{H}_2}^2 = \|G_1\|_{\mathrm{H}_2}^2 + \|G_2\|_{\mathrm{H}_2}^2.$$

Fact 12.22.17. Let $G_1, G_2 \in \mathbb{R}^{m \times m}(s)$ be strictly proper. Then,

$$\left\| \begin{bmatrix} G_1 \\ G_2 \end{bmatrix} \right\|_{\mathrm{H}_2} = \left\| \begin{bmatrix} G_1 & G_2 \end{bmatrix} \right\|_{\mathrm{H}_2}.$$

Fact 12.22.18. Let $G(s) \triangleq \frac{\alpha}{s+\beta}$, where $\beta > 0$. Then,

$$\|G\|_{\mathrm{H}_2} = \frac{|\alpha|}{\sqrt{2\beta}}.$$

Fact 12.22.19. Let $G(s) \triangleq \frac{\alpha_1 s + \alpha_0}{s^2 + \beta_1 s + \beta_0}$, where $\beta_0, \beta_1 > 0$. Then,

$$\|G\|_{\mathrm{H}_2} = \sqrt{\frac{\alpha_0^2}{2\beta_0\beta_1} + \frac{\alpha_1^2}{2\beta_1}}.$$

Fact 12.22.20. Let $G_1(s) = \frac{\alpha_1}{s+\beta_1}$ and $G_2(s) = \frac{\alpha_2}{s+\beta_2}$, where $\beta_1, \beta_2 > 0$. Then,

$$\|G_1 G_2\|_{\mathrm{H}_2} \leq \|G_1\|_{\mathrm{H}_2} \|G_2\|_{\mathrm{H}_2}$$

if and only if $\beta_1 + \beta_2 \geq 2$.

Remark: The H_2 norm is not submultiplicative.

12.23 Facts on the Riccati Equation

Fact 12.23.1. Assume that (A, B) is stabilizable, and assume that $\mathcal{H}$ defined by (12.16.8) has an imaginary eigenvalue λ. Then, every Jordan block of $\mathcal{H}$ associated with λ has even order.

Proof: Let P be a solution of (12.16.4), and let $\mathcal{J}$ denote the Jordan form of $A - \Sigma P$. Then, there exists a nonsingular $2n \times 2n$ block-diagonal matrix $\mathcal{S}$ such that $\hat{\mathcal{H}} \triangleq \mathcal{S}^{-1}\mathcal{H}\mathcal{S} = \begin{bmatrix} \mathcal{J} & \hat{\Sigma} \\ 0 & -\mathcal{J}^{\mathrm{T}} \end{bmatrix}$, where $\hat{\Sigma}$ is positive semidefinite. Next, let $\mathcal{J}_\lambda \triangleq \lambda I_r + N_r$ be a Jordan block of $\mathcal{J}$ associated with λ, and consider the submatrix of $\lambda I - \hat{\mathcal{H}}$ consisting of the rows and columns of $\lambda I - \mathcal{J}_\lambda$ and $\lambda I + \mathcal{J}_\lambda^{\mathrm{T}}$. Since (A, B) is stabilizable, it follows that the rank of this submatrix is $2r - 1$. Hence, every Jordan block of $\mathcal{H}$ associated with λ has even order.

Remark: Canonical forms for symplectic and Hamiltonian matrices are discussed in [898].

Fact 12.23.2. Let $A, B \in \mathbb{C}^{n \times n}$, assume that A and B are positive definite, let $S \in \mathbb{C}^{n \times n}$ satisfy $A = S^*S$, and define

$$X \triangleq S^{-1}(SBS^*)^{1/2}S^{-*}.$$

Then, X satisfies $XAX = B$.

Proof: See [701, p. 52].

Fact 12.23.3. Let $A, B \in \mathbb{C}^{n \times n}$, and assume that the $2n \times 2n$ matrix

$$\begin{bmatrix} A & -2I \\ 2B - \frac{1}{2}A^2 & A \end{bmatrix}$$

is simple. Then, there exists a matrix $X \in \mathbb{C}^{n \times n}$ satisfying

$$X^2 + AX + B = 0.$$

Proof: See [1369].

Fact 12.23.4. Let $A, B \in \mathbb{F}^{n \times n}$, and assume that A and B are positive semidefinite. Then, the following statements hold:

i) If A is positive definite, then $X = A \# B$ is the unique positive-definite solution of
$$XA^{-1}X - B = 0.$$

ii) If A is positive definite, then $X = \frac{1}{2}[-A + A\#(A + 4B)]$ is the unique positive-definite solution of
$$XA^{-1}X + X - B = 0.$$

iii) If A is positive definite, then $X = \frac{1}{2}[A + A\#(A + 4B)]$ is the unique positive-definite solution of
$$XA^{-1}X - X - B = 0.$$

iv) If B is positive definite, then $X = A \# B$ is the unique positive-definite solution of
$$XB^{-1}X = A.$$

v) If A is positive definite, then $X = \frac{1}{2}[A + A\#(A + 4BA^{-1}B)]$ is the unique positive-definite solution of
$$BX^{-1}B - X + A = 0.$$

vi) If A is positive definite, then $X = \frac{1}{2}[-A + A\#(A + 4BA^{-1}B)]$ is the unique positive-definite solution of
$$BX^{-1}B - X - A = 0.$$

vii) If $0 < A \leq B$, then $X = \frac{1}{2}[A + A\#(4B - 3A)]$ is the unique positive-definite solution of
$$XA^{-1}X - X - (B - A) = 0.$$

viii) If $0 < A \le B$, then $X = \frac{1}{2}[-A + A\#(4B - 3A)]$ is the unique positive-definite solution of

$$XA^{-1}X + X - (B - A) = 0.$$

ix) If $0 < A < B$, $X(0)$ is positive definite, and $X(t)$ satisfies

$$\dot{X} = -XA^{-1}X + X + (B - A),$$

then

$$\lim_{t \to \infty} X(t) = \frac{1}{2}[A + A\#(4B - 3A)].$$

x) If $0 < A < B$, $X(0)$ is positive definite, and $X(t)$ satisfies

$$\dot{X} = -XA^{-1}X - X + (B - A),$$

then

$$\lim_{t \to \infty} X(t) = \frac{1}{2}[A + A\#(4B - 3A)].$$

xi) If $0 < A < B$, $X(0)$ and $Y(0)$ are positive definite, $X(t)$ satisfies

$$\dot{X} = -XA^{-1}X + X + (B - A)$$

and $Y(t)$ satisfies

$$\dot{Y} = -YA^{-1}Y - Y + (B - A),$$

then

$$\lim_{t \to \infty} X(t)\#Y(t) = A\#(B - A).$$

Proof: See [936].

Remark: $A\#B$ is the geometric mean of A and B. See Fact 8.10.43.

Remark: The solution X given by *vii*) is the *golden mean* of A and B. In the scalar case with $A = 1$ and $B = 2$, the solution X of $X^2 - X - 1 = 0$ is the *golden ratio* $\frac{1}{2}(1 + \sqrt{5})$. See Fact 4.11.13.

Fact 12.23.5. Let $P_0 \in \mathbb{R}^{n \times n}$, assume that P_0 is positive definite, let $V \in \mathbb{R}^{n \times n}$ be positive semidefinite, and, for all $t \ge 0$, let $P(t) \in \mathbb{R}^{n \times n}$ satisfy

$$\dot{P}(t) = A^{\mathrm{T}}P(t) + P(t)A + P(t)VP(t),$$
$$P(0) = P_0.$$

Then, for all $t \ge 0$,

$$P(t) = e^{tA^{\mathrm{T}}}\left[P_0^{-1} - \int_0^t e^{\tau A}Ve^{\tau A^{\mathrm{T}}}\,\mathrm{d}\tau\right]^{-1}e^{tA}.$$

Remark: $P(t)$ satisfies a Riccati differential equation.

Fact 12.23.6. Let $G_c \sim \left[\begin{array}{c|c} A_c & B_c \\ \hline C_c & 0 \end{array}\right]$ denote an nth-order dynamic controller for the standard control problem. If G_c minimizes $\|\tilde{\mathcal{G}}\|_2$, then G_c is given by

$$A_c \triangleq A + BC_c - B_cC - B_cDC_c,$$
$$B_c \triangleq \left(QC^T + V_{12}\right)V_2^{-1},$$
$$C_c \triangleq -R_2^{-1}(B^TP + R_{12}^T),$$

where P and Q are positive-semidefinite solutions to the algebraic Riccati equations

$$\hat{A}_R^TP + P\hat{A}_R - PBR_2^{-1}B^TP + \hat{R}_1 = 0,$$
$$\hat{A}_EQ + Q\hat{A}_E^T - QC^TV_2^{-1}CQ + \hat{V}_1 = 0,$$

where $\hat{A}_R$ and $\hat{R}_1$ are defined by

$$\hat{A}_R \triangleq A - BR_2^{-1}R_{12}^T, \quad \hat{R}_1 \triangleq R_1 - R_{12}R_2^{-1}R_{12}^T,$$

and $\hat{A}_E$ and $\hat{V}_1$ are defined by

$$\hat{A}_E \triangleq A - V_{12}V_2^{-1}C, \quad \hat{V}_1 \triangleq V_1 - V_{12}V_2^{-1}V_{12}^T.$$

Furthermore, the eigenvalues of the closed-loop system are given by

$$\text{mspec}\left(\left[\begin{array}{cc} A & BC_c \\ B_cC & A_c + B_cDC_c \end{array}\right]\right) = \text{mspec}(A + BC_c) \cup \text{mspec}(A - B_cC).$$

Fact 12.23.7. Let $G_c \sim \left[\begin{array}{c|c} A_c & B_c \\ \hline C_c & 0 \end{array}\right]$ denote an nth-order dynamic controller for the discrete-time standard control problem. If G_c minimizes $\|\tilde{\mathcal{G}}\|_2$, then G_c is given by

$$A_c \triangleq A + BC_c - B_cC - B_cDC_c,$$
$$B_c \triangleq \left(AQC^T + V_{12}\right)\left(V_2 + CQC^T\right)^{-1},$$
$$C_c \triangleq -\left(R_2 + B^TPB\right)^{-1}\left(R_{12}^T + B^TPA\right),$$

and the eigenvalues of the closed-loop system are given by

$$\text{mspec}\left(\left[\begin{array}{cc} A & BC_c \\ B_cC & A_c + B_cDC_c \end{array}\right]\right) = \text{mspec}(A + BC_c) \cup \text{mspec}(A - B_cC).$$

Now, assume in addition that $D = 0$ and $G_c \sim \left[\begin{array}{c|c} A_c & B_c \\ \hline C_c & D_c \end{array}\right]$. Then,

$$A_c \triangleq A + BC_c - B_cC - BD_cC,$$
$$B_c \triangleq \left(AQC^T + V_{12}\right)\left(V_2 + CQC^T\right)^{-1} + BD_c,$$
$$C_c \triangleq -\left(R_2 + B^TPB\right)^{-1}\left(R_{12}^T + B^TPA\right) - D_cC,$$
$$D_c \triangleq \left(R_2 + B^TPB\right)^{-1}\left[B^TPAQC^T + R_{12}^TQC^T + B^TPV_{12}\right]\left(V_2 + CQC^T\right)^{-1},$$

and the eigenvalues of the closed-loop system are given by

$$\text{mspec}\left(\left[\begin{array}{cc} A + BD_cC & BC_c \\ B_cC & A_c \end{array}\right]\right) = \text{mspec}(A + BC_c) \cup \text{mspec}(A - B_cC).$$

In both cases, P and Q are positive-semidefinite solutions to the discrete-time algebraic Riccati equations

$$P = \hat{A}_{\mathrm{R}}^{\mathrm{T}} P \hat{A}_{\mathrm{R}} - \hat{A}_{\mathrm{R}}^{\mathrm{T}} PB(R_2 + B^{\mathrm{T}} PB)^{-1} B^{\mathrm{T}} P \hat{A}_{\mathrm{R}} + \hat{R}_1,$$

$$Q = \hat{A}_{\mathrm{E}} Q \hat{A}_{\mathrm{E}}^{\mathrm{T}} - \hat{A}_{\mathrm{E}} Q C^{\mathrm{T}} (V_2 + C Q C^{\mathrm{T}})^{-1} C Q \hat{A}_{\mathrm{E}}^{\mathrm{T}} + \hat{V}_1,$$

where $\hat{A}_{\mathrm{R}}$ and $\hat{R}_1$ are defined by

$$\hat{A}_{\mathrm{R}} \triangleq A - B R_2^{-1} R_{12}^{\mathrm{T}}, \quad \hat{R}_1 \triangleq R_1 - R_{12} R_2^{-1} R_{12}^{\mathrm{T}},$$

and $\hat{A}_{\mathrm{E}}$ and $\hat{V}_1$ are defined by

$$\hat{A}_{\mathrm{E}} \triangleq A - V_{12} V_2^{-1} C, \quad \hat{V}_1 \triangleq V_1 - V_{12} V_2^{-1} V_{12}^{\mathrm{T}}.$$

Proof: See [633].

12.24 Notes

Linear system theory is treated in [265, 1179, 1368, 1485]. Time-varying linear systems are considered in [375, 1179], while discrete-time systems are emphasized in [678]. The PBH test is due to [673]. Spectral factorization results are given in [345]. Stabilization aspects are discussed in [439]. Observable asymptotic stability and controllable asymptotic stability were introduced and used to analyze Lyapunov equations in [1238]. Zeros are treated in [23, 491, 809, 813, 968, 1101, 1184, 1209]. Matrix-based methods for linear system identification are developed in [1397], while stochastic theory is considered in [648].

Solutions of the LQR problem under weak conditions are given in [558]. Solutions of the Riccati equation are considered in [576, 869, 872, 889, 890, 999, 1151, 1468, 1475, 1480]. Proposition 12.16.16 is based on Theorem 3.6 of [1490, p. 79]. A variation of Theorem 12.18.1 is given without proof by Theorem 7.2.1 of [771, p. 125].

There are numerous extensions to the results given in this chapter relating to various generalizations of (12.16.4). These generalizations include the case in which R_1 is indefinite [575, 1472, 1474] as well as the case in which Σ is indefinite [1196]. The latter case is relevant to H_∞ optimal control theory [192]. Additional extensions include the Riccati inequality $A^{\mathrm{T}} P + PA + R_1 - P\Sigma P \geq 0$ [1143, 1195, 1196, 1197], the discrete-time Riccati equation [9, 679, 764, 889, 1143, 1479], and fixed-order control [759].

Bibliography

[1] A. Abdessemed and E. B. Davies, "Some Commutator Estimates in the Schatten Classes," *J. London Math. Soc.*, Vol. s2-39, pp. 299–308, 1989. (Cited on page 641.)

[2] R. Ablamowicz, "Matrix Exponential via Clifford Algebras," *J. Nonlin. Math. Phys.*, Vol. 5, pp. 294–313, 1998. (Cited on page 737.)

[3] H. Abou-Kandil, G. Freiling, V. Ionescu, and G. Jank, *Matrix Riccati Equations in Control and Systems Theory.* Basel: Birkhauser, 2003. (Cited on page xvii.)

[4] S. Abramovich, J. Baric, M. Matic, and J. Pecaric, "On Van De Lune-Alzer's Inequality," *J. Math. Ineq.*, Vol. 1, pp. 563–587, 2007. (Cited on page 32.)

[5] L. Aceto and D. Trigiante, "The Matrices of Pascal and Other Greats," *Amer. Math. Monthly*, Vol. 108, pp. 232–245, 2001. (Cited on pages 387, 396, 491, and 737.)

[6] S. Afriat, "Orthogonal and Oblique Projectors and the Characteristics of Pairs of Vector Spaces," *Proc. Cambridge Phil. Soc.*, Vol. 53, pp. 800–816, 1957. (Cited on page 228.)

[7] R. Agaev and P. Chebotarev, "On the Spectra of Nonsymmetric Laplacian Matrices," *Lin. Alg. Appl.*, Vol. 399, pp. 157–168, 2005. (Cited on page 777.)

[8] R. P. Agarwal, *Difference Equations and Inequalities: Theory, Methods, and Applications*, 2nd ed. New York: Marcel Dekker, 2000. (Cited on page 793.)

[9] C. D. Ahlbrandt and A. C. Peterson, *Discrete Hamiltonian Systems: Difference Equations, Continued Fractions, and Riccati Equations.* Dordrecht: Kluwer, 1996. (Cited on page 879.)

[10] E. Ahn, S. Kim, and Y. Lim, "An Extended Lie-Trotter Formula and Its Application," *Lin. Alg. Appl.*, Vol. 427, pp. 190–196, 2007. (Cited on page 749.)

[11] A. C. Aitken, *Determinants and Matrices*, 9th ed. Edinburgh: Oliver and Boyd, 1956. (Cited on page xix.)

[12] M. Aivazis, *Group Theory in Physics—Problems and Solutions.* Singapore: World Scientific, 1991. (Cited on page 211.)

[13] M. Al-Ahmar, "An Identity of Jacobi," *Amer. Math. Monthly*, Vol. 103, pp. 78–79, 1996. (Cited on page 207.)

[14] A. A. Albert and B. Muckenhoupt, "On Matrices of Trace Zero," *Michigan Math. J.*, Vol. 4, pp. 1–3, 1957. (Cited on page 341.)

[15] A. E. Albert, "Conditions for Positive and Nonnegative Definiteness in Terms of Pseudoinverses," *SIAM J. Appl. Math.*, Vol. 17, pp. 434–440, 1969. (Cited on page 595.)

[16] A. E. Albert, *Regression and the Moore-Penrose Pseudoinverse.* New York: Academic Press, 1972. (Cited on pages 415, 418, 421, 427, and 678.)

[17] J. M. Aldaz, "A Refinement of the Inequality between Arithmetic and Geometric Means," *J. Math. Ineq.*, Vol. 2, pp. 473–477, 2007. (Cited on page 62.)

[18] R. Aldrovandi, *Special Matrices of Mathematical Physics: Stochastic, Circulant and Bell Matrices.* Singapore: World Scientific, 2001. (Cited on pages xvii, 299, and 390.)

[19] M. Aleksiejczuk and A. Smoktunowicz, "On Properties of Quadratic Matrices," *Mathematica Pannonica*, Vol. 112, pp. 239–248, 2000. (Cited on page 429.)

[20] A. Y. Alfakih, "On the Nullspace, the Rangespace and the Characteristic Polynomial of Euclidean Distance Matrices," *Lin. Alg. Appl.*, Vol. 416, pp. 348–354, 2006. (Cited on page 629.)

[21] M. Alic, P. S. Bullen, J. E. Pecaric, and V. Volenec, "On the Geometric-Arithmetic Mean Inequality for Matrices," *Mathematical Communications*, Vol. 2, pp. 125–128, 1997. (Cited on page 58.)

[22] M. Alic, B. Mond, J. E. Pecaric, and V. Volenec, "Bounds for the Differences of Matrix Means," *SIAM J. Matrix Anal. Appl.*, Vol. 18, pp. 119–123, 1997. (Cited on page 510.)

[23] H. Aling and J. M. Schumacher, "A Nine-Fold Decomposition for Linear Systems," *Int. J. Contr.*, Vol. 39, pp. 779–805, 1984. (Cited on page 879.)

[24] G. Alpargu and G. P. H. Styan, "Some Remarks and a Bibliography on the Kantorovich Inequality," in *Multidimensional Statistical Analysis and Theory of Random Matrices*, A. K. Gupta and V. L. Girko, Eds. Utrecht: VSP, 1996, pp. 1–13. (Cited on page 551.)

[25] R. C. Alperin, "The Matrix of a Rotation," *College Math. J.*, Vol. 20, p. 230, 1989. (Cited on page 215.)

[26] C. Alsina and R. Ger, "On Some Inequalities and Stability Results Related to the Exponential Function," *J. Ineq. Appl.*, Vol. 2, pp. 373–380, 1998. (Cited on page 37.)

[27] C. Alsina and R. B. Nelsen, "On Candido's Identity," *Math. Mag.*, Vol. 80, pp. 226–228, 2007. (Cited on page 35.)

[28] S. L. Altmann, *Rotations, Quaternions, and Double Groups.* New York: Oxford University Press, 1986. (Cited on pages xvii, 249, and 251.)

[29] S. L. Altmann, "Hamilton, Rodrigues, and the Quaternion Scandal," *Math. Mag.*, Vol. 62, pp. 291–308, 1989. (Cited on pages 214 and 215.)

[30] H. Alzer, "A Lower Bound for the Difference between the Arithmetic and Geometric Means," *Nieuw. Arch. Wisk.*, Vol. 8, pp. 195–197, 1990. (Cited on page 56.)

[31] S. Amghibech, "Problem 11296," *Amer. Math. Monthly*, Vol. 114, p. 452, 2007. (Cited on page 536.)

[32] B. D. O. Anderson, "Orthogonal Decomposition Defined by a Pair of Skew-Symmetric Forms," *Lin. Alg. Appl.*, Vol. 8, pp. 91–93, 1974. (Cited on pages xvii and 375.)

[33] B. D. O. Anderson, "Weighted Hankel-Norm Approximation: Calculation of Bounds," *Sys. Contr. Lett.*, Vol. 7, pp. 247–255, 1986. (Cited on page 552.)

[34] B. D. O. Anderson, E. I. Jury, and M. Mansour, "Schwarz Matrix Properties for Continuous and Discrete Time Systems," *Int. J. Contr.*, Vol. 23, pp. 1–16, 1976. (Cited on page 772.)

[35] B. D. O. Anderson and J. B. Moore, "A Matrix Kronecker Lemma," *Lin. Alg. Appl.*, Vol. 15, pp. 227–234, 1976. (Cited on page 511.)

[36] G. Anderson, M. Vamanamurthy, and M. Vuorinen, "Monotonicity Rules in Calculus," *Amer. Math. Monthly*, Vol. 113, pp. 805–816, 2006. (Cited on pages 25, 30, and 32.)

[37] T. W. Anderson and I. Olkin, "An Extremal Problem for Positive Definite Matrices," *Lin. Multilin. Alg.*, Vol. 6, pp. 257–262, 1978. (Cited on page 504.)

[38] W. N. Anderson, "Shorted Operators," *SIAM J. Appl. Math.*, Vol. 20, pp. 520–525, 1971. (Cited on pages 582 and 583.)

[39] W. N. Anderson and R. J. Duffin, "Series and Parallel Addition of Matrices," *J. Math. Anal. Appl.*, Vol. 26, pp. 576–594, 1969. (Cited on page 582.)

[40] W. N. Anderson, E. J. Harner, and G. E. Trapp, "Eigenvalues of the Difference and Product of Projections," *Lin. Multilin. Alg.*, Vol. 17, pp. 295–299, 1985. (Cited on page 366.)

[41] W. N. Anderson and M. Schreiber, "On the Infimum of Two Projections," *Acta Sci. Math.*, Vol. 33, pp. 165–168, 1972. (Cited on pages 421, 422, and 506.)

[42] W. N. Anderson and G. E. Trapp, "Shorted Operators II," *SIAM J. Appl. Math.*, Vol. 28, pp. 60–71, 1975. (Cited on pages 486, 582, 583, and 595.)

[43] W. N. Anderson and G. E. Trapp, "Inverse Problems for Means of Matrices," *SIAM J. Alg. Disc. Meth.*, Vol. 7, pp. 188–192, 1986. (Cited on page 510.)

[44] W. N. Anderson and G. E. Trapp, "Symmetric Positive Definite Matrices," *Amer. Math. Monthly*, Vol. 95, pp. 261–262, 1988. (Cited on page 522.)

[45] T. Ando, "Concavity of Certain Maps on Positive Definite Matrices and Applications to Hadamard Products," *Lin. Alg. Appl.*, Vol. 26, pp. 203–241, 1979. (Cited on pages 482, 589, 591, 594, 596, and 752.)

[46] T. Ando, "Inequalities for M-Matrices," *Lin. Multilin. Alg.*, Vol. 8, pp. 291–316, 1980. (Cited on page 588.)

[47] T. Ando, "On the Arithmetic-Geometric-Harmonic-Mean Inequalities for Positive Definite Matrices," *Lin. Alg. Appl.*, Vol. 52/53, pp. 31–37, 1983. (Cited on page 509.)

[48] T. Ando, "On Some Operator Inequalities," *Math. Ann.*, Vol. 279, pp. 157–159, 1987. (Cited on pages 509 and 510.)

[49] T. Ando, "Majorizations and Inequalities in Matrix Theory," *Lin. Alg. Appl.*, Vol. 199, pp. 17–67, 1994. (Cited on pages 520, 539, 640, 645, 657, 671, 756, and 761.)

[50] T. Ando, "Majorization Relations for Hadamard Products," *Lin. Alg. Appl.*, Vol. 223–224, pp. 57–64, 1995. (Cited on page 589.)

[51] T. Ando, "Matrix Young Inequalities," *Oper. Theory Adv. Appl.*, Vol. 75, pp. 33–38, 1995. (Cited on page 671.)

[52] T. Ando, "Problem of Infimum in the Positive Cone," in *Analytic and Geometric Inequalities and Applications*, T. M. Rassias and H. M. Srivastava, Eds. Dordrecht: Kluwer, 1999, pp. 1–12. (Cited on pages 506 and 583.)

[53] T. Ando, "Lowner Inequality of Indefinite Type," *Lin. Alg. Appl.*, Vol. 385, pp. 73–80, 2004. (Cited on pages 503 and 868.)

[54] T. Ando and R. Bhatia, "Eigenvalue Inequalities Associated with the Cartesian Decomposition," *Lin. Multilin. Alg.*, Vol. 22, pp. 133–147, 1987. (Cited on page 566.)

[55] T. Ando and F. Hiai, "Log-Majorization and Complementary Golden-Thompson Type Inequalities," *Lin. Alg. Appl.*, Vol. 197/198, pp. 113–131, 1994. (Cited on pages 512, 749, 753, and 761.)

[56] T. Ando and F. Hiai, "Hölder Type Inequalities for Matrices," *Math. Ineq. Appl.*, Vol. 1, pp. 1–30, 1998. (Cited on pages 485, 506, and 507.)

[57] T. Ando, F. Hiai, and K. Okubo, "Trace Inequalities for Multiple Products of Two Matrices," *Math. Ineq. Appl.*, Vol. 3, pp. 307–318, 2000. (Cited on pages 524 and 526.)

[58] T. Ando, R. A. Horn, and C. R. Johnson, "The Singular Values of a Hadamard Product: A Basic Inequality," *Lin. Multilin. Alg.*, Vol. 21, pp. 345–365, 1987. (Cited on page 674.)

[59] T. Ando, C.-K. Li, and R. Mathias, "Geometric Means," *Lin. Alg. Appl.*, Vol. 385, pp. 305–334, 2004. (Cited on page 510.)

[60] T. Ando and X. Zhan, "Norm Inequalities Related to Operator Monotone Functions," *Math. Ann.*, Vol. 315, pp. 771–780, 1999. (Cited on pages 566, 647, and 763.)

[61] T. Andreescu and D. Andrica, *Complex Numbers from A to Z*. Boston: Birkhauser, 2006. (Cited on pages 76, 77, 168, 169, 171, 172, and 173.)

[62] T. Andreescu and Z. Feng, *103 Trigonometry Problems*. Boston: Birkhauser, 2004. (Cited on page 172.)

[63] G. E. Andrews and K. Eriksson, *Integer Partitions*. Cambridge: Cambridge University Press, 2004. (Cited on page 391.)

[64] E. Andruchow, G. Corach, and D. Stojanoff, "Geometric Operator Inequalities," *Lin. Alg. Appl.*, Vol. 258, pp. 295–310, 1997. (Cited on pages 632 and 646.)

[65] E. Angel, *Interactive Computer Graphics*, 3rd ed. Reading: Addison-Wesley, 2002. (Cited on pages xvii and 215.)

[66] Anonymous, "Like the Carlson's inequality," http://www.mathlinks.ro/viewtopic.php?t=151210. (Cited on page 46.)

[67] A. C. Antoulas and D. C. Sorensen, "Lyapunov, Lanczos, and Inertia," *Lin. Alg. Appl.*, Vol. 326, pp. 137–150, 2001. (Cited on page 867.)

[68] J. D. Aplevich, *Implicit Linear Systems*. Berlin: Springer, 1991. (Cited on pages xvii and 331.)

[69] T. Apostol, *Introduction to Analytic Number Theory*. New York: Springer, 1998. (Cited on page 491.)

[70] T. M. Apostol, "Some Explicit Formulas for the Exponential Matrix," *Amer. Math. Monthly*, Vol. 76, pp. 289–292, 1969. (Cited on page 737.)

[71] T. M. Apostol, Ed., *Mathematical Analysis*, 2nd ed. Reading: Addison Wesley, 1974. (Cited on pages xxxv, 14, 68, 74, 693, 695, and 696.)

[72] H. Araki, "On an Inequality of Lieb and Thirring," *Lett. Math. Phys.*, Vol. 19, pp. 167–170, 1990. (Cited on pages 528 and 639.)

[73] H. Araki and S. Yamagami, "An Inequality for Hilbert-Schmidt Norm," *Comm. Math. Phys.*, Vol. 81, pp. 89–96, 1981. (Cited on page 640.)

[74] J. Araujo and J. D. Mitchell, "An Elementary Proof that Every Singular Matrix Is a Product of Idempotent Matrices," *Amer. Math. Monthly*, Vol. 112, pp. 641–645, 2005. (Cited on page 383.)

[75] A. Arimoto, "A Simple Proof of the Classification of Normal Toeplitz Matrices," *Elec. J. Lin. Alg.*, Vol. 9, pp. 108–111, 2002. (Cited on page 390.)

[76] T. Arponen, "A Matrix Approach to Polynomials," *Lin. Alg. Appl.*, Vol. 359, pp. 181–196, 2003. (Cited on page 737.)

[77] V. Arsigny, P. Fillard, X. Pennec, and N. Ayache, "Geometric Means in a Novel Vector Space Structure on Symmetric Positive-Definite Matrices," *SIAM J. Matrix Anal. Appl.*, Vol. 29, pp. 328–347, 2007. (Cited on page 753.)

[78] M. Artin, *Algebra*. Englewood Cliffs: Prentice-Hall, 1991. (Cited on pages 245 and 252.)

[79] M. Artzrouni, "A Theorem on Products of Matrices," *Lin. Alg. Appl.*, Vol. 49, pp. 153–159, 1983. (Cited on page 786.)

[80] A. Arvanitoyeorgos, *An Introduction to Lie Groups and the Geometry of Homogeneous Spaces*. Providence: American Mathematical Society, 2003. (Cited on page 252.)

[81] H. Aslaksen, "SO(2) Invariants of a Set of 2×2 Matrices," *Math. Scand.*, Vol. 65, pp. 59–66, 1989. (Cited on page 283.)

[82] H. Aslaksen, "Laws of Trigonometry on SU(3)," *Trans. Amer. Math. Soc.*, Vol. 317, pp. 127–142, 1990. (Cited on pages 165, 198, and 284.)

[83] H. Aslaksen, "Quaternionic Determinants," *Mathematical Intelligencer*, Vol. 18, no. 3, pp. 57–65, 1996. (Cited on pages 249 and 251.)

[84] H. Aslaksen, "Defining Relations of Invariants of Two 3×3 Matrices," *J. Algebra*, Vol. 298, pp. 41–57, 2006. (Cited on page 283.)

[85] B. A. Asner, "On the Total Nonnegativity of the Hurwitz Matrix," *SIAM J. Appl. Math.*, Vol. 18, pp. 407–414, 1970. (Cited on pages 765 and 770.)

[86] Y.-H. Au-Yeung, "A Note on Some Theorems on Simultaneous Diagonalization of Two Hermitian Matrices," *Proc. Cambridge Phil. Soc.*, Vol. 70, pp. 383–386, 1971. (Cited on pages 554 and 558.)

[87] Y.-H. Au-Yeung, "Some Inequalities for the Rational Power of a Nonnegative Definite Matrix," *Lin. Alg. Appl.*, Vol. 7, pp. 347–350, 1973. (Cited on page 503.)

[88] Y.-H. Au-Yeung, "On the Semi-Definiteness of the Real Pencil of Two Hermitian Matrices," *Lin. Alg. Appl.*, Vol. 10, pp. 71–76, 1975. (Cited on page 396.)

[89] K. M. R. Audenaert, "A Norm Compression Inequality for Block Partitioned Positive Semidefinite Matrices," *Lin. Alg. Appl.*, Vol. 413, pp. 155–176, 2006. (Cited on page 652.)

[90] K. M. R. Audenaert, "On a Norm Compression Inequality for $2 \times N$ Partitioned Block Matrices," *Lin. Alg. Appl.*, Vol. 428, pp. 781–795, 2008. (Cited on page 653.)

[91] K. M. R. Audenaert, "On the Araki-Lieb-Thirring Inequality," *Int. J. Inform. Sys. Sci.*, Vol. 4, pp. 78–83, 2008. (Cited on page 528.)

[92] J. S. Aujla, "Some Norm Inequalities for Completely Monotone Functions," *SIAM J. Matrix Anal. Appl.*, Vol. 22, pp. 569–573, 2000. (Cited on page 566.)

[93] J. S. Aujla and J.-C. Bourin, "Eigenvalue Inequalities for Convex and Log-Convex Functions," *Lin. Alg. Appl.*, Vol. 424, pp. 25–35, 2007. (Cited on pages 504, 574, 591, 639, and 649.)

[94] J. S. Aujla and F. C. Silva, "Weak Majorization Inequalities and Convex Functions," *Lin. Alg. Appl.*, Vol. 369, pp. 217–233, 2003. (Cited on pages 565 and 566.)

[95] J. S. Aujla and H. L. Vasudeva, "Inequalities Involving Hadamard Product and Operator Means," *Math. Japonica*, Vol. 42, pp. 265–272, 1995. (Cited on pages 586, 592, and 595.)

[96] B. Aupetit and J. Zemanek, "A Characterization of Normal Matrices by Their Exponentials," *Lin. Alg. Appl.*, Vol. 132, pp. 119–121, 1990. (Cited on page 762.)

[97] M. Avriel, *Nonlinear Programming: Analysis and Methods.* Englewood Cliffs: Prentice-Hall, 1976, reprinted by Dover, Mineola, 2003. (Cited on page 555.)

[98] O. Axelsson, *Iterative Solution Methods.* Cambridge: Cambridge University Press, 1994. (Cited on page xviii.)

[99] L. E. Azar, "On Some Extensions of Hardy-Hilbert's Inequality and Applications," *J. Ineq. Appl.*, pp. 1–14, 2008, article ID 546829. (Cited on page 69.)

[100] J. C. Baez, "Symplectic, Quaternionic, Fermionic," http://math.ucr.edu/home /baez/symplectic.html. (Cited on pages 247 and 250.)

[101] J. C. Baez, "The Octonions," *Bull. Amer. Math. Soc.*, Vol. 39, pp. 145–205, 2001. (Cited on page 249.)

[102] O. Bagdasar, *Inequalities and Applications.* Cluj-Napoca: Babes-Bolyai University, 2006, Bachelor's Degree Thesis, http://rgmia.vu.edu.au/monographs/index.html. (Cited on pages 24, 40, 47, and 59.)

[103] Z. Bai and G. Golub, "Bounds for the Trace of the Inverse and the Determinant of Symmetric Positive Definite Matrices," *Ann. Numer. Math.*, Vol. 4, pp. 29–38, 1997. (Cited on page 523.)

[104] D. W. Bailey and D. E. Crabtree, "Bounds for Determinants," *Lin. Alg. Appl.*, Vol. 2, pp. 303–309, 1969. (Cited on page 294.)

[105] H. Bailey and Y. A. Rubinstein, "A Variety of Triangle Inequalities," *College Math. J.*, Vol. 31, pp. 350–355, 2000. (Cited on page 171.)

[106] A. Baker, *Matrix Groups: An Introduction to Lie Group Theory.*   New York: Springer, 2001. (Cited on pages 198, 239, 248, 252, 379, 724, and 748.)

[107] J. K. Baksalary and O. M. Baksalary, "Nonsingularity of Linear Combinations of Idempotent Matrices," *Lin. Alg. Appl.*, Vol. 388, pp. 25–29, 2004. (Cited on pages 219 and 221.)

[108] J. K. Baksalary and O. M. Baksalary, "Particular Formulae for the Moore-Penrose Inverse of a Columnwise Partitioned Matrix," *Lin. Alg. Appl.*, Vol. 421, pp. 16–23, 2007. (Cited on page 429.)

[109] J. K. Baksalary, O. M. Baksalary, and X. Liu, "Further Relationships between Certain Partial Orders of Matrices and their Squares," *Lin. Alg. Appl.*, Vol. 375, pp. 171–180, 2003. (Cited on page 130.)

[110] J. K. Baksalary, O. M. Baksalary, and T. Szulc, "Properties of Schur Complements in Partitioned Idempotent Matrices," *Lin. Alg. Appl.*, Vol. 397, pp. 303–318, 2004. (Cited on page 425.)

[111] J. K. Baksalary, O. M. Baksalary, and G. Trenkler, "A Revisitation of Formulae for the Moore-Penrose Inverse of Modified Matrices," *Lin. Alg. Appl.*, Vol. 372, pp. 207–224, 2003. (Cited on page 412.)

[112] J. K. Baksalary, K. Nordstrom, and G. P. H. Styan, "Lowner-Ordering Antitonicity of Generalized Inverses of Hermitian Matrices," *Lin. Alg. Appl.*, Vol. 127, pp. 171–182, 1990. (Cited on pages 337, 503, and 580.)

[113] J. K. Baksalary and F. Pukelsheim, "On the Lowner, Minus, and Star Partial Orderings of Nonnegative Definite Matrices," *Lin. Alg. Appl.*, Vol. 151, pp. 135–141, 1990. (Cited on pages 514, 576, and 579.)

[114] J. K. Baksalary, F. Pukelsheim, and G. P. H. Styan, "Some Properties of Matrix Partial Orderings," *Lin. Alg. Appl.*, Vol. 119, pp. 57–85, 1989. (Cited on pages 130, 579, and 589.)

[115] J. K. Baksalary and G. P. H. Styan, "Generalized Inverses of Bordered Matrices," *Lin. Alg. Appl.*, Vol. 354, pp. 41–47, 2002. (Cited on page 429.)

[116] O. M. Baksalary, "An Inequality Involving a Product of Two Orthogonal Projectors," *IMAGE*, Vol. 41, p. 44, 2008. (Cited on page 532.)

[117] O. M. Baksalary, A. Mickiewicz, and G. Trenkler, "Rank of a Nonnegative Definite Matrix," *IMAGE*, Vol. 39, pp. 27–28, 2007. (Cited on page 578.)

[118] O. M. Baksalary and G. Trenkler, "Rank of a Generalized Projector," *IMAGE*, Vol. 39, pp. 25–27, 2007. (Cited on page 192.)

[119] O. M. Baksalary and G. Trenkler, "Characterizations of EP, Normal, and Hermitian Matrices," *Lin. Multilin. Alg.*, Vol. 56, pp. 299–304, 2008. (Cited on pages 194, 195, 344, 407, 436, 437, and 438.)

[120] K. Ball, E. Carlen, and E. Lieb, "Sharp Uniform Convexity and Smoothness Inequalities for Trace Norms," *Invent. Math.*, Vol. 115, pp. 463–482, 1994. (Cited on pages 625 and 643.)

[121] C. S. Ballantine, "Products of Positive Semidefinite Matrices," *Pac. J. Math.*, Vol. 23, pp. 427–433, 1967. (Cited on pages 382 and 596.)

[122] C. S. Ballantine, "Products of Positive Definite Matrices II," *Pac. J. Math.*, Vol. 24, pp. 7–17, 1968. (Cited on page 596.)

[123] C. S. Ballantine, "Products of Positive Definite Matrices III," *J. Algebra*, Vol. 10, pp. 174–182, 1968. (Cited on page 596.)

[124] C. S. Ballantine, "A Note on the Matrix Equation $H = AP + PA^*$," *Lin. Alg. Appl.*, Vol. 2, pp. 37–47, 1969. (Cited on pages 341 and 869.)

[125] C. S. Ballantine, "Products of Positive Definite Matrices IV," *Lin. Alg. Appl.*, Vol. 3, pp. 79–114, 1970. (Cited on page 596.)

[126] C. S. Ballantine, "Products of EP Matrices," *Lin. Alg. Appl.*, Vol. 12, pp. 257–267, 1975. (Cited on page 192.)

[127] C. S. Ballantine, "Some Involutory Similarities," *Lin. Multilin. Alg.*, Vol. 3, pp. 19–23, 1975. (Cited on page 383.)

[128] C. S. Ballantine, "Products of Involutory Matrices I," *Lin. Multilin. Alg.*, Vol. 5, pp. 53–62, 1977. (Cited on page 383.)

[129] C. S. Ballantine, "Products of Idempotent Matrices," *Lin. Alg. Appl.*, Vol. 19, pp. 81–86, 1978. (Cited on page 383.)

[130] C. S. Ballantine and C. R. Johnson, "Accretive Matrix Products," *Lin. Multilin. Alg.*, Vol. 3, pp. 169–185, 1975. (Cited on page 770.)

[131] S. Banerjee, "Revisiting Spherical Trigonometry with Orthogonal Projectors," *College Math. J.*, Vol. 35, pp. 375–381, 2004. (Cited on page 175.)

[132] J. Bang-Jensen and G. Gutin, *Digraphs: Theory, Algorithms and Applications.* New York: Springer, 2001. (Cited on page xvii.)

[133] W. Bani-Domi and F. Kittaneh, "Norm Equalities and Inequalities for Operator Matrices," *Lin. Alg. Appl.*, Vol. 429, pp. 57–67, 2008. (Cited on pages 650 and 651.)

[134] R. B. Bapat, "Two Inequalities for the Perron Root," *Lin. Alg. Appl.*, Vol. 85, pp. 241–248, 1987. (Cited on pages 73 and 305.)

[135] R. B. Bapat, "On Generalized Inverses of Banded Matrices," *Elec. J. Lin. Alg.*, Vol. 16, pp. 284–290, 2007. (Cited on page 237.)

[136] R. B. Bapat and M. K. Kwong, "A Generalization of $A \circ A^{-1} \geq I$," *Lin. Alg. Appl.*, Vol. 93, pp. 107–112, 1987. (Cited on page 586.)

[137] R. B. Bapat and T. E. S. Raghavan, *Nonnegative Matrices and Applications.* Cambridge: Cambridge University Press, 1997. (Cited on page 299.)

[138] R. B. Bapat and B. Zheng, "Generalized Inverses of Bordered Matrices," *Elec. J. Lin. Alg.*, Vol. 10, pp. 16–30, 2003. (Cited on pages 134 and 429.)

[139] I. Y. Bar-Itzhack, D. Hershkowitz, and L. Rodman, "Pointing in Real Euclidean Space," *J. Guid. Contr. Dyn.*, Vol. 20, pp. 916–922, 1997. (Cited on pages 212 and 738.)

[140] E. J. Barbeau, *Polynomials.* New York: Springer, 1989. (Cited on pages 34, 43, 46, 51, 52, 171, 778, and 779.)

[141] E. R. Barnes and A. J. Hoffman, "On Bounds for Eigenvalues of Real Symmetric Matrices," *Lin. Alg. Appl.*, Vol. 40, pp. 217–223, 1981. (Cited on page 295.)

[142] S. Barnett, "A Note on the Bezoutian Matrix," *SIAM J. Appl. Math.*, Vol. 22, pp. 84–86, 1972. (Cited on page 279.)

[143] S. Barnett, "Congenial Matrices," *Lin. Alg. Appl.*, Vol. 41, pp. 277–298, 1981. (Cited on pages 387 and 396.)

[144] S. Barnett, "Inversion of Partitioned Matrices with Patterned Blocks," *Int. J. Sys. Sci.*, Vol. 14, pp. 235–237, 1983. (Cited on page 234.)

[145] S. Barnett, *Polynomials and Linear Control Systems*. New York: Marcel Dekker, 1983. (Cited on pages 396 and 871.)

[146] S. Barnett, *Matrices in Control Theory*, revised ed. Malabar: Krieger, 1984. (Cited on page xvii.)

[147] S. Barnett, "Leverrier's Algorithm: A New Proof and Extensions," *SIAM J. Matrix Anal. Appl.*, Vol. 10, pp. 551–556, 1989. (Cited on page 307.)

[148] S. Barnett, *Matrices: Methods and Applications*. Oxford: Clarendon, 1990. (Cited on pages 158, 386, 396, 411, 430, 499, and 679.)

[149] S. Barnett and P. Lancaster, "Some Properties of the Bezoutian for Polynomial Matrices," *Lin. Multilin. Alg.*, Vol. 9, pp. 99–110, 1980. (Cited on page 279.)

[150] S. Barnett and C. Storey, *Matrix Methods in Stability Theory*. New York: Barnes and Noble, 1970. (Cited on pages xvii, 147, 396, 494, and 771.)

[151] W. Barrett, "A Theorem on Inverses of Tridiagonal Matrices," *Lin. Alg. Appl.*, Vol. 27, pp. 211–217, 1979. (Cited on page 238.)

[152] W. Barrett, "Hermitian and Positive Definite Matrices," in *Handbook of Linear Algebra*, L. Hogben, Ed. Boca Raton: Chapman & Hall/CRC, 2007, pp. 8-1-8-12. (Cited on page 592.)

[153] W. Barrett, C. R. Johnson, and P. Tarazaga, "The Real Positive Definite Completion Problem for a Simple Cycle," *Lin. Alg. Appl.*, Vol. 192, pp. 3–31, 1993. (Cited on page 493.)

[154] J. Barria and P. R. Halmos, "Vector Bases for Two Commuting Matrices," *Lin. Multilin. Alg.*, Vol. 27, pp. 147–157, 1990. (Cited on page 348.)

[155] H. Bart, I. Gohberg, M. A. Kaashoek, and A. C. M. Ran, "Schur Complements and State Space Realizations," *Lin. Alg. Appl.*, Vol. 399, pp. 203–224, 2005. (Cited on page 873.)

[156] H. Baruh, *Analytical Dynamics*. Boston: McGraw-Hill, 1999. (Cited on pages 214, 249, and 741.)

[157] A. Barvinok, *A Course in Convexity*. Providence: American Mathematical Society, 2002. (Cited on pages 47, 51, 120, 124, 547, 562, and 696.)

[158] S. Barza, J. Pecaric, and L.-E. Persson, "Carlson Type Inequalities," *J. Ineq. Appl.*, Vol. 2, pp. 121–135, 1998. (Cited on page 64.)

[159] F. L. Bauer, J. Stoer, and C. Witzgall, "Absolute and Monotonic Norms," *Numer. Math.*, Vol. 3, pp. 257–264, 1961. (Cited on page 680.)

[160] D. S. Bayard, "An Optimization Result with Application to Optimal Spacecraft Reaction Wheel Orientation Design," in *Proc. Amer. Contr. Conf.*, Arlington, VA, June 2001, pp. 1473–1478. (Cited on page 364.)

[161] M. S. Bazaraa, H. D. Sherali, and C. M. Shetty, *Nonlinear Programming*, 2nd ed. New York: Wiley, 1993. (Cited on pages 305, 684, and 686.)

[162] N. Bebiano and J. da Providencia, "On the Determinant of Certain Strictly Dissipative Matrices," *Lin. Alg. Appl.*, Vol. 83, pp. 117–128, 1986. (Cited on page 534.)

[163] N. Bebiano, J. da Providencia, and R. Lemos, "Matrix Inequalities in Statistical Mechanics," *Lin. Alg. Appl.*, Vol. 376, pp. 265–273, 2004. (Cited on pages xvii, 752, and 754.)

[164] N. Bebiano, R. Lemos, and J. da Providencia, "Inequalities for Quantum Relative Entropy," *Lin. Alg. Appl.*, Vol. 401, pp. 159–172, 2005. (Cited on pages xvii and 753.)

[165] N. Bebiano and M. E. Miranda, "On a Recent Determinantal Inequality," *Lin. Alg. Appl.*, Vol. 201, pp. 99–102, 1994. (Cited on page 365.)

[166] E. F. Beckenbach and R. Bellman, *Inequalities*. Berlin: Springer, 1965. (Cited on pages 84 and 598.)

[167] R. I. Becker, "Necessary and Sufficient Conditions for the Simultaneous Diagonability of Two Quadratic Forms," *Lin. Alg. Appl.*, Vol. 30, pp. 129–139, 1980. (Cited on pages 396 and 554.)

[168] W. Beckner, "Inequalities in Fourier Analysis," *Ann. Math.*, Vol. 102, p. 159182, 1975. (Cited on page 643.)

[169] L. W. Beineke and R. J. Wilson, Eds., *Topics in Algebraic Graph Theory*. Cambridge: Cambridge University Press, 2005. (Cited on page xvii.)

[170] T. N. Bekjan, "On Joint Convexity of Trace Functions," *Lin. Alg. Appl.*, Vol. 390, pp. 321–327, 2004. (Cited on page 549.)

[171] P. A. Bekker, "The Positive Semidefiniteness of Partitioned Matrices," *Lin. Alg. Appl.*, Vol. 111, pp. 261–278, 1988. (Cited on pages 531, 532, and 595.)

[172] J. G. Belinfante, B. Kolman, and H. A. Smith, "An Introduction to Lie Groups and Lie Algebras with Applications," *SIAM Rev.*, Vol. 8, pp. 11–46, 1966. (Cited on page 793.)

[173] G. R. Belitskii and Y. I. Lyubich, *Matrix Norms and Their Applications*. Basel: Birkhauser, 1988. (Cited on page 680.)

[174] R. Bellman, *Introduction to Matrix Analysis*, 2nd ed. New York: McGraw-Hill, 1960, reprinted by SIAM, Philadelphia, 1995. (Cited on pages 158, 306, 499, and 748.)

[175] R. Bellman, "Some Inequalities for the Square Root of a Positive Definite Matrix," *Lin. Alg. Appl.*, Vol. 1, pp. 321–324, 1968. (Cited on page 506.)

[176] A. Ben-Israel, "A Note on Partitioned Matrices and Equations," *SIAM Rev.*, Vol. 11, pp. 247–250, 1969. (Cited on page 429.)

[177] A. Ben-Israel, "The Moore of the Moore-Penrose Inverse," *Elect. J. Lin. Alg.*, Vol. 9, pp. 150–157, 2002. (Cited on page 438.)

[178] A. Ben-Israel and T. N. E. Greville, *Generalized Inverses: Theory and Applications*, 2nd ed. New York: Springer, 2003. (Cited on pages 406, 410, 429, 434, 435, 438, and 558.)

[179] B. Ben Taher and M. Rachidi, "Some Explicit Formulas for the Polynomial Decomposition of the Matrix Exponential and Applications," *Lin. Alg. Appl.*, Vol. 350, pp. 171–184, 2002. (Cited on page 737.)

[180] A. Ben-Tal and A. Nemirovski, *Lectures on Modern Convex Optimization*. Philadelphia: SIAM, 2001. (Cited on pages 119 and 178.)

[181] A. T. Benjamin and J. J. Quinn, *Proofs That Really Count: The Art of Combinatorial Proof*. Washington, DC: Mathematical Association of America, 2003. (Cited on pages 12, 19, and 304.)

[182] L. Berg, "Three Results in Connection with Inverse Matrices," *Lin. Alg. Appl.*, Vol. 84, pp. 63–77, 1986. (Cited on page 141.)

[183] C. Berge, *The Theory of Graphs*. London: Methuen, 1962, reprinted by Dover, Mineola, 2001. (Cited on page xvii.)

[184] L. D. Berkovitz, *Convexity and Optimization in $\mathbb{R}^n$*. New York: Wiley, 2002. (Cited on pages 178 and 697.)

[185] A. Berman, M. Neumann, and R. J. Stern, *Nonnegative Matrices in Dynamic Systems*. New York: Wiley, 1989. (Cited on pages 299, 774, and 776.)

[186] A. Berman and R. J. Plemmons, *Nonnegative Matrices in the Mathematical Sciences*. New York: Academic Press, 1979, reprinted by SIAM, Philadelphia, 1994. (Cited on pages 190, 252, 301, 303, 774, 775, and 776.)

[187] A. Berman, R. S. Varga, and R. C. Ward, "ALPS: Matrices with Nonpositive Off-Diagonal Entries," *Lin. Alg. Appl.*, Vol. 21, pp. 233–244, 1978. (Cited on page 775.)

[188] D. S. Bernstein, "Inequalities for the Trace of Matrix Exponentials," *SIAM J. Matrix Anal. Appl.*, Vol. 9, pp. 156–158, 1988. (Cited on pages 756 and 759.)

[189] D. S. Bernstein, "Some Open Problems in Matrix Theory Arising in Linear Systems and Control," *Lin. Alg. Appl.*, Vol. 162–164, pp. 409–432, 1992. (Cited on page 769.)

[190] D. S. Bernstein and S. P. Bhat, "Lyapunov Stability, Semistability, and Asymptotic Stability of Matrix Second-Order Systems," *ASME Trans. J. Vibr. Acoustics*, Vol. 117, pp. 145–153, 1995. (Cited on pages 368, 774, and 793.)

[191] D. S. Bernstein and S. P. Bhat, "Nonnegativity, Reducibility, and Semistability of Mass Action Kinetics," in *Proc. Conf. Dec. Contr.*, Phoenix, AZ, December 1999, pp. 2206–2211. (Cited on page 776.)

[192] D. S. Bernstein and W. M. Haddad, "LQG Control with an H_∞ Performance Bound: A Riccati Equation Approach," *IEEE Trans. Autom. Contr.*, Vol. 34, pp. 293–305, 1989. (Cited on page 879.)

[193] D. S. Bernstein, W. M. Haddad, D. C. Hyland, and F. Tyan, "Maximum Entropy-Type Lyapunov Functions for Robust Stability and Performance Analysis," *Sys. Contr. Lett.*, Vol. 21, pp. 73–87, 1993. (Cited on page 374.)

[194] D. S. Bernstein and D. C. Hyland, "Compartmental Modeling and Second-Moment Analysis of State Space Systems," *SIAM J. Matrix Anal. Appl.*, Vol. 14, pp. 880–901, 1993. (Cited on pages 252, 774, and 777.)

[195] D. S. Bernstein and W. So, "Some Explicit Formulas for the Matrix Exponential," *IEEE Trans. Autom. Contr.*, Vol. 38, pp. 1228–1232, 1993. (Cited on page 737.)

[196] D. S. Bernstein and C. F. Van Loan, "Rational Matrix Functions and Rank-1 Updates," *SIAM J. Matrix Anal. Appl.*, Vol. 22, pp. 145–154, 2000. (Cited on page 138.)

[197] K. V. Bhagwat and R. Subramanian, "Inequalities between Means of Positive Operators," *Math. Proc. Camb. Phil. Soc.*, Vol. 83, pp. 393–401, 1978. (Cited on pages 476, 506, and 749.)

[198] S. P. Bhat and D. S. Bernstein, "Average-Preserving Symmetries and Energy Equipartition in Linear Hamiltonian Systems," *Math. Contr. Sig. Sys.*, preprint. (Cited on page 250.)

[199] S. P. Bhat and D. S. Bernstein, "Nontangency-Based Lyapunov Tests for Convergence and Stability in Systems Having a Continuum of Equilibria," *SIAM J. Contr. Optim.*, Vol. 42, pp. 1745–1775, 2003. (Cited on pages 793 and 870.)

[200] R. Bhatia, *Perturbation Bounds for Matrix Eigenvalues*. Essex: Longman Scientific and Technical, 1987. (Cited on page 657.)

[201] R. Bhatia, *Matrix Analysis*. New York: Springer, 1997. (Cited on pages 176, 177, 201, 355, 356, 476, 479, 482, 484, 485, 486, 526, 528, 560, 561, 565, 573, 595, 596, 615, 634, 636, 639, 641, 645, 646, 647, 657, 658, 663, 752, 761, and 763.)

[202] R. Bhatia, "Linear Algebra to Quantum Cohomology: The Story of Alfred Horn's Inequalities," *Amer. Math. Monthly*, Vol. 108, pp. 289–318, 2001. (Cited on pages 562, 565, and 596.)

[203] R. Bhatia, "Infinitely Divisible Matrices," *Amer. Math. Monthly*, Vol. 113, pp. 221–235, 2006. (Cited on pages 241, 489, 491, and 584.)

[204] R. Bhatia, *Perturbation Bounds for Matrix Eigenvalues*. Philadelphia: SIAM, 2007. (Cited on pages 296, 570, 630, 657, and 658.)

[205] R. Bhatia, *Positive Definite Matrices*. Princeton: Princeton University Press, 2007. (Cited on pages 485, 489, 490, 491, 494, 502, 509, 752, and 755.)

[206] R. Bhatia and C. Davis, "More Matrix Forms of the Arithmetic-Geometric Mean Inequality," *SIAM J. Matrix Anal. Appl.*, Vol. 14, pp. 132–136, 1993. (Cited on page 646.)

[207] R. Bhatia and C. Davis, "A Cauchy-Schwarz Inequality for Operators with Applications," *Lin. Alg. Appl.*, Vol. 223/224, pp. 119–129, 1995. (Cited on pages 639 and 646.)

[208] R. Bhatia and D. Drissi, "Generalized Lyapunov Equations and Positive Definite Functions," *SIAM J. Matrix Anal. Appl.*, Vol. 27, pp. 103–114, 2005. (Cited on page 492.)

[209] R. Bhatia and L. Elsner, "Higher Order Logarithmic Derivatives of Matrices in the Spectral Norm," *SIAM J. Matrix Anal. Appl.*, Vol. 25, pp. 662–668, 2003. (Cited on page 758.)

[210] R. Bhatia and J. Holbrook, "Frechet Derivatives of the Power Function," *Indiana University Math. J.*, Vol. 49, pp. 1155–1173, 2000. (Cited on pages 26 and 566.)

[211] R. Bhatia and J. Holbrook, "Riemannian Geometry and Matrix Geometric Means," *Lin. Alg. Appl.*, Vol. 413, pp. 594–618, 2006. (Cited on page 755.)

[212] R. Bhatia and F. Kittaneh, "Norm Inequalities for Partitioned Operators and an Application," *Math. Ann.*, Vol. 287, pp. 719–726, 1990. (Cited on page 650.)

[213] R. Bhatia and F. Kittaneh, "On the Singular Values of a Product of Operators," *SIAM J. Matrix Anal. Appl.*, Vol. 11, pp. 272–277, 1990. (Cited on pages 504, 646, and 671.)

[214] R. Bhatia and F. Kittaneh, "Norm Inequalities for Positive Operators," *Lett. Math. Phys.*, Vol. 43, pp. 225–231, 1998. (Cited on page 647.)

[215] R. Bhatia and F. Kittaneh, "Cartesian Decompositions and Schatten Norms," *Lin. Alg. Appl.*, Vol. 318, pp. 109–116, 2000. (Cited on pages 643 and 644.)

[216] R. Bhatia and F. Kittaneh, "Notes on Matrix Arithmetic-Geometric Mean Inequalities," *Lin. Alg. Appl.*, Vol. 308, pp. 203–211, 2000. (Cited on pages 638 and 639.)

[217] R. Bhatia and F. Kittaneh, "Clarkson Inequalities with Several Operators," *Bull. London Math. Soc.*, Vol. 36, pp. 820–832, 2004. (Cited on page 643.)

[218] R. Bhatia and F. Kittaneh, "Commutators, Pinchings, and Spectral Variation," *Oper. Matrices*, Vol. 2, pp. 143–151, 2008. (Cited on pages 641 and 642.)

[219] R. Bhatia and F. Kittaneh, "The Matrix Arithmetic-Geometric Mean Inequality," *Lin. Alg. Appl.*, Vol. 428, pp. 2177–2191, 2008. (Cited on pages 567, 636, and 647.)

[220] R. Bhatia and K. R. Parthasarathy, "Positive Definite Functions and Operator Inequalities," *Bull. London Math. Soc.*, Vol. 32, pp. 214–228, 2000. (Cited on pages 490, 646, 647, and 763.)

[221] R. Bhatia and P. Rosenthal, "How and Why to Solve the Operator Equation $AX - XB = Y$," *Bull. London Math. Soc.*, Vol. 29, pp. 1–21, 1997. (Cited on pages 349 and 365.)

[222] R. Bhatia and P. Semrl, "Orthogonality of Matrices and Some Distance Problems," *Lin. Alg. Appl.*, Vol. 287, pp. 77–85, 1999. (Cited on page 626.)

[223] R. Bhatia and X. Zhan, "Norm Inequalities for Operators with Positive Real Part," *J. Operator Theory*, Vol. 50, pp. 67–76, 2003. (Cited on pages 535 and 644.)

[224] R. Bhattacharya and K. Mukherjea, "On Unitary Similarity of Matrices," *Lin. Alg. Appl.*, Vol. 126, pp. 95–105, 1989. (Cited on page 346.)

[225] S. P. Bhattacharyya, H. Chapellat, and L. Keel, *Robust Control: The Parametric Approach.* Englewood Cliffs: Prentice-Hall, 1995. (Cited on page 764.)

[226] M. R. Bicknell, "The Lambda Number of a Matrix: The Sum of Its n^2 Cofactors," *Amer. Math. Monthly*, Vol. 72, pp. 260–264, 1965. (Cited on page 155.)

[227] N. Biggs, *Algebraic Graph Theory*, 2nd ed. Cambridge: Cambridge University Press, 2000. (Cited on pages xvii and 367.)

[228] P. Binding, B. Najman, and Q. Ye, "A Variational Principle for Eigenvalues of Pencils of Hermitian Matrices," *Integ. Equ. Oper. Theory*, Vol. 35, pp. 398–422, 1999. (Cited on page 396.)

[229] K. Binmore, *Fun and Games: A Text on Game Theory.* Lexington: D. C. Heath and Co., 1992. (Cited on page xvii.)

[230] A. Bjorck, *Numerical Methods for Least Squares Problems.* Philadelphia: SIAM, 1996. (Cited on page 678.)

[231] S. Blanes and F. Casas, "On the Convergence and Optimization of the Baker-Campbell-Hausdorff Formula," *Lin. Alg. Appl.*, Vol. 378, pp. 135–158, 2004. (Cited on page 749.)

[232] S. Blanes, F. Casas, J. A. Oteo, and J. Ros, "Magnus and Fer Expansions for Matrix Differential Equations: The Convergence Problem," *J. Phys. A: Math. Gen.*, Vol. 31, pp. 259–268, 1998. (Cited on page 743.)

[233] E. D. Bloch, *Proofs and Fundamentals: A First Course in Abstract Mathematics.* Boston: Birkhauser, 2000. (Cited on pages 11 and 84.)

[234] V. Blondel and J. N. Tsitsiklis, "When Is a Pair of Matrices Mortal?" *Inform. Proc. Lett.*, Vol. 63, pp. 283–286, 1997. (Cited on page 786.)

[235] V. Blondel and J. N. Tsitsiklis, "The Boundedness of All Products of a Pair of Matrices Is Undecidable," *Sys. Contr. Lett.*, Vol. 41, pp. 135–140, 2000. (Cited on page 786.)

[236] L. M. Blumenthal, *Theory and Applications of Distance Geometry.* Oxford: Oxford University Press, 1953. (Cited on page 174.)

[237] W. Boehm, "An Operator Limit," *SIAM Rev.*, Vol. 36, p. 659, 1994. (Cited on page 785.)

[238] A. W. Bojanczyk and A. Lutoborski, "Computation of the Euler Angles of a Symmetric 3×3 Matrix," *SIAM J. Matrix Anal. Appl.*, Vol. 12, pp. 41–48, 1991. (Cited on page 289.)

[239] B. Bollobas, *Modern Graph Theory.* New York: Springer, 1998. (Cited on page xvii.)

[240] J. V. Bondar, "Comments on and Complements to *Inequalities: Theory of Majorization and Its Applications*," *Lin. Alg. Appl.*, Vol. 199, pp. 115–129, 1994. (Cited on pages 49 and 66.)

[241] A. Borck, *Numerical Methods for Least Squares Problems.* Philadelphia: SIAM, 1996. (Cited on page 438.)

[242] J. Borwein, D. Bailey, and R. Girgensohn, *Experimentation in Mathematics: Computational Paths to Discovery.* Natick: A K Peters, 2004. (Cited on pages 19, 20, 22, 23, 62, 174, and 300.)

[243] J. M. Borwein and A. S. Lewis, *Convex Analysis and Nonlinear Optimization.* New York: Springer, 2000. (Cited on pages 119, 120, 178, 305, 363, 364, 484, 506, 693, and 696.)

[244] A. J. Bosch, "The Factorization of a Square Matrix into Two Symmetric Matrices," *Amer. Math. Monthly*, Vol. 93, pp. 462–464, 1986. (Cited on pages 382 and 396.)

[245] A. J. Bosch, "Note on the Factorization of a Square Matrix into Two Hermitian or Symmetric Matrices," *SIAM Rev.*, Vol. 29, pp. 463–468, 1987. (Cited on pages 382 and 396.)

[246] A. Bottcher and D. Wenzel, "How Big Can the Commutator of Two Matrices Be and How Big Is it Typically?" *Lin. Alg. Appl.*, Vol. 403, pp. 216–228, 2005. (Cited on page 641.)

[247] A. Bottcher and D. Wenzel, "The Frobenius Norm and the Commutator," *Lin. Alg. Appl.*, Vol. 429, pp. 1864–1885, 2008. (Cited on page 641.)

[248] O. Bottema, Z. Djordjovic, R. R. Janic, D. S. Mitrinovic, and P. M. Vasic, *Inequalities: Theory of Majorization and Its Applications*. Groningen: Wolters-Noordhoff, 1969. (Cited on page 172.)

[249] T. L. Boullion and P. L. Odell, *Generalized Inverse Matrices*. New York: Wiley, 1971. (Cited on page 438.)

[250] J.-C. Bourin, "Some Inequalities for Norms on Matrices and Operators," *Lin. Alg. Appl.*, Vol. 292, pp. 139–154, 1999. (Cited on pages 528, 530, 632, and 666.)

[251] J.-C. Bourin, "Convexity or Concavity Inequalities for Hermitian Operators," *Math. Ineq. Appl.*, Vol. 7, pp. 607–620, 2004. (Cited on page 530.)

[252] J.-C. Bourin, "Hermitian Operators and Convex Functions," *J. Ineq. Pure Appl. Math.*, Vol. 6, no. 5, pp. 1–6, 2005, Article 139. (Cited on pages 530 and 531.)

[253] J.-C. Bourin, "Reverse Inequality to Araki's Inequality Comparison of $A^p Z^p Z^p$ and $(AZA)^p$," *Math. Ineq. Appl.*, Vol. 232, pp. 373–378, 2005. (Cited on pages 511 and 512.)

[254] J.-C. Bourin, "Matrix Versions of Some Classical Inequalities," *Lin. Alg. Appl.*, Vol. 407, pp. 890–907, 2006. (Cited on pages 532 and 656.)

[255] J.-C. Bourin, "Reverse Rearrangement Inequalities via Matrix Technics," *J. Ineq. Pure Appl. Math.*, Vol. 7, no. 2, pp. 1–6, 2006, Article 43. (Cited on pages 67 and 504.)

[256] J.-C. Bourin and Y. Seo, "Reverse Inequality to Golden-Thompson Type Inequalities: Comparison of e^{A+B} and $e^A e^B$," *Lin. Alg. Appl.*, Vol. 426, pp. 312–316, 2007. (Cited on page 754.)

[257] K. Bourque and S. Ligh, "Matrices Associated with Classes of Arithmetical Functions," *J. Number Theory*, Vol. 45, pp. 367–376, 1993. (Cited on page 491.)

[258] S. Boyd, "Entropy and Random Feedback," in *Open Problems in Mathematical Systems and Control Theory*, V. D. Blondel, E. D. Sontag, M. Vidyasagar, and J. C. Willems, Eds. New York: Springer, 1998, pp. 71–74. (Cited on page 633.)

[259] S. Boyd and L. Vandenberghe, *Convex Optimization*. Cambridge: Cambridge University Press, 2004. (Cited on pages xvii and 178.)

[260] J. L. Brenner, "A Bound for a Determinant with Dominant Main Diagonal," *Proc. Amer. Math. Soc.*, Vol. 5, pp. 631–634, 1954. (Cited on page 294.)

[261] J. L. Brenner, "Expanded Matrices from Matrices with Complex Elements," *SIAM Rev.*, Vol. 3, pp. 165–166, 1961. (Cited on pages 166 and 251.)

[262] J. L. Brenner and J. S. Lim, "The Matrix Equations $A = XYZ$ and $B = ZYX$ and Related Ones," *Canad. Math. Bull.*, Vol. 17, pp. 179–183, 1974. (Cited on page 378.)

[263] J. W. Brewer, "Kronecker Products and Matrix Calculus in System Theory," *IEEE Trans. Circ. Sys.*, Vol. CAS-25, pp. 772–781, 1978, correction: CAS-26, p. 360, 1979. (Cited on page 458.)

[264] L. Brickman, "On the Field of Values of a Matrix," *Proc. Amer. Math. Soc.*, Vol. 12, pp. 61–66, 1961. (Cited on page 548.)

[265] R. Brockett, *Finite Dimensional Linear Systems*. New York: Wiley, 1970. (Cited on page 879.)

[266] R. W. Brockett, "Using Feedback to Improve System Identification," in *Control of Uncertain Systems*, B. A. Francis, M. C. Smith, and J. C. Willems, Eds. Berlin: Springer, 2006, pp. 45–65. (Cited on pages 562 and 871.)

[267] G. Brown, "Convexity and Minkowski's Inequality," *Amer. Math. Monthly*, Vol. 112, pp. 740–742, 2005. (Cited on page 72.)

[268] E. T. Browne, *Introduction to the Theory of Determinants and Matrices*. Chapel Hill: University of North Carolina Press, 1958. (Cited on page 595.)

[269] R. Bru, J. J. Climent, and M. Neumann, "On the Index of Block Upper Triangular Matrices," *SIAM J. Matrix Anal. Appl.*, Vol. 16, pp. 436–447, 1995. (Cited on page 374.)

[270] R. A. Brualdi, *Combinatorial Matrix Classes*. Cambridge: Cambridge University Press, 2006. (Cited on page xvii.)

[271] R. A. Brualdi, "Combinatorial Matrix Theory," in *Handbook of Linear Algebra*, L. Hogben, Ed. Boca Raton: Chapman & Hall/CRC, 2007, pp. 27-1–27-12. (Cited on page 300.)

[272] R. A. Brualdi and D. Cvetkovic, *A Combinatorial Approach to Matrix Theory and Its Applications*. Boca Raton: CRC Press, 2009. (Cited on pages xvii, 178, and 285.)

[273] R. A. Brualdi and J. Q. Massey, "Some Applications of Elementary Linear Algebra in Combinatorics," *College Math. J.*, Vol. 24, pp. 10–19, 1993. (Cited on page 142.)

[274] R. A. Brualdi and S. Mellendorf, "Regions in the Complex Plane Containing the Eigenvalues of a Matrix," *Amer. Math. Monthly*, Vol. 101, pp. 975–985, 1994. (Cited on pages 293 and 772.)

[275] R. A. Brualdi and H. J. Ryser, *Combinatorial Matrix Theory*. Cambridge: Cambridge University Press, 1991. (Cited on pages xvii, 137, 142, and 550.)

[276] R. A. Brualdi and H. Schneider, "Determinantal Identities: Gauss, Schur, Cauchy, Sylvester, Kronecker, Jacobi, Binet, Laplace, Muir, and Cayley," *Lin. Alg. Appl.*, Vol. 52/53, pp. 769–791, 1983. (Cited on page 140.)

[277] D. Buckholtz, "Inverting the Difference of Hilbert Space Projections," *Amer. Math. Monthly*, Vol. 104, pp. 60–61, 1997. (Cited on page 228.)

[278] D. Buckholtz, "Hilbert Space Idempotents and Involutions," *Proc. Amer. Math. Soc.*, Vol. 128, pp. 1415–1418, 1999. (Cited on pages 366 and 367.)

[279] P. S. Bullen, *A Dictionary of Inequalities*. Essex: Longman, 1998. (Cited on pages 32, 38, 46, 60, 62, 70, 84, 600, and 626.)

[280] P. S. Bullen, *Handbook of Means and Their Inequalities*. Dordrecht: Kluwer, 2003. (Cited on pages 25, 28, 29, 30, 32, and 84.)

[281] P. S. Bullen, D. S. Mitrinovic, and P. M. Vasic, *Means and Their Inequalities*. Dordrecht: Reidel, 1988. (Cited on pages 41, 56, 84, and 622.)

[282] F. Bullo, J. Cortes, and S. Martinez, *Distributed Control of Robotic Networks*. Princeton: Princeton University Press, 2009. (Cited on pages xvii and 301.)

[283] A. Bultheel and M. Van Barel, *Linear Algebra, Rational Approximation and Orthogonal Polynomials.* Amsterdam: Elsevier, 1997. (Cited on page 307.)

[284] F. Burns, D. Carlson, E. V. Haynsworth, and T. L. Markham, "Generalized Inverse Formulas Using the Schur-Complement," *SIAM J. Appl. Math*, Vol. 26, pp. 254–259, 1974. (Cited on page 429.)

[285] P. J. Bushell and G. B. Trustrum, "Trace Inequalities for Positive Definite Matrix Power Products," *Lin. Alg. Appl.*, Vol. 132, pp. 173–178, 1990. (Cited on page 571.)

[286] N. D. Cahill, J. R. D'Errico, D. A. Narayan, and J. Y. Narayan, "Fibonacci Determinants," *College Math. J.*, Vol. 33, pp. 221–225, 2002. (Cited on page 304.)

[287] B. E. Cain, "Inertia Theory," *Lin. Alg. Appl.*, Vol. 30, pp. 211–240, 1980. (Cited on pages 336 and 868.)

[288] D. Callan, "When Is 'Rank' Additive?" *College Math. J.*, Vol. 29, pp. 145–147, 1998. (Cited on page 128.)

[289] E. Camouzis and G. Ladas, *Dynamics of Third-Order Rational Difference Equations with Open Problems and Conjectures.* New York: CRC Press, 2007. (Cited on pages 778 and 793.)

[290] S. L. Campbell, *Singular Systems.* London: Pitman, 1980. (Cited on page 438.)

[291] S. L. Campbell and C. D. Meyer, *Generalized Inverses of Linear Transformations.* London: Pitman, 1979, reprinted by Dover, Mineola, 1991. (Cited on pages 411, 420, 429, 435, 438, and 793.)

[292] S. L. Campbell and N. J. Rose, "Singular Perturbation of Autonomous Linear Systems," *SIAM J. Math. Anal.*, Vol. 10, pp. 542–551, 1979. (Cited on page 767.)

[293] L. Cao and H. M. Schwartz, "A Decomposition Method for Positive Semidefinite Matrices and its Application to Recursive Parameter Estimation," *SIAM J. Matrix Anal. Appl.*, Vol. 22, pp. 1095–1111, 2001. (Cited on pages 460 and 503.)

[294] E. A. Carlen and E. H. Lieb, "A Minkowski Type Trace Inequality and Strong Subadditivity of Quantum Entropy," *Amer. Math. Soc. Transl.*, Vol. 189, pp. 59–62, 1999. (Cited on pages 484 and 485.)

[295] D. Carlson, "Controllability, Inertia, and Stability for Tridiagonal Matrices," *Lin. Alg. Appl.*, Vol. 56, pp. 207–220, 1984. (Cited on page 770.)

[296] D. Carlson, "What Are Schur Complements Anyway?" *Lin. Alg. Appl.*, Vol. 74, pp. 257–275, 1986. (Cited on page 596.)

[297] D. Carlson, "On the Controllability of Matrix Pairs (A, K) with K Positive Semidefinite, II," *SIAM J. Matrix Anal. Appl.*, Vol. 15, pp. 129–133, 1994. (Cited on page 867.)

[298] D. Carlson, E. V. Haynsworth, and T. L. Markham, "A Generalization of the Schur Complement by Means of the Moore-Penrose Inverse," *SIAM J. Appl. Math.*, Vol. 26, pp. 169–175, 1974. (Cited on pages 424, 485, and 596.)

[299] D. Carlson and R. Hill, "Generalized Controllability and Inertia Theory," *Lin. Alg. Appl.*, Vol. 15, pp. 177–187, 1976. (Cited on page 868.)

[300] D. Carlson, C. R. Johnson, D. C. Lay, and A. D. Porter, Eds., *Linear Algebra Gems: Assets for Undergraduate Mathematics.* Washington, DC: Mathematical Association of America, 2002. (Cited on page 178.)

[301] D. Carlson, C. R. Johnson, D. C. Lay, A. D. Porter, A. E. Watkins, and W. Watkins, Eds., *Resources for Teaching Linear Algebra*. Washington, DC: Mathematical Association of America, 1997. (Cited on page 178.)

[302] P. Cartier, "Mathemagics, A Tribute to L. Euler and R. Feynman," in *Noise, Oscillators and Algebraic Randomness*, M. Planat, Ed. New York: Springer, 2000, pp. 6–67. (Cited on page 793.)

[303] D. I. Cartwright and M. J. Field, "A Refinement of the Arithmetic Mean-Geometric Mean Inequality," *Proc. Amer. Math. Soc.*, Vol. 71, pp. 36–38, 1978. (Cited on pages 56 and 793.)

[304] N. Castro-Gonzalez and E. Dopazo, "Representations of the Drazin Inverse for a Class of Block Matrices," *Lin. Alg. Appl.*, Vol. 400, pp. 253–269, 2005. (Cited on page 429.)

[305] H. Caswell, *Matrix Population Models: Construction, Analysis, and Interpretation*, 2nd ed. Sunderland: Sinauer Associates, 2000. (Cited on page xvii.)

[306] F. S. Cater, "Products of Central Collineations," *Lin. Alg. Appl.*, Vol. 19, pp. 251–274, 1978. (Cited on page 380.)

[307] Y. Chabrillac and J.-P. Crouzeix, "Definiteness and Semidefiniteness of Quadratic Forms Revisited," *Lin. Alg. Appl.*, Vol. 63, pp. 283–292, 1984. (Cited on page 554.)

[308] N. N. Chan and M. K. Kwong, "Hermitian Matrix Inequalities and a Conjecture," *Amer. Math. Monthly*, Vol. 92, pp. 533–541, 1985. (Cited on page 503.)

[309] H. Chapellat, M. Mansour, and S. P. Bhattacharyya, "Elementary Proofs of Some Classical Stability Criteria," *IEEE Trans. Educ.*, Vol. 33, pp. 232–239, 1990. (Cited on page 764.)

[310] G. Chartrand, *Graphs and Digraphs*, 4th ed. Boca Raton: Chapman & Hall, 2004. (Cited on page xvii.)

[311] G. Chartrand and L. Lesniak, *Graphs and Digraphs*, 4th ed. Boca Raton: Chapman & Hall/CRC, 2004. (Cited on page xvii.)

[312] F. Chatelin, *Eigenvalues of Matrices*. New York: Wiley, 1993. (Cited on page xviii.)

[313] J.-J. Chattot, *Computational Aerodynamics and Fluid Dynamics*. Berlin: Springer, 2002. (Cited on page xvii.)

[314] N. A. Chaturvedi, N. H. McClamroch, and D. S. Bernstein, "Asymptotic Smooth Stabilization of the Inverted 3D Pendulum," *IEEE Trans. Autom. Contr.*, 2009. (Cited on page 204.)

[315] J.-P. Chehab and M. Raydan, "Geometrical Properties of the Frobenius Condition Number for Positive Definite Matrices," *Lin. Alg. Appl.*, Vol. 429, pp. 2089–2097, 2008. (Cited on page 523.)

[316] V.-S. Chellaboina and W. M. Haddad, "Is the Frobenius Matrix Norm Induced?" *IEEE Trans. Autom. Contr.*, Vol. 40, pp. 2137–2139, 1995. (Cited on page 636.)

[317] V.-S. Chellaboina and W. M. Haddad, "Solution to 'Some Matrix Integral Identities'," *SIAM Rev.*, Vol. 39, pp. 763–765, 1997. (Cited on page 840.)

[318] V.-S. Chellaboina, W. M. Haddad, D. S. Bernstein, and D. A. Wilson, "Induced Convolution Operator Norms of Linear Dynamical Systems," *Math. Contr. Sig. Sys.*, Vol. 13, pp. 216–239, 2000. (Cited on page 680.)

[319] B. M. Chen, Z. Lin, and Y. Shamash, *Linear Systems Theory: A Structural Decomposition Approach.* Boston: Birkhauser, 2004. (Cited on pages xvii, 123, 124, 354, 363, and 396.)

[320] C. T. Chen, "A Generalization of the Inertia Theorem," *SIAM J. Appl. Math.*, Vol. 25, pp. 158–161, 1973. (Cited on page 867.)

[321] C.-T. Chen, *Linear System Theory and Design.* New York: Holt, Rinehart, Winston, 1984. (Cited on pages xvii and 772.)

[322] H. Chen, "A Unified Elementary Approach to Some Classical Inequalities," *Int. J. Math. Educ. Sci. Tech.*, Vol. 31, pp. 289–292, 2000. (Cited on pages 56 and 68.)

[323] J. Chen, "On Bounds of Matrix Eigenvalues," *Math. Ineq. Appl.*, Vol. 10, pp. 723–726, 2007. (Cited on page 654.)

[324] J.-Q. Chen, J. Ping, and F. Wang, *Group Representation for Physicists*, 2nd ed. Singapore: World Scientific, 2002. (Cited on page 244.)

[325] L. Chen and C. S. Wong, "Inequalities for Singular Values and Traces," *Lin. Alg. Appl.*, Vol. 171, pp. 109–120, 1992. (Cited on pages 525 and 671.)

[326] S. Chen, "Inequalities for M-Matrices and Inverse M-Matrices," *Lin. Alg. Appl.*, Vol. 426, pp. 610–618, 2007. (Cited on page 588.)

[327] C.-M. Cheng, "Cases of Equality for a Singular Value Inequality for the Hadamard Product," *Lin. Alg. Appl.*, Vol. 177, pp. 209–231, 1992. (Cited on page 674.)

[328] C.-M. Cheng, R. A. Horn, and C.-K. Li, "Inequalities and Equalities for the Cartesian Decomposition of Complex Matrices," *Lin. Alg. Appl.*, Vol. 341, pp. 219–237, 2002. (Cited on pages 486, 535, and 566.)

[329] H.-W. Cheng and S. S.-T. Yau, "More Explicit Formulas for the Matrix Exponential," *Lin. Alg. Appl.*, Vol. 262, pp. 131–163, 1997. (Cited on page 737.)

[330] S. Cheng and Y. Tian, "Moore-Penrose Inverses of Products and Differences of Orthogonal Projectors," *Acta Scientiarum Math.*, Vol. 69, pp. 533–542, 2003. (Cited on pages 416, 419, and 425.)

[331] S. Cheng and Y. Tian, "Two Sets of New Characterizations for Normal and EP Matrices," *Lin. Alg. Appl.*, Vol. 375, pp. 181–195, 2003. (Cited on pages 192, 194, 406, 407, and 434.)

[332] M.-T. Chien and M. Neumann, "Positive Definiteness of Tridiagonal Matrices via the Numerical Range," *Elec. J. Lin. Alg.*, Vol. 3, pp. 93–102, 1998. (Cited on page 238.)

[333] M.-D. Choi, "Tricks or Treats with the Hilbert Matrix," *Amer. Math. Monthly*, Vol. 90, pp. 301–312, 1983. (Cited on page 234.)

[334] M. D. Choi, T. Y. Lam, and B. Reznick, "Sums of Squares of Real Polynomials," in *K Theory and Algebraic Geometry*, B. Jacob and A. Rosenberg, Eds. Providence: American Mathematical Society, 1995, pp. 103–126. (Cited on page 46.)

[335] M.-D. Choi and P. Y. Wu, "Convex Combinations of Projections," *Lin. Alg. Appl.*, Vol. 136, pp. 25–42, 1990. (Cited on pages 226 and 395.)

[336] J. Chollet, "Some Inequalities for Principal Submatrices," *Amer. Math. Monthly*, Vol. 104, pp. 609–617, 1997. (Cited on page 522.)

[337] A. Choudhry, "Extraction of nth Roots of 2×2 Matrices," *Lin. Alg. Appl.*, Vol. 387, pp. 183–192, 2004. (Cited on page 381.)

[338] M. T. Chu, R. E. Funderlic, and G. H. Golub, "A Rank-One Reduction Formula and Its Application to Matrix Factorizations," *SIAM Rev.*, Vol. 37, pp. 512–530, 1995. (Cited on pages 412 and 425.)

[339] M. T. Chu, R. E. Funderlic, and G. H. Golub, "Rank Modifications of Semidefinite Matrices Associated with a Secant Update Formula," *SIAM J. Matrix Anal. Appl.*, Vol. 20, pp. 428–436, 1998. (Cited on page 503.)

[340] X.-G. Chu, C.-E. Zhang, and F. Qi, "Two New Algebraic Inequalities with $2n$ Variables," *J. Ineq. Pure Appl. Math.*, Vol. 8, no. 4, pp. 1–6, 2007, Article 102. (Cited on page 19.)

[341] J. Chuai and Y. Tian, "Rank Equalities and Inequalities for Kronecker Products of Matrices with Applications," *Appl. Math. Comp.*, Vol. 150, pp. 129–137, 2004. (Cited on pages 447 and 451.)

[342] N. L. C. Chui and J. M. Maciejowski, "Realization of Stable Models with Subspace Methods," *Automatica*, Vol. 32, pp. 1587–1595, 1996. (Cited on page 783.)

[343] F. R. K. Chung, *Spectral Graph Theory.* Providence: American Mathematical Society, 2000. (Cited on page xvii.)

[344] A. Cizmesija and J. Pecaric, "Classical Hardy's and Carleman's Inequalities and Mixed Means," in *Survey on Classical Inequalities*, T. M. Rassias, Ed. Dordrecht: Kluwer, 2000, pp. 27–65. (Cited on pages 63 and 64.)

[345] D. J. Clements, B. D. O. Anderson, A. J. Laub, and J. B. Matson, "Spectral Factorization with Imaginary-Axis Zeros," *Lin. Alg. Appl.*, Vol. 250, pp. 225–252, 1997. (Cited on page 879.)

[346] R. E. Cline, "Representations for the Generalized Inverse of a Partitioned Matrix," *SIAM J. Appl. Math.*, Vol. 12, pp. 588–600, 1964. (Cited on page 428.)

[347] R. E. Cline and R. E. Funderlic, "The Rank of a Difference of Matrices and Associated Generalized Inverses," *Lin. Alg. Appl.*, Vol. 24, pp. 185–215, 1979. (Cited on pages 128, 129, 131, and 420.)

[348] M. J. Cloud and B. C. Drachman, *Inequalities with Applications to Engineering.* New York: Springer, 1998. (Cited on page 84.)

[349] E. S. Coakley, F. M. Dopico, and C. R. Johnson, "Matrices for Orthogonal Groups Admitting Only Determinant One," *Lin. Alg. Appl.*, Vol. 428, pp. 796–813, 2008. (Cited on page 246.)

[350] J. E. Cohen, "Spectral Inequalities for Matrix Exponentials," *Lin. Alg. Appl.*, Vol. 111, pp. 25–28, 1988. (Cited on page 757.)

[351] J. E. Cohen, S. Friedland, T. Kato, and F. P. Kelly, "Eigenvalue Inequalities for Products of Matrix Exponentials," *Lin. Alg. Appl.*, Vol. 45, pp. 55–95, 1982. (Cited on page 759.)

[352] D. K. Cohoon, "Sufficient Conditions for the Zero Matrix," *Amer. Math. Monthly*, Vol. 96, pp. 448–449, 1989. (Cited on page 137.)

[353] D. Constales, "A Closed Formula for the Moore-Penrose Generalized Inverse of a Complex Matrix of Given Rank," *Acta Math. Hung.*, Vol. 80, pp. 83–88, 1998. (Cited on page 405.)

[354] J. C. Conway and D. A. Smith, *On Quaternions and Octonions: Their Geometry, Arithmetic, and Symmetry.* Natick: A. K. Peters, 2003. (Cited on pages 51, 52, 245, and 249.)

[355] G. Corach, H. Porta, and L. Recht, "An Operator Inequality," *Lin. Alg. Appl.*, Vol. 142, pp. 153–158, 1990. (Cited on page 646.)

[356] M. J. Corless and A. E. Frazho, *Linear Systems and Control: An Operator Perspective.* New York: Marcel Dekker, 2003. (Cited on page xvii.)

[357] E. B. Corrachano and G. Sobczyk, Eds., *Geometric Algebra with Applications in Science and Engineering.* Boston: Birkhauser, 2001. (Cited on pages 204 and 249.)

[358] P. J. Costa and S. Rabinowitz, "Matrix Differentiation Identities," *SIAM Rev.*, Vol. 36, pp. 657–659, 1994. (Cited on page 704.)

[359] R. W. Cottle, "Quartic Barriers," *Comp. Optim. Appl.*, Vol. 12, pp. 81–105, 1999. (Cited on pages 546, 556, and 764.)

[360] T. M. Cover and J. A. Thomas, "Determinant Inequalities via Information Theory," *SIAM J. Matrix Anal. Appl.*, Vol. 9, pp. 384–392, 1988. (Cited on pages 485, 536, and 556.)

[361] T. M. Cover and J. A. Thomas, *Elements of Information Theory*, 2nd ed. New York: Wiley, 2006. (Cited on pages xvii, 485, 536, and 541.)

[362] D. E. Crabtree, "The Characteristic Vector of the Adjoint Matrix," *Amer. Math. Monthly*, Vol. 75, pp. 1127–1128, 1968. (Cited on page 373.)

[363] C. R. Crawford and Y. S. Moon, "Finding a Positive Definite Linear Combination of Two Hermitian Matrices," *Lin. Alg. Appl.*, Vol. 51, pp. 37–48, 1983. (Cited on page 554.)

[364] T. Crilly, "Cayley's Anticipation of a Generalised Cayley-Hamilton Theorem," *Historia Mathematica*, Vol. 5, pp. 211–219, 1978. (Cited on page 284.)

[365] M. D. Crossley, *Essential Topology.* New York: Springer, 2005. (Cited on page 212.)

[366] C. G. Cullen, "A Note on Normal Matrices," *Amer. Math. Monthly*, Vol. 72, pp. 643–644, 1965. (Cited on page 374.)

[367] C. G. Cullen, *Matrices and Linear Transformations*, 2nd ed. Reading: Addison-Wesley, 1972, reprinted by Dover, Mineola, 1990. (Cited on pages 689 and 703.)

[368] W. J. Culver, "On the Existence and Uniqueness of the Real Logarithm of a Matrix," *Proc. Amer. Math. Soc.*, Vol. 17, pp. 1146–1151, 1966. (Cited on page 793.)

[369] R. F. Curtain and H. J. Zwart, *An Introduction to Infinite-Dimensional Linear Systems Theory.* New York: Springer, 1995. (Cited on page 758.)

[370] M. L. Curtis, *Matrix Groups*, 2nd ed. New York: Springer, 1984. (Cited on pages 214, 248, and 252.)

[371] D. Cvetkovic, P. Rowlinson, and S. Simic, *Eigenspaces of Graphs.* Cambridge: Cambridge University Press, 1997. (Cited on page xvii.)

[372] P. J. Daboul and R. Delbourgo, "Matrix Representations of Octonions and Generalizations," *J. Math. Phys.*, Vol. 40, pp. 4134–4150, 1999. (Cited on page 249.)

[373] G. Dahlquist, "On Matrix Majorants and Minorants, with Application to Differential Equations," *Lin. Alg. Appl.*, Vol. 52/53, pp. 199–216, 1983. (Cited on page 774.)

[374] R. D'Andrea, "Extension of Parrott's Theorem to Nondefinite Scalings," *IEEE Trans. Autom. Contr.*, Vol. 45, pp. 937–940, 2000. (Cited on page 668.)

[375] H. D'Angelo, *Linear Time-Varying Systems: Analysis and Synthesis.* Boston: Allyn and Bacon, 1970. (Cited on page 879.)

[376] J. P. D'Angelo, *Inequalities from Complex Analysis.* Washington, DC: The Mathematical Association of America, 2002. (Cited on pages 77 and 620.)

[377] F. M. Dannan, "Matrix and Operator Inequalities," *J. Ineq. Pure. Appl. Math.*, Vol. 2, no. 3/34, pp. 1–7, 2001. (Cited on pages 528 and 537.)

[378] R. Datko and V. Seshadri, "A Characterization and a Canonical Decomposition of Hurwitzian Matrices," *Amer. Math. Monthly*, Vol. 77, pp. 732–733, 1970. (Cited on page 773.)

[379] B. N. Datta, *Numerical Linear Algebra and Applications.* Pacific Grove: Brooks/Cole, 1995. (Cited on page xvii.)

[380] B. N. Datta, "Stability and Inertia," *Lin. Alg. Appl.*, Vol. 302–303, pp. 563–600, 1999. (Cited on page 867.)

[381] B. N. Datta, *Numerical Methods for Linear Control Systems.* San Diego, CA: Elsevier Academic Press, 2003. (Cited on page xvii.)

[382] J. Dattorro, *Convex Optimization and Euclidean Distance Geometry.* Palo Alto: Meboo Publishing, 2005. (Cited on pages xvii, 128, 137, 352, and 705.)

[383] I. Daubechies and J. C. Lagarias, "Sets of Matrices all Infinite Products of Which Converge," *Lin. Alg. Appl.*, Vol. 162, pp. 227–263, 1992. (Cited on page 786.)

[384] P. J. Davis, *The Mathematics of Matrices*, 2nd ed. Boston: Ginn, 1965. (Cited on page xix.)

[385] P. J. Davis, *Circulant Matrices*, 2nd ed. New York: Chelsea, 1994. (Cited on page 390.)

[386] R. J. H. Dawlings, "Products of Idempotents in the Semigroup of Singular Endomorphisms of a Finite-Dimensional Vector Space," *Proc. Royal Soc. Edinburgh*, Vol. 91A, pp. 123–133, 1981. (Cited on page 383.)

[387] J. Day, W. So, and R. C. Thompson, "Some Properties of the Campbell-Baker-Hausdorff Formula," *Lin. Multilin. Alg.*, Vol. 29, pp. 207–224, 1991. (Cited on page 720.)

[388] J. Day, W. So, and R. C. Thompson, "The Spectrum of a Hermitian Matrix Sum," *Lin. Alg. Appl.*, Vol. 280, pp. 289–332, 1998. (Cited on page 565.)

[389] P. W. Day, "Rearrangement Inequalities," *Canad. J. Math.*, Vol. 24, pp. 930–943, 1972. (Cited on page 67.)

[390] C. de Boor, "An Empty Exercise," *SIGNUM*, Vol. 25, pp. 2–6, 1990. (Cited on page 178.)

[391] P. P. N. de Groen, "A Counterexample on Vector Norms and the Subordinate Matrix Norms," *Amer. Math. Monthly*, Vol. 97, pp. 406–407, 1990. (Cited on page 636.)

[392] F. R. de Hoog, R. P. Speed, and E. R. Williams, "On a Matrix Identity Associated with Generalized Least Squares," *Lin. Alg. Appl.*, Vol. 127, pp. 449–456, 1990. (Cited on page 413.)

[393] W. de Launey and J. Seberry, "The Strong Kronecker Product," *J. Combinatorial Thy., Series A*, Vol. 66, pp. 192–213, 1994. (Cited on page 458.)

[394] J. de Pillis, "Transformations on Partitioned Matrices," *Duke Math. J.*, Vol. 36, pp. 511–515, 1969. (Cited on pages 533 and 543.)

[395] J. de Pillis, "Inequalities for Partitioned Semidefinite Matrices," *Lin. Alg. Appl.*, Vol. 4, pp. 79–94, 1971. (Cited on page 543.)

[396] J. E. De Pillis, "Linear Operators and their Partitioned Matrices," *Notices Amer. Math. Soc.*, Vol. 14, p. 636, 1967. (Cited on page 543.)

[397] L. G. de Pillis, "Determinants and Polynomial Root Structure," *Int. J. Math. Educ. Sci. Technol.*, Vol. 36, pp. 469–481, 2005. (Cited on page 277.)

[398] E. de Souza and S. P. Bhattacharyya, "Controllability, Observability and the Solution of $AX - XB = C$," *Lin. Alg. Appl.*, Vol. 39, pp. 167–188, 1981. (Cited on page 870.)

[399] E. de Souza and S. P. Bhattarcharyya, "Controllability, Observability, and the Solution of $AX - XB = -C$," *Lin. Alg. Appl.*, Vol. 39, pp. 167–188, 1981. (Cited on page 870.)

[400] P. N. De Souza and J.-N. Silva, *Berkeley Problems in Mathematics*, 3rd ed. New York: Springer, 2004. (Cited on pages 122, 127, 163, and 503.)

[401] L. M. DeAlba and C. R. Johnson, "Possible Inertia Combinations in the Stein and Lyapunov Equations," *Lin. Alg. Appl.*, Vol. 222, pp. 227–240, 1995. (Cited on pages 470, 868, and 869.)

[402] H. P. Decell, "An Application of the Cayley-Hamilton Theorem to Generalized Matrix Inversion," *SIAM Rev.*, Vol. 7, pp. 526–528, 1965. (Cited on page 408.)

[403] N. Del Buono, L. Lopez, and T. Politi, "Computation of Functions of Hamiltonian and Skew-Symmetric Matrices," *Math. Computers Simulation*, Vol. 79, pp. 1284–1297, 2008. (Cited on page 240.)

[404] J. W. Demmel, *Applied Numerical Linear Algebra.* Philadelphia: SIAM, 1997. (Cited on page xviii.)

[405] C. Y. Deng, "The Drazin Inverses of Sum and Difference of Idempotents," *Lin. Alg. Appl.*, Vol. 430, pp. 1282–1291, 2009. (Cited on page 433.)

[406] E. D. Denman and A. N. Beavers, "The Matrix Sign Function and Computations in Systems," *Appl. Math. Computation*, Vol. 2, pp. 63–94, 1976. (Cited on page 381.)

[407] C. R. DePrima and C. R. Johnson, "The Range of $A^{-1}A^*$ in $GL(n, C)$," *Lin. Alg. Appl.*, Vol. 9, pp. 209–222, 1974. (Cited on page 377.)

[408] C. A. Desoer and H. Haneda, "The Measure of a Matrix as a Tool to Analyze Computer Algorithms for Circuit Analysis," *IEEE Trans. Circ. Thy.*, Vol. 19, pp. 480–486, 1972. (Cited on page 758.)

[409] D. W. DeTemple, "A Geometric Look at Sequences that Converge to Euler's Constant," *College Math. J.*, Vol. 37, pp. 128–131, 2006. (Cited on page 21.)

[410] E. Deutsch, "Matricial Norms," *Numer. Math.*, Vol. 16, pp. 73–84, 1970. (Cited on page 649.)

[411] E. Deutsch, "Matricial Norms and the Zeros of Polynomials," *Lin. Alg. Appl.*, Vol. 3, pp. 483–489, 1970. (Cited on pages 779 and 780.)

[412] E. Deutsch and M. Mlynarski, "Matricial Logarithmic Derivatives," *Lin. Alg. Appl.*, Vol. 19, pp. 17–31, 1978. (Cited on page 758.)

[413] P. J. Dhrymes, *Mathematics for Econometrics*, 3rd ed. New York: Springer, 2000. (Cited on page xvii.)

[414] L. Dieci, "Real Hamiltonian Logarithm of a Symplectic Matrix," *Lin. Alg. Appl.*, Vol. 281, pp. 227–246, 1998. (Cited on pages 724 and 751.)

[415] R. Diestel, *Graph Theory*, 3rd ed. Berlin: Springer, 2006. (Cited on page xvii.)

[416] L. L. Dines, "On the Mapping of Quadratic Forms," *Bull. Amer. Math. Soc.*, Vol. 47, pp. 494–498, 1941. (Cited on pages 547 and 548.)

[417] J. Ding, "Perturbation of Systems of Linear Algebraic Equations," *Lin. Multilin. Alg.*, Vol. 47, pp. 119–127, 2000. (Cited on page 677.)

[418] J. Ding, "Lower and Upper Bounds for the Perturbation of General Linear Algebraic Equations," *Appl. Math. Lett.*, Vol. 14, pp. 49–52, 2001. (Cited on page 677.)

[419] J. Ding and W. C. Pye, "On the Spectrum and Pseudoinverse of a Special Bordered Matrix," *Lin. Alg. Appl.*, Vol. 331, pp. 11–20, 2001. (Cited on page 286.)

[420] A. Dittmer, "Cross Product Identities in Arbitrary Dimension," *Amer. Math. Monthly*, Vol. 101, pp. 887–891, 1994. (Cited on page 204.)

[421] G. M. Dixon, *Division Algebras: Octonions, Quaternions, Complex Numbers and the Algebraic Design of Physics*. Dordrecht: Kluwer, 1994. (Cited on page 249.)

[422] J. D. Dixon, "How Good Is Hadamard's Inequality for Determinants?" *Canad. Math. Bull.*, Vol. 27, pp. 260–264, 1984. (Cited on page 540.)

[423] T. E. Djaferis and S. K. Mitter, "Algebraic Methods for the Study of Some Linear Matrix Equations," *Lin. Alg. Appl.*, Vol. 44, pp. 125–142, 1982. (Cited on page 870.)

[424] D. Z. Djokovic, "Product of Two Involutions," *Arch. Math.*, Vol. 18, pp. 582–584, 1967. (Cited on page 383.)

[425] D. Z. Djokovic, "On Some Representations of Matrices," *Lin. Multilin. Alg.*, Vol. 4, pp. 33–40, 1976. (Cited on pages 163 and 345.)

[426] D. Z. Djokovic and O. P. Lossers, "A Determinant Inequality," *Amer. Math. Monthly*, Vol. 83, pp. 483–484, 1976. (Cited on page 166.)

[427] D. Z. Djokovic and J. Malzan, "Products of Reflections in the Unitary Group," *Proc. Amer. Math. Soc.*, Vol. 73, pp. 157–160, 1979. (Cited on page 380.)

[428] D. Z. Dokovic, "On the Product of Two Alternating Matrices," *Amer. Math. Monthly*, Vol. 98, pp. 935–936, 1991. (Cited on page 375.)

[429] D. Z. Dokovic, "Unitary Similarity of Projectors," *Aequationes Mathematicae*, Vol. 42, pp. 220–224, 1991. (Cited on pages 342, 343, and 347.)

[430] D. Z. Dokovic, F. Szechtman, and K. Zhao, "An Algorithm that Carries a Square Matrix into Its Transpose by an Involutory Congruence Transformation," *Elec. J. Lin. Alg.*, Vol. 10, pp. 320–340, 2003. (Cited on page 340.)

[431] V. Dolotin and A. Morozov, *Introduction to Non-Linear Algebra*. Singapore: World Scientific, 2007. (Cited on page 458.)

[432] W. F. Donoghue, *Monotone Matrix Functions and Analytic Continuation*. New York: Springer, 1974. (Cited on pages 489, 490, 594, and 595.)

[433] F. M. Dopico and C. R. Johnson, "Complementary Bases in Symplectic Matrices and a Proof That Their Determinant Is One," *Lin. Alg. Appl.*, Vol. 419, pp. 772–778, 2006. (Cited on page 239.)

[434] F. M. Dopico, C. R. Johnson, and J. M. Molera, "Multiple LU Factorizations of a Singular Matrix," *Lin. Alg. Appl.*, Vol. 419, pp. 24–36, 2006. (Cited on page 378.)

[435] C. Doran and A. Lasenby, *Geometric Algebra for Physicists*. Cambridge: Cambridge University Press, 2005. (Cited on pages 204 and 249.)

[436] L. Dorst, D. Fontijne, and S. Mann, *Geometric Algebra for Computer Science: An Object-Oriented Approach to Geometry*. Amsterdam: Elsevier, 2007. (Cited on pages 249 and 620.)

[437] R. G. Douglas, "On Majorization, Factorization, and Range Inclusion of Operators on Hilbert Space," *Proc. Amer. Math. Soc.*, Vol. 17, pp. 413–415, 1966. (Cited on page 474.)

[438] P. G. Doyle and J. L. Snell, *Random Walks and Electric Networks*. Washington, DC: Mathematical Association of America, 1984. (Cited on page xvii.)

[439] V. Dragan and A. Halanay, *Stabilization of Linear Systems*. Boston: Birkhauser, 1999. (Cited on page 879.)

[440] S. S. Dragomir, "A Survey on Cauchy-Bunyakovsky-Schwarz Type Discrete Inequalities," *J. Ineq. Pure Appl. Math.*, Vol. 4, no. 3, pp. 1–142, 2003, Article 3. (Cited on pages 39, 63, 68, 70, 71, 72, 73, and 77.)

[441] S. S. Dragomir, *Discrete Inequalities of the Cauchy-Bunyakovsky-Schwarz Type*. Hauppauge: Nova Science Publishers, 2004. (Cited on pages 39, 63, 68, 70, 71, 72, 73, and 77.)

[442] S. S. Dragomir, "Some Reverses of the Generalised Triangle Inequality in Complex Inner Product Spaces," *Lin. Alg. Appl.*, Vol. 402, pp. 245–254, 2005. (Cited on page 77.)

[443] S. S. Dragomir and C. J. Goh, "A Counterpart of Jensen's Discrete Inequality for Differentiable Convex Mappings and Applications in Information Theory," *Math. Comput. Modelling*, Vol. 24, pp. 1–11, 1996. (Cited on page 65.)

[444] S. S. Dragomir and C. J. Goh, "Some Bounds on Entropy Measures in Information Theory," *Appl. Math. Lett.*, Vol. 10, pp. 23–28, 1997. (Cited on page 65.)

[445] S. S. Dragomir, C. E. Pearce, and J. Pecaric, "Some New Inequalities for the Logarithmic Map, with Applications to Entropy and Mutual Information," *Kyungpook Math. J.*, Vol. 41, pp. 115–125, 2001. (Cited on pages 65 and 70.)

[446] M. P. Drazin, "A Note on Skew-Symmetric Matrices," *Math. Gaz.*, Vol. 36, pp. 253–255, 1952. (Cited on page 282.)

[447] D. Drissi, "Sharp Inequalities for Some Operator Means," *SIAM J. Matrix Anal. Appl.*, Vol. 28, pp. 822–828, 2006. (Cited on page 41.)

[448] D. Drivaliaris, S. Karanasios, and D. Pappas, "Factorizations of EP Operators," *Lin. Alg. Appl.*, Vol. 429, pp. 1555–1567, 2008. (Cited on page 414.)

[449] R. Drnovsek, H. Radjavi, and P. Rosenthal, "A Characterization of Commutators of Idempotents," *Lin. Alg. Appl.*, Vol. 347, pp. 91–99, 2002. (Cited on page 217.)

[450] S. W. Drury and G. P. H. Styan, "Normal Matrix and a Commutator," *IMAGE*, Vol. 31, p. 24, 2003. (Cited on page 200.)

[451] H. Du, C. Deng, and Q. Li, "On the Infimum Problem of Hilbert Space Effects," *Science in China: Series A Math.*, Vol. 49, pp. 545–556, 2006. (Cited on page 506.)

[452] K. Du, "Norm Inequalities Involving Hadamard Products," *Lin. Multilin. Alg.*, preprint. (Cited on page 676.)

[453] F. Dubeau, "Linear Algebra and the Sums of Powers of Integers," *Elec. J. Lin. Alg.*, Vol. 17, pp. 577–596, 2008. (Cited on page 20.)

[454] J. J. Duistermaat and J. A. C. Kolk, Eds., *Multidimensional Real Analysis I: Differentiation.* Cambridge: Cambridge University Press, 2004. (Cited on pages 696, 700, and 748.)

[455] I. Duleba, "On a Computationally Simple Form of the Generalized Campbell-Baker-Hausdorff-Dynkin Formula," *Sys. Contr. Lett.*, Vol. 34, pp. 191–202, 1998. (Cited on page 743.)

[456] G. E. Dullerud and F. Paganini, *A Course in Robust Control Theory: A Convex Approach*, 2nd ed. New York: Springer, 1999. (Cited on page xvii.)

[457] D. S. Dummit and R. M. Foote, *Abstract Algebra*, 3rd ed. New York: Wiley, 2004. (Cited on pages 242, 244, 245, 390, 391, and 396.)

[458] C. F. Dunkl and K. S. Williams, "A Simple Inequality," *Amer. Math. Monthly*, Vol. 71, pp. 53–54, 1964. (Cited on page 619.)

[459] H. Dym, *Linear Algebra in Action.* Providence: American Mathematical Society, 2006. (Cited on pages 61, 140, 142, 145, 173, 306, 336, 337, 338, 365, 366, 429, 474, 503, 558, 561, 578, 648, 668, 695, 697, 698, 701, 766, and 868.)

[460] A. Edelman and G. Strang, "Pascal Matrices," *Amer. Math. Monthly*, Vol. 111, pp. 189–197, 2004. (Cited on page 491.)

[461] C. H. Edwards, Jr., *Advanced Calculus of Several Variables.* New York: Academic Press, 1973. (Cited on page 688.)

[462] O. Egecioglu, "Parallelogram-Law-Type Identities," *Lin. Alg. Appl.*, Vol. 225, pp. 1–12, 1995. (Cited on page 77.)

[463] H. G. Eggleston, *Convexity.* Cambridge: Cambridge University Press, 1958. (Cited on page 178.)

[464] L. Elsner, D. Hershkowitz, and H. Schneider, "Bounds on Norms of Compound Matrices and on Products of Eigenvalues," *Bull. London Math. Soc.*, Vol. 32, pp. 15–24, 2000. (Cited on pages 20 and 454.)

[465] L. Elsner and K. D. Ikramov, "Normal Matrices: An Update," *Lin. Alg. Appl.*, Vol. 285, pp. 291–303, 1998. (Cited on pages 194, 714, and 762.)

[466] L. Elsner, C. R. Johnson, and J. A. D. Da Silva, "The Perron Root of a Weighted Geometric Mean of Nonnegative Matrices," *Lin. Multilin. Alg.*, Vol. 24, pp. 1–13, 1988. (Cited on page 457.)

[467] L. Elsner and M. H. C. Paardekooper, "On Measures of Nonnormality of Matrices," *Lin. Alg. Appl.*, Vol. 92, pp. 107–124, 1987. (Cited on page 194.)

[468] L. Elsner and P. Rozsa, "On Eigenvectors and Adjoints of Modified Matrices," *Lin. Multilin. Alg.*, Vol. 10, pp. 235–247, 1981. (Cited on page 154.)

[469] L. Elsner and T. Szulc, "Convex Sets of Schur Stable and Stable Matrices," *Lin. Multilin. Alg.*, Vol. 48, pp. 1–19, 2000. (Cited on page 293.)

[470] A. Engel, *Problem-Solving Strategies.* New York: Springer, 1998. (Cited on pages 34, 35, 37, 42, 44, 45, 46, 48, 49, 50, 51, 54, 57, 64, 66, and 171.)

[471] K. Engo, "On the BCH Formula in so(3)," *Numer. Math. BIT*, Vol. 41, pp. 629–632, 2001. (Cited on page 749.)

[472] K. Erdmann and M. J. Wildon, *Introduction to Lie Algebras.* New York: Springer, 2006. (Cited on page 252.)

[473] J. A. Erdos, "On Products of Idempotent Matrices," *Glasgow Math. J.*, Vol. 8, pp. 118–122, 1967. (Cited on page 383.)

[474] R. Eriksson, "On the Measure of Solid Angles," *Math. Mag.*, Vol. 63, pp. 184–187, 1990. (Cited on page 175.)

[475] J.-C. Evard and F. Uhlig, "On the Matrix Equation $f(X) = A$," *Lin. Alg. Appl.*, Vol. 162–164, pp. 447–519, 1992. (Cited on page 793.)

[476] S. Fallat and M. J. Tsatsomeros, "On the Cayley Transform of Positivity Classes of Matrices," *Elec. J. Lin. Alg.*, Vol. 9, pp. 190–196, 2002. (Cited on page 498.)

[477] K. Fan, "Generalized Cayley Transforms and Strictly Dissipative Matrices," *Lin. Alg. Appl.*, Vol. 5, pp. 155–172, 1972. (Cited on page 498.)

[478] K. Fan, "On Real Matrices with Positive Definite Symmetric Component," *Lin. Multilin. Alg.*, Vol. 1, pp. 1–4, 1973. (Cited on pages 498 and 536.)

[479] K. Fan, "On Strictly Dissipative Matrices," *Lin. Alg. Appl.*, Vol. 9, pp. 223–241, 1974. (Cited on pages 498 and 534.)

[480] M. Fang, "Bounds on Eigenvalues of the Hadamard Product and the Fan Product of Matrices," *Lin. Alg. Appl.*, Vol. 425, pp. 7–15, 2007. (Cited on page 457.)

[481] Y. Fang, K. A. Loparo, and X. Feng, "Inequalities for the Trace of Matrix Product," *IEEE Trans. Autom. Contr.*, Vol. 39, pp. 2489–2490, 1994. (Cited on page 364.)

[482] R. W. Farebrother, "A Class of Square Roots of Involutory Matrices," *IMAGE*, Vol. 28, pp. 26–28, 2002. (Cited on page 438.)

[483] R. W. Farebrother, J. Gross, and S.-O. Troschke, "Matrix Representation of Quaternions," *Lin. Alg. Appl.*, Vol. 362, pp. 251–255, 2003. (Cited on page 251.)

[484] R. W. Farebrother and I. Wrobel, "Regular and Reflected Rotation Matrices," *IMAGE*, Vol. 29, pp. 24–25, 2002. (Cited on page 380.)

[485] D. R. Farenick, M. Krupnik, N. Krupnik, and W. Y. Lee, "Normal Toeplitz Matrices," *SIAM J. Matrix Anal. Appl.*, Vol. 17, pp. 1037–1043, 1996. (Cited on page 390.)

[486] A. Fassler and E. Stiefel, *Group Theoretical Methods and Their Applications*. Boston: Birkhauser, 1992. (Cited on page 252.)

[487] A. E. Fekete, *Real Linear Algebra*. New York: Marcel Dekker, 1985. (Cited on pages 204 and 217.)

[488] B. Q. Feng, "Equivalence Constants for Certain Matrix Norms," *Lin. Alg. Appl.*, Vol. 374, pp. 247–253, 2003. (Cited on page 640.)

[489] B. Q. Feng and A. Tonge, "Equivalence Constants for Certain Matrix Norms, II," *Lin. Alg. Appl.*, Vol. 420, pp. 388–399, 2007. (Cited on pages 630 and 640.)

[490] R. Fenn, *Geometry*. New York: Springer, 2001. (Cited on pages 175, 204, 213, 248, 249, and 304.)

[491] P. G. Ferreira and S. P. Bhattacharyya, "On Blocking Zeros," *IEEE Trans. Autom. Contr.*, Vol. AC-22, pp. 258–259, 1977. (Cited on page 879.)

[492] J. H. Ferziger and M. Peric, *Computational Methods for Fluid Dynamics*, 3rd ed. Berlin: Springer, 2002. (Cited on page xvii.)

[493] M. Fiedler, "A Note on the Hadamard Product of Matrices," *Lin. Alg. Appl.*, Vol. 49, pp. 233–235, 1983. (Cited on pages 585 and 589.)

[494] M. Fiedler, *Special Matrices and Their Applications in Numerical Mathematics*. Dordrecht: Martinus Nijhoff, 1986, reprinted by Dover, Mineola, 2008. (Cited on pages xvii, 277, 279, 299, 359, 388, 427, 454, and 770.)

[495] M. Fiedler, "Some Applications of Matrices and Graphs in Euclidean Geometry," in *Handbook of Linear Algebra*, L. Hogben, Ed. Boca Raton: Chapman & Hall/CRC, 2007, pp. 66-1–66-15. (Cited on pages 174 and 585.)

[496] M. Fiedler and T. L. Markham, "A Characterization of the Moore-Penrose Inverse," *Lin. Alg. Appl.*, Vol. 179, pp. 129–133, 1993. (Cited on pages 161 and 410.)

[497] M. Fiedler and T. L. Markham, "An Observation on the Hadamard Product of Hermitian Matrices," *Lin. Alg. Appl.*, Vol. 215, pp. 179–182, 1995. (Cited on page 587.)

[498] M. Fiedler and T. L. Markham, "Some Results on the Bergstrom and Minkowski Inequalities," *Lin. Alg. Appl.*, Vol. 232, pp. 199–211, 1996. (Cited on page 515.)

[499] M. Fiedler and V. Ptak, "A New Positive Definite Geometric Mean of Two Positive Definite Matrices," *Lin. Alg. Appl.*, Vol. 251, pp. 1–20, 1997. (Cited on pages 509 and 510.)

[500] J. A. Fill and D. E. Fishkind, "The Moore-Penrose Generalized Inverse for Sums of Matrices," *SIAM J. Matrix Anal. Appl.*, Vol. 21, pp. 629–635, 1999. (Cited on page 413.)

[501] P. A. Fillmore, "On Similarity and the Diagonal of a Matrix," *Amer. Math. Monthly*, Vol. 76, pp. 167–169, 1969. (Cited on page 341.)

[502] P. A. Fillmore, "On Sums of Projections," *J. Funct. Anal.*, Vol. 4, pp. 146–152, 1969. (Cited on page 394.)

[503] P. A. Fillmore and J. P. Williams, "On Operator Ranges," *Adv. Math.*, Vol. 7, pp. 254–281, 1971. (Cited on page 474.)

[504] C. H. Fitzgerald and R. A. Horn, "On Fractional Hadamard Powers of Positive Definite Matrices," *J. Math. Anal. Appl.*, Vol. 61, pp. 633–642, 1977. (Cited on page 584.)

[505] H. Flanders, "Methods of Proof in Linear Algebra," *Amer. Math. Monthly*, Vol. 63, pp. 1–15, 1956. (Cited on page 233.)

[506] H. Flanders, "On the Norm and Spectral Radius," *Lin. Multilin. Alg.*, Vol. 2, pp. 239–240, 1974. (Cited on page 660.)

[507] H. Flanders, "An Extremal Problem in the Space of Positive Definite Matrices," *Lin. Multilin. Alg.*, Vol. 3, pp. 33–39, 1975. (Cited on page 504.)

[508] W. Fleming, *Functions of Several Variables*, 2nd ed. New York: Springer, 1987. (Cited on page 543.)

[509] T. M. Flett, *Differential Analysis*. Cambridge: Cambridge University Press, 1980. (Cited on page 705.)

[510] S. Foldes, *Fundamental Structures of Algebra and Discrete Mathematics*. New York: Wiley, 1994. (Cited on page 242.)

[511] J. Foley, A. van Dam, S. Feiner, and J. Hughes, *Computer Graphics Principles and Practice*, 2nd ed. Reading: Addison-Wesley, 1990. (Cited on pages xvii and 215.)

[512] E. Formanek, "Polynomial Identities and the Cayley-Hamilton Theorem," *Mathematical Intelligencer*, Vol. 11, pp. 37–39, 1989. (Cited on pages 161, 233, and 283.)

[513] E. Formanek, *The Polynomial Identities and Invariants of $n \times n$ Matrices*. Providence: American Mathematical Society, 1991. (Cited on pages 161, 233, and 283.)

[514] L. R. Foulds, *Graph Theory Applications*. New York: Springer, 1992. (Cited on page xvii.)

[515] B. A. Francis, *A Course in H_∞ Control Theory*. New York: Springer, 1987. (Cited on page xvii.)

[516] J. Franklin, *Matrix Theory*. Englewood Cliffs: Prentice-Hall, 1968. (Cited on page xix.)

[517] M. Frazier, *An Introduction to Wavelets through Linear Algebra*. New York: Springer, 1999. (Cited on page xvii.)

[518] S. Friedland, "A Note on a Determinantal Inequality," *Lin. Alg. Appl.*, Vol. 141, pp. 221–222, 1990. (Cited on page 659.)

[519] S. Friedland, D. Hershkowitz, and S. M. Rump, "Positive Entries of Stable Matrices," *Elec. J. Lin. Alg.*, Vol. 12, pp. 17–24, 2005. (Cited on page 777.)

[520] S. Friedland and A. Torokhti, "Generalized Rank-Constrained Matrix Approximation," *SIAM J. Matrix Anal. Appl.*, Vol. 29, pp. 656–659, 2007. (Cited on pages 672
and 679.)

[521] P. A. Fuhrmann, *A Polynomial Approach to Linear Algebra.* New York: Springer,
1996. (Cited on pages 208, 209, 277, 279, 280, 307, 336, and 520.)

[522] J. I. Fujii and M. Fujii, "Kolmogorov's Complexity for Positive Definite Matrices,"
Lin. Alg. Appl., Vol. 341, pp. 171–180, 2002. (Cited on page 552.)

[523] J. I. Fujii, M. Fujii, T. Furuta, and R. Nakamoto, "Norm Inequalities Equivalent
to Heinz Inequality," *Proc. Amer. Math. Soc.*, Vol. 118, pp. 827–830, 1993. (Cited
on page 646.)

[524] J.-I. Fujii, S. Izumino, and Y. Seo, "Determinant for Positive Operators and
Specht's Theorem," *Scientiae Mathematicae*, Vol. 1, pp. 307–310, 1998. (Cited
on page 57.)

[525] M. Fujii, E. Kamei, and R. Nakamoto, "On a Question of Furuta on Chaotic
Order," *Lin. Alg. Appl.*, Vol. 341, pp. 119–129, 2002. (Cited on pages 507, 513,
and 575.)

[526] M. Fujii, E. Kamei, and R. Nakamoto, "On a Question of Furuta on Chaotic Order,
II," *Math. J. Okayama University*, Vol. 45, pp. 123–131, 2003. (Cited on page 513.)

[527] M. Fujii and F. Kubo, "Buzano's Inequality and Bounds for Roots of Algebraic
Equations," *Proc. Amer. Math. Soc.*, Vol. 117, pp. 359–361, 1993. (Cited on
pages 74, 620, and 780.)

[528] M. Fujii, S. H. Lee, Y. Seo, and D. Jung, "Reverse Inequalities on Chaotically
Geometric Mean via Specht Ratio," *Math. Ineq. Appl.*, Vol. 6, pp. 509–519, 2003.
(Cited on page 37.)

[529] M. Fujii, S. H. Lee, Y. Seo, and D. Jung, "Reverse Inequalities on Chaotically
Geometric Mean via Specht Ratio, II," *J. Ineq. Pure Appl. Math.*, Vol. 4/40, pp.
1–8, 2003. (Cited on page 57.)

[530] M. Fujii and R. Nakamoto, "Rota's Theorem and Heinz Inequalities," *Lin. Alg.
Appl.*, Vol. 214, pp. 271–275, 1995. (Cited on page 646.)

[531] M. Fujii, Y. Seo, and M. Tominaga, "Golden-Thompson Type Inequalities Related
to a Geometric Mean via Specht's Ratio," *Math. Ineq. Appl.*, Vol. 5, pp. 573–582,
2002. (Cited on page 752.)

[532] A. T. Fuller, "Conditions for a Matrix to Have Only Characteristic Roots with
Negative Real Parts," *J. Math. Anal. Appl.*, Vol. 23, pp. 71–98, 1968. (Cited on
pages 447, 454, 458, and 765.)

[533] W. Fulton and J. Harris, *Representation Theory.* New York: Springer, 2004.
(Cited on page 244.)

[534] S. Furuichi, "Matrix Trace Inequalities on the Tsallis Entropies," *J. Ineq. Pure
Appl. Math.*, Vol. 9, no. 1, pp. 1–7, 2008, Article 1. (Cited on page 527.)

[535] S. Furuichi, K. Kuriyama, and K. Yanagi, "Trace Inequalities for Products of Matrices," *Lin. Alg. Appl.*, Vol. 430, pp. 2271–2276, 2009. (Cited on page 526.)

[536] T. Furuta, "$A \geq B \geq 0$ Assures $(B^r A^p B^r)^{1/q} \geq B^{(p+2r)/q}$ for $r \geq 0$, $p \geq 0$, $q \geq 1$ with $(1+2r)q \geq p+2r$," *Proc. Amer. Math. Soc.*, Vol. 101, pp. 85–88, 1987. (Cited on page 476.)

[537] T. Furuta, "Norm Inequalities Equivalent to Löwner-Heinz Theorem," *Rev. Math. Phys.*, Vol. 1, pp. 135–137, 1989. (Cited on pages 573 and 639.)

[538] T. Furuta, "Two Operator Functions with Monotone Property," *Proc. Amer. Math. Soc.*, Vol. 111, pp. 511–516, 1991. (Cited on page 507.)

[539] T. Furuta, "A Note on the Arithmetic-Geometric-Mean Inequality for Every Unitarily Invariant Matrix Norm," *Lin. Alg. Appl.*, Vol. 208/209, pp. 223–228, 1994. (Cited on page 646.)

[540] T. Furuta, "Extension of the Furuta Inequality and Ando-Hiai Log-Majorization," *Lin. Alg. Appl.*, Vol. 219, pp. 139–155, 1995. (Cited on pages 507 and 513.)

[541] T. Furuta, "Generalizations of Kosaki Trace Inequalities and Related Trace Inequalities on Chaotic Order," *Lin. Alg. Appl.*, Vol. 235, pp. 153–161, 1996. (Cited on page 529.)

[542] T. Furuta, "Operator Inequalities Associated with Hölder-McCarthy and Kantorovich Inequalities," *J. Ineq. Appl.*, Vol. 2, pp. 137–148, 1998. (Cited on page 511.)

[543] T. Furuta, "Simple Proof of the Concavity of Operator Entropy $f(A) = -A \log A$," *Math. Ineq. Appl.*, Vol. 3, pp. 305–306, 2000. (Cited on page 485.)

[544] T. Furuta, *Invitation to Linear Operators: From Matrices to Bounded Linear Operators on a Hilbert Space.* London: Taylor and Francis, 2001. (Cited on pages 26, 27, 227, 476, 477, 479, 501, 507, 511, 513, 552, 562, 575, and 623.)

[545] T. Furuta, "Spectral Order $A \succ B$ if and only if $A^{2p-r} \geq \left(A^{-r/2} B^p A^{-r/2}\right)^{(2p-r)/(p-r)}$ for all $p > r \geq 0$ and Its Application," *Math. Ineq. Appl.*, Vol. 4, pp. 619–624, 2001. (Cited on pages 513 and 575.)

[546] T. Furuta, "The Hölder-McCarthy and the Young Inequalities Are Equivalent for Hilbert Space Operators," *Amer. Math. Monthly*, Vol. 108, pp. 68–69, 2001. (Cited on page 552.)

[547] T. Furuta, "Convergence of Logarithmic Trace Inequalities via Generalized Lie-Trotter Formulae," *Lin. Alg. Appl.*, Vol. 396, pp. 353–372, 2005. (Cited on pages 749 and 753.)

[548] T. Furuta, "Concrete Examples of Operator Monotone Functions Obtained by an Elementary Method without Appealing to Löwner Integral Representation," *Lin. Alg. Appl.*, Vol. 429, pp. 972–980, 2008. (Cited on page 514.)

[549] F. Gaines, "A Note on Matrices with Zero Trace," *Amer. Math. Monthly*, Vol. 73, pp. 630–631, 1966. (Cited on pages 199, 200, and 341.)

[550] A. Galantai, *Projectors and Projection Methods.* Dordrecht: Kluwer, 2004. (Cited on pages 221, 252, 343, 344, 358, 366, 627, and 698.)

[551] A. Galantai, "Subspaces, Angles and Pairs of Orthogonal Projections," *Lin. Multilin. Alg.*, Vol. 56, pp. 227–260, 2008. (Cited on pages 122, 227, 228, 343, 344, 358, 366, 367, and 416.)

[552] J. Gallier and D. Xu, "Computing Exponentials of Skew-Symmetric Matrices and Logarithms of Orthogonal Matrices," *Int. J. Robotics Automation*, Vol. 17, pp. 1–11, 2002. (Cited on page 214.)

[553] L. Gangsong and Z. Guobiao, "Inverse Forms of Hadamard Inequality," *SIAM J. Matrix Anal. Appl.*, Vol. 23, pp. 990–997, 2002. (Cited on page 563.)

[554] F. R. Gantmacher, *The Theory of Matrices.* New York: Chelsea, 1959, Vol. I. (Cited on pages xix and 595.)

[555] F. R. Gantmacher, *The Theory of Matrices.* New York: Chelsea, 1959, Vol. II. (Cited on pages xix, 331, 346, and 793.)

[556] D. J. H. Garling, *Inequalities: A Journey into Linear Analysis.* Cambridge: Cambridge University Press, 2007. (Cited on pages 36, 58, 59, 64, 69, 82, 176, 624, and 630.)

[557] J. Garloff and D. G. Wagner, "Hadamard Products of Stable Polynomials Are Stable," *J. Math. Anal. Appl.*, Vol. 202, pp. 797–809, 1996. (Cited on page 765.)

[558] T. Geerts, "A Necessary and Sufficient Condition for Solvability of the Linear-Quadratic Control Problem without Stability," *Sys. Contr. Lett.*, Vol. 11, pp. 47–51, 1988. (Cited on page 879.)

[559] I. M. Gelfand, M. M. Kapranov, and A. V. Zelevinsky, *Discriminants, Resultants, and Multidimensional Determinants.* Boston: Birkhauser, 1994. (Cited on page 458.)

[560] M. G. Genton, "Classes of Kernels for Machine Learning: A Statistics Perspective," *J. Machine Learning Res.*, Vol. 2, pp. 299–312, 2001. (Cited on page 489.)

[561] A. George and K. D. Ikramov, "Common Invariant Subspaces of Two Matrices," *Lin. Alg. Appl.*, Vol. 287, pp. 171–179, 1999. (Cited on pages 373, 391, and 392.)

[562] P. Gerdes, *Adventures in the World of Matrices.* Hauppauge: Nova Publishers, 2007. (Cited on page 390.)

[563] A. Gerrard and J. M. Burch, *Introduction to Matrix Methods in Optics.* New York: Wiley, 1975. (Cited on page xvii.)

[564] A. Gheondea, S. Gudder, and P. Jonas, "On the Infimum of Quantum Effects," *J. Math. Phys.*, Vol. 46, pp. 1–11, 2005, paper 062102. (Cited on page 506.)

[565] F. A. A. Ghouraba and M. A. Seoud, "Set Matrices," *Int. J. Math. Educ. Sci. Technol.*, Vol. 30, pp. 651–659, 1999. (Cited on page 252.)

[566] M. I. Gil, "On Inequalities for Eigenvalues of Matrices," *Lin. Alg. Appl.*, Vol. 184, pp. 201–206, 1993. (Cited on page 354.)

[567] R. Gilmore, *Lie Groups, Lie Algebras, and Some of Their Applications.* New York: Wiley, 1974, reprinted by Dover, Mineola, 2005. (Cited on page 252.)

[568] R. Gilmore, *Lie Groups, Physics, and Geometry: An Introduction for Physicists, Engineers, and Chemists.* Cambridge: Cambridge University Press, 2008. (Cited on page 252.)

[569] P. R. Girard, *Quaternions, Clifford Algebras and Relativistic Physics.* Boston: Birkhauser, 2007. (Cited on pages 204, 213, 214, and 246.)

[570] M. L. Glasser, "Exponentials of Certain Hilbert Space Operators," *SIAM Rev.*, Vol. 34, pp. 498–500, 1992. (Cited on page 750.)

[571] C. Godsil and G. Royle, *Algebraic Graph Theory*. New York: Springer, 2001. (Cited on page xvii.)

[572] S. K. Godunov, *Modern Aspects of Linear Algebra*. Providence: American Mathematical Society, 1998. (Cited on pages xix, 229, and 767.)

[573] I. Gohberg, P. Lancaster, and L. Rodman, *Matrix Polynomials*. New York: Academic Press, 1982. (Cited on pages 256, 307, and 396.)

[574] I. Gohberg, P. Lancaster, and L. Rodman, *Invariant Subspaces of Matrices with Applications*. New York: Wiley, 1986. (Cited on pages 366, 396, and 698.)

[575] I. Gohberg, P. Lancaster, and L. Rodman, "On Hermitian Solutions of the Symmetric Algebraic Riccati Equation," *SIAM J. Contr. Optim.*, Vol. 24, pp. 1323–1334, 1986. (Cited on pages 860, 861, and 879.)

[576] I. Gohberg, P. Lancaster, and L. Rodman, *Indefinite Linear Algebra and Applications*. Boston: Birkhauser, 2005. (Cited on pages 396 and 879.)

[577] M. A. Golberg, "The Derivative of a Determinant," *Amer. Math. Monthly*, Vol. 79, pp. 1124–1126, 1972. (Cited on pages 703 and 743.)

[578] M. Goldberg, "Mixed Multiplicativity and l_p Norms for Matrices," *Lin. Alg. Appl.*, Vol. 73, pp. 123–131, 1986. (Cited on page 640.)

[579] M. Goldberg, "Equivalence Constants for l_p Norms of Matrices," *Lin. Multilin. Alg.*, Vol. 21, pp. 173–179, 1987. (Cited on page 640.)

[580] M. Goldberg, "Multiplicativity Factors and Mixed Multiplicativity," *Lin. Alg. Appl.*, Vol. 97, pp. 45–56, 1987. (Cited on page 640.)

[581] M. Goldberg and G. Zwas, "On Matrices Having Equal Spectral Radius and Spectral Norm," *Lin. Alg. Appl.*, Vol. 8, pp. 427–434, 1974. (Cited on page 661.)

[582] H. Goller, "Shorted Operators and Rank Decomposition Matrices," *Lin. Alg. Appl.*, Vol. 81, pp. 207–236, 1986. (Cited on page 583.)

[583] G. H. Golub and C. F. Van Loan, *Matrix Computations*, 3rd ed. Baltimore: Johns Hopkins University Press, 1996. (Cited on pages xviii, 672, and 710.)

[584] N. C. Gonzalez, J. J. Koliha, and Y. Wei, "Integral Representation of the Drazin Inverse," *Elect. J. Lin. Alg.*, Vol. 9, pp. 129–131, 2002. (Cited on page 745.)

[585] F. M. Goodman, *Algebra: Abstract and Concrete*, 2nd ed. Englewood Cliffs: Prentice-Hall, 2003. (Cited on page 245.)

[586] L. E. Goodman and W. H. Warner, *Statics*. Belmont: Wadsworth Publishing Company, 1964, reprinted by Dover, Mineola, 2001. (Cited on page 204.)

[587] N. Gordon and D. Salmond, "Bayesian Pattern Matching Technique for Target Acquisition," *J. Guid. Contr. Dyn.*, Vol. 22, pp. 68–77, 1999. (Cited on page 151.)

[588] A. Goroncy and T. Rychlik, "How Deviant Can You Be? The Complete Solution," *Math. Ineq. Appl.*, Vol. 9, pp. 633–647, 2006. (Cited on page 56.)

[589] W. Govaerts and J. D. Pryce, "A Singular Value Inequality for Block Matrices," *Lin. Alg. Appl.*, Vol. 125, pp. 141–145, 1989. (Cited on page 668.)

[590] W. Govaerts and B. Sijnave, "Matrix Manifolds and the Jordan Structure of the Bialternate Matrix Product," *Lin. Alg. Appl.*, Vol. 292, pp. 245–266, 1999. (Cited on pages 454 and 458.)

[591] R. Gow, "The Equivalence of an Invertible Matrix to Its Transpose," *Lin. Alg. Appl.*, Vol. 8, pp. 329–336, 1980. (Cited on pages 340 and 384.)

[592] R. Gow and T. J. Laffey, "Pairs of Alternating Forms and Products of Two Skew-Symmetric Matrices," *Lin. Alg. Appl.*, Vol. 63, pp. 119–132, 1984. (Cited on pages 375 and 384.)

[593] A. Graham, *Kronecker Products and Matrix Calculus with Applications*. Chichester: Ellis Horwood, 1981. (Cited on page 458.)

[594] F. A. Graybill, *Matrices with Applications in Statistics*, 2nd ed. Belmont: Wadsworth, 1983. (Cited on page xvii.)

[595] J. Grcar, "Linear Algebra People," http://seesar.lbl.gov/ccse/people/grcar/index.html. (Cited on page 178.)

[596] J. F. Grcar, "A Matrix Lower Bound," Lawrence Berkeley National Laboratory, Report LBNL-50635, 2002. (Cited on pages 613, 614, and 615.)

[597] W. L. Green and T. D. Morley, "Operator Means and Matrix Functions," *Lin. Alg. Appl.*, Vol. 137/138, pp. 453–465, 1990. (Cited on pages 509 and 582.)

[598] D. H. Greene and D. E. Knuth, *Mathematics for the Analysis of Algorithms*, 3rd ed. Boston: Birkhauser, 1990. (Cited on page 19.)

[599] W. H. Greub, *Multilinear Algebra*. New York: Springer, 1967. (Cited on page 458.)

[600] W. H. Greub, *Linear Algebra*. New York: Springer, 1981. (Cited on page xix.)

[601] T. N. E. Greville, "Notes on the Generalized Inverse of a Matrix Product," *SIAM Rev.*, Vol. 8, pp. 518–521, 1966. (Cited on page 415.)

[602] T. N. E. Greville, "Solutions of the Matrix Equation $XAX = X$ and Relations between Oblique and Orthogonal projectors," *SIAM J. Appl. Math*, Vol. 26, pp. 828–832, 1974. (Cited on pages 228, 416, and 417.)

[603] R. Grone, C. R. Johnson, E. M. Sa, and H. Wolkowicz, "Normal Matrices," *Lin. Alg. Appl.*, Vol. 87, pp. 213–225, 1987. (Cited on pages 194, 205, and 372.)

[604] J. Gross, "A Note on a Partial Ordering in the Set of Hermitian Matrices," *SIAM J. Matrix Anal. Appl.*, Vol. 18, pp. 887–892, 1997. (Cited on pages 576, 578, and 579.)

[605] J. Gross, "Some Remarks on Partial Orderings of Hermitian Matrices," *Lin. Multilin. Alg.*, Vol. 42, pp. 53–60, 1997. (Cited on pages 576 and 578.)

[606] J. Gross, "More on Concavity of a Matrix Function," *SIAM J. Matrix Anal. Appl.*, Vol. 19, pp. 365–368, 1998. (Cited on page 583.)

[607] J. Gross, "On Oblique Projection, Rank Additivity and the Moore-Penrose Inverse of the Sum of Two Matrices," *Lin. Multilin. Alg.*, Vol. 46, pp. 265–275, 1999. (Cited on page 416.)

[608] J. Gross, "On the Product of Orthogonal Projectors," *Lin. Alg. Appl.*, Vol. 289, pp. 141–150, 1999. (Cited on pages 226 and 417.)

[609] J. Gross, "The Moore-Penrose Inverse of a Partitioned Nonnegative Definite Matrix," *Lin. Alg. Appl.*, Vol. 321, pp. 113–121, 2000. (Cited on page 429.)

[610] J. Gross and G. Trenkler, "On the Product of Oblique Projectors," *Lin. Multilin. Alg.*, Vol. 44, pp. 247–259, 1998. (Cited on page 221.)

[611] J. Gross and G. Trenkler, "Nonsingularity of the Difference of Two Oblique Projectors," *SIAM J. Matrix Anal. Appl.*, Vol. 21, pp. 390–395, 1999. (Cited on pages 219, 220, and 221.)

[612] J. Gross and G. Trenkler, "Product and Sum of Projections," *Amer. Math. Monthly*, Vol. 111, pp. 261–262, 2004. (Cited on page 366.)

[613] J. Gross, G. Trenkler, and S.-O. Troschke, "The Vector Cross Product in $\mathbb{C}$," *Int. J. Math. Educ. Sci. Tech.*, Vol. 30, pp. 549–555, 1999. (Cited on page 204.)

[614] J. Gross, G. Trenkler, and S.-O. Troschke, "Quaternions: Further Contributions to a Matrix Oriented Approach," *Lin. Alg. Appl.*, Vol. 326, pp. 205–213, 2001. (Cited on page 251.)

[615] J. Gross and S.-O. Troschke, "Some Remarks on Partial Orderings of Nonnegative Definite Matrices," *Lin. Alg. Appl.*, Vol. 264, pp. 457–461, 1997. (Cited on pages 576 and 579.)

[616] J. Gross and J. Yellen, *Graph Theory and Its Applications*, 2nd ed. Boca Raton: Chapman & Hall/CRC, 2005. (Cited on page xvii.)

[617] I. Grossman and W. Magnus, *Groups and Their Graphs*. Washington, DC: The Mathematical Association of America, 1992. (Cited on page 244.)

[618] L. C. Grove and C. T. Benson, *Finite Reflection Groups*, 2nd ed. New York: Springer, 1996. (Cited on page 245.)

[619] K. Guan and H. Zhu, "A Generalized Heronian Mean and Its Inequalities," *University Beogradu Publ. Elektrotehn. Fak.*, Vol. 17, pp. 1–16, 2006, http://pefmath.etf.bg.ac.yu/to%20appear/rad528.pdf. (Cited on page 41.)

[620] S. Gull, A. Lasenby, and C. Doran, "Imaginary Numbers Are not Real: The Geometric Algebra of Spacetime," *Found. Phys.*, Vol. 23, pp. 1175–1201, 1993. (Cited on pages 204, 249, and 620.)

[621] A. K. Gupta and D. K. Nagar, *Matrix Variate Distributions*. Boca Raton: CRC, 1999. (Cited on page xvii.)

[622] K. Gurlebeck and W. Sprossig, *Quaternionic and Clifford Calculus for Physicists and Engineers*. New York: Chichester, 1997. (Cited on pages 249, 251, and 740.)

[623] L. Gurvits, "Stability of Discrete Linear Inclusion," *Lin. Alg. Appl.*, Vol. 231, pp. 47–85, 1995. (Cited on page 786.)

[624] K. E. Gustafson, "Matrix Trigonometry," *Lin. Alg. Appl.*, Vol. 217, pp. 117–140, 1995. (Cited on page 633.)

[625] K. E. Gustafson and D. K. M. Rao, *Numerical Range*. New York: Springer, 1997. (Cited on page 546.)

[626] W. H. Gustafson, P. R. Halmos, and H. Radjavi, "Products of Involutions," *Lin. Alg. Appl.*, Vol. 13, pp. 157–162, 1976. (Cited on page 383.)

[627] P. W. Gwanyama, "The HM-GM-QM-QM Inequalities," *College Math. J.*, Vol. 35, pp. 47–50, 2004. (Cited on page 57.)

[628] W. M. Haddad and D. S. Bernstein, "Robust Stabilization with Positive Real Uncertainty: Beyond the Small Gain Theorem," *Sys. Contr. Lett.*, Vol. 17, pp. 191–208, 1991. (Cited on page 539.)

[629] W. M. Haddad and D. S. Bernstein, "Controller Design with Regional Pole Constraints," *IEEE Trans. Autom. Contr.*, Vol. 37, pp. 54–69, 1992. (Cited on page 871.)

[630] W. M. Haddad and V. Chellaboina, "Stability and Dissipativity Theory for Nonnegative Dynamical Systems: A Unified Analysis Framework for Biological and Physiological Systems," *Nonlinear Anal. Real World Appl.*, Vol. 6, pp. 35–65, 2005. (Cited on page 776.)

[631] W. M. Haddad and V. Chellaboina, *Nonlinear Dynamical Systems and Control.* Princeton: Princeton University Press, 2008. (Cited on page xvii.)

[632] W. M. Haddad, V. Chellaboina, and S. G. Nersesov, *Thermodynamics: A Dynamical Systems Approach.* Princeton: Princeton University Press, 2005. (Cited on pages 252 and 777.)

[633] W. M. Haddad, V. Kapila, and E. G. Collins, "Optimality Conditions for Reduced-Order Modeling, Estimation, and Control for Discrete-Time Linear Periodic Plants," *J. Math. Sys. Est. Contr.*, Vol. 6, pp. 437–460, 1996. (Cited on page 879.)

[634] W. W. Hager, "Updating the Inverse of a Matrix," *SIAM Rev.*, Vol. 31, pp. 221–239, 1989. (Cited on page 178.)

[635] W. Hahn, *Stability of Motion.* Berlin: Springer, 1967. (Cited on page 793.)

[636] E. Hairer, C. Lubich, and G. Wanner, *Geometric Numerical Integration: Structure-Preserving Algorithms for Ordinary Differential Equations.* Berlin: Springer, 2002. (Cited on page 743.)

[637] M. Hajja, "A Method for Establishing Certain Trigonometric Inequalities," *J. Ineq. Pure Appl. Math.*, Vol. 8, no. 1, pp. 1–11, 2007, Article 29. (Cited on page 171.)

[638] A. Hall, "Conditions for a Matrix Kronecker Lemma," *Lin. Alg. Appl.*, Vol. 76, pp. 271–277, 1986. (Cited on page 511.)

[639] B. C. Hall, *Lie Groups, Lie Algebras, and Representations: An Elementary Introduction.* New York: Springer, 2003. (Cited on pages 239, 718, 720, 722, 723, 724, 746, 750, and 793.)

[640] G. T. Halliwell and P. R. Mercer, "A Refinement of an Inequality from Information Theory," *J. Ineq. Pure Appl. Math.*, Vol. 5, no. 1, pp. 1–3, 2004, Article 3. (Cited on pages 30 and 74.)

[641] P. R. Halmos, *Finite-Dimensional Vector Spaces.* Princeton: Van Nostrand, 1958, reprinted by Springer, New York, 1974. (Cited on pages 341, 371, and 411.)

[642] P. R. Halmos, *A Hilbert Space Problem Book*, 2nd ed. New York: Springer, 1982. (Cited on pages 346, 374, 421, and 422.)

[643] P. R. Halmos, "Bad Products of Good Matrices," *Lin. Multilin. Alg.*, Vol. 29, pp. 1–20, 1991. (Cited on page 382.)

[644] P. R. Halmos, *Problems for Mathematicians Young and Old*. Washington, DC: Mathematical Association of America, 1991. (Cited on pages 122, 305, 496, and 747.)

[645] P. R. Halmos, *Linear Algebra Problem Book*. Washington, DC: Mathematical Association of America, 1995. (Cited on pages 98, 219, and 374.)

[646] M. Hamermesh, *Group Theory and its Application to Physical Problems*. Reading: Addison-Wesley, 1962, reprinted by Dover, Mineola, 1989. (Cited on page 244.)

[647] J. H. Han and M. D. Hirschhorn, "Another Look at an Amazing Identity of Ramanujan," *Math. Mag.*, Vol. 79, pp. 302–304, 2006. (Cited on page 139.)

[648] E. J. Hannan and M. Deistler, *The Statistical Theory of Linear Systems*. New York: Wiley, 1988. (Cited on page 879.)

[649] F. Hansen, "Extrema for Concave Operator Mappings," *Math. Japonica*, Vol. 40, pp. 331–338, 1994. (Cited on page 531.)

[650] F. Hansen, "Operator Inequalities Associated with Jensen's Inequality," in *Survey on Classical Inequalities*, T. M. Rassias, Ed. Dordrecht: Kluwer, 2000, pp. 67–98. (Cited on page 531.)

[651] F. Hansen, "Some Operator Monotone Functions," *Lin. Alg. Appl.*, Vol. 430, pp. 795–799, 2009. (Cited on page 40.)

[652] A. J. Hanson, *Visualizing Quaternions*. Amsterdam: Elsevier, 2006. (Cited on pages 249 and 251.)

[653] G. Hardy, J. E. Littlewood, and G. Polya, *Inequalities*. Cambridge: Cambridge University Press, 1988. (Cited on page 84.)

[654] J. M. Harris, J. L. Hirst, and M. J. Missinghoff, *Combinatorics and Graph Theory*, 2nd ed. New York: Springer, 2008. (Cited on page xvii.)

[655] L. A. Harris, "Factorizations of Operator Matrices," *Lin. Alg. Appl.*, Vol. 225, pp. 37–41, 1995. (Cited on pages 520 and 521.)

[656] L. A. Harris, "The Inverse of a Block Matrix," *Amer. Math. Monthly*, Vol. 102, pp. 656–657, 1995. (Cited on page 161.)

[657] W. A. Harris, J. P. Fillmore, and D. R. Smith, "Matrix Exponentials–Another Approach," *SIAM Rev.*, Vol. 43, pp. 694–706, 2001. (Cited on page 737.)

[658] G. W. Hart, *Multidimensional Analysis: Algebras and Systems for Science and Engineering*. New York: Springer, 1995. (Cited on pages xvii and 252.)

[659] D. J. Hartfiel, *Nonhomogeneous Matrix Products*. Singapore: World Scientific, 2002. (Cited on pages xvii and 785.)

[660] R. Hartwig, X. Li, and Y. Wei, "Representations for the Drazin Inverse of a 2×2 Block Matrix," *SIAM J. Matrix Anal. Appl.*, Vol. 27, pp. 757–771, 2006. (Cited on page 429.)

[661] R. E. Hartwig, "Resultants and the Solutions of $AX - XB = -C$," *SIAM J. Appl. Math.*, Vol. 23, pp. 104–117, 1972. (Cited on page 870.)

[662] R. E. Hartwig, "Block Generalized Inverses," *Arch. Rat. Mech. Anal.*, Vol. 61, pp. 197–251, 1976. (Cited on page 429.)

[663] R. E. Hartwig, "A Note on the Partial Ordering of Positive Semi-Definite Matrices," *Lin. Multilin. Alg.*, Vol. 6, pp. 223–226, 1978. (Cited on pages 576, 579, and 580.)

[664] R. E. Hartwig, "How to Partially Order Regular Elements," *Math. Japonica*, Vol. 25, pp. 1–13, 1980. (Cited on page 129.)

[665] R. E. Hartwig and I. J. Katz, "On Products of EP Matrices," *Lin. Alg. Appl.*, Vol. 252, pp. 339–345, 1997. (Cited on page 418.)

[666] R. E. Hartwig and M. S. Putcha, "When Is a Matrix a Difference of Two Idempotents?" *Lin. Multilin. Alg.*, Vol. 26, pp. 267–277, 1990. (Cited on pages 351 and 367.)

[667] R. E. Hartwig and M. S. Putcha, "When Is a Matrix a Sum of Idempotents?" *Lin. Multilin. Alg.*, Vol. 26, pp. 279–286, 1990. (Cited on page 395.)

[668] R. E. Hartwig and K. Spindelböck, "Matrices for which A^* and A^+ Commute," *Lin. Multilin. Alg.*, Vol. 14, pp. 241–256, 1984. (Cited on pages 344, 406, 407, and 436.)

[669] R. E. Hartwig and G. P. H. Styan, "On Some Characterizations of the "Star" Partial Ordering for Matrices and Rank Subtractivity," *Lin. Alg. Appl.*, Vol. 82, pp. 145–161, 1989. (Cited on pages 129, 130, 422, and 579.)

[670] R. E. Hartwig, G. Wang, and Y. Wei, "Some Additive Results on Drazin Inverse," *Lin. Alg. Appl.*, Vol. 322, pp. 207–217, 2001. (Cited on page 431.)

[671] D. A. Harville, *Matrix Algebra from a Statistician's Perspective.* New York: Springer, 1997. (Cited on pages xvii, 216, 219, 405, 411, 413, 417, 420, 535, 557, and 705.)

[672] J. Hauke and A. Markiewicz, "On Partial Orderings on the Set of Rectangular Matrices," *Lin. Alg. Appl.*, Vol. 219, pp. 187–193, 1995. (Cited on page 577.)

[673] M. L. J. Hautus, "Controllability and Observability Conditions of Linear Autonomous Systems," *Proc. Koniklijke Akademic Van Wetenshappen*, Vol. 72, pp. 443–448, 1969. (Cited on page 879.)

[674] J. Havil, *Gamma: Exploring Euler's Constant.* Princeton: Princeton University Press, 2003. (Cited on page 21.)

[675] T. Haynes, "Stable Matrices, the Cayley Transform, and Convergent Matrices," *Int. J. Math. Math. Sci.*, Vol. 14, pp. 77–81, 1991. (Cited on page 783.)

[676] E. V. Haynsworth, "Applications of an Inequality for the Schur Complement," *Proc. Amer. Math. Soc.*, Vol. 24, pp. 512–516, 1970. (Cited on pages 485, 521, and 596.)

[677] E. Hecht, *Optics*, 4th ed. Reading: Addison Wesley, 2002. (Cited on page xvii.)

[678] C. Heij, A. Ran, and F. van Schagen, Eds., *Introduction to Mathematical Systems Theory: Linear Systems, Identification and Control.* Basel: Birkhauser, 2007. (Cited on page 879.)

[679] C. Heij, A. Ran, and F. van Schagen, *Introduction to Mathematical Systems Theory: Linear Systems, Identification and Control.* Basel: Birkhauser, 2007. (Cited on page 879.)

[680] G. Heinig, "Matrix Representations of Bezoutians," *Lin. Alg. Appl.*, Vol. 223/224, pp. 337–354, 1995. (Cited on page 279.)

[681] U. Helmke and P. A. Fuhrmann, "Bezoutians," *Lin. Alg. Appl.*, Vol. 122–124, pp. 1039–1097, 1989. (Cited on page 279.)

[682] B. W. Helton, "Logarithms of Matrices," *Proc. Amer. Math. Soc.*, Vol. 19, pp. 733–738, 1968. (Cited on page 793.)

[683] H. V. Henderson, F. Pukelsheim, and S. R. Searle, "On the History of the Kronecker Product," *Lin. Multilin. Alg.*, Vol. 14, pp. 113–120, 1983. (Cited on page 458.)

[684] H. V. Henderson and S. R. Searle, "On Deriving the Inverse of a Sum of Matrices," *SIAM Rev.*, Vol. 23, pp. 53–60, 1981. (Cited on pages 157 and 178.)

[685] H. V. Henderson and S. R. Searle, "The Vec-Permutation Matrix, The Vec Operator and Kronecker Products: A Review," *Lin. Multilin. Alg.*, Vol. 9, pp. 271–288, 1981. (Cited on page 458.)

[686] J. Herman, R. Kucera, and J. Simsa, *Equations and Inequalities.* New York: Springer, 2000. (Cited on pages 19, 20, 26, 27, 29, 35, 39, 40, 43, 44, 45, 46, 47, 48, 50, 51, 54, 57, and 66.)

[687] D. Hershkowitz, "Positive Semidefinite Pattern Decompositions," *SIAM J. Matrix Anal. Appl.*, Vol. 11, pp. 612–619, 1990. (Cited on page 522.)

[688] D. Hestenes, *Space-Time Algebra.* New York: Gordon and Breach, 1966. (Cited on page 249.)

[689] D. Hestenes, *New Foundations for Classical Mechanics.* Dordrecht: Kluwer, 1986. (Cited on pages 204 and 249.)

[690] D. Hestenes and G. Sobczyk, *Clifford Algebra to Geometric Calculus: A Unified Language for Mathematics and Physics.* Dordrecht: D. Riedel, 1984. (Cited on pages 204 and 249.)

[691] F. Hiai and H. Kosaki, Eds., *Means of Hilbert Space Operators.* Berlin: Springer, 2003. (Cited on pages 32, 38, and 489.)

[692] F. Hiai and D. Petz, "The Golden-Thompson Trace Inequality Is Complemented," *Lin. Alg. Appl.*, Vol. 181, pp. 153–185, 1993. (Cited on pages 507 and 753.)

[693] F. Hiai and X. Zhan, "Submultiplicativity vs Subadditivity for Unitarily Invariant Norms," *Lin. Alg. Appl.*, Vol. 377, pp. 155–164, 2004. (Cited on page 637.)

[694] D. J. Higham and N. J. Higham, *Matlab Guide*, 2nd ed. Philadelphia: SIAM, 2005. (Cited on page 178.)

[695] N. J. Higham, "Newton's Method for the Matrix Square Root," *Math. Computation*, Vol. 46, pp. 537–549, 1986. (Cited on page 381.)

[696] N. J. Higham, "Computing Real Square Roots of a Real Matrix," *Lin. Alg. Appl.*, Vol. 88/89, pp. 405–430, 1987. (Cited on page 381.)

[697] N. J. Higham, "Matrix Nearness Problems and Applications," in *Applications of Matrix Theory*, M. J. C. Gover and S. Barnett, Eds. Oxford: Oxford University Press, 1989, pp. 1–27. (Cited on pages 666, 768, and 769.)

[698] N. J. Higham, "Estimating the Matrix p-Norm," *Numer. Math.*, Vol. 62, pp. 539–555, 1992. (Cited on pages 626, 630, and 631.)

[699] N. J. Higham, *Accuracy and Stability of Numerical Algorithms*, 2nd ed. Philadelphia: SIAM, 2002. (Cited on pages xviii, 234, 241, 357, 360, 626, 629, 630, 631, 632, 661, 666, 673, and 680.)

[700] N. J. Higham, "Cayley, Sylvester, and Early Matrix Theory," *Lin. Alg. Appl.*, Vol. 428, pp. 39–43, 2008. (Cited on page 284.)

[701] N. J. Higham, *Functions of Matrices: Theory and Computation.* Philadelphia: SIAM, 2008. (Cited on pages xviii, 78, 81, 381, 393, 394, 640, 690, 698, 699, 722, 742, 747, 751, 759, 761, 793, and 876.)

[702] G. N. Hile and P. Lounesto, "Matrix Representations of Clifford Algebras," *Lin. Alg. Appl.*, Vol. 128, pp. 51–63, 1990. (Cited on page 249.)

[703] R. D. Hill, R. G. Bates, and S. R. Waters, "On Centrohermitian Matrices," *SIAM J. Matrix Anal. Appl.*, Vol. 11, pp. 128–133, 1990. (Cited on page 242.)

[704] R. D. Hill and E. E. Underwood, "On the Matrix Adjoint (Adjugate)," *SIAM J. Alg. Disc. Meth.*, Vol. 6, pp. 731–737, 1985. (Cited on page 154.)

[705] C.-J. Hillar, "Advances on the Bessis-Moussa-Villani Trace Conjecture," *Lin. Alg. Appl.*, Vol. 426, pp. 130–142, 2007. (Cited on page 530.)

[706] L. O. Hilliard, "The Case of Equality in Hopf's Inequality," *SIAM J. Alg. Disc. Meth.*, Vol. 8, pp. 691–709, 1987. (Cited on page 306.)

[707] D. Hinrichsen, E. Plischke, and F. Wirth, "Robustness of Transient Behavior," in *Unsolved Problems in Mathematical Systems and Control Theory*, V. D. Blondel and A. Megretski, Eds. Princeton: Princeton University Press, 2004. (Cited on page 767.)

[708] D. Hinrichsen and A. J. Pritchard, *Mathematical Systems Theory I: Modelling, State Space Analysis, Stability and Robustness.* Berlin: Springer, 2005. (Cited on pages 347, 355, 356, 609, 676, 701, 758, 762, and 778.)

[709] M. W. Hirsch and S. Smale, *Differential Equations, Dynamical Systems, and Linear Algebra.* San Diego: Academic Press, 1974. (Cited on pages xvii and 339.)

[710] M. W. Hirsch, S. Smale, and R. L. Devaney, *Differential Equations, Dynamical Systems and an Introduction to Chaos*, 2nd ed. New York: Elsevier, 2003. (Cited on page xvii.)

[711] O. Hirzallah, "Inequalities for Sums and Products of Operators," *Lin. Alg. Appl.*, Vol. 407, pp. 32–42, 2005. (Cited on page 671.)

[712] O. Hirzallah and F. Kittaneh, "Matrix Young Inequalities for the Hilbert-Schmidt Norm," *Lin. Alg. Appl.*, Vol. 308, pp. 77–84, 2000. (Cited on page 671.)

[713] O. Hirzallah and F. Kittaneh, "Commutator Inequalities for Hilbert-Schmidt Norm," *J. Math. Anal. Appl.*, Vol. 268, pp. 67–73, 2002. (Cited on pages 42 and 77.)

[714] O. Hirzallah and F. Kittaneh, "Non-Commutative Clarkson Inequalities for Unitarily Invariant Norms," *Pac. J. Math.*, Vol. 202, pp. 363–369, 2002. (Cited on pages 642 and 643.)

[715] O. Hirzallah and F. Kittaneh, "Norm Inequalities for Weighted Power Means of Operators," *Lin. Alg. Appl.*, Vol. 341, pp. 181–193, 2002. (Cited on page 552.)

[716] O. Hirzallah and F. Kittaneh, "Inequalities for Sums and Direct Sums of Hilbert Space Operators," *Lin. Alg. Appl.*, Vol. 424, pp. 71–82, 2007. (Cited on pages 670 and 671.)

[717] A. Hmamed, "A Matrix Inequality," *Int. J. Contr.*, Vol. 49, pp. 363–365, 1989. (Cited on page 595.)

[718] J. B. Hoagg, J. Chandrasekar, and D. S. Bernstein, "On the Zeros, Initial Undershoot, and Relative Degree of Lumped-Parameter Structures," *ASME J. Dyn. Sys. Meas. Contr.*, Vol. 129, pp. 493–502, 2007. (Cited on page 238.)

[719] K. Hoffman and R. Kunze, *Linear Algebra*, 2nd ed. Englewood Cliffs: Prentice-Hall, 1971. (Cited on page xix.)

[720] L. Hogben, Ed., *Handbook of Linear Algebra*. Boca Raton: Chapman & Hall/CRC, 2007. (Cited on page xvii.)

[721] R. R. Holmes and T. Y. Tam, "Group Representations," in *Handbook of Linear Algebra*, L. Hogben, Ed. Boca Raton: Chapman & Hall/CRC, 2007, pp. 68-1–68-11. (Cited on page 244.)

[722] O. Holtz, "On Convergence of Infinite Matrix Products," *Elec. J. Lin. Alg.*, Vol. 7, pp. 178–181, 2000. (Cited on page 786.)

[723] O. Holtz, "Hermite-Biehler, Routh-Hurwitz, and Total Positivity," *Lin. Alg. Appl.*, Vol. 372, pp. 105–110, 2003. (Cited on page 764.)

[724] Y. Hong and R. A. Horn, "The Jordan Canonical Form of a Product of a Hermitian and a Positive Semidefinite Matrix," *Lin. Alg. Appl.*, Vol. 147, pp. 373–386, 1991. (Cited on page 382.)

[725] Y. P. Hong and C.-T. Pan, "A Lower Bound for the Smallest Singular Value," *Lin. Alg. Appl.*, Vol. 172, pp. 27–32, 1992. (Cited on page 664.)

[726] K. J. Horadam, *Hadamard Matrices and Their Applications*. Princeton: Princeton University Press, 2007. (Cited on page 391.)

[727] A. Horn, "Doubly Stochastic Matrices and the Diagonal of a Rotation Matrix," *Amer. J. Math.*, Vol. 76, pp. 620–630, 1954. (Cited on page 562.)

[728] R. A. Horn and C. R. Johnson, *Matrix Analysis*. Cambridge: Cambridge University Press, 1985. (Cited on pages 151, 201, 205, 233, 276, 295, 297, 298, 299, 300, 306, 307, 319, 342, 348, 374, 377, 378, 392, 394, 454, 470, 474, 490, 522, 541, 555, 558, 561, 562, 563, 564, 585, 591, 592, 595, 600, 604, 605, 607, 628, 629, 633, 636, 658, 659, and 667.)

[729] R. A. Horn and C. R. Johnson, "Hadamard and Conventional Submultiplicativity for Unitarily Invariant Norms on Matrices," *Lin. Multilin. Alg.*, Vol. 20, pp. 91–106, 1987. (Cited on page 634.)

[730] R. A. Horn and C. R. Johnson, *Topics in Matrix Analysis*. Cambridge: Cambridge University Press, 1991. (Cited on pages 176, 177, 209, 302, 355, 356, 373, 380, 444, 447, 457, 474, 486, 517, 546, 547, 555, 560, 615, 616, 617, 633, 634, 649, 660, 663, 664, 670, 674, 675, 703, 705, 718, 719, 722, 743, 756, 759, 761, 771, 776, and 871.)

[731] R. A. Horn and R. Mathias, "An Analog of the Cauchy-Schwarz Inequality for Hadamard Products and Unitarily Invariant Norms," *SIAM J. Matrix Anal. Appl.*, Vol. 11, pp. 481–498, 1990. (Cited on pages 649, 661, and 675.)

[732] R. A. Horn and R. Mathias, "Cauchy-Schwarz Inequalities Associated with Positive Semidefinite Matrices," *Lin. Alg. Appl.*, Vol. 142, pp. 63–82, 1990. (Cited on pages 593, 632, and 638.)

[733] R. A. Horn and R. Mathias, "Block-Matrix Generalizations of Schur's Basic Theorems on Hadamard Products," *Lin. Alg. Appl.*, Vol. 172, pp. 337–346, 1992. (Cited on page 458.)

[734] R. A. Horn and I. Olkin, "When Does $A^*A = B^*B$ and Why Does One Want to Know?" *Amer. Math. Monthly*, Vol. 103, pp. 470–482, 1996. (Cited on page 348.)

[735] R. A. Horn and G. G. Piepmeyer, "Two Applications of the Theory of Primary Matrix Functions," *Lin. Alg. Appl.*, Vol. 361, pp. 99–106, 2003. (Cited on page 705.)

[736] R. A. Horn and F. Zhang, "Basic Properties of the Schur Complement," in *The Schur Complement and Its Applications*, F. Zhang, Ed. New York: Springer, 2004, pp. 17–46. (Cited on pages 362, 430, 507, 539, and 542.)

[737] B. G. Horne, "Lower Bounds for the Spectral Radius of a Matrix," *Lin. Alg. Appl.*, Vol. 263, pp. 261–273, 1997. (Cited on page 661.)

[738] H.-C. Hou and H.-K. Du, "Norm Inequalities of Positive Operator Matrices," *Integ. Eqns. Operator Theory*, Vol. 22, pp. 281–294, 1995. (Cited on pages 516, 573, and 668.)

[739] S.-H. Hou, "A Simple Proof of the Leverrier-Faddeev Characteristic Polynomial Algorithm," *SIAM Rev.*, Vol. 40, pp. 706–709, 1998. (Cited on page 307.)

[740] A. S. Householder, *The Theory of Matrices in Numerical Analysis*. New York: Blaisdell, 1964, reprinted by Dover, New York, 1975. (Cited on pages xviii, 412, 669, and 695.)

[741] A. S. Householder, "Bezoutiants, Elimination and Localization," *SIAM Rev.*, Vol. 12, pp. 73–78, 1970. (Cited on page 279.)

[742] A. S. Householder and J. A. Carpenter, "The Singular Values of Involutory and of Idempotent Matrices," *Numer. Math.*, Vol. 5, pp. 234–237, 1963. (Cited on page 358.)

[743] R. Howe, "Very Basic Lie Theory," *Amer. Math. Monthly*, Vol. 90, pp. 600–623, 1983. (Cited on pages 252 and 793.)

[744] J. M. Howie, *Complex Analysis*. New York: Springer, 2003. (Cited on page 84.)

[745] R. A. Howland, *Intermediate Dynamics: A Linear Algebraic Approach*. New York: Springer, 2006. (Cited on page xvii.)

[746] P.-F. Hsieh and Y. Sibuya, *Basic Theory of Ordinary Differential Equations*. New York: Springer, 1999. (Cited on pages xvii and 339.)

[747] G.-D. Hu and G.-H. Hu, "A Relation between the Weighted Logarithmic Norm of a Matrix and the Lyapunov Equation," *Numer. Math. BIT*, Vol. 40, pp. 606–610, 2000. (Cited on page 767.)

[748] S. Hu-yun, "Estimation of the Eigenvalues of AB for $A > 0$ and $B > 0$," *Lin. Alg. Appl.*, Vol. 73, pp. 147–150, 1986. (Cited on pages 138 and 572.)

[749] R. Huang, "Some Inequalities for Hadamard Products of Matrices," *Lin. Multilin. Alg.*, Vol. 56, pp. 543–554, 2008. (Cited on pages 661, 675, and 676.)

[750] R. Huang, "Some Inequalities for the Hadamard Product and the Fan Product of Matrices," *Lin. Alg. Appl.*, Vol. 428, pp. 1551–1559, 2008. (Cited on page 457.)

[751] T.-Z. Huang and L. Wang, "Improving Bounds for Eigenvalues of Complex Matrices Using Traces," *Lin. Alg. Appl.*, Vol. 426, pp. 841–854, 2007. (Cited on pages 56 and 655.)

[752] X. Huang, W. Lin, and B. Yang, "Global Finite-Time Stabilization of a Class of Uncertain Nonlinear Systems," *Automatica*, Vol. 41, pp. 881–888, 2005. (Cited on page 42.)

[753] P. C. Hughes, *Spacecraft Attitude Dynamics*. New York: Wiley, 1986. (Cited on pages xvii and 741.)

[754] M. Huhtanen, "Real Linear Kronecker Product Operations," *Lin. Alg. Appl.*, Vol. 418, pp. 347–361, 2006. (Cited on page 77.)

[755] Q. Hui and W. M. Haddad, "$\mathcal{H}_2$ Optimal Semistable Stabilisation for Linear Discrete-Time Dynamical Systems with Applications to Network Consensus," *Int. J. Contr.*, Vol. 82, pp. 456–469, 2009. (Cited on page 782.)

[756] S. Humphries and C. Krattenthaler, "Trace Identities from Identities for Determinants," *Lin. Alg. Appl.*, Vol. 411, pp. 328–342, 2005. (Cited on page 143.)

[757] C. H. Hung and T. L. Markham, "The Moore-Penrose Inverse of a Partitioned Matrix," *Lin. Alg. Appl.*, Vol. 11, pp. 73–86, 1975. (Cited on page 429.)

[758] J. J. Hunter, "Generalized Inverses and Their Application to Applied Probability Problems," *Lin. Alg. Appl.*, Vol. 45, pp. 157–198, 1982. (Cited on page 438.)

[759] D. C. Hyland and D. S. Bernstein, "The Optimal Projection Equations for Fixed-Order Dynamic Compensation," *IEEE Trans. Autom. Contr.*, Vol. AC-29, pp. 1034–1037, 1984. (Cited on page 879.)

[760] D. C. Hyland and E. G. Collins, "Block Kronecker Products and Block Norm Matrices in Large-Scale Systems Analysis," *SIAM J. Matrix Anal. Appl.*, Vol. 10, pp. 18–29, 1989. (Cited on page 458.)

[761] N. H. Ibragimov, *Elementary Lie Group Analysis and Ordinary Differential Equations*. Chichester: Wiley, 1999. (Cited on page 793.)

[762] Y. Ikebe and T. Inagaki, "An Elementary Approach to the Functional Calculus for Matrices," *Amer. Math. Monthly*, Vol. 93, pp. 390–392, 1986. (Cited on page 705.)

[763] K. D. Ikramov, "A Simple Proof of the Generalized Schur Inequality," *Lin. Alg. Appl.*, Vol. 199, pp. 143–149, 1994. (Cited on pages 653 and 655.)

[764] V. Ionescu, C. Oar, and M. Weiss, *Generalized Riccati Theory and Robust Control*. Chichester: Wiley, 1999. (Cited on pages xvii and 879.)

[765] I. Ipsen and C. Meyer, "The Angle between Complementary Subspaces," *Amer. Math. Monthly*, Vol. 102, pp. 904–911, 1995. (Cited on pages 122, 228, 358, 366, and 416.)

[766] A. Iserles, "Solving Linear Ordinary Differential Equations by Exponentials of Iterated Commutators," *Numer. Math.*, Vol. 45, pp. 183–199, 1984. (Cited on page 743.)

[767] A. Iserles, H. Z. Munthe-Kaas, S. P. Norsett, and A. Zanna, "Lie-Group Methods," *Acta Numerica*, Vol. 9, pp. 215–365, 2000. (Cited on pages 204, 739, 743, and 748.)

[768] T. Ito, "Mixed Arithmetic and Geometric Means and Related Inequalities," *J. Ineq. Pure Appl. Math.*, Vol. 9, no. 3, pp. 1–21, 2008, Article 65. (Cited on page 46.)

[769] Y. Ito, S. Hattori, and H. Maeda, "On the Decomposition of a Matrix into the Sum of Stable Matrices," *Lin. Alg. Appl.*, Vol. 297, pp. 177–182, 1999. (Cited on page 773.)

[770] S. Izumino, H. Mori, and Y. Seo, "On Ozeki's Inequality," *J. Ineq. Appl.*, Vol. 2, pp. 235–253, 1998. (Cited on pages 71 and 551.)

[771] D. H. Jacobson, D. H. Martin, M. Pachter, and T. Geveci, *Extensions of Linear Quadratic Control Theory*. Berlin: Springer, 1980. (Cited on page 879.)

[772] A. Jeffrey, *Handbook of Mathematical Formulas and Integrals*. San Diego: Academic Press, 1995. (Cited on pages 81, 82, 84, and 705.)

[773] C. Jeffries, V. Klee, and P. van der Driessche, "Quality Stability of Linear Systems," *Lin. Alg. Appl.*, Vol. 87, pp. 1–48, 1987. (Cited on page 777.)

[774] A. Jennings and J. J. McKeown, *Matrix Computation*, 2nd ed. New York: Wiley, 1992. (Cited on page xviii.)

[775] G. A. Jennings, *Modern Geometry with Applications*. New York: Springer, 1997. (Cited on page 175.)

[776] S. T. Jensen and G. P. H. Styan, "Some Comments and a Bibliography on the Laguerre-Samuelson Inequality with Extensions and Applications in Statistics and Matrix Theory," in *Analytic and Geometric Inequalities and Applications*, T. M. Rassias and H. M. Srivastava, Eds. Dordrecht: Kluwer, 1999, pp. 151–181. (Cited on pages 56 and 499.)

[777] J. Ji, "Explicit Expressions of the Generalized Inverses and Condensed Cramer Rules," *Lin. Alg. Appl.*, Vol. 404, pp. 183–192, 2005. (Cited on page 141.)

[778] G. Jia and J. Cao, "A New Upper Bound of the Logarithmic Mean," *J. Ineq. Pure Appl. Math.*, Vol. 4, no. 4, pp. 1–4, 2003, Article 80. (Cited on pages 30 and 41.)

[779] C.-C. Jiang, "On Products of Two Hermitian Operators," *Lin. Alg. Appl.*, Vol. 430, pp. 553–557, 2009. (Cited on page 340.)

[780] Y. Jiang, W. W. Hager, and J. Li, "The Geometric Mean Decomposition," *Lin. Alg. Appl.*, Vol. 396, pp. 373–384, 2005. (Cited on page 344.)

[781] D. Jocic and F. Kittaneh, "Some Perturbation Inequalities for Self-Adjoint Operators," *J. Operator Theory*, Vol. 31, pp. 3–10, 1994. (Cited on page 645.)

[782] C. R. Johnson, "An Inequality for Matrices Whose Symmetric Part Is Positive Definite," *Lin. Alg. Appl.*, Vol. 6, pp. 13–18, 1973. (Cited on page 536.)

[783] C. R. Johnson, "Inequalities for a Complex Matrix whose Real Part Is Positive Definite," *Trans. Amer. Math. Soc.*, Vol. 212, pp. 149–154, 1975. (Cited on page 534.)

[784] C. R. Johnson, "The Inertia of a Product of Two Hermitian Matrices," *J. Math. Anal. Appl.*, Vol. 57, pp. 85–90, 1977. (Cited on page 336.)

[785] C. R. Johnson, "Closure Properties of Certain Positivity Classes of Matrices under Various Algebraic Operations," *Lin. Alg. Appl.*, Vol. 97, pp. 243–247, 1987. (Cited on page 765.)

[786] C. R. Johnson, "A Gersgorin-Type Lower Bound for the Smallest Singular Value," *Lin. Alg. Appl.*, Vol. 112, pp. 1–7, 1989. (Cited on page 664.)

[787] C. R. Johnson and L. Elsner, "The Relationship between Hadamard and Conventional Multiplication for Positive Definite Matrices," *Lin. Alg. Appl.*, Vol. 92, pp. 231–240, 1987. (Cited on page 587.)

[788] C. R. Johnson, M. Neumann, and M. J. Tsatsomeros, "Conditions for the Positivity of Determinants," *Lin. Multilin. Alg.*, Vol. 40, pp. 241–248, 1996. (Cited on page 238.)

[789] C. R. Johnson and M. Newman, "A Surprising Determinantal Inequality for Real Matrices," *Math. Ann.*, Vol. 247, pp. 179–186, 1980. (Cited on page 143.)

[790] C. R. Johnson and P. Nylen, "Monotonicity Properties of Norms," *Lin. Alg. Appl.*, Vol. 148, pp. 43–58, 1991. (Cited on page 680.)

[791] C. R. Johnson, K. Okubo, and R. Beams, "Uniqueness of Matrix Square Roots," *Lin. Alg. Appl.*, Vol. 323, pp. 51–60, 2001. (Cited on page 381.)

[792] C. R. Johnson and L. Rodman, "Convex Sets of Hermitian Matrices with Constant Inertia," *SIAM J. Alg. Disc. Meth.*, Vol. 6, pp. 351–359, 1985. (Cited on pages 335 and 337.)

[793] C. R. Johnson and R. Schreiner, "The Relationship between AB and BA," *Amer. Math. Monthly*, Vol. 103, pp. 578–582, 1996. (Cited on page 378.)

[794] C. R. Johnson and H. M. Shapiro, "Mathematical Aspects of the Relative Gain Array $A \circ A^{-\mathrm{T}}$," *SIAM J. Alg. Disc. Meth.*, Vol. 7, pp. 627–644, 1986. (Cited on pages 456 and 585.)

[795] C. R. Johnson and R. L. Smith, "Linear Interpolation Problems for Matrix Classes and a Transformational Characterization of M-Matrices," *Lin. Alg. Appl.*, Vol. 330, pp. 43–48, 2001. (Cited on pages 200, 212, and 738.)

[796] C. R. Johnson and T. Szulc, "Further Lower Bounds for the Smallest Singular Value," *Lin. Alg. Appl.*, Vol. 272, pp. 169–179, 1998. (Cited on page 664.)

[797] C. R. Johnson and F. Zhang, "An Operator Inequality and Matrix Normality," *Lin. Alg. Appl.*, Vol. 240, pp. 105–110, 1996. (Cited on page 552.)

[798] M. Jolly, "On the Calculus of Complex Matrices," *Int. J. Contr.*, Vol. 61, pp. 749–755, 1995. (Cited on page 705.)

[799] T. F. Jordan, *Quantum Mechanics in Simple Matrix Form*. New York: Wiley, 1986, reprinted by Dover, Mineola, 2005. (Cited on page 251.)

[800] E. A. Jorswieck and H. Boche, "Majorization and Matrix Monotone Functions in Wireless Communications," *Found. Trends Comm. Inform. Theory*, Vol. 3, pp. 553–701, 2006. (Cited on page xvii.)

[801] E. A. Jorswieck and H. Boche, "Performance Analysis of MIMO Systems in Spatially Correlated Fading Using Matrix-Monotone Functions," *IEICE Trans. Fund.*, Vol. E89-A, pp. 1454–1472, 2006. (Cited on pages xvii and 176.)

[802] A. Joseph, A. Melnikov, and R. Rentschler, Eds., *Studies in Memory of Issai Schur*. Boston: Birkhauser, 2002. (Cited on page 596.)

[803] D. Joyner, *Adventures in Group Theory*. Baltimore: Johns Hopkins University Press, 2002. (Cited on pages 139, 231, 245, 246, and 252.)

[804] E. I. Jury, *Inners and Stability of Dynamic Systems*, 2nd ed. Malabar: Krieger, 1982. (Cited on pages 454, 458, and 778.)

[805] W. J. Kaczor and M. T. Nowak, *Problems in Mathematical Analysis II*. American Mathematical Society, 2001. (Cited on pages 26, 30, 32, 37, and 72.)

[806] R. V. Kadison, "Order Properties of Bounded Self-Adjoint Operators," *Proc. Amer. Math. Soc.*, Vol. 2, pp. 505–510, 1951. (Cited on page 505.)

[807] A. Kagan and P. J. Smith, "A Stronger Version of Matrix Convexity as Applied to Functions of Hermitian Matrices," *J. Ineq. Appl.*, Vol. 3, pp. 143–152, 1999. (Cited on page 531.)

[808] J. B. Kagstrom, "Bounds and Perturbation Bounds for the Matrix Exponential," *Numer. Math. BIT*, Vol. 17, pp. 39–57, 1977. (Cited on page 767.)

[809] T. Kailath, *Linear Systems*. Englewood Cliffs: Prentice-Hall, 1980. (Cited on pages 259, 307, 331, 386, and 879.)

[810] D. Kalman and J. E. White, "Polynomial Equations and Circulant Matrices," *Amer. Math. Monthly*, Vol. 108, pp. 821–840, 2001. (Cited on page 390.)

[811] T. R. Kane, P. W. Likins, and D. A. Levinson, *Spacecraft Dynamics*. New York: McGraw-Hill, 1983. (Cited on page xvii.)

[812] I. Kaplansky, *Linear Algebra and Geometry: A Second Course*. New York: Chelsea, 1974, reprinted by Dover, Mineola, 2003. (Cited on pages xix and 346.)

[813] N. Karcanias, "Matrix Pencil Approach to Geometric System Theory," *Proc. IEE*, Vol. 126, pp. 585–590, 1979. (Cited on pages 396 and 879.)

[814] S. Karlin and F. Ost, "Some Monotonicity Properties of Schur Powers of Matrices and Related Inequalities," *Lin. Alg. Appl.*, Vol. 68, pp. 47–65, 1985. (Cited on page 457.)

[815] E. Kaszkurewicz and A. Bhaya, *Matrix Diagonal Stability in Systems and Computation*. Boston: Birkhauser, 2000. (Cited on page 793.)

[816] M. Kato, K.-S. Saito, and T. Tamura, "Sharp Triangle Inequality and Its Reverse in Banach Space," *Math. Ineq. Appl.*, Vol. 10, pp. 451–460, 2007. (Cited on page 623.)

[817] T. Kato, "Spectral Order and a Matrix Limit Theorem," *Lin. Multilin. Alg.*, Vol. 8, pp. 15–19, 1979. (Cited on page 506.)

[818] T. Kato, *Perturbation Theory for Linear Operators*. Berlin: Springer, 1980. (Cited on pages 657 and 680.)

[819] H. Katsuura, "Generalizations of the Arithmetic-Geometric Mean Inequality and a Three Dimensional Puzzle," *College Math. J.*, Vol. 34, pp. 280–282, 2003. (Cited on pages 51, 58, and 59.)

[820] M. Kauderer, *Symplectic Matrices: First Order Systems and Special Relativity*. Singapore: World Scientific, 1994. (Cited on page xvii.)

[821] S. Kayalar and H. L. Weinert, "Oblique Projections: Formulas, Algorithms, and Error Bounds," *Math. Contr. Sig. Sys.*, Vol. 2, pp. 33–45, 1989. (Cited on page 344.)

[822] J. Y. Kazakia, "Orthogonal Transformation of a Trace Free Symmetric Matrix into One with Zero Diagonal Elements," *Int. J. Eng. Sci.*, Vol. 26, pp. 903–906, 1988. (Cited on page 341.)

[823] N. D. Kazarinoff, *Analytic Inequalities.* New York: Holt, Rinehart, and Winston, 1961, reprinted by Dover, Mineola, 2003. (Cited on pages 40 and 600.)

[824] M. G. Kendall, *A Course in the Geometry of n Dimensions.* London: Griffin, 1961, reprinted by Dover, Mineola, 2004. (Cited on page 199.)

[825] C. Kenney and A. J. Laub, "Controllability and Stability Radii for Companion Form Systems," *Math. Contr. Sig. Sys.*, Vol. 1, pp. 239–256, 1988. (Cited on pages 357 and 780.)

[826] C. Kenney and A. J. Laub, "Rational Iteration Methods for the Matrix Sign Function," *SIAM J. Matrix Anal. Appl.*, Vol. 12, pp. 273–291, 1991. (Cited on page 699.)

[827] H. Kestelman, "Eigenvectors of a Cross-Diagonal Matrix," *Amer. Math. Monthly*, Vol. 93, p. 566, 1986. (Cited on page 371.)

[828] N. Keyfitz, *Introduction to the Mathematics of Population.* Reading: Addison-Wesley, 1968. (Cited on pages xvii and 299.)

[829] W. Khalil and E. Dombre, *Modeling, Identification, and Control of Robots.* New York: Taylor and Francis, 2002. (Cited on page xvii.)

[830] C. G. Khatri, "A Note on Idempotent Matrices," *Lin. Alg. Appl.*, Vol. 70, pp. 185–195, 1985. (Cited on page 408.)

[831] C. G. Khatri and S. K. Mitra, "Hermitian and Nonnegative Definite Solutions of Linear Matrix Equations," *SIAM J. Appl. Math.*, Vol. 31, pp. 579–585, 1976. (Cited on page 421.)

[832] S. Kim and Y. Lim, "A Converse Inequality of Higher Order Weighted Arithmetic and Geometric Means of Positive Definite Operators," *Lin. Alg. Appl.*, Vol. 426, pp. 490–496, 2007. (Cited on pages 58 and 510.)

[833] C. King, "Inequalities for Trace Norms of 2×2 Block Matrices," *Comm. Math. Phys.*, Vol. 242, pp. 531–545, 2003. (Cited on page 652.)

[834] C. King and M. Nathanson, "New Trace Norm Inequalities for 2×2 Blocks of Diagonal Matrices," *Lin. Alg. Appl.*, Vol. 389, pp. 77–93, 2004. (Cited on page 643.)

[835] M. K. Kinyon, "Revisited: $\arctan 1 + \arctan 2 + \arctan 3 = \pi$," *College Math. J.*, Vol. 37, pp. 218–219, 2006. (Cited on page 83.)

[836] F. Kittaneh, "Inequalities for the Schatten p-Norm III," *Commun. Math. Phys.*, Vol. 104, pp. 307–310, 1986. (Cited on page 640.)

[837] F. Kittaneh, "Inequalities for the Schatten p-Norm. IV," *Commun. Math. Phys.*, Vol. 106, pp. 581–585, 1986. (Cited on page 640.)

[838] F. Kittaneh, "On Zero-Trace Matrices," *Lin. Alg. Appl.*, Vol. 151, pp. 119–124, 1991. (Cited on page 341.)

[839] F. Kittaneh, "A Note on the Arithmetic-Geometric-Mean Inequality for Matrices," *Lin. Alg. Appl.*, Vol. 171, pp. 1–8, 1992. (Cited on page 646.)

[840] F. Kittaneh, "Norm Inequalities for Fractional Powers of Positive Operators," *Lett. Math. Phys.*, Vol. 27, pp. 279–285, 1993. (Cited on pages 566 and 671.)

[841] F. Kittaneh, "Singular Values of Companion Matrices and Bounds on Zeros of Polynomials," *SIAM J. Matrix Anal. Appl.*, Vol. 16, pp. 333–340, 1995. (Cited on pages 357 and 780.)

[842] F. Kittaneh, "Norm Inequalities for Certain Operator Sums," *J. Funct. Anal.*, Vol. 143, pp. 337–348, 1997. (Cited on pages 568, 569, and 645.)

[843] F. Kittaneh, "Some Norm Inequalities for Operators," *Canad. Math. Bull.*, Vol. 42, pp. 87–96, 1999. (Cited on page 646.)

[844] F. Kittaneh, "Commutator Inequalities Associated with the Polar Decomposition," *Proc. Amer. Math. Soc.*, Vol. 130, pp. 1279–1283, 2001. (Cited on pages 637, 641, 646, and 660.)

[845] F. Kittaneh, "Norm Inequalities for Sums of Positive Operators," *J. Operator Theory*, Vol. 48, pp. 95–103, 2002. (Cited on pages 568, 645, 668, and 669.)

[846] F. Kittaneh, "Bounds for the Zeros of Polynomials from Matrix Inequalities," *Arch. Math.*, Vol. 81, pp. 601–608, 2003. (Cited on pages 361 and 780.)

[847] F. Kittaneh, "A Numerical Radius Inequality and an Estimate for the Numerical Radius of the Frobenius Companion Matrix," *Studia Mathematica*, Vol. 158, pp. 11–17, 2003. (Cited on pages 633 and 781.)

[848] F. Kittaneh, "Norm Inequalities for Sums and Differences of Positive Operators," *Lin. Alg. Appl.*, Vol. 383, pp. 85–91, 2004. (Cited on pages 567, 568, 569, and 660.)

[849] F. Kittaneh, "Norm Inequalities for Sums of Positive Operators II," *Positivity*, Vol. 10, pp. 251–260, 2006. (Cited on pages 645 and 649.)

[850] F. Kittaneh, "Bounds for the Zeros of Polynomials from Matrix Inequalities–II," *Lin. Multilin. Alg.*, Vol. 55, pp. 147–158, 2007. (Cited on page 780.)

[851] F. Kittaneh and H. Kosaki, "Inequalities for the Schatten p-Norm. V," *Publications of RIMS Kyoto University*, Vol. 23, pp. 433–443, 1987. (Cited on pages 640 and 642.)

[852] A.-L. Klaus and C.-K. Li, "Isometries for the Vector (p, q) Norm and the Induced (p, q) Norm," *Lin. Multilin. Alg.*, Vol. 38, pp. 315–332, 1995. (Cited on page 640.)

[853] J. A. Knox and H. J. Brothers, "Novel Series-Based Approximations to e," *College Math. J.*, Vol. 30, pp. 269–275, 1999. (Cited on page 28.)

[854] C. T. Koch and J. C. H. Spence, "A Useful Expansion of the Exponential of the Sum of Two Non-Commuting Matrices, One of Which is Diagonal," *J. Phys. A: Math. Gen.*, Vol. 36, pp. 803–816, 2003. (Cited on page 743.)

[855] D. Koks, *Explorations in Mathematical Physics*. New York: Springer, 2006. (Cited on pages 249 and 338.)

[856] J. J. Koliha, "A Simple Proof of the Product Theorem for EP Matrices," *Lin. Alg. Appl.*, Vol. 294, pp. 213–215, 1999. (Cited on page 418.)

[857] J. J. Koliha and V. Rakocevic, "Invertibility of the Sum of Idempotents," *Lin. Multilin. Alg.*, Vol. 50, pp. 285–292, 2002. (Cited on page 219.)

[858] J. J. Koliha and V. Rakocevic, "Invertibility of the Difference of Idempotents," *Lin. Multilin. Alg.*, Vol. 51, pp. 97–110, 2003. (Cited on page 221.)

[859] J. J. Koliha and V. Rakocevic, "The Nullity and Rank of Linear Combinations of Idempotent Matrices," *Lin. Alg. Appl.*, Vol. 418, pp. 11–14, 2006. (Cited on page 219.)

[860] J. J. Koliha, V. Rakocevic, and I. Straskraba, "The Difference and Sum of Projectors," *Lin. Alg. Appl.*, Vol. 388, pp. 279–288, 2004. (Cited on pages 219, 220, 221, and 222.)

[861] N. Komaroff, "Bounds on Eigenvalues of Matrix Products with an Applicaton to the Algebraic Riccati Equation," *IEEE Trans. Autom. Contr.*, Vol. 35, pp. 348–350, 1990. (Cited on page 363.)

[862] N. Komaroff, "Rearrangement and Matrix Product Inequalities," *Lin. Alg. Appl.*, Vol. 140, pp. 155–161, 1990. (Cited on pages 67 and 570.)

[863] V. Komornik, "Another Short Proof of Descartes's Rule of Signs," *Amer. Math. Monthly*, Vol. 113, pp. 829–830, 2006. (Cited on page 763.)

[864] R. H. Koning, H. Neudecker, and T. Wansbeek, "Block Kronecker Products and the vecb Operator," *Lin. Alg. Appl.*, Vol. 149, pp. 165–184, 1991. (Cited on page 458.)

[865] T. Koshy, *Fibonacci and Lucas Numbers with Applications*. New York: Wiley, 2001. (Cited on pages xvii, 238, and 304.)

[866] J. Kovac-Striko and K. Veselic, "Trace Minimization and Definiteness of Symmetric Pencils," *Lin. Alg. Appl.*, Vol. 216, pp. 139–158, 1995. (Cited on page 396.)

[867] O. Krafft, "An Arithmetic-Harmonic-Mean Inequality for Nonnegative Definite Matrices," *Lin. Alg. Appl.*, Vol. 268, pp. 243–246, 1998. (Cited on page 582.)

[868] C. Krattenthaler, "Advanced Determinant Calculus: A Complement," *Lin. Alg. Appl.*, Vol. 411, pp. 68–166, 2005. (Cited on page xvii.)

[869] W. Kratz, *Quadratic Functionals in Variational Analysis and Control Theory*. New York: Wiley, 1995. (Cited on page 879.)

[870] E. Kreindler and A. Jameson, "Conditions for Nonnegativeness of Partitioned Matrices," *IEEE Trans. Autom. Contr.*, Vol. AC-17, pp. 147–148, 1972. (Cited on page 595.)

[871] R. Kress, H. L. de Vries, and R. Wegmann, "On Nonnormal Matrices," *Lin. Alg. Appl.*, Vol. 8, pp. 109–120, 1974. (Cited on page 654.)

[872] V. Kucera, "On Nonnegative Definite Solutions to Matrix Quadratic Equations," *Automatica*, Vol. 8, pp. 413–423, 1972. (Cited on page 879.)

[873] A. Kufner, L. Maligranda, and L.-E. Persson, "The Prehistory of the Hardy Inequality," *Amer. Math. Monthly*, Vol. 113, pp. 715–732, 2006. (Cited on pages 64 and 69.)

[874] J. B. Kuipers, *Quaternions and Rotation Sequences: A Primer with Applications to Orbits, Aerospace, and Virtual Reality*. Princeton: Princeton University Press, 1999. (Cited on pages xvii, 249, and 251.)

[875] M. R. S. Kulenovic and G. Ladas, *Dynamics of Second Order Rational Difference Equations with Open Problems and Conjectures*. New York: Chapman & Hall/CRC Press, 2002. (Cited on page 793.)

[876] S. Kurepa, "A Note on Logarithms of Normal Operators," *Proc. Amer. Math. Soc.*, Vol. 13, pp. 307–311, 1962. (Cited on page 756.)

[877] K. Kwakernaak and R. Sivan, *Linear Optimal Control Systems.* New York: Wiley, 1972. (Cited on page xvii.)

[878] M. Kwapisz, "The Power of a Matrix," *SIAM Rev.*, Vol. 40, pp. 703–705, 1998. (Cited on page 782.)

[879] M. K. Kwong, "Some Results on Matrix Monotone Functions," *Lin. Alg. Appl.*, Vol. 118, pp. 129–153, 1989. (Cited on pages 242 and 506.)

[880] I. I. Kyrchei, "Analogs of the Adjoint Matrix for Generalized Inverses and Corresponding Cramer Rules," *Lin. Multilin. Alg.*, Vol. 56, pp. 453–469, 2008. (Cited on page 141.)

[881] K. R. Laberteaux, "Hermitian Matrices," *Amer. Math. Monthly*, Vol. 104, p. 277, 1997. (Cited on page 195.)

[882] T. J. Laffey, "Products of Skew-Symmetric Matrices," *Lin. Alg. Appl.*, Vol. 68, pp. 249–251, 1985. (Cited on page 384.)

[883] T. J. Laffey, "Products of Matrices," in *Generators and Relations in Groups and Geometries*, A. Barlotti, E. W. Ellers, P. Plaumann, and K. Strambach, Eds. Dordrecht: Kluwer, 1991, pp. 95–123. (Cited on page 384.)

[884] T. J. Laffey and S. Lazarus, "Two-Generated Commutative Matrix Subalgebras," *Lin. Alg. Appl.*, Vol. 147, pp. 249–273, 1991. (Cited on pages 347 and 348.)

[885] T. J. Laffey and E. Meehan, "A Refinement of an Inequality of Johnson, Loewy and London on Nonnegative Matrices and Some Applications," *Elec. J. Lin. Alg.*, Vol. 3, pp. 119–128, 1998. (Cited on page 306.)

[886] J. C. Lagarias and Y. Wang, "The Finiteness Conjecture for the Generalized Spectral Radius of a Set of Matrices," *Lin. Alg. Appl.*, Vol. 214, pp. 17–42, 1995. (Cited on page 786.)

[887] S. Lakshminarayanan, S. L. Shah, and K. Nandakumar, "Cramer's Rule for Non-Square Matrices," *Amer. Math. Monthly*, Vol. 106, p. 865, 1999. (Cited on page 141.)

[888] P. Lancaster, *Lambda-Matrices and Vibrating Systems.* Oxford: Pergamon, 1966, reprinted by Dover, Mineola, 2002. (Cited on page xvii.)

[889] P. Lancaster and L. Rodman, "Solutions of the Continuous and Discrete Time Algebraic Riccati Equations: A Review," in *The Riccati Equation*, S. Bittanti, J. C. Willems, and A. Laub, Eds. New York: Springer, 1991, pp. 11–51. (Cited on page 879.)

[890] P. Lancaster and L. Rodman, *Algebraic Riccati Equations.* Oxford: Clarendon, 1995. (Cited on pages xvii and 879.)

[891] P. Lancaster and L. Rodman, "Canonical Forms for Hermitian Matrix Pairs under Strict Equivalence and Congruence," *SIAM Rev.*, Vol. 47, pp. 407–443, 2005. (Cited on pages 331 and 554.)

[892] P. Lancaster and M. Tismenetsky, *The Theory of Matrices*, 2nd ed. Orlando: Academic Press, 1985. (Cited on pages 349, 371, 447, 615, 635, 680, 867, and 868.)

[893] L. Larson, *Problem Solving Through Problems.* New York: Springer, 1983. (Cited on pages 27, 32, 36, 38, 42, 44, 45, 46, 47, 48, 49, 53, 168, and 171.)

[894] L. Larsson, L. Maligranda, J. Pecaric, and L.-E. Persson, *Multiplicative Inequalities of Carlson Type and Interpolation.* Singapore: World Scientific, 2006. (Cited on page 64.)

[895] J. Lasenby, A. N. Lasenby, and C. J. L. Doran, "A Unified Mathematical Language for Physics and Engineering in the 21st Century," *Phil. Trans. Math. Phys. Eng. Sci.*, Vol. 358, pp. 21–39, 2000. (Cited on pages 204, 249, and 620.)

[896] J. B. Lasserre, "A Trace Inequality for Matrix Product," *IEEE Trans. Autom. Contr.*, Vol. 40, pp. 1500–1501, 1995. (Cited on page 363.)

[897] A. J. Laub, *Matrix Analysis for Scientists and Engineers.* Philadelphia: SIAM, 2004. (Cited on pages xix, 122, 331, 332, and 369.)

[898] A. J. Laub and K. Meyer, "Canonical Forms for Symplectic and Hamiltonian Matrices," *Celestial Mechanics*, Vol. 9, pp. 213–238, 1974. (Cited on page 876.)

[899] C. Laurie, B. Mathes, and H. Radjavi, "Sums of Idempotents," *Lin. Alg. Appl.*, Vol. 208/209, pp. 175–197, 1994. (Cited on page 395.)

[900] J. L. Lavoie, "A Determinantal Inequality Involving the Moore-Penrose Inverse," *Lin. Alg. Appl.*, Vol. 31, pp. 77–80, 1980. (Cited on page 430.)

[901] C. L. Lawson, *Solving Least Squares Problems.* Englewood Cliffs: Prentice-Hall, 1974, reprinted by SIAM, Philadelphia, 1995. (Cited on page 438.)

[902] J. D. Lawson and Y. Lim, "The Geometric Mean, Matrices, Metrics, and More," *Amer. Math. Monthly*, Vol. 108, pp. 797–812, 2001. (Cited on pages 474 and 509.)

[903] P. D. Lax, *Linear Algebra.* New York: Wiley, 1997. (Cited on pages 173, 282, 502, and 657.)

[904] S. R. Lay, *Convex Sets and Their Applications.* New York: Wiley, 1982. (Cited on pages 101, 178, and 697.)

[905] B. Leclerc, "On Identities Satisfied by Minors of a Matrix," *Adv. Math.*, Vol. 100, pp. 101–132, 1993. (Cited on page 140.)

[906] K. J. LeCouteur, "Representation of the Function $\mathrm{Tr}(\exp(A\text{-}\lambda B))$ as a Laplace Transform with Positive Weight and Some Matrix Inequalities," *J. Phys. A: Math. Gen.*, Vol. 13, pp. 3147–3159, 1980. (Cited on pages 571 and 748.)

[907] A. Lee, "On S-Symmetric, S-Skewsymmetric, and S-Orthogonal Matrices," *Periodica Math. Hungar.*, Vol. 7, pp. 71–76, 1976. (Cited on page 340.)

[908] A. Lee, "Centrohermitian and Skew-Centrohermitian Matrices," *Lin. Alg. Appl.*, Vol. 29, pp. 205–210, 1980. (Cited on pages 242 and 410.)

[909] J. M. Lee and D. A. Weinberg, "A Note on Canonical Forms for Matrix Congruence," *Lin. Alg. Appl.*, Vol. 249, pp. 207–215, 1996. (Cited on page 396.)

[910] S. H. Lehnigk, *Stability Theorems for Linear Motions.* Englewood Cliffs: Prentice-Hall, 1966. (Cited on page 793.)

[911] T.-G. Lei, C.-W. Woo, and F. Zhang, "Matrix Inequalities by Means of Embedding," *Elect. J. Lin. Alg.*, Vol. 11, pp. 66–77, 2004. (Cited on pages 498, 531, 532, 542, 551, 555, 556, 559, and 561.)

[912] F. S. Leite, "Bounds on the Order of Generation of so(n, r) by One-Parameter Subgroups," *Rocky Mountain J. Math.*, Vol. 21, pp. 879–911, 1183–1188, 1991. (Cited on pages 380 and 739.)

[913] E. Leonard, "The Matrix Exponential," *SIAM Rev.*, Vol. 38, pp. 507–512, 1996. (Cited on page 710.)

[914] G. Letac, "A Matrix and Its Matrix of Reciprocals Both Positive Semi-Definite," *Amer. Math. Monthly*, Vol. 82, pp. 80–81, 1975. (Cited on page 584.)

[915] J. S. Lew, "The Cayley Hamilton Theorem in n Dimensions," *Z. Angew. Math. Phys.*, Vol. 17, pp. 650–653, 1966. (Cited on pages 283 and 284.)

[916] A. S. Lewis, "Convex Analysis on the Hermitian Matrices," *SIAM J. Optim.*, Vol. 6, pp. 164–177, 1996. (Cited on pages 363 and 364.)

[917] D. C. Lewis, "A Qualitative Analysis of S-Systems: Hopf Bifurcations," in *Canonical Nonlinear Modeling*, E. O. Voit, Ed. New York: Van Nostrand Reinhold, 1991, pp. 304–344. (Cited on page 456.)

[918] C.-K. Li, "Positive Semidefinite Matrices and Tensor Product," *IMAGE*, Vol. 41, p. 44, 2008. (Cited on page 590.)

[919] C.-K. Li and R.-C. Li, "A Note on Eigenvalues of Perturbed Hermitian Matrices," *Lin. Alg. Appl.*, Vol. 395, pp. 183–190, 2005. (Cited on page 296.)

[920] C.-K. Li and R. Mathias, "The Determinant of the Sum of Two Matrices," *Bull. Austral. Math. Soc.*, Vol. 52, pp. 425–429, 1995. (Cited on pages 669 and 670.)

[921] C.-K. Li and R. Mathias, "The Lidskii-Mirsky-Wielandt Theorem–Additive and Multiplicative Versions," *Numer. Math.*, Vol. 81, pp. 377–413, 1999. (Cited on page 657.)

[922] C.-K. Li and R. Mathias, "Extremal Characterizations of the Schur Complement and Resulting Inequalities," *SIAM Rev.*, Vol. 42, pp. 233–246, 2000. (Cited on pages 479, 485, 586, and 596.)

[923] C.-K. Li and R. Mathias, "Inequalities on Singular Values of Block Triangular Matrices," *SIAM J. Matrix Anal. Appl.*, Vol. 24, pp. 126–131, 2002. (Cited on page 357.)

[924] C.-K. Li and S. Nataraj, "Some Matrix Techniques in Game Theory," *Math. Ineq. Appl.*, Vol. 3, pp. 133–141, 2000. (Cited on page xvii.)

[925] C.-K. Li and E. Poon, "Additive Decomposition of Real Matrices," *Lin. Multilin. Alg.*, Vol. 50, pp. 321–326, 2002. (Cited on page 394.)

[926] C.-K. Li and L. Rodman, "Some Extremal Problems for Positive Definite Matrices and Operators," *Lin. Alg. Appl.*, Vol. 140, pp. 139–154, 1990. (Cited on pages 505 and 506.)

[927] C.-K. Li and H. Schneider, "Orthogonality of Matrices," *Lin. Alg. Appl.*, Vol. 347, pp. 115–122, 2002. (Cited on page 626.)

[928] J.-L. Li and Y.-L. Li, "On the Strengthened Jordan's Inequality," *J. Ineq. Appl.*, pp. 1–8, 2007, article ID 74328. (Cited on page 32.)

[929] Q. Li, "Commutators of Orthogonal Projections," *Nihonkai. Math. J.*, Vol. 15, pp. 93–99, 2004. (Cited on page 637.)

[930] X. Li and Y. Wei, "A Note on the Representations for the Drazin Inverse of 2×2 Block Matrices," *Lin. Alg. Appl.*, Vol. 423, pp. 332–338, 2007. (Cited on page 429.)

[931] E. H. Lieb, "Convex Trace Functions and the Wigner-Yanase-Dyson Conjecture," *Adv. Math.*, Vol. 11, pp. 267–288, 1973. (Cited on pages 484, 549, 596, and 754.)

[932] E. H. Lieb and M. Loss, *Analysis.* Providence: American Mathematical Society, 2001. (Cited on page 625.)

[933] E. H. Lieb and M. B. Ruskai, "Some Operator Inequalities of the Schwarz Type," *Adv. Math.*, Vol. 12, pp. 269–273, 1974. (Cited on pages 521 and 596.)

[934] E. H. Lieb and R. Seiringer, "Equivalent Forms of the Bessis-Moussa-Villani Conjecture," *J. Stat. Phys.*, Vol. 115, pp. 185–190, 2004. (Cited on page 530.)

[935] E. H. Lieb and W. E. Thirring, "Inequalities for the Moments of the Eigenvalues of the Schrodinger Hamiltonian and Their Relation to Sobolev Inequalities," in *Studies in Mathematical Physics*, E. H. Lieb, B. Simon, and A. Wightman, Eds. Princeton: Princeton University Press, 1976, pp. 269–303. (Cited on page 571.)

[936] Y. Lim, "The Matrix Golden Mean and its Applications to Riccati Matrix Equations," *SIAM J. Matrix Anal. Appl.*, Vol. 29, pp. 54–66, 2006. (Cited on page 877.)

[937] C.-S. Lin, "On Halmos's Sharpening of Reid's Inequality," *Math. Reports Acad. Sci. Canada*, Vol. 20, pp. 62–64, 1998. (Cited on page 552.)

[938] C.-S. Lin, "Inequalities of Reid Type and Furuta," *Proc. Amer. Math. Soc.*, Vol. 129, pp. 855–859, 2000. (Cited on page 553.)

[939] C.-S. Lin, "Heinz-Kato-Furuta Inequalities with Bounds and Equality Conditions," *Math. Ineq. Appl.*, Vol. 5, pp. 735–743, 2002. (Cited on page 620.)

[940] C.-S. Lin, "On Operator Order and Chaotic Operator Order for Two Operators," *Lin. Alg. Appl.*, Vol. 425, pp. 1–6, 2007. (Cited on pages 512 and 575.)

[941] T.-P. Lin, "The Power Mean and the Logarithmic Mean," *Amer. Math. Monthly*, Vol. 81, pp. 879–883, 1974. (Cited on page 40.)

[942] W.-W. Lin, "The Computation of the Kronecker Canonical Form of an Arbitrary Symmetric Pencil," *Lin. Alg. Appl.*, Vol. 103, pp. 41–71, 1988. (Cited on page 396.)

[943] H. Linden, "Numerical Radii of Some Companion Matrices and Bounds for the Zeros of Polynomials," in *Analytic and Geometric Inequalities and Applications*, T. M. Rassias and H. M. Srivastava, Eds. Dordrecht: Kluwer, 1999, pp. 205–229. (Cited on pages 780 and 781.)

[944] L. Lipsky, *Queuing Theory: A Linear Algebraic Approach*, 2nd ed. New York: Springer, 2008. (Cited on page xvii.)

[945] B. Liu and H.-J. Lai, *Matrices in Combinatorics and Graph Theory.* New York: Springer, 2000. (Cited on page xvii.)

[946] H. Liu and L. Zhu, "New Strengthened Carleman's Inequality and Hardy's Inequality," *J. Ineq. Appl.*, pp. 1–7, 2007, article ID 84104. (Cited on page 28.)

[947] J. Liu and J. Wang, "Some Inequalities for Schur Complements," *Lin. Alg. Appl.*, Vol. 293, pp. 233–241, 1999. (Cited on pages 485 and 596.)

[948] J. Liu and Q. Xie, "Inequalities Involving Khatri-Rao Products of Positive Semidefinite Hermitian Matrices," *Int. J. Inform. Sys. Sci.*, Vol. 4, pp. 30–40–135, 2008. (Cited on page 458.)

[949] R.-W. Liu and R. J. Leake, "Exhaustive Equivalence Classes of Optimal Systems with Separable Controls," *SIAM Rev.*, Vol. 4, pp. 678–685, 1966. (Cited on page 201.)

[950] S. Liu, "Matrix Results on the Khatri-Rao and Tracy-Singh Products," *Lin. Alg. Appl.*, Vol. 289, pp. 267–277, 1999. (Cited on pages 458, 586, 589, and 593.)

[951] S. Liu, "Several Inequalities Involving Khatri-Rao Products of Positive Semidefinite Matrices," *Lin. Alg. Appl.*, Vol. 354, pp. 175–186, 2002. (Cited on page 458.)

[952] S. Liu and H. Neudecker, "Several Matrix Kantorovich-Type Inequalities," *Math. Anal. Appl.*, Vol. 197, pp. 23–26, 1996. (Cited on pages 63 and 551.)

[953] S. Liu and G. Trenkler, "Hadamard, Khatri-Rao, Kronecker and Other Matrix Products," *Int. J. Inform. Sys. Sci.*, Vol. 4, pp. 160–177, 2008. (Cited on page 458.)

[954] E. Liz, "A Note on the Matrix Exponential," *SIAM Rev.*, Vol. 40, pp. 700–702, 1998. (Cited on page 710.)

[955] R. Loewy, "An Inertia Theorem for Lyapunov's Equation and the Dimension of a Controllability Space," *Lin. Alg. Appl.*, Vol. 260, pp. 1–7, 1997. (Cited on page 867.)

[956] D. O. Logofet, *Matrices and Graphs: Stability Problems in Mathematical Ecology.* Boca Raton: CRC Press, 1993. (Cited on page xvii.)

[957] R. Lopez-Valcarce and S. Dasgupta, "Some Properties of the Matrix Exponential," *IEEE Trans. Circ. Sys. Analog. Dig. Sig. Proc.*, Vol. 48, pp. 213–215, 2001. (Cited on page 746.)

[958] M. Loss and M. B. Ruskai, Eds., *Inequalities: Selecta of Elliott H. Lieb.* New York: Springer, 2002. (Cited on pages 549 and 754.)

[959] P. Lounesto, *Clifford Algebras and Spinors*, 2nd ed. Cambridge: Cambridge University Press, 2001. (Cited on pages 204 and 249.)

[960] D. G. Luenberger, *Optimization by Vector Space Methods.* New York: Wiley, 1969. (Cited on page xvii.)

[961] D. G. Luenberger, *Introduction to Linear and Nonlinear Programming*, 2nd ed. Reading: Addison-Wesley, 1984. (Cited on page 555.)

[962] M. Lundquist and W. Barrett, "Rank Inequalities for Positive Semidefinite Matrices," *Lin. Alg. Appl.*, Vol. 248, pp. 91–100, 1996. (Cited on page 134.)

[963] M. Lundquist and W. Barrett, "Rank Inequalities for Positive Semidefinite Matrices," *Lin. Alg. Appl.*, Vol. 248, pp. 91–100, 1996. (Cited on pages 488 and 541.)

[964] H. Lutkepohl, *Handbook of Matrices.* Chichester: Wiley, 1996. (Cited on page xix.)

[965] E.-C. Ma, "A Finite Series Solution of the Matrix Equation $AX - XB = C$," *SIAM J. Appl. Math.*, Vol. 14, pp. 490–495, 1966. (Cited on page 870.)

[966] Y. Ma, S. Soatto, J. Kosecka, and S. S. Sastry, *An Invitation to 3-D Vision.* New York: Springer, 2004. (Cited on page xvii.)

[967] C. C. MacDuffee, *The Theory of Matrices*. New York: Chelsea, 1956. (Cited on pages 447, 451, 454, and 458.)

[968] A. G. J. Macfarlane and N. Karcanias, "Poles and Zeros of Linear Multivariable Systems: A Survey of the Algebraic, Geometric, and Complex-Variable Theory," *Int. J. Contr.*, Vol. 24, pp. 33–74, 1976. (Cited on page 879.)

[969] D. S. Mackey, N. Mackey, and F. Tisseur, "Structured Tools for Structured Matrices," *Elec. J. Lin. Alg.*, Vol. 10, pp. 106–145, 2003. (Cited on page 252.)

[970] J. H. Maddocks, "Restricted Quadratic Forms, Inertia Theorems, and the Schur Complement," *Lin. Alg. Appl.*, Vol. 108, pp. 1–36, 1988. (Cited on pages 338 and 554.)

[971] J. R. Magnus, "A Representation Theorem for $(\operatorname{tr} A^p)^{1/p}$," *Lin. Alg. Appl.*, Vol. 95, pp. 127–134, 1987. (Cited on pages 523, 525, and 530.)

[972] J. R. Magnus, *Linear Structures*. London: Griffin, 1988. (Cited on pages xvii, 428, and 458.)

[973] J. R. Magnus and H. Neudecker, *Matrix Differential Calculus with Applications in Statistics and Econometrics*. Chichester: Wiley, 1988. (Cited on pages xvii, 405, 421, 425, 429, 438, 458, 523, 525, 553, and 705.)

[974] W. Magnus, "On the Exponential Solution of Differential Equations for a Linear Operator," *Commun. Pure Appl. Math.*, Vol. VII, pp. 649–673, 1954. (Cited on page 743.)

[975] K. N. Majindar, "On Simultaneous Hermitian Congruence Transformations of Matrices," *Amer. Math. Monthly*, Vol. 70, pp. 842–844, 1963. (Cited on page 559.)

[976] L. Maligranda, "Simple Norm Inequalities," *Amer. Math. Monthly*, Vol. 113, pp. 256–260, 2006. (Cited on pages 622, 623, and 624.)

[977] A. N. Malyshev and M. Sadkane, "On the Stability of Large Matrices," *J. Comp. Appl. Math.*, Vol. 102, pp. 303–313, 1999. (Cited on page 767.)

[978] O. Mangasarian, *Nonlinear Programming*. New York: McGraw-Hill, 1969, reprinted by Krieger, Malabar, 1982. (Cited on page xvii.)

[979] S. M. Manjegani, "Hölder and Young Inequalities for the Trace of Operators," *Positivity*, Vol. 11, pp. 239–250, 2007. (Cited on page 525.)

[980] H. B. Mann, "Quadratic Forms with Linear Constraints," *Amer. Math. Monthly*, Vol. 50, pp. 430–433, 1943. (Cited on page 555.)

[981] L. E. Mansfield, *Linear Algebra with Geometric Application*. New York: Marcel-Dekker, 1976. (Cited on page xix.)

[982] M. Marcus, "An Eigenvalue Inequality for the Product of Normal Matrices," *Amer. Math. Monthly*, Vol. 63, pp. 173–174, 1956. (Cited on page 363.)

[983] M. Marcus, *Finite Dimensional Multilinear Algebra, Part I*. Marcel Dekker, 1973. (Cited on pages 454 and 458.)

[984] M. Marcus, *Finite Dimensional Multilinear Algebra, Part II*. Marcel Dekker, 1975. (Cited on page 458.)

[985] M. Marcus, "Two Determinant Condensation Formulas," *Lin. Multilin. Alg.*, Vol. 22, pp. 95–102, 1987. (Cited on page 148.)

[986] M. Marcus and S. M. Katz, "Matrices of Schur Functions," *Duke Math. J.*, Vol. 36, pp. 343–352, 1969. (Cited on page 543.)

[987] M. Marcus and N. A. Khan, "A Note on the Hadamard Product," *Canad. Math. J.*, Vol. 2, pp. 81–83, 1959. (Cited on pages 444, 458, and 590.)

[988] M. Marcus and H. Minc, *A Survey of Matrix Theory and Matrix Inequalities*. Boston: Prindle, Weber, and Schmidt, 1964, reprinted by Dover, New York, 1992. (Cited on pages xix, 60, 84, 252, 354, 598, and 655.)

[989] M. Marcus and W. Watkins, "Partitioned Hermitian Matrices," *Duke Math. J.*, Vol. 38, pp. 237–249, 1971. (Cited on pages 495, 532, 533, 543, and 564.)

[990] M. Margaliot and G. Langholz, "The Routh-Hurwitz Array and Realization of Characteristic Polynomials," *IEEE Trans. Autom. Contr.*, Vol. 45, pp. 2424–2427, 2000. (Cited on page 772.)

[991] T. L. Markham, "An Application of Theorems of Schur and Albert," *Proc. Amer. Math. Soc.*, Vol. 59, pp. 205–210, 1976. (Cited on page 595.)

[992] T. L. Markham, "Oppenheim's Inequality for Positive Definite Matrices," *Amer. Math. Monthly*, Vol. 93, pp. 642–644, 1986. (Cited on page 588.)

[993] G. Marsaglia and G. P. H. Styan, "Equalities and Inequalities for Ranks of Matrices," *Lin. Multilin. Alg.*, Vol. 2, pp. 269–292, 1974. (Cited on pages 131, 178, 424, and 426.)

[994] J. E. Marsden and R. S. Ratiu, *Introduction to Mechanics and Symmetry: A Basic Exposition of Classical Mechanical Systems*, 2nd ed. New York: Springer, 1994. (Cited on page 214.)

[995] J. E. Marsden and T. S. Ratiu, *Introduction to Mechanics and Symmetry*. New York: Springer, 1994. (Cited on page xvii.)

[996] A. W. Marshall and I. Olkin, *Inequalities: Theory of Majorization and Its Applications*. New York: Academic Press, 1979. (Cited on pages 49, 66, 67, 84, 171, 172, 175, 176, 177, 178, 201, 355, 356, 364, 365, 454, 485, 486, 561, 595, 596, and 679.)

[997] A. W. Marshall and I. Olkin, "Matrix Versions of the Cauchy and Kantorovich Inequalities," *Aequationes Math.*, Vol. 40, pp. 89–93, 1990. (Cited on page 500.)

[998] A. W. Marshall, I. Olkin, and B. Arnold, *Inequalities: Theory of Majorization and Its Applications*, 2nd ed. New York: Springer, 2009. (Cited on page 178.)

[999] K. Martensson, "On the Matrix Riccati Equation," *Inform. Sci.*, Vol. 3, pp. 17–49, 1971. (Cited on page 879.)

[1000] A. M. Mathai, *Jacobians of Matrix Transformations and Functions of Matrix Argument*. Singapore: World Scientific, 1997. (Cited on page 705.)

[1001] R. Mathias, "The Spectral Norm of a Nonnegative Matrix," *Lin. Alg. Appl.*, Vol. 131, pp. 269–284, 1990. (Cited on page 674.)

[1002] R. Mathias, "Evaluating the Frechet Derivative of the Matrix Exponential," *Numer. Math.*, Vol. 63, pp. 213–226, 1992. (Cited on page 748.)

[1003] R. Mathias, "Matrices with Positive Definite Hermitian Part: Inequalities and Linear Systems," *SIAM J. Matrix Anal. Appl.*, Vol. 13, pp. 640–654, 1992. (Cited on pages 484 and 563.)

[1004] R. Mathias, "An Arithmetic-Geometric-Harmonic Mean Inequality Involving Hadamard Products," *Lin. Alg. Appl.*, Vol. 184, pp. 71–78, 1993. (Cited on page 492.)

[1005] R. Mathias, "A Chain Rule for Matrix Functions and Applications," *SIAM J. Matrix Anal. Appl.*, Vol. 17, pp. 610–620, 1996. (Cited on page 705.)

[1006] R. Mathias, "Singular Values and Singular Value Inequalities," in *Handbook of Linear Algebra*, L. Hogben, Ed. Boca Raton: Chapman & Hall/CRC, 2007, pp. 17-1–17-16. (Cited on page 362.)

[1007] M. Matic, C. E. M. Pearce, and J. Pecaric, "Shannon's and Related Inequalities in Information Theory," in *Survey on Classical Inequalities*, T. M. Rassias, Ed. Dordrecht: Kluwer, 2000, pp. 127–164. (Cited on pages 65 and 74.)

[1008] T. Matsuda, "An Inductive Proof of a Mixed Arithmetic-Geometric Mean Inequality," *Amer. Math. Monthly*, Vol. 102, pp. 634–637, 1995. (Cited on page 63.)

[1009] V. Mazorchuk and S. Rabanovich, "Multicommutators and Multianticommutators of Orthogonal Projectors," *Lin. Multilin. Alg.*, Vol. 56, pp. 639–646, 2008. (Cited on page 637.)

[1010] J. E. McCarthy, "Pick's Theorem—What's the Big Deal?" *Amer. Math. Monthly*, Vol. 110, pp. 36–45, 2003. (Cited on page 494.)

[1011] J. M. McCarthy, *Geometric Design of Linkages*. New York: Springer, 2000. (Cited on page 793.)

[1012] J. P. McCloskey, "Characterizations of *r*-Potent Matrices," *Math. Proc. Camb. Phil. Soc.*, Vol. 96, pp. 213–222, 1984. (Cited on page 231.)

[1013] A. R. Meenakshi and C. Rajian, "On a Product of Positive Semidefinite Matrices," *Lin. Alg. Appl.*, Vol. 295, pp. 3–6, 1999. (Cited on page 583.)

[1014] C. L. Mehta, "Some Inequalities Involving Traces of Operators," *J. Math. Phys.*, Vol. 9, pp. 693–697, 1968. (Cited on page 549.)

[1015] Y. A. Melnikov, *Influence Functions and Matrices*. New York: Marcel Dekker, 1998. (Cited on page xvii.)

[1016] P. R. Mercer, "The Dunkl-Williams Inequality in an Inner Product Space," *Math. Ineq. Appl.*, Vol. 10, pp. 447–450, 2007. (Cited on pages 621 and 623.)

[1017] J. K. Merikoski, H. Sarria, and P. Tarazaga, "Bounds for Singular Values Using Traces," *Lin. Alg. Appl.*, Vol. 210, pp. 227–254, 1994. (Cited on page 661.)

[1018] R. Merris, "Laplacian Matrices of Graphs: A Survey," *Lin. Alg. Appl.*, Vol. 198, pp. 143–176, 1994. (Cited on pages 301 and 550.)

[1019] R. Merris, *Multilinear Algebra*. Amsterdam: Gordon and Breach, 1997. (Cited on pages 458 and 590.)

[1020] R. Merris and S. Pierce, "Monotonicity of Positive Semidefinite Hermitian Matrices," *Proc. Amer. Math. Soc.*, Vol. 31, pp. 437–440, 1972. (Cited on pages 507 and 516.)

[1021] C. D. Meyer, "The Moore-Penrose Inverse of a Bordered Matrix," *Lin. Alg. Appl.*, Vol. 5, pp. 375–382, 1972. (Cited on page 429.)

[1022] C. D. Meyer, "Generalized Inverses and Ranks of Block Matrices," *SIAM J. Appl. Math*, Vol. 25, pp. 597–602, 1973. (Cited on pages 390 and 429.)

[1023] C. D. Meyer, *Matrix Analysis and Applied Linear Algebra*. Philadelphia: SIAM, 2000. (Cited on pages 174, 190, 191, 231, 265, 305, 409, and 784.)

[1024] C. D. Meyer and N. J. Rose, "The Index and the Drazin Inverse of Block Triangular Matrices," *SIAM J. Appl. Math.*, Vol. 33, pp. 1–7, 1977. (Cited on pages 374, 429, and 435.)

[1025] J.-M. Miao, "General Expressions for the Moore-Penrose Inverse of a 2×2 Block Matrix," *Lin. Alg. Appl.*, Vol. 151, pp. 1–15, 1991. (Cited on page 429.)

[1026] L. Mihalyffy, "An Alternative Representation of the Generalized Inverse of Partitioned Matrices," *Lin. Alg. Appl.*, Vol. 4, pp. 95–100, 1971. (Cited on page 429.)

[1027] K. S. Miller, *Some Eclectic Matrix Theory*. Malabar: Krieger, 1987. (Cited on pages 276, 451, 452, and 557.)

[1028] G. A. Milliken and F. Akdeniz, "A Theorem on the Difference of the Generalized Inverses of Two Nonnegative Matrices," *Commun. Statist. Theory Methods*, Vol. 6, pp. 73–79, 1977. (Cited on page 579.)

[1029] G. V. Milovanovic, D. S. Mitrinovic, and T. M. Rassias, *Topics in Polynomials: Extremal Problems, Inequalities, Zeros*. Singapore: World Scientific, 1994. (Cited on pages 778 and 781.)

[1030] G. V. Milovanovic and T. M. Rassias, "Inequalities for Polynomial Zeros," in *Survey on Classical Inequalities*, T. M. Rassias, Ed. Dordrecht: Kluwer, 2000, pp. 165–202. (Cited on pages 700, 778, 779, and 781.)

[1031] N. Minamide, "An Extension of the Matrix Inversion Lemma," *SIAM J. Alg. Disc. Meth.*, Vol. 6, pp. 371–377, 1985. (Cited on page 413.)

[1032] H. Miranda and R. C. Thompson, "A Trace Inequality with a Subtracted Term," *Lin. Alg. Appl.*, Vol. 185, pp. 165–172, 1993. (Cited on page 595.)

[1033] L. Mirsky, *An Introduction to Linear Algebra*. Oxford: Clarendon, 1972, reprinted by Dover, Mineola, 1990. (Cited on pages xix and 211.)

[1034] P. Misra and R. V. Patel, "A Determinant Identity and its Application in Evaluating Frequency Response Matrices," *SIAM J. Matrix Anal. Appl.*, Vol. 9, pp. 248–255, 1988. (Cited on page 874.)

[1035] D. S. Mitrinovic, J. E. Pecaric, and A. M. Fink, *Classical and New Inequalities in Analysis*. Dordrecht: Kluwer, 1993. (Cited on pages 25, 77, 84, 295, 619, 620, 623, and 624.)

[1036] D. S. Mitrinovic, J. E. Pecaric, and V. Volenec, *Recent Advances in Geometric Inequalities*. Dordrecht: Kluwer, 1989. (Cited on page 171.)

[1037] B. Mityagin, "An Inequality in Linear Algebra," *SIAM Rev.*, Vol. 33, pp. 125–127, 1991. (Cited on pages 504 and 595.)

[1038] M. Moakher, "Means and Averaging in the Group of Rotations," *SIAM J. Matrix Anal. Appl.*, Vol. 24, pp. 1–16, 2002. (Cited on pages 207, 394, 739, 755, and 759.)

[1039] M. Moakher, "A Differential Geometric Approach to the Geometric Mean of Symmetric Positive-Definite Matrices," *SIAM J. Matrix Anal. Appl.*, Vol. 26, pp. 735–747, 2005. (Cited on pages 510, 703, and 755.)

[1040] C. Moler and C. F. Van Loan, "Nineteen Dubious Ways to Compute the Exponential of a Matrix, Twenty-Five Years Later," *SIAM Rev.*, Vol. 45, pp. 3–49, 2000. (Cited on pages 759 and 793.)

[1041] B. Mond, "Generalized Inverse Extensions of Matrix Inequalities," *Lin. Alg. Appl.*, Vol. 2, pp. 393–399, 1969. (Cited on page 551.)

[1042] B. Mond and J. E. Pecaric, "Reverse Forms of a Convex Matrix Inequality," *Lin. Alg. Appl.*, Vol. 220, pp. 359–364, 1995. (Cited on page 502.)

[1043] B. Mond and J. E. Pecaric, "Inequalities for the Hadamard Product of Matrices," *SIAM J. Matrix Anal. Appl.*, Vol. 19, pp. 66–70, 1998. (Cited on page 591.)

[1044] B. Mond and J. E. Pecaric, "On Inequalities Involving the Hadamard Product of Matrices," *Elec. J. Lin. Alg.*, Vol. 6, pp. 56–61, 2000. (Cited on pages 590 and 592.)

[1045] V. V. Monov, "On the Spectrum of Convex Sets of Matrices," *IEEE Trans. Autom. Contr.*, Vol. 44, pp. 1009–1012, 1992. (Cited on page 293.)

[1046] A. Morawiec, *Orientations and Rotations: Computations in Crystallographic Textures.* New York: Springer, 2003. (Cited on pages 249 and 740.)

[1047] T. Moreland and S. Gudder, "Infima of Hilbert Space Effects," *Lin. Alg. Appl.*, Vol. 286, pp. 1–17, 1999. (Cited on page 506.)

[1048] T. Mori, "Comments on 'A Matrix Inequality Associated with Bounds on Solutions of Algebraic Riccati and Lyapunov Equation'," *IEEE Trans. Autom. Contr.*, Vol. 33, p. 1088, 1988. (Cited on page 595.)

[1049] Y. Moschovakis, *Notes on Set Theory*, 2nd ed. New York: Springer, 2005. (Cited on page 15.)

[1050] T. Muir, *The Theory of Determinants in the Historical Order of Development.* New York: Dover, 1966. (Cited on page 178.)

[1051] W. W. Muir, "Inequalities Concerning the Inverses of Positive Definite Matrices," *Proc. Edinburgh Math. Soc.*, Vol. 19, pp. 109–113, 1974–75. (Cited on pages 485 and 596.)

[1052] I. S. Murphy, "A Note on the Product of Complementary Principal Minors of a Positive Definite Matrix," *Lin. Alg. Appl.*, Vol. 44, pp. 169–172, 1982. (Cited on page 541.)

[1053] R. M. Murray, Z. Li, and S. S. Sastry, *A Mathematical Introduction to Robotic Manipulation.* Boca Raton: CRC, 1994. (Cited on pages xvii and 793.)

[1054] I. Najfeld and T. F. Havel, "Derivatives of the Matrix Exponential and Their Computation," *Adv. Appl. Math.*, Vol. 16, pp. 321–375, 1995. (Cited on page 748.)

[1055] R. Nakamoto, "A Norm Inequality for Hermitian Operators," *Amer. Math. Monthly*, Vol. 110, pp. 238–239, 2003. (Cited on page 762.)

[1056] Y. Nakamura, "Any Hermitian Matrix is a Linear Combination of Four Projections," *Lin. Alg. Appl.*, Vol. 61, pp. 133–139, 1984. (Cited on page 396.)

[1057] A. W. Naylor and G. R. Sell, *Linear Operator Theory in Engineering and Science.* New York: Springer, 1986. (Cited on pages 77, 84, 620, 623, 683, 684, 685, and 705.)

[1058] T. Needham, *Visual Complex Analysis.* Oxford: Oxford University Press, 1997. (Cited on page 84.)

[1059] C. N. Nett and W. M. Haddad, "A System-Theoretic Appropriate Realization of the Empty Matrix Concept," *IEEE Trans. Autom. Contr.*, Vol. 38, pp. 771–775, 1993. (Cited on page 178.)

[1060] M. G. Neubauer, "An Inequality for Positive Definite Matrices with Applications to Combinatorial Matrices," *Lin. Alg. Appl.*, Vol. 267, pp. 163–174, 1997. (Cited on page 540.)

[1061] M. F. Neuts, *Matrix-Geometric Solutions in Stochastic Models*. Baltimore: Johns Hopkins University Press, 1981, reprinted by Dover, Mineola, 1994. (Cited on page xvii.)

[1062] R. W. Newcomb, "On the Simultaneous Diagonalization of Two Semi-Definite Matrices," *Quart. Appl. Math.*, Vol. 19, pp. 144–146, 1961. (Cited on page 595.)

[1063] M. Newman, "Lyapunov Revisited," *Lin. Alg. Appl.*, Vol. 52/53, pp. 525–528, 1983. (Cited on page 336.)

[1064] M. Newman, W. So, and R. C. Thompson, "Convergence Domains for the Campbell-Baker-Hausdorff Formula," *Lin. Multilin. Alg.*, Vol. 24, pp. 301–310, 1989. (Cited on page 720.)

[1065] D. W. Nicholson, "Eigenvalue Bounds for of $AB + BA$ for A, B Positive Definite Matrices," *Lin. Alg. Appl.*, Vol. 24, pp. 173–183, 1979. (Cited on page 572.)

[1066] C. Niculescu and L.-E. Persson, *Convex Functions and Their Applications: A Contemporary Approach*. New York: Springer, 2006. (Cited on pages 23, 24, 38, 40, 46, 59, 60, 62, 69, 143, 531, 620, 700, and 702.)

[1067] C. P. Niculescu, "Convexity According to the Geometric Mean," *Math. Ineq. Appl.*, Vol. 3, pp. 155–167, 2000. (Cited on pages 43, 59, 171, and 781.)

[1068] C. P. Niculescu and F. Popovici, "A Refinement of Popoviciu's Inequality," *Bull. Math. Soc. Sci. Math. Roumanie*, Vol. 49(97), pp. 285–290, 2006. (Cited on page 24.)

[1069] M. A. Nielsen and I. L. Chuang, *Quantum Computation and Quantum Information*. Cambridge: Cambridge University Press, 2000. (Cited on page xvii.)

[1070] M. Niezgoda, "Laguerre-Samuelson Type Inequalities," *Lin. Alg. Appl.*, Vol. 422, pp. 574–581, 2007. (Cited on pages 56 and 499.)

[1071] K. Nishio, "The Structure of a Real Linear Combination of Two Projections," *Lin. Alg. Appl.*, Vol. 66, pp. 169–176, 1985. (Cited on page 217.)

[1072] B. Noble and J. W. Daniel, *Applied Linear Algebra*, 3rd ed. Englewood Cliffs: Prentice-Hall, 1988. (Cited on pages xix and 178.)

[1073] K. Nomakuchi, "On the Characterization of Generalized Inverses by Bordered Matrices," *Lin. Alg. Appl.*, Vol. 33, pp. 1–8, 1980. (Cited on page 429.)

[1074] K. Nordstrom, "Some Further Aspects of the Lowner-Ordering Antitonicity of the Moore-Penrose Inverse," *Commun. Statist. Theory Meth.*, Vol. 18, pp. 4471–4489, 1989. (Cited on pages 503 and 580.)

[1075] J. Nunemacher, "Which Real Matrices Have Real Logarithms?" *Math. Mag.*, Vol. 62, pp. 132–135, 1989. (Cited on page 793.)

[1076] H. Ogawa, "An Operator Pseudo-Inversion Lemma," *SIAM J. Appl. Math.*, Vol. 48, pp. 1527–1531, 1988. (Cited on page 413.)

[1077] I. Olkin, "An Inequality for a Sum of Forms," *Lin. Alg. Appl.*, Vol. 52–53, pp. 529–532, 1983. (Cited on page 502.)

[1078] J. M. Ortega, *Matrix Theory, A Second Course.* New York: Plenum, 1987. (Cited on page xix.)

[1079] B. Ortner and A. R. Krauter, "Lower Bounds for the Determinant and the Trace of a Class of Hermitian Matrices," *Lin. Alg. Appl.*, Vol. 236, pp. 147–180, 1996. (Cited on page 523.)

[1080] S. L. Osburn and D. S. Bernstein, "An Exact Treatment of the Achievable Closed-Loop H$_2$ Performance of Sampled-Data Controllers: From Continuous-Time to Open-Loop," *Automatica*, Vol. 31, pp. 617–620, 1995. (Cited on page 748.)

[1081] A. Ostrowski and H. Schneider, "Some Theorems on the Inertia of General Matrices," *J. Math. Anal. Appl.*, Vol. 4, pp. 72–84, 1962. (Cited on pages 867 and 868.)

[1082] A. M. Ostrowski, "On Some Metrical Properties of Operator Matrices and Matrices Partitioned into Blocks," *J. Math. Anal. Appl.*, Vol. 2, pp. 161–209, 1961. (Cited on page 649.)

[1083] J. A. Oteo, "The Baker-Campbell-Hausdorff Formula and Nested Commutator Identities," *J. Math. Phys.*, Vol. 32, pp. 419–424, 1991. (Cited on page 743.)

[1084] D. V. Ouellette, "Schur Complements and Statistics," *Lin. Alg. Appl.*, Vol. 36, pp. 187–295, 1981. (Cited on pages 515, 535, and 596.)

[1085] D. A. Overdijk, "Skew-Symmetric Matrices in Classical Mechanics," Eindhoven University, Memorandum COSOR 89-23, 1989. (Cited on page 204.)

[1086] C. C. Paige and M. Saunders, "Towards a Generalized Singular Value Decomposition," *SIAM J. Numer. Anal.*, Vol. 18, pp. 398–405, 1981. (Cited on page 344.)

[1087] C. C. Paige, G. P. H. Styan, B. Y. Wang, and F. Zhang, "Hua's Matrix Equality and Schur Complements," *Int. J. Inform. Sys. Sci.*, Vol. 4, pp. 124–135, 2008. (Cited on pages 336, 520, 539, and 553.)

[1088] C. C. Paige and M. Wei, "History and Generality of the CS Decomposition," *Lin. Alg. Appl.*, Vol. 208/209, pp. 303–326, 1994. (Cited on page 344.)

[1089] H. Palanthandalam-Madapusi, D. S. Bernstein, and R. Venugopal, "Dimensional Analysis of Matrices: State-Space Models and Dimensionless Units," *IEEE Contr. Sys. Mag.*, Vol. 27, no. December, pp. 100–109, 2007. (Cited on pages 252 and 744.)

[1090] H. Palanthandalam-Madapusi, T. Van Pelt, and D. S. Bernstein, "Matrix Pencils and Existence Conditions for Quadratic Programming with a Sign-Indefinite Quadratic Equality Constraint," *J. Global Optimization*, preprint. (Cited on page 548.)

[1091] H. Palanthandalam-Madapusi, T. Van Pelt, and D. S. Bernstein, "Parameter Consistency and Quadratically Constrained Errors-in-Variables Least-Squares Identification," preprint. (Cited on page 548.)

[1092] F. C. Paliogiannis, "On Commuting Operator Exponentials," *Proc. Amer. Math. Soc.*, Vol. 131, pp. 3777–3781, 2003. (Cited on page 747.)

[1093] B. P. Palka, *An Introduction to Complex Function Theory.* New York: Springer, 1991. (Cited on page 84.)

[1094] C. V. Pao, "Logarithmic Derivatives of a Square Matrix," *Lin. Alg. Appl.*, Vol. 6, pp. 159–164, 1973. (Cited on page 758.)

[1095] J. G. Papastavridis, *Tensor Calculus and Analytical Dynamics.* Boca Raton: CRC, 1998. (Cited on page xvii.)

[1096] J. G. Papastavridis, *Analytical Mechanics.* Oxford: Oxford University Press, 2002. (Cited on pages xvii and 496.)

[1097] F. C. Park, "Computational Aspects of the Product-of-Exponentials Formula for Robot Kinematics," *IEEE Trans. Autom. Contr.*, Vol. 39, pp. 643–647, 1994. (Cited on page 793.)

[1098] P. Park, "On the Trace Bound of a Matrix Product," *IEEE Trans. Autom. Contr.*, Vol. 41, pp. 1799–1802, 1996. (Cited on page 364.)

[1099] D. F. Parker, *Fields, Flow and Waves: An Introduction to Continuum Models.* London: Springer, 2003. (Cited on page 396.)

[1100] P. C. Parks, "A New Proof of the Routh-Hurwitz Stability Criterion Using the Second Method of Liapunov," *Proc. Camb. Phil. Soc.*, Vol. 58, pp. 694–702, 1962. (Cited on page 772.)

[1101] R. V. Patel, "On Blocking Zeros in Linear Multivariable Systems," *IEEE Trans. Autom. Contr.*, Vol. AC-31, pp. 239–241, 1986. (Cited on page 879.)

[1102] R. V. Patel and M. Toda, "Trace Inequalities Involving Hermitian Matrices," *Lin. Alg. Appl.*, Vol. 23, pp. 13–20, 1979. (Cited on pages 532, 533, and 759.)

[1103] C. Pearcy, "A Complete Set of Unitary Invariants for Operators Generating Finite W^*-Algebras of Type I," *Pac. J. Math.*, Vol. 12, pp. 1405–1414, 1962. (Cited on page 346.)

[1104] M. C. Pease, *Methods of Matrix Algebra.* New York: Academic Press, 1965. (Cited on page 252.)

[1105] J. Pecaric and R. Rajic, "The Dunkl-Williams Inequality with n Elements in Normed Linear Spaces," *Math. Ineq. Appl.*, Vol. 10, pp. 461–470, 2007. (Cited on page 623.)

[1106] J. E. Pecaric, S. Puntanen, and G. P. H. Styan, "Some Further Matrix Extensions of the Cauchy-Schwarz and Kantorovich Inequalities, with Some Statistical Applications," *Lin. Alg. Appl.*, Vol. 237/238, pp. 455–476, 1996. (Cited on page 551.)

[1107] J. E. Pecaric, S. Puntanen, and G. P. H. Styan, "Some Further Matrix Extensions of the Cauchy-Schwarz and Kantorovich Inequalities, with Some Statistical Applications," *Lin. Alg. Appl.*, Vol. 237/238, pp. 455–476, 1996. (Cited on page 674.)

[1108] S. Perlis, *Theory of Matrices.* Reading: Addison-Wesley, 1952, reprinted by Dover, New York, 1991. (Cited on pages 178, 255, 256, 259, 307, and 396.)

[1109] T. Peter, "Maximizing the Area of a Quadrilateral," *College Math. J.*, Vol. 34, pp. 315–316, 2003. (Cited on page 172.)

[1110] I. R. Petersen and C. V. Hollot, "A Riccati Equation Approach to the Stabilization of Uncertain Systems," *Automatica*, Vol. 22, pp. 397–411, 1986. (Cited on page 553.)

[1111] D. Petz and R. Temesi, "Means of Positive Numbers and Matrices," *SIAM J. Matrix Anal. Appl.*, Vol. 27, pp. 712–720, 2005. (Cited on pages 510 and 514.)

[1112] L. A. Pipes, "Applications of Laplace Transforms of Matrix Functions," *J. Franklin Inst.*, Vol. 285, pp. 436–451, 1968. (Cited on pages 736 and 737.)

[1113] A. O. Pittenger and M. H. Rubin, "Convexity and Separation Problem of Quantum Mechanical Density Matrices," *Lin. Alg. Appl.*, Vol. 346, pp. 47–71, 2002. (Cited on page xvii.)

[1114] I. Polik and T. Terlaky, "A Survey of the S-Lemma," *SIAM Rev.*, Vol. 49, pp. 371–418, 2007. (Cited on page 548.)

[1115] T. Politi, "A Formula for the Exponential of a Real Skew-Symmetric Matrix of Order 4," *Numer. Math. BIT*, Vol. 41, pp. 842–845, 2001. (Cited on pages 737 and 740.)

[1116] D. S. G. Pollock, "Tensor Products and Matrix Differential Calculus," *Lin. Alg. Appl.*, Vol. 67, pp. 169–193, 1985. (Cited on page 705.)

[1117] B. T. Polyak, "Convexity of Quadratic Transformations and Its Use in Control and Optimization," *J. Optim. Theory Appl.*, Vol. 99, pp. 553–583, 1998. (Cited on pages 547 and 548.)

[1118] B. Poonen, "A Unique $(2k+1)$-th Root of a Matrix," *Amer. Math. Monthly*, Vol. 98, p. 763, 1991. (Cited on page 381.)

[1119] B. Poonen, "Positive Deformations of the Cauchy Matrix," *Amer. Math. Monthly*, Vol. 102, pp. 842–843, 1995. (Cited on page 491.)

[1120] D. Popa and I. Rasa, "Inequalities Involving the Inner Product," *J. Ineq. Pure Appl. Math.*, Vol. 8, no. 3, pp. 1–4, 2007, Article 86. (Cited on page 620.)

[1121] V. M. Popov, *Hyperstability of Control Systems.* Berlin: Springer, 1973. (Cited on pages xvii and 793.)

[1122] G. J. Porter, "Linear Algebra and Affine Planar Transformations," *College Math. J.*, Vol. 24, pp. 47–51, 1993. (Cited on page 215.)

[1123] B. H. Pourciau, "Modern Multiplier Rules," *Amer. Math. Monthly*, Vol. 87, pp. 433–452, 1980. (Cited on page 697.)

[1124] C. R. Pranesachar, "Ratio Circum-to-In-Radius," *Amer. Math. Monthly*, Vol. 114, p. 648, 2007. (Cited on page 171.)

[1125] V. V. Prasolov, *Problems and Theorems in Linear Algebra.* Providence: American Mathematical Society, 1994. (Cited on pages xix, 155, 162, 199, 200, 209, 225, 232, 249, 277, 279, 282, 295, 299, 340, 341, 349, 358, 364, 366, 371, 372, 374, 377, 381, 383, 392, 430, 454, 485, 521, 535, 537, 542, 558, 559, 587, 629, 645, 656, 658, 666, 703, and 724.)

[1126] U. Prells, M. I. Friswell, and S. D. Garvey, "Use of Geometric Algebra: Compound Matrices and the Determinant of the Sum of Two Matrices," *Proc. Royal Soc. Lond. A*, Vol. 459, pp. 273–285, 2003. (Cited on page 454.)

[1127] J. S. Przemieniecki, *Theory of Matrix Structural Analysis.* New York: McGraw-Hill, 1968. (Cited on page xvii.)

[1128] P. J. Psarrakos, "On the mth Roots of a Complex Matrix," *Elec. J. Lin. Alg.*, Vol. 9, pp. 32–41, 2002. (Cited on page 381.)

[1129] N. J. Pullman, *Matrix Theory and Its Applications: Selected Topics.* New York: Marcel Dekker, 1976. (Cited on page 793.)

[1130] S. Puntanen and G. P. H. Styan, "Historical Introduction: Issai Schur and the Early Development of the Schur Complement," in *The Schur Complement and Its Applications*, F. Zhang, Ed. New York: Springer, 2004, pp. 1–16. (Cited on page 423.)

[1131] F. Qi, "Inequalities between the Sum of Squares and the Exponential of Sum of a Nonnegative Sequence," *J. Ineq. Pure Appl. Math.*, Vol. 8, no. 3, pp. 1–5, 2007, Article 78. (Cited on page 66.)

[1132] F. Qi and L. Debnath, "Inequalities for Power-Exponential Functions," *J. Ineq. Pure. Appl. Math.*, Vol. 1, no. 2/15, pp. 1–5, 2000. (Cited on page 42.)

[1133] C. Qian and J. Li, "Global Finite-Time Stabilization by Output Feedback for Planar Systems without Observation Linearization," *IEEE Trans. Autom. Contr.*, Vol. 50, pp. 885–890, 2005. (Cited on page 42.)

[1134] R. X. Qian and C. L. DeMarco, "An Approach to Robust Stability of Matrix Polytopes through Copositive Homogeneous Polynomials," *IEEE Trans. Autom. Contr.*, Vol. 37, pp. 848–852, 1992. (Cited on page 293.)

[1135] L. Qiu, B. Bernhardsson, A. Rantzer, E. J. Davison, P. M. Young, and J. C. Doyle, "A Formula for Computation of the Real Stability Radius," *Automatica*, Vol. 31, pp. 879–890, 1995. (Cited on page 769.)

[1136] L. Qiu and X. Zhan, "On the Span of Hadamard Products of Vectors," *Lin. Alg. Appl.*, Vol. 422, pp. 304–307, 2007. (Cited on page 455.)

[1137] J. F. Queiro and A. Kovacec, "A Bound for the Determinant of the Sum of Two Normal Matrices," *Lin. Multilin. Alg.*, Vol. 33, pp. 171–173, 1993. (Cited on page 365.)

[1138] V. Rabanovich, "Every Matrix Is a Linear Combination of Three Idempotents," *Lin. Alg. Appl.*, Vol. 390, pp. 137–143, 2004. (Cited on page 395.)

[1139] H. Radjavi, "Decomposition of Matrices into Simple Involutions," *Lin. Alg. Appl.*, Vol. 12, pp. 247–255, 1975. (Cited on page 380.)

[1140] H. Radjavi and P. Rosenthal, *Simultaneous Triangularization.* New York: Springer, 2000. (Cited on page 392.)

[1141] H. Radjavi and J. P. Williams, "Products of Self-Adjoint Operators," *Michigan Math. J.*, Vol. 16, pp. 177–185, 1969. (Cited on pages 347, 380, and 410.)

[1142] V. Rakocevic, "On the Norm of Idempotent Operators in a Hilbert Space," *Amer. Math. Monthly*, Vol. 107, pp. 748–750, 2000. (Cited on pages 228 and 367.)

[1143] A. C. M. Ran and R. Vreugdenhil, "Existence and Comparison Theorems for Algebraic Riccati Equations for Continuous- and Discrete-Time Systems," *Lin. Alg. Appl.*, Vol. 99, pp. 63–83, 1988. (Cited on page 879.)

[1144] A. Rantzer, "On the Kalman-Yakubovich-Popov Lemma," *Sys. Contr. Lett.*, Vol. 28, pp. 7–10, 1996. (Cited on page 348.)

[1145] C. R. Rao and S. K. Mitra, *Generalized Inverse of Matrices and Its Applications*. New York: Wiley, 1971. (Cited on pages 438, 582, and 595.)

[1146] C. R. Rao and M. B. Rao, *Matrix Algebra and Its Applications to Statistics and Econometrics*. Singapore: World Scientific, 1998. (Cited on pages xvii, 304, and 458.)

[1147] J. V. Rao, "Some More Representations for the Generalized Inverse of a Partitioned Matrix," *SIAM J. Appl. Math.*, Vol. 24, pp. 272–276, 1973. (Cited on page 429.)

[1148] U. A. Rauhala, "Array Algebra Expansion of Matrix and Tensor Calculus: Part 1," *SIAM J. Matrix Anal. Appl.*, Vol. 24, pp. 490–508, 2003. (Cited on page 458.)

[1149] P. A. Regalia and S. K. Mitra, "Kronecker Products, Unitary Matrices and Signal Processing Applications," *SIAM Rev.*, Vol. 31, pp. 586–613, 1989. (Cited on page 458.)

[1150] M. W. Reinsch, "A Simple Expression for the Terms in the Baker-Campbell-Hausdorff Series," *J. Math. Physics*, Vol. 41, pp. 2434–2442, 2000. (Cited on page 749.)

[1151] T. J. Richardson and R. H. Kwong, "On Positive Definite Solutions to the Algebraic Riccati Equation," *Sys. Contr. Lett.*, Vol. 7, pp. 99–104, 1986. (Cited on pages 863 and 879.)

[1152] D. S. Richeson, *Euler's Gem: The Polyhedron Formula and the Birth of Topology*. Princeton: Princeton University Press, 2008. (Cited on page 16.)

[1153] A. N. Richmond, "Expansions for the Exponential of a Sum of Matrices," in *Applications of Matrix Theory*, M. J. C. Gover and S. Barnett, Eds. Oxford: Oxford University Press, 1989, pp. 283–289. (Cited on page 751.)

[1154] K. S. Riedel, "A Sherman-Morrison-Woodbury Identity for Rank Augmenting Matrices with Application to Centering," *SIAM J. Matrix Anal. Appl.*, Vol. 13, pp. 659–662, 1992. (Cited on page 413.)

[1155] J. R. Ringrose, *Compact Non-Self-Adjoint Operators*. New York: Van Nostrand Reinhold, 1971. (Cited on page 680.)

[1156] R. S. Rivlin, "Further Remarks on the Stress Deformation Relations for Isotropic Materials," *J. Rational Mech. Anal.*, Vol. 4, pp. 681–702, 1955. (Cited on pages 283 and 284.)

[1157] J. W. Robbin, *Matrix Algebra Using MINImal MATlab*. Wellesley: A. K. Peters, 1995. (Cited on pages 84, 125, 178, 229, 252, 307, 309, 348, 377, 378, and 703.)

[1158] J. W. Robbin and D. A. Salamon, "The Exponential Vandermonde Matrix," *Lin. Alg. Appl.*, Vol. 317, pp. 225–226, 2000. (Cited on page 456.)

[1159] D. W. Robinson, "Nullities of Submatrices of the Moore-Penrose Inverse," *Lin. Alg. Appl.*, Vol. 94, pp. 127–132, 1987. (Cited on page 237.)

[1160] P. Robinson and A. J. Wathen, "Variational Bounds on the Entries of the Inverse of a Matrix," *IMA J. Numerical Anal.*, Vol. 12, pp. 463–486, 1992. (Cited on page 523.)

[1161] R. T. Rockafellar, *Convex Analysis*. Princeton: Princeton University Press, 1990. (Cited on pages 178, 684, 693, 697, and 705.)

[1162] R. T. Rockafellar and R. J. B. Wets, *Variational Analysis.* Berlin: Springer, 1998. (Cited on pages 693 and 705.)

[1163] L. Rodman, "Products of Symmetric and Skew Symmetric Matrices," *Lin. Multilin. Alg.*, Vol. 43, pp. 19–34, 1997. (Cited on page 384.)

[1164] G. S. Rogers, *Matrix Derivatives.* New York: Marcel Dekker, 1980. (Cited on page 705.)

[1165] C. A. Rohde, "Generalized Inverses of Partitioned Matrices," *SIAM J. Appl. Math.*, Vol. 13, pp. 1033–1035, 1965. (Cited on page 429.)

[1166] J. Rohn, "Computing the Norm $||A||_{\infty,1}$ Is NP-Hard," *Lin. Multilin. Alg.*, Vol. 47, pp. 195–204, 2000. (Cited on page 630.)

[1167] O. Rojo, "Further Bounds for the Smallest Singular Value and the Spectral Condition Number," *Computers Math. Appl.*, Vol. 38, pp. 215–228, 1999. (Cited on page 662.)

[1168] O. Rojo, "Inequalities Involving the Mean and the Standard Deviation of Nonnegative Real Numbers," *J. Ineq. Appl.*, Vol. 2006, pp. 1–15, 2006, article ID 43465. (Cited on page 56.)

[1169] J. Rooin, "On Ky Fan's Inequality and its Additive Analogues," *Math. Ineq. Appl.*, Vol. 6, pp. 595–604, 2003. (Cited on page 60.)

[1170] T. G. Room, "The Composition of Rotations in Euclidean Three-Space," *Amer. Math. Monthly*, Vol. 59, pp. 688–692, 1952. (Cited on page 740.)

[1171] D. J. Rose, "Matrix Identities of the Fast Fourier Transform," *Lin. Alg. Appl.*, Vol. 29, pp. 423–443, 1980. (Cited on page 390.)

[1172] K. H. Rosen, Ed., *Handbook of Discrete and Combinatorial Mathematics.* Boca Raton: CRC, 2000. (Cited on pages xvii and xix.)

[1173] H. H. Rosenbrock, *State-Space and Multivariable Theory.* New York: Wiley, 1970. (Cited on page 836.)

[1174] M. Rosenfeld, "A Sufficient Condition for Nilpotence," *Amer. Math. Monthly*, Vol. 103, pp. 907–909, 1996. (Cited on pages xvii and 233.)

[1175] A. Rosoiu, "Triangle Radii Inequality," *Amer. Math. Monthly*, Vol. 114, p. 640, 2007. (Cited on page 171.)

[1176] W. Rossmann, *Lie Groups: An Introduction Through Linear Groups.* Oxford: Oxford University Press, 2002. (Cited on page 252.)

[1177] U. G. Rothblum, "Nonnegative Matrices and Stochastic Matrices," in *Handbook of Linear Algebra*, L. Hogben, Ed. Boca Raton: Chapman & Hall/CRC, 2007, pp. 9-1–9-25. (Cited on pages 298, 299, 306, and 438.)

[1178] J. J. Rotman, *An Introduction to the Theory of Groups*, 4th ed. New York: Springer, 1999. (Cited on pages 245 and 391.)

[1179] W. J. Rugh, *Linear System Theory*, 2nd ed. Upper Saddle River: Prentice Hall, 1996. (Cited on pages 274, 275, 743, 816, 829, and 879.)

[1180] A. M. Russell and C. J. F. Upton, "A Class of Positive Semidefinite Matrices," *Lin. Alg. Appl.*, Vol. 93, pp. 121–126, 1987. (Cited on page 493.)

[1181] D. L. Russell, *Mathematics of Finite-Dimensional Control Systems.* New York: Marcel Dekker, 1979. (Cited on page 866.)

[1182] A. Saberi, P. Sannuti, and B. M. Chen, *H_2 Optimal Control.* Englewood Cliffs: Prentice-Hall, 1995. (Cited on page xvii.)

[1183] N. Sabourova, *Real and Complex Operator Norms.* Lulea, Sweden: Lulea University of Technology, 2007, Ph.D. Dissertation, http://epubl.ltu.se/1402-1757/2007/09/LTU-LIC-0709-SE.pdf. (Cited on page 609.)

[1184] M. K. Sain and C. B. Schrader, "The Role of Zeros in the Performance of Multi-input, Multioutput Feedback Systems," *IEEE Trans. Educ.*, Vol. 33, pp. 244–257, 1990. (Cited on page 879.)

[1185] J. Sandor, "On Certain Inequalities for Means, II," *J. Math. Anal. Appl.*, Vol. 199, pp. 629–635, 1996. (Cited on page 40.)

[1186] J. Sandor, "Inequalities for Generalized Convex Functions with Applications I," *Studia University Babes-Bolyai, Mathematica*, Vol. XLIV, pp. 93–107, 1999. (Cited on page 24.)

[1187] J. Sandor, "On Certain Limits Related to the Number *e*," *Libertas Mathematica*, Vol. XX, pp. 155–159, 2000. (Cited on page 28.)

[1188] J. Sandor, "Inequalities for Generalized Convex Functions with Applications II," *Studia University Babes-Bolyai, Mathematica*, Vol. XLVI, pp. 79–91, 2001. (Cited on page 24.)

[1189] J. Sandor, "On an Inequality of Ky Fan," *Int. J. Math. Educ. Sci. Tech.*, Vol. 32, pp. 133–160, 2001. (Cited on page 60.)

[1190] J. Sandor and L. Debnath, "On Certain Inequalities Involving the Constant *e* and Their Applications," *J. Math. Anal. Appl.*, Vol. 249, pp. 569–582, 2000. (Cited on pages 28, 29, and 73.)

[1191] R. A. Satnoianu, "General Power Inequalities between the Sides and the Circumscribed and Inscribed Radii Related to the Fundamental Triangle Inequality," *Math. Ineq. Appl.*, Vol. 5, pp. 745–751, 2002. (Cited on page 171.)

[1192] D. H. Sattinger and O. L. Weaver, *Lie Groups and Algebras with Applications to Physics, Geometry, and Mechanics.* New York: Springer, 1986. (Cited on pages 162, 187, 720, 724, and 793.)

[1193] A. H. Sayed, *Fundamentals of Adaptive Filtering.* New York: Wiley, 2003. (Cited on page xvii.)

[1194] H. Schaub, P. Tsiotras, and J. L. Junkins, "Principal Rotation Representations of Proper $N \times N$ Orthogonal Matrices," *Int. J. Eng. Sci.*, Vol. 33, pp. 2277–2295, 1995. (Cited on page 738.)

[1195] C. W. Scherer, "The Solution Set of the Algebraic Riccati Equation and the Algebraic Riccati Inequality," *Lin. Alg. Appl.*, Vol. 153, pp. 99–122, 1991. (Cited on page 879.)

[1196] C. W. Scherer, "The Algebraic Riccati Equation and Inequality for Systems with Uncontrollable Modes on the Imaginary Axis," *SIAM J. Matrix Anal. Appl.*, Vol. 16, pp. 1308–1327, 1995. (Cited on pages 860 and 879.)

[1197] C. W. Scherer, "The General Nonstrict Algebraic Riccati Inequality," *Lin. Alg. Appl.*, Vol. 219, pp. 1–33, 1995. (Cited on page 879.)

[1198] P. Scherk, "On the Decomposition of Orthogonalities into Symmetries," *Proc. Amer. Math. Soc.*, Vol. 1, pp. 481–491, 1950. (Cited on page 379.)

[1199] C. Schmoeger, "Remarks on Commuting Exponentials in Banach Algebras," *Proc. Amer. Math. Soc.*, Vol. 127, pp. 1337–1338, 1999. (Cited on page 747.)

[1200] C. Schmoeger, "Remarks on Commuting Exponentials in Banach Algebras, II," *Proc. Amer. Math. Soc.*, Vol. 128, pp. 3405–3409, 2000. (Cited on page 747.)

[1201] C. Schmoeger, "On Normal Operator Exponentials," *Proc. Amer. Math. Soc.*, Vol. 130, pp. 697–702, 2001. (Cited on page 747.)

[1202] C. Schmoeger, "On Logarithms of Linear Operators on Hilbert Spaces," *Demonstratio Math.*, Vol. XXXV, pp. 375–384, 2002. (Cited on pages 714, 715, and 756.)

[1203] C. Schmoeger, "On the Operator Equation $e^A = e^B$," *Lin. Alg. Appl.*, Vol. 359, pp. 169–179, 2003. (Cited on page 749.)

[1204] H. Schneider, "Olga Taussky-Todd's Influence on Matrix Theory and Matrix Theorists," *Lin. Multilin. Alg.*, Vol. 5, pp. 197–224, 1977. (Cited on page 293.)

[1205] B. Scholkopf and A. J. Smola, *Learning with Kernels: Support Vector Machines, Regularization, Optimization, and Beyond.* Cambridge: MIT Press, 2001. (Cited on page 489.)

[1206] D. Scholz and M. Weyrauch, "A Note on the Zassenhaus Product Formula," *J. Math. Phys.*, Vol. 47, pp. 033 505/1–7, 2006. (Cited on page 751.)

[1207] J. R. Schott, "Kronecker Product Permutation Matrices and Their Application to Moment Matrices of the Normal Distribution," *J. Multivariate Analysis*, Vol. 87, pp. 177–190, 2003. (Cited on page 19.)

[1208] J. R. Schott, *Matrix Analysis for Statistics*, 2nd ed. New York: Wiley, 2005. (Cited on pages xvii, 121, 229, 236, 391, 411, 413, 414, 415, 419, 420, 428, 449, 470, 486, 536, 557, 564, 565, 569, 570, 571, 585, 586, 587, 589, 605, 656, 678, and 696.)

[1209] C. B. Schrader and M. K. Sain, "Research on System Zeros: A Survey," *Int. J. Contr.*, Vol. 50, pp. 1407–1433, 1989. (Cited on page 879.)

[1210] B. S. W. Schroder, *Ordered Sets: An Introduction.* Boston: Birkhauser, 2003. (Cited on page 84.)

[1211] A. J. Schwenk, "Tight Bounds on the Spectral Radius of Asymmetric Nonnegative Matrices," *Lin. Alg. Appl.*, Vol. 75, pp. 257–265, 1986. (Cited on page 456.)

[1212] S. R. Searle, *Matrix Algebra Useful for Statistics.* New York: Wiley, 1982. (Cited on page xvii.)

[1213] P. Sebastian, "On the Derivatives of Matrix Powers," *SIAM J. Matrix Anal. Appl.*, Vol. 17, pp. 640–648, 1996. (Cited on page 705.)

[1214] P. Sebastiani, "On the Derivatives of Matrix Powers," *SIAM J. Matrix Anal. Appl.*, Vol. 17, pp. 640–648, 1996. (Cited on page 703.)

[1215] G. A. F. Seber, *A Matrix Handbook for Statisticians.* New York: Wiley, 2008. (Cited on pages 74, 121, 127, 151, 167, 168, 198, 224, 226, 231, 236, 241, 252, 293, 296, 408, 409, 414, 431, 535, 550, 579, 588, 627, 665, and 775.)

[1216] J. M. Selig, *Geometric Fundamentals of Robotics*, 2nd ed. New York: Springer, 2005. (Cited on pages xvii, 213, 249, and 252.)

[1217] D. Serre, *Matrices: Theory and Applications.* New York: Springer, 2002. (Cited on pages 187, 239, 283, and 427.)

[1218] J.-P. Serre and L. L. Scott, *Linear Representations of Finite Groups* . New York: Springer, 1996. (Cited on pages 244 and 245.)

[1219] C. Shafroth, "A Generalization of the Formula for Computing the Inverse of a Matrix," *Amer. Math. Monthly*, Vol. 88, pp. 614–616, 1981. (Cited on pages 151 and 207.)

[1220] W. M. Shah and A. Liman, "On the Zeros of a Class of Polynomials," *Math. Ineq. Appl.*, Vol. 10, pp. 733–744, 2007. (Cited on page 778.)

[1221] H. Shapiro, "A Survey of Canonical Forms and Invariants for Unitary Similarity," *Lin. Alg. Appl.*, Vol. 147, pp. 101–167, 1991. (Cited on page 346.)

[1222] H. Shapiro, "Notes from Math 223: Olga Taussky Todd's Matrix Theory Course, 1976–1977," *Mathematical Intelligencer*, Vol. 19, no. 1, pp. 21–27, 1997. (Cited on page 383.)

[1223] R. Shaw and F. I. Yeadon, "On $(a \times b) \times c$," *Amer. Math. Monthly*, Vol. 96, pp. 623–629, 1989. (Cited on page 204.)

[1224] S.-Q. Shen and T.-Z. Huang, "Several Inequalities for the Largest Singular Value and the Spectral Radius of Matrices," *Math. Ineq. Appl.*, Vol. 10, pp. 713–722, 2007. (Cited on pages 299, 457, 458, 675, and 676.)

[1225] G. E. Shilov, *Linear Algebra.* Englewood Cliffs: Prentice-Hall, 1971, reprinted by Dover, New York, 1977. (Cited on page xix.)

[1226] M. D. Shuster, "A Survey of Attitude Representations," *J. Astron. Sci.*, Vol. 41, pp. 439–517, 1993. (Cited on pages 249 and 738.)

[1227] M. D. Shuster, "Attitude Analysis in Flatland: The Plane Truth," *J. Astronautical Sci.*, Vol. 52, pp. 195–209, 2004. (Cited on page 214.)

[1228] D. D. Siljak, *Large-Scale Dynamic Systems: Stability and Structure.* New York: North-Holland, 1978. (Cited on pages xvii, 252, and 774.)

[1229] F. C. Silva and R. Simoes, "On the Lyapunov and Stein Equations," *Lin. Alg. Appl.*, Vol. 420, pp. 329–338, 2007. (Cited on page 869.)

[1230] F. C. Silva and R. Simoes, "On the Lyapunov and Stein Equations, II," *Lin. Alg. Appl.*, Vol. 426, pp. 305–311, 2007. (Cited on page 869.)

[1231] S. F. Singer, *Symmetry in Mechanics: A Gentle, Modern Introduction.* Boston: Birkhauser, 2001. (Cited on page xvii.)

[1232] R. E. Skelton, T. Iwasaki, and K. Grigoriadis, *A Unified Algebraic Approach to Linear Control Design.* London: Taylor and Francis, 1998. (Cited on pages xvii and 793.)

[1233] D. M. Smiley and M. F. Smiley, "The Polygonal Inequalities," *Amer. Math. Monthly*, Vol. 71, pp. 755–760, 1964. (Cited on page 620.)

[1234] O. K. Smith, "Eigenvalues of a Symmetric 3×3 Matrix," *Comm. ACM*, Vol. 4, p. 168, 1961. (Cited on page 289.)

[1235] R. A. Smith, "Matrix Calculations for Lyapunov Quadratic Forms," *J. Diff. Eqns.*, Vol. 2, pp. 208–217, 1966. (Cited on page 870.)

[1236] A. Smoktunowicz, "Block Matrices and Symmetric Perturbations," *Lin. Alg. Appl.*, Vol. 429, pp. 2628–2635, 2008. (Cited on pages 200 and 649.)

[1237] R. Snieder, *A Guided Tour of Mathematical Methods for the Physical Sciences*, 2nd ed. Cambridge: Cambridge University Press, 2004. (Cited on page 404.)

[1238] J. Snyders and M. Zakai, "On Nonnegative Solutions of the Equation $AD + DA' = C$," *SIAM J. Appl. Math.*, Vol. 18, pp. 704–714, 1970. (Cited on pages 793, 818, 819, and 879.)

[1239] W. So, "Equality Cases in Matrix Exponential Inequalities," *SIAM J. Matrix Anal. Appl.*, Vol. 13, pp. 1154–1158, 1992. (Cited on pages 194, 524, 714, 747, 749, 756, and 761.)

[1240] W. So, "The High Road to an Exponential Formula," *Lin. Alg. Appl.*, Vol. 379, pp. 69–75, 2004. (Cited on page 755.)

[1241] W. So and R. C. Thompson, "Product of Exponentials of Hermitian and Complex Symmetric Matrices," *Lin. Multilin. Alg.*, Vol. 29, pp. 225–233, 1991. (Cited on page 755.)

[1242] W. So and R. C. Thompson, "Singular Values of Matrix Exponentials," *Lin. Multilin. Alg.*, Vol. 47, pp. 249–258, 2000. (Cited on pages 560 and 757.)

[1243] E. D. Sontag, *Mathematical Control Theory: Deterministic Finite-Dimensional Systems*, 2nd ed. New York: Springer, 1998. (Cited on pages xvii and 866.)

[1244] E. D. Sontag, "Passivity Gains and the 'Secant Condition' for Stability," *Sys. Contr. Lett.*, Vol. 55, pp. 177–183, 2006. (Cited on page 773.)

[1245] A. R. Sourour, "A Factorization Theorem for Matrices," *Lin. Multilin. Alg.*, Vol. 19, pp. 141–147, 1986. (Cited on page 383.)

[1246] A. R. Sourour, "Nilpotent Factorization of Matrices," *Lin. Multilin. Alg.*, Vol. 31, pp. 303–308, 1992. (Cited on page 383.)

[1247] E. Spiegel, "Sums of Projections," *Lin. Alg. Appl.*, Vol. 187, pp. 239–249, 1993. (Cited on page 395.)

[1248] M. Spivak, *A Comprehensive Introduction to Differential Geometry*, 3rd ed. Houston: Publish or Perish, 1999. (Cited on pages 249 and 684.)

[1249] R. P. Stanley, *Enumerative Combinatorics, Volume 1*, 2nd ed. Cambridge: Cambridge University Press, 2000. (Cited on page 12.)

[1250] W.-H. Steeb, *Matrix Calculus and Kronecker Product with Applications and C++ Programs*. Singapore: World Scientific, 2001. (Cited on page 458.)

[1251] W.-H. Steeb and F. Wilhelm, "Exponential Functions of Kronecker Products and Trace Calculation," *Lin. Multilin. Alg.*, Vol. 9, pp. 345–346, 1981. (Cited on page 755.)

[1252] J. M. Steele, *The Cauchy-Schwarz Master Class*. Washington, DC: Mathematical Association of America, 2004. (Cited on page 84.)

[1253] R. F. Stengel, *Flight Dynamics*. Princeton: Princeton University Press, 2004. (Cited on page xvii.)

[1254] C. Stepniak, "Ordering of Nonnegative Definite Matrices with Application to Comparison of Linear Models," *Lin. Alg. Appl.*, Vol. 70, pp. 67–71, 1985. (Cited on page 579.)

[1255] H. J. Stetter, *Numerical Polynomial Algebra*. Philadelphia: SIAM, 2004. (Cited on page xviii.)

[1256] G. W. Stewart, *Introduction to Matrix Computations*. New York: Academic Press, 1973. (Cited on page xviii.)

[1257] G. W. Stewart, "On the Perturbation of Pseudo-Inverses, Projections and Linear Least Squares Problems," *SIAM Rev.*, Vol. 19, pp. 634–662, 1977. (Cited on page 678.)

[1258] G. W. Stewart, *Matrix Algorithms Volume I: Basic Decompositions*. Philadelphia: SIAM, 1998. (Cited on page xviii.)

[1259] G. W. Stewart, "On the Adjugate Matrix," *Lin. Alg. Appl.*, Vol. 283, pp. 151–164, 1998. (Cited on pages xxxvi and 178.)

[1260] G. W. Stewart, *Matrix Algorithms Volume II: Eigensystems*. Philadelphia: SIAM, 2001. (Cited on page xviii.)

[1261] G. W. Stewart and J. Sun, *Matrix Perturbation Theory*. Boston: Academic Press, 1990. (Cited on pages 331, 332, 344, 367, 391, 392, 554, 595, 625, 634, 635, 657, 672, 673, 680, 697, and 698.)

[1262] E. U. Stickel, "Fast Computation of Matrix Exponential and Logarithm," *Analysis*, Vol. 5, pp. 163–173, 1985. (Cited on page 774.)

[1263] E. U. Stickel, "An Algorithm for Fast High Precision Computation of Matrix Exponential and Logarithm," *Analysis*, Vol. 10, pp. 85–95, 1990. (Cited on pages 774 and 793.)

[1264] L. Stiller, "Multilinear Algebra and Chess Endgames," in *Games of No Chance*, R. Nowakowski, Ed. Berkeley: Mathematical Sciences Research Institute, 1996, pp. 151–192. (Cited on page xvii.)

[1265] J. Stoer, "On the Characterization of Least Upper Bound Norms in Matrix Space," *Numer. Math*, Vol. 6, pp. 302–314, 1964. (Cited on page 636.)

[1266] J. Stoer and C. Witzgall, *Convexity and Optimization in Finite Dimensions I*. Berlin: Springer, 1970. (Cited on pages 178 and 697.)

[1267] K. B. Stolarsky, "Generalizations of the Logarithmic Mean," *J. Math. Anal. Appl.*, Vol. 48, pp. 87–92, 1975. (Cited on pages 40 and 41.)

[1268] M. G. Stone, "A Mnemonic for Areas of Polygons," *Amer. Math. Monthly*, Vol. 93, pp. 479–480, 1986. (Cited on page 173.)

[1269] G. Strang, *Linear Algebra and Its Applications*, 3rd ed. San Diego: Harcourt, Brace, Jovanovich, 1988. (Cited on pages xvii and xix.)

[1270] G. Strang, "The Fundamental Theorem of Linear Algebra," *Amer. Math. Monthly*, Vol. 100, pp. 848–855, 1993. (Cited on page 178.)

[1271] G. Strang, "Every Unit Matrix is a LULU," *Lin. Alg. Appl.*, Vol. 265, pp. 165–172, 1997. (Cited on page 379.)

[1272] G. Strang and K. Borre, *Linear Algebra, Geodesy, and GPS*. Wellesley-Cambridge Press, 1997. (Cited on page xvii.)

[1273] G. Strang and T. Nguyen, "The Interplay of Ranks of Submatrices," *SIAM Rev.*, Vol. 46, pp. 637–646, 2004. (Cited on pages 135 and 237.)

[1274] S. Strelitz, "On the Routh-Hurwitz Problem," *Amer. Math. Monthly*, Vol. 84, pp. 542–544, 1977. (Cited on page 765.)

[1275] R. S. Strichartz, "The Campbell-Baker-Hausdorff-Dynkin Formula and Solutions of Differential Equations," *J. Funct. Anal.*, Vol. 72, pp. 320–345, 1987. (Cited on page 743.)

[1276] T. Strom, "On Logarithmic Norms," *SIAM J. Numer. Anal.*, Vol. 12, pp. 741–753, 1975. (Cited on page 758.)

[1277] J. Stuelpnagel, "On the Parametrization of the Three-Dimensional Rotation Group," *SIAM Rev.*, Vol. 6, pp. 422–430, 1964. (Cited on page 738.)

[1278] G. P. H. Styan, "On Lavoie's Determinantal Inequality," *Lin. Alg. Appl.*, Vol. 37, pp. 77–80, 1981. (Cited on page 430.)

[1279] R. P. Sullivan, "Products of Nilpotent Matrices," *Lin. Multilin. Alg.*, Vol. 56, pp. 311–317, 2008. (Cited on page 383.)

[1280] K. N. Swamy, "On Sylvester's Criterion for Positive-Semidefinite Matrices," *IEEE Trans. Autom. Contr.*, Vol. AC-18, p. 306, 1973. (Cited on page 595.)

[1281] P. Szekeres, *A Course in Modern Mathematical Physics: Groups, Hilbert Space and Differential Geometry*. Cambridge: Cambridge University Press, 2004. (Cited on page 249.)

[1282] G. Szep, "Simultaneous Triangularization of Projector Matrices," *Acta Math. Hung.*, Vol. 48, pp. 285–288, 1986. (Cited on page 392.)

[1283] T. Szirtes, *Applied Dimensional Analysis and Modeling*. New York: McGraw-Hill, 1998. (Cited on page xvii.)

[1284] H. Takagi, T. Miura, T. Kanzo, and S.-E. Takahasi, "A Reconsideration of Hua's Inequality," *J. Ineq. Appl.*, Vol. 2005, pp. 15–23, 2005. (Cited on pages 56 and 621.)

[1285] Y. Takahashi, S.-E. Takahasi, and S. Wada, "A General Hlawka Inequality and Its Reverse Inequality," *Math. Ineq. Appl.*, Vol. 12, pp. 1–10, 2009. (Cited on page 620.)

[1286] Y. Tao, "More Results on Singular Value Inequalities of Matrices," *Lin. Alg. Appl.*, Vol. 416, pp. 724–729, 2006. (Cited on page 567.)

[1287] K. Tapp, *Matrix Groups for Undergraduates*. Providence: American Mathematical Society, 2005. (Cited on pages 214, 245, 247, 248, 249, and 700.)

[1288] O. Taussky, "Positive-Definite Matrices and Their Role in the Study of the Characteristic Roots of General Matrices," *Adv. Math.*, Vol. 2, pp. 175–186, 1968. (Cited on page 396.)

[1289] O. Taussky, "The Role of Symmetric Matrices in the Study of General Matrices," *Lin. Alg. Appl.*, Vol. 5, pp. 147–154, 1972. (Cited on page 396.)

[1290] O. Taussky, "How I Became a Torchbearer for Matrix Theory," *Amer. Math. Monthly*, Vol. 95, pp. 801–812, 1988. (Cited on page 178.)

[1291] O. Taussky and J. Todd, "Another Look at a Matrix of Mark Kac," *Lin. Alg. Appl.*, Vol. 150, pp. 341–360, 1991. (Cited on page 360.)

[1292] O. Taussky and H. Zassenhaus, "On the Similarity Transformation between a Matrix and Its Transpose," *Pac. J. Math.*, Vol. 9, pp. 893–896, 1959. (Cited on page 396.)

[1293] W. Tempelman, "The Linear Algebra of Cross Product Operations," *J. Astron. Sci.*, Vol. 36, pp. 447–461, 1988. (Cited on page 204.)

[1294] G. ten Have, "Structure of the nth Roots of a Matrix," *Lin. Alg. Appl.*, Vol. 187, pp. 59–66, 1993. (Cited on page 381.)

[1295] R. E. Terrell, "Solution to 'Exponentials of Certain Hilbert Space Operators'," *SIAM Rev.*, Vol. 34, pp. 498–500, 1992. (Cited on page 750.)

[1296] R. E. Terrell, "Matrix Exponentials," *SIAM Rev.*, Vol. 38, pp. 313–314, 1996. (Cited on page 750.)

[1297] R. C. Thompson, "On Matrix Commutators," *J. Washington Acad. Sci.*, Vol. 48, pp. 306–307, 1958. (Cited on page 199.)

[1298] R. C. Thompson, "A Determinantal Inequality for Positive Definite Matrices," *Canad. Math. Bull.*, Vol. 4, pp. 57–62, 1961. (Cited on page 543.)

[1299] R. C. Thompson, "Some Matrix Factorization Theorems," *Pac. J. Math.*, Vol. 33, pp. 763–810, 1970. (Cited on page 384.)

[1300] R. C. Thompson, "Dissipative Matrices and Related Results," *Lin. Alg. Appl.*, Vol. 11, pp. 155–169, 1975. (Cited on page 534.)

[1301] R. C. Thompson, "A Matrix Inequality," *Comment. Math. Univ. Carolinae*, Vol. 17, pp. 393–397, 1976. (Cited on page 539.)

[1302] R. C. Thompson, "Matrix Type Metric Inequalities," *Lin. Multilin. Alg.*, Vol. 5, pp. 303–319, 1978. (Cited on pages 75, 76, and 640.)

[1303] R. C. Thompson, "Proof of a Conjectured Exponential Formula," *Lin. Multilin. Alg.*, Vol. 19, pp. 187–197, 1986. (Cited on page 755.)

[1304] R. C. Thompson, "Special Cases of a Matrix Exponential Formula," *Lin. Alg. Appl.*, Vol. 107, pp. 283–292, 1988. (Cited on page 755.)

[1305] R. C. Thompson, "Convergence Proof for Goldberg's Exponential Series," *Lin. Alg. Appl.*, Vol. 121, pp. 3–7, 1989. (Cited on page 743.)

[1306] R. C. Thompson, "Pencils of Complex and Real Symmetric and Skew Matrices," *Lin. Alg. Appl.*, Vol. 147, pp. 323–371, 1991. (Cited on page 396.)

[1307] R. C. Thompson, "High, Low, and Quantitative Roads in Linear Algebra," *Lin. Alg. Appl.*, Vol. 162–164, pp. 23–64, 1992. (Cited on page 178.)

[1308] Y. Tian, "Eight Expressions for Generalized Inverses of a Bordered Matrix," *Lin. Multilin. Alg.*, preprint. (Cited on page 429.)

[1309] Y. Tian, "EP Matrices Revisited," preprint. (Cited on pages 192 and 339.)

[1310] Y. Tian, "The Moore-Penrose Inverses of $m \times n$ Block Matrices and Their Applications," *Lin. Alg. Appl.*, Vol. 283, pp. 35–60, 1998. (Cited on pages 422 and 429.)

[1311] Y. Tian, "Matrix Representations of Octonions and Their Applications," *Adv. Appl. Clifford Algebras*, Vol. 10, pp. 61–90, 2000. (Cited on page 249.)

[1312] Y. Tian, "Commutativity of EP Matrices," *IMAGE*, Vol. 27, pp. 25–27, 2001. (Cited on pages 417 and 418.)

[1313] Y. Tian, "How to Characterize Equalities for the Moore-Penrose Inverse of a Matrix," *Kyungpook Math. J.*, Vol. 41, pp. 1–15, 2001. (Cited on pages 165, 219, 406, and 409.)

[1314] Y. Tian, "Some Equalities for Generalized Inverses of Matrix Sums and Block Circulant Matrices," *Archivum Mathematicum (BRNO)*, Vol. 37, pp. 301–306, 2001. (Cited on pages 160, 166, 422, 423, and 431.)

[1315] Y. Tian, "Some Inequalities for Sums of Matrices," *Scientiae Mathematicae Japonicae*, Vol. 54, pp. 355–361, 2001. (Cited on page 511.)

[1316] Y. Tian, "How to Express a Parallel Sum of k Matrices," *J. Math. Anal. Appl.*, Vol. 266, pp. 333–341, 2002. (Cited on page 582.)

[1317] Y. Tian, "The Minimal Rank of the Matrix Expression $A - BX - YC$," *Missouri J. Math. Sci.*, Vol. 14, pp. 40–48, 2002. (Cited on page 424.)

[1318] Y. Tian, "Upper and Lower Bounds for Ranks of Matrix Expressions Using Generalized Inverses," *Lin. Alg. Appl.*, Vol. 355, pp. 187–214, 2002. (Cited on page 425.)

[1319] Y. Tian, "A Range Equality for Idempotent Matrix," *IMAGE*, Vol. 30, pp. 26–27, 2003. (Cited on page 216.)

[1320] Y. Tian, "A Range Equality for the Difference of Orthogonal Projectors," *IMAGE*, Vol. 30, p. 36, 2003. (Cited on page 415.)

[1321] Y. Tian, "On Mixed-Type Reverse Order Laws for the Moore-Penrose Inverse of a Matrix Product," *Int. J. Math. Math. Sci.*, Vol. 58, pp. 3103–3116, 2004. (Cited on pages 415, 416, and 428.)

[1322] Y. Tian, "Rank Equalities for Block Matrices and Their Moore-Penrose Inverses," *Houston J. Math.*, Vol. 30, pp. 483–510, 2004. (Cited on page 423.)

[1323] Y. Tian, "Using Rank Formulas to Characterize Equalities for Moore-Penrose Inverses of Matrix Products," *Appl. Math. Comp.*, Vol. 147, pp. 581–600, 2004. (Cited on page 415.)

[1324] Y. Tian, "A Range Equality for the Commutator with Two Involutory Matrices," *IMAGE*, Vol. 35, pp. 32–33, 2005. (Cited on page 231.)

[1325] Y. Tian, "A Range Equality for the Kronecker Product of Matrices," *IMAGE*, Vol. 34, pp. 26–27, 2005. (Cited on page 447.)

[1326] Y. Tian, "Similarity of Two Block Matrices," *IMAGE*, Vol. 34, pp. 27–32, 2005. (Cited on page 350.)

[1327] Y. Tian, "The Schur Complement in an Orthogonal Projector," *IMAGE*, Vol. 35, pp. 34–36, 2005. (Cited on page 424.)

[1328] Y. Tian, "Two Characterizations of an EP Matrix," *IMAGE*, Vol. 34, p. 32, 2005. (Cited on page 406.)

[1329] Y. Tian, "Inequalities Involving Rank and Kronecker Product," *Amer. Math. Monthly*, Vol. 113, p. 851, 2006. (Cited on page 447.)

[1330] Y. Tian, "On the Inverse of a Product of Orthogonal Projectors," *Amer. Math. Monthly*, Vol. 113, pp. 467–468, 2006. (Cited on page 417.)

[1331] Y. Tian, "Two Commutativity Equalities for the Regularized Tikhonov Inverse," *IMAGE*, Vol. 41, pp. 37–39, 2008. (Cited on page 500.)

[1332] Y. Tian, "Two Equalities for the Moore-Penrose Inverse of a Row Block Matrix," *IMAGE*, Vol. 41, pp. 39–42, 2008. (Cited on page 428.)

[1333] Y. Tian and S. Cheng, "Some Identities for Moore-Penrose Inverses of Matrix Products," *Lin. Multilin. Alg.*, Vol. 52, pp. 405–420, 2004. (Cited on page 414.)

[1334] Y. Tian and S. Cheng, "The Moore-Penrose Inverse for Sums of Matrices under Rank Additivity Conditions," *Lin. Multilin. Alg.*, Vol. 53, pp. 45–65, 2005. (Cited on pages 419, 422, and 584.)

[1335] Y. Tian and Y. Liu, "Extremal Ranks of Some Symmetric Matrix Expressions with Applications," *SIAM J. Matrix Anal. Appl.*, Vol. 28, pp. 890–905, 2006. (Cited on page 424.)

[1336] Y. Tian and G. P. H. Styan, "Universal Factorization Inequalities for Quaternion Matrices and their Applications," *Math. J. Okayama University*, Vol. 41, pp. 45–62, 1999. (Cited on page 251.)

[1337] Y. Tian and G. P. H. Styan, "How to Establish Universal Block-Matrix Factorizations," *Elec. J. Lin. Alg.*, Vol. 8, pp. 115–127, 2001. (Cited on pages 146 and 251.)

[1338] Y. Tian and G. P. H. Styan, "Rank Equalities for Idempotent and Involutory Matrices," *Lin. Alg. Appl.*, Vol. 335, pp. 101–117, 2001. (Cited on pages 219, 220, 221, 228, and 405.)

[1339] Y. Tian and G. P. H. Styan, "A New Rank Formula for Idempotent Matrices with Applications," *Comment. Math. University Carolinae*, Vol. 43, pp. 379–384, 2002. (Cited on pages 134 and 218.)

[1340] Y. Tian and G. P. H. Styan, "When Does $\mathrm{rank}(ABC) = \mathrm{rank}(AB) + \mathrm{rank}(BC) - \mathrm{rank}(B)$ Hold?" *Int. J. Math. Educ. Sci. Tech.*, Vol. 33, pp. 127–137, 2002. (Cited on pages 127, 134, 220, 225, and 426.)

[1341] Y. Tian and G. P. H. Styan, "Rank Equalities for Idempotent Matrices with Applications," *J. Comp. Appl. Math.*, Vol. 191, pp. 77–97, 2006. (Cited on pages 219, 220, and 228.)

[1342] Y. Tian and Y. Takane, "Schur Complements and Banachiewicz-Schur Forms," *Elec. J. Lin. Alg.*, Vol. 13, pp. 405–418, 2005. (Cited on page 429.)

[1343] T. Toffoli and J. Quick, "Three-Dimensional Rotations by Three Shears," *Graph. Models Image Process.*, Vol. 59, pp. 89–96, 1997. (Cited on page 379.)

[1344] M. Tominaga, "A Generalized Reverse Inequality of the Cordes Inequality," *Math. Ineq. Appl.*, Vol. 11, pp. 221–227, 2008. (Cited on page 572.)

[1345] A. Tonge, "Equivalence Constants for Matrix Norms: A Problem of Goldberg," *Lin. Alg. Appl.*, Vol. 306, pp. 1–13, 2000. (Cited on page 640.)

[1346] G. E. Trapp, "Hermitian Semidefinite Matrix Means and Related Matrix Inequalities—An Introduction," *Lin. Multilin. Alg.*, Vol. 16, pp. 113–123, 1984. (Cited on page 509.)

[1347] L. N. Trefethen and D. Bau, *Numerical Linear Algebra*. Philadelphia: SIAM, 1997. (Cited on page xviii.)

[1348] L. N. Trefethen and M. Embree, *Spectra and Pseudospectra*. Princeton: Princeton University Press, 2005. (Cited on page 758.)

[1349] D. Trenkler, G. Trenkler, C.-K. Li, and H. J. Werner, "Square Roots and Additivity," *IMAGE*, Vol. 29, p. 30, 2002. (Cited on page 504.)

[1350] G. Trenkler, "A Trace Inequality," *Amer. Math. Monthly*, Vol. 102, pp. 362–363, 1995. (Cited on page 524.)

[1351] G. Trenkler, "The Moore-Penrose Inverse and the Vector Product," *Int. J. Math. Educ. Sci. Tech.*, Vol. 33, pp. 431–436, 2002. (Cited on pages 204 and 427.)

[1352] G. Trenkler, "A Matrix Related to an Idempotent Matrix," *IMAGE*, Vol. 31, pp. 39–40, 2003. (Cited on page 409.)

[1353] G. Trenkler, "On the Product of Orthogonal Projectors," *IMAGE*, Vol. 31, p. 43, 2003. (Cited on pages 227 and 417.)

[1354] G. Trenkler, "An Extension of Lagrange's Identity to Matrices," *Int. J. Math. Educ. Sci. Technol.*, Vol. 35, pp. 245–249, 2004. (Cited on page 68.)

[1355] G. Trenkler, "A Property for the Sum of a Matrix A and its Moore-Penrose Inverse A^+," *IMAGE*, Vol. 35, pp. 38–40, 2005. (Cited on page 406.)

[1356] G. Trenkler, "Factorization of a Projector," *IMAGE*, Vol. 34, pp. 33–35, 2005. (Cited on page 383.)

[1357] G. Trenkler, "On the Product of Orthogonal Projectors," *IMAGE*, Vol. 35, pp. 42–43, 2005. (Cited on page 504.)

[1358] G. Trenkler, "Projectors and Similarity," *IMAGE*, Vol. 34, pp. 35–36, 2005. (Cited on pages 409 and 429.)

[1359] G. Trenkler, "Property of the Cross Product," *IMAGE*, Vol. 34, pp. 36–37, 2005. (Cited on page 204.)

[1360] G. Trenkler, "A Range Equality for Idempotent Hermitian Matrices," *IMAGE*, Vol. 36, pp. 34–35, 2006. (Cited on page 226.)

[1361] G. Trenkler, "Necessary and Sufficient Conditions for $A + A^*$ to be a Nonnegative Definite Matrix," *IMAGE*, Vol. 37, pp. 28–30, 2006. (Cited on page 577.)

[1362] G. Trenkler, "Another Property for the Sum of a Matrix A and its Moore-Penrose Inverse A^+," *IMAGE*, Vol. 39, pp. 23–25, 2007. (Cited on page 406.)

[1363] G. Trenkler, "Characterization of EP-ness," *IMAGE*, Vol. 39, pp. 30–31, 2007. (Cited on page 406.)

[1364] G. Trenkler and S. Puntanen, "A Multivariate Version of Samuelson's Inequality," *Lin. Alg. Appl.*, Vol. 410, pp. 143–149, 2005. (Cited on pages 56 and 499.)

[1365] G. Trenkler and D. Trenkler, "The Sherman-Morrison Formula and Eigenvalues of a Special Bordered Matrix," *Acta Math. University Comenianae*, Vol. 74, pp. 255–258, 2005. (Cited on page 286.)

[1366] G. Trenkler and D. Trenkler, "The Vector Cross Product and 4×4 Skew-Symmetric Matrices," in *Recent Advances in Linear Models and Related Areas*, Shalabh and C. Heumann, Eds. Berlin: Springer, 2008, pp. 95–104. (Cited on pages 204, 287, and 741.)

[1367] G. Trenkler and H. J. Werner, "Partial Isometry and Idempotent Matrices," *IMAGE*, Vol. 29, pp. 30–32, 2002. (Cited on page 224.)

[1368] H. L. Trentelman, A. A. Stoorvogel, and M. L. J. Hautus, *Control Theory for Linear Systems*. New York: Springer, 2001. (Cited on pages xvii and 879.)

[1369] P. Treuenfels, "The Matrix Equation $X^2 - 2AX + B = 0$," *Amer. Math. Monthly*, Vol. 66, pp. 145–146, 1959. (Cited on page 876.)

[1370] B. Tromborg and S. Waldenstrom, "Bounds on the Diagonal Elements of a Unitary Matrix," *Lin. Alg. Appl.*, Vol. 20, pp. 189–195, 1978. (Cited on page 207.)

[1371] J. A. Tropp, A. C. Gilbert, and M. J. Strauss, "Algorithms for Simultaneous Sparse Approximation, Part I: Greedy Pursuit," *Sig. Proc.*, Vol. 86, pp. 572–588, 2006. (Cited on page 635.)

[1372] M. Tsatsomeros, J. S. Maybee, D. D. Olesky, and P. Van Den Driessche, "Nullspaces of Matrices and Their Compounds," *Lin. Multilin. Alg.*, Vol. 34, pp. 291–300, 1993. (Cited on page 369.)

[1373] N.-K. Tsing and F. Uhlig, "Inertia, Numerical Range, and Zeros of Quadratic Forms for Matrix Pencils," *SIAM J. Matrix Anal. Appl.*, Vol. 12, pp. 146–159, 1991. (Cited on pages 396 and 554.)

[1374] P. Tsiotras, "Further Passivity Results for the Attitude Control Problem," *IEEE Trans. Autom. Contr.*, Vol. 43, pp. 1597–1600, 1998. (Cited on page 204.)

[1375] P. Tsiotras, J. L. Junkins, and H. Schaub, "Higher-Order Cayley Transforms with Applications to Attitude Representations," *J. Guid. Contr. Dyn.*, Vol. 20, pp. 528–534, 1997. (Cited on page 212.)

[1376] S. H. Tung, "On Lower and Upper Bounds of the Difference between the Arithmetic and the Geometric Mean," *Math. Comput.*, Vol. 29, pp. 834–836, 1975. (Cited on page 56.)

[1377] D. A. Turkington, *Matrix Calculus and Zero-One Matrices*. Cambridge: Cambridge University Press, 2002. (Cited on page 458.)

[1378] R. Turkmen and D. Bozkurt, "A Note for Bounds of Norms of Hadamard Product of Matrices," *Math. Ineq. Appl.*, Vol. 9, pp. 211–217, 2006. (Cited on page 674.)

[1379] H. W. Turnbull, *The Theory of Determinants, Matrices and Invariants*. London: Blackie, 1950. (Cited on page 178.)

[1380] G. M. Tuynman, "The Derivation of the Exponential Map of Matrices," *Amer. Math. Monthly*, Vol. 102, pp. 818–820, 1995. (Cited on page 748.)

[1381] F. Tyan and D. S. Bernstein, "Global Stabilization of Systems Containing a Double Integrator Using a Saturated Linear Controller," *Int. J. Robust Nonlin. Contr.*, Vol. 9, pp. 1143–1156, 1999. (Cited on page 869.)

[1382] M. Uchiyama, "Norms and Determinants of Products of Logarithmic Functions of Positive Semi-Definite Operators," *Math. Ineq. Appl.*, Vol. 1, pp. 279–284, 1998. (Cited on pages 574 and 761.)

[1383] M. Uchiyama, "Some Exponential Operator Inequalities," *Math. Ineq. Appl.*, Vol. 2, pp. 469–471, 1999. (Cited on page 755.)

[1384] F. E. Udwadia and R. E. Kalaba, *Analytical Dynamics: A New Approach.* Cambridge: Cambridge University Press, 1996. (Cited on page xvii.)

[1385] F. Uhlig, "A Recurring Theorem About Pairs of Quadratic Forms and Extensions: A Survey," *Lin. Alg. Appl.*, Vol. 25, pp. 219–237, 1979. (Cited on pages 396, 554, 558, and 559.)

[1386] F. Uhlig, "On the Matrix Equation $AX = B$ with Applications to the Generators of a Controllability Matrix," *Lin. Alg. Appl.*, Vol. 85, pp. 203–209, 1987. (Cited on page 126.)

[1387] F. Uhlig, "Constructive Ways for Generating (Generalized) Real Orthogonal Matrices as Products of (Generalized) Symmetries," *Lin. Alg. Appl.*, Vol. 332–334, pp. 459–467, 2001. (Cited on page 379.)

[1388] F. A. Valentine, *Convex Sets.* New York: McGraw-Hill, 1964. (Cited on page 178.)

[1389] M. Van Barel, V. Ptak, and Z. Vavrin, "Bezout and Hankel Matrices Associated with Row Reduced Matrix Polynomials, Barnett-Type Formulas," *Lin. Alg. Appl.*, Vol. 332–334, pp. 583–606, 2001. (Cited on page 279.)

[1390] R. van der Merwe, *Sigma-Point Kalman Filters for Probabilistic Inference in Dynamic State-Space Models.* Portland: Oregon Health and Science University, 2004, Ph.D. Dissertation. (Cited on page 741.)

[1391] P. Van Dooren, "The Computation of Kronecker's Canonical Form of a Singular Pencil," *Lin. Alg. Appl.*, Vol. 27, pp. 103–140, 1979. (Cited on page 396.)

[1392] S. Van Huffel and J. Vandewalle, *The Total Least Squares Problem: Computational Aspects and Analysis.* Philadelphia: SIAM, 1987. (Cited on page 679.)

[1393] C. F. Van Loan, "Computing Integrals Involving the Matrix Exponential," *IEEE Trans. Autom. Contr.*, Vol. AC-23, pp. 395–404, 1978. (Cited on page 748.)

[1394] C. F. Van Loan, "How Near Is a Stable Matrix to an Unstable Matrix," *Contemporary Math.*, Vol. 47, pp. 465–478, 1985. (Cited on page 769.)

[1395] C. F. Van Loan, *Computational Frameworks for the Fast Fourier Transform.* Philadelphia: SIAM, 1992. (Cited on pages xvii and 390.)

[1396] C. F. Van Loan, "The Ubiquitous Kronecker Product," *J. Comp. Appl. Math.*, Vol. 123, pp. 85–100, 2000. (Cited on page 458.)

[1397] P. Van Overschee and B. De Moor, *Subspace Identification for Linear Systems: Theory, Implementation, Applications.* Dordrecht: Kluwer, 1996. (Cited on page 879.)

[1398] R. Vandebril and M. Van Barel, "A Note on the Nullity Theorem," *J. Comp. Appl. Math.*, Vol. 189, pp. 179–190, 2006. (Cited on page 135.)

[1399] R. Vandebril, M. Van Barel, and N. Mastronardi, *Matrix Computations and Semiseparable Matrices.* Baltimore: Johns Hopkins University Press, 2008. (Cited on pages 135, 136, and 140.)

[1400] V. S. Varadarajan, *Lie Groups, Lie Algebras, and Their Representations.* New York: Springer, 1984. (Cited on pages 252, 720, 748, 749, and 793.)

[1401] J. M. Varah, "A Lower Bound for the Smallest Singular Value of a Matrix," *Lin. Alg. Appl.*, Vol. 11, pp. 3–5, 1975. (Cited on page 664.)

[1402] A. I. G. Vardulakis, *Linear Multivariable Control: Algebraic Analysis and Synthesis Methods.* Chichester: Wiley, 1991. (Cited on pages xvii and 307.)

[1403] R. S. Varga, *Matrix Iterative Analysis.* Englewood Cliffs: Prentice-Hall, 1962. (Cited on pages xviii and 299.)

[1404] R. S. Varga, *Gersgorin and His Circles.* New York: Springer, 2004. (Cited on page 293.)

[1405] P. R. Vein, "A Short Survey of Some Recent Applications of Determinants," *Lin. Alg. Appl.*, Vol. 42, pp. 287–297, 1982. (Cited on page 178.)

[1406] R. Vein and P. Dale, *Determinants and Their Applications in Mathematical Physics.* New York: Springer, 1999. (Cited on pages xvii and 178.)

[1407] D. Veljan, "The Sine Theorem and Inequalities for Volumes of Simplices and Determinants," *Lin. Alg. Appl.*, Vol. 219, pp. 79–91, 1995. (Cited on page 174.)

[1408] D. Veljan and S. Wu, "Parametrized Klamkin's Inequality and Improved Euler's Inequality," *Math. Ineq. Appl.*, Vol. 11, pp. 729–737, 2008. (Cited on pages 46 and 171.)

[1409] J. Vermeer, "Orthogonal Similarity of a Real Matrix and Its Transpose," *Lin. Alg. Appl.*, Vol. 428, pp. 382–392, 2008. (Cited on page 377.)

[1410] K. Veselic, "On Real Eigenvalues of Real Tridiagonal Matrices," *Lin. Alg. Appl.*, Vol. 27, pp. 167–171, 1979. (Cited on page 359.)

[1411] K. Veselic, "Estimating the Operator Exponential," *Lin. Alg. Appl.*, Vol. 280, pp. 241–244, 1998. (Cited on page 769.)

[1412] K. Veselic, "Bounds for Exponentially Stable Semigroups," *Lin. Alg. Appl.*, Vol. 358, pp. 309–333, 2003. (Cited on page 769.)

[1413] W. J. Vetter, "Matrix Calculus Operations and Taylor Expansions," *SIAM Rev.*, Vol. 15, pp. 352–369, 1973. (Cited on page 458.)

[1414] M. Vidyasagar, "On Matrix Measures and Convex Liapunov Functions," *J. Math. Anal. Appl.*, Vol. 62, pp. 90–103, 1978. (Cited on page 758.)

[1415] G. Visick, "A Weak Majorization Involving the Matrices $A \circ B$ and AB," *Lin. Alg. Appl.*, Vol. 223/224, pp. 731–744, 1995. (Cited on page 589.)

[1416] G. Visick, "Majorizations of Hadamard Products of Matrix Powers," *Lin. Alg. Appl.*, Vol. 269, pp. 233–240, 1998. (Cited on pages 589 and 594.)

[1417] G. Visick, "A Quantitative Version of the Observation that the Hadamard Product Is a Principal Submatrix of the Kronecker Product," *Lin. Alg. Appl.*, Vol. 304, pp. 45–68, 2000. (Cited on pages 585, 591, 592, and 593.)

[1418] G. Visick, "Another Inequality for Hadamard Products," *IMAGE*, Vol. 29, pp. 32–33, 2002. (Cited on page 585.)

[1419] S.-W. Vong and X.-Q. Jin, "Proof of Bottcher and Wenzel's Conjecture," *Oper. Matrices*, Vol. 2, pp. 435–442, 2008. (Cited on page 641.)

[1420] S. Wada, "On Some Refinement of the Cauchy-Schwarz Inequality," *Lin. Alg. Appl.*, Vol. 420, pp. 433–440, 2007. (Cited on pages 66 and 70.)

[1421] G. G. Walter and M. Contreras, *Compartmental Modeling with Networks.* Boston: Birkhauser, 1999. (Cited on pages xvii and 777.)

[1422] B. Wang and F. Zhang, "Some Inequalities for the Eigenvalues of the Product of Positive Semidefinite Hermitian Matrices," *Lin. Alg. Appl.*, Vol. 160, pp. 113–118, 1992. (Cited on pages 570, 571, and 672.)

[1423] B.-Y. Wang and M.-P. Gong, "Some Eigenvalue Inequalities for Positive Semidefinite Matrix Power Products," *Lin. Alg. Appl.*, Vol. 184, pp. 249–260, 1993. (Cited on page 571.)

[1424] B.-Y. Wang, B.-Y. Xi, and F. Zhang, "Some Inequalities for Sum and Product of Positive Semidefinite Matrices," *Lin. Alg. Appl.*, Vol. 293, pp. 39–49, 1999. (Cited on pages 580 and 673.)

[1425] B.-Y. Wang and F. Zhang, "A Trace Inequality for Unitary Matrices," *Amer. Math. Monthly*, Vol. 101, pp. 453–455, 1994. (Cited on page 209.)

[1426] B.-Y. Wang and F. Zhang, "Trace and Eigenvalue Inequalities for Ordinary and Hadamard Products of Positive Semidefinite Hermitian Matrices," *SIAM J. Matrix Anal. Appl.*, Vol. 16, pp. 1173–1183, 1995. (Cited on pages 571 and 587.)

[1427] B.-Y. Wang and F. Zhang, "Schur Complements and Matrix Inequalities of Hadamard Products," *Lin. Multilin. Alg.*, Vol. 43, pp. 315–326, 1997. (Cited on page 595.)

[1428] C.-L. Wang, "On Development of Inverses of the Cauchy and Hölder Inequalities," *SIAM Rev.*, Vol. 21, pp. 550–557, 1979. (Cited on pages 63 and 71.)

[1429] D. Wang, "The Polar Decomposition and a Matrix Inequality," *Amer. Math. Monthly*, Vol. 96, pp. 517–519, 1989. (Cited on page 476.)

[1430] G. Wang, Y. Wei, and S. Qiao, *Generalized Inverses: Theory and Computations.* Beijing/New York: Science Press, 2004. (Cited on pages 141, 219, 412, and 438.)

[1431] J.-H. Wang, "The Length Problem for a Sum of Idempotents," *Lin. Alg. Appl.*, Vol. 215, pp. 135–159, 1995. (Cited on page 395.)

[1432] L.-C. Wang and C.-L. Li, "On Some New Mean Value Inequalities," *J. Ineq. Pure Appl. Math.*, Vol. 8, no. 3, pp. 1–8, 2007, Article 87. (Cited on page 63.)

[1433] Q.-G. Wang, "Necessary and Sufficient Conditions for Stability of a Matrix Polytope with Normal Vertex Matrices," *Automatica*, Vol. 27, pp. 887–888, 1991. (Cited on page 295.)

[1434] X. Wang, "A Simple Proof of Descartes's Rule of Signs," *Amer. Math. Monthly*, Vol. 111, pp. 525–526, 2004. (Cited on page 763.)

[1435] Y. W. Wang and D. S. Bernstein, "Controller Design with Regional Pole Constraints: Hyperbolic and Horizontal Strip Regions," *AIAA J. Guid. Contr. Dyn.*, Vol. 16, pp. 784–787, 1993. (Cited on page 871.)

[1436] Y. W. Wang and D. S. Bernstein, "L_2 Controller Synthesis with L_∞-Bounded Closed-Loop Impulse Response," *Int. J. Contr.*, Vol. 60, pp. 1295–1306, 1994. (Cited on page 756.)

[1437] A. J. B. Ward and F. Gerrish, "A Constructive Proof by Elementary Transformations of Roth's Removal Theorems in the Theory of Matrix Equations," *Int. J. Math. Educ. Sci. Tech.*, Vol. 31, pp. 425–429, 2000. (Cited on page 349.)

[1438] J. Warga, *Optimal Control of Differential and Functional Equations*. New York: Academic Press, 1972. (Cited on pages 685 and 686.)

[1439] W. E. Waterhouse, "A Determinant Identity with Matrix Entries," *Amer. Math. Monthly*, Vol. 97, pp. 249–250, 1990. (Cited on pages 148 and 149.)

[1440] W. Watkins, "Convex Matrix Functions," *Proc. Amer. Math. Soc.*, Vol. 44, pp. 31–34, 1974. (Cited on pages 537 and 590.)

[1441] W. Watkins, "Generating Functions," *Coll. Math. J.*, Vol. 18, pp. 195–211, 1987. (Cited on page 304.)

[1442] W. Watkins, "A Determinantal Inequality for Correlation Matrices," *Lin. Alg. Appl.*, Vol. 104, pp. 59–63, 1988. (Cited on page 586.)

[1443] G. S. Watson, G. Alpargu, and G. P. H. Styan, "Some Comments on Six Inequalities Associated with the Inefficiency of Ordinary Least Squares with One Regressor," *Lin. Alg. Appl.*, Vol. 264, pp. 13–54, 1997. (Cited on pages 63 and 72.)

[1444] J. R. Weaver, "Centrosymmetric (Cross-Symmetric) Matrices, Their Basic Properties, Eigenvalues, and Eigenvectors," *Amer. Math. Monthly*, Vol. 92, pp. 711–717, 1985. (Cited on pages 242 and 342.)

[1445] J. H. Webb, "An Inequality in Algebra and Geometry," *College Math. J.*, Vol. 36, p. 164, 2005. (Cited on page 49.)

[1446] R. Webster, *Convexity*. Oxford: Oxford University Press, 1994. (Cited on page 178.)

[1447] E. Wegert and L. N. Trefethen, "From the Buffon Needle Problem to the Kreiss Matrix Theorem," *Amer. Math. Monthly*, Vol. 101, pp. 132–139, 1994. (Cited on page 786.)

[1448] J. Wei and E. Norman, "Lie Algebraic Solution of Linear Differential Equations," *J. Math. Phys.*, Vol. 4, pp. 575–581, 1963. (Cited on page 743.)

[1449] J. Wei and E. Norman, "On Global Representations of the Solutions of Linear Differential Equations as a Product of Exponentials," *Proc. Amer. Math. Soc.*, Vol. 15, pp. 327–334, 1964. (Cited on page 743.)

[1450] M. Wei, "Reverse Order Laws for Generalized Inverses of Multiple Matrix Products," *Lin. Alg. Appl.*, Vol. 293, pp. 273–288, 1999. (Cited on page 415.)

[1451] Y. Wei, "A Characterization and Representation of the Drazin Inverse," *SIAM J. Matrix Anal. Appl.*, Vol. 17, pp. 744–747, 1996. (Cited on page 431.)

[1452] Y. Wei, "Expressions for the Drazin Inverse of a 2×2 Block Matrix," *Lin. Multilin. Alg.*, Vol. 45, pp. 131–146, 1998. (Cited on page 429.)

[1453] G. H. Weiss and A. A. Maradudin, "The Baker-Hausdorff Formula and a Problem in Crystal Physics," *J. Math. Phys.*, Vol. 3, pp. 771–777, 1962. (Cited on page 743.)

[1454] E. M. E. Wermuth, "Two Remarks on Matrix Exponentials," *Lin. Alg. Appl.*, Vol. 117, pp. 127–132, 1989. (Cited on pages 747 and 762.)

[1455] E. M. E. Wermuth, "A Remark on Commuting Operator Exponentials," *Proc. Amer. Math. Soc.*, Vol. 125, pp. 1685–1688, 1997. (Cited on page 747.)

[1456] H.-J. Werner, "On the Matrix Monotonicity of Generalized Inversion," *Lin. Alg. Appl.*, Vol. 27, pp. 141–145, 1979. (Cited on page 579.)

[1457] H. J. Werner, "On the Product of Orthogonal Projectors," *IMAGE*, Vol. 32, pp. 30–37, 2004. (Cited on pages 227, 409, and 416.)

[1458] H. J. Werner, "Solution 34-4.2 to 'A Range Equality for the Commutator with Two Involutory Matrices'," *IMAGE*, Vol. 35, pp. 32–33, 2005. (Cited on page 217.)

[1459] J. R. Wertz and W. J. Larson, Eds., *Space Mission Analysis and Design*. Dordrecht: Kluwer, 1999. (Cited on page 175.)

[1460] P. Wesseling, *Principles of Computational Fluid Dynamics*. Berlin: Springer, 2001. (Cited on page xvii.)

[1461] J. R. Westlake, *A Handbook of Numerical Matrix Inversion and Solution of Linear Equations*. New York: Wiley, 1968. (Cited on page xviii.)

[1462] N. A. Wiegmann, "Normal Products of Matrices," *Duke Math. J.*, Vol. 15, pp. 633–638, 1948. (Cited on page 374.)

[1463] Z. Wiener, "An Interesting Matrix Exponent Formula," *Lin. Alg. Appl.*, Vol. 257, pp. 307–310, 1997. (Cited on page 785.)

[1464] E. P. Wigner and M. M. Yanase, "On the Positive Semidefinite Nature of a Certain Matrix Expression," *Canad. J. Math.*, Vol. 16, pp. 397–406, 1964. (Cited on page 502.)

[1465] R. M. Wilcox, "Exponential Operators and Parameter Differentiation in Quantum Physics," *J. Math. Phys.*, Vol. 8, pp. 962–982, 1967. (Cited on pages xviii and 750.)

[1466] J. B. Wilker, J. S. Sumner, A. A. Jagers, M. Vowe, and J. Anglesio, "Inequalities Involving Trigonometric Functions," *Amer. Math. Monthly*, Vol. 98, pp. 264–266, 1991. (Cited on page 32.)

[1467] J. H. Wilkinson, *The Algebraic Eigenvalue Problem*. London: Oxford University Press, 1965. (Cited on page xviii.)

[1468] J. C. Willems, "Least Squares Stationary Optimal Control and the Algebraic Riccati Equation," *IEEE Trans. Autom. Contr.*, Vol. AC-16, pp. 621–634, 1971. (Cited on page 879.)

[1469] D. A. Wilson, "Convolution and Hankel Operator Norms for Linear Systems," *IEEE Trans. Autom. Contr.*, Vol. AC-34, pp. 94–97, 1989. (Cited on page 680.)

[1470] P. M. H. Wilson, *Curved Spaces: From Classical Geometries to Elementary Differential Geometry*. Cambridge: Cambridge University Press, 2008. (Cited on pages 175 and 214.)

[1471] H. K. Wimmer, "Inertia Theorems for Matrices, Controllability, and Linear Vibrations," *Lin. Alg. Appl.*, Vol. 8, pp. 337–343, 1974. (Cited on pages 867 and 870.)

[1472] H. K. Wimmer, "The Algebraic Riccati Equation without Complete Controllability," *SIAM J. Alg. Disc. Math.*, Vol. 3, pp. 1–12, 1982. (Cited on page 879.)

[1473] H. K. Wimmer, "On Ostrowski's Generalization of Sylvester's Law of Inertia," *Lin. Alg. Appl.*, Vol. 52/53, pp. 739–741, 1983. (Cited on page 337.)

[1474] H. K. Wimmer, "The Algebraic Riccati Equation: Conditions for the Existence and Uniqueness of Solutions," *Lin. Alg. Appl.*, Vol. 58, pp. 441–452, 1984. (Cited on page 879.)

[1475] H. K. Wimmer, "Monotonicity of Maximal Solutions of Algebraic Riccati Equations," *Sys. Contr. Lett.*, Vol. 5, pp. 317–319, 1985. (Cited on pages 861 and 879.)

[1476] H. K. Wimmer, "Extremal Problems for Hölder Norms of Matrices and Realizations of Linear Systems," *SIAM J. Matrix Anal. Appl.*, Vol. 9, pp. 314–322, 1988. (Cited on pages 505 and 642.)

[1477] H. K. Wimmer, "Linear Matrix Equations, Controllability and Observability, and the Rank of Solutions," *SIAM J. Matrix Anal. Appl.*, Vol. 9, pp. 570–578, 1988. (Cited on page 870.)

[1478] H. K. Wimmer, "On the History of the Bezoutian and the Resultant Matrix," *Lin. Alg. Appl.*, Vol. 128, pp. 27–34, 1990. (Cited on page 279.)

[1479] H. K. Wimmer, "Normal Forms of Symplectic Pencils and the Discrete-Time Algebraic Riccati Equation," *Lin. Alg. Appl.*, Vol. 147, pp. 411–440, 1991. (Cited on page 879.)

[1480] H. K. Wimmer, "Lattice Properties of Sets of Semidefinite Solutions of Continuous-Time Algebraic Riccati Equations," *Automatica*, Vol. 31, pp. 173–182, 1995. (Cited on page 879.)

[1481] A. Witkowski, "A New Proof of the Monotonicity of Power Means," *J. Ineq. Pure Appl. Math.*, Vol. 5, no. 1, pp. 1–2, 2004, Article 6. (Cited on pages 25 and 30.)

[1482] J. Wittenburg and L. Lilov, "Decomposition of a Finite Rotation into Three Rotations about Given Axes," *Multibody Sys. Dyn.*, Vol. 9, pp. 353–375, 2003. (Cited on page 740.)

[1483] H. Wolkowicz and G. P. H. Styan, "Bounds for Eigenvalues Using Traces," *Lin. Alg. Appl.*, Vol. 29, pp. 471–506, 1980. (Cited on pages 56, 351, 361, and 362.)

[1484] H. Wolkowicz and G. P. H. Styan, "More Bounds for Eigenvalues Using Traces," *Lin. Alg. Appl.*, Vol. 31, pp. 1–17, 1980. (Cited on page 361.)

[1485] W. A. Wolovich, *Linear Multivariable Systems*. New York: Springer, 1974. (Cited on page 879.)

[1486] M. J. Wonenburger, "A Decomposition of Orthogonal Transformations," *Canad. Math. Bull.*, Vol. 7, pp. 379–383, 1964. (Cited on page 383.)

[1487] M. J. Wonenburger, "Transformations Which are Products of Two Involutions," *J. Math. Mech.*, Vol. 16, pp. 327–338, 1966. (Cited on page 383.)

[1488] C. S. Wong, "Characterizations of Products of Symmetric Matrices," *Lin. Alg. Appl.*, Vol. 42, pp. 243–251, 1982. (Cited on page 382.)

[1489] E. Wong, Ed., *Stochastic Processes in Information and Dynamical Systems*. New York: McGraw-Hill, 1971. (Cited on page 489.)

[1490] W. M. Wonham, *Linear Multivariable Control: A Geometric Approach*, 2nd ed. New York: Springer, 1979. (Cited on pages xvii, 285, 710, 865, and 879.)

[1491] C.-F. Wu, "On Some Ordering Properties of the Generalized Inverses of Nonnegative Definite Matrices," *Lin. Alg. Appl.*, Vol. 32, pp. 49–60, 1980. (Cited on page 579.)

[1492] P. Y. Wu, "Products of Nilpotent Matrices," *Lin. Alg. Appl.*, Vol. 96, pp. 227–232, 1987. (Cited on page 383.)

[1493] P. Y. Wu, "Products of Positive Semidefinite Matrices," *Lin. Alg. Appl.*, Vol. 111, pp. 53–61, 1988. (Cited on pages 382 and 596.)

[1494] P. Y. Wu, "The Operator Factorization Problems," *Lin. Alg. Appl.*, Vol. 117, pp. 35–63, 1989. (Cited on pages 382, 383, 384, and 770.)

[1495] P. Y. Wu, "Sums of Idempotent Matrices," *Lin. Alg. Appl.*, Vol. 142, pp. 43–54, 1990. (Cited on pages 394 and 395.)

[1496] S. Wu, "A Further Generalization of Aczel's Inequality and Popoviciu's Inequality," *Math. Ineq. Appl.*, Vol. 10, pp. 565–573, 2007. (Cited on page 70.)

[1497] S. Wu, "A Sharpened Version of the Fundamental Triangle Inequality," *Math. Ineq. Appl.*, Vol. 11, pp. 477–482, 2008. (Cited on page 171.)

[1498] S. Wu, "Note on a Conjecture of R. A. Satnoianu," *Math. Ineq. Appl.*, Vol. 12, pp. 147–151, 2009. (Cited on page 65.)

[1499] Y.-D. Wu, Z.-H. Zhang, and V. Lokesha, "Sharpening on Mircea's Inequality," *J. Ineq. Pure Appl. Math.*, Vol. 8, no. 4, pp. 1–6, 2007, Article 116. (Cited on page 171.)

[1500] Z.-G. Xiao and Z.-H. Zhang, "The Inequalities $G \leq L \leq I \leq A$ in n Variables," *J. Ineq. Pure. Appl. Math.*, Vol. 4, no. 2/39, pp. 1–6, 2003. (Cited on page 59.)

[1501] C. Xu, "Bellman's Inequality," *Lin. Alg. Appl.*, Vol. 229, pp. 9–14, 1995. (Cited on page 526.)

[1502] C. Xu, Z. Xu, and F. Zhang, "Revisiting Hua-Marcus-Bellman-Ando Inequalities on Contractive Matrices," *Lin. Alg. Appl.*, Vol. 430, pp. 1499–1508, 2009. (Cited on pages 76, 157, 520, 539, and 620.)

[1503] H. Xu, "Two Results About the Matrix Exponential," *Lin. Alg. Appl.*, Vol. 262, pp. 99–109, 1997. (Cited on pages 714 and 769.)

[1504] A. Yakub, "Symmetric Inequality," *Amer. Math. Monthly*, Vol. 114, p. 649, 2007. (Cited on page 49.)

[1505] T. Yamazaki, "An Extension of Specht's Theorem via Kantorovich Inequality and Related Results," *Math. Ineq. Appl.*, Vol. 3, pp. 89–96, 2000. (Cited on page 57.)

[1506] T. Yamazaki, "Further Characterizations of Chaotic Order via Specht's Ratio," *Math. Ineq. Appl.*, Vol. 3, pp. 259–268, 2000. (Cited on page 575.)

[1507] B. Yang, "On a New Inequality Similar to Hardy-Hilbert's Inequality," *Math. Ineq. Appl.*, Vol. 6, pp. 37–44, 2003. (Cited on page 69.)

[1508] B. Yang and T. M. Rassias, "On the Way of Weight Coefficient and Research for the Hilbert-Type Inequalities," *Math. Ineq. Appl.*, Vol. 6, pp. 625–658, 2003. (Cited on page 69.)

[1509] X. Yang, "Necessary Conditions of Hurwitz Polynomials," *Lin. Alg. Appl.*, Vol. 359, pp. 21–27, 2003. (Cited on pages 764 and 765.)

[1510] X. Yang, "Some Necessary Conditions for Hurwitz Stability," *Automatica*, Vol. 40, pp. 527–529, 2004. (Cited on page 765.)

[1511] Z. P. Yang and X. X. Feng, "A Note on the Trace Inequality for Products of Hermitian Matrix Power," *J. Ineq. Pure. Appl. Math.*, Vol. 3, no. 5/78, pp. 1–12, 2002. (Cited on pages 524 and 526.)

[1512] S. F. Yau and Y. Bresler, "A Generalization of Bergstrom's Inequality and Some Applications," *Lin. Alg. Appl.*, Vol. 161, pp. 135–151, 1992. (Cited on pages 588 and 592.)

[1513] D. M. Young, *Iterative Solution of Large Linear Systems*. New York: Academic Press, 1971, reprinted by Dover, Mineola, 2003. (Cited on page xviii.)

[1514] N. J. Young, "Matrices which Maximise any Analytic Function," *Acta Math. Hung.*, Vol. 34, pp. 239–243, 1979. (Cited on page 665.)

[1515] R. Zamfir, "Refining Some Inequalities," *J. Ineq. Pure Appl. Math.*, Vol. 9, no. 3, pp. 1–5, 2008, Article 77. (Cited on page 781.)

[1516] A. Zanna and H. Z. Munthe-Kaas, "Generalized Polar Decompositions for the Approximation of the Matrix Exponential," *SIAM J. Matrix Anal. Appl.*, Vol. 23, pp. 840–862, 2002. (Cited on page 742.)

[1517] X. Zhan, "Inequalities for the Singular Values of Hadamard Products," *SIAM J. Matrix Anal. Appl.*, Vol. 18, pp. 1093–1095, 1997. (Cited on page 674.)

[1518] X. Zhan, "Inequalities for Unitarily Invariant Norms," *SIAM J. Matrix Anal. Appl.*, Vol. 20, pp. 466–470, 1998. (Cited on page 492.)

[1519] X. Zhan, "Singular Values of Differences of Positive Semidefinite Matrices," *SIAM J. Matrix Anal. Appl.*, Vol. 22, pp. 819–823, 2000. (Cited on pages 567 and 647.)

[1520] X. Zhan, "Span of the Orthogonal Orbit of Real Matrices," *Lin. Multilin. Alg.*, Vol. 49, pp. 337–346, 2001. (Cited on page 394.)

[1521] X. Zhan, *Matrix Inequalities*. New York: Springer, 2002. (Cited on pages 177, 356, 477, 485, 492, 560, 561, 562, 573, 582, 583, 584, 589, 591, 592, 594, 595, 603, 638, 671, and 752.)

[1522] X. Zhan, "On Some Matrix Inequalities," *Lin. Alg. Appl.*, Vol. 376, pp. 299–303, 2004. (Cited on page 567.)

[1523] F. Zhang, *Linear Algebra: Challenging Problems for Students*. Baltimore: Johns Hopkins University Press, 1996. (Cited on pages 142, 199, and 200.)

[1524] F. Zhang, "Quaternions and Matrices of Quaternions," *Lin. Alg. Appl.*, Vol. 251, pp. 21–57, 1997. (Cited on page 251.)

[1525] F. Zhang, "A Compound Matrix with Positive Determinant," *Amer. Math. Monthly*, Vol. 105, p. 958, 1998. (Cited on page 166.)

[1526] F. Zhang, *Matrix Theory: Basic Results and Techniques*. New York: Springer, 1999. (Cited on pages 136, 137, 235, 340, 373, 390, 502, 511, 518, 524, 525, 538, 595, and 619.)

[1527] F. Zhang, "Schur Complements and Matrix Inequalities in the Lowner Ordering," *Lin. Alg. Appl.*, Vol. 321, pp. 399–410, 2000. (Cited on page 539.)

[1528] F. Zhang, "Matrix Inequalities by Means of Block Matrices," *Math. Ineq. Appl.*, Vol. 4, pp. 481–490, 2001. (Cited on pages 497, 516, 542, 560, 578, 586, 587, 592, and 593.)

[1529] F. Zhang, "Inequalities Involving Square Roots," *IMAGE*, Vol. 29, pp. 33–34, 2002. (Cited on page 503.)

[1530] F. Zhang, "Block Matrix Techniques," in *The Schur Complement and Its Applications*, F. Zhang, Ed. New York: Springer, 2004, pp. 83–110. (Cited on pages 498, 531, 532, 542, 551, 555, 556, 559, and 561.)

[1531] F. Zhang, "A Matrix Identity on the Schur Complement," *Lin. Multilin. Alg.*, Vol. 52, pp. 367–373, 2004. (Cited on page 430.)

[1532] L. Zhang, "A Characterization of the Drazin Inverse," *Lin. Alg. Appl.*, Vol. 335, pp. 183–188, 2001. (Cited on page 431.)

[1533] Q.-C. Zhong, "*J*-Spectral Factorization of Regular Para-Hermitian Transfer Matrices," *Automatica*, Vol. 41, pp. 1289–1293, 2005. (Cited on page 223.)

[1534] Q.-C. Zhong, *Robust Control of Time-delay Systems*. London: Springer, 2006. (Cited on page 223.)

[1535] K. Zhou, *Robust and Optimal Control*. Upper Saddle River: Prentice-Hall, 1996. (Cited on pages xvii, 307, 664, 668, and 860.)

[1536] L. Zhu, "A New Simple Proof of Wilker's Inequality," *Math. Ineq. Appl.*, Vol. 8, pp. 749–750, 2005. (Cited on page 32.)

[1537] L. Zhu, "On Wilker-Type Inequalities," *Math. Ineq. Appl.*, Vol. 10, pp. 727–731, 2005. (Cited on page 32.)

[1538] G. Zielke, "Some Remarks on Matrix Norms, Condition Numbers, and Error Estimates for Linear Equations," *Lin. Alg. Appl.*, Vol. 110, pp. 29–41, 1988. (Cited on pages 629 and 677.)

[1539] S. Zlobec, "An Explicit Form of the Moore-Penrose Inverse of an Arbitrary Complex Matrix," *SIAM Rev.*, Vol. 12, pp. 132–134, 1970. (Cited on page 400.)

[1540] D. Zwillinger, *Standard Mathematical Tables and Formulae*, 31st ed. Boca Raton: Chapman & Hall/CRC, 2003. (Cited on pages 82, 171, 175, 277, and 391.)

Author Index

Index

H

O

Milton Keynes UK
Ingram Content Group UK Ltd.
UKHW010841140924
448309UK00009B/335